Trends in Quorum Sensing and Quorum Quenching

Trends in Quorum Sensing and Quorum Quenching

New Perspectives and Applications

Edited by
V. Ravishankar Rai
Jamuna A Bai

CRC Press is an imprint of the
Taylor & Francis Group, an **informa** business

CRC Press
Taylor & Francis Group
52 Vanderbilt Avenue,
New York, NY 10017

CRC Press is an imprint of Taylor & Francis Group, an Informa business

Printed on acid-free paper

International Standard Book Number-13: 978-0-367-22428-8 (Hardback)

Visit the Taylor & Francis Web site at
http://www.taylorandfrancis.com

and the CRC Press Web site at
http://www.crcpress.com

Contents

Preface

Trends in Quorum Sensing and Quorum Quenching: New Perspectives and Applications focuses on the evolution and our current understanding of quorum sensing mechanisms in bacteria and the potential application of quorum sensing inhibitors in clinical and industrial settings. Discovered in the 1960s and 1970s, quorum sensing has garnered an increasing interest in the research community. Elucidating the quorum sensing mechanism in bacteria has revealed previously unknown coordinated group behavior in bacteria. Studies on cell-cell signaling or intercellular communication started with the understanding of bioluminescence in marine vibrios, fruiting body development in myxobacteria, and competence in pneumococci. Research on quorum sensing further advanced when it was realized that it had a central and crucial role in regulation of bacterial pathogenicity. The discovery of quorum sensing inhibitory compound furanones from red seaweed *Delisea pulchra* and the characterization of quorum quenching enzyme, the AiiA lactonase from *Bacillus*, indicated the novel strategy that could be used to combat and control bacterial infections.

This book has two major sections with key topics. Section one deals with advances and perspectives on molecular mechanism of QS in bacteria. The topics covered include influence of quorum sensing on bacterial central metabolism; novel quorum sensing signaling molecules and detection techniques; molecular insights in the role of QS in clinical pathogens, foodborne bacteria, agriculturally important bacteria, industrial relevant bacteria and its application in metabolic engineering; the evolution and role of QS in biofilm formation and development; and QS in regulating morphology and metabolic pathways in eukaryotic microbes (fungi).

The second section focuses on trends in the development and application of quorum sensing inhibitors. The emphasis is on the mechanism and types of QS inhibitors; evolution of quorum quenching in bacteria; application of metagenomics tools for the identification of novel quorum quenching genes and enzymes; bioprospecting of bacteria, fungi, actinomycetes and endophytes from rhizospheres and marine ecosystems for novel QS inhibitors; design of QS inhibitors based on nanotechnology; etc. The potential application of QS inhibitors as—anti-infectives and therapeutics (quorum quenching monoclonal antibodies and AHL acylase nanoparticles), novel intervention techniques in the food industry (sanitizers for food contact surfaces and as preservatives), anti-biofouling agents with commercial and industrial applications, infection control strategies in aquaculture, and as biocontrol agents for plant pathogens is discussed.

The book is comprehensive and detailed in nature, covering all the important aspects and highly relevant topics in quorum sensing and quorum quenching in bacteria. Special focus is given on exploring quorum sensing inhibitors from microbes and flora inhabiting biodiversity rich regions including tropical rain forests, various tropical soils, and oceans. Graduate students, researchers and academicians from the field of Medical Microbiology, Pharmaceutical Biology, Genetics and Food Biotechnology will find the book an invaluable tool.

Editors

Prof. V. Ravishankar Rai received his MSc (1980) and PhD (1989) from the University of Mysore, India. Currently, he is working at the Department of Studies in Microbiology, University of Mysore, Mysore. His current research and publications in food microbiology, microbial quorum sensing, microbial influenced corrosion, and nanotechnology has been well received by the international scientific committee. His series of edited books with reputed publishers such as CRC Press and Wiley Publications—*Biotechnology: Concepts and Applications* (2009), *Microbial Food Safety and Preservation Techniques* (2014), *Beneficial Microbes in Fermented and Functional Foods* (2014), *Advances in Food Biotechnology* (2015), *Food Safety and Protection* (2016), and *Nanotechnology Applications in the Food Industry* (2018)—are comprehensive in nature and have contributions from international experts in the field. Prof. Rai has received awards from UNESCO Biotechnology Action Council Programme (Visiting Fellow, 1996), UGC Indo-Israel Culture Exchange Programme (1998), DBT Overseas Fellowship (2008), Indo-Hungarian Educational Exchange Programme fellowship (2011), INSA—bilateral exchange fellowship (2015), Incoming Fellowship (2017) from Cardiff University, UK, and an invitation from Mauritius Research Council, Mauritius (2018) to conduct collaborative research with renowned scientists from international universities. He has been awarded the Bilateral Exchange Fellowship by the Indian National Academy of Sciences to visit Germany (2020) and Cambridge-Hamied Visiting Lecture Scheme (2020) to visit University of Cambridge, UK.

Dr. Jamuna A Bai has completed her MSc and PhD in Microbiology from the University of Mysore, India. She is working as an Assistant Professor in JSS Academy of Higher Education and Research, Mysore. She has previously worked as a Researcher in UGC sponsored University with the Potential Excellence Project, University of Mysore, India and as ICMR Senior Research Fellow. She has carried research work on food safety, role of quorum sensing, and biofilms in food-related bacteria and developing quorum-sensing inhibitors. Her research interests also include antimicrobial application of functionalized nanomaterials and peptides against pathogenic bacteria.

Contributors

J. Fernando Ayala-Zavala
Centro de Investigación en Alimentación y Desarrollo, A.C. (CIAD, AC)
Carretera Gustavo Enrique Astiazarán
Hermosillo, Mexico

Jamuna A Bai
Department of Studies in Microbiology
University of Mysore
Mysore, India

Dimitra C. Banti
Department of Food Technology
Alexander Technological Educational Institute of Thessaloni
Thessaloníki, Greece

Jorge Barriuso
Centro de Investigaciones Biológicas
Consejo Superior de Investigaciones Científicas
Madrid, Spain

José E. Belizário
Department of Pharmacology
Institute of Biomedical Sciences of University of São Paulo
São Paulo, Brazil

A. Thalia Bernal-Mercado
Centro de Investigación en Alimentación y Desarrollo, A.C. (CIAD, AC)
Carretera Gustavo Enrique Astiazarán
Hermosillo, Mexico

Elisa V. Bertini
Planta Piloto de Procesos Industriales Microbiológicos
Tucumán, Argentina

Gustavo Bodelón
Department of Physical Chemistry and Biomedical Research Centre (CINBIO)
Universidade de Vigo
Vigo, Spain

Greicy Kelly Bonifacio Pereira
Faculdade de Medicina de Ribeirão Preto
Universidade de São Paulo
São Paulo, Brazil

Deisy Guimarães Carneiro
Department of Microbiology
Universidade Federal de Viçosa (UFV)
Viçosa, Brazil

Lucía I. Castellanos de Figueroa
Department of Microbiology
Faculty of Biochemistry, Chemistry and Pharmacy
National University of Tucumán
Tucumán, Argentina

Israel Castillo-Juárez
Department of Botany
Postgraduate College
Texcoco, Mexico

Adrián Cazares
Institute of Infection and Global Health
University of Liverpool
Liverpool, UK

Christiane Chbib
Department of Pharmaceutical Sciences
College of Pharmacy
Larkin University
Miami, Florida

Ana Ćirić
Institute for Biological Research "Siniša Stanković"- National Institute of Republic of Serbia
University of Belgrade
Belgrade, Serbia

Humberto Cortes-López
Department of Botany
Postgraduate College
Texcoco, Mexico

Maryam Dadashi
Department of Oral Biology
Rady Faculty of Health Sciences
University of Manitoba
Winnipeg, Manitoba, Canada

Felipe Alves de Almeida
Department of Nutrition
Universidade Federal de Juiz de Fora (UFJF)
Governador Valadares, Brazil

Iñigo de la Fuente
Centro de Investigaciones Biológicas
Consejo Superior de Investigaciones Científicas
Madrid, Spain

Sarah De Marchi-Lourenco
Departament of Physical Chemistry and
Biomedical Research Centre (CINBIO)
Universidade de Vigo
Vigo, Spain

Stephen K. Dolan
Department of Biochemistry
University of Cambridge
Cambridge, United Kingdom

Joel Faintuch
Laboratory of Genetics
Butantan Institute
São Paulo, Brazil

Zhaolu Feng
Graduate School at Shenzhen
Tsinghua University
Shenzhen, China

G.C.P. Fernando
Faculty of Agricultural Sciences
Deparment of Livestock Production
Sabaragamuwa University of Sri Lanka
Belihuloya, Sri Lanka

Rodolfo García-Contreras
Faculty of Medicine
Department of Microbiology and Parasitology
Universidad Nacional Autónoma de México
Mexico City, Mexico

Chenchen Gao
China-Norway Joint Lab on Fish Gut Microbiota
Feed Research Institute
Chinese Academy of Agricultural Sciences
Beijing, China

M. Melissa Gutierrez-Pacheco
Centro de Investigación en Alimentación y Desarrollo, A.C. (CIAD, AC)
Carretera Gustavo Enrique Astiazarán
Hermosillo, Mexico

Anton Hartmann
Faculty of Biology, Host-Microbe Interactions
Ludwig-Maximilians-Universität München
Munich, Germany

Imran Hashmi
Institute of Environmental Sciences and Engineering (IESE)
School of Civil and Environmental Engineering (SCEE)
National University of Sciences and Technology (NUST)
Islamabad, Pakistan

Juhász János
Faculty of Information Technology and Bionics
Pázmány Péter Catholic University
Budapest, Hungary

T.S.P. Jayaweera
Faculty of Agricultural Sciences
Deparment of Livestock Production
Sabaragamuwa University of Sri Lanka
Belihuloya, Sri Lanka

Sunny C. Jiang
Civil and Environmental Engineering
University of California Irvine
Irvine, California

Petrović Jovana
Institute for Biological Research “Siniša Stanković”- National Institute of Republic of Serbia
University of Belgrade
Belgrade, Serbia

Obaroakpo Joy Ujiroghene
Department of Food Science and Technology
Auchi Polytechnic
Auchi, Nigeria

Martha Juárez-Rodríguez
Department of Botany
Postgraduate College
Texcoco, Mexico

Ioannis D. Kampouris
Alexander Technological Educational Institute of Thessaloni
Department of Food Technology
Thessaloníki, Greece

Sher Jamal Khan
Institute of Environmental Sciences and Engineering (IESE)
School of Civil and Environmental Engineering (SCEE)
National University of Sciences and Technology (NUST)
Islamabad, Pakistan

Christina Kuttler
Zentrum Mathematik
Technical University of Munich
Garching, Germany

Mariano J. Lacosegliaz
Pilot Plant for Microbiological Industrial Processes (PROIMI-CONICET)
Tucumán, Argentina

Ana Carolina del V. Leguina
Pilot Plant for Microbiological Industrial Processes (PROIMI-CONICET)
Tucumán, Argentina

Tianle Li
Graduate School at Shenzhen
Tsinghua University
Shenzhen, China

Balázs Ligeti
Institute of Medical Microbiology
Semmelweis University
Budapest, Hungary

Emília Maria França Lima
Food Research Center
Department of Food and Experimental Nutrition
Universidade de São Paulo (USP)
São Paulo, Brazil

Luis M. Liz-Marzán
CIC biomaGUNE and CIBER-BBN
Donostia-San Sebastián, Spain

Jing Lu
Institute of Food Science and Technology
Chinese Academy of Agricultural Science
Beijing, China

Jesus M. Luna-Solorza
Centro de Investigación en Alimentación y Desarrollo, A.C. (CIAD, AC)
Carretera Gustavo Enrique Astiazarán
Hermosillo, Mexico

Jiaping Lv
Institute of Food Science and Technology
Chinese Academy of Agricultural Science
Beijing, China

Ivanov Marija
Institute for Biological Research "Siniša Stanković"- National Institute of Republic of Serbia
University of Belgrade
Belgrade, Serbia

Kostić Marina
Institute for Biological Research "Siniša Stanković"- National Institute of Republic of Serbia
University of Belgrade"
Belgrade, Serbia

Soković Marina
Institute for Biological Research "Siniša Stanković"- National Institute of Republic of Serbia
University of Belgrade
Belgrade, Serbia

Andrea Muras
Facultade de Bioloxía – CIBUS
Grupo de Acuicultura e Biotecnoloxía
Dpt. Microbioloxía e Parasitoloxía
Universidade de Santiago de Compostela
Santiago, Spain

Filomena Nazzaro
ISA CNR, Institute of Food Science
Avellino, Italy

Carlos G. Nieto-Peñalver
Faculty of Biochemistry, Chemistry and Pharmacy
Microbiology Department
National University of Tucumán
Tucumán, Argentina

Hyun-Suk Oh
Department of Environmental Engineering
Seoul National University of Science and Technology
Seoul, South Korea

Ana Otero
Facultade de Bioloxía – CIBUS
Grupo de Acuicultura e Biotecnoloxía
Dpt. Microbioloxía e Parasitoloxía
Universidade de Santiago de Compostela
Santiago, Spain

Soumya Palliyil
Scottish Biologics Facility
University of Aberdeen
Aberdeen, United Kingdom

Xiaoyang Pang
Institute of Food Science and Technology
Chinese Academy of Agricultural Science
Beijing, China

Shabila Parveen
Institute of Environmental Sciences and Engineering (IESE)
School of Civil and Environmental Engineering (SCEE)
National University of Sciences and Technology (NUST)
Islamabad, Pakistan

Isabel Pastoriza-Santos
Departament of Physical Chemistry
Biomedical Research Centre (CINBIO)
Universidade de Vigo
Vigo, Spain

Jorge Pérez-Juste
Departament of Physical Chemistry
Biomedical Research Centre (CINBIO)
Universidade de Vigo
Vigo, Spain

Judith Pérez-Velázquez
Departamento de Matemáticas y Mecánica
Instituto de Matemáticas Aplicadas y Sistemas
Universidad Nacional Autónoma de México
México City, México

Uelinton Manoel Pinto
Food Research Center
Department of Food and Experimental Nutrition
Universidade de São Paulo (USP)
São Paulo, Brazil

Alejandra Ponce
Facultad de Ingeniería
Grupo de Investigación en Ingeniería en Alimentos
Universidad Nacional de Mar del Plata y Consejo Nacional de Investigaciones Científicas y Técnicas (CONICET)
Mar del Plata, Argentina

Sándor Pongor
Faculty of Information Technology and Bionics
Pázmány Péter Catholic University
Budapest, Hungary

Chao Ran
Key Laboratory for Feed Biotechnology of the Ministry of Agriculture
Feed Research Institute
Chinese Academy of Agricultural Sciences
Beijing, China

J.L.P.C. Randika
Faculty of Agricultural Sciences
Deparment of Livestock Production
Sabaragamuwa University of Sri Lanka
Belihuloya, Sri Lanka

V. Ravishankar Rai
Department of Studies in Microbiology
University of Mysore
Mysore, India

Carolina Ripolles-Avila
Veterinary Faculty
Human Nutrition and Food Science
Universitat Autònoma de Barcelona
Cerdanyola del Vallès, Spain

José Juan Rodríguez-Jerez
Veterinary Faculty
Human Nutrition and Food Science
Universitat Autònoma de Barcelona
Cerdanyola del Vallès, Spain

Michael Rothballer
Helmholtz Zentrum München
German Research Center for Environmental Health
Department of Environmental Sciences
Institute of Network Biology
Neuherberg, Germany

H.A.D. Ruwandeepika
Faculty of Agricultural Sciences
Deparment of Livestock Production
Sabaragamuwa University of Sri Lanka
Belihuloya, Sri Lanka

Petros Samaras
Department of Food Technology
Alexander Technological Educational Institute of Thessaloni
Thessaloníki, Greece

Rafael Silva-Rocha
Faculdade de Medicina de Ribeirão Preto
Universidade de São Paulo
São Paulo, Brazil

Marcelo Sircili
Department of Gastroenterology
University of São Paulo School of Medical
São Paulo, Brazil

Marcos Soto-Hernández
Department of Botany
Postgraduate College
Texcoco, Mexico

Sujatha Subramoni
Singapore Centre for Environmental Life Sciences Engineering
Nanyang Technological University
Singapore

Marcos Sulca-Lopez
Department of Pharmacology
Institute of Biomedical Sciences of University of São Paulo
São Paulo, Brazil

Yuepeng Sun
Graduate School at Shenzhen
Tsinghua University
Shenzhen, China

Chuan Hao Tan
The School of Materials Science and Engineering
Nanyang Technological University
Singapore

Tsegay Teame
China-Norway Joint Lab on Fish Gut Microbiota
Feed Research Institute
Chinese Academy of Agricultural Sciences
Beijing, China

Stephen Trigg
Department of Biochemistry
University of Cambridge
Cambridge, United Kingdom

Leda K. Tse
Civil & Environmental Engineering
University of California Irvine
Irvine, California

Barbara Tomadoni
Grupo de Materiales Compuestos Termoplásticos (CoMP)
Instituto de Ciencia y Tecnología de Materiales (INTEMA)
Universidad Nacional de Mar del Plata y Consejo Nacional de Investigaciones Científicas y Técnicas (CONICET)
Buenos Aires, Argentina

Maria Cristina Dantas Vanetti
Department of Microbiology
Universidade Federal de Viçosa (UFV)
Viçosa, Brazil

Erika Lorena Giraldo Vargas
Department of Microbiology
Universidade Federal de Viçosa (UFV)
Viçosa, Brazil

Francisco J. Vazquez-Armenta
Centro de Investigación en Alimentación y Desarrollo, A.C. (CIAD, AC)
Carretera Gustavo Enrique Astiazarán
Hermosillo, Mexico

Martin Welch
Department of Biochemistry
University of Cambridge
Cambridge, United Kingdom

Cauã Antunes Westmann
Faculdade de Medicina de Ribeirão Preto
Universidade de São Paulo
São Paulo, Brazil

Guangxue Wu
Graduate School at Shenzhen
Tsinghua University
Shenzhen, China

Min Wu
Department of Biomedical Sciences
University of North Dakota
Grand Forks, North Dakota

Rui Xia
China-Norway Joint Lab on Fish Gut Microbiota
Feed Research Institute
Chinese Academy of Agricultural Sciences
Beijing, China

Lan Yang
Institute of Food Science and Technology
Chinese Academy of Agricultural Science
Beijing, China

Yalin Yang
Key Laboratory for Feed Biotechnology of the Ministry of Agriculture
Feed Research Institute
Chinese Academy of Agricultural Sciences
Beijing, China

Fengli Zhang
Key Laboratory for Feed Biotechnology of the Ministry of Agriculture
Feed Research Institute
Chinese Academy of Agricultural Sciences
Beijing, China

Hongling Zhang
China-Norway Joint Lab on Fish Gut Microbiota
Feed Research Institute
Chinese Academy of Agricultural Sciences
Beijing, China

Shuwen Zhang
Institute of Food Science and Technology
Chinese Academy of Agricultural Science
Beijing, China

Zhen Zhang
China-Norway Joint Lab on Fish Gut Microbiota
Feed Research Institute
Chinese Academy of Agricultural Sciences
Beijing, China

Chuanmin Zhou
Department of Biomedical Sciences
University of North Dakota
Grand Forks, North Dakota

Zhigang Zhou
China-Norway Joint Lab on Fish Gut Microbiota
Feed Research Institute
Chinese Academy of Agricultural Sciences
Beijing, China

1

Expanding Roles and Regulatory Networks of LadS/RetS in Pseudomonas aeruginosa

Chuanmin Zhou, Maryam Dadashi, and Min Wu

CONTENTS

1.1 The Two-Component Systems

Gram-negative opportunistic pathogen *Pseudomonas aeruginosa* is a severe host pathogen, found widely in nature, exposing in dynamic environmental conditions. Of note, the two-component system (TCS) is important for sensing those environmental challenges which in turn modulate a number of gene expressions (Stock, Robinson, and Goudreau 2000). Typically, TCS is coupled with a sensor histidine protein kinase and a response regulator protein. Histidine protein kinase is responsible for detecting extracellular signals, regulating the downstream effectors in response to the stimuli through phosphorylated response regulator protein (Stock, Robinson, and Goudreau 2000). To date, over 100 TCS genes have been found in *P. aeruginosa* (Rodrigue et al. 2000, Stover et al. 2000).

1.2 Discovery of RetS and LadS

Hybrid sensor kinase RetS (regulator of exopolysaccharide and Type III secretion) was first described in *P. aeruginosa* in 2004, which encoded 942 amino acids. This protein not only contains N-terminal cleaved signal sequences, a large periplasmic domain, and seven transmembrane domains (associated with environmental signal transduction), but also possesses TCS-like histidine kinase and response regulator domains in tandem, revealing that other TCS regulators may exist (Laskowski, Osborn, and Kazmierczak 2004, Goodman et al. 2004). RetS orthologs were also found in *Pseudomonas putida*, *Pseudomonas fluorescens*, *Pseudomonas syringae*, and *Azotobacter vinelandii* (Goodman et al. 2004). In 2005, another hybrid sensor kinase named LadS (lost adherence sensor) was noticed in *P. aeruginosa* PAO1 genome, which showed an opposite role of RetS, promoting biofilm formation and inhibiting T3SS activation (Ventre et al. 2006).

Domain analysis of LadS amino acids showed that LadS contained 795 amino acids with similar domains seen in RetS, including N-terminal cleaved signal sequences, a large periplasmic domain, seven transmembrane domains, as well as a histidine kinase and a response regulator domain (Ventre et al. 2006). These transmembrane domains in LadS and RetS were also observed in a number of other carbohydrate binding proteins (Ventre et al. 2006, Anantharaman and Aravind 2003). In particular, these domains exhibited 35% sequence identity, suggesting that this periplasmic sensor may respond to similar but not 100% identical environmental signals through its unique transmembrane domains (Ventre et al. 2006) (Figures 1.1 and 1.2).

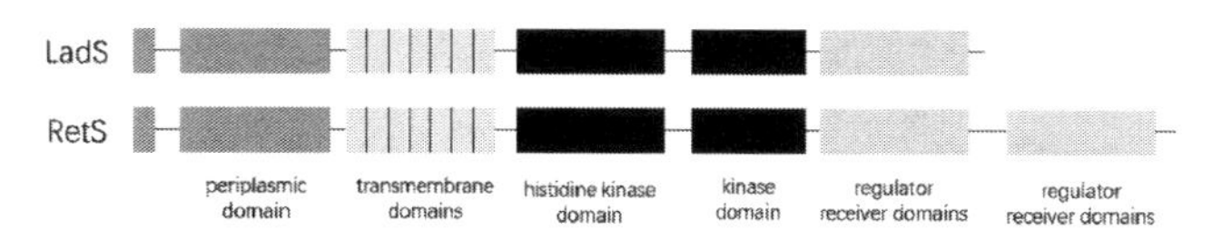

FIGURE 1.1 Domain organization of LadS and RetS hybrid sensors.

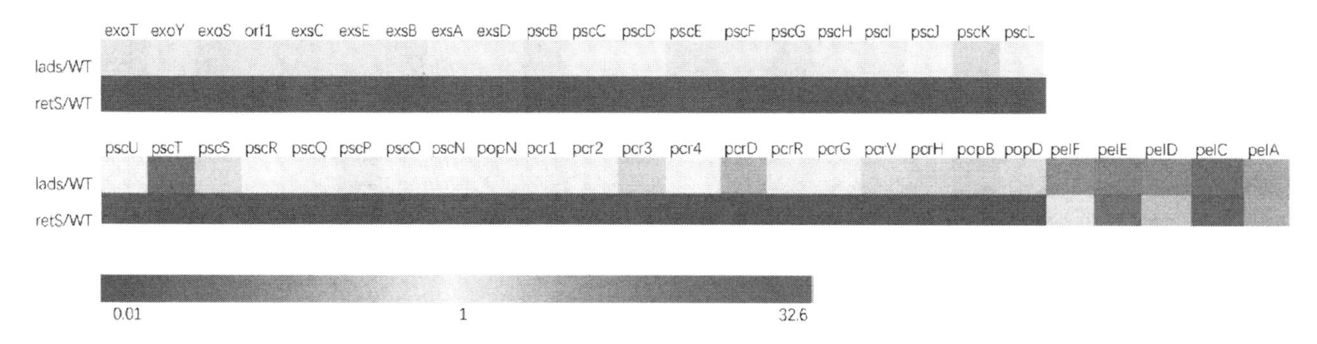

FIGURE 1.2 Pel and TTSS expression levels in Δ*ladS* and Δ*retS* and WT *P. aeruginosa*. (From Goodman, A. L. et al., *Dev. Cell.*, 7, 745–754, 2004; Ventre, I. et al., *Proc. Natl. Acad. Sci. USA*, 103, 171–176, 2006.)

1.3 Opposing Roles of RetS and LadS

RetS was characterized as a global pleiotropic regulatory protein, and the expression levels of almost 400 genes were significantly altered in a *retS* mutant strain (Goodman et al. 2004). RetS was necessary for the transcription of the T3SS operons under low calcium and host cell contact conditions. Deletion of *retS* significantly prohibited the activation of T3SS (Laskowski, Osborn, and Kazmierczak 2004). However, no DNA binding motifs were identified in RetS protein domains, indicating that it might modulate T3SS function indirectly (Laskowski, Osborn, and Kazmierczak 2004). The *retS* deletion strain exhibited robust biofilm formation by promoting the expression of biofilm related genes *pel* and *psl* (Goodman et al. 2004). A mutant with a deletion of *ladS* behaved in an opposite manner (Ventre et al. 2006). Transcriptome analysis of *ladS* mutant strain compared to WT strain showed that 79 genes were significantly affected, including that *pel* and *psl* gene expression were repressed and T3SS were activated in *ladS* mutant strain (Ventre et al. 2006). Compared to transcriptome analysis in *retS* mutant strain, 49% were oppositely regulated in *ladS* mutant strain, indicating that LadS and RetS signaling transduction pathways are antagonistic (Ventre et al. 2006).

RetS promotes the formation of heterodimers with GacS, reducing the GacS autophosphorylation. Gene screening of transposon insertions showed that RetS may modulate GacS/GacA (Goodman et al. 2004), whereas LadS showed an opposite activity by promoting the activation of GacS/GacA pathway (Ventre et al. 2006). Interestingly, LadS did not interact with GacS and hence may upregulate the GacS/GacA pathway through a phosphor-relay mechanism, resulting in phosphotransfer to the HPT domain of GacS, which in turn promoted chronic infection (Chambonnier et al. 2016). Furthermore, another histidine kinase PA1611 showed a similar role of GacS by interacting with RetS to modulate the GacS/GacA pathway (Kong et al. 2013). The phenotype of *retS* mutant strain was completely blocked in *gacS* mutant strain, indicating that RetS works through the GacS/GacA pathway (Ventre et al. 2006). Phosphorylated GacA in turn directly modulated the expression of small noncoding regulatory RNA (sRNA) *rsmY/rsmZ*. RsmA was a global post-transcriptional regulator, influencing expression of over 500 genes by binding to targeted mRNAs, which was inhibited by sRNA *rsmY/rsmZ*. Different pathogens including *P. aeruginosa*, *Escherichia coli*, *Legionella pneumophila*, *Vibrio cholera*, and *Salmonella typhimurium*, are found expressing *rsmA*. Finally, deletion of *rsmA* also showed similar phenotypes to *retS* deletion mutant strain, exhibiting activation of biofilm and T3SS, which in turn contributes to acute infection (Coggan and Wolfgang 2012).

Although the *retS* deletion mutant exhibited increased attachment to host cells, it showed less cytotoxicity in eukaryotes and less virulence in a pneumonia mouse model, indicating that the *retS* mutant strain was unable to respond to environmental signals (Laskowski, Osborn, and Kazmierczak 2004, Goodman et al. 2004). Deletion of *ladS* showed hyper cytotoxicity compared to the WT strain, and the phenotype of *ladS/retS* double mutant strain was similar to *retS* mutant, showing no cytotoxicity, indicating that LadS may function at the upstream of RetS in response to input signals (Ventre et al. 2006) (Figure 1.3).

The stimuli triggering RetS and LadS activity remain largely uncharacterized. Recent research revealed that deletion of *ladS* causes *P. aeruginosa* calcium-blind through genetic, biochemical, and proteomic study (Broder, Jaeger, and Jenal 2016). LadS detected calcium, while did not respond to other divalent cations (Mg^{2+}, Fe^{2+}, Zn^{2+}, Mn^{2+} or Cu^{2+}), through its DISMED2 domain, promoting chronic infection. The presence of an additional helix inhibited the binding of carbohydrate, which in turn promoted the binding of calcium. In addition, deletion of *gacA*, *gacS*, and *rsmA* also showed calcium unresponsive, whereas *rsmY* or *rsmZ* single deletion remained calcium sensitive. These studies suggest that calcium signaling plays a key role in host-*P. aeruginosa* interaction by facilitating acute-to-chronic infection transformation.

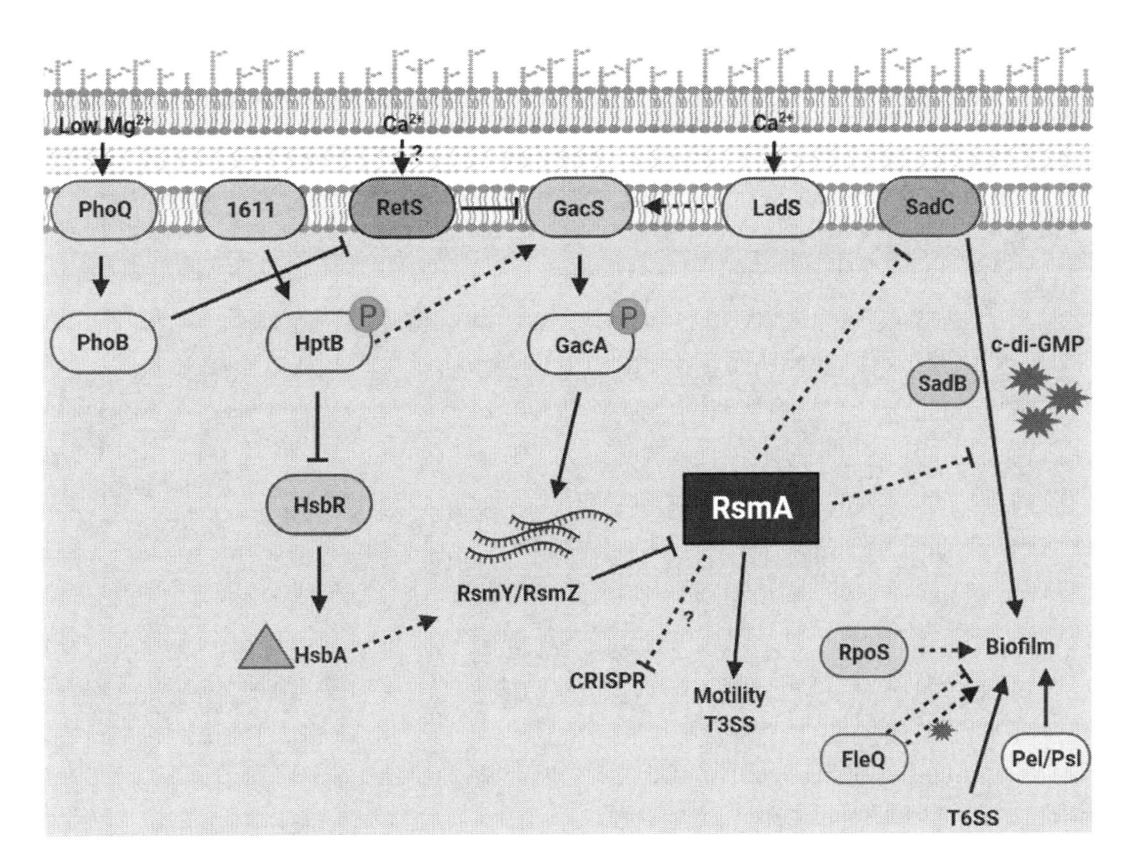

FIGURE 1.3 Schematic representation of RsmA signaling in *P. aeruginosa*.

1.4 Function of LadS/RetS Is Dependent on Small RNAs

Researchers have studied the mechanism of QS systems for decades. To date, LasI/LasR, RhlI/RhlR QS systems are found important for encoding AHLs, and more than 10% of *P. aeruginosa* genes are affected by AHLs. It is noticed that production of *las* and *rhl* dependent AHLs are positively modulated by two sRNAs, *rsmY* and *rsmZ*, and negatively by RsmA. Most QS-dependent genes are repressed by post-transcriptional RsmA effector by bind to its mRNA motif. Considering that LadS/RetS are in the upstream of RsmA, LadS and RetS also showed important roles in influencing QS. Importantly, QS showed important roles associated with acute and chronic infections, indicating that QS is also necessary for acute-to-chronic infection transformation. In addition, RetS contributes in transcriptional repression of sigma factor RpoS, Pel, Psl, and FleQ. FleQ is a repressor of Pel in the absence of c-di-GMP. When c-di-GMP is available, FleQ activates the Pel operon. It was shown that the two-component system PhoQ/B directly interact with RetS. TCS PhoQ/B is a Mg^{2+} sensing system. When exogenous Mg^{2+} content is low, the PhoQ/B represses RetS and promotes biofilm formation.

1.5 Other Hybrid Sensor Kinases in Gac/Rsm Pathway

When planktonic cells hit a proper surface, they form biofilms stepwise. Several factors participate in establishment of *Pseudomonas* biofilms. Below we looked into some role players of biofilm formation. c-di-GMP is a secondary messenger, and its abundance in the cell decides the transition between motility and sessility of the bacterial cells. Biofilms of *P. aeruginosa* have 75–110 pmol mg^{-1} c-di-GMP in total cell extracts compared to planktonic cells, which bear merely 30 pmol mg^{-1} (Basu Roy and Sauer 2014). High levels of c-di-GMP are the hallmark of a biofilm forming lifestyle, which is modulated by diguanylate cyclases like SadC and phosphodiesterases (Merritt et al. 2007).

Recent study indicated that another hybrid sensor kinase, PA1611, modulated genes of acute and chronic infection, which played an important role in downregulation of T3SS and upregulation of biofilm formation (Kong et al. 2013). PA1611 showed similar function to LadS. However, PA1611 did not show a Lads dependent manner. In addition, PA1611 associated with RetS which was similar to GacS showing phosphorelay independent, causing PA1611 shared similar protein domains with GacS

(Kong et al. 2013). PA1611 is capable of influencing the Gac/Rsm pathway by promoting the phosphorylation of HptB (Kong et al. 2013). Phosphorylated HptB phosphorylates HsbR, which further phosphorylates HsbA. HsbA is an anti-anti-sigma factor. HsbA indirectly and positively modulates the expression of RsmY (Bordi et al. 2010), resulting in modulation of RsmA function. Further, two other hybrid sensor kinases, PA1976 and PA2824, are also involved in phosphorylating HptB like PA1611 (Lin et al. 2006, Hsu et al. 2008). Additional research showed that deletion of *hptB* and *retS* led to similar phenotypes. However, HptB signaling only controlled the expression of *rsmY*, which is dependent on the σ^{28} dependent genes (Bordi et al. 2010).

1.6 Other sRNAs as Regulators

1.6.1 P27 sRNA

Polynucleotide phosphorylase (PNPase) is an RNA processing enzyme. It modulates several virulence factors by destabilizing RsmY and RsmZ (Chen et al. 2019). PNPase mutant cells have an increased level of rsmY/Z (Chen et al. 2016). Also, rhamnolipid production is defective in this mutant, leading to lower biofilm formation. Rhamnolipids are regulated by RhlI-RhlR QS system, and PNPase modulates the translation of RhlI by sRNA P27. P27 sRNA directly binds to 5′-untranslated region (UTR) of RhlI mRNA by recruiting Hfq and represses RhlI translation. Mutations in P27 or 5′UTR of RhlI result in unpairing of these two RNAs and RhlI expression restoration (Chen et al. 2019). These results indicate that P27 sRNA may play critical roles in RhlI QS through Hfq-mediated signaling.

1.6.2 PhrS sRNA

In addition to Las and Rhl QS molecules, *Pseudomonas* quinolone signal (PQS), a member of 4-hydroxy-2-alkylquinolines compounds, links the Las and Rhl systems. PQS controls its own expression by inducing *pqsABCDE* operon when it binds to PqsR responsive regulator (Sonnleitner et al. 2011). *pqsABCDE* encodes for intermediate molecules called HHQ, which converts to PQS by LasR-dependent PqsH. PqsR also regulates PqsE and aids RhlR to respond to (C4-HSL) molecules (Farrow et al. 2008).

ANR is an oxygen-responsive regulator. When the oxygen content of the cell is low, ANR induces oxygen limiting-dependent sRNA PhrS. PhrS is shown to activate the PqsR and found to be the first sRNA, linking the QS system with oxygen availability (Sonnleitner et al. 2011).

1.6.3 RsmV and RsmW sRNAs

In *Pseudomonas*, sRNAs play a pivotal role in cellular response to signal molecules (Jakobsen et al. 2017). When RsmY and RsmZ are abundant in the cell, RsmA is sequestered and the production of AHL-based molecules is increased. The sRNA RsmV, which shares sequence similarity to RsmY and RsmZ, is capable of sequestering RsmA and is under control of RhlR. A predicated RhlR binding site is found upstream of RsmV (Janssen et al. 2018).

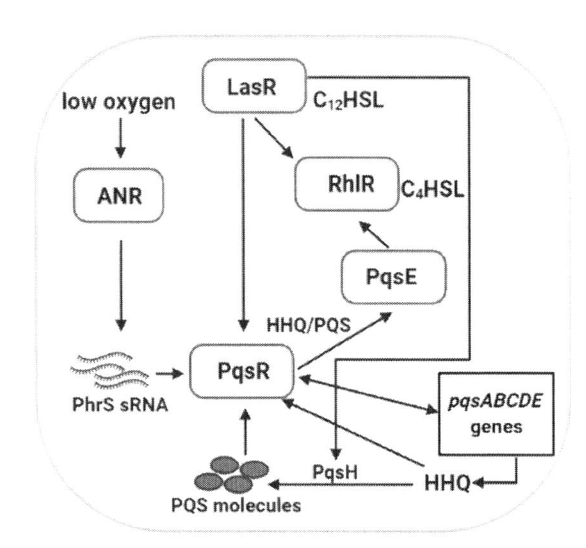

FIGURE 1.4 Schematic representation of sRNA and QS axis.

RsmW indeed is an additional sRNA whose expression is increased during stationary phase in minimal media resembling a biofilm forming environment (Janssen et al. 2018). RsmW sequesters RsmA in a lower efficacy than RsmY and RsmZ. RsmV and RsmW are not under positive regulation of GacA/S system. The transcription regulation of RsmV and RsmW has remained to be elucidated (Janssen et al. 2018) (Figure 1.4).

1.7 Is RetS a Calcium Sensitive Histidine Kinase?

It should be mentioned that RetS also contains a DISMED2 domain, but the RetS input signals are still elusive except that the activation signal was noticed in lyzed kin cells (LeRoux et al. 2015). It remains elusive whether calcium binds to the DISMED2 domain of RetS to activate its kinase activity. Although transcription of *retS* was not modulated in response to calcium, RetS might be activated under low calcium conditions (Laskowski, Osborn, and Kazmierczak 2004). Broder et al., reported that deletion of *retS* led to a calcium blind (Broder, Jaeger, and Jenal 2016). Also, *P. aeruginosa* challenged with lower to higher calcium concentrations promoted the transformation of acute infection to chronic status, indicating that low calcium might activate RetS dependent acute infection. Also, deletion of *ladS* showed lower chronic infection levels compared to the WT counterpart in response to calcium treatment. Hence, this raises the possibility that RetS also acts as a calcium responsive kinase and interacts with LadS. The histidine kinase RetS/LadS are responsible for lower or higher calcium concentration stimulation. Nevertheless, it is unknown whether LadS/RetS play a role in detecting other unknown exogenous signals, which is worth further investigating.

1.8 Can RetS and LadS Interact with CRISPR-Cas Systems?

Bacteria possess multiple defense systems against the invading bacteriophages (Mohanraju et al. 2016, Koonin, Makarova, and Wolf 2017, Forsberg and Malik 2018). Among these, clustered regularly interspaced short palindromic repeats (CRISPR), first described in 1987 (Ishino et al. 1987), was found as a heritable immunity system in 2007 (Barrangou et al. 2007). To date, 2 classes and 6 types CRISPR-CRISPR-associated (Cas) systems, based on the characteristic of Cas proteins, have been identified in various bacteria and archaea (Makarova et al. 2015, Koonin, Makarova, and Zhang 2017). Class 1 CRISPR-Cas systems rely on multiple CRISPR-Cas protein effector complexes, while Class 2 CRISPR-Cas systems are dependent on a single CRISPR-Cas effector protein. CRISPR and their Cas proteins function as prokaryotic adaptive immunity by targeting acquired mobile genetic elements (MGEs) against invasion of bacteriophages or plasmids (Marraffini 2015, Makarova et al. 2015, Mohanraju et al. 2016). Approximately 45% bacteria and 84% archaea were found containing CRISPR-Cas systems (Grissa, Vergnaud, and Pourcel 2007). Generally, CRISPR-Cas adaptive immunity processes function through three sequential phases. First, acquiring short DNA sequences occurs in CRISPR arrays upon the bacteriophage invasion (Levy et al. 2015, Yosef, Goren, and Qimron 2012, Makarova et al. 2015). Then, the integrated CRISPR arrays containing the recently acquired foreign genetic substance are transcribed and processed into crRNAs (Deltcheva et al. 2011, Haurwitz et al. 2010). Finally, Cas proteins and crRNAs are assembled together to degrade the invading complementary nucleic acids (Brouns et al. 2008).

Most CRISPR-Cas regulators were found to target *cas* promoters (Patterson, Yevstigneyeva, and Fineran 2017). Quorum sensing (QS) was found to modulate different CRISPR-Cas systems (I-E, I-F and III-A) in *Serratia* sp. ATCC39006 and *P. aeruginosa* by influencing the expression of *cas* promoters (Patterson et al. 2016, Hoyland-Kroghsbo et al. 2017). Additionally, cAMP receptor protein (CRP) and histone-like nucleoid structuring proteins (H-NS) play pleiotropic roles in regulating CRISPR-Cas systems (Agari et al. 2010, Shinkai et al. 2007, Patterson et al. 2015, Pul et al. 2010, Westra et al. 2010, Medina-Aparicio et al. 2011).

We showed that CRISPR-Cas targets endogenous RNA to regulate the master QS molecule, LasR and impacting the host response in a TLR4-mediated manner (Li et al. 2016). Because the regulation is reciprocal, we recently identified some QS regulating signals, such as CdpR, can also modulate the activity of CRISPR-Cas (Lin et al. 2019). This line of research is recently broadened to small RNAs that regulated CRISPR-Cas (Lin et al., manuscript under revision). Due to the necessity for precise regulation of CRISPR-Cas activity to respond to foreign invasion but avoiding potential autoimmunity or toxicity, the CRISPR-Cas is tightly controlled by a complex network in the prokaryotes (Figure 1.5).

The *Pseudomonas* species has a well-characterized type I CRISPR-Cas system, which lead us to speculate that LadS/RetS may interact with the adaptive immune system to coordinate virulence. Thus far, it remains unclear whether calcium and LadS/RetS are capable of modulating CRISPR-Cas systems. RsmA, the downstream regulator of LadS/RetS, is a global post-transcriptional regulator through targeting 5' untranslated regions (UTR) or early coding regions of targeting mRNA motif (ANGGA). CRISPR mRNA blast in *P. aeruginosa* showed that a conserved RsmA binding site exists in the *cas1* mRNA 5' UTR, which may be worthwhile to study in the future.

Despite the function of CRISPR-Cas systems in adaptive immunity and its biotechnological application being well studied

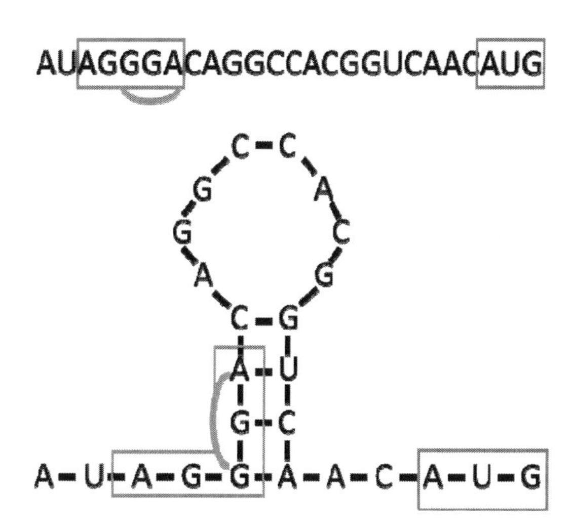

FIGURE 1.5 Predicted RsmA binding sites using RNA fold for *cas* promoter. GGA motif; long box, start codon; short box, predicted RBS (proteins primarily bind to the sequence motif A(N)GGA in single-stranded mRNA regions).

(Barrangou et al. 2007, Barrangou and Doudna 2016, Hsu, Lander, and Zhang 2014), limited research has discussed the regulation of CRISPR-Cas systems in bacteria (Patterson, Yevstigneyeva, and Fineran 2017). It is still unknown whether RsmA takes part in endogenous regulation of CRISPR. Furthermore, it remains to be discovered whether CRISPR function is influenced by environmental signals and calcium through a LadS/RetS/Gac/Rsm cascade. We speculate that the LadS/RetS/Gac/Rsm axis may also interact with or regulate CRISPR-Cas immunity.

1.9 Drug Targets

Disruption of small molecules (i.e., QS molecules) is considered as an antimicrobial strategy (Jakobsen et al. 2017). QS inhibitors are found in many herbal extracts such as Ajoene presenting in garlic. It is shown that, ajoene could block the production of rhamnolipid, which is regulated by QS. Ajoene helped polymorphonuclear neutrophils (PMNs) phagocytize biofilms more efficiently. Moreover, the biofilms were more susceptible to antibiotics, like tobramycin, showed milder pulmonary infection in mice treated with ajoene. Ajoene exerts its effect through modulation of RsmY and RsmZ (Jakobsen et al. 2017).

Cells in biofilm state are 1,000 times more resistant to antimicrobial therapy. By manipulating the c-di-GMP content of the cell and switching the bacteria to planktonic lifestyle, the susceptibility of the bacterial cells to antibiotics will increase (Valentini and Filloux 2016). Therefore, either direct regulating QS signaling or modulating biofilm may affect the clinical control of bacterial infection.

1.10 Concluding Remarks

Although it is known that LadS/RetS are essential for the regulation of *P. aeruginosa* infection status, its roles in sensing other exogenous signals and interacting with other endogenous signal pathways are still elusive. We are just a beginning to unravel the detailed mechanisms of LadS/RetS. Several questions remain to be discovered: Is RetS responsible for detecting exogenous environmental signals? What is the structure mechanism of LadS/RetS in detecting calcium and signal transduction? Are there any other new co-factors that interact with LadS/RetS signal transduction? Are LadS/RetS capable of being served as novel drug targets for infection modulation? Are LadS/RetS taking part in modulation of CRSIPR system? If so, how do LadS/RetS and CRSIPR systems interact with each other? Are there other new exogenous signals involved in modulating CRSIPR system? Discovery of homologous calcium sensors in other bacteria as well as answering and dissecting those possible interconnections would enhance our understanding of the functioning mechanisms of this unique immunity system and bacterial virulence, meanwhile providing better tools to enable accurate gene-editing or transcription regulation.

REFERENCES

Agari, Y., K. Sakamoto, M. Tamakoshi, T. Oshima, S. Kuramitsu, and A. Shinkai. 2010. "Transcription profile of *Thermus thermophilus* CRISPR systems after phage infection." *J Mol Biol* no. 395 (2):270–281. doi:10.1016/j.jmb.2009.10.057.

Anantharaman, V., and L. Aravind. 2003. "Application of comparative genomics in the identification and analysis of novel families of membrane-associated receptors in bacteria." *BMC Genomics* no. 4 (1):34. doi:10.1186/1471-2164-4-34.

Barrangou, R., and J. A. Doudna. 2016. "Applications of CRISPR technologies in research and beyond." *Nat Biotechnol* no. 34 (9):933–941. doi:10.1038/nbt.3659.

Barrangou, R., C. Fremaux, H. Deveau, M. Richards, P. Boyaval, S. Moineau, D. A. Romero, and P. Horvath. 2007. "CRISPR provides acquired resistance against viruses in prokaryotes." *Science* no. 315 (5819):1709–1712. doi:10.1126/science.1138140.

Basu Roy, A., and K. Sauer. 2014. "Diguanylate cyclase NicD-based signalling mechanism of nutrient-induced dispersion by *Pseudomonas* aeruginosa." *Mol Microbiol* no. 94 (4):771–793. doi:10.1111/mmi.12802.

Bordi, C., M. C. Lamy, I. Ventre, E. Termine, A. Hachani, S. Fillet, B. Roche, S. Bleves, V. Mejean, A. Lazdunski, and A. Filloux. 2010. "Regulatory RNAs and the HptB/RetS signalling pathways fine-tune *Pseudomonas aeruginosa* pathogenesis." *Mol Microbiol* no. 76 (6):1427–1443. doi:10.1111/j.1365-2958.2010.07146.x.

Broder, U. N., T. Jaeger, and U. Jenal. 2016. "LadS is a calcium-responsive kinase that induces acute-to-chronic virulence switch in *Pseudomonas aeruginosa*." *Nat Microbiol* no. 2:16184. doi:10.1038/nmicrobiol.2016.184.

Brouns, S. J., M. M. Jore, M. Lundgren, E. R. Westra, R. J. Slijkhuis, A. P. Snijders, M. J. Dickman, K. S. Makarova, E. V. Koonin, and J. van der Oost. 2008. "Small CRISPR RNAs guide antiviral defense in prokaryotes." *Science* no. 321 (5891):960–964. doi:10.1126/science.1159689.

Chambonnier, G., L. Roux, D. Redelberger, F. Fadel, A. Filloux, M. Sivaneson, S. de Bentzmann, and C. Bordi. 2016. "The hybrid histidine kinase LadS forms a multicomponent signal transduction system with the GacS/GacA two-component system in *Pseudomonas* aeruginosa." *PLoS Genet* no. 12 (5):e1006032. doi:10.1371/journal.pgen.1006032.

Chen, R., X. Wei, Z. Li, Y. Weng, Y. Xia, W. Ren, X. Wang, Y. Jin, F. Bai, Z. Cheng, S. Jin, and W. Wu. 2019. "Identification of a small RNA that directly controls the translation of the quorum sensing signal synthase gene rhlI in *Pseudomonas aeruginosa*." *Environ Microbiol*. doi:10.1111/1462-2920.14686.

Chen, R., Y. Weng, F. Zhu, Y. Jin, C. Liu, X. Pan, B. Xia, Z. Cheng, S. Jin, and W. Wu. 2016. "Polynucleotide phosphorylase regulates multiple *Virulence* factors and the stabilities of small RNAs RsmY/Z in *Pseudomonas aeruginosa*." *Front Microbiol* no. 7:247. doi:10.3389/fmicb.2016.00247.

Coggan, K. A., and M. C. Wolfgang. 2012. "Global regulatory pathways and cross-talk control *pseudomonas aeruginosa* environmental lifestyle and virulence phenotype." *Curr Issues Mol Biol* no. 14 (2):47–70.

Deltcheva, E., K. Chylinski, C. M. Sharma, K. Gonzales, Y. Chao, Z. A. Pirzada, M. R. Eckert, J. Vogel, and E. Charpentier. 2011. "CRISPR RNA maturation by trans-encoded small RNA and host factor RNase III." *Nature* no. 471 (7340):602–607. doi:10.1038/nature09886.

Farrow, J. M., 3rd, Z. M. Sund, M. L. Ellison, D. S. Wade, J. P. Coleman, and E. C. Pesci. 2008. "PqsE functions independently of PqsR-*Pseudomonas* quinolone signal and enhances the rhl quorum-sensing system." *J Bacteriol* no. 190 (21):7043–7051. doi:10.1128/JB.00753-08.

Forsberg, K. J., and H. S. Malik. 2018. "Microbial genomics: The expanding universe of bacterial defense systems." *Curr Biol* no. 28 (8):R361–R364. doi:10.1016/j.cub.2018.02.053.

Goodman, A. L., B. Kulasekara, A. Rietsch, D. Boyd, R. S. Smith, and S. Lory. 2004. "A signaling network reciprocally regulates genes associated with acute infection and chronic persistence in *Pseudomonas aeruginosa*." *Dev Cell* no. 7 (5):745–754. doi:10.1016/j.devcel.2004.08.020.

Grissa, I., G. Vergnaud, and C. Pourcel. 2007. "CRISPRFinder: A web tool to identify clustered regularly interspaced short palindromic repeats." *Nucleic Acids Res* no. 35 (Web Server issue):W52–W57. doi:10.1093/nar/gkm360.

Haurwitz, R. E., M. Jinek, B. Wiedenheft, K. Zhou, and J. A. Doudna. 2010. "Sequence- and structure-specific RNA processing by a CRISPR endonuclease." *Science* no. 329 (5997):1355–1358. doi:10.1126/science.1192272.

Hoyland-Kroghsbo, N. M., J. Paczkowski, S. Mukherjee, J. Broniewski, E. Westra, J. Bondy-Denomy, and B. L. Bassler. 2017. "Quorum sensing controls the *Pseudomonas aeruginosa* CRISPR-Cas adaptive immune system." *Proc Natl Acad Sci U S A* no. 114 (1):131–135. doi:10.1073/pnas.1617415113.

Hsu, J. L., H. C. Chen, H. L. Peng, and H. Y. Chang. 2008. "Characterization of the histidine-containing phosphotransfer protein B-mediated multistep phosphorelay system in *Pseudomonas aeruginosa* PAO1." *J Biol Chem* no. 283 (15):9933–9944. doi:10.1074/jbc.M708836200.

Hsu, P. D., E. S. Lander, and F. Zhang. 2014. "Development and applications of CRISPR-Cas9 for genome engineering." *Cell* no. 157 (6):1262–1278. doi:10.1016/j.cell.2014.05.010.

Ishino, Y., H. Shinagawa, K. Makino, M. Amemura, and A. Nakata. 1987. "Nucleotide sequence of the IAP gene, responsible for alkaline phosphatase isozyme conversion in *Escherichia coli*, and identification of the gene product." *J Bacteriol* no. 169 (12):5429–5433.

Jakobsen, T. H., A. N. Warming, R. M. Vejborg, J. A. Moscoso, M. Stegger, F. Lorenzen, M. Rybtke, J. B. Andersen, R. Petersen, P. S. Andersen, T. E. Nielsen, et al. 2017. "A broad range quorum sensing inhibitor working through sRNA inhibition." *Sci Rep* no. 7 (1):9857. doi:10.1038/s41598-017-09886-8.

Janssen, K. H., M. R. Diaz, C. J. Gode, M. C. Wolfgang, and T. L. Yahr. 2018. "RsmV, a Small Noncoding Regulatory RNA in *Pseudomonas aeruginosa* That Sequesters RsmA and RsmF from Target mRNAs." *J Bacteriol* no. 200 (16): e00277–e00218. doi:10.1128/JB.00277-18.

Kong, W., L. Chen, J. Zhao, T. Shen, M. G. Surette, L. Shen, and K. Duan. 2013. "Hybrid sensor kinase PA1611 in *Pseudomonas aeruginosa* regulates transitions between acute and chronic infection through direct interaction with RetS." *Mol Microbiol* no. 88 (4):784–797. doi:10.1111/mmi.12223.

Koonin, E. V., K. S. Makarova, and F. Zhang. 2017. "Diversity, classification and evolution of CRISPR-Cas systems." *Curr Opin Microbiol* no. 37:67–78. doi:10.1016/j.mib.2017.05.008.

Koonin, E. V., K. S. Makarova, and Y. I. Wolf. 2017. "Evolutionary genomics of defense systems in *Archaea* and Bacteria." *Annu Rev Microbiol* no. 71:233–261. doi:10.1146/annurev-micro-090816-093830.

Laskowski, M. A., E. Osborn, and B. I. Kazmierczak. 2004. "A novel sensor kinase-response regulator hybrid regulates type III secretion and is required for virulence in *Pseudomonas aeruginosa*." *Mol Microbiol* no. 54 (4):1090–1103. doi:10.1111/j.1365-2958.2004.04331.x.

LeRoux, M., R. L. Kirkpatrick, E. I. Montauti, B. Q. Tran, S. B. Peterson, B. N. Harding, J. C. Whitney, A. B. Russell, B. Traxler, Y. A. Goo, D. R. Goodlett et al. 2015. "Kin cell lysis is a danger signal that activates antibacterial pathways of *Pseudomonas aeruginosa*." *Elife* no. 4. doi:10.7554/eLife.05701.

Levy, A., M. G. Goren, I. Yosef, O. Auster, M. Manor, G. Amitai, R. Edgar, U. Qimron, and R. Sorek. 2015. "CRISPR adaptation biases explain preference for acquisition of foreign DNA." *Nature* no. 520 (7548):505–510. doi:10.1038/nature14302.

Li, R., L. Fang, S. Tan, M. Yu, X. Li, S. He, Y. Wei, G. Li, J. Jiang, and M. Wu. 2016. "Type I CRISPR-Cas targets endogenous genes and regulates virulence to evade mammalian host immunity." *Cell Res* no. 26 (12):1273–1287. doi:10.1038/cr.2016.135.

Lin, C. T., Y. J. Huang, P. H. Chu, J. L. Hsu, C. H. Huang, and H. L. Peng. 2006. "Identification of an HptB-mediated multi-step phosphorelay in *Pseudomonas aeruginosa* PAO1." *Res Microbiol* no. 157 (2):169–175. doi:10.1016/j.resmic.2005.06.012.

Lin, P., Q. Pu, G. Shen, R. Li, K. Guo, C. Zhou, H. Liang, J. Jiang, and M. Wu. 2019. "CdpR inhibits CRISPR-Cas adaptive immunity to lower anti-viral defense while avoiding self-reactivity." *iScience* no. 13:55–68. doi:10.1016/j.isci.2019.02.005.

Makarova, K. S., Y. I. Wolf, O. S. Alkhnbashi, F. Costa, S. A. Shah, S. J. Saunders, R. Barrangou, S. J. Brouns, E. Charpentier, D. H. Haft, P. Horvath, et al. 2015. "An updated evolutionary classification of CRISPR-Cas systems." *Nat Rev Microbiol* no. 13 (11):722–736. doi:10.1038/nrmicro3569.

Marraffini, L. A. 2015. "CRISPR-Cas immunity in prokaryotes." *Nature* no. 526 (7571):55–61. doi:10.1038/nature15386.

Medina-Aparicio, L., J. E. Rebollar-Flores, A. L. Gallego-Hernandez, A. Vazquez, L. Olvera, R. M. Gutierrez-Rios, E. Calva, and I. Hernandez-Lucas. 2011. "The CRISPR/Cas immune system is an operon regulated by LeuO, H-NS, and leucine-responsive regulatory protein in *Salmonella enterica* serovar Typhi." *J Bacteriol* no. 193 (10):2396–2407. doi:10.1128/JB.01480-10.

Merritt, J. H., K. M. Brothers, S. L. Kuchma, and G. A. O'Toole. 2007. "SadC reciprocally influences biofilm formation and swarming motility via modulation of exopolysaccharide production and flagellar function." *J Bacteriol* no. 189 (22):8154–8164. doi:10.1128/JB.00585-07.

Mohanraju, P., K. S. Makarova, B. Zetsche, F. Zhang, E. V. Koonin, and J. van der Oost. 2016. "Diverse evolutionary roots and mechanistic variations of the CRISPR-Cas systems." *Science* no. 353 (6299):aad5147. doi:10.1126/science.aad5147.

Patterson, A. G., J. T. Chang, C. Taylor, and P. C. Fineran. 2015. "Regulation of the Type I-F CRISPR-Cas system by CRP-cAMP and GalM controls spacer acquisition and interference." *Nucleic Acids Res* no. 43 (12):6038–6048. doi:10.1093/nar/gkv517.

Patterson, A. G., S. A. Jackson, C. Taylor, G. B. Evans, G. P. C. Salmond, R. Przybilski, R. H. J. Staals, and P. C. Fineran. 2016. "Quorum sensing controls adaptive immunity through the regulation of multiple CRISPR-Cas systems." *Mol Cell* no. 64 (6):1102–1108. doi:10.1016/j.molcel.2016.11.012.

Patterson, A. G., M. S. Yevstigneyeva, and P. C. Fineran. 2017. "Regulation of CRISPR-Cas adaptive immune systems." *Curr Opin Microbiol* no. 37:1–7. doi:10.1016/j.mib.2017.02.004.

Pul, U., R. Wurm, Z. Arslan, R. Geissen, N. Hofmann, and R. Wagner. 2010. "Identification and characterization of *E. coli* CRISPR-cas promoters and their silencing by H-NS." *Mol Microbiol* no. 75 (6):1495–1512. doi:10.1111/j.1365-2958.2010.07073.x.

Rodrigue, A., Y. Quentin, A. Lazdunski, V. Mejean, and M. Foglino. 2000. "Two-component systems in *Pseudomonas aeruginosa*: Why so many?" *Trends Microbiol* no. 8 (11):498–504.

Shinkai, A., S. Kira, N. Nakagawa, A. Kashihara, S. Kuramitsu, and S. Yokoyama. 2007. "Transcription activation mediated by a cyclic AMP receptor protein from *Thermus thermophilus* HB8." *J Bacteriol* no. 189 (10):3891–901. doi:10.1128/JB.01739-06.

Sonnleitner, E., N. Gonzalez, T. Sorger-Domenigg, S. Heeb, A. S. Richter, R. Backofen, P. Williams, A. Huttenhofer, D. Haas, and U. Blasi. 2011. "The small RNA PhrS stimulates synthesis of the *Pseudomonas aeruginosa* quinolone signal." *Mol Microbiol* no. 80 (4):868–885. doi:10.1111/j.1365-2958.2011.07620.x.

Stock, A. M., V. L. Robinson, and P. N. Goudreau. 2000. "Two-component signal transduction." *Annu Rev Biochem* no. 69:183–215. doi:10.1146/annurev.biochem.69.1.183.

Stover, C. K., X. Q. Pham, A. L. Erwin, S. D. Mizoguchi, P. Warrener, M. J. Hickey, F. S. Brinkman, W. O. Hufnagle, D. J. Kowalik, M. Lagrou, R. L. Garber, et al. 2000. "Complete genome sequence of *Pseudomonas aeruginosa* PAO1, an opportunistic pathogen." *Nature* no. 406 (6799):959–964. doi:10.1038/35023079.

Valentini, M., and A. Filloux. 2016. "Biofilms and Cyclic di-GMP (c-di-GMP) Signaling: Lessons from *Pseudomonas aeruginosa* and Other Bacteria." *J Biol Chem* no. 291 (24):12547–12555. doi:10.1074/jbc.R115.711507.

Ventre, I., A. L. Goodman, I. Vallet-Gely, P. Vasseur, C. Soscia, S. Molin, S. Bleves, A. Lazdunski, S. Lory, and A. Filloux. 2006. "Multiple sensors control reciprocal expression of *Pseudomonas aeruginosa* regulatory RNA and virulence genes." *Proc Natl Acad Sci U S A* no. 103 (1):171–176. doi:10.1073/pnas.0507407103.

Westra, E. R., U. Pul, N. Heidrich, M. M. Jore, M. Lundgren, T. Stratmann, R. Wurm, A. Raine, M. Mescher, L. Van Heereveld, M. Mastop et al. 2010. "H-NS-mediated repression of CRISPR-based immunity in *Escherichia coli* K12 can be relieved by the transcription activator LeuO." *Mol Microbiol* no. 77 (6):1380–1393. doi:10.1111/j.1365-2958.2010.07315.x.

Yosef, I., M. G. Goren, and U. Qimron. 2012. "Proteins and DNA elements essential for the CRISPR adaptation process in *Escherichia coli*." *Nucleic Acids Res* no. 40 (12):5569–5576. doi:10.1093/nar/gks216.

2

Autoinducer-1 Quorum Sensing Communication Mechanism in Gram-Negative Bacteria

Maria Cristina Dantas Vanetti, Deisy Guimarães Carneiro, Felipe Alves de Almeida, Erika Lorena Giraldo Vargas, Emília Maria França Lima, and Uelinton Manoel Pinto

CONTENTS

2.1 Introduction

Bacteria have extraordinary ability to survive and grow in basically every niche in the environment. Microbial life strategies should consider stress tolerance such as high or low pH, temperature, osmolarity, nutrient availability, antimicrobials, and population density that lead to competitive relationships. In order to respond to changes in their immediate environment, bacterial cells must be able to alter the cellular pathways to survival or resume growth. The response and adaptation to diverse environmental conditions are related to reception and processing signals present outside their borders. This adaptation process is mainly mediated by a striking combination of transcriptional regulatory networks, which allow bacteria to sense and convert physical or chemical extracellular stimuli into a specific response that results in altered gene expression and enzyme activities. Whereas some of these alterations are reversible and disappear when the stress is over, others are maintained and can even be passed on to surviving bacteria.

The bacterial response could be related to individual cells, but in a community, bacteria are able to interact and regulate, in a coordinated way, their response to environmental changes through the sophisticated mechanism of cellular communication called quorum sensing (QS) (Bassler and Losick 2006; Fuqua, Winans, and Greenberg 1994). This signaling process allows communication between cells leading to differential gene expression in response to changes in population density and allows bacteria to act as a group.

The ability of bacteria to communicate and to present social interactions like a multi-cellular organism has provided significant benefits to bacterial populations in host colonization, formation of biofilm, defense against competitors, and adaptation to changing environments (Li and Tian 2012). Bacteria are not limited to communication within their own species but are capable of "listening in" and "broadcasting to" unrelated species to intercept messages and coerce cohabitants into behavioral modifications, either for the good of the population or for the benefit of one species over another (Atkinson and Williams 2009). The perception that bacteria are social organisms has produced new insights into bacterial physiology and gene regulation from the point of view of population and evolutionary biology (Goo et al. 2015).

The mechanism of QS is mediated by diffusible signaling molecules called autoinducers (AIs) synthesized throughout the growth of the bacteria and released into the surrounding medium. At low population densities, the production and secretion of QS signal molecules proceed at a basal level. As population density increases, the signal molecules accumulate above the threshold in the external environment and bind to and activate receptors inside bacterial cells, and collectively induce the expression of specific target genes to activate behaviors that are beneficial under the particular condition encountered. Therefore, one of the common and often observed consequences of the QS is gene regulation in the increased synthesis of the proteins involved in the signaling molecule production. The higher the concentration of signaling molecules, the greater the signaling protein synthesis, which leads to positive feedback loop. This is the reason for the term autoinducers; the signaling molecule initiates the synthesis of the protein responsible for its own production (Geethanjali et al. 2019).

Over the years, since its introduction as a cell density dependent mechanism, the use of the term quorum sensing has evolved to become the general description of the signaling production process and response at the level of gene expression. However, many biotic and abiotic environmental factors can influence the chemical gradients of these signaling molecules. These factors include the spatial distribution of signaling molecule-producing cells, the rate at which the signaling molecule is produced and diffused, and the stability of the signaling molecule. This has led to the proposition of new terms, including diffusion sensing, confinement-induced QS, and efficiency sensing, to describe these genetic and biochemical processes (Platt and Fuqua 2010). These new names emphasize different adaptive functions of regulation by the QS mechanism, related to a specific subset of factors that influence the concentration of signal molecules in the environment. However, the use of a different term for each adaptive function may complicate the understanding of the QS process rather than clarify it. Thus, it is important to remember that the ecological context of QS regulation, as the process itself, is complex and influenced by multiple aspects of natural environments (Platt and Fuqua 2010).

The QS signaling molecules, also known as AIs, are chemically diverse, and many bacteria synthesize and utilize multiple signaling molecules from the same or different classes that constitute a regulatory hierarchy. Most signaling molecules are small organic molecules (<1000 Da) or small peptides with five to 20 amino acids in length (Williams 2007). Multiple QS signals have been identified in bacteria, and the most common ones are *N*-acyl-homoserine lactones (AHLs, AI-1) in Gram-negative bacteria, oligopeptides (AIPs) in Gram-positive bacteria, and furanosyl borate diester (AI-2) in both Gram-negative and Gram-positive bacteria. Other signals, such as auto-inducer-3 (AI-3), 2-heptyl-3-hydroxy-4(1*H*)-quinolone (*Pseudomonas* quinolone signal [PQS]) and its precursor 2-heptyl-4(1*H*)-hydroxyquinoline (HHQ), *cis*-11-methyl-2-dodecenoic acid (diffusible signal factor [DSF]) (LaSarre and Federle 2013), 3-hydroxypalmitic acid methyl ester (3-OH PAME), indole (Lee et al. 2015), diketopiperazines (DKP), and others have been detected in a limited number of bacteria or suggested as a signal molecule for bacterial communication. A number of other extracellular bacterial metabolites, including compounds with antibiotic activity, have the potential to function as signal molecules. However, it is important to differentiate between a true signal molecule involved in cell-to-cell communication and other metabolites.

AHLs are the most studied QS signaling molecules, also known as AI-1, and are produced by Gram-negative bacteria. The AHLs are neutral lipid molecules normally produced by proteins homologous to LuxI from the lactone fraction of S-adenosyl methionine (SAM) and, in most cases, the acyl chain is obtained from intermediates of the fatty acid biosynthesis pathway (Papenfort and Bassler 2016). The length of the acyl chain can range from four to 18 carbons and contains possibly a 3-oxo or 3-hydroxy function (Churchill and Chen 2011; Galloway et al. 2011; Yajima 2014). Short side-chain AHLs are directly released out of the cell upon synthesis while long side-chain AHLs are actively secreted to the environment (Liu et al. 2018). This diversity in the AHLs is recognized by different and compatible LuxR proteins promoting specificity to intraspecies-specific cell-cell communication in bacteria (Husain et al. 2019).

Although all known QS mechanisms differ in the regulatory components and molecular mechanisms, they are dependent on three basic principles: first, secretion of signaling molecules (AIs); second, detection of AIs by the receptors existing in the cytoplasm or in the membrane; and third, activation of gene expression necessary for cooperative behaviors (Figure 2.1).

Several bacterial phenotypes have been described as being controlled by the QS, among them bioluminescence, competence, biofilm, metabolism, cell differentiation, sporulation, surface motility, toxin production, expression of virulence genes, and others (Table 2.1). Therefore, the mechanism of communication by QS plays a critical role for survival and colonization both in symbiotic and pathogenic host-bacterial interactions. The use of QS to regulate processes associated with virulence increases the pathogen's prospects for survival, because the coordinated attack against the host will only be done when the bacterial population reaches high population density, increasing the probability of successfully overcoming host defenses.

While it is advantageous that the regulation of some phenotypes is quorum-dependent, the communication incurs a cost in terms of signal production; therefore, the communication has only been maintained throughout the evolution because this transfer of information gives benefits to both parties, signaling and receiving bacteria (Diggle et al. 2007; Keller and Surette 2006). The QS regulators LuxI and LuxR arose early in the evolution of the Proteobacteria and subsequently diverged within each group of organisms (Gray and Garey 2001). The construction of phylogenetic trees indicates that duplication and horizontal gene transfer have played an important role in the distribution of the system across bacterial species (Lerat and Moran 2004). The inducer/receptor elements in the LuxI/R systems evolved together and maintained their paired functional relationship, but loss and exchange of elements occurred in several γ-Proteobacteria lineages (Lerat and Moran 2004). In a systematic survey for LuxR QS-domains in sequenced bacterial genomes included in the InterPro database, it was identified that 40%–70% have a complete QS system depending on taxa, while the remaining species have only LuxR solos or orphans (Subramoni, Florez Salcedo, and Suarez-Moreno 2015). It is believed that in some bacteria belonging to the Enterobacteriaceae family, such as *Escherichia coli* and *Salmonella*, a deletion event has removed the *luxI* homolog after their divergence from *Pantoea* and *Erwinia* genera, leaving only the LuxR homolog known as SdiA (Sabag-Daigle and Ahmer 2012).

The mechanism of QS was first described in the regulation of bioluminescence in *Vibrio fischeri* (Nealson, Platt, and Hastings 1970), now *Aliivibrio fischeri*, a marine bacterium found in a Hawaiian squid known for its striking bioluminescence. The luciferase operon in *A. fischeri* is regulated by two proteins, LuxI, responsible for the production of the AHL, and LuxR protein, which is activated by this auto-inducer to increase the transcription of the luciferase operon (Engebrecht and Silverman 1984).

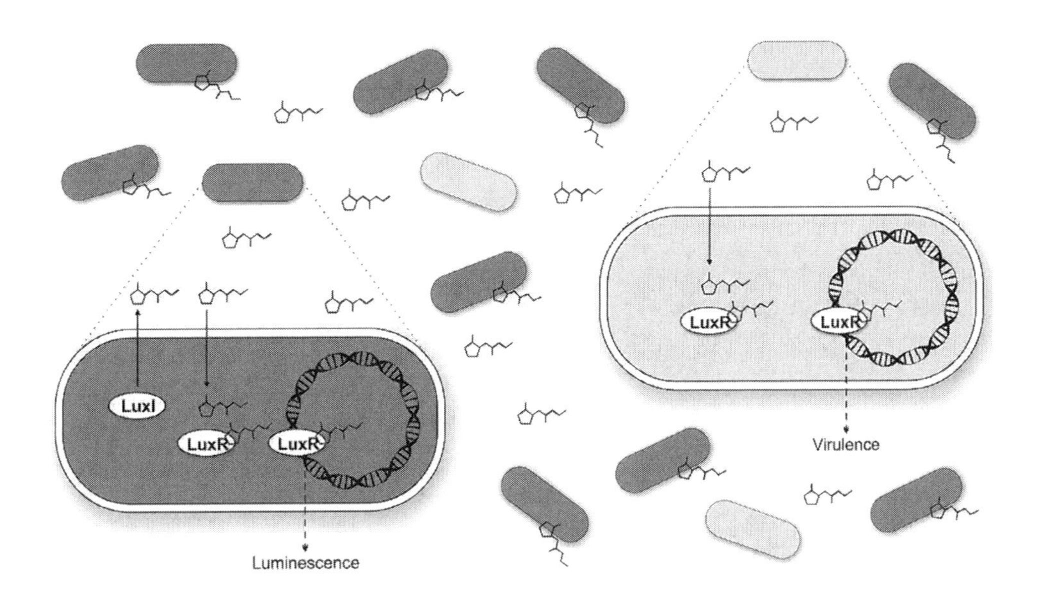

FIGURE 2.1 LuxIR signaling circuit. The LuxIR QS circuit, represented in dark gray cells, is composed of an AHL synthase LuxI that synthesizes the AIs which are exported to the exterior of the cell. When these AIs reach a threshold concentration, they are internalized, bind to the response regulatory protein LuxR and bind to the DNA regulating the expression of target genes. Some bacteria (light gray) lack the LuxI protein and do not synthesize AHL. However, they possess the LuxR homologue and are able to recognize and respond to molecules produced by other bacteria.

TABLE 2.1

Phenotypes Regulated by Auto-Inducer-1 Quorum Sensing Mechanisms

Bacteria	Phenotype	Regulation[a]	References
Bacteria with complete auto-inducer-1 quorum sensing mechanisms			
A. fischeri MAV	Bioluminescence	+	Nealson et al. (1970)
A. fischeri B-61	Bioluminescence	+	Eberhard et al. (1981)
A. fischeri MJ-1	Bioluminescence	+	Engebrecht and Silverman (1984)
A. fischeri ES114	Bioluminescence	+	Lupp et al. (2003)
	Persistence in squid	+	
	Motility	−	Lupp and Ruby (2005)
A. salmonicida LFI1238	Biofilm formation in polystyrene	−	Hansen et al. (2015)
V. cholerae O1	Resuscitation of cell in the viable but non-cultivable state	+	Bari et al. (2013)
V. harveyi BB120	Bioluminescence	+	Henke and Bassler (2004)
	Type III secretion system	+	
	Metalloprotease production	+	
P. aeruginosa PAO1	Elastase production	+	Jones et al. (1993)
	Elastase production	+	Brint and Ohman (1995)
	Pyocyanin production	+	
	Protease production	+	
	Rhamnolipid production	+	
	Rhamnolipid production	+	Ochsner and Reiser (1995)
	Lung infection	+	Wu et al. (2001)
P. fluorescens B52	Biofilm formation in glass	+	Allison et al. (1998)
P. fluorescens NCIMB 10586	Mupirocin biosynthesis	+	El-sayed, Hothersall, and Thomas (2001)
P. fluorescens 395	Protease production	+	Liu, Wang, and Griffiths (2007)

(*Continued*)

TABLE 2.1 (*Continued*)

Phenotypes Regulated by Auto-Inducer-1 Quorum Sensing Mechanisms

Bacteria	Phenotype	Regulation[a]	References
P. fluorescens 07A	Growth	Ø	Pinto et al. (2010)
	Proteolytic activity	Ø	
P. putida IsoF	Biofilm formation	+/Ø	Steidle et al. (2002)
P. syringae B728a	Epiphytic fitness	+	Quiñones, Pujol, and Lindow (2004)
B. cepacia H111	Swarming motility	+	Huber et al. (2001)
	Biofilm formation	+	
B. glumae BGR1	Production of excreted oxalate	+	Goo et al. (2012)
B. thailandensis E264	Production of excreted oxalate	+	Goo et al. (2012)
A. hydrophila SSU	Type VI secretion system	+	Khajanchi et al. (2009)
	Metalloprotease production	+	
	Biofilm formation	+	
A. hydrophila ATCC 7966	Proteolytic activity on casein	+	Ponce-Rossi et al. (2016)
	Proteolytic activity on gelatin	+	
	Amylolytic activity	+	
	Lipolytic activity	+	
	β-hemolytic activity	+	
	Biofilm formation on stainless steel	+	
C. violaceum ATCC 31532	Violacein production	+/–	McClean et al. (1997), Chen et al. (2011)
	Chitinase production	+	Chernin et al. (1998)
	Swarming motility	+	Oca-Mejía et al. (2014)
	Cell aggregation	+	
	Biofilm formation in glass	+	
	Oxidative stress resistance	+	
	Exoprotease production	+	
C. violaceum ATCC 12472	Violacein production	+/–	Morohoshi et al. (2008), Chen et al. (2011)
	Chitinase production	+	Liu et al. (2013)
	Biofilm formation in glass	+	
S. liquefaciens MG1	Surfactant production	+	Lindum et al. (1998)
	Growth	Ø	Givskov et al. (1998)
	Cell elongation	Ø	
	Cell flagellation	Ø	
	Swarming motility	+	
S. plymuthica RVH1	Nuclease production	+	Van Houdt, Givskov, and Michiels (2007)
	Chitinase production	+	
	Protease production	+	
	Butanediol fermentation	+	
S. proteamaculans B5a	Lipolytic activity	+	Christensen et al. (2003)
	Proteolytic activity	+	
	Chitinolytic activity	+	
Serratia sp. ATCC 39006	Carbapenem antibiotic production	+	Thomsonet al. (2000)
	Prodigiosin pigment production	+	
A. tumefaciens	Ti plasmid copy number	+	Cho, Pinto, and Winans (2009)
	Ti plasmid conjugation	+	
	Tumorigenesis in plants	+	

(*Continued*)

TABLE 2.1 (*Continued*)

Phenotypes Regulated by Auto-Inducer-1 Quorum Sensing Mechanisms

Bacteria	Phenotype	Regulation[a]	References
P. carotovorum SCRI193	Elastase production	+	Jones et al. (1993)
P. carotovorum EC153	PCWDE production	+	Chatterjee et al. (2005)
	Virulence	+	
Bacteria with incomplete auto-inducer-1 quorum sensing mechanisms			
E. coli	Cell division	+	Ahmer et al. (1998)
E. coli MG1655	Growth	Ø	Van Houdt et al. (2006)
	Acid resistance	+	
E. coli K-12	Biofilm formation in polystyrene	–	Lee et al. (2007)
	Acid resistance	+	
	Quinolones resistance	+/Ø	Dyszel et al. (2010)
	Acid resistance	+/Ø	
	Cell division	+/Ø	
EHEC	Calve fitness	+	Dziva et al. (2004), Hughes et al. (2010)
	Quinolones resistance	+/Ø	Dyszel et al. (2010)
	Acid resistance	+/Ø	
	Cell division	+/Ø	
	Adherence to HEp-2 cells *in vitro*	–	Sharma and Bearson (2013)
aEPEC ONT:H25	Biofilm formation	–	Culler et al. (2018)
	Motility	–	Culler et al. (2018)
Salmonella Typhimurium ATCC 14028	Cell division	Ø	Ahmer et al. (1998)
	Invasion to HEp-2 cells *in vitro*	+	Nesse et al. (2011)
Salmonella Enteritidis PT4 578	Growth	Ø	Campos-Galvão et al. (2016)
	Biofilm formation in polystyrene	+	
	Growth	Ø	Almeida et al. (2017b)
	Initial adhesion in polystyrene	Ø	
	Biofilm formation in polystyrene	+	
	Swarming motility	Ø	
	Twitching motility	Ø	
	Level of thiol	+	Almeida et al. (2018)
Salmonella Typhi ST_8	Adherence to HeLa cells *in vitro*	+	Liu et al. (2014)
	Biofilm formation in polystyrene	+	
	Survival in rabbit serum *in vitro*	+	
	Survival in guinea pig serum *in vitro*	+	

[a] Positive = +, Negative = –, no effect = Ø.
PCWDE = plant cell wall-degrading enzymes.

Different homologues of the LuxI-LuxR mechanism have been identified in other bacteria and, in all these mechanisms, its components have been observed to be performing the same functions such that the AHLs are the AIs that bind to the LuxR-type proteins that regulate different phenotypic characteristics. Although this mechanism of communication was considered exclusive to some marine vibrios for many years, the presence of homologues of a complete QS (*luxI*/*luxR*) mechanism has also been demonstrated in many Gram-negative bacteria capable of producing specific AHLs, including *Agrobacterium*, *Aeromonas*, *Acinetobacter*, *Brucella*, *Burkholderia*, *Chromobacterium*, *Citrobacter*, *Enterobacter*, *Erwinia*, *Hafnia*, *Nitrosomonas*,

Obesumbacterium, *Pantoea*, *Pseudomonas*, *Rahnella*, *Ralstonia*, *Rhodobacter*, *Rhizobium*, *Serratia*, *Vibrio*, and *Yersinia* (Smith and Iglewski 2003).

2.2 Bacteria with Complete Auto-Inducer-1 Quorum Sensing Mechanism

2.2.1 QS in *Aliivibrio* and *Vibrio*

Aliivibrio and *Vibrio* are moderately halophilic bacteria, inhabitants of marine and estuarine environments, and several species are related to infections in humans and in animals reared in aquaculture (Urbanczyk et al. 2007). The cell signaling system dependent of AI-1 in marine *Aliivibrio* and *Vibrio* comprises a LuxIR-type, although they received different denominations, such as LuxMN in *V. harveyi* (Cao and Meighen 1989), LuxIR in *A. fischeri* (Liu et al. 2018), and VanIR in *Vibrio anguillarum* (Milton et al. 2001). Different AHLs have been noted among the various strains of the same species. Based on current literature, a total of 32 AHLs-producing marine *Aliivibrio* and *Vibrio* species have already been identified and 23 different AHLs were definitely classified, including 10 short side-chain and 13 long side-chain AHLs (Liu et al. 2018). Marine species of *Aliivibrio* and *Vibrio* produce many types of long side-chain AHLs, such as C14-HSL (Girard et al. 2017) and, different from those found in terrestrial bacteria, AHLs such as C7-HSL, 3-OH-C9-HSL, 3-oxo-C9-HSL, 3-OH-C11-HSL, and 3-oxo-C11-HSL are also detected (Rasmussen et al. 2014).

The mechanism of cell communication based on the LuxIR system of *A. fischeri* is the paradigm of Gram-negative QS systems; however, it is not found in all vibrios. In luminescent *V. harveyi*, QS positively regulates phenotypes such as bioluminescence (Freeman and Bassler 1999), metalloprotease, siderophore, and exopolysaccharide production (Henke and Bassler 2004). In *V. anguillarum*, AHLs regulate biofilm formation, metalloprotease, and siderophore production (Milton 2006). AHLs also participate in the regulation of marine *Vibrio* pathogenicity via other virulence-related proteins. For example, ToxR, a classic *Vibrio* virulence factor encoded by the virulence-related gene *toxR*, is directly regulated by AHLs. ToxR was first discovered in *V. cholerae*, and subsequent studies showed that homologous genes of *toxR* also exist in many pathogenic *Vibrio* species such as *V. parahaemolyticus*, *V. vulnificus*, and *V. alginolyticus* (Liu et al. 2018).

In 2004, the regulation in QS system by sRNAsQrr (quorum regulatory RNAs) was identified in *Vibrio* and was the first demonstration of a role for sRNAs in QS in Gram-negative bacteria (Lenz et al., 2004). The sRNAsQrr are classified as *trans*-sRNAs and act to activate or repress translation of target mRNAs via unique base-pairing to the 5′ UTR in conjunction with the RNA chaperone Hfq (Shao and Bassler, 2012).

In low population density, LuxO is phosphorylated and together with the σ^{54} activates the expression of sRNAsQrr (Lilley and Bassler 2000; Waters and Bassler 2006). Five QrrsRNAs were identified and characterized in *V. harveyi* and, with Hfq, they strongly repress translation of the master QS regulator LuxR by occupying the ribosome binding site or mediating degradation of the *luxR* mRNA (Bejerano-Sagie and Xavier 2007; Feng et al. 2015; Tu and Bassler 2007). Without the bound LuxR protein, the *lux* operon is not expressed, resulting in inhibition of luminescence (Miyamoto et al. 1996). The regulation by sRNA is important because it provides a fine-tuning to the bioluminescence mechanism due to the highly dynamic nature of sRNAs (Rutherford et al. 2015). The regulation of AHL-QS by sRNAs has also been described in *Sinorhizobium meliloti* and *P. aeruginosa* (Gao et al. 2015; Malgaonkar and Nair 2019).

2.2.2 QS in *Pseudomonas*

Members of the family Pseudomonaceae are ubiquitous Gram-negative bacteria, comprising many genera and several hundred species. Most species have great metabolic and physiological versatility, which explains their presence in several environmental niches, including soil and fresh water, and have the ability to undergo transitions to become important and dangerous pathogens. Some species cause disease in plants, like *Pseudomonas syringae*, and a few cause serious diseases in humans, as in the case of *Pseudomonas aeruginosa* (Venturi 2006).

Several *Pseudomonas* produce AIs such as AHL by QS system, which control important functions including pathogenicity, biofilm formation, and production of a variety of extracellular metabolites and enzymes. The most extensive studies on QS have been performed on *P. aeruginosa*, making this bacterium a well-known study model. Additionally, many AHL QS systems, as well the involved genes, have been reported in other species such as *P. aureofaciens*, *P. chlororaphis*, *P. putida*, *P. fluorescens*, and *P. syringae*, demonstrating the comprehensiveness of QS in the genus (Chen et al. 2019; Venturi 2006; Martins et al. 2014; Pang et al. 2019; Barbarossa et al. 2010).

Pseudomonas aeruginosa is a nonspore-forming bacterium that presents motility through a single polar flagellum, and it has a bacillar shape. It is an opportunist pathogen that causes severe infections and diseases in both plants and animals, and is a problematic human pathogen since it causes serious infections mainly in hospitalized and immunocompromised individuals, such as those with cancer or AIDS (Azam and Khan 2019; Schütz and Empting 2018; Lee and Zhang 2014). Some infections caused by *P. aeruginosa* are hospital-acquired infections (HAIs), hospital-acquired pneumonia (HAP), and gastrointestinal, bone, and skin infections, besides representing a major cause of morbidity and mortality in burn patients and those with cystic fibrosis (CF) (Azam and Khan 2019; Schütz and Empting 2018; Pang et al. 2019). The bacterial genome is relatively large, which provides metabolic versatility and high adaptability to environmental changes (Pang et al. 2019). This may explain the variety of virulence mechanisms employed during *P. aeruginosa* infections, like the ability to form a biofilm matrix, motility, iron scavenging, and cytotoxicity capabilities, and many of these virulence factors are regulated by QS (Lee and Zhang 2014; Schütz and Empting 2018). In addition, the infections may be aggravated by the intrinsic and acquired antibiotic resistance of this pathogen, including formation of multidrug-tolerant persister cells in biofilm settings which generate chronic diseases that cannot be eradicated with antibiotic treatment (D'Angelo et al. 2018; Pang et al. 2019). *P. aeruginosa* became the main model in anti-virulence strategy studies for multiple reasons: first, the

TABLE 2.2

QS Mechanisms in *P. aeruginosa*

Autoinducer (AI)	AI Abbreviation	AI Synthesis	AI Receptor Protein	References
N-(3-oxododecanoyl)-homoserine lactone	OdDHL (3-oxo C12-HSL)	LasI[a]	LasR[b]	Pearson et al. (1994)
N-butyrylhomoserine lactone	BHL (C4-HSL)	RhlI[1]	RhlR[b]	Ochsner and Reiser (1995), Pearson et al. (1995)
2-heptyl-3-hydroxy-4-quinolone/2-heptyl-4-hydroxyquinoline	PQS/HHQ	PqsABCDE	MvfR (PqsR)[c]	Pesci et al. (1999), D'Angelo et al. (2018)
2-(2-hydroxyphenyl)-thiazole-4-carbaldehyde	IQS	AmbBCDE	IqsR	Lee et al. (2013)

[a] LuxI homologue.

[b] LuxR homologue, based in LuxI-LuxR QS mechanism of *A. fisheri*. Source: Lee and Zhang (2014).

[c] MvfR "Multiple virulence factor Regulator" and PqsR are the same protein that historically have been named differently by several research groups.

QS network is well characterized; second, QS regulates the expression of multiple virulence-related factors; and third, this bacterium is important to the medical community due to antibiotic resistance (LaSarre and Federle 2013).

In *P. aeruginosa*, a complex QS network consisting of four interconnected systems, i.e., *las*, *rhl*, *pqs*, and *iqs* is found (Table 2.2). These systems collectively control group behaviors and the expression of virulence determinants, such as proteolytic and lipolytic enzymes, swarming motility, toxin production, tolerance to stress, and biofilm formation (D'Angelo et al. 2018; Turan et al. 2017; Choudhary and Schmidt-Dannert 2010; Azam and Khan 2019; Quecan et al. 2019).

Two different QS LuxI-LuxR mechanisms exist in *P. aeruginosa*: LasI-LasR and RhlI-RhR (Table 2.2). LasI is a synthase that produces an extracellularly diffusible AHL signal molecule called *N*-(3-oxododecanoyl)-L-homoserine lactone (3-oxo-C12-HSL), which is recognized by the transcriptional regulator LasR that directs the expression of various genes. Likewise, RhlI produces the *N*-butyryl-L-homoserine lactone (C4-HSL) signaling molecule that can bind to its RhlR cognate transcription regulator. The transcriptional regulators LasR and RhlR are activated when sufficient levels of 3-oxo-C12-HSL and C4-HSL are present as a result of the high population density, and regulate the production of multiple virulence factors (Smith and Iglewski 2003).

The QS cascade in *P. aeruginosa* is organized in a hierarchical way, starting with the *las* QS system. More recently, a circular model, in which multiple feedback loops control QS gene expression, has been proposed by the Rahme's group (Maura et al. 2016). Activation of the *las*QS system (LasR-OdDHL) stimulates the transcription of the other QS systems, *rhl*, *pqs*, and *iqs* (Lee and Zhang 2014). The *rhl* system is under control of *las* and *pqs*, and several virulence factors regulated by QS are predominantly activated by the RhlR-BHL complex, demonstrating its importance for the bacterium. On the other hand, RhlR represses PQS signal production by interfering with *mvfR* (*pqsR*) and *pqsABCDE* expression. Thus, a reduced activity of *pqs* QS system may be due to a negative effect of *rhl* QS system (D'Angelo et al. 2018). MvfR autoinducers, PQS, and its precursor HHQ can bind and activate the transcription of the MvfR (PqsR) regulon (Schütz and Empting 2018). However, PQS has additional functions, such as iron chelation, and it is 100 times more potent than HHQ. The *iqs*QS system has been recently identified (Lee et al. 2013) as strictly dependent on LasI-LasR under rich media conditions, and disruption of LasI or LasR completely abolishes *ambBCDE* expression and the IQS production (Lee and Zhang 2014; D'Angelo et al. 2018; Schütz and Empting 2018).

The necessity to develop new strategies to combat infections caused by *P. aeruginosa* is urgent. This concern was highlighted in a recent World Health Organization report in which this pathogen was classified into the most critical group (priority 1) for which new antibiotics are urgently needed, due to antibiotic resistance (D'Angelo et al. 2018; WHO 2017). Thus, the understanding of the QS inhibition mechanism by gene expression may be the key to develop this new treatment generation.

Differently from *P. aeruginosa*, *P. fluorescens* is not generally known as a pathogen in humans. It is especially considered for its role in soil and the rhizosphere as well as in food spoilage, especially refrigerated raw products. However, it does possess functional traits that provide it with the capability to grow in mammalian hosts (Scales et al. 2014).

P. fluorescens is reported to have a significant ability to form biofilms and is one of the most important spoilage bacteria of refrigerated foods. Both traits can be regulated by QS (Zhang et al. 2019). In addition, biofilm formation is a serious problem for the food industry. The structure formation can be accelerated by the bacterium's ability to use swarming motility to colonize nutrient-rich environments, facilitating colony spreading. There is little information about *P. fluorescens* biofilm formation on mammalian surfaces, but, whether in humans or plant cells, this structure is very important for successful long-term colonization (Scales et al. 2014; Martins et al. 2014).

Two QS systems have been described for some strains of this species. First, a LuxI-LuxR homologue pair was discovered in *P. fluorescens* NCIMB 10586 and was termed *mupI-mupR* system due to its regulation of mupirocin (pseudomonic acid) biosynthesis, an important and potent polyketide antibiotic (El-sayed, Hothersall, and Thomas 2001; Scales et al. 2014). Second, the *hdtS* system, a new class of AHL synthase, was discovered in *P. fluorescens* F113. The HdtS enzyme synthesizes at least three signaling molecules: (i) *N*-(3-hydroxy-7-*cis*-tetradecenoyl) homoserine lactone (3-OH-C14:1-HSL), (ii) *N*-decanoylhomoserine lactone (C10-HSL), and (iii) *N*-hexanoylhomoserine lactone (C6-HSL). HdtS is not a member of the LuxI family and received this name since it directs the synthesis of AHLs with acyl side chains of six (hexa-), ten (deca-) and fourteen (tetradeca-) carbons in length (Laue et al. 2000).

Among Gram-negative psychotrophic bacteria, *P. fluorescens* constitutes the major raw milk deteriorative species due to proteolysis and lipolysis. There is some speculation whether the secretion of these enzymes is regulated by AHLs. There have been studies showing that protease and lipase production by milk isolates of *P. fluorescens* is not regulated by AI-1 type QS system (Pinto et al. 2010; Martins et al. 2014). On the other hand, other works have shown AI-1 QS regulation in *P. fluorescens* strains isolated from milk and fish (Tang Rong et al. 2019; Liu, Wang, and Griffiths 2007). Considering the genetic versatility of the species, it is likely that there are different regulatory systems controlling the production of these enzymes, depending on the strain.

Pseudomonas putida has importance at the ecological level, since it promotes the growth of plants, as well as inhibits plant pathogens and contributes to the degradation of toxic organic compounds (Barbarossa et al. 2010). In this microorganism, QS is mediated by AHL autoinducer molecules. Given that a large proportion of root-colonizing bacteria produces AIs, these interactions appear to be important in controlling many populations within the rhizosphere community, *P. putida* being a highly attractive candidate for agricultural and environmental uses (Steidle et al. 2002). Examples of QS systems homologous to the LuxI-LuxR are present in this bacterium (Barbarossa et al. 2010). This is the case for PhzR-PhzI, which utilizes *N*-hexanoyl-homoserine lactone (C6-HSL) to control the synthesis of phenazine antibiotics (Wood et al. 1997), and the PpuI-PpuR in *P. putida* IsoF, which uses at least four kinds of AHLs to influence biofilm structural development (Steidle et al. 2002).

Pseudomonas syringae is a bacterium studied mainly for its role as a plant pathogen, and it has the capability to control virulence through the QS signaling system (Pérez-Velázquez et al. 2015; Venturi 2006). This microorganism possesses a LuxI-LuxR homolog pair, called AhlI-AhlR. The genes *ahlI* and *ahlR* (Quiñones, Pujol, and Lindow 2004), also called *psyI* and *psyR* (Nakatsu et al. 2019; Ichinose et al. 2016), are responsible for encoding AHL protein synthase and the AHL transcriptional factor. The molecules *N*-(3-oxo-hexanoly)-L-homoserine lactone (OHHL) and *N*-hexanoly-L-homoserine lactone (HHL) are the main AIs detected in this bacterium (Nakatsu et al. 2019; Quiñones, Pujol, and Lindow 2004). For instance, *P. syringae* pv. *syringae* strain B728a is a plant pathogen that causes brown spots in beans and produces and responds to 3-oxo-C6-HSL in a cell-density dependent manner (Quiñones, Pujol, and Lindow 2004; Venturi 2006). However, this system is not generalized in *P. syringae*, since some isolates such as *P. syringae* pv. *tomato* DV3000 do not produce AHLs (Nakatsu et al. 2019). The authors suggest that ancestors of *P. syringae* had produced AHL, but the production might have become inconvenient for successful infection, and then, most strains have lost the signal production through mutations in *psyI* or *psyR* genes (Nakatsu et al. 2019).

2.2.3 QS in *Chromobacterium violaceum*

Another example of a Gram-negative bacteria with a complete QS mechanism is *C. violaceum*, an opportunistic human pathogen that can cause fatal sepsis, skin lesions, and liver and lung abscesses in immunocompromised individuals (De Lamo Marin et al. 2007; Jitmuang 2008; Yang and Li 2011). The *C. violaceum* QS consists of the LuxI/LuxR homologues called CviI/CviR, which produce and respond to AHLs of different acyl lengths (McClean et al. 1997). This bacterium produces violacein, a water-insoluble purple pigment with antibacterial activity which is synthesized from tryptophan by the products of the *vioABCD* operon (August et al. 2000). The production of violacein is regulated by QS and is the most well-studied phenotype in *C. violaceum* (McClean et al. 1997; Stauff and Bessler 2011). For this reason, and because it is an easily observable phenotype, *C. violaceum* wild type and mutants, with interruptions in the QS mechanism, have been used as tools in several studies related to bacterial communication mechanisms as well as inhibition of the mechanism of QS (QSI) by natural and synthetic products (Adonizio et al. 2006; Steindler and Venturi 2007). An example of a biomonitor commonly used is *C. violaceum* CV026, a mutant strain derived from wild type *C. violaceum* ATCC 31532, which is unable to produce AHL but retains the ability to respond to exogenous AHLs (McClean et al. 1997).

Other phenotypes studied in *C. violaceum* that are under the regulation of QS include biofilm formation, cell aggregation, chitinase production, and exoprotease production (Oca-Mejía et al. 2014; Chernin et al. 1998).

2.2.4 QS in *Aeromonas hydrophila*

Aeromonas hydrophila is a Gram-negative opportunistic pathogen that is capable of infecting a wide variety of hosts, which include terrestrial and aquatic animals in addition to humans. Its pathogenicity typically includes minor skin infections or gastroenteritis in humans. Furthermore, *A. hydrophila* is present in raw milk and is an important spoilage bacterium due to its ability to grow and to present proteolytic activity in chilled foods. *A. hydrophila* has the homologous *luxRI* genes termed *ahyRI*, and the QS mechanism is mediated by C4-HSL and C6-HSL. Phenotypes regulated by QS in *A. hydrophila* included biofilm formation (Ponce-Rossi et al. 2016), proteolytic activity related with serine protease and metalloprotease (Khajanchi et al. 2009; Martins et al. 2018; Ponce-Rossi et al. 2016; Swift et al. 1999), and virulence (Khajanchi et al. 2009).

The effect of QS on the virulence of *A. hydrophila* was demonstrated using mutants in *ahyI* and, or *ahyR* genes and therefore incapable of synthesizing and/or detecting AHLs. A double mutant *ΔahyRI* of *A. hydrophila* SSU presented reduced protease production, less biofilm formation, and attenuation of virulence in mice (Khajanchi et al. 2009). Mutants of *A. hydrophila* AH-1N in *ahyI* or *ahyR* genes were used to challenge burbot (*Lotalota*) and resulted in higher survival of larvae when compared to challenge with the wild type (Natrah et al. 2012). The addition of the signal molecule C4-HSL restored the virulence of the QS mutant. These results with mutants *A. hydrophila* in QS are examples of models for studying this mechanism applied to other bacterial genera and also to elucidate when and how cellular communication is involved in the regulation of important phenotypes in pathogenesis. This knowledge can be useful in the development of specific and promising strategies to block this communication for the control of pathogens.

2.2.5 QS in *Serratia*

Species of *Serratia* are widely dispersed in the environment, including soil, water, plant surfaces, and the gastrointestinal tract of several animals, including humans. Many species are part of the spoilage microbiota of diverse foods, and some have been related to outbreaks and opportunistic infections (Doulgeraki et al. 2012; Mahlen 2011). A range of AHLs and genes for production and regulation have been described in *Serratia*, and four systems have already been studied: SmaI/SmaR in *Serratia* sp. ATCC 39006, SwrI/SwrR in *S. marcescens* MG1, SplI/SplR in *S. plymuthica*, and SprI/SprR in *S. proteamaculans* (Van Houdt, Givskov, and Michiels 2007). Not all *Serratia* species have a homologous LuxIR system and produce AHL. In addition, there is considerable strain-dependent variation in both the ability to synthesize AHLs and in the nature of the AHL produced (Wei and Lai 2006).

Several phenotypes have been described as being regulated by QS in *Serratia*, such as virulence, biofilm formation and sloughing, butanediol fermentation, biosynthesis of antibiotic, and production of lipase, protease, chitinase, and the prodigiosin pigment (Christensen et al. 2003; Rice et al. 2005; Thomsonet al. 2000; Van Houdt et al. 2006). In *S. marcescens* strain SS-1, the QS system SpnIR is located in a mobile transposon, and that means lateral gene transfer may play an important role in the transfer of QS units between different bacterial genera and species (Wei et al. 2006). This may have significant implications for the diversity of the mechanism, and the acquisition of such a mobile QS system may allow the bacterium to bypass a possible specific disruption of their native QS system as long as the new signal-receptor complex is capable of activating target gene expression (Defoirdt, Boon, and Bossier, 2010).

2.2.6 QS in *Burkholderia*

The genus *Burkholderia* is very diverse and contains more than 30 species that occupy different niches, having agricultural, biotechnological, and clinical importance (Coenye and Vandamme 2003). All species of *Burkholderia* investigated encode at least one QS system that relies on AHL signal molecules for coordinated gene expression and is usually referred to as a CepI/CepR system (Eberl 2006). This system is present in the collectively called *Burkholderia cepacia* (Bcc) complex, which includes at least nine species recognized as problematic opportunistic pathogens in patients with cystic fibrosis and in immunocompromised individuals (Mahenthiralingam, Urban, and Goldberg 2005).

The CepI/CepR system positively regulates different functions, such as the production of exoproteases, siderophores, swarming motility, and biofilm production, besides contributing to the virulence of *Bcc* complex (Venturi et al. 2004). The detection of AHL in sputum and mucopurulent respiratory secretions in patients with cystic fibrosis provides clinical evidence of the occurrence of the mechanism during infection (Chambers et al. 2005; Middleton et al. 2002). In addition, the analysis of sequential strains of cystic fibrosis patients, obtained several years apart, indicates that the QS genes are maintained and expressed during chronic infections (McKeon et al. 2011).

A more complex QS system than in other *Burkholderia* species with more than one LuxI/LuxR homologue and numerous AHL-signaling molecules was described in the *Bptm* group, which includes the non-pathogenic soil saprophyte *Burkholderia thailandensis* and the pathogens *Burkholderia pseudomallei* and *Burkholderia mallei*, the causative agents of melioidosis and glanders, respectively (Breck et al. 2009; Ulrich 2004; Ulrich et al. 2004). In *B. thailandensis* and *B. pseudomallei*, AHL QS systems are described as QS-1, QS-2, and QS-3 made up of AHL synthase/AHL receptor pairs BpsI1/BpsR1, BpsI2/BpsR2, and BpsI3/BpsR3 respectively, besides two additional solo AHL receptors (R4 and R5).

As in the *Bcc* complex, the QS mechanism is also related to virulence, biofilm formation, and production of biomolecules that provide fitness advantages in the *Bptm* group (Mott, Panchal, and Rajamani 2017). Furthermore, recent findings about the mechanisms of QS have drawn attention because of their influence on the regulation of physiology and microbial metabolism, in order to provide strategies for competitiveness and at the same time perpetuating the species through cooperative behaviors (Abisado et al. 2018; Majerczyk et al. 2014a, 2014b). In *B. pseudomallei* and *B. thailandensis*, QS induces the production and excretion of oxalate in the stationary phase, which becomes a shared resource with the whole population and protects cells from self-intoxication and killing as a result of ammonia production (Goo et al. 2012).

2.2.7 QS in Gram-Negative Phytopathogens

Phytopathogenic bacteria benefit from QS mechanisms to control gene expression related to virulence and colonization of hosts (Von Bodman, Bauer, and Coplin 2003). Important phytopathogens such as *Agrobacterium tumefaciens*, *Pantoea stewartia*, *Pectobacterium carotovora*, *P. syringae*, *P. aeruginosa*, *Ralstonia solanacearum*, *Xanthomonas campestres*, among others use QS to fine tune plant-microbe interactions. These pathogens have developed abilities to colonize the ryzosphere or aerial surfaces of plants in order to circumvent plant defenses and cause diseases by using a plethora of virulence factors.

Agrobacterium tumefaciens is an alpha-proteobacterium that belongs to the Rhizobiaceae family, which includes plant pathogens and nitrogen-fixing microbes (Slater et al. 2009; Wood et al. 2001). The bacterium is found in soil and can cause crown gall disease in dicotyledonous plants at wounded sites (Winans 1992). The illness is usually non-fatal and is characterized by the growth of tumors which can reduce crop productivity (Escobar and Dandekar 2003).

The tumor inducing principle is as a piece of DNA that is transferred from the bacteria to the plant cells and is linked to the presence of the so-called tumor inducing plasmid (Ti plasmid) (Chilton et al. 1977). Research with Ti plasmids has impacted many different fields including plant biology, agriculture, biotechnology, and molecular biology (Binns 2002; Escobar and Dandekar 2003).

The Ti-plasmid is a large circular replicon that carries the transferred DNA (also known as transforming or T-DNA) and most genes required for tumorigenesis (Pinto, Pappas, and Winans 2012; White and Winans 2007). The T-DNA carries a set of genes responsible for plant cell proliferation and another set of genes required for the synthesis of opines which support bacterial growth (Zhu et al. 2000). The Ti-plasmid also codes for the transport and catabolism of opines produced in the tumors. In fact, Ti-plasmids are usually classified according to the type of opines that are encoded in the T-DNA.

Infection starts when bacterial cells containing the Ti plasmid encounter a plant wounded site, which releases compounds such as amino acids, organic acids, and sugars that activate the transfer of the T-DNA from bacterial to plant cells. Once the T-DNA is transported to the nucleus, it can integrate into the plant genome and initiate expression of the tumor inducing and opine synthase genes (Escobar and Dandekar 2003; Pappas 2008; Zhu et al. 2000).

Interestingly, *luxI* and *luxR* homologues known as *traI* and *traR* are found within the Ti plasmid and regulate plasmid copy number, Ti plasmid conjugation, and entry exclusion as well as increased tumorigenesis in plants infected with bacteria containing these plasmids (Cho, Pinto, and Winans 2009; Fuqua and Winans 1994; Pinto, Pappas, and Winans 2012). The crystal structure of TraR complexed with 3-oxo-octanoyl-L-homoserine lactone bound to the *tra* box DNA has been solved by two groups (Zhang et al. 2002; Vannini et al. 2002). The protein binds DNA as a dimer and both the N-terminal and C-terminal domains contribute to protein dimerization (Pinto and Winans 2009). Several studies have further confirmed and extended the structural predictions, broadening the understanding of TraR transcription activation and making it a pivotal model for the LuxR family of transcriptional regulators.

Pectobacterium (previously classified as *Erwinia*) is a genus of Gram-negative phytopathogenic bacteria that belongs to the Enterobacteriaceae family (Davidsson et al. 2013). These bacteria can cause soft rot and degradation of plant cell wall polysaccharides in commercially important plants such as those destined for food (especially crop potatoes) and for ornamental purposes (Joshi et al. 2016; Park et al. 2012). In fact, pectobacteria encode for a large number of plant cell wall-degrading enzymes (PCWDEs) which are under control of QS (Barnard et al. 2007). The PCWDEs are usually cellulases, hemicellulases, pectinases, and proteinases, mainly secreted by type II secretion systems (Davidsson et al. 2013).

The three main species that can cause soft rot are *P. carotovorum*, *P. atrosepticum*, and *P. parmentieri*. The species *P. carotovorum* is further divided into *P. carotovorum* subsp. *brasiliense* (*Pcb*), *P. carotovorum* subsp. *carotovorum* (*Pcc*), and *P. carotovorum* subsp. *odoriferum* (*Pco*) (Li et al. 2018). One *luxI* homologue and two or more *luxR* homologues are found in these organisms. Signaling in *Pectobacterium* is usually mediated by 3-oxo-hexanoyl homoserine lactone (3OC6HSL), 3-oxo-octanoyil homoserine lactone (3OC8HSL), and autoinducer-2(AI-2), which regulate the expression of PCWDEs, contributing to the soft rot phenotype (Põllumaa, Alamäe, and Mäe 2012).

Signaling molecules vary according to the subspecies and the strain type, as well as the *luxI/luxR* pair. For instance, different homologues have been found, and for the case of *Pcc* strain SCC3193 they are named ExpI/ExpR1/ExpR2. On the other hand, in strain *Pcc* EC153 they have been named AhlI/ExpR; while in *Pcc* ATCC390048, the homologues are CarI/CarR and ExpR1/VirR. These QS systems are responsible for controlling the production of PCWDEs and virulence factors, in addition to the production of carbapenem antibiotics in *Pcc* ATCC390048 (Põllumaa, Alamäe, and Mäe 2012). A pioneer work by Dong et al. (2001) demonstrated that inactivation of QS through enzymatic hydrolysis of AHLs rendered transgenic tobacco plants resistant to *Pcc* infections, bringing interesting insights into the role of QS inactivation in controlling bacterial infections.

2.3 Bacteria with Incomplete Auto-Inducer-1 Quorum Sensing Mechanisms

Some Proteobacteria belonging to the Enterobacteriaceae family, such as *Escherichia coli* and *Salmonella*, presents incomplete AI-1 QS mechanism. These bacteria do not have the *luxI*-homologous gene encoding AI-1 synthase; consequently, there is no synthesis of AHLs. However, a LuxR homolog, known as SdiA (cell division inhibition suppressor), which shows an amino acid sequence similar to that of the LuxR-type transcriptional activators, is present and allows the detection of signal molecules produced by other bacterial species leading to the regulation of gene expression (Michael et al. 2001; Dyszel et al. 2010; Smith and Ahmer 2003; Smith et al. 2008). In *Salmonella* and *E. coli* some phenotypes regulated by QS have been described.

2.3.1 QS in *E. coli*

Escherichia coli is a very diverse bacterial species belonging to the family Enterobacteriaceae, which comprises Gram-negative bacilli, and inhabits the lower gastrointestinal tract of humans and other animals. This bacterium is a paradigm of bacterial versatility and comprises harmless commensal as well as different pathogenic variants with the ability to cause either intestinal or extraintestinal diseases (Leimbach, Hacker, and Dobrindt 2013). *E. coli* strains regulate their virulence gene expression in response to a variety of environmental factors and can use QS to modulate gene expression. Different intercellular signaling systems have been identified: the LuxR homolog SdiA, the LuxS/AI-2 system, an AI-3 system, and a signaling system mediated by indole.

The SdiA protein played a role in the regulation of *ftsQAZ* cell division genes in *E. coli* (Ahmer et al. 1998), and increases of up to four-fold in the *ftsQAZ* expression were reported (Dyszel et al. 2010). In contrast, SdiA repressed the expression of virulence factors in enterohemorrhagic *E. coli* O157:H7 (EHEC) (Kanamaru et al. 2000), conferred multidrug resistance and increased levels of AcrAB (Dyszel et al. 2010; Rahmati et al. 2002), and increased the acid tolerance of *E. coli* upon exposure to AHLs (Van Houdt et al. 2006). SdiA also decreases early *E. coli* biofilm formation 51-fold, enhances acid resistance, and is required to reduce *E. coli* biofilm formation in the presence of AHLs (Lee et al. 2007). The quinolone resistance, expression of *acrAB* and *ftsQAZ* were not increased by chromosomal *sdiA* and/or AHL in *E. coli* K-12 or EHEC (Dyszel et al. 2010). However, using plasmid-encoded *sdiA* a two-fold change in response to some antibiotics was observed, an increase of up to two-fold in *acrA* expression and four-fold in *ftsQAZ* expression (Dyszel et al. 2010).

The *sdiA* mutants of atypical enteropathogenic *E. coli* (aEPEC) were capable of forming thicker biofilm structures and showed increased motility when compared to the wild type and complemented strains (Culler et al. 2018). These authors also demonstrated increased *csgA*, *csgD*, and *fliC* transcription on mutant strains. Biofilm formation, as well as *csgD*, *csgA*, and *fimA* transcription decreased on wild type strains by the addition of AHL. These results indicate that SdiA participates on the regulation of these phenotypes in aEPEC and that AHL addition enhances the repressor effect of this receptor on the transcription of biofilm and motility related genes (Culler et al. 2018). In *sdiA* mutant of EHEC, the expression of the glutamate decarboxylase acid-resistance system genes (*gad* genes) was dramatically decreased, even in the absence of AHLs and, consequently, this mutant was less resistant to acidic environments than wild-type of EHEC (Dyszel et al. 2010; Hughes et al. 2010).

The rumen of cattle harbors AHLs, and these chemical signals can be sensed in part through SdiA to modulate gene expression in EHEC, leading to successful colonization of these animals (Hughes et al. 2010). The presence of SdiA seems to be essential in EHEC colonization of the bovine intestine since *sdiA* transposon insertion mutants were not recovered or were recovered at low levels in the feces of old Friesian bull calves (Dziva et al. 2004). The *sdiA* mutant was detected in feces of only one of the four calves at low levels (10^2 CFU/g feces) from days 19 to 27 post-inoculation, whereas the fecal shedding of the wild-type strain persisted at approximately four-logs in all four calves. AHLs activated expression of the *gad* genes and repressed expression of the locus of enterocyte effacement (LEE) of EHEC (Hughes et al. 2010). Of note, the arginine acid-resistance system (*adi*) was not regulated by SdiA or the addition of AHLs (Hughes et al. 2010). Sharma and Bearson (2013) confirmed that SdiA represses *ler*, which encodes a positive transcriptional regulator of LEE in response to AHLs and reduces adherence of EHEC to HEp-2 cells.

Phenotypes regulated by SdiA protein in the absence of AHLs in *Salmonella* and *E. coli* have been reported (Dyszel et al. 2010; Hughes et al. 2010; Nguyen et al. 2015; Smith and Ahmer 2003). For instance, SdiA of EHEC is constitutively activated by the binding of molecule 1-octanoyl-*rac*-glycerol (OCL) in the absence of AHLs (Nguyen et al. 2015). The OCL is a monoglycerol present in prokaryotes and eukaryotes and is used as an energy source and substrate for the synthesis of membrane and a signaling molecule (Alvarez and Steinbüchel 2002; Liu et al. 2012). However, the activation of SdiA from EHEC by AHLs conferred greater stability and affinity to DNA, albeit not affecting *sdiA* gene transcription (Nguyen et al. 2015). Additionally, these authors observed conformational changes of EHEC SdiA protein complexed with different ligands such as: OCL in the absence of AHLs; *N*-(3-oxo-hexanoyl)-L-homoserine lactone (3-oxo-C6-HSL); and *N*-(3-oxo-octanoyl)-L-homoserine lactone (3-oxo-C8-HSL).

2.3.2 QS in *Salmonella*

Salmonella is a genus of rod-shaped, Gram-negative, facultative anaerobic bacteria belonging to the family Enterobacteriaceae, and it is considered the main foodborne bacterial pathogen. It is an important cause of gastrointestinal diseases worldwide, and complications can lead to death. In this pathogen, the communication by QS can be mediated by three types of AIs, called AI-1, AI-2, and AI-3 (Ahmer 1998; Hughes and Sperandio, 2008). In *Salmonella*, SdiA was described for the first time by Ahmer et al. (1998) and they showed that, in *Salmonella enterica* serovar Typhimurium, this protein is able to partially activate the promoter two of the *ftsQAZ* operon and suppress *ftsZ* responsible for cell filamentation, unlike what occurs in *E. coli*. The AHLs regulate the expression of the *rck* operon (resistant to complement killing), which codes for *pefI*, *srgD*, *srgA*, *srgB*, *rck*, and *srgC* genes, and it is found in plasmids influencing virulence of *Salmonella* Typhimurium (Ahmer et al. 1998; Michael et al. 2001; Smith and Ahmer 2003; Soares and Ahmer 2011).

Genes related to virulence such as *hilA*, *invA*, and *invF* present on the pathogenicity island PAI-1, and genes involved in the formation of biofilm by *Salmonella* Enteritidis were more expressed in the presence of exogenous AHLs (Campos-Galvão et al. 2016). A global analysis carried out on the influence of AHL on proteins of *Salmonella* Enteritidis showed that the abundance of proteins involved in translation (PheT), transport (PtsI), metabolic processes (TalB, PmgI, Eno and PykF), and response to stress (HtpG and Adi) increased while the abundance of other proteins related to translation (RplB, RplE, RpsB, and Tsf), transport (OmpA, OmpC, and OmpD), and metabolic processes (GapA) decreased in the presence of AI-1 (Almeida et al. 2017a). It was hypothesized that these changes observed in cells in the middle of logarithmic phase in presence of AHL are correlated with those into the early stationary phase of growth, without AHL. In other organisms, the effect of AHLs in anticipating the stationary phase responses was confirmed by global analysis, such as the transcriptome of *P. aeruginosa* (Schuster et al. 2003) and *B. thailandensis* (Majerczyk et al. 2014a), as well as the metabolomes of *Burkholderia glumae*, *B. pseudomallei*, and *B. thailandensis* (Goo et al. 2012).

The suggestion that QS signal anticipated a stationary phase response in *Salmonella* was reinforced when cells were cultivated in anaerobic condition in the presence of *N*-dodecanoyl-homoserine lactone (C12-HSL), and the fatty acid profiles were altered and similar to those of cells at late stationary phase (Almeida et al. 2017a). The presence of C12-HSL increased the abundance of thiol related proteins such as Tpx, Q7CR42, Q8ZP25, YfgD, AhpC, NfsB, YdhD, and TrxA, as well as the levels of free cellular thiol in late log phase, suggesting that these cells have greater potential to resist oxidative stress (Almeida et al. 2018). Additionally, the LuxS protein which synthesizes the AI-2 signaling molecule was differentially abundant in the presence of C12-HSL. The increased abundance of NfsB protein in the presence of C12-HSL suggested that the cells may be susceptible to the action of nitrofurans or that AHLs present some toxicity. Overall, the presence of C12-HSL altered important pathways related to oxidative stress and stationary phase response in *Salmonella* (Almeida et al. 2018).

The role of AHLs in *Salmonella* pathogenicity was suggested when *N*-hexanoyl homoserine lactone (C6-HSL) and *N*-octanoylhomoserine lactone (C8-HSL) increased the invasion of HEp-2 cells by *Salmonella* Typhimurium at 37°C (Nesse et al. 2011) whereas C8-HSL increased adhesion of *S. enterica* serovar Typhi containing plasmid pRST98, which harbors the virulence gene *rck*, to HeLa cells after 1 h at 37°C in the presence of 5% CO_2 gas (Liu et al. 2014). Biofilm formation by *S. enterica* serovar Enteritidis PT4 578 in polyestyrene surface was positively regulated by C12-HSL in anaerobiose, even though no growth changes were observed in planktonic cells (Almeida et al. 2017b; Campos-Galvão et al. 2016). *N*-butyrilhomoserine lactone (C4-HSL) and C6-HSL also increased biofilm formation by *Salmonella* Typhimurium on polyestyrene (Aswathanarayan and Vittal 2016).

On the other hand, a cell free supernatant (CFS) rich in AHLs, AI-2, and other unknown compounds of *Y. enterocolitica* and *Serratia proteamaculans* altered growth of different phage types of *Salmonella* Enteritidis and *Salmonella* Typhimurium under aerobiose (Dourou et al. 2011). Similarly, the CFS of *P. aeruginosa* containing AHLs and different metabolites decreased growth of nine serovars of *S. enterica* (Wang et al. 2013). Conversely, the CFS of *Hafnia alvei* containing AHLs, as well as the addition of synthetic *N*-3-oxo-hexanoyl homoserine lactone (3-oxo-C6-HSL) to the growth medium in aerobioses did not influence biofilm formation by *Salmonella* Typhimurium (Blana et al. 2017). It is noteworthy that in these studies, the CFS of different bacteria contained metabolites other than AHLs which might have interfered in the detection of the AI's subtle effects in the cells.

2.4 Quorum Quenching of Autoinducer-1

A large variety of microorganisms have adopted the QS communication system leading to the expression of specific genes in order to coordinate certain basic cellular functions in response to changes in the environment (Fuqua, Parsek, and Greenberg 2001; Whitehead et al. 2001; Federle and Bassler 2003). Some phenotypes regulated by AI-1 of QS mechanisms are shown in Table 2.1.

Such coordinate functions in the bacterial population can provide competitive advantages to microorganisms to remain in ecological niches. Likewise, a microorganism's ability to neutralize QS signaling from its competitors can also significantly increase its competitive strength in the ecosystem (Zhang and Dong 2004). Besides the ability of many bacteria to produce and utilize AHL-based communication systems, inhibitors and quorum quenching (QQ) enzymes have been identified from different sources, including both prokaryotic and eukaryotic organisms (Hentzer et al. 2003; Zhang 2003; Zhang and Dong 2004; Uroz, Dessaux, and Oger 2009). Therefore, the interruption of this communication system or the activity of the QS mechanism in bacteria can lead to the attenuation of the microbial virulence (Whiteley, Lee, and Greenberg 1999; De Kievit and Iglewski 2000; Smith and Iglewski 2003).

Natural and synthetic compounds with QQ action have gained interest as potential attractive strategies for controlling bacterial pathogenesis. One of the first mechanisms most studied is related to chemical compounds quorum sensing inhibitors (QSI) acting as antagonists and interfering with the transcriptional regulator structure. Plants are potential sources of antimicrobials with QSI activity due to the production of a broad spectrum of secondary metabolites, such as phenolic compounds, alkaloids, terpernoids, and polyacetylenes, among other classes (Givskov et al. 1996; Zhang et al. 2002; Vattem et al. 2007; Sybiya Vasantha Packiavathy et al. 2012). Another mechanism that strongly interferes or even eliminates the functions regulated by QS is related to the production of enzymes capable of degrading the AHL signal molecules (Uroz, Dessaux, and Oger 2009). The potential for biological decomposition of these signals is interesting because other bacteria sharing the same local environment as quorum-sensitive bacteria could gain a competitive advantage by degrading acyl-HSL signals. Four types of enzymes have been shown to possess an ability to degrade QS signals-AHLs: lactonases and decarboxylases hydrolyze lactone ring, whereas acylase and deaminase cleave the acyl side chain (Kalia 2013).

2.5 Concluding Remarks

There is a growing interest in QS mechanism in the bacterial world and its implications for biotechnology, medicine, ecology, and agriculture. The great challenge of QS studies is expected to be an insightful observation and practical application of chemical signaling that occurs in the natural world. Scientists need to develop intelligent and sophisticated strategies to study such an intricate network of interactions at the crossroads of chemistry, physics, and biology. Most of the research that led to the current understanding of QS used well-controlled pure culture and mixed culture in laboratories, but the borders need to be broadened to understand the impact and exploitation of QS in complex microbial communities.

REFERENCES

Abisado, Rhea G., Saida Benomar, Jennifer R. Klaus, Ajai A. Dandekar, and Josephine R. Chandler. 2018. "Bacterial Quorum Sensing and Microbial Community Interactions." *mBio* 9, no. 3: 1–14. https://doi.org/10.1128/mBio.02331-17.

Adonizio, Allison L., Kelsey Downum, Bradley C. Bennett, and Kalai Mathee. 2006. "Anti-Quorum Sensing Activity of Medicinal Plants in Southern Florida." *Journal of Ethnopharmacology* 105, no. 3: 427–435. https://doi.org/10.1016/j.jep.2005.11.025.

Ahmer, Brian M. M., Jeroen van Reeuwijk, Cynthia D. Timmers, Peter J. Valentine, and Fred Heffron. 1998. "*Salmonella* Typhimurium Encodes an SdiA Homolog, a Putative Quorum Sensor of the LuxR Family, That Regulates Genes on the Virulence Plasmid." *Journal of Bacteriology* 180, no. 5: 1185–1193. http://www.ncbi.nlm.nih.gov/pubmed/9495757.

Allison, David G., Begoña Ruiz, Carmen San Jose, Almudena Jaspe, and Peter Gilbert. 1998. "Extracellular Products as Mediators of the Formation and Detachment of *Pseudomonas fluorescens* Biofilms." *FEMS Microbiology Letters* 167, no. 2: 179–184. https://doi.org/10.1016/S0378-1097(98)00386-3.

Almeida, Felipe Alves, Deisy Carneiro, Tiago Mendez, Edvaldo Barros, Uelinton Pinto, Leandro Oliveira, and Maria Cristina Dantas Vanetti. 2018. "*N*-Dodecanoyl-Homoserine Lactone Influences the Levels of Thiol and Proteins Related to Oxidation-Reduction Process in *Salmonella*." *PLoS ONE* 13, no. 10: 1–31. https://doi.org/10.1371/journal.pone.0204673.

Almeida, Felipe Alves, Natan de Jesus Pimentel-Filho, Lanna Clícia Carrijo, Cláudia Braga Pereira Bento, Maria Cristina Baracat-Pereira, Uelinton Manoel Pinto, Leandro Licursi de Oliveira, and Maria Cristina Dantas Vanetti. 2017a. "Acyl

Homoserine Lactone Changes the Abundance of Proteins and the Levels of Organic Acids Associated with Stationary Phase in *Salmonella* Enteritidis." *Microbial Pathogenesis* 102: 148–159. https://doi.org/10.1016/j.micpath.2016.11.027.

Almeida, Felipe Alves, Natan de Jesus Pimentel-Filho, Uelinton Manoel Pinto, Hilário Cuquetto Mantovani, Leandro Licursi de Oliveira, and Maria Cristina Dantas Vanetti. 2017b. "Acyl Homoserine Lactone-Based Quorum Sensing Stimulates Biofilm Formation by *Salmonella* Enteritidis in Anaerobic Conditions." *Archives of Microbiology* 199, no. 3: 475–486. https://doi.org/10.1007/s00203-016-1313-6.

Alvarez, H. M., and A. Steinbüchel. 2002. "Triacylglycerols in Prokaryotic Microorganisms." *Applied Microbiology and Biotechnology* 60, no. 4: 367–376. https://doi.org/10.1007/s00253-002-1135-0.

Aswathanarayan, Jamuna B., and Ravishankar R. Vittal. 2016. "Effect of Small Chain *N*-Acyl Homoserine Lactone Quorum Sensing Signals on Biofilms of Food-Borne Pathogens." *Journal of Food Science and Technology* 53, no. 9: 3609–3614. https://doi.org/10.1007/s13197-016-2346-1.

Atkinson, Steve, and Paul Williams. 2009. "Quorum Sensing and Social Networking in the Microbial World." *Journal of the Royal Society Interface* 6, no. 40: 959–978. https://doi.org/10.1098/rsif.2009.0203.

August, Paul R., Trudy H. Grossman, Charles Minor, M. P. Draper, Ian A. MacNeill, John M. Pemberton, K. M. Call, Denis Holt, and Marcia S. Osburne. 2000. "Sequence Analysis and Functional Characterization of the Violacein Biosynthetic Pathway from *Chromobacterium violaceum*." *Journal of Molecular Microbiology and Biotechnology* 2, no. 4: 513–519.

Azam, Mohd W., and Asad U. Khan. 2019. "Updates on the Pathogenicity Status of *Pseudomonas aeruginosa*." *Drug Discovery Today* 24, no. 1: 350–359. https://doi.org/10.1016/j.drudis.2018.07.003.

Bai, A. J., and Rai V. H. 2016. "Effect of Small Chain *N*-Acyl Homoserine Lactone Quorum Sensing Signals on Biofilms of Food-Borne Pathogens." *Journal of Food Science and Technology*, 53, no. 9: 3609–3614. https://doi.org/10.1007/s13197-016-2346-1.

Barbarossa, M. V., C. Kuttler, A. Fekete, and M. Rothballer. 2010. "A Delay Model for Quorum Sensing of *Pseudomonas putida*." *BioSystems* 102: 148–156. https://doi.org/10.1016/j.biosystems.2010.09.001.

Bari, S. M. Nayeemul, M. Kamruzzaman Roky, M. Mohiuddin, M. Kamruzzaman, John J. Mekalanos, and Shah M. Faruque. 2013. "Quorum-Sensing Autoinducers Resuscitate Dormant *Vibrio cholerae* in Environmental Water Samples." *Proceedings of the National Academy of Sciences* 110, no. 24: 9926–9931. https://doi.org/10.1073/pnas.1307697110.

Barnard, A. M., S. D. Bowden, T. Burr, S. J. Coulthurst, R. E. Monson, and G. P. Salmond. 2007. "Quorum Sensing, Virulence and Secondary Metabolite Production in Plant Soft-Rotting Bacteria." *Philosophical Transactions of the Royal Society B: Biological Sciences* 29, no. 362: 1165–1183. 10.1098/rstb.2007.2042.

Bassler, Bonnie Lynn, and Richard Losick. 2006. "Bacterially Speaking." *Cell* 125, no. 2: 237–246. https://doi.org/10.1016/j.cell.2006.04.001.

Bejerano-Sagie, Michal, and Karina Bivar Xavier. 2007. "The Role of Small RNAs in Quorum Sensing." *Current Opinion in Microbiology* 10, no. 2: 189–198. https://doi.org/10.1016/j.mib.2007.03.009.

Binns, A. N. 2002. "T-DNA of *Agrobacterium tumefaciens*: 25 Years and Counting." *Trends in Plant Science* 7, no. 5: 231–233.

Blana, Vasiliki, Aliki Georgomanou, and Efstathios Giaouris. 2017. "Assessing Biofilm Formation by *Salmonella* Enterica Serovar Typhimurium on Abiotic Substrata in the Presence of Quorum Sensing Signals Produced by Hafnia Alvei." *Food Control* 80: 83–91. https://doi.org/10.1016/j.foodcont.2017.04.037.

Brint, J. Mark, and Dennis E. Ohman. 1995. "Synthesis of Multiple Exoproducts in *Pseudomonas aeruginosa* Is under the Control of RhlR-RhlI, Another Set of Regulators in Strain PAO1 with Homology to the Autoinducer-Responsive LuxR-LuxI Family." *Journal of Bacteriology* 177, no. 24: 7155–7163. https://doi.org/10.1128/jb.177.24.7155-7163.1995.

Campos-Galvão, Maria Emilene Martino, Andrea Oliveira Barros Ribon, Elza Fernandes Araújo, and Maria Cristina Dantas Vanetti, Campos-Galvão, Maria Emilene Martino, Andrea Oliveira Barros Ribon, Elza Fernandes Araújo, and Maria Cristina Dantas Vanetti. 2016. "Changes in the *Salmonella* Enterica Enteritidis Phenotypes in Presence of Acyl Homoserine Lactone Quorum Sensing Signals." *Journal of Basic Microbiology* 56, no. 5: 493–501. https://doi.org/10.1002/jobm.201500471.

Cao, Jen G., Edward Meighen.1989. "Purification and Structural Identification of an Autoinducer for the Luminescence System of *Vibrio harveyi*." *Journal of Biological Chemistry* 264, no. 36: 21670–21676.

Chambers, Catherine E., Michelle B. Visser, Ute Schwab, and Pamela A. Sokol. 2005. "Identification of Chambers, Catherine E., Michelle B. Visser, Ute Schwab, and Pamela A. Sokol. 2005. "Identification of *N*-Acylhomoserine Lactones in Mucopurulent Respiratory Secretions from Cystic Fibrosis Patients." *FEMS Microbiology Letters* 244, no. 2: 297–304. https://doi.org/10.1016/j.femsle.2005.01.055.

Chatterjee, Asita, Yaya Cui, Hiroaki Hasegawa, Nathan Leigh, Vaishali Dixit, and Arun K. Chatterjee. 2005. "Comparative Analysis of Two Classes of Quorum-Sensing Signaling Systems That Control Production of Extracellular Proteins and Secondary Metabolites in *Erwinia carotovora* Subspecies." *Journal of Bacteriology* 187, no. 23: 8026–8038. https://doi.org/10.1128/JB.187.23.8026-8038.2005.

Chen, Guozhou, Lee R. Swem, Danielle L. Swem, Devin L. Stauff, Colleen T. O'Loughlin, Philip D. Jeffrey, Bonnie L. Bassler, and Frederick M. Hughson. 2011. "A Strategy for Antagonizing Quorum Sensing." *Molecular Cell* 42, no. 2: 199–209. https://doi.org/10.1016/j.molcel.2011.04.003.

Chen, R., Eric Déziel, Marie-Christine Groleau, Amy L. Schaefer, and Everett P. Greenberg. 2019. "Social Cheating in a *Pseudomonas aeruginosa* Quorum-Sensing Variant." *Proceedings of the National Academy of Sciences* 116, no. 14: 7021–7026. https://doi.org/10.1073/pnas.1819801116

Chernin, Leonid S., Michael K. Winson, Jacquelyn M. Thompson, Shoshan Haran, Barrie W. Bycroft, Ilan Chet, Paul Williams, and Gordon S. A. B. Stewart. 1998. "Chitinolytic Activity in *Chromobacterium violaceum*: Substrate Analysis and Regulation by Quorum Sensing." *Journal of Bacteriology* 180, no. 17: 4435–4441.

Chilton, M. D., Drummond, M. H., Merio, D. J., Sciaky, D., Montoya, A. L., Gordon, M. P., and Nester, E. W. 1977. "Stable Incorporation of Plasmid DNA into Higher Plant Cells: The Molecular Basis of Crown Gall Tumorigenesis." *Cell* 11, no. 2: 263–271. https://doi.org/10.1016/0092-8674(77)90043-5

Cho, Hongbaek, Uelinton M. Pinto, and Stephen C. Winans. 2009. "Transsexuality in the Rhizosphere: Quorum Sensing Reversibly Converts *Agrobacterium tumefaciens* from Phenotypically Female to Male." *Journal of Bacteriology* 191, no. 10: 3375–3383. https://doi.org/10.1128/JB.01608-08.

Choudhary, Swati, and Claudia Schmidt-Dannert. 2010. "Applications of Quorum Sensing in Biotechnology." *Applied Microbiology and Biotechnology* 86, no. 5: 1267–1279. https://doi.org/10.1007/s00253-010-2521-7.

Christensen, Allan B., Kathrin Riedel, Leo Eberl, Lars R. Flodgaard, Søren Molin, Lone Gram and Michael Givskov. 2003. "Quorum-Sensing-Directed Protein Expression in *Serratia Proteamaculans* B5a." *Microbiology* 149, no. 2: 471–483. https://doi.org/10.1099/mic.0.25575-0.

Churchill, Mair. E. A., Chen, L. 2011. "Structural Basis of Acyl-Homoserine Lactone-Dependent Signaling." *Chemical Review* 111, no. 1: 68–85.

Coenye, Tom, and Peter Vandamme. 2003. "Diversity and Significance of *Burkholderia* Species Occupying Diverse Ecological Niches." *Environmental Microbiology* 5, no. 9: 719–729. https://doi.org/10.1046/j.1462-2920.2003.00471.x.

Culler, Hebert F., Samue C. F. Couto, Juliana S. Higa, Renato M. Ruiz, Min J. Yang, Vanessa Bueris, Marcia R. Franzolin and Marcelo P. Sircili. 2018. "Role of SdiA on Biofilm Formation by Atypical Enteropathogenic *Escherichia coli*." *Genes* 9, no. 5: 253. https://doi.org/10.3390/genes9050253.

D'Angelo, Francesca, Valerio Baldelli, Nigel Halliday, Paolo Pantalone, Fabio Polticelli, Ersilia Fiscarelli, Paul Williams, Paolo Visca, Livia Leoni, and Giordano Rampioni. 2018. "Identification of FDA-Approved Drugs as Antivirulence Agents Targeting the *pqs* Quorum-Sensing System of *Pseudomonas aeruginosa*." *Antimicrobial Agents and Chemotherapy* 62, no. 11. https://doi.org/10.1128/AAC.01296-18.

Davidsson, Pär R., Tarja Kariola, Outi Niemi, and E. Tapio Palva. 2013. "Pathogenicity of and Plant Immunity to Soft Rot Pectobacteria." *Frontiers in Plant Science* 4: 191. doi:10.3389/fpls.2013.00191

De Keersmaecker, Sigrid C. J., Kathleen Sonck, and Jos Vanderleyden. 2006. "Let LuxS Speak up in AI-2 Signaling." *Trends in Microbiology* 14, no. 3: 114–119. https://doi.org/10.1016/j.tim.2006.01.003.

De Kievit, Teresa, and Barbara Iglewski. 2000. "Bacterial quorum sensing in pathogenic relationships." *Infection and Immunity* 68, no. 9: 4839–4849. https://doi.org/10.1128/IAI.68.9.4839-4849.2000.

De Lamo Marin, Sandra, Yang Xu, Michael M. Meijler, and Kim D. Janda. 2007. "Antibody Catalyzed Hydrolysis of a Quorum Sensing Signal Found in Gram-Negative Bacteria." *Bioorganic & Medicinal Chemistry Letters* 17, no. 6: 1549–1552. https://doi.org/10.1016/j.bmcl.2006.12.118.

Defoirdt, Tom, Nico Boon, and Peter Bossier. 2010. "Can Bacteria Evolve Resistance to Quorum Sensing Disruption?" *PLoS Pathogens* 6, no. 7: e1000989. https://doi.org/10.1371/journal.ppat.1000989.

Diggle, Stephen P., Andy Gardner, Stuart A. West, and Ashleigh S. Griffin. 2007. "Evolutionary Theory of Bacterial Quorum Sensing: When Is a Signal Not a Signal?" *Philosophical Transactions of the Royal Society B: Biological Sciences* 362, no. 1483: 1241–1249. https://doi.org/10.1098/rstb.2007.2049.

Dong, Yi-Hu, Lian-Hui Wang, Jin-Ling Xu, Hai-Bao Zhang, Xi-Fen Zhang, and Lian-Hui Zhang. 2001. "Quenching Quorum-Sensing-Dependent Bacterial Infection by an *N*-Acyl Homoserine Lactonase." *Nature* 411: 813–817. https://doi.org/10.1038/35081101.

Doulgeraki, Agapi I., Danilo Ercolini, Francesco Villani, and George John E. Nychas. 2012. "Spoilage Microbiota Associated to the Storage of Raw Meat in Different Conditions." *International Journal of Food Microbiology* 157, no. 2: 130–141. https://doi.org/10.1016/j.ijfoodmicro.2012.05.020.

Dourou, Dimitra, Mohammed S. Ammor, Panagiotis N. Skandamis, and George-John E. Nychas. 2011. "Growth of *Salmonella* Enteritidis and *Salmonella* Typhimurium in the Presence of Quorum Sensing Signalling Compounds Produced by Spoilage and Pathogenic Bacteria." *Food Microbiology* 28, no. 5: 1011–1018. https://doi.org/10.1016/j.fm.2011.02.004.

Dyszel, Jessica L., Jitesh A. Soares, Matthew C. Swearingen, Amber Lindsay, Jenee N. Smith, and Brian M. M. Ahmer. 2010. "*E. coli* K-12 and EHEC Genes Regulated by SdiA." *PLoS ONE* 5, no. 1: e8946. https://doi.org/10.1371/journal.pone.0008946.

Dziva, Francis, Pauline M. van Diemen, Mark P. Stevens, Amanda J. Smith, and Timothy S. Wallis.2004. "Identification of *Escherichia coli* O157: H7 Genes Influencing Colonization of the Bovine Gastrointestinal Tract Using Signature-Tagged Mutagenesis." *Microbiology* 150, no. 11: 3631–3645. https://doi.org/10.1099/mic.0.27448-0.

Eberhard, Anatol, Al Burlingame, C. Eberhard, G. L. Kenyon, Kenneth H. Nealson, and N. J. Oppenheimer. 1981. "Structural Identification of Autoinducer of *Photobacterium fischeri* Luciferase." *Biochemistry* 20, no. 9: 2444–2449. https://doi.org/10.1021/bi00512a013.

Eberl, Leo. 2006. "Quorum Sensing in the Genus *Burkholderia*." *International Journal of Medical Microbiology* 296, nos. 2–3: 103–110. https://doi.org/10.1016/j.ijmm.2006.01.035.

El-sayed, A. Kassem, Joanne Hothersall, and Christopher M. Thomas. 2001. "Quorum-Sensing-Dependent Regulation of Biosynthesis of the Polyketide Antibiotic Mupirocin in *Pseudomonas fluorescens* NCIMB 10586." *Microbiology* 147: 2127–2139. https://doi.org/10.1099/00221287-147-8-2127.

Engebrecht, Joanne, and Silverman, Michael 1984. "Identification of Genes and Gene Products Necessary for Bacterial Bioluminescence." *Proceedings of the National Academy of Sciences USA* 81, no. 13: 4154–4158. https://doi.org/10.1073/pnas.81.13.4154.

Escobar, M. A., and A. M. Dandekar. 2003. "*Agrobacterium tumefaciens* as an Agent of Disease." *Trends in Plant Science* 8, no. 8: 380–386. https://doi.org/10.1016/S1360-1385(03)00162-6.

Federle, Michael, and Bonnie Lynn Bassler. 2003. "Interspecies communication in bacteria." *The Journal of Clinical Ivestigation* 112, no. 9: 1291–1299. doi:10.1172/JCI200320195.

Feng, Lihui, Steven T. Rutherford, Kai Papenfort, John D. Bagert, Julia C. van Kessel, David A. Tirrell, Ned S. Wingreen, and Bonnie L. Bassler. 2015. "A Qrr Noncoding RNA Deploys Four Different Regulatory Mechanisms to Optimize Quorum-Sensing Dynamics." *Cell* 160, no. 1–2: 228–240. https://doi.org/10.1016/j.cell.2014.11.051.

Freeman, Jeremy A., and Bonnie Lynn Bassler. 1999. "A Genetic Analysis of the Function of LuxO, a Two-Component Response Regulator Involved in Quorum Sensing in *Vibrio harveyi*." *Molecular Microbiology* 31, no. 2: 665–677. https://doi.org/10.1046/j.1365-2958.1999.01208.x.

Fuqua, Clay, and Stephen Carlyle Winans. 1994. "A LuxR-LuxI Type Regulatory System Activates Agrobacterium Ti Plasmid Conjugal Transfer in the Presence of a Plant Tumor Metabolite." *Journal of Bacteriology* 176, no. 10: 2796–2806.

Fuqua, Clay, Matthew R. Parsek, and Everett P. Greenberg. 2001. "Regulation of Gene Expression by Cell-to-Cell Communication: Acyl-Homoserine Lactone Quorum Sensing." *Annual Review of Genetics* 35, no. 1: 439–468. https://doi.org/10.1146/annurev.genet.35.102401.090913.

Fuqua, Clay, Stephen Carlyle Winans, and Everett P. Greenberg. 1994. "Quorum Sensing in Bacteria: The LuxR-LuxI Family of Cell Density-Responsive Transcriptional Regulators." *Journal of Bacteriology* 176, no. 2: 269–275. https://doi.org/10.1128/jb.176.2.269-275.1994.

Galloway, Warren R. J. D., James T. Hodgkinson, Steven D. Bowden, Martin Welch, and David R. Spring. 2011. "Quorum Sensing in Gram-Negative Bacteria: Small-Molecule Modulation of AHL and AI-2 Quorum Sensing Pathways." *Chemical Reviews* 111, no. 1: 28–67. https://doi.org/10.1021/cr100109t.

Gao, Mengsheng, Ming Tang, Lois Guerich, Isai Salas-Gonzalez, and Max Teplitski. 2015. "Modulation of *Sinorhizobium meliloti* Quorum Sensing by Hfq-Mediated Post-Transcriptional Regulation of ExpR." *Environmental Microbiology Reports* 7, no. 1: 148–154. https://doi.org/10.1111/1758-2229.12235.

Geethanjali, Dinesh K. V., N. Raghu, T. S. Gopenath, S. Veerana Gowda, K. W. Ong, M. S. Ranjith, A. Gnanasekaran, M. Karthikeyan, B. Roy, B. Pugazhandhi, et al. 2019. "Quorum Sensing: A Molecular Cell Communication in Bacterial Cells." *Journal of Biomedical Sciences* 5, no. 2: 23–34. https://doi.org/10.3126/jbs.v5i2.23635.

Girard, Léa, Élodie Blanchet, Laurent Intertaglia, Julia Baudart, Didier Stien, Marcelino Suzuki, Philippe Lebaron, Raphaël Lami. 2017. "Characterization of *N*-Acyl Homoserine Lactones in *Vibrio tasmaniensis* LGP32 by a Biosensor-Based UHPLC-HRMS/MS Method." *Sensors* 17, no. 4:1–13. https://doi.org/10.3390/s17040906.

Givskov, Michael, Jörgen Östling, Leo Eberl, Peter W. Lindum, Allan B. Christensen, Gunna Christiansen, Søren Molin, and Staffan Kjelleberg. 1998. "Two Separate Regulatory Systems Participate in Control of Swarming Motility of *Serratia liquefaciens* MG1." *Journal of Bacteriology* 180, no. 3: 742–745. https://www.ncbi.nlm.nih.gov/pmc/articles/PMC106947/.

Givskov, Michael, Rocky De Nys, Michael Manefield, Lone Gram, Ria Maximilien, Leo Eberl, Søren Molin, Peter D. Steinberg, and Staffan Kjelleberg. 1996. "Eukaryotic interference with homoserine lactone-mediated prokaryotic signalling." *Journal of Bacteriology* 178, no. 22: 6618–6622. doi:10.1128/jb.178.22.6618-6622.1996.

Goo, Eunhye, C. D. Majerczyk, J. H. An, J. R. Chandler, Y.-S. Seo, H. Ham, J. Y. Lim, et al. 2012. "Bacterial Quorum Sensing, Cooperativity, and Anticipation of Stationary-Phase Stress." *Proceedings of the National Academy of Sciences* 109, no. 48: 19775–197780. https://doi.org/10.1073/pnas.1218092109.

Goo, Eunhye, Jae Hyung An, Yongsung Kang, and Ingyu Hwang. 2015. "Control of Bacterial Metabolism by Quorum Sensing." *Trends in Microbiology* 23, no. 9: 567–576. https://doi.org/10.1016/j.tim.2015.05.007.

Gray, Kendall M., and James R. Garey. 2001. "The Evolution of Bacterial LuxI and LuxR Quorum Sensing Regulators." *Microbiology* 147, no. 8: 2379–2387. https://doi.org/10.1099/00221287-147-8-2379.

Hansen, Hilde, Amit Anand Purohit, Hanna-Kirsti S. Leiros, Jostein A. Johansen, Stefanie J. Kellermann, Ane Mohn Bjelland, and Nils Peder Willassen. 2015. "The Autoinducer Synthases LuxI and AinS Are Responsible for Temperature-Dependent AHL Production in the Fish Pathogen *Aliivibrio salmonicida*." *BMC Microbiology* 15, no. 1: 69. https://doi.org/10.1186/s12866-015-0402-z.

Henke, Jennifer M., and Bonnie Lynn Bassler. 2004. "Three Parallel Quorum-Sensing Systems Regulate Gene Expression in *Vibrio harveyi*." *Journal of Bacteriology* 186, no. 20: 6902–6914. https://doi.org/10.1128/JB.186.20.6902-6914.2004.

Hentzer, Morten, Hong Wu, Jens B. Andersen, Kathrin Riedel, Thomas B. Rasmussen, Niels Bagge, Naresh Kumar, Mark A. Schembri, Zhijun Song, Peter Kristoffersen, et al. 2003. "Attenuation of *Pseudomonas aeruginosa* Virulence by Quorum Sensing Inhibitors." *The EMBO Journal* 22, no. 15: 3803–3815. https://doi.org/10.1093/emboj/cdg366.

Huber, Birgit, Kathrin Riedel, Morten Hentzer, Arne Heydorn, Astrid Gotschlich, Michael Givskov, Søren Molin, and Leo Eberl. 2001. "The *cep* Quorum-Sensing System of *Burkholderia cepacia* H111 Controls Biofilm Formation and Swarming Motility." *Microbiology* 147, no. 9: 2517–2528. https://doi.org/10.1099/00221287-147-9-2517.

Hughes, David T., and Vanessa Sperandio. 2008. "Inter-Kingdom Signalling: Communication between Bacteria and Their Hosts." *Nature Reviews Microbiology* 6, no. 2: 111–120. https://doi.org/10.1038/nrmicro1836.

Hughes, David T., Darya A. Terekhova, Linda Liou, Carolyn J. Hovde, Jason W. Sahl, Arati V. Patankar, Juan E. Gonzalez, Thomas S. Edrington, David A. Rasko, and Vanessa Sperandio. 2010. "Chemical Sensing in Mammalian Host-Bacterial Commensal Associations." *Proceedings of the National Academy of Sciences* 107, no. 21: 9831–9836. https://doi.org/10.1073/pnas.1002551107.

Husain, Fohad Mabood, Nasser A. Shabib, Saba Noor, Rais Ahmad Khan, Mohammad S havez Khan, Firoz Ahmad Ansari, Mohd Shahnawaz Khan, Altaf Khan, Iqbal Ahmad. 2019. "Current Strategy to Target Bacterial Quorum Sensing and Virulence by Phytocompounds." *New Look to Phytomedicine* 301–329. https://doi.org/10.1016/b978-0-12-814619-4.00012.

Ichinose, Y., Takahiro Sawada, Hidenori Matsui, Mikihiro Yamamoto, Kazuhiro Toyoda, Yoshiteru Noutoshi, and Fumiko Taguchi. 2016. "Motility-Mediated Regulation of Virulence In *Pseudomonas syringae*." *Physiological and Molecular Plant Pathology* 95: 50–54. https://doi.org/10.1016/j.pmpp.2016.02.005.

Janssens, Joost C. A., Hans Steenackers, Stijn Robijns, Edith Gellens, Jeremy Levin, Hui Zhao, Kim Hermans, David De Coster, Tine L. Verhoeven, Kathleen Marchal, et al. 2008. "Brominated Furanones Inhibit Biofilm Formation by *Salmonella enterica* serovar Typhimurium." *Applied and Environmental Microbiology* 74, no. 21: 6639–6648. https://doi.org/10.1128/AEM.01262-08.

Jitmuang, Anupop. 2008. "Human *Chromobacterium violaceum* Infection in Southeast Asia: Case Reports and Literature Review." *Southeast Asian Journal of Tropical Medicine and Public Health* 39, no. 3: 452–460.

Jones, S., B. Yu, N. J. Bainton, M. Birdsall, B. W. Bycroft, S. R. Chhabra, A. J. Cox, P. Golby, P. J. Reeves, S. Stephens, et al. 1993. "The Lux Autoinducer Regulates the Production of Exoenzyme Virulence Determinants in *Erwinia carotovora* and *Pseudomonas aeruginosa*." *The EMBO Journal* 12, no. 6: 2477–2482. http://www.ncbi.nlm.nih.gov/pubmed/8508773.

Joshi, Janak Raj, Netaly Khazanov, Hanoch Senderowitz, Saul Burdman, Alexander Lipsky, and Iris Yedidia. 2016. "Plant Phenolic Volatiles Inhibit Quorum Sensing in Pectobacteria and Reduce Their Virulence by Potential Binding to ExpI and ExpR Proteins." *Scientific Reports* 6: 38126. https://doi.org/10.1038/srep38126.

Kalia, Vipin Chandra. 2013. "Quorum sensing inhibitors: an overview." *Biotechnology Advances* 31, no. 2: 224–245. doi:10.1016/j.biotechadv.2012.10.004.

Kanamaru, Kyoko, Kengo Kanamaru, Ichiro Tatsuno, Toru Tobe, and Chihiro Sasakawa. 2000. "SdiA, an *Escherichia coli* Homologue of Quorum-Sensing Regulators, Controls the Expression of Virulence Factors in Enterohaemorrhagic *Escherichia coli* O157:H7." *Molecular Microbiology* 38, no. 4: 805–816. https://doi.org/10.1046/j.1365-2958.2000.02171.x.

Khajanchi, Bijay K., J. Sha, Elena V. Kozlova, Tatiana E. Erova, G. Suarez, Johanna C. Sierra, Vsevolod L. Popov, Amy J. Horneman, and Ashok K. Chopra. 2009. "N-acylhomoserine lactones involved in quorum sensing control the type VI secretion system, biofilm formation, protease production, and in vivo virulence in a clinical isolate of *Aeromonas hydrophila*". *Microbiology* 155, no. 11: 3518–3531.

Kay, Elisabeth, Cornelia Reimmann, and Dieter Haas. 2006. "Small RNAs in Bacterial Cell-Cell Communication." *Microbe Magazine* 1, no. 2: 63–69. https://doi.org/10.1128/microbe.1.63.1.

Keller, Laurent, and Michael G. Surette. 2006. "Communication in Bacteria: An Ecological and Evolutionary Perspective." *Nature Reviews Microbiology* 4, no. 4: 249–258. https://doi.org/10.1038/nrmicro1383.

LaSarre, Breah, and Michael J. Federle. 2013. "Exploiting Quorum Sensing to Confuse Bacterial Pathogens." *Microbiology and Molecular Biology Reviews* 77, no. 1: 73–111. https://doi.org/10.1128/MMBR.00046-12.

Laue, Bridget E., Yan Jiang, Siri Ram Chhabra, Sinead Jacob, Gordon S. A. B. Stewart, A Hardman, J. Allan Downie, Fergal O'Gara, and Paul Williams. 2000. "The Biocontrol Strain *Pseudomonas fluorescens* F113 Produces the Rhizobium Small Bacteriocin, *N*-(3-Hydroxy-7-Cis-Tetradecenoyl) Homoserine Lactone, via HdtS, a Putative Novel *N*-Acylhomoserine Lactone Synthase." *Microbiology* 146: 2469–2480. https://doi.org/10.1099/00221287-146-10-2469.

Lee, Jasmine, and Lianhui Zhang. 2014. "The Hierarchy Quorum Sensing Network in *Pseudomonas aeruginosa*." *Protein Cell* 6, no. 1: 26–41. https://doi.org/10.1007/s13238-014-0100-x.

Lee, Jasmine, Jien Wu, Yinyue Deng, Jing Wang, Chao Wang, Jianhe Wang, Changqing Chang, Yihu Dong, Paul Williams, and Lian-Hui Zhang. 2013. "A Cell-Cell Communication Signal Integrates Quorum Sensing and Stress Response." *Nature Chemical Biology* 9, no. 5: 339–343. https://doi.org/10.1038/nchembio.1225.

Lee, Jintae, Arul Jayaraman, and Thomas K. Wood. 2007. "Indole Is an Inter-Species Biofilm Signal Mediated by SdiA." *BMC Microbiology* 7, no. 1: 42. https://doi.org/10.1186/1471-2180-7-42.

Lee Jin-Hyung, Wood, Thomas Keith, Lee, Jintae. 2015. "Roles of indole as an interspecies and interkingdom signaling molecule." *Trends in Microbiology* 23, no. 11:707–718. http://doi.org/10.1016/j.tim.2015.08.001.

Leimbach, Andreas, Jörg Hacker, and Ulrich Dobrindt. 2013. "*E. coli* as an All-Rounder: The Thin Line Between Commensalism and Pathogenicity." *Current Topics in Microbiology and Immunology* 358, 3–32. https://doi.org/10.1007/82_2012_303.

Lenz, Derrick H., Kenny C. Mok, Brendan N. Lilley, Rahul V. Kulkarni, Ned S. Wingreen, and Bonnie Lynn Bassler. 2004. "The Small RNA Chaperone Hfq and Multiple Small RNAs Control Quorum Sensing in *Vibrio harveyi* and *Vibrio cholerae*." *Cell* 118, no. 1: 69–82. https://doi.org/10.1016/j.cell.2004.06.009.

Lerat, Emmanuelle, and Nancy A. Moran. 2004. "The Evolutionary History of Quorum-Sensing Systems in Bacteria." *Molecular Biology and Evolution* 21, no. 5: 903–913. https://doi.org/10.1093/molbev/msh097.

Li, Xiaoying, Yali Ma, Shuqing Liang, Yu Tian, Sanjun Yin, Sisi Xie, and Hua Xie. 2018. "Comparative Genomics of 84 Pectobacterium Genomes Reveals the Variations Related to a Pathogenic Lifestyle." *BMC Genomics* 19, no. 1: 889. https://doi.org/10.1186/s12864-018-5269-6.

Li, Yung-Hua, and Xiaolin Tian. 2012. "Quorum Sensing and Bacterial Social Interactions in Biofilms." *Sensors* 12, no. 3: 2519–2538. https://doi.org/10.3390/s120302519.

Li, Yung-Hua, Nan Tang, Marcelo B. Aspiras, Peter C. Y. Lau, Janet H. Lee, Richard P. Ellen, and Dennis G. Cvitkovitch. 2002. "A Quorum-Sensing Signaling System Essential for Genetic Competence in *Streptococcus mutans* Is Involved in Biofilm Formation." *Journal of Bacteriology* 184, no. 10: 2699–2708. https://doi.org/10.1128/JB.184.10.2699-2708.2002.

Lilley, Brendan N., and Bonnie L. Bassler. 2000. "Regulation of Quorum Sensing in *Vibrio harveyi* by LuxO and Sigma-54." *Molecular Microbiology* 36, no. 4: 940–954. https://doi.org/10.1046/j.1365-2958.2000.01913.x.

Lindum, P. W., U. Anthoni, C. Christophersen, L. Eberl, S. Molin, and M. Givskov. 1998. "*N*-Acyl-L-Homoserine Lactone Autoinducers Control Production of an Extracellular Lipopeptide Biosurfactant Required for Swarming Motility of *Serratia liquefaciens* MG1." *Journal of Bacteriology* 180, no. 23:6384–6388.

Liu, Jianfei, Fu, Kaifei, Wu, Chenglin. Li, Fei, Zhou, Lijun. 2018. "In-Group" Communication in Marine *Vibrio*: A Review of *N*-Acyl Homoserine Lactones-Driven Quorum Sensing." *Frontiers in Cellular and Infection Microbiology* 8: 139–146. https://doi.org/10.3389/fcimb.2018.00139.

Liu, M., H. Wang, and M. W. Griffiths. 2007. "Regulation of Alkaline Metalloprotease Promoter by N-Acyl Homoserine Lactone Quorum Sensing in *Pseudomonas fluorescens*." *Journal of Applied Microbiology* 103, no. 6: 2174–2184. 10.1111/j.1365-2672.2007.03488.x.

Liu, Qin, Rodrigo M. P. Siloto, Richard Lehner, Scot J. Stone, and Randall J. Weselake. 2012. "Acyl-CoA:Diacylglycerol Acyltransferase: Molecular Biology, Biochemistry and Biotechnology." *Progress in Lipid Research* 51, no. 4: 350–377. https://doi.org/10.1016/j.plipres.2012.06.001.

Liu, Zhanjun, Weishan Wang, Ying Zhu, Qianhong Gong, Wengong Yu, and Xinzhi Lu. 2013. "Antibiotics at Subinhibitory Concentrations Improve the Quorum Sensing Behavior of *Chromobacterium violaceum*." *FEMS Microbiology Letters* 341, no. 1: 37–44. https://doi.org/10.1111/1574-6968.12086.

Liu, Zhen, Fengxia Que, Li Liao, Min Zhou, Lixiang You, Qing Zhao, Yuanyuan Li, Hua Niu, Shuyan Wu, and Rui Huang. 2014. "Study on the Promotion of Bacterial Biofilm Formation by a *Salmonella* Conjugative Plasmid and the Underlying Mechanism." *PLoS ONE* 9, no. 10: e109808. https://doi.org/10.1371/journal.pone.0109808.

Lupp, Claudia, and Edward G. Ruby. 2005. "*Vibrio fischeri* Uses Two Quorum-Sensing Systems for the Regulation of Early and Late Colonization Factors." *Journal of Bacteriology* 187, no. 11: 3620–3629. https://doi.org/10.1128/JB.187.11.3620-3629.2005.

Lupp, Claudia, Mark Urbanowski, Everett P. Greenberg, and Edward G. Ruby. 2003. "The *Vibrio fischeri* Quorum-Sensing Systems Ain and Lux Sequentially Induce Luminescence Gene Expression and Are Important for Persistence in the Squid Host." *Molecular Microbiology* 50, no. 1: 319–331. https://doi.org/10.1046/j.1365-2958.2003.t01-1-03585.x.

Mahenthiralingam, Eshwar, Teresa A. Urban, and Joanna B. Goldberg. 2005. "The Multifarious, Multireplicon *Burkholderia cepacia* Complex." *Nature Reviews Microbiology* 3, no. 2: 144–156. https://doi.org/10.1038/nrmicro1085.

Mahlen, Steven D. 2011. "*Serratia* Infections: From Military Experiments to Current Practice." *Clinical Microbiology Reviews* 24, no. 4: 755–791. https://doi.org/10.1128/CMR.00017-11.

Majerczyk, Charlotte D., Mitchell J. Brittnacher, Michael A. Jacobs, Christopher D. Armour, Matthew C. Radey, Richard Bunt, Hillary S. Hayden, Ryland Bydalek, and E. Peter Greenberg. 2014b. "Cross-Species Comparison of the *Burkholderia pseudomallei*, *Burkholderia thailandensis*, and *Burkholderia mallei* Quorum-Sensing Regulons." *Journal of Bacteriology* 196, no. 22: 3862–3871. https://doi.org/10.1128/JB.01974-14.

Majerczyk, Charlotte, Mitchell Brittnacher, Michael Jacobs, Christopher D. Armour, Mathew Radey, Emily Schneider, Somsak Phattarasokul, Richard Bunt, and Everett P. Greenberg. 2014a. "Global Analysis of the *Burkholderia thailandensis* Quorum Sensing-Controlled Regulon." *Journal of Bacteriology* 196, no. 7: 1412–1424. https://doi.org/10.1128/JB.01405-13.

Malgaonkar, Anuja, and Mrinalini Nair. 2019. "Quorum Sensing in *Pseudomonas aeruginosa* Mediated by RhlR Is Regulated by a Small RNA PhrD." *Scientific Reports* 9, no. 1: 1–11. https://doi.org/10.1038/s41598-018-36488-9.

Martins, Maurilio L., Uelinton M. Pinto, Katharina Riedel, Maria C. D. Vanetti, Hilário C. Mantovani, and Elza F. de Araújo. 2014. "Lack of AHL-Based Quorum Sensing in *Pseudomonas fluorescens* Isolated from Milk." *Brazilian Journal of Microbiology* 45, no. 3: 1039–1046. https://doi.org/10.1590/S1517-83822014000300037.

Martins, Maurilio L., Uelinton M. Pinto, Katharina Riedel, and Maria C. D. Vanetti. 2018. "Quorum Sensing and Spoilage Potential of Psychrotrophic Enterobacteriaceae Isolated from Milk." *BioMed Research International* 2018: 1–13. https://doi.org/doi:10.1155/2018/2723157.

Maura, Damien, Ronen Hazan, Tomoe Kitao, Alicia E. Ballok, and Laurence G. Rahme. 2016. "Evidence for Direct Control of Virulence and Defense Gene Circuits by the *Pseudomonas aeruginosa* Quorum Sensing Regulator, MvfR." *Scientific Reports* 6: 34083. https://doi.org/10.1038/srep34083.

McClean, Kay H., Michael K. Winson, Leigh Fish, Adrian Taylor, Siri Ram Chhabra, Miguel Camara, Mavis Daykin, et al. 1997. "Quorum Sensing and *Chromobacterium violaceum*: Exploitation of Violacein Production and Inhibition for the Detection of N-Acylhomoserine Lactones." *Microbiology* 143: 3703–3711. https://doi.org/10.1099/00221287-143-12-3703.

McKeon, Suzanne A., David T. Nguyen, Duber F. Viteri, James E. A. Zlosnik, and Pamela A. Sokol. 2011. "Functional Quorum Sensing Systems Are Maintained during Chronic *Burkholderia cepacia* Complex Infections in Patients with Cystic Fibrosis." *The Journal of Infectious Diseases* 203, no. 3: 383–392. https://doi.org/10.1093/infdis/jiq054.

Michael, Bindhu, Jenee N. Smith, Simon Swift, Fred Heffron, and Brian M. M. Ahmer. 2001. "SdiA of *Salmonella enterica* Is a LuxR Homolog That Detects Mixed Microbial Communities." *Journal of Bacteriology* 183, no. 19: 5733–5742. https://doi.org/10.1128/JB.183.19.5733-5742.2001.

Middleton, Barry, Helen C. Rodgers, Miguel Cámara, Alan J. Knox, Paul Williams, and Andrea Hardman. 2002. "Direct Detection of *N*-Acylhomoserine Lactones in Cystic Fibrosis Sputum." *FEMS Microbiology Letters* 207, no. 1: 1–7. https://doi.org/10.1111/j.1574-6968.2002.tb11019.x.

Milton, Debra L., Victoria J. Chalker, David Kirke, Andrea Hardman, Miguel Cámara, Williams Paul. 2001. "The LuxM homologue VanM from *Vibrio anguillarum* directs the synthesis of *N*-(3-hydroxy hexanoyl) homoserine lactone and N-hexanoylhomoserine lactone." *Journal of Bacteriology* 183, no. 12: 3537–3547. https://doi.org/10.1128/JB.183.12.3537-3547.2001.

Miyamoto, Carol M., Jaidip Chatterjee, Elana Swartzman, Rose Szittner, and Edward A. Meighen. 1996. "The Role of the Lux Autoinducer in Regulating Luminescence in *Vibrio harveyi*; Control of LuxR Expression." *Molecular Microbiology* 19, no. 4: 767–775. https://doi.org/10.1046/j.1365-2958.1996.417948.x.

Morohoshi, Tomohiro, Masashi Kato, Katsumasa Fukamachi, Norihiro Kato, and Tsukasa Ikeda. 2008. "*N*-Acylhomoserine Lactone Regulates Violacein Production in *Chromobacterium violaceum* Type Strain ATCC 12472." *FEMS Microbiology Letters* 279, no. 1: 124–130. https://doi.org/10.1111/j.1574-6968.2007.01016.x.

Mott, Tiffany, Rekha G. Panchal, and Sathish Rajamani. 2017. "Quorum Sensing in *Burkholderia pseudomallei* and Other *Burkholderia* Species." *Current Tropical Medicine Reports* 4, no. 4: 199–207. https://doi.org/10.1007/s40475-017-0127-1.

Nakatsu, Yukiko., Hidenori Matsui, Mikihiro Yamamoto, Yoshiteru Noutoshi, Kazuhiro Toyoda, and Yuki Ichinose. 2019. "Quorum-Dependent Expression of RsmX and RsmY, Small Non-Coding RNAs, in *Pseudomonas syringae*." *Microbiological Research* 223–225: 72–78. https://doi.org/10.1016/j.micres.2019.04.004.

Natrah, F. M. I., M. Iftakharul Alam, Sushant Pawar, A. Shiri Harzevili, Nancy Nevejan, Nico Boon, Patrick Sorgeloos, Peter Bossier, and Tom Defoirdt. 2012. "The Impact of Quorum Sensing on the Virulence of *Aeromonas hydrophila* and *Aeromonas salmonicida* towards Burbot (*Lota lota* L.) Larvae." *Veterinary Microbiology* 159, nos. 1–2: 77–82. https://doi.org/10.1016/j.vetmic.2012.03.014.

Nealson, Kenneth, Platt, Terry and Hastings, Woodland. 1970. "Cellular Control of the Synthesis and Activity of the Bacterial Luminescent Systems." *Journal of Bacteriology* 104, no. 1: 313–322.

Nesse, Live L., Kristin Berg, Lene K. Vestby, Ingrid Olsaker, and Berit Djønne. 2011. "*Salmonella* Typhimurium Invasion of HEp-2 Epithelial Cells *in vitro* Is Increased by *N*-Acylhomoserine Lactone Quorum Sensing Signals." *Acta Veterinaria Scandinavica* 53, no. 1: 44. https://doi.org/10.1186/1751-0147-53-44.

Nguyen, Y., Nam X. Nguyen, Jamie L. Rogers, Jun Liao, John B. MacMillan, Youxing Jiang, and Vanessa Sperandio. 2015. "Structural and Mechanistic Roles of Novel Chemical Ligands on the SdiA Quorum-Sensing Transcription Regulator." *mBio* 6, no. 2. https://doi.org/10.1128/mBio.02429-14.

Oca-Mejía, Marielba Montes de, Israel Castillo-Juárez, Mariano Martínez-Vázquez, Marcos Soto-Hernandez, and Rodolfo García-Contreras. 2014. "Influence of Quorum Sensing in Multiple Phenotypes of the Bacterial Pathogen *Chromobacterium violaceum*." *Pathogens and Disease* 73, no. 2: 1–4. https://doi.org/10.1093/femspd/ftu019.

Ochsner, U. A., and J. Reiser. 1995. "Autoinducer-Mediated Regulation of Rhamnolipid Biosurfactant Synthesis in *Pseudomonas aeruginosa*." *Proceedings of the National Academy of Sciences of USA* 92, no. 14: 6424–6428. https://doi.org/10.1073/pnas.92.14.6424.

Pang, Zheng, Renee Raudonis, Bernard R. Glick, Tong-jun Lin, and Zhenyu Cheng. 2019. "Antibiotic Resistance in *Pseudomonas aeruginosa*: Mechanisms and Alternative Therapeutic Strategies." *Biotechnology Advances* 37, no. 1: 177–192. https://doi.org/10.1016/j.biotechadv.2018.11.013.

Papenfort, Kai, and Bonnie Lynn Bassler. 2016. "Quorum Sensing Signal–Response Systems in Gram-Negative Bacteria." *Nature Reviews Microbiology* 14, no. 9: 576–588. https://doi.org/10.1038/nrmicro.2016.89.

Pappas, Katherine M. 2008. "Cell-Cell Signaling and the *Agrobacterium tumefaciens* Ti Plasmid Copy Number Fluctuations." *Plasmid* 60, no. 2: 89–107. https://doi.org/10.1016/j.plasmid.2008.05.003.

Park, T. H., B. S. Choi, A. Y. Choi, I. Y. Choi, S. Heu, and B. S. Park. 2012. "Genome Sequence of *Pectobacterium carotovorum* Subsp. *carotovorum* Strain PCC21, a Pathogen Causing Soft Rot in Chinese Cabbage." *Journal of Bacteriology* 194, no. 22: 6345–6346. 10.1128/JB.01583-12.

Pearson, James P., Kendall M. Gray, Luciano Passador, Kenneth D. Tucker, Anatol Eberhard, Barbara H. Iglewski, and Everett P. Greenberg. 1994. "Structure of the Autoinducer Required for Expression of *Pseudomonas aeruginosa* Virulence Genes." *Proceedings of the National Academy of Sciences of USA* 91, no. 1: 197–201. https://doi.org/10.1073/pnas.91.1.197.

Pearson, James P., Luciano Passador, Barbara H. Iglewski, and Everett P. Greenberg. 1995. "A Second *N*-Acylhomoserine Lactone Signal Produced by *Pseudomonas aeruginosa*." *Proceedings of the National Academy of Sciences of USA* 92, no. 5: 1490–1494. https://doi.org/10.1073/pnas.92.5.1490.

Pérez-Velázquez, Judith, Beatriz Quiñones, Burkhard A. Hense, and Christina Kuttler. 2015. "A Mathematical Model to Investigate Quorum Sensing Regulation and Its Heterogeneity in *Pseudomonas Syringae* on Leaves." *Ecological Complexity* 21: 128–141. https://doi.org/10.1016/j.ecocom.2014.12.003.

Pesci, Everett. C., Jared. B. J. Milbank, James. P. Pearson, Susan McKnight, Andrew S. Kende, Everett P. Greenberg, and Barbara H. Iglewski. 1999. "Quinolone Signaling in the Cell-to-Cell Communication System of *Pseudomonas aeruginosa*." *Proceedings of the National Academy of Sciences of USA* 96, no. 20: 11229–11234. https://doi.org/10.1073/pnas.96.20.11229.

Pinto, Uelinton M., and Stephen C. Winans. 2009. "Dimerization of the Quorum-Sensing Transcription Factor TraR Enhances Resistance to Cytoplasmic Proteolysis." *Molecular Microbiology* 73, no. 1: 32–42. https://doi.org/10.1111/j.1365-2958.2009.06730.x.

Pinto, Uelinton M., Esther D. Costa, Hilario C. Mantovani, and M. C. D. Vanetti. 2010. "The Proteolytic Activity of *Pseudomonas fluorescens* 07a Isolated from Milk Is Not Regulated by Quorum Sensing Signals." *Brazilian Journal of Microbiology* 41, no. 1: 91–96. https://doi.org/10.1590/S1517-83822010000100015.

Pinto, Uelinton M., Katherine M. Pappas, and Stephen C. Winans. 2012. "The ABCs of Plasmid Replication and Segregation." *Nature Reviews Microbiology*: 755–765. https://doi.org/10.1038/nrmicro2882.

Platt, Thomas Gene, and Clay Fuqua. 2010. "What's in a Name? The Semantics of Quorum Sensing." *Trends in Microbiology* 18, no. 9: 383–387. https://doi.org/10.1016/j.tim.2010.05.003.

Põllumaa, Lee, Tiina Alamäe, and Andres Mäe. 2012. "Quorum Sensing and Expression of Virulence in Pectobacteria." *Sensors* 12, no. 3: 3327–3349. doi:10.3390/s120303327.

Ponce-Rossi, Adriana R., Uelinton M. Pinto, Andrea O. B. Ribon, Denise M. S. Bazzolli, and Maria Criatin Dantas Vanetti. 2016. "Quorum Sensing Regulated Phenotypes in *Aeromonas hydrophila* ATCC 7966 Deficient in AHL Production." *Annals of Microbiology* 66, no. 3: 1117–1126. https://doi.org/10.1007/s13213-016-1196-4.

Quecan, Beatriz X. V., José T. C. Santos, Milagros L. C. Rivera, Neuza M. A. Hassimotto, Felipe A. Almeida, and Uelinton M. Pinto. 2019. "Effect of Quercetin Rich Onion Extracts on Bacterial Quorum Sensing." *Frontiers in Microbiology* 10, no. 867. https://doi.org/10.3389/fmicb.2019.00867.

Quiñones, Beatriz, Catherine J. Pujol, and Steven E. Lindow. 2004. "Regulation of AHL Production and Its Contribution to Epiphytic Fitness in *Pseudomonas syringae*." *Molecular Plant-Microbe Interactions* 17, no. 5: 521–531. https://doi.org/10.1094/mpmi.2004.17.5.521.

Rahmati, Sonia, Shirley Yang, Amy L. Davidson, and E. Lynn Zechiedrich. 2002. "Control of the AcrAB Multidrug Efflux Pump by Quorum-Sensing Regulator SdiA." *Molecular Microbiology* 43, no. 3: 677–685. https://doi.org/10.1046/j.1365-2958.2002.02773.x.

Rasmussen, Thomas B., Kristian Fog Nielsen, Henrique Machado, Jette Melchiorsen, Lone Gram, Eva Sonnenschein. 2014."Global and Phylogenetic Distribution of Quorum Sensing Signals, Acyl Homoserine Lactones, in the Family of *Vibrionaceae*." *Marine Drugs* 12: 5527–5546. https://doi.org/10.3390/md12115527.

Rice, S. A., K. S. Koh, S. Y. Queck, M. Labbate, K. W. Lam, and S. Kjelleberg. 2005. "Biofilm Formation and Sloughing in *Serratia marcescens* Are Controlled by Quorum Sensing and Nutrient Cues." *Journal of Bacteriology* 187, no. 10: 3477–3485. https://doi.org/10.1128/JB.187.10.3477-3485.2005.

Rutherford, Steven T., Julie S. Valastyan, Thibaud Taillefumier, Ned S. Wingreen, and Bonnie L. Bassler. 2015. "Comprehensive Analysis Reveals How Single Nucleotides Contribute to Noncoding RNA Function in Bacterial Quorum Sensing." *Proceedings of the National Academy of Sciences* 112, no. 44: E6038–E6047. https://doi.org/10.1073/pnas.1518958112.

Sabag-Daigle, Anice, and Brian M. M. Ahmer. 2012. "ExpI and PhzI Are Descendants of the Long Lost Cognate Signal Synthase for SdiA." *PLoS ONE* 7, no. 10: e47720. https://doi.org/10.1371/journal.pone.0047720.

Scales, Brittan S., Robert P. Dickson, John J. LiPuma, and Gary B. Huffnagle. 2014. "Microbiology, Genomics, and Clinical Significance of the *Pseudomonas fluorescens* Species Complex, an Unappreciated Colonizer of Humans." *Clinical Microbiology Reviews* 27, no. 4: 927–948. https://doi.org/10.1128/CMR.00044-14.

Schuster, Martin, Phoebe C. Lostroh, Tomoo Ogi, and Everett P. Greenberg. 2003. "Identification, Timing, and Signal Specificity of *Pseudomonas aeruginosa* Quorum-Controlled Genes: A Transcriptome Analysis." *Journal of Bacteriology* 185, no. 7: 2066–2079. https://doi.org/10.1128/JB.185.7.2066-2079.2003.

Schütz, Christian, and Martin Empting. 2018. "Targeting the *Pseudomonas* Quinolone Signal Quorum Sensing System for the Discovery of Novel Anti-Infective Pathoblockers." *Beilstein Journal of Organic Chemistry* 14: 2627–2645. https://doi.org/10.3762/bjoc.14.241.

Shao, Yi, and Bonnie Lynn Bassler. 2012. "Quorum-Sensing Non-Coding Small RNAs Use Unique Pairing Regions to Differentially Control MRNA Targets." *Molecular Microbiology* 83, no. 3: 599–611. https://doi.org/10.1111/j.1365-2958.2011.07959.x.

Sharma, V. K., and S. M. D. Bearson. 2013. "Evaluation of the Impact of Quorum Sensing Transcriptional Regulator SdiA on Long-Term Persistence and Fecal Shedding of *Escherichia coli* O157:H7 in Weaned Calves." *Microbial Pathogenesis* 57: 21–26. https://doi.org/10.1016/j.micpath.2013.02.002.

Slater, Steven C., Barry S. Goldman, Brad Goodner, João C. Setubal, Stephen K. Farrand, Eugene W. Nester, Thomas J. Burr, et al. 2009. "Genome Sequences of Three Agrobacterium Biovars Help Elucidate the Evolution of Multichromosome Genomes in Bacteria." *Journal of Bacteriology* 191, no. 8: 2501–2511. https://doi.org/10.1128/JB.01779-08.

Smith, Jenée N., and Brian M. M. Ahmer. 2003. "Detection of Other Microbial Species by *Salmonella*: Expression of the SdiA Regulon." *Journal of Bacteriology* 185, no. 4: 1357–1366. https://doi.org/10.1128/JB.185.4.1357-1366.2003.

Smith, Jenee N., Jessica L. Dyszel, Jitesh A. Soares, Ellermeier, C. Altier, Sara D. Lawhon, Garry Adams, V. Konjufca, R. Slauch Curtiss, M. James, Brian M. M. Ahmer. 2008. "SdiA, an N-acylhomoserine lactone receptor, becomes active during the transit of *Salmonella enterica* through the gastrointestinal tract of turtles". *PLoS* One 3, no. 7: e2826. https://doi.org/10.1371/journal.pone.0002826.

Smith, Roger S., and Barbara H. Iglewski. 2003. "*P. aeruginosa* Quorum-Sensing Systems and Virulence." *Current Opinion in Microbiology* 6: 56–60. https://doi.org/10.1016/S1369-5274(03)00008-0.

Soares, Jitesh A., and Brian M. M. Ahmer. 2011. "Detection of Acyl-Homoserine Lactones by *Escherichia* and *Salmonella*." *Current Opinion in Microbiology* 14, no. 2: 188–193. https://doi.org/10.1016/j.mib.2011.01.006.

Stauff, Devin L., and Bonnie L. Bassler. 2011. "Quorum Sensing in *Chromobacterium violaceum*: DNA Recognition and Gene Regulation by the CviR Receptor." *Journal of Bacteriology* 193, no. 15: 3871–3878. https://doi.org/10.1128/JB.05125-11.

Steidle, A., Marie Allesen-holm, Kathrin Riedel, Gabriele Berg, Michael Givskov, Søren Molin, and Leo Eberl. 2002. "Identification and Characterization of an N-Acylhomoserine Lactone-Dependent Quorum-Sensing System in *Pseudomonas putida* Strain IsoF." *Applied and Environmental Microbiology* 68, no. 12: 6371–6382. doi: 10.1128/aem.68.12.6371-6382.2002.

Steindler, Laura, and Vittorio Venturi. 2007. "Detection of Quorum-Sensing *N*-Acyl Homoserine Lactone Signal Molecules by Bacterial Biosensors." *FEMS Microbiology Letters* 266, no. 1: 1–9. https://doi.org/10.1111/j.1574-6968.2006.00501.x.

Subramoni, Sujatha, Diana Vanessa Florez Salcedo, and Zulma R. Suarez-Moreno. 2015. "A Bioinformatic Survey of Distribution, Conservation, and Probable Functions of LuxR Solo Regulators in Bacteria." *Frontiers in Cellular and Infection Microbiology* 5: 1–17. https://doi.org/10.3389/fcimb.2015.00016.

Sybiya Vasantha Packiavathy, Issac Abraham, Palani Agilandeswari, Khadar Syed Musthafa, Shunmugiah Karutha Pandian, and Arumugam Veera Ravi. 2012. "Antibiofilm and quorum sensing inhibitory potential of *Cuminum cyminum* and its secondary metabolite methyl eugenol against gram negative bacterial pathogens." *Food Research International* 45, no. 1: 85–92. doi:10.1016/j.foodres.2011.10.022.

Tang Rong, Junli Zhu, Lifang Feng, Jianrong Li, and Xiaoxiang Liu. 2019. "Characterization of LuxI/LuxR and Their Regulation Involved in Biofilm Formation and Stress Resistance in Fish Spoilers *Pseudomonas fluorescens*." *International Journal of Food Microbiology* 297: 60–71. https://doi.org/10.1016/j.ijfoodmicro.2018.12.011.

Thomson, N. R., M. A. Crow, S. J. McGowan, A. Cox, and G. P. C. Salmond. 2000. "Biosynthesis of Carbapenem Antibiotic and Prodigiosin Pigment in *Serratia* Is under Quorum Sensing Control." *Molecular Microbiology* 36, no. 3: 539–556. https://doi.org/10.3109/00498258509045329.

Tu, Kimberly C., and Bonnie L. Bassler. 2007. "Multiple Small RNAs Act Additively to Integrate Sensory Information and Control Quorum Sensing in *Vibrio harveyi*." *Genes & Development* 21, no. 2: 221–233. https://doi.org/10.1101/gad.1502407.

Turan, N. Bakaraki, Dotse S. Chormey, Çağdaş Büyükpınar, Güleda O. Engin, and Sezgin Bakirdere. 2017. "Quorum Sensing: Little Talks for an Effective Bacterial Coordination." *TrAC–Trends in Analytical Chemistry* 91: 1–11. https://doi.org/10.1016/j.trac.2017.03.007.

Ulrich, Ricky L. 2004. "Role of Quorum Sensing in the Pathogenicity of *Burkholderia pseudomallei*." *Journal of Medical Microbiology* 53, no. 11: 1053–1064. https://doi.org/10.1099/jmm.0.45661-0.

Ulrich, Ricky L., David DeShazer, Harry B. Hines, and Jeffrey A. Jeddeloh. 2004. "Quorum Sensing: A Transcriptional Regulatory System Involved in the Pathogenicity of *Burkholderia mallei*." *Infection and Immunity* 72, no. 11: 6589–6596. https://doi.org/10.1128/IAI.72.11.6589-6596.2004.

Urbanczyk, Henryk, Jennifer C. Ast, Melissa J. Higgins, Jeremy Carson, and Paul V. Dunlap. 2007. "Reclassification of *Vibrio fischeri*, *Vibrio logei*, *Vibrio salmonicida* and *Vibrio wodanis* as *Aliivibrio fischeri* gen. nov., comb. nov., *Aliivibrio logei* comb. nov., *Aliivibrio salmonicida* comb. nov. and *Aliivibrio wodanis* comb. nov." *International Journal of Systematic and Evolutionary Microbiology* 57, no. 12: 2823–2829. https://doi.org/10.1099/ijs.0.65081-0.

Uroz, Stéphane, Yves Dessaux, and Phil Oger. 2009. "Quorum sensing and quorum quenching: the yin and yang of bacterial communication." *ChemBioChem* 10, no. 2: 205–16. doi:10.1002/cbic.200800521.

Van Houdt, Rob, Abram Aertsen, Pieter Moons, Kristof Vanoirbeek, and Chris W. Michiels. 2006. "*N*-Acyl-L-Homoserine Lactone Signal Interception by *Escherichia coli*." *FEMS Microbiology Letters* 256, no. 1: 83–89. https://doi.org/10.1111/j.1574-6968.2006.00103.x.

Van Houdt, Rob, Michael Givskov, and Chris W. Michiels. 2007. "Quorum Sensing in *Serratia*." *FEMS Microbiology Reviews* 31, no. 4: 407–424. https://doi.org/10.1111/j.1574-6976.2007.00071.x.

Van Houdt, Rob, Pieter Moons, M. Hueso Buj, and Chris W. Michiels. 2006. "*N*-Acyl-L-Homoserine Lactone Quorum Sensing Controls Butanediol Fermentation in *Serratia plymuthica* RVH1 and *Serratia marcescens* MG1." *Journal of Bacteriology* 188, no. 12: 4570–4572. https://doi.org/10.1128/JB.00144-06.

Vannini, A., C. Volpari, C. Gargioli, E. Muraglia, R. Cortese, R. De Francesco, P. Neddermann, and S. D. Marco. 2002. "The Crystal Structure of the Quorum Sensing Protein TraR Bound to Its Autoinducer and Target DNA." *The EMBO Journal* 21, no. 17: 4393–4401. https://doi.org/10.1093/emboj/cdf459.

Vattem, D. A., K. Mihalik, S. H. Crixell, and R. J. C. McLean. 2007. "Dietary Phytochemicals as Quorum Sensing Inhibitors." *Fitoterapia* 78, no. 4: 302–310. https://doi.org/10.1016/j.fitote.2007.03.009.

Venturi, Vittorio, Arianna Friscina, Iris Bertani, Giulia Devescovi, and Claudio Aguilar. 2004. "Quorum Sensing in the *Burkholderia cepacia* Complex." *Research in Microbiology* 155, no. 4: 238–244. https://doi.org/10.1016/j.resmic.2004.01.006.

Venturi, Vittorio. 2006. "Regulation of Quorum Sensing in *Pseudomonas*." *FEMS Microbiology Reviews* 30, no. 2: 274–291. https://doi.org/10.1111/j.1574-6976.2005.00012.x.

Von Bodman, S. B., W. D. Bauer, and D. L. Coplin. 2003. "Quorum Sensing in Plant-Pathogenic Bacteria." *Annual Review of Phytopathology* 41: 455–842. 10.1146/annurev.phyto.41.052002.095652.

Wang, Hu-Hu, Ke-Ping Ye, Qiu-Qin Zhang, Yang Dong, Xing-Lian Xu, and Guang-Hong Zhou. 2013. "Biofilm Formation of Meat-Borne *Salmonella enterica* and Inhibition by the Cell-Free Supernatant from *Pseudomonas aeruginosa*." *Food Control* 32, no. 2: 650–658. https://doi.org/10.1016/j.foodcont.2013.01.047.

Waters, Christopher M., and Bonnie L. Bassler. 2006. "The *Vibrio harveyi* Quorum-Sensing System Uses Shared Regulatory Components to Discriminate between Multiple Autoinducers." *Genes & Development* 20, no. 19: 2754–2767. https://doi.org/10.1101/gad.1466506.

Wei, Jun Rong, and Hsin Chih Lai. 2006. "*N*-Acylhomoserine Lactone-Dependent Cell-to-Cell Communication and Social Behavior in the Genus *Serratia*." *International Journal of Medical Microbiology* 296, nos. 2–3: 117–124. https://doi.org/10.1016/j.ijmm.2006.01.033.

Wei, Jun-Rong, Y. H. Tsai, Y. T. Horng , P.C. Soo, S. C. Hsieh, P. R. Hsueh, J. T. Horng, P. Williams, and H. C. Lai. 2006. "A Mobile Quorum-Sensing System in *Serratia marcescens*." *Journal of Bacteriology* 188, no. 4: 1518–1525. https://doi.org/10.1128/JB.188.4.1518-1525.2006.

White, C. E., and S. C. Winans. 2007. "The Quorum-Sensing Transcription Factor TraR Decodes Its DNA Binding Site by Direct Contacts with DNA Bases and by Detection of DNA Flexibility." *Molecular Microbiology* 64, no. 1: 245–256. https://doi.org/10.1111/j.1365-2958.2007.05647.x.

Whitehead, Neil A, Anne M. L. Barnard, Holly Slater, Natalie J. L. Simpson, and George P. C. Salmond. 2001. "Quorum sensing in gram-negative bacteria." *FEMS Microbiology Reviews* 25: 365–404.

Whiteley, Marvin, Kimberly Lee, and E. Greenberg. 1999. "Identification of genes controlled by quorum sensing in *Pseudomonas aeruginosa*." *Clinical Infectious Diseases* 96, no. 24: 13904–13909. doi:10.1073/pnas.96.24.13904.

WHO. 2017. "WHO Publishes List of Bacteria for Which New Antibiotics Are Urgently Needed." Accessed April 25, 2019. https://www.who.int/news-room/detail/27-02-2017-who-publishes-list-of-bacteria-for-which-new-antibiotics-are-urgently-needed.

Williams, Paul. 2007. "Quorum Sensing, Communication and Cross-Kingdom Signalling in the Bacterial World." *Microbiology* 153: 3923–3938. https://doi.org/10.1099/mic.0.2007/012856-0.

Winans, Stephen C. 1992. "Two-Way Chemical Signaling in Agrobacterium-Plant Interactions." *Microbiology and Molecular Biology Reviews* 56: 12–31.

Wood, D. W., F. Gong, M. M. Daykin, P. Willians, and L. S. Pierson. 1997. "*N*-Acyl-Homoserine Lactone-Mediated Regulation of Phenazine Gene Expression by *Pseudomonas aureofaciens* 30–84 in the Wheat Rhizosphere" *Journal of Bacteriology* 179, no. 24: 7663–7670. doi:10.1128/jb.179.24.7663-7670.1997.

Wood, Derek W., Joao C. Setubal, Rajinder Kaul, Dave E. Monks, Joao P. Kitajima, Vagner K. Okura, Yang Zhou, Lishan Chen, Gwendolyn E. Wood, Nalvo F. Almeida Jr. et al. 2001. "The Genome of the Natural Genetic Engineer *Agrobacterium tumefaciens* C58." *Science* 294, no. 5550: 2317–2323. doi:10.1126/science.1066804.

Wu, Hong, Zhijun Song, Michael Givskov, Gerd Doring, Dieter Worlitzsch, Kalai Mathee, Jørgen Rygaard, and Niels Høiby. 2001. "*Pseudomonas aeruginosa* Mutations in *lasI* and *rhlI* Quorum Sensing Systems Result in Milder Chronic Lung Infection." *Microbiology* 147, no. 5: 1105–1113.

Yajima, Arata. 2014. "Recent Progress in the Chemistry and Chemical Biology of Microbial Signaling Molecules: Quorum-Sensing Pheromones and Microbial Hormones." *Tetrahedron Letters* 55, no. 17: 2773–2780. https://doi.org/10.1016/j.tetlet.2014.03.051.

Yang, Ching Huei, and Li, Yi Hwei. 2011. "*Chromobacterium violaceum* Infection: A Clinical Review of an Important but Neglected Infection." *Journal of the Chinese Medical Association* 74, no. 10: 435–441. https://doi.org/10.1016/j.jcma.2011.08.013.

Zhang, Lian-hui. 2003. "Quorum quenching and proactive host defense." Trends in Plant Science 8, no. 5: 238–244. doi:10.1016/S1360-1385(03)00063-3.

Zhang, Lian-hui, and Yi-hu Dong. 2004. "Microreview quorum sensing and signal interference : diverse." *Molecular Cell* 53: 1563–1571. doi:10.1111/j.1365-2958.2004.04234.x.

Zhang, Rong G., Terina Pappas, Jennifer L. Brace, Paula C. Miller, Tim Oulmassov, John M. Molyneaux, John C. Anderson, James K. Bashkin, Stephen C. Wlnans, and Andrzej Joachimiak. 2002. "Structure of a Bacterial Quorum-Sensing Transcription Factor Complexed with Pheromone and DNA." *Nature* 417, no. 6892: 971–974. https://doi.org/10.1038/nature00833.

Zhang, Ying, Jie Kong, Yunfei Xie, Yahui Guo, Hang Yu, Yuliang Cheng, He Qian, Rui Shi, and Weirong Yao. 2019. "Quorum-Sensing Inhibition by Hexanal in Biofilms Formed by *Erwinia carotovora* and *Pseudomonas fluorescens*." *LWT – Food Science and Technology* 109: 145–152. https://doi.org/10.1016/j.lwt.2019.04.023.

Zhu, J., P. M. Oger, B. Schrammeijer, P. J. Hooykaas, S. K. Farrand, and S. C. Winans. 2000. "The Bases of Crown Gall Tumorigenesis." *Journal of Bacteriology* 182, no. 14: 3885– 3895. doi: 10.1128/jb.182.14.3885-3895.2000.

3

Toward a Systematic Genomic Survey of Bacterial Quorum Sensing Genes: Cross Cutting Regulatory and Genomic Concepts

Juhász János, Sándor Pongor, and Balázs Ligeti

CONTENTS

3.1 Introduction

Quorum sensing (QS) is a general mechanism that allows cells of some bacteria to sense the proximity of each other via the environmental concentration of a QS signal molecule that members of the same species release into the environment. When this concentration is above a certain level, certain cellular functions will be activated in each cell so that the population will behave in a synchronized fashion. This synchronized behavior will then allow a bacterial community to carry out tasks that a single cell cannot.

The underlying mechanism is autocrine signaling whereby the signal produced by a cell is sensed by both the same cell and the target cells (in this case, members of the same strain). Such a mechanism taking place between microbial cells was first described in a seminal paper by Tomasz (1965) which stated that a population-level response can be triggered by a secreted chemical signal. The phenomenon got wider attention as ground-breaking work started to appear on bacterial luminescence (Nealson, Platt, and Hastings 1970; Greenberg, Hastings, and Ulitzur 1979; Engebrecht, Nealson, and Silverman 1983; Engebrecht and Silverman 1984). Finally the term quorum sensing was coined in the 1990s (Fuqua, Winans, and Greenberg 1994). As of today, quorum sensing signals have been described in most major molecular classes, carbohydrates, amino acids, peptides, etc. (Table 3.1). Importantly, each class of signaling molecule contains several members differing in minor molecular details such as the length and type of molecular side-chains attached to the same general molecular scaffold. For such a mechanism to be called a true case of QS signaling, it is supposed that bacteria need three general characteristics: (i) an ability to secrete a signaling molecule, an autoinducer (see below), (ii) an ability to detect the change in concentration of signaling molecules, and (iii) an ability to regulate gene transcription as a response.

With the advent of next generation sequencing, the number of known bacterial genomes is dramatically increasing. One would expect a concomitant increase in the number of known QS systems, but this does not appear to be the case. First, a large proportion of genes in the known genomes is not annotated. The annotated parts include mostly the general housekeeping genes of bacteria, or in other words, those gene families that are included in databases such as COG (Tatusov, Koonin, and Lipman 1997; Tatusov et al. 2000; Galperin et al. 2015) or PFAM (Sonnhammer, Eddy, and Durbin 1997; El-Gebali et al. 2019), etc. However, QS genes are not part of the core genome; they most often belong to what is termed "shell genome," which remains largely unannotated in current sequence databases. The long term goal of our project is to develop annotation tools that would identify and automatically classify QS systems in newly determined genomes. The first step in this direction, summarized in this chapter, is to describe and classify the ingredients and the structure of QS operons, and to overview the types of variability we can expect.

TABLE 3.1
Types of Quorum Sensing Signals

Small Signaling Molecules	
Acil-homoserine lactones (AHLs, HAI-1)	
Quinolones	
Alpha-pyrones (PPYs)	
Dialkylresorcinols (DARs), cyclohexanediones (CHDs)	
Cyclodipeptides	
Gamma-butyrolactones (CHB)	
Furanosyl-borate diesters (AI-2)	
Alpha-hydroxyketones (AHKs), e.g., CAI-1, LAI-1	
Fatty acids and derivatives, e.g., DSFs, PAME, MAME	
Epinephrine, norepinephrin-like molecules (AI-3)	
Peptide Signals	
Synthetized as pre-form, active after modifications (Thoendel et al. 2011)	
Cleavage	Short hydrophobic peptide signals, genes e.g., HSP2, HSP3, ComS (CIP, XIP signal)
	RNPP family, genes e.g., PhrC (CSF signal), NprX/NprRB, PapR, PgrQ (iCF10 signal)
	Gly-Gly motif peptides, genes e.g., ComC (CSP signal), SilCR
Circularization	Agr-type cyclic pheromones (AIP), genes e.g., ArgD, FrsD, LamD
Amino acid modification	ComX
	Small antimicrobial peptides/lantibiotics, genes e.g., NisA (nisin signal), SpaS

3.2 Regulatory Foundations

The heart of a QS system is an autoinduction cycle which contains three fundamental elements (Figure 3.1): (1) An intracellular signal synthase enzyme that produces a QS signal S such as a acyl-homoserine lactone molecule, used by many Gram-negative bacteria. (2) The signal is detected by a receptor-regulator protein R which will upregulate the production of the synthase I. The result is a classical feed forward loop that can in principle increase signal production without limits, which is not a probable event in nature. For this overproduction not to happen we need (3) a further, downregulating mechanism which will decrease the intensity of signal production above a certain signal concentration. The result is an incoherent feed forward loop (IFFL), well known in nature, originally discovered in ecological networks (Stone, Simberloff, and Artzy-Randrup 2019), later identified as a reoccurring motif in various regulatory networks (Mangan and Alon 2003). IFFLs are regulatory circuits in which one element can both activate and inhibit another element (Alon 2007; Milo et al. 2002). The result is a set of complex behavior patterns (Whitehead et al. 2001; Rogozin et al. 2002; Ahlgren et al. 2011), the most important being a stable and bounded output, which ensures robustness against fluctuations in the input signal levels whence. The second noteworthy property of IFFL networks is adaptation: under appropriate circumstances the circuits will react to the change rather than to the level of the signal (Novák and Tyson 2008; Tyson and Novák 2010). This can even lead to fold change detection (Goentoro et al. 2009), when identical short response is given by the network for unequal fold changes in the input. A third property is that IFFLs can work as cock-and-fire systems, which ensures that a process happens only once in a cell cycle (Csikász-Nagy et al. 2009). Theoretically, this property may also allow bacteria to react to changes rather than the level of the population density.

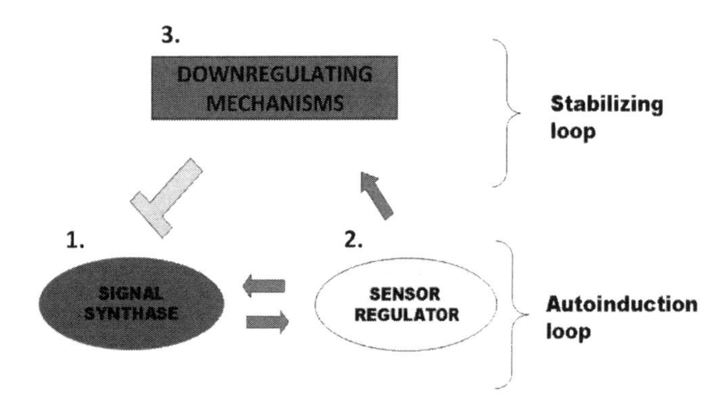

FIGURE 3.1 Generalized structure of the quorum sensing mechanism. The heart of QS is the autoinduction loop which upregulates signal synthesis (**1.**) in response to signal concentration. This upregulation via sensor and regulator proteins (**2.**) needs to be stabilized by a negative feedback mechanism (**3.**) (stabilizing loop), which can be a gene encoding a repressor protein or signal degrading enzyme, etc. The existence of a negative feedback is a logical necessity, but the underlying molecular mechanisms are frequently unknown.

The stabilizing loop can be formed, for instance, by a single repressor gene, co-transcribed by one of the in QS system genes of the autoinduction circuit. For instance, the master regulator TraM can form a non-functional heterodimer with the TraR regulator in *Agrobacterium tumefaciens* (Chen et al. 2004; Qin et al. 2004). The DNA-binding negative regulator RsaL acts as a homo-dimer by binding on the bi-directional rsaL-luxI promoter (Venturi et al. 2011). RsaM is another small regulatory protein believed to be acting in a similar way to RsaL in many *Burkholderia* species (Mattiuzzo et al. 2011). Other avenues of downregulation include enzymes that degrade the signal, sRNA-based mechanisms, and ultimately resource limitation wherein a reduction cell density would decrease QS signal concentration itself. Finally we have to point out that the paths of the QS circle are of different nature. I → R upregulation is via chemical signaling, based on the QS signal S. R → I interaction is a classical transcriptional upregulation with R as a transcriptional factor.

There are interesting variations also in the regulatory mechanism of the autoinduction loop itself (Gelencsér et al. 2012). For instance, this mechanism in Gram-negatives is often thought to follow the luxR/luxI scenario in which the dimeric LuxR protein forms a regulatory complex with N-AHL, but there are a growing number of cases where this happens in a different way. For example, a few members of the LuxR family are able to fold, dimerize, bind DNA, and regulate transcription in the absence of AHLs. Moreover, these proteins are antagonized by their cognate AHLs in the sense that some of the receptors/DNA complexes are disrupted by AHLs in vitro (Tsai and Winans 2010).

How does this simple regulatory setup sense the environment? From the regulatory point of view, the answer is simple. The path

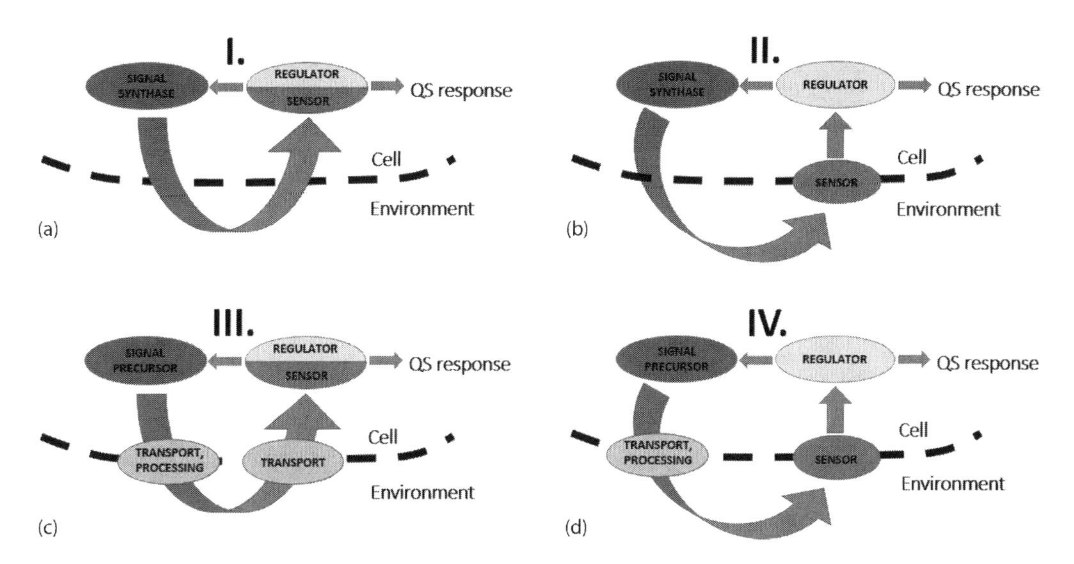

FIGURE 3.2 Types of QS systems. A QS system is able to sense the environment because the signal (wide, curved arrow) can leave the cell and then reenter the cell in some form. In many Gram-negative QS systems this happens via diffusion (a). In more complex systems the signal leaves via diffusion but the environmental signal acts on a transmembrane sensor (b). Alternatively, signal secretion and reentry are both mediated by active transport (c). Finally, signal secretion can be mediated by active transport while the effect of the environmental signal can be carried out by a transmembrane receptor (sensor) (d).

of the autoinducer loop mediated by S is coupled to the environment. In AHL systems of Gram-negative bacteria, S can freely diffuse in and out the cell via the semipermeable cell membrane (Figure 3.2a). As a result, concentration of S inside and outside the cell will be in equilibrium, so R will respond directly to the environmental concentration of S, due to this simple environmental coupling. It is customary to use the terms *public* and *private* to environmental (intercellular) and intracellular parts of this system, respectively. For instance, cytosolic signals that do not cross the membranes are parts of private signaling.

The situation is more complex in Gram-positive bacteria the cell membrane of which does not allow free movement of S between the cytosol and the medium outside of the cell. These bacteria use active transport for pumping out the signal. The return path can be either active transport, such as in short peptide signaling (Figure 3.2c), or a more complex two-component system which is analogous to transmembrane receptor-based signaling seen in eukaryotes (Figure 3.2d). Here the signal first binds to the extracellular part of a transmembrane receptor, and the intracellular part of the receptor, a histidine kinase will phosphorylate a transcription factor. The transcription factor will be activated upon phosphorylation and will mediate the upregulation of the QS system by binding to the respective promoter(s) on the genome. In other words, the S-mediated, environmentally coupled path is divided into three distinct sections, the first and the last of which are internal (private) while the central one is entirely in the environment. This system can be apparently more finely tuned than those based on signal diffusion simply because the inside and outside concentrations do not need to be in equilibrium, and autoinduction is brought about via a freely tuneable public pathway connected via the environment. Some Gram-negative species apply a related mechanism wherein a small signal molecule leaves the cell by diffusion and binds to transmembrane receptors of a two-component system (Figure 3.2b).

3.3 Mechanistic Classification of QS Systems

The above brief overview allows us now to list the functional steps required for QS signaling (Table 3.2 and Figure 3.3) which allows one to identify broad mechanism types I-IV informally described in the previous section. For instance, type I designates QS systems wherein the signal is a small molecule and both inbound and outbound traffic is carried out by diffusion. Type II includes QS systems wherein the small molecular signal leaves the cell diffusion, but it does not reenter the cell; rather it binds to a transmembrane receptor which triggers a cascade that will ultimately activate both signal production and the downstream target genes. It is interesting to note that QS systems consist of elements that are used in many other molecular mechanisms. For instance, homodimeric transcription factors (TFs) are known to drive a very large number of gene regulation events. In Gram-negative QS systems the TFs are regulated by signal binding that occurs according to at least five known mechanisms. In Gram-positive systems, TF dimerization is triggered by phosphorylation.

Another characteristic feature is the modular architecture of QS proteins which has an impact on how such proteins and their genes can be identified in bacterial genomes. As an example, the sensor-regulator protein R of AHL systems used in Gram-negative bacteria consists of two domains, a DNA-binding domain DB and a signal-binding (autoinducer binding) domain AIB.

The DNA-binding domain, which contains a helix-turn-helix (HTH) DNA binding motif, is a very common structural motif. The autoinducer-binding domain is specific for AHL systems. When searching for R homologs we find a large variety of hits to various helix-turn-helix proteins, with only a few also containing regions similar to AIB sequences. When searching a comprehensive protein sequence database for proteins with both HTH

TABLE 3.2

A Possible Classification Schema and Typical Components of Bacterial Quorum Sensing Systems

	Small Molecular Signals		Peptide Signals	
QS Mechanism	**Inward Signal Transport: Diffusion**	**Signal Binding to Receptor: Extracellular**	**Inward Signal Transport: Active Transport**	**Signal Binding to Receptor: Extracellular**
	Type I	**Type II**	**Type III**	**Type IV**
Typically occurs in:	Gram-negatives		Gram-positives	
Signal (autoinducer, AI) synthesis	Secondary metabolite synthesis		Protein (peptide) synthesis	
Synthetized form of signal	Small molecule		Propeptide signal precursor	
Outward transport	Diffusion		Signal modification, processing (cleavage); Active transport	Signal modification, processing (cleavage, circularization, amino acid modification); Active transport
Active, extracellular form of signal	Small molecule		Small peptide	Peptide
Inward signal transport	Diffusion	Transmembrane receptor binding	Active transport	Transmembrane receptor binding
Intracellular signal	Small molecule	Phosphorylation signal	Small peptide	Phosphorylation signal
Signal transduction mechanism	Gene expression regulation via DNA binding proteins			
Negative feedback control	Various: repressor proteins, RNA molecules, metabolite or resource depletion			

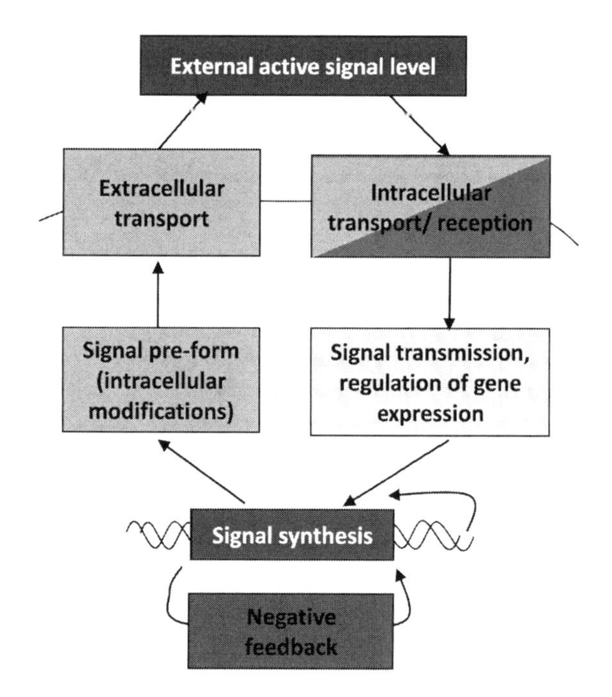

FIGURE 3.3 Functional components of the QS regulatory mechanism. The steps outlined in Figures 3.1 and 3.2 can be combined into a simple scheme shown in this figure

and AIB similarities, we will find a variety of domain architectures, only one of them corresponding to canonical R proteins, the other, much less frequent ones containing domains in inverted order, several copies of one or the other domain and/or additional domains. One of the possibilities for computational identification is to limit protein length to a value corresponding to validated R sequences, which can be extracted from a length distribution of known R proteins.

The situation is no less complicated in Gram-positive QS systems.

3.4 Gene Topologies

It has been apparent since the first sequencing results that QS genes are located in a conserved manner with respect to each other, i.e., they are part of conserved topologies. The two genes of the AHL system can be located in four different ways with respect to each other, and in addition, there can be overlaps between reading frames. A census of AHL genes identified 27 topologies that differed not only in the mutual orientation of the QS genes but also in the genes found in between the synthase and the sensor-regulator genes (Gelencsér et al. 2012). The interspersed genes included negative regulatory TFs, some of which were shown to play a stabilizing role outlined in Figure 3.1. Others were involved in DNA mobilization, signal degradation, etc. so one could suppose that the interspersed genes are tightly co-regulated with the QS genes and are part of the target genes to be activated at higher cell densities. A further, interesting type of AHL topologies are solo sensor regulator genes (Venturi and Ahmer 2015; Hudaiberdiev et al. 2015). These genes may respond to the same signal as the canonical QS system found in the same cell; in this case they can provide a link between the QS system and target genes. Clearly, the signal concentration within the cell is in equilibrium with the environment, but it can regulate a variety of solo sensor-regulator genes, thereby switching on or off a variety of genes. (Alternatively, solos may respond to other signals coming from the environment, but this case is outside the scope of this chapter.) The evolution of the AHL QS genes has been studied in a few bacterial species, and the general conclusion is that the QS topologies seem to be linked with bacterial life-style, for instance in *Burkholderia* the *pseudomallei* group and *PBE* group cluster separately and each cluster has a characteristic topology (Choudhary et al. 2013). Also, the QS proteins (both the AHL synthase I and the sensor-regulator R) form orthologous groups according to topology, for instance amino acid sequences of R genes present in different strains but in the same topology group are more closely related to each other than to other R genes present in the same strain.

A typical Gram-positive QS system is ComQXPA known primarily in *Bacilli* group 1. It consists of four tandemly oriented genes with little variation among various occurrences. A systematic survey revealed, however, that there are characteristic overlap patterns which form characteristic topology groups whose members cluster together by protein sequence similarity (Dogsa et al. 2014), which is analogous to the situation seen in AHL genes.

3.5 Brief Description of Selected Quorum Sensing Systems

As mentioned above, QS systems can be grouped according to two criteria (Table 3.2) (Rutherford and Bassler 2012; Hawver, Jung, and Ng 2016). One of these is the nature of the signal, which is connected to its ability to leave the cell. Different small molecules can leave the bacterial via diffusion (Figure 3.2a and b) while specific active transport mechanisms are needed for the outward transport of peptide signals (Figure 3.2c and d). The other criterion that helps to classify quorum sensing systems is the signal mechanism of action. Some active signals can enter the cell either via diffusion or through transporter protein complexes and bind to transcriptional factors or start signaling cascades from the cytoplasm (Figure 3.2a and c). In contrast, other signals cannot cross the cell wall and membrane but bind to the extracellular segments of specific transmembrane receptors and initialize the quorum sensing signal transduction cascade from outside (Figure 3.2b and d). This classification system specifies the four main types of QS circuits. Each type contains multiple QS systems that can be grouped into subcategories based on their genes, proteins, and the structure of their signaling molecules (Table 3.3). In the following, we briefly describe some of these subcategories.

3.5.1 Type I. Systems: Small Molecular Signal, Inward Signal Diffusion via Diffusion

Acyl-homosenine lactones (**AHL**, AI-1) were the first identified quorum sensing signal molecules. The first QS circuit was described in *Vibrio fischeri*, and it is responsible for controlling its bioluminescence in a cell concentration dependent manner (Nealson, Platt, and Hastings 1970). The circuit consists of two proteins: a signal synthase, called LuxI, and response regulator, called LuxR, which has a signal binding and a DNA binding domain (W. C. Fuqua and Winans 1994; Churchill and Chen 2011). AHL systems are widespread in Gram-negative bacteria; they are extensively studied (Fuqua and Greenberg 2002). Many luxR and luxI homologs were found, and they form different gene topologies (Gelencsér et al. 2012) and can regulate a variety of processes (for example swarming in *Pseudomonas aeruginosa* [Köhler et al. 2000]). The system also can contains negative regulators that modify (e.g., RsaL in *P. aeruginosa* [De Kievit et al. 1999] or RsaM in *Burkholderia cenocepacia* [Michalska et al. 2014]) or degrade the AHL signals (e.g., MupX in *Pseudomonas fluorescens* [Hothersall et al. 2011]).

Many bacteria have multiple quorum sensing systems that can work either in a hierarchical or parallel way (Hawver, Jung, and Ng 2016). *P. aeruginosa*, for example, operates four QS circuits from which two are not based on AHL signals. One of these is the **quinolone** QS system. Its signal molecule is 2-heptyl-3-hydroxy-4-quinolone (PQS), which promotes biofilm production and virulence (Papaioannou, Utari, and Quax 2013; Hodgkinson et al. 2010; Chan, Liu, and Chang 2015). QS quinolones are synthesized by an operon of genes: PhnA, B produce the signal precursors, PqsA, B, C, D are responsible for the signal synthesis, PqsE can have role both in synthesis and QS response, and PqsH and PqsL modify the signal. PqsR (or in its more common name MvfR) is the signal sensor and response regulator (Papaioannou, Utari, and Quax 2013). PmpR can bind to PqsR and act as a negative regulator of the system (Liang et al. 2008). Even though quinolone signaling is best known in *P. aeruginosa*, its homologs were found in some *Burkholderia* species where similar trpE, G, and hmqA, B, C, D, E, F, G genes products are responsible for signal production (Arslan 2012; Chapalain et al. 2017; Vial et al. 2008).

IQS, the forth *P. aeruginosa* QS system, is less known than the previous ones. Its signal molecule is 2-(2-hydroxyphenyl)-thiazole-4-carbaldehyde, which is thought to be produced by the ambBCDE non ribosomal peptide synthase gene cluster (Lee and Zhang 2015), while others suggest that this is a byproduct from siderophore biosynthesis (Ye et al. 2014). The signal receptor and response regulator IqsR is still not known for the system (Sun et al. 2016). The circuit is connected to las and rhl (AHL type) QS systems, plays role in virulence of the bacteria, and also reacts to phosphate limitation stress (Lee et al. 2013).

3-OH-PAME (3-hydroxypalmitic acid methyl ester) and 3-OH-MAME (methyl 3-hydroxymyristate) are fatty acid derivatives and known autoregulators that control the virulence of the plant pathogen, *Ralstonia solanacearum* (Hikichi et al. 2017). The **Phc QS** circuit contains three genes. PhcB is the signal synthase (a putative methyltransferase), PhcS is the receptor, a histidine kinase that binds the signal and phosphorylates PhcR, the response regulator that can modify the expression of virulence genes (Kai et al. 2015).

AHL quorum sensing systems are widespread in Gram-negative bacteria, but not every LuxR has its LuxI pair (Hudaiberdiev

TABLE 3.3

Examples for Different Quorum Sensing Systems

Type (from Table 3.2)	System	Signal Molecule	Synthesis Gene	Receptor and Regulatory Gene	Typical Species
I.	AHL (N-AHL, AI-1)	acyl-homoserine lactones	luxI; lasI	luxR; lasR	Widespread system, e.g., *Vibrio fischeri*; *Pseudomonas aeruginosa*,
I.	Quinolone signaling	quinolones (AQS, HHQ, PQS, HMAQs)	phnA, B; pqsA, B, C, D, E, (pqsH, L); trpE, G, hmqA, B, C, D, E, F, G	pqsR (mvfR)	*Pseudomonas aeruginosa*; *Burkholderia* spp.
I.	IQS	2-(2-hydroxyphenyl)-thiazole-4-carbaldehyde	ambB, C, D, E	iqsR (unknown)	*Pseudomonas aeruginosa*
I.	phc QS	3-OH-PAME, 3-OH-MAME	phcB	phcS → phcR	*Ralstonia solanacearum*
I.	Photopyrone QS (PPYs)	alpha-pyrones	ppyS	pluR	*Photorhabdus luminescens*
I.	Dialkylresorcinols (DARs), cyclohexanediones (CHDs)	Dialkylresorcinols (DARs), cyclohexanediones (CHDs)	darA, B, C	pauR	*Photorhabdus asymbiotica*
I.	GBLs (gamma-butyrolactones, A-factor)	CHB	scgA, X	scgR	*Streptomyces chattanoogensis* Gram-positive!
II.	alpha-hydroxyketones (AHK) QS	CAI-1 ((Z)-3-aminoundec-2-en-4-one) LAI-2	cqsA; lqsA	cqsS → … → luxR; lqsS, lqsT → lqsR	Vibrionaceae (*Vibrio harveyi*); *Legionella pneumophila*
II.	HAI-1 QS	*N*-(3-hydroxybutyryl)-HSL	luxM	luxN → … → luxR	*Vibrio harveyi*
II.	Diffusible signal factor: (B/C/I)DSF	cis-2-unsaturated fatty acids	rpfF	rpfC(,H) → rpfG → clp	*Burkholderia* spp.; *Xanthomonas citri*; *Pseudomonas aeruginosa*
II.	AI-3	epinephrine, norepinephrine-like molecule	(unknown)	qseC → qseB; qseE → qseF	*Escherichia coli*
II. III.	AI-2 (autoinducer 2, QS-2, DPD)	furanosyl-borate diesters	luxS	luxP → luxQ → … → luxR; lsrA, B,C, D → lsrR	Very frequent in both Gram-positives and Gram-negatives, e.g., *Vibrio harveyi*; *Escherichia coli*
III. Peptide signal	Short hydrophobic peptide (SHP) signaling	SHP, CIP, XIP	hsp2,3; comS	rgg2,3; comR (rgg4)	*Streptococcus* spp., *Lactobacillus* spp., *Listeria* spp.
III. Peptide signal	RNPP family	CSF; PapR-AIP; NprX/NprRB-AIP; iCF10, cCF12; PhrA-AIP	phrC; nprX; papR; ccfA, prgQ; phrA	rapB, C; nprR; plcR; prgX; trpA	*Bacillus subtilis*, *B. cereus*, *B. thuringiensis*, *Enterococcus faecalis*; *Streptococcus pneumoniae D39*
IV. Peptide signal	ComQXPA	ComX	comX	comP → comA	*Bacillus subtilis*
IV. Peptide signal	Arg-type cyclic pheromones	AIP (I-IV); GBAP; Lam-AIP	argD; frsD; lamD	argC → argA; frsC → frsA; lamC → lamA, lamK → lamR	*Streptococcus aureus*; *Enterococcus faecalis*; *Lactobacillus plantarum*
IV. Peptide signal	Gly-Gly peptides (class II. bacteriocin AMP related)	e.g., CSP; Blp signal; Sil signal	comC; blpC; silCR	comD → comE; blpH → blpR; silD → silA	e.g., *Streptococcus pneumonia*; *Streptococcus pyogenes*
IV. Peptide signal	lantibiotics (class I. small AMPs)	e.g., nisin	nisA	nisK → nisR	e.g., *Lactobacillus lactis*

Note: "Gene1" → "gene2" stands for "receptor gene" → "response regulator gene" in the fifth column. More details are found in the text.

et al. 2015). The role of these solo LuxRs is not always clear. Sometimes they can sense AHL signals from other circuits (Choudhary et al. 2013) while in other cases they are binding not AHL type molecular autoinducers (which can be compound even with antimicrobial or toxic properties). An example for that is the PluR receptor of *Photorhabdus luminescens* (Brachmann et al. 2013). It is responsible for enhancing the expression of cell clumping factors after binding its signal. The signals for PluR receptors are photopyrones (**PPYs**, alpha-pyrones) produced by PpyS, a ketosynthase (Brachmann et al. 2013). A similar system was found in *Photorhabdus asymbiotica*, where the PauR, soloLuxR controls cell clumping. Its signal is a dialkylresorcinol (**DAR**) molecule synthesized by Dar proteins (e.g., DarA, B, C) (Brameyer, Bode, and Heermann 2015).

Some other molecules, for example diketopiperazines (**DKPs**, cyclic dipeptides), are also able to bind LuxR homologs and tune (usually inhibit) quorum sensing functions. They can interact with quinolone and arg (Gram-positive) system receptors as well in a wide variety of bacteria, for example in *Bacillus* (Wang et al. 2010), *Streptomyces* (Li et al. 2006), or *Pseudomonas* (Holden et al. 1999) species. Their structure shows similarities with mammalian thyrotropin-releasing hormone (Holden et al. 2002) and plant hormone auxin (Ortiz-Castro et al. 2011), which makes them potential candidates of interspecies or even interkingdom communication. Diketopiperazine biosynthesis has multiple pathways. One of them is based on NRPS (nonribosomal peptide sythetase) enzyme complexes which can be dedicated for the signal or produce it as a side product (De Carvalho and Abraham 2012). The other mechanism is using specific t-RNA dependent CDPS (cyclodipeptide synthase) systems (those are sharing similarities with AlbC) (Giessen and Marahiel 2014, 2015). In spite of these it is not known whether DKPs have some own QS effect or they are only inhibitors, regulators of other systems.

Small diffusible molecules are common autoinducers of Gram-negative bacteria, but they can have a similar role in some Gram-positive species. Gamma-butyrolactones (**GBLs**, A-factors) are the signals of an AHL-like quorum sensing system in some *Streptomyces* (Takano 2006), for example in *S. chattanoogensis* (CHB system) (Du et al. 2011). The QS circuit has two signal synthase genes: scgA and scgX, and a signal receptor and response regulator part: scgR. The GBL mediated QS systems can effect cell metabolism, nutrient utilization, and morphological differentiation or antibiotic production (Du et al. 2011).

3.5.2 Type II. Systems: Small Molecular Signal, Extracellular Signal Binding

Furanosyl-borate diester (autoinducer-2, **AI-2** or QS-2) is produced by nearly the half of bacterial species (both Gram-negative and Gram-positive bacteria) (Federle 2009). AI-2 is synthetized from S-adenosil methionine (SAM), which has an important role in activated methyl circle (Winzer et al. 2002). Thus, unlike other quorm sensing systems, AI-2 is directly connected to the cell metabolism (Vendeville et al. 2005). These properties make it a valuable candidate for interspecies (Federle 2009) and host-microbe communication (Ismail et al. 2016). It was first described in the context of QS in *Vibrio harveyi* where it regulates bioluminescence (Bassler, Wright, and Silverman 1994). LuxS enzyme synthesizes AI-2 from SAM. AI-2 leaves the cell via diffusion and binds to the extracellular segment of LuxP transmembrane receptor. In high enough AI-2 concentration, their complex with intracellular membrane bounded LuxQ dephosphorylates LuxU, that dephosphorylates LuxO that enables LuxR functions by disabling the production of regulatory RNA molecules (Waters and Bassler 2006).

In other bacteria, for example mainly in Enterobacteriaceae, Pasteurellaceae or Bacillaceae, AI-2 re-enters into the cytosol through the Lsr ABC-transporter complex. There it is phosphorylated by LsrK and can bind and inactivate LsrR transcriptional suppressor, and the quorum regulated gene products can express (Rezzonico, Smits, and Duffy 2012). This pathway is more complex than simple AHL circuits and highlights that the capability of AI-2 production is not enough for QS; a (sometimes quite complex) sensory machinery is also needed. AI-2 regulated processes are usually connected to biofilm development or pathogenity and were found in many species that are endosymbiotes or living in microbial communities with high cell density (Rezzonico, Smits, and Duffy 2012) which can make them universal, inter-species signals.

Vibrio harveyi has two additional QS systems that parallel control the LuxU-LuxO pathway with AI-2 (Waters and Bassler 2006). One of them uses *N*-(3-hydroxybutyryl)-HSL (**HAI-1**) signal produced by LuxM, and its transmembrane receptor (that is connected with LuxQ) is called LuxN. This signal is specific to *V. harveyi*, so it can be used for intra-species communication. The third signal of *V. harveyi* is (*Z*)-3-aminoundec-2-en-4-one (CAI-1), an alpha-hydroxyketone (AHK) with the synthase CqsA and the transmembrane receptor CqsS. It is specific to Vibrionaceae and can represent an inter-genus communication signal (Waters and Bassler 2006). **AHKs** are used as signaling molecules in other bacteria as well, for example in *Legionella* species, where *L. pneumophila* is the best known example (Hochstrasser and Hilbi 2017; Schell et al. 2016). In the Legionella QS system, the signal, 3-hydroxypentadecane-4-one (*Leginella* autoinducer-1, LAI-1) is produced by LqsA and inhibits the phosphorylation of two homologous transmembrane receptor kinases LqsS and LqsT. The inactivated kinases cannot phosphorylate LqsR that split into two monomers from its dimer form and change the cells' metabolism from replication to transmissive (pathogenic) phase (Schell et al. 2016).

Diffusible signaling factors (**DSFs**) are cis-2-unsaturated fatty acids and known QS autoinducers of *Xanthomonas* species, for example *X. campestris* or *X. citri* (Guo et al. 2012). The system has three main elements: RpfF, RpfC, and RpfG. RpfF is responsible for DSF synthesis and can work with maximal intensity when it is not bounded to RpfC. RpfC undergoes conformational changes when its extracellular domain binds DSF. The results of these changes are releasing RpfF and activating RpfG (via phosphorylation). Active RpfG decreases c-di-GMP level that affects the activity of Clp transcription factor. Clp without c-di-GMP enhances virulence gene transcription but cannot repress RpfB production. RpfB is the negative regulator of the circuit; it modifies and inactivates DSF molecules (Guo et al. 2012; Zhou et al. 2017). Similar systems were found, for example, in *Burkholderia*, *Cronobacter* species (Deng et al. 2012), and *P. aeruginosa* (Amari, Marques, and Davies 2013) suggesting the importance of this kind of less known but intensively studied signaling mechanism.

Autoinducer-3 (**AI-3**) is an epinephrine/norepinephrine-like molecule identified in *Escherichia coli* (Walters and Sperandio 2006). Two transmembrane receptors QseC and QseE can bind it. They autophosphorylate in that form and phosphorylate QseB and QseF response regulators respectively (Clarke and Sperandio 2005). Active QseB promotes flagella biosynthesis, motility, enterocyte effacement (LEE), and toxin production, while QseF affects mainly LEE genes (Kendall and Sperandio 2014). The bacterial synthesis of AI-3 is not known; maybe is connected to SAM (LuxS) or tyrosine metabolism, but its receptors can bind external epinephrine or norepinephrine suggesting its potential importance in host-bacteria signaling (Walters, Sircili, and Sperandio 2006).

3.5.3 Type III. Systems: Peptide Signal, Inward Signal Transport via Active Transport

Peptide signals (which are typical in Gram-positive bacteria) cannot cross the cell membrane via diffusion. Their transmission requires active transport mechanisms, and they reach their active form during or after that process.

Short hydrophobic peptide (**SHP**) quorum sensing signals are common in *Firmicutes*. It is well known in *Streptococcus* species for example *S. pyogenes*, but was also found in different *Lactobacillus* and *Listeria* spp. (Jimenez and Federle 2014). Shp genes (hsp2, hsp3) code the pre-peptides. These small genes are hard to find and annotate; nevertheless, they were identified based on their location near their receptor genes (rgg2, rgg3 respectively), but in the reverse DNA strand (Lasarre, Aggarwal, and Federle 2013; Ibrahim et al. 2007). Eep enzyme cleaves (Cook and Federle 2014) PptAB complex transports the signal to the extracellular space (extracellular signal processing is also possible) (Chang and Federle 2016). Oligopeptide transporters (Opp), for example AmiACDEF complexes, import the mature SHP signals that can bind to Rgg2 or Rgg3 receptors in the cytoplasm that facilitate biofilm formation and lysozyme resistance as response regulators (Federle 2012). Rgg receptors have homologs in other bacteria, for example RovS in *S. agalactiae*, MutR in *S. mutans*, or LasX in *Lactobacillus sakei* (Lasarre, Aggarwal, and Federle 2013). PepO enzyme can degrade mature SHPs, and it was identified recently as a negative regulator of these systems (Wilkening, Chang, and Federle 2016).

The other family of short hydrophobic peptide signals are XIP (CIP) double-tryptophan containing pheromones. They are less conserved than other HSPs and can be clustered into multiple groups in *Streptococci* (for example to *S. mutans*, *S. pyogenic*, *S. bovis* or *S. salivarius* groups) (Shanker et al. 2016). ComS genes code the precursors of XIPs. The maturation, export, and import of these signals are similar to other HSPs, and their receptors are ComR proteins (also called Rgg4) which are encoded upstream to comS in the same strand (Federle 2012). Activated ComR facilitates genetic competence via regulating ComX (Li and Tian 2012).

The **RNPP** family was named after its four typical members: RapR, NprR, PlcR, and PrgX. They are all signal receptors and response regulators in the cytosol with similar structure. They all have tetratricopeptide repeat domains in their C-terminal for peptide signal binding, and most of them possess N-terminal HTH domains for interacting with the DNA (Lasarre, Aggarwal, and Federle 2013; Hawver, Jung, and Ng 2016). The precursors of their signals are usually synthesized from genes adjacent to the receptor (Monnet and Gardan 2015). These pre-forms undergo cleavage steps during extracellular transport (connected to the Sec system) and sometimes in the extracellular space also, until they reach their active state. Active signal peptides enter the cell through Opp complexes and bind to their receptors (Lasarre, Aggarwal, and Federle 2013).

RapB and RapC are two phosphatases that can bind CSF (competence and sporulation factor) in *Bacillus subtilis* and promote sporulation and inhibit competence respectively. The precursor of CSF is coded by phrC gene (Waters and Bassler 2006; Hawver, Jung, and Ng 2016) and is processed by extracellular Epr, Vpr, and Subtilisin enzymes (Hoover et al. 2015). NprR inhibits sporulation (binds Spo0A) without its ligand in *B. cereus* and *B. thuringiensis*. Binding with small signaling peptide produced from NprX (or NprRB), it releases Spo0A (making sporulation possible) and activates the transcription of genes involved in necrotrophism (Perchat, Talagas, Poncet, et al. 2016). PlcR is another transcription factor of virulence related genes in *B. cereus* and *B. thuringiensis*. Its peptide signal is derived from PapR precursor and is cleaved by NprB (Slamti et al. 2014). These three systems share the same origin (Perchat, Talagas, Zouhir, et al. 2016), and they work together in the gene regulatory network responsible for *B. thuringiensis* infection (Slamti et al. 2014). PrgX is a transcriptional repressor of plasmid conjugation in *Enterococcus faecalis*. It can bind two different peptide signals: chromosomally encoded cCF10, which induces conjugation (decrease PrgX function), and plasmid encoded iCF10 that inhibits it (by enhancing PrgX efficiency) (Lasarre, Aggarwal, and Federle 2013). CcfA is the precursor of cCF10 and PrgQ is the precursor of iCF10. The signals are processed (cleaved) in a similar way by Eep and re-enter to the cell through Opp and PrgZ transporter complexes (Cook and Federle 2014). This complex mechanism that controls plasmid uptake in the function of cell and plasmid concentration has a significant role in antibiotic-resistance gene transfer (Chen et al. 2017; Chatterjee et al. 2013). A relatively new member of the RNPP family is TprA of *Streptococcus pneumonia D39*. It is a negative regulator of lantibiotics production and it can be inactivated by its signal pheromone. The precursor of its pheromone is synthesized from phrA gene in the absence of glucose, which is near tprA but in the other DNA strand (Hoover et al. 2015).

3.5.4 Type IV. Systems: Peptide Signal, Extracellular Signal Binding

The **ComQXPA** quorum sensing system regulates the genetic competence in *Bacillus subtilis* (Comella and Grossman 2005), but was also found in other bacteria, for example in *Desulfosporosinus* and *Syntrophobotulus* species (Dogsa et al. 2014). The four genes of the system—comQ, comX, comP, and comA—form an operon and are located in the same strand of the bacterial chromosome. ComX gene encodes the signal peptide precursor which is extremely variable but contains a tryptophan amino acid between 5 and 10 positions from its C terminus. This tryptophan undergoes posttranslational modification by the ComQ membrane transporter that attaches an isoprenyl unit to it.

The mature signal leaves the cell via ComQ and can bind to the extracellular segment of its special receptor ComP. The activated ComP can transmit this signal to the transcriptional response regulator ComA (via phosphorylation) which induces the expression of competence genes (Okada et al. 2005).

A similar signaling mechanism was found in *Staphylococcus aureus* controlling toxin production and spreading (Thoendel et al. 2011). The QS genes here are the arg genes and the signaling molecules are the **Arg-type cyclic pheromones** (or AIPs, autoinducing peptides). ArgD encodes the pre-form of the signal, and ArgB is responsible for modifying it with a trilactone ring and exporting to the extracellular space. Therefore, the posttranslational modification that activates the signal is circularization. The active AIP can bind to its ArgC transmembrane receptor, which then phosphorylates ArgA, the DNA binding response regulator of the system (Waters and Bassler 2006). There are different kinds of AIPs with different circular structures, and all have their own ArgC receptor to activate (binding to other ArgCs is possible, but these complexes cannot activate ArgA) (Waters and Bassler 2006). Homologs of the Arg system were found in many different bacteria (Kuipers et al. 1998), for example in *Enterococcus faecalis* (Cook and Federle 2014; Gobbetti et al. 2007) or in *Lactobacillus plantarum* (Fujii et al. 2008). The Fsr system of *E. faecalis* is responsible for biofilm formation and virulence through the GBAP pheromone which FsrD precursor is in frame with the FsrB transporter and processing enzyme (Thoendel et al. 2011). *L. plantarum* has two Arg-like receptors, response regulator pairs LamC, A and LamK, R which controls cell shape, adherence, and viability (Fujii et al. 2008).

Double-glycine (Gly-Gly) **motif** peptides are another class of QS signals (Cook and Federle 2014). Their pre-form contains a leading sequence with Gly-Gly motif which is recognized by the machinery that cleaves this leading segment and transports the mature signal to the extracellular space (Havarstein, Holo, and Nes 1994; Havarstein, Diep, and Nes 1995). Competence stimulation peptide (CSP) of *Streptococcus mutans* is a typical example for this signaling (Petersen, Fimland, and Scheie 2006). ComC codes the CSP precursor that is cut and transported by the ComAB complex. ComD receptor binds the signal with its extracellular domain and phosphorylates ComE in the cytoplasm. Active ComE enhances the transcription of comX and other competence related genes (Petersen, Fimland, and Scheie 2006). Another example for Gly-Gly motif peptide signaling is the Sil (*Streptococcus* invasion locus) system in *Streptococcus pyogenes* and other *Group G Streptococci* which has a role in host infection (Cook and Federle 2014). SilCR is the signal precursor, SilDE is the complex responsible for cleavage and transport, SilB is the receptor, and SilA is the response regulator of this circuit (Jimenez and Federle 2014).

Many small antimicrobial peptides (AMPs, bacteriocin) are also regulating their production in a QS based manner in order to eliminate their competitors rapidly and sufficiently (and prevent developing immunity or wasting AMP molecules) (Miller et al. 2017). The AMPs are not the QS signals in most systems, but there are small peptide pheromones (similar to the AMPs) that are sensed via specific cell membrane receptor kinases that are connected with response regulators. Peptide pheromone precursor, sensor, response regulator genes are usually in the same cluster with AMP, AMP immunity, processing, and secretion system genes in the bacterial genome (Michiel Kleerebezema 2001).

Many class II AMP pheromones have precursors with the Gly-Gly motif described above, for example carnobacteriocins, plantaricin A, sakacin A, sakacin P, or enterocine A (Michiel Kleerebezema 2001), but Blp system is probably the simplest example for bacteriocin based QS. It was found in many *Streptococcus* species, and it is similar to the CSP system (cross-talk is possible between them) (Miller et al. 2017). The precursor of the signal is transcribed from blpC gene BlpAB complex modifies and secretes the mature signal which binds to the BlpH receptor and phosphorylates the BlpR response regulator (Blomqvist, Steinmoen, and Havarstein 2006).

In contrast to class II AMPs, **class I AMPs** (or **lantibiotics**) undergo posttranslational amino acid modifications in addition to cleavage before secretion. Some of them have dual antimicrobial and QS pheromone activity (for example nisin or subtilin) while others use AMP-like pheromones for QS (Michiel Kleerebezema 2001). Nisin is a peptide antibiotic of *Lactococcus lactis* (Kuipers et al. 1998) and the best described member of this QS family. In its gene cluster, nisA codes the nisin precursor, nisB, nisC the enzymes for intracellular, and nisP the enzyme for extracellular modifications. NisT is the outward membrane transporter (ABC translocator), and NisK is the transmembrane receptor that binds the signal and transmit its message to NisR response regulator via phosphorylation. NisF, E, G, I proteins are responsible for the nisin immunity of the producing cell (Gobbetti et al. 2007; Quadri 2002).

3.5.5 Outside the Bacterial Kingdom

Besides bacteria, quorum sensing-like control mechanisms were found also in **eukaryotic** cells. Some fungal species (e.g., *Candida albicans, Saccharomyces cerevisiae*) also use signaling molecules to estimate cell density (Sprague and Winans 2006; Lee et al. 2007; Suarez-Moreno et al. 2010). Moreover, density dependent behavior was discovered recently in some **phages** of the *Bacillus* and *Vibrio* genera where the mechanism is supposed to control the lysis-lysogeny decision (Erez et al. 2017; Silpe and Bassler 2019).

3.6 Toward an Automated Identification of QS Genes in Genomic Data

Given the amount of genomic data available, automated identification and annotation of QS genes is an important and timely effort. As shown by the overview above, the QS systems widely differ in their structures, regulatory mechanisms, the sequences of their components, as well as in the signals they utilize. In view of this variability, there is no *ab initio* method to find QS systems in genomic data, rather one has to use known examples for finding new occurrences. Even though there are useful bacterial databases of QS genes such as Sigmol for Gram-negatives bacteria (Rajput, Kaur, and Kumar 2016) and QuorumPeps for Gram-positives (Wynendaele et al. 2013), identification of example sequence sets needs a critical

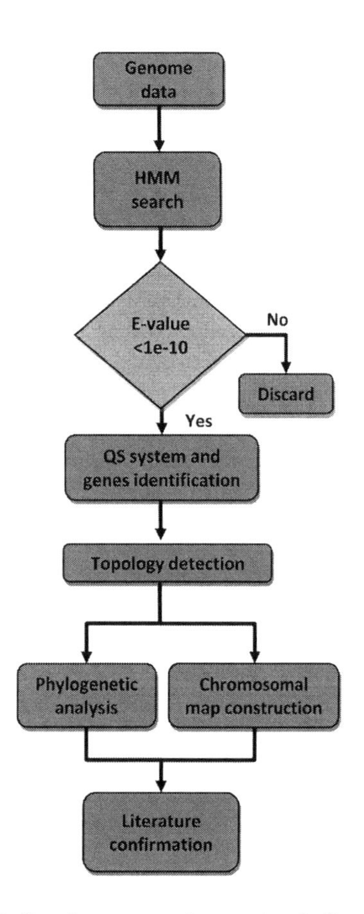

FIGURE 3.4 Outline of an automated annotation pipeline for QS genes.

review of the literature and of the public protein sequence and domain databases. An automated sequence annotation pipeline (Figure 3.4) can start from validated multiple alignments for all proteins in a given QS system, and use a reliable similarity search program, such as HMMER (Eddy 1998, 2011) to pick up potential QS protein homologs. If these putative homologs are within a certain distance within the chromosome, topology identification is initiated wherein the orientation, intergenic distances, overlaps, etc. are identified. For this to be efficient, typical properties of the QS operon in question need to be determined (size of the operon, length of the reading frames, intergenic distances). All these properties are variable within a given QS system, so in practice only probabilistic models can be built. The biggest challenges include the size and complexity of QS operons, meaning, for instance, that QS genes separated by a large number of interspersed genes are especially difficult to analyze. Another problem is the abundance of truncated operons. Operons with one gene missing or containing a truncated QS protein may or may not be functional as QS systems – such cases require experimental validation but at least a careful and critical analysis of the pertinent literature.

Therefore the advantage of the described *in silico* QS system prediction methodology is in its speed and low price compared to *in vitro* studies and in its high-throughput nature due to the use of the rapidly growing publicly available databases. The crucial point of the approach is the construction of multiple alignment profiles. For these one needs known examples of the genes, indicating entirely new systems cannot be identified with it, only new occurrences and closer relatives of already described ones. It is also important to highlight that *in silico* methods give predictions that have to be proven via *in vitro* and *in vivo* experiments.

In spite of these, automated searching in genomic data has proven its worth in identifying new quorum sensing operon candidates for different QS systems. It was used, for instance, to describe multiple, sometimes quite complex AHL operon topologies (Gelencsér et al. 2012; Choudhary et al. 2013), and many solo LuxR genes (Hudaiberdiev et al. 2015). ComQXPA gene overlap patterns and their distribution in *Bacilli* and other bacteria were also found with these techniques (Dogsa et al. 2014). *In silico* scans showed that Dar genes are common in bacteria (Brameyer, Bode, and Heermann 2015), but also helped to found new AIP homologs in *Lactobacillus plantarum* (Sturme et al. 2007). Mining microbial genomes is becoming a fast developing field with approaches that also could be applied to QS genes (Ziemert, Alanjary, and Weber 2016). Finding QS genes in metagenomics samples is feasible with the currently available search tools, and recent studies highlighted the wide distribution of some QS systems in metagenomics samples (Kimura 2014; Barriuso and Martínez 2018).

One of the bottlenecks of finding new QS systems arises from the need for experimental validation. Members of known QS signal families can be tested by a battery of QS sensor cells (Verbeke et al. 2017), but the signal molecule needs to be isolated and its structure has to be confirmed by classical techniques such as mass (or NMR, IR) spectrometry (Brelles-Mariño and Bedmar 2001; Taminiau et al. 2002; Okutsu et al. 2015). Peptide signals are somewhat different since the preprotein carrying them can be recognized relatively easily from their sequence (Rajput, Gupta, and Kumar 2015). Nevertheless, the processing sites need to be confirmed, which calls for MS sequencing methods, and the modified residues and the peptide cyclization events may present additional problems (Verbeke et al. 2017).

In conclusion, we hope that new genomic approaches will become more and more widespread and help to uncover the true diversity, prevalence, and phylogenetic distribution of bacterial quorum sensing systems, taking us closer to understanding microbial communication.

ACKNOWLEDGEMENTS

The research project has been supported by the Hungarian government grants OTKA 120650 (Mikrobiom bioinformatics: Computational analysis of complex bacterial communities); EFOP-3.6.2-16-2017-00013 (4. Integration of high-throughput biological data); EFOP-3.6.3-VEKOP-16-2017-00002 (2. Nonconventional computing and modeling approaches) which were supported by the European Union and co-financed by the European Social Fund as well as by grant ED_17-1-2017-0009 of the National Bionics Program sponsored by Hungarian Ministry of Technology and Innovation.

REFERENCES

Ahlgren, Nathan A., Caroline S. Harwood, Amy L. Schaefer, Eric Giraud, and E. Peter Greenberg. 2011. "Aryl-Homoserine Lactone Quorum Sensing in Stem-Nodulating Photosynthetic Bradyrhizobia." *Proceedings of the National Academy of Sciences of the United States of America* 108 (17): 7183–88. doi:10.1073/pnas.1103821108.

Alon, Uri. 2007. "Network Motifs: Theory and Experimental Approaches." *Nature Reviews Genetics* 8 (6): 450–61. doi:10.1038/nrg2102.

Amari, Diana T., Cláudia N. H. Marques, and David G. Davies. 2013. "The Putative Enoyl-Coenzyme A Hydratase DspI is Required for Production of the *Pseudomonas aeruginosa* Biofilm Dispersion Autoinducer Cis-2-Decenoic Acid." *Journal of Bacteriology* 195 (20): 4600–4610. doi:10.1128/JB.00707-13.

Arslan, Evrim. 2012. "Quantification and Comparison of Quorum Sensing Response to Various Quinolone Molecules in *Pseudomonas aeruginosa* and *Burkholderia thailandensis*." doi:10.7282/T3GH9GQ7.

Barriuso, Jorge, and María J. Martínez. 2018. "In Silico Analysis of the Quorum Sensing Metagenome in Environmental Biofilm Samples." *Frontiers in Microbiology* 9: 1243. doi:10.3389/fmicb.2018.01243.

Bassler, Bonnie L., Miriam Wright, and Michael R. Silverman. 1994. "Multiple Signalling Systems Controlling Expression of Luminescence in Vibrio Harveyi: Sequence and Function of Genes Encoding a Second Sensory Pathway." *Molecular Microbiology* 13 (2): 273–86. http://www.ncbi.nlm.nih.gov/pubmed/7984107.

Blomqvist, Trinelise, Hilde Steinmoen, and Leiv Sigve Havarstein. 2006. "Pheromone-Induced Expression of Recombinant Proteins in *Streptococcus thermophilus*." *Archives of Microbiology* 186 (6): 465–73. doi:10.1007/s00203-006-0162-0.

Brachmann, Alexander O., Sophie Brameyer, Darko Kresovic, Ivana Hitkova, Yannick Kopp, Christian Manske, Karin Schubert, Helge B. Bode, and Ralf Heermann. 2013. "Pyrones as Bacterial Signaling Molecules." *Nature Chemical Biology* 9 (9): 573–81. doi:10.1038/nchembio.1295.

Brameyer, Sophie, Helge B. Bode, and Ralf Heermann. 2015. "Languages and Dialects: Bacterial Communication beyond Homoserine Lactones." *Trends in Microbiology* 23 (9): 521–23. doi:10.1016/j.tim.2015.07.002.

Brelles-Mariño, Graciela, and Eulogio J. Bedmar. 2001. "Detection, Purification and Characterisation of Quorum-Sensing Signal Molecules in Plant-Associated Bacteria." *Journal of Biotechnology* 91 (2–3): 197–209. doi:10.1016/S0168-1656(01)00330-3.

Chan, Kok-Gan, Yi-Chia Liu, and Chien-Yi Chang. 2015. "Inhibiting *N*-Acyl-Homoserine Lactone Synthesis and Quenching *Pseudomonas quinolone* Quorum Sensing to Attenuate Virulence." *Frontiers in Microbiology* 6. doi:10.3389/FMICB.2015.01173.

Chang, Jennifer C., and Michael J. Federle. 2016. "PptAB Exports Rgg Quorum-Sensing Peptides in *Streptococcus*." *PloS One* 11 (12): e0168461. doi:10.1371/journal.pone.0168461.

Chapalain, Annelise, Marie-Christine Groleau, Servane Le Guillouzer, Aurélie Miomandre, Ludovic Vial, Sylvain Milot, and Eric Déziel. 2017. "Interplay between 4-Hydroxy-3-Methyl-2-Alkylquinoline and *N*-Acyl-Homoserine Lactone Signaling in a *Burkholderia cepacia* Complex Clinical Strain." *Frontiers in Microbiology* 8: 1021. doi:10.3389/fmicb.2017.01021.

Chatterjee, Anushree, Laura C. C. Cook, Che-Chi Shu, Yuqing Chen, Dawn A. Manias, Doraiswami Ramkrishna, Gary M. Dunny, and Wei-Shou Hu. 2013. "Antagonistic Self-Sensing and Mate-Sensing Signaling Controls Antibiotic-Resistance Transfer." *Proceedings of the National Academy of Sciences of the United States of America* 110 (17): 7086–90. doi:10.1073/pnas.1212256110.

Chen, Guozhou, James W. Malenkos, Mee-Rye Cha, Clay Fuqua, and Lingling Chen. 2004. "Quorum-Sensing Antiactivator TraM Forms a Dimer That Dissociates to Inhibit TraR." *Molecular Microbiology* 52 (6): 1641–51. doi:10.1111/j.1365-2958.2004.04110.x.

Chen, Yuqing, Arpan Bandyopadhyay, Briana K. Kozlowicz, Heather A. H. Haemig, Albert Tai, Wei-Shou Hu, and Gary M. Dunny. 2017. "Mechanisms of Peptide Sex Pheromone Regulation of Conjugation in *Enterococcus faecalis*." *Microbiology Open* 6 (4). doi:10.1002/mbo3.492.

Choudhary, Kumari Sonal, Sanjarbek Hudaiberdiev, Zsolt Gelencsér, Bruna Gonçalvescoutinho, Vittorio Venturi, and Sándor Pongor. 2013. "The Organization of the Quorum Sensing luxI/R Family Genes in *Burkholderia*." *Open Access International Journal of Molecular Science* 14: 14. doi:10.3390/ijms140713727.

Churchill, Mair E. A., and Lingling Chen. 2011. "Structural Basis of Acyl-Homoserine Lactone-Dependent Signaling." *Chemical Reviews* 111 (1): 68–85. doi:10.1021/cr1000817.

Clarke, Marcie B., and Vanessa Sperandio. 2005. "Events at the Host-Microbial Interface of the Gastrointestinal Tract III. Cell-to-Cell Signaling among Microbial Flora, Host, and Pathogens: There is a Whole Lot of Talking Going on." *American Journal of Physiology-Gastrointestinal and Liver Physiology* 288 (6): G1105–9. doi:10.1152/ajpgi.00572.2004.

Comella, Natalia, and Alan D. Grossman. 2005. "Conservation of Genes and Processes Controlled by the Quorum Response in Bacteria: Characterization of Genes Controlled by the Quorum-Sensing Transcription Factor ComA in *Bacillus subtilis*." *Molecular Microbiology* 57 (4): 1159–74. doi:10.1111/j.1365-2958.2005.04749.x.

Cook, Laura C., and Michael J. Federle. 2014. "Peptide Pheromone Signaling in *Streptococcus* and *Enterococcus*." *FEMS Microbiology Reviews* 38 (3): 473–92. doi:10.1111/1574-6976.12046.

Csikász-Nagy, Attila, Orsolya Kapuy, Attila Tóth, Csaba Pál, Lars Juhl Jensen, Frank Uhlmann, John J Tyson, and Béla Novák. 2009. "Cell Cycle Regulation by Feed-Forward Loops Coupling Transcription and Phosphorylation." *Molecular Systems Biology* 5. : 236. doi:10.1038/msb.2008.73.

De Carvalho, Maria Peres, and Wolf-Rainer Abraham. 2012. "Antimicrobial and Biofilm Inhibiting Diketopiperazines." *Current Medicinal Chemistry* 19 (21): 3564–77. http://www.ncbi.nlm.nih.gov/pubmed/22709011.

De Kievit, Teri R., Patrick C. Seed, Jonathon Nezezon, Luciano Passador, and Barbara H. Iglewski. 1999. "RsaL, a Novel Repressor of Virulence Gene Expression in *Pseudomonas aeruginosa*." *Journal of Bacteriology* 181 (7): 2175–84. http://www.ncbi.nlm.nih.gov/pubmed/10094696.

Deng, Yinyue, Nadine Schmid, Chao Wang, Jianhe Wang, Gabriella Pessi, Donghui Wu, Jasmine Lee, et al. 2012. "Cis-2-Dodecenoic Acid Receptor RpfR Links Quorum-Sensing Signal Perception with Regulation of Virulence through Cyclic Dimeric Guanosine Monophosphate Turnover." *Proceedings of the National Academy of Sciences of the United States of America* 109 (38): 15479–84. doi:10.1073/pnas.1205037109.

Dogsa, Iztok, Kumari Sonal Choudhary, Ziva Marsetic, Sanjarbek Hudaiberdiev, Roberto Vera, Sándor Pongor, and Ines Mandic-Mulec. 2014. "ComQXPA Quorum Sensing Systems May Not Be Unique to *Bacillus subtilis*: A Census in Prokaryotic Genomes." *PloS One* 9 (5): e96122. doi:10.1371/journal.pone.0096122.

Du, Yi-Ling, Xue-Ling Shen, Pin Yu, Lin-Quan Bai, and Yong-Quan Li. 2011. "Gamma-Butyrolactone Regulatory System of *Streptomyces chattanoogensis* Links Nutrient Utilization, Metabolism, and Development." *Applied and Environmental Microbiology* 77 (23): 8415–26. doi:10.1128/AEM.05898-11.

Eddy, Sean R. 1998. "Profile Hidden Markov Models." *Bioinformatics* 14 (9): 755–63. doi:10.1093/bioinformatics/14.9.755.

Eddy, Sean R. 2011. "Accelerated Profile HMM Searches." *PLoS Computational Biology* 7 (10): e1002195. doi:10.1371/journal.pcbi.1002195.

El-Gebali, Sara, Jaina Mistry, Alex Bateman, Sean R. Eddy, Aurélien Luciani, Simon C. Potter, Matloob Qureshi, et al. 2019. "The Pfam Protein Families Database in 2019." *Nucleic Acids Research* 47 (D1): D427–32. doi:10.1093/nar/gky995.

Engebrecht, Johanne, Nealson, Kenneth H and Michael Silverman. 1983. "Bacterial Bioluminescence: Isolation and Genetic Analysis of Functions from Vibrio Fischeri." *Cell* 32: 773–81.

Engebrecht, Johanne, and Michael Silverman. 1984. "Identification of Genes and Gene Products Necessary for Bacterial Bioluminescence." *Proceeding of National Academy of Sciences USA* 81: 4154–58.

Erez, Zohar, Ida Steinberger-Levy, Maya Shamir, Shany Doron, Avigail Stokar-Avihail, Yoav Peleg, Sarah Melamed, et al. 2017. "Communication between Viruses Guides Lysis–lysogeny Decisions." *Nature* 541 (7638): 488–93. doi:10.1038/nature21049.

Federle, Michael. 2012. "Pathogenic *Streptococci* Speak, but What Are They Saying?" *Virulence* 3 (1): 92–94. doi:10.4161/viru.3.1.18652.

Federle, Michael J. 2009. "Autoinducer-2-Based Chemical Communication in Bacteria: Complexities of Interspecies Signaling." *Contributions to Microbiology* 16: 18–32. doi:10.1159/000219371.

Fujii, Toshio, Colin Ingham, Jiro Nakayama, Marke Beerthuyzen, Ryoko Kunuki, Douwe Molenaar, Mark Sturme, Elaine Vaughan, Michiel Kleerebezem, and Willem de Vos. 2008. "Two Homologous Agr-like Quorum-Sensing Systems Cooperatively Control Adherence, Cell Morphology, and Cell Viability Properties in *Lactobacillus plantarum* WCFS1." *Journal of Bacteriology* 190 (23): 7655–65. doi:10.1128/JB.01489-07.

Fuqua, Clay, and E. Peter Greenberg. 2002. "Listening in on Bacteria: Acyl-Homoserine Lactone Signalling." *Nature Reviews Molecular Cell Biology* 3 (9): 685–95. doi:10.1038/nrm907.

Fuqua, Claiborne W., and Stephen C. Winans. 1994. "A LuxR-LuxI Type Regulatory System Activates Agrobacterium Ti Plasmid Conjugal Transfer in the Presence of a Plant Tumor Metabolite." *Journal of Bacteriology* 176 (10): 2796–2806. http://www.ncbi.nlm.nih.gov/pubmed/8188582.

Fuqua, Claiborne W., Stephen C. Winans, and Peter E. Greenberg. 1994. "Quorum Sensing in Bacteria: The LuxR-LuxI Family of Cell Density-Responsive Transcriptional Regulators." *Journal of Bacteriology* 176 (2): 269–75. http://www.ncbi.nlm.nih.gov/pubmed/8288518.

Galperin, Michael Y., Kira S. Makarova, Yuri I. Wolf, and Eugene V. Koonin. 2015. "Expanded Microbial Genome Coverage and Improved Protein Family Annotation in the COG Database." *Nucleic Acids Research* 43 (D1): D261–69. doi:10.1093/nar/gku1223.

Gelencsér, Zsolt, Kumari Sonal Choudhary, Bruna Goncalves Coutinho, Sanjarbek Hudaiberdiev, Borisz Galbáts, Vittorio Venturi, and Sándor Pongor. 2012. "Classifying the Topology of AHL-Driven Quorum Sensing Circuits in Proteobacterial Genomes." *Sensors* 12: 5432–44.

Giessen, Tobias W., and Mohamed A. Marahiel. 2014. "The tRNA-Dependent Biosynthesis of Modified Cyclic Dipeptides." *International Journal of Molecular Sciences* 15 (8): 14610. doi:10.3390/IJMS150814610.

Giessen, Tobias W., and Mohamed A. Marahiel. 2015. "Rational and Combinatorial Tailoring of Bioactive Cyclic Dipeptides." *Frontiers in Microbiology*. doi:10.3389/fmicb.2015.00785.

Gobbetti, Marco, Maria De Angelis, Raffaella Di Cagno, Fabio Minervini, and Antonio Limitone. 2007. "Cell-Cell Communication in Food Related Bacteria." *International Journal of Food Microbiology* 120 (1–2): 34–45. doi:10.1016/j.ijfoodmicro.2007.06.012.

Goentoro, Lea, Oren Shoval, Marc W. Kirschner, and Uri Alon. 2009. "The Incoherent Feedforward Loop Can Provide Fold-Change Detection in Gene Regulation." *Molecular Cell* 36 (5): 894–99. doi:10.1016/j.molcel.2009.11.018.

Greenberg, Peter E., John Woodland Hastings, and Shimon Ulitzur. 1979. "Induction of Luciferase Synthesis in Beneckea Harveyi by Other Marine Bacteria." *Archives of Microbiology* 120 (2): 87–91. doi:10.1007/BF00409093.

Guo, Yinping, Yanping Zhang, Jian-Liang Li, and Nian Wang. 2012. "Diffusible Signal Factor-Mediated Quorum Sensing Plays a Central Role in Coordinating Gene Expression of *Xanthomonas Citri* Subsp. *citri*." *Molecular Plant-Microbe Interactions* 25 (2): 165–79. doi:10.1094/MPMI-07-11-0184.

Havarstein, Leiv Sigve, Dzung Bao Diep, and Ingolf F. Nes. 1995. "A Family of Bacteriocin ABC Transporters Carry out Proteolytic Processing of Their Substrates Concomitant with Export." *Molecular Microbiology* 16 (2): 229–40. http://www.ncbi.nlm.nih.gov/pubmed/7565085.

Havarstein, Leiv Sigve, Helge Holo, and Ingolf F. Nes 1994. "The Leader Peptide of Colicin V Shares Consensus Sequences with Leader Peptides That Are Common among Peptide Bacteriocins Produced by Gram-Positive Bacteria." *Microbiology* 140 (9): 2383–89. doi:10.1099/13500872-140-9-2383.

Hawver, Lisa A., Sarah A. Jung, and Wai Leung Ng. 2016. "Specificity and Complexity in Bacterial Quorum-Sensing Systemsa." *FEMS Microbiology Reviews* 40 (5): 738–52. doi:10.1093/femsre/fuw014.

Hikichi, Yasufumi, Yuka Mori, Shiho Ishikawa, Kazusa Hayashi, Kouhei Ohnishi, Akinori Kiba, and Kenji Kai. 2017. "Regulation Involved in Colonization of Intercellular Spaces of Host Plants in Ralstonia Solanacearum." *Frontiers in Plant Science* 8: 967. doi:10.3389/fpls.2017.00967.

Hochstrasser, Ramon, and Hubert Hilbi. 2017. "Intra-Species and Inter-Kingdom Signaling of Legionella Pneumophila." *Frontiers in Microbiology* 8: 79. doi:10.3389/fmicb.2017.00079.

Hodgkinson, James, Steven D. Bowden, Warren R. J. D. Galloway, David R. Spring, and Martin Welch. 2010. "Structure-Activity Analysis of the *Pseudomonas quinolone* Signal Molecule." *Journal of Bacteriology* 192 (14): 3833–37. doi:10.1128/JB.00081-10.

Holden, Matthew T.G., Siri Ram Chhabra, Rocky De Nys, Paul Stead, Nigel J. Bainton, Philip J. Hill, Mike Manefield, et al. 1999. "Quorum-Sensing Cross Talk: Isolation and Chemical Characterization of Cyclic Dipeptides from *Pseudomonas aeruginosa* and Other Gram-Negative Bacteria." *Molecular Microbiology* 33 (6): 1254–66. http://www.ncbi.nlm.nih.gov/pubmed/10510239.

Holden, Matthew T.G., Siri Ram Chhabra, Rocky De Nys, Paul Stead, Nigel J. Bainton, Philip J. Hill, Mike Manefield, et al. 2002. "Quorum-Sensing Cross Talk: Isolation and Chemical Characterization of Cyclic Dipeptides from *Pseudomonas aeruginosa* and Other Gram-Negative Bacteria." *Molecular Microbiology* 33 (6): 1254–66. doi:10.1046/j.1365-2958.1999.01577.x.

Hoover, Sharon E., Amilcar J. Perez, Ho-Ching T. Tsui, Dhriti Sinha, David L. Smiley, Richard D. Dimarchi, Malcolm E. Winkler, and Beth A. Lazazzera. 2015. "A New Quorum Sensing System (TprA/PhrA) for *Streptococcus pneumoniae* D39 That Regulates a Lantibiotic Biosynthesis Gene Cluster HHS Public Access." *Mol Microbiol* 97 (2): 229–43. doi:10.1111/mmi.13029.

Hothersall, Joanne, Annabel C. Murphy, Zafar Iqbal, Genevieve Campbell, Elton R. Stephens, Ji'en Wu, Helen Cooper, et al. 2011. "Manipulation of Quorum Sensing Regulation in *Pseudomonas fluorescens* NCIMB 10586 to Increase Mupirocin Production." *Applied Microbiology and Biotechnology* 90 (3): 1017–26. doi:10.1007/s00253-011-3145-2.

Hudaiberdiev, Sanjarbek, Kumari S. Choudhary, Roberto Vera Alvarez, Zsolt Gelencsér, Balázs Ligeti, Doriano Lamba, and Sándor Pongor. 2015. "Census of Solo LuxR Genes in Prokaryotic Genomes." *Frontiers in Cellular and Infection Microbiology* 5: 20. doi:10.3389/fcimb.2015.00020.

Ibrahim, Mariam, Pierre Nicolas, Philippe Bessières, Alexander Bolotin, Véronique Monnet, and Rozenn Gardan. 2007. "A Genome-Wide Survey of Short Coding Sequences in *Streptococci*." *Microbiology* 153 (11): 3631–44. doi:10.1099/mic.0.2007/006205-0.

Ismail, Anisa S., Julie S. Valastyan, Bonnie L. Bassler Correspondence, and Bonnie L. Bassler. 2016. "A Host-Produced Autoinducer-2 Mimic Activates Bacterial Quorum Sensing." *Cell Host & Microbe* 19: 470–80. doi:10.1016/j.chom.2016.02.020.

Jimenez, Juan Cristobal, and Michael J. Federle. 2014. "Quorum Sensing in Group A *Streptococcus*." *Frontiers in Cellular and Infection Microbiology* 4: 1–17. doi:10.3389/fcimb.2014.00127.

Kai, Kenji, Hideyuki Ohnishi, Mika Shimatani, Shiho Ishikawa, Yuka Mori, Akinori Kiba, Kouhei Ohnishi, Mitsuaki Tabuchi, and Yasufumi Hikichi. 2015. "Methyl 3-Hydroxymyristate, a Diffusible Signal Mediating *phc* Quorum Sensing in *Ralstonia Solanacearum*." *ChemBioChem* 16 (16): 2309–18. doi:10.1002/cbic.201500456.

Kendall, Melissa M., and Vanessa Sperandio. 2014. "Cell-to-Cell Signaling in *Escherichia Coli* and Salmonella." *EcoSal Plus* 6 (1). doi:10.1128/ecosalplus. ESP-0002-2013.

Kimura, Nobutada. 2014. "Metagenomic Approaches to Understanding Phylogenetic Diversity in Quorum Sensing." *Virulence* 5 (3): 433. doi:10.4161/VIRU.27850.

Kohler, Thilo, Lasta Kocjancic Curty, Fracsisco Barja, Christian Van Delden, and Jean-Claude Pechere. 2000. "Swarming of *Pseudomonas aeruginosa* is Dependent on Cell-to-Cell Signaling and Requires Flagella and Pili." *Journal of Bacteriology* 182 (21): 5990–96. doi:10.1128/jb.182.21.5990-5996.2000.

Kuipers, Oscar P., Pascalle G. G. a de Ruyter, Michiel Kleerebezem, and Willem M. de Vos. 1998. "Quorum Sensing-Controlled Gene Expression in Lactic Acid Bacteria." *Journal of Biotechnology* 64 (1): 15–21. doi:10.1016/S0168-1656(98)00100-X.

Lasarre, Breah, Chaitanya Aggarwal, and Michael J. Federle. 2013. "Antagonistic Rgg Regulators Mediate Quorum Sensing via Competitive DNA Binding in *Streptococcus pyogenes*." *mBio* 3 (6). doi:10.1128/mBio.00333-12.

Lee, Hyeseung, Yun C. Chang, Glenn Nardone, and Kyung J. Kwon-Chung. 2007. "TUP1 Disruption in *Cryptococcus neoformans* Uncovers a Peptide-Mediated Density-Dependent Growth Phenomenon That Mimics Quorum Sensing." *Molecular Microbiology* 64 (3): 591–601. doi:10.1111/j.1365-2958.2007.05666.x.

Lee, Jasmine, Jien Wu, Yinyue Deng, Jing Wang, Chao Wang, Jianhe Wang, Changqing Chang, Yihu Dong, Paul Williams, and Lian-Hui Zhang. 2013. "A Cell-Cell Communication Signal Integrates Quorum Sensing and Stress Response." *Nature Chemical Biology* 9 (5): 339–43. doi:10.1038/nchembio.1225.

Lee, Jasmine, and Lianhui Zhang. 2015. "The Hierarchy Quorum Sensing Network in *Pseudomonas aeruginosa*." *Protein & Cell* 6 (1): 26–41. doi:10.1007/s13238-014-0100-x.

Li, Xiancui, Sergey Dobretsov, Ying Xu, Xiang Xiao, Oi Shing Hung, and Pei-Yuan Qian. 2006. "Antifouling Diketopiperazines Produced by a Deep-Sea Bacterium, *Streptomyces* Fungicidicus." *Biofouling* 22 (3–4): 201–8. http://www.ncbi.nlm.nih.gov/pubmed/17290864.

Li, Yung-Hua, and Xiaolin Tian. 2012. "Quorum Sensing and Bacterial Social Interactions in Biofilms." *Sensors (Basel, Switzerland)* 12 (3): 2519–38. doi:10.3390/s120302519.

Liang, Haihua, Lingling Li, Zhaolin Dong, Michael G. Surette, and Kangmin Duan. 2008. "The YebC Family Protein PA0964 Negatively Regulates the *Pseudomonas aeruginosa* Quinolone Signal System and Pyocyanin Production." *Journal of Bacteriology* 190 (18): 6217–27. doi:10.1128/JB.00428-08.

Mangan, Scott A., and Uri Alon. 2003. "Structure and Function of the Feed-Forward Loop Network Motif." *Proceedings of the National Academy of Sciences* 100 (21): 11980–85. doi:10.1073/pnas.2133841100.

Mattiuzzo, Maura, Iris Bertani, Sara Ferluga, Laura Cabrio, Joseph Bigirimana, Corrado Guarnaccia, Sandor Pongor, Henri Maraite, and Vittorio Venturi. 2011. "The Plant Pathogen *Pseudomonas fuscovaginae* Contains Two Conserved Quorum Sensing Systems Involved in Virulence and Negatively Regulated by RsaL and the Novel Regulator RsaM." *Environmental Microbiology* 13 (1): 145–62. doi:10.1111/j.1462-2920.2010.02316.x.

Michalska, Karolina, Gekleng Chhor, Shonda Clancy, Robert Jedrzejczak, Gyorgy Babnigg, Stephen C. Winans, and Andrzej Joachimiak. 2014. "RsaM: A Transcriptional Regulator of *Burkholderia* Spp. with Novel Fold." *FEBS Journal* 281 (18): 4293–4306. doi:10.1111/febs.12868.

Michiel Kleerebezema, Luis E. Quadric. 2001. "Peptide Pheromone Dependent Regulation of Antimicrobial Peptide Production in Gram Positive Bacteria: A Case of Multicellular Behavior." *Peptides* 22 (10): 1579–96.

Miller, Eric, Morten Kjos, Monica Abrudan, Ian S Roberts, Jan-Willem Veening, and Daniel Rozen. 2017. "Crosstalk and Eavesdropping among Quorum Sensing Peptide Signals That Regulate Bacteriocin Production in *Streptococcus pneumoniae*." *bioRxiv*. doi:10.1101/087247.

Milo, Ron, Shai S. Shen-Orr, Shalev Itzkovitz, Nadav Kashtan, Dimitri Chklovskii, and Uri Alon. 2002. "Network Motifs: Simple Building Blocks of Complex Networks." *Science (New York, NY)* 298 (5594): 824–27. doi:10.1126/science.298.5594.824.

Monnet, Véronique, and Rozenn Gardan. 2015. "Quorum-Sensing Regulators in Gram-Positive Bacteria: 'cherchez Le Peptide.'" *Molecular Microbiology* 97 (2): 181–84. doi:10.1111/mmi.13060.

Nealson, Kenneth H., Terry Platt, and John Woodland Hastings. 1970. "Cellular Control of the Synthesis and Activity of the Bacterial Luminescent System." *Journal of Bacteriology* 104: 313–22.

Novák, Béla, and John J. Tyson. 2008. "Design Principles of Biochemical Oscillators." *Nature Reviews Molecular Cell Biology* 9 (12): 981–91. doi:10.1038/nrm2530.

Okada, Masahiro, Isao Sato, Soo Jeong Cho, Hidehisa Iwata, Toshihiko Nishio, David Dubnau, and Youji Sakagami. 2005. "Structure of the *Bacillus subtilis* Quorum-Sensing Peptide Pheromone ComX." *Nature Chemical Biology* 1 (1): 23–24. doi:10.1038/nchembio709.

Okutsu, Noriya, Tomohiro Morohoshi, Xiaonan Xie, Norihiro Kato, and Tsukasa Ikeda. 2015. "Characterization of *N*-Acyl-Homoserine Lactones Produced by Bacteria Isolated from Industrial Cooling Water Systems." *Sensors (Basel, Switzerland)* 16 (1). doi:10.3390/s16010044.

Ortiz-Castro, Randy, César Díaz-Pérez, Miguel Martínez-Trujillo, Rosa E. del Río, Jesús Campos-García, and José López-Bucio. 2011. "Transkingdom Signaling Based on Bacterial Cyclodipeptides with Auxin Activity in Plants." *Proceedings of the National Academy of Sciences of the United States of America* 108 (17): 7253–58. doi:10.1073/pnas.1006740108.

Papaioannou, Evelina, Putri Dwi Utari, and Wim J. Quax. 2013. "Choosing an Appropriate Infection Model to Study Quorum Sensing Inhibition in *Pseudomonas* Infections." *International Journal of Molecular Sciences* 14 (9): 19309–40. doi:10.3390/ijms140919309.

Perchat, Stéphane, Antoine Talagas, Sandrine Poncet, Noureddine Lazar, Inès Li de la Sierra-Gallay, Michel Gohar, Didier Lereclus, and Sylvie Nessler. 2016. "How Quorum Sensing Connects Sporulation to Necrotrophism in *Bacillus thuringiensis*." *PLoS Pathogens* 12 (8): e1005779. doi:10.1371/journal.ppat.1005779.

Perchat, Stéphane, Antoine Talagas, Samira Zouhir, Sandrine Poncet, Laurent Bouillaut, Sylvie Nessler, and Didier Lereclus. 2016. "NprR, a Moonlighting Quorum Sensor Shifting from a Phosphatase Activity to a Transcriptional Activator." *Microbial Cell (Graz, Austria)* 3 (11): 573–75. doi:10.15698/mic2016.11.542.

Petersen, Fernanda Cristina, Gunnar Fimland, and Anne Aamdal Scheie. 2006. "Purification and Functional Studies of a Potent Modified Quorum-Sensing Peptide and a Two-Peptide Bacteriocin in *Streptococcus mutans*." *Molecular Microbiology* 61 (5): 1322–34. doi:10.1111/j.1365-2958.2006.05312.x.

Qin, Yinping, Audra J. Smyth, Shengchang Su, and Stephen K. Farrand. 2004. "Dimerization Properties of TraM, the Antiactivator That Modulates TraR-Mediated Quorum-Dependent Expression of the Ti Plasmid Tra Genes." *Molecular Microbiology* 53 (5): 1471–85. doi:10.1111/j.1365-2958.2004.04216.x.

Quadri, Luis E. N. 2002. "Regulation of Antimicrobial Peptide Production by Autoinducer-Mediated Quorum Sensing in Lactic Acid Bacteria." *Antonie van Leeuwenhoek* 82 (1–4): 133–45. http://www.ncbi.nlm.nih.gov/pubmed/12369185.

Rajput, Akanksha, Amit Kumar Gupta, and Manoj Kumar. 2015. "Prediction and Analysis of Quorum Sensing Peptides Based on Sequence Features." Edited by Lukasz Kurgan. *PLoS One* 10 (3): e0120066. doi:10.1371/journal.pone.0120066.

Rajput, Akanksha, Karambir Kaur, and Manoj Kumar. 2016. "SigMol: Repertoire of Quorum Sensing Signaling Molecules in Prokaryotes." *Nucleic Acids Research* 44 (D1): D634–39. doi:10.1093/nar/gkv1076.

Rezzonico, Fabio, Theo H. M. Smits, and Brion Duffy. 2012. "Detection of AI-2 Receptors in Genomes of Enterobacteriaceae Suggests a Role of Type-2 Quorum Sensing in Closed Ecosystems." *Sensors* 12 (5): 6645–65. doi:10.3390/s120506645.

Rogozin, Igor B., Kira S. Makarova, Janos Murvai, Eva Czabarka, Yuri I. Wolf, Roman L. Tatusov, Laszlo A. Szekely, and Eugene V. Koonin. 2002. "Connected Gene Neighborhoods in Prokaryotic Genomes." *Nucleic Acids Research* 30 (10): 2212–23. doi:10.1093/nar/30.10.2212.

Rutherford, Steven T., and Bonnie L. Bassler. 2012. "Bacterial Quorum Sensing: Its Role in Virulence and Possibilities for Its Control." *Cold Spring Harbor Perspectives in Medicine* 2 (11): a012427–a012427. doi:10.1101/cshperspect.a012427.

Schell, Ursula, Sylvia Simon, Tobias Sahr, Dominik Hager, Michael F. Albers, Aline Kessler, Felix Fahrnbauer, et al. 2016. "The A-Hydroxyketone LAI-1 Regulates Motility, Lqs-Dependent Phosphorylation Signalling and Gene Expression of *Legionella pneumophila*." *Molecular Microbiology* 99 (4): 778–93. doi:10.1111/mmi.13265.

Shanker, Erin, Donald A. Morrison, Antoine Talagas, Sylvie Nessler, Michael J. Federle, and Gerd Prehna. 2016. "Pheromone Recognition and Selectivity by ComR Proteins among *Streptococcus* Species." *PLOS Pathogens* 12 (12): e1005979. doi:10.1371/journal.ppat.1005979.

Silpe, Justin E., and Bonnie L. Bassler. 2019. "A Host-Produced Quorum-Sensing Autoinducer Controls a Phage Lysis-Lysogeny Decision." *Cell* 176 (1–2): 268–80.e13. doi:10.1016/J.CELL.2018.10.059.

Slamti, Leyla, Stéphane Perchat, Eugénie Huillet, and Didier Lereclus. 2014. "Quorum Sensing in *Bacillus thuringiensis* is Required for Completion of a Full Infectious Cycle in the Insect." *Toxins* 6 (8): 2239–55. doi:10.3390/toxins6082239.

Sonnhammer, Erik L.L., Sean R. Eddy, and Richard Durbin. 1997. "Pfam: A Comprehensive Database of Protein Domain Families Based on Seed Alignments." *Proteins* 28 (3): 405–20. http://www.ncbi.nlm.nih.gov/pubmed/9223186.

Sprague, George F., and Stephen C. Winans. 2006. "Eukaryotes Learn How to Count: Quorum Sensing by Yeast." *Genes & Development* 20 (9): 1045–49. doi:10.1101/gad.1432906.

Stone, Lewi, Daniel Simberloff, and Yael Artzy-Randrup. 2019. "Network Motifs and Their Origins." *PLoS Computational Biology* 15 (4): e1006749. doi:10.1371/journal.pcbi.1006749.

Sturme, Mark H.J., Christof Francke, Roland J. Siezen, Willem M. de Vos, and Michiel Kleerebezem. 2007. "Making Sense of Quorum Sensing in Lactobacilli: A Special Focus on *Lactobacillus plantarum* WCFS1." *Microbiology* 153 (12): 3939–47. doi:10.1099/mic.0.2007/012831-0.

Suarez-Moreno, Zulma Rocio, Ádám Kerényi, Sándor Pongor, and Vittorio Venturi. 2010. "Multispecies Microbial Communities. Part I: Quorum Sensing Signaling in Bacterial and Mixed Bacterial-Fungal Communitis." *Mikologia Lekarska* 17 (2): 108–12.

Sun, Shuang, Lian Zhou, Kaiming Jin, Haixia Jiang, and Ya-Wen He. 2016. "Quorum Sensing Systems Differentially Regulate the Production of Phenazine-1-Carboxylic Acid in the Rhizobacterium *Pseudomonas aeruginosa* PA1201." *Scientific Reports* 6 (1): 30352. doi:10.1038/srep30352.

Takano, Eriko. 2006. "G-Butyrolactones: *Streptomyces* Signalling Molecules Regulating Antibiotic Production and Differentiation." *Current Opinion in Microbiology* 9: 287–94. doi:10.1016/j.mib.2006.04.003.

Taminiau, Bernard, Mavis Daykin, Simon Swift, Maria-Laura Boschiroli, Anne Tibor, Pascal Lestrate, Xavier De Bolle, David O'Callaghan, Paul Williams, and Jean-Jacques Letesson. 2002. "Identification of a Quorum-Sensing Signal Molecule in the Facultative Intracellular Pathogen *Brucella melitensis*." *Infection and Immunity* 70 (6): 3004–11. doi:10.1128/iai.70.6.3004-3011.2002.

Tatusov, Roman L., Michael Y. Galperin, Darren A. Natale, and Eugene V. Koonin. 2000. "The COG Database: A Tool for Genome-Scale Analysis of Protein Functions and Evolution." *Nucleic Acids Research* 28 (1): 33–36. doi:10.1093/nar/28.1.33.

Tatusov, Roman L., Eugene V. Koonin, and David J. Lipman. 1997. "A Genomic Perspective on Protein Families." *Science (New York, NY)* 278 (5338): 631–37. http://www.ncbi.nlm.nih.gov/pubmed/9381173.

Thoendel, Matthew, Jeffrey S. Kavanaugh, Caralyn E. Flack, and Alexander R. Horswill. 2011. "Peptide Signaling in the *Staphylococci*." *Chemical Reviews* 111 (1): 117–51. doi:10.1021/cr100370n.

Tomasz, Alexander. 1965. "Control of the Competent State in *Pneumococcus* by a Hormone-Like Cell Product: An Example for a New Type of Regulatory Mechanism in Bacteria." *Nature* 208 (5006): 155–59. doi:10.1038/208155a0.

Tsai, Ching-Sung, and Stephen C. Winans. 2010. "LuxR-Type Quorum-Sensing Regulators That are Detached from Common Scents." *Molecular Microbiology* 77 (5): 1072–82. doi:10.1111/j.1365-2958.2010.07279.x.

Tyson, John J., and Béla Novák. 2010. "Functional Motifs in Biochemical Reaction Networks." *Annual Review of Physical Chemistry* 61 (1): 219–40. doi:10.1146/annurev.physchem.012809.103457.

Vendeville, Agnès, Klaus Winzer, Karin Heurlier, Christoph M. Tang, and Kim R. Hardie. 2005. "Making 'Sense' of Metabolism: Autoinducer-2, LuxS and Pathogenic Bacteria." *Nature Reviews Microbiology* 3: 383–96.

Venturi, Vittorio, and Brian M. M. Ahmer. 2015. "Editorial: LuxR Solos are Becoming Major Players in Cell-Cell Communication in Bacteria." *Frontiers in Cellular and Infection Microbiology* 5: 89. doi:10.3389/fcimb.2015.00089.

Venturi, Vittorio, Giordano Rampioni, Sándor Pongor, and Livia Leoni. 2011. "The Virtue of Temperance: Built-in Negative Regulators of Quorum Sensing in *Pseudomonas*." *Molecular Microbiology* 82 (5): 1060–70. doi:10.1111/j.1365-2958.2011.07890.x.

Verbeke, Frederick, Severine De Craemer, Nathan Debunne, Yorick Janssens, Evelien Wynendaele, Christophe Van de Wiele, and Bart De Spiegeleer. 2017. "Peptides as Quorum Sensing Molecules: Measurement Techniques and Obtained Levels In Vitro and In Vivo." *Frontiers in Neuroscience* 11: 183. doi:10.3389/fnins.2017.00183.

Vial, Ludovic, François Lépine, Sylvain Milot, Marie-Christine Groleau, Valérie Dekimpe, Donald E Woods, and Eric Déziel. 2008. "*Burkholderia pseudomallei*, *B. Thailandensis*, and *B. Ambifaria* Produce 4-Hydroxy-2-Alkylquinoline Analogues with a Methyl Group at the 3 Position That is Required for Quorum-Sensing Regulation." *Journal of Bacteriology* 190 (15): 5339–52. doi:10.1128/JB.00400-08.

Walters, Matthew, Marcelo P. Sircili, and Vanessa Sperandio. 2006. "AI-3 Synthesis is Not Dependent on luxS in *Escherichia Coli*." *Journal of Bacteriology* 188 (16): 5668–81. doi:10.1128/JB.00648-06.

Walters, Matthew, and Vanessa Sperandio. 2006. "Autoinducer 3 and Epinephrine Signaling in the Kinetics of Locus of Enterocyte Effacement Gene Expression in Enterohemorrhagic *Escherichia Coli*." *Infection and Immunity* 74 (10): 5445–55. doi:10.1128/IAI.00099-06.

Wang, Guanghua, Shikun Dai, Minjie Chen, Houbuo Wu, Lianwu Xie, Xiongming Luo, and Xiang Li. 2010. "Two Diketopiperazine Cyclo(pro-Phe) Isomers from Marine Bacteria *Bacillus subtilis* sp. 13-2." *Chemistry of Natural Compounds* 46 (4): 583–85. doi:10.1007/s10600-010-9680-8.

Waters, Christopher M., and Bonnie L. Bassler. 2006. "The Vibrio Harveyi Quorum-Sensing System Uses Shared Regulatory Components to Discriminate between Multiple Autoinducers." *Genes & Development* 20 (19): 2754–67. doi:10.1101/gad.1466506.

Whitehead, Neil A., Anne M. L. Barnard, Holly Slater, Natalie J. L. Simpson, and George P. C. Salmond. 2001. "Quorum-Sensing in Gram-Negative Bacteria." *FEMS Microbiology Reviews* 25 (4): 365–404. doi:10.1111/j.1574-6976.2001.tb00583.x.

Wilkening, Reid V., Jennifer C. Chang, and Michael J. Federle. 2016. "PepO, a CovRS-Controlled Endopeptidase, Disrupts *Streptococcus pyogenes* Quorum Sensing." *Molecular Microbiology* 99 (1): 71–87. doi:10.1111/mmi.13216.

Winzer, Klaus, Kim R. Hardie, Nicola Burgess, Neil Doherty, David Kirke, Matthew T. G. Holden, Rob Linforth, et al. 2002. "LuxS: Its Role in Central Metabolism and the in Vitro Synthesis of 4-Hydroxy-5-Methyl-3(2H)-Furanone." *Microbiology* 148 (4): 909–22.

Wynendaele, Evelien, Antoon Bronselaer, Joachim Nielandt, Matthias D'Hondt, Sofie Stalmans, Nathalie Bracke, Frederick Verbeke, Christophe Van De Wiele, Guy De Tré, and Bart De Spiegeleer. 2013. "Quorumpeps Database: Chemical Space, Microbial Origin and Functionality of Quorum Sensing Peptides." *Nucleic Acids Research* 41 (D1): D655–59. doi:10.1093/nar/gks1137.

Ye, Lumeng, Pierre Cornelis, Karel Guillemyn, Steven Ballet, and Ole Hammerich. 2014. "Structure Revision of *N*-Mercapto-4-Formylcarbostyril Produced by *Pseudomonas fluorescens* G308 to 2-(2-Hydroxyphenyl)thiazole-4-Carbaldehyde [aeruginaldehyde]." *Natural Product Communications* 9 (6): 789–94. http://www.ncbi.nlm.nih.gov/pubmed/25115080.

Zhou, Lian, Lian-Hui Zhang, Miguel Cámara, and Ya-Wen He. 2017. "The DSF Family of Quorum Sensing Signals: Diversity, Biosynthesis, and Turnover." *Trends in Microbiology* 25 (4): 293–303. doi:10.1016/j.tim.2016.11.013.

Ziemert, Nadine, Mohammad Alanjary, and Tilmann Weber. 2016. "The Evolution of Genome Mining in Microbes—A Review." *Natural Product Reports* 33 (8): 988–1005. doi:10.1039/C6NP00025H.

4

Old Acquaintances in a New Role: Regulation of Bacterial Communication Systems by Fatty Acids

Humberto Cortes-López, Martha Juárez-Rodríguez, Rodolfo García-Contreras, Marcos Soto-Hernández, and Israel Castillo-Juárez

CONTENTS

4.1 Introduction

Lipids are one of the essential molecules for life, and for this reason, their chemical diversity and biological functions have been widely studied. Specifically, fatty acids (FA) are lipids that form a structural part of cells, perform energy storage functions, and can also act as signal transduction molecules (Soto, Calatrava-Morales and López-Lara 2019). FA are classified according to the presence and number of double bonds in their carbon chain. Saturated fatty acids (SAFA) contain no double bonds, monounsaturated fatty acids (MUFA) contain one, and polyunsaturated fatty acids (PUFA) contain more than one double bond. For PUFA, the position of the double bonds is also important; for this reason, they are also classified as omega 6 and omega 3 according to the position of the terminal double bond, counting from the last aliphatic carbon (omega carbon). In addition, PUFA can be classified as *cis* or *trans*, depending on whether hydrogen is bound to the same or is on the opposite side of the double bond (Garrett and Grisham 2010). The property of FA as signaling molecules has been studied in eukaryotic cells, but their role as communication molecules in prokaryotes has only recently been explored, and we are currently in the process of elucidating its role in various biological and ecological phenomena.

FA are one of the most ubiquitous components of cell membranes and are common compounds in terrestrial and aquatic organisms. Approximately 3,700 million years ago, microorganisms dominated all the environments of the Earth and determined the development of multicellular organisms (Nutman et al. 2016). Although they have existed during all this time, it is only now, barely 300 years after their discovery by A.V. Leeuwenhoek, that we have begun to know them and to discover their diversity and complexity, as well as their participation in different ecological processes.

Initially, the main interest for studying microorganisms was to obtain industrial products and to understand their role as causes of diseases. It was believed that, compared with eukaryotes, they used unsophisticated mechanisms to perform their biological functions. However, this perspective has been changing because of growing knowledge of how they regulate their cell-cell communication and how they determine the general state of multicellular organisms.

In the 1970s, a striking phenomenon that aroused the curiosity of scientists was production of bioluminescence by some bacterial species of the genus *Vibrio*. Even more interesting was that these bacteria produced light only when there was an agglomeration of a certain number of bacterial cells (Zavilgelsky and Shakulov 2018). How unicellular microorganisms perceive that their numbers are sufficient (microbial quorum) to activate the genes and produce homogeneous bioluminescence? This is possible because bacteria and archaea are social microorganisms that communicate to perform various functions and exhibit multicellular behavior. QS is the mechanism through which communication takes place (Muñoz-Cazares et al. 2017). Currently, we know that many bacterial functions are regulated by QS, including the production of virulence factors, which are responsible for causing damage to the host (Castillo-Juárez et al. 2015). Bacterial QS communication is a complex phenomenon designed to promote multicellular behavior of unicellular organisms, in which there is coordination at a population level in time and space for the expression of certain phenotypes (Muñoz-Cazares et al. 2018).

Different QS systems (QSS) have been described in recent decades and have been found to be regulated by specific autoinducers (AI). For FA-type AI, there is a new expanding family called DSF ("diffusible signal factor"), whose chemical nature was unknown when it was discovered. We now know that DSF plays an important role in controlling biological functions, such as virulence and production of pigments and proteins, as well as resistance to antibiotics (Ryan et al. 2015). DSF was subsequently identified as α, β-unsaturated FA, produced mainly by a variety of phytopathogenic bacteria. However, recently, other genera have been found that are pathogenic for humans and respond to this AI, for example *Burkholderia cenocepacia*, *Pseudomonas aeruginosa*, and *Mycobacterium* sp. The DSF of *B. cenocepacia* is also of critical importance for the maintenance of ecology through communication between species (*Xanthomonas campestris*) and between domains (*Candida albicans*) (Ryan et al. 2015).

Also, FA with the ability to regulate the QS of bacterial pathogens have been identified in foods such as vegetables and meats. The implications of these discoveries are yet to be known and we are convinced that they will have a major impact on the way we understand the role of lipids with AI properties in the global metabolism of living organisms.

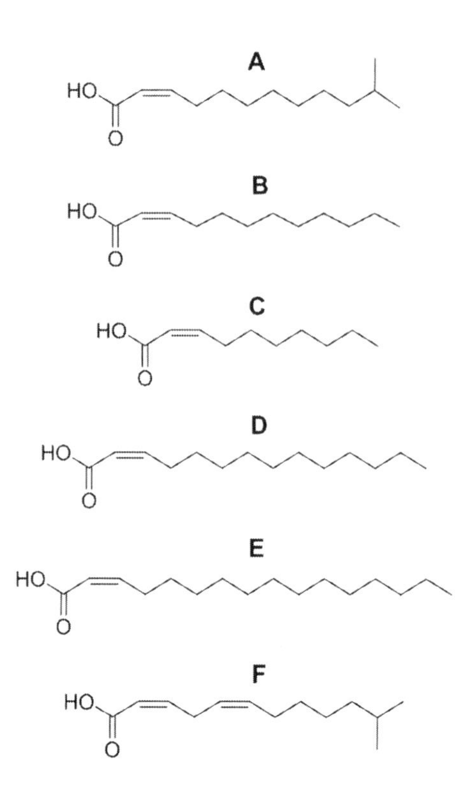

FIGURE 4.1 Main autoinducers of "diffusible signal factor" family. **A:** DSF (*cis*-11-methyl-dodecenoic acid); **B:** BDSF (*cis*-2 dodecenoic acid); **C:** CDA (*cis*-2 decenoic acid); **D:** XfDSF1 (*cis*-2 tetradecenoic acid); **E:** XfDSF2 (*cis*-2 hexadecenoic acid); and **F:** CDSF (*cis,cis*-11-methyl dodeca-2,5-dienoic acid).

4.2 Bacterial Language Based on Fatty Acids

AIs are produced by microbial cells and released, accumulated, and detected by specific receptors when cell density increases to induce collective gene expression (Castillo-Juárez et al. 2015). The best characterized AIs are those of the N-acyl homoserin lactone (AHL) family used by Gram-negative bacteria and some archaea, as well as the cyclic peptides in Gram positive bacteria (Muñoz-Cazares et al. 2017).

Type 2 autoinducer (AI-2) is not a particular structure but consists of a set of molecules derived from 4,5-dihydroxy-2,3-pentanedione (DPD). This AI has the ability to induce both Gram negative and Gram positive bacteria and has been identified in more than 500 bacterial species, the reason it is recognized as a universal signaling molecule (Castillo-Juárez et al. 2015).

DSFs are a recently discovered AI family. They were initially described in phytopathogenic bacteria species such as *X. campestris* (Wang et al. 2004). Currently, the QSS that use this type of AI have been described in other bacterial groups and have been shown to be involved in inter-domain communication (Boon et al. 2008). These AIs are *cis*-2-unsaturated fatty acids of different chain lengths and branching patterns present in bacteria that are pathogenic to plants and animals (Dow 2017). Some members of the DSF family have been described, such as *cis*-2-dodecenoic acid (BDSF) in *B. cenocepacia*, *cis*-2 tetradecenoic acid (XfDSF1), and *cis*-2-hexadecanoic acid (XfDSF2) in *Xylella fastidiosa*, *cis*-2-decenoic acid (CDA) in *P. aeruginosa* and *cis,cis*-11-methyl dodeca-2,5-dienoic acid (CDSF) in *X. oryzae* (Figure 4.1) (Ryan et al. 2015; Dow 2017). According to Zhou and colleagues (2017), QSS using DSF can be classified into three groups based on their genomic context.

4.2.1 DSF Group I

This group is represented by the *X. campestris* bacterium in which the gene *RpfF* encodes the enzyme enoyl-coA-hydratase (superfamily of the crotonases) involved in the synthesis of DSF (Figure 4.1a). It should be noted that the RpfF enzyme is the only member of the crotonases with desaturase/thioesterase activity, which uses fatty acyl ACP (acyl carrier protein) substrates for DSF synthesis (Ryan et al. 2015). The *RpfB* gene codes for the free fatty acid CoA ligase, which can be used for both biosynthesis and degradation of phospholipids (Figure 4.2a).

DSF detection and signal transduction involve a two-component system, where RpfC is the receptor and RpfG is the regulator. Once RpfC (transmembrane domain) recognizes the DSF signal, autophosphorylation occurs, which triggers a phosphorus cascade mechanism, passing through the histidine kinase A domain, a receptor domain, a histidine phosphotransferase domain, and finally, phosphotransfer to RpfG. This activates phosphodiesterase activity, which degrades cyclic-di-GMP (c-di-GMP) and releases the global transcription factor Clp that controls the expression of virulence genes (Zhou et al. 2017).

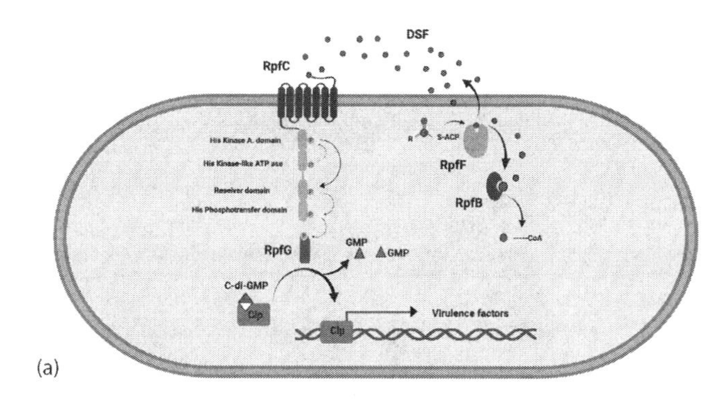

(a)

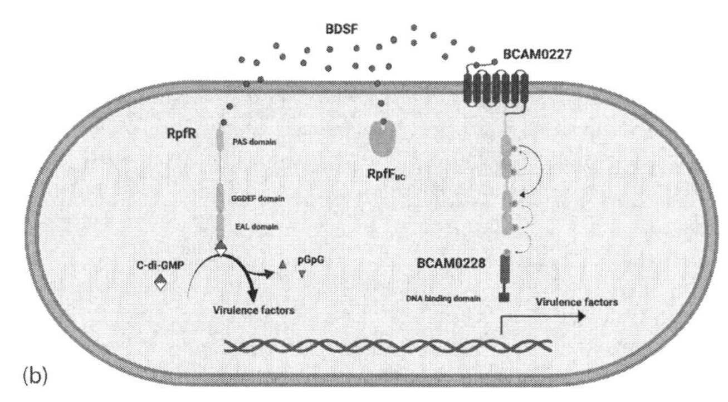

(b)

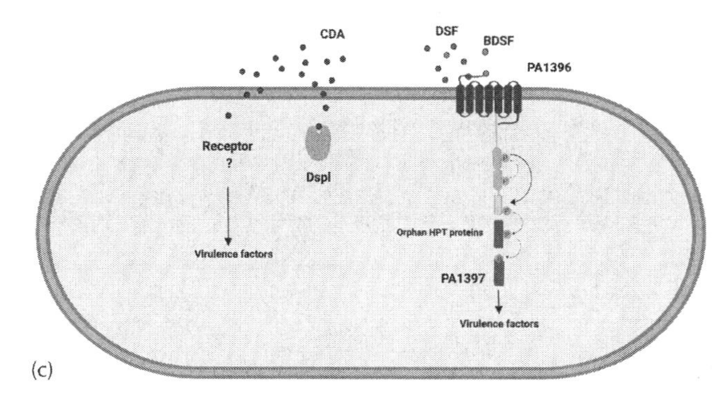

(c)

FIGURE 4.2 DSF quorum sensing system in (a): *X. campestris*, (b): *B. cenocepacia*, and (c): *P. aeruginosa*.

4.2.2 DSF Group II

In this group is *B. cenocepacia* with a QSS called RpfF/RpfR. The synthesis of BDSF (Figure 4.1b) is carried out by an enzyme DfsA (or $\mathrm{RpfF_{Bc}}$, homologous to RpfF), while the receptor protein designated RpfR contains the PAS, GGDEF, and EAL domains that regulate the levels of c-di-GMP through phosphodiesterase activity (Ryan et al. 2015; Dow 2017).

Additionally, this bacterium has a BDSF sensor system that involves the BCAM0227 histidine kinase (constituted by two transmembrane α-helices and a large periplasmic domain), which triggers a phosphorus cascade mechanism, ending with its transference to the binding regulator of DNA (BCAM0228) that activates the expression of virulence genes (Figure 4.2b) (McCarthy et al. 2010).

4.2.3 DSF Group III

This group is represented by *P. aeruginosa*, in which CDA (Figure 4.1c) is synthesized by enzymes encoded in groups of genes homologous to *dspI* (*rpfF-rpfC-rpfG* or *rpfF-rpfR*); however, the identity of the receptor and the mechanism of transduction are still unknown (Ryan et al. 2015). Interestingly, this bacterium is capable of being induced with DSF and BDSF by means of a sensor-kinase protein called PA1396, which contains an input domain of 5 transmembrane helices, very similar to that of RpfC from *X. campestris*. This interspecies induction increases tolerance to antibiotics but also reduces the production of some virulence factors in *P. aeruginosa* (Figure 4.2c) (Ryan et al. 2008; Deng et al. 2013).

4.3 Quenching Bacterial Communication by Fatty Acids

Disruption of QS is known as quorum quenching (QQ), a newly proposed approach that has the potential to combat bacterial infections in humans and plants (Castillo-Juarez et al. 2017). To interfere with the microbial communities with which they interact, many eukaryotic organisms produce and secrete compounds that mimic bacterial AIs (Muñoz-Cazares et al. 2018). Similarly, over millions of years, prokaryotes have generated a language that allows them to interact between microbial communities and also with multicellular organisms. Many of these interactions between unicellular and multicellular microorganisms determine ecological relationships, as well as physiological processes.

Higher plants have been identified as one of the main sources of molecules that regulate QS and virulence (Muñoz-Cazares et al. 2017). In the case of FAs with QQ capacity (FA-QQ), although the number of reports is still scarce, most of the FAs described are present in human diets, having been identified in plants, animals, and microorganisms.

So far, the identified FA-QQ are characterized by their acid character and by a length of mainly saturated and monounsaturated hydrocarbon chains of C-12 to C-18 (Figure 4.3). These FA, through QS modulation may offer a strategy to control bacterial infections or to stimulate communication in beneficial microorganisms.

4.3.1 QQ-Saturated Fatty Acids

Palmitic acid (3C) and ***myristic acid*** (3B) are the most reported FA-QQ to date (Table 4.1). Several studies have identified palmitic acid as the main FA in different samples and bioactive fractions. These FA have been identified in samples of poultry meat, oil seeds, and microorganisms (Widmer et al. 2007; Inoue, Shingaki and Fukui 2008; Santhakumari et al. 2017; Pérez-López et al. 2017).

Of the genus *Vibrio*, there are several species that are pathogenic for fish and humans, as well as bioluminescent species that are used as biosensor strains to evaluate QQ properties of various metabolites. In this bacterial genus it is well established that both virulence and light production are controlled by several QSS, of which the one is regulated by AI-2.

In *Vibrio harveyi* BB170 strain that uses AI-2 as QSS, **palmitic acid** (3C), **linoleic acid** (3L), and **oleic acid** (3J) inhibit light production without affecting bacterial growth (Widmer et al. 2007). It is suggested that the acidic property of FAs is involved in the QQ mechanism, possibly by inhibiting the

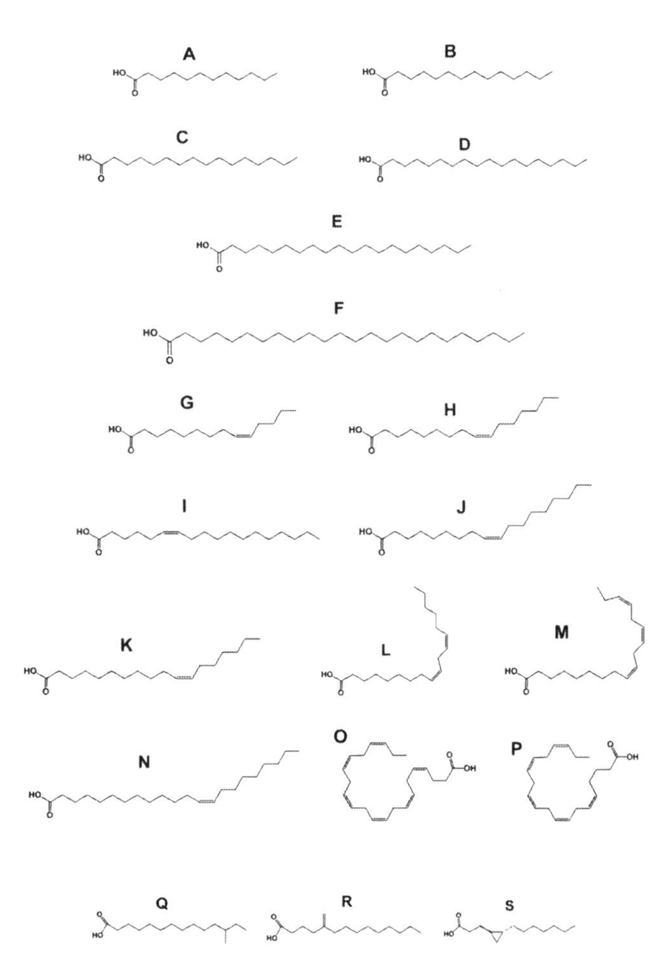

FIGURE 4.3 Fatty acids with QQ activity. **A:** lauric; **B:** myristic; **C:** palmitic; **D:** stearic; **E:** arachidic; **F:** lignoceric; **G:** myristoleic; **H:** palmitoleic; **I:** petroselinic; **J:** oleic; **K:** vaccenic; **L:** linoleic; **M:** linolenic; **N:** erucic; **O:** DHA; **P:** EPA; **Q:** anteiso-C15:0; **R:** pitinoic and **S:** lyngbyoic.

activity of Lux synthase and probably by interfering with AI-2 transport systems (Widmer et al. 2007).

This same effect was observed with other *Vibrio* species (*V. harveyi* MTCC 3438, *V. parahaemolyticus* ATCC 17802, *V. vulnificus* MTCC 1145, and *V. alginolyticus* ATCC 17749) in which **palmitic acid** (3C), isolated from the cyanobacterium *Synechococcus elongatus*, has a strong inhibitory effect on biofilm formation, as well as on exopolysaccharide and light production. The antibiofilm property of **palmitic acid** (3C) is very important because the formation of biofilms plays a major role in causing multiple diseases. Biofilms can make bacteria up to 1000 times more resistant to antimicrobials (Santhakumari et al. 2016). The mechanism of action suggested involves a decrease in the transcription of the LuxS and LuxR proteins because treatment with palmitic acid decreases *V. harveyi* transcription levels from 0.6 to 1.2 times, respectively (Santhakumari et al. 2017).

Chromobacterium violaceum is an opportunistic pathogenic bacterium that infects animals and is one of the main QQ biosensors (Pérez-López et al. 2017). Recently, it was found that amaranth (*Amaranthus hypochondriacus* L.), chia (*Salvia hispanica* L.), and sunflower (*Helianthus annuus* L.) oils inhibit the production of violacein pigment and exoprotease activity without affecting growth. In tests with commercial compounds it was clearly determined that the FA-QQ effect is present only in **palmitic acid** (3C), **myristic acid** (3B), **lauric acid** (3A), and **stearic acid** (3D), while PUFA exhibit bactericidal properties attributed to the presence of unsaturation in the structure. The results obtained in the docking analysis suggest that the mechanism of action of SAFA is related by competition to the AI by binding to the active site of the CviR receptor (Pérez-López et al. 2017).

Swarming is a type of motility regulated by QS where a differentiation toward vegetative hyperflagellate cells favors multicellular behavior and allows displacement on solid surfaces (Nickzad, Lépine and Déziel 2015). *Proteus mirabilis* is a pathogenic bacterium of the urinary tract and the main infectious agent of permanent catheters (Armbruster, Mobley and Pearson 2018). **Palmitic acid** (3C), **myristic acid** (3B), **lauric acid** (3A), and **stearic acid** (3D) inhibit swarming and reduce hemolysin activity, a virulence factor that produces lysis of erythrocytes. The suggested target for the first three FAs is the RsbA protein (homologous to the sensor proteins of the two-component systems), whereas stearic acid seems to have an independent mechanism. On the other hand, **oleic acid** (3J) has an antagonistic effect and favors swarming, while **myristic acid** (3B) slightly stimulates biofilm formation and exopolysaccharide production (Liaw, Lai and Wang 2004).

P. aeruginosa is an opportunistic pathogen that causes respiratory and urinary tract infections in immunocompromised hosts. It also infects plants such as lettuce, tobacco, onions, and others (Abd-Alla and Bashandy 2012). The virulence of *P. aeruginosa* involves the secretion of various bacterial QSS-regulated products such as pyocyanin, proteases, lipases, rhamnolipids, among others (Soto-Aceves et al. 2019). Also, it can produce biofilms, swarming, and several types of bacterial secretion systems. In this regard, it has been reported that **myristic acid** (3B) and **linoleic acid** (3L) are produced in onion bulbs in response to *P. aeruginosa* infection. Similarly, the exogenous application of **myristic acid** (3B) to cultures significantly inhibits pyocyanin production, as well as the activity of protease, lipase, and polygalacturonase (Abd-Alla and Bashandy 2012).

Finally, SAFA with QQ properties in marine species of the genus *Streptomyces* have also been reported. **Arachidic acid** (3E) and **lignoceric acid** (3F) produced by *Streptomyces griseoincarnatus* HK 12 inhibit biofilm formation in *P. aeruginosa*. According to the molecular docking analysis, a mechanism of action is suggested by competition with the receptor protein LasI (Kamarudheen and Rao 2019).

4.3.2 QQ-Unsaturated Fatty Acids

A. baumannii causes nosocomial infections worldwide and has a strong capacity to develop resistance to antimicrobials (Castillo-Juárez et al. 2017). Also, it is one of the bacterial agents classified as "critical" (priority 1) by the World Health Organization for which the development of new treatment strategies is urgently required. This bacterium also forms biofilms regulated by QSS that use acyl homoserine lactones (AHL). **Palmitoleic** (3H) and **myristoleic** (3G) reduce biofilm formation and motility in *A. baumannii* ATCC 1797. The mechanism involved is a decrease in the expression of the abaR regulator of the LuxIR type QSS, as well as autoinducer production. However, it is possible that FA have

TABLE 4.1

Fatty Acids with Anti-virulence and Quorum Quenching Properties

Fatty Acids	IUPAC	Source	QSS	Strain	Phenotypes	Suggested Target	References
SAFA							
Lauric	Dodecanoic acid 12:0	Commercial compound	RsbA	*P. mirabilis*		RsbA protein	Liaw et al. (2004)
		Seed oils	CviI/ CviR	*C. violaceum*	Violacein and protease	Competition for autoinducer binding sites (CviR)	Pérez-López et al. (2017)
Myristic	Tetradecanoic acid 14:0	Infected onion bulb and commercial compound		*P. aeruginosa*	Pyocyanin, protease, lipase, and polygalacturonase		Abd-Alla and Bashandy (2012)
		Commercial compound	RsbA	*P. mirabilis*		RsbA protein	Liaw et al. (2004)
		Seed oils	CviI/ CviR	*C. violaceum*	Violacein and protease	Competition for autoinducer binding sites (CviR)	Pérez-López et al. (2017)
		Commercial compound		*P. aeruginosa*	Swarming		Inoue et al. (2008)
Palmitic	Hexadecanoic acid 16:0	Poultry meat and commercial compound	AI-2	*V. harveyi* BB170	Inhibition of bioluminescence	Lux S and AI-2 transport systems	Windmer et al. (2007)
		S. elongatus	AI-2	*Vibrios* sp.	Biofilm, bioluminescence, exopolysaccharides and hydrophobicity	Interference in the initial stages of adhesion	Santhakumari et al. (2016, 2017)
		Commercial compound	RsbA	*P. mirabilis*		RsbA protein	Liaw et al. (2004)
		Seed oils	CviI/ CviR	*C. violaceum*	Violacein and protease	Competition for autoinducer binding sites (CviR)	Pérez-López et al. (2018)
		Commercial compound		*P. aeruginosa*	Swarming	Flagella and surface polysaccharides	Inoue et al. (2008)
Stearic	Octadecanoic acid 18:0	Poultry meat and commercial compound	AI-2	*V. harveyi* BB170	Bioluminescence	Lux S and AI-2 transport systems	Windmer et al. (2007)
		Commercial compound	RsbA	*P. mirabilis*		Mechanism independent of the RsbA protein	Liaw et al. (2004)
		Seed oils	CviI/ CviR	*C. violaceum*	Violacein and protease	Competition for autoinducer binding sites (CviR)	Pérez-López et al. (2018)
		Commercial compound		*P. aeruginosa*	Swarming		Inoue et al. (2008)
Arachidic	Eicosanoic acid 20:0	*S. griseoincarnatus*	LasI	*P. aeruginosa*	Biofilm		Kamarudheen and Rao (2019)
Lignoceric	Tetracosanoic acid 24:0			*S. aureus*			
MUFA							
Myristoleic	9-tetradecenoic acid $14{:}1^{\Delta9}$	Commercial compound		*P. aeruginosa*	Swarming		Inoue et al. (2008)
Palmitoleic	9-hexadecenoic acid $16{:}1^{\Delta9}$		LuxIR	*V. cholerae* *A. baumannii*	Biofilm and motility	AbaR regulator. Modification of interface from air-liquid physical material	Nicol et al. (2018)

(Continued)

TABLE 4.1 (*Continued*)

Fatty Acids with Anti-virulence and Quorum Quenching Properties

Fatty Acids	IUPAC	Source	QSS	Strain	Phenotypes	Suggested Target	References
Petroselinic	*cis*-6-octadecenoic acid $18{:}1^{\Delta 6}$	Commercial compound		*S. marcescens*	Prodigiosin, protease, biofilm, exopolysaccharide, swimming, and swarming	Downregulating genes associated with QS systems	Ramanathan et al. (2018)
Oleic	*cis*-9-octadecenoic acid $18{:}1^{\Delta 9}$	*S. maltophilia* BJ01		*C. violaceum, P. aeruginosa*	Violacein and biofilm		Singh et al. (2013)
		Poultry meat and commercial compound	AI-2	*V. harveyi* BB170	Bioluminescence	Lux S and AI-2 transport systems	Windmer et al. (2007)
Vaccenic	*cis*-11-octadecenoic acid $18{:}1^{\Delta 11}$	Commercial compound		*P. aeruginosa*	Swarming		Inoue et al. (2008)
PUFA							
Linoleic	*cis*-9, *cis*-12-octadecadienoic acid $18{:}1^{\Delta 9,12}$	Poultry meat and commercial compound	AI-2	*V. harveyi* BB170	Bioluminescence	Lux S and AI-2 transport systems	Windmer et al. (2007)
		Commercial compound		*P. aeruginosa*	Swarming		Inoue et al. (2008)
Linolenic	*cis*-9,12,13-octadecatrienoic acid $18{:}1^{\Delta 9,12,15}$	Commercial compound	Las Rhl Pqs	*P. aeruginosa*	Biofilm, swarming, and growth inhibition	Downregulating genes associated with QS systems	Chanda et al. (2017)
Erucic	*cis*-13-docosenoic acid $22{:}1^{\Delta 13}$	*S. griseoincarnatus*	LasI	*P. aeruginosa* *S. aureus*	Biofilm		Kamarudheen and Rao (2019)
DHA	*cis*-4,7,10,13,16,19-docosahexaenoic acid $22{:}6^{\Delta 4,7,10,13,16,19}$	Herring oil		*S. aureus*	Biofilm and hemolytic activity	*hla* gene of α-hemolysin	Kim et al. (2018)
EPA	*cis*-5,8,11,14,17-eicosapentaenoic acid $20{:}5^{\Delta 5,8,11,14,17}$						
Other FA							
Anteiso-C15:0	12-methyltetradecanoic acid 15:0	Commercial compound		*P. aeruginosa*	Swarming, swimming, and biofilm	Flagella and surface polysaccharides	Inoue et al. (2008)
BDFS	*cis*-2-dodecenoic $12{:}1^{\Delta 2}$	Commercial compound		*P. aeruginosa*		LasR, PqsR y RhlR. Reduce 3-oxo-C12-HSL, PQS y C4-HSL	Deng et al. (2013)
Pitinoic Acid A	5-methylene decanoic acid $11{:}1^{\Delta 5}$	*Lyngbya* sp.		*P. aeruginosa*	Pyocyanin, elastase and LasB		Montaser et al. (2013)
Lyngbyoic	3-[(1R,2R)-2-heptylcyclopropyl propanoic acid	*Lyngbya* cf. majuscula	LasR	*P. aeruginosa*	Pyocyanin, elastase and LasB	Competition for autoinducer (AHL) binding sites	Kwan et al. (2011)

additional effects on QQ, such as the ability to modify material interface properties that act in the initial adhesion process (Nicol et al. 2018).

Vibrio cholerae regulates the production of virulence factors, such as cholera toxin, motility, protease production and biofilm formation. **Palmitoleic acid** (3H) and **myristoleic acid** (3G) reduce production of the ToxT protein, which is responsible for activating the transcription of genes that encode the cholera toxin and toxin-regulated pilus (Childers et al. 2011). The proposed mechanism is that FA interfere with the N-terminal dimerization of ToxT (necessary for transcriptional activation and pathogenesis) and prevent interaction with DNA (Lowden et al. 2010; Childers et al. 2011; Plecha and Withey 2015).

Other MUFA capable of reducing biofilm formation in *P. aeruginosa* are **oleic acid** (3J) and **erucic acid** (3N). The first is produced by the bacterium *Stenotrophomonas maltophilia* BJ01 isolated from the rhizosphere of *Cyperus laevigatus* L., while erucic acid is found in the marine actinobacteria *Streptomyces griseoincarnatus* HK 12 (Singh et al. 2013; Kamarudheen and Rao 2019). It should be noted that **oleic acid** (3J) also inhibits the production of violacein in *C. violaceum* CV026 (Singh et al. 2013). It has been proposed that **erucic acid** (3N) acts through competition in the binding with LasI receptor (Kamarudheen and Rao 2019).

Serratia marcescens is an opportunistic pathogenic bacterium of humans that causes nosocomial infections. It regulates the production of virulence factors by QSS with short chain AHL. Swarming motility, serrawatin biosurfactant production, prodigiosin (red pigment), proteases, and biofilm formation are some of the phenotypes QS regulated. **Petroselinic acid** (3I) inhibits prodigiosin, proteases, and exopolysaccharide production, as well as swarming and biofilm formation. The mechanism of action suggested involves the negative regulation of genes controlled by QS, such as *bsmB*, *fimA*, *fimC*, and *flhD* (Ramanathan et al. 2018).

Staphylococcus aureus is a Gram-positive bacterium pathogenic for humans that regulates the formation of biofilm and other virulence factors by QSS using autoinducer peptides (Castillo-Juárez et al. 2015). So far, the only reported PUFA with QQ properties are the **DHA** (3O) and **EPA** (3P) identified as the main constituent of herring oil. The oil, as well as PUFA, significantly decreases biofilm formation and hemolytic capacity of *S. aureus* strains in red blood cells and reduces the expression of the *hla* gene of α-hemolysin (Kim et al. 2018).

4.3.3 Other Fatty Acids with QQ Properties

One of the first studies focused on analyzing the effect of FA on bacterial motility in *P. aeruginosa* was carried out by Inoue, Shingaki, and Fukui (2008). They found that several SAFA [**myristic** (3B), **palmitic** (3C), and **stearic** (3D)] and MUFA [**myristoleic** (3G), **palmitoleic** (3H), **oleic** (3J), **vacenic** (3K), and **linoleic** (3L)] inhibit swarming in *P. aeruginosa* PAO1. The most effective was **12-methyltetradecanoic acid** (iso and anteiso) (3Q), which also reduced formation of biofilms. Although the mechanism of action suggested involves the partial repression of flagella and decreasing moisture in the bacterial colony due to inhibition of surface polysaccharides (exopolysaccharides and lipopolysaccharides), studies have focused on determining the site of action in the QSS (Inoue, Shingaki and Fukui 2008).

BDSF (1B) is produced by *B. cenocepacia* to control the production of virulence factors (Boon et al. 2008). However, in *P. aeruginosa* it inhibits virulence factor production, biofilm formation, and type 3 secretion system (T3SS). Also, it reduces transcription of the *lasR*, *pqsR*, and *rhlR* genes and production of AI 3-oxo-C12-HSL, PQS, and C4-HSL (Deng et al. 2013). T3SS is a protein syringe that uses bacteria to secrete effector molecules into host cells. Although it is not yet clear whether the T3SS are regulated by QS in *P. aeruginosa*, a similar effect is obtained with some non-esterified FA (oleate, myristate, and palmitate) on the T3SS of *Salmonella enterica* serovar Typhimurium. External addition of these FAs represses expression of *hilA*, which encodes the transcriptional activator of T3SS structural genes (Golubeva et al. 2016).

On the other hand, marine cyanobacteria are a group of microorganisms in which several metabolites with QQ properties have been identified. **Pitinoic acid A** (3R), is a FA with an exomethyl group isolated from *Lyngbya* sp. that inhibits pyocyanin production and elastase activity (LasB) in *P. aeruginosa*. Also, it reduces the transcription levels of the genes encoding *lasB* and the biosynthetic member of pyocyanin (Montaser, Paul and Luesch 2013). Another is **lyngbyoic acid** (3S), which is one of the main metabolites produced by *Lyngbya* c.f *majuscule* and contains a cyclopropane in its structure. This metabolite interferes mainly with the LasR system of *P. aeruginosa*, reducing the production of pyocyanin and elastase (lasB), as well as the genes responsible for transcription (Kwan et al. 2011). Interestingly, this effect is contrary to lauric acid, which increases the production of pyocyanin and LasB. The proposed mechanism of action indicates an interaction of lynbyoic acid with the LasR receptor, although it has been suggested that other mechanisms may be involved (Kwan et al. 2011).

4.3.4 QQ-Fatty Acid in Vivo Studies

Currently, one of the limitations for the application of QQ molecules is the lack of *in vivo* studies that validate their effectiveness in counteracting bacterial infections. However, in the case of QQ-FA, some studies have already reported positive results in reducing damage and increasing survival in different animal models.

Infection by *Vibrio* species in aquatic organisms is a serious problem for aquaculture, as it leads to significant economic losses in this industry (Osunla and Okoh 2017). Recently, it was demonstrated that treating brine shrimp larvae (*Artemia franciscana*) with **palmitic acid** (3C) increases survival and reduces colonization of several *Vibrio* species (*harveyi*, *parahaemolyticus*, *vulnificus*, and *alginolyticus*) in larval intestine (Santhakumari et al. 2017). Similarly, **BDSF** (1B) in a zebrafish model reduced *P. aeruginosa* PA14 virulence and increased animal survival by 20% 6 days post-infection (Deng et al. 2013), while **DHA** (3O) and **EPA** (3P) prolonged survival of the nematode *Caenorhabditis elegans* infected with *S. aureus* (Kim et al. 2018).

Recently, participation of FA-rich oils in the survival of animals was reported using a murine model. Oral administration of oilseed (chia, sunflower, and amaranth) prolonged survival of animals infected with *C. violaceum*. However, it should be noted that the effect of SAFA [**lauric** (3A), **myristic** (3B), **palmitic** (3C), and **stearic** (3D)] to which the QQ activity is attributed was not evaluated. It is suggested that this positive effect may be due to a synergism with PUFAs (also present in oils) that exhibit a bactericidal effect on the bacteria *in vitro* (Pérez-López et al. 2017).

4.4 Fatty Acids-Autoinducer, Beyond Intra-Species Bacterial Communication

DSFs were originally identified as intra-species signals to AIs that regulate biofilm formation, virulence, and motility (Soto, Calatrava-Morales and López-Lara 2019; Liu et al. 2018). However, recent reports indicate that they also have a critical importance in maintenance of ecological interactions among different species, genera, kingdoms, and domains (Ryan et al. 2015).

Because microorganisms in nature develop in polymicrobial communities, it is necessary to develop interaction mechanisms that allow them to excel and establish themselves. Thus, those bacteria that produce DSF have an advantage since DSF modulate the behavior of other microorganisms present in these communities (Soto, Calatrava-Morales and López-Lara 2019; Ryan et al. 2015; Deng et al. 2013; Twomey et al. 2012).

QSS regulation by DSF can result in the stimulation or inhibition of establishment and survival of the microorganisms in their community. An example of positive interaction is that which develops between *P. aeruginosa* with *S. maltophilia* and *B. cenocepacia*. This relationship is observed especially in patients with cystic fibrosis where these three bacteria establish and cause nosocomial infection, the most predominant being *P. aeruginosa* (Ryan et al. 2015; Twomey et al. 2012; Ryan et al. 2008).

BDSF (1B) has the ability to regulate virulence factors in other bacterial genera, such as *X. campestris* pv. campestris in which exogenous addition of BDSF restores dispersion of biofilms and production of virulence factors in mutants deficient in **DSF** (1A). Similarly, it interacts with other kingdoms such as *C. albicans*, in which **BDSF** (1B) inhibits the formation of its germ tube (Boon et al. 2008). It also modulates virulence, resistance to antibiotics, and persistence in *P. aeruginosa* (Deng et al. 2013).

Myristic acid (3B) and related FAs play an important role in the communication between *Pseudomonas* species. These FAs are present in the cultures of *P. aeruginosa* with *P. putida* where they activate the density-dependent expression of the *ddcA* gene, which is responsible for the colonization of seed and root by *P. putida* (Fernández-Piñar et al. 2012).

Also, the combination of ajoene with tobramycin potentiates the bactericidal activity of the antibiotic by 80%. This phenomenon is possible because QS inhibition prevents the correct formation of biofilms, causing bacteria to be more vulnerable to the effect of antibiotics and the immune system (Christensen et al. 2012). It has been reported that FA can also reduce bacterial resistance to antibiotics (adjuvant effect) by a mechanism that involves QSS regulation. AIs of the FA type modulate bacterial resistance to antibiotics through intra-species signaling but may also affect their tolerance. It has been reported that DFS (1A) and its structurally related molecules induce antibiotic susceptibility. Exogenous addition of DSF (1A) to *Bacillus cereus* cultures increases susceptibility to gentamicin, and this adjuvant effect can occur in other species such as *Bacillus thuringiensis*, *S. aureus*, *Mycobacterium smegmatis*, *Neisseria subflava*, and *P. aeruginosa* (Deng et al. 2014). Also, the CDA (1C) produced by *P. aeruginosa* induces dispersion of biofilms in multiple types of bacteria, in addition to having a function as an adjuvant when combined with antimicrobials that favor dispersion of biofilms (Rahmani-Badi et al. 2015). Interestingly, **linolenic acid** (3M) has an adjuvant effect, which reduces production of virulence factors by interference in the QSS of *P. aeruginosa* and increases sensitivity to tobramycin (Chanda et al. 2017).

4.5 Implications of Fatty-Acid-Rich Diets in Bacterial Communication and Health Human

Edible fats and oils are an essential part of the animal diet because they play an important role in nutrition. Although current opinions argue that dietary fat consumption should be reduced, it is also true that some FAs have an important role in maintaining health. In recent years, research on FA consumption has become an important topic. The PUFA series omega-3 and omega-6 have been attributed with properties beneficial to health, while the SAFA or *trans*-MUFA have been linked to the development of various diseases (Figure 4.4).

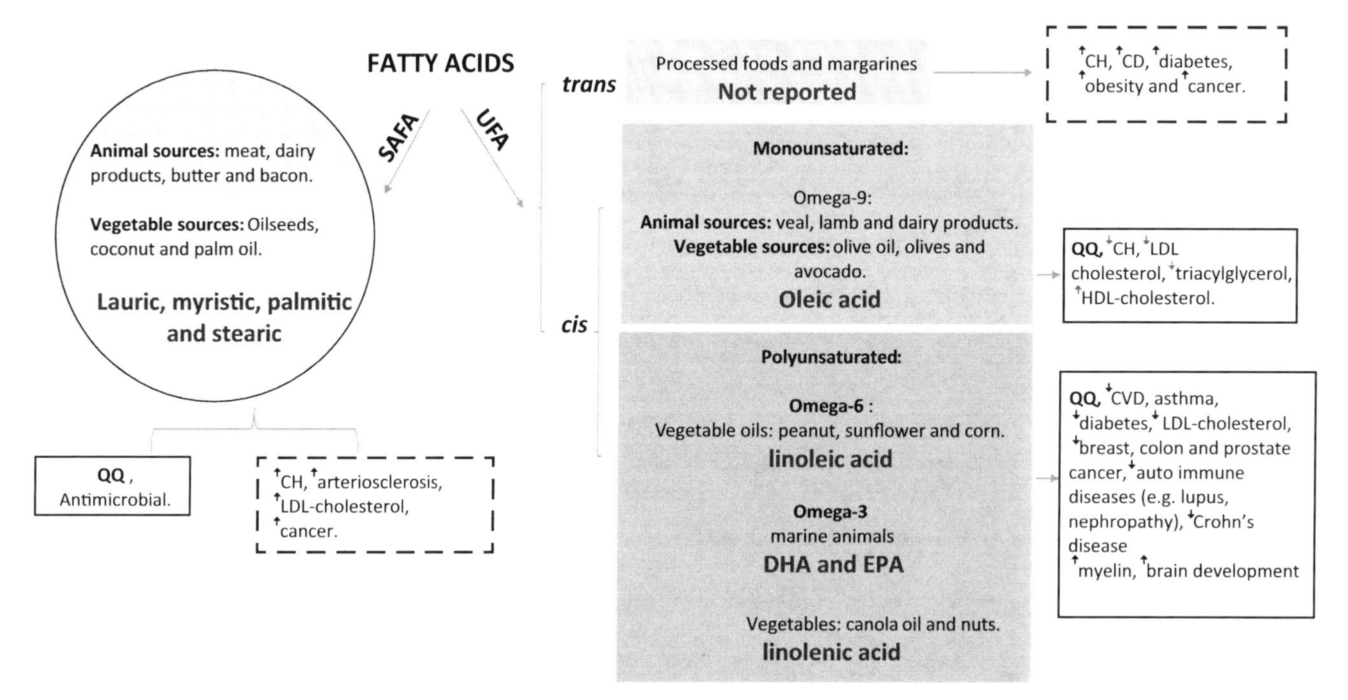

FIGURE 4.4 Main sources of FA and representative QS inhibitors. QQ: quorum quenching activity; SAFA: saturated fatty acids; UFA: unsaturated fatty acids; MUFA: monounsaturated fatty acids and PUFA: polyunsaturated fatty acids; EPA: eicosapentaenoic acid; DHA: docosahexaenoic acid. Harmful effects (dotted boxes) and beneficial effects (solid boxes) for health. CH: Coronary heart disease; CD: Cognitive disorders and Alzheimer´s; CVD: cardiovascular disease; LDL: low-density lipoprotein (bad cholesterol); and HDL: high-density lipoprotein (good cholesterol).

FAs have been reported to possess several pharmacological properties such as anticancer, anti-inflammatory, analgesic, cytotoxic, apoptotic, and bactericidal (Soni et al. 2008). However, as we have already mentioned, it has been discovered that mainly SAFA and MUFA (some related to harmful effects on human health) have a role in the regulation of bacterial communication (Xiang, Tan and Macia 2018) (Figure 4.4). Interest in the role of FA as modulating molecules of QS is reflected in the recent increase in publications related to this topic.

It has been proposed that in chronic infections (often opportunistic in immunocompromised patients), a diet rich in QQ compounds may be responsible for resistance or susceptibility to this type of infection (Givskov 2012). An analysis of 69 food products and common plants showed that garlic extract has QQ properties, tested both *in vitro* and in a murine model of infection with *P. aeruginosa* (Bjarnsholt et al. 2005). Interestingly, because garlic extract is more potent than one of the purified compounds (ajoene), synergism with other compounds seems to be very important. It has been suggested that it is unlikely that a food contains a quantity of QQ molecule sufficient to eradicate established infections. To achieve the positive effect obtained in the murine model (if extrapolated to an adult human), it would require a daily intake of 50 garlic bulbs for several days (Givskov 2012). Thus, it is suggested that QQ compounds could have a preventive antimicrobial mechanism of action, which would prevent the establishment of infections, rather than their elimination once they have been established.

The content of QQ molecules in medicinal and edible plants has been fully demonstrated (Muñoz-Cazares et al. 2017). Thus, it is clear that a diet rich in vegetables and food supplements of plant origin would have beneficial effects on health (Givskov 2012). However, the QS-regulating FA (QQ, QS stimulators, or adjuvants) are present in more diverse sources, including meat (Widmer et al. 2007; Kim et al. 2018).

What has been the role of diets rich in regulatory FA on QS in the developed world? FA produced by bacteria represent the greatest communication mechanism of the microbiome present in the human intestine (Xiang, Tan and Macia 2018). But it is only recently that we have begun to understand the scope of DSF symbiotic relationships, whether mutualistic or pathogenic, on a large scale, as well as their role in cell signaling that modulates metabolic activities of natural microbial communities.

4.6 Concluding Remarks

SAFA and MUFA have regulatory properties of QSS in pathogenic bacteria of both humans and plants, but little has been explored of its effect on the beneficial microbiota associated with these organisms. Similarly, the number of bacterial genera where the effect of FA has been investigated is low, when compared to the total diversity present in the microbiome of animals and plants. However, we are in the process of understanding the role of FA in the regulation of microbial communication, as well as the beneficial and negative effects it generates in the social networks of those microbial populations that have the capacity to perceive them.

So far, the advantage of using FA inhibitors of QS on pathogenic bacteria is clear. Nonetheless, we do not know the effect on symbiotic bacteria that use QS to regulate the production of substances beneficial for health. Also, studies on the effect of FA with QQ properties on human microbiomes are necessary.

On the other hand, FA may have other anti-virulence targets such as SST3, or those related to the alteration of membrane fluidity. However, their great potential application for use as food supplements or as adjuvants for the control of infections resistant to antibiotics is clear.

ACKNOWLEDGMENTS

C.L.-H and J -R. M. research is supported by the CONACYT PhD Grant 273770 and 907614 respectively. C.J.-I. research is supported by Cátedras-CONACYT program. This work was supported by grants from Scientific Development Projects for Solving National Problems/CONACYT Mexico no. 2015-01-402.

REFERENCES

Abd-Alla, Mohamed H., and Shymaa R. Bashandy. 2012. "Production of Quorum Sensing Inhibitors in Growing Onion Bulbs Infected with *Pseudomonas aeruginosa* E (HQ324110)." *ISRN Microbiology* 2012: 1–7. https://doi.org/10.5402/2012/161890.

Armbruster, Chelsie E., Harry L. T. Mobley, and Melanie M. Pearson. 2018. "Pathogenesis of *Proteus mirabilis* Infection." *EcoSal Plus* 8 (1): 1–123. https://doi.org/10.1128/ecosalplus.esp-0009-2017.

Bjarnsholt, Thomas, Peter Østrup Jensen, and Thomas B. Rasmussen, et al. 2005. "Garlic Blocks Quorum Sensing and Promotes Rapid Clearing of Pulmonary *Pseudomonas aeruginosa* Infections." *Microbiology* 151 (12): 3873–80. https://doi.org/10.1099/mic.0.27955-0.

Boon, Calvin, Yinyue Deng, and Lian Hui Wang, et al. 2008. "A Novel DSF-like Signal from *Burkholderia cenocepacia* Interferes with *Candida albicans* Morphological Transition." *ISME Journal* 2 (1): 27–36. https://doi.org/10.1038/ismej.2007.76.

Castillo-Juárez, Israel, Luis Esau López-Jacome, and Gloria Soberón-Chávez, et al. 2017. "Exploiting Quorum Sensing Inhibition for the Control of *Pseudomonas aeruginosa* and *Acinetobacter baumannii* Biofilms." *Current Topics in Medicinal Chemistry* 17 (17): 1915–27. https://doi.org/10.2174/1568026617666170105144104.

Castillo-Juárez, Israel, Toshinari Maeda, and Edna Ayerim Mandujano-Tinoco, et al. 2015. "Role of Quorum Sensing in Bacterial Infections." *World Journal of Clinical Cases* 3 (7): 575–98. https://doi.org/10.12998/wjcc.v3.i7.575.

Chanda, Warren, Thomson Patrick Joseph, and Arshad Ahmed Padhiar, et al. 2017. "Combined Effect of Linolenic Acid and Tobramycin on *Pseudomonas aeruginosa* Biofilm Formation and Quorum Sensing." *Experimental and Therapeutic Medicine* 14 (5): 4328–38. https://doi.org/10.3892/etm.2017.5110.

Childers, Brandon M., Xiaohang Cao, and Gregor G. Weber, et al. 2011. "N-Terminal Residues of the *Vibrio cholerae* Virulence Regulatory Protein ToxT Involved in Dimerization and Modulation by Fatty Acids." *Journal of Biological Chemistry* 286 (32): 28644–55. https://doi.org/10.1074/jbc.M111.258780.

Christensen, Louise D., Maria Van Gennip, and Tim H. Jakobsen, et al. 2012. "Synergistic Antibacterial Efficacy of Early Combination Treatment with Tobramycin and Quorum-Sensing Inhibitors against *Pseudomonas aeruginosa* in an Intraperitoneal Foreign-Body Infection Mouse Model." *Journal of Antimicrobial Chemotherapy* 67 (5): 1198–1206. https://doi.org/10.1093/jac/dks002.

Deng, Yinyue, Amy Lim, and Jasmine Lee, et al. 2014. "Diffusible Signal Factor (DSF) Quorum Sensing Signal and Structurally Related Molecules Enhance the Antimicrobial Efficacy of Antibiotics against Some Bacterial Pathogens." *BMC Microbiology* 14 (1): 1–9. https://doi.org/10.1186/1471-2180-14-51.

Deng, Yinyue, Calvin Boon, and Shaohua Chen, et al. 2013. "*Cis*-2-Dodecenoic Acid Signal Modulates Virulence of *Pseudomonas aeruginosa* through Interference with Quorum Sensing Systems and T3SS." *BMC Microbiology* 13 (1): 1–11. https://doi.org/10.1186/1471-2180-13-231.

Dow, Jhon Maxwell. 2017. "Diffusible Signal Factor-Dependent Quorum Sensing in Pathogenic Bacteria and Its Exploitation for Disease Control." *Journal of Applied Microbiology* 122 (1): 2–11. https://doi.org/10.1111/jam.13307.

Fernández-Piñar, Regina, Manuel Espinosa-Urgel, and Jean Frederic Dubern, et al. 2012. "Fatty Acid-Mediated Signalling between Two *Pseudomonas* Species." *Environmental Microbiology Reports* 4 (4): 417–23. https://doi.org/10.1111/j.1758-2229.2012.00349.x.

Garrett, Reginald H., and Charless M. Grisham. 2010. "Pressure Vessel Design Manual." *Front Matter.* https://doi.org/10.1016/B978-0-12-387000-1.01001-9.

Givskov, Michael. 2012. "Beyond Nutrition: Health-Promoting Foods by Quorum-Sensing Inhibition." *Future Microbiology* 7 (9): 1025–28. https://doi.org/10.2217/fmb.12.84.

Golubeva, Yekaterina A., Jeremy R. Ellermeier, and Jessica E. Cott Chubiz, et al. 2016 "Intestinal Long-Chain Fatty Acids Act as a Direct Signal to Modulate Expression of the *Salmonella* Pathogenicity Island 1 Type III Secretion System." *mBio* 7 (1): 1–9. https://doi.org/10.1128/mBio.02170-15.

Inoue, Tetsuyoshi, Ryuji Shingaki, and Kazuhiro Fukui. 2008. "Inhibition of Swarming Motility of *Pseudomonas aeruginosa* by Branched-Chain Fatty Acids." *FEMS Microbiology Letters* 281 (1): 81–6. https://doi.org/10.1111/j.1574-6968.2008.01089.x.

Kamarudheen, Neethu, and K. V. Bhaskara Rao. 2019. "Fatty Acyl Compounds from Marine *Streptomyces griseoincarnatus* Strain HK12 against Two Major Bio-Film Forming Nosocomial Pathogens; an *in Vitro* and *in Silico* Approach." *Microbial Pathogenesis* 127: 121–30. https://doi.org/10.1016/j.micpath.2018.11.050.

Kim, Suhyun, S. Jordan Kerns, and Marika Ziesack, et al. 2018. "Quorum Sensing Can Be Repurposed to Promote Information Transfer between Bacteria in the Mammalian Gut." *ACS Synthetic Biology* 7 (9): 2270–81. https://doi.org/10.1021/acssynbio.8b00271.

Kwan, Jason Christopher, Theresa Meickle, and Dheran Ladwa, et al. 2011. "Lyngbyoic Acid, a 'Tagged' Fatty Acid from a Marine Cyanobacterium, Disrupts Quorum Sensing in *Pseudomonas aeruginosa*." *Molecular BioSystems* 7 (4): 1205–16. https://doi.org/10.1039/c0mb00180e.

Liaw, Shwu Jen, Hsin Chih Lai, and Won Bo Wang. 2004. "Modulation of Swarming and Virulence by Fatty Acids through the RsbA Protein in *Proteus mirabilis*." *Infection and Immunity* 72 (12): 6836–45. https://doi.org/10.1128/IAI.72.12.6836-6845.2004.

Liu, Li, Tao Li, and Xing Jun Cheng, et al. 2018. "Structural and Functional Studies on *Pseudomonas aeruginosa* DspI: Implications for Its Role in DSF Biosynthesis." *Scientific Reports* 8 (1): 1–11. https://doi.org/10.1038/s41598-018-22300-1.

Lowden, M. J., K. Skorupski, and M. Pellegrini, et al. 2010. "Structure of *Vibrio cholerae* ToxT Reveals a Mechanism for Fatty Acid Regulation of Virulence Genes." *Proceedings of the National Academy of Sciences* 107 (7): 2860–65. https://doi.org/10.1073/pnas.0915021107.

McCarthy, Yvonne, Liang Yang, and Kate B. Twomey, et al. 2010. "A Sensor Kinase Recognizing the Cell-Cell Signal BDSF (*cis*-2-dodecenoic acid) Regulates Virulence in *Burkholderia cenocepacia*." *Molecular Microbiology* 77 (5): 1220–36. https://doi.org/10.1111/j.1365-2958.2010.07285.x.

Montaser, Rana, Valerie J. Paul, and Hendrik Luesch. 2013. "Modular Strategies for Structure and Function Employed by Marine Cyanobacteria: Characterization and Synthesis of Pitinoic Acids." *Organic Letters* 15 (16): 4050–53. https://doi.org/10.1021/ol401396u.

Muñoz-Cazares, Naybi, Rodolfo García-Contreras, and Macrina Pérez-López, et al. 2017. "Phenolic Compounds with Anti-Virulence Properties." *Phenolic Compounds: Biological Activity*, 139–67. https://doi.org/10.5772/66367.

Muñoz-Cazares, Naybi, Rodolfo García-Contreras, and Marcos Soto-Hernández, et al. 2018. "Natural Products with Quorum Quenching-Independent Antivirulence Properties." *Studies in Natural Products Chemistry* 57: 327–51. https://doi.org/10.1016/B978-0-444-64057-4.00010-7.

Nickzad, Arvin, François Lépine, and Eric Déziel. 2015. "Quorum Sensing Controls Swarming Motility of *Burkholderia glumae* through Regulation of Rhamnolipids." *PLoS ONE* 10 (6): 1–10. https://doi.org/10.1371/journal.pone.0128509.

Nicol, Marion, Stéphane Alexandre, and Jean Baptiste Luizet, et al. 2018. "Unsaturated Fatty Acids Affect Quorum Sensing Communication System and Inhibit Motility and Biofilm Formation of *Acinetobacter baumannii*." *International Journal of Molecular Sciences* 19 (1): 1–10. https://doi.org/10.3390/ijms19010214.

Nutman, Allen P., Vickie C. Bennett, and Clark R.L. Friend, et al. 2016. "Rapid Emergence of Life Shown by Discovery of 3,700-Million-Year-Old Microbial Structures." *Nature* 537 (7621): 535–38. https://doi.org/10.1038/nature19355.

Osunla, Charles A., and Anthony I. Okoh. 2017. "Vibrio Pathogens: A Public Health Concern in Rural Water Resources in Sub-Saharan Africa." *International Journal of Environmental Research and Public Health* 14 (10): 1–27. https://doi.org/10.3390/ijerph14101188.

Pérez-López, Macrina, Rodolfo García-Contreras, and Marcos Soto-Hernández, et al. 2017. "Antiquorum Sensing Activity of Seed Oils from Oleaginous Plants and Protective Effect During Challenge with *Chromobacterium violaceum*." *Journal of Medicinal Food* 21 (4): 356–63. https://doi.org/10.1089/jmf.2017.0080.

Plecha, Sarah C., and Jeffrey H. Withey. 2015. "Mechanism for Inhibition of *Vibrio cholerae* ToxT Activity by the Unsaturated Fatty Acid Components of Bile." *Journal of Bacteriology* 197 (10): 1716–25. https://doi.org/10.1128/jb.02409-14.

Rahmani-Badi, Azadeh, Shayesteh Sepehr, and Hossein Fallahi, et al. 2015. "Dissection of the *Cis*-2-Decenoic Acid Signaling Network in *Pseudomonas aeruginosa* Using Microarray Technique." *Frontiers in Microbiology* 6 (APR): 1–13. https://doi.org/10.3389/fmicb.2015.00383.

Ramanathan, Srinivasan, Durgadevi Ravindran, and Kannappan Arunachalam, et al. 2018. "Inhibition of Quorum Sensing-Dependent Biofilm and Virulence Genes Expression in Environmental Pathogen *Serratia marcescens* by Petroselinic Acid." *Antonie Van Leeuwenhoek, International Journal of General and Molecular Microbiology* 111 (4): 501–15. https://doi.org/10.1007/s10482-017-0971-y.

Ryan, Robert P., Shi qi An, John H. Allan and Yvonne McCarthy, et al. 2015. "The DSF Family of Cell–Cell Signals: An Expanding Class of Bacterial Virulence Regulators." *PLoS Pathogens* 11 (7): 1–14. https://doi.org/10.1371/journal.ppat.1004986.

Ryan, Robert P., Yvonne Fouhy, and Belen Fernandez Garcia, et al. 2008. "Interspecies Signalling via the *Stenotrophomonas maltophilia* Diffusible Signal Factor Influences Biofilm Formation and Polymyxin Tolerance in *Pseudomonas aeruginosa*." *Molecular Microbiology* 68 (1): 75–86. https://doi.org/10.1111/j.1365-2958.2008.06132.x.

Santhakumari, Sivasubramanian, Arunachalam Kannappan, and Shunmugiah Karutha Pandian, et al. 2016. "Inhibitory Effect of Marine Cyanobacterial Extract on Biofilm Formation and Virulence Factor Production of Bacterial Pathogens Causing Vibriosis in Aquaculture." *Journal of Applied Phycology* 28 (1): 313–24. https://doi.org/10.1007/s10811-015-0554-0.

Santhakumari, Sivasubramanian, Nizam Mohamed Nilofernisha, and Jeyaraj Godfred Ponraj, et al. 2017. "*In Vitro* and *In Vivo* Exploration of Palmitic Acid from *Synechococcus elongatus* as an Antibiofilm Agent on the Survival of *Artemia Franciscana* against Virulent Vibrios." *Journal of Invertebrate Pathology* 150: 21–31. https://doi.org/10.1016/j.jip.2017.09.001.

Singh, Vijay Kumar, Kumari Kavita, and Rathish Prabhakaran, et al. 2013. "*Cis*-9-Octadecenoic Acid from the Rhizospheric Bacterium *Stenotrophomonas maltophilia* BJ01 Shows Quorum Quenching and Anti-Biofilm Activities." *Biofouling* 29 (7): 855–67. https://doi.org/10.1080/08927014.2013.807914.

Soni, Kamelash A., Palmy Jesudhasan, and Martha Cepeda, et al. 2008. "Identification of Ground Beef–Derived Fatty Acid Inhibitors of Autoinducer-2–Based Cell Signaling." *Journal of Food Protection* 71 (1): 134–38. https://doi.org/10.4315/0362-028x-71.1.134.

Soto, María J., N. Calatrava-Morales, and Isabel M. López-Lara. 2019. "Functional Roles of Non-Membrane Lipids in Bacterial Signaling." *Biogenesis of Fatty Acids, Lipids and Membranes*, 273–89. https://doi.org/10.1007/978-3-319-50430-8_16.

Soto-Aceves, Martín P., Miguel Cocotl-Yañez, and Enrique Merino, et al. 2019. "Inactivation of the Quorum-Sensing Transcriptional Regulators LasR or RhlR Does Not Suppress the Expression of Virulence Factors and the Virulence of *Pseudomonas aeruginosa* PAO1." *Microbiology* 165 (4): 425–32. https://doi.org/10.1099/mic.0.000778.

Twomey, Kate B., Oisin J. O'Connell, and Yvonne McCarthy, et al. 2012. "Bacterial *Cis*-2-Unsaturated Fatty Acids Found in the Cystic Fibrosis Airway Modulate Virulence and Persistence of *Pseudomonas aeruginosa*." *ISME Journal* 6 (5): 939–50. https://doi.org/10.1038/ismej.2011.167.

Wang, Lian Hui, Yawen He, and Yunfeng Gao, et al. 2004. "A Bacterial Cell-Cell Communication Signal with Cross-Kingdom Structural Analogues." *Molecular Microbiology* 51 (3): 903–12. https://doi.org/10.1046/j.1365-2958.2003.03883.x.

Widmer, K. W., K. A. Soni, and M. E. Hume, et al. 2007. "Identification of Poultry Meat-Derived Fatty Acids Functioning as Quorum Sensing Signal Inhibitors to Autoinducer-2 (AI-2)." *Journal of Food Science* 72 (9): 363–68. https://doi.org/10.1111/j.1750-3841.2007.00527.x.

Xiang, Michelle S.W., Jian K. Tan, and Laurence Macia. 2018. Fatty Acids, Gut Bacteria, and Immune Cell Function. *The Molecular Nutrition of Fats*. https://doi.org/10.1016/b978-0-12-811297-7.00011-1.

Zavilgelsky, G. B., and R. S. Shakulov. 2018. "Mechanisms and Origin of Bacterial Biolumenescence." *Molecular Biology* 52 (6): 812–22. https://doi.org/10.1134/s0026893318060183.

Zhou, Lian, Lian Hui Zhang, and Miguel Cámara, et al. 2017. "The DSF Family of Quorum Sensing Signals: Diversity, Biosynthesis, and Turnover." *Trends in Microbiology* 25 (4): 293–303. https://doi.org/10.1016/j.tim.2016.11.013.

5

Analysis of Quorum Sensing by Surface-Enhanced Raman Scattering Spectroscopy

Gustavo Bodelón, Sarah De Marchi-Lourenco, Jorge Pérez-Juste, Isabel Pastoriza-Santos, and Luis M. Liz-Marzán

CONTENTS

5.1 Introduction

Quorum sensing regulates the production of an ample repertoire of extracellular chemical compounds such as exopolysaccharides, enzymes, and bioactive metabolites that can act as signals, cues, and virulence determinants, which are known to greatly influence the interaction of microbes with their environment. These biomolecules are believed to mediate intra- and interspecies, as well as interkingdom interactions (Rumbaugh and Kaufmann 2012), which play important roles in the ecology of the microorganisms (O'Brien and Wright 2011, Davies 2013), and influence human health and disease (Rutherford and Bassler 2012). This entails an even greater significance as the vast majority of microbial life exists in complex communities, often associated with higher organisms, in ways that are still poorly understood. Although it is assumed that microbial populations use quorum sensing processes for coordinating social behaviors driving colonization and survival, very little is known with respect to the impact of chemical exchange processes on the competitive fitness of microbes in the host and the natural environment (Liu, Qin, and Defoirdt 2018). Furthermore, a better understanding of how quorum sensing is used by microbes to interact with their hosts could assist novel strategies to improve human health and combat infectious diseases (Defoirdt 2018). Traditionally, the characterization of quorum sensing systems has been carried out by analytical techniques that rely on the isolation or purification of the effectors and signals involved, as well as by the use of quorum sensing mutant strains and genetically modified biosensors (Fletcher et al. 2014). These approaches have greatly expanded our understanding of how microbial cells communicate. However, the capability to noninvasively monitor the dynamics of quorum sensing in natural microbial populations could provide new fundamental insights into their function and impact in the ecology and social behavior of microbes. To achieve this challenging task, techniques such as imaging mass spectrometry (Stasulli and Shank 2016, Dunham et al. 2017), electrochemical sensing (Connell et al. 2014, Bellin et al. 2014), and surface-enhanced Raman scattering (SERS) spectroscopy (Bodelón et al. 2018) have been recently applied for noninvasive detection and visualization of the molecules involved in microbial communication processes.

We commence this chapter with a brief introduction of SERS, describing the enhancement mechanisms and the strategies for SERS detection. Next, we highlight recent advances in the application of SERS for the study of quorum sensing and, with the aim to facilitate new opportunities and synergies for the study of microbial communication, we conclude the chapter with a summary of perspectives focusing the integration of SERS with microfabricated devices, which have recently revolutionized our understanding of the social behavior of bacteria at the microscale.

5.2 Surface-Enhanced Raman Scattering Spectroscopy

SERS spectroscopy is an analytical technique that combines the rich molecular fingerprint information provided by Raman spectroscopy with the high sensitivity provided by the outstanding optical properties of metallic nanostructures (Schlucker 2014). In SERS, the inherently weak Raman scattering signals of molecules located on or nearby metallic nanostructures, often made

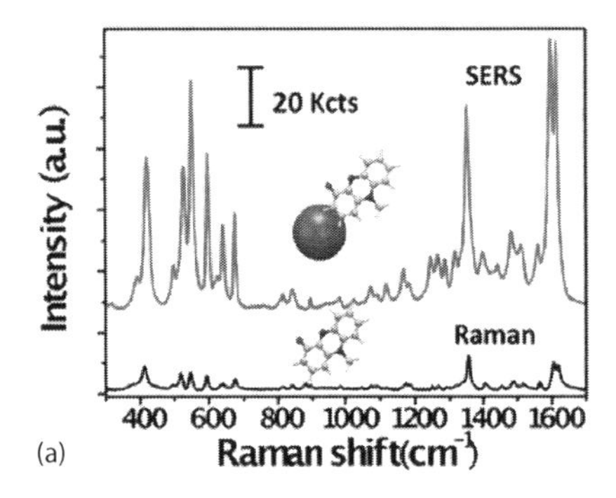

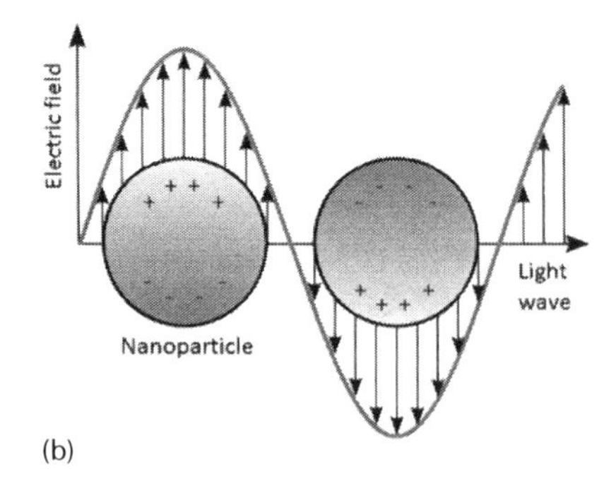

FIGURE 5.1 Principles of SERS and LSPR. (a) Raman (down) and SERS (up) spectra of pyocyanin bound to a plasmonic Au nanoparticle (shiny sphere). (Reproduced with permission from Bodelón, G. et al., *Front. Cell. Infect. Microbiol.*, 8, 2018.) (b) Schematic representation of the localized surface plasmon resonance for a spherical metallic nanoparticle. (Reproduced with permission from Hammond, J.L. et al., *Biosensors*, 4, 172–188, 2014. Copyright 2014, MDPI.)

of gold (Au) or silver (Ag), can be amplified by several orders of magnitude, even reaching single molecule detection under favorable conditions (Figure 5.1a) (Le Ru and Etchegoin 2012). Since its first observation by Fleischmann and coworkers in the 1970s (Fleischmann, Hendra, and Mcquillan 1974), SERS has matured into a powerful analytical tool that has been used in many different research fields, ranging from materials science to biomedicine. As a vibrational spectroscopy, the signals observed in SERS arise from the structure of the molecule and are thus chemically specific (i.e., molecular fingerprints). Major features of SERS include an outstanding multiplexing capability for simultaneous detection of target analytes due to the narrow width of the vibrational Raman bands, possibility for quantification, and high photostability that allows repetitive signal measurements to increase the reliability of the analysis. Other advantages of SERS for biological studies comprise its nondestructive nature, it requires little or no sample preparation that allows on-line analysis, and amenability to miniaturization in lab-on-a-chip (LoC) systems.

The amplification of Raman scattering signals observed in SERS is mainly attributed to two possible mechanisms, an electromagnetic enhancement (Stiles et al. 2008), and to a lesser extent, a chemical enhancement (Jensen, Aikens, and Schatz 2008). The electromagnetic enhancement is strictly related with the optical response of the metallic nanostructure, that is, with the localized surface plasmon resonance (LSPR). The LSPR results in the interactions between the incident light and conduction electrons that collectively oscillate in resonance with the frequency of the incident light (Figure 5.1b) (Schlucker 2014). The regions of intense electromagnetic enhancement, also known as hot spots, which predominantly contribute to the SERS effect, are primarily concentrated at vertices, tips, or edges of metal nanoparticles, as well as in interparticle gaps. The Raman scattering of molecules in these hot spots can be dramatically enhanced with theoretical enhancement factors which can be as high as 10^{11} (Camden et al. 2008). The intensity of the electromagnetic enhancement, which determines the intensity of the measured Raman signal, strongly depends on the nanoparticle size, shape, and composition, the dielectric constant of the surrounding medium, as well as the interparticle distance (Willets and Van Duyne 2007). The chemical enhancement in SERS involves mainly charge transfer mechanisms between the molecule and the metal, enabling new resonance Raman processes that otherwise would not exist in the molecule alone (Moskovits 1985, LeRu and Etchegoin 2009). This phenomenon is analyte-specific and much weaker than the electromagnetic enhancement. The total SERS enhancement factor is the combined contribution of the electromagnetic and chemical enhancement mechanisms, which provides an estimation of the efficiency by which a plasmonic substrate amplifies the Raman signal of a molecule. For molecules bearing chromophores, the magnitude of the enhancement may be further increased if the excitation wavelength of the laser line coincides with an electronic transition of the analyte, which is known as surface-enhanced resonance Raman scattering (SERRS) (Hildebrandt and Stockburger 1984). Under optimal conditions SERS/SERRS can achieve femtomolar (10^{-14} mol·L^{-1}) and attomolar (10^{-18} mol·L^{-1}) levels of detection (Chang et al. 2016, Yang et al. 2016).

In practical terms, silver and gold nanoparticles are the most frequently used optical transducers for SERS, because their LSPR frequencies can be tuned within the visible and near-IR (NIR) region. A crucial aspect for SERS is the plasmonic nanostructured substrate used for the enhancement of the Raman scattering signal. The ideal substrate should possess a large surface area, high number of hot spots, high enhancement factor, reproducibility, and long-term stability. Over the last decade, diverse approaches have been applied to fabricate highly sensitive and reproducible plasmonic substrates, which have been recently discussed elsewhere (Jahn et al. 2016, Mosier-Boss 2017, Hamon and Liz-Marzán 2018).

In general, two distinct modalities can be used for bioanalytical detection by SERS: direct detection (i.e., label-free) and indirect detection using SERS labels (i.e., SERS-tags). In the case of direct SERS, the signal arises upon interaction of the analyte with the plasmonic nanoparticle (Figure 5.2a). Plasmonic colloids or assemblies of plasmonic nanoparticles over a surface (i.e., nanostructured plasmonic substrates) are typically used as optical transducers in this modality. The use of plasmonic colloids often involves the aggregation of nanoparticles to generate hot spots via an external agent (e.g., salt, electrolyte, etc.). However, the SERS measurements obtained by means of this approach are usually characterized by low reproducibility. On the other hand, rationally designed plasmonic nanoparticle assemblies offer the possibility to control nanoparticle clustering and hot spots, thereby leading to improved sensitivity and reproducibility of the SERS measurements. Although label-free SERS has been extensively applied for the molecular detection of a wide variety of biomolecules (Zheng et al. 2018), the

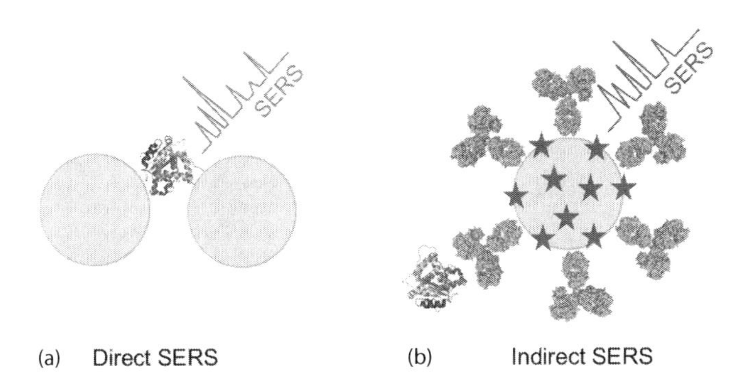

FIGURE 5.2 Modalities of SERS detection. (a) Direct (i.e., label-free) detection of a protein interacting with plasmonic nanospheres. (b) Indirect detection of the protein by means of SERS-tags, composed by a plasmonic nanosphere encoded with Raman-active molecules (stars) and functionalized with specific antibodies.

exposure of the plasmonic transducer to a biological environment may result in matrix interferences that may hinder SERS measurements (see challenges and outlook section). The indirect approach comprises the detection of the target analyte employing plasmonic nanoparticles labeled with Raman reporter molecules, which are additionally functionalized with selected targeting ligands (e.g., antibodies) (Figure 5.2b) (Wang, Yan, and Chen 2013). This approach using SERS-labels excludes the possibility of the spectral contribution from other (bio)molecules that may be present in the medium, thereby enhancing the selectivity of the detection. This method has been applied to the selective detection and imaging of diverse biomolecules (Wang, Yan, and Chen 2013, Shan et al. 2018). Nevertheless, the large size of SERS-labels, compared to conventional organic fluorophores, may be a potential disadvantage making them less attractive for measurements requiring high spatial resolution.

5.3 Applications of SERS to Quorum Sensing Studies

5.3.1 SERS Detection of Autoinducer Molecules

All quorum sensing systems utilize secreted signaling molecules, known as autoinducers (AIs), which regulate their own expression by feedback loop mechanisms (Ng and Bassler 2009). In general, the most prominent AIs can be classified into the following classes: *N*-acyl-L-homoserine lactones (AHLs) utilized by Gram-negative bacteria; oligopeptide signals used by Gram-positive bacteria; and furanosyl borate diester, as a universal signal for interspecies communication used by both Gram-negative and Gram-positive bacteria. Other relevant AIs that go beyond these classes include 2-heptyl-3-hydroxy-4-quinolone (PQS) and its precursor molecule 2-heptyl-4-quinolone (HHQ), autoinducer-3 (AI-3) (LaSarre and Federle 2013), among others (Papenfort et al. 2017).

Direct detection of purified AHLs by SERS has been achieved by means of colloidal Ag nanoparticles (Pearman et al. 2007, Claussen et al. 2013). One study detected up to seven types of commercial AHLs in water, showing that the SERS spectra of the different AHLs are very similar with just subtle differences, which may impair their differentiation by SERS (Pearman et al. 2007), especially in biological media. Claussen and collaborators used a similar approach to detect *N*-dodecanoyl-DL-homoserine lactones (C12-AHLs) in spiked ABtGcasa culture medium, at a biologically relevant concentration of 0.2 nM (Claussen et al. 2013). Although the aforementioned studies suggest the potential of SERS for direct detection of AHLs in bacterial cultures, this has not been achieved to date, most likely due to their low Raman cross section and the SERS contribution of bacterial cells and extracellular exoproducts that can hinder the interaction of the autoinducer molecules with the metal surface. Indeed, as reported by Claussen and coworkers, many of the ABtGcasa Raman spectral peaks are similar to those exhibited by C12-AHLs, hindering the identification of autoinducer SERS signatures (Claussen et al. 2013). In another study, continuous Ag films with nanosized structures, fabricated by a laser deposition technique, demonstrated their suitability for SERS detection of *N*-butyryl-L-homoserine lactone (Culhane et al. 2017). Owing to the inherently low enhancement factor of the system (ca. 10^6), the sensitivity of the detection method was lower than that of previous reports using Ag nanoparticle suspensions (Pearman et al. 2007, Claussen et al. 2013).

With the exception of AHLs, the use of SERS for detection of the other classes of quorum sensing autoinducers mentioned above remains to be shown. Interestingly, it should be noted that Raman scattering spectroscopy was applied to assess the production and spatiotemporal distribution of PQS and HHQ quinolone AIs across dried biofilms of the human opportunistic pathogen *Pseudomonas aeruginosa* at different growth stages (Lanni et al. 2014, Baig et al. 2015). This result suggests that the Raman-active molecules PQS and HHQ could also be detected by SERS.

5.3.2 SERS Detection of Bioactive Metabolites

Besides AIs, quorum sensing regulates the expression of a wide variety of bioactive small diffusible molecules, which can act as signals, cues, and virulence factors, influencing interactions between microbial cells and with their hosts (Rumbaugh and Kaufmann 2012, O'Brien and Wright 2011, Davies 2013, Rutherford and Bassler 2012). Pyocyanin (5-*N*-methyl-1-hydroxyphenazine) is a redox-active, nitrogen-containing heterocyclic compound excreted by *P. aeruginosa* that has been recognized as an intercellular signaling molecule of the quorum sensing network of this organism (Dietrich et al. 2006). Pyocyanin is also a virulence factor that is toxic to cells due to its capacity to produce reactive oxygen species, which are believed to play a major role in oxidative damage elicited by *P. aeruginosa* in the respiratory epithelium (Rada and Leto 2013). Moreover, this phenazine has been involved in the dysregulation of the host immune mechanisms (Hall et al. 2016), and abrogating its biosynthesis leads to attenuation of *P. aeruginosa* virulence in various infection models (Lau et al. 2004). In the context of human airway infections, pyocyanin has been detected in the sputum of cystic fibrosis (CF) patients, and its accumulation negatively correlates with lung function and CF disease progression. Thus, pyocyanin has been suggested as a predictive indicator of disease state in adult patients chronically infected with *P. aeruginosa* (Hunter et al. 2012).

In clinical samples, the concentration of pyocyanin is traditionally determined by UV-vis absorption spectroscopy and high performance liquid chromatography (HPLC) after chloroform extraction from the biological matrix. With the aim to implement a fast and simple diagnostic tool to assess the presence of pyocyanin in sputum samples, Wu and coworkers reported the detection of this molecule by SERS employing substrates of Ag nanorod arrays (Wu et al. 2014) with a limit of detection (LOD) of 5 ppb. Despite the necessity of a pretreatment step (i.e., chloroform extraction), this approach offers the possibility to analyze multiple samples rapidly and overcomes the need for large volumes of samples and reagents. Bodelón et al. (2016) explored three types of rationally designed plasmonic nanostructured substrates, with the aim of detecting and imaging pyocyanin in live *P. aeruginosa* biofilms by SERS. These sensing platforms comprised a plasmonic component (Au nanoparticles) embedded in a porous matrix that acts as a molecular sieve to restrict the interaction of the metal nanoparticle with high molecular weight interfering biomolecules present in the biofilm. Taking advantage that pyocyanin exhibits a broad absorption band centered at 690 nm, SERRS was used to enhance the sensitivity and selectivity of pyocyanin detection (Figure 5.3a and b) versus other types of phenazines that may be produced during the biofilm growth (Figure 5.3c). In this framework, the analysis of cell-free stationary-phase cultures obtained from wild-type *P. aeruginosa* PA14 bacteria (WT) evidenced a SERRS signature identical to that of pure pyocyanin (PYO), whereas no pyocyanin signal was detected in a sample obtained from the phenazine-null isogenic mutant strain (Δ*phz*) (Mavrodi et al. 2001) (Figure 5.3c). The recorded SERS fingerprint is pyocyanin-specific since it was not detected in stationary-phase cultures of *P. aeruginosa* PA14 mutant strains Δ*phzM* and Δ*phzS*, unable to produce this phenazine (Mavrodi et al. 2001) (Figure 5.3d). Moreover, the measurement under resonance Raman conditions facilitated the selective detection of pyocyanin over the rest of the phenazines that may be produced by PA14 *P. aeruginosa* (Mavrodi et al. 2001), which lack the 550–900 nm absorption band (Figures 5.3e and f).

Bodelón et al. reported that the use of poly-N-isopropylacrylamide (pNIPAM) hydrogel loaded with Au nanorods (Au@pNIPAM), devised as a highly porous platform with enhanced diffusivity, allowed bacterial proliferation as a colony on its surface, as well as detection of pyocyanin secretion into the underlying sensor (Figure 5.4). Notably, the use of a 785 nm NIR laser enabled to detect this metabolite at biologically relevant concentrations in spiked Au@pNIPAM hydrogels implanted subcutaneously in mice (Figure 5.4). This result suggested that pyocyanin could be used as a biomarker for noninvasive monitoring of quorum sensing and screening potential antimicrobial drugs in animal models of infections. In view of these results, pre-colonized plasmonic hydrogels with natural populations of *P. aeruginosa* could be potentially used as implantable materials to assess anti-virulence therapies in experimental

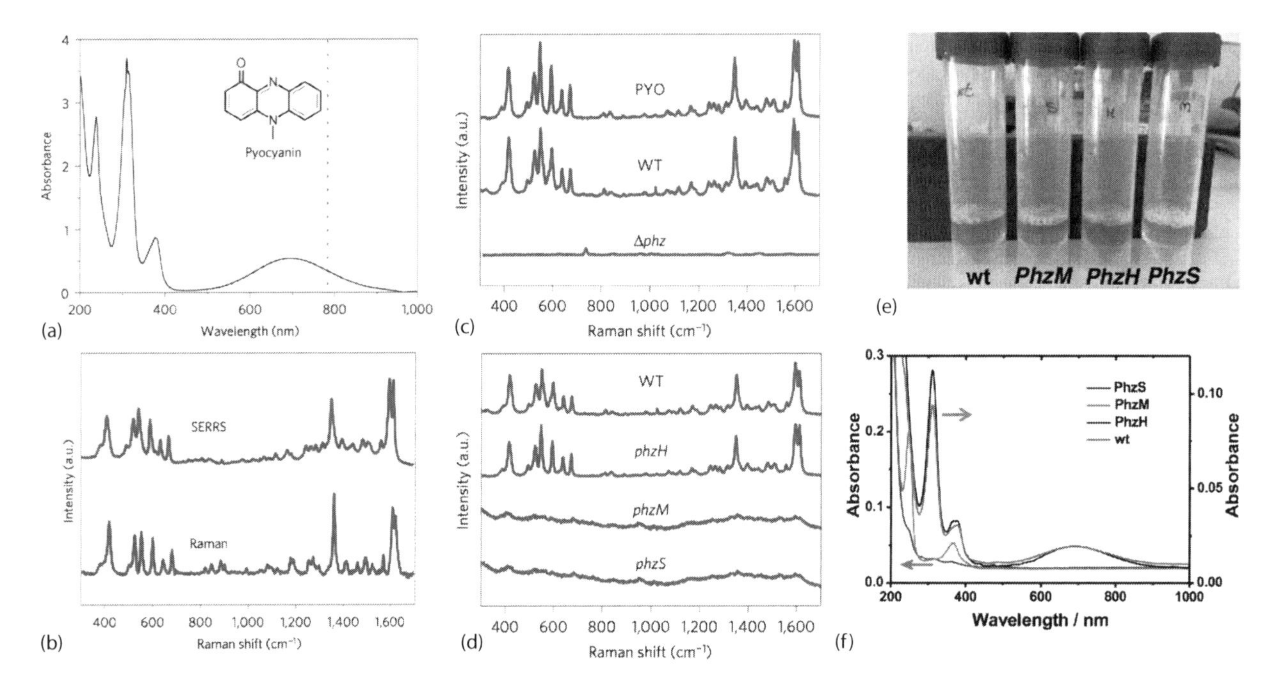

FIGURE 5.3 SERS detection of pyocyanin produced by *P. aeruginosa* grown in liquid culture. (a) UV–visible–NIR absorption spectrum of an aqueous pyocyanin solution (10^{-4} M) and molecular structure of pyocyanin (inset). The dotted line indicates 785 nm, corresponding to the excitation wavelength used for recording resonance Raman (RR) and SERRS spectra. (b) Resonance Raman spectrum of pyocyanin measured in solid state and SERRS spectrum of pyocyaninaqueous solution (1 μM) recorded using a poly-N-isopropylacrylamide (pNIPAM) hydrogeldoped with Au nanorods (Au@pNIPAM). Resonance Raman measurement was carried out with a 50× objective, a maximum power of 54.22 kW cm^{-2}, and an acquisition time of 10 s. SERRS measurements were carried out with a 20× objective, a maximum power of 4.24 kW cm^{-2} and an acquisition time of 10 s. (c) SERRS spectra of commercial pyocyanin (PYO) and of pyocyanin produced by the wild-type PA14 (WT) and phenazine-null *phz1/2* (Δ*phz*) strains. (d) SERRS spectra of pyocyanin produced by wild-type and different phenazine mutant strains (PhzH, PhzS, and PhzM). (e) Photographs of supernatant samples obtained from liquid cultures of WT, PhzH, PhzS, and PhzM mutants, as labeled. (f) UV-visible-NIR absorption spectra of samples containing different phenazines; pyocyanin (wt and PhzH), 1-hydroxyphenazine (1-HO-PHZ, wt, PhzM, and PhzH), and phenazine-1-carboxamide (PCN, wt, PhzS, and PhzM). All SERRS measurements were performed with a 785 nm laser line employing a 20× objective, maximum power between 1.72 kW cm^{-2}, and an acquisition time of 10 s (intensity at 418 cm^{-1}) employing Au@pNIPAM hydrogels. (Reproduced with permission from Bodelón, G. et al., *Nat. Mat.*, 15, 1203, 2016. Copyright 2016, Springer Nature.)

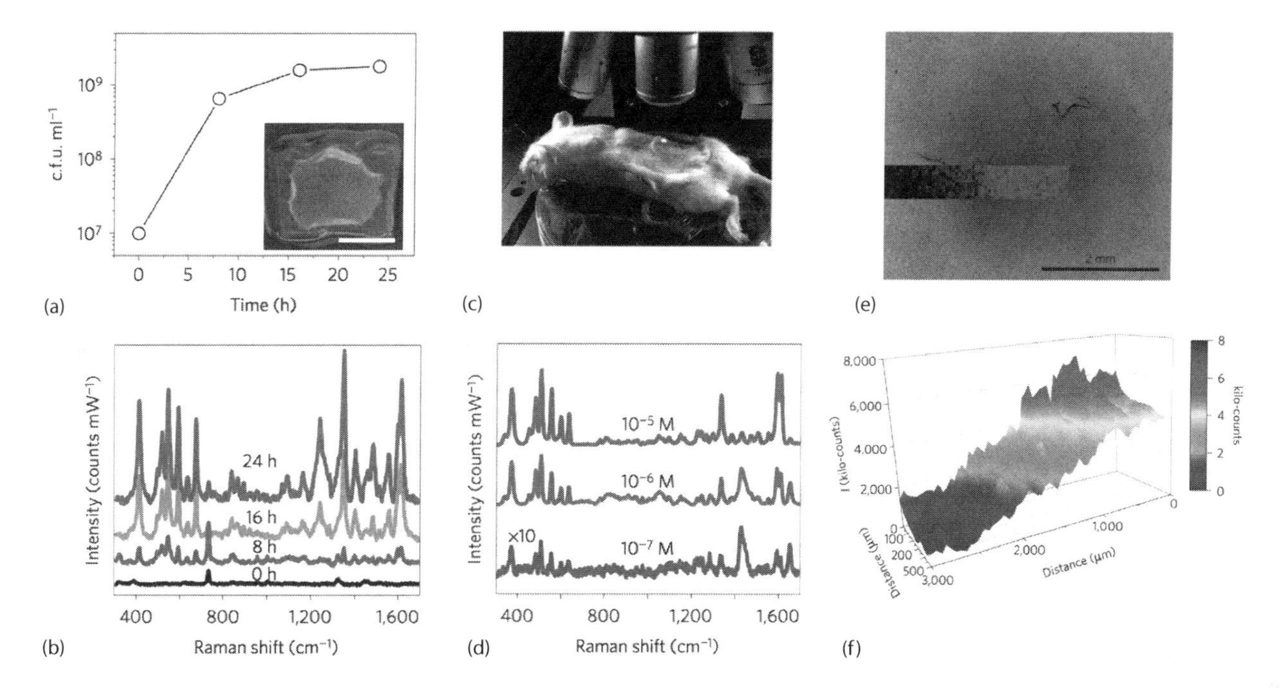

FIGURE 5.4 *In situ* SERS detection and imaging of pyocyanin excreted by *P. aeruginosa* biofilms grown on plasmonic nanostructured substrates. (a) Graphical representation of viable bacteria (c.f.u. mL^{-1}) quantified over time. The inset shows an image of the colony-biofilm grown on Au@pNIPAM (scale bar, 0.5 cm). (b) SERRS spectra recorded at the indicated times. Measurements of colony-biofilms were done using a 785 nm laser line for 10 s and using a maximum power of 0.91 kW cm^{-2} employing a 20× objective. (c) Photograph showing the Raman experimental set-up for detection of pyocyanin in subcutaneous implants in mice. (d) Under-skin SERRS spectra of pyocyanin spiked at the indicated concentrations on Au@pNIPAM hydrogel. SERRS measurements of pyocyanin-spiked hydrogels were performed using a 785 nm laser line for 10 s using a maximum power of 24.45 kW cm^{-2} employing a 10× objective. For clarity, the spectra noted with ×10 have been multiplied by a factor of 10. (e) Optical image of bacterial biofilm (dark central region) grown on Au@TiO_2 substrate captured with the Raman microscope and superimposed pyocyanin SERRS mapping (418 cm^{-1}) acquired with excitation laser wavelength of 785 nm, 5× objective and a laser power of 0.94 mW for 10 s. (f) Graphical representation of the SERRS intensity mapping shown in (e). (Reproduced with permission from Bodelón, G. et al., *Nat. Mat.*, 15, 1203, 2016. Copyright 2016, Springer Nature.)

animal models of infection by SERRS (Bodelón et al. 2016). In this work, the authors demonstrated the detection of pyocyanin produced by *P. aeruginosa* biofilms grown on mesoporous TiO_2 thin films, over a sub-monolayer of Au nanospheres (Au@TiO_2) with high spatial resolution (ca. 20 μm), over micrometer-large surface areas (Figure 5.4e and f). In addition, it was shown that plasmonic substrates based on mesoporous silica-coated supercrystal arrays of Au nanorods (Au@SiO_2), bearing an extremely high SERS efficiency, enabled the detection of pyocyanin at very early stages of biofilm formation, as well as imaging of the phenazine produced by clusters of bacterial cells colonizing micron-sized plasmonic features (ca. 25 μm^2). This highly sensitive approach, combining purpose-designed nanostructured hybrid materials and SERS, not only allowed the detection and visualization of pyocyanin with very high limits of detection (between 10^{-9} and 10^{-14} M depending on the plasmonic substrate), but also provided an efficient and versatile tool for noninvasive chemical detection of quorum sensing communication in live microbial communities, based on SERS-active diffusible molecules (Bodelón et al. 2016).

Analytical methods toward the chemical profiling of cellular metabolites that require extraction of the biomolecules often ignore their spatial distribution, thereby missing important biological information. In this context, it has been shown that SERS can be successfully applied to directly visualize the production and spatial distributions of quorum sensing-regulated bioactive metabolites released by interacting bacterial populations (Bodelón et al. 2017, De Marchi et al. 2019). Bodelón et al. (2017) focused their study in the simultaneous detection of pyocyanin and violacein produced by interacting colonies of *P. aeruginosa* PA14 and *Chromobacterium violaceum* CV026, grown as a coculture on agar-based hybrid plasmonic nanostructures (Au@agar) (Figure 5.5a). These plasmonic substrates comprised a multilayer film of Au nanospheres on glass covered by a layer (c.a. 440 μm of thickness) of LB agar. The rationale behind this co-culture approach is that it enables the identification of the interacting organisms as individual colonies, as well as spatial control over the interacting microbial species, which is often a crucial factor to elicit a stable system. Significantly, the CV026 mutant strain of *C. violaceum* cannot generate its own AHL signals, but can respond to compatible AHLs bearing short acyl chains, such as *N*-butyryl-L-homoserine lactones (C4-AHLs) excreted by *P. aeruginosa* (McClean et al. 1997), thereby resulting in the expression of quorum sensing-regulated phenotypes, including violacein biosynthesis. This study demonstrated simultaneous SERS detection and imaging of pyocyanin and violacein produced by the interacting bacterial species (Figure 5.5), as well as the inhibition of pyocyanin biosynthesis in coculture. Interestingly, the analysis of the spatial distributions of both metabolites showed that the levels of pyocyanin and violacein detected were inversely proportional within the confrontation zone (Figure 5.5c–e), suggesting a putative role of violacein in the down-regulation of the phenazine. Indeed, analysis of

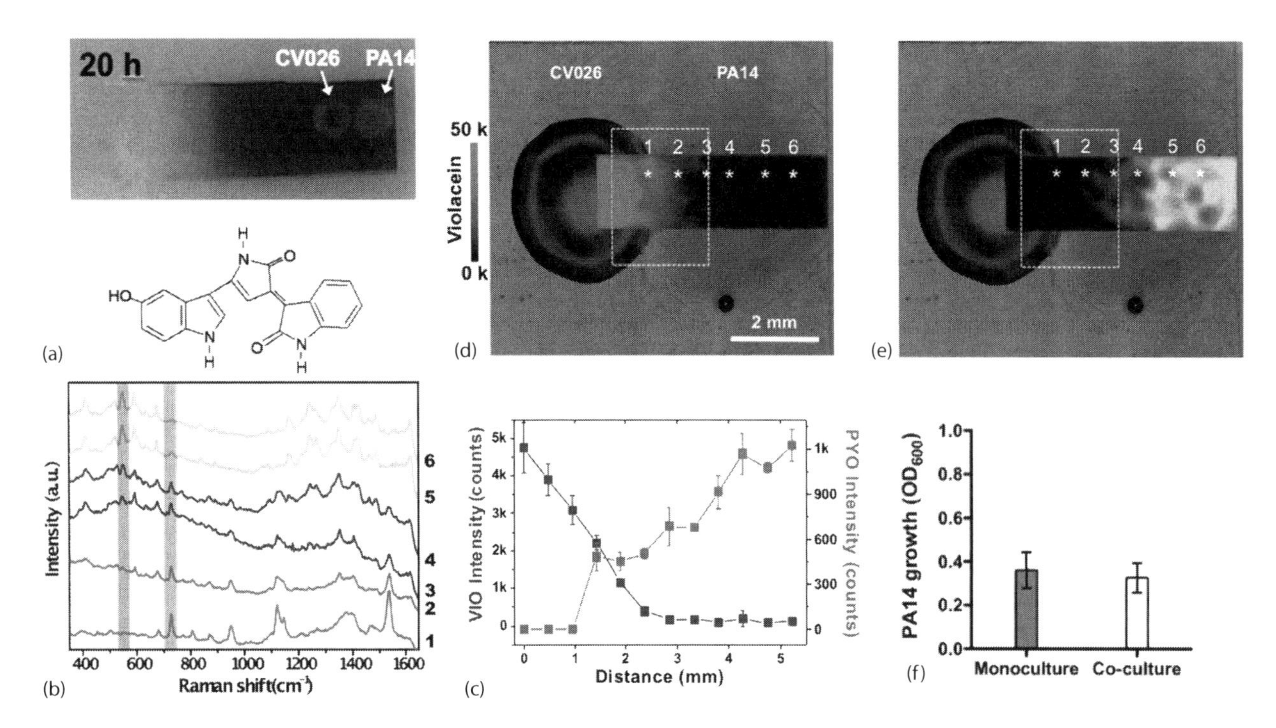

FIGURE 5.5 SERS detection and imaging of interspecies quorum sensing. (a) Photographs of *C. violaceum* CV026 and *P. aeruginosa* PA14 colonies cocultured on Au@agar taken at 20 hours, and molecular structure of violacein. (b) SERS spectra recorded at the points indicated with asterisks in (d) and (e). The left and right bars indicate violacein (727 cm^{-1}) and pyocyanin (544 cm^{-1}) specific bands. (c) SERS intensities of violacein (727 cm^{-1}) and pyocyanin (544 cm^{-1}) measured in (d) and (e) and plotted as a function of distance. The error bars show the standard deviation from three different measurements. All measurements were acquired with an excitation laser wavelength of 785 nm, 5× objective, and a laser power of 12.21 kW cm^{-2} for 10 s. (d) SERS mapping of violacein (727 cm^{-1}) and (e) SERRS mapping of pyocyanin (544 cm^{-1}) in coculture at 20 h. The dashed squares indicate the confrontation zone. (f) Growth of *P. aeruginosa* in monoculture and in coculture with CV026. Error bars indicate the standard deviation of biological triplicates. (Reprinted with permission from Bodelón, G. et al., *Nat. Mat.*, 15, 1203, 2016. Copyright 2017 American Chemical Society.)

gene expression by quantitative real-time PCR indicated that violacein partially repressed the *P. aeruginosa phzS* gene responsible for the last step of pyocyanin biosynthesis. Since the growth of *P. aeruginosa* in monoculture and coculture was very similar (Figure 5.5f), the differential production of pyocyanin was not attributed to growth defects. In light of these results, the authors suggested that the promiscuous quorum sensing CviR receptor of CV026 (McClean et al. 1997) can sense *P. aeruginosa* C4-AHLs and subsequently produce violacein, which in turn down-regulates the expression of the toxic phenazine. The authors hypothesized that violacein is involved in a defensive mechanism of *C. violaceum*, in the chemical interplay with *P. aeruginosa* (Bodelón et al. 2017). Thus, the capability to noninvasively detect and visualize secreted bioactive metabolites by SERS is important to provide meaningful insights into their true biological roles.

In the context of polymicrobial diseases, De Marchi et al. (2019) reported a SERS-based approach for the detection and imaging of indole and pyocyanin, secreted by *Escherichia coli* MG1655 and *P. aeruginosa* PA14, respectively, within a mixed community grown on Au@agar. In this study, the signaling capabilities of indole in the dual-species bacterial consortium were investigated (Figure 5.6a–d), showing by SERS that indole promoted the inhibition of pyocyanin biosynthesis, most likely due to inhibition of the quorum sensing network of *P. aeruginosa* by a yet unknown mechanism (Lee et al. 2009, Chu et al. 2012). Remarkably, the analysis of *E. coli* and *P. aeruginosa* expressing green and red fluorescent proteins, respectively, enabled the authors to visualize the presence of both bacterial species in the mixed culture (Figure 5.6e–j), as well as to quantify bacterial numbers, demonstrating that the decrease in pyocyanin biosynthesis could not be attributed to an impaired growth of *P. aeruginosa* (Figure 5.6k).

It is widely recognized that environmental cues and nutrient availability (e.g., carbon source) have a high impact on bacterial quorum sensing signaling, virulence, and biofilm formation. In this context, Polisetti et al., evaluated by SERS the production of pyocyanin as a function of the carbon source in growing pellicle biofilms (biofilms that form at the air-liquid interface) of two different *P. aeruginosa* strains. To this end, pellicle biofilms of a cystic fibrosis lung isolate (FRD1) and a laboratory strain (PAO1C) were grown in the presence of glucose or glutamate and probed with colloidal Ag nanoparticles. The SERS analysis demonstrated that, whereas the PAO1C is competent to synthesize pyocyanin from glutamate, but not glucose, the FRD1 lung isolate can biosynthesize pyocyanin from either carbon source, indicating strain-level differences in carbon metabolism (Polisetti et al. 2017). Owing to its inherent limitations (e.g., heterogeneous distribution of the local electromagnetic field across the nanostructured material), quantitative analysis by SERS can be quite challenging. The use of internal standards has been proposed to improve the accuracy and precision of quantitative SERS measurements (Bell and Sirimuthu 2008). In this context, Guo and coworkers designed a multifunctional plasmonic sensing

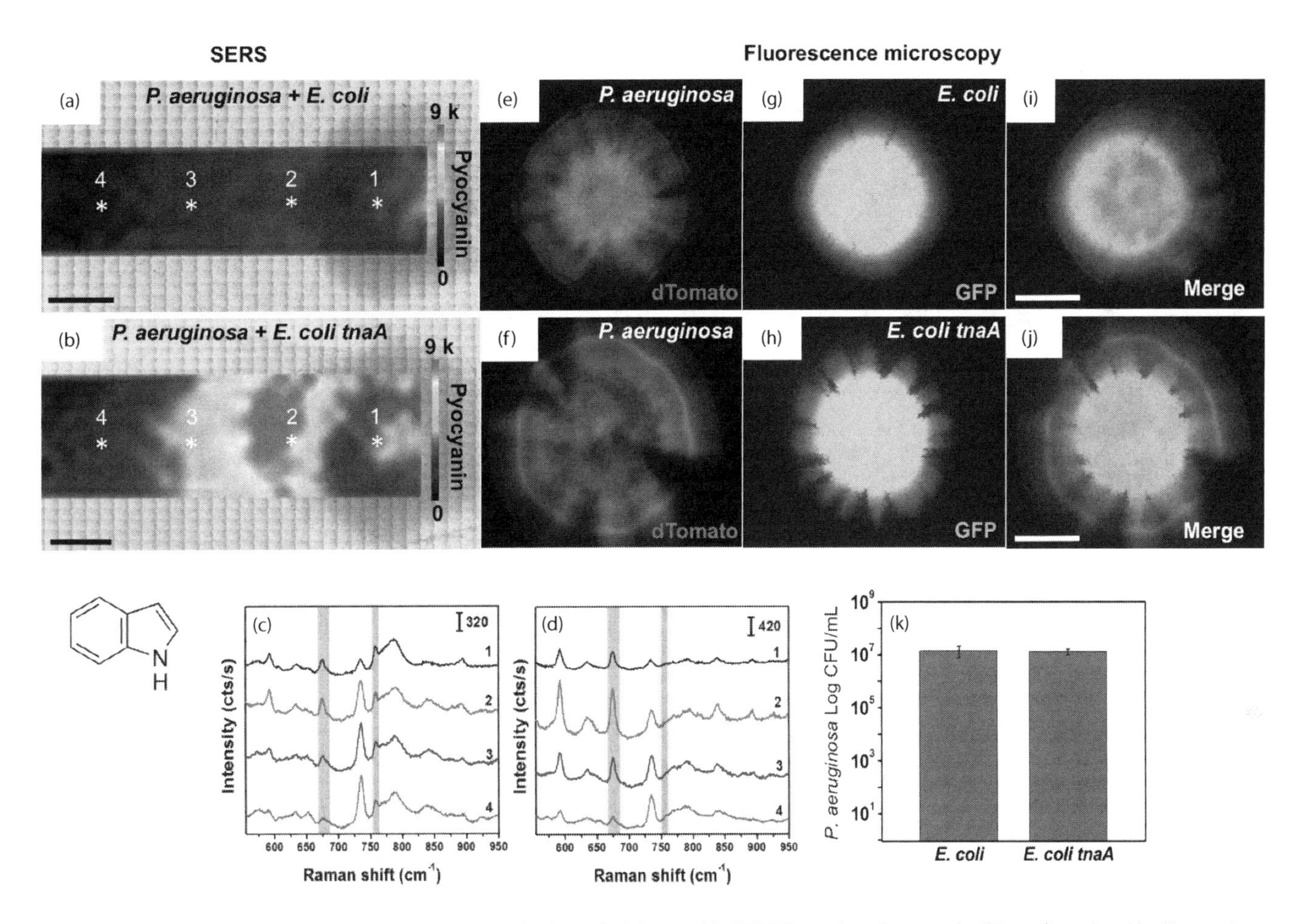

FIGURE 5.6 SERS detection and imaging of pyocyanin inhibition by indole. (a and b) SERRS mapping of pyocyanin (675 cm^{-1}) produced by *P. aeruginosa* expressing dTomato cocultured with GFP-expressing *E. coli* (a) and *E. coli tnaA* (b). Scale bar is 1 mm. Below is represented the molecular structure of indole. (c and d) SERS spectra recorded at the points indicated with asterisks in (a) and (b). The bars at the right and left handsides indicate indole (760 cm^{-1}) and pyocyanin (544 cm^{-1}) specific bands respectively. All measurements were performed with a 785 nm excitation laser line, 20× objective, and a laser power of 4 mW. (e–j) Fluorescence microscopy images of dTomato (e and f) and GFP (g and h) produced by *P. aeruginosa* in coculture with *E. coli* and *E. coli tnaA*. (i and j) Merged images of (e), (g) and (f), (h), respectively. Scale bar is 1 mm. (k) Quantification of *P. aeruginosa* growth in coculture with the indicated *E. coli* strains. The values represent the average of three independent experiments, each counted in at least triplicate. (Reprinted from *Appl. Mat. Today*, De Marchi, S. et al., Surface-enhanced Raman scattering (SERS) imaging of bioactive metabolites in mixed bacterial populations, 207–215, doi:10.1016/j.apmt.2018.12.005, Copyright 2018, with permission from Elsevier.)

device that consists of gold nanostars encapsulated between two pieces of hexagonal boron nitride layers functionalized with 4-mercaptobenzoic acid as an internal standard for SERS calibration during quantification (Guo et al. 2018). This plasmonic substrate with self-calibration properties was successfully used for direct detection and quantification of pyocyanin secreted by cultures of *P. aeruginosa*, as well as assessment of its space-time distribution (Guo et al. 2018).

5.3.3 SERS Detection of Volatile Organic Compounds

The study of cell-to-cell interactions has primarily focused on soluble bioactive metabolites, but bacteria also excrete into their headspace a wide variety of volatile organic compounds (VOCs), which are known to influence the growth of neighboring microbes and eukaryotic hosts (Audrain et al. 2015, Schmidt et al. 2015). These small molecules are typically characterized by low molecular mass and low boiling point, and, together with a lipophilic character, these characteristics support evaporation and diffusion. Since the production of VOCs is associated with cell density, it has been speculated that they may be regulated by quorum sensing (Schmidt et al. 2015). This holds true for hydrogen cyanide and 2′-aminoacetophenone (2-AA), whose biosynthesis is under quorum sensing control in *Chromobacterium* and *Pseudomonas* species (Chernin et al. 1998, Pessi and Haas 2000, Que et al. 2013). In addition, it has been reported that VOCs can affect quorum sensing in bacteria (Schulz et al. 2010, Chernin et al. 2011, Ahmad, Viljoen, and Chenia 2015), and fungi (Hornby et al. 2001, Martins et al. 2007). For example, rhizobacterial volatiles produced by *Pseudomonas fluorescens* and *Serratia plymuthica* act as inhibitors of intercellular communication mediated by AHLs in strains of *Agrobacterium*, *Chromobacterium*, *Pectobacterium*, and *Pseudomonas*. VOCs emitted by *S. plymuthica* led to a significant suppression of transcription of AHL synthase genes in the above mentioned bacterial species (Chernin et al. 2011). The ecological role of volatiles in microbial communication and competition has been overlooked for a long time, probably due to the requirement of expensive and sophisticated chemical-detection technologies (Schmidt et al. 2015). Currently, gas chromatography-mass spectrometry (GC-MS) is the gold standard for VOCs detection and quantification. The detection of trace amounts in real time is limited by the necessity of pre-concentration steps (e.g., solid phase microextraction), which are typically followed by thermal desorption for chemical analysis. However, the significant relevance of microbial VOCs, along with current interest in microbiome research, is attracting attention to the development of novel approaches for their rapid and *in situ* detection. In this context, monitoring volatiles may be used as a potential indicator of microbial activity (e.g., quorum sensing), and one important area of VOC research is the identification of novel biomarkers for disease diagnosis and early detection of infections in exhaled breath and open wounds (Bos, Sterk, and Schultz 2013). Recently, SERS has emerged as an alternative tool for the detection of microbial volatiles because it allows direct measurements on gas state without pretreatment steps. Lauridsen et al. (2017) used gold-coated silicon nanopillars for direct SERS detection of hydrogen cyanide emission from liquid cultures of *P. aeruginosa* strains isolated from the airways of CF patients employing an air sampling pump (Figure 5.7a and b). The production of this compound in *P. aeruginosa* is under the control of the quorum sensing transcriptional regulator *lasR* (Pessi and Haas 2000). Significantly, patho-adaptive mutations in this gene commonly emerge in chronically infected CF patients, which may explain why hydrogen cyanide cannot be detected in all patients chronically infected with *P. aeruginosa* (Lauridsen et al. 2017). Aiming to investigate whether mutants in *lasR* were still producing volatiles, the selected strains consisted of early isolates obtained prior to chronicity of the infection and the corresponding late isolates obtained from the same chronically infected patient. The results of this study showed that the volatile was detected by SERS at ppb concentrations in gases from *in vitro* cultures of all tested early *P. aeruginosa* isolates, as well as in half of the late strains with mutated *lasR*. This result suggested the possibility to use SERS detection of this molecule as an early indicator of *P. aeruginosa* infection. The authors hypothesized that the production of hydrogen cyanide in strains with malfunctioning *lasR* could be explained by changes in the regulation of other quorum sensing signaling systems (e.g., RhlR-RhlI) and hydrogen cyanide production (Lauridsen et al. 2017). Shown in Figure 5.7c are SERS

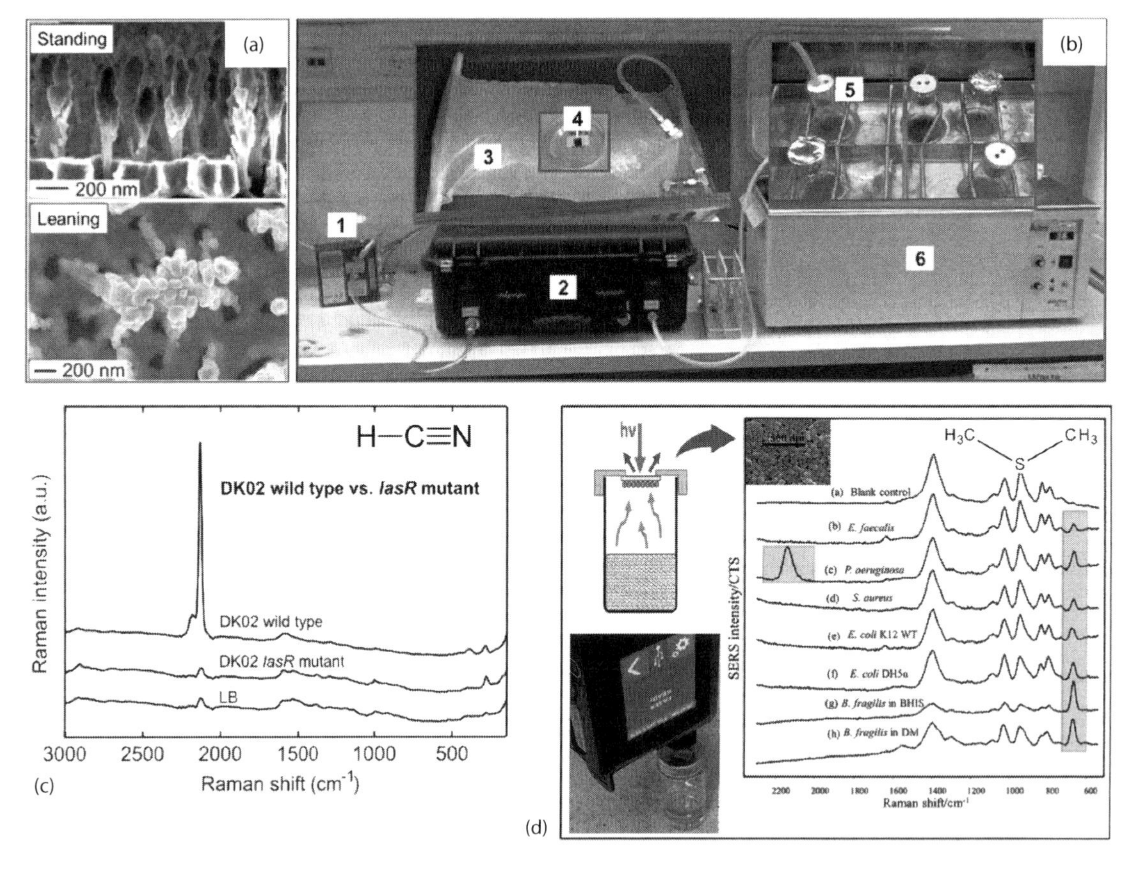

FIGURE 5.7 SERS detection of bacterial VOCs. (a) SEM image of the Au-coated Si nanopillars SERS substrate, before (standing) and after (leaning) drying for analyte detection. (b) The pump (1) induces a vacuum inside the vacuum chamber (2), leading to inflation of the bag (3) and exposure of the SERS substrate (4) to volatiles created by the bacterial culture (5) in the water bath (6). (c) SERS spectra of emissions from overnight cultures of the *P. aeruginosa* DK02 lineage and *lasR* mutated strain. Both wild types show intense hydrogen cyanide signals at 2135 cm^{-1}. The spectrum of LB emissions is included as reference. Measurements were performed with a 1064 nm excitation laser line. (Reprinted with permission from Lauridsen, R.K. et al., *Sci. Rep.*, 7, 45264, 2017.) (d) SERS spectra of the headspace of (a) blank control, (b) *E. faecalis* ATCC 10541, (c) *P. aeruginosa* OA1, (d) *S. aureus* Cowan I, (e) *E. coli* K12 WT, (f) *E. coli* DH5α, (g) *B. fragilis* NCTC 9343 cultivated in brain heart infusion broth supplemented (BHIS), and (h) *B. fragilis* NCTC 9343 cultivated in defined medium (DM). The SERS spectra were recorded with a 785 nm laser line and an excitation power of 20 mW. 1 s accumulations were used for experiments with the Ag substrate and 20 s for Au substrates. Inset: SEM image of the SERS substrate. The top image is a scheme of the detection principle and configuration for headspace SERS measurements. (Reprinted with permission from Kelly, J. et al., *Angew Chem. Int. Ed. Engl.*, 57, 15686–15690, 2018.)

spectra corresponding to the C≡N stretching peak at 2135 cm^{-1} for hydrogen cyanide detected from an overnight culture of wild type *P. aeruginosa* DK02 strain. In another recent study, Kelly et al. (2018) developed a SERS approach for *in situ* detection of dimethyl disulphide (DMDS) in the headspace of liquid cultures of several bacterial species including *Enterococcus faecalis*, *P. aeruginosa*, *Staphylococcus aureus*, *E. coli*, and *Bacteroides fragilis*, employing a portable Raman instrument (Figure 5.7d). The SERS substrate comprised a densely packed monolayer of Ag or Au nanoparticles deposited onto a quartz window that was fixed onto the cap of a culture vial. Interestingly, the authors reported that Au nanoparticle films yielded lower absolute SERS signals but much faster kinetics than the Ag substrates, so that the time required to achieve a detectable signal was significantly shorter. The measured SERS spectra exhibited a spectral band assigned to the ν(C–S) vibration of surface-bound methyl sulfide, $Ag–S–CH_3$. The presence of DMDS was further confirmed by GC-MS. It should be mentioned that besides this volatile, the authors could also identify a SERS peak characteristic of hydrogen cyanide in the SERS spectrum from *P. aeruginosa* (Figure 5.7d).

DeJong et al. (2017) applied Au-coated nanostructured arrays for the differentiation of *E. coli*, *Enterobacter cloacae*, and *Serratia marcescens* by means of the specific SERS signatures of excreted VOCs into the headspace of the bacterial cultures. Interestingly, the authors reported the identification of SERS spectral bands that were assigned to indole, a molecule previously detected in gas phase in the headspace of diverse bacterial species (Effmert et al. 2012). Despite these recent advancements, direct gas detection remains a challenge due to poor affinity to the plasmonic surface and VOCs low concentrations. A novel approach for SERS detection of VOCs consists in the integration of metal–organic frameworks (MOFs) with a plasmonic nanostructure. MOFs are adsorbent materials characterized by high surface area, enhanced gas adsorption, thermal stability, and tailorable pore sizes (Liu et al. 2018), which can confine gas molecules directly at the metal surfaces, even in the absence of specific gas-metal affinity (Koh et al. 2018). This provides a pre-concentration effect that has been shown to greatly enhance the SERS performance, up to *in situ* femtomolar detection and quantification of toxic volatiles, such as toluene and chloroform (Koh et al. 2018), as well as gaseous aldehydes, as indicators of lung cancer (Qiao et al. 2018). The application of such hybrid SERS sensors may open new avenues for the detection of quorum sensing-regulated VOCs and other microbial volatiles.

5.3.4 SERS Detection of Other Quorum Sensing-Regulated Phenotypes

The crowded environment in biofilms facilitates chemical exchange processes and quorum sensing, which are believed to play an important role during biofilm biogenesis (Parsek and Greenberg 2005). SERS has been applied for noninvasive chemical analysis of biofilms, and to evaluate the spatial bio distribution and relative abundance of the components of the extracellular polymeric matrix, which has been recently reviewed elsewhere (Ivleva, Kubryk, and Niessner 2017, Kelestemur, Avci, and Culha 2018). In general, spectral signatures of biofilm constituents such as DNA, RNA, proteins, lipids, and carbohydrates can be measured by incubating biofilms with plasmonic nanoparticle suspensions. This strategy has enabled scientists to qualitatively analyze the chemical composition and monitor changes in biofilms at different growth stages (Ivleva et al. 2008, 2010, Ramya et al. 2010, Chao and Zhang 2012, Efeoglu and Culha 2013), as well as to monitor the fouling process in nanofiltration membranes (Cui et al. 2015), or to assess the development of a dual-species biofilm on a mixed cellulose ester membrane surface (Chen, Cui, and Zhang 2015). In one study, Chao and Zhang applied SERS to study chemical variation in single-species biofilms of *E. coli*, *Pseudomonas putida*, and *Bacillus subtilis* from initial attachment to mature biofilms. The authors claimed that the increase in intensity of a Raman band at 730 cm^{-1} during biofilm growth, which was assigned to nucleic acids (i.e., adenine-related compounds), could be attributed to the accumulation of extracellular DNA (eDNA). In this context, eDNA is used by microbial cells as a nutrient source, for genetic exchange, and as a structural component during biofilm formation. It has been shown that in certain bacterial species the release of this major structural component of the biofilm matrix comes from programmed lysis of a bacterial subpopulation in response to quorum sensing (Ibáñez de Aldecoa, Zafra, and González-Pastor 2017). The capability to monitor the production of eDNA noninvasively by SERS could provide new insights of the role of this extracellular polymeric substance during biofilm biogenesis.

Recently, SERS has been widely applied for the study and characterization of proteins, predominantly in purified samples (Feliu et al. 2017). However, the direct identification of specific cellular proteins by SERS is a challenging task due to their common biochemical composition and the intrinsic complexity of the biological matrix (e.g., bacterial outer-membrane, extracellular medium, biofilm, etc.). In this regard, the direct detection of specific bacterial surface proteins by SERS has been recently reported (Witkowska et al. 2018). Witkowska et al., reported the identification of SERS signatures corresponding to proteins in the cell envelope of *Listeria monocytogenes* termed CadA1/CadA2, and BcrB/BcrC that determine the resistance to $Cd^{2}+$ and benzalkonium chloride, respectively. In this study, SERS signals were detected upon treatment of the bacterial cells with $Cd^{2}+$ or benzalkonium chloride, which are known to be inducers of the aforementioned resistance proteins. These highly toxic compounds can also change the gene expression of other cell membrane components, and therefore, the SERS features detected cannot be solely attributed to the proteins. It would therefore be exciting to evaluate the SERS detection of Gram-positive autoinducer peptides, as well as other quorum sensing regulated proteins, such as toxins and adhesion molecules, which has not been shown to date.

5.4 Perspectives for the SERS Analysis of Microbial Communication at the Microscale

Understanding how microbial communities communicate and function is a huge challenge that requires interdisciplinary approaches. The original concept of quorum sensing focused on the relevance of a process that coordinates the behavior of large groups of cells. Theoretical considerations soon realized that depending on the local concentration of quorum sensing signals, small groups of cells could also initiate quorum sensing. Physical models predicted that, besides cell density, this process would be influenced by a combination of other factors such as spatial cell organization, clustering, diffusion rates, and flow conditions (Hense and Schuster 2015). In this context, modern analytical tools are needed to provide new insights on the chemistry that microbial cells use to communicate and interact with other microbes, as well as to manipulate their hosts. By means of microfabrication it is now possible to realize materials with micron-scale features that match the physical dimensions of microbial cells and their microenvironment, thereby enabling us to "zoom in" into the microbial world and providing detailed insight into the complex lives of microbes (Weibel, Diluzio, and Whitesides 2007, Wessel et al. 2013, Hol and Dekker 2014, Wu and Dekker 2016). By allowing precise control and manipulation of physical and chemical conditions, microtechnology and nanotechnology tools can reproduce relevant characteristics of natural microbial habitats at different scales, with an unprecedented level of versatility and quantification. This has recently enabled researchers to address long-lasting questions regarding cellular processes such as motility, growth, cell-fate, evolution, horizontal gene transfer, and quorum sensing (Figure 5.8).

The seminal studies of Boedicker with *P. aeruginosa* and Carnes with *S. aureus* demonstrated that small groups of bacterial cells and even single cells could activate quorum sensing mechanisms when physically confined in ultra-small volumes (Boedicker, Vincent, and Ismagilov 2009, Carnes et al. 2010). Recently, Gao et al. (2016) developed a method to encapsulate bacteria in hydrogel microcapsules to investigate the effect of cell clustering on quorum sensing, showing that cell aggregates of approximately 25 μm size exhibited much stronger quorum sensing capacity than smaller clusters or uniform cell populations. In static conditions, diffusion becomes a key driver of microbial community dynamics. Microfabricated arrays of poly(ethylene glycol) diacrylate (PEGDA) 1.5-mm wide chambers were used to quantitatively address how gradients of quorum sensing signals and secondary metabolites influence cell growth and biofilm formation between adjacent, but physically separated, *P. aeruginosa* populations (Flickinger et al. 2011). This study showed that *N*-(3-oxododecanoyl)-L-homoserine lactone secreted by a growing biofilm formed a gradient in the hydrogel that induced quorum sensing in *P. aeruginosa* cells situated 8 mm away, stimulating the growth of neighboring cells in close proximity (less than 3 mm away from the nascent biofilm). The authors suggested that this observation

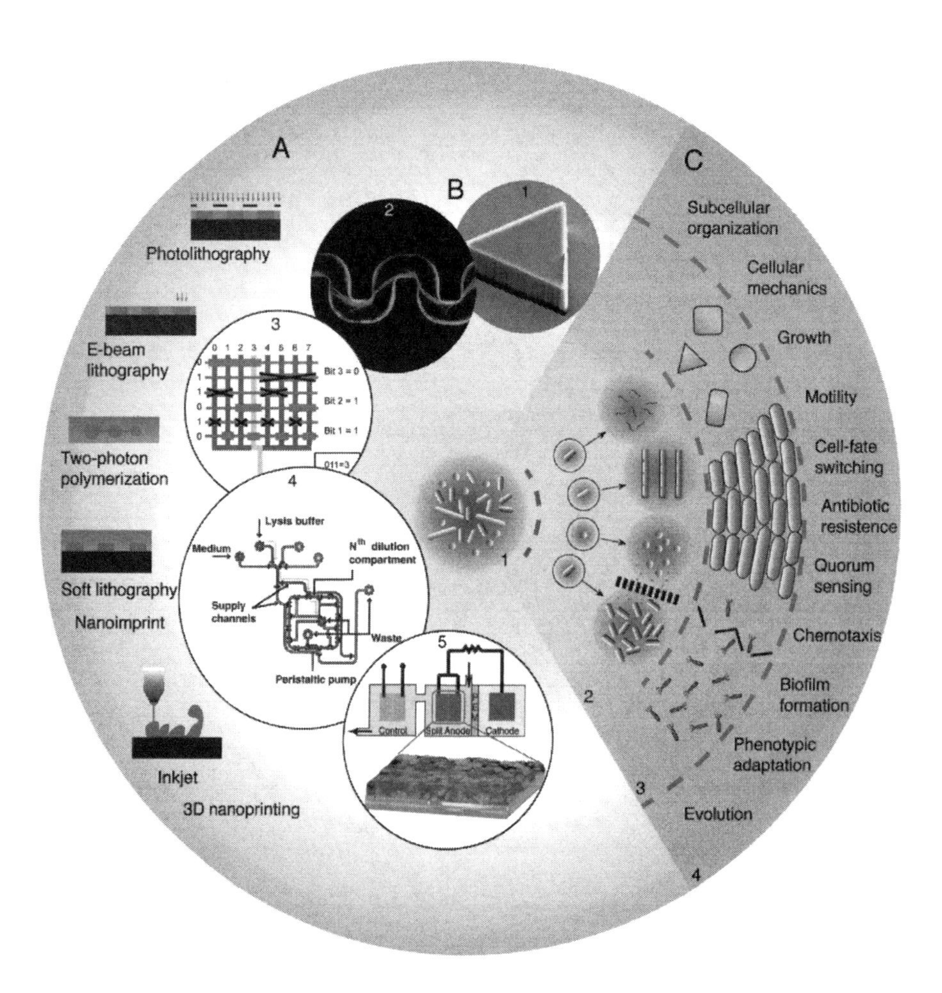

FIGURE 5.8 Microfabricated devices for microbiological applications. (a) Illustration of basic microfabrication processes viewed at cross sections. (b) Images in circles show microscale structures and (c) their multiplex applications. (Wu, F., and Dekker, C., *Chem. Soc. Rev.*, 45, 268–280, 2016. Reproduced by permission of The Royal Society of Chemistry.)

supports the "efficiency sensing" theory (Hense et al. 2007), by which quorum sensing in natural environments is affected by three key parameters: cell density, mass transfer, and spatial distribution. In this context, the Whiteley group assembled bacterial populations within 3D-printed "traps" to investigate how diffusion and spatial parameters influence quorum sensing behaviors (Connell et al. 2013, 2014, Darch et al. 2018). Microfluidics has significantly contributed to elucidate many fundamental aspects of microbial behavior offering a means to experimentally explore the role of a dynamic environment and assess the impact of fluids, spatial structure, and chemical gradients on microbial interactions (Rusconi, Garren, and Stocker 2014, Yawata et al. 2016). In this context, Kim and collaborators (2016) investigated quorum sensing in *S. aureus* and *Vibrio cholerae* biofilms under different flow regimes using a microfluidic approach bearing geometric and topographic features, so as to mimic the dynamics and surface complexity of host environments such as tooth cavities. The authors identified conditions where flow, which carries away signaling molecules, repressed quorum sensing, and other situations in which quorum sensing was active despite flow, for instance, within the crevices of a groove-like surface, where flow reduction leads to autoinducer accumulation and activation of quorum sensing. This study highlights that, under flow and in complex environments, even genetically identical cells can exhibit remarkably different social behaviors (Kim et al. 2016). Thus, understanding the consequences of nonuniform quorum sensing responses in natural environments might be the key for successful deployment of anti-quorum sensing strategies. Furthermore, the ability to engineer patterns of individual bacteria (Hochbaum and Aizenberg 2010) or cellular aggregates (Jahed et al. 2017) on surfaces with spatial control offers new, exciting avenues for the study of cell-to-cell communication phenomena especially at initial stages during biofilm formation. Thus, the manipulation of microbial environments at the microscale offers new opportunities to investigate quorum sensing, as well as the chemical exchange processes that modulate the social behavior of bacteria. This approach would greatly benefit from highly sensitive detection methods capable of measuring target biomolecules *in situ*, at very low concentrations and in extremely small volumes, a capability for which SERS stands out. In recent years, the rapid progress of nanotechnology has triggered significant advancements in reliable SERS substrates, as well as SERS-enabled LoC systems that can quantitatively or qualitatively monitor bio/chemical responses in real time (Huang et al. 2015, Jahn et al. 2017, Kant and Abalde-Cela 2018). Several strategies have been implemented for SERS detection of biomolecules in microfabricated and LoC systems including: (1) colloidal plasmonic nanoparticles, (2) SERS labels, (3) 3D nanostructured substrates, and (4) hybrid materials based on plasmonic polymer nanocomposites.

1. **Colloidal plasmonic nanoparticles.** These plasmonic transducers have been used in microfluidics under two distinct modalities: continuous flow and droplet-based approaches (Kant and Abalde-Cela 2018). Compared with conventional SERS measurements on open settings, the capability to operate within a continuous flow regime and generate homogeneous mixing conditions within microfluidics has demonstrated to afford quantitative SERS-based analysis (Jahn et al. 2017). This strategy has been applied for the identification of *E. coli* bacteria at the strain level (Walter et al. 2011), or differentiation of methicillin-resistant *S. aureus* (Lu et al. 2013). Nguyen et al. fabricated chemically assembled oligomers of Au nanoparticles with controlled nanometer gap spacing as SERS sensors for the detection of pyocyanin in a microfluidic device (Figure 5.9a). The authors reported longitudinal monitoring and quantification of pyocyanin concentration, down to the nanomolar range, excreted by *P. aeruginosa* biofilms grown in flow channels. This strategy avoided the need to grow biofilms directly on SERS substrates or to directly probe the microbial populations with plasmonic nanoparticles (Nguyen et al. 2018). The droplet-based approach relies on the combination of the sample and nanoparticles into aqueous droplets surrounded by an immiscible liquid phase. The microdroplets contribute to enhance mixing in a reproducible way, leading to more uniform SERS signals (Jahn et al. 2017). In a recent study, Zukovskaja and coworkers (2017) employed silver nanoparticles in a segmented flow cell for detection of pyocyanin in saliva at a clinically relevant range, with no need for sample processing. Mühlig et al. (2016) demonstrated the identification of six species of *Mycobacterium tuberculosis* and nontuberculous mycobacteria in a droplet-based platform by direct SERS, focusing on the vibrational signals of the cell-wall component mycolic acid specific for each species. It is important to note that free mycolic acid is known to participate in the maturation of the biofilm pellicle in this organism (Ojha et al. 2008). Nevertheless, the role of biofilms in the pathogenesis of tuberculosis remains unclear (Esteban and Garcia-Coca 2017).
2. **SERS labels.** The use of SERS-tags can increase both the selectivity and the multiplexing capabilities of the detection. Pazos-Perez et al. reported the simultaneous detection of four microbial species by SERS-tags functionalized with specific antibodies, employing a millifluidic system (Figure 5.9b). This device enabled the screening and quantification of several bacterial species in serum and blood, with sensitivity down to the single colony-forming unit (Pazos-Perez et al. 2016). The combination of antibody-functionalized magnetic nanoparticles and SERS-tags in microdroplet platforms have enabled a wash-free immunoassay for the detection of the quorum sensing-regulated virulence factor antigen fraction 1 (F1) from *Yersinia pestis* (Figure 5.9c). The reported limit of detection of the F1 antigen was 59.6 pg/mL, a value approximately 2 orders of magnitude more sensitive than conventional enzyme-linked immunosorbent assays. Additional advantages of this fully automated system were reduced sample consumption and rapid assay times (Choi et al. 2017).

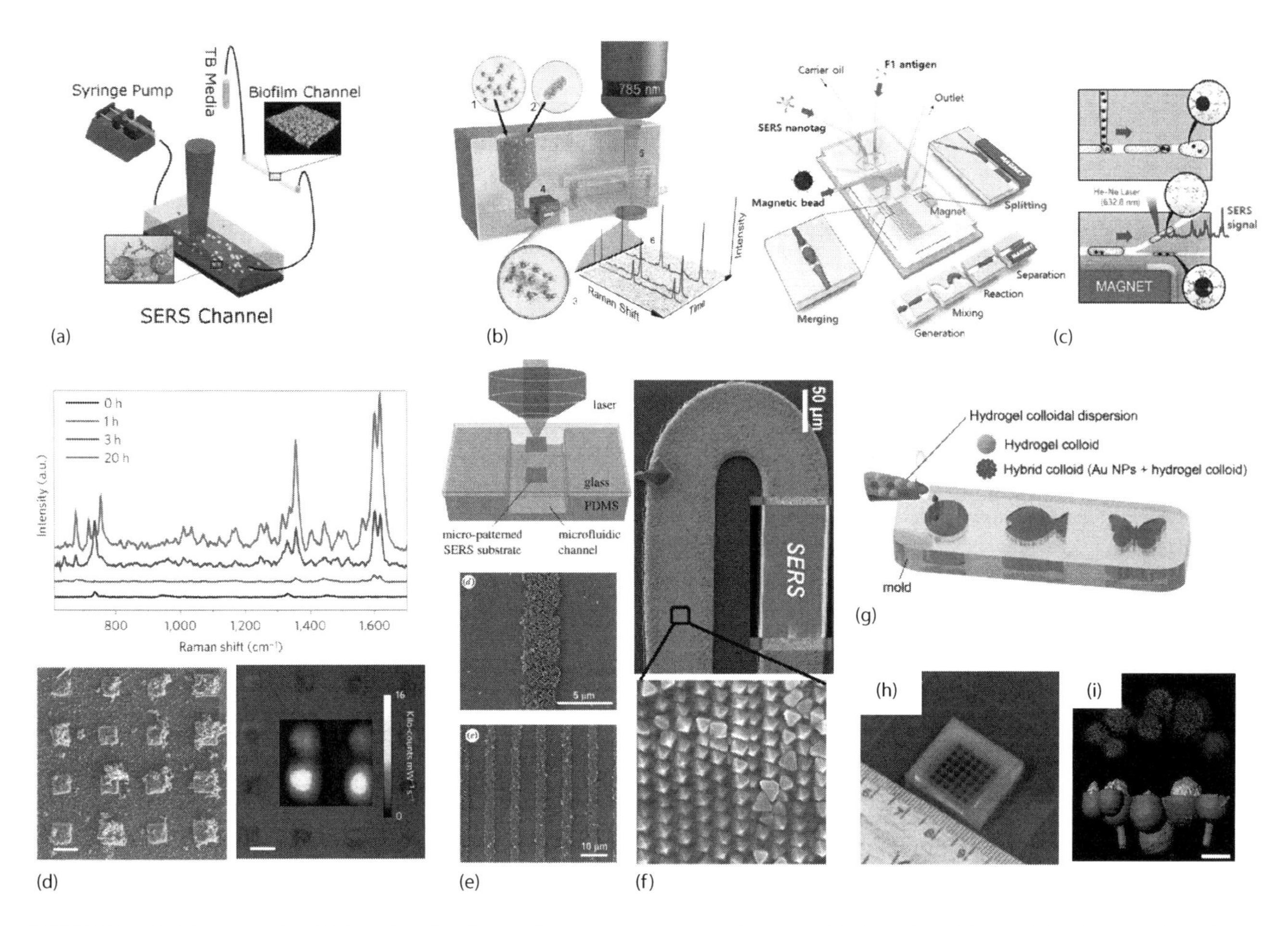

FIGURE 5.9 SERS-active microfabricated devices for applications in quorum sensing. (a to c) Schematics of the LoC-SERS setups for (a) in line measurement of pyocyanin produced by *P. aeruginosa* biofilms employing chemically assembled oligomers of Au nanoparticles (Reprinted with permission from Nguyen, C.Q. et al., *ACS Appl.,Mater. Interfaces.*, 10, 12364–12373, 2018. Copyright 2018 American Chemical Society.), (b) identification of bacterial species employing SERS-tags (Reproduced with permission from Pazos-Perez, N. et al., *Sci. Rep.*, 6, 29014, 2016), and (c) automatic immunoassay for the quantification of F1 antigen of *Yersinia pestis* by antibody-functionalized magnetic nanoparticles and SERS-tags. (Reprinted with permission from Choi, N. et al., *Anal. Chem.*, 89, 8413–8420, 2017. Copyright 2017 American Chemical Society.) (d) *In situ* detection and imaging of pyocyanin secreted by *P. aeruginosa* biofilms grown on micropatterned Au@SiO_2 supercrystals. Scale bars, 5 µm. (Reproduced with permission from Bodelón, G. et al. Copyright 2016 Springer Nature.) (e) Images of patterned gold nanoparticle arrays in microfluidic chip fabricated by lift-up lithography. (Reproduced with permission from Wu, Y. et al., *R. Soc. Open Sci.*, 5, 172034, 2018.) (f) Evaporation-based microfluidic cell used for controlled assembly of Au octahedral nanoparticles and SERS detection. (Reprinted with permission from Gómez-Graña, S. et al., *Chem. Mat.*, 27, 8310–8317, 2015. Copyright 2015 American Chemical Society.) (g) Schematic illustrating of the fabrication process of colloidal plasmonic hydrogel architectures. (Reproduced with permission from Song, J.E., and Cho, E.C., *Sci. Rep.*, 6, 34622, 2016.) (h) Image of structured PEGDA hydrogel. (Reprinted with permission from Flickinger, S.T. et al., *J. Am. Chem. Soc.*, 133, 5966–5975, 2011. Copyright 2011 American Chemical Society.) (i) Cut-out views of confocal fluorescence isosurfaces illustrating six physically segregated *P. aeruginosa* populations organized in three dimensions within a series of spheroid cavities (2–15 pL in volume) obtained by gelatin-based micro-3D printing. (Reproduced with permission from Connell, J.L. et al., *Proc. Natl. Acad. Sci. USA.*, 110, 18380–18385, 2013.)

3. **3D nanostructured substrates.** Plasmonic platforms fabricated by top-down or bottom-up approaches allow a precise generation of uniform hot-spot arrays, which are essential to achieving high reproducibility and sensitivity. Among various nanostructured SERS substrates, 3D plasmonic supercrystals offer a high densityof hot spots and a large surface area for the adsorption of the analyte. In the context of quorum sensing, plasmonic supercrystals of Au@SiO_2 bearing a highly sensitive detection of pyocyanin by SERS (LOD 10^{-14} M) enabled the detection of this molecule in low-density *P. aeruginosa* liquid cultures, as well as SERS imaging of quorum sensing communication triggered by bacterial aggregates colonizing the plasmonic supercrystals (Figure 5.9d) (Bodelón et al. 2016). It is noteworthy that the mesoporous silica infiltrated within the highly ordered Au nanorods structure increases the "plasmonically active space," thereby contributing to the extremely high electromagnetic enhancement factor of this plasmonic substrate (Bodelón et al. 2016). The integration of SERS substrates within continuous flow cells may be realized by either *in-situ* fabrication of metallic micro/nanostructures inside a microfluidic channel, or by

the construction of microfluidic channels on top of preformed plasmonic substrates. Alternatively, SERS substrates could be assembled inside the microfluidic cells and shaped into designed patterns by subsequent lift-up soft lithography (Figure 5.9e) (Wu et al. 2018). Gomez-Graña et al. reported a microfluidic-assisted 3D assembly of Au octahedra to form supercrystals over extended areas, which led to uniform and SERS-active nanostructures with high electromagnetic enhancement factor (Figure 5.9f) (Gómez-Graña et al. 2015). Recent reviews provide excellent overviews regarding the integration of nanostructures in such platforms for direct SERS sensing, as well as the challenges associated with specific applications (Huang et al. 2015, Jahn et al. 2017, Kant and Abalde-Cela 2018).

4. **Hybrid materials based on plasmonic polymer nanocomposites.** The cross-linked polymer networks of hydrogels have been used as tunable materials, amenable to manipulation of controlled hydrogel architecture and structure across multiple length scales for growing cells and tissues (Zhang and Khademhosseini 2017). The integration of plasmonic nanoparticles within polymeric matrices such as gels and hydrogels may lead to nanocomposite hybrids with unique features (e.g., increased robustness, responsiveness, and flexibility), as well as enhanced optical performance for the fabrication of SERS sensors (Pastoriza-Santos et al. 2018). Owing to their high moldability, plasmonic hydrogels with predetermined 3D architectures can be easily formed by heating an aqueous mixture of hydrogel colloids and gold nanoparticles (Figure 5.9g). Since these two colloids have thermo- and ionic-sensitive functions, the as prepared multifunctional architectures can respond to both stimuli with reversible, on-off, responses (Song and Cho 2016). Hybrid plasmonic gels and hydrogels with different compositions have been recently applied for SERS detection of quorum sensing. For example, as mentioned above, pNIPAM hydrogels mixed with Au nanoparticles were used to grow colony-biofilms of *P. aeruginosa* and direct, *in situ* SERS detection of quorum sensing pyocyanin that diffuse to the underlying optical sensor without significant interferences from the biofilm matrix (Figure 5.4) (Bodelón et al. 2016). This study suggests the feasibility of implementing SERS-active plasmonic nanoparticles within hydrogels, with rationally designed architectures and functionalities such as multiarray chambers (Figure 5.9h) and cages (Figure 5.9i), which may be used to investigate quorum sensing communication and other chemical exchange processes between microbial populations. Encapsulation of living cells in hydrogel-based capsules and alginate beads is an approach that has been widely used for the study of quorum sensing and bacterial virulence in mice models of infection (Li et al. 2017). As mentioned in the previous section, pre-colonized SERS-active hydrogels with bacterial pathogens could be used as implantable materials in experimental animal models to investigate quorum sensing and to assess anti-virulence therapies by SERS (Bodelón et al. 2016). Furthermore, taking advantage that agar has been successfully used in soft lithography techniques to fabricate microdevices for patterning bacterial cells, we envision that a plasmonic sensing component could be incorporated in similar micro platforms to study quorum sensing communication between patterned bacterial populations. Additionally, encapsulation and growth of bacterial populations in agarose or alginate microspheres doped with plasmonic nanoparticles could represent an alternative high throughput method to rapidly screen by SERS, cells that display quorum sensing phenotypes of interest.

In Table 5.1 we summarize the biomolecules and quorum sensing phenotypes mentioned in this chapter, which have been detected by SERS.

TABLE 5.1

Quorum Sensing Biomolecules and Phenotypes Detected by SERS

Class	Biomolecule	Bacterial Species	References
AIs	AHLs	Gram-negative	Pearman et al. (2007), Claussen et al. (2013), Culhane et al. (2017)
Other bioactive metabolites	Pyocyanin	*P. aeruginosa*	Bodelón et al. (2016, 2017), De Marchi et al. (2019), Polisetti et al. (2017), Guo et al. (2018), Nguyen et al. (2018), Zukovskaja et al. (2017)
	Violacein	*C. violaceum*	Bodelón et al. (2017)
	Indole	*E. coli*	De Marchi et al. (2019)
VOCs	Hydrogen cyanide	*P. aeruginosa*	Kelly et al. (2018)
	Dimethyl disulphide	Gram-negative and Gram-positive	Kelly et al. (2018)
	Indole	*E. coli*	DeJong et al. (2017)
Biofilm/EPS[a]/ secreted proteins	Biofilm/EPS[a]/secreted proteins	Gram-negative and Gram-positive	Cui et al. (2015), Chen, Cui, and Zhang (2015), Mühlig et al. (2016)
	Virulence factor antigen fraction 1	*Yersinia pestis*	Choi et al. (2017)

[a] Extracellular polymeric substances.

5.5 Challenges and Outlook

Novel analytical techniques have provided access to study quorum sensing molecules, as discussed in this review, as well as others that are yet to be characterized. The works reviewed here illustrate how SERS can be successfully applied to the detection of quorum sensing-regulated phenotypes. The ability to detect trace amounts of quorum sensing signals, and to visualize these chemical substances *in situ* is fundamental to provide new knowledge into their function. In this context, revealing the spatiotemporal dependencies required for the intercellular communication processes and chemical interactions shaping microbial communities can prove breeding ground to yield ecological insight and valuable information for antimicrobial drug prospecting. In addition, SERS can be an effective approach for the evaluation of quorum quenching strategies and the screening of synthetic chemical compounds and natural products targeting SERS-active biomolecules (e.g., pyocyanin). The understanding of SERS mechanisms, together with the ability to engineer sensitive and reliable plasmonic nanostructures (Hamon and Liz-Marzán 2018), has allowed researchers to apply this technique in the analytical field with great success (Zong et al. 2018). However, direct SERS detection and imaging in complex biological environments faces many challenges due to inherent limitations of this surface-sensitive technique. For instance, the detection of molecules with no affinity to the SERS-active surface, or with low Raman cross section, such as quorum sensing AHLs, is an important roadblock that limits its practical application. Another difficulty that must be overcome is the interference from biomolecules present in biological media. In this respect, two recent reviews have focused on the various strategies that can be followed for the selective capture and confinement of molecules with no specific affinity toward metals near SERS-active surfaces, as well as to increase the selectivity of detection (Wang et al. 2017, Koh et al. 2018). In general, such strategies rely on the use of surface-grafted chemical agents and biomolecules (e.g., antibodies, aptamers) for selective capture of analytes. Smart materials such as plasmonic hydrogels have been successfully used for trapping target molecules at hotspots (Alvarez-Puebla et al. 2009, Abalde-Cela et al. 2011). Furthermore, the interactions between molecules and the metallic surface may be enhanced by tuning electrostatic (attraction/repulsion) and the dielectric (hydrophilic/hydrophobic) properties of the material (Abalde-Cela et al. 2010). In a recent study, Kim and collaborators devised charged microgels-containing aggregates of gold nanoparticles for direct SERS detection of oppositely charged, small molecules adsorbed through electrostatic attraction (Kim et al. 2018). By employing SERS-active-positively charged microgels, the authors detected negatively charged fipronil sulfone, a pesticide metabolite, spiked in an egg without any pretreatment of the samples. Interestingly, the limiting mesh size of the hydrogel excluded large proteins from accessing the plasmonic sensor, which contributed to further increasing the sensitivity of the detection method (Kim et al. 2018). A similar concept of plasmonic hybrid materials with "molecular filtering" functionality was developed in a quorum sensing study for reducing the spectral contribution of biomolecules present in the biological medium (Bodelón et al. 2016). Another strategy to reduce the spectral complexity of the biological medium involves measuring under resonance Raman conditions. The Raman cross section of the analyte is an important parameter to be considered. In general, heterocyclic molecules-containing aromatic rings feature high Raman scattering activities. Remarkably, many chromophore-containing molecules that may be regulated by quorum sensing are characterized by potential Raman-active features. As mentioned above, the direct SERS analysis of biological systems often generates spectra that are characterized by the contribution of many different vibrational modes from biomolecules, making their interpretation a challenge for most adsorbates. In order to maximally exploit the capabilities of SERS for biochemical identification, it is necessary to know in advance the Raman/SERS fingerprint of the target analyte. In this context, the spectral fingerprints of numerous types of biomolecules have been reported. Nevertheless, a comprehensive SERS database of biochemical compounds is still needed for band assignment. Furthermore, due to spectral overlap of biomolecules, unambiguous band assignment requires the understanding of the true biochemical origins of the SERS vibrational signatures. For example, direct SERS detection of bacterial components (e.g., proteins, signaling molecules, etc.) requires the analysis of isogenic mutant strains deficient in the target biomolecule, in order to unequivocally ascertain the biochemical origins of the observed SERS features. The correct interpretation of complex vibrational spectra demands data processing, as well as statistical treatments such as multivariate data analysis, to accurately differentiate vibrational signatures of target analytes from other molecular interferences, which is favored by the high resolution of the SERS spectrum. The virtually unlimited multiplexing capability offered by SERS-tags, which overcomes a major limitation of fluorescence, renders these plasmonic materials an appealing tool when aiming at multiplex detection. However, their size is usually too large compared with the targeted biomolecules, which becomes an important drawback that limits their application when high spatial resolution is demanded. An excellent review describing in detail the strengths and disadvantages of SERS-tags for bioanalysis can be found elsewhere (Wang, Yan, and Chen 2013).

A major challenge toward obtaining reliable measurements is represented by the performance of the SERS-active substrates. Ideally, the preparation of these structures should be simple, while at the same time the signal enhancement has to be homogenous throughout the plasmonic substrate. The ability to engineer plasmonic transducers with tunable optical properties, large SERS enhancement factors, and appropriate surface functionalization through colloidal chemistry has made it possible to overcome previous limitations to produce SERS-active substrates with high sensitivity, stability, and reproducibility (Hamon and Liz-Marzán 2018). Although various SERS substrates, typically prepared by physical deposition methods, are commercially available (Mosier-Boss 2017), standard manufacturing of highly sensitive and reproducible SERS-active products for commercial use is still under development.

Owing to current technological limitations for *in vivo* studies, many researchers have turned to chemical engineering for the fabrication of microscale devices for culturing microbial cells that emulate the characteristics of the microbial microenvironments. Such devices not only have the capability to transform the study of microbial physiology and quorum sensing, but also hold great potential for many practical applications including drug discovery and diagnosis. The success of this emerging field requires the incorporation of sensitive analytical tools able to detect trace amounts of target biomolecules, an application for which SERS holds a great potential. Adapting SERS-active nanoprobes with other functional materials in microfabricated devices offers attractive strategies for numerous applications in microbiology and quorum sensing research. Although the body of literature focusing on the use of miniaturized SERS platforms for investigating microbial chemical communication is still very scarce, the examples provided here may illuminate new opportunities to explore the chemical exchange processes governing the social behavior of bacteria. To conclude, despite some limitations, SERS has shown great potential as an alternative analytical technique to investigate the chemistry underpinning the complex social lives of microbes. Upon successful implementation with microfabricated cellular environments this analytical tool can significantly expand our knowledge in quorum sensing and ultimately bring about a deeper and more complete understanding of the microbial world. Finally, SERS can be used for multimodal analysis, when coupled to other techniques such as fluorescence microscopy (De Marchi et al. 2019), stable-isotope labeling (Ivleva, Kubryk, and Niessner 2017), and surface-assisted laser desorption/ionization mass spectrometry (Nitta et al. 2014). We hope these insights can stimulate further scientific advancements not only in SERS and the quorum sensing research field, but also in other microbiology and cell biology disciplines, where the objective of trace molecule detection is highly sought after for environmental, industrial, and biomedical applications.

REFERENCES

Abalde-Cela, S., B. Auguie, M. Fischlechner, W. T. S. Huck, R. A. Alvarez-Puebla, L. M. Liz-Marzan, and C. Abell. 2011. "Microdroplet fabrication of silver-agarose nanocomposite beads for SERS optical accumulation." *Soft Matter* 7 (4):1321–25. doi:10.1039/c0sm00601g.

Abalde-Cela, S., P. Aldeanueva-Potel, C. Mateo-Mateo, L. Rodriguez-Lorenzo, R. A. Alvarez-Puebla, and L. M. Liz-Marzan. 2010. "Surface-enhanced Raman scattering biomedical applications of plasmonic colloidal particles." *J R Soc Interface* 7 (4):S435–50. doi:10.1098/rsif.2010.0125.focus.

Ahmad, A., A. M. Viljoen, and H. Y. Chenia. 2015. "The impact of plant volatiles on bacterial quorum sensing." *Lett Appl Microbiol* 60 (1):8–19. doi:10.1111/lam.12343.

Alvarez-Puebla, R. A., R. Contreras-Caceres, I. Pastoriza-Santos, J. Perez-Juste, and L. M. Liz-Marzan. 2009. "Au@pNIPAM colloids as molecular traps for surface-enhanced, spectroscopic, ultra-sensitive analysis." *Angew Chem Int Ed Engl* 48 (1):138–43. doi:10.1002/anie.200804059.

Audrain, B., M. A. Farag, C. M. Ryu, and J. M. Ghigo. 2015. "Role of bacterial volatile compounds in bacterial biology." *FEMS Microbiol Rev* 39 (2):222–33. doi:10.1093/femsre/fuu013.

Baig, Nameera F., S. J. B. Dunham, N. Morales-Soto, J. D. Shrout, J. V. Sweedler, and P. W. Bohn. 2015. "Multimodal chemical imaging of molecular messengers in emerging *Pseudomonas aeruginosa* bacterial communities." *Analyst* 140 (19):6544–52. doi:10.1039/c5an01149c.

Bell, S. E., and N. M. Sirimuthu. 2008. "Quantitative surface-enhanced Raman spectroscopy." *Chem Soc Rev* 37 (5):1012–24. doi:10.1039/b705965p.

Bellin, D. L., H. Sakhtah, J. K. Rosenstein, P. M. Levine, J. Thimot, K. Emmett, L. E. Dietrich, and K. L. Shepard. 2014. "Integrated circuit-based electrochemical sensor for spatially resolved detection of redox-active metabolites in biofilms." *Nat Commun* 5:3256. doi:10.1038/ncomms4256.

Bodelón, G., V. Montes-Garcia, C. Costas, I. Perez-Juste, J. Perez-Juste, I. Pastoriza-Santos, and L. M. Liz-Marzan. 2017. "Imaging bacterial interspecies chemical interactions by surface-enhanced Raman scattering." *ACS Nano* 11 (5):4631–40. doi:10.1021/acsnano.7b00258.

Bodelón, G., V. Montes-Garcia, J. Perez-Juste, and I. Pastoriza-Santos. 2018. "Surface-enhanced Raman scattering spectroscopy for label-free analysis of *P. aeruginosa* quorum sensing." *Front Cell Infect Microbiol* 8. doi:10.3389/fcimb.2018.00143.

Bodelón, G., V. Montes-Garcia, V. Lopez-Puente, E. H. Hill, C. Hamon, M. N. Sanz-Ortiz, S. Rodal-Cedeira et al. 2016. "Detection and imaging of quorum sensing in *Pseudomonas aeruginosa* biofilm communities by surface-enhanced resonance Raman scattering." *Nat Mat* 15 (11):1203. doi:10.1038/Nmat4720.

Boedicker, J. Q., M. E. Vincent, and R. F. Ismagilov. 2009. "Microfluidic confinement of single cells of bacteria in small volumes initiates high-density behavior of quorum sensing and growth and reveals its variability." *Angew Chem Int Ed Engl* 48 (32):5908–11. doi:10.1002/anie.200901550.

Bos, L. D., P. J. Sterk, and M. J. Schultz. 2013. "Volatile metabolites of pathogens: A systematic review." *PLoS Pathog* 9 (5):e1003311. doi:10.1371/journal.ppat.1003311.

Camden, J. P., J. A. Dieringer, Y. M. Wang, D. J. Masiello, L. D. Marks, G. C. Schatz, and R. P. Van Duyne. 2008. "Probing the structure of single-molecule surface-enhanced Raman scattering hot spots." *J Am Chem Soc* 130 (38):12616. doi:10.1021/ja8051427.

Carnes, E. C., D. M. Lopez, N. P. Donegan, A. Cheung, H. Gresham, G. S. Timmins, and C. J. Brinker. 2010. "Confinement-induced quorum sensing of individual *Staphylococcus aureus* bacteria." *Nat Chem Biol* 6 (1):41–5. doi:10.1038/nchembio.264.

Chang, H. J., H. M. Kang, E. Ko, B. H. Jun, H. Y. Lee, Y. S. Lee, and D. H. Jeang. 2016. "PSA detection with femtomolar sensitivity and a broad dynamic range using SERS nanoprobes and an area-scanning method." *ACS Sensors* 1 (6):645–49. doi:10.1021/acssensors.6b00053.

Chao, Y. Q., and T. Zhang. 2012. "Surface-enhanced Raman scattering (SERS) revealing chemical variation during biofilm formation: From initial attachment to mature biofilm." *Anal Bioanal Chem* 404 (5):1465–75. doi:10.1007/s00216-012-6225-y.

Chen, P. Y., L. Cui, and K. S. Zhang. 2015. "Surface-enhanced Raman spectroscopy monitoring the development of dual-species biofouling on membrane surfaces." *J Membr Sci* 473:36–44. doi:10.1016/j.memsci.2014.09.007.

Chernin, L. S., M. K. Winson, J. M. Thompson, S. Haran, B. W. Bycroft, I. Chet, P. Williams, and G. S. Stewart. 1998. "Chitinolytic activity in *Chromobacterium violaceum*: Substrate analysis and regulation by quorum sensing." *J Bacteriol* 180 (17):4435–41.

Chernin, L., N. Toklikishvili, M. Ovadis, S. Kim, J. Ben-Ari, I. Khmel, and A. Vainstein. 2011. "Quorum-sensing quenching by rhizobacterial volatiles." *Environ Microbiol Rep* 3 (6):698–704. doi:10.1111/j.1758-2229.2011.00284.x.

Choi, N., J. Lee, J. Ko, J. H. Jeon, G. E. Rhie, A. J. deMello, and J. Choo. 2017. "Integrated SERS-based microdroplet platform for the automated immunoassay of F1 antigens in *Yersinia pestis*." *Anal Chem* 89 (16):8413–20. doi:10.1021/acs.analchem.7b01822.

Chu, W., T. R. Zere, M. M. Weber, T. K. Wood, M. Whiteley, B. Hidalgo-Romano, E. Valenzuela, Jr., and R. J. McLean. 2012. "Indole production promotes *Escherichia coli* mixed-culture growth with *Pseudomonas aeruginosa* by inhibiting quorum signaling." *Appl Environ Microbiol* 78 (2):411–9. doi:10.1128/AEM.06396-11.

Claussen, A., S. Abdali, R. W. Berg, M. Givskov, and T. Sams. 2013. "Detection of the quorum sensing signal molecule *N*-dodecanoyl-DL-homoserine lactone below 1 nanomolar concentrations using surface enhanced Raman spectroscopy." *Curr Phys Chem* 3 (2):199–210. doi:10.2174/1877946811303020010.

Connell, J. L., E. T. Ritschdorff, M. Whiteley, and J. B. Shear. 2013. "3D printing of microscopic bacterial communities." *Proc Natl Acad Sci U S A* 110 (46):18380–5. doi:10.1073/pnas.1309729110.

Connell, J. L., J. Kim, J. B. Shear, A. J. Bard, and M. Whiteley. 2014. "Real-time monitoring of quorum sensing in 3D-printed bacterial aggregates using scanning electrochemical microscopy." *Proc Natl Acad Sci U S A* 111 (51):18255–60. doi:10.1073/pnas.1421211111.

Cui, L., P. Chen, B. Zhang, D. Zhang, J. Li, F. L. Martin, and K. Zhang. 2015. "Interrogating chemical variation via layer-by-layer SERS during biofouling and cleaning of nanofiltration membranes with further investigations into cleaning efficiency." *Water Res* 87:282–91. doi:10.1016/j.watres.2015.09.037.

Culhane, K., K. Jiang, A. Neumann, and A. O. Pinchuk. 2017. "Laser-fabricated plasmonic nanostructures for surface-enhanced Raman spectroscopy of bacteria quorum sensing molecules." *MRS Adv* 2 (42):2287–94. doi:10.1557/adv.2017.98.

Darch, S. E., O. Simoska, M. Fitzpatrick, J. P. Barraza, K. J. Stevenson, R. T. Bonnecaze, J. B. Shear, and M. Whiteley. 2018. "Spatial determinants of quorum signaling in a *Pseudomonas aeruginosa* infection model." *Proc Natl Acad Sci U S A* 115 (18):4779–84. doi:10.1073/pnas.1719317115.

Davies, J. 2013. "Specialized microbial metabolites: Functions and origins." *J Antibiot* 66 (7):361–4. doi:10.1038/ja.2013.61.

De Marchi, S., G. Bodelón, L. Vazquez-Iglesias, L. M. Liz-Marzan, J. Perez-Juste, and I. Pastoriza-Santos. 2019. "Surface-enhanced Raman scattering (SERS) imaging of bioactive metabolites in mixed bacterial populations." *Appl Mat Today*:207–15. doi:10.1016/j.apmt.2018.12.005.

Defoirdt, T. 2018. "Quorum-sensing systems as targets for antivirulence therapy." *Trends Microbiol* 26 (4):313–28. doi:10.1016/j.tim.2017.10.005.

DeJong, C. S., D. I. Wang, A. Polyakov, A. Rogacs, S. J. Simske, and V. Shkolnikov. 2017. "Bacterial detection and differentiation via direct volatile organic compound sensing with surface enhanced Raman spectroscopy." *Chemistryselect* 2 (27):8431–35. doi:10.1002/slct.201701669.

Dietrich, L. E., A. Price-Whelan, A. Petersen, M. Whiteley, and D. K. Newman. 2006. "The phenazine pyocyanin is a terminal signalling factor in the quorum sensing network of *Pseudomonas aeruginosa*." *Mol Microbiol* 61 (5):1308–21. doi:10.1111/j.1365-2958.2006.05306.x.

Dunham, S. J., J. F. Ellis, B. Li, and J. V. Sweedler. 2017. "Mass spectrometry imaging of complex microbial communities." *ACC Chem Res* 50 (1):96–104. doi:10.1021/acs.accounts.6b00503.

Efeoglu, E., and M. Culha. 2013. "*In situ*-monitoring of biofilm formation by using surface-enhanced Raman scattering." *Appl Spectrosc* 67 (5):498–505. doi:10.1366/12-06896.

Effmert, U., J. Kalderas, R. Warnke, and B. Piechulla. 2012. "Volatile mediated interactions between bacteria and fungi in the soil." *J Chem Ecol* 38 (6):665–703. doi:10.1007/s10886-012-0135-5.

Esteban, J., and M. Garcia-Coca. 2017. "Mycobacterium biofilms." *Front Microbiol* 8:2651. doi:10.3389/fmicb.2017.02651.

Feliu, N., M. Hassan, E. Garcia Rico, D. Cui, W. Parak, and R. Alvarez-Puebla. 2017. "SERS quantification and characterization of proteins and other biomolecules." *Langmuir* 33 (38):9711–30. doi:10.1021/acs.langmuir.7b01567.

Fleischmann, M., P. J. Hendra, and A. J. Mcquillan. 1974. "Raman-spectra of pyridine adsorbed at a silver electrode." *Chem Phys Let* 26 (2):163–166. doi:10.1016/0009-2614(74)85388-1.

Fletcher, M., M. Camara, D. A. Barrett, and P. Williams. 2014. "Biosensors for qualitative and semiquantitative analysis of quorum sensing signal molecules." *Methods Mol Biol* 1149:245–54. doi:10.1007/978-1-4939-0473-0_20.

Flickinger, S. T., M. F. Copeland, E. M. Downes, A. T. Braasch, H. H. Tuson, Y. J. Eun, and D. B. Weibel. 2011. "Quorum sensing between *Pseudomonas aeruginosa* biofilms accelerates cell growth." *J Am Chem Soc* 133 (15):5966–75. doi:10.1021/ja111131f.

Gao, M., H. Zheng, Y. Ren, R. Lou, F. Wu, W. Yu, X. Liu, and X. Ma. 2016. "A crucial role for spatial distribution in bacterial quorum sensing." *Sci Rep* 6:34695. doi:10.1038/srep34695.

Gómez-Graña, S., C. Fernández-López, L. Polavarapu, J.-B. Salmon, J. Leng, I. Pastoriza-Santos, and J. Pérez-Juste. 2015. "Gold nanooctahedra with tunable size and microfluidic-induced 3D assembly for highly uniform SERS-active supercrystals." *Chem Mat* 27 (24):8310–17. doi:10.1021/acs.chemmater.5b03620.

Guo, J., Y. Liu, Y. Chen, J. Li, and H. Ju. 2018. "A multifunctional SERS sticky note for real-time quorum sensing tracing and inactivation of bacterial biofilms." *Chem Sci* 9 (27):5906–11. doi:10.1039/c8sc02078g.

Hall, S., C. McDermott, S. Anoopkumar-Dukie, A. J. McFarland, A. Forbes, A. V. Perkins, A. K. Davey et al. 2016. "Cellular effects of pyocyanin, a secreted virulence factor of *Pseudomonas aeruginosa*." *Toxins* 8 (8). doi:10.3390/toxins8080236.

Hammond, J. L., N. Bhalla, S. D. Rafiee, and P. Estrela. 2014. "Localized surface plasmon resonance as a biosensing platform for developing countries." *Biosensors* 4 (2):172–88. doi:10.3390/bios4020172.

Hamon, C., and L. M. Liz-Marzán. 2018. "Colloidal design of plasmonic sensors based on surface enhanced Raman scattering." *J Colloid Interface Sci* 512:834–843. doi:10.1016/j.jcis.2017.10.117.

Hense, B. A., and M. Schuster. 2015. "Core principles of bacterial autoinducer systems." *Microbiol Mol Biol Rev* 79 (1):153–69. doi:10.1128/MMBR.00024-14.

Hense, B. A., C. Kuttler, J. Muller, M. Rothballer, A. Hartmann, and J. U. Kreft. 2007. "Does efficiency sensing unify diffusion and quorum sensing?" *Nat Rev Microbiol* 5 (3):230–9. doi:10.1038/nrmicro1600.

Hildebrandt, P., and M. Stockburger. 1984. "Surface-enhanced resonance Raman-spectroscopy of rhodamine-6g adsorbed on colloidal silver." *J Phys Chem* 88 (24):5935–44. doi:10.1021/j150668a038.

Hochbaum, A. I., and J. Aizenberg. 2010. "Bacteria pattern spontaneously on periodic nanostructure arrays." *Nano Lett* 10 (9):3717–21. doi:10.1021/nl102290k.

Hol, F. J., and C. Dekker. 2014. "Zooming in to see the bigger picture: Microfluidic and nanofabrication tools to study bacteria." *Science* 346 (6208):1251821. doi:10.1126/science.1251821.

Hornby, J. M., E. C. Jensen, A. D. Lisec, J. J. Tasto, B. Jahnke, R. Shoemaker, P. Dussault, and K. W. Nickerson. 2001. "Quorum sensing in the dimorphic fungus Candida albicans is mediated by farnesol." *Appl Environ Microbiol* 67 (7):2982–92. doi:10.1128/Aem.67.7.2982-2992.2001.

Huang, J. A., Y. L. Zhang, H. Ding, and H. B. Sun. 2015. "SERS-enabled lab-on-a-chip systems." *Adv Opt Mat* 3 (5):618–33. doi:10.1002/adom.201400534.

Hunter, R. C., V. Klepac-Ceraj, M. M. Lorenzi, H. Grotzinger, T. R. Martin, and D. K. Newman. 2012. "Phenazine content in the cystic fibrosis respiratory tract negatively correlates with lung function and microbial complexity." *Am J Respir Cell Mol Biol* 47 (6):738–45. doi:10.1165/rcmb.2012-0088OC.

Ibáñez de Aldecoa, A. L., O. Zafra, and J. E. González-Pastor. 2017. "Mechanisms and regulation of extracellular DNA release and its biological roles in microbial communities." *Front Microbiol* 8. doi:10.3389/fmicb.2017.01390.

Ivleva, N. P., M. Wagner, H. Horn, R. Niessner, and C. Haisch. 2008. "*In situ* surface-enhanced Raman scattering analysis of biofilm." *Anal Chemistry* 80 (22):8538–44. doi:10.1021/ac801426m.

Ivleva, N. P., P. Kubryk, and R. Niessner. 2017. "Raman microspectroscopy, surface-enhanced Raman scattering microspectroscopy, and stable-isotope Raman microspectroscopy for biofilm characterization." *Anal Bioanal Chem* 409 (18):4353–75. doi:10.1007/s00216-017-0303-0.

Ivleva, N. P., M. Wagner, H. Horn, R. Niessner, and C. Haisch. 2010. "Raman microscopy and surface-enhanced Raman scattering (SERS) for *in situ* analysis of biofilms." *J Biophotonics* 3 (8–9):548–56. doi:10.1002/jbio.201000025.

Jahed, Z., H. Shahsavan, M. S. Verma, J. L. Rogowski, B. B. Seo, B. Zhao, T. Y. Tsui, F. X. Gu, and M. R. Mofrad. 2017. "Bacterial networks on hydrophobic micropillars." *ACS Nano* 11 (1):675–83. doi:10.1021/acsnano.6b06985.

Jahn, I. J., O. Žukovskaja, X. S. Zheng, K. Weber, T. W. Bocklitz, D. Cialla-May, and J. Popp. 2017. "Surface-enhanced Raman spectroscopy and microfluidic platforms: Challenges, solutions and potential applications." *Analyst* 142 (7):1022–47. doi:10.1039/c7an00118e.

Jahn, M., S. Patze, I. J. Hidi, R. Knipper, A. I. Radu, A. Mühlig, S. Yuksel et al. 2016. "Plasmonic nanostructures for surface enhanced spectroscopic methods." *Analyst* 141 (3):756–93. doi:10.1039/c5an02057c.

Jensen, L., C. M. Aikens, and G. C. Schatz. 2008. "Electronic structure methods for studying surface-enhanced Raman scattering." *Chem Soc Rev* 37 (5):1061–73. doi:10.1039/b706023h.

Kant, K., and S. Abalde-Cela. 2018. "Surface-enhanced Raman scattering spectroscopy and microfluidics: Towards ultrasensitive label-free sensing." *Biosensors* 8 (3). doi:10.3390/bios8030062.

Kelestemur, S., E. Avci, and M. Culha. 2018. "Raman and surface-enhanced Raman scattering for biofilm characterization." *Chemosensors* 6 (1). doi:10.3390/chemosensors6010005.

Kelly, J., R. Patrick, S. Patrick, and S. E. J. Bell. 2018. "Surface-enhanced Raman spectroscopy for the detection of a metabolic product in the headspace above live bacterial cultures." *Angew Chem Int Ed Engl* 57 (48):15686–90. doi:10.1002/anie.201808185.

Kim, D. J., S. G. Park, D. H. Kim, and S. H. Kim. 2018. "SERS-active-charged microgels for size- and charge-selective molecular analysis of complex biological samples." *Small* 14 (40):e1802520. doi:10.1002/smll.201802520.

Kim, M. K., F. Ingremeau, A. Zhao, B. L. Bassler, and H. A. Stone. 2016. "Local and global consequences of flow on bacterial quorum sensing." *Nat Microbiol* 1:15005. doi:10.1038/nmicrobiol.2015.5.

Koh, C. S. L., H. K. Lee, X. Han, H. Y. F. Sim, and X. Y. Ling. 2018. "Plasmonic nose: Integrating the MOF-enabled molecular preconcentration effect with a plasmonic array for recognition of molecular-level volatile organic compounds." *Chem Commun* 54 (20):2546–49. doi:10.1039/c8cc00564h.

Lanni, E. J., R. N. Masyuko, C. M. Driscoll, S. J. B. Dunham, J. D. Shrout, P. W. Bohn, and J. V. Sweedler. 2014. "Correlated imaging with C60-SIMS and confocal Raman microscopy: Visualization of cell-scale molecular distributions in bacterial biofilms." *Anal Chem* 86 (21):10885–91. doi:10.1021/ac5030914.

LaSarre, B., and M. J. Federle. 2013. "Exploiting quorum sensing to confuse bacterial pathogens." *Microbiol Mol Biol Rev* 77 (1):73–111. doi:10.1128/MMBR.00046-12.

Lau, G. W., D. J. Hassett, H. Ran, and F. Kong. 2004. "The role of pyocyanin in *Pseudomonas aeruginosa* infection." *Trends Mol Med* 10 (12):599–606. doi:10.1016/j.molmed.2004.10.002.

Lauridsen, R. K., L. M. Sommer, H. K. Johansen, T. Rindzevicius, S. Molin, L. Jelsbak, S. B. Engelsen, and A. Boisen. 2017. "SERS detection of the biomarker hydrogen cyanide from *Pseudomonas aeruginosa* cultures isolated from cystic fibrosis patients." *Sci Rep* 7:45264. doi:10.1038/srep45264.

Le Ru, E. C., and P. G. Etchegoin. 2012. "Single-molecule surface-enhanced Raman spectroscopy." *Annu Rev Phys Chem* 63:65–87. doi:10.1146/annurev-physchem-032511-143757.

Lee, J., C. Attila, S. L. Cirillo, J. D. Cirillo, and T. K. Wood. 2009. "Indole and 7-hydroxyindole diminish *Pseudomonas aeruginosa* virulence." *Microb Biotechnol* 2 (1):75–90. doi:10.1111/j.1751-7915.2008.00061.x.

LeRu, E. C., and P. G. Etchegoin. 2009. *Principles of Surface-Enhanced Raman Spectroscopy: And Related Plasmonic Effects*, Elsevier, Amsterdam, the Netherlands, pp. 1–663.

Li, P., M. Muller, M. W. Chang, M. Frettloh, and H. Schonherr. 2017. "Encapsulation of autoinducer sensing reporter bacteria in reinforced alginate-based microbeads." *ACS Appl Mater Interfaces* 9 (27):22321–331. doi:10.1021/acsami.7b07166.

Liu, D., J. Wan, G. Pang, and Z. Tang. 2018. "Hollow metal-organic-framework micro/nanostructures and their derivatives: Emerging multifunctional materials." *Adv Mater*:e1803291. doi:10.1002/adma.201803291.

Liu, Y., Q. Qin, and T. Defoirdt. 2018. "Does quorum sensing interference affect the fitness of bacterial pathogens in the real world?" *Environ Microbiol*. doi:10.1111/1462-2920.14446.

Lu, X., D. R. Samuelson, Y. Xu, H. Zhang, S. Wang, B. A. Rasco, J. Xu, and M. E. Konkel. 2013. "Detecting and tracking nosocomial methicillin-resistant Staphylococcus aureus using a microfluidic SERS biosensor." *Anal Chem* 85 (4):2320–7. doi:10.1021/ac303279u.

Martins, M., M. Henriques, J. Azeredo, S. M. Rocha, M. A. Coimbra, and R. Oliveira. 2007. "Morphogenesis control in *Candida albicans* and *Candida dubliniensis* through signaling molecules produced by planktonic and biofilm cells." *Eukaryotic Cell* 6 (12):2429–36. doi:10.1128/Ec.00252-07.

Mavrodi, D. V., R. F. Bonsall, S. M. Delaney, M. J. Soule, G. Phillips, and L. S. Thomashow. 2001. "Functional analysis of genes for biosynthesis of pyocyanin and phenazine-1-carboxamide from *Pseudomonas aeruginosa* PAO1." *J Bacteriol* 183 (21):6454–65. doi:10.1128/JB.183.21.6454-6465.2001.

McClean, K. H., M. K. Winson, L. Fish, A. Taylor, S. R. Chhabra, M. Camara, M. Daykin et al. 1997. "Quorum sensing and *Chromobacterium violaceum*: Exploitation of violacein production and inhibition for the detection of *N*-acylhomoserine lactones." *Microbiology* 143:3703–11. doi:10.1099/00221287-143-12-3703.

Mosier-Boss, P. A. 2017. "Review of SERS substrates for chemical sensing." *Nanomaterials* 7 (6). doi:10.3390/nano7060142.

Moskovits, M. 1985. "Surface-enhanced spectroscopy." *Rev Mod Phys* 57 (3):783–826. doi:10.1103/RevModPhys.57.783.

Mühlig, A., T. Bocklitz, I. Labugger, S. Dees, S. Henk, E. Richter, S. Andres et al. 2016. "LOC-SERS: A promising closed system for the identification of mycobacteria." *Anal Chem* 88 (16):7998–8004. doi:10.1021/acs.analchem.6b01152.

Ng, W. L., and B. L. Bassler. 2009. "Bacterial quorum-sensing network architectures." *Annu Rev Genet* 43:197–222. doi:10.1146/annurev-genet-102108-134304.

Nguyen, C. Q., W. J. Thrift, A. Bhattacharjee, S. Ranjbar, T. Gallagher, M. Darvishzadeh-Varcheie, R. N. Sanderson et al. 2018. "Longitudinal monitoring of biofilm formation via robust surface-enhanced Raman scattering quantification of *Pseudomonas aeruginosa*-produced metabolites." *ACS Appl Mater Interfaces* 10 (15):12364–73. doi:10.1021/acsami.7b18592.

Nitta, S., A. Yamamoto, M. Kurita, R. Arakawa, and H. Kawasaki. 2014. "Gold-decorated titania nanotube arrays as dual-functional platform for surface-enhanced Raman spectroscopy and surface-assisted laser desorption/ionization mass spectrometry." *ACS Appl Mater Interfaces* 6 (11):8387–95. doi:10.1021/am501291d.

O'Brien, J., and G. D. Wright. 2011. "An ecological perspective of microbial secondary metabolism." *Curr Opin Biotechnol* 22 (4):552–8. doi:10.1016/j.copbio.2011.03.010.

Ojha, A. K., A. D. Baughn, D. Sambandan, T. Hsu, X. Trivelli, Y. Guerardel, A. Alahari, L. Kremer, W. R. Jacobs, Jr., and G. F. Hatfull. 2008. "Growth of *Mycobacterium tuberculosis* biofilms containing free mycolic acids and harbouring drug-tolerant bacteria." *Mol Microbiol* 69 (1):164–74. doi:10.1111/j.1365-2958.2008.06274.x.

Papenfort, K., J. E. Silpe, K. R. Schramma, J. P. Cong, M. R. Seyedsayamdost, and B. L. Bassler. 2017. "A *Vibrio cholerae* autoinducer-receptor pair that controls biofilm formation." *Nat Chem Biol* 13 (5):551–7. doi:10.1038/nchembio.2336.

Parsek, M. R., and E. P. Greenberg. 2005. "Sociomicrobiology: The connections between quorum sensing and biofilms." *Trends Microbiol* 13 (1):27–33. doi:10.1016/j.tim.2004.11.007.

Pastoriza-Santos, I., C. Kinnear, J. Perez-Juste, P. Mulvaney, and L. M. Liz-Marzan. 2018. "Plasmonic polymer nanocomposites." *Nat Rev Mat* 3 (10):375–91. doi:10.1038/s41578-018-0050-7.

Pazos-Perez, N., E. Pazos, C. Catala, B. Mir-Simon, S. Gomez-de Pedro, J. Sagales, C. Villanueva et al. 2016. "Ultrasensitive multiplex optical quantification of bacteria in large samples of biofluids." *Sci Rep* 6:29014. doi:10.1038/srep29014.

Pearman, W. F., M. Lawrence-Snyder, S. M. Angel, and A. W. Decho. 2007. "Surface-enhanced Raman spectroscopy for in situ measurements of signaling molecules (autoinducers) relevant to bacteria quorum sensing." *Appl Spectrosc* 61 (12):1295–300. doi:10.1366/000370207783292244.

Pessi, G., and D. Haas. 2000. "Transcriptional control of the hydrogen cyanide biosynthetic genes hcnABC by the anaerobic regulator ANR and the quorum-sensing regulators LasR and RhlR in Pseudomonas aeruginosa." *J Bacteriol* 182 (24):6940–9.

Polisetti, S., N. F. Baig, N. Morales-Soto, J. D. Shrout, and P. W. Bohn. 2017. "Spatial mapping of pyocyanin in *Pseudomonas aeruginosa* bacterial communities using surface enhanced Raman scattering." *Appl Spectrosc* 71 (2):215–23. doi:10.1177/0003702816654167.

Qiao, X., B. Su, C. Liu, Q. Song, D. Luo, G. Mo, and T. Wang. 2018. "Selective surface enhanced Raman scattering for quantitative detection of lung cancer biomarkers in superparticle@MOF structure." *Adv Mater* 30 (5). doi:10.1002/adma.201702275.

Que, Y. A., R. Hazan, B. Strobel, D. Maura, J. He, M. Kesarwani, P. Panopoulos et al. 2013. "A quorum sensing small volatile molecule promotes antibiotic tolerance in bacteria." *PLoS One* 8 (12):e80140. doi:10.1371/journal.pone.0080140.

Rada, B., and T. L. Leto. 2013. "Pyocyanin effects on respiratory epithelium: Relevance in *Pseudomonas aeruginosa* airway infections." *Trends Microbiol* 21 (2):73–81. doi:10.1016/j.tim.2012.10.004.

Ramya, S., R. P. George, R. V. S. Rao, and R. K. Dayal. 2010. "Detection of algae and bacterial biofilms formed on titanium surfaces using micro-Raman analysis." *Appl Surf Sci* 256 (16):5108–15. doi:10.1016/j.apsusc.2010.03.079.

Rumbaugh, K. P., and G. F. Kaufmann. 2012. "Exploitation of host signaling pathways by microbial quorum sensing signals." *Curr Opin Microbiol* 15 (2):162–8. doi:10.1016/j.mib.2011.12.003.

Rusconi, R., M. Garren, and R. Stocker. 2014. "Microfluidics expanding the frontiers of microbial ecology." *Annu Rev Biophys* 43:65–91. doi:10.1146/annurev-biophys-051013-022916.

Rutherford, S. T., and B. L. Bassler. 2012. "Bacterial quorum sensing: Its role in virulence and possibilities for its control." *Cold Spring Harb Perspect Med* 2 (11). doi:10.1101/cshperspect.a012427.

Schlucker, S. 2014. "Surface-enhanced Raman spectroscopy: Concepts and chemical applications." *Angew Chem Int Ed Engl* 53 (19):4756–95. doi:10.1002/anie.201205748.

Schmidt, R., V. Cordovez, W. de Boer, J. Raaijmakers, and P. Garbeva. 2015. "Volatile affairs in microbial interactions." *ISME J* 9 (11):2329–35. doi:10.1038/ismej.2015.42.

Schulz, S., J. S. Dickschat, B. Kunze, I. Wagner-Dobler, R. Diestel, and F. Sasse. 2010. "Biological activity of volatiles from marine and terrestrial bacteria." *Mar Drugs* 8 (12):2976–87. doi:10.3390/md8122976.

Shan, B. B., Y. H. Pu, Y. F. Chen, M. L. Liao, and M. Li. 2018. "Novel SERS labels: Rational design, functional integration and biomedical applications." *Coord Chem Rev* 371:11–37. doi:10.1016/j.ccr.2018.05.007.

Song, J. E., and E. C. Cho. 2016. "Dual-responsive and multi-functional plasmonic hydrogel valves and biomimetic architectures formed with hydrogel and gold nanocolloids." *Sci Rep* 6:34622. doi:10.1038/srep34622.

Stasulli, N. M., and E. A. Shank. 2016. "Profiling the metabolic signals involved in chemical communication between microbes using imaging mass spectrometry." *FEMS Microbiol Rev* 40 (6):807–13. doi:10.1093/femsre/fuw032.

Stiles, P. L., J. A. Dieringer, N. C. Shah, and R. R. Van Duyne. 2008. "Surface-enhanced Raman spectroscopy." *Annu Rev Anal Chem* 1:601–26. doi:10.1146/annurev.anchem.1.031207.112814.

Walter, A., A. Marz, W. Schumacher, P. Rosch, and J. Popp. 2011. "Towards a fast, high specific and reliable discrimination of bacteria on strain level by means of SERS in a microfluidic device." *Lab Chip* 11 (6):1013–21. doi:10.1039/c0lc00536c.

Wang, F., S. Cao, R. Yan, Z. Wang, D. Wang, and H. Yang. 2017. "Selectivity/specificity improvement strategies in surface-enhanced Raman spectroscopy analysis." *Sensors* 17 (11). doi:10.3390/s17112689.

Wang, Y., B. Yan, and L. Chen. 2013. "SERS tags: Novel optical nanoprobes for bioanalysis." *Chem Rev* 113 (3):1391–428. doi:10.1021/cr300120g.

Weibel, D. B., W. R. Diluzio, and G. M. Whitesides. 2007. "Microfabrication meets microbiology." *Nat Rev Microbiol* 5 (3):209–18. doi:10.1038/nrmicro1616.

Wessel, A. K., L. Hmelo, M. R. Parsek, and M. Whiteley. 2013. "Going local: Technologies for exploring bacterial microenvironments." *Nat Rev Microbiol* 11 (5):337–48. doi:10.1038/nrmicro3010.

Willets, K. A., and R. P. Van Duyne. 2007. "Localized surface plasmon resonance spectroscopy and sensing." *Annu Rev Phys Chem* 58:267–97. doi:10.1146/annurev.physchem.58.032806.104607.

Witkowska, E., D. Korsak, A. Kowalska, A. Janeczek, and A. Kaminska. 2018. "Strain-level typing and identification of bacteria – A novel approach for SERS active plasmonic nanostructures." *Anal Bioanal Chem*. doi:10.1007/s00216-018-1153-0.

Wu, F., and C. Dekker. 2016. "Nanofabricated structures and microfluidic devices for bacteria: From techniques to biology." *Chem Soc Rev* 45 (2):268–80. doi:10.1039/c5cs00514k.

Wu, X., J. Chen, X. Li, Y. Zhao, and S. M. Zughaier. 2014. "Culture-free diagnostics of *Pseudomonas aeruginosa* infection by silver nanorod array based SERS from clinical sputum samples." *Nanomedicine* 10 (8):1863–70. doi:10.1016/j.nano.2014.04.010.

Wu, Y., Y. Jiang, X. Zheng, S. Jia, Z. Zhu, B. Ren, and H. Ma. 2018. "Facile fabrication of microfluidic surface-enhanced Raman scattering devices via lift-up lithography." *R Soc Open Sci* 5 (4):172034. doi:10.1098/rsos.172034.

Yang, S., X. Dai, B. B. Stogin, and T. S. Wong. 2016. "Ultrasensitive surface-enhanced Raman scattering detection in common fluids." *Proc Natl Acad Sci U S A* 113 (2):268–73. doi:10.1073/pnas.1518980113.

Yawata, Y., J. Nguyen, R. Stocker, and R. Rusconi. 2016. "Microfluidic studies of biofilm formation in dynamic environments." *J Bacteriol* 198 (19):2589–95. doi:10.1128/JB.00118-16.

Zhang, Y. S., and A. Khademhosseini. 2017. "Advances in engineering hydrogels." *Science* 356 (6337). doi:10.1126/science.aaf3627.

Zheng, X. S., I. J. Jahn, K. Weber, D. Cialla-May, and J. Popp. 2018. "Label-free SERS in biological and biomedical applications: Recent progress, current challenges and opportunities." *Spectrochim Acta A Mol Biomol Spectrosc* 197:56–77. doi:10.1016/j.saa.2018.01.063.

Zong, C., M. Xu, L. J. Xu, T. Wei, X. Ma, X. S. Zheng, R. Hu, and B. Ren. 2018. "Surface-enhanced Raman spectroscopy for bioanalysis: Reliability and challenges." *Chem Rev* 118 (10):4946–80. doi:10.1021/acs.chemrev.7b00668.

Zukovskaja, O., I. J. Jahn, K. Weber, D. Cialla-May, and J. Popp. 2017. "Detection of *Pseudomonas aeruginosa* metabolite pyocyanin in water and saliva by employing the SERS technique." *Sensors* 17 (8). doi:10.3390/s17081704.

6

Quorum Sensing in Pseudomonas aeruginosa*: From Gene and Metabolic Networks to Bacterial Pathogenesis*

Stephen K. Dolan, Cauã Antunes Westmann, Greicy Kelly Bonifacio Pereira, Stephen Trigg, Rafael Silva-Rocha, and Martin Welch

CONTENTS

6.1 Background

The opportunistic pathogen *Pseudomonas aeruginosa* has a profound ability to thrive in numerous ecological niches, including animal, plant, and human hosts. Within the human host environment, *P. aeruginosa* has been isolated from both gastrointestinal and urinary tracts, burn wounds, and the respiratory system. Notably, *P. aeruginosa* has demonstrated competence in colonizing the surfaces of medical equipment, such as catheters and ventilators. The capacity to inhabit and persist in multiple disparate environments confers an exceptional talent for host infection and transmission.

Generally, *P. aeruginosa* presents few problems to healthy individuals. However, in elderly and immunocompromised patients, *P. aeruginosa* poses a considerable threat to health and is one of the leading nosocomial pathogens. *P. aeruginosa* is notorious for its intrinsic high level of tolerance to antibiotic treatment due to a number of efflux pumps and resistance genes [1]. Furthermore, *P. aeruginosa* can produce a diverse range of toxins and virulence factors, allowing it to damage surrounding tissue and expand colonization within the host. *P. aeruginosa* also possesses a highly adaptive metabolism that allows utilization of a range of carbon sources and energy-generating pathways. In addition to its flexible physiology, the outer membrane of *P. aeruginosa* comprises a tough physical barrier to host immune responses, as well as chemotherapeutic agents; effects are amplified when the organism switches to a biofilm-forming mode of growth.

Given the ability of *P. aeruginosa* to flourish in a variety of environments, it is important to understand the regulatory mechanisms underlying bacterial cell-cell communication and how any downstream effects may be coordinated. Communication between individual members of the bacterial population is mediated by a system known as quorum sensing (QS), a cell-density-based intracellular communication system which plays a key role in regulating bacterial pathogenicity and biofilm formation by controlling a battery of virulence genes. The concept of QS, first proposed in the 1990s, was based on population density dependent regulatory mechanisms found in *Vibrio fischeri* [2]. In *P. aeruginosa*, QS is mediated by two chemically distinct classes of signaling molecules, the *N*-acyl-homoserine lactones and the 4-quinolones. Typically, as a bacterial population increases in number, QS signaling molecules will accumulate until reaching a threshold concentration. When combined with other critical factors, such as the distribution and diffusion rate of these signaling molecules, this accumulation results in the transcription of genes under QS regulation.

Due to the importance of *P. aeruginosa* as a nosocomial human pathogen, it has become a model system for examining the functionality of QS. Recent investigations have unravelled a multilayered and hierarchical QS network in this organism, comprising at least four distinct signaling systems, diverging from the conventional linear signaling model.

6.2 QS Systems in *P. aeruginosa*

As illustrated in Figure 6.1, *P. aeruginosa* encodes a variety of distinct but interconnected QS signaling pathways. These are classified into the *las*, *rhl*, *pqs*, and *iqs* systems.

las: The *las* system consists of a transcriptional activator, LasR, and an acyl-homoserine lactone (AHL) synthase, LasI, which catalyzes synthesis of the autoinducer molecule *N*-(3-oxododecanoyl)-L-homoserine lactone (3-oxo-C_{12}-HSL, also known as PAI-1 or OdDHL) [3,4]. This signaling molecule binds to LasR, and the OdDHL-LasR complex then stimulates (or, more rarely, represses) the expression of target genes by recognizing the *las* box, a conserved dyad symmetry DNA motif that is found upstream of many genes known to be directly regulated by this system [5–7].

rhl: A second autoinducer system was discovered in *P. aeruginosa* that interlinks with the previously characterized *rhlABR* gene cluster, which encodes a rhamnosyl transferase (RhlAB) and a regulatory protein (RhlR) required for rhamnolipid biosynthesis [8]. The LuxI type autoinducer synthetase, RhlI, which synthesises the autoinducer molecule *N*-butyryl-L-homoserine lactone (C_4-HSL, also known as PAI-2 or BHL), was found just downstream of this cluster [8–10]. As with the LasR-LasI system, the RhlR-RhlI regulatory system was also shown to be essential for optimal production of elastase, suggesting cross-communication between these two distinct QS systems [8]. RhlR-dependent transcriptional regulation was shown to be mediated by its binding to a specific sequence in the *rhlAB* regulatory region that occurs both in the presence and absence of its autoinducer, BHL. However, the RhlR-BHL complex activates transcription, whereas the ligand-free form acts as a transcriptional repressor; therefore, both forms finely tune the expression of QS-regulated genes [11]. RhlR was recently shown to mediate a distinct transcriptional program even in the absence of RhlI, suggesting that RhlR may also respond to a different, unknown ligand. Additionally, a Δ*rhlI* mutant displays full virulence in animals whereas the Δ*rhlR* mutant is attenuated [12].

pqs: The *P. aeruginosa* intracellular QS repertoire was increased further in complexity by the discovery of a third signaling molecule. The addition of spent culture supernatant from *P. aeruginosa* wild-type PAO1 cultures to PAO-R1 (*lasR*⁻) resulted in significant activation of transcription from the *lasB* promoter, as shown by the induction of a *lasB*′-*lacZ* fusion. This transcriptional activation could not be mimicked by the addition of OdDHL or BHL, suggesting the existence of an uncharacterized QS signaling molecule. HPLC fractionation of the spent media extract resulted in the identification of this new activatory molecule as an alkylquinolone. This *Pseudomonas* quinolone signal (PQS), which is capable of inducing *lasB* expression in *P. aeruginosa*, was shown to be 2-heptyl-3-hydroxy-4-quinolone [13]. The genes required for PQS biosynthesis were subsequently identified and shown to consist of *pqsABCD*, *phnAB*, *pqsH*, and a LysR family regulator encoded by *pqsR* [14,15]. The first step of PQS biosynthesis is mediated by

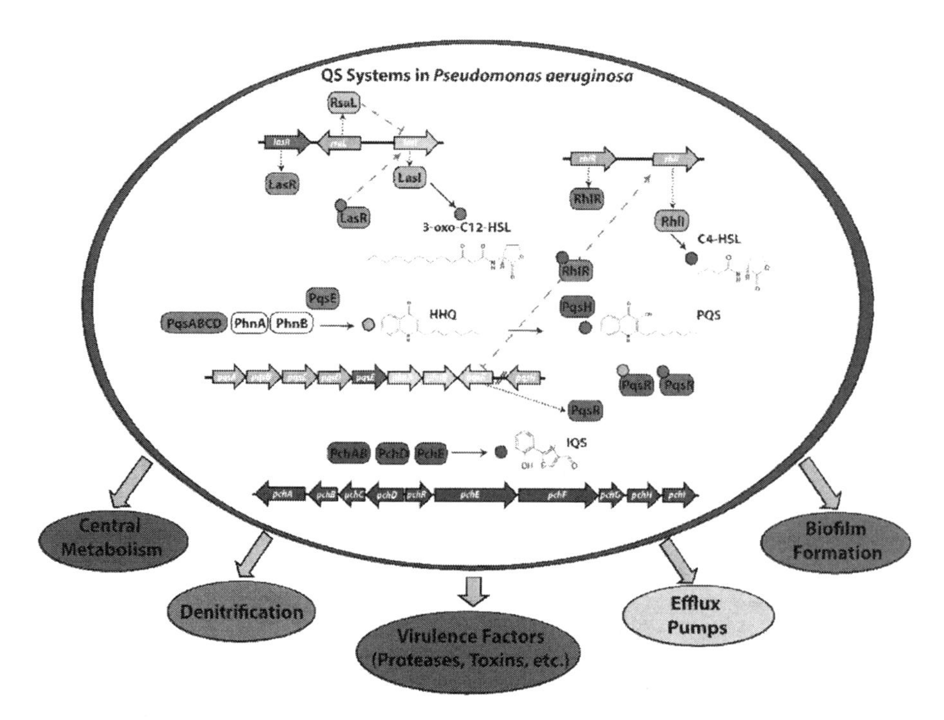

FIGURE 6.1 A schematic overview of the *P. aeruginosa* quorum sensing network, their respective signaling molecules, and downstream targets. 3-oxo-C_{12}-HSL = *N*-(3-oxododecanoyl)-L-homoserine lactone or OdDHL, C_4-HSL = *N*-butyryl-L-homoserine lactone or BHL, PQS = 2-heptyl-3-hydroxy-4-quinolone, IQS = 2-(2-hydroxyphenyl)-thiazole-4-carbaldehyde.

the anthranilate-CoA ligase, PqsA, which activates the anthranilate made by PhnAB to form anthraniloyl-CoA. This is subsequently processed by the 3-oxoacyl-ACP synthases, PqsB, PqsC, and PqsD, to generate 2-heptyl-4-quinolone (HHQ) [14,16,17]. HHQ is then converted into PQS by PqsH, a putative flavin-dependent monooxygenase [18,19]. The PQS system has been shown regulate the production of virulence determinants including elastase, rhamnolipids, LecA, and pyocyanin (a blue-green phenazine pigment that contributes to *P. aeruginosa*'s distinctive hue) [20,21].

The PQS biosynthetic cluster also contains a fifth ORF, *pqsE*. The role of PqsE remained a mystery due to its puzzling phenotype. Mutations in *pqsE* do not affect PQS biosynthesis, but these mutants fail to respond to PQS, or to produce PQS regulated factors such as pyocyanin. However, overexpression of PqsE led to enhanced pyocyanin and rhamnolipid production, phenotypes which are strongly-dependent on PQS signaling [14,20,22]. PqsE was finally characterized as a thioesterase, which cleaves the biosynthetic intermediate 2-aminobenzoylacetyl-CoA, resulting in 2-aminobenzoylacetate, the precursor of HHQ, and 2-aminoacetophenone [23]. Some redundancy is built into this pathway, as the promiscuous thioesterase TesB can sustain HHQ production in the absence of *psqE*, thus clarifying how the *pqsE* mutant can still synthesize HHQ and PQS. It is possible that PqsE plays a role in balancing the level of secondary metabolites derived from the 2-N-alkyl-4(1H)-quinolone biosynthetic pathway [23]. However, blocking the thioesterase activity of PqsE still did not block pyocyanin production, indicating that the secondary metabolite regulatory activity of PqsE is independent of this activity. However, PqsE is clearly not a transcriptional regulator [24].

iqs: Recently, a new QS system was identified, termed IQS after its proposed function in the integration of QS signals and the stress response [25]. Blockage of the *iqs* system disrupts *pqs*- and *rhl*-mediated QS signals. The IQS system is controlled by the *las* signaling system in nutrient-replete conditions and can also be activated by phosphate limitation. IQS was first proposed to be 2-(2-hydroxyphenyl)-thiazole-4-carbaldehyde, which is a non-ribosomal peptide encoded by the gene cluster *ambBCDE* [25]. When this operon is disrupted, the production of PQS and BHL are significantly decreased. The production of the virulence factors pyocyanin, elastase, and rhamnolipids were also concomitantly decreased, leading to reduced virulence. However, recent data has strongly suggested that IQS is actually a product of the operon required for biosynthesis of the siderophore pyochelin [26,27]. It has been known for several years that mutations in the genes *lasI* or *lasR* are frequently detected in *P. aeruginosa* clinical isolates, despite the central importance of this QS system [28,29]. The IQS system may be key to this, as it has been shown to override the *las* system during phosphate depletion, a scenario commonly encountered during infection scenarios [25].

6.3 Integration of QS Signals

As shown in Figure 6.1, the *las* system is thought to sit at the top of the QS signaling hierarchy. This system is auto-activated by OdDHL, resulting in the formation of a LasR-OdDHL complex, which in turn activates the transcription of *rhlR*, *rhlI*, *lasI* (as part of a positive feedback loop), and several other virulence-associated genes [30,31]. This stepwise activation cascade is triggered when a population "quorum" is attained.

The *las* system stimulates expression of the *rhl* subsystem. Here, RhlR dimerizes upon forming a complex with BHL (RhlR-BHL), activating the expression of its own regulon and *rhlI* [32,33]. LasR-OdDHL also positively regulates the expression of *pqsR* and *pqsH*, which encode the transcriptional activator of the HHQ biosynthetic operon *pqsABCDE*, and the enzyme that converts HHQ to the more bioactive derivative PQS, respectively [13,34]. Despite the commonly held view of a clear *las*-driven hierarchy, RhlR-BHL activates most QS-dependent virulence factors, thus the *rhl* system can be considered the "beating heart" of the QS system. However, PQS also enhances the transcription of *rhlI*, thereby stimulating BHL production and subsequent expression of the *rhl* QS regulon, indirectly modulating *rhl*-dependent phenotypes. In an elegant feedback loop, the expression of *pqsR* and *pqsABCDE* are inhibited by RhlR/BHL, thus the ratio between OdDHL and BHL is key to fine tuning of *pqs* signaling [15]. Adding a further layer of complexity to QS signal integration, recent work by Carloni et al. has shown that the small RNA, reaL, regulates synthesis of PQS via a positive post-transcriptional effect on *pqsC* [35]. ReaL is negatively regulated by LasR, and overexpression of reaL results in a hypervirulent phenotype. This highlights a novel mode—a QS regulation that links the *pqs* and *las* systems.

There are some exceptions to LasR-mediated control. Deletion of *lasR* results in delayed PQS production instead of an abolished PQS system [20]. LasR dependency can also be mitigated by activating the *rhl* QS system and downstream virulence factors [36]. The dominance of *las* over IQS can also be reversed in conditions of phosphate starvation, even in a *lasI* or *lasR* mutant [25]. It is possible that there are more conditions under which these systems are activated, highlighting the importance of understanding the environmental context of an infection when using small-molecule QS inhibitors [37].

A recent examination of 580 *lasR* coding variants from *P. aeruginosa* CF isolates identified 173 unique *lasR* variants, 116 of which were either missense or nonsense mutations. Remarkably, in some LasR-null isolates, genes that are LasR-dependent in laboratory strains were still expressed. RhlR activity was retained in approximately half of these LasR-null isolates. This suggests that *in vivo*, *lasR* mutants may be far from QS-silent, which has been the consensus to date. It appears that these isolates have instead "rewired" their QS circuitry to make RhlR signaling independent of the canonical hierarchy—possibly through PqsE involvement [38]. RhlR can also be activated in LasR-null mutants under conditions of stress or following activation of the stringent response [39]. Furthermore it is known that 2-alkyl-quinolone signals can activate the RhlR-I system in stationary phase [20]. Considered together, these data suggest that in certain circumstances it is possible that the inactivation of

LasR could actually provide a metabolic advantage—the cost of which can be minimized by the rewiring of other QS systems to maintain the activation of downstream virulence factors such as elastase and pyocyanin [40].

It has been noted that in studied examples of QS systems, the gene required for signal production is positively autoregulated by the QS molecule itself. This feedback induction has been described as an essential component of these systems, but the precise reason for this remains unknown. To examine this, the QS signal synthase gene promoter was replaced with an arabinose-inducible promoter system, resulting in a QS signal which was produced at a constant rate across the population, regardless of the population density. This work showed that positive autoregulation is not required for a strong QS response, but instead enhances the response synchrony. This synchrony may be particularly important in circumstances where population cooperation is essential for success [41].

In the multi-signal QS system of *P. aeruginosa*, the three signaling pathways (*las*, *rhl*, and *pqs*) differ mainly in the QS molecules that they use, since both the *las* and *rhl* pathways use *N*-acylhomoserine lactones, and *pqs* uses 2-alkyl-4-quinolones (AQs) [42,43]. Rampioni et al. studied the interplay between components of the *pqs* system during QS in *P. aeruginosa*. Due to the intrinsic complexity of these systems, the authors constructed a series of mutants where the expression of *pqsE*, a key effector of the pathway, was controlled by an exogenous signal (IPTG) and where the *pqsA*, *pqsH*, and *pqsL* genes were deleted (a "Δ4AQ strain"). Using this approach, it was possible to differentiate the effects of AQ molecules from those of PqsE. Rampioni et al., demonstrated that PqsE controls the expression of 145 genes in the Δ4AQ strain, including a subset of genes related to pyocyanin, an important virulence factor of *P. aeruginosa*. PqsE was also found to be involved in the transcriptional regulation of genes related to biofilm formation and of the multidrug efflux pumpMexHI-OpmD [43].

From the analysis of *las*, *rhl*, and *pqs*, it is clear that the regulatory circuitry for QS in *P. aeruginosa* features intricate interplay between the cognate QS signaling molecules, expression of virulence determinants, and the multidrug resistance mechanisms of this bacterium.

6.4 Targets of the QS System in *P. aeruginosa*

6.4.1 Efflux Pumps

As a Gram-negative pathogen, *P. aeruginosa* has an asymmetric bilayer outer membrane (OM) composed of glycolipid lipopolysaccharides (LPS) and glycerol phospholipids [44]. Gram-negative cell walls can resist high turgor pressure, high temperatures, and pH variation, and are elastic enough to expand their normal surface area multifold. In addition, the OM serves as a barrier to toxic compounds such as antibiotics [45]. In many cases, this drug-resistance property is amplified by the presence of cell envelope-spanning multidrug efflux transporters. Among the efflux pumps, the resistance-nodulation-division efflux systems (RND type transporters) arguably provide the most potent protection to the organism, as they often transport several different types of substrate [44,46]. *P. aeruginosa* encodes no less than 12 RND-family transporters, many of which have been shown to play key roles in resistance to xenobiotics and antimicrobial agents. Additionally, there is evidence to suggest that some of these pumps also impact on aspects of *P. aeruginosa* physiology, such as QS, biofilm formation, and cell motility [47]. For example, the MexAB-OprM efflux pump translocates *N*-acyl-homoserine lactones [48]. Another significant RND transporter system, MexEF-OprN, transportsHHQ, the direct precursor of PQS [49]. In later work, Wolloscheck et al., analyzed the substrate specificities and efflux capacities of the MexGHI-OpmD and MexEF-OprN pumps, which have also been implicated in the efflux of QS signals. For example, MexG has been shown by Hodgkinson et al. to bind PQS [50]. Wolloscheck et al., found that overexpression of MexGHI-OpmD provides resistance to fluoroquinolone antibiotics, and is more efficient than MexEF-OprN in protecting *P. aeruginosa* against toxic phenazines [47]. Interestingly, the authors also showed that the interaction of MexG with MexHI-OpmD reduced efflux activity of fluoroquinolones in a substrate-dependent manner. Finally, the same authors concluded that the endogenous activity of MexGHI-OpmD establishes the steady-state intra- and extracellular concentration of virulence factor pyocyanin, another signaling factor in the QS network of *P. aeruginosa*, previously shown by Dietrich et al. to be capable of up-regulating QS-controlled genes during stationary phase [51].

These conclusions may be significant because they highlight that not only does QS signaling play a significant role in the expression of these efflux systems, but it also alters the activities of these pumps, resulting in a significant impact on the intra- and extra-cellular levels of QS molecules. Additionally, this complex regulatory network will respond differentially to changes in the environment, as each component of the QS system in *P. aeruginosa* has a particular level of sensitivity to nutrient-limited conditions [37].

6.5 Biofilm Formation

The regulation of biofilm formation by QS in *P. aeruginosa* was first described twenty years ago [52], when it was shown that a *lasI* mutant develops flat, undifferentiated biofilms which are sensitive to sodium dodecyl sulfate. This phenotype could be complemented by the addition of OdDHL. In subsequent years, this finding has been reinforced and there is now an established link between biofilm formation and QS in the literature [53–56]. Some studies have debated this link, although these discrepancies may be related to a lack of standardized methodologies to examine biofilm formation [57–60].

How might QS impinge on a multifactorial phenotype like biofilm formation in *P. aeruginosa*? One insightful study has shown that a tyrosine phosphatase (TpbA) acts as a bridge between QS and biofilm formation. The gene encoding TpbA is under QS control, and expression of *tpbA* in the flat, undifferentiated biofilm cells of a *lasI* mutant is diminished, although no diminution is observed in a *rhlI* mutant. TpbA responds to OdDHL by negatively regulating the level of the intracellular secondary messenger 3,5-cyclic diguanylate (c-di-GMP), resulting in decreased biofilm formation. However, and as *tpbA* expression was not completely abolished in the *lasI* mutant, additional, uncharacterized factors are likely to play a role in the regulation of this phosphatase [61].

The full extent of the link between QS and biofilm formation remains to be elucidated. A recent study revealed that RhlR can control biofilm development independently of its canonical autoinducer, BHL. This suggests that a currently unknown alternative ligand is also capable of activating RhlR expression or activity [12].

6.6 Central Metabolism

The literature is replete with studies describing how QS controls the synthesis of key virulence factors in *P. aeruginosa*. Several omics-based studies carried out on *P. aeruginosa* QS mutants have indicated that QS regulates several primary metabolic nodes in this organism (Figure 6.1). As many of these metabolic pathways are either essential or have pathway redundancy, few studies have moved into a detailed characterization of these metabolic QS targets in *P. aeruginosa*.

The activation of QS was shown to result in global metabolic changes in *P. aeruginosa*. A detailed 'omic analysis of a *lasI rhlI* double mutant compared with the wild-type identified alterations in a third of all intracellular metabolites, including TCA cycle intermediates, amino acids, fatty acids, and polyamines. The TCA cycle metabolites citrate, malate, succinate, and fumarate were shown to be more abundant in a *lasI rhlI* mutant compared with the wild-type. This altered metabolic profile may be explained by a simple case of supply and demand. The synthesis of secreted of QS molecules would require a reallocation of cellular resources to the respective pathways—depleting TCA cycle intermediates. In support of this hypothesis, the *lasI rhlI* mutant had a small but consistent growth advantage compared with the wild-type [62].

A clear selection pressure to abandon a fully functional QS system has been noted in clinical scenarios. Loss-of-function mutations of *lasR* are a common occurrence during early stage chronic infection of the CF airway [28]. It was shown that although these mutations confer reduced expression of virulence genes, they also confer a growth advantage in the presence of certain amino acids such as phenylalanine, isoleucine, and tyrosine [29].

The influence of metabolism on quorum sensing is further emphasised by the increased production of QS molecules in the presence of precursor metabolites [63]. When cultured in the presence of the aromatic amino acids phenylalanine and tyrosine, which are found at high levels in CF sputum, *P. aeruginosa* produces increased levels of PQS, thus underscoring the importance of the nutritional environment as a cue for QS responses. The exact *in vivo* consequence of enhanced PQS production remains to be elucidated, but it may confer a significant advantage during lung colonization and/or polymicrobial infection [64]. The close integration of QS signaling with nutritional cues is also seen with rhamnolipids. Although the *rhlAB* operon (which encodes the enzymes for rhamnolipid biosynthesis) is primarily controlled by QS, depletion of available carbon abruptly stops transcription. This may be a strategy to reduce the bioconversion of carbon-rich exo-metabolites during carbon starvation. However, when extracellular iron is depleted, the growth rate slows and *rhlAB* expression increases. This may be due to the reduced growth rate during iron limitation, slowing the carbon demand for biomass [65]. These data suggest that QS signals and nutrient availability are integrated in this organism, possibly as a form of metabolic prudence, maintaining an individual's fitness in the face of a cooperating population [66–68].

6.7 Anaerobic Metabolism: Denitrification

Anaerobic growth of *P. aeruginosa* is also strongly influenced by QS. *P. aeruginosa* is thought to survive in the oxygen-limited or anaerobic CF airway mucus by respiratory NO_3^- reduction, or denitrification. This pathway uses nitric oxides as terminal electron acceptors to support anaerobic respiration [69]. Complete denitrification occurs through four sequential steps to reduce nitrate (NO_3^-) to dinitrogen (N_2) via nitrite (NO_2^-), nitric oxide (NO^{α}), and nitrous oxide (N_2O). Each step of the pathway is catalyzed by the denitrification enzymes nitrate reductase (NAR), nitrite reductase (NIR), nitric oxide reductase (NOR), and nitrous oxide reductase (NOS) [70,71]. A key point is that denitrification produces potentially toxic reactive nitrogen intermediates including nitric oxide and nitrous acid [72]. The first hint that QS is in control of this pathway was demonstrated through the examination of anaerobic *P. aeruginosa* biofilms. When growing as anaerobic biofilm, *rhl*-deficient mutants were effectively committing "metabolic suicide" due to the build-up of toxic NO, which is normally removed by downstream enzymes in the denitrification pathway. This work showed that *rhl* mediated regulation of NOR is necessary to keep NO levels low during anaerobic respiration [73].

The influence of QS on *P. aeruginosa* denitrification under anaerobic conditions was further clarified by a detailed *in vivo* and *in vitro* analysis [74]. This work showed that the *las* and *rhl* quorum-sensing systems repress denitrification, and that regulation by *las* is dependent on *rhl*. The authors then explored if PQS could also influence denitrification in this organism—this QS molecule was shown to affect denitrifying enzyme activities primarily due to its iron chelating activity. When $FeCl_3$ was added to the medium alongside PQS, this fully reversed the PQS suppression of NO_3^- reduction and N_2O production [75]. A recent comparative transcriptomic study carried out on a QS-deficient clinical isolate (J215) compared to *P. aeruginosa* wild-type grown in 1% oxygen showed that the denitrification machinery was expressed at a higher rate in this mutant, supporting the previous findings that QS may influence denitrification [76].

6.8 Using Systems Biology to Understand QS

While QS has become a very popular concept, it is important to highlight that it is just the tip of the iceberg in the modulation of bacterial group behavior. Currently, QS systems are viewed as highly-compact signal integrators, where the concentration of the autoinducer is the combined response to many inputs, such as nutritional factors, host factors, population size, host diffusion rate, and spatial distribution [77–82]. Moreover, this environmental information is processed in parallel among many processing layers until it is finally transferred to autoinducer molecules. For example, a myriad of environmental signals modulate, from transcriptional to post-translational levels, the expression and dynamics of enzymes responsible for producing/degrading QS molecules, which will subsequently control the autoinducer concentration and ultimately the community behavior [80,83].

To gain a holistic understanding of QS—and to further rewire its components for novel applications—two nonexclusive main frameworks have been employed: Systems Biology and Synthetic Biology. Most contributions from Systems Biology in understanding QS systems on a global scale are based on analysis and integration of data from metabolic, transcriptional, and protein-protein networks. A bird's-eye view of these network architecture(s) allows for meaningful insights regarding its topology, hierarchy, connectivity, robustness, and evolvability. It also allows us to focus in on the network and explore the dynamics of the desired subsystem, such as regulatory motifs or signaling pathways, which can then be modeled and experimentally tested to reveal novel properties or constraints for QS systems.

In this context, the analysis of the *P. aeruginosa* regulatory system has revealed a gene regulatory network of 690 genes and 1020 regulatory interactions, constituting the third largest regulatory network reported in bacteria [84]. This network is also enriched for self-activation interactions for transcription factors, which is rather uncommon in most bacterial systems, with some components having extensive control over the network; for example, LasR is the most global regulator with influence on 15% of the genes in the network. This emphasizes the importance of QS systems in *P. aeruginosa* gene regulation. As previously mentioned, there is some regulatory hierarchy between the four QS systems of *P. aeruginosa*, in which LasR induces the expression and activation of the *rhl*, *pqs*, and *iqs*, while *rhl* and *pqs* systems can also regulate each other [80,83].

The global analysis of transcriptional, metabolic, and protein-protein responses to QS systems are also important resources for constructing an overview of complex QS effects. Recently, Rampioni et al. analyzed PQS system effects on *P. aeruginosa*, uncovering specific regulons controlled by this system [43]. Another study assessed the global effects of anti-activators (proteins which sequester QS regulators), showing additive effects of anti-activation on QS gene expression, likely via LasR and RhlR systems [85].

The study of network subsets and motifs can also provide relevant insights regarding QS dynamics [86,87]. The analysis of a common and simple motif in the *P. aeruginosa* regulatory network, positive autoregulation, has shown that while it is not essential, it enhances the synchrony of individual responses in a population, suggesting that autoregulation allows an adaptive balance between coherent and less structured responses [41]. The exploration of a more complex motif, such as the multi-output type 1 incoherent feedforward loop (IFFL-1) that governs OdDHL production by means of LasR and RsaL dynamics, has been shown to split the QS regulon into two distinct sub-regulons with different robustness with respect to LasR fluctuations, enhancing phenotypic plasticity [88].

Mathematical models of QS systems are also a complementary approach to network analysis, depicting more general aspects of the system's dynamics and emergent phenomena [89]. The first model of QS in *P. aeruginosa* was of the *las* systems [90]. One of the major contributions of this study was highlighting the existence of a fold bifurcation structure in this system for the concentration of the response regulator, LasR, in response to cell concentration. Over the years these systems have been constantly modified, ranging from qualitative models [91] to exhaustive rule-based approaches [92]. Recent work has also explored the QS system by considering nonlinear positive and negative feedback loops associated with the production of the synthase LasI and the regulator RsaL [93]. These nonlinear effects create the possibility of novel dynamical behaviors in the model such as pulse generation and the emergence of a quorum memory for triggering rhamnolipid production in specific conditions [93].

6.9 QS Systems as Biological Parts for Synthetic Biology

While Systems Biology usually adopts a top-down approach for exploring biological systems, Synthetic Biology tends to adopt a bottom-up approach. Thus, both fields can be integrated into an iterative cycle for insightful discoveries. The Synthetic Biology framework can also utilize information gathered from a system to rewire it toward novel functions, achieving both engineering principles and an understanding of novel emergent rules. Using a "building to understand" methodology, a recent study has evaluated the crosstalk between LuxR/I and LasR/I systems and found that QS crosstalk can be dissected into signal and promoter crosstalk. When studied in the context of a synthetic positive feedback gene network, QS crosstalk leads to distinct dynamic behaviors: signal crosstalk significantly decreases the circuit's induction range for bistability, but promoter crosstalk causes transposon insertions into the regulator gene and yields trimodal responses due to a combination of mutagenesis and noise-induced state transitions [94].

The *las*/*rhl* systems of *P. aeruginosa* have also been employed in a number of different applications in synthetic biology, ranging from biological biocomputation to engineering microbial consortia. Brenner et al. used the *las* and *rhl* networks to build two strains of *E. coli* that form a biofilm together only once both populations reach a threshold density [95]. Therefore, rewriting QS networks allows an elegant way to engineer complex community behavior in bacteria trigged by an autonomous signal (such as cell density). Balagaddé et al. used components from the *lux* and *las* networks to engineer a predator–prey relationship between *E. coli* strains [96]. This type of approach can be used both to study basic ecological interactions using minimal microbial systems, and to create oscillation-like dynamics in artificial microbial communities. Tamsir et al., used *las* and *rhl* networks to link circuits expressed in multiple cell populations, implementing complex Boolean expressions through different spatial arrangements on agar plates [97]. In this system, complex logic operations can be performed at a population level using simple NOR gates interconnected from cell to cell using QS molecules. Many advances have also been achieved in synthetic cell communication using *las* and *rhl* systems. Rampioni et al., have developed a giant vesicle-based "synthetic cell" capable of producing QS signaling molecules that can be recognized by *P. aeruginosa* [98]. Kylilis et al. have characterized *rhl*, *lux*, *tra*, *las*, *cin*, and *rpa* QS systems at single cell level and with non-cognate AHL inducers, generating the largest characterized library of QS devices with mapped chemical crosstalk interactions [99]. Using a similar approach, Scott et al. engineered two systems, *rpa* and *tra*, to add to the extensively used *lux* and *las* systems [100]. By testing the cross-talk between all systems, they characterized new inducible systems for versatile control of engineered communities. In an even more drastic approach, Wen and colleagues have reverse-engineered the *las* QS system to create a cell-free biosensor of AHL molecules for diagnostic purposes [101]. This remarkable approach holds the potential to exploit the components of bacterial QS networks for novel biomedical applications.

6.10 Conclusions and Perspectives

In conclusion, recent advances in the study of *P. aeruginosa* QS systems have shifted the classical concept of population density sensors to a far more complex and sophisticated one: they are highly compressed environmental signal integrators with multiple outputs. The autoinducer concentration ultimately embodies intra-, inter-, and extracellular information, deeply modulating other layers of organization such as transcriptional and metabolic networks in a bilateral process. QS modules are also hierarchical yet flexible, and as in most adaptive systems the interactions between their components are highly dynamic, producing a range of emergent behaviors such as bistable and oscillatory ones depending on the overall information provided. Exploring these systems from a holistic viewpoint thus requires a combination of high-throughput technologies, mathematical models, and insightful analysis. In this context, Systems and Synthetic Biology can provide a unique and synergetic approach for both unravelling the global effects of QS systems in bacterial behavior and redirecting the obtained knowledge to more applied frameworks, such as drug-targeting methods and the engineering of novel relationships between microorganisms in biotechnological/medical applications [100–103].

While several works have demonstrated the intricate interplay between QS systems, virulence factors, resistance determinants, and biofilms, we still lack a comprehensive overview of these processes. Unravelling the tight control of these QS systems and elucidating these networks may be the key to controlling *P. aeruginosa* infections. A more complete understanding may result in more precisely targeted antimicrobials. For example, in a lung cell assay, a *lasRrhlR* double mutant actually causes more cell death than the wild-type strain—which may be due to a disruption of these overlapping networks—particularly the expression of virulence factors which are regulated by the antagonistic action of RhlR and LasR [104]. In line with this rationale, a quadruple Δ*lasI*, Δ*lasR*, Δ*rhlI*, Δ*rhlR* mutant remains virulent in a mouse lung infection model [105]. Therefore, new systems approaches are still required to fully understand the way these mechanisms coordinate the physiology of *P. aeruginosa* and their impact in the infection progression.

ACKNOWLEDGEMENTS

This work was supported by BBSRC/FAPESP grant BB/R005435/1.

REFERENCES

1. Lambert, P.A. Mechanism of antibiotic resistance in *Pseudomonas aeruginosa*. *J. R. Soc. Med.* **2002**, *95*, S22–26.
2. Fuqua, W.C.; Winans, S.C.; Greenberg, E.P. Quorum sensing in bacteria: The LuxR-LuxI family of cell density-responsive transcriptional regulators. *J. Bacteriol.* **1994**, *176*, 269–275.
3. Gambello, M.J.; Iglewski, B.H. Cloning and characterization of the *Pseudomonas aeruginosa* lasR gene, a transcriptional activator of elastase expression. *J. Bacteriol.* **1991**, *173*, 3000–9.
4. Pearson, J.P.; Gray, K.M.; Passador, L.; Tucker, K.D.; Eberhard, A.; Iglewski, B.H.; Greenberg, E.P.; Ramsey, B.W.; Speert, D.P.; Moskowitz, S.M.; et al. Structure of the autoinducer required for expression of *Pseudomonas aeruginosa* virulence genes. *Proc. Natl. Acad. Sci.* **1994**, *91*, 197–201.
5. Anderson, R.M.; Zimprich, C.A.; Rust, L. A second operator is involved in *Pseudomonas aeruginosa* elastase (lasB) activation. *J. Bacteriol.* **1999**, *181*, 6264–70.
6. Rust, L.; Pesci, E.C.; Iglewski, B.H. Analysis of the *Pseudomonas aeruginosa* elastase (lasB) regulatory region. *J. Bacteriol.* **1996**, *178*, 1134–40.
7. Whiteley, M.; Greenberg, E.P. Promoter specificity elements in *Pseudomonas aeruginosa* quorum-sensing-controlled genes. *J. Bacteriol.* **2001**, *183*, 5529–34.
8. Ochsner, U.A.; Reiser, J. Autoinducer-mediated regulation of rhamnolipid biosurfactant synthesis in *Pseudomonas aeruginosa*. *Proc. Natl. Acad. Sci. USA*. **1995**, *92*, 6424–8.
9. Pearson, J.P.; Passador, L.; Iglewski, B.H.; Greenberg, E.P. A second *N*-acylhomoserine lactone signal produced by *Pseudomonas aeruginosa*. *Proc. Natl. Acad. Sci. USA* **1995**, *92*, 1490–4.
10. Brint, J.M.; Ohman, D.E. Synthesis of multiple exoproducts in *Pseudomonas aeruginosa* is under the control of RhlR-RhlI, another set of regulators in strain PAO1 with homology to the autoinducer-responsive LuxR-LuxI family. *J. Bacteriol.* **1995**, *177*, 7155–63.
11. Medina, G.; Juárez, K.; Valderrama, B.; Soberón-Chávez, G. Mechanism of *Pseudomonas aeruginosa* RhlR transcriptional regulation of the rhlAB promoter. *J. Bacteriol.* **2003**, *185*, 5976–83.
12. Mukherjee, S.; Moustafa, D.; Smith, C.D.; Goldberg, J.B.; Bassler, B.L. The RhlR quorum-sensing receptor controls *Pseudomonas aeruginosa* pathogenesis and biofilm development independently of its canonical homoserine lactone autoinducer. *PLOS Pathog.* **2017**, *13*, e1006504.
13. Pesci, E.C.; Milbank, J.B.; Pearson, J.P.; McKnight, S.; Kende, A.S.; Greenberg, E.P.; Iglewski, B.H. Quinolone signaling in the cell-to-cell communication system of *Pseudomonas aeruginosa*. *Proc. Natl. Acad. Sci. USA*. **1999**, *96*, 11229–34.
14. Gallagher, L.A.; McKnight, S.L.; Kuznetsova, M.S.; Pesci, E.C.; Manoil, C. Functions required for extracellular quinolone signaling by *Pseudomonas aeruginosa*. *J. Bacteriol.* **2002**, *184*, 6472–80.
15. Cao, H.; Krishnan, G.; Goumnerov, B.; Tsongalis, J.; Tompkins, R.; Rahme, L.G. A quorum sensing-associated virulence gene of *Pseudomonas aeruginosa* encodes a LysR-like transcription regulator with a unique self-regulatory mechanism. *Proc. Natl. Acad. Sci.* **2001**, *98*, 14613–18.
16. Coleman, J.P.; Hudson, L.L.; McKnight, S.L.; Farrow, J.M.; Calfee, M.W.; Lindsey, C.A.; Pesci, E.C. *Pseudomonas aeruginosa* PqsA is an anthranilate-coenzyme A ligase. *J. Bacteriol.* **2008**, *190*, 1247–55.

17. Déziel, E.; Lépine, F.; Milot, S.; He, J.; Mindrinos, M.N.; Tompkins, R.G.; Rahme, L.G. Analysis of *Pseudomonas aeruginosa* 4-hydroxy-2-alkylquinolines (HAQs) reveals a role for 4-hydroxy-2-heptylquinoline in cell-to-cell communication. *Proc. Natl. Acad. Sci. USA*. **2004**, *101*, 1339–44.
18. Deziel, E.; Lepine, F.; Milot, S.; He, J.; Mindrinos, M.N.; Tompkins, R.G.; Rahme, L.G. Analysis of *Pseudomonas aeruginosa* 4-hydroxy-2-alkylquinolines (HAQs) reveals a role for 4-hydroxy-2-heptylquinoline in cell-to-cell communication. *Proc. Natl. Acad. Sci.* **2004**, *101*, 1339–44.
19. Schertzer, J.W.; Boulette, M.L.; Whiteley, M. More than a signal: Non-signaling properties of quorum sensing molecules. *Trends Microbiol.* **2009**, *17*, 189–95.
20. Diggle, S.P.; Winzer, K.; Chhabra, S.R.; Worrall, K.E.; Cámara, M.; Williams, P. The *Pseudomonas aeruginosa* quinolone signal molecule overcomes the cell density-dependency of the quorum sensing hierarchy, regulates rhl-dependent genes at the onset of stationary phase and can be produced in the absence of LasR. *Mol. Microbiol.* **2003**, *50*, 29–43.
21. Pesci, E.C.; Milbank, J.B.; Pearson, J.P.; McKnight, S.; Kende, A.S.; Greenberg, E.P.; Iglewski, B.H.; Plotnikova, J.; Tan, M.-W.; Tsongalis, J. Quinolone signaling in the cell-to-cell communication system of *Pseudomonas aeruginosa*. *Proc. Natl. Acad. Sci. USA*. **1999**, *96*, 11229–34.
22. Farrow, J.M.; Sund, Z.M.; Ellison, M.L.; Wade, D.S.; Coleman, J.P.; Pesci, E.C. PqsE functions independently of PqsR-*Pseudomonas* quinolone signal and enhances the rhl quorum-sensing system. *J. Bacteriol.* **2008**, *190*, 7043–51.
23. Drees, S.L.; Fetzner, S. PqsE of *Pseudomonas aeruginosa* acts as pathway-specific thioesterase in the biosynthesis of alkylquinolone signaling molecules. *Chem. Biol.* **2015**, *22*, 611–618.
24. Zender, M.; Witzgall, F.; Drees, S.L.; Weidel, E.; Maurer, C.K.; Fetzner, S.; Blankenfeldt, W.; Empting, M.; Hartmann, R.W. Dissecting the multiple roles of PqsE in *Pseudomonas aeruginosa* virulence by discovery of small tool compounds. *ACS Chem. Biol.* **2016**, *11*, 1755–1763.
25. Lee, J.; Wu, J.; Deng, Y.; Wang, J.; Wang, C.; Wang, J.; Chang, C.; Dong, Y.; Williams, P.; Zhang, L.-H. A cell-cell communication signal integrates quorum sensing and stress response. *Nat. Chem. Biol.* **2013**, *9*, 339–343.
26. Ye, L.; Cornelis, P.; Guillemyn, K.; Ballet, S.; Hammerich, O. Structure revision of N-mercapto-4-formylcarbostyril produced by *Pseudomonas* fluorescens G308 to 2-(2-hydroxyphenyl) thiazole-4-carbaldehyde [aeruginaldehyde]. *Nat. Prod. Commun.* **2014**, *9*, 789–94.
27. Rojas Murcia, N.; Lee, X.; Waridel, P.; Maspoli, A.; Imker, H.J.; Chai, T.; Walsh, C.T.; Reimmann, C. The *Pseudomonas aeruginosa* antimetabolite L-2-amino-4-methoxy-trans-3-butenoic acid (AMB) is made from glutamate and two alanine residues via a thiotemplate-linked tripeptide precursor. *Front. Microbiol.* **2015**, *6*, 170.
28. Smith, E.E.; Buckley, D.G.; Wu, Z.; Saenphimmachak, C.; Hoffman, L.R.; D'Argenio, D.A.; Miller, S.I.; Ramsey, B.W.; Speert, D.P.; Moskowitz, S.M.; et al. Genetic adaptation by *Pseudomonas aeruginosa* to the airways of cystic fibrosis patients. *Proc. Natl. Acad. Sci.* **2006**, *103*, 8487–92.
29. D'Argenio, D.A.; Wu, M.; Hoffman, L.R.; Kulasekara, H.D.; Déziel, E.; Smith, E.E.; Nguyen, H.; Ernst, R.K.; Larson Freeman, T.J.; Spencer, D.H.; et al. Growth phenotypes of *Pseudomonas aeruginosa* lasR mutants adapted to the airways of cystic fibrosis patients. *Mol. Microbiol.* **2007**, *64*, 512–533.
30. Kiratisin, P.; Tucker, K.D.; Passador, L. LasR, a transcriptional activator of *Pseudomonas aeruginosa* virulence genes, functions as a multimer. *J. Bacteriol.* **2002**, *184*, 4912–9.
31. Pesci, E.C.; Pearson, J.P.; Seed, P.C.; Iglewski, B.H. Regulation of las and rhl quorum sensing in *Pseudomonas aeruginosa*. *J. Bacteriol.* **1997**, *179*, 3127–32.
32. Ventre, I.; Ledgham, F.; Prima, V.; Lazdunski, A.; Foglino, M.; Sturgis, J.N. Dimerization of the quorum sensing regulator RhlR: Development of a method using EGFP fluorescence anisotropy. *Mol. Microbiol.* **2003**, *48*, 187–98.
33. Winson, M.K.; Camara, M.; Latifi, A.; Foglino, M.; Chhabra, S.R.; Daykin, M.; Bally, M.; Chapon, V.; Salmond, G.P.; Bycroft, B.W.; et al. Multiple *N*-acyl-L-homoserine lactone signal molecules regulate production of virulence determinants and secondary metabolites in *Pseudomonas aeruginosa*. *Proc. Natl. Acad. Sci. USA*. **1995**, *92*, 9427–31.
34. McKnight, S.L.; Iglewski, B.H.; Pesci, E.C. The *Pseudomonas* quinolone signal regulates rhl quorum sensing in *Pseudomonas aeruginosa*. *J. Bacteriol.* **2000**, *182*, 2702–8.
35. Carloni, S.; Macchi, R.; Sattin, S.; Ferrara, S.; Bertoni, G. The small RNA ReaL: A novel regulatory element embedded in the *Pseudomonas aeruginosa* quorum sensing networks. *Environ. Microbiol.* **2017**, *19*, 4220–37.
36. Dekimpe, V.; Deziel, E. Revisiting the quorum-sensing hierarchy in *Pseudomonas aeruginosa*: The transcriptional regulator RhlR regulates LasR-specific factors. *Microbiology* **2009**, *155*, 712–23.
37. Welsh, M.A.; Blackwell, H.E. Chemical genetics reveals environment-specific roles for quorum sensing circuits in *Pseudomonas aeruginosa*. *Cell Chem. Biol.* **2016**, *23*, 361–369.
38. Feltner, J.B.; Wolter, D.J.; Pope, C.E.; Groleau, M.-C.; Smalley, N.E.; Greenberg, E.P.; Mayer-Hamblett, N.; Burns, J.; Déziel, E.; Hoffman, L.R.; et al. LasR variant cystic fibrosis isolates reveal an adaptable quorum-sensing hierarchy in *Pseudomonas aeruginosa*. *mBio* **2016**, *7*, e01513–16.
39. Van Delden, C.; Pesci, E.C.; Pearson, J.P.; Iglewski, B.H. Starvation selection restores elastase and rhamnolipid production in a *Pseudomonas aeruginosa* quorum-sensing mutant. *Infect. Immun.* **1998**, *66*, 4499–502.
40. Wilder, C.N.; Allada, G.; Schuster, M. Instantaneous within-patient diversity of *Pseudomonas aeruginosa* quorum-sensing populations from cystic fibrosis lung infections. *Infect. Immun.* **2009**, *77*, 5631–9.
41. Scholz, R.L.; Greenberg, E.P. Positive autoregulation of an acyl-homoserine lactone quorum-sensing circuit synchronizes the population response. *mBio* **2017**, *8*, e01079–17.
42. Williams, P.; Cámara, M. Quorum sensing and environmental adaptation in *Pseudomonas aeruginosa*: A tale of regulatory networks and multifunctional signal molecules. *Curr. Opin. Microbiol.* **2009**, *12*, 182–191.
43. Rampioni, G.; Falcone, M.; Heeb, S.; Frangipani, E.; Fletcher, M.P.; Dubern, J.-F.; Visca, P.; Leoni, L.; Cámara, M.; Williams, P. Unravelling the genome-wide contributions of specific 2-Alkyl-4-Quinolones and PqsE to quorum sensing in *Pseudomonas aeruginosa*. *PLOS Pathog.* **2016**, *12*, e1006029.
44. Blair, J.M.A.; Webber, M.A.; Baylay, A.J.; Ogbolu, D.O.; Piddock, L.J. V. Molecular mechanisms of antibiotic resistance. *Nat. Rev. Microbiol.* **2015**, *13*, 42–51.
45. Beveridge, T.J. Structures of Gram-negative cell walls and their derived membrane vesicles. *J. Bacteriol.* **1999**, *181*, 4725–33.

46. Li, X.-Z.; Plésiat, P.; Nikaido, H. The challenge of efflux-mediated antibiotic resistance in Gram-negative bacteria. *Clin. Microbiol. Rev.* **2015**, *28*, 337–418.
47. Wolloscheck, D.; Krishnamoorthy, G.; Nguyen, J.; Zgurskaya, H.I. Kinetic control of quorum sensing in *Pseudomonas aeruginosa* by multidrug efflux pumps. *ACS Infect. Dis.* **2018**, *4*, 185–95.
48. Minagawa, S.; Inami, H.; Kato, T.; Sawada, S.; Yasuki, T.; Miyairi, S.; Horikawa, M.; Okuda, J.; Gotoh, N. RND type efflux pump system MexAB-OprM of *Pseudomonas aeruginosa* selects bacterial languages, 3-oxo-acyl-homoserine lactones, for cell-to-cell communication. *BMC Microbiol.* **2012**, *12*, 70.
49. Lamarche, M.G.; Déziel, E. MexEF-OprN efflux pump exports the *Pseudomonas* quinolone signal (PQS) precursor HHQ (4-hydroxy-2-heptylquinoline). *PLoS One* **2011**, *6*, e24310.
50. Hodgkinson, J.T.; Gross, J.; Baker, Y.R.; Spring, D.R.; Welch, M. A new *Pseudomonas* quinolone signal (PQS) binding partner: MexG. *Chem. Sci.* **2016**, *7*, 2553–62.
51. Dietrich, L.E.P.; Price-Whelan, A.; Petersen, A.; Whiteley, M.; Newman, D.K. The phenazine pyocyanin is a terminal signalling factor in the quorum sensing network of *Pseudomonas aeruginosa*. *Mol. Microbiol.* **2006**, *61*, 1308–21.
52. Davies, D.G.; Parsek, M.R.; Pearson, J.P.; Iglewski, B.H.; Costerton, J.W.; Greenberg, E.P. The involvement of cell-to-cell signals in the development of a bacterial biofilm. *Science.* **1998**, *280*, 295–8.
53. Sakuragi, Y.; Kolter, R. Quorum-sensing regulation of the biofilm matrix genes (pel) of *Pseudomonas aeruginosa*. *J. Bacteriol.* **2007**, *189*, 5383–86.
54. Rice, S.A.; Givskov, M.; Høiby, N.; Parsek, M.R.; Riedel, K.; Heydorn, A.; Eberl, L.; Andersen, J.B.; Hentzer, M.; Molin, S.; et al. Inhibition of quorum sensing in *Pseudomonas aeruginosa* biofilm bacteria by a halogenated furanone compound. *Microbiology* **2002**, *148*, 87–102.
55. Allesen-Holm, M.; Barken, K.B.; Yang, L.; Klausen, M.; Webb, J.S.; Kjelleberg, S.; Molin, S.; Givskov, M.; Tolker-Nielsen, T. A characterization of DNA release in *Pseudomonas aeruginosa* cultures and biofilms. *Mol. Microbiol.* **2006**, *59*, 1114–1128.
56. Barken, K.B.; Pamp, S.J.; Yang, L.; Gjermansen, M.; Bertrand, J.J.; Klausen, M.; Givskov, M.; Whitchurch, C.B.; Engel, J.N.; Tolker-Nielsen, T. Roles of type IV pili, flagellum-mediated motility and extracellular DNA in the formation of mature multicellular structures in *Pseudomonas aeruginosa* biofilms. *Environ. Microbiol.* **2008**, *10*, 2331–2343.
57. Heydorn, A.; Ersbøll, B.; Kato, J.; Hentzer, M.; Parsek, M.R.; Tolker-Nielsen, T.; Givskov, M.; Molin, S. Statistical analysis of *Pseudomonas aeruginosa* biofilm development: Impact of mutations in genes involved in twitching motility, cell-to-cell signaling, and stationary-phase sigma factor expression. *Appl. Environ. Microbiol.* **2002**, *68*, 2008–17.
58. Dhevan, V.; Griswold, J.A.; Burrowes, B.H.; Colmer-Hamood, J.A.; Williams, S.C.; Carty, N.L.; Hamood, A.N.; Hammond, A.; Schaber, J.A. Diversity of biofilms produced by quorum-sensing-deficient clinical isolates of *Pseudomonas aeruginosa*. *J. Med. Microbiol.* **2007**, *56*, 738–48.
59. Purevdorj, B.; Costerton, J.W.; Stoodley, P. Influence of hydrodynamics and cell signaling on the structure and behavior of *Pseudomonas aeruginosa* biofilms. *Appl. Environ. Microbiol.* **2002**, *68*, 4457–64.
60. Parsek, M.R.; Greenberg, E.P. Sociomicrobiology: The connections between quorum sensing and biofilms. *Trends Microbiol.* **2005**, *13*, 27–33.
61. Ueda, A.; Wood, T.K. Connecting quorum sensing, c-di-GMP, Pel polysaccharide, and biofilm formation in *Pseudomonas aeruginosa* through tyrosine phosphatase TpbA (PA3885). *PLoS Pathog.* **2009**, *5*, e1000483.
62. Davenport, P.W.; Griffin, J.L.; Welch, M. Quorum sensing is accompanied by global metabolic changes in the opportunistic human pathogen *Pseudomonas aeruginosa*. *J. Bacteriol.* **2015**, *197*, 2072–82.
63. Farrow, J.M.; Pesci, E.C. Two distinct pathways supply anthranilate as a precursor of the *Pseudomonas* quinolone signal. *J. Bacteriol.* **2007**, *189*, 3425–33.
64. Palmer, K.L.; Aye, L.M.; Whiteley, M. Nutritional cues control *Pseudomonas aeruginosa* multicellular behavior in cystic fibrosis sputum. *J. Bacteriol.* **2007**, *189*, 8079–87.
65. Boyle, K.E.; Monaco, H.; van Ditmarsch, D.; Deforet, M.; Xavier, J.B. Integration of metabolic and quorum sensing signals governing the decision to cooperate in a bacterial social trait. *PLOS Comput. Biol.* **2015**, *11*, e1004279.
66. de Vargas Roditi, L.; Boyle, K.E.; Xavier, J.B. Multilevel selection analysis of a microbial social trait. *Mol. Syst. Biol.* **2014**, *9*, 684–84.
67. Xavier, J.B.; Kim, W.; Foster, K.R. A molecular mechanism that stabilizes cooperative secretions in *Pseudomonas aeruginosa*. *Mol. Microbiol.* **2011**, *79*, 166–79.
68. Boyle, K.E.; Monaco, H.T.; Deforet, M.; Yan, J.; Wang, Z.; Rhee, K.; Xavier, J.B. Metabolism and the evolution of social behavior. *Mol. Biol. Evol.* **2017**, *34*, 2367–79.
69. Worlitzsch, D.; Tarran, R.; Ulrich, M.; Schwab, U.; Cekici, A.; Meyer, K.C.; Birrer, P.; Bellon, G.; Berger, J.; Weiss, T.; et al. Effects of reduced mucus oxygen concentration in airway *Pseudomonas* infections of cystic fibrosis patients. *J. Clin. Invest.* **2002**, *109*, 317–25.
70. Henry, Y.; Bessières, P. Denitrification and nitrite reduction: *Pseudomonas aeruginosa* nitrite-reductase. *Biochimie.* **1984**, *66*, 259–89.
71. Cutruzzolà, F.; Frankenberg-Dinkel, N. Origin and impact of nitric oxide in *Pseudomonas aeruginosa* biofilms. *J. Bacteriol.* **2016**, *198*, 55–65.
72. Fang, F.C. Perspectives series: Host/pathogen interactions. Mechanisms of nitric oxide-related antimicrobial activity. *J. Clin. Invest.* **1997**, *99*, 2818–25.
73. Yoon, S.S.; Hennigan, R.F.; Hilliard, G.M.; Ochsner, U.A.; Parvatiyar, K.; Kamani, M.C.; Allen, H.L.; DeKievit, T.R.; Gardner, P.R.; Schwab, U.; et al. *Pseudomonas aeruginosa* anaerobic respiration in biofilms: Relationships to cystic fibrosis pathogenesis. *Dev. Cell.* **2002**, *3*, 593–603.
74. Toyofuku, M.; Nomura, N.; Fujii, T.; Takaya, N.; Maseda, H.; Sawada, I.; Nakajima, T.; Uchiyama, H. Quorum sensing regulates denitrification in *Pseudomonas aeruginosa* PAO1. *J. Bacteriol.* **2007**, *189*, 4969–72.
75. Toyofuku, M.; Nomura, N.; Kuno, E.; Tashiro, Y.; Nakajima, T.; Uchiyama, H. Influence of the *Pseudomonas* quinolone signal on denitrification in *Pseudomonas aeruginosa*. *J. Bacteriol.* **2008**, *190*, 7947–56.
76. Hammond, J.H.; Dolben, E.F.; Smith, T.J.; Bhuju, S.; Hogan, D.A. Links between Anr and quorum sensing in *Pseudomonas aeruginosa* biofilms. *J. Bacteriol.* **2015**, *197*, 2810–2820.
77. Redfield, R.J. Is quorum sensing a side effect of diffusion sensing? *Trends Microbiol.* **2002**, *10*, 365–70.

78. Hense, B.A.; Kuttler, C.; Müller, J.; Rothballer, M.; Hartmann, A.; Kreft, J.-U. Does efficiency sensing unify diffusion and quorum sensing? *Nat. Rev. Microbiol.* **2007**, *5*, 230–39.
79. Cornforth, D.M.; Popat, R.; McNally, L.; Gurney, J.; Scott-Phillips, T.C.; Ivens, A.; Diggle, S.P.; Brown, S.P. Combinatorial quorum sensing allows bacteria to resolve their social and physical environment. *Proc. Natl. Acad. Sci. USA.* **2014**, *111*, 4280–4.
80. Lee, J.; Zhang, L. The hierarchy quorum sensing network in *Pseudomonas aeruginosa*. *Protein Cell.* **2014**, *6*, 26–41.
81. Hense, B.A.; Schuster, M. Core principles of bacterial autoinducer systems. *Microbiol. Mol. Biol. Rev.* **2015**, *79*, 153–69.
82. Welsh, M.A.; Blackwell, H.E. Chemical probes of quorum sensing: From compound development to biological discovery. *FEMS Microbiol. Rev.* **2016**, *40*, 774–94.
83. Balasubramanian, D.; Schneper, L.; Kumari, H.; Mathee, K. A dynamic and intricate regulatory network determines *Pseudomonas aeruginosa* virulence. *Nucleic Acids Res.* **2013**, *41*, 1–20.
84. Galán-Vásquez, E.; Luna, B.; Martínez-Antonio, A. The regulatory network of *Pseudomonas aeruginosa*. *Microb. Inform. Exp.* **2011**, *1*, 3.
85. Asfahl, K.L.; Schuster, M. Additive effects of quorum sensing anti-activators on *Pseudomonas aeruginosa* virulence traits and transcriptome. *Front. Microbiol.* **2018**, *8*, 2654.
86. Milo, R.; Shen-Orr, S.; Itzkovitz, S.; Kashtan, N.; Chklovskii, D.; Alon, U. Network motifs: Simple building blocks of complex networks. *Science.* **2002**, *298*, 824–7.
87. Alon, U. Network motifs: Theory and experimental approaches. *Nat. Rev. Genet.* **2007**, *8*, 450–461.
88. Bondí, R.; Longo, F.; Messina, M.; D'Angelo, F.; Visca, P.; Leoni, L.; Rampioni, G. The multi-output incoherent feedforward loop constituted by the transcriptional regulators LasR and RsaL confers robustness to a subset of quorum sensing genes in *Pseudomonas aeruginosa*. *Mol. Biosyst.* **2017**, *13*, 1080–89.
89. Pérez-Velázquez, J.; Gölgeli, M.; García-Contreras, R. Mathematical modelling of bacterial quorum sensing: A review. *Bull. Math. Biol.* **2016**, *78*, 1585–1639.
90. Dockery, J.; Keener, J.P. A mathematical model for quorum sensing in *Pseudomonas aeruginosa*. *Bull. Math. Biol.* **2001**, *63*, 95–116.
91. Viretta, A.U.; Fussenegger, M. Modeling the quorum sensing regulatory network of human-pathogenic *Pseudomonas aeruginosa*. *Biotechnol. Prog.* **2004**, *20*, 670–78.
92. Schaadt, N.S.; Steinbach, A.; Hartmann, R.W.; Helms, V. Rule-based regulatory and metabolic model for quorum sensing in *P. aeruginosa*. *BMC Syst. Biol.* **2013**, *7*.
93. Alfiniyah, C.; Bees, M.A.; Wood, A.J. Pulse generation in the quorum machinery of *Pseudomonas aeruginosa*. *Bull. Math. Biol.* **2017**, *79*, 1360–89.
94. Wu, F.; Menn, D.J.; Wang, X. Quorum-sensing crosstalk-driven synthetic circuits: From unimodality to trimodality. *Chem. Biol.* **2014**, *21*, 1629–38.
95. Brenner, K.; Karig, D.K.; Weiss, R.; Arnold, F.H. Engineered bidirectional communication mediates a consensus in a microbial biofilm consortium. *Proc. Natl. Acad. Sci.* **2007**, *104*, 17300–304.
96. Balagaddé, F.K.; Song, H.; Ozaki, J.; Collins, C.H.; Barnet, M.; Arnold, F.H.; Quake, S.R.; You, L. A synthetic *Escherichia coli* predator–prey ecosystem. *Mol. Syst. Biol.* **2008**, *4*, 187.
97. Tamsir, A.; Tabor, J.J.; Voigt, C.A. Robust multicellular computing using genetically encoded NOR gates and chemical "wires." *Nature.* **2011**, *469*, 212–15.
98. Rampioni, G.; D'Angelo, F.; Messina, M.; Zennaro, A.; Kuruma, Y.; Tofani, D.; Leoni, L.; Stano, P. Synthetic cells produce a quorum sensing chemical signal perceived by *Pseudomonas aeruginosa*. *Chem. Commun.* **2018**, *54*, 2090–93.
99. Kylilis, N.; Stan, G.-B.; Polizzi, K.M. Tools for engineering coordinated system behaviour in synthetic microbial consortia. *bioRxiv.* **2017**, 231597.
100. Scott, S.R.; Hasty, J. Quorum sensing communication modules for microbial consortia. *ACS Synth. Biol.* **2016**, *5*, 969–77.
101. Wen, K.Y.; Cameron, L.; Chappell, J.; Jensen, K.; Bell, D.J.; Kelwick, R.; Kopniczky, M.; Davies, J.C.; Filloux, A.; Freemont, P.S. A cell-free biosensor for detecting quorum sensing molecules in *P. aeruginosa* -infected respiratory Samples. *ACS Synth. Biol.* **2017**, *6*, 2293–301.
102. Moore, J.D.; Rossi, F.M.; Welsh, M.A.; Nyffeler, K.E.; Blackwell, H.E. A comparative analysis of synthetic quorum sensing modulators in *Pseudomonas aeruginosa*: New insights into mechanism, active efflux susceptibility, phenotypic response, and next-generation ligand design. *J. Am. Chem. Soc.* **2015**, *137*, 14626–639.
103. Pérez-Pérez, M.; Jorge, P.; Pérez Rodríguez, G.; Pereira, M.O.; Lourenço, A. Quorum sensing inhibition in *Pseudomonas aeruginosa* biofilms: New insights through network mining. *Biofouling.* **2017**, *33*, 128–142.
104. Parsek, M.R.; Greenberg, E.P.; Iglewski, B.; Greenberg, E.; Ausubel, F.; Bassler, B.L. Acyl-homoserine lactone quorum sensing in Gram-negative bacteria: A signaling mechanism involved in associations with higher organisms. *Proc. Natl. Acad. Sci.* **2000**, *97*, 8789–793.
105. Lazenby, J.J.; Griffin, P.E.; Kyd, J.; Whitchurch, C.B.; Cooley, M.A. A quadruple knockout of lasIR and rhlIR of *Pseudomonas aeruginosa* PAO1 that retains wild-type twitching motility has equivalent infectivity and persistence to PAO1 in a mouse model of lung infection. *PLoS One.* **2013**, *8*, e60973.

7

Quorum Sensing in Vibrios

H.A.D. Ruwandeepika, G.C.P. Fernando, J.L.P.C. Randika, and T.S.P. Jayaweera

CONTENTS

7.1 Introduction

The Vibrionaceae is a large family of Gram-negative gammaproteobacteria having morphology of short rods which are often curved or comma-shaped. They are ubiquitous in nature, occupying a large variety of ecological niches including human and animal gastrointestinal tracts, on the surface of chitinous organisms, aquatic environments, such the water column, in association with hosts (both pathogenic and symbiotic), and even in extreme habitats as biofilms (Thompson et al. 2004). According to the Bergey's Manual, the family Vibrionaceae is composed of eight genera: *Vibrio* (65 species), *Allomonas* (1 species), *Catenococcus* (1 species), *Enterovibrio* (2 species), *Grimontia* (1 species), *Listonella* (2 species), *Photobacterium* (8 species), and *Salinivibrio* (1 species). There are new additions to the vibrionacae family with the use of molecular techniques, hence the taxonomic improvements. On the basis of concatenated 16S rRNA, recA, and rpoA gene sequences and phenotypic data, the Vibrionaceae family has been split into four families: Enterovibrionaceae (comprising the genera Enterovibrio and Grimontia), Photobacteriaceae (comprising the genus Photobacterium), and Salinivibrionaceae (comprising the genus Salinivibrio) (Thompson et al. 2004). Recently Swabe et al. (2013) reported that there are 142 species in the Vibrionaceae family and they are classified into seven genera of *Aliivibrio, Echinimonas, Enterovibrio, Grimontia, Photobacterium, Salinivibrio*, and *Vibrio* (Swabe et al. 2013). Reconstructing the evolutionary history of the genus (Swabe et al. 2007), defined 14 different clades of Vibrio using multilocus sequence analysis of nine genes. It was further updated by Swabe and his group in 2013, and supertree reconstruction was done using the basis of split decomposition analysis defining eight new clades to the family. Fourteen clades that previously defined were Photobacterium clade, the Salinivibrio clade, the Splendidus clade, the Nereis clade, the Orientalis clade, the Coraliilyticus clade, the Scophthalmi clade, the Diazotrophicus clade, the Cholerae clade, the Anguillarum clade, the Vulnificus clade, the Halioticoli clade, the Fischeri clade, and the Harveyi clade (or Vibrio core group) (Swabe et al. 2007) and Damselae, Mediterranei, Pectenicida, Phosphoreum, Profundum, Porteresiae, Rosenbergii, and Rumoiensis are the eight new clades (Swabe et al. 2013). Vibrios possess biological and genetic plasticity which aid them to live in diverse environments having high tolerability and adaptability to the surroundings (Swabe et al. 2007, 2013). In the event of evolution and speciation of Vibrios, there can be different genomic events such as mutations, chromosomal rearrangements, loss of genes by deletion, and gene acquisitions through duplication or horizontal transfer (Gogarten et al. 2002; Henke and Bassler 2004c). Horizontal gene transfer is one of the effective mechanisms in introducing new phenotypes into the genome of bacteria. Several studies revealed the horizontal gene transfer involved in the distribution of virulence genes among vibrios (McCarter 1999; Bai et al. 2008; Snoussi et al. 2008; Ruwandeepika et al. 2010).

The family Vibrio is a versatile group of organisms containing few numbers of human pathogens, and only 12 species are known to be of clinical significance, that is *Vibrio cholerae*, *V. parahaemolyticus*, *V. mimicus*, *V. hollisae*, *V. fluvialis*, *V. furnissii*, *V. vulnificus*, *V. alginolyticus*, *V. damselae*, *V. metschnikovii*, *V. cincinnatiensis*, and *V. harveyi* (Lin et al. 1993; Kim et al. 2003a; Milton 2006) and also the pathogens of aquatic organisms such as *Vibrio harveyi*, *V. campbellii*, and *V. alginolyticus*, which cause multibillion-dollar losses to the aquaculture industry (Austin and Zhang 2006). Not all these vibrios are pathogenic to human or animals and some of them are important in nutrient regeneration. Some vibrios are involved in chitin degeneration exhibiting enzymatic degradation abilities which helps for the balance of the ecosystem (Defoirdt et al. 2010). Some of the vibrios have symbiotic relationships and are localized in light organs of fish and squid (Williams et al. 2007). Some vibrios have the ability to biodegradade the polycyclic aromatic hydrocarbons in polluted marine sediments (Kostka et al. 2011). Having an immense role in the ecosystem, vibrios live in diverse and dramatically changing environments which can be stimulatory or inhibitory for them. The genomic diversity within the vibrios might help for the survival within the different ecosystems, and survival depends on the efficient adaptation of cells to changes in environmental conditions. Vibrios have shown different adaptive mechanisms such as starvation adaptation, the viable but nonculturable (VBNC) response, biofilm formation, virulence factor secretion, bioluminescence, antibiotic production, sporulation, Type Three Secretion System, quorum sensing, etc. (Henke and Bassler 2004a; Ruwandeepika et al. 2015; Papenfort and Bassler 2016). Quorum sensing is the bacterial cell to cell communication, or sensing each other through signaling molecules, which help bacteria to collectively change behavior by altering gene expressions in response to the cell density in the surrounding environment. Quorum sensing plays an important role in both pathogenic and symbiotic bacteria-host interactions (Boyer 2009). It is reported that many pathogenic bacteria use quorum sensing for their virulence, and it regulates the expression of virulence genes which help them to combact the host defenses (Geske et al. 2007; Jayaraman and Wood 2008; Raina et al. 2009; Ruwandeepika et al. 2011, 2015).

7.2 Quorum Sensing

Quorum sensing (QS) is a process of the bacterial cell to cell communication which involves the production, detection, and response to extracellular signaling molecules which are small diffusible molecules termed autoinducers (AIs). Quorum sensing was first discovered and described in the early 1970s in two luminous marine bacterial species, *V. fischeri* and *V. harveyi* (Ruby 1996; Hastings and Greenberg 1999; Miller and Bassler 2001). Since then, QS circuits have been identified in many bacteria including in Gram-negative and Gram-positive species (Bassler 1999; Rutherford and Bassler 2012). This mechanism coordinates the expression of certain genes in response to the concentration of signal molecules (Henke and Bassler 2004a, 2004b). Quorum sensing bacteria produce and release chemical signal molecules (autoinducers) which concentration is proportional to cell density (Waters and Bassler 2005). When the concentration of autoinducers reaches a threshold level, it stimulates a cascade of reactions and it upregulates or downregulates the expression of several genes. Gram-positive and Gram-negative bacteria use quorum sensing communication circuits to regulate many physiological activities including symbiosis, virulence, competence, conjugation, antibiotic production, motility, sporulation, and biofilm formation. Gram-negative bacteria use acylated homoserine lactones as autoinducers, and the quorum sensing information is often integrated by small RNAs (sRNAs) (Papenfort and Vogel 2010; Papenfort and Bassler 2016). Gram-positive bacteria use processed oligo-peptides and two-component systems, which consist of membrane-bound sensor kinase receptors and cytoplasmic transcription factors to communicate, which alter the gene expression (Miller and Bassler 2001; Papenfort and Bassler 2016). These signal molecules are synthesized intracellularly and are secreted outside the cells either passively or actively. When autoinducers accumulate above the threshold level, i.e., required for detection, cognate receptors bind the autoinducers and trigger signal transduction cascades that result in population-wide changes in gene expression, and the organisms act collectively.

7.2.1 Quorum Sensing in *Vibrio harveyi*

The quorum-sensing system of *V. harveyi* has been studied intensively, and it uses a three-channel quorum sensing system mediated by different autoinducers, namely, HAI-1 (Harveyi autoinducer 1) (Cao and Meighen 1989), AI-2 (Auto inducer 2) (Chen et al. 2002), and CAI-1 (Cholera autoinducer 1) (Higgins et al. 2007) for the control of expression of genes responsible for bioluminescence and other traits (Henke and Bassler 2004c). HAI-1 (a homoserine lactone), a species-specific molecule, is synthesized by LuxM (Bassler et al. 1993) while AI-2 is a furanosyl borate diester synthesized by the LuxS enzyme, and CAI-1 ((S)-3-hydroxytridecan-4-one) (Higgins et al. 2007) is a genus-specific signal produced by CqsA. The autoinducers are detected at the cell surface by distinct membrane-bound histidine sensor kinase proteins which feed a common phosphorylation/dephosphorylation signal transduction cascade (Taga and Bassler 2003). LuxN detects HAI-1, LuxQ responds to AI-2 via the periplasmic protein LuxP, and CqsS detects CAI-1 (Henke and Bassler 2004c; Miller et al. 2002). These three sensory proteins deliver phosphate to LuxO via LuxU (Lilley and Bassler 2000; Freeman and Bassler 1999a, 1999b) in the absence of autoinducers. Phosphorylated LuxO is active and in conjunction with $\sigma 54$ activates the expression of genes encoding small regulatory RNAs. These together with the RNA chaperone Hfq destabilize the mRNA encoding the transcriptional regulator protein LuxR (Lilley and Bassler 2000; Lenz et al. 2004). At high signal molecule concentration, LuxO is dephosphorylated (sensors are switched from kinase mode to phosphate mode in the presence of their cognate signal) and dephosphorylated LuxO is inactive and consequently, the expression of luxR takes place (Freeman and Bassler 1999a).

LuxR is a transcriptional regulator protein which acts as the master regulator in the quorum sensing system of *V. harveyi* (Henke and Bassler 2004c). It directly activates the Lux operon whereas the majority of other genes are controlled indirectly (Waters and Bassler 2006). The expression of several virulence factors is regulated by quorum sensing. LuxR has been shown to activate the expression of an extracellular toxin, metalloprotease, and a siderophore and to inactivate the expression of a type III secretion system and chitinase (Lilley and Bassler 2000; Henke and Bassler 2004a; Ruwandeepika et al. 2011, 2015). Also the

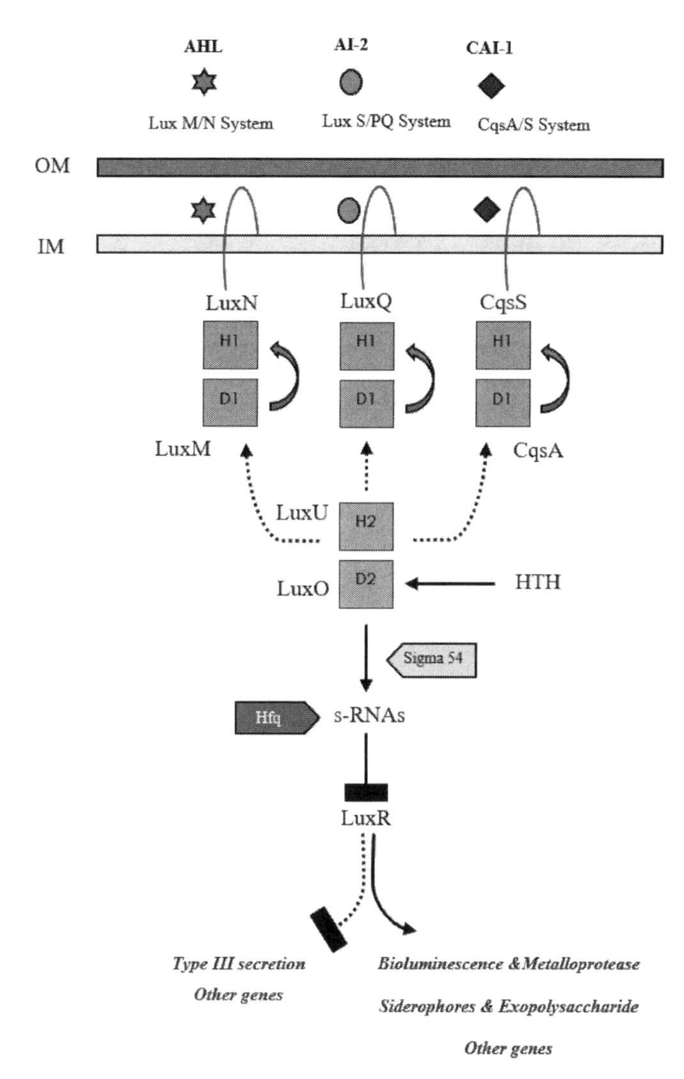

FIGURE 7.1 Model of the *Vibrio harveyi* quorum sensing circuits. For a detailed account of the model, refer to the corresponding text. At low cell populations, less or no signal molecules are made by LuxM, LuxS, and CqsA which are detected via LuxN, LuxP/Q, and CqsS. At low cell densities the AI molecules are auto phosphorylated and LuxO is activated through transfer of phosphates via LuxU that will combinedly trigger a sRNA activation sequence with sigma-54. sRNAs together with Hfq chaperone result in destabilization of mRNAs encoding for LuxR. At high cell populations, the receptors will switch from kinases to phosphatases which dephosphorylate LuxO, triggering a reverse cascade of events. Transportation systems are not depicted in the figure. Dashed arrows indicate phosphoryl group transfers. Solid single arrows indicate positive inducing effects and solid lines with a bar at the end indicate negative inducing effects.

quorum sensing has been shown to regulate virulence of luminescent vibrios toward different hosts *in vivo* (Figure 7.1).

7.2.2 Quorum Sensing in *Vibrio fischeri*

Bioluminescence is one of the manifestations of the quorum sensing mechanism, and *V. fischeri* was the first organism that was discovered as the luminescence producing organism (Nealson et al. 1970). *V. fischeri* is a marine symbiotic organism, and it forms a symbiont with Hawaiian bobtail squid (Euprymna scolopes) (Milton 2006; Abisado et al. 2018). The bacteria dwell in the crypts of the light organ of the Hawaiian bobtail squid, and the squid uses the bioluminescence of the bacteria both as an antipredation tactic as well as a method of attracting prey. During the night time, the squid's reflector is focused downwards and the lens system and the aperture of the light organ are regulated as per the intensity of light received from the top surface light-sensitive cells, which aids in mimicking the moon or starlight. Usually, this is used to reduce the formation of a shadow when the squid hunts in the open shallow waters during the night time. The colonization of bobtail squid's crypts occurs hours after hatching, where it triggers a series of developmental changes in both the squid and the bacteria, which stays in symbiosis with the squid to its death (Milton 2006).

Quorum sensing based on the lux system was first described in *V. fischeri* as the principal process regulating luminescence expression. The quorum sensing mechanism of *V. fischeri* is actually two quorum sensing mechanisms, i.e., ain and lux, using acyl homoserine lactones as signaling molecules.

The proteins required for the bioluminescent activities are encoded within the luxICDABEG/Luciferase operon. Light is emitted during the oxidation reaction that converts long chain fatty acids (RCHO) to flavin mononucleotide ($FMNH_2$) and oxygen into aliphatic acids (RCOOH) and FMN via an enzyme, luciferase (Whiteley et al. 2017). This enzyme has two subunits, α and β, which are respectively transcribed of luxA and luxB genes operon. A reductase complex composed of proteins expressed from luxCDE genes synthesizes the substrate RCHO, and LuxG converts FMN to $FMNH_2$. Although multiple QS systems are involved with regulation of luxICDABEG operon, the LuxI-LuxR system is identified as the predominant quorum sensing circuit on control where LuxI (encoded in luxI gene) is the synthase of the autoinducer N-(3-oxo-hexanoyl)-L-homoserine lactone (3-oxo-C6-HSL), and it is identified to bind and activate its cognate receptor or the transcription factor LuxR (encoded in luxR gene). This bound complex regulates the luxICDABEG as a dimer upstream, which recruits the RNA polymerase to trigger the transcription of the operon (Schaefer et al. 1996) *V. fischeri*, AinS-AinR, and LuxS-LuxP/Q are responsible for regulating the bioluminescent mechanism in an indirect upstream basis through the regulation of modulating the transcription of the luxR gene. AinS is the synthase for N-octanoyl-homoserine-lactone (C8-HSL) that is activating the histidine kinase AinR. Lux S synthesizes the autoinducer 2 (AI-2) and the AI2 bind with periplasmic porin protein LuxP to form a complex with the LuxQ (histidine kinase in the inner membrane). AI-2 binding is known to elevate the rotational shift of the corresponding dimer of LuxQ that inhibits kinase activity (Milton 2006). AinR and LuxP/Q act in parallel by controlling the phosphorylation stage of LuxU, which is an intermediate component of the phosphorylation cycle and also a controller of the LuxO response regulator. When the cell density is below the threshold limits, the AHL signal is weak and the histidine kinases initiates the process of phosphorylating the LuxO that activates the transcription of qrr1 that transcribes for a sRNA, which is known to post-translationally repress the LitR transcription factor through an RNA chaperone Hfq. This is reversed at high cell density stabilizing the litR transcript mRNA. Produced LitR will enhance luxR expression and thereby elevate the light production. Apart from the bioluminescence, certain functions like production of early and late colonization factors is also regulated via a mechanism of quorum sensing. This is identified as regulated by the AinS-AinR QS system.

The two signal systems, AinS-AinR and LuxI-LuxR, synergistically induce a phosphorelay, likely via LuxO to relieve repression of LitR, a *V. harveyi* LuxR homologue (Miyamoto et al. 2003; Lupp and Ruby 2004). Repression of LitR is anticipated to occur via destabilization of litR mRNA similar to that of luxR in *V. harveyi*. LitR activates the LuxI-LuxR system linking the AinS-AinR and LuxS-LuxP/Q systems to the LuxI-LuxR system. The QS system of *V. fischeri* is largely significant from others of the Vibrionaceae, and the LuxI-LuxR system is only present in the Aliivibrio clade. When considering the number of qrr genes, *V. fischeri* has only one qrr gene, but in *V. harveyi* there are multiple qrr sRNAs that encode the expression of LitR homologues. At the same time, it has been identified that the QS networks are converged and function in cross connection among different autoinducers. C8-HSL, which is produced by the AinS synthase system, can also directly bind to LuxR, producing a LuxR-C8-HSL complex. This is not quite stable as LuxR-3-oxo-C6-HSL that activates the transcription cascade of lux genes. At higher concentrations, C8-HSL may therefore function as a competitive inhibitor which suppresses bioluminescence (Lupp et al. 2003). At low cell densities, luminescence gene regulation is repressed through LuxO, a negative modulator of the transcriptional regulator LitR. At the intermediate cell densities achieved in laboratory culture, the AinS-synthesized signal has two effects: induction of luminescence gene expression through direct interaction with LuxR and inactivation of LuxO; the second effect results in increased transcription of litR. LitR positively regulates transcription of *V. fischeri* luxR, thereby connecting the two quorum-sensing systems ain and lux and allowing the sequential induction of luminescence gene expression with increasing cell density. Inactivation of the *V. fischeri* ain system results in a luminescence defect and affects colonization competence. The *V. fischeri* LuxS signal AI-2 was shown to be involved in the regulation of luminescence expression and colonization competence as well; however, compared to that of ain and lux quorum sensing, the impact of AI-2 is recognized as quantitatively small (Miyamoto et al. 2003). The QS systems also consist of feedback loops that regulate the overall dynamics of the system. The autoinducer 3-oxo-C6-HSL promotes its own synthesis as the luxI gene is upregulated by the LuxR/3-oxo-C6-HSL. In the same manner it has been discovered that the C8-HSL also has a positive feedback loop involved in its QS circuit as ainS expression is regulated through LitR. Another biological activity mediated via QS in *V. fischeri* is its flagellar motility (Lupp and Ruby 2004). This occurrs in part by the LuxO-Qrr1-LitR cascade where the motility is enhanced under lower cell densities that repress the bioluminescence. As stated in prior phosphorylation of LuxO/expression of qrr1 leads to enhanced motility. Recently an epistasis experiment was done which revealed that litR is the transcription factor that mediates the effect on motility by LuxO and Qrr1. AinS has also been recognized as an acs expression regulator that encodes acetyl-CoA synthetase and as a result can control acetate metabolism.

Very recently it was identified that the LuxP/Q complex has an influence on biofilm formation of *V. fischeri* that is a process mediated by an 18-gene symbiosis polysaccharide (syp) locus. Using a model, reearchers suggest that two systems are arranged to allow the induction or repression of colonization factors important during specific temporal phases of the symbiotic relationship. In their model, the fact that the ain system is functional and essential at the early stages of colonization is focused, where they have proven the lux system is neither fully induced nor required at the time of early colonization, suggesting that the ain system is operative at a lower threshold cell density than the lux system. This provides evidence for a stepwise pattern in their control during the onset of symbiosis (Lupp and Ruby 2005). The sequential nature of quorum-sensing induction is further supported by a previous finding that in symbiotic strains of *V. fischeri*, the effect of the ain system on luminescence is apparent at the relatively moderate bacterial concentrations occurring in culture, whereas the lux system is the predominant inducer of luminescence expression at the very high cell densities achieved only within the squid light organ (Lupp and Ruby 2004; Turovskiy et al. 2007) (Figure 7.2).

7.2.3 Quorum Sensing in *Vibrio cholerae*

V. cholerae is a natural inhabitant of brackish riverine, estuarine, and coastal waters and few strains are known to be human pathogens, causing life-threatening diarrhea (Almagro-Moreno and Taylor 2013). It is the causative agent of cholera, which is a devastating, severely dehydrating diarrheal disease, and the cholera epidemic is mainly by the O1 serogroup two biotypes known as classical and El Tor (Dziejman et al. 2002; Tsou and Zhu 2010) whereas a new serogroup, O139, has been reported an association with cholera epidemics (Chowdhury et al. 2015). Pathogenic *V. cholera* produces several kinds of virulence factors, mainly the cholera toxin (CT), toxin-coregulated pilus (TCP). Cholera toxin (CT), a factor which is responsible for the profuse watery diarrhea, is an ADP-ribosylating protein that induces the secretion of massive amounts of fluid from intestinal epithelial cells, whereas toxin-coregulated pilus (TCP) is necessary for efficient colonization (Childers and Klose 2007). Hemolysin encoded by hlyA, a hemagglutinin (HA)/protease encoded by hapA, and a multifunctional autoprocessing RTX toxin are some of the toxins produced by the organism hlyA having a vital role in lethality (Olivier et al. 2007). In addition, they have type three and four secretion systems playing roles in virulence. Virulence regulation is regulated by two inner membrane-localized regulators, ToxR and TcpP, required to transcribe the toxT gene, which encodes the master regulator of virulence in choleragenic *V. cholerae*. ToxT is a transcriptional regulator required for the expression of the ctxAB operon, which encodes the two subunits of CT, and the tcp operon, which encodes TCP (Childers and Klose 2007). *V. cholera* is known to use the QS mechanism to regulate the virulence factor production, biofilm formation, Type VI secretion, and competence development, which are important for survival and adaptation inside and outside of its human hosts (Jung et al. 2015). *V. cholerae* must shift between two distinct environmental conditions, free-floating in the environment and attachment to a host, which in turn requires switching between two phases through differential gene expression patterns. These shifts have been identified to occur under the aid of both quorum sensing and a secondary regulatory system, 3′5′-cyclic diguanylic acid (c-di-GMP) signaling cascade. Two main QS systems have been identified in *V. cholarae* which are of greater similarity to that of *V. harveyi*. Namely, these are the LuxS-LuxP/Q system and the CqsA/S system (Miller et al. 2002). Though it is not fully elucidated, a recent study has proposed two additional autoinducer sensors in *V. cholerae*, namely VpsS and CqsR, but the signals sensed

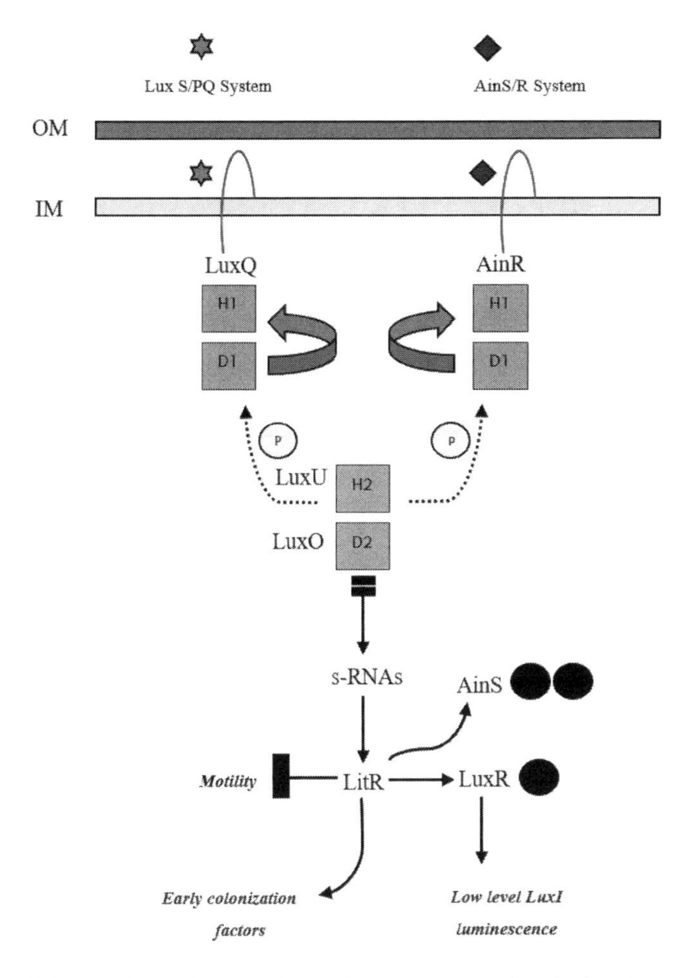

FIGURE 7.2 Model of the *Vibrio fischeri* quorum sensing circuits. For a detailed account of the model, refer to the corresponding text. The model for low cell population is predicted to function similar to that of *Vibrio harveyi*. Thus, only the models for intermediate and high cell population density are shown to indicate differences between species. The low cell densities LuxI will only produce low levels of the QS signals, N-(3-oxohexanoyl)-L-homoserine lactone (3-oxo-C6-HSL), yet at higher cell densities the signaling molecules will reach their critical concentrations and binds with the LuxR causing the DNA binding domain to unravel. The LuxR-AHL complex directly activates transcription of the *luxICDABEG* genes and activates genes required for the late stage of colonization; this results an exponential increase in 3-OXO-C6-HSL leading to elevated bioluminescence. LuxS synthesizes an AI-2 signal that is similar to that of *Vibrio harveyi* and sensed by LuxP and LuxQ. AinS synthase produces N-octanoyl-L-homoserine lactone (C8-HSL) that is sensed by AinR, a *Vibrio harveyi* LuxN homologue. These two signal systems synergistically induce phosphorelays via LuxO to relieve repression of LitR, a *Vibrio harveyi* LuxR homologue. Repression of LitR occur via destabilization of litR mRNA similar to that of luxR in *Vibrio harveyi*. LitR activates the LuxI/R system linking the AinS/R and LuxS/PQ systems to the LuxI/R system. The AinS/R system also starts a low-level induction of lux gene when 3-OXO-C6-HSL is limiting as the AinS signal C8-HSL binds and activates LuxR. Explanation of symbols is given in a diagram key stated with reference to *Vibrio harveyi* QS circuits.

by these sensors are presumably different from the two canonical autoinducers, CAI-1 and AI-2 (Jung et al. 2015). Autoinducer 2 (AI-2) is a furanosyl borate diester whose presence in the cell is detected by LuxP/Q; the cognate receptor and the second system occupies the cholera autoinducer 1 (CAI-1), which is a hydroxylated alkyl ketone, where the cognate receptor is CqsS. CAI-1 is unique to the Vibrio family, where AI-2 is identified in various groups of bacteria (Rutherford and Bassler 2012; Jung et al. 2015). The LuxM/N system which is available in *V. harveyi* is not available in the *V. cholerae* genome, but the CqsA/S system and LuxS-LuxP/Q system function similarly to that of *V. harveyi* in that the sensory information is channeled in parallel through LuxU to the transcriptional activator LuxO (Milton 2006). A *V. cholerae* autoinducer-receptor pair which controls biofilm formation and toxin production has been identified recently (Papenfort et al. 2017). The autoinducer 3,5-dimethylpyrazin-2-ol (DPO) (made from threonine and alanine) binds and activates a transcription factor, VqmA, where the VqmA-DPO complex activates the expression of vqmR, which encodes a small regulatory RNA. VqmR represses genes required for biofilm formation and toxin production (Papenfort et al. 2017).

At lower cell population densities, the AI molecules function as kinases that phosphorylate LuxO, together with Sigma 54, conjugated with LuxP, and activate expression qrr genes encoding for four sRNAs that destabilize hapR mRNA repressing the expression of HapR, a LuxR homologue (Jobling and Holmes 1997; Lenz et al. 2004). In the absence of HapR, the ctx and tcp genes are expressed alongside polysaccharide genes (vps) required for biofilm formation (Miller et al. 2002; Hammer and Bassler 2003), and this also results in the upregulation of the extracellular proteases HapA and PrtV. At higher cell population densities, LuxO is dephosphorylated and inactivated and repression of hapR is terminated, which leads to downregulation of virulence factor expression and biofilm formation, causing bacteria to disperse. Recently it was identified that HapR exerts a second type of control on biofilms formation by binding to vpsT inhibiting the expression of biofilm genes. vpsT is produced under the trigger of c-di-GMP that controls the switch of bacteria between motile and sessile states. c-di-GMP is produced when two GTP molecules cyclize with the loss of two phosphate groups through a process carried out by proteins with GGDEF motifs such as QrgB.

HapR has an influence on this system because it represses the expression of QrgB. Thus, when cell density HapR concentrations are low, more c-di-GMP is produced. The second type of regulation that affects c-di-GMP levels has been identified as the protein VieA, controlled by the promoter V1086, which contains an EAL domain that acts as a c-di-GMP phosphodiesterase, that could degrade the molecule and reduce its overall cellular concentration that catastrophically repress the exopolysaccharides needed for biofilm formation. At the same time, HapR is identified to represses expression of AphA, the key activator of tcpPH (Kovacikova and Skorupski 2002). Thus, TcpP expression is repressed, resulting in repression of ToxT expression and consequently repression of virulence genes. HapR also represses biofilm formation while activating expression of the hemagglutination protease. Biofilm formation is found to increase the acid resistance of *V. cholerae* and thereby a critical factor which aids in entry into the host through the severely acidic stomach environment (Zhu and Mekalanos 2003). Once in the intestinal environment, quorum-sensing signals within the biofilm repress vps, ctx, and tcp expression. The bacterium detaches from the biofilm and colonizes the intestines, where quorum-sensing signals are low and CT and TCP are expressed (Figure 7.3).

7.2.4 Quorum Sensing in *Vibrio parahaemolyticus*

V. parahaemolyticus is a natural inhabitant of the marine ecosystems which causes a seafood-borne human pathogen causing

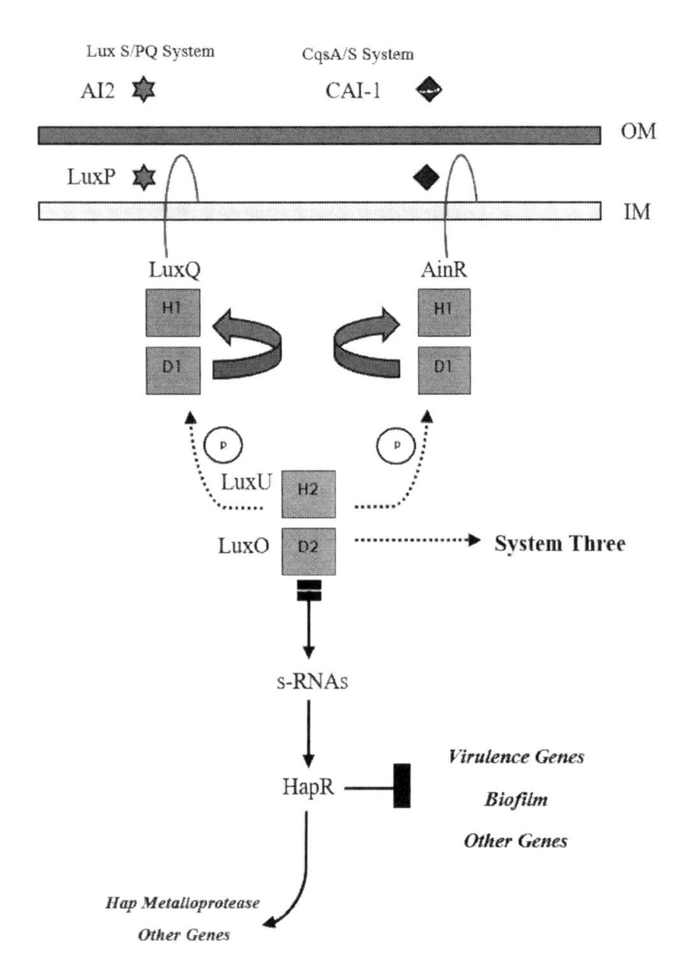

FIGURE 7.3 Model of the *Vibrio cholerae* quorum sensing circuits. For a detailed account of the model, refer to the text. The model for low cell population density is predicted to function similar to that of *Vibrio harveyi*. Thus, only the model at high cell population density is shown for indicating differences between species. The LuxM/N system which is available in *Vibrio harveyi* is not available in *Vibrio cholerae's* genome, but the CqsA/S system and LuxS/PQ system function similarly to that of *Vibrio harveyi* in that the sensory information is channeled in parallel through LuxU to the transcriptional activator LuxO. The third system is believed to channel sensory information through LuxO, and all three systems are believed to converge at LuxO. At lower cell population densities which are well below the critical limits, LuxO, together with s54, activates expression of four sRNAs that destabilize hapR mRNA, repressing the expression of HapR, a LuxR homologue. At the absence of HapR, the *ctx* and *tcp* genes are expressed alongside polysaccharide genes (*vps*) required for biofilm formation. Explanation of symbols is given in a diagram key stated with reference to *Vibrio harveyi* QS circuits.

gastroenteritis (Su and Liu 2007; Broberg et al. 2011). Virulence of the *V. parahaemolyticus* is due to the presence of virulence factors and also the virulence mechanisms such as thermostable direct hemolysin (TDH), the TDH-related hemolysin, and type III secretion systems (Makino et al. 2003). TDH genes and the T3SS2 genes are clustered together as a pathogenicity island Vp-PAI, and this pathogenicity island and the T3SS1 are known to be regulated by the quorum sensing mechanism (Zhang et al. 2019). There is complex regulatory network of QS regulators AphA and OpaR and an AraC-type transcriptional regulator QsvR to control the expression of T3SS1, Vp-PAI genes. The quorum sensing mechanism of the *V. parahaemolyticus* is very similar to *V. harveyi*. It produces three types of autoinducers, termed harveyi autoinducer 1 (HAI-1), autoinducer 2 (AI-2), and cholerae autoinducer 1 (CAI-1), and receptor proteins (histidine kinases) LuxN, LuxP/LuxQ, and CqsS detect the autoinducers HAI-1, AI-2, and CAI-1, respectively (Zhang et al. 2012). Autoinducers HAI-1, AI-2, and CAI-1 are synthesized by LuxM, LuxS, and CqsA, respectively (Henke and Bassler 2004c; Rutherford and Bassler 2012). At low cell density confirmation, concentrations of autoinducers are low and the kinase activity predominates, and a phosphate group from the receptor is transferred to LuxU and the receptor proteins and LuxU initiate a phosphorylation cascade and phosphorylate, LuxO. At low cell density, the sigma 54-dependent response regulator LuxO is active, and it activates the transcription of Qrr sRNAs (Qrr1–5), post-transcriptionally promoting the QS master regulator AphA, while OpaR, the other master regulator, is repressed (McCarter 1998; Tu et al. 2008; Sun et al. 2012; Kalburge et al. 2017; Lin et al. 1993; Zhang et al. 2019). At high cell density, concentration of AIs are higher and the LuxO is dephosphorylated and Qrr1-5 are not transcribed, resulting low levels of Qrr1–5. Hence there is expression of OpaR and but not AphA. OpaR activates transcription of qrr1–5 but represses the transcription of aphA, opaR, and other virulence genes, as well as biofilm-related genes. It has been shown that the AphA and OpaR control hundreds of target genes during QS signal transduction in *V. parahaemolyticus* OpaR; the quorum sensing master regulator in *V. parahemolyticus* positively regulates the capsule polysaccharide (CPS) production and type VI secretion system-2 (T6SS-2) and negatively regulates motility, biofilm, and T3SS-1 and T6SS-1 production (Sun et al. 2012; Zhang et al. 2012; Kalburge et al. 2017) (Figure 7.4).

7.2.5 Quorum Sensing in *Vibrio anguillarum*

V. anguillarum is the causative agent of vibriosis, which is fatal haemorrhagic septicaemia affecting many aquatic organisms such as finfish, crustaceans, and mollusks leading to significant economic losses in the aquaculture industry (Frans et al. 2011; Austin and Austin 2012). These bacteria produce many virulence factors, but the role they play during the infection is not fully elucidated. Some virulence factors such as iron uptake system involving the siderophore anguibactin, chemotactic, and exopolysaccharide production are reported to be the factors related to the pathogenicity, and in addition, haemolysin, lipase, and protease are also present though their role in virulence is unknown (Croxatto et al. 2007; Yang et al. 2007; Li et al. 2014).

A direct relationship between the quorum sensing mechanism of *Vibrio anguillarum* and its virulence is not described well (Milton et al. 1997; Li et al. 2018). *V. anguillarum* is known to have two quorum sensing systems, a classical acylhomoserine lactone (AHL) system involving the signal synthase/receptor pair VanI/VanR and a three-channel system which is found in many vibrios (Milton 2006; Li et al. 2014, 2018). AHL is produced by both the pathogenic and environmental isolates, and Van T is the homologous of LuxR in *V. harveyi*. Two systems are homologous to the LuxM/N and LuxS/PQ systems of *V. harveyi*. The VanM synthase produces N-hexanoyl-L-homoserine lactone and N-(3-hydroxyhexanoyl)-L-homoserine lactone, which are sensed by the hybrid sensor kinase VanN, while VanS produces AI-2 molecules are sensed by VanP and the hybrid sensor kinase VanQ. CAI-1 is synthesized by the CqsA synthase, which is sensed by the hybrid sensor kinase CqsS (Milton et al. 2001;

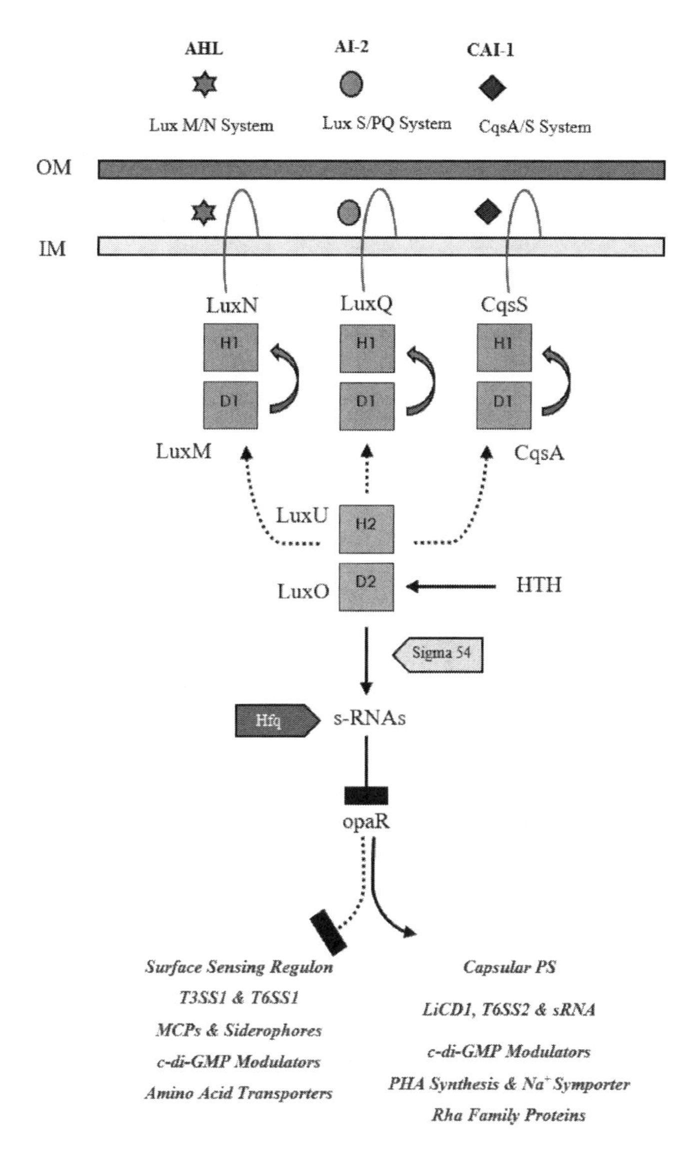

FIGURE 7.4 Model of the *Vibrio parahaemolyticus* quorum sensing circuits. For a detailed account of the model, refer to the text. It is not known if the low cell population model functions similarly to that of *Vibrio harveyi*. However, for simplicity of discussion, it will be predicted to function similarly. Thus, only the model at high cell population density is shown for indicating differences between species. At a low cell density, when the concentrations of autoinducer molecules are also low, sensor histidine kinases phosphorylate the Sigma 54-dependent LuxO regulator via the small histidine phosphorelay protein LuxU. LuxO—P induces transcription of small quorum-regulatory RNAs (Qrrs), and the Qrrs destabilize the mRNA for the central output regulator of the system. At a high cell density in the presence of autoinducers, the histidine kinases become LuxO phosphatases, resulting in an inactive form of LuxO. The Qrrs are no longer transcribed, and the mRNA for the central output regulator is then translated. The output regulon in the case of *Vibrio parahaemolyticus* is the colony opacity termed OpaR. For a detailed account of the model, refer to the text. Explanation of symbols is given in a diagram key stated with reference to *Vibrio harveyi* QS circuits.

Croxatto et al. 2002, 2004). Signal transduction happens through a phosphorylation cascade and regulates the VanT. VanT of *V. anguillarum*, which is a LuxR homologue of *V. harveyi*, acts as a transcriptional regulator, and it positively regulates metalloproteases, pigment production, serine, and biofilm production and negatively regulates type IV secretion system (Croxatto et al. 2002). Another QS mechanism which is similar to the *V. fischeri* LuxI/R system exists in *V. anguillarum*, and VanI synthesises 3-oxo-C10-HSL, which likely binds VanR, a transcriptional activator of VanI. At low cell densities there is low concentration of signal molecules, and this autophosphorylate three hybrid sensor kinases, VanN, VanQ, and CqsS, autophosphorylate. It initiates a phosphorylation cascade to transmit a phosphoryl group to the phosphotransferase VanU, which phosphorylates and activates the σ54-dependent response regulator VanO. Phosphorylated VanO is activated and with the alternative sigma factor σ54, induces expression of small regulatory RNA (sRNA) Qrrl. The Qrr, together with the RNA chaperone Hfq, destabilizes vanT mRNA. It represses the expression of the quorum-sensing transcriptional regulator VanT (Croxatto et al. 2004; Weber et al. 2008). At high cell density, the threshold for the signal molecules is reached, and VanT expression is induced. Binding of the signal molecules to the cognate sensor kinases VanN, VanQ, and CqsS inhibits kinase activity, allowing phosphatase activity to predominate (Table 7.1).

At higher cell densities, a certain threshold of signal molecules is reached and the expression of the vanT is induced. Three types of signal molecules are produced and are sensed, and linking of the signal molecules to the sensor kinases VanN, VanQ, and CqsS inhibits kinase activity, allowing phosphatase activity to predominate. This will result in dephosphorylation and inactivation of VanO. Therefore the, qrrl is not expressed and VanT expression is induced, leading to gene regulation within the quorum-sensing regulon (Croxatto et al. 2004; Weber et al. 2008, 2009; Denkin and Nelson 2004; Higgins et al. 2007) (Figure 7.5).

7.2.6 Quorum Sensing in *Vibrio vulnificus*

V. vulnificus is a natural habitat of aquatic ecosystems mostly in estuarine environments and is an opportunistic pathogen which causes septicemia and gastroenteritis following ingestion or localized infections following contamination of a wound. *V. vulnificus* is composed of three main biotypes: biotype 1 (associated with human disease), biotype 2 (prevails in eel infections), and biotype 3 (found in human vibriosis). This organism possesses many virulence factors such as cytolysin, siderophores, exoenzymes, production of capsular polysaccharide (for survival of the pathogen within the host), extracellular metalloprotease (VVP), haemolysin, protease, hyaluronidase, elastase, chondroitin sulphatase, phospholipase, and mucina.

Though there are not many reports, as in many bacteria, *V. vulnificus* also regulate their phenotypes through a signaling pathway (quorum sensing). In *V. vulnificus*, there are no homologues of AHL synthases or AHL signaling molecules. It is lacking systems like LuxM/N, CqsA/S, LuxI homologue of *V. fischeri* (Kim et al. 2003), and it has a LuxS homologue with AI2 autoinducer activity (Kawase et al. 2004), *V. harveyi* homologues to the LuxS/PQ system, LuxU, LuxO, and the LuxR transcriptional regulator, SmcR, and five small RNAs (Lenz et al. 2004; McDougald et al. 2000; Shao and Hor 2001; Chen et al. 2003). Regulatory gene (SmcR) is reported to play an important role in starvation adaptation and in the regulation of many stationary-phase-regulated genes, including some virulence factors. LuxS and SmcR regulate the metalloprotease, VvpE expression positively and cytolysin VvhA expression negatively (Linkous and Oliver 1999; Shao

TABLE 7.1
Quorum Sensing Systems and Regulated Virulence Genes/Mechanisms in Vibrios

Organism	QS Signal	QS-Regulated Mechanisms/Virulence Factors	References
Vibrio cholerae	AI-2, CAI-1	Toxin-coregulated pilus, exotoxin, cholera toxin, biofilm formation, proteases, hemolysin	Tsou and Zhu (2010); Jung et al. (2015)
Vibrio campbellii	HAI-1, AI-2, CAI-1	Bioluminescence, Toxin T1, Metalloprotease, Siderophore, Chitinase A, Extracellular polysaccharide, Biofilm Formation, Lethality Phospholipase, TTSS (Type three secretion system)	Defoirdt et al. (2006); Haldar et al. (2010); Ruwandeepika et al., (2011)
Vibrio alginolyticus	AI-2	Biosynthesis of Flagella, Proteases, EPS, Biofilm Formation	Rui et al. (2008); Wong et al. (2012); Liu et al. (2012); Ye et al. (2008)
Vibrio anguillarum	HHL, AI-2, CAI-1	Biofilm formation, empA Métalloprotéases, Serine and Glycine Proteases, Melanine Pigment, Hemagglutinine Proteases, Hemolytic Activity, Extracellular Polysaccharides, Lipase	Milton (2006); Croxatto et al. (2002); Frans et al. (2011)
Vibrio fischeri	3OC6-HSL, C8-HSL AI-2	Bioluminescence, Biofilm formation, EPS Secretion, Colonization Patterns	Pérez and Hagen (2010); Pallaval and Sheela (2019)
Vibrio harveyi	3OHC4-HSL, AI-2, CAI-1	Bioluminescence, Biofilm formation, EPS Secretion, Colonization Patterns, Virulence Protein Production, Bioluminescence, Metalloprotease, Siderophore, chitinase, phospholipases, TTSS	Defoirdt et al. (2006); Cao and Meighen (1989); Bassler et al. (1993); Ball et al. (2017)
Vibrio vulnificus	C4-HSL, 3OC6-HSL, 3OHC6-HSL, AI-2	VvpE Metalloprotease, VvhA Cytolysin	Milton (2006)
Vibrio fluvialis	AHL, AI-2, CAI-1	Metalloprotease, vfqR Expression	Kalburge et al. (2017); Wang et al. (2013)
Vibrio parahaemolyticus	AHL, AI-2, CAI-1	Surface Sensing Regulon, T3SS1 and T6SS1, Siderophores, c-di-GMP Modulators, Amino Acid Transporters, Capsular PS, T6SS2 and sRNA, PHA Synthesis and Na Symporter Activity, hemolysins	Kalburge et al. (2017); Gode-Potratz and McCarter (2011)

and Hor 2001; Kim et al. 2003). Murine macrophages challenged with *V. vulnificus* secrete proinflammatory mediators, tumor necrosis factor-alpha, and nitric oxide, which has a role in septic shock and is known to be regulated by both LuxS and SmcR proteins (Shin et al. 2004) (Figure 7.6).

7.2.7 Quorum Sensing in *Vibrio alginolyticus*

V. alginolyticus is one of the most invasive organisms and is a highly fatal fish pathogen in aquaculture, causing vibriosis in the large yellow croaker, sea bream, grouper, and Kuruma prawn and shellfish species, and also a human pathogen causing gastroenteritis, conjunctivitis, and otitis (Chien et al. 2002; Liu et al. 2004; Rossez et al. 2015). The virulence mechanism in *V. alginolyticus* is yet fully unraveled, but it is known to produce two kinds of extracellular proteases (ECP) (alkaline serine protease and collagenase), possesses polar and lateral flagella, and produces hydroxamate type siderophores and three hemolysins: Thermostable Direct Hemolysin (TDH), Thermostable Related Hemolysin (TRH), and Thermolabile Direct Hemolysin (TLH) (Xie et al. 2005; Wong et al. 2012). As with other vibrios, quorum sensing regulates the production of some virulence factors and virulence related genes in *V. alginolyticus* sharing a similar mechanism as in *V. harveyi* (Rui et al. 2008; Tian et al. 2008; Liu et al. 2011). The LuxO-LuxR regulatory circuit is characterized to be closely related to the virulence (Rui et al. 2008), and luxT (TetR family regulator) has been identified as a new regulator playing multiple roles in regulating ECP production and motility in *V. alginolyticus*, which is different from *V. harveyi* and *V. vulnificus* (Liu et al. 2012). Another regulator VqsA has been identified recently regulating the QS by binding to the promoters of aphA and luxR to regulate the expression of aphA negatively and luxR positively (Gao et al. 2018). VqsA plays an essential role in QS-regulated phenotypes, such as type VI secretion system 2 (T6SS2)-dependent interbacterial competition, exotoxin production biofilm formation (Gao et al. 2018), and serine proteases (Zhang et al. 2019) (Figure 7.7).

7.2.8 Quorum Sensing in *Vibrio fluvialis*

V. fluvialis has been considered as an emerging food-borne human pathogen commonly found in coastal surroundings such as marine and estuarine environments. It has a close phenotypic identity either with *V. cholerae* or *Aeromonas* spp. (Ramamurthy et al. 2014). This organism is isolated from different sources including, human diarrheal cases, aquatic environments, and treated wastewater effluent system, and there are reports of food poisoning caused by this organism associated with extra-intestinal infections (Huq et al. 1980; Yoshii et al. 1987; Levine and Griffin 1993; Igbinosa et al. 2009). *V. fluvialis* produces several virulence factors such as enterotoxin-like substance, lipase, protease (metaloproteaes), cytotoxin, and hemolysin (VFH) (Miyoshi et al. 2002; Kothary et al. 2003; Igbinosa and Okoh 2010). Though there are different virulence factors discovered,

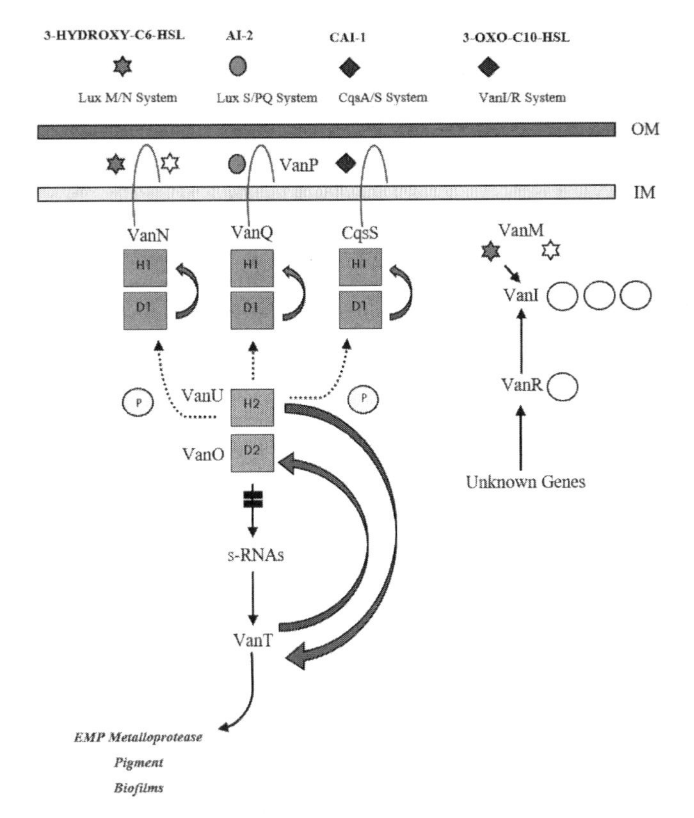

FIGURE 7.5 Model of the *Vibrio anguillarum* quorum sensing circuits. For a detailed account of the model, refer to the text. It is not known if the low cell population model functions similarly to that of *Vibrio harveyi*. However, for simplicity of discussion, it will be predicted to function similarly. Thus, only the model at high cell population density is shown for indicating differences between species. Four QS systems are available in *Vibrio anguillarum*. An AHL synthase VanM produces N-hexanoyl-L-homoserine lactone (C6-HSL) and N-(3-hydroxyhexanoyl)-L-homoserine lactone (3-hydroxy-C6-HSL) for the LuxM/N that is detected by VanO and the signal transmission is converged at VanU, causing slight repression of VanT expression via VanO. The VanT mRNA appears to be comparatively stable at lower cell densities and not triggered as the cell densities increase, which suggests that putative sRNA does not stabilize VanT mRNA, yet it may inhibit translation of VanT mRNA by other means. Recent findings show that the systems LuxS/PQ and LuxM/N function differently from those of other *Vibrio* spp. VanO represses VanT expression while VanU activates the VanT. VanN and VanO both repress the VanT expression in the absence of VanU, indicating that these might be bypassing the VanU during signal transmission. VanT has a autoregulatory mechanism where it represses the VanT and initiates VanOU. The third system is quite similar to the *Vibrio fischeri* LuxI/R QS circuit where VanI synthesizes N-(3-oxodecanoyl)-L-homoserine lactone (3-oxo-C10-HSL), which likely binds VanR, a transcriptional activator of VanI, and the fourth predicted CqsA/S signal system is yet to be extensively exploited. Explanation of symbols is given in a diagram key stated with reference to *Vibrio harveyi* QS circuits.

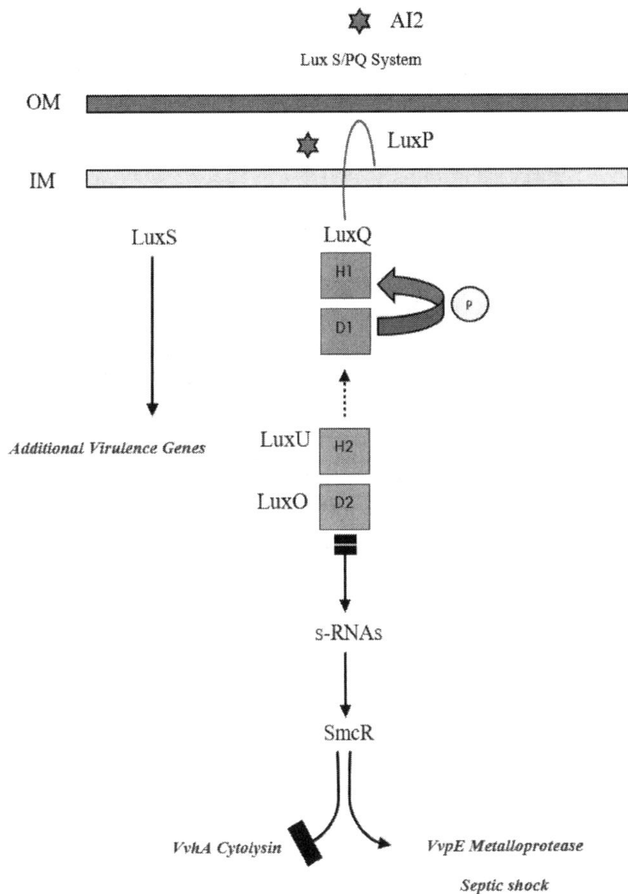

FIGURE 7.6 Model of the *Vibrio vulnificus* quorum sensing circuits. For a detailed description of the model, refer to the text. The model for low cell population density functions similar to that of *Vibrio harveyi*. Thus, only the model at high cell population density is shown for indicating differences between species. A LuxS homologue with AI-2 autoinducer activity and LuxS/PQ system, LuxU, LuxO, and the LuxR transcriptional regulator, SmcR, are available in *V. vulnificus* as seen in *Vibrio harveyi*. LuxS and SmcR positively regulate the metalloprotease VvpE expression and negatively regulate cytolysin VvhA expression where both proteins also stimulate the secretion of the proinflammatory mediators, tumor necrosis factor-alpha, and nitric oxide from murine macrophages challenged with *V. vulnificus*, suggesting a vital role in septic shock. But it has been identified that LuxS is the essential component of virulence generating an excess loading of iron and believed to regulate the other virulence genes independent of the SmcR pathway. Explanation of symbols is given in a diagram key stated with reference to *Vibrio harveyi* QS circuits.

research on virulence regulation mechanisms are hitherto lacking, and it is yet to be covered. However, there are a few reports on the existence of quorum sensing mechanism in *V. fluvialis*, and it has been shown that some potential virulence factors (extracellular protease and hemolysin) are regulated by quorum sensing systems (Wang et al. 2013). *V. fluvialis* produces three types of autoinducers consequential for their QS system in a cell density-dependent manner, namely CAI-1, AI-2, and AHL QS systems (Yang et al. 2011; Wang et al. 2013), and three types of autoinducers are synthesized in cell density-dependent manner similar with other vibrios. The CAI-1/AI-2 QS system of *V. fluvialis* negatively regulates the AHL QS system, unlike some other Vibrio species which have positive regulation. The *V. fluvialis* genome draft sequence analysis identified 50% and 38% LuxI and LuxR homologs respectively at the amino acid level (Wang et al. 2013) and annotated them as VfqI (*V. fluvialis* QS I) and VfqR (*V. fluvialis* QS R). Both VfqR and VfqI are required for vfqI expression and thus the AHL production. Wang and colleagues suggested that the AHLs produced by VfqI serve as the ligand for VfqR, thereby modulating its activity, possibly in a similar manner as the LuxR-family proteins. Homologs of *V. cholerae* CqsA (CAI-1 synthase), LuxS (AI-2 synthase), LuxO (σ^{54}-dependent regulator), and HapR (QS master regulator, a

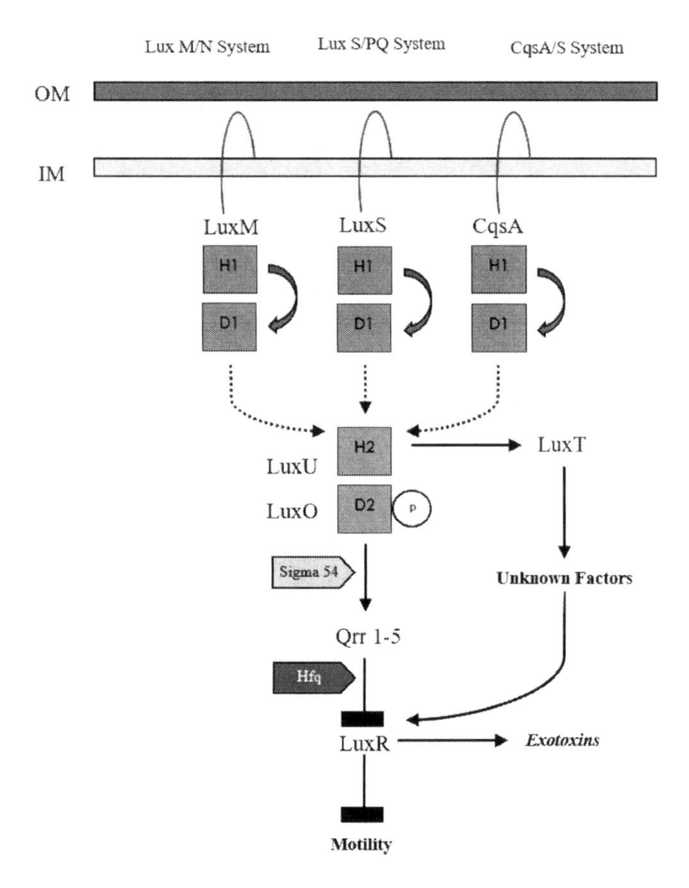

FIGURE 7.7 Model of the *Vibrio alginolyticus* quorum sensing circuits. For a detailed account of the model, refer to the text. It is not known if the low cell population model functions similarly to that of *Vibrio harveyi*. However, for simplicity of discussion, it will be predicted to function similarly. Thus, only the model at high cell population density is shown for indicating differences between species. The QS signaling AIs are produced via LuxM, LuxS, and CqsA, which will function in a very similar manner to that of *V. harveyi* phosphorylation cascade operates under the influence of sigma-54, sRNAs, and Hfq, but with additional pathway involving LuxT is known to activate a set of genes which are of unknown functionality. Explanation of symbols is given in a diagram key stated with reference to *Vibrio harveyi* QS circuits.

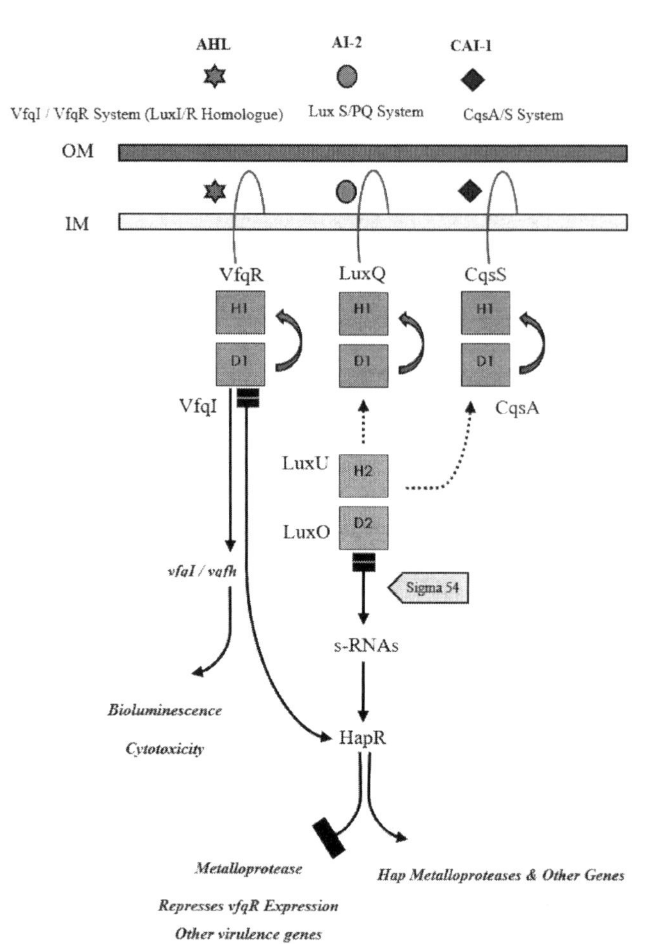

FIGURE 7.8 Model of the *Vibrio fluvialis* quorum sensing circuits. For a detailed account of the model, refer to the text. Three QS circuits have been proposed to exist in *V. fluvialis* where two of these are homologues of LuxS/PQ system and CqsA/S system mediated via a LuxO sigma 54-dependent regulator. LuxO has a negative regulatory effect on *hapR* expression similar to the systems in *V. cholerae* and *V. harveyi*. HapR negatively regulates the expression of vfqR by repressing its function, and cross-talks between the different QS circuits is not yet fully clarified. HapR positively affects Hap metalloprotease production while other metalloprotease functions are repressed and *vfqR* expression is also reduced. vfqI/vfqR are directly activated by AHL 3-OXO-C10-HSL where they positively affect cytotoxic effects and bioluminescence.

TetR-family transcriptional regulator similar to LuxR of *V. harveyi*) have been identified in the *V. flivialis* genome (Wang et al. 2013). Further it is found that the deletion in cqsA and luxS stopped CAI-1 and AI-2 production, respectively, whereas the CAI-1 and AI-2 production were not affected by deletion of LuxO or HapR. For the activation of hapR, either CAI-1 or AI-2 is sufficient, and further it was revealed that the expression of hapR was higher in the luxO mutant, indicating that LuxO has a negative regulatory effect on hapR expression, similar to the autoinducer systems of *V. cholerae* and *V. harveyi*. In *V. fluvialis*, the AHL QS system does not regulate the CAI-1/AI-2 system, but the deletion of hapR greatly affected the AHL QS system where in hapR mutants, AHL production was significantly increased relative to the wild type. Production of some virulence factors is known to be controlled by quorum sensing in *V. fluvialis* as well, regulating the protease production by CAI-1/AI-2 QS system in *V. fluvialis* and also regulating the hemolysin production. The hemolysin gene (vhf) expression is activated by the AHL-VfqR QS system and is repressed by the CAI-1/AI-2-HapR QS system, and further it showed that the VfqR-AHL QS system regulates the production of other extracellular factors contributing to cytotoxic activity (Wang et al. 2013) (Figure 7.8).

7.3 Genomic and Transcriptomic Approaches of Vibrio QS

Transcriptional and post-transcriptional regulatory networks play crucial roles in maintaining the homeostasis in bacteria and adapting to the changing growth conditions, and the regulatory small RNAs (sRNAs) are the vital elements of these regulations. They are the main effectors in signal transduction cascades (quorum sensing) of many bacteria including vibrios (Waters and

Storz 2009). The Qrr sRNAs (quorum regulatory RNAs) are Hfq-dependent trans-encoded sRNAs that control the quorum sensing in vibrio (Lenz et al. 2004). Closely related Qrr sRNAs are found in quorum sensing circuits in vibrios, and they act as positive and negative regulators controlling multiple target mRNAs. Five homologues sRNAs are reported associated with the vibrio quorum sensing circuits controlling 20 mRNA targets (Feng et al. 2015). Qrr sRNAs uses four mechanisms in controlling the targets such as catalytic degradation, coupled degradation, sequestration, and activation-induced degradation. AphA (transcription factor) is a master regulator of quorum sensing that operates at low cell density (LCD) in *V. harveyi* and *V. cholera*, whereas the LuxR (*V. harveyi*)/HapR (*V. cholerae*) is the master regulator that operates at high cell density (HCD). The Qrr sRNAs positively control the production of the low-cell-density master regulator AphA, and they repress the production of the high-cell-density master regulator LuxR. At LCD, Qrr sRNAs activate production of AphA achieving, maximal AphA production and AphA together with sRNAs repress production of LuxR/HapR. In contrast, at high cell density LuxR/HapR represses aphA, achieving maximal LuxR/HapR production (Tu and Bassler 2007; Rutherford et al. 2011; Feng et al. 2015). LuxO (central QS response regulator) negatively autoregulates its own transcription irrespective of its phosphorylation state, and the Qrr sRNAs posttranscriptionally control LuxO production by pairing with and repressing the luxO mRNA transcript. The Qrr sRNAs feedback repress the genes encoding one of the quorum-sensing synthases-receptor pairs, LuxMN, and the gene encoding the transcriptional factor LuxO (Teng et al. 2011; Tu et al. 2010). These transcriptomes are studied with transcriptomics technologies. Studies on transcriptomes began in the 1990s, and with advances in technology, these studies have been improved. Microarrays (quantify a set of predetermined sequences) and RNA sequencing (RNA-Seq; high-throughput sequencing to capture all sequences) are the two key techniques that are used in the transcriptome studies (Lowe et al. 2017; Nelson 2002). Microarrays measure the abundances of a defined set of transcripts via their hybridization to an array of complementary probes (Pozhitkov et al. 2007). The high-throughput sequencing technologies have made it possible to obtain detailed transcriptomic profiles of several bacterial pathogens including vibrios (Nydam et al. 2014). Different technologies, even single molecule sequencing, are available such as Solexa (sequencing by synthesis), SOLiD (sequencing by oligo ligation detection), 454 (first commercially successful next generation system), nanopore sequencing, and SMRT sequencing (Lowe et al. 2017).

7.4 Symbiosis of *Vibrio fischeri* and Euprymna Scolopes

V. fischeri forms a symbiont with Hawaiian bobtail squid (Euprymna scolopes), which is a nocturnal hunter that dwells hidden under the sandy bottoms of Hawaiian shallow coastal water during the day time (Ruby and McFall-Ngai 1999; Visick et al. 2000; Nyholm and McFall-Ngai 2004). The symbiotic strains of bacteria dwell in the crypts of the squid, but *V. fischeri* is also found as free forms in marine ecosystems. The colonization of bobtail squid's crypts occurs hours after hatching, where it triggers a series of developmental changes in both the squid and the bacteria. In the squid's embryo, the mucus secretion from the ciliated epithelial cells is stimulated at the pore openings of the light organ, extending the chance of bacterial contact with the squid. This mucus provides an environment for the *V. fischeri* to aggregate and out compete other microbes, where it travels down toward internal light organs and colonizes from the inside. After about 12 h of infection, *V. fischeri* induces ciliated epithelial cell regression and enlargement of the cell lining of the crypts, and stimulates actin production in the ducts of the light organ to decrease the circumference of the ducts, prohibiting entry to other colonizers (Visick and McFall-Ngai 2000). The light organ later matures to form ink sac, lens, and reflectors. The *V. fischeri* itself goes through a series of changes where it gains the abilities of entry, colonization, and persistence in the crypts. At the stage of persistence, the bacteria lose its flagella and show a noticeable reduction in the size; decelerated growth rate after initial growth burst has been observed. At certain critical cellular densities, the bacteria will collectively exert bioluminescence to minimize the production of oxygen radicals by the squid's enzymes.

For this state to occur, it is essential to have lux gene expression and optimal levels of nutrients supplied by the host tissues. Hawaiian bobtail squid uses the bioluminescence of the bacteria both as an antipredation tactic as well as a method of attracting prey. During the night time, the squid's reflector is focused downwards and the lens system and the opening of the light organ is regulated as per the intensity of light received from the top mimicking the moonlight or starlight. Usually this is used to reduce the occurrence of a shadow when the squid hunts in the open shallow waters. Usually it takes an overnight for the bacterial culture in the crypts to reach its critical cell density, and at the first light around 90% of the dying bacterial culture is vented out from the crypts while filling the crypts with fresh sea water. In *V. fischeri*, it utilizes hierarchal regulatory cascade, which responds to the cell density that sequentially induces bioluminescence associated genes, luxICDABEG and luxR, which is different from the luxR of the *V. harveyi*. The underlying mechanism at the bottom of the hierarchy is the QS paradigm for most Gram-negative bacteria, which is the LuxI/R system that is not found in *V. cholerae*, *V. vulnificus*, or *V. harveyi*. The LuxI/R system is found to regulate the function of four genes of unknown activity plus the lux genes. The low cell densities LuxI will only produce low levels of the QS signals, N-(3-oxohexanoyl)-L-homoserine lactone (3-oxo-C6-HSL), yet at higher cell densities the signaling molecules will reach their critical concentrations and bind with the LuxR, causing the DNA binding domain to unravel (Miller and Bassler 2001; Neil et al. 2001).

The LuxR-AHL complex directly activates transcription of the luxICDABEG genes and activates genes required for the late stage of colonization in the squid; this results in an exponential increase in 3-OXO-C6-HSL, leading to elevated bioluminescence. In *V. fischeri*, the intercellular signaling cascade does cause a phosphorelay cascade that generates a transcriptional repressor as for *V. harveyi* and *V. cholerae*. LuxS synthesizes an AI-2 signal that is similar to that of *V. harveyi* and sensed by LuxP and LuxQ (Lupp and Ruby 2004; Ruby 1996). AinS synthase produces N-octanoyl-L-homoserine lactone (C8-HSL) that is sensed by AinR, a *V. harveyi* LuxN homologue (Cao and Meighen 1989;

Gilson et al. 1995). These two signal systems synergistically induce a phosphorelay via LuxO to relieve repression of LitR, a *V. harveyi* LuxR homologue (Fidopiastis et al. 2002; Miyamoto et al. 2003; Lupp and Ruby 2004). Repression of LitR is anticipated to occur via destabilization of litR mRNA similar to that of luxR in *V. harveyi*. LitR activates the LuxI/R system linking the AinS/R and LuxS/PQ systems to the LuxI/R system (Fidopiastis et al. 2002). The AinS/R system also starts a low-level induction of lux gene when 3-OXO-C6-HSL is limiting as the AinS signal C8-HSL binds and activates LuxR (Lupp et al. 2003).

A positive feedback loop is expected to occur as LitR activates AinS, and this regulation cascade triggers bioluminescence in the light organ of the squid (Visick et al. 2000; Lupp et al. 2003; Lupp and Ruby 2004, 2005). At low cell population densities in the marine waters, bioluminescence is repressed and at moderately high cell population densities in lab isolates and early stages of squid colonization, 3-OXO-C6-HSL is limiting. However, C8-HSL induces a low level of luminescence by binding LuxR. In the squid, the AinS/R system is not essential for luminescence. However, it regulates the expression of early colonization factors and motility of the bacteria by controlling the disjointing of the flagella (Lupp and Ruby 2004, 2005). At high cell population densities in the squid light organ, 3-oxo-C6-HSL is in high amounts and the LuxI/R system induces luminescence and expression of late colonization factors, and regulation of the lux genes also involves additional regulatory factors not regulated by quorum sensing (Ulitzur et al. 1997).

7.5 Virulence of Vibrios and Quorum Sensing

In general, the infectious cycle of pathogenic bacteria consists of entry of the pathogen, establishment and multiplication, and avoidance of host defenses, causing damage. These different steps involve the expression of virulence factors, that is, gene products that allow the pathogens to infect and damage the host. They are known to produce an array of virulence factors including adhesion factors (pilli, flagellae, chitin binding proteins), extracellular polysaccharides, and lytic enzymes (hemolysins, proteases, lipases). Lytic enzymes are produced by many pathogenic bacteria and often play a central role in pathogenesis, and these enzymes cause damage to host tissues, thereby allowing the pathogen to obtain nutrients and to spread through the tissues (Ruwandeepika et al. 2011). Virulence factors are often costly metabolic products; therefore, the production is under strict regulatory control, including cell-to-cell communication (quorum sensing) and ToxR, and also influenced by host factors. *In vitro* and *in vivo* studies have shown that the quorum sensing regulates the expression of different virulence factors in *V. harveyi*, including an extracellular toxin, metalloprotease, siderophore, type III secretion system, chitinase, and three phospholipase genes (Ruwandeepika et al. 2010, 2011; Natrah et al. 2011).

Whereas in *V. vulnificus* cytotoxicity is an important virulence determinant in the pathogenesis and two cytotoxins, RTX (encoded by rtxA1) and cytolysin/hemolysin (encoded by vvhA), and also extracellular protease called elastase, (one of the major virulence factors), encoded by vvpE are known to regulate by the QS (Kim et al. 2003). The LuxS QS system plays an important role in regulating the expression of virulence factors in *V. alginolyticus*. In *V. alginolyticus*, extracellular alkaline serine protease, Asp, is closely related to the quorum sensing system, and its production is in a cell-density-dependent manner. The *V. alginolyticus* luxS quorum-sensing system plays an essential role in the pathogenicity of the organism by regulating the protease and EPS production, biofilm development, and flagella formation. The LysR family regulator VqsA regulates QS, and VqsA regulates the expression of exotoxins and other QS-controlled virulence factors i.e., type VI secretion system 2 (T6SS2)-dependent interbacterial competition, biofilm formation, exotoxin production, and *in vivo* virulence of *V. alginolyticus* (Gao et al. 2018). The *V. cholerae* QS mechanism controls virulence factor production and biofilm formation (Jemielita et al. 2018). In *V. cholerae*, there are two main virulence factors: cholera toxin (CT) and the toxin-coregulated pilus (TCP), encoded respectively by the ctx gene and tcp gene cluster, which are indirectly activated by AphA. At low cell densities, sRNAs are transcribed in the presence of phosphorylated LuxO. sRNAs destabilize the hapR mRNA but promote the expression of AphA. At high cell densities, HapR is available for expression and inhibits aphA transcription, resulting in the cessation of virulence factor formation. AphA promotes and HapR represses the production of four components required for biofilm formation, i.e., vibrio polysaccharide (VPS) and three matrix proteins, RbmA, Bap1, and RbmC. *V. cholerae* forms biofilms at LCD and disperses from them at HCD (Yang et al. 2010; Singh et al. 2017; Teschler et al. 2015). Major virulence factors of *V. parahaemolyticus* are thermostable direct hemolysin (TDH), the TDH-related hemolysin, and two distinct type III secretion systems (T3SS1 and T3SS2). The tdh genes and the T3SS2 gene cluster constitute a pathogenicity island (Vp-PAI located on the chromosome II). Expression of T3SS1 and Vp-PAI is regulated in a quorum sensing (QS)-dependent manner. It is reported that the three QS regulators AphA and OpaR and an AraC-type transcriptional regulator (QsvR) form a complex regulatory network to control the expression of T3SS1 and Vp-PAI genes. At low cell density (LCD), Vp-PAI expression is repressed and T3SS1 genes are induced by AphA, which directly binds (an operator region of) the exsBAD-vscBCD operon. At high cell density (HCD), the bacterium turns off T3SS1 expression by replacing AphA with OpaR, triggering the induction of Vp-PAI (Zhang et al. 2019).

7.6 Recent Findings on Quorum Sensing of Vibrios

Bacteria use a chemical communication process called quorum sensing to monitor cell density and to alter behavior in response to fluctuations in population numbers. Although the key event was discovered associated to *V. fischeri*, multitudinous other QS-regulated biochemical processes, various ways of inhibitory mechanisms and compounds with potential to initiate/terminate those cascades have been discovered in various other Vibrio species (Plener et al. 2015). In a recent study, it was found that in *V. harveyi*, the ratios of kinase activities to phosphatase activities in its QS system and, hence, the extent of phosphorylation of LuxU/LuxO are important not only for the signaling output but also for the degree of noise in the system. This unveiled that the pools of phosphorylated LuxU/LuxO per cell directly determine the amounts of sRNAs produced and, consequently, the copy number of LuxR, generating heterogeneous quorum-sensing activation at the single-cell level. By this it is possible to conclude that the ability to drive the heterogeneous expression of QS-regulated genes in *V. harveyi* is an inherent feature of the architecture of the QS cascade (Plener et al. 2015). Another parallel study has revealed that in *V. harveyi* the process of osmotic stress response is also under QS regulation, and it was shown that LuxR activates expression of the glycine betaine operon betIBA-proXWV, which enhances growth recovery under osmotic stress conditions. BetI, an autorepressor of the *V. harveyi* betIBA-proXWV operon, activates the expression of genes encoding regulatory small RNAs that control quorum-sensing transitions (Kessel et al. 2015).

As per the study, the process of connecting quorum-sensing and glycine betaine pathways presumably enables *V. harveyi* to tune its execution of collective behaviors to its tolerance to stress (Kessel et al. 2015). Apart from the aforementioned, recently it was identified that QS has an influence on the choice of anti-phage defense strategy in *V. anguillarum*, which is a key driver of bacterial diversity and evolution (Tan et al. 2015). Its results demonstrate that *V. anguillarum* employs quorum-sensing information to choose between two complementary antiphage defense strategies. Further, the prevalence of nonmutational defense mechanisms in a strain of *V. anguillarum* suggests highly flexible adaptations to KVP40 phage infection pressure, possibly allowing the long-term coexistence of phage and host (Tan et al. 2015). This finding is important since detailed understanding of the mechanisms of phage protection in pathogenic bacteria is vital for the development of successful phage-based control of bacterial pathogens. Recently another two QS systems associated with *V. cholerae* were discovered, and these are associated with phosphotransfer protein LuxU: CqsR, VpsS. VpsS is known to be cytosolic and its cognate autoinducer is still unknown, but this VpsS is anticipated to contribute to QS via a nitric oxide-dependent pathway through an association with NosP (a nitric oxide responsive hemoprotein where VcNosP is encoded by Vc1444, where the product of gene is known as VpsV that inhibits the autophosphorylation of VpsS and thus phosphate flow to LuxU). CqsR is known to function as an upstream modulator of LuxO function, but the signaling effects are yet to be explored (Jung et al. 2015). The function of CqsR is regulated by the common metabolite ethanolamine since it has the ability of binding to the CACHE domain of CqsR *in vitro* and inducing HCD QS response via CqsR kinase inhibition in *V. cholerae* models (Jung et al. 2015). Another associated study where the biofilm formation and the vpsL expression by RpoN activators was assessed has unveiled that NtrC (one of the seven bEBPs—bacterial enhancer binding proteins) functions as a vps expression and biofilm formation aid in *V. cholerae*. This discovery is very important since it is necessary to identify the aspects of colonization of cholera pathogen in human tissues via biofilm formation (Chen et al. 2003). Very recently it was discovered that in *V. fischeri*, calcium ions function as a signal promoter that aids the QS circuit for biofilm formation; it was further revealed that calcium ions are capable of inducing both the syp-dependent and bacterial cellulose synthesis-dependent biofilms at the level of transcription of these loci. This finding has enabled manipulation of surface conditions of ionic compositions to regulate attachment of vibrio and other bacterial species of significance to various surfaces, and the dependency of calcium induced biofilm formation only at the sufficient presence of SypF's Hpt domain opened a new pathway where it led to discovery of another two sensor kinases, RscS and HahK. The process of multicellular aggregate formation/multicellularity of *V. cholerae* has been also identified to have a QS regulation. This is dissimilar from surface biofilm formation, occurs in cell suspensions at a rapid scale without the need of a cell division, and aids in successful transit between marine/saline niche and the human host. This process is known to upregulate via a HapR involved activation cascade that affects negatively on biofilm formation at high cell densities (Jemielita et al. 2018). Further, it is unveiled that Type VI secretion systems in *V. cholerae* also have a QS regulation, which aids the cholera pathogen to secrete compounds that have the potential to combat the phagocytic mechanisms of the hosts tissues. Two mechanisms of regulation were identified by this study that are involved with Qrr sRNAs where they repress the expression of the large type VI secretion system cluster through base pairing and its repression of HapR, the activator of two type VI secretion clusters (Shao and Bassler 2014). A study published in *Nature* magazine demonstrates that *V. cholerae* has gained its resistance to the environmental bacteriophages via another QS-regulated process. The demonstration carried out using the AI synthase mutated *V. cholarae* strains revealed that they were more susceptible to environmental bacteriophages compared to its parent bacteria (Hoque et al. 2016). Apart from the major vibrio, species like *V. fluvialis* (VfqR mediated pathogenesis cascade) (Wang et al. 2013), *V. tubiashii* (VtpR system), and the hydrophobicity response to biofilm propagation in *V. parahaemolyticus* were also found to have a QS regulation (Lee et al. 2016). A study carried out to assess the correlation between the ureolytic activity and QS-regulated biofilm formation capabilities of selected isolates of *V. campbellii* and *V. harveyi* resulted in evidence of direct connection of urea utilization process and QS (Defoirdt et al. 2017). Unconventional vibrio species like *V. variabilis* have demonstrated processes regulated by QS as well. A study done on the aforementioned bacteria led to identification of three unique N-acyl homoserine lactones which were reported to be first QS profiles in the *V. variabilis* T01, yet the functionalities are to be exploited. This expands the QS members of vibrio family (Mohamad et al. 2016). Very recently, the modulatory functions by integrated QsvR on *V. parahaemolyticus* were assessed, revealing the Vp-PAI genes and the characters encoded by them (Zhang et al. 2018). There is ongoing research associated with vibrio QS under various aspects that targets toward anti-virulence therapy, quorum quenching compound identification, and diverse other medical as well food industrial applications.

7.7 Diversity of Quorum Sensing in Vibrios

The QS mechanisms in genus vibrio have a large degree of variation and diversity even though the components involved are very similar in both biochemical and functional aspects, which is reflective of the pliancy of the genomic makeup of the vibrios. Vibrios are found in diverse environmental conditions ranging from animal tissues to aquatic ecosystems with both freshwater and saline nature. Almost every vibrio species identified with QS so far have circuitry of quorum sensing involved with LuxO master regulatory system with varying numbers of signal transmissions converging on the activator LuxO. Slight deviations of this mechanism have been observed in recent studies in *V. fluvialis* and *V. cholerae* (Wang et al. 2013), expanding the possibility of alternate pathways and diversifications. Single circuit convergence through LuxO is only seen in *V. vulnificus*, and most of the other species have two or more parallel circuits converging on LuxO. *V. cholerae* and *V. harveyi* have three parallel circuits of QS that converge on LuxO, where the systems have been identified as homologous. Evidence of a fourth QS in *V. cholerae* was recently discovered which operates through an alternate parallel pathway instead of the LuxO system. *V. fischeri* contains two parallel QS circuits (Lupp and Ruby 2005) that converge on Lux O. *V. alginolyticus* Rui et al. (2008) and *V. parahaemolyticus* (Kalburge et al. 2017) also have three QS circuits that converge on LuxO homologue master regulator. In *V. anguillarum*, the QS signal transmits through a VanO system, which is a homologue of the LuxO/U regulator. The specialty of *V. anguillarum* is that that there is a LuxI/R signaling cascade linked alongside the LuxO master regulator. These QS circuits are involved with a variety of phosphotransferase proteins and different small regulatory RNAs that aid in maintaining fastened regulation and diversified cell responses. The complexity of these systems is yet to be exploited, and topically it could be stated that these QS systems mediate the survival/colonization and responses against host/environment triggered conditions, radiation protection, competition, adaptations toward physiochemical fluctuations (pH, salinity, temperature, nutrients), and other seasonal variations, which are of positive/negative effect for the free living vibrio bacteria while providing protection from tissue iron limitation, host immune responses, and oxidative stress to the animal tissue associated vibrios which are of pathogenic nature. The tool QS with diversified circuit arrangements is of great aid for vibrios to survive in competitive environmental conditions that are of a large diversity ranging up to 50 plus species of individuals belonging to the same genus. Apart from the generally explained QS systems, a vast range of mutation triggered alterations have been identified in various vibrio species, and more are yet to be discovered. These are mainly due to inductions of the environment, and another part comes from inter/intra specie communications and genetic material transactions. However, it is noteworthy that the genus vibrio has a comparatively larger diversity of QS to that of other genera of bacteria.

7.8 Quorum Quenching in Vibrios

Development of antibiotics resistance, more importantly, multidrug resistance, is an emerging as well as a critical global issue. Some of the pathogenic bacteria have shown defeating abilities against the currently available antibacterials, hence this is becoming a greater threat in therapeutics. Thus scientists are in search of alternatives to antibiotics for combatting this emerging threat, and it seems that the inhibition of the quorum sensing mechanism is one of the more focused areas in this regard. QS inhibition (quorum quenching) is garnering a greater interest as there is a less risk of developing resistance due to the lack of selective pressure on the organisms. Quorum quenching molecules are diverse in nature, comprised of enzymes and chemical compounds, which are derivatives of natural compounds or synthetic compounds; they have various modes of action as well as different targets (Defoirdt 2018). Quorum quenching molecules disrupt the quorum sensing in the process of synthesis, diffusion, accumulation, and perception of the QS signals, and they alter the cascade of gene expressions which are needed for the virulence of the organism. QS inhibitors act on the QS system by inhibiting the synthesis of autoinducers, degrading autoinducers, interfering with autoinducer receptors, or inhibiting the autoinducer/receptor complex formation. As with many other bacteria, the virulence of the vibrio species is said to be controlled by the quorum sensing mechanism, and quorum quenching acts vice versa, suppressing the virulence of Vibrios including *V. harveyi*, *V. campbelli*, *V. vulnificus*, *V. cholerae*, and *V. parahemolyticus* (Defoirdt et al. 2006; Ruwandeepika et al. 2010). Among the QS signaling molecules of Vibrios, acyl-homoserine lactones (AHLs) are a well-known and studied class of autoinducers which are specific in nature. AI-2 is the universal language used for intra- and inter-species communication, and beside these two kinds of molecules, CA1 is also involved in QS. Inhibition of all the types of QS molecules imposes an equal impact on the QS mechanism; however, most studies have been conducted intensively to investigate the inhibition of AHLs pathway. Though there are many QS inhibitors identified and claimed, brominated furanones are the most intensively studied QS inhibitors, and these compounds have been reported to disrupt QS in various Gram-negative bacteria including Vibrios. It has been reported that furanones inhibit the cell to communication by interfering with auto inducers AI-1 as well as the AI-2 (Ren et al. 2004). In addition to furanone, compounds such as furanones and analogs, bismuth porphyrin complexes, glycosylation reagents of glycosylated flavonoids, glycomonoterpenols, heavy metals, and nanomaterials are known to disrupt the QS mechanism. Use of the (bromo) alkylmaleic anhydrides, Brominated thiophenones like Qstatin (1-(5-bromothiophene-2-sulfonyl)-1H-pyrazole), 2,6-Di-tert-butyl-4-methylphenol, brominated furanones, 9H-fluorene vinyl ether derivative like SAM 461, cinnamaldehyde derivatives, and phenylboronic acid compounds have been reported as quorum sensing inhibitors (Steenackers et al. 2010). The red marine alga *Delisea pulchra* produces halogenated furanones as antagonists for AHL-mediated quorum sensing. Also, the higher plants such as pea, rice, soybean, tomato, etc. and micro algae influence AHL-mediated quorum (Defoirdt et al. 2004). Several marine micro-algae such as *Chlorella saccharophila* CCAP211/48, *Nannochloropsis* CCAP849/9, *Isochrysis* sp. CCAP927/14, *Tetraselmis suecica* CCAP66/4, *T. striata* SAG41.85, and *T. tetrathele* SAG161-2C are capable of inhibiting the QS-dependent responses of reporter strains *E. coli* JB523 and *V. harveyi* JMH612 (Natrah et al. 2011). Traditional medicines, such as fruits, herbs, and medicinal plants, and some food additives are known to be potential sources of QS inhibitory compounds. Cinnamaldehyde is widely used as a flavoring substance in the food industry and is found to inhibit both AHL and AI-2-dependent QS in *V. harveyi* (Niu et al. 2006). Also, the mobility shift assays revealed that cinnamaldehyde decreases the DNA-binding ability of LuxR of *V. harveyi*. Cinnamaldehyde and its analogs increase the survival of the nematode *Caenorhabditis elegans*, brine shrimp, giant freshwater prawn, and *Macrobrachium rosenbergii* infected with pathogenic *Vibrio* spp. (Brackman et al. 2008, 2011; Defoirdt 2018; Pande et al. 2013). Coumarin has been reported as an inhibitor of the quorum sensing of *V. splendidus*, and it downregulates the expression of virulence genes of *V. splendidus* (Zhang et al. 2017). There are some reports that inhibition of QS by marine microorganisms, *Halobacillus salinus*, produces two compounds, namely N-(2-phenylethyl)-isobutyramide and 3-methyl-N-(2-phenylethyl)-butyramide, which are capable of inhibiting violacein biosynthesis of *Chromobacterium violaceum* CV026, *Bacillus cereus*, and *Marinobacter* sp. SK-3 produce diketopiperazines (DKPs) which inhibit AHL-dependent QS. Actinobacteria produce piericidin, which inhibits violacein biosynthesis in *C. violaceum* CV026 (Teasdale et al. 2009, 2011; Ooka et al. 2013). Brominated thiophenones, sulphur analogues of the brominated furanones, also have been synthesized, and these compounds are found to be more active than the corresponding furanones (Defoirdt et al. 2012; Yang et al. 2015). Enzymetic involvement in QS mechanisms is also a known fact there are three classes of enzymes: lactonases, acylases, and oxidoreductases. AHL lactonases are metalloproteins which hydrolyze the ester bond of the homoserine lactone ring, whereas the AHL acylases hydrolyze the acyl-amide bond between the acyl tail and lactone ring of AHLs in a nonreversible manner, resulting in the release of a fatty acid chain and a homoserine lactone moiety. The enzyme oxidoreductases do not degrade the AHL but modify it to an inactive form by oxidizing or reducing the acyl side chain (Zhang and Li 2016). Nonenzymatic signal inactivation and sequestration by monoclonal antibodies (MAbs) also have been investigated (Kaufmann et al. 2006; Zhang and Li 2016). The antibody sequestration of AHLs is therapeutically beneficial as it inhibits the activation of QS cascades and subsequently the production of bacterial virulence factors, thereby preventing the host cell cytotoxicity induced by the AHL molecules. QS inhibitor Honaucin A, isolated from cyanobacterium (Leptolyngbya), acts as an inhibitor against QS of *V. harveyi* and *V. fischeri*; moreover the two honaucin A derivates (4'-bromohonaucin A and 4-iodohonaucin A) have shown a remarkable QS inhibitory activity of honaucin A (Choi et al. 2012). Other molecules such as trifluoromethyl analog of (4S)-4,5-dihydroxy-2,3-pentanedione (DPD), alkyl substituted DPD analogs, pyrogallol-type compounds, phenothiazine, sulfone compounds, and arylboronic acids are reported to act as an agonist on AI-2-mediated quorum sensing (Frezza et al. 2006; Lowery et al. 2008; Ni et al. 2008, 2009; Li et al. 2008; Peng et al. 2009).

7.9 Conclusion

Quorum sensing is one of the major events in vibrios to regulate vital functions of them in order to maintain their cellular function needed for surveillance, survival, and adaptation to their changing environments. It is a process of communication among the bacteria through chemical signals by which bacteria track changes in their cell numbers and collectively alter expression of several genes. Though almost all the vibrios share the common components in QS (similarities at genetic level), the regulatory mechanisms differ to each other. There are several processes controlled by QS such as bioluminescence, sporulation, competence, antibiotic production, biofilm formation, and virulence factor secretion. QS influences the virulence of many pathogenic organisms, including vibrios controlling the expressions of virulence genes. Hence strategies have been developed to control infectious diseases caused by vibrios by disrupting the QS, i.e., the quorum quenching; moreover, this is an alternative strategy to get rid of antibiotic resistant bacteria. There are many more unknown factors associated with QS to be identified, characterized, and explored.

REFERENCES

Abisado, Rhea G., Saida Benomar, Jennifer R. Klaus, Ajai A. Dandekar, and Josephine R. Chandler. 2018. "Bacterial quorum sensing and microbial community interactions." *mBio* 9, no. 3: e02331–e02317. doi:10.1128/mBio.02331-17.

Almagro-Moreno, Salvador, and Ronald K. Taylor. 2013. "Cholera: Environmental reservoirs and impact on disease transmission." *Microbiology Spectrum* 1, no. 2. 1–12. doi:10.1128/microbiolspec. OH-0003–2012.

Austin, Brian, and X-H. Zhang. 2006. "*Vibrio harveyi*: significant pathogen of marine vertebrates and invertebrates." *Letters in Applied Microbiology* 43, no. 2: 119–124. https://doi.org/10.1111/j.1472-765X.2006.01989.x.

Austin, Brian, Dawn A. Austin. 2012. *Bacterial Fish Pathogens*. 5th ed. Springer, New York.

Bai, Fangfang, Lingling Pang, Zizhong Qi, Jixiang Chen, Brian Austin, and Xiao-Hua Zhang. 2008. "Distribution of five Vibrio virulence-related genes among Vibrio harveyi isolates." *The Journal of General and Applied Microbiology* 54, no. 1: 71–78. https://doi.org/10.2323/jgam.54.71.

Ball, Alyssa S., Ryan R. Chaparian, and Julia C. van Kessel. 2017. "Quorum sensing gene regulation by LuxR/HapR master regulators in Vibrios." *Journal of Bacteriology*. doi:10.1128/jb.00105-17.

Bassler, Bonnie L. 1999. "How bacteria talk to each other: Regulation of gene expression by quorum sensing." *Current Opinion in Microbiology* 2, no. 6: 582–587. https://doi.org/10.1016/S1369-5274(99)00025-9.

Bassler, Bonnie L., Miriam Wright, Richard E. Showalter, and Michael R. Silverman. 1993. "Intercellular signalling in *Vibrio harveyi*: Sequence and function of genes regulating expression of luminescence." *Molecular Microbiology* 9, no. 4: 773–786. https://doi.org/10.1111/j.1365-2958.1993.tb01737.x.

Boyer, Mickaël, and Florence Wisniewski-Dye. 2009. "Cell–cell signalling in bacteria: Not simply a matter of quorum." *FEMS Microbiology Ecology* 70, no. 1: 1–19. https://doi.org/10.1111/j.1574-6941.2009.00745.x.

Brackman, Gilles, Shari Celen, Ulrik Hillaert, Serge van Calenbergh, Paul Cos, Louis Maes, Hans J. Nelis, and Tom Coenye. 2011. "Structure-activity relationship of cinnamaldehyde analogs as inhibitors of AI-2 based quorum sensing and their effect on virulence of *Vibrio* spp." *PLoS One*. doi:10.1371/journal.pone.0016084.

Brackman, Gilles, Tom Defoirdt, Carol Miyamoto, Peter Bossier, Serge Van Calenbergh, Hans Nelis, and Tom Coenye. 2008. "Cinnamaldehyde and cinnamaldehyde derivatives reduce virulence in *Vibrio* spp. by decreasing the DNA-binding activity of the quorum sensing response regulator LuxR." *BMC Microbiology*. doi:10.1186/1471-2180-8-149.

Broberg, Christopher A., Thomas J. Calder, and Kim Orth. 2011. "Vibrio parahaemolyticus cell biology and pathogenicity determinants." *Microbes and Infection* 13, no. 12–13: 992–1001.

Cao, Jie-Gang, and E. A. Meighen. 1989. "Purification and structural identification of an autoinducer for the luminescence system of *Vibrio harveyi*." *Journal of Biological Chemistry* 264, no. 36: 21670–21676. doi:10.1016/j.micinf.2011.06.013.

Chen, Chung-Yung, Keh-Ming Wu, Yo-Cheng Chang, Chuan-Hsiung Chang, Hui-Chi Tsai, Tsai-Lien Liao, Yen-Ming Liu et al. 2003. "Comparative genome analysis of *Vibrio vulnificus*, a marine pathogen." *Genome Research* 13, no. 12: 2577–2587. doi:10.1101/gr.1295503.

Chen, Xin, Stephan Schauder, Noelle Potier, Alain Van Dorsselaer, István Pelczer, Bonnie L. Bassler, and Frederick M. Hughson. 2002. "Structural identification of a bacterial quorum-sensing signal containing boron." *Nature* 415, no. 6871: 545.doi. 10.1038/4155a.

Chien, Jason W., Jin-Yuan Shih, Po Ren Hsueh, Phillip C. Yang, and K.T. Luh. 2002. "*Vibrio alginolyticus* as the cause of pleural empyema and bacteremia in an immunocompromised patient." *European Journal of Clinical Microbiology & Infectious Diseases* 21, no. 5: 401–403.doi:10.1007?s10096-002-0726-0.

Childers, Brandon M., and Karl E. Klose. 2007. "Regulation of virulence in *Vibrio cholerae*: The ToxR regulon." *Future Microbiology* 335–344. doi:10.2217/17460913.2.3.335.

Choi, Hyukjae, Samantha J. Mascuch, Francisco A. Villa, Tara Byrum, Margaret E. Teasdale, Jennifer E. Smith, Linda B. Preskitt, David C. Rowley, Lena Gerwick, and William H. Gerwick. 2012. "Honaucins A-C, potent inhibitors of inflammation and bacterial quorum sensing: Synthetic derivatives and structure-activity relationships." *Chemistry and Biology*. doi:10.1016/j.chembiol.2012.03.014.

Chowdhury, Fahima, Alison E. Mather, Yasmin Ara Begum, Muhammad Asaduzzaman, Nabilah Baby, Salma Sharmin, Rajib Biswas et al. 2015. "*Vibrio cholerae* serogroup O139: Isolation from cholera patients and asymptomatic household family members in Bangladesh between 2013 and 2014." *PLoS Neglected Tropical Diseases* 9, no. 11: e0004183. doi: 10.1371/journal.pntd.0004183.

Croxatto, Antony, Johan Lauritz, Chang Chen, and Debra L. Milton. 2007. "*Vibrio anguillarum* colonization of rainbow trout integument requires a DNA locus involved in exopolysaccharide transport and biosynthesis." *Environmental Microbiology* 9, no. 2: 370–382. doi:10.1111/j.1462-2920.2006.01147.x.

Croxatto, Antony, John Pride, Andrea Hardman, Paul Williams, Miguel Cámara, and Debra L. Milton. 2004. "A distinctive dual-channel quorum-sensing system operates in *Vibrio anguillarum*." *Molecular Microbiology* 52, no. 6: 1677–1689. doi:10.1111/j.1365-2958.2004.04083.x.

Croxatto, Antony, Victoria J. Chalker, Johan Lauritz, Jana Jass, Andrea Hardman, Paul Williams, Miguel Cámara, and Debra L. Milton. 2002. "VanT, a homologue of *Vibrio harveyi* LuxR, regulates serine, metalloprotease, pigment, and biofilm production in *Vibrio anguillarum*." *Journal of Bacteriology* 184, no. 6: 1617–1629. doi:10.1128/jb.184.6.1617-1629.2002.

Defoirdt, Tom, H. A. Darshanee Ruwandeepika, Indrani Karunasagar, Nico Boon, and Peter Bossier. 2010. "Quorum sensing negatively regulates chitinase in *Vibrio harveyi*." *Environmental Microbiology Reports* 2, no. 1: 44–49. doi: 10.1111/j.1758-2229.2009.00043.

Defoirdt, Tom, Nico Boon, Peter Bossier, and Willy Verstraete. 2004. "Disruption of bacterial quorum sensing: An unexplored strategy to fight infections in aquaculture." *Aquaculture*. doi: 10.1016/j.aquaculture.2004.06.031.

Defoirdt, Tom, Roselien Crab, Thomas K. Wood, Patrick Sorgeloos, Willy Verstraete, and Peter Bossier. 2006. "Quorum sensing-disrupting brominated furanones protect the gnotobiotic brine shrimp artemia franciscana from pathogenic *Vibrio harveyi, Vibrio campbellii*, and *Vibrio parahaemolyticus* isolates." *Applied and Environmental Microbiology*. doi:10.1128/AEM.00753-06.

Defoirdt, Tom, Siegfried E. Vlaeminck, Xiaoyan Sun, Nico Boon, and Peter Clauwaert. 2017. "Ureolytic activity and its regulation in *Vibrio campbellii* and *Vibrio harveyi* in relation to nitrogen recovery from human urine." *Environmental Science and Technology*. doi:10.1021/acs.est.7b03829.

Defoirdt, Tom, Tore Benneche, Gilles Brackman, Tom Coenye, Patrick Sorgeloos, and Anne Aamdal Scheie. 2012. "A quorum sensing-disrupting brominated thiophenone with a promising therapeutic potential to treat luminescent Vibriosis." *PLoS ONE* doi:10.1371/journal.pone.0041788.

Defoirdt, Tom. 2018. "Quorum-sensing systems as targets for antivirulence therapy." *Trends in Microbiology*. doi:10.1016/j.tim.2017.10.005.

Denkin, Steven M., and David R. Nelson. 2004. "Regulation of *Vibrio anguillarum* empA metalloprotease expression and its role in virulence." *Applied and Environmental Microbiology* 70, no. 7: 4193–4204. doi:10.1128/AEM.70.7.4193–4204.2004.

Dziejman, Michelle, Emmy Balon, Dana Boyd, Clare M. Fraser, John F. Heidelberg, and John J. Mekalanos. 2002. "Comparative genomic analysis of *Vibrio cholerae*: Genes that correlate with cholera endemic and pandemic disease." *Proceedings of the National Academy of Sciences* 99, no. 3: 1556–1561. doi:10.1073/pnas.042667999.

Feng, Lihui, Steven T. Rutherford, Kai Papenfort, John D. Bagert, Julia C. Van Kessel, David A. Tirrell, Ned S. Wingreen, and Bonnie L. Bassler. 2015. "A Qrr noncoding RNA deploys four different regulatory mechanisms to optimize quorum-sensing dynamics." *Cell* doi:10.1016/j.cell.2014.11.051.

Fidopiastis, Pat M., Carol M. Miyamoto, Michael G. Jobling, Edward A. Meighen, and Edward G. Ruby. 2002. "LitR, a new transcriptional activator in *Vibrio fischeri*, regulates luminescence and symbiotic light organ colonization." *Molecular Microbiology* doi:10.1046/j.1365-2958.2002.02996. x.

Frans, Ingeborg, Chris W. Michiels, Peter Bossier, K. A. Willems, Bart Lievens, and Hans Rediers. 2011. "*Vibrio anguillarum* as a fish pathogen: Virulence factors, diagnosis and prevention." *Journal of Fish Diseases* 34, no. 9: 643–661. doi:10.1111/j.1365-2761.2011. 01279.x.

Freeman, Jeremy A., and Bonnie L. Bassler. 1999a. "A genetic analysis of the function of LuxO, a two-component response regulator involved in quorum sensing in *Vibrio harveyi*." *Molecular Microbiology* 31, no. 2: 665–677. https://doi.org/10.1046/j.1365-2958.1999.01208.x.

Freeman, Jeremy A., and Bonnie L. Bassler. 1999b. "Sequence and function of LuxU: A two-component phosphorelay protein that regulates quorum sensing in *Vibrio harveyi*." *Journal of Bacteriology* 181, no. 3: 899–906.

Frezza, Marine, Damien Balestrino, Laurent Soulère, Sylvie Reverchon, Yves Queneau, Christiane Forestier, and Alain Doutheau. 2006. "Synthesis and biological evaluation of the trifluoromethyl analog of (4S)-4,5-Dihydroxy-2,3-Pentanedione (DPD)." *European Journal of Organic Chemistry*. doi:10.1002/ejoc.200600416.

Gao, Xiating, Xuetong Wang, Qiaoqiao Mao, Rongjing Xu, Xiaohui Zhou, Yue Ma, Qin Liu, Yuanxing Zhang, and Qiyao Wang. 2018. "VqsA, a novel LysR-type transcriptional regulator, coordinates quorum sensing (QS) and is controlled by QS to regulate virulence in the pathogen *Vibrio alginolyticus*." *Applied and Environmental Microbiology* 84, no. 12: e00444–e00418. doi:10.1128/AEM.00444-18.

Geske, Grant D., Jennifer C. O'Neill, David M. Miller, Margrith E. Mattmann, and Helen E. Blackwell. 2007. "Modulation of bacterial quorum sensing with synthetic ligands: Systematic evaluation of N-acylated homoserine lactones in multiple species and new insights into their mechanisms of action." *Journal of the American Chemical Society* 129, no. 44: 13613–13625. doi:10.1021/ja074135h.

Gilson, Lynne., Alan Kuo, and Paul V. Dunlap. 1995. "AinS and a new family of autoinducer synthesis proteins." *Journal of Bacteriology*. doi:10.1128/jb.177.23.6946-6951.1995.

Gode-Potratz, Cindy J., and Linda L. McCarter. 2011. "Quorum sensing and silencing in Vibrio parahaemolyticus." *Journal of Bacteriology*. doi:10.1128/JB.00432-11.

Gogarten, J. Peter, W. Ford Doolittle, and Jeffrey G. Lawrence. 2002. "Prokaryotic evolution in light of gene transfer." *Molecular Biology and Evolution* 19, no. 12: 2226–2238. doi:10.1093/oxfordjournals.molbev.a004046.

Haldar, S., S. B. Neogi, K. Kogure, S. Chatterjee, N. Chowdhury, A. Hinenoya, M. Asakura, and Shinji Yamasaki. 2010. "Development of a haemolysin gene-based multiplex PCR for simultaneous detection of *Vibrio campbellii*, *Vibrio harveyi* and *Vibrio parahaemolyticus*." *Letters in Applied Microbiology*. doi:10.1111/j.1472-765X.2009.02769.x.

Hammer, Brian K., and Bonnie L. Bassler. 2003. "Quorum sensing controls biofilm formation in *Vibrio cholerae*." *Molecular Microbiology* 50, no. 1: 101–104. https://doi.org/10.1046/j.1365-2958.2003.03688.x.

Hastings, J. Woody, and E. Peter Greenberg. 1999. "Quorum sensing: The explanation of a curious phenomenon reveals a common characteristic of bacteria." *Journal of Bacteriology* 181, no. 9: 2667–2668.

Henke, Jennifer M., and Bonnie L. Bassler. 2004a. "Quorum sensing regulates type III secretion in *Vibrio harveyi* and *Vibrio parahaemolyticus*." *Journal of Bacteriology* 186, no. 12: 3794–3805. doi:10.1128/JB.186.12.3794-3805.2004

Henke, Jennifer M., and Bonnie L. Bassler. 2004b. "Bacterial social engagements." *Trends in Cell Biology* 14, no. 11: 648–656. doi:10.1016/j.tcb.2004.09.012.

Henke, Jennifer M., and Bonnie L. Bassler. 2004c. "Three parallel quorum-sensing systems regulate gene expression in *Vibrio harveyi*." *Journal of Bacteriology* 186, no. 20: 6902–6914. doi:10.1128/JB.186.20.6902-6914.

Higgins, Douglas A., Megan E. Pomianek, Christina M. Kraml, Ronald K. Taylor, Martin F. Semmelhack, and Bonnie L. Bassler. 2007. "The major *Vibrio cholerae* autoinducer and its role in virulence factor production." *Nature* 450, no. 7171: 883.

Hoque, M. Mozammel, Iftekhar Bin Naser, S. M. Nayeemul Bari, Jun Zhu, John J. Mekalanos, and Shah M. Faruque. 2016. "Quorum regulated resistance of *Vibrio cholerae* against environmental bacteriophages." *Scientific Reports*. doi:10.1038/srep37956.

Huq, M. I., A. K. Alam, D. J. Brenner, and G. K. Morris. 1980. "Isolation of Vibrio-like group, EF-6, from patients with diarrhea." *Journal of Clinical Microbiology* 11, no. 6: 621–624.

Igbinosa, Etinosa O., and Anthony I. Okoh. 2010. "Vibrio fluvialis: An unusual enteric pathogen of increasing public health concern." *International Journal of Environmental Research and Public Health*. doi:10.3390/ijerph7103628.

Igbinosa, Etinosa O., Larry C. Obi, and Anthony I. Okoh. 2009. "Occurrence of potentially pathogenic Vibrios in final effluents of a wastewater treatment facility in a rural community of the eastern cape province of South Africa." *Research in Microbiology*. doi:10.1016/j.resmic.2009.08.007.

Jayaraman, Arul, and Thomas K. Wood. 2008. "Bacterial quorum sensing: Signals, circuits, and implications for biofilms and disease." *Annual Review of Biomedical Engineering* 10: 145–167. doi:10.1146/annurev.bioeng.10.061807.160536.

Jemielita, Matthew, Ned S Wingreen, and Bonnie L Bassler. 2018. "Quorum sensing controls *Vibrio cholerae* multicellular aggregate formation." *ELife*. doi:10.7554/elife.42057.

Jobling, Michael G., and Randall K. Holmes. 1997. "Characterization of hapR, a positive regulator of the *Vibrio cholerae* HA/protease gene hap, and its identification as a functional homologue of the *Vibrio harveyi* luxR gene." *Molecular Microbiology* 26, no. 5: 1023–1034. https://doi.org/10.1046/j.1365-2958.1997.6402011.x.

Jung, Sarah A., Christine A. Chapman, and Wai-Leung Ng. 2015. "Quadruple quorum-sensing inputs control *Vibrio cholerae* virulence and maintain system robustness." *PLoS Pathogens* 11, no. 4: e1004837. https://doi.org/10.1371/journal.ppat.1004837.

Kalburge, Sai Siddarth, Megan R. Carpenter, Sharon Rozovsky, and E. Fidelma Boyd. 2017. "Quorum sensing regulators are required for metabolic fitness in *Vibrio parahaemolyticus*." *Infection and Immunity* 85, no. 3: e00930–e00916. doi:10.1128/IAI.00930-16.

Kaufmann, Gunnar F., Rafaella Sartorio, Sang Hyeup Lee, Jenny M. Mee, Laurence J. Altobell, David P. Kujawa, Emily Jeffries, Bruce Clapham, Michael M. Meijler, and Kim D. Janda. 2006. "Antibody interference with N-Acyl homoserine lactone-mediated bacterial quorum sensing." *Journal of the American Chemical Society*. doi:10.1021/ja0578698.

Kawase, Tomoka, Shin-ichi Miyoshi, Zafar Sultan, and Sumio Shinoda. 2004. "Regulation system for protease production in *Vibrio vulnificus*." *FEMS Microbiology Letters* 240, no. 1: 55–59. doi:10.1016/j.femsle.2004.09.023.

Kessel, Julia C. van, Steven T. Rutherford, Jian Ping Cong, Sofia Quinodoz, James Healy, and Bonnie L. Basslera. 2015. "Quorum sensing regulates the osmotic stress response in *Vibrio harveyi*." *Journal of Bacteriology*. doi:10.1128/JB.02246-14.

Kim, Soo Young, Shee Eun Lee, Young Ran Kim, Choon Mee Kim, Phil Youl Ryu, Hyon E. Choy, Sun Sik Chung, and Joon Haeng Rhee. "Regulation of *Vibrio vulnificus* virulence by the LuxS quorum-sensing system." *Molecular Microbiology* 48, no. 6 (2003a): 1647–1664. https://doi.org/10.1046/j.1365-2958.2003.03536.x.

Kim, Young Ran, Shee Eun Lee, Choon Mee Kim, Soo Young Kim, Eun Kyoung Shin, Dong Hyeon Shin, Sun Sik Chung et al. 2003. "Characterization and pathogenic significance of

Vibrio vulnificus antigens preferentially expressed in septicemic patients." *Infection and Immunity* 71, no. 10: 5461–5471. doi:10.1128/iai.71.10.5461-5471.2003.

Kostka, Joel E., Om Prakash, Will A. Overholt, Stefan J. Green, Gina Freyer, Andy Canion, Jonathan Delgardio, Nikita Norton, Terry C. Hazen, and Markus Huettel. 2011. "Hydrocarbon-degrading bacteria and the bacterial community response in Gulf of Mexico beach sands impacted by the Deepwater Horizon oil spill." *Applied and Environmental Microbiology* 77, no. 22: 7962–7974. doi:10.1128/AEM.05402-11.

Kothary, Mahendra H., Heather Lowman, Barbara A. McCardell, and Ben D. Tall. 2003. "Purification and characterization of enterotoxigenic El Tor-like hemolysin produced by *Vibrio fluvialis*." *Infection and Immunity*. doi:10.1128/IAI.71.6.3213-3220.2003.

Kovacikova, Gabriela, and Karen Skorupski. 2002. "Regulation of virulence gene expression in *Vibrio cholerae* by quorum sensing: HapR functions at the aphA promoter." *Molecular Microbiology* 46, no. 4: 1135–1147. https://doi.org/10.1046/j.1365-2958.2002.03229.x.

Lee, Jin Hyung, Yong Guy Kim, Giyeon Gwon, Thomas K. Wood, and Jintae Lee. 2016. "Halogenated indoles eradicate bacterial persister cells and biofilms." *AMB Express*. doi:10.1186/s13568-016-0297-6.

Lenz, Derrick H., Kenny C. Mok, Brendan N. Lilley, Rahul V. Kulkarni, Ned S. Wingreen, and Bonnie L. Bassler. 2004. "The small RNA chaperone Hfq and multiple small RNAs control quorum sensing in *Vibrio harveyi* and *Vibrio cholerae*." *Cell* 118, no. 1: 69–82. doi:10.1016/j.cell.2004.06.009.

Levine, W. C., and P. M. Griffin. 1993. "Vibrio infections on the gulf coast: Results of first year of regional surveillance." *Journal of Infectious Diseases*. doi:10.1093/infdis/167.2.479.

Li, Minyong, Nanting Ni, Han Ting Chou, Chung Dar Lu, Phang C. Tai, and Binghe Wang. 2008. "Structure-based discovery and experimental verification of novel AI-2 quorum sensing inhibitors against *Vibrio harveyi*." *ChemMedChem*. doi:10.1002/cmdc.200800076.

Li, Xuan, Kristof Dierckens, Peter Bossier, and Tom Defoirdt. 2018. "The impact of quorum sensing on the virulence of *Vibrio anguillarum* towards gnotobiotic sea bass (Dicentrarchus labrax) larvae." *Aquaculture Research* 49, no. 11: 3686–3689. https://doi.org/10.1111/are.13821.

Li, Xuan, Qian Yang, Kristof Dierckens, Debra L. Milton, and Tom Defoirdt. 2014. "RpoS and indole signaling control the virulence of *Vibrio anguillarum* towards gnotobiotic sea bass (Dicentrarchus labrax) larvae." *PloS one* 9, no. 10: e111801. https://doi.org/10.1371/journal.pone.0111801.

Lilley, Brendan N., and Bonnie L. Bassler. 2000. "Regulation of quorum sensing in *Vibrio harveyi* by LuxO and sigma-54." *Molecular Microbiology* 36, no. 4: 940–954. https://doi.org/10.1046/j.1365-2958.2000.01913.x.

Lin, Z., K. Kumagai, K. Baba, J. J. Mekalanos, and M. Nishibuchi. 1993. "*Vibrio parahaemolyticus* has a homolog of the *Vibrio cholerae* toxRS operon that mediates environmentally induced regulation of the thermostable direct hemolysin gene." *Journal of Bacteriology* 175, no. 12: 3844–3855. doi:10.1128/jb.175.12.3844-3855.1993.

Linkous, Debra A., and James D. Oliver. 1999. "Pathogenesis of *Vibrio vulnificus*." *FEMS Microbiology Letters* 174, no. 2: 207–214. doi:10.1111/j.1574-6968.1999.tb13570.x.

Liu, Chun-Hung, Winton Cheng, Jung-Ping Hsu, and Jiann-Chu Chen. 2004 "*Vibrio alginolyticus* infection in the white shrimp *Litopenaeus vannamei* confirmed by polymerase chain reaction and 16S rDNA sequencing." *Diseases of Aquatic Organisms* 61, no. 1–2: 169–174. doi:10.3354/dao061169.

Liu, Huan, Dan Gu, Xiaodan Cao, Qin Liu, Qiyao Wang, and Yuanxing Zhang. 2012. "Characterization of a new quorum sensing regulator luxT and its roles in the extracellular protease production, motility, and virulence in fish pathogen *Vibrio alginolyticus*." *Archives of Microbiology* 194, no. 6: 439–452. doi:10.1007/s00203-011-0774-x.

Liu, Huan, Qiyao Wang, Qin Liu, Xiaodan Cao, Cunbin Shi, and Yuanxing Zhang. 2011. "Roles of Hfq in the stress adaptation and virulence in fish pathogen *Vibrio alginolyticus* and its potential application as a target for live attenuated vaccine." *Applied Microbiology and Biotechnology* 91, no. 2: 353–364. doi:10.1007/s00253-011-3286-3.

Lowe, Rohan, Neil Shirley, Mark Bleackley, Stephen Dolan, and Thomas Shafee. 2017. "Transcriptomics technologies." *PLoS Computational Biology*. doi:10.1371/journal.pcbi.1005457.

Lowery, Colin A., Junguk Park, Gunnar F. Kaufmann, and Kim D. Janda. 2008. "An unexpected switch in the modulation of AI-2-based quorum sensing discovered through synthetic 4,5-Dihydroxy-2,3-Pentanedione analogues." *Journal of the American Chemical Society*. doi:10.1021/ja802353j.

Lupp, Claudia, and Edward G. Ruby. 2004. "Vibrio fischeri LuxS and AinS: Comparative study of two signal synthases." *Journal of Bacteriology* 186, no. 12: 3873–3881. doi:10.1128/JB.186.12.3873-3881.2004.

Lupp, Claudia, and Edward G. Ruby. 2005. "*Vibrio fischeri* uses two quorum-sensing systems for the regulation of early and late colonization factors." *Journal of Bacteriology* 187, no. 11: 3620–3629. doi:10.1128/JB.187.11.3620-3629.

Lupp, Claudia, Mark Urbanowski, E. Peter Greenberg, and Edward G. Ruby. 2003. "The *Vibrio fischeri* quorum-sensing systems ain and lux sequentially induce luminescence gene expression and are important for persistence in the squid host." *Molecular Microbiology* 50, no. 1: 319–331. https://doi.org/10.1046/j.1365-2958.2003.t01-1-03585.x.

Makino, Kozo, Kenshiro Oshima, Ken Kurokawa, Katsushi Yokoyama, Takayuki Uda, Kenichi Tagomori, Yoshio Iijima et al. 2003. "Genome sequence of *Vibrio parahaemolyticus*: A pathogenic mechanism distinct from that of V cholerae." *The Lancet* 361, no. 9359: 743–749. doi:10.1016/S0140-6736(03)12659-1.

McCarter, Linda L. 1998. "OpaR, a homolog of Vibrio harveyi LuxR, controls opacity of *Vibrio parahaemolyticus*." *Journal of Bacteriology* 180, no. 12: 3166–3173.

McCarter, Linda. 1999. "The multiple identities of Vibrio parahaemolyticus." *Journal of Molecular Microbiology and Biotechnology* 1, no. 1: 51–57.

McDougald, Diane, Scott A. Rice, and Staffan Kjelleberg. 2000. "The marine pathogen *Vibrio vulnificus* encodes a putative homologue of the *Vibrio harveyi* regulatory gene, luxR: A genetic and phylogenetic comparison." *Gene* 248, no. 1–2: 213–221. https://doi.org/10.1016/S0378-1119(00)00117-7.

Miller, Melissa B., and Bonnie L. Bassler. 2001. "Quorum sensing in bacteria." *Annual Reviews in Microbiology* 55, no. 1: 165–199. doi:10.1146/annurev.micro.55.1.165.

Miller, Melissa B., Karen Skorupski, Derrick H. Lenz, Ronald K. Taylor, and Bonnie L. Bassler. 2002. "Parallel quorum sensing systems converge to regulate virulence in *Vibrio cholerae*." *Cell* 110, no. 3: 303–314. https://doi.org/10.1016/S0092-8674(02)00829-2.

Milton, Debra L. 2006. "Quorum sensing in Vibrios: Complexity for diversification." *International Journal of Medical Microbiology* 296, no. 2–3: 61–71. doi:10.1016/j.ijmm.2006.01.044.

Milton, Debra L., Andrea Hardman, Miguel Camara, Siri Ram Chhabra, Barrie W. Bycroft, G. S. Stewart, and Paul Williams. 1997. "Quorum sensing in *Vibrio anguillarum*: characterization of the vanI/vanR locus and identification of the autoinducer N-(3-oxodecanoyl)-L-homoserine lactone." *Journal of Bacteriology* 179, no. 9: 3004–3012. doi:10.1128/jb.179.9.3004-3012.1997.

Milton, Debra L., Victoria J. Chalker, David Kirke, Andrea Hardman, Miguel Cámara, and Paul Williams. 2001. "The LuxM homologue vanM from *Vibrio anguillarum* directs the synthesis of N-(3-Hydroxyhexanoyl) homoserine Lactone and N-Hexanoylhomoserine Lactone." *Journal of Bacteriology* 183, no. 12: 3537–3547. doi:10.1128/JB.183.12.3537-3547.2001.

Miyamoto, Carol M., Paul V. Dunlap, Edward G. Ruby, and Edward A. Meighen. 2003. "LuxO controls luxR expression in *Vibrio harveyi*: Evidence for a common regulatory mechanism in Vibrio." *Molecular Microbiology* 48, no. 2: 537-548. https://doi.org/10.1046/j.1365-2958.2003.03453.x.

Miyoshi, Shin Ichi, Yuka Sonoda, Hiroko Wakiyama, Md Monzur Rahman, Ken Ichi Tomochika, Sumio Shinoda, Shigeo Yamamoto, and Kazuo Tobe. 2002. "An exocellular thermolysin-like metalloprotease produced by *Vibrio fluvialis*: Purification, characterization, and gene cloning." *Microbial Pathogenesis*. doi:10.1006/mpat.2002.0520.

Mohamad, Nur Izzati, Tan Guan Sheng Adrian, Wen Si Tan, Nina Yusrina Muhamad Yunos, Pui Wan Tan, Wai Fong Yin, and Kok Gan Chan. 2016. "*Vibrio variabilis* T01: A tropical marine bacterium exhibiting unique N-Acyl homoserine lactone production." *Frontiers in Life Science*. doi:10.1080/21553769.2015.1066716.

Natrah, F. M I, Mireille Mardel Kenmegne, Wiyoto Wiyoto, Patrick Sorgeloos, Peter Bossier, and Tom Defoirdt. 2011. "Effects of micro-algae commonly used in aquaculture on acyl-homoserine lactone quorum sensing." *Aquaculture*. doi:10.1016/j.aquaculture.2011.04.038.

Nealson, Kenneth H., Terry Platt, and J. Woodland Hastings. 1970. "Cellular control of the synthesis and activity of the bacterial luminescent system." *Journal of Bacteriology* 104, no. 1: 313–322.

Neil, A. Whitehead, Swift, Simon, J. Allan Downie, Anne M. L. Barnard, George P. C. Salmond, and Paul Williams. 2001. "Quorum sensing as a population-density-dependent determinant of bacterial physiology." *Advances in Microbial Physiology*. doi:10.1016/S0065-2911(01)45005-3.

Nelson, N. J. 2002. "Microarrays have arrived: Gene expression tool matures." *JNCI Journal of the National Cancer Institute*. doi:10.1093/jnci/93.7.492.

Ni, Nanting, Gaurav Choudhary, Minyong Li, and Binghe Wang. 2008. "Pyrogallol and its analogs can antagonize bacterial quorum sensing in Vibrio harveyi." *Bioorganic and Medicinal Chemistry Letters*. doi:10.1016/j.bmcl.2008.01.081.

Ni, Nanting, Gaurav Choudhary, Minyong Li, and Binghe Wang. 2009. "A new phenothiazine structural scaffold as inhibitors of bacterial quorum sensing in *Vibrio harveyi*." *Biochemical and Biophysical Research Communications*. doi:10.1016/j.bbrc.2009.02.157.

Niu, C., S. Afre, and E. S. Gilbert. 2006. "Subinhibitory concentrations of cinnamaldehyde interfere with quorum sensing." *Letters in Applied Microbiology*. doi:10.1111/j.1472-765X.2006.02001. x.

Nydam, Seth D., Devendra H. Shah, and Douglas R. Call. 2014. "Transcriptome analysis of *Vibrio parahaemolyticus* in type III secretion system 1 inducing conditions." *Frontiers in Cellular and Infection Microbiology*. doi:10.3389/fcimb.2014.00001.

Nyholm, Spencer V., and Margaret J. McFall-Ngai. 2004. "The winnowing: Establishing the squid—Vibrios symbiosis." *Nature Reviews Microbiology*. doi:10.1038/nrmicro957.

Olivier, Verena, G. Kenneth Haines, Yanping Tan, and Karla J. Fullner Satchell. 2007. "Hemolysin and the multifunctional autoprocessing RTX toxin are virulence factors during intestinal infection of mice with *Vibrio cholerae* El Tor O1 strains." *Infection and Immunity* 75, no. 10: 5035–5042. doi:10.1128/IAI.00506-07.

Ooka, Kazuhiro, Atsushi Fukumoto, Tomoe Yamanaka, Kanako Shimada, Ryo Ishihara, Yojiro Anzai, and Fumio Kato. 2013. "Piericidins, novel quorum-sensing inhibitors against chromobacterium violaceum CV026, from *streptomyces* sp. TOHO-Y209 and TOHO-O348." *Open Journal of Medicinal Chemistry*. doi:10.4236/ojmc.2013.34012.

Pallaval Veera Bramhachari, and G. Mohana Sheela. 2019. "*Vibrio fischeri* symbiotically synchronizes bioluminescence in marine animals via quorum sensing mechanism." In *Implication of Quorum Sensing System in Biofilm Formation and Virulence*. doi:10.1007/978-981-13-2429-1_13.

Pande, Gde Sasmita Julyantoro, Anne Aamdal Scheie, Tore Benneche, Mathieu Wille, Patrick Sorgeloos, Peter Bossier, and Tom Defoirdt. 2013. "Quorum sensing-disrupting compounds protect larvae of the giant freshwater prawn *Macrobrachium rosenbergii* from *Vibrio harveyi* infection." *Aquaculture*. doi:10.1016/j.aquaculture.2013.05.015.

Papenfort, Kai, and Bonnie L. Bassler. 2016. "Quorum sensing signal–response systems in Gram-negative bacteria." *Nature Reviews Microbiology* 14, no. 9: 576. doi:10.1038/nrmicro.2016.89.

Papenfort, Kai, and Jörg Vogel. 2010. "Regulatory RNA in bacterial pathogens." *Cell Host & Microbe* 8, no. 1: 116–127. doi.org/10.1016/j.chom.2010.06.008.

Papenfort, Kai, Justin E. Silpe, Kelsey R. Schramma, Jian-Ping Cong, Mohammad R. Seyedsayamdost, and Bonnie L. Bassler. 2017. "A *Vibrio cholerae* autoinducer–receptor pair that controls biofilm formation." *Nature Chemical Biology* 13, no. 5: 551. doi:10.1038/nchembio.2336.

Peng, Hanjing, Yunfeng Cheng, Nanting Ni, Minyong Li, Gaurav Choudhary, Han Ting Chou, Chung Dar Lu, Phang C. Tai, and Binghe Wang. 2009. "Synthesis and evaluation of new antagonists of bacterial quorum sensing in *Vibrio harveyi*." *ChemMedChem*. doi:10.1002/cmdc.200900180.

Pérez, Pablo Delfino, and Stephen J. Hagen. 2010. "Heterogeneous response to a quorum-sensing signal in the luminescence of individual *Vibrio fischeri*." *PLoS ONE*. doi:10.1371/journal.pone.0015473.

Plener, Laure, Nicola Lorenz, Matthias Reiger, Tiago Ramalho, Ulrich Gerland, and Kirsten Jung. 2015. "The phosphorylation flow of the *Vibrio harveyi* quorum-sensing cascade determines levels of phenotypic heterogeneity in the population." *Journal of Bacteriology*. doi:10.1128/JB.02544-14.

Pozhitkov, Alex E., Diethard Tautz, and Peter A. Noble. 2007. "Oligonucleotide microarrays: Widely applied—Poorly understood." *Briefings in Functional Genomics and Proteomics*. doi:10.1093/bfgp/elm014.

Raina, Sheetal, Daniela De Vizio, Mark Odell, Mark Clements, Sophie Vanhulle, and Tajalli Keshavarz. 2009. "Microbial quorum sensing: A tool or a target for antimicrobial therapy?" *Biotechnology and Applied Biochemistry* 54, no. 2: 65–84. doi:10.1042/BA20090072.

Ramamurthy, Thandavarayan, Goutam Chowdhury, Gururaja P. Pazhani, and Sumio Shinoda. 2014. "*Vibrio fluvialis*: An emerging human pathogen." *Frontiers in Microbiology*. doi:10.3389/fmicb.2014.00091.

Ren, Dacheng, Laura A. Bedzyk, Rick W. Ye, Stuart M. Thomas, and Thomas K. Wood. 2004. "Differential gene expression shows natural brominated furanones interfere with the autoinducer-2 bacterial signaling system of *Escherichia coli*." *Biotechnology and Bioengineering*. doi:10.1002/bit.20259.

Rossez, Yannick, Eliza B. Wolfson, Ashleigh Holmes, David L. Gally, and Nicola J. Holden. 2015. "Bacterial flagella: Twist and stick, or dodge across the kingdoms." *PLoS Pathogens* 11, no. 1: e1004483. doi.org/10.1371/journal.ppat.1004483.

Ruby, Edward G. 1996. "Lessons from a cooperative, bacterial-animal association: The *Vibrio fischeri*–Euprymna scolopes light organ symbiosis." *Annual Reviews in Microbiology* 50, no. 1: 591–624. doi:10.1146/annurev.micro.50.1.591.

Ruby, Edward G., and Margaret J. McFall-Ngai. 1999. "Oxygen-utilizing reactions and symbiotic colonization of the squid light organ by *Vibrio fischeri*." *Trends in Microbiology*. doi:10.1016/S0966-842X(99)01588-7.

Rui, H., Q. Liu, Y. Ma, Q. Wang, and Y. Zhang. 2008. "Roles of LuxR in extracellular alkaline serine protease A expression, extracellular polysaccharide production and mobility of *Vibrio alginolyticus*." *FEMS Microbiology Letters* 285, no. 2: 155–162. doi:10.1111/j.1574-6968.2008.01185.x.

Rutherford, Steven T., and Bonnie L. Bassler. 2012. "Bacterial quorum sensing: Its role in virulence and possibilities for its control." *Cold Spring Harbor Perspectives in Medicine* 2, no. 11: a012427. doi:10.1101/cshperspect.a012427.

Rutherford, Steven T., Julia C. Van Kessel, Yi Shao, and Bonnie L. Bassler. 2011. "AphA and LuxR/HapR reciprocally control quorum sensing in Vibrios." *Genes and Development*. doi:10.1101/gad.2015011.

Ruwandeepika, H. A. D., Tom Defoirdt, P. P. Bhowmick, M. Shekar, Peter Bossier, and Indrani Karunasagar. 2010. "Presence of typical and atypical virulence genes in Vibrio isolates belonging to the Harveyi clade." *Journal of Applied Microbiology* 109, no. 3: 888–899. doi.org/10.1111/j.1365-2672.2010.04715.x.

Ruwandeepika, HA Darshanee, Indrani Karunasagar, Peter Bossier, and Tom Defoirdt. 2015. "Expression and quorum sensing regulation of type III secretion system genes of *Vibrio harveyi* during infection of gnotobiotic brine shrimp." *PLoS One* 10, no. 12: e0143935. doi.org/10.1371/journal.pone.0143935.

Ruwandeepika, HA Darshanee, Patit Paban Bhowmick, Indrani Karunasagar, Peter Bossier, and Tom Defoirdt. 2011. "Quorum sensing regulation of virulence gene expression in Vibrio harveyi in vitro and in vivo during infection of gnotobiotic brine shrimp larvae." *Environmental Microbiology Reports* 3, no. 5: 597–602. doi.org/10.1111/j.1758-2229.2011. 00268.x.

Schaefer, Amy L., Brian L. Hanzelka, Anatol Eberhard, and E. P. Greenberg. 1996. "Quorum sensing in *Vibrio fischeri*: Probing autoinducer-LuxR interactions with autoinducer analogs." *Journal of Bacteriology* 178, no. 10: 2897–2901. doi:10.1128/jb.178.10.2897-2901.

Schaefer, Amy L., Dale L. Val, Brian L. Hanzelka, John E. Cronan Jr, and E. P. Greenberg. 1996. "Generation of cell-to-cell signals in quorum sensing: Acyl homoserine lactone synthase activity of a purified *Vibrio fischeri* LuxI protein." *Proceedings of the National Academy of Sciences of the United States of America* 93, no. 18: 9505. doi:10.1073/pnas.93.18.9505.

Shao, Chung-Ping, and Lien-I. Hor. 2001 "Regulation of metalloprotease gene expression in *Vibrio vulnificus* by a *Vibrio harveyi* LuxR homologue." *Journal of Bacteriology* 183, no. 4: 1369–1375. doi:10.1128/JB.183.4.1369-1375.

Shao, Yi, and Bonnie L. Bassler. 2014. "Quorum regulatory small RNAs repress type VI secretion in *Vibrio cholerae*." *Molecular Microbiology*. doi:10.1111/mmi.12599.

Shin, Na-Ri, Deog-Yong Lee, Sung Jae Shin, Kun-Soo Kim, and Han-Sang Yoo. 2004. "Regulation of proinflammatory mediator production in RAW264. 7 macrophages by *Vibrio vulnificus* luxS and smcR." *FEMS Immunology & Medical Microbiology* 41, no. 2: 169–176. doi.org/10.1016/j.femsim.2004.03.001.

Singh, Praveen K., Sabina Bartalomej, Raimo Hartmann, Hannah Jeckel, Lucia Vidakovic, Carey D. Nadell, and Knut Drescher. 2017. "*Vibrio cholerae* combines individual and collective sensing to trigger biofilm dispersal." *Current Biology*. doi:10.1016/j.cub.2017.09.041.

Snoussi, Mejdi, Emira Noumi, Donatella Usai, Leonardo Antonio Sechi, Stefania Zanetti, and Amina Bakhrouf. 2008. "Distribution of some virulence related-properties of Vibrio alginolyticus strains isolated from Mediterranean seawater (Bay of Khenis, Tunisia): Investigation of eight *Vibrio cholerae* virulence genes." *World Journal of Microbiology and Biotechnology* 24, no. 10: 2133–2141. doi:10.1007/s11274-008-9719-1.

Steenackers, Hans P., Jeremy Levin, Joost C. Janssens, Ami De Weerdt, Jan Balzarini, Jos Vanderleyden, Dirk E. De Vos, and Sigrid C. De Keersmaecker. 2010. "Structure-activity relationship of brominated 3-Alkyl-5-Methylene-2(5H)-furanones and alkylmaleic anhydrides as inhibitors of salmonella biofilm formation and quorum sensing regulated bioluminescence in *Vibrio harveyi*." *Bioorganic and Medicinal Chemistry*. doi:10.1016/j.bmc.2010.05.055.

Su, Yi-Cheng, and Chengchu Liu. 2007. "Vibrio parahaemolyticus: A concern of seafood safety." *Food Microbiology* 24, no. 6: 549–558. doi.org/10.1016/j.fm.2007.01.005.

Sun, Fengjun, Yiquan Zhang, Li Wang, Xiaojuan Yan, Yafang Tan, Zhaobiao Guo, Jingfu Qiu, Ruifu Yang, Peiyuan Xia, and Dongsheng Zhou. 2012. "Molecular characterization of direct target genes and cis-acting consensus recognized by quorum-sensing regulator AphA in Vibrio parahaemolyticus." *PLoS One* 7, no. 9: e44210. doi.org/10.1371/journal.pone.0044210.

Swabe, Tomoo, Kumiko Kita-Tsukamoto, and Fabiano L. Thompson. 2007. "Inferring the evolutionary history of Vibrios by means of multilocus sequence analysis." *Journal of Bacteriology* 189, no. 21: 7932–7936. doi.org/10.1111/j.1758-2229.2011. 00268.x.

Swabe, Tomoo, Yoshitoshi Ogura, Yuta Matsumura, Feng Gao, A. K. M. Amin, Sayaka Mino, Satoshi Nakagawa et al. 2013. "Updating the Vibrio clades defined by multilocus sequence phylogeny: Proposal of eight new clades, and the description of *Vibrio tritonius* sp. nov." *Frontiers in Microbiology* 4: 414. doi.org/10.3389/fmicb.2013.00414.

Taga, Michiko E., and Bonnie L. Bassler. 2003. "Chemical communication among bacteria." *Proceedings of the National Academy of Sciences* 100, no. suppl 2: 14549–14554. doi.org/10.1073/pnas.1934514100.

Tan, Demeng, Sine Lo Svenningsen, and Mathias Middelboe. 2015. "Quorum sensing determines the choice of antiphage defense strategy in Vibrio anguillarum." *MBio*. doi:10.1128/mbio.00627-15.

Teasdale, Margaret E., Jiayuan Liu, Joselynn Wallace, Fatemeh Akhlaghi, and David C. Rowley. 2009. "Secondary metabolites produced by the marine bacterium halobacillus salinus that inhibit quorum sensing-controlled phenotypes in Gram-negative bacteria." *Applied and Environmental Microbiology*. doi:10.1128/AEM.00632-08.

Teasdale, Margaret E., Kellye A. Donovan, Stephanie R. Forschner-Dancause, and David C. Rowley. 2011. "Gram-positive marine bacteria as a potential resource for the discovery of quorum sensing inhibitors." *Marine Biotechnology*. doi:10.1007/s10126-010-9334-7.

Teng, Shu Wen, Jessica N. Schaffer, Kimberly C. Tu, Pankaj Mehta, Wenyun Lu, N. P. Ong, Bonnie L. Bassler, and Ned S. Wingreen. 2011. "Active regulation of receptor ratios controls integration of quorum-sensing signals in *Vibrio harveyi*." *Molecular Systems Biology*. doi:10.1038/msb.2011.30.

Teschler, Jennifer K., David Zamorano-Sánchez, Andrew S. Utada, Christopher J. A. Warner, Gerard C. L. Wong, Roger G. Linington, and Fitnat H. Yildiz. 2015. "Living in the matrix: Assembly and control of *Vibrio cholerae* biofilms." *Nature Reviews Microbiology*. doi:10.1038/nrmicro3433.

Thompson, Fabiano L., Tetsuya Iida, and Jean Swings. 2004. "Biodiversity of Vibrios." *Microbiology and Molecular Biology Reviews* 68, no. 3: 403–431. doi:10.1128/MMBR.68.3.403-431.

Tian, Yang, Qiyao Wang, Qin Liu, Yue Ma, Xiaodan Cao, Lingyu Guan, and Yuanxing Zhang. 2008. "Involvement of LuxS in the regulation of motility and flagella biogenesis in *Vibrio alginolyticus*." *Bioscience, Biotechnology, and Biochemistry* 72, no. 4: 1063–1071. doi.org/10.1271/bbb.70812.

Tsou, Amy M., and Jun Zhu. 2010. "Quorum sensing negatively regulates hemolysin transcriptionally and posttranslationally in *Vibrio cholerae*." *Infection and Immunity* 78, no. 1: 461–467. doi:10.1128/IAI.00590-09.

Tu, Kimberly C., and Bonnie L. Bassler. 2007. "Multiple Small RNAs Act additively to integrate sensory information and control quorum sensing in *Vibrio harveyi*." *Genes and Development*. doi:10.1101/gad.1502407.

Tu, Kimberly C., Christopher M. Waters, Sine L. Svenningsen, and Bonnie L. Bassler. 2008. "A small-RNA-mediated negative feedback loop controls quorum-sensing dynamic in *Vibrio harveyi*." *Molecular Microbiology* 70, no. 4: 896–907. doi.org/10.1111/j.1365-2958.2008. 06452.x.

Tu, Kimberly C., Tao Long, Sine L. Svenningsen, Ned S. Wingreen, and Bonnie L. Bassler. 2010. "Negative feedback loops involving small regulatory RNAs precisely control the *Vibrio harveyi* quorum-sensing response." *Molecular Cell*. doi:10.1016/j.molcel.2010.01.022.

Turovskiy, Yevgeniy, Dimitri Kashtanov, Boris Paskhover, and Michael L. Chikindas. 2007. "Quorum sensing: Fact, fiction, and everything in between." *Advances in Applied Microbiology* 62: 191–234. doi.org/10.1016/S0065-2164(07)62007-3.

Ulitzur, Shimon, Abdul Matin, Cress Fraley, and Edward Meighen. 1997. "H-NS protein represses transcription of the Lux systems of *Vibrio fischeri* and other luminous bacteria cloned into Escherichia coli." *Current Microbiology*. doi:10.1007/s002849900265.

Visick, Karen L., Jamie Foster, Judith Doino, Margaret McFall-Ngai, and Edward G. Ruby. 2000. "*Vibrio fischeri* Lux genes play an important role in colonization and development of the host light organ." *Journal of Bacteriology*. doi:10.1128/JB.182.16.4578-4586.2000.

Visick, Karen L., and Margaret J. McFall-Ngai. 2000. "An exclusive contract: Specificity in the *Vibrio fischeri*-euprymna scolopes partnership." *Journal of Bacteriology*. doi:10.1128/JB.182.7.1779-1787.2000.

Wang, Yunduan, Hui Wang, Weili Liang, Amanda J. Hay, Zengtao Zhong, Biao Kan, and Jun Zhu. 2013. "Quorum sensing regulatory cascades control *Vibrio fluvialis* pathogenesis." *Journal of Bacteriology*. doi:10.1128/JB.00508-13.

Waters, Christopher M., and Bonnie L. Bassler. 2005. "Quorum sensing: Cell-to-cell communication in bacteria." *Annual Review of Cell and Developmental Biology*. 21: 319–346. doi:10.1146/annurev.cellbio.21.012704.131001.

Waters, Christopher M., and Bonnie L. Bassler. 2006. "The *Vibrio harveyi* quorum-sensing system uses shared regulatory components to discriminate between multiple autoinducers." *Genes & Development* 20, no. 19: 2754–2767. doi/10.1101/gad.1466506.

Waters, Lauren S., and Gisela Storz. 2009. "Regulatory RNAs in bacteria." *Cell*. doi:10.1016/j.cell.2009.01.043.

Weber, Barbara, Antony Croxatto, Chang Chen, and Debra L. Milton. 2008. "RpoS induces expression of the *Vibrio anguillarum* quorum-sensing regulator VanT." *Microbiology* 154, no. 3: 767–780. doi:10.1099/mic.0.2007/014167-0.

Weber, Barbara, Medisa Hasic, Chang Chen, Sun Nyunt Wai, and Debra L. Milton. 2009. "Type VI secretion modulates quorum sensing and stress response in *Vibrio anguillarum*." *Environmental Microbiology* 11, no. 12: 3018–3028. https://doi.org/10.1111/j.1462-2920.2009.02005.x.

Whiteley, Marvin, Stephen P. Diggle, and E. Peter Greenberg. 2017. "Progress in and promise of bacterial quorum sensing research." *Nature* 551, no. 7680: 313. doi:10.1038/nature24624.

Williams, Paul, Klaus Winzer, Weng C. Chan, and Miguel Camara. 2007. "Look who's talking: Communication and quorum sensing in the bacterial world." *Philosophical Transactions of the Royal Society B: Biological Sciences* 362, no. 1483: 1119–1134. doi.org/10.1098/rstb.2007.2039.

Wong, Sze Ki, Xiao-Hua Zhang, and Norman YS Woo. 2012. "Vibrio alginolyticus thermolabile hemolysin (TLH) induces apoptosis, membrane vesiculation and necrosis in sea bream erythrocytes." *Aquaculture* 330: 29–36. doi.org/10.1016/j.aquaculture.2011.12.012.

Xie, Z-Y., C-Q. Hu, C. Chen, L-P. Zhang, and C-H. Ren. 2005. "Investigation of seven *Vibrio virulence* genes among Vibrio alginolyticus and *Vibrio parahaemolyticus* strains from the coastal mariculture systems in Guangdong, China." *Letters in Applied Microbiology* 41, no. 2: 202–207. doi.org/10.1111/j.1472-765X.2005.01688.x.

Yang, Hui, Jixiang Chen, Guanpin Yang, Xiao-Hua Zhang, and Yun Li. 2007. "Mutational analysis of the zinc metalloprotease EmpA of *Vibrio anguillarum*." *FEMS Microbiology Letters* 267, no. 1: 56–63. doi.org/10.1111/j.1574-6968.2006.00533.x.

Yang, Menghua, Erin M. Frey, Zhi Liu, Rima Bishar, and Jun Zhu. 2010. "The virulence transcriptional activator AphA enhances biofilm formation by *Vibrio cholerae* by activating expression of the biofilm regulator VpsT." *Infection and Immunity*. doi:10.1128/IAI.00429-09.

Yang, Q., Y. Han, and X. H. Zhang. 2011. "Detection of quorum sensing signal molecules in the family Vibrionaceae." *Journal of Applied Microbiology*. doi:10.1111/j.1365-2672.2011.04998. x.

Yang, Qian, Anne Aamdal Scheie, Tore Benneche, and Tom Defoirdt. 2015. "Specific quorum sensing-disrupting activity (AQSI) of thiophenones and their therapeutic potential." *Scientific Reports*. doi:10.1038/srep18033.

Ye, J., Y. Ma, Q. Liu, D. L. Zhao, Q. Y. Wang, and Y. X. Zhang. 2008. "Regulation of *Vibrio alginolyticus* virulence by the LuxS quorum-sensing system." *Journal of Fish Diseases*. doi: 10.1111/j.1365-2761.2007.00882. x.

Yoshii, Yukiko, Hiroji Nishino, Katsusuke Satake, and Kaoru Umeyama. 1987. "Isolation of *Vibrio fluvialis*, an unusual pathogen in acute suppurative cholangitis." *The American Journal of Gastroenterology*. doi:10.1111/j.1572-0241. 1987. tb06044. x.

Zhang, Hui, Zhenquan Yang, Yan Zhou, Hongduo Bao, Ran Wang, Tingwu Li, Maoda Pang, Lichang Sun, and Xiaohui Zhou. 2018. "Application of a phage in decontaminating *Vibrio parahaemolyticus* in oysters." *International Journal of Food Microbiology*. doi:10.1016/j.ijfoodmicro.2018.03.027.

Zhang, Jun, Yuan Hao, Kaiyu Yin, Qiaoqiao Mao, Rongjing Xu, Yuanxing Zhang, Yue Ma, and Qiyao Wang. 2019. "VqsA controls exotoxin production by directly binding to the promoter of asp in the pathogen *Vibrio alginolyticus*." *FEMS Microbiology Letters*. doi.org/10.1093/femsle/fnz056.

Zhang, Shanshan, Ningning Liu, Weikang Liang, Qingxi Han, Weiwei Zhang, and Chenghua Li. 2017. "Quorum sensing-disrupting coumarin suppressing virulence phenotypes in *Vibrio splendidus*." *Applied Microbiology and Biotechnology*. doi:10.1007/s00253-016-8009-3.

Zhang, Weiwei, and Chenghua Li. 2016. "Exploiting quorum sensing interfering strategies in gram-negative bacteria for the enhancement of environmental applications." *Frontiers in Microbiology*. doi:10.3389/fmicb.2015.01535.

Zhang, Yiquan, Linghui Hu, Yue Qiu, George Osei-Adjei, Hao Tang, Ying Zhang, Rui Zhang et al. 2019. "QsvR integrates into quorum sensing circuit to control *Vibrio parahaemolyticus* virulence." *Environmental Microbiology* 21, no. 3: 1054–1067. doi:10.1111/1462-2920.14524.

Zhang, Yiquan, Yefeng Qiu, Yafang Tan, Zhaobiao Guo, Ruifu Yang, and Dongsheng Zhou. 2012. "Transcriptional regulation of opaR, qrr2–4 and aphA by the master quorum-sensing regulator OpaR in *Vibrio parahaemolyticus*." *PLoS One* 7, no. 4: e34622. doi.org/10.1371/journal.pone.0034622.

Zhu, Jun, and John J. Mekalanos. 2003. "Quorum sensing-dependent biofilms enhance colonization in *Vibrio cholerae*." *Developmental Cell* 5, no. 4: 647–656. doi.org/10.1016/S1534-5807(03)00295-8.

8

Quorum Sensing in Lactic Acid Bacteria

Xiaoyang Pang, Jiaping Lv, Shuwen Zhang, Jing Lu, Obaroakpo Joy Ujiroghene, and Lan Yang

CONTENTS

8.1 Introduction

Lactic acid bacteria (LAB) is a generic term for bacteria that produce large amounts of lactic acid from fermentable carbohydrates [1]. These bacteria are widely distributed in nature and are rich in species diversity. They are not only ideal materials for research classification, biochemistry, genetics, molecular biology, and genetic engineering, but also have important academic value in theory, and are applied in important fields closely related to human life such as industry, agriculture, animal husbandry, food, and medicine [2–5]. The value is also extremely high. Therefore, it has received great attention. Lactic acid bacteria is a group of quite complex bacteria, currently divided into at least 18 genera, a total of more than 200 species [6]. Many lactic acid bacteria have been widely used in the dairy industry, especially as a starter for yogurt production [7–9].

8.2 Quorum Sensing in Lactic Acid Bacteria

In recent years, it has been found that the important characteristics of lactic acid bacteria, such as biofilm formation ability [10,11] and intestinal colonization ability [12,13], are affected and regulated by quorum sensing (QS) systems. The *Lactobacillus* QS system is cell-density-dependent, and can be divided into intraspecific and interspecific information exchanges [14,15]. Intraspecific information exchange takes auto-inducible peptide (AIPs) as signal molecules. The precursor AIPs are transcribed and modified to form mature AIPs [16]. However, it cannot pass through cell wall freely and therefore needs to be transported by an ABC system or other membrane channel proteins in order to reach the extracellular level [17]. When a certain threshold concentration is reached, AIPs activates a two-component system (TCS) and then regulates the expression of the target genes [18]. On the other hand, interspecific signaling is a signal communication mode between different bacterial species, with AI-2 as signaling molecules. A large number of studies have shown that AI-2/LuxS QS regulatory mechanisms exist in Gram-positive and Gram-negative bacteria; capable of sensing environmental changes by AI-2 signaling molecules [19–21]. They also play an important role in intraspecific and interspecific information exchanges between different bacteria species [22]. It has been proven that the luxS gene involved in QS regulates many physiological activities of lactic acid bacteria. According to the existing reports, there are two ways to synthesize AI-2. The first way of AI-2 synthesis is by a four-step enzymatic reaction, using methionine as the starting material. Firstly, methionine produces S-adenosylmethionine (SAM) catalyzed by S-adenosylmethionine synthase (MetK). Secondly, SAM is used as a methyl donor to produce S-adenosylhomocysteine (SAH) under the action of methyltransferase (MT). Thirdly, SAH is hydrolyzed by S-adenosylhomocysteine nucleosidase (Pfs) to S-nucleoside homocysteine (SRH) and adenine; while SRH is hydrolyzed by S-adenosylhomocysteine nucleosidase (Pfs), Homocysteine, and 4,5-dihydroxy-2,3-pentanedione (DPD) are then produced by S-protein. Lastly, the former further produces

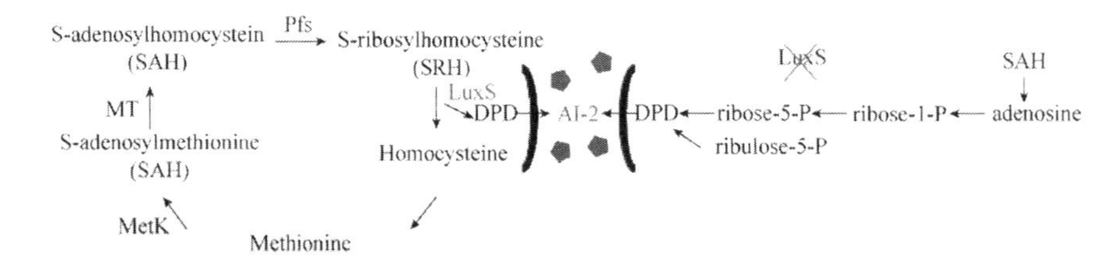

FIGURE 8.1 AI-2 synthesis pathway.

methionine into the methyl cycle, while DPD rearranges to AI-2. In this route, Pfs may contribute more to AI-2 production than LuxS. Another way of AI-2 synthesis involves a semi-biosynthetic pathway derived from thermophilic bacteria. In the absence of LuxS and presence of Pfs, SAH was cleaved onto adenosine and homocysteine by SAH hydrolase. Adenosine was transformed into ribose-1-phosphate by nucleoside phosphorylase. Phosphosemutase isomerized SAH into ribose-5-phosphate, and was further transformed into DPD and AI-2 by thermal induction. However, some studies have suggested that ribose-5-phosphate could also be converted to AI-2 in the absence of LuxS (Figure 8.1).

It has been found that the QS system of lactic acid bacteria is involved in regulating various physiological activities of bacteria. The bacterial quorum sensing system includes signal molecule AIPs and TCS. A typical TCS consists of a histidine protein kinase (HPK) and a cytoplasmic response regulator (RR). The reported *Lactobacillus* quorum sensing signal molecules are listed in Table 8.1.

8.3 Role of QS in the Biofilm Formation of Lactic Acid Bacteria

Biofilm is a highly organized microbial colony aggregate formed by bacteria adhering to inert objects or living organisms, and is encapsulated by extracellular matrix secreted by the bacteria itself. It is a self-protective mode developed by bacteria in order to resist adverse environments. In nature, most bacteria exist in the form of biofilms, except a few which exist in the form of planktons. As probiotics, the key factors for optimizing its functions and promoting health depend on its viability and the formation of biofilms, which can subsequently improve its acid and osmotic resistance.

TABLE 8.1

Lactic Acid Bacteria Quorum Sensing Signal Molecule

Sequence	Trivial Name	Molecular Formula	Species Origin	Receptor	Function	References
MLDYFLFEN	AIPII	$C_{53}H_{57}N_{12}O_{17}S_1$	*Lacbobacillus bulgaricus*	Hpk	Autolysis	[23]
CVGIW, thiolacton linkage between C1 and W5	LamD	$C_{27}H_{38}N_6O_5S_1$	*Lactobacillus plantarum*	LamC	Biofilm-forming	[18]
EQLSFTSIGILQLLTIGTRSCWFFYCRY	PltA	$C_{156}H_{233}N_{37}O_{41}S_2$	*Lactobacillus plantarum*	PltK	Antibacterial	[18]
TPGGFDIISGGPHVAQDVLNAIKDFFK	IP-TX	$C_{131}H_{200}N_{33}O_{38}$	*Lactobacillus sakei*	StxK	Bacteriocin production	[24,25]
KSSAYSLQMGATAIKQVKKLFKKWGW	Plantaricin A, PlnA	$C_{140}H_{223}N_{36}O_{34}S_1$	*Lactobacillus plantarum*	PlnB	Antibacterial	[26]
MAGNSSNFIHKIKQIFTHR	IP-673	$C_{99}H_{157}N_{31}O_{26}S_1$	*Lactobacillus sakei*	SppK	Bacteriocin production	[27]
TNRNYGKPNKDIGTCIWSGFRHC	Orf4	$C_{115}H_{176}N_{37}O_{33}S_2$	*Lactobacillus sakei*	SapK	Bacteriocin production	[24,28]
EGIIVIVVG	SHP1358(15-23)	$C_{42}H_{75}N_9O_{12}$	*Streptococcus thermophilus*	Rgg1358	Unknown	[29]
AILPYFAGCL	SHP0316(18-24), ComS	$C_{52}H_{78}N_{10}O_{12}S$	*Streptococcus thermophilus*	ComR	Responsible for comX induction in species from the salivarius group	[30]
I(Dhb)AI(Dha)LA(Abu)PGAK(Abu) GALMGANMK(Abu)A(Abu) AHASIHV(Dha)K, cyclization between (Ala3-S-Ala7), (Abu8-S-Ala11), (Abu13-S-Ala19), (Abu23-S-Ala26) and (Abu25-S-Ala28)	Nisin A	$C_{143}H_{230}N_{42}O_{37}S_7$	*Lactococcus lactis*	NisK	Antibacterial function	[31]

The formation of biofilms by lactic acid bacteria originates from the gradual transformation of bacteria from reversible adhesion to irreversible binding and accumulation on the surface of the object. It is controlled by QS and the interaction between bacteria and cells. Directly through their biofilms, the bacteria make contact and their physiological activity is changed when compared to their planktonic form. Compared with the planktonic bacteria, those on the surface of the biofilm secrete more organic matrix to the outside of the cell in order to improve the adhesion of the bacteria and further improve their ability to form biofilm on the surface of the object.

8.3.1 Relationship between QS System and Bacterial Adhesion

The improvement of cell adhesion which contributes to bacteria accumulation is a key factor in promoting the formation and maturation of biofilm. It has been demonstrated in some studies as the relationship between biofilm formation ability and cell adhesion.

It has been shown that the ability of *L. rhamnosus* to form biofilm increased by 20% with the addition of mucin and absence of glucose in MRS medium; *L. plantarum* and *L. fermentum* formed biofilm under high concentration of bile and mucus; and Lebeer et al., obtained a mutant strain of *L. rhamnosus*, containing genes related to the synthesis of extracellular polysaccharide which improved the ability of biofilm formation when compared with the wild type [32]. There have been reports on the signal molecules of the QS system that can affect the adhesion of lactic acid bacteria. Buck et al. found out that the ability of gene deletion mutants to adhere to intestinal epithelial cells of *L. acidophilus* luxS decreased by 58% when compared with the wild-type strain [33], and luxS is a key enzyme gene for the molecular synthesis of AI-2. Sturme et al., found a homologous lamBDCA system with the QS system of *L. plantarum* WCFS1 (agrBDCA system), involved in the regulation of biofilm formation in *S. aureus* [18,34]. In the process of regulation of biofilm formation by the agrBDCA system, AgrD acts as a precursor to produce octapeptide-type AIPs with a unique thioester ring structure and is extracellularly transported via the AgrB membrane protein. The concentration of AIP gradually increases as the cell density increases, binds and phosphorylates the histidine kinase receptor AgrC, and then activates the response regulator AgrA. The activated AgrA synergizes with the global regulator SarA, resulting in the transcription of intracellular RNAII (associated with QS amplification) and RNAIII (related to exoprotein production). In the lamBDCA system of *L. plantarum* WCFS1, LamD, a self-induced peptide precursor, is processed into a LamD-derived peptide by LamB processing protein, both of which are expressed by the lamBD gene. The LamD-derived peptide is a cyclic thiolactone pentapeptide similar to the AIP structure of the *agr* system and is named LamD558. It was found that the Lam system is involved in the regulation of the surface adhesion of *L. plantarum*, indicating that the Agr-like QS system (lamBDCA system) in *Lactobacillus* has a non-negligible role in improving cell adhesion. Consequently, some evidence can be provided by the QS system for the formation of biofilm by lactic acid bacteria.

8.3.2 Mechanisms of Formation of QS-Induced Biofilm

The QS system can mediate the formation of biofilm by lactic acid bacteria. A study by Lebeer found that the ability of mutant strains with luxS gene deletion of *Lactobacillus rhamnosus* to form biofilm decreased. Adding AI-2 precursor molecules or wild-type strains could partially complement the formation of biofilm by mutant strains, although not reversed to its original level. The signal molecule of *Lactobacillus* autoinducible peptide is produced at a low level during the growth of *Lactobacillus*. When the cell density reaches a certain threshold, it binds to the receptor kinase HPK in TCS to make it self-phosphorylated. The activation response regulator (RR) is then transferred by the phosphoric acid group, thus resulting in a dynamic range of self-induction. The formation of biofilm by lactic acid bacteria mediated by the QS system plays an important role in the growth and dissemination of the biofilm. Growth period refers to the period of irreversible adherence and accumulation of bacteria. This stage is characterized by the secretion of extracellular viscous matrix, large amounts of adherence and aggregation of lactic acid bacteria, perception of cell density by signal molecules, and transcription of specific RNA in the cells after reaching the threshold, thereby promoting the maturation of biofilm. After biofilm maturation, the metabolic wastes of the inherent bacteria can only be removed through interstitial water channels. When a large number of bacteria in a limited space adhere to the biofilm, accumulation of metabolic wastes occurs as well as intraspecific or interspecific competition. In this process, the partial exfoliation in the biofilm of bacteria is regulated by the cell QS system by activating specific gene expression under adverse environmental factors. Consequently, the bacteria are reversed to their planktonic form, thus preventing bacterial overgrowth, metabolic waste accumulation, and nutrient deficiency.

8.4 Role of QS in Autolysis of *Lactobacillus bulgaricus*

Lactic acid bacteria (LAB) are common starters for the production of yogurt and other dairy products, and their autolysis attracts increasing attention [35–37]. During cheese production, the role of ripening is critical in determining the flavor and texture of the final product, which constitute the basis for cheese product differentiation [38]. Cheese production is a costly process due to its lengthy procedure (from about 3 weeks to >2 years) [39]. Attempts to speed up ripening include: temperature increase [40], starter culture adjustment [41,42], and enzyme supplementation [40]. The lysis of starter strains during ripening releases cytoplasmic enzymes that degrades amino acids into cheese [43]. Such enzymes are believed to promote the degradation of peptides as well as the removal of bitter ones [44–46]. It is therefore important to initiate the autolysis of LAB starter during this process. During yogurt production, LAB autolysis lowers the live cell count of starters [47]. This demonstrates the significant role of LAB autolysis in food production, thereby unveiling the prime importance of its underlying mechanisms. Bacteria usually sense and react to various changes in their environment via a two-component system (TCS) [48]. TCS are found in most bacteria, primarily as essential environmental sensors and cell signaling effectors [49–51]. TCS typically consist of a sensor (membrane-bound histidine protein kinase, HPK) and a signaling effector (response regulator, RR) [49–52].

Genome sequencing predicts multiple TCS in LAB [50], while several others remained uncharacterized. TCS were shown to be associated with the production of bacteriocins [53,54]. The TCS LBA1524/LBA1525 of *L. acidophilus* was implicated in acid tolerance [55], while LBA1430/LBA1431 was implicated in bile tolerance [56]. The TCS lamBDCA system of *Lactobacillus plantarum* is likely to affect commensal host–microbe interactions, since a lamA mutant adheres to surfaces [34]. Autolysis of lactic acid bacteria usually occurs at high cell density [57–59]; TCS enables bacteria to sense, respond, and adapt to a wide range of environments, stressors, and growth conditions [60–62]. The whole genome sequence of *L. bulgaricus* BAA-365 was scanned using the Hmmer software for HisKA, HATPase-c, and Response-reg domains. A total of 7 HPKs and 7 RRs were identified, as shown in Table 8.2.

In our previous research, we demonstrated that *N*-acetylmuramidase has a critical function in the autolysis of *L. bulgaricus* [47], as one of the major degraders of the cell wall. However, our interest lies in the specific protein responsible for the transfer of autolytic signals to N-acetylmuramidase. Two component systems (TCS) are bacterial components that sense the surrounding environment, and LAB autolysis usually occurs at high cell density. In order to study the correlation between cell autolysis and TCS in our research, molecular biology was used to knock out the TCS-encoding gene of an unknown function of *Lactobacillus bulgaricus.* The results showed that the rates of autolysis were remarkably lower for the LBUL_RS00115 gene mutant, in comparison with the wild type strain BAA-365 suggesting that LBUL_RS00115 (coding gene of WP_011677872.1) contributes to the autolysis of *L. bulgaricus* (Figure 8.2). Furthermore, we found a direct interaction with the inclusion of a phosphorelay between WP_011677872.1 and WP_011677871.1. The interaction was characterized by yeast two-hybrid analysis. The above results demonstrated that the TCS of P_011677872.1/ WP_011677871.1 is related to the cell autolysis in L. *bulgaricus*, confirming our previous assumptions. However, a further investigation is necessary to ascertain if the response regulator of this TCS can directly regulate the N-acetylmuramidase gene. The regulatory system of WP_011677872.1/WP_011677871.1 in *L. bulgaricus* could be a novel target for controlling cell autolysis. *N*-acetylmuramidase is involved in other metabolic processes *in vivo*, such as bacterial division, and less impact on bacteria is produced by regulating TCS than *N*-acetylmuramidase.

Conclusively, *Lactobacillus bulgaricus* secretes a signal peptide molecule to the surrounding environment during its growth. When the concentration of the signal peptide in the surrounding environment reaches or exceeds the threshold, the histidine protein kinase vicK located on the cell wall initiates phosphorylation reaction, which further phosphorylates the feedback regulated protein vicR. The phosphorylated vicR binds to RNA polymerase located at lytM, ssaA, and atlA. The transcriptional initiation sites up-regulate the expression of these three genes, and the expressed peptidoglycan hydrolase accelerates cell wall rupture (Figure 8.3).

TABLE 8.2

Functional Prediction of *L. bulgaricus* BAA-365 TCS System

HK Protein No.	RR Protein No.	HK/RR Order	Sequence Homology (HK/RR, Identities/ Identities)	Function Prediction
WP_011678182.1	WP_011678181.1	RH	YP_194286/YP_194287 44%/67% L. aci	Bile tolerance
WP_011677912.1	WP_003620636.1	RH	NP_814923/NP_814922 49%/75% E. fae	Vancomycin resistance
WP_011543754.1	WP_003613569.1	RH	WP_011374912.1/ WP_011374913.1 42%/66% L. sak	Anaerobic regulation
WP_011678447.1	WP_003618182.1	HR	YP_194374 /YP_194375 62%/91% L. aci	Protein hydrolysis, Acid resistance
WP_003621126.1	WP_003624707.1	HR	WP_011373988.1/ WP_011373987.1 55%/82% L. sak	Vancomycin resistance, Anaerobic regulation
WP_011543855.1	WP_003620064.1	RH	WP_005726711.1/ WP_060785076.1 95%/87% L. cri	Unknown
WP_011677872.1	WP_011677871.1	RH	WP_009557837.1/ WP_009557836.1 83%/88% L. equ	Unknown

Note: L. aci, *Lactobacillus acidophilus*; E. fae, *Enterococcus faecalis*; L. sak, *Lactobacillus sakei*; L. cri, *Lactobacillus crispatus*; L. equ, *Lactobacillus equicursoris*.

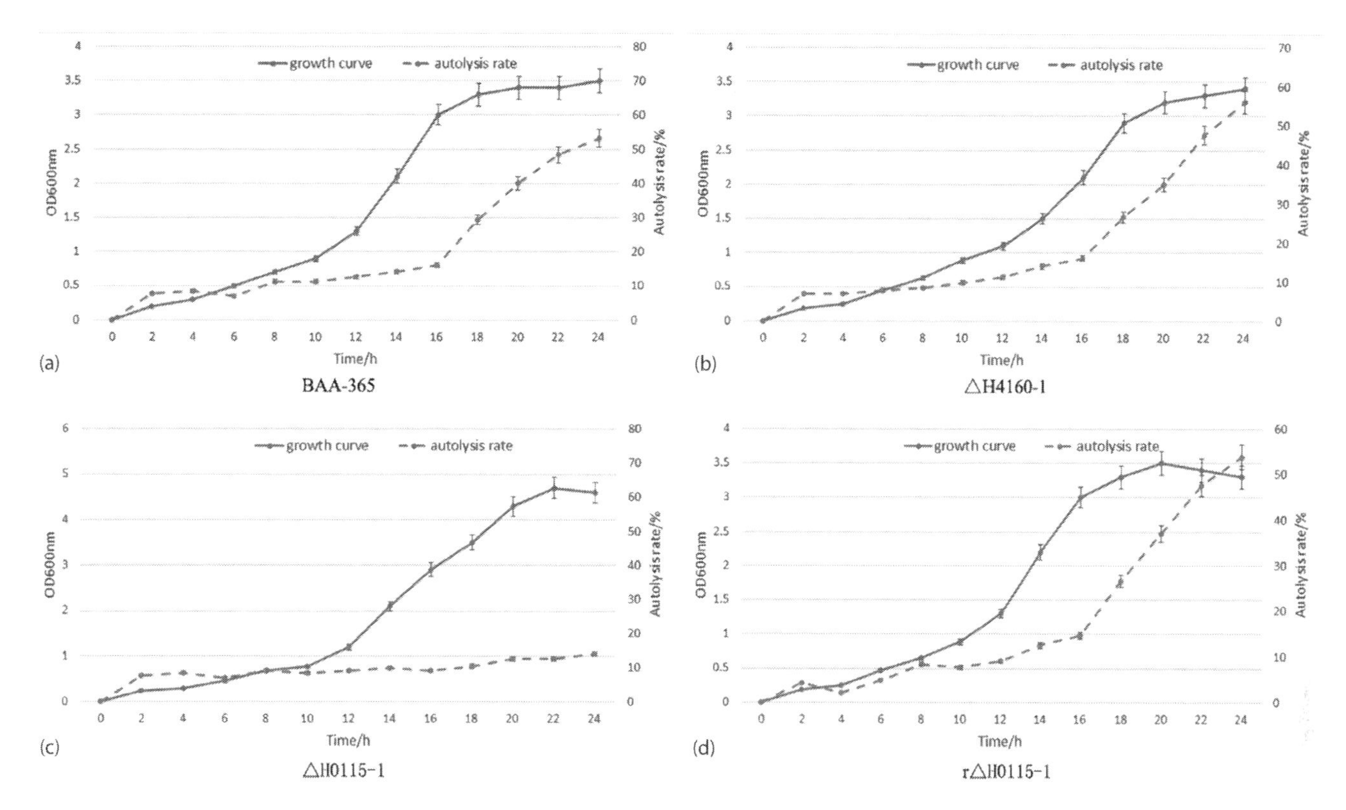

FIGURE 8.2 Autolysis data (a) BAA-365 (b) ΔH4160-1 (c) ΔH0115-1 and (d) rΔH0115-1.

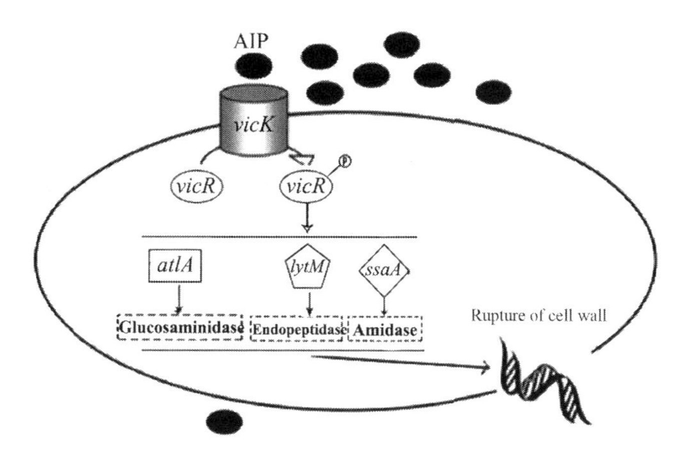

FIGURE 8.3 *Lactobacillus bulgaricus* regulates autolysis through the vicKR pathway.

8.5 Role of QS in Lantibiotics Production of *Lactococcus lactis*

Lantibiotics production by a number of lactic acid bacteria is regulated in a cell-density-dependent manner [63]. Production of these antibiotics is generally initiated at mid- to end-logarithmic growth phase to reach a maximum at the beginning of the stationary phase. In many cases, it has been shown that this mode of regulation involves secreted peptides that act as communication molecules (peptide pheromones) [31]. These peptide pheromones accumulate in the environment during growth, and when a certain threshold concentration is reached, high level lantibiotics production is triggered, which is mediated by interaction of the peptide pheromone with its cognate receptor (histidine kinase sensor protein), resulting in transmembrane signal transduction leading to activation of lantibiotics production. This mode of regulation groups these lantibiotics production systems among the peptide pheromone-mediated quorum sensing systems known in lactic acid bacteria. These systems are involved in cell-density-dependent regulation of a large variety of phenotypes.

Nisin is the most common lantibiotics produced by lactic acid bacteria. The interest in this lantibiotic molecule was greatly stimulated by the broad spectrum of its antimicrobial activity and its increasing application as a natural preservative in the food industry. The biosynthesis of Nisin is completed by the

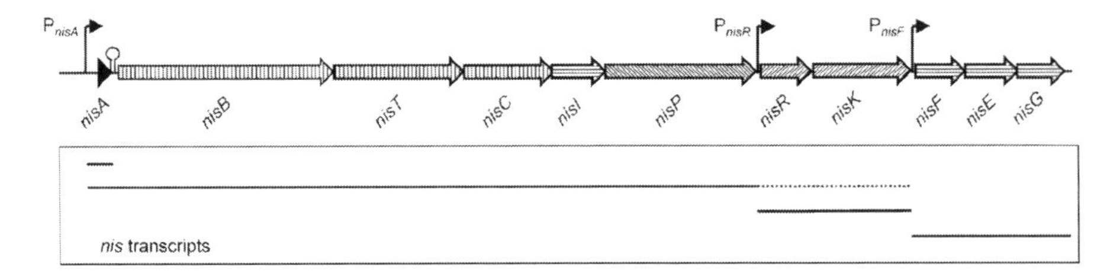

FIGURE 8.4 Organization of Nisin biosynthetic gene clusters.

gene cluster nisABTCIPRKFEG (Figure 8.4), which includes the structural gene encoding the Nisin precursor (nisA), the post-transcriptional modification gene of the Nisin precursor (nisB, nisC), the transmembrane transport gene (nisT), the protease gene of cutting leader sequence (nisP), encoding the immunoprotein gene (nisI, nisF, nisE, nisG) and the Nisin biosynthesis regulatory gene (nisK, nisR).

The synthesis of Nisin is regulated by a two-component signal transduction system consisting of nisK and nisR. When the histidine protein kinase NisK or the response regulatory protein NisR is disrupted, Nisin cannot be synthesized. The quorum sensing regulation process of Nisin synthesis is shown in Figure 8.5. First, the Nisin precursor is formed in the cell. The Nisin precursor is modified by NisB and NisC, NisP is cleaved, and NisT is transported to the outside of the cell. Nisin is accumulated as a signal molecule to a certain threshold. It is recognized by the two-component regulatory system NisK and NisR, which makes the bacteriocin synthesis gene expression. Due to the presence of immune proteins, the bacteria themselves are not inhibited by Nisin.

8.6 Role of QS in Acid Stress Response in LAB

The main research on the environmental stress of lactic acid bacteria is acid stress. Moslehi-Jenabian et al. [64] found that the survival rate of lactic acid bacteria gradually adapting to acid stress environment is higher than that of direct exposure to the acid environment. The concentration of the signal molecule AI-2 is lower than that directly exposed to the acid environment. At the same time, the secretion of the signal molecule AI-2 increases with the decreasing pH value. The transcription levels of Lux S gene in *L. rhamnosus* GG and *L. acidophilus* NCFM increased

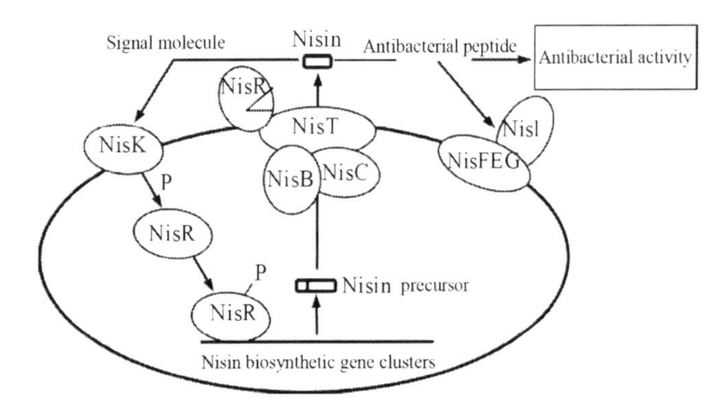

FIGURE 8.5 Regulation of Nisin synthesis by quorum sensing in LAB.

significantly at 1 h under acidic stress, and there were differences in transcription levels between different strains. This may be due to the involvement of *LuxS* gene in the self-protection process of the strain. Rogers et al. [65] found that when stimulated by low doses of penicillin, the *LuxS* gene is over expressed by *Streptococcus pneumoniae* for self-protection. In the study of *L. rhamnosus* GG, the *LuxS* gene mutant showed a significant decrease in acid resistance, which also indicates that the *LuxS* gene plays a key role in the process of adapting to acid stress in *L. rhamnosus* GG. The above results indicate that the QS system mediated by AI-2 may be involved in the acidic stress response of lactic acid bacteria.

8.7 Conclusion

The QS system is an important regulatory mechanism for the interaction of lactic acid bacteria with the environment. In recent years, lactic acid bacteria genome research has been widely carried out, which provides a possibility for systematically expounding the physiological and metabolic mechanisms of lactic acid bacteria at the molecular level. The bioinformatics method is used to predict and analyze the genome of known lactic acid bacteria. The QS system has been found in *Lactobacillus bulgaricus*, *Lactobacillus acidophilus*, and other lactic acid bacteria, but there are still many problems to be further explored. For example: Does AI-2 only regulate the behavior of the bacteria itself, or participate in the exchange of other bacteria in the environment? The in-depth study of the lactic acid bacteria QS system is helpful to clarify the mechanism of lactic acid bacteria adapting to environmental factors.

ACKNOWLEDGMENTS

This chapter and the studies described herein by the author's group were supported by grants from the National Natural Science Foundation of China (31871833). Thanks to Dr. JOY for editing the English language of the manuscript.

REFERENCES

1. Usui, Y., et al., Effects of long-term intake of a yogurt fermented with *Lactobacillus delbrueckii* subsp. *bulgaricus* 2038 and *Streptococcus thermophilus* 1131 on mice. *Int Immunol*, 2018. **30**(7): pp. 319–31.
2. Chen, H., et al., Effects of six substances on the growth and freeze-drying of *Lactobacillus delbrueckii* subsp. *bulgaricus*. *Acta Sci Pol Technol Aliment*, 2017. **16**(4): pp. 403–412.

3. Weiss, N., U. Schillinger, and O. Kandler, *Lactobacillus lactis, Lactobacillus leichmannii* and *Lactobacillus bulgaricus*, subjective synonyms of *Lactobacillus delbrueckii*, and description of *Lactobacillus delbrueckii* subsp. *lactis* comb. nov. and *Lactobacillus delbrueckii* subsp. *bulgaricus* comb. nov. *Syst Appl Microbiol*, 1983. **4**(4): pp. 552–7.
4. Dellaglio, F., et al., *Lactobacillus delbrueckii* subsp. *indicus* subsp. nov., isolated from Indian dairy products. *Int J Syst Evol Microbiol*, 2005. **55**(Pt 1): pp. 401–4.
5. Tanigawa, K. and K. Watanabe, Multilocus sequence typing reveals a novel subspeciation of *Lactobacillus delbrueckii*. *Microbiology*, 2011. **157**(Pt 3): pp. 727–38.
6. Adimpong, D.B., et al., *Lactobacillus delbrueckii* subsp. *jakobsenii* subsp. nov., isolated from dolo wort, an alcoholic fermented beverage in Burkina Faso. *Int J Syst Evol Microbiol*, 2013. **63**(Pt 10): pp. 3720–6.
7. Li, B., et al., Soluble *Lactobacillus delbrueckii* subsp. *bulgaricus* 92059 PrtB proteinase derivatives for production of bioactive peptide hydrolysates from casein. *Appl Microbiol Biotechnol*, 2019. **103**(6): pp. 2731–43.
8. Wang, C., Y. Cui, and X. Qu, Identification of proteins regulated by acid adaptation related two component system HPK1/RR1 in *Lactobacillus delbrueckii* subsp. *bulgaricus*. *Arch Microbiol*, 2018. **200**(9): pp. 1381–93.
9. Yin, X., et al., Proteomes of *Lactobacillus delbrueckii* subsp. *bulgaricus* LBB.B5 incubated in milk at optimal and low temperatures. *MSystems*, 2017. **2**(5): pp. e00027–17.
10. Liu, L., et al., Overexpression of luxS promotes stress resistance and biofilm formation of *Lactobacillus paraplantarum* L-ZS9 by regulating the expression of multiple genes. *Front Microbiol*, 2018. **9**: p. 2628.
11. Wasfi, R., et al., Probiotic *Lactobacillus* sp. inhibit growth, biofilm formation and gene expression of caries-inducing *Streptococcus mutans*. *J Cell Mol Med*, 2018. **22**(3): pp. 1972–83.
12. Pezzulo, A.A., et al., Expression of human paraoxonase 1 decreases superoxide levels and alters bacterial colonization in the gut of *Drosophila melanogaster*. *PLoS One*, 2012. **7**(8): p. e43777.
13. Medellin-Pena, M.J. and M.W. Griffiths, Effect of molecules secreted by *Lactobacillus acidophilus* strain La-5 on *Escherichia coli* O157:H7 colonization. *Appl Environ Microbiol*, 2009. **75**(4): pp. 1165–72.
14. Park, H., et al., Autoinducer-2 properties of kimchi are associated with lactic acid bacteria involved in its fermentation. *Int J Food Microbiol*, 2016. **225**: pp. 38–42.
15. Wilson, C.M., et al., Transcriptional and metabolomic consequences of LuxS inactivation reveal a metabolic rather than quorum-sensing role for LuxS in *Lactobacillus reuteri* 100–23. *J Bacteriol*, 2012. **194**(7): pp. 1743–6.
16. Di Cagno, R., et al., Proteomics of the bacterial cross-talk by quorum sensing. *J Proteomics*, 2011. **74**(1): pp. 19–34.
17. Lubkowicz, D., et al., Reprogramming probiotic *Lactobacillus reuteri* as a Biosensor for *Staphylococcus aureus* derived AIP-I detection. *ACS Synth Biol*, 2018. **7**(5): pp. 1229–1237.
18. Sturme, M.H., et al., Making sense of quorum sensing in lactobacilli: A special focus on *Lactobacillus plantarum* WCFS1. *Microbiology*, 2007. **153**(Pt 12): pp. 3939–47.
19. Duarte, V.D.S., et al., The complete genome sequence of *Trueperella pyogenes* UFV1 reveals a processing system involved in the quorum-sensing signal response. *Genome Announc*, 2017. **5**(29): pp. e00639–17.
20. Park, H., et al., Autoinducer-2 quorum sensing influences viability of *Escherichia coli* O157:H7 under osmotic and in vitro gastrointestinal stress conditions. *Front Microbiol*, 2017. **8**: p. 1077.
21. Zhao, J., et al., Production, detection and application perspectives of quorum sensing autoinducer-2 in bacteria. *J Biotechnol*, 2018. **268**: pp. 53–60.
22. Waheed, H., et al., Insights into quorum quenching mechanisms to control membrane biofouling under changing organic loading rates. *Chemosphere*, 2017. **182**: pp. 40–47.
23. Pang, X., et al., Identification of quorum sensing signal molecule of *Lactobacillus delbrueckii* subsp. *bulgaricus*. *J Agric Food Chem*, 2016. **64**(49): pp. 9421–27.
24. van Belkum, M.J., et al., Structure function relationship of inducer peptide pheromones involved in bacteriocin production in *Carnobacterium maltaromaticum* and *Enterococcus faecium*. *Microbiology*, 2007. **153**(Pt 11): pp. 3660–6.
25. Vaughan, A., V.G. Eijsink, and D. Van Sinderen, Functional characterization of a composite bacteriocin locus from malt isolate *Lactobacillus sakei* 5. *Appl Environ Microbiol*, 2003. **69**(12): pp. 7194–203.
26. Diep, D.B., H.V. LS, and I.F. Nes, A bacteriocin-like peptide induces bacteriocin synthesis in *Lactobacillus plantarum* C11. *Mol Microbiol*, 1995. **18**(4): pp. 631–39.
27. Brurberg, M.B., I.F. Nes, and V.G. Eijsink, Pheromone-induced production of antimicrobial peptides in *Lactobacillus*. *Mol Microbiol*, 1997. **26**(2): pp. 347–60.
28. Diep, D.B., et al., The synthesis of the bacteriocin sakacin A is a temperature-sensitive process regulated by a pheromone peptide through a three-component regulatory system. *Microbiology*, 2000. **146**(Pt 9): pp. 2155–60.
29. Fleuchot, B., et al., Rgg proteins associated with internalized small hydrophobic peptides: A new quorum-sensing mechanism in streptococci. *Mol Microbiol*, 2011. **80**(4): pp. 1102–19.
30. Fontaine, L., et al., A novel pheromone quorum-sensing system controls the development of natural competence in *Streptococcus thermophilus* and *Streptococcus salivarius*. *J Bacteriol*, 2010. **192**(5): pp. 1444–54.
31. Kleerebezem, M., et al., Quorum sensing by peptide pheromones and two-component signal-transduction systems in Gram-positive bacteria. *Mol Microbiol*, 1997. **24**(5): pp. 895–904.
32. Lebeer, S., et al., Functional analysis of luxS in the probiotic strain *Lactobacillus rhamnosus* GG reveals a central metabolic role important for growth and biofilm formation. *J Bacteriol*, 2007. **189**(3): pp. 860–71.
33. Buck, B.L., M.A. Azcarate-Peril, and T.R. Klaenhammer, Role of autoinducer-2 on the adhesion ability of *Lactobacillus acidophilus*. *J Appl Microbiol*, 2009. **107**(1): pp. 269–79.
34. Sturme, M.H., et al., An agr-like two-component regulatory system in *Lactobacillus plantarum* is involved in production of a novel cyclic peptide and regulation of adherence. *J Bacteriol*, 2005. **187**(15): pp. 5224–35.
35. Cibik, R. and M.P. Chapot-chartier, autolysis of dairy leuconostocs and detection of peptidoglycan hydrolases by renaturing SDS-PAGE. *J Appl Microbiol*, 2000. **89**(5): pp. 862–9.
36. Ortakci, F., et al., Late blowing of cheddar cheese induced by accelerated ripening and ribose and galactose supplementation in presence of a novel obligatory heterofermentative nonstarter *Lactobacillus wasatchensis*. *J Dairy Sci*, 2015. **98**(11): pp. 7460–72.

37. Ouzari, H., A. Cherif, and D. Mora, Autolytic phenotype of *Lactococcus lactis* strains isolated from traditional Tunisian dairy products. *J Appl Microbiol*, 2002. **92**(5): pp. 812–20.
38. Lazzi, C., et al., Can the development and autolysis of lactic acid bacteria influence the cheese volatile fraction? The case of Grana Padano. *Int J Food Microbiol*, 2016. **233**: pp. 20–8.
39. Sondergaard, L., et al., Impact of NaCl reduction in Danish semi-hard Samsoe cheeses on proliferation and autolysis of DL-starter cultures. *Int J Food Microbiol*, 2015. **213**: pp. 59–70.
40. Fox, P.F., et al., Acceleration of cheese ripening. *Antonie Van Leeuwenhoek*, 1996. **70**(2–4): pp. 271–97.
41. Garbowska, M., A. Pluta, and A. Berthold-Pluta, Dipeptidase activity and growth of heat-treated commercial dairy starter culture. *Appl Biochem Biotechnol*, 2015. **175**(5): pp. 2602–15.
42. Williams, A.G., et al., Glutamate dehydrogenase activity in lactobacilli and the use of glutamate dehydrogenase-producing adjunct *Lactobacillus* spp. cultures in the manufacture of cheddar cheese. *J Appl Microbiol*, 2006. **101**(5): pp. 1062–75.
43. Xu, Y. and J. Kong, Construction and potential application of controlled autolytic systems for *Lactobacillus casei* in cheese manufacture. *J Food Prot*, 2013. **76**(7): pp. 1187–93.
44. Visweswaran, G.R., et al., Expression of prophage-encoded endolysins contributes to autolysis of *Lactococcus lactis*. *Appl Microbiol Biotechnol*, 2017. **101**(3): pp. 1099–1110.
45. Collins, Y.F., P.L. McSweeney, and M.G. Wilkinson, Evidence of a relationship between autolysis of starter bacteria and lipolysis in cheddar cheese during ripening. *J Dairy Res*, 2003. **70**(1): pp. 105–13.
46. Valence, F., et al., Autolysis and related proteolysis in Swiss cheese for two *Lactobacillus helveticus* strains. *J Dairy Res*, 2000. **67**(2): pp. 261–71.
47. Pang, X., et al., Gene knockout and overexpression analysis revealed the role of N-acetylmuramidase in autolysis of *Lactobacillus delbrueckii* subsp. *bulgaricus* ljj-6. *PLoS One*, 2014. **9**(8): pp. e104829.
48. Cui, Y., et al., A two component system is involved in acid adaptation of *Lactobacillus delbrueckii* subsp. *bulgaricus*. *Microbiol Res*, 2012. **167**(5): pp. 253–61.
49. El-Sharoud, W.M., Two-component signal transduction systems as key players in stress responses of lactic acid bacteria. *Sci Prog*, 2005. **88**(Pt 4): pp. 203–28.
50. Thevenard, B., et al., Characterization of *Streptococcus thermophilus* two-component systems: In silico analysis, functional analysis and expression of response regulator genes in pure or mixed culture with its yogurt partner, *Lactobacillus delbrueckii* subsp. *bulgaricus*. *Int J Food Microbiol*, 2011. **151**(2): pp. 171–81.
51. Zuniga, M., C.L. Gomez-Escoin, and F. Gonzalez-Candelas, Evolutionary history of the OmpR/IIIA family of signal transduction two component systems in *Lactobacillaceae* and *Leuconostocaceae*. *BMC Evol Biol*, 2011. **11**: p. 34.
52. Borland, S., et al., Genome-wide survey of two-component signal transduction systems in the plant growth-promoting bacterium *Azospirillum*. *BMC Genomics*, 2015. **16**: p. 833.
53. Marx, P., M. Meiers, and R. Bruckner, Activity of the response regulator CiaR in mutants of *Streptococcus pneumoniae* R6 altered in acetyl phosphate production. *Front Microbiol*, 2014. **5**: p. 772.
54. Roces, C., et al., Isolation of *Lactococcus* lactis mutants simultaneously resistant to the cell wall-active bacteriocin Lcn972, lysozyme, nisin, and bacteriophage c2. *Appl Environ Microbiol*, 2012. **78**(12): pp. 4157–63.
55. Azcarate-Peril, M.A., et al., Microarray analysis of a two-component regulatory system involved in acid resistance and proteolytic activity in *Lactobacillus acidophilus*. Appl Environ Microbiol, 2005. **71**(10): pp. 5794–804.
56. Pfeiler, E.A., M.A. Azcarate-Peril, and T.R. Klaenhammer, Characterization of a novel bile-inducible operon encoding a two-component regulatory system in *Lactobacillus acidophilus*. *J Bacteriol*, 2007. **189**(13): pp. 4624–34.
57. Hong, W., et al., Nontypeable *Haemophilus* influenzae inhibits autolysis and fratricide of *Streptococcus pneumoniae* in vitro. *Microbes Infect*, 2014. **16**(3): pp. 203–13.
58. Chu, X., et al., Role of Rot in bacterial autolysis regulation of *Staphylococcus aureus* NCTC8325. *Res Microbiol*, 2013. **164**(7): pp. 695–700.
59. Kovacs, Z., et al., Effect of cell wall integrity stress and RlmA transcription factor on asexual development and autolysis in *Aspergillus nidulans*. *Fungal Genet Biol*, 2013. **54**: pp. 1–14.
60. Yu, S., et al., Comparative genomic analysis of two-component signal transduction systems in probiotic *Lactobacillus casei*. *Indian J Microbiol*, 2014. **54**(3): pp. 293–301.
61. Faralla, C., et al., Analysis of two-component systems in group B *Streptococcus* shows that RgfAC and the novel FspSR modulate virulence and bacterial fitness. *MBio*, 2014. **5**(3): pp. e00870–14.
62. Straube, R., Reciprocal regulation as a source of ultrasensitivity in two-component systems with a bifunctional sensor kinase. *PLoS Comput Biol*, 2014. **10**(5): pp. e1003614.
63. Twomey, D., et al., Lantibiotics produced by lactic acid bacteria: Structure, function and applications. *Antonie Van Leeuwenhoek*, 2002. **82**(1–4): pp. 165–85.
64. Moslehi-Jenabian, S., K. Gori, and L. Jespersen, AI-2 signalling is induced by acidic shock in probiotic strains of *Lactobacillus* spp. *Int J Food Microbiol*, 2009. **135**(3): pp. 295–302.
65. Rogers, P.D., et al., Gene expression profiling of the response of *Streptococcus pneumoniae* to penicillin. *J Antimicrob Chemother*, 2007. **59**(4): pp. 616–26.

9

Role of N*-acyl-homoserine Lactone QS Signals in Bacteria-Plant Interactions*

Anton Hartmann and Michael Rothballer

CONTENTS

9.1 Introduction

Plants have, like any other eukaryotic organism, micro-organisms (fungi, bacteria, and archaea) associated at their outer and inner surfaces (Mendes et al. 2013). There is accumulating evidence that these interacting organisms can be regarded as a hologenome assembly, altogether contributing to improved plant health and well-adapted performance at varying environmental conditions (Berendsen et al. 2012). It is proposed that the hologenomes are also relevant in evolutionary aspects (Zilber-Rosenberg and Rosenberg 2008). In the holobiont concept, the plant microbiome is recognized as an integral part of the plant (Sánchez-Canizares et al. 2017). Within this plant microbiome, the root microbiome, associated with the plant´s rhizosphere/root compartment, is a very important player (Bakker et al. 2013). In the pioneer times of soil microbiology, the term rhizosphere was coined referring to the soil/root interface where the roots attract and nourish beneficial, but also harmful, microbes (Hiltner 1904, Hartmann et al. 2008). The highly diverse plant-associated microbial community is able to contribute essential traits to the well-being of the plant throughout its life cycle. The interactions of plants with microorganisms can lead to symbiosis with fungi in mycorrhiza and with bacteria, like rhizobia in legumes. Root-associated and endophytic microbes provide nutrients and water to the plant and protection against pathogens, while they are nourished with mainly carbon substrates from the plant (Sànchez-Canizares et al. 2017). However, the interaction can also result in pathogenesis, and the entire plant microbiome may contribute to which extent the attack by virulent pathogens can finally damage the development of a plant. These different types of microbe-plant interactions in the rhizosphere and endosphere of plants were recently summarized (Mendes et al. 2013, Brink 2016, Kandel et al. 2017, and Hartmann et al. 2019).

Several mechanisms are known which regulate the interaction of microbes and plants. Of major importance is the innate immune response on the plant side. Specific microbial-associated molecular patterns (MAMPs), which are called pathogen-associated molecular patterns (PAMPs) in the case of pathogen-plant interactions (Zipfel and Robatzek 2010), are recognized by the plant with high sensitivity. These PAMPs are specific molecular structures, usually proteins (e.g., flagellum) or exopolysaccharides on the surface of the microbes, which are recognized by a genetically determined response leading to cascades of defense reactions on the plant side. Saprophytic or symbiotic bacteria, which do not provoke plant defense reactions or manage to escape from them, can enter and establish an endophytic or symbiotic interaction (Alqueres et al. 2013, Gonzalez and Marketon 2003,

Reinhold-Hurek et al. 2015). In the initial phase of this interaction, when it is decided which microbial partner is friend or foe, molecular signals play a crucial role.

Bacteria are known to use a high diversity of small diffusible molecules (Rajput et al. 2016) to sense the density of their population and the quality of their environment (Hense et al. 2007). These molecules are so-called autoinducers (AI), since they are produced at very low amounts in the non-induced state but by an autoinducer mechanism, where the produced substance induces its own synthesis in a feed-back loop; the production can be quickly multiplied. A key role is played by the cellular regulator protein R, which forms the R-receptor-AI-complex with the autoinducer molecules. This complex binds to the promotor of the AI-biosynthesis gene (Ng and Bassler 2009), upregulating its expression resulting in an increase of autoinducer synthesis. Core principles of AI systems are described by Hense and Schuster (2015). The general purpose of AI-systems is the control of costly cooperative behaviors, such as secreted public goods. Costly behavior or the initiation and establishment of a completely new life phase, such as becoming a pathogen, symbiont, or beneficial endophyte to a eukaryotic organism, requires a pre-assessment of efficiency using cheap AI signals. The determining factors were explained by Hense et al. (2007) with a hybrid "push-pull" model. "Push" factors are cell density, diffusion, and spatial clustering which determine when a behavior becomes effective. On the other hand, "pull" factors are often stress situations, like starvation, which reduce the activation threshold for the initiation of a changed behavior or initiation of a different life phase (Figure 9.1). Furthermore, AI-controlled behavior offers a homoeostatic and cooperative lifestyle supported by accessory

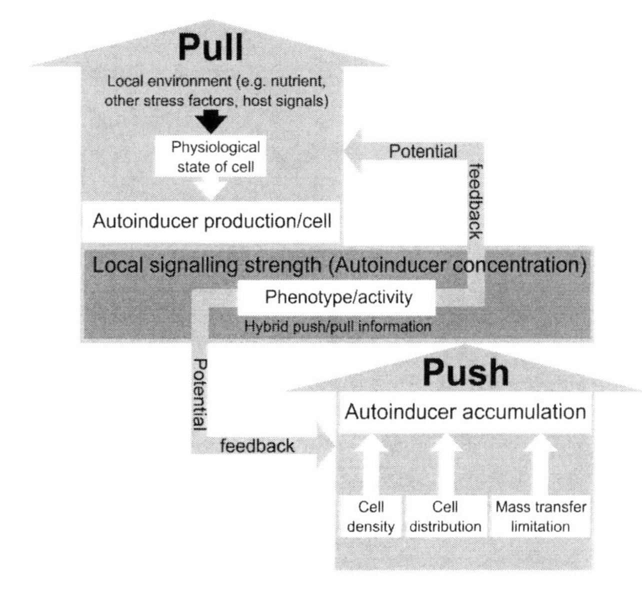

FIGURE 9.1 Scheme of the hybrid push/pull model. Autoinducer regulation systems integrate information on the cell's demand (pull) with the potential cooperative strength of the regulated activity (push). The resulting local autoinducer concentration triggers gene expression based on this integrated information. Via a feedback loop, the regulated phenotype can influence both the pull information (e.g., if produced exoenzymes increased the availability of nutrients) and the push information (e.g., if an altered migratory behavior changed cell density or distribution). (Reproduced from Hense, B. A. et al., *Sensors* 12, 4156–4171, 2012.)

mechanisms. In addition, negative regulators contribute to a better balancing of the bacterial quorum sensing system as they fine tune the synthesis of public goods after an induced state is reached and allow prompt responses to environmental changes (Gupta and Schuster 2013). Because of this extraordinary importance of AI-systems for optimizing cell activities especially for single cell organisms, it is not surprising that quorum sensing using AI-systems is present within a wide taxonomic range. This includes Archaea (Zhang et al. 2012), but also eukaryotes, like fungi and yeasts (Hogan 2006, Sprague and Winans 2006, Padder et al. 2018), algae (Hallmann 2011), animals (Burnard et al. 2008), and possibly even viruses, where host cell lysis may depend on virus concentration (Moussa et al. 2012).

In Gram-negative bacteria, the most widely distributed QS-AI-systems are based on *N*-acyl-homoserine lactones (AHL, or AI-1), which vary in the length of the acyl residue (C4 to C14) having a hydrogen, hydroxyl-residue, or even a carbonylfunction at the C3-atom. Furthermore, furanosyl-borate ester or alternative autoinducer AI-2 (Rajamani and Sayrae 2018) are widespread in Gram-negative and Gram-positive positive bacteria as well as the quinolone-type autoinducer A2 (Fletcher et al. 2007). In Gram-positive but also some Gram-negative bacteria, small peptides have the functions of AIs (Holden et al. 1999, La Rosa et al. 2015). Certainly, many more small diffusible molecules may turn out as AIs, which are still to be discovered.

In the case of surface colonization in the environment e.g., on plant roots, the AI-system reaches its induced state already within micro-colonies, as has been experimentally shown *in situ* at the rhizoplane of tomato and wheat for the AHL-production of *Serratia liquefaciens* and *Pseudomonas putida* (Gantner et al. 2006) and is usually fully upregulated throughout whole root colonizing biofilms. However, within bacterial colonies or biofilms, there is a spatial heterogeneity of AI-production, which may reflect a division of work within the biofilm or ensure a quicker response to changing conditions. This was demonstrated and modeled in colonies of AHL-producing bacteria (Hense et al. 2012). In addition, autoinducers act as biological timers of behavior in *Vibrio harveyi*, where AI-1 and AI-2 systems operate in a coordinated but time shifted fashion (Anetzberger et al. 2012).

The big question, if plants recognize the AHL production of colonizing bacteria, was answered recently. Certainly, in the case of AHL-producing pathogens, chemical signaling between plants and plant-pathogenic bacteria does occur by several means to fight back the attack (Venturi and Fuqua 2013, see below). In the case of beneficial and symbiotic microbes, it has been shown for several cases that QS-AIs also play an important role in these interactions (see below). The role of AHLs in the *Rhizobium*-Legume symbiosis has been recently reviewed in detail (Calatrava-Morales et al. 2018) and will not be treated further in this chapter. In the systemic phase of the interaction of AHL-producing bacteria with plants, AHLs are known as important mediators of beneficial systemic plant responses (Hartmann and Schikora 2012). These are preceded by a specific signaling cascade (Schenk and Schikora 2015).

In this chapter, the current knowledge about the involvement of bacterial QS autoinducers of different plant pathogenic and beneficial/symbiotic bacteria is summarized. We focus on *N*-acyl-homoserine lactones (AHLs) of Gram-negative-bacteria, because there is the most detailed knowledge about the bacteria-plant interactive system, based on transcriptome studies of the involved bacterial and plant partners, as well as powerful chemical analytics of AHL molecules, which will be the central point of the next chapter.

9.2 Analytical Approaches for *N*-acyl-homoserine Lactone and Other QS Autoinducers

9.2.1 Biosensor-Based Analyses

For the analysis of AI compounds, including AHLs (also called AI-1), alternative autoinducer AI-2 (Rajamani and Sayre 2018), quinolone type A2 (Fletcher et al. 2007), peptides (Holden et al. 1999, La Rosa et al. 2015), and DSF factor (Deng et al. 2011) biosensor assays are commonly used because they facilitate straightforward *in situ* detection (Fekete et al. 2010a). In order to use biosensors not only for detection of AIs on plate assays but also for identification, extracts of culture supernatants have to be separated by thin-layer chromatography according to Shaw et al. (1997). To be applicable as an AI biosensor, a bacterial strain must contain a functional AI receptor gene but has to be unable to produce any AI itself. In the case of *N*-acyl-homoserine lactones, mutants in the biosynthesis gene (*I*-gene) or bacterial strains naturally unable to produce AHLs were used as AHL-biosensors. To construct a biosensor, marker genes like luciferase (*luxA*), Green Fluorescent Protein (*gfp*), Red Fluorescent Protein (*rfp*), or others are fused to specific promotors of AHL biosynthesis genes. Steidle et al. (2001) developed for the first time different fluorescent biosensors for a range of short-length (C4) to long-length (C14) acyl chain AHLs using two AHL negative mutants, one of *Pseudomonas putida* IsoF, called F117, and another one of *Serratia liquefaciens* MG1, called MG44. Particular QS biosensor constructs are usually most sensitive for a narrow group of AHLs, as indicated. In general, there is the obstacle that only known AHLs can be identified with this approach. The AHL biosensor strains of Steidle et al. (2001) allowed the *in situ* demonstration of AHL-production in the rhizosphere/rhizoplane of tomato roots in unsterile soil (Figure 9.2). Detailed signaling distances of these *N*-acyl-homoserine lactones, produced by fluorescence labeled *S. liquefaciens* MG1 and *P. putida* IsoF in the rhizosphere of wheat plants, were detected by differentially fluorescence labeled AHL biosensors. This allowed the calculation of calling distances with bio-imaging (Gantner et al. 2006, Dazzo et al. 2015). Furthermore, Ma et al. (2016) screened successfully for AHL using the biosensor technique in the rhizosphere of mangrove plants. A sufficient overview of available biosensors including a list with specificities, reporter system, references, etc. can be found in Steindler and Venturi (2007).

Biosensor-based assays for PQS, HHQ, and related 2-alkyl-4-quinolone quorum sensing signal compounds have also been established (Fletcher et al. 2007). Rajamani and Sayre (2018) developed and successfully applied a LuxP-FRET *in vitro* protein-based bio-monitoring assay for furanosyl borate diester (AI-2). However, false identifications using bacterial biosensors are possible, especially when new compounds do not have a correct reference compound or

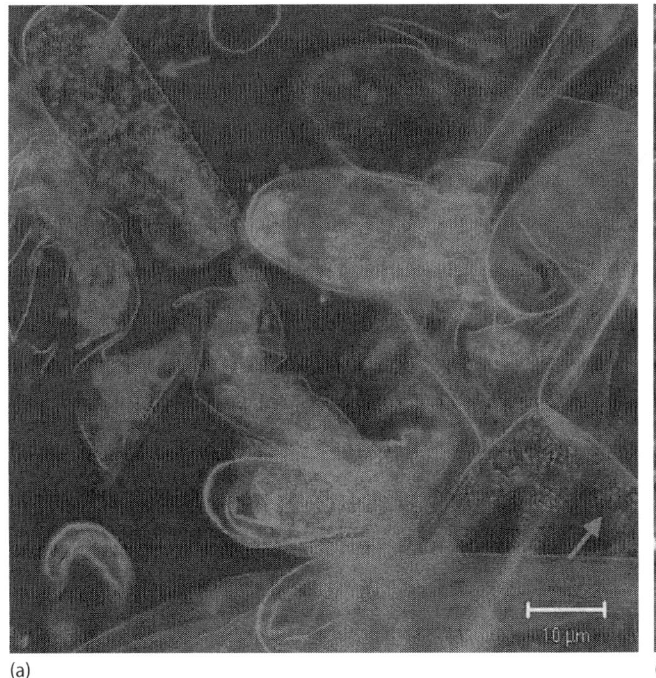

(a)

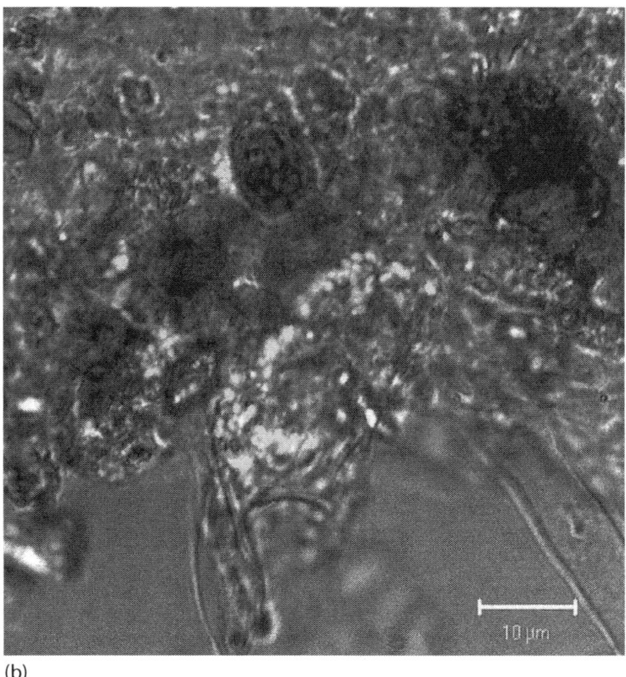

(b)

FIGURE 9.2 AHL-mediated cell-cell communication in the tomato rhizosphere. Axenically grown tomato plants were inoculated with a mixture of a Gfp-based AHL monitor strain and an AHL-producing tomato isolate. (a) Results of a colonization experiment using a mixture of the monitor strain *Serratia liquefaciens* MG44(pJBA132) and a DsRed-tagged derivative of *Rahnella aquatilis* T13 (dark arrow). The light arrow indicates the position of activated sensor cells. (b) The DsRed-tagged monitor strain *Pseudomonas putida* KS35(F117::Tn5-LAS) was also stimulated when it was used for the inoculation of tomato plants grown in nonsterile soil, indicating the production of AHL molecules by the indigenous rhizosphere community. (Reprinted with permission from Steidle, A. et al., *Appl. Environ. Microbiol.* 67, 5761–5770, 2001. Copyright American Society for Microbiology.)

the chromatography was not able to resolve the compound from a similar but not identical running reference (Holden et al. 1999). Also, false-positive cases were reported which can, for example, be generated from medium constituents, like diketopiperazines (Holden et al. 1999). Therefore, a verification of the initial biosensor screening is necessary using different high resolving chemical or immunological methods (Fekete et al. 2010b, Rothballer et al. 2018).

9.2.2 High Resolving UPLC-MS Analyses

The combination of ultrahigh performance liquid chromatography coupled to mass spectrometry (UPLC-MS) is currently the most sensitive quantitative method of analyzing the complete spectrum of AHLs and *N*-acyl-homoserines (HSs), as the first degradation product after lactonase cleavage of AHLs (Li et al. 2006, Fekete et al. 2007, Rothballer et al. 2018). The method was developed for quantitative analysis of AHLs/HSs even in small sample culture volumes and low analyte concentration (Buddrus-Schiemann et al. 2014). Much care has to be taken with matrix effects, which can disturb a reliable performance. Either acidified ethyl acetate extracts, reconstituted in acetate-acetonitrile (50:50% v/v), or eluent samples from solid phase extraction are applied to an UHPLC column, using, for example, a Waters Acquity BEH C18 column and as eluant A acetonitirile and formic acid (10% and 0.1% resp.) in deionized water, and eluant B 100% acetonitrile. The column is coupled to an electrospray ionization quadrupole time-of-flight mass spectrometer (ESI-QToF-MS). For the simultaneous quantitative profiling of *N*-acyl-homoserine lactone and the additional 2-alkyl-4(14)-quinolone families in *Pseudomonas aeruginosa* PAO, also LC-MS/MS in positive electrospray mode (Ortori et al. 2011) was used.

9.2.3 Fourier Transform Ion Cyclotron Resonance-Mass Spectrometry (FTICR-MS) and Other Analyses

This ultra-high resolving mass spectrometry resolves the separated masses to a precision of 1/100 mass units. This allows for each resolved mass a prediction of the composition of elements. Thus, in a background of a high diversity of unknown compounds, the screening and identification of applied AHL compounds was possible in extracts of plant samples (Götz et al. 2007, von Rad et al. 2008). Other analytical approaches for identifying AHL molecules employed partial filling micellar electrokinetic chromatography, followed by electrospray ionization-ion trap mass spectrometry (Frommberger et al. 2003). Also, nano-liquid chromatography was used as robust set-up for on-column sample preparation for AHL-identification by MS (Frommberger et al. 2004). Alternatively, AHLs can be hydrolyzed at alkaline conditions and identified after anion-exchange solid-phase extraction by capillary zone electrophoresis-mass spectrometry (Frommberger et al. 2005). Since almost all AHL-compounds can be purchased nowadays, the set-up of specific analytical approaches is rather easily possible.

9.2.4 Immunochemical Analysis Using Monoclonal Antibodies

An alternative method for separation and quantification of AHLs and HSs is the application of monoclonal antibodies (mAbs) raised for particular *N*-acyl-homoserine lactones (Chen et al. 2010a). Using the ELISA technique, specific AHLs could be identified and quantified in cultures of *Burkholderia cepacia* (Chen et al. 2010b) and plant sap inoculated with AHL producing bacteria. Since the mAbs also had some affinity to HSs, the quantification of HSs was possible after quantitative acid lactonolysis of the AHLs in the samples. In this way, the quantities of AHLs and HSs in samples could be identified (Buddrus-Schiemann et al. 2014). Monoclonal antibodies against 3-oxo-C12-HSL and C4-HSL, the major AHLs of *Pseudomonas aeruginosa*, were raised and characterized (Kaufmann et al. 2006). It could be shown that the application of these AHL-mAbs inhibited the expression of virulence by *P. aeruginosa* and thus succeeded to block infection. Similar results were obtained by Palliyil et al. (2014), who also showed that monoclonal antibodies can be used for immunomodulation of AHL-molecules.

9.3 Plant Responses to Pure AHL Compounds

It has been shown in several studies that some plants have lactonases, which effectively cleave AHL compounds. The majority of legumes have lactonases and thus AHLs have no major systemic effects on the plants. In contrast, many plants like *Arabidopsis*, tomato, wheat, and barley have no lactonases or other hydrolases attacking the AHL molecule specifically, as was shown by Sieper et al. (2014). However, it has been reported that plants produce compounds, which disturb the binding and interaction of AHL signals with plants (Teplitski et al. 2000, Bauer and Mathesius 2004). Interestingly, not only plants, but also human tissues have different activities to modify and degrade AHL-compounds (Teplitski et al. 2011). Very recently it was shown for various barley cultivars that they respond very differently to AHL compounds (Shrestha et al. 2019). While in some cultivars AHL application could stimulate the resistance towards pathogenic fungi, there was no response in other cultivars. Therefore, even within the same plant species genetic differences in the QS signal perception and response occurs on the level of cultivars.

It has been known for more than twenty years that eukaryotic organisms respond to homoserine-lactone mediated prokaryotic signaling (Givskov et al. 1996). It was also demonstrated by Fray et al. (1999) that plants being genetically modified to produce AHLs interfere with rhizosphere bacteria. Degrassi et al. (2007) showed that *Oryza sativa* (rice) plants contain molecules that activate different AHL biosensors and are degraded by an AiiA lactonase. Furthermore, plants genetically modified by introducing AHL biosynthesis genes were shown to produce AHLs in the phylosphere and rhizosphere and release it even into the soil. Two transgenic tomato lines producing either short or a long chain AHLs differentially modulated the beneficial effects of inoculated *Burkholderia graminis* strains M12 and M14, which themselves produced different amounts of long and short chain AHLs (Barriuso et al. 2008). One major conclusion from this study was that the presence of a constantly high concentration of plant derived AHLs interferes with AHL-mediated beneficial mechanisms of the bacteria rendering them ineffective. Likewise, tobacco plants genetically engineered to produce or degrade AHLs showed alterations of induced systemic resistance to *Serratia marcescens* 90–166 (Ryu et al. 2013), demonstrating again the importance of AHLs in bacteria plant interactions.

9.3.1 Early Plant Responses to AHL Compounds

To investigate the interaction of QS active rhizobacteria with host plants, pure AI compounds have to be applied in axenic systems to avoid the influence of any other bacterial molecules or other possibly interfering bacterial activities. Furthermore, the influence of the plant on the stability of the AHL compounds is of key importance. The application of 10 μM C8- and C12-homoserine lactone resulted in immediate nitric oxide (NO) increase in the root tissues. This NO generation is linked to cytosolic Ca^{2+} increase through the action of calmodulin or a calmodulin-like protein (Ma et al. 2008, Zhao et al. 2015). This response cascade has an immediate impact on the stimulation of lateral root formation (Rankl 2017). It has also been shown in *Arabidopsis* that AHL induces intracellular calcium elevation (Song et al. 2011). Ortiz-Castro et al. (2008) had already shown that AHL application initiates stimulatory effects on post-embryonic root development in *Arabidopsis*. Also, von Rad et al. (2008) demonstrated that C6- and C8-AHLs stimulated root growth and demonstrated a specifically altered expression profile in the roots and shoots in AHL treated *Arabidopsis* seedlings. In addition, they found that small and medium length acyl chain AHLs are taken up by *Arabidopsis* using the FTICR-MS-approach (von Rad et al. 2008). Bai et al. (2012) reported that in mung bean, *N*-3-oxo-C12-homoserine lactone activates auxin-induced adventitious root formation via hydrogen peroxide and nitric oxide dependent cyclic GMP signaling.

9.3.2 Uptake and Fate Within Different Plants

Using radio-labeled AHL compounds with different *N*-acyl chain length, Sieper et al. (2014) applied axenic pitman chambers where root tip and cut root base were located in tightly sealed separate compartments. By applying radio-labeled AHLs into the compartment with the root tips, the transport of radiolabeled AHLs could be followed through the root to the basal end of the segment. It was found that in barley roots, AHLs with short acyl chain lengths (C4 and C6) were transported much faster through the root as compared to AHLs with longer acyl chain lengths (C10 and C12). The uptake of the AHLs could be inhibited by blocking ATP-dependent ABC transporters with orthovanadate. When symplastic transport was blocked by addition of potassium chloride, the AHLs were not translocated to the basal end of the root, indicating that functional AHL transport works via the central cylinder. In barley seedlings, a chiral discrimination of AHL molecules to some extent favoring L- instead of D-isomers was also demonstrated during uptake into the shoot (Götz et al. 2007). Compared to barley, the legume yam bean (*Pachyrhizus erosus*), which was also studied, showed more degradation of the added AHL compounds probably due to the release of lactonases by the plant roots.

9.3.3 Systemic Plant Responses to AHLs

Since in *Arabidopsis*, barley, wheat, and tomato AHLs are not degraded, they can be taken up by the plants, as was shown by several methodological approaches in *Arabidopsis* and barley (von Rad et al. 2008, Götz et al. 2007), and ELISA with AHL-specific monoclonal antibodies. In *Arabidopsis thaliana*, C6- and C8-HSL changed the phytohormone balance in roots, causing morphological changes and an increase in root length (von Rad et al. 2008). Accordingly, the RNA profiles of roots and shoots after C6- and C8-AHL application showed a priming of phytohormone-linked gene expression (von Rad et al. 2008). Furthermore, it was demonstrated that the AHL-stimulated root growth in *Arabidopsis* is dependent on two G-protein coupled receptor candidates, Cand2 and Cand7, encoded by the GCR1- and GPA1-genes (Liu et al. 2012, Jin et al. 2012). Schikora et al. (2011) studied the response of *Arabidopsis thaliana* and barley (*Hordeum vulgare*) to the application of oxo-C14-HSL. They observed increased resistance of *Arabidopsis* toward the biotrophic fungus *Golovinomyces orontii*, of barley to *Blumeria graminis* f.sp. *hordei*, and also increased resistance to the hemibiotrophic bacterium *Pseudomonas syringae* pv. *tomato* DC 3000. In the signaling chain, the protein kinases AttMPK3 and AttMPG6 were further stimulated by AHLs when challenged with the elicitor flg22, and the expression of the mitogen-activated transcription factors WRKY22 and WRKY29 were increased. It was concluded that the AHL-induced resistance is MPK6-dependent. *N*-acyl-homoserine lactones with long side chains prime plants also for cell reinforcement and use the salicylic acid/oxylipin pathway for induced resistance to bacterial pathogens (Schenk et al. 2014). A positive influence of bacterial AHLs on growth parameters, pigments, anti-oxidative capacities, and the xenobiotic phase II detoxification enzymes was found after application of short- to medium-length AHLs in barley (Götz-Rösch et al. 2015). According to the better uptake of more water soluble AHLs with shorter acyl-hydrocarbon chain, the modulation of anti-oxidative enzymes and detoxification enzymes is higher. As a systemic response upon application of C8- and C12-HSL to barley roots, the phytohormone salicylic acid accumulated in barley leaves, while jasmonic acid and jasmonic acid-isoleucine remained unaffected. Additionally, only C12-HSL induced abscisic acid accumulation (Rankl et al. 2016). RNA seq based transcriptome analysis revealed that C8-HSL induced gene transcripts involved in cell metabolism and partly in defense, while C12-HSL induced differentially mainly defense genes (Rankl 2017). The activity of the enzyme phenylalanine ammonium lyase increased upon AHL application and reached its peak in barley leaves 12h after AHL application. Despite this enhanced enzyme activity, the flavonoid content (lutonarin and saponarin) was not increased. A 24h to 96h treatment with C8- and C12-HSL of barley seedlings caused a systemic reduction of the biotrophic pathogen *Xanthomonas translucens* pv. *cerealis*. Thus, AHLs stimulate a SA-dependent induced systemic resistance (ISR) in barley via defense gene priming, while the root-accumulated NO could be a possible second messenger leading to SA-accumulation in barley leaves (Rankl 2017). In tobacco, a salicylic acid induced resistance was found toward *Erwinia carotovora* subsp. *carotovora* (Palva et al. 1994). Since this is a plant response toward a plant pathogen, this parallels to some extent the response of plants toward non-pathogenic, AHL-producing rhizobacteria, which have an immune-stimulatory, priming effect.

9.4 Role of *rhizobacterial* AHL Production for Bacteria-Plant Interactions

9.4.1 Role of QS in Pathogen-Host Interactions

The role of quorum sensing in the acquirement of virulence by plant and human pathogens and possibilities of its control has been studied in many cases and reviewed in detail (von Bodman et al. 2003, Rutherford and Bassler 2012). The understanding of the central role of QS signaling and regulation of virulence development and interaction with their host is especially interesting, because human pathogens like *Pseudomonas aeruginosa* and *Burkholderia cepacia* harbor QS systems very similar to strains of the same species colonizing plants. Bacteria closely related to human pathogens are found rather often associated with roots, and closely related strains even show plant growth promotion and biotechnologically valuable traits (Berg at al. 2005). Since many pathogenic bacteria are resistant to conventional antibiotics, there is a need for alternative approaches avoiding resistance development (De Kievit and Iglewski 2000). In this context, the quenching of the central quorum sensing regulation by several means is a hot topic, which is, however, not in the focus of this chapter.

There are very interesting observations about species interactions based on quorum sensing signaling of pathogens and neighboring commensalic bacteria within plants. In the olive and oleander knot, the knot endophytes *Erwinia toletana* and *Pseudomonas agglomerans* support the virulence development of the pathogen *Pseudomonas savastanoi* pv. *savastanoi* based on QS signaling (Hosni et al. 2011, Caballo-Ponce et al. 2018). The use of the same AHL signaling compound by both co-localized bacteria resulted in a more aggressive disease development. QS increased the carbohydrate metabolism by, for example, increasing the expression of the aldolase gene (*garL*) in *E. toletana*. This QS-based commensalism presents an example of *in vivo* interaction based interspecies AHL signaling. Another case study of QS-based cross talk among epiphytic bacteria was reported by Dulla and Lindow (2009). In this study, epiphytic bacteria modulated the behavior of the pathogen *Pseudomonas syringae* pv. *syringae*. 3-oxo-C6 HSL biosynthesis was up-regulated ten times resulting in the biosynthesis of extracellular polysaccharides, motility, and virulence factors in *Ps. syringae*. However, this causes a reduction of virulence, because the acceleration of virulence development was too early in a premature development phase of *Ps. syringae*. Similar findings were reported by Quiñones et al. (2005) for *Ps. syringae* py. *tabaci* growing in transgenic tobacco plants that produced AHL. Based on these observations, the AHL-producing commensals may be used to reduce the virulence of *Ps. syringae*. In other studies, a tight regulatory interaction of QS regulation and intracellular metabolism was found for *Pseudomonas syringae* py. *tabaci* 11528 (Cha et al. 2008). The global regulator for iron uptake (Fur) not only regulates iron homoeostasis, but also keeps the expression of the AHL regulatory system (*psyI* and *psyR*) active. Thus, Fur deletion mutants show decreased swarming and other activities, which are dependent on AHL-QS.

Compared to our understanding of AHL-mediated QS in virulence acquirement by plant pathogens, the knowledge about the role of AHLs in beneficial interactions of rhizobacteria with plants is rather limited.

9.4.2 Role of QS in Beneficial Rhizobacteria-Plant Interactions

Serratia liquefaciens MG1 and *Pseudomonas putida* IsoF applied to roots of tomato plants conferred systemic resistance against the attack of *Alternaria alternata* on the leaves (Schuhegger et al. 2006). An increased level of salicylic acid (SA) and ethylene (ET) was found in roots and shoots of inoculated tomato plants as well as the expression of defense genes, like PR1a and chitinase. In treatments with AHL-deficient mutants, changes in SA and ET levels alongside biocontrol effects were lost. Similar results were obtained from *Serratia plymuthica* HRO-C48, which is an efficient biocontrol agent against damping off-disease in cucumber caused by *Pythium apahnidermatum* and in bean and tomato against *Botrytis cinera* induced grey mold. Transconjugant strains expressing the *aiiA* gene resulting in lactonase degradation of any produced AHL failed efficient root colonization and biocontrol activities (Pang et al. 2009). In *Arabidopsis thaliana* Col-0, the biocontrol rhizosphere bacterium *Chryseobacterium balustrinum* could induce a similar systemic response with increased SA-, ET-, and PR1-expression as *Pseudomonas syringae* pv. *tomato* DC3000 (Ramos-Solano et al. 2008). Although QS and AHLs were not explicitly mentioned, it is quite possible that they are involved too. Another example for the importance of AHL-mediated interactions with plants has been provided by Gao et al. (2007), who found severely changed protein patterns in *Sinorhizobium meliloti* after introduction of the lactonase gene *aiiA* accompanied by a reduced capacity to initiate nodules on the plant host *Medicago truncatula*.

The potent plant growth promoting fungus *Piriformospora indica*, which has a wide application potential in agriculture for improving salt and stress resistance in plants, and induction of systemic resistance to a wide range of pathogens or other beneficial traits (Varma et al. 2012), turned out to harbor an endofungal bacterium, which could not be cured (Sharma et al. 2008). The bacterium was identified as *Rhizobium radiobacter* F4, very closely related to the plant pathogen *R. radiobacter* C58 (formerly *Agrobacterium tumefaciens* C58), except major deletions in the tumor inducing (pTi) and the accessor (pAT) plasmids (Glaeser et al. 2016). Inoculation of barley seedlings with *R. radiobacter* F4 resulted in systemic colonization of the host plant by plant growth promotion and systemic resistance acquirement to powdery mildew fungus *Blumeria graminis*, similar to *P. indica* inoculation. Like the pathogen *R. radiobacter* C58, *R. radiobacter* F4 also produced *N*-hydroxy- and oxo-acyl-homoserine lactone molecules with chain lengths C8, C10, and C12 (Li 2010). Transconjugants with an *aiiA* lactonase gene showed a strongly reduced colonization of barley roots by *R. radiobacter* F4 and diminished systemic beneficial effects (Glaeser et al. unpublished results). Thus, at least part of the PGP-effects of the fungus *P. indica* could be based on the QS activity of the endofungal bacterium, which *P. indica* may have acquired to improve an effective interaction with host plants.

More evidence for the importance of AHL signaling in beneficial plant-microbe interactions comes from the wheat rhizosphere isolate *Acidovorax radicis* N35. It is an endophyte of barley and wheat and has plant growth promoting effects on both host plants (Li et al. 2011). One major type of AHL (3-OH-C10 HSL, as determined by FTICR-MS) was found to be produced by *A. radicis* N35 (Li 2010). Deletion mutations in the *araI* biosynthesis gene had reduced root colonization abilities and PGPR activity (Han et al. 2016). Most interestingly, RNAseq analysis of expressed plant genes in barley seedlings under the influence of the *araI*-mutant showed higher expression of flavonoid biosynthesis genes (like chalcone synthase or chalcone-flavonone isomerase) and higher saponarin and lutonarin content in the shoots (Han et al. 2016). This is in contrast to plants inoculated with the *A. radicis* N35 wildtype, which showed no increased expression of the flavonoid biosynthesis pathway and had lower flavonoid contents. It appeared that the plant exhibited an attenuated flavonoid defense response against the AHL-producing wildtype compared to the AHL-deficient mutant.

Azospirillum spp. are PGPR, which are widely applied in agriculture in Brazil and Argentina. It was shown by Vial et al. (2006) that AHL production was rather variable in *Azospirillum* and mostly present in *Azospirillum lipoferum* strains, while lacking in *A. brasilense*. Most recently it has been shown that in the very frequently applied strain *A. brasilense* Ab-V5 (closely related to the type strain Sp7), only an AHL receptor (LuxR-solo) is present. There is no AHL production since the *luxI* homologous biosynthesis gene is missing (Fukami et al. 2018). However, the stimulation of plant growth was lost using a transconjugant with an AiiA-lactonase as inoculum. Therefore, it can be concluded that in the interaction with the host plant, AHL mimic compounds of plant origin may have been involved, steering the QS regulation of the inoculant *A. brasilense* Ab-V5. It has been suggested by Patel et al. (2013) that LuxR-solos evolved to respond to different molecules from host organisms. Most interestingly, molecules which activate different AHL biosensors were found in rice, which are sensitive to the specific AiiA lactonase. This again supports the suggestion that there are signal molecules in plants which can interact with bacterial LuxR-solos (Degrassi et al. 2007). In human-pathogenic *E. coli* (EHEC), SidA, a LuxR-solo type transcription factor, was found which binds 1-octanoyl-rac-glycerol, a ubiquitous metabolite in all organisms (Nguyen et al. 2015). Also, in *Pseudomonas aeruginosa*, the orphan (=solo) quorum sensing regulator gene *qscR* was described having a relaxed specificity toward QS signals and might therefore widen its interaction potential with the host but also other microbial co-inhabitants in its environment (Lee et al. 2006, Chugani et al. 2014). These are examples of evolving communication systems in bacteria-host interactions.

Gluconacetobacter diazotrophicus is an endophytic diazotroph which can be found in many plants, including sugarcane and rice. It has recently been shown that *G. diazotrophicus* PAL3 mitigates drought stress in rice (*Oryza sativa*) (Filgueiras et al. 2019). This endophyte harbors an AHL quorum sensing system. It was demonstrated that upon inoculation, QS was activated via the expression of the *luxI* gene. Beneficial plant effects were increased growth under salt stress conditions and activation of the PR 1 and PR 10 gene expression, known to lead to defense reactions. In LuxR mutants of *Gluconacetobacter diazotrophicus* PAL3, having no functional QS system anymore, the endophytic colonization of rice roots was lost and the expression of important genes for host-interaction, like exopolysaccharide production and biofilm formation, was abolished (Hofmann and Baldani, unpublished results, Meneses et al. 2011, Alqueres et al. 2013). For sugarcane, it has been reported by Arencibia et al. (2006) that *Gluconacetobacter diazotrophicus* elicits defense response against a pathogenic bacterium *Xanthomonas albilineans*. This could be the result of induced systemic resistance as consequence of QS-AHL signaling. Finally, also the cross talk between sugarcane and the diazotrophic *Burkholderia* sp. strain Q208 leading to plant growth promotion and increased resistance could at least partly be based on QS-AHL signal interaction, because the corresponding gene set was detected in its genome sequence (Paungfoo-Lonhienne et al. 2016). Researchers at EMBRAPA-Agrobiologia, Rio de Janeiro, Brazil, developed an inoculation mixture with five diazotrophic strains (*Gluconacetobacter diazotrophicus*, *Nitrobacter amazonense*, *Herbaspirillum seropedicae*, *Herbaspirillum rubrisubalbicans*, and *Burkholderia tropica*). All five strains harbor genes for quorum sensing systems, based on AHLs, which could very well be responsible for the reliable inoculation effect on sugarcane. As it was shown for a plant pathogen system (Hosni et al. 2011), *in situ* exchange and cooperative action using AHL signaling may also occur in beneficial bacteria-plant interactions. It was further concluded by Hernández-Reyes (2014) that AHL-producing commensal bacteria are able to protect plants against plant and human pathogens.

9.5 Application Aspects

Since the QS autoinducer systems are key regulators of the interaction of bacteria with plants, there are several possibilities for applications. One important approach to fight back virulence outbreaks of pathogenic bacteria is the application of quorum quenching AI-degrading bacteria or molecules interfering with the fine-tuned auto-induction process (LaSarre and Federle 2013, Chen et al. 2013, Defoirdt 2018). In the agricultural context, genetically engineered crop plants could be applied, which express lactonases to counteract the establishment of a QS-controlled virulence of pathogenic bacteria (Ryu et al. 2013).

On the other hand, QS also regulates the interaction of beneficial bacteria with plants (Hartmann et al. 2014). As QS signaling is a key interactive trait, one should make use and apply combinations of, for example, AHL-producing bacteria as bacterial consortia for plant inoculation to ensure the establishment on the target host plant and their positive activities. In this context, it is important to check on the plant side which cultivars do perceive and support AHL-producing bacteria. While the AHL-producing bacteria themselves have measures to cope with plant defense efforts (like reactive oxygen species), as was shown for endophytic bacteria by Alqueres et al. (2013), chemically synthesized AHL or compounds simulating these signals could help to optimize the acceptance of other inoculated bacteria by the plant. PGPR strains with LuxR solos could be good candidates for successful plant inoculation, if the plant signals match the bacterial LuxR-receptor as in the case of *A. brasilense* Ab-V5 (Fukami et al. 2018).

9.6 Summary and Concluding Remarks

The growing knowledge of biological and ecological principles should now be transferred into agricultural practice. In a multidisciplinary approach, the "Phytobiome concept" is going to be developed as a global project, spanning efforts of research institutes with commercial companies and farmers. An important aspect in this multidisciplinary endeavor is the understanding and application of microbial interactions within plant holobionts (Harsani et al. 2018) and the communication level of the organisms involved. Thus, the quorum sensing of microorganisms, among other interactive systems, and the possible use and manipulation of it to control plant pathogens and to support beneficial microbes and symbionts should be regarded as an important issue for agro-biotechnological developments. It may contribute to a sustainable future agriculture in the frame of the "Phytobiome concept" (Leach et al. 2017).

ACKNOWLEDGEMENTS

This chapter is dedicated to Dr. Michael (Mike) Schmid, intermittent head of the Research Unit Microbe-Plant Interactions, and Dr. Burkhard A. Hense, group leader at the Institute of Computational Biology, both at Helmholtz Zentrum München, Neuherberg, Germany. Both unexpectedly died in 2016/2017 at the peak of their scientific productivity. Burkhard was pioneering biosystemic and mathematical modeling aspects of bacterial quorum sensing and Mike was a leading expert in bacterial molecular biology and phylogeny. Their contributions to the development in this research field are highly appreciated.

REFERENCES

Alqueres, S., C. Menses, L. M. Rouws et al. 2013. The bacterial superoxide dismutase and glutathione reductase are crucial for endophytic colonization of rice roots by *Gluconacetobacter diazotrophicus* strain PAL5. *Mol. Plant Microbe Interact.* 26:937–945.

Anetzberger, C., M. Reiger, A. Fekete et al. 2012. Autoinducers act as biological timers in *Vibrio harveyi*. *PLoS One* 7:e48310.

Arencibia, A. D., F. Vinagre, Y. Estevez et al. 2006. *Gluconacetobacter diazotrophicus* elicits a sugarcane defense response against a pathogenic bacterium *Xanthomonas albilineans*. *Plant Signal Behav.* 1:265–273.

Bai, X., C. D. Todd, R. Desikan, Y. Yang, and X. Hu. 2012. *N*-3-oxo-decanoyl-L-homoserine lactone activates auxin-induced adventitious root formation via hydrogen peroxide- and nitric oxide-dependent cyclic GMP signaling in mung bean. *Plant Physiol.* 158:725–736.

Bakker, P. A. H. M., R. L. Berendsen, R. F. Doombos, P. C. A. Wintermans, and C. M. J. Pieterse. 2013. The rhizosphere revisited: Root microbiome. *Front. Plant Sci.* 4:165.

Barriuso, J., B. R. Solano, R. G. Fray, M. Cámara, A. Hartmann, F. J. Gutierrez Manero. 2008. Transgenic tomato plants alter quorum sensing in plant growth-promoting rhizobacteria. *Plant Biotech. J.* 6: 442–452.

Bauer, W. D., and U. Mathesius. 2004. Plant responses to bacterial quorum sensing signals. *Curr. Opin. Plant Biol.* 7:429–433.

Berendsen, R. L., C. M. J. Pieterse, and P. A. H. M. Bakker. 2012. The rhizosphere microbiome and plant health. *Cell* 17:478–486.

Berg, G., L. Eberl, and A. Hartmann. 2005. The rhizosphere as a reservoir for opportunistic human pathogenic bacteria. *Environ. Microbiol.* 7: 1673–1685.

Brink, S. C. 2016. Unlocking the secrets of the rhizosphere. *Trends Plant Sci.* 21:169–170.

Buddrus-Schiemann, K., M. Rieger, M. Mühlbauer et al. 2014. Analysis of *N*-acyl-homoserine lactone dynamics in continuous culture of *Pseudomonas putida* IsoF using ELISA and UPLC/qTOF-MS-related measurements and mathematical models. *Anal. Bioanal. Chem.* 406: 6373–6383.

Burnard, D., R. E. Gozlan, and S. W. Griffiths. 2008. The role of pheromones in freshwater fishes. *J. Fish Biol.* 73:1–16.

Caballo-Ponce, E., X. Meng, G. Uzelac et al. 2018. Quorum sensing in *Pseudomonas savastanoi* pv. *savastanoi* and *Erwinia toletana*: Role in virulence and interspecies interaction in the olive knot. *Appl. Env. Microbiol.* 84:e-00950-18.

Calatrava-Morales, N., M. Intosh, and M. I. Soto. 2018. Regulation mediated by *N*-acyl-homoserine lactone quorum sensing signals in the *Rhizobium*-Legume symbiosis. *Genes* 29:263.

Cha, J. Y., J. S. Lee, J. I. Oh, J. W. Choi, and H. S. Baik. 2008. Functional analysis of the role of *Fur* in the virulence of *Pseudomonas syringae* pv. *tabaci* 11528: *Fur* controls expression of genes involved in quorum sensing. *Biochem. Biophys. Res. Commun.* 366:281–287.

Chen, F., Y. Gao, X. Chen, Z. Yu, and X. Li. 2013. Quorum-quenching enzymes and their application in degrading signal molecules to block quorum-sensing-dependent infections. *Int. J. Mol. Sci.* 14:17477–17500.

Chen, X., E. Kremmer, M. F. Gouzy et al. 2010a. Development and characterization of rat monoclonal antibodies for *N*-acylated homoserine lactones. *Anal. Bioanal. Chem.* 398: 2655–2667.

Chen, X., K. Buddrus-Schiemann, M. Rothballer, P. Krämer, and A. Hartmann. 2010b. Detection of quorum sensing molecules in *Burkholderia cepacia* culture supernatants with enzyme-linked immunosorbent assays. *Anal. Bioanal. Chem.* 398: 2669–2676.

Chugani, S., and Greenberg, E. P. 2014. An evolving perspective on the *Pseudomonas aeruginosa* orphan quorum sensing regulator *QscR*. *Front Cell Infect. Microbiol.* 4:152.

Dazzo, F. B., Y. Yanni, A. Jones, and A. Elsadany. 2015. CMEIAS bio-image informatics that define the landscape ecology of immature microbial biofilms developed on plant rhizoplane surfaces. *AIMS Bioengineering* 2:469–486.

De Kievit, T. R., and B. H. Iglewski. 2000. Bacterial quorum sensing in pathogenic relationships. *Infect. Immun.* 68: 4839–4849.

Defoirdt, T. 2018. Quorum sensing systems as targets for antivirulence therapy. *Trends Microbiol.* 26:313–328.

Degrassi, G., G. Devescovi, R. Solis, L. Steindler, and V. Venturi. 2007. *Oryza sativa* rice plants contain molecules that activate different quorum sensing *N*-acyl-homoserine lactone biosensors and are sensitive to the specific AiiA lactonase. *FEMS Microbiol. Lett.* 269:213–220.

Deng, Y., J. Wu, F. Tao, and L. H. Zhang (2011) Listening to a new language: DSF-based quorum sensing in Gram-negative bacteria. *Chem. Rev.* 111:160–173.

Dulla, G. F. J., and Lindow S. E. 2009. Acyl-homoserine lactone-mediated cross talk among epiphytic bacteria modulates behavior of *Pseudomonas syringae* on leaves. *ISME J.* 3:825–834.

Fekete, A., M. Frommberger, M. Rothballer et al. 2007. Identification of bacterial N-acyl-homoserine lactones (AHL) using ultra performance liquid chromatography (UPLC™), ultrahigh resolution mass spectrometry and *in situ* biosensor constructs. *Anal. Bioanal. Chem.* 387: 455–467.

Fekete, A., C. Kuttler, M. Rothballer et al. 2010a. Dynamic regulation of AHL-production and degradation in *Pseudomonas putida* IsoF. *FEMS Microbiol. Ecol.* 72: 22–34.

Fekete, A., M. Rothballer, A. Hartmann, and P. Schmitt-Kopplin. 2010b. Identification of bacterial autoinducers. In: *Bacterial Signaling* (eds.), R. Krämer, Kirsten Jung, pp. 95–111. WILEY-VCH Verlag GmbH & Co, Weinheim, Germany.

Filgueiras, L., R. Silva, J. Almeida et al. 2019. *Gluconacetobacter diazotrophicus* mitigates drought stress in *Oryza sativa* (L.). *Plant Soil* 2019:1–17.

Fletcher, M. P., S. P. Diggle, M. Camara, and P. Williams. 2007. Biosensor-based assays for PQS, HHQ and related 2-alkyl-4-quinolone quorum sensing signal molecules. *Nat. Protoc.* 2:1254–1262.

Fray, R. G., J. P. Throup, and M. Daykin. 1999. Plants genetically modified to produce AHLs communicate with bacteria. *Nat. Biotechnol.* 17:1017–1020.

Frommberger, M., N. Hertkorn, M. Englmann et al. 2005. Analysis of *N*-acyl-homoserine lactones after alkaline hydrolysis and anion-exchange solid-phase extraction by capillary zone electrophoresis-mass spectrometry. *Electrophoresis* 26:1523–1532.

Frommberger, M., P. Schmitt-Kopplin, F. Menzinger et al. 2003. Analysis of *N*-acyl-L-homoserine lactones produced by *Burkholderia cepacia* with partial filling micellar electrokinetic chromatography–electrospray ionization-ion trap mass spectrometry. *Electrophoresis* 24:3067–3074.

Frommberger, M., P. Schmitt-Kopplin, G. Pin et al. 2004. A simple and robust set-up for on-column sample preconcentration – nano-liquid chromatography – electrospray ionization mass spectrometry for the analysis of *N*-acyl-homoserine lactones. *Anal. Bioanal. Chem.* 378:1014–1020.

Fukami, J., J. L. F. Abrantes, P. del Cerro et al. 2018. Revealing strategies of quorum sensing in *Azospirillum brasilense* strains Ab-V5 and Ab-V6. *Arch Microbiol.* 200:47–56.

Gantner, S., M. Schmid, C. Duerr et al. 2006. *In situ* spatial scale of calling distances and population density-dependent *N*-acyl-homoserine lactone mediated communication by rhizobacteria colonized on plant roots. *FEMS Microbiol. Ecol.* 56:188–194.

Gao, M., H. Chen, A. Eberhard et al. 2007. Effects of AiiA-mediated quorum quenching in *Sinorhizobium meliloti* on quorum sensing signals, proteome pattern, and symbiotic interaction. *Mol. Plant Microbe Interact.* 20:843–856.

Givskov, M., R. de Nys, M. Manefield et al. 1996. Eukaryotic interference with homoserine lactone-mediated prokaryotic signaling. *J. Bacteriol.* 178:6618–6622.

Glaeser, S. P., J. Imani, I. Alabid et al. 2016. Non-pathogenic *Rhizobium radiobacter* RrF4 deploys plant-beneficial activity independent of its host *Piriformospora indica*. *ISME J.* 10:871–884.

Gonzalez, J. E., and M. M. Marketon. 2003. Quorum sensing in nitrogen-fixing rhizobia. *Microbiol. Mol. Biol. Rev.* 67:574–592.

Götz, C., A. Fekete, I. Gebefügi et al. 2007. Uptake, degradation and chiral discrimination of *N*-acyl-D/L-homoserine lactones by barley (*Hordeum vulgare*) and yam bean (*Pachyrhizus erosus*) plants. *Anal. Bioanal. Chem.* 389:1447–1457.

Götz-Rösch, C., T. Riedel, P. Schmitt-Kopplin et al. 2015. Influence of bacterial *N*-acyl-homoserine lactones on growth parameters, pigments, anti-oxidative capacities and the xenobiotic phase II detoxification enzymes in barley and yam bean. *Front. Plant Sci.* 6:205.

Gupta, R., M. Schuster. 2013. Negative regulation of bacterial quorum sensing tunes public goods cooperation. *ISME J.* 7:2159–2168.

Hallmann, A. 2011. Evolution of reproductive development in the volvocine algae. *Sex Plant Reprod.* 42:97–112.

Han, S., D. Li, E. Trost et al. 2016. Systemic responses of barley to the 3-hydroxy-decanoyl-homoserine lactone producing plant beneficial endophyte *Acidovorax radicis* N35. *Front. Plant Sci.* 7:1868.

Harsani, M. A., P. Durán, S. Hacquard. 2018. Microbial interactions within the plant holobiont. *Microbiome* 6:58.

Hartmann, A., D. Fischer, L. Kinzel et al. 2019. Assessment of the structural and functional diversities of plant microbiota: Achievements and challenges – A review. *J. Adv. Res.* 19:3–13.

Hartmann, A., M. Rothballer, and M. Schmid. 2008. Lorenz Hiltner, a pioneer in rhizosphere microbial ecology and soil bacteriology research. *Plant Soil* 312:7–14.

Hartmann, A., M. Rothballer, B. A. Hense, and P. Schröder. 2014. Bacterial quorum sensing compounds are important modulators of microbe-plant interactions. *Front. Plant Sci.* 5: 131.

Hartmann, A., and A. Schikora. 2012. Quorum sensing of bacteria and trans-kingdom interactions of *N*-acyl-homoserine lactones with eukaryotes. *J. Chem. Ecol.* 38:704–713.

Hense, B. A., C. Kuttler, J. Müller et al. 2007. Does efficiency sensing unify diffusion and quorum sensing? *Nature Rev. Microbiol.* 5:230–239.

Hense, B. A., J. Müller, C. Kuttler, A. Hartmann. 2012. Spatial heterogeneity of auto-inducer regulation systems. *Sensors* 12: 4156–4171.

Hense, B. A., and M. Schuster. 2015. Core principles of bacterial autoinducer systems. *Microbiol. Molec. Biol. Rev.* 79:153–169.

Hernández-Reyes, C., S.T. Schenk, C. Neumann, K.-H. Kogel, and A. Schikora. 2014. *N*-acyl-homoserine lactones producing bacteria protect plants against plant and human pathogens. *Microb. Biotechnol.* 7:580–588.

Hiltner, L. 1904. Über neuere Erfahrungen und Probleme auf dem Gebiete der Bodenbakteriologie unter besonderer Berücksichtigung der Brache. *Arb. Dtsch. Landwirtsch Gesellsch.* 98:59–78.

Hogan, D. A. 2006. Talking to themselves: Autoregulation and quorum sensing in fungi. Eukaryot. *Cell* 5:613–619.

Holden, M. T. G., S. R. Chhabra, R. de Nys et al. 1999. Quorum-sensing cross talk: Isolation and chemical characterization of cyclic dipeptides from *Pseudomonas aeruginosa* and other Gram-negative bacteria. *Mol. Microbiol.* 33:1254–1266.

Hosni, T., C. Moretti, G. Devescovi et al. 2011. Sharing of quorum sensing signals and role of interspecies communities in a bacterial plant disease. *ISME J.* 5:1857–1870.

Jin, G., F. Liu, H. Ma et al. 2012. Two G-protein-coupled-receptor candidates, *Cand2* and *Cand7*, are involved in *Arabidopsis* root growth mediated by the bacterial quorum-sensing signals *N*-acyl-homoserine lactones. *Biochem. Biophys. Res. Commun.* 417:991–995.

Kandel, S. L., P. M. Joubert, and S. L. Doty. 2017. Bacterial endophytes: Colonization and distribution within plants. *Microorganisms* 5:77.

Kaufmann, G. F., R. Sartorio, S. H. Lee et al. 2006. Antibody interference with *N*-acyl-homoserine lactone-mediated bacterial quorum sensing. *J. Am. Chem. Soc.* 128:2802–2803.

La Rosa, S. L., M. Solheim, D. B. Diep, I. F. Nes, and D. A. Brede. 2015. Bioluminescence based biosensors for quantitative detection of enterococcal peptide-pheromone activity reveal inter-strain telesensing *in vivo* during polymicrobial systemic infection. *Sci. Rep.* 5:8339.

LaSarre, B., and M. J. Federle. 2013. Exploiting quorum sensing to confuse bacterial pathogens, *Microbiol. Mol. Biol. Rev.* 77: 73–111.

Leach, J. E., L. R. Triplett, C. T. Argueso, and P. Trivedi. 2017. Communication in the phytobiome. *Cell* 169:587–596.

Lee, J. H., Y. Lequette, and E. P. Greenberg 2006. Activity of purified QscR, a *Pseudomonas aeruginosa* orphan quorum-sensing transcription factor. *Mol. Microbiol.* 59:602–609.

Li, D. 2010. Phenotypic variation and molecular signalling in the interactions of the rhizosphere bacteria *Acidovorax radicis* N35 and *Rhizobium radiobacter* F4 with roots. PhD-thesis, Faculty of Biology, Ludwig-Maximilian-University, München, Germany.

Li, D., M. Rothballer, M. Schmid, J. Esperschütz, and A. Hartmann. 2011. *Acidovorax radicis* sp. nov., a rhizosphere bacterium isolated from wheat roots. *Int. J. Syst. Evol. Microbiol.* 61: 2589–2594.

Li, X., A. Fekete, M. Englmann et al. 2006. Development of a solid phase extraction – ultra pressure liquid chromatography method for the determination of *N*-acyl-homoserine lactones from bacterial supernatants. *J. Chromatography A* 1134:186–193.

Liu, F., Z. Bian, Z. Jia et al. 2012. The GCR1 and GPA1 participate in promotion of *Arabidopsis* primary root elongation induced by *N*-acyl-homoserine lactones, the bacterial quorum-sensing signals. *Mol. Plant-Microbe Interact.* 25:677–683.

Ma, W., A. Smigel, Y.-C. Tsai, J. Braam, G. A. Berkowitz. 2008. Innate immunity signaling: Cytosolic Ca^{2+} elevation is linked to downstream nitric oxide generation through the action of calmodulin or a calmodulin-like protein. *Plant Physiol.* 148:818–828.

Ma, Z. P., Y. M. Lao, H. Jin et al. 2016. Diverse profiles of AI-1 type quorum sensing molecules in cultivable bacteria from mangrowe (*Kandelia obovata*) rhizosphere environment. *Front. Microbiol.* 7:1957.

Mendes, R., P. Garbeva, P., and J. Raaijmakers. 2013. The rhizosphere microbiome: Significance of plant beneficial, plant pathogenic, and human pathogenic microorganisms. *FEMS Microbiol. Rev.* 37:634–663.

Meneses, C. H., L. M. F. Rouws, J. L. Simoes-Araujo, M. S. Vidal, and J. I. Baldani. 2011. Exopolysaccharide production is required for biofilm formation and plant colonization by the nitrogen-fixing endophyte *Gluconacetobacter diazotrophicus* PAL5. *Mol. Plant-Microbe Interact.* 24:1448–14458.

Moussa, S. H., V. Kuznetsov, T. A. Tran, J. C. Sacchettini, and R. Young. 2012. Protein determinants of phage T4 lysis inhibition. *Protein Sci.* 21:571–582.

Ng, W. L., and B. L. Bassler. 2009. Quorum sensing network architectures. *Annu. Rev. Genet.* 43:197–222.

Nguyen, Y., N. X. Nguyen, J. L. Rogers et al. 2015. Structural and mechanistic roles of novel chemical ligands on the SdiA quorum-sensing transcription regulator. *mBio* 6:e02429-14.

Ortiz-Castro, R., M. Martinez-Trujillo, and J. López-Bucio. 2008. *N*-acyl-L-homoserine lactones: A class of bacterial quorum-sensing signals alter post-embryonic root development in *Arabidopsis thaliana*. *Plant Cell Environ.* 31:1597–1509.

Ortori, C. A., J. F. Dubern, S. R. Chhabra et al. 2011. Simultaneous quantitative profiling of *N*-acyl-homoserine lactones and 2-alkyl-4(14)-quinolone families of quorum-sensing signaling molecules by LC-MS/MS. *Anal. Bioanal. Chem.* 399:839–850.

Padder, S. A., R. Prasad, and A. H. Shah. 2018. Quorum sensing: A less known mode of communication among fungi. *Microbiol. Res.* 210:51–58.

Palliyil, S., C. Downham, I. Broadbent, K. Charlton, and A. J. Porter. 2014. High-sensitivity monoclonal antibodies specific for homoserine lactones protect mice from lethal *Pseudomonas aeruginosa* infections. *Appl. Environ. Microbiol.* 80:462–469.

Palva, T. K., M. Hurtig, P. Saindrenan, and E. T. Palva. 1994. Salicylic acid induced resistance to *Erwinia carotovora* subsp. *carotovora* in tobacco. *Mol. Plant-Microbe Interact.* 7:356–363.

Pang, Y. D., X. G. Liu, Y. X. Ma et al. 2009. Induction of systemic resistance, root colonization, and biocontrol activities of the rhizosphere strain of *Serratia plymuthica* are dependent on *N*-acyl-homoserine lactones. *Eur. J. Plant Pathol.* 124:261–268.

Patel, H. K., Z. R. Suárez-Moreno, G. Degrassi et al. 2013. Bacterial *LuxR* solos have evolved to respond to different molecules including signals from plants. *Front. Plant Sci.* 4:447.

Paungfoo-Lonhienne, C., T. G. A. Lonhienne, Y. K. Yeoh et al. 2016. Crosstalk between sugarcane and a plant-growth promoting *Burkholderia* species. *Sci. Rep.* 6:37389.

Quiñones, B., G. Dulla, and S. E. Lindow. 2005. Quorum sensing regulates exopolysaccharide production, motility and virulence in *Pseudomonas syringae*. *Mol. Plant-Microbe Interact.* 18:682–693.

Rajamani, S., and R. Sayre. 2018. Biosensors for the detection and quantification of AI-2 class quorum sensing compounds. *Methods Mol. Biol.* 1673:73–88.

Rajput, A., K. Kaur, and M. Kumar. 2016. SigMol: Repertoire of quorum sensing signaling molecules in prokaryotes. *Nucleic Acids Res.* 44:D634–D639.

Ramos-Solano, B., J. Barriuso, P. de la Iglesia et al. 2008. Systemic disease protection elicited by plant-growth promoting *rhizobacterial* strains: Relationship between metabolic responses, systemic disease protection and biotic elicitors. *Phytopathol.* 98:451–457.

Rankl, S. C. 2017. Inter-kingdom signaling: The role of homoserine lactones in early responses and resistance in barley (*Hordeum vulgare* L.). Dissertation Technische Universität München, Germany.

Rankl, S. C., B. Gunsé, T. Sieper et al. 2016. Microbial *N*-acyl-homoserine lactones (AHLs) are effectors of root morphological changes in barley. *Plant Sci.* 253:130–140.

Reinhold-Hurek, B., W. Buenger, C. S. Burbano, M. Sabale, and T. Hurek. 2015. Roots shaping their microbiome: Global hot spots for microbial activity. *Annu. Rev. Phytopathol.* 53:403–424.

Rothballer, M., J. Uhl, J. Kunze, P. Schmitt-Kopplin, and A. Hartmann. 2018. Detection of the bacterial quorum sensing signaling molecules *N*-acyl-homoserine lactones and *N*-acyl-homoserine with enzyme-linked immunosorbent assay (ELISA) and via ultrahigh performance liquid chromatography coupled to mass spectrometry (UPLC-MS). In: *Quorum Sensing: Methods and Protocols*. (eds.), L. Leoni, G. Rampioni, Springer series Methods in Molecular Biology 1673:61–72, Springer Nature.

Rutherford, S. T., and B. L. Bassler. 2012. Bacterial quorum sensing: Its role in virulence and possibilities for its control. *Cold Spring Harb. Perspect. Med.* 2:a012427.

Ryu, C. M., H. K. Choi, C. H. Lee et al. 2013. Modulation of quorum sensing in acyl-homoserine lactone-producing or – degrading tobacco plants leads to alteration of induced systemic resistance elicited by the rhizobacterium *Serratia marcescens* 90–166. *Plant Pathol.* 160:413–420.

Sánchez-Canizares, C., B. Jorrin, P. S. Poole, and A. Tkacz. 2017. Understanding the holobiont: The interdependence of plants and their microbiome. *Curr. Opin. Microbiol.* 38:188–196.

Schenk, S. T., and A. Schikora. 2015. AHL-priming functions via oxylipin and salicylic acid. *Front. Plant Sci.* 5:784

Schenk, S. T., C. Hernández-Reyes, B. Samans et al. 2014. *N*-acyl-homoserine lactone primes plants for cell wall reinforcement and induces resistance to bacterial pathogens via the salicylic acid/oxylipin pathway. *Plant Cell* 26:2708–2723.

Schikora, A., S. T. Schenk, E. Stein et al. 2011. *N*-acyl-homoserine lactone confers resistance towards biotrophic and hemibiotrophic pathogens via altered activation of AtMPK6. *Plant Physiol.* 157:1407–1418.

Schuhegger, R., A. Ihring, S. Gantner et al. 2006. Induction of systemic resistance in tomato plants by *N*-acyl-homoserine lactone–producing rhizosphere bacteria. *Plant Cell Environm.* 29: 909-918.

Sharma, M., M. Schmid, M. Rothballer et al. 2008. Detection and identification of mycorrhiza helper bacteria intimately associated with representatives of the order *Sebacinales*. *Cell. Microbiol.* 10: 2235–2246.

Shaw, P. D., G. Ping, S. L. Daly et al. 1997. Detecting and characterizing *N*-acyl-homoserine lactone signal molecules by thin-layer chromatography. *Proc. Natl. Acad. Sci. USA* 94:6036–6041.

Shrestha, A., A. Elhady, S. Adss et al. 2019. Genetic differences in barley govern the responsiveness to *N*-acyl-homoserine lactones. *Phytobiomes J.* 3(3):191–202.

Sieper, T, S. Forczek, M. Matucha, P. Krämer, A. Hartmann, P. Schröder. 2014. *N*-acyl-homoserine lactone uptake and systemic transport in barley rest upon active parts of the plant. *New Phytol.* 201:545–555.

Song, S., Z. Jia, J. Xu, Z. Zhang, Z. Bian 2011. *N*-butyryl-homoserine lactone, a bacterial quorum-sensing signaling molecule, induces intracellular calcium elevation in *Arabidopsis* root cells. *Biochem. Biophys. Res. Commun.* 414:355–360.

Sprague, G. F., and S. C. Winans. 2006. Eukaryotes learn to count: Quorum sensing by yeasts. *Genes Dev.* 20:1045–1049.

Steidle, A., K. Sigl, R. Schuhegger et al. 2001. Visualization of *N*-acyl-homoserine lactone-mediated cell-cell communication between bacteria colonizing the tomato rhizosphere. *Appl. Environ. Microbiol.* 67:5761–5770.

Steindler, L., and V. Venturi. 2007. Detection of QS-*N*-acyl-homoserine lactone signal molecules by bacterial biosensors. *FEMS Microbiol. Lett.* 266:1–9.

Teplitski, M., J. B. Robinson, and W. D. Bauer. 2000. Plants secrete substances that mimic bacterial *N*-acyl-homoserine lactone signal activities and affect population density-dependent behaviors in associated bacteria. *Mol. Plant-Microbe Interact.* 13:637–648.

Teplitski, M., U. Mathesius, K. P. Rumbaugh. 2011. Perception and degradation of *N*-acyl-homoserine lactone quorum sensing signals by mammalian and plant cells. *Chem. Rev.* 111:100–116.

Varma, A., M. Bakshi, B. Lou, A. Hartmann, and R. Oelmueller. 2012. *Piriformospora indica*: A novel plant growth-promoting mycorrhizal fungus. *Agric. Res.* 1:117–131.

Venturi, V., and C. Fuqua. (2013) Chemical signalling between plants and plant-pathogenic bacteria. *Annu. Rev. Phytopathol.* 51:17–37.

Vial, L., C. Cuny, K. Gluchoff-Fiasson et al. 2006. *N*-acyl-homoserine lactone-mediated quorum sensing in *Azospirillum*: An exception rather than a rule. *FEMS Microbiol. Ecol.* 58:155–168.

von Bodman, S. B., W. D. Bauer, and D. L. Coplin. 2003. Quorum sensing in plant-pathogenic bacteria. *Annu. Rev. Phytopathol.* 41:455–482.

von Rad, U., I. Klein, P. I. Dobrev et al. 2008. The response of *Arabidopsis thaliana* to *N*-hexanoyl-DL-homoserine lactone, a bacterial quorum sensing molecule produced in the rhizosphere. *Planta* 229:73–85.

Zhang, G., F. Zhang, G. Ding et al. 2012. Acyl-homoserine lactone-based quorum sensing in a methanogenic archaeon. *ISME J.* 6:1336–1344.

Zhao, Q., C. Zhang, Z. Jia et al. 2015. Involvement of calmodulin in the regulation of primary root elongation by *N*-3-oxo-hexanoyl-homoserine lactone in *Arabidopsis thaliana*. *Front. Plant Sci.* 5:807.

Zilber-Rosenberg, I., and E. Rosenberg. 2008. Role of micro-organisms in the evolution of animals and plants: The hologenome theory of evolution. *FEMS Microbiol. Rev.* 32:723–735.

Zipfel, C., and S. Robatzek. 2010. Pathogen-associated molecular pattern-triggered-immunity: Veni, vidi…? *Plant Physiol.* 154: 551–554.

10

Quorum Sensing and the Environment: Open Questions in Plant-Associated Bacteria

Ana Carolina del V. Leguina, Elisa V. Bertini, Mariano J. Lacosegliaz, Lucía I. Castellanos de Figueroa, and Carlos G. Nieto-Peñalver

CONTENTS

10.1 Introduction

As a general term, the environment comprises the biotic (animals, plants, and microorganisms) and abiotic (physical and chemical) factors influencing an organism and a community. The modification of the behavior and fitness of a microorganism is due to the reprogramming in gene expression in response to fluctuations in biotic and abiotic components of the environment. These factors put the microorganisms in circumstances that are far from the well-controlled *in vitro* conditions utilized in the laboratory for their characterization. The environment alters the microbe physiology directly through devoted microbial systems for sensing, responding and adapting to these changes. The environment also modifies the activity of microbial regulatory mechanisms not connected with the sensing of the environment. Among these, quorum sensing (QS) systems, which mainly provide information about the cell density, can also be altered. The environment can directly influence the activity of the QS systems and their regulation through the signal molecule, its production or reception, or indirectly through other cellular components. The understanding of how these different environmental factors modulate the microbial communication is indispensable for the exploitation of the beneficial microbes and for a better fight against the phytopathogens.

10.2 Quorum Sensing in Plant Pathogenic Bacteria

A wide variety of plants are susceptible to the attack of pectinolytic bacteria that cause soft rotting by means of an array of plant-cell wall degrading enzymes (PCWDEs), including pectate lyases, pectin lyases, cellulases, and proteases, that macerate the vegetal tissues producing the characteristic rotting. In 1993, Palva and collaborators reported that *expI* gene from *Pectobacterium carotovorum*, one of the best characterized members of this group of bacteria, could complement a *luxI* mutation in *V. fischeri* (Pirhonen et al. 1993) (Figure 10.1). Together with the biochemical data, this evidence showed that the same signal molecule, 3OC6-HSL, was produced by the phytopathogen, suggesting for the first time that acyl homoserine lactones (AHLs) could be conserved for genetic regulation.

The solving of the chemical structure of signal molecule produced by another plant-associated bacterium, *A. fabrum* (formerly *A. tumefaciens*), confirmed that a wide group of bacterial species shared the regulatory mechanism based on AHL signals. *Agrobacterium* spp. easily acquire the oncogenic Ti plasmid by a conjugative pilus (Lin, Binns, and Lynn 2008; Lang and Faure

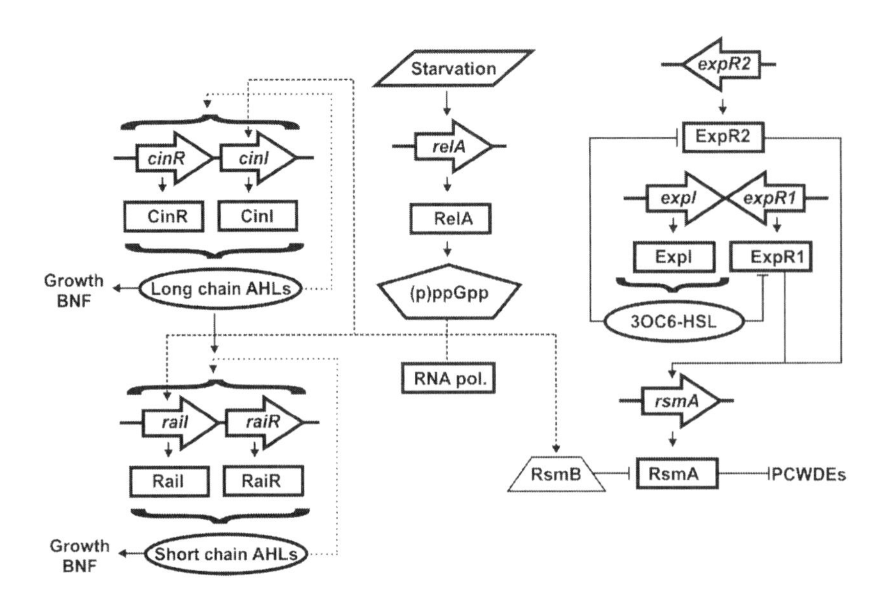

FIGURE 10.1 QS systems in a plant-beneficial and a phytopathogenic bacterium and the interconnection with the stringent response. Left, in *R. etli* CNPAF512 the CinR/CinI system coded in *cinR* and *cinI* is involved in the synthesis of AHLs with long acyl chains. Located at the top of a hierarchical regulatory cascade, this system induces the activity of the RaiR/RaiI system coded in *raiR* and *raiI*, which produces AHLs with short acyl chains. Both systems are involved in the regulation of growth and BNF. Positive autoregulatory loops have been described in both systems. Right, the QS system in *Pca* is composed of ExpI and ExpR1, coded in *expI* and *expR1*, and the second AHL receptor ExpR2, coded in *expR2*. In contrast to other systems, when C6-HSL is bound by ExpR1 or ExpR2, the induction of the *rsmA* transcription by the LuxR homologs is no longer exerted. RsmA destabilizes mRNAs that code for PCWDEs, among others. RsmB is a non-coding RNA that sequesters RsmA. Center, the SR induced under starvation conditions leads to the production of (p)ppGpp by RelA. This alarmone modifies the activity of the RNA polymerase. The SR increases the expression of *cinI* and *raiI* in *R. etli*, and the expression of *rsmB* in *Pca*. Intersecting lines denote repression; arrows denote induction. Dashed lines are related to SR, and dotted lines indicate autoinduction.

2014; Zhu et al. 2000). When a pTi-harboring bacterium detects sugars and phenolic compounds produced from a wounded plant, the bacterial T-DNA is transferred to a plant cell via a type IV secretion system. Once integrated, T-DNA directs the production of the phytohormones responsible for the characteristic crown gall tumors. T-DNA also permits the production of opines that serve as nutrients for the bacterium in the tumor. A subgroup of opines, the conjugative opines allow the activity of the QS system, and the conjugation of pTi to a new bacterial cell (Lin, Binns, and Lynn 2008; Lang and Faure 2014; Zhu et al. 2000). In 1991, Zhang and Kerr reported the presence of a compound in the culture supernatants of *A. tumefaciens* that increased the conjugation frequency (Zhang and Kerr 1991). The chemical structure of the conjugal factor 3OC8-HSL was remarkably similar to the autoinducer 3OC6-HSL of *V. fischeri* and *P. carotovorum* (Zhang et al. 1993). This report represented a hallmark in the confirmation of the hypothesis about a common regulatory mechanism based on common signal molecules in bacteria. The characterizations of the *A. tumefaciens* 3OC8-HSL started to shed light on the influence of the environment on QS: the authors documented the loss of QS activity when 3OC8-HSL was alkalinized or autoclaved at 121°C (Zhang et al. 1993).

In pathogenic bacteria, the QS mechanism has been analyzed not only as a means for understanding part of the regulation of the pathogenesis, as in *A. fabrum*, but also as a putative target for combating the infection. After *P. carotovorum* (Pirhonen et al. 1993; Chatterjee et al. 2010), AHL-based QS was described in other soft rotting bacteria, including *Pantoea* (Ramachandran et al. 2014; Beck von Bodman and Farrand 1995) and *Erwinia* (Venturi et al. 2004). As shown in Table 10.1, in this group of microorganisms QS is mainly related to the production of PCWDEs and antibiotics. Among the pathogenic species of the genus *Pseudomonas* (*Ps.*), only *Ps. fuscovaginae* (Uzelac et al. 2017), *Ps. syringae* pv. *syringae* (*Pss*) (Quiñones, Pujol, and Lindow 2004; Quiñones, Dulla, and Lindow 2005; Scott and Lindow 2016), and *Ps. syringae* pv. *tabaci* (Cheng et al. 2017; Taguchi et al. 2006) have been studied in detail. *Burkholderia* is a genus of both beneficial and pathogenic bacteria for plants. In the latter group, AHL-based QS has been described in *B. glumae* (Kang et al. 2017; Devescovi et al. 2007; Nickzad and Déziel 2016), *B. plantarii* (Solis et al. 2006), and *B. gladioli* (Kim et al. 2014), in which QS has been related to plant pathogenicity. *Ralstonia solanacearum* possesses an AHL-based QS system (SolI/SolR) but, to date, the better described signaling mechanism is based on 3-hydroxy palmitic acid methyl ester (3-OH PAME) (Flavier, Clough, et al. 1997; Flavier, Ganova-Raeva, et al. 1997). *Xylella fastidiosa* and *Xanthomonas* spp. also utilize lipidic molecules, collectively known as diffusible signal factors (DSF), like 2-*cis*-11-methyldodecenoic acid or 2-*cis*-tetradecenoic acid. QS in *X. fastidiosa* is involved in the biofilm formation on the walls of vector insects' foreguts, which is required for the invasion of the host plant (Beaulieu et al. 2013; Newman et al. 2004). Exopolysaccharide production and extracellular enzymes have been related to QS in *X. campestris* pv. *campestris* (*Xcc*) (He et al. 2006; Barber et al. 1997).

TABLE 10.1

Quorum Sensing in Plant-Associated Bacteria

Microorganism	QS System	QS Molecule	QS-regulated Phenotype	References
Plant Pathogenic Bacteria				
P. wasabiae	ExpI/ExpR1/ExpR2	3OC8-, C8-, 3OC6-HSL	Pectate lyase, polygalacturonase, protease, cellulase	Valente et al. (2017); A. Chatterjee et al. (2005)
P. carotovorum	ExpI/ExpR1/ExpR2	3OC8-, 3OC6-HSL	Pectate lyase, polygalacturonase, protease, cellulase, motility	A. Chatterjee et al. (2010); Pirhonen et al. (1993)
P. atrosepticum	ExpI/ExpR1/ExpR2	3OC6-HSL	PCWDEs, secondary metabolism, signal transduction, lipid metabolism	Barnard et al. (2007); Bowden et al. (2013)
A. fabrum	TraI/TraR	3OC8-HSL	Conjugation	Lang and Faure (2014); L. Zhang et al. (1993)
E. amylovora	Unknown	3OC6-, 3OHC6-HSL	Unknown	Venturi et al. (2004)
Pantoea stewartii subsp. *stewartii*	EsaI/EsaR	3OC6-HSL	Exopolysaccharide production, motility, stress response	Ramachandran et al. (2014); Beck von Bodman and Farrand (1995)
Ps. fuscovaginae	PfsI/PfsR PfvI/PfvR	C10-, C12-, 3OC10-, 3OC12-HSL	Plant virulence	Uzelac et al. (2017)
Ps. syringae pv. *tabaci*	PsyI/PsyR	C6-, 3OC6-HSL	Motility, chemotaxis, pilus, extracellular polysaccharides, secretion systems, two-component system	Cheng et al. (2017); Taguchi et al. (2006)
Ps. syringae pv. *syringae*	AhlI/AhlR	3OC6-HSL	Exopolysaccharide, motility, epiphytic fitness	Quiñones, Pujol, and Lindow (2004); Quiñones, Dulla, and Lindow (2005); Scott and Lindow (2016)
B. glumae	TofI/TofR	C6-, C8-HSL	LipA lipase, toxoflavin, rhamnolipids, bacterial osmolality	Kang et al. (2017); Devescovi et al. (2007)
B. plantarii	PlaI/PlaR	C6-, C8-HSL	Plant virulence	Solis et al. (2006)
B. gladioli	bgla_2g11050 (*luxI*) bgla_2g11070 (*luxR*)	Uncharacterized AHLs	Motility, toxin production, oxalogenesis, polyketide biosynthesis	Kim et al. (2014)
R. solanacearum	SolI/SolR PhcB/PhcS	C6-, C8-HSL 3-OH PAME	?? Exopolysaccharide I, endoglucanase, pectin methyl esterase	Flavier, Clough, et al. (1997); Flavier, Ganova-Raeva, et al. (1997)
Xylella fastidiosa	RpfF/RpfC	2-*cis*-tetradecenoic acid	Biofilm formation, hemagglutinin-like proteins	Beaulieu et al. (2013); Newman et al. (2004)
Xanthomonas campestris pv. *campestris*	RpfF/RpfC	2-*cis*-11-methyldodecenoic acid	Exopolysaccharide production, extracellular enzymes	He et al. (2006); Barber et al. (1997)
Plant Beneficial Bacteria				
Rhizobium etli	CinI/CinR RaiI/RaiR TraI/TraR	Uncharacterized long- and shortchain AHLs 3OC8-, 3OHC8-HSL	Growth, nitrogen fixation, conjugation, swarming	Daniels et al. (2002, 2006); Rosemeyer et al. (1998); Tun-Garrido et al. (2003)
Bradyrhizobium spp.	BjaI/BjaR1 BraI/BraR	Isovaleryl-HSL Cinnamoyl-HSL Several AHLs	Unknown Unknown Cell aggregation, biofilm formation, motility	Ahlgren et al. (2011); Lindemann et al. (2011); Nievas et al. (2012)
M. extorquens	TslI MsaI/MsaR MlaI/MlaR	Unknown C6-, C8- C14:1-, C14:2-HSL	Exopolysaccharides Exopolysaccharides Unknown	Nieto Peñalver, Cantet, et al. (2006); Nieto Peñalver, Morin, et al. (2006)
G. diazotrophicus	GDI2836 (*luxI*) GDI2838 (*luxR*)	C6-, C8-, C10-, C12-, C14-, 3OC10-, 3OC12-, 3OC14-HSL	Plant colonization	Bertini et al. (2014); Nieto Peñalver, Bertini, and Castellanos de Figueroa (2012)
A. lipoferum	AlpI/AlpR	3OC6-, C6-, 3OC8-, 3OHC8- C8-HSL	Pectinase activity, siderophore, auxin production	Boyer et al. (2008)
Ps. capeferrum	PpuI/PpuR/PpoR	3OC6-, 3OC8-, 3OC10-, 3OC12-HSL	Pyoverdine, biofilm formation	Rampioni et al. (2012); Bertani and Venturi (2004)
Ps. chlororaphis	PhzI/PhzR CsaR/CsaI	C6-HSL	Pyrrolnitrin, phenazine production	Zhang and Pierson (2001); Maddula et al. (2006)
Paraburkholderia phytophirmans	XenI2/XenR2 BraI/BraR	3OHC8-HSL 3OC14-, 3OHC14-HSL	Biofilm formation Plant colonization	Coutinho et al. (2013); Zúñiga et al. (2017)

Note: The list of both pathogenic and beneficial bacteria is not exhaustive.

10.3 Quorum Sensing in Plant Beneficial Bacteria

Research on plant beneficial bacteria has profound economical and agronomical importance. This group of microorganisms allows increases in crop yields with reduced amounts of agrochemicals, redounding in a positive ecological impact. In general, plant growth-promoting bacteria (PGPBs) exert their positive effects by direct or indirect mechanisms. Direct mechanisms include biological nitrogen fixation (BNF), the solubilization of nutrients, and the production of the phytohormones auxins, cytokinins, gibberellins, abscisic acid, and ethylene (Gray and Smith 2005). Indirect growth promotion is due to antibiotics that inhibit pathogens, siderophores that render iron unavailable, hydrolytic enzymes, or the induction of systemic resistance of the host (Gray and Smith 2005).

AHL-based QS in PGPBs has been mainly studied in symbiotic nitrogen-fixing bacteria. *Rhizobium*, *Sinorhizobium*, *Mesorhizobium*, *Ensifer*, *Bradyrhizobium*, and related genera of Alphaproteobacteria possess complex QS systems with usually more than one AHL synthase and several AHL receptors (Table 10.1 and Figure 10.1). *Bradyrhizobium* spp. also synthesize AHL-related molecules with branched or aromatic chains: isovaleryl-HSL and cinnamoyl-HSL (Nievas et al. 2012; Ahlgren et al. 2011; Lindemann et al. 2011). QS systems in these PGPBs are involved in different aspects of the BNF, nodulation, motility, production of exopolysaccharides, and biofilm formation. Noteworthy, regulation of plasmid transfer by QS is frequently described in rhizobia, which are closely related to *A. fabrum* (Table 10.1). QS has also been described in the free-living diazotrophic *Gluconacetobacter diazotrophicus* (Bertini et al. 2014; Nieto Peñalver, Bertini, and Castellanos de Figueroa 2012) and *Azospirillum lipoferum* (Boyer et al. 2008). Proteomic characterizations of these species suggest that their QS systems are related to relevant functions for the endophytic and rhizospheric colonization, respectively (Bertini et al. 2014; Boyer and Wisniewski-Dyé 2009). QS regulation in non-diazotrophic PGPBs, including *Ps. capeferrum* (formerly, *Ps. putida*) (Rampioni et al. 2012; Subramoni and Venturi 2009), *Ps. chlororaphis* (Zhang and Pierson 2001; Maddula et al. 2006) (Figure 10.2), *Methylobacterium extorquens* (Nieto Peñalver, Cantet, et al. 2006; Nieto Peñalver, Morin, et al. 2006), and *Paraburkholderia phytophirmans* (Coutinho et al. 2013; Zúñiga et al. 2017), is related to the production of antibiotics and biofilm formation, features highly relevant for survival and plant colonization.

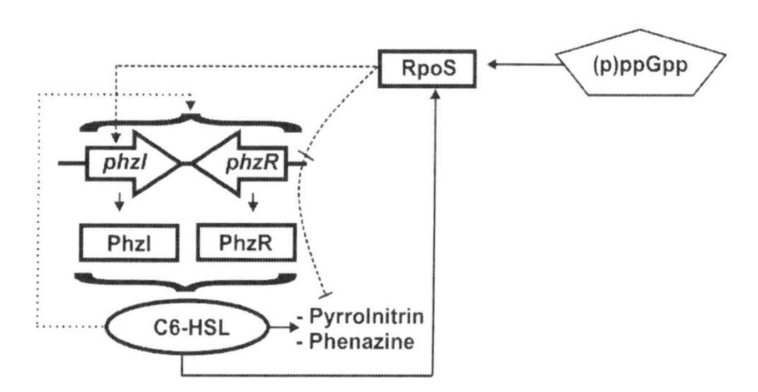

FIGURE 10.2 QS system in *Ps. chlororaphis* PA23 and its interconnection with the GSR. In PA23 the QS system is composed of the *csaR/csaI* (not depicted) and the *phzR/phzI* homologs. The production of C6-HSL depends on the latter, in which a characteristic positive autoregulatory loop has been described. The PhzR/PhzI system is involved in the regulation of pyrrolnitrin and phenazine production. This QS system also shows a positive effect on RpoS production. RpoS is the key transcriptional factor of the GSR that induces and represses *phzI* and *phzR*, respectively. The RpoS effect on pyrrolnitrin is exerted through the repression of *phzR*. The SR triggers the production of (p)ppGpp that, in turn, increases the synthesis of RpoS. Intersecting lines denote repression; arrows denote induction. Dashed lines are related to GSR, and dotted lines indicate autoinduction.

10.4 QS and pH

During the first characterizations of the QS signal molecule in *A. fabrum* B6, Zhang and Kerr obtained spent supernatants from media supplemented with opines that increased the conjugation frequency in the bacterium. The researchers evidenced a complete loss of activity when a supernatant sample was alkalinized with NH_4OH (Zhang and Kerr 1991). Later in 2002, Paul Williams and colleagues published a fundamental report describing the effect of pH on the stability of AHLs (Yates et al. 2002). The authors showed that the treatment with NaOH for only 15 min was sufficient to produce a lactonolysis of the molecules, with the concomitant loss of QS activity. Noteworthy, the process was reversible under acidic conditions, with a complete recovery of the chemical structure and the QS activity of the molecule (Yates et al. 2002).

Under alkaline conditions, the loss of QS activity of AHLs is not only dependent on the homoserine lactone ring, but also on the acyl-side chain: a homoserine lactone ring is completely opened when the pH rises from 1 to 2, but remains closed when it is esterified with an acyl chain. As the chain length increases, AHLs tend to be more resistant to alkaline conditions (Yates et al. 2002). To date, the shortest AHL described is C4-HSL, produced by *Ps. aeruginosa*. It is possible that shorter molecules (e.g., C2- or C3-HSL) are not utilized as QS molecules due to the relatively rapid inactivation. The substitution at the third carbon in the acyl chain has a strong influence on the stability of the AHLs: the carbonyl group of 3-oxo-substituted AHLs reduces the stability of the molecules due to its electronegativity, which reduces the electron-donating capacity of the lactone, rendering the ring more unstable (Yates et al. 2002).

The influence of pH on the QS activity of AHLs has a practical interest. For instance, when analyzing the presence of a QS system in a bacterial isolate, it is possible to find a negative result due to the lactonolysis of the AHLs produced by an alkaline pH of the sample. However, AHL lactonolysis has been scarcely analyzed *in situ*. Decho and colleagues showed that, according to the kinetic of AHL lactonolysis under laboratory conditions, pH-dependent inactivation of AHLs could be of relevance in certain environments (Decho et al. 2009). To date, this chemical inactivation has not been evidenced *in planta*, but it is interesting to note that pH of xylem and phloem saps can vary seasonally (Thomas and Eamus 2002) and under drought conditions (Gloser et al. 2016). It is plausible that these modifications of the pH homeostasis in the host plant can alter the QS activity of endophytic microorganisms. In the rhizosphere, the pH-dependent lactonolysis of AHLs could be restricted to alkaline soils, which are not preferred for agriculture exploitation. In addition, the presence of organic anions can transiently and locally alkalinize certain soils attaining pH values above 8, which are high enough to inactivate AHLs (Rukshana et al. 2014).

AHL lactonolysis for plant-associated bacteria could also be a consequence of the microbial metabolism itself. *M. extorquens* is a ubiquitous bacterium that colonizes the surface of aerial parts and the inner tissues of plants. Its capacity for utilizing the methanol released as a waste product from the stomata and the production of phytohormones allows the classification of *M. extorquens* as a PGPB. In addition, the plant colonization by *M. extorquens* hinders the attack by phytopathogens. *M. extorquens* is a facultative methylotroph that also utilizes other carbon sources, including organic acids like succinic acid. In contrast to methylotrophic growth conditions, the depletion of the organic acid tends to alkalize the growth medium when *M. extorquens* utilizes succinate as the carbon source (Nieto Peñalver, Morin, et al. 2006). As a consequence, the AHL QS molecules produced by the bacterium are lactonyzed. To note, it is not the production but the stability of the AHLs that is affected (Nieto Peñalver, Morin, et al. 2006). In plants, organic acids are important components of the xylem and phloem saps, and of root exudates (Sasse, Martinoia, and Northen 2018; Gloser et al. 2016). It is then expected that there will be alkalinization in the surroundings of endophytic or rhizospheric bacteria utilizing organic acids and, in consequence, an inactivation of the AHLs. The microbial metabolism of these carbon sources would imply that in a microniche of the plant, bacteria require a larger quorum of cells to attain a threshold concentration of AHLs and to regulate their physiology in a concerted manner.

10.5 QS and Nutrients

It is expected that alteration in the levels of the precursors for the QS signal synthesis, produced by the availability of nutrients in the environment, will modify the profile of QS molecules. To date, this hypothesis has only been evaluated for the AHL-related QS molecules produced by *Rhodopseudomonas palustris* and bradyrhizobia: while *R. palustris p*-coumaroyl-HSL is only produced when the precursor *p*-coumarate is present, cinnamate is not required for the synthesis of *Bradyrhizobium* cinnamoyl-HSL (Schaefer et al. 2008; Ahlgren et al. 2011). Nutrients, nevertheless, have an important influence on the activity of QS systems. For instance, in the free-living diazotrophic bacterium *G. diazotrophicus*, an endophytic PGPB of several important crops, the proportions of AHLs changes according to the composition of the growth medium. *G. diazotrophicus* produces short- and long-chain AHLs, totaling eight different molecules (Nieto Peñalver, Bertini, and Castellanos de Figueroa 2012). When growing in a synthetic growth medium, *G. diazotrophicus* reduces 30% the levels of long-chain AHLs, in comparison to a complex medium (Nieto Peñalver, Bertini, and Castellanos de Figueroa 2012).

A number of mechanisms behind the nutritional influence on QS have been partially characterized in different microorganisms. One of these is the carbon catabolite repression (CCR), which determines the preferred carbon source utilized in bacteria. The role of CCR on QS has been analyzed in deeper detains in *V. harveyi* and *V. cholera*, in which the cAMP receptor protein (CRP) induces directly the QS system by binding to the *luxR* promoter (Chatterjee et al. 2002), or indirectly by stabilizing the main autoinducer synthase *cqsA* mRNA (Liang et al. 2008). CRP also binds to the promoter of *ainS*, which codes for the alternative autoinducer synthase AinS (Lyell et al. 2013). The role of CCR on QS regulation in plant-associated bacteria remains largely unexplored. In the phytopathogen *E. chrysanthemi*, cAMP-CRP induces the expression of *expR* and represses that of *expI*, which code the AHL receptor and AHL synthase, respectively (Reverchon et al. 1998). This dual behavior is explained by the DNA binding site for CRP located inside the *expR* coding sequence, and also upstream of *expI* in *E. chrysanthemi* (Reverchon et al. 1998). Considering that CCR is an important mechanism for global control of the microbial physiology, it is expected that its influence on QS will not be restricted to *Vibrio* spp. and *E. chrysanthemi*. Interestingly, rhizobia present an alternative carbon catabolite repression, named Reverse CCR, which allows the preferred utilization of organic acids over carbohydrates (Iyer et al. 2016). Whether Reverse CCR alters the QS regulation in rhizobia needs to be evaluated. In addition, as discussed above, the utilization of organic acids alkalinizes the surrounding environment of a bacterium, producing the pH-dependent lactonolysis of AHLs. It is then possible that CCR in general, and Reverse CCR in particular, alters indirectly the QS activity.

The stringent response (SR) has also been analyzed with respect to QS regulation. The SR, characterized by the accumulation of the alarmones ppGpp and pppGpp [collectively known as (p)ppGpp], is triggered by the deprivation of nutrients, in particular amino acids (Hauryliuk et al. 2015). The lack of amino acids causes the apparition of uncharged tRNA in the ribosomes, which leads to the synthesis of (p)ppGpp by the GTP pyrophosphokinase RelA protein. Alarmone reprogramming of gene transcription is accomplished after binding to the β and β′ subunits the RNA polymerase (Hauryliuk et al. 2015). In addition, DksA protein boosts the effect of (p)ppGpp by binding to the RNA polymerase, which modifies the structure of the enzyme increasing the activity of ppGpp (Hauryliuk et al. 2015). The link with QS was unveiled in the diazotrophic *R. etli*. In this bacterium, the SR is indispensable for the symbiotic establishment with the plant host and the BNF (Calderón-Flores et al. 2005). It was showed that a mutation in RelA-homolog encoding *rsh* (previously known as *relA*) causes a lower expression of the *luxI*-homologs *raiI* and *cinI*, lower levels of AHLs, and defective symbiosis with *Phaseolus vulgaris* in *R. etli* CNPAF512 (Moris et al. 2005; Vercruysse et al. 2011) (Figure 10.1). To date the mechanism of the QS induction by *rsh* has not been elucidated. Noteworthy, the effect of the SR on QS seems not to be conserved, even for closely related microorganisms. For instance, while a *relA* mutation causes a repression of *sinR* in *Ensifer meliloti* Rm2011 (Krol and Becker 2011), the SR increase the expression of *attM* in *A. fabrum* K588, which codes for the AttM lactonase (see below), lowering the levels of the AHL 3OC8-HSL (Zhang, Wang, and Zhang 2004).

SR and QS are connected not only in rhizobia but also in other plant-associated bacteria. *Ps. chlororaphis* PA23 produces hydrogen cyanide and the antibiotics pyrrolnitrin and phenazine under the control of several regulatory mechanisms including QS (see also below). In PA23 QS is influenced by and forms a regulatory network together with the SR for controlling the production of antibiotics (Selin et al. 2014). Through RpoS (see also below), the SR causes a repression of *phzR* and an induction of *phzI* (Selin et al. 2014) (Figure 10.2). A similar interconnection has been described in the phytopathogen *Pectobacterium atrosepticum* (*Pca*) (Bowden et al. 2013). In *Pca*, QS and RelA are involved in regulation of virulence factors, including PCWDEs required in soft rotting (Figure 10.1). At low cell densities, the *Pca* LuxR homolog ExpR2 represses posttranscriptionally the production of the enzymes through RsmA protein, which interact with the PCWDE mRNAs destabilizing them (Põllumaa, Alamäe, and Mäe 2012). As the 3OC6-HSL concentration increases, ExpR2 binds the AHLs and inhibits the expression of *rsmA* with the concomitant translation of PCWDE mRNAs. *rsmB* is a noncoding regulatory RNA induced by (p)ppGpp that sequesters RsmA, forming in this way a regulatory network between the stringent response and QS (Bowden et al. 2013).

10.6 QS and General Stress

Under stress conditions, including temperature, acid, UV, oxidative, or osmotic stress, different adaptive and defense mechanisms are triggered in bacteria to deal with the specific stressful situation. Bacteria can also display a more global response that simultaneously protects from different stresses. The key player of this general stress response (GSR) in *E. coli* is the alternative RpoS transcriptional factor, which interacts with the RNA polymerase modifying the expression of about 500 genes (Battesti, Majdalani, and Gottesman 2011). *Ps. aeruginosa* is one of the first microorganisms where the connection between QS and RpoS was established. *rhlI*, which codes the AHL synthase RhlI, is repressed by RpoS in *Ps. aeruginosa*, though to date it is not clear whether this effect is direct or indirect through other regulators. At the same time, RpoS and QS form a regulatory network with overlapping regulons: a large list of genes is regulated in a concerted manner by both RpoS and the QS regulators RhlR and/or LasR (Schuster et al. 2004). Among plant-associated bacteria, QS and RpoS have been analyzed in deeper detail in *Ps. chlororaphis*. As mentioned above, RpoS is also involved in the production of pyrrolnitrin and phenazine antibiotics in *Ps. chlororaphis* PA23. However, RpoS action is indirect and exerted through the Phz QS system (Table 10.1) in PA23 (Figure 10.2). While the expression of *phzI* is induced, *phzR* is repressed by RpoS. At the same time, PhzR and C6-HSL both have a positive effect on *rpoS* expression (Maddula et al. 2006) (Figure 10.2). In contrast, a *Ps. chlororaphis* PCL1391 *rpoS* mutant has decreased levels of C6-HSL, but QS has no effect on *rpoS* (Girard et al. 2006). It is possible that these differences on the regulation between PA23 and PCL1391 strains are due to a lack of conservation in the architecture of the regulatory networks. However, it is important to note that the experiments on PA23 were conducted on Lennox Luria–Bertani medium, while Modified Vogel–Bonner salts #1 medium was employed for PCL1391, which could also explain, at least partially, the differences reported. *Ps. capeferrum* (formerly, *Ps. putida*) WCS358 is a PGPB where the link between QS and RpoS was also revealed. In opposite to *Ps. chlororaphis*, RpoS has a negative effect on the expression of the *luxI*-homolog *ppuI* expression and has no effect on the *luxR*-homolog *ppuR*. In WCS358, RpoS has also no effect on the QS repressor *rsaL* (Bertani and Venturi 2004).

Alphaproteobacteria lack RpoS homologs, and the GSR depends on extracytoplasmic function σ factors of the ECF15 and EcfG-like subgroups (Francez-Charlot et al. 2015). Among the six identified in *M. extorquens* AM1, σ^{EcfG1} is normally repressed by the anti-σ factor NepR. Under stress conditions, the anti-anti-σ factor PhyR is phosphorylated by histidine kinases, which allows the PhyR-NepR interaction, releasing σ^{EcfG1}. σ^{EcfG2} is also a main component of the GSR, but its activity is constitutive and not repressed by NepR (Francez-Charlot et al. 2015). Transcriptome analyses under ethanol stress in σ^{EcfG1} and σ^{EcfG2} mutants showed that *msaI* (Table 10.1) was repressed and induced, respectively, in comparison to the wild type, demonstrating an influence of these σ factors on the QS in *M. extorquens* AM1. In contrast, no significant differences were determined for the *mlaI-mlaR* QS system (Francez-Charlot et al. 2016). On the other side, microarrays performed at early stationary growth phase in *S. meliloti* showed no influence of RpoE2 on QS (Sauviac et al. 2007). In *S. meliloti*, RpoE2 is the extracytoplasmic function σ factor involved in the GSR, and is part of a regulatory mechanism similar to that described in *M. extorquens* (Sauviac et al. 2007).

10.7 QS and Heavy Metals

Heavy metals (e.g., iron, copper, cadmium, chromium, gold, and silver) are defined as metallic compounds, including metalloids, with a relatively high density (more than 5 g/cm^3) compared to water (Tchounwou et al. 2012). Microorganisms require some of these metals as micro-nutrients for different biological activities: in cell membrane, protein and DNA structures, and in enzymatic activities. At the same time, they possess different mechanisms that protect from the excessive metal concentrations: reduced uptake; efflux; sequestration in the extracellular, periplasmic, or intracellular space; or chemical modifications that render the metal less toxic (Macomber and Hausinger 2011).

The first link between metals and QS came to light with the discovery of VqsR in *Ps. aeruginosa*, an AraC-type transcriptional regulator that increases QS activity. Transcriptomic analysis performed by Juhas and colleagues showed the overlapping of QS and VqsR regulons, and that the expression of genes regulated by iron requires VqsR (Juhas et al. 2004). Shortly after, it was evidenced that the concentration of iron in the culture medium has a profound influence on the expression of the luxR-homolog *lasR* in *Ps. aeruginosa* (Kim et al. 2005). Kim and colleagues analyzed the *lasR* expression when *Ps. aeruginosa* was cultured in media supplemented with replete (7 mg/L) and low (0.6 mg/L) iron concentrations, finding that at low iron concentrations *lasR* was expressed at a higher level (Kim et al. 2005). Results obtained later with the plant pathogen *Pss* were in contrast with the observations in *Ps. aeruginosa*. Dulla and colleagues evidenced the inhibition of QS in *Pss* B728a when the strain was cultured together with different epiphytic isolates of bacteria. A screening of a Tn5 mutant library of the inhibitory isolate *Pseudomonas* sp. 114 showed that the iron uptake from the culture medium was responsible for QS inhibition of *Pss* B728a (Dulla, Krasileva, and Lindow 2010).

To note, metals can also reduce the QS activity. For instance, in contrast to *Pss* B728a, Ni^{2+}, Cd^{2+}, and Co^{2+} repress the QS system of *B. multivorans* ATCC 17616 (Vega et al. 2014). It is possible, as Vega and colleagues suggested, that competition between these metals and native metal cofactors in ATCC 17616 is responsible for QS inhibition (Vega et al. 2014). However, the analysis of the inhibitory effect of Cd^{2+} on the QS system of the model bacterium *Chromobacterium violaceum* showed that, at least for this metal and this microorganism, the inhibition could follow a non-competitive mechanism (Thornhill et al. 2017). The effect of a metal seems then to depend not only on the specific metal but also on the analyzed bacterium.

The modification of the QS activity by metals can be explained by metal response regulators connected with the QS systems. However, at least two other explanations are plausible. Through different mechanisms, the presence of metals increases the production of reactive oxygen species (ROSs; e.g., peroxides, superoxide, hydroxyl radical, and singlet oxygen) that, at high doses, produce an oxidative stress for the cell. Under these conditions, certain cysteine residues of QS proteins can be oxidized, reducing their activities. This has been shown for the Cys^{201} and Cys^{203} located in the DNA binding domain and Cys79; in the AHL binding domain of LasR, which form a disulfide bond when the protein is exposed to H_2O_2 (Kafle et al. 2016). In addition, ROSs can also affect AHLs, which generates molecules with unsaturated acyl chains or with oxidized substituents, which are bound with less affinity by the corresponding AHL receptor (Frey et al. 2010). Another explanation for the effect on QS could be the direct interaction among metals and AHLs. Recently, it was shown that AHLs form extracellular complexes with Cu^{2+} and Ag^{+}, rendering the signal molecule unavailable for the bacterium (McGivney et al. 2018).

To date, the effect of metals has not been further evaluated in plant-associated bacteria. However, the presence of these chemical elements can be of high relevance for QS under certain conditions. For instance, in metal contaminated soils, the exposition of rhizospheric bacteria to high concentrations of these pollutants may alter the QS regulation interfering with the plant colonization. At the same time, the uptake of the metal by the plant could interfere with the signal mechanism of endophytic bacteria.

10.8 QS and Soil Components

Among the different amendments utilized to increase the agricultural output, biochar (i.e., charcoal of biological origin) provides several benefits for crops reducing the tensile strength of soils, improving water storage capabilities, and increasing the pool of organic carbon, the soil pH, and exchangeable cations (Novak et al. 2012; Chan et al. 2007). However, the high AHL adsorption capacity of biochar functions as a natural quencher for microbial communication in soil (Gao et al. 2016; Masiello et al. 2013). This quenching effect on QS signaling is dependent, at least partially, on the temperature of pyrolysis utilized for the biochar preparation, since higher temperatures produce higher surface areas in the biochar. In consequence, the higher the pyrolysis temperatures utilized for the biochar preparation, the lowers the amount of biochar required to quench the QS signaling (Masiello et al. 2013). The presence of biochar also increases the pH of the medium with the subsequent AHL lactonolysis (Gao et al. 2016).

Clays, naturally present in soils or utilized as amendments in agriculture to improve crop productivity, can also potentially disrupt the microbial signaling. Through nonelectrostatic forces (i.e., Van der Waals and hydrophobic interactions), 3OC12-HSL is adsorbed to colloidal clays, in particular montmorillonite, kaolinite, and goethite, which are important components of montmorillonitic, kaolinitic, and oxisol soils, respectively (Liu, Chen, and Chen 2015). In agreement with the behavior of clays, alfisol and oxisol soils adsorb chemically AHLs in a differential manner depending on the acyl-side chain (Sheng et al. 2018). Indeed, C12-HSL is more absorbed than C6-HSL. Alfisol shows faster and higher sorption capacities than oxisol, due to the different physicochemical properties (Sheng et al. 2018). To date, it is not known at which extent the AHL adsorption interferes with the microbial communication in soils, though it is expected that both plant-beneficial and -pathogenic microorganisms might be affected, in a level that will depend on the specific AHL utilized in the signaling.

Soil components can also interfere with microbial signaling, not by quenching the signals but by acting directly in the bacterial cell. It has been known since a long time ago that humic substances have important modulator activities of soil biological properties. However, only recently was the impact of humic substances on QS shown. Yuan and colleagues found that the water-soluble fraction of humic acids represses the main components of the QS system of *S. meliloti* composed of the *luxI* homolog *sinI* and the two *luxR* homologs *sinR* and *expR* (Xu et al. 2018). Although to date the exact mechanism of this repression is not completely understood, humic substances seem to increase the repression exerted by the QsrR repressor on *expR* and *sinR*. In consequence the expression of *sinI* is compromised (Xu et al. 2018). QS in *S. meliloti* is related to important functions for the interactions with the host plant, including the repression of flagellar motility and the increase in the production of specific exopolysaccharides (Calatrava-Morales, McIntosh, and Soto 2018). Surprisingly, the nodulation of the model leguminous *Medicago sativa* by *S. meliloti* and the weight of nodules are higher in the presence of humic substances (Xu et al. 2018). These observations indicate that, beyond the repression of *sinI*, *sinR*, and *expR*, humic substances alter other aspects of the regulation of the BNF process, may be also related with QS, not known to date. To note, *S. meliloti* possesses at least four other *luxR* homologs (identified as SMc00658, SMc00877, SMc00878, and SMc04032) whose functions in the complete architecture of the QS system are very poorly understood (Calatrava-Morales, McIntosh, and Soto 2018).

10.9 QS and Natural Products

The first biological compounds characterized as QS inhibitor (QSI) were the halogenated furanones produced by the red algae *Delisea pulchra*, which accelerate the degradation of the AHL receptor in the cell (Manefield et al. 2002). Since then, a vast quantity of compounds produced by plants and microorganisms has been described mainly as QS inhibitors, though few examples exist of QS induction. For instance, 3-hydroxy-oxindole produced by *R. solanacearum* can induce QS activity in LuxR-harboring bacteria (Delaspre et al. 2007); also *Chlamydomonas reinhardtii* produces unidentified compounds with stimulatory QS activity (Teplitski et al. 2004).

The exact mechanisms by which natural QSIs alter the QS activity have not been elucidated, though computational analyses utilizing these molecules for docking suggest that the corresponding LuxR homologs are the main targets in the QS interference (see for instance Ravichandran et al. 2018). It is, at least, surprising that it has a common inhibitory mechanism of LuxR-homolog inhibition considering the wide diversity of chemical structures that have been described for natural QSIs (Asfour 2018). Indeed, certain evidence suggests that signal production can also be potentially altered by QSIs (Norizan, Yin, and Chan 2013; Truchado et al. 2012).

Beyond the molecular mechanisms of QS inhibition by natural QSIs, it is interesting to consider the potential effects of these compounds in the niche of a microorganism. Plant-associated bacteria are continuously exposed to molecules produced by the host plant as part of the defense system or the secondary metabolism. In addition, microorganisms are found most of the time in nature as part of a complex community that also produces a large variety of biologically active molecules. It is then plausible that the QS systems of plant-associated bacteria will be altered, both positively and negatively, by these compounds. To date, this hypothesis has only been evaluated *in vitro*. For instance, phenolic coumarins utilized at concentrations of 30 μM produce an inhibition of 50% in the QS systems of *C. violaceum* and *Ps. aeruginosa in vitro* (D'Almeida et al. 2017). At first sight, it seems that mixtures of natural compounds tend to inhibit rather than to induce QS systems. However, this conclusion could be biased due to the fact that the study of natural products has been traditionally directed to the search of QS inhibition.

10.10 QS and the Microbial Metabolism of AHLs

The first characterization of AHLs as nutrient sources was obtained from *Variovorax paradoxus*, an aerobic Betaproteobacterium commonly found in soils. Utilizing ^{14}C-labeled C6-HSL, Leadbetter and Greenberg showed that *V. paradoxus* VAI-C could metabolize the QS molecule. Indeed, growth yield analysis with different AHLs showed the VAI-C biomass formation from the QS molecules. Those results clearly suggested the ecological implications on QS: the presence of an AHL-metabolizing microorganism interferes with the signaling systems (Leadbetter and Greenberg 2000). Similar results were obtained later with *Pseudomonas* sp. PAI-A (Huang et al. 2003). PAI-A showed doubling times of 16.5 h and 25.0 h in culture media supplemented with C12-HSL or 3OC12-HSL, respectively, in contrast to VAI-C that utilized equally well all the AHLs with shorter doubling time (3.5 h) (Leadbetter and Greenberg 2000).

Wang and Leadbetter described the complete mineralization of AHLs to CO_2 in soils, due to microbial activity (Wang and Leadbetter 2005). This AHL mineralization is constitutive and not inducible by the presence of the QS molecules (Wang and Leadbetter 2005). It is then possible that the microbial metabolism of AHLs is frequent, at least in certain environments. In agreement with this hypothesis, it is important to consider that, as Safari and colleagues discussed elsewhere (Safari et al. 2014), lactone-containing compounds are part of central metabolic pathways in microorganisms, including 6-phospho D-glucono-1,5-lactone in the Entner-Doudoroff pathway, the oxidative branch of the pentose phosphate pathway, and formaldehyde oxidation, where lactone-hydrolyzing enzymes are involved. It is also possible that the microbial ruptures the amide bond in the AHL by amidases (see below), rendering free fatty acids that are metabolized through the β-oxidation pathway. AHLs can also be metabolized through yet unexplored pathways, as those related to AHL decarboxylation (Sheng et al. 2017).

Assuming nM concentrations for AHLs *in situ* (e.g., *in planta*) (Nievas et al. 2012), it is expected that a QS molecule will not be the preferred nutrient. However, its metabolism by certain microorganisms cannot be ruled out. In addition, it is unlikely that the AHL metabolism implies the absolute interruption of the signaling, since the molecules are constantly synthesized. However, a higher density of cells would be required in order to attain a threshold concentration of AHLs. At the same time, if a bacterium produces more than one QS signal, and each molecule metabolizes at different rates, their relative proportion could also change altering in consequence the QS regulation.

10.11 QS and Microbial Transformation of AHLs

The corresponding AHL receptors show an exquisite specificity for these molecules, even when small variations (e.g., a carbonyl or a hydroxyl substituent) are present. For instance, the specificity of *C. violaceum* CviR is 30 folds higher for C6-HSL than for C4-HSL (McClean et al. 1997); *A. fabrum* TraR is 21 folds more sensitive for 3OC8-HSL than for C8-HSL (Zhu et al. 1998). It is then expected that there may be an alteration of the QS communication if the AHL structure is modified. The cytochrome P450 CYP102A1 (UniProt P14779) from *Bacillus megaterium* has the capacity for oxidizing *N*-fatty acyl amino acids at ω-1, ω-2 and ω-3. Analysis with purified CYP102A1 showed the capacity to oxidize C12-HSL producing ω-1, ω-2, and ω-3 hydroxylated derivatives that are still detected by an AHL receptor with a QS bioassay but with less affinity (Chowdhary et al. 2007). CYP102A1 also oxidizes dodecanoic acid released from C12-HSL by amidase enzymes, and C12-homoserine obtained after the action of lactonase enzymes (Chowdhary et al. 2007). It is not known whether CYP102A1 can oxidize other AHLs and, maybe more important, how this AHL-oxidizing capacity of *B. megaterium* modifies the AHL-mediated signaling in a polymicrobial community. Noteworthy, CYP102A1 orthologs are also present in plant-associated bacteria, though the role in QS has not been evaluated. It is interesting to highlight the presence of a CYP102A1 orthologs in *Bradyrhizobium* spp., which produce AHL-related QS molecules (Lindemann et al. 2011; Ahlgren et al. 2011). For instance, the *B. diazoefficiens* USDA 110 ortholog (UniProt Q89R90) shows an identity of 47% and a similarity of 62% with CYP102A1. It is plausible that the utilization of an alternative QS signal molecule protects these microorganisms from its enzymatic oxidation by the CYP102A1 ortholog.

3-oxo substituted AHLs can also be enzymatically reduced. BpiB09 is a NADP-dependent short-chain dehydrogenase/reductase obtained from a soil metagenomic library that reduces 3OC12-HSL to 3OHC12-HSL. When BpiB09 is expressed in *Ps. aeruginosa*, its QS system is quenched with a reduced pathogenicity (Bijtenhoorn et al. 2011). Its dehydrogenase/reductase specificity has not been reported; however, it is possible that BpiB09 has a broad spectrum of activity on AHLs, since the BpiB09 expression in *C. violaceum*, which utilizes C6-HSL as QS signal, also decreases violacein production (Bijtenhoorn et al. 2011). These results with *C. violaceum* open the question about other quenching mechanisms of BpiB09, since the third position in the acyl-side chain of C6-HSL cannot be further reduced.

10.12 QS and the Enzymatic Hydrolysis of the Signals

The enzymatic inactivation of QS signals, known as quorum quenching (QQ), is one of the most studied aspects of the environment in relation to QS. The first enzyme to be characterized was the AiiA lactonase produced by *Bacillus* sp. 240B1 (Dong et al. 2000). AiiA is a Zn^{2+}-dependent protein that belongs to the metallo-β-lactamase superfamily. When expressed in *P. carotovorum*, AiiA reduced PCWDE activities and the pathogenesis on Chinese cabbage (Dong et al. 2000). Close homologs to AiiA are widely present in different *Bacillus* species (Dong et al. 2002). With a low degree of specificity, AiiA homologs catalyze the reversible hydrolysis of the AHL homoserine lactone ring generating the corresponding acyl homoserine (Grandclément et al. 2016).

In contrast to lactonases, amidase/acylase enzymes catalyze the irreversible cleavage of the AHL acyl-side chain. When *V. paradoxus* VAI-C was cultured with AHLs (see above), only the acyl moiety was utilized, allowing the hypothesis that an enzyme was acting on the amide bond (Leadbetter and Greenberg 2000). The amide cleavage of AHLs was later confirmed with the amidase/acylase Aac (formerly AiiD) from *Ralstonia* sp. XJ12B, a protein of the Penicil_Amidase superfamily (Lin et al. 2003). Aac is related to aculeacin A acylase from *Actinoplanes utahensis* involved in the hydrolysis of the amide bonds of aculeacin-A and aliphatic penicillins, though Aac shows no activity with these substrates (Lin et al. 2003).

3-OH PAME can also be hydrolyzed and inactivated by hydrolase enzymes. Firstly identified in *Ideonella* sp. 0-0013 (Shinohara, Nakajima, and Uehara 2007), the capacity for 3-OH PAME hydrolysis was also later determined in *Stenotrophomonas maltophilia*, *Ps. aeruginosa*, and *Rhodococcus corynebacterioides* (Achari and Ramesh 2015). A recent soil metagenome analysis identified novel 3-OH PAME hydrolases that suppress the QS regulation in *R. solanacearum* (Lee et al. 2018).

As shown in Table 10.2, QQ enzymes have been found in soil and plant-associated bacteria that produce the inactivated signal molecule

TABLE 10.2
QQ Enzymes in Plant Associated Bacteria

Host	QQ Enzyme	Type	Substrates	References
Plant pathogenic bacteria				
A. fabrum	AiiB	Lactonase	Broad spectrum	Carlier et al. (2003)
	Blc	Lactonase	Broad spectrum	
R. solanacearum	Aac	Amidase/acylase	Broad spectrum, mainly long-chain AHLs	Chen et al. (2009)
Ps. syringae pv. *syringae*	HacB (Psyr 4858)	Amidase/acylase	Broad spectrum	Shepherd and Lindow (2009)
	HacA (Psyr 1971)	Amidase/acylase	Long chain AHLs	
Plant beneficial bacteria				
Bacillus spp.[ab]	AiiA	Lactonase	Broad spectrum	Dong et al. (2002)
Microbacterium testaceum[a]	AiiM	Lactonase	Broad spectrum, mainly long-chain 3-oxo substituted AHLs	Wang et al. (2010)
Ochrobactrum spp.[ab]	AidH	Lactonase	C6, 3OC8[c]	Mei et al. (2010)
	AiiO	Amidase/acylase	Broad spectrum, mainly long chain AHLs	Czajkowski et al. (2011)
Rhizobium sp.	QsdR1	Lactonase	3OC8-HSL	Krysciak et al. (2011)
	DlhR	Lactonase	3OC8-HSL	
Chryseobacterium sp.[a]	AidC	Lactonase	Broad spectrum	Wang et al. (2012)
Rhodococcus erythropolis[a]	QsdA	Lactonase	Broad spectrum	Uroz et al. (2005, 2008)
	Unidentified enzyme	Oxidoreductase	3OC8–3OC14[c]	
	Unidentified enzyme	Amidase/acylase	Broad spectrum	
Enterobacter asburiae[a]	AiiA	Lactonase	C4, 3OC12[c]	Rajesh and Rai (2014)
Solibacillus silvestris[a]	AhlS	Lactonase	C6, 3OC6, C10, 3OC10[c]	Morohoshi et al. (2012)
Arthrobacter sp.[ab]	AhlD	Lactonase	C4, C6, 3OC6, C8, 3OC12[c]	Park et al. (2003)
Lysinibacillus sp.[ab]	AdeH	Lactonase	C4, C6, 3OC6, C8, 3OC8[c]	Garge and Nerurkar (2016)
Streptomyces sp.[ab]	AhlM	Amidase/acylase	Broad spectrum, mainly long chain AHLs	Park et al. (2005)
Comamonas sp.[a]	Unidentified enzyme	Amidase/acylase	Broad spectrum	Uroz et al. (2007)
Burkholderia sp.	Unidentified enzyme	Oxidoreductase	Broad spectrum	Chan et al. (2011)

Note: The section of plant beneficial bacteria includes microorganisms isolated from plants whose positive effects on the growth of the host have not been reported but, at the same time, are not known as phytopathogens.
[a] Reported in a soil isolate.
[b] Not reported as phytopathogen.
[c] Results reported only in the indicated substrates.

(e.g., *Pss* and *R. solanacearum*) and in species that do not produce them (e.g., *Bacillus* spp.). It has been largely speculated about their biological role, in particular with respect to QS. For AHL-producing bacteria, QQ enzymes can have a role in the recycling and/or fine tuning of QS, as with the AiiB and BlcC lactonases of *A. fabrum* (Lang and Faure 2014), or with the QuiP amidase of *Ps. aeruginosa* (Huang et al. 2006). The degradation of AHLs in non-AHL-producers could protect against toxic effects of AHLs. For instance, the lactonase from *Bacillus* spp. confers protection against tetramic acid-derived compounds spontaneously produced from 3OC12-HSL that shows antibacterial and siderophore-like activities (Kaufmann et al. 2005). In agreement, this protection enhances the root colonization of *B. thuringiensis* in pepper plants (Park et al. 2008). The characterization of the enzymatic inactivation of DSF, described in several species but studied in more detail in *Pseudomonas* sp. strain G, showed interesting aspects of the QQ phenomenon. In this bacterium, the disruption of *carAB*, involved in carbamoylphosphate production, abolishes the degradation of DSF (Newman et al. 2008). Carbamoylphosphate is a precursor for the synthesis of arginine and pyrimidine. The lack of connection between these pathways in *Pseudomonas* sp. strain G and DSF-dependent QS suggests that, at least for certain cases, the inactivation of signal molecules could be an incidental but not less real and important situation in nature.

10.13 QS and the Fungal Partners

In the last years the alteration of bacterial QS by fungi has started to be revealed. Uroz and Heinonsalo performed the first attempt for detecting QQ activities in these microorganisms, through a screening of filamentous fungi obtained from forest soils (Uroz and Heinonsalo 2008). Utilizing C6-HSL or 3OC6-HSL as sole carbon source, the authors documented reversible inactivations (i.e., lactonase-like activities) with an isolate of *Meliniomyces variabilis* and a mycorrhizal Basiodiomycete related to *Tylospora* sp. The inactivation by an Ascomycete isolate, closely related to *Phialocephala* sp., was irreversible, suggesting an amidase/acylase-like activity. In contrast, another 12 isolates were not able to inactivate these AHLs (Uroz and Heinonsalo 2008). The presence of a second isolate of *M. variabilis* in this group suggests that the enzyme/s involved in the degradation of AHLs could belong to strain-specific metabolic pathways and that is/are not part of the central metabolism.

Trichosporon loubieri WW1C is a basiodiomycete yeast that is able to growth in a synthetic culture medium supplemented with 3OC6-HSL (Wong et al. 2013). This observation led to the hypothesis that WW1C was capable of degrading AHLs, which was confirmed by analyzing supernatants of a buffered solution of AHLs with resting yeast cells. WW1C shows reversible QQ activity on AHLs with short to medium acyl-side chain, while AHLs with a C10 chain or longer are not inactivated (Wong et al. 2013). *Rhodotorula* (*Rh.*) *mucilaginosa* B2 and *Rh. mucilaginosa* B3 are also basiodiomycetes that, similar to WW1C, inactivate reversibly C6-HSL, 3OC6-HSL, and 3OHC6-HSL (Ghani et al. 2014). Unfortunately, to date the QQ spectrum of B2 and B3 is not known.

The coprophilous fungus *Coprinopsis cinerea* harbors two intracellular AHL lactonases, FAII1 and FAII2, which present activity on AHLs with short and long acyl-side chains (Stöckli et al. 2017). FAII1 and FAII2 are related to AiiA and AiiB from *B. thuringiensis* and *A. fabrum*, respectively. It is supposed that FAII1- and FAII2-mediated lactonolysis of AHLs could interfere with the QS signaling among competitor and antagonizing bacteria. To date, this hypothesis has not been evaluated *in vivo*. Noteworthy, the presence of FAII1 and FAII2 orthologs is restricted to the class Agaricomycetes with a saprotrophic lifestyle (Stöckli et al. 2017).

It is important to highlight that not only lactonase-, but also acylase-like activities are present in fungi. A recent large survey of QQ activities showed that both the reversible and the irreversible AHL inactivation is widely present in ascomycete and basidiomycete yeasts. Isolates obtained from two dissimilar geographical locations (i.e., Antarctica and the inner tissues of sugarcane) evidenced the enzymatic inactivation of AHLs in both reversible and irreversible manners (Leguina et al. 2018). The grouping of the yeasts according to their enzymatic QQ activities in clusters and subclusters showed that it is mostly not possible to correlate the profiles of AHL inactivation with the niche or the geographical origin of the isolates (Figure 10.3).

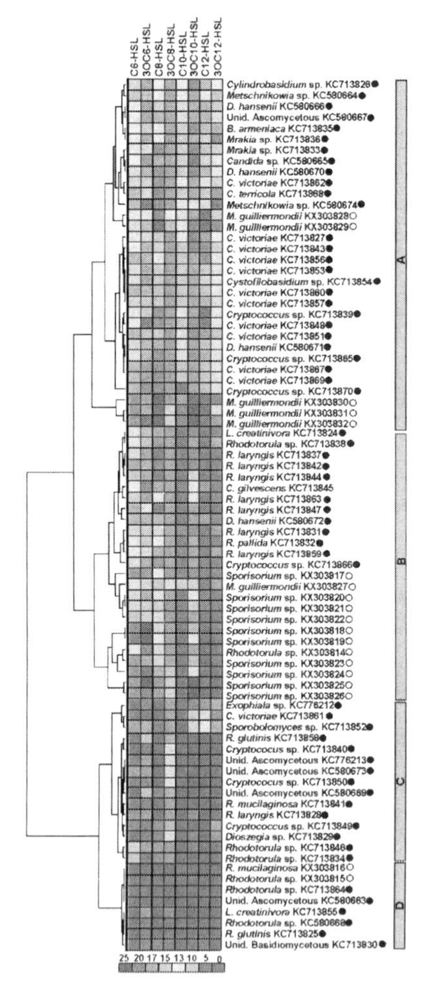

FIGURE 10.3 Clusters of yeast isolates according to their AHL-inactivation profiles. Each isolate is identified according to the corresponding GenBank accession number. White and black circles indicate the endophytic or Antarctic origin of the isolate, respectively. Two main clusters divided in two (A and B) and three (C, D and E) subclusters were identified. Reference bar indicates the corresponding AHL concentration expressed in μM. For a color figure, and a better interpretation of it, the reader is referred to the source article. (Reprinted from *Fungal Biol.*, 122, Leguina, A.C.D.V. et al., Inactivation of bacterial quorum sensing signals *N*-acyl homoserine lactones is widespread in yeasts, 52–62, Copyright 2018, with permission from Elsevier.)

However, a group of yeasts that inactivated all the AHLs assayed (with the exception of 3OC8-HSL) was composed exclusively by isolates from Antarctica (see subcluster C in Figure 10.3) (Leguina et al. 2018). It is remarkable the fungal inactivation of QS molecules with C10 and C12 acyl chains in subclusters B, C, and D (Figure 10.3). This tendency could explain the limited number of QQ positive isolates from forest soil obtained by Uroz and Heinonsalo, who utilized C6-HSL and 3OC6-HSL for the screening of forest fungi (Uroz and Heinonsalo 2008). To date it is not known whether the AHLs are utilized as a nutrient source by the yeasts. It is also not known whether the fungal enzymes are part of other metabolic pathways, and in consequence, the inactivation of these molecules is incidental (Safari et al. 2014), or whether they play a role in a putative AHL detoxification, as suggested for lactonases and amidases/acylases (Kaufmann et al. 2005). Beyond those hypotheses, it is important to note how these findings shed light on the relatively little known effect of these microorganisms on the bacterial communication. Indeed, the AHL concentration in a coculture of a QQ yeast with an AHL-producing bacterium is lower, compared to a bacterial pure culture (Leguina et al. 2018). It is then plausible that in the environment, the presence of yeasts modulate the cell density required to trigger the bacterial QS regulation.

10.14 Advantages and Limitations of Potential Biotechnological Application of QS Mediated Processes of Plant-Associated Bacteria in Agriculture

The constantly growing global population, the increasing demand for food, and the many environmental issues that the world has to face nowadays have driven the search of more eco-friendly biotechnological approaches to enhance productivity in agriculture. Since QS regulates processes involved in plant growth-promoting activities in bacteria, its stimulation using biotic or abiotic factors could have potential benefits for the host. For instance, promoting the QS system of beneficial bacteria could help avoiding the economic loss caused by phytopathogens in crops. In the same sense, intensification of the QS system activities in symbiotic nitrogen-fixing rhizobia may have economical and ecological benefits by means of an increase in nitrogen fixation and a reduced utilization of fertilizers (Galloway et al. 2012). A strategy proposed for taking advantage of QS has been the expression of an AHL synthase gene in plants. Successful examples are tobacco plants harboring the *expI* gene of *Pca* that drives the synthesis of 3OC6-HSL (Figure 10.1), which showed enhanced resistance to infection by *Pca*. Similar results were obtained when exogenous C6-HSL was added to wild-type tobacco plants (Mäe et al. 2001). In agriculture, another promising approach is to manipulate phytopathogens using QS as a target. For instance, genetically engineered *Lysobacter enzymogenes* expressing the *Bacillus* AiiA lactonase significantly decreased bacterial soft rot symptoms of plants infected by *Pca* (Qian et al. 2010). It has also been demonstrated that transgenic plants expressing the AiiA lactonase are protected against Gram-negative phytopathogens (Dong et al. 2001).

However, at the same time both silencing and stimulating QS systems in bacteria could elicit non-targeted interactions that could interfere with beneficial traits of other strains present in the plant environment. For instance, the search of the interference of the QS system of a phytopathogen through the introduction of a QQ microorganism can be, at the same time, detrimental for PGPBs whose beneficial activities are regulated by this regulatory mechanism (Molina et al. 2003). Moreover, there is a chance that bacteria could evolve resistance to QQ strategies (Defoirdt, Boon, and Bossier 2010). With regard to transgenic techniques, their applicability is limited, as to this day there are biosafety issues to be solved before being totally accepted in agricultural practice. Exploiting QS as a sustainable way to increase crop yields and control plant diseases in agriculture is a very promising strategy. Nevertheless, more information about the efficiency and success of these methods in the field is needed to fully estimate their applicability.

10.15 Conclusions and Open Questions

Quorum sensing is an important regulatory system in bacteria, including those associated with plants. As other mechanisms, QS receives inputs others than the cell density from the environment. In a context where more than one factor can act on a bacterial population at the same time, the complexity of the QS regulation could be highly significant. However, the question about the QS activity *in situ* in the context of fluctuating environmental conditions remains open. For instance, how do conditions negative for the QS communication (e.g., alkaline pH and the presence of lactonase activity) function together? Can two conditions with opposite activities on QS neutralize each other? Is it possible to manipulate the QS through certain environmental conditions for a better utilization of a PGPB? How can all these findings be exploited against phytopathogens? Although significant progress has been made *in vitro* for the understanding of QS in both pathogenic and beneficial bacteria, the next step toward the environmental effects has begun.

ACKNOWLEDGMENT

This work was supported by the Consejo Nacional de Investigaciones Científicas y Técnicas (CONICET, PIP 946), Agencia Nacional de Promoción Científica y Tecnológica (PICT 2016 N° 0532; PICT 2016 N° 2013), and Secretaría de Ciencia, Arte e Innovación Tecnológica from the Universidad Nacional de Tucumán (PIUNT D609).

We apologize to all the authors whose articles, due to space limits, could not be referenced in this work.

REFERENCES

Achari, G. A., and R. Ramesh. 2015. Characterization of bacteria degrading 3-hydroxy palmitic acid methyl ester (3OH-PAME), a quorum sensing molecule of *Ralstonia solanacearum*. *Letters in Applied Microbiology* 60:447–55.

Ahlgren, N. A., C. S. Harwood, A. L. Schaefer, E. Giraud, and E. P. Greenberg. 2011. Aryl-homoserine lactone quorum sensing in stem-nodulating photosynthetic bradyrhizobia. *Proceedings of the National Academy of Sciences of the United States of America* 108:7183–88.

Asfour, H. Z. 2018. Anti-quorum sensing natural compounds. *Journal of Microscopy and Ultrastructure* 6:1–10.

Barber, C. E., J. L. Tang, J. X. Feng, M. Q. Pan, T. J. Wilson, H. Slater, J. M. Dow, P. Williams, and M. J. Daniels. 1997. A novel regulatory system required for pathogenicity of *Xanthomonas campestris* is mediated by a small diffusible signal molecule. *Molecular Microbiology* 24:555–66.

Barnard, A. M., S. D. Bowden, T. Burr, S. J. Coulthurst, R. E. Monson, and G. P. Salmond. 2007. Quorum sensing, virulence and secondary metabolite production in plant soft-rotting bacteria. *Philosophical Transactions of the Royal Society B: Biological Sciences* 362:1165–83.

Battesti, A., N. Majdalani, and S. Gottesman. 2011. The RpoS-mediated general stress response in *Escherichia coli*. *Annual Review of Microbiology* 65:189–213.

Beaulieu, E. D., M. Ionescu, S. Chatterjee, K. Yokota, D. Trauner, and S. Lindow. 2013. Characterization of a diffusible signaling factor from *Xylella fastidiosa*. *mBio* 4:e00539–12.

Beck von Bodman, S., and S. K. Farrand. 1995. Capsular polysaccharide biosynthesis and pathogenicity in *Erwinia stewartii* require induction by an *N*-acylhomoserine lactone autoinducer. *Journal of Bacteriology* 177:5000–5008.

Bertani, I., and V. Venturi. 2004. Regulation of the *N*-Acyl homoserine lactone-dependent quorum-sensing system in rhizosphere *Pseudomonas putida* WCS358 and cross-talk with the stationary-phase RpoS Sigma factor and the global regulator GacA. *Applied and Environmental Microbiology* 70:5493–502.

Bertini, E. V, C. G. Nieto Peñalver, A. C. Leguina, V. P. Irazusta, and L. I. Castellanos de Figueroa. 2014. *Gluconacetobacter diazotrophicus* PAL5 possesses an active quorum sensing regulatory system. *Antonie van Leeuwenhoek* 106:497–506.

Bijtenhoorn, P., H. Mayerhofer, J. Müller-Dieckmann, C. Utpatel, C. Schipper, C. Hornung, M. Szesny, et al. 2011. A novel metagenomic short-chain dehydrogenase/reductase attenuates *Pseudomonas aeruginosa* biofilm formation and virulence on *Caenorhabditis elegans*. *PloS One* 6:e26278.

Bowden, S. D., A. Eyres, J. C. S. Chung, R. E. Monson, A. Thompson, G. P. C. Salmond, D. R. Spring, and M. Welch. 2013. Virulence in *Pectobacterium atrosepticum* is regulated by a coincidence circuit involving quorum sensing and the stress alarmone, (p)ppGpp. *Molecular Microbiology* 90:457–71.

Boyer, M., R. Bally, S. Perrotto, C. Chaintreuil, and F. Wisniewski-Dyé. 2008. A quorum-quenching approach to identify quorum-sensing-regulated functions in *Azospirillum lipoferum*. *Research in Microbiology* 159:699–708.

Boyer, M., and F. Wisniewski-Dyé. 2009. Cell-cell signalling in bacteria: Not simply a matter of quorum. *FEMS Microbiology Ecology* 70:1–19.

Calatrava-Morales, N., M. McIntosh, and M. J. Soto. 2018. Regulation mediated by *N*-acyl homoserine lactone quorum sensing signals in the *Rhizobium*-legume symbiosis. *Genes* 9:263.

Calderón-Flores, A., G. Du Pont, A. Huerta-Saquero, H. Merchant-Larios, L. Servín-González, and S. Durán. 2005. The stringent response is required for amino acid and nitrate utilization, nod factor regulation, nodulation, and nitrogen fixation in *Rhizobium etli*. *Journal of Bacteriology* 187:5075–83.

Carlier, A., S. Uroz, B. Smadja, R. Fray, X. Latour, Y. Dessaux, and D. Faure. 2003. The Ti plasmid of *Agrobacterium tumefaciens* harbors an *attM*-paralogous gene, *aiiB*, also encoding *N*-acyl homoserine lactonase activity. *Applied and Environmental Microbiology* 69:4989–93.

Chan, K.-G., S. Atkinson, K. Mathee, C.-K. Sam, S. R. Chhabra, M. Cámara, C.-L. Koh, and P. Williams. 2011. Characterization of *N*-acylhomoserine lactone-degrading bacteria associated with the *Zingiber officinale* (ginger) rhizosphere: Co-existence of quorum quenching and quorum sensing in *Acinetobacter* and *Burkholderia*. *BMC Microbiology* 11:51.

Chan, K. Y., L. Van Zwieten, I. Meszaros, A. Downie, and S. Joseph. 2007. Agronomic values of greenwaste biochar as a soil amendment. *Australian Journal of Soil Research* 45:629–34.

Chatterjee, A., Y. Cui, P. Chakrabarty, and A. K. Chatterjee. 2010. Regulation of motility in *Erwinia carotovora* subsp. *carotovora*: Quorum-sensing signal controls FlhDC, the global regulator of flagellar and exoprotein genes, by modulating the production of RsmA, an RNA-binding protein. *Molecular Plant-Microbe Interactions* 23:1316–23.

Chatterjee, A., Y. Cui, H. Hasegawa, N. Leigh, V. Dixit, and A. K. Chatterjee. 2005. Comparative analysis of two classes of quorum-sensing signaling systems that control production of extracellular proteins and secondary metabolites in *Erwinia carotovora* subspecies. *Journal of Bacteriology* 187:8026–38.

Chatterjee, J., C. M. Miyamoto, A. Zouzoulas, B. F. Lang, N. Skouris, and E. A. Meighen. 2002. MetR and CRP bind to the *Vibrio harveyi lux* promoters and regulate luminescence. *Molecular Microbiology* 46:101–11.

Chen, C.-N., C.-J. Chen, C.-T. Liao, and C.-Y. Lee. 2009. A probable aculeacin A acylase from the *Ralstonia solanacearum* GMI1000 is *N*-acyl-homoserine lactone acylase with quorum-quenching activity. *BMC Microbiology* 9:89.

Cheng, F., A. Ma, J. Luo, X. Zhuang, and G. Zhuang. 2017. *N*-acylhomoserine lactone-regulation of genes mediating motility and pathogenicity in *Pseudomonas syringae* pathovar *tabaci* 11528. *MicrobiologyOpen* 6:e00440.

Chowdhary, P. K., N. Keshavan, H. Q. Nguyen, J. A. Peterson, J. E. González, and D. C. Haines. 2007. *Bacillus megaterium* CYP102A1 oxidation of acyl homoserine lactones and acyl homoserines. *Biochemistry* 46:14429–37.

Coutinho, B., B. Mitter, C. Talbi, A. Sessitsch, E. Bedmar, N. Halliday, E. James, M. Cámara, and V. V. 2013. Regulon studies and in planta role of the BraI/R quorum-sensing system in the plant-beneficial *Burkholderia* cluster. *Applied and Environmental Microbiology* 79:4421–32.

Czajkowski, R., D. Krzyzanowska, J. Karczewska, S. Atkinson, J. Przysowa, E. Lojkowska, P. Williams, and S. Jafra. 2011. Inactivation of AHLs by *Ochrobactrum* sp. A44 depends on the activity of a novel class of AHL acylase. *Environmental Microbiology Reports* 3:59–68.

D'Almeida, R. E., R. R. D. I. Molina, C. M. Viola, M. C. Luciardi, C. Nieto Peñalver, A. Bardón, and M. E. Arena. 2017. Comparison of seven structurally related coumarins on the inhibition of quorum sensing of *Pseudomonas aeruginosa* and *Chromobacterium violaceum*. *Bioorganic Chemistry* 73:37–42.

Daniels, R., S. Reynaert, H. Hoekstra, C. Verreth, J. Janssens, K. Braeken, M. Fauvart, et al. 2006. Quorum signal molecules as biosurfactants affecting swarming in *Rhizobium etli*. *Proceedings of the National Academy of Sciences of the United States of America* 103:14965–70.

Daniels, R., D. E. De Vos, J. Desair, G. Raedschelders, E. Luyten, V. Rosemeyer, C. Verreth, E. Schoeters, J. Vanderleyden, and J. Michiels. 2002. The *cin* quorum sensing locus of *Rhizobium etli* CNPAF512 affects growth and symbiotic nitrogen fixation. *The Journal of Biological Chemistry* 277:462–68.

Decho, A. W., P. T. Visscher, J. Ferry, T. Kawaguchi, L. He, K. M. Przekop, R. S. Norman, and R. P. Reid. 2009. Autoinducers extracted from microbial mats reveal a surprising diversity of *N*-acylhomoserine lactones (AHLs) and abundance changes that may relate to diel pH. *Environmental Microbiology* 11:409–20.

Defoirdt, T., N. Boon, and P. Bossier. 2010. Can bacteria evolve resistance to quorum sensing disruption? *PLoS Pathogens* 6:e1000989.

Delaspre, F., C. G. Nieto Peñalver, O. Saurel, P. Kiefer, E. Gras, A. Milon, C. Boucher, S. Genin, and J. A. Vorholt. 2007. The *Ralstonia solanacearum* pathogenicity regulator HrpB induces 3-hydroxy-oxindole synthesis. *Proceedings of the National Academy of Sciences of the United States of America* 104:15870–75.

Devescovi, G., J. Bigirimana, G. Degrassi, L. Cabrio, J. J. LiPuma, J. Kim, I. Hwang, and V. Venturi. 2007. Involvement of a quorum-sensing-regulated lipase secreted by a clinical isolate of *Burkholderia glumae* in severe disease symptoms in rice. *Applied and Environmental Microbiology* 73:4950–58.

Dong, Y.-H., A. R. Gusti, Q. Zhang, J.-L. Xu, and L.-H. Zhang. 2002. Identification of quorum-quenching *N*-acyl homoserine lactonases from *Bacillus* species. *Applied and Environmental Microbiology* 68:1754–59.

Dong, Y. H., L. H. Wang, J. L. Xu, H. B. Zhang, X. F. Zhang, and L. H. Zhang. 2001. Quenching quorum-sensing-dependent bacterial infection by an *N*-acyl homoserine lactonase. *Nature* 411:813–17.

Dong, Y. H., J. L. Xu, X. Z. Li, and L. H. Zhang. 2000. AiiA, an enzyme that inactivates the acylhomoserine lactone quorum-sensing signal and attenuates the virulence of *Erwinia carotovora*. *Proceedings of the National Academy of Sciences of the United States of America* 97:3526–31.

Dulla, G. F. J., K. V. Krasileva, and S. E. Lindow. 2010. Interference of quorum sensing in *Pseudomonas syringae* by bacterial epiphytes that limit iron availability. *Environmental Microbiology* 12:1762–74.

Flavier, A. B., S. J. Clough, M. A. Schell, and T. P. Denny. 1997. Identification of 3-hydroxypalmitic acid methyl ester as a novel autoregulator controlling virulence in *Ralstonia solanacearum*. *Molecular Microbiology* 26:251–59.

Flavier, A. B., L. M. Ganova-Raeva, M. A. Schell, and T. P. Denny. 1997. Hierarchical autoinduction in *Ralstonia solanacearum*: Control of acyl-homoserine lactone production by a novel autoregulatory system responsive to 3-hydroxypalmitic acid methyl ester. *Journal of Bacteriology* 179:7089–97.

Francez-Charlot, A., J. Frunzke, J. Zingg, A. Kaczmarczyk, and J. A. Vorholt. 2016. Multiple σEcfG and NepR proteins are involved in the general stress response in *Methylobacterium extorquens*. *PLoS One* 11:e0152519.

Francez-Charlot, A., A. Kaczmarczyk, H.-M. Fischer, and J. A. Vorholt. 2015. The general stress response in *Alphaproteobacteria*. *Trends in Microbiology* 23:164–71.

Frey, R. L., L. He, Y. Cui, A. W. Decho, T. Kawaguchi, P. L. Ferguson, and J. L. Ferry. 2010. Reaction of *N*-acylhomoserine lactones with hydroxyl radicals: Rates, products, and effects on signaling activity. *Environmental Science & Technology* 44:7465–69.

Galloway, W. R. J. D., J. T. Hodgkinson, S. Bowden, M. Welch, and D. R. Spring. 2012. Applications of small molecule activators and inhibitors of quorum sensing in Gram-negative bacteria. *Trends in Microbiology* 20:449–58.

Gao, X., H.-Y. Cheng, I. Del Valle, S. Liu, C. A. Masiello, and J. J. Silberg. 2016. Charcoal disrupts soil microbial communication through a combination of signal sorption and hydrolysis. *ACS Omega* 1:226–33.

Garge, S. S., and A. S. Nerurkar. 2016. Attenuation of quorum sensing regulated virulence of *Pectobacterium carotovorum* subsp. *carotovorum* through an AHL lactonase produced by *Lysinibacillus* sp. Gs50. *PloS One* 11:e0167344.

Ghani, N. A., J. Sulaiman, Z. Ismail, X.-Y. Chan, W.-F. Yin, and K.-G. Chan. 2014. *Rhodotorula mucilaginosa*, a quorum quenching yeast exhibiting lactonase activity isolated from a tropical shoreline. *Sensors (Basel)* 14:6463–73.

Girard, G., E. T. van Rij, B. J. J. Lugtenberg, and G. V Bloemberg. 2006. Regulatory roles of *psrA* and *rpoS* in phenazine-1-carboxamide synthesis by *Pseudomonas chlororaphis* PCL1391. *Microbiology* 152:43–58.

Gloser, V., H. Korovetska, A. I. Martín-Vertedor, M. Hájíčková, Z. Prokop, S. Wilkinson, and W. Davies. 2016. The dynamics of xylem sap pH under drought: A universal response in herbs? *Plant and Soil* 409:259–72.

Grandclément, C., M. Tannières, S. Moréra, Y. Dessaux, and D. Faure. 2016. Quorum quenching: Role in nature and applied developments. *FEMS Microbiology Reviews* 40:86–116.

Gray, E. J., and D. L. Smith. 2005. Intracellular and extracellular PGPR: Commonalities and distinctions in the plant-bacterium signaling processes. *Soil Biology and Biochemistry* 37:395–412.

Hauryliuk, V., G. C. Atkinson, K. S. Murakami, T. Tenson, and K. Gerdes. 2015. Recent functional insights into the role of (p) ppGpp in bacterial physiology. *Nature Reviews Microbiology* 13:298–309.

He, Y.-W., M. Xu, K. Lin, Y.-J. A. Ng, C.-M. Wen, L.-H. Wang, Z.-D. Liu, et al. 2006. Genome scale analysis of diffusible signal factor regulon in *Xanthomonas campestris* pv. *campestris*: Identification of novel cell-cell communication-dependent genes and functions. *Molecular Microbiology* 59:610–22.

Huang, J. J., J.-I. Han, L.-H. Zhang, and J. R. Leadbetter. 2003. Utilization of acyl-homoserine lactone quorum signals for growth by a soil pseudomonad and *Pseudomonas aeruginosa* PAO1. *Applied and Environmental Microbiology* 69:5941–49.

Huang, J. J., A. Petersen, M. Whiteley, and J. R. Leadbetter. 2006. Identification of QuiP, the product of gene PA1032, as the second acyl-homoserine lactone acylase of *Pseudomonas aeruginosa* PAO1. *Applied and Environmental Microbiology* 72:1190–97.

Iyer, B., M. S. Rajput, R. Jog, E. Joshi, K. Bharwad, and S. Rajkumar. 2016. Organic acid mediated repression of sugar utilization in rhizobia. *Microbiological Research* 192:211–20.

Juhas, M., L. Wiehlmann, B. Huber, D. Jordan, J. Lauber, P. Salunkhe, A. S. Limpert, et al. 2004. Global regulation of quorum sensing and virulence by VqsR in *Pseudomonas aeruginosa*. *Microbiology* 150:831–41.

Kafle, P., A. N. Amoh, J. M. Reaves, E. G. Suneby, K. A. Tutunjian, R. L. Tyson, and T. L. Schneider. 2016. Molecular insights into the impact of oxidative stress on the quorum-sensing regulator protein LasR. *The Journal of Biological Chemistry* 291:11776–86.

Kang, Y., E. Goo, J. Kim, and I. Hwang. 2017. Critical role of quorum sensing-dependent glutamate metabolism in homeostatic osmolality and outer membrane vesiculation in *Burkholderia glumae*. *Scientific Reports* 7:44195.

Kaufmann, G. F., R. Sartorio, S.-H. Lee, C. J. Rogers, M. M. Meijler, J. A. Moss, B. Clapham, A. P. Brogan, T. J. Dickerson, and K. D. Janda. 2005. Revisiting quorum sensing: Discovery of additional chemical and biological functions for 3-oxo-*N*-acylhomoserine lactones. *Proceedings of the National Academy of Sciences of the United States of America* 102:309–14.

Kim, E.-J., W. Wang, W.-D. Deckwer, and A.-P. Zeng. 2005. Expression of the quorum-sensing regulatory protein LasR is strongly affected by iron and oxygen concentrations in cultures of *Pseudomonas aeruginosa* irrespective of cell density. *Microbiology* 151:1127–38.

Kim, S., J. Park, O. Choi, J. Kim, and Y.-S. Seo. 2014. Investigation of quorum sensing-dependent gene expression in *Burkholderia gladioli* BSR3 through RNA-seq analyses. *Journal of Microbiology and Biotechnology* 24:1609–21.

Krol, E., and A. Becker. 2011. PpGpp in *Sinorhizobium meliloti*: Biosynthesis in response to sudden nutritional downshifts and modulation of the transcriptome. *Molecular Microbiology* 81:1233–54.

Krysciak, D., C. Schmeisser, S. Preuß, J. Riethausen, M. Quitschau, S. Grond, and W. R. Streit. 2011. Involvement of multiple loci in quorum quenching of autoinducer I molecules in the nitrogen-fixing symbiont *Rhizobium* (*Sinorhizobium*) sp. strain NGR234. *Applied and Environmental Microbiology* 77:5089–99.

Lang, J., and D. Faure. 2014. Functions and regulation of quorum-sensing in *Agrobacterium tumefaciens*. *Frontiers in Plant Science* 5:14.

Leadbetter, J. R., and E. P. Greenberg. 2000. Metabolism of acyl-homoserine lactone quorum-sensing signals by *Variovorax paradoxus*. *Journal of Bacteriology* 182:6921–26.

Lee, M. H., R. Khan, W. Tao, K. Choi, S. Y. Lee, J. W. Lee, E. C. Hwang, and S.-W. Lee. 2018. Soil metagenome-derived 3-hydroxypalmitic acid methyl ester hydrolases suppress extracellular polysaccharide production in *Ralstonia solanacearum*. *Journal of Biotechnology* 270:30–38.

Leguina, A. C. D. V., C. Nieto, H. F. Pajot, E. V Bertini, W. Mac Cormack, L. I. Castellanos de Figueroa, and C. G. Nieto-Peñalver. 2018. Inactivation of bacterial quorum sensing signals *N*-acyl homoserine lactones is widespread in yeasts. *Fungal Biology* 122:52–62.

Liang, W., S. Z. Sultan, A. J. Silva, and J. A. Benitez. 2008. Cyclic AMP post-transcriptionally regulates the biosynthesis of a major bacterial autoinducer to modulate the cell density required to activate quorum sensing. *FEBS Letters* 582:3744–50.

Lin, Y.-H., A. N. Binns, and D. G. Lynn. 2008. The initial steps in *Agrobacterium tumefaciens* pathogenesis: Chemical biology of host recognition. In *Agrobacterium: From Biology to Biotechnology*, eds. T. Tzfira and V. Citovsky, 221–41. New York: Springer New York.

Lin, Y.-H., J.-L. Xu, J. Hu, L.-H. Wang, S. L. Ong, J. R. Leadbetter, and L.-H. Zhang. 2003. Acyl-homoserine lactone acylase from *Ralstonia* strain XJ12B represents a novel and potent class of quorum-quenching enzymes. *Molecular Microbiology* 47:849–60.

Lindemann, A., G. Pessi, A. L. Schaefer, M. E. Mattmann, Q. H. Christensen, A. Kessler, H. Hennecke, H. B. Blackwell, E. P. Greenberg, and C. S. Harwood. 2011. Isovaleryl-homoserine lactone, an unusual branched-chain quorum-sensing signal from the soybean symbiont *Bradyrhizobium japonicum*. *Proceedings of the National Academy of Sciences of the United States of America* 108:16765–70.

Liu, P., X. Chen, and W. Chen. 2015. Adsorption of *N*-acyl-homoserine lactone onto colloidal minerals presents potential challenges for quorum sensing in the soil environment. *Geomicrobiology Journal* 32:602–8.

Lyell, N. L., D. M. Colton, J. L. Bose, M. P. Tumen-Velasquez, J. H. Kimbrough, and E. V Stabb. 2013. Cyclic AMP receptor protein regulates pheromone-mediated bioluminescence at multiple levels in *Vibrio fischeri* ES114. *Journal of Bacteriology* 195:5051–63.

Macomber, L., and R. P. Hausinger. 2011. Mechanisms of nickel toxicity in microorganisms. *Metallomics* 3:1153–62.

Maddula, V. S. R. K., Z. Zhang, E. A. Pierson, and L. S. Pierson. 2006. Quorum sensing and phenazines are involved in biofilm formation by *Pseudomonas chlororaphis* (*aureofaciens*) strain 30-84. *Microbial Ecology* 52:289–301.

Mäe, A., M. Montesano, V. Koiv, and E. T. Palva. 2001. Transgenic plants producing the bacterial pheromone *N*-acyl-homoserine lactone exhibit enhanced resistance to the bacterial phytopathogen *Erwinia carotovora*. *Molecular Plant-Microbe Interactions* 14:1035–42.

Manefield, M., T. B. Rasmussen, M. Henzter, J. B. Andersen, P. Steinberg, S. Kjelleberg, and M. Givskov. 2002. Halogenated furanones inhibit quorum sensing through accelerated LuxR turnover. *Microbiology* 148:1119–27.

Masiello, C. A., Y. Chen, X. Gao, S. Liu, H.-Y. Cheng, M. R. Bennett, J. A. Rudgers, D. S. Wagner, K. Zygourakis, and J. J. Silberg. 2013. Biochar and microbial signaling: Production conditions determine effects on microbial communication. *Environmental Science & Technology* 47:11496–503.

McClean, K. H., M. K. Winson, L. Fish, A. Taylor, S. R. Chhabra, M. Camara, M. Daykin, et al. 1997. Quorum sensing and *Chromobacterium violaceum*: Exploitation of violacein production and inhibition for the detection of *N*-acylhomoserine lactones. *Microbiology* 143:3703–11.

McGivney, E., K. E. Jones, B. Weber, A. M. Valentine, J. M. VanBriesen, and K. B. Gregory. 2018. Quorum sensing signals form complexes with Ag+ and Cu^{2+} cations. *ACS Chemical Biology* 13:894–99.

Mei, G.-Y., X.-X. Yan, A. Turak, Z.-Q. Luo, and L.-Q. Zhang. 2010. AidH, an alpha/beta-hydrolase fold family member from an *Ochrobactrum* sp. strain, is a novel *N*-acylhomoserine lactonase. *Applied and Environmental Microbiology* 76:4933–42.

Molina, L., F. Constantinescu, L. Michel, C. Reimmann, B. Duffy, and G. Défago. 2003. Degradation of pathogen quorum-sensing molecules by soil bacteria: A preventive and curative biological control mechanism. *FEMS Microbiology Ecology* 45:71–81.

Moris, M., K. Braeken, E. Schoeters, C. Verreth, S. Beullens, J. Vanderleyden, and J. Michiels. 2005. Effective symbiosis between *Rhizobium etli* and *Phaseolus vulgaris* requires the alarmone ppGpp. *Journal of Bacteriology* 187:5460–69.

Morohoshi, T., Y. Tominaga, N. Someya, and T. Ikeda. 2012. Complete genome sequence and characterization of the *N*-acylhomoserine lactone-degrading gene of the potato leaf-associated *Solibacillus silvestris*. *Journal of Bioscience and Bioengineering* 113:20–25.

Newman, K. L., R. P. P. Almeida, A. H. Purcell, and S. E. Lindow. 2004. Cell-cell signaling controls *Xylella fastidiosa* interactions with both insects and plants. *Proceedings of the National Academy of Sciences of the United States of America* 101:1737–42.

Newman, K. L., S. Chatterjee, K. A. Ho, and S. E. Lindow. 2008. Virulence of plant pathogenic bacteria attenuated by degradation of fatty acid cell-to-cell signaling factors. *Molecular Plant-Microbe Interactions* 21:326–34.

Nickzad, A., and E. Déziel. 2016. Adaptive significance of quorum sensing-dependent regulation of rhamnolipids by integration of growth rate in *Burkholderia glumae*: A trade-off between survival and efficiency. *Frontiers in Microbiology* 7:1215.

Nieto Peñalver, C. G., E. V Bertini, and L. I. Castellanos de Figueroa. 2012. Identification of *N*-acyl homoserine lactones produced by *Gluconacetobacter diazotrophicus* PAL5 cultured in complex and synthetic media. *Archives of Microbiology* 194:615–22.

Nieto Peñalver, C. G., F. Cantet, D. Morin, D. Haras, and J. A. Vorholt. 2006. A plasmid-borne truncated luxI homolog controls quorum-sensing systems and extracellular carbohydrate production in *Methylobacterium extorquens* AM1. *Journal of Bacteriology* 188:7321–24.

Nieto Peñalver, C. G., D. Morin, F. Cantet, O. Saurel, A. Milon, and V. J. A. 2006. *Methylobacterium extorquens* AM1 produces a novel type of acyl-homoserine lactone with a double unsaturated side chain under methylotrophic growth conditions. *FEBS Letters* 580:561–67.

Nievas, F., P. Bogino, F. Sorroche, and W. Giordano. 2012. Detection, characterization, and biological effect of quorum-sensing signaling molecules in peanut-nodulating bradyrhizobia. *Sensors (Basel)* 12:2851–73.

Norizan, S. N. M., W.-F. Yin, and K.-G. Chan. 2013. Caffeine as a potential quorum sensing inhibitor. *Sensors (Basel)* 13:5117–29.

Novak, J. M., W. J. Busscher, D. W. Watts, J. E. Amonette, J. A. Ippolito, I. M. Lima, J. Gaskin, et al. 2012. Biochars impact on soil-moisture storage in an ultisol and two aridisols. *Soil Science* 177:310–20.

Park, S.-J., S.-Y. Park, C.-M. Ryu, S.-H. Park, and J.-K. Lee. 2008. The role of AiiA, a quorum-quenching enzyme from *Bacillus thuringiensis*, on the rhizosphere competence. *Journal of Microbiology and Biotechnology* 18:1518–21.

Park, S.-Y., H.-O. Kang, H.-S. Jang, J.-K. Lee, B.-T. Koo, and D.-Y. Yum. 2005. Identification of extracellular *N*-acylhomoserine lactone acylase from a *Streptomyces* sp. and its application to quorum quenching. *Applied and Environmental Microbiology* 71:2632–41.

Park, S. Y., S. J. Lee, T. K. Oh, J. W. Oh, B. T. Koo, D. Y. Yum, and J. K. Lee. 2003. AhlD, an *N*-acylhomoserine lactonase in *Arthrobacter* sp., and predicted homologues in other bacteria. *Microbiology* 149:1541–50.

Pirhonen, M., D. Flego, R. Heikinheimo, and E. T. Palva. 1993. A small diffusible signal molecule is responsible for the global control of virulence and exoenzyme production in the plant pathogen *Erwinia carotovora*. *The EMBO Journal* 12:2467–76.

Põllumaa, L., T. Alamäe, and A. Mäe. 2012. Quorum sensing and expression of virulence in pectobacteria. *Sensors (Basel)* 12:3327–49.

Qian, G.-L., J.-Q. Fan, D.-F. Chen, Y.-J. Kang, B. Han, B.-S. Hu, and F.-Q. Liu. 2010. Reducing *Pectobacterium* virulence by expression of an *N*-acyl homoserine lactonase gene P_{lpp}-*aiiA* in *Lysobacter enzymogenes* strain OH11. *Biological Control* 52:17–23.

Quiñones, B., G. Dulla, and S. E. Lindow. 2005. Quorum sensing regulates exopolysaccharide production, motility, and virulence in *Pseudomonas syringae*. *Molecular Plant-Microbe Interactions* 18:682–93.

Quiñones, B., C. J. Pujol, and S. E. Lindow. 2004. Regulation of AHL production and its contribution to epiphytic fitness in *Pseudomonas syringae*. *Molecular Plant-Microbe Interactions* 17:521–31.

Rajesh, P. S., and V. Ravishankar Rai. 2014. Quorum quenching activity in cell-free lysate of endophytic bacteria isolated from *Pterocarpus santalinus* Linn., and its effect on quorum sensing regulated biofilm in *Pseudomonas aeruginosa* PAO1. *Microbiological Research* 169:561–69.

Ramachandran, R., A. K. Burke, G. Cormier, R. V. Jensen, and A. M. Stevens. 2014. Transcriptome-based analysis of the *Pantoea stewartii* quorum-sensing regulon and identification of EsaR direct targets. *Applied and Environmental Microbiology* 80:5790–800.

Rampioni, G., I. Bertani, C. R. Pillai, V. Venturi, E. Zennaro, and L. Leoni. 2012. Functional characterization of the quorum sensing regulator RsaL in the plant-beneficial strain *Pseudomonas putida* WCS358. *Applied and Environmental Microbiology* 78:726–34.

Ravichandran, V., L. Zhong, H. Wang, G. Yu, Y. Zhang, and A. Li. 2018. Virtual screening and biomolecular interactions of CviR-based quorum sensing inhibitors against *Chromobacterium violaceum*. *Frontiers in Cellular and Infection Microbiology* 8:292.

Reverchon, S., M. L. Bouillant, G. Salmond, and W. Nasser. 1998. Integration of the quorum-sensing system in the regulatory networks controlling virulence factor synthesis in *Erwinia chrysanthemi*. *Molecular Microbiology* 29:1407–18.

Rosemeyer, V., J. Michiels, C. Verreth, and J. Vanderleyden. 1998. *luxI*- and *luxR*-homologous genes of *Rhizobium etli* CNPAF512 contribute to synthesis of autoinducer molecules and nodulation of *Phaseolus vulgaris*. *Journal of Bacteriology* 180:815–21.

Rukshana, F., C. R. Butterly, J.-M. Xu, J. A. Baldock, and C. Tang. 2014. Organic anion-to-acid ratio influences pH change of soils differing in initial pH. *Journal of Soils and Sediments* 14:407–14.

Safari, M., R. Amache, E. Esmaeilishirazifard, and T. Keshavarz. 2014. Microbial metabolism of quorum-sensing molecules acyl-homoserine lactones, γ-heptalactone and other lactones. *Applied Microbiology and Biotechnology* 98:3401–12.

Sasse, J., E. Martinoia, and T. Northen. 2018. Feed your friends: do plant exudates shape the root microbiome? *Trends in Plant Science* 23:25–41.

Sauviac, L., H. Philippe, K. Phok, and C. Bruand. 2007. An extracytoplasmic function sigma factor acts as a general stress response regulator in *Sinorhizobium meliloti*. *Journal of Bacteriology* 189:4204–16.

Schaefer, A. L., E. P. Greenberg, C. M. Oliver, Y. Oda, J. J. Huang, G. Bittan-Banin, C. M. Peres, et al. 2008. A new class of homoserine lactone quorum-sensing signals. *Nature* 454:595–99.

Schuster, M., A. C. Hawkins, C. S. Harwood, and E. P. Greenberg. 2004. The *Pseudomonas aeruginosa* RpoS regulon and its relationship to quorum sensing. *Molecular Microbiology* 51:973–85.

Scott, R. A., and S. E. Lindow. 2016. Transcriptional control of quorum sensing and associated metabolic interactions in *Pseudomonas syringae* strain B728a. *Molecular Microbiology* 99:1080–98.

Selin, C., J. Manuel, W. G. D. Fernando, and T. de Kievit. 2014. Expression of the *Pseudomonas chlororaphis* strain PA23 Rsm system is under control of GacA, RpoS, PsrA, quorum sensing and the stringent response. *Biological Control* 69:24–33.

Sheng, H., M. Harir, L. A. Boughner, X. Jiang, P. Schmitt-Kopplin, R. Schroll, and F. Wang. 2017. *N*-acyl-homoserine lactone dynamics during biofilm formation of a 1,2,4-trichlorobenzene mineralizing community on clay. *The Science of the Total Environment* 605–606:1031–38.

Sheng, H., F. Wang, C. Gu, R. Stedtfeld, Y. Bian, G. Liu, W. Wu, and X. Jiang. 2018. Sorption characteristics of *N*-acyl homserine lactones as signal molecules in natural soils based on the analysis of kinetics and isotherms. *RSC Advances* 8:9364–74.

Shepherd, R. W., and S. E. Lindow. 2009. Two dissimilar *N*-acyl-homoserine lactone acylases of *Pseudomonas syringae* influence colony and biofilm morphology. *Applied and Environmental Microbiology* 75:45–53.

Shinohara, M., N. Nakajima, and Y. Uehara. 2007. Purification and characterization of a novel esterase (beta-hydroxypalmitate methyl ester hydrolase) and prevention of the expression of virulence by *Ralstonia solanacearum*. *Journal of Applied Microbiology* 103:152–62.

Solis, R., I. Bertani, G. Degrassi, G. Devescovi, and V. Venturi. 2006. Involvement of quorum sensing and RpoS in rice seedling blight caused by *Burkholderia plantarii*. *FEMS Microbiology Letters* 259:106–12.

Stöckli, M., C.-W. Lin, R. Sieber, D. F. Plaza, R. A. Ohm, and M. Künzler. 2017. *Coprinopsis cinerea* intracellular lactonases hydrolyze quorum sensing molecules of Gram-negative bacteria. *Fungal Genetics and Biology* 102:49–62.

Subramoni, S., and V. Venturi. 2009. PpoR is a conserved unpaired LuxR solo of *Pseudomonas putida* which binds *N*-acyl homoserine lactones. *BMC Microbiology* 15:125.

Taguchi, F., Y. Ogawa, K. Takeuchi, T. Suzuki, K. Toyoda, T. Shiraishi, and Y. Ichinose. 2006. A homologue of the 3-oxoacyl-(acyl carrier protein) synthase III gene located in the glycosylation island of *Pseudomonas syringae* pv. *tabaci* regulates virulence factors via *N*-acyl homoserine lactone and fatty acid synthesis. *Journal of Bacteriology* 188:8376–84.

Tchounwou, P. B., C. G. Yedjou, A. K. Patlolla, and D. J. Sutton. 2012. Heavy metal toxicity and the environment. *Experientia Supplementum* 101:133–64.

Teplitski, M., H. Chen, S. Rajamani, M. Gao, M. Merighi, R. T. Sayre, J. B. Robinson, B. G. Rolfe, and W. D. Bauer. 2004. *Chlamydomonas reinhardtii* secretes compounds that mimic bacterial signals and interfere with quorum sensing regulation in bacteria. *Plant Physiology* 134:137–46.

Thomas, D. S., and D. Eamus. 2002. Seasonal patterns of xylem sap pH, xylem abscisic acid concentration, leaf water potential and stomatal conductance of six evergreen and deciduous Australian savanna tree species. *Australian Journal of Botany* 50:229–36.

Thornhill, S. G., M. Kumar, L. M. Vega, and R. J. C. McLean. 2017. Cadmium ion inhibition of quorum signalling in *Chromobacterium violaceum*. *Microbiology* 163:1429–35.

Truchado, P., J.-A. Giménez-Bastida, M. Larrosa, I. Castro-Ibáñez, J. C. Espín, F. A. Tomás-Barberán, M. T. García-Conesa, and A. Allende. 2012. Inhibition of quorum sensing (QS) in *Yersinia enterocolitica* by an orange extract rich in glycosylated flavanones. *Journal of Agricultural and Food Chemistry* 60:8885–94.

Tun-Garrido, C., P. Bustos, V. González, and S. Brom. 2003. Conjugative transfer of p42a from Rhizobium etli CFN42, which is required for mobilization of the symbiotic plasmid, is regulated by quorum sensing. *Journal of Bacteriology* 185:1681–92.

Uroz, S., S. R. Chhabra, M. Cámara, P. Williams, P. Oger, and Y. Dessaux. 2005. *N*-acylhomoserine lactone quorum-sensing molecules are modified and degraded by *Rhodococcus erythropolis* W2 by both amidolytic and novel oxidoreductase activities. *Microbiology* 151:3313–22.

Uroz, S., and J. Heinonsalo. 2008. Degradation of *N*-acyl homoserine lactone quorum sensing signal molecules by forest root-associated fungi. *FEMS Microbiology Ecology* 65:271–78.

Uroz, S., P. Oger, S. R. Chhabra, M. Cámara, P. Williams, and Y. Dessaux. 2007. *N*-acyl homoserine lactones are degraded via an amidolytic activity in *Comamonas* sp. strain D1. *Archives of Microbiology* 187:249–56.

Uroz, S., P. M. Oger, E. Chapelle, M.-T. Adeline, D. Faure, and Y. Dessaux. 2008. A *Rhodococcus qsdA*-encoded enzyme defines a novel class of large-spectrum quorum-quenching lactonases. *Applied and Environmental Microbiology* 74:1357–66.

Uzelac, G., H. K. Patel, G. Devescovi, D. Licastro, and V. Venturi. 2017. Quorum sensing and RsaM regulons of the rice pathogen *Pseudomonas fuscovaginae*. *Microbiology* 163:765–77.

Valente, R. S., P. Nadal-Jimenez, A. F. P. Carvalho, F. J. D. Vieira, and K. B. Xavier. 2017. Signal integration in quorum sensing enables cross-species induction of virulence in *Pectobacterium wasabiae*. *mBio* 8:e00398–17.

Vega, L. M., J. Mathieu, Y. Yang, B. H. Pyle, R. J. C. McLean, and P. J. J. Alvarez. 2014. Nickel and cadmium ions inhibit quorum sensing and biofilm formation without affecting viability in *Burkholderia multivorans*. *International Biodeterioration & Biodegradation* 91:82–87.

Venturi, V., C. Venuti, G. Devescovi, C. Lucchese, A. Friscina, G. Degrassi, C. Aguilar, and U. Mazzucchi. 2004. The plant pathogen *Erwinia amylovora* produces acyl-homoserine lactone signal molecules in vitro and in planta. *FEMS Microbiology Letters* 241:179–83.

Vercruysse, M., M. Fauvart, A. Jans, S. Beullens, K. Braeken, L. Cloots, K. Engelen, K. Marchal, and J. Michiels. 2011. Stress response regulators identified through genome-wide transcriptome analysis of the (p)ppGpp-dependent response in *Rhizobium etli*. *Genome Biology* 12:R17.

Wang, W.-Z., T. Morohoshi, M. Ikenoya, N. Someya, and T. Ikeda. 2010. AiiM, a novel class of *N*-acylhomoserine lactonase from the leaf-associated bacterium *Microbacterium testaceum*. *Applied and Environmental Microbiology* 76:2524–30.

Wang, W.-Z., T. Morohoshi, N. Someya, and T. Ikeda. 2012. AidC, a novel *N*-acylhomoserine lactonase from the potato root-associated cytophaga-flavobacteria-bacteroides (CFB) group bacterium *Chryseobacterium* sp. strain StRB126. *Applied and Environmental Microbiology* 78:7985–92.

Wang, Y.-J., and J. R. Leadbetter. 2005. Rapid acyl-homoserine lactone quorum signal biodegradation in diverse soils. *Applied and Environmental Microbiology* 71:1291–99.

Wong, C.-S., C.-L. Koh, C.-K. Sam, J. W. Chen, Y. M. Chong, W.-F. Yin, and K.-G. Chan. 2013. Degradation of bacterial quorum sensing signaling molecules by the microscopic yeast *Trichosporon loubieri* isolated from tropical wetland waters. *Sensors (Basel)* 13:12943–57.

Xu, Y.-Y., J.-S. Yang, C. Liu, E.-T. Wang, R.-N. Wang, X.-Q. Qiu, B.-Z. Li, W.-F. Chen, and H.-L. Yuan. 2018. Water-soluble humic materials regulate quorum sensing in *Sinorhizobium meliloti* through a novel repressor of *expR*. *Frontiers in Microbiology* 9:3194.

Yates, E. A., B. Philipp, C. Buckley, S. Atkinson, S. R. Chhabra, R. E. Sockett, M. Goldner, and Y. Dessaux. 2002. *N*-acylhomoserine lactones undergo lactonolysis in a pH-, temperature-, and acyl chain length-dependent manner during growth of *Yersinia pseudotuberculosis* and *Pseudomonas aeruginosa*. *Infection and Immunity* 70:5635–46.

Zhang, H.-B., C. Wang, and L.-H. Zhang. 2004. The quormone degradation system of *Agrobacterium tumefaciens* is regulated by starvation signal and stress alarmone (p)ppGpp. *Molecular Microbiology* 52:1389–1401.

Zhang, L. H., and A. Kerr. 1991. A diffusible compound can enhance conjugal transfer of the Ti plasmid in *Agrobacterium tumefaciens*. *Journal of Bacteriology* 173:1867–72.

Zhang, L., P. J. Murphy, A. Kerr, and M. E. Tate. 1993. *Agrobacterium* conjugation and gene regulation by *N*-acyl-L-homoserine lactones. *Nature* 362:446–48.

Zhang, Z., and L. S. Pierson. 2001. A second quorum-sensing system regulates cell surface properties but not phenazine antibiotic production in *Pseudomonas aureofaciens*. *Applied and Environmental Microbiology* 67:4305–15.

Zhu, J., J. W. Beaber, M. I. Moré, C. Fuqua, A. Eberhard, and S. C. Winans. 1998. Analogs of the autoinducer 3-oxooctanoyl-homoserine lactone strongly inhibit activity of the TraR protein of *Agrobacterium tumefaciens*. *Journal of Bacteriology* 180:5398–405.

Zhu, J., P. M. Oger, B. Schrammeijer, P. J. Hooykaas, S. K. Farrand, and S. C. Winans. 2000. The bases of crown gall tumorigenesis. *Journal of Bacteriology* 182:3885–95.

Zúñiga, A., R. Donoso, D. Ruiz, G. Ruz, and B. González. 2017. Quorum-sensing systems in the plant growth-promoting bacterium *Paraburkholderia phytofirmans* PsJN exhibit cross-regulation and are involved in biofilm formation. *Molecular Plant-Microbe Interactions* 30:557–65.

11

In Silico Mining of Quorum Sensing *Genes in Genomes and Metagenomes for Ecological and Evolutionary Studies*

Iñigo de la Fuente and Jorge Barriuso

CONTENTS

11.1 *Quorum Sensing* in Bacteria and Fungi

In the late 1960s Nealson and colleagues discovered that the fluorescence in the marine bacterium *Vibrio fischeri* was regulated by an autoinduction mechanism dependent on the population density of the microbial population (Nealson, Platt, & Hastings, 1970). Since then it is known that microorganisms communicate between individuals, allowing the populations to synchronize their behaviors. This coordination is based in a complex cell to cell signal system known as *quorum sensing* (QS), which is mediated by small diffusible molecules. These molecules act as autoinductors that spread out and accumulate in the media during the exponential growth phase of the microbial population (Bandara, Lam, Jin, & Samaranayake, 2012). The individuals within the same population are able of sense and respond to the concentration of QS molecules in the environment, and when a threshold concentration is reached, coordinate a specific genetic expression (Miller & Bassler, 2001). These mechanisms are involved in the regulation of very important processes in bacteria such as pathogenesis, symbiosis, competence, conjugation, nutrient uptake, morphological differentiation, secondary metabolite production, and biofilms formation (Bandara, Lam, Jin, & Samaranayake, 2012).

QS mechanisms have been deeply studied in bacteria, where different systems exist. The first mechanism described was the one from *V. fischeri* that was shown to be common among Gram-negative bacteria, and particularly among proteobacteria. This mechanism (AI-1) is regulated by a two-component system composed by a synthase (LuxI) able to produce the QS molecule N-acyl homoserine lactone (AHL), and a receptor (LuxR) that can detect the AHLs from the external media and directly bind DNA to regulate the expression of certain genes (Greenberg, Fuqua, & Winans, 1994). Other two-component system mediated by AHLs was discovered in Gram-negative bacteria later, but in this case the reception of the QS molecule is carried out by a membrane histidine kinase from the LuxN family (i.e., AinR or LuxN). In addition, other AHLs-synthases have been described such as those from the LuxM or HdtS families (Laue et al., 2000; Milton, 2006). Finally, another QS system was described in the Gram-negative bacterium *Pseudomonas aeruginosa* and related species. In this system, the QS molecules are alkyl-hydroxy-quinolones synthesizes by the pqsABCDEH gene cluster (Diggle et al., 2007).

Later on, QS mechanisms were described in Gram-positive bacteria; these are usually mediated by small peptides that act as QS molecules. In this case, a pre-pro-peptide is produced, exported, and processed by extracellular proteases leading to the final signal peptide (Dunny & Leonard, 1997). For example, in *Streptococcus pneumoniae* this system is encoded by the ComACDE cassette. Another QS molecule in Gram-positive bacteria is the γ-butyrolactone, known to be used by certain species from the genus *Streptomyces* (Polkade, Mantri, Patwekar, & Jangid, 2016).

More recently, a QS system common to Gram- positive and Gram-negative bacteria has been found, the autoinducer-2 (AI-2). This is mediated by a furanosyl borate diester, which is synthesized by an enzyme from the LuxS family, and the signal transduction is carried out by a LuxP protein (Hardie & Heurlier, 2008).

It is worthy to mention that some bacteria, like *P. aeruginosa* or *Vibrio harveyi*, possess more than one QS system to communicate with their own siblings and with other bacteria from the community. It has been found that many bacteria use both AI-1 and AI-2 QS systems (Mok, Wingreen, & Bassler, 2003).

On the other hand, other organisms have evolved strategies to inactivate the QS molecules from other microorganisms in a phenomenon called quorum quenching (Zhu & Kaufmann, 2013).

QS phenomenon in eukaryotes was reported much later, particularly in fungi, in the pathogenic yeast *Candida albicans*, where the sesquiterpene alcohol—farnesol—was described as the first QS molecule in an eukaryotic organism (Hornby et al., 2001). QS has also been found in other fungi (De Salas, Martínez, & Barriuso, 2015), but little is known yet about the molecular components of their QS systems. In fungi, QS mechanisms have been described to regulate processes such as sporulation, secondary metabolite production, morphological transition, and enzyme secretion (Barriuso, 2015). Moreover, communication between bacteria and eukaryotes have been described (Barriuso, Hogan, Keshavarz, & Martínez, 2018).

11.2 Genome Mining

In 1995, the first genome fully sequenced ever was published, the one from the bacterium *Haemophilus influenzae* (Fleischmann et al., 1995). The published protocol was based on whole-genome shotgun, which consists of the sequencing of short DNA fragments digested by restriction enzymes and their reassembling by bioinformatic means. This method showed the way for a new revolution in the biological sciences, and the fast development of the different nucleic acid sequencing technologies alongside the conclusion of the Human Genome Project followed in 2003 (for a review: Moraes & Góes, 2016). Later on, the second generation of sequencing (also known as Next Generation Sequencing, NGS) was established, with different platforms of massive parallel sequencing (i.e., 454 pyro-sequencing or Illumina). With these technologies the genomes of a huge number of different organisms could be studied in a very short period of time and with decreasing cost and effort. Thousands of genomes and microbiomes of a vast variety of ecological niches are daily becoming available to the scientific community, as whole culture collections are currently being sequenced (Loman & Pallen, 2015) and new technologies like single cell genomics and metagenomics generate massive data to be analyzed (Ziemert, Alanjary, & Weber, 2016).

This vast amount of data being is usually uploaded to the different databases where it could be safely and freely manipulated, stored, and analyzed. Most of the sequencing projects are deposited in several public databases. Since the early 1980s, the International Nucleotide Sequence Database Collaboration (INSDC, http://www.insdc.org (Karsch-Mizrachi, Takagi, & Cochrane, 2018)) has become one of the largest consortiums focused on establishing formats, protocols, and storage of DNA related data and metadata. The mains partners of INSDC are the following: DNA Data Bank of Japan (DDBJ; http://www.ddbj.nig.ac.jp/ (Mashima et al., 2016)), the European Nucleotide Archive (ENA, www.ebi.ac.uk/ena (Silvester et al., 2015)), GenBank (http://www.ncbi.nlm.nih.gov/genbank (Benson et al., 2018)) at the National Center for Biotechnology Information (NCBI), and IMG/M (Markowitz et al., 2014) from the Joint Genome Institute (DOE-JGI).

Moreover, the advances in bioinformatics methodologies are allowing the discovery of new genes and proteins in these datasets (Barriuso, Prieto, & Martínez, 2013). In the words of Nicholas Negroponte, from MIT (Biology is the New Digital): "And it is true that biology is not only an experimental science anymore: Sequencing and experimental procedures generate thousands and thousands of terabytes of data that has to be stored, classified, analyzed, and discussed using *ad hoc* computational methods in order to be able to use its full potential for human and environmental wellbeing."

Traditional genome screening methods such as mutagenesis (Nolan et al., 2000) or the use of transposons (van Opijnen & Camilli, 2013) were soon left behind after the apparition of the new sequencing techniques. In this sense, the sequencing of the genome of bacteria, such as those from the genus *Sreptomyces*, has exploded in the field of drug discovery, even to the point of being considered exhausted as a source of new antibiotics (Bachmann, Van Lanen, & Baltz, 2014; Challis & Hopwood, 2003). However, after the sequencing of the full genomes of two of the main natural product producer strains, *Streptomyces avermitilis* and *Streptomyces coelicolor*, the scientific community started to understand the true potential behind of the genomic dataset. These *Streptomyces* contained on average 30 secondary metabolite gene clusters, from which only three were known before the sequencing. From here the concept of genome mining started developing as a way to isolate natural products (terpenoids, alkaloids, aromatic compounds, saccharides, nonribosomal peptides…) based on genetic information of the organism. Soon, after realizing that even well studied organisms still had potential for discovery of new products, genome mining was extended to other types of organism, such as anaerobic bacteria (Behnken & Hertweck, 2012), myxobacteria (Herrmann, Fayad, & Müller, 2017), or cyanobacteria (Calteau et al., 2014).

The term "genome mining" derives from the idea of "data mining," which can be understood as the interpretation of big amounts of data sets (also commonly known as "big data") in order to discover trends, relationships, or patterns that will eventually lead to new knowledge (Fayyad, Piatetsky-Shapiro, & Padhraic, 1996). In this case, the whole genome of the organisms is the source of the big data sets, which is decoded by bioinformatic technics and tools to get new catalysts, biomolecules, or different products (Paterson et al., 2017). Using genome mining methodologies, the gene cluster encoding compounds such as polyketides, non-ribosomal peptides, terpenes, bacteriocins, nucleosides, beta-lactams, butyrolactones, siderophores, QS metabolites, and others have been identified and characterized. As an example of a work based in a procedure as previously described, Barriuso and colleagues (2013) screened different fungal genomes published at JGI database to find new lipases. They used a methodology based on the search of conserved motifs and in sequence homology; after that they selected several enzymes and predicted their catalytic properties based on 3D models of the structure of the candidate proteins.

On the other hand, it is known that 99% of the microorganisms are not able to grow *in vitro* in lab conditions; among this unculturable fraction is included the so-called Microbial Dark Matter, which includes the microorganisms that are suspected to live in any microbiome but are out of reach by current analytical tools and methods (Rinke et al., 2013). To overcome this problem, metagenomic approaches are utilized. According to Garza and Dutilh (2015), metagenomics can be defined as the study of the genetic material recovered from a sample and subjected to random fragmentation and sequencing. In this way, we may be able to have a much deeper view of the actual microbial ecology in different niches and microbiotas (Handelsman & Handelsman, 2004). Furthermore, this opens a huge window of opportunities for searching and mining new organisms, enzymes, metabolites, and molecules (Piel, 2011).

11.3 Metagenome Mining

11.3.1 Classics Screening

The traditional approach for a metagenomic analysis begins with the isolation of DNA from environmental samples. Next is the construction of metagenomic libraries cloning DNA fragments into a convenient vector (plasmid, cosmid, fosmid, bacterial artificial chromosome, etc.), and finally the screening of the libraries using different methodologies to look for the desired genes, proteins, or activities (Kimura, 2014). Typically, the clones in the library are screened for an enzymatic activity, and the positive ones subjected to DNA extraction and sequencing to identify the coding genes. However, this classical metagenomic library construction has still some limitations to overcome. The hit rate for a positive clone is technically challenging; statistically for a small insert library (less 10 kb), 10^5–10^6 clones are needed to get a single hit. Factors like the selected vector, the heterologous gene expression host organism, or the assay method may have a strong influence (Henne, Daniel, & Schmitz, 1999). A perfect example of the potential of the metagenomic screening is shown in the work published by Popovic et al. (2017). Here, 16 different environmental samples were analyzed (from soil, sea water, waste water) with a variety of conditions (3°C–50°C, high salt concentrations, contaminated grounds or water), and over one million clones screened to find for esterase activities. They were able to identify 714 positive clones that rendered 80 final active enzymes of 17 different families. Furthermore, the new enzymes showed different optimal conditions and substrates according to different environmental adaptations. On the other hand, an example of a metagenomic study to identify QS genes into soil samples was provided by Williamson and colleagues (2005). They took soil samples from non-permafrost ground in Alaska. The samples were subjected to DNA extraction and fragmentation and then cloned in to bacterial artificial chromosomes and fosmids, constructing eight metagenomic libraries containing more than 50,000 clones between 1 and 190 kb. To identify metagenomic clones that produce biologically active QS molecules, they designed a rapid high throughput "intracellular" screen based in a host cell containing a biosensor for compounds that induce bacterial QS. With this methodology, 11 clones resulted positive, and one of them carried a LuxI homologue gene that directs the synthesis of an AHL (Williamson et al., 2005). Other examples of *in vitro* metagenome mining for different QS molecules is the work by Hao and collaborators (Hao, Winans, Glick, & Charles, 2010), in which they built a *Agrobacterium tumefaciens*, strain HC103 (pJZ381), as biosensor for new QS inducers in metagenomic libraries. The system was based on the activation of traC-lacZ fusion gene in presence of QS inducer synthases. In case of activation, and with the X-Gal source, the positive colonies would become blue. They prove this method using metagenomic libraries based in activated sludge and soil samples, isolating three LuxI/LuxR-like systems closely related to those previously found in *Alphaproteobacteria*, *Betaproteobacteria*, and *Gammaproteobacteria*.

11.3.2 *In Silico* Screening

On the other hand, *in silico* approaches avoid these drawbacks and offer great opportunities to exploit the hidden information that the microbial dark matter still has to offer. The last trend in metagenome mining is none other than the analysis *in silico*, which is getting more and more attention these days. In contrast to the classical *in vivo* and *in vitro* assays, this method is purely based in computational tools, showing some clear advantages: (i) there is no need of buying and/or transforming any genomic or metagenomic libraries, (ii) a huge amount of data is publicly available in the databases, and the amount is increasing day by day, and (iii) this methodology requires lower experimental effort and less time and money investment (Fayyad et al., 1996).

Overall, the bioinformatic analysis tends to follow the same pattern, with some differences to adapt to each specific study. The searches usually take into account gene or protein sequence similarity, presence of conservative motifs, or known structural domains. In addition, searches may be refined using detailed sequence and phylogenetic analysis, or 3D-modeling (Barriuso et al., 2013).

Among the public databases containing metagenomic sequences, there is the Integrated Microbial Genomes & Microbiomes data management system: IMG/M (https://img.jgi.doe.gov/) that provides user friendly analytical tools with huge potential. This is a service dependent on the Department of Energy (DOE) of the United States, and together with the Genomes OnLine Database (GOLD) (Mukherjee et al., 2018) and GenBank (Benson et al., 2018), they include genomes from archaea, bacteria, eukarya; genomes from uncultured organisms represented by single-cell amplified genomes (SAGs); metagenome-assembled genomes (MAGs); metagenomes from environmental, engineered, and host-associated microbiomes; and metatranscriptome datasets, plasmids, and viruses (Chen et al., 2018).

By December 2018, the IMG contained more than 70,000 publicly available genomes from archaea, bacteria, and eukaryotic origin, as well as 8,390 viral genomes, 1,190 sequenced plasmids, and more than 24,000 metagenome datasets from various origins. In the later cases, this means more than 54 billion metagenome genes. The open access of these datasets varies depending of their origin; while all the genomes imported from NCBI are public, the external submitted data remains private for 18 months for genomes and 24 months for the metagenomes (Chen et al., 2018). Finally, IMG also contains two important side sites for management and analysis of genomic information: IMG/ABC (Hadjithomas et al., 2017) and IMG/VR (Paez-Espino et al., 2017). IMG/ABC is a specialized site that predicts biosynthetic gene clusters (BGCs) and contains more than a million BGCs predicted from different genomes and metagenomes. In the case of IMG/VR, it currently hosts around 700,000 isolated viruses and metagenomic viral contigs, providing the statistical tools for their analysis as well as their sequence, clustering, and host information. As previously mentioned, for the *in silico* analysis, databases are key for information management and analysis. The earlier cited IMG/M data management system is one of the most used systems, as it offers a comprehensive set of gene analysis tools already available in their own interface. Once signed in (free for both students or researchers), the user can select a different set of genomes or metagenomes according to their interest, add then to a workspace, and use the native options specialized in statistical analysis and in finding different genes, functions, and scaffolds within the workspace. A basic outlook of how to use this tool is provided by Chen et al. (2018). Once a workspace is built, several methods can be used in order to find new genes

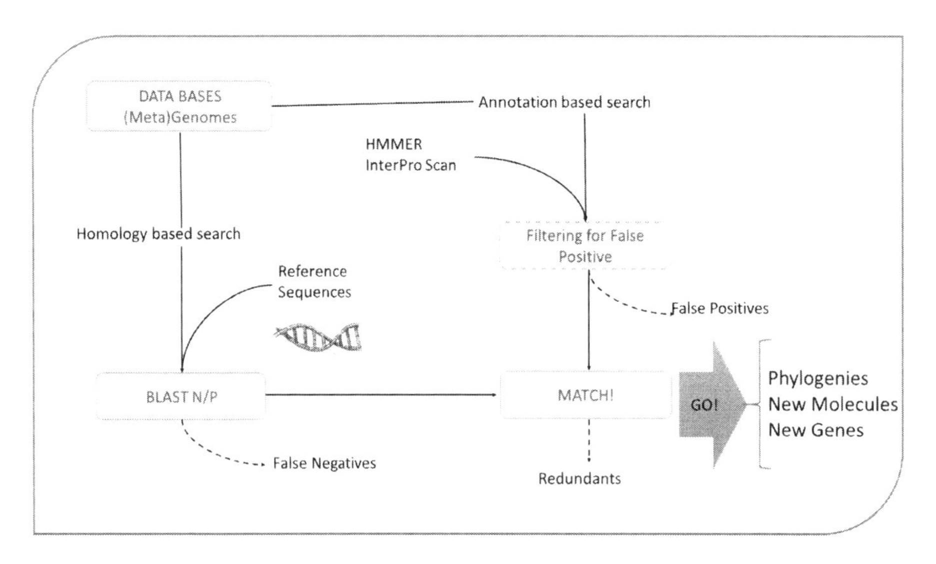

FIGURE 11.1 Flow chart of two different strategies for the *in silico* mining of genomes and metagenomes.

encoding different proteins. In this chapter we will focus on two simple yet effective strategies (Figure 11.1). First, use the "search by term" option, filter by "Gene Product Name," and then apply different queries related to your molecule of interest (if considering QS molecules, words like "autoinducer," "Quorum," or "Lux_" would be good examples). This method will retrieve all genes that, based on the JGI notation, contain the inquired term. The annotation of these genes is automatic in most of the cases, so some false positive results may be encountered. An easy way to filter false positives is to compare the conserved motifs of the results with the reference sequence's motifs, using algorithms as InterPro Scan (Jones et al., 2014; Mitchell et al., 2018) and discarding those non-coherent with the expected results. The second strategy is based on sequence homology: Beginning with a reference sequence from known proteins, a BlastP can be run against all the pull of proteins formed from the (meta)genome stored in the workspace. Cut-off value of $1e^{-10}$ is recommended in the case of proteins, as it usually retrieves matches with sequence identity higher than 20%, avoiding false positives. A downside of this method is the need of a reference protein, which may be a problem because of low variety of reference sequences. Moreover, homology-based searches can be applied to other sequences, such as promoter sequences like the ones described in Schuster, Urbanowski, and Greenberg (2004).

The IMG annotation pipeline is mostly focused in prokaryotic gene finders (Chen et al., 2018), and once the metagenomic data have been submitted, the protein-coding genes go through different analysis based on diverse bioinformatic tools. First, signal peptides and transmembrane regions are predicted using TMHMM (Moller, Croning, & Apweiler, 2002) and SignalIP (Emanuelsson, Brunak, von Heijne, & Nielsen, 2007). Next, protein family assignments are made by COG (Galperin, Makarova, Wolf, & Koonin, 2015), as well as functional annotation using Hidden Markov Models (HMMER 3.1 (Finn et al., 2015)), and the results are compared from Pfam-A (Finn et al., 2016) and TIGRfam (Haft et al., 2013) are made. Then, the different genes are processed by a personalized version of InterProScan5 (Jones et al., 2014) to be assigned into a InterPro families. Finally, LAST (Kiełbasa, Wan, Sato, Horton, & Frith, 2011) is used to associate proteins with the KEGG Orthologt (KO) terms (Kanehisa, Sato, Kawashima, Furumichi, & Tanabe, 2016).

On the other hand, another tool that has had a huge impact in genomes and metagenome mining is the "antibiotics and secondary metabolite analysis shell": AntiSMASH (Medema et al., 2011). This tool offers a comprehensive pipeline to identify the biosynthetic loci of secondary metabolites, for both bacteria and fungi, comparing to their closest relatives from a database harboring all known gene clusters, integrating and cross-linking previously available secondary specific gene analysis tools in one interactive interface (Blin et al., 2017).

In spite of all its potential, *in silico* genome mining is far from being the promised land. As a drawback, it has to be considered that the method based in sequence screening relies on homology between the reference and different sequences or conserved motifs, which means that this approach is not able to find new structures with the same function. Another matter to consider is the fact that, due to the low sequencing depth, the relative abundance of the original DNA samples of the bacterial DNA is usually predominant against other eukaryote or fungi, which may increase the difficulty of finding new or less known sequences of secondary metabolism or proteins (Barriuso et al., 2013; Bolduc et al. 2012).

11.4 Opportunities in Ecological and Evolutionary Studies

Another important topic in which the new dimension of computational biology may cause a revolution is in the evolutive and ecological study of the QS mechanisms. The use of QS systems in microbial populations has been considered a big evolutionary advantage to adapt to changing environments. It has even been suggested that these systems could be considered as a neo-Darwinian mechanism of evolution, with a primordial role in the development of the first multicellular organisms (Miller & Bassler, 2001). Considering that the

TABLE 11.1

Overview of QS Genes and Proteins Identified by *In Silico* Genomes and Metagenome Mining

Genome/Metagenome	QS System	Accession Num.	Gene	Method
***Pseudoalteromonas luteoviolacea* S4054**	LuxI/R	–	AHL synthase	ANTISmash
***Paracoccus* sp. S4493**	LuxI/R	–	AHL synthase	ANTISmash
***Phaeobacter inhibens* DSM17395**	LuxI/R	–	AHL synthase	ANTISmash
***Loktanella* sp. S4079**	LuxI/R	–	AHL synthase	ANTISmash
***Ruegeria mobilis* S1942**	LuxI/R	–	AHL synthase	ANTISmash
***Ruegeria mobilis* S1926**	LuxI/R	–	AHL synthase	ANTISmash
Fissure Spring Biofilm (FS08)	HdtS	03DRAFT_10002733	LptA-like protein	JGI term + blastp search 3D modeling
Pozzo dei Cristalli (PC08)	HdtS	66DRAFT_100008891	LptA-like protein	JGI term + blastp search 3D modeling
Fissure Spring Biofilm (FS08)	HdtS	03DRAFT_10008623	LptA-like protein	JGI term + blastp search 3D modeling
Fissure Spring Biofilm (FS08)	HdtS	03DRAFT_10009017	LptA-like protein	JGI term + blastp search 3D modeling
Grotta Sulfurea Biofilms (GS10)	HdtS	10DRAFT_10083017	LptA-like protein	JGI term + blastp search 3D modeling
Fissure Spring Biofilm (FS08)	HdtS	03DRAFT_10095361	LptA-like protein	JGI term + blastp search 3D modeling

Source: Machado, H. et al., *BMC Genomics*, 16, 1–12, 2015; Barriuso, J. et al., *FEMS Microbiology Reviews*, 42, 627–638, 2018.

culture-independent genomic analysis of microbial communities is already a fact, this has contributed to a deeper understanding of the microbial communities and the biodiversity in different environments (Prifti & Zucker, 2015). The rich and freely available databases and bioinformatic tools are making the discovery of new genes and proteins a new trend (Barriuso & Martínez, 2018; Barriuso et al., 2013).

Using *in silico* metagenomic approaches, it is now clear the LuxI/LuxR QS system is not limited to *Alpha*, *Beta*, *Gamma*, or *Epsilon-proteobacteria*, but it is distributed among other kinds of bacteria such as those from the genus Nitrospira (Kimura, 2014; Barriuso et al., 2018) or methanogenic archaeon (Zhang et al., 2012). The phylogenetic analysis the QS proteins have clustered the protein families according to groups defined by the 16S rRNA (Gray & Garey, 2001), overall showing an ancient origin of the LuxI/LuxR system in the *Proteobacteria* phylum. Other works purely focused in *in silico* approaches have been recently published. In the study developed by Machado and colleagues, they checked the genomes of 21 different marine species, *Alpha*- and *Gamma-proteobacteria*, taken from a sea water prospection. Considering that the classical activity assay may have not been enough to disclose the full potential of the different species, they decided to sequence and analyze them by bioinformatic means (Table 11.1). After full genome annotation using RAST (Glass & Meyer, 2011), they tested the samples for genetic clusters related to different secondary metabolites (chitin metabolism, iron scavenging, and QS system) using AntiSMASH (Blin et al., 2017) software, and showing three strains with AHL related gene clusters that were not recognized in the *in vitro* assays. This may be explained due to the fact that the AHL may be under the detection range of the biomonitor experiment or the growing conditions may not be adequate for the production of QS molecules (Machado, Sonnenschein, Melchiorsen, & Gram, 2015).

While genome mining may unlock the potential of the different species, a closer look at the microbial communities and the role of the QS within then can give a wider view of the ecological importance of this mechanism. The study by Barriuso & Martínez (2018) is an example of the use of the different approaches described above to look for bacterial QS proteins in environmental or biotechnological relevant biofilms available in the JGI database. First, they used the "search by term" with more than 30 different terms, followed by a "homology-based search" (with 37 different reference sequence). Then, they performed phylogenetic analysis and 3D-modeling to select the protein candidate (Figure 11.1). After multivariable analysis of the abundance and diversity of the QS proteins in each habitat, they concluded that the number of hits in each sample is influenced by number of reads in the sequenced sample and by the ecological abundance and the complexity of the microbial community. This work shows the complex interaction between the different species of each community. In addition, they described a group of six putative new HdtS-like proteins that did not match with any reference sequence (Barriuso & Martínez, 2018) (Table 11.1).

11.5 Conclusions

The study of the components of the QS systems in microorganisms will be essential in the future to understand the interactions among the complex communities of microbes and as a source of molecules with biotechnological and clinical implications. Taking into account the unculturable nature of the microbial communities, the amount of data generated by NGS technologies, and the bioinformatic tools and approaches that are being developed on a daily basis, the potential of the *in silico* data mining is huge. However, wet lab experiments cannot be denied as important sources of information and knowledge, as previous experience is what enables the actual *in silico* revolution. In this sense, in most cases, the searches are based on homologies with proteins or genes from which there is previous knowledge.

ACKNOWLEDGMENTS

This work was supported by the Spanish Project BIO2015-73697-JIN from MEICOMP and co-financed with FEDER Funds.

REFERENCES

Bachmann, B. O., Van Lanen, S. G., & Baltz, R. H. (2014). Microbial genome mining for accelerated natural products discovery: Is a renaissance in the making? *Journal of Industrial Microbiology & Biotechnology*, *41*(2), 175–184. https://doi.org/10.1007/s10295-013-1389-9.Microbial.

Bandara, H. M. H. N., Lam, O. L. T., Jin, L. J., & Samaranayake, L. (2012). Microbial chemical signaling: A current perspective. *Critical Reviews in Microbiology*, *38*(3), 217–249. https://doi.org/10.3109/1040841X.2011.652065.

Barriuso, J. (2015). Quorum sensing mechanisms in fungi. *AIMS Microbiology*, *1*(1), 37–47. https://doi.org/10.3934/microbiol.2015.1.37.

Barriuso, J., & Martínez, M. J. (2018). *In silico* analysis of the quorum sensing metagenome in environmental biofilm samples. *Frontiers in Microbiology*, *9* (June), 1–8. https://doi.org/10.3389/fmicb.2018.01243.

Barriuso, J., Hogan, D. A., Keshavarz, T., & Martínez, M. J. (2018). Role of quorum sensing and chemical communication in fungal biotechnology and pathogenesis. *FEMS Microbiology Reviews*, *42*(5), 627–638. https://doi.org/10.1093/femsre/fuy022.

Barriuso, J., Prieto, A., & Martínez, M. J. (2013). Fungal genomes mining to discover novel sterol esterases and lipases as catalysts. *BMC Genomics*, *14*(1). https://doi.org/10.1186/1471-2164-14-712.

Behnken, S., & Hertweck, C. (2012). Anaerobic bacteria as producers of antibiotics. *Applied Microbiology and Biotechnology*, *96*(1), 61–67. https://doi.org/10.1007/s00253-012-4285-8.

Benson, D. A., Cavanaugh, M., Clark, K., Karsch-Mizrachi, I., Ostell, J., Pruitt, K. D., & Sayers, E. W. (2018). GenBank. *Nucleic Acids Research*, *46*(D1), D41–D47. https://doi.org/10.1093/nar/gkx1094.

Blin, K., Wolf, T., Chevrette, M. G., Lu, X., Schwalen, C. J., Kautsar, S. A., … Medema, M. H. (2017). AntiSMASH 4.0 - improvements in chemistry prediction and gene cluster boundary identification. *Nucleic Acids Research*, *45*(W1), W36–W41. https://doi.org/10.1093/nar/gkx319.

Bolduc, B., Shaughnessy, D. P., Wolf, Y. I., Koonin, E. V., & Roberto, F. F., Young, M. (2012). Identification of novel positive-strand RNA viruses by *Metagenomic* analysis of archaea-dominated yellowstone hot springs. *Journal of Virology*, *13*(2), 401–407. https://doi.org/10.1128/JVI.07196-11.

Calteau, A., Fewer, D. P., Latifi, A., Coursin, T., Laurent, T., Jokela, J., … Gugger, M. (2014). Phylum-wide comparative genomics unravel the diversity of secondary metabolism in Cyanobacteria. *BMC Genomics*, *15*(1), 1–14. https://doi.org/10.1186/1471-2164-15-977.

Chen, I.-M. A., Chu, K., Palaniappan, K., Pillay, M., Ratner, A., Huang, J., … Kyrpides, N. C. (2018). IMG/M v.5.0: An integrated data management and comparative analysis system for microbial genomes and microbiomes. *Nucleic Acids Research*, 1–12. https://doi.org/10.1093/nar/gky901.

De Salas, F., Martínez, M. J., & Barriuso, J. (2015). *Quorum-sensing* mechanisms mediated by farnesol in *Ophiostoma piceae*: Effect on secretion of sterol esterase. *Applied and Environmental Microbiology*, *81*(13), 4351–4357. https://doi.org/10.1128/AEM.00079-15.

Diggle, S. P., Matthijs, S., Wright, V. J., Fletcher, M. P., Chhabra, S. R., Lamont, I. L., … Williams, P. (2007). The *Pseudomonas aeruginosa* 4-quinolone signal molecules HHQ and PQS play multifunctional roles in quorum sensing and iron entrapment. *Chemistry and Biology*, *14*(1), 87–96. https://doi.org/10.1016/j.chembiol.2006.11.014.

Dunny, G. M., & Leonard, B. A. (1997). Cell-cell communication in gram-positive bacteria. *Annual Review of Microbiology*, *51*, 527–564. https://doi.org/10.1146/annurev.micro.51.1.527.

Emanuelsson, O., Brunak, S., von Heijne, G., & Nielsen, H. (2007). Locating proteins in the cell using TargetP, SignalP and related tools. *Nature Protocols*, *2*(4), 953–971. https://doi.org/10.1038/nprot.2007.131.

Fayyad, U., Piatetsky-Shapiro, G., & Padhraic, S. (1996). From data mining to knowledge discovery in databases. *AI Magazine*. https://doi.org/10.1007/978-3-319-18032-8_50.

Finn, R. D., Clements, J., Arndt, W., Miller, B. L., Wheeler, T. J., Schreiber, F., … Eddy, S. R. (2015). HMMER web server: 2015 Update. *Nucleic Acids Research*, *43*(W1), W30–W38. https://doi.org/10.1093/nar/gkv397.

Finn, R. D., Coggill, P., Eberhardt, R. Y., Eddy, S. R., Mistry, J., Mitchell, A. L., … Bateman, A. (2016). The Pfam protein families database: Towards a more sustainable future. *Nucleic Acids Research*, *44*(D1), D279–D285. https://doi.org/10.1093/nar/gkv1344.

Fleischmann, R. D., Adams, M. D., White, O., Clayton, R. A., Kirkness, E. F., Kerlavage, A. R., … Venter, J. C. (1995). Whole-genome random sequencing and assembly of *Haemophilus influenzae* Rd. *Science*, *269*(5223), 496–512. https://doi.org/10.1126/science.7542800.

Galperin, M. Y., Makarova, K. S., Wolf, Y. I., & Koonin, E. V. (2015). Expanded microbial genome coverage and improved protein family annotation in the COG database. *Nucleic Acids Research*, *43*(D1), D261–D269. https://doi.org/10.1093/nar/gku1223.

Garza, D. R., & Dutilh, B. E. (2015). From cultured to uncultured genome sequences: Metagenomics and modeling microbial ecosystems. *Cellular and Molecular Life Sciences*, *72*(22), 4287–4308. https://doi.org/10.1007/s00018-015-2004-1.

Glass, E.M., & Meyer, F. (2011). The metagenomics RAST server: A public resource for the automatic phylogenetic and functional analysis of metagenomes. In *Handbook of Molecular Microbial Ecology I*, F.J. de Bruijn (ed.). https://doi.org/10.1002/9781118010518.ch37.

Gray, K. M., & Garey, J. R. (2001). The evolution of bacterial LuxI and LuxR quorum sensing regulators. *Microbiology*, *147*(8), 2379–2387. https://doi.org/10.1099/00221287-147-8-2379.

Greenberg, E. P., Fuqua, W. C., & Winans, S. C. (1994). MINIREVIEW quorum sensing in bacteria: The LuxR-LuxI family of cell density-responsive transcriptional regulators. *Journal of Bacteriology*, *176*(2), 269–275. https://doi.org/10.1128/jb.176.2.269-275.1994.

Hadjithomas, M., Chen, I. M. A., Chu, K., Huang, J., Ratner, A., Palaniappan, K., ... Ivanova, N. N. (2017). IMG-ABC: New features for bacterial secondary metabolism analysis and targeted biosynthetic gene cluster discovery in thousands of microbial genomes. *Nucleic Acids Research*, *45*(D1), D560–D565. https://doi.org/10.1093/nar/gkw1103.

Haft, D. H., Selengut, J. D., Richter, R. A., Harkins, D., Basu, M. K., & Beck, E. (2013). TIGRFAMs and genome properties in 2013. *Nucleic Acids Research*, *41*(D1), 387–395. https://doi.org/10.1093/nar/gks1234.

Handelsman, J., & Handelsman, J. (2004). Metagenomics: Application of genomics to uncultured microorganisms. *Microbiology and Molecular Biology Reviews*, *68*(4), 669–685. https://doi.org/10.1128/MBR.68.4.669.

Hao, Y., Winans, S. C., Glick, B. R., & Charles, T. C. (2010). NIH Identification and characterization of new LuxR/LuxI-type quorum sensing systems from metagenomic libraries. *Environmental Microbiology*, *12*(1), 105–117. https://doi.org/10.1111/j.1462-2920.2009.02049.x. Identification.

Hardie, K. R., & Heurlier, K. (2008). Establishing bacterial communities by "word of mouth": LuxS and autoinducer 2 in biofilm development. *Nature Reviews Microbiology*, *6*(8), 635–643. https://doi.org/10.1038/nrmicro1916.

Henne, A., Daniel, R., & Schmitz, R. A. (1999). Construction of environmental DNA libraries in *Escherichia coli* and screening for the presence of genes conferring utilization of 4-Hydroxybutyrate, *65*(9), 3901–3907.

Herrmann, J., Fayad, A. A., & Müller, R. (2017). Natural products from myxobacteria: Novel metabolites and bioactivities. *Natural Product Reports*, *34*(2), 135–160. https://doi.org/10.1039/c6np00106h.

Challis, G. L., & Hopwood, D. A. (2003). Synergy and contingency as driving forces for the evolution of multiple secondary metabolite production by *Streptomyces* species. *Proceedings of the National Academy of Sciences*, *100*, 14555–14561. https://doi.org/10.1073/pnas.1934677100.

Hornby, J. M., Jensen, E. C., Lisec, A. D., Tasto, J. J., Jahnke, B., Shoemaker, R., ... Nickerson, A. W. (2001). Quorum sensing in the Dimorphic Fungus *Candida albicans* is mediated by farnesol. *Applied and Environmental Microbiology*, *67*(7), 98–99. https://doi.org/10.1128/AEM.67.7.2982.

Jones, P., Binns, D., Chang, H. Y., Fraser, M., Li, W., McAnulla, C., ... Hunter, S. (2014). InterProScan 5: Genome-scale protein function classification. *Bioinformatics*, *30*(9), 1236–1240. https://doi.org/10.1093/bioinformatics/btu031.

Kanehisa, M., Sato, Y., Kawashima, M., Furumichi, M., & Tanabe, M. (2016). KEGG as a reference resource for gene and protein annotation. *Nucleic Acids Research*, *44*(D1), D457–D462. https://doi.org/10.1093/nar/gkv1070.

Karsch-Mizrachi, I., Takagi, T., & Cochrane, G. (2018). The international nucleotide sequence database collaboration. *Nucleic Acids Research*, *46*(D1), D48–D51. https://doi.org/10.1093/nar/gkx1097.

Kiełbasa, S. M., Wan, R., Sato, K., Horton, P., & Frith, M. C. (2011). Adaptive seeds tame genomic sequence comparison. *Genome Research*, *21*(3), 487–493. https://doi.org/10.1101/gr.113985.110.

Kimura, N. (2014). Metagenomic approaches to understanding phylogenetic diversity in quorum sensing. *Virulence*, *5*(3), 433–442. https://doi.org/10.4161/viru.27850.

Laue, B. E., Jiang, Y., Chhabra, S. R., Jacob, S., Stewart, G. S. A. B., Hardman, A., ... Williams, P. (2000). The biocontrol strain *Pseudomonas fluorescens* F113 produces the Rhizobium small bacteriocin, lactone, via HdtS, a putative novel N-acylhomoserine lactone synthase. *Microbiology*, *146*(10), 2469–2480.

Loman, N. J., & Pallen, M. J. (2015). Twenty years of bacterial genome sequencing. *Nature Reviews Microbiology*, 1–9. https://doi.org/10.1038/nrmicro3565.

Machado, H., Sonnenschein, E. C., Melchiorsen, J., & Gram, L. (2015). Genome mining reveals unlocked bioactive potential of marine Gram-negative bacteria. *BMC Genomics*, *16*(1), 1–12. https://doi.org/10.1186/s12864-015-1365-z.

Markowitz, V. M., Chen, I. M. A., Chu, K., Szeto, E., Palaniappan, K., Pillay, M., ... Kyrpides, N. C. (2014). IMG/M 4 version of the integrated metagenome comparative analysis system. *Nucleic Acids Research*, *42*(D1), 568–573. https://doi.org/10.1093/nar/gkt919.

Mashima, J., Kodama, Y., Kosuge, T., Fujisawa, T., Katayama, T., Nagasaki, H., ... Takagi, T. (2016). DNA data bank of Japan (DDBJ) progress report. *Nucleic Acids Research*, *44*(D1), D51–D57. https://doi.org/10.1093/nar/gkv1105.

Medema, M. H., Blin, K., Cimermancic, P., De Jager, V., Zakrzewski, P., Fischbach, M. A., ... Breitling, R. (2011). AntiSMASH: Rapid identification, annotation and analysis of secondary metabolite biosynthesis gene clusters in bacterial and fungal genome sequences. *Nucleic Acids Research*, *39*(SUPPL. 2), 339–346. https://doi.org/10.1093/nar/gkr466.

Miller, M., & Bassler, B. (2001). Quorum sensing in bacteria. *Annual Review of Microbiology*, *55*, 165–169.

Milton, D. L. (2006). Quorum sensing in vibrios: Complexity for diversification. *International Journal of Medical Microbiology*, *296*(2–3), 61–71. https://doi.org/10.1016/j.ijmm.2006.01.044.

Mitchell, A. L., Attwood, T. K., Babbitt, P. C., Blum, M., Bork, P., Bridge, A., ... Finn, R. D. (2018). InterPro in 2019: Improving coverage, classification and access to protein sequence annotations. *Nucleic Acids Research*, *47*(November 2018), 351–360. https://doi.org/10.1093/nar/gky1100.

Mok, K. C., Wingreen, N. S., & Bassler, B. L. (2003). Vibrio harveyi quorum sensing: A coincidence detector for two autoinducers controls gene expression. *EMBO Journal*, *22*(4), 870–881. https://doi.org/10.1093/emboj/cdg085.

Moller, S., Croning, M. D. R., & Apweiler, R. (2002). Evaluation of methods for the prediction of membrane spanning regions, *Bioinformatics*, *18*(1), 218.

Moraes, F., & Góes, A. (2016). A decade of human genome project conclusion: Scientific diffusion about our genome knowledge. *Biochemistry and Molecular Biology Education*, *44*(3), 215–223. https://doi.org/10.1002/bmb.20952.

Mukherjee, S., Stamatis, D., Bertsch, J., Ovchinnikova, G., Katta, H. Y., Mojica, A., ... Reddy, T. (2018). Genomes OnLine database (GOLD) v.7: Updates and new features. *Nucleic Acids Research*, *47*(D1), 1–11. https://doi.org/10.1093/nar/gky977.

Nealson, K. H., Platt, T., & Hastings, J. W. (1970). Cellular control of the synthesis and activity of the bacterial *Luminescent. Microbiology*, *104*(1), 313–322.

Nolan, P. M., Peters, J., Strivens, M., Rogers, D., Hagan, J., Spurr, N., ... Hunter, J. (2000). Mutagenesis programme for gene function studies in the mouse. *Nature Genetics*, *25*(August), 440–443.

Paez-Espino, D., Chen, I. M. A., Palaniappan, K., Ratner, A., Chu, K., Szeto, E., ... Kyrpides, N. C. (2017). IMG/VR: A database of cultured and uncultured DNA viruses and retroviruses. *Nucleic Acids Research*, *45*(D1), D457–D465. https://doi.org/10.1093/nar/gkw1030.

Paterson, J., Jahanshah, G., Li, Y., Wang, Q., Mehnaz, S., & Gross, H. (2017). The contribution of genome mining strategies to the understanding of active principles of PGPR strains. *FEMS Microbiology Ecology*, *93*(3), 1–12. https://doi.org/10.1093/femsec/fiw249.

Piel, J. (2011). Approaches to capturing and designing biologically active small molecules produced by uncultured microbes. *Annual Review of Microbiology*, *65*(1), 431–453. https://doi.org/10.1002/humu.22781.

Polkade, A. V., Mantri, S. S., Patwekar, U. J., & Jangid, K. (2016). Quorum sensing: An under-explored phenomenon in the phylum Actinobacteria. *Frontiers in Microbiology*, *7*(FEB), 1–13. https://doi.org/10.3389/fmicb.2016.00131.

Popovic, A., Hai, T., Tchigvintsev, A., Hajighasemi, M., Nocek, B., Khusnutdinova, A. N., ... Yakunin, A. F. (2017). Activity screening of environmental metagenomic libraries reveals novel carboxylesterase families. *Scientific Reports*, *7*(December 2016), 1–15. https://doi.org/10.1038/srep44103.

Prifti, E., & Zucker, J.-D. (2015). The new science of *Metagenomics* and the challenges of its use in both developed and developing countries. *Socio-Ecological Dimensions of Infectious Diseases in Southeast Asia*, 191–216. https://doi.org/10.1007/978-981-287-527-3_12.

Rinke, C., Schwientek, P., Sczyrba, A., Ivanova, N. N., Anderson, I. J., Cheng, J. F., ... Woyke, T. (2013). Insights into the phylogeny and coding potential of microbial dark matter. *Nature*, *499*(7459), 431–437. https://doi.org/10.1038/nature12352.

Schuster, M., Urbanowski, M. L., & Greenberg, E. P. (2004). Promoter specificity in *Pseudomonas aeruginosa* quorum sensing revealed by DNA binding of purified LasR. *Proceedings of the National Academy of Sciences*, *101*(45), 2–8.

Silvester, N., Alako, B., Amid, C., Cerdeño-Tárraga, A., Cleland, I., Gibson, R., ... Cochrane, G. (2015). Content discovery and retrieval services at the European nucleotide archive. *Nucleic Acids Research*, *43*(D1), D23–D29. https://doi.org/10.1093/nar/gku1129.

van Opijnen, T., & Camilli, A. (2013). Transposon insertion sequencing: A new tool for systems-level analysis of microorganisms. *Nature Reviews Microbiology*, *11*(7), 43. *Nature Reviews Microbiology*, *11*(7), 435–442. https://doi.org/10.1038/nrmicro3033.

Williamson, L. L., Borlee, B. R., Schloss, P. D., Guan, C., Allen, H. K., & Handelsman, J. (2005). Intracellular screen to identify metagenomic clones that induce or inhibit a quorum-sensing biosensor. *Applied and Environmental Microbiology*, *71*(10), 6335–6344. https://doi.org/10.1128/AEM.71.10.6335-6344.2005.

Zhang, G., Zhang, F., Ding, G., Li, J., Guo, X., Zhu, J., ... Dong, X. (2012). Acyl homoserine lactone-based quorum sensing in a methanogenic archaeon. *ISME Journal*, *6*(7), 1336–1344. https://doi.org/10.1038/ismej.2011.203.

Zhu, J., & Kaufmann, G. F. (2013). Quo vadis quorum quenching? *Current Opinion in Pharmacology*, *13*(5), 688–698. https://doi.org/10.1016/j.coph.2013.07.003.

Ziemert, N., Alanjary, M., & Weber, T. (2016). The evolution of genome mining in microbes—a review. *Natural Product Reports*, *33*(8), 988–1005. https://doi.org/10.1039/c6np00025h.

12

Bacterial Quorum Sensing in Multispecies Communities: The Presence, Functions and Applications

Chuan Hao Tan, Sujatha Subramoni, and Hyun-Suk Oh

CONTENTS

12.1 Introduction

Koch's definition of a bacterial pathogen has led to the long-time perception that bacteria are living as self-sufficient individuals with a strictly unicellular lifestyle. This single-celled concept has resulted in great success in identification of many host pathogens and subsequent discovery of antibiotics against the associated acute bacterial infections. However, antimicrobials derived from such models often fail in treating chronic infectious diseases of a biofilm origin, a multicellular lifestyle of bacteria that is commonly found in nature. Indeed, the discovery of biofilm as the dominant form of microbial life has spurred intensive research interest in the last 40 years and caused a paradigm shift in our understanding of the norm of bacteria as independent individuals to multicellular entities with capability of coordinating group behaviors for the benefits of the population. Recent technological advancements in sequencing and computational biology as well as cell imaging have further unveiled that many of these biofilms, either natural or engineered, are composed of diverse microbial species including bacteria, fungi, and archaea (Tan et al. 2017, Flemming and Wuertz 2019). These microorganisms are spatially segregated by extracellular matrices in a highly organized manner, forming complex, structured microbial communities (Flemming and Wingender 2010). Studies using simplified model communities indicate that social interactions that occur within and between species, either cooperative or competitive, are deterministic of the community structure, organization, and function (Elias and Banin 2012, Rendueles and Ghigo 2012, Abisado et al. 2018). Quorum sensing (QS) is arguably one of the most important means of social interaction that has evolved for group behavior coordination in the community (Abisado et al. 2018).

QS was first discovered in late 1960s as a form of bacterial cell-cell communication system to orchestrate group activities, including bioluminescence (Engebrecht et al. 1983), virulence factor expression (Rutherford and Bassler 2012), secondary metabolite production (Barnard et al. 2007), and biofilm formation and dispersal (Solano et al. 2014). Cooperative behavior is achieved by pegging the expression of a group of related genes to cell density, which can be inferred from the accumulated extracellular signaling molecules secreted by the cells. It is believed that the QS-mediated cooperative behaviors have evolved because of the population-wide benefits that allow QS organisms to thrive in diverse environments. Among other things, QS enables cells to withstand stresses (Goo et al. 2012, Leung and Lévesque 2012), to acquire nutrients from the environment (Darch et al. 2012, Drescher et al. 2014), and to facilitate host infections (Smith and Iglewski 2003, Qazi et al. 2001, Sully et al. 2014). In addition to the intraspecies cooperative behaviors, QS has been proposed to mediate interspecies cooperation

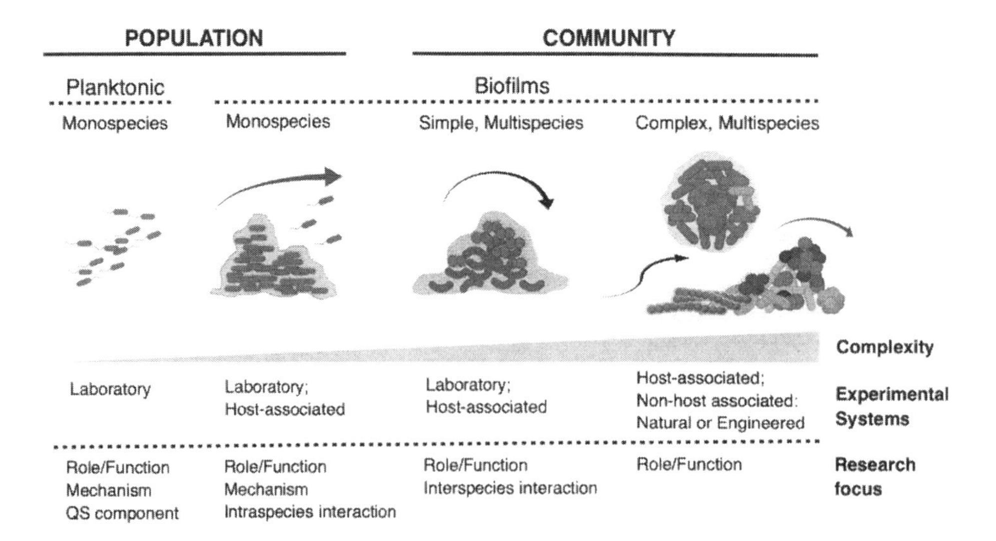

FIGURE 12.1 The experimental systems used in quorum sensing studies and the associated research focus. The experimental systems can be categorized based on the species composition and if the bacteria were in planktonic or highly structured biofilm lifestyle. In this chapter, we concentrate on studies using the biofilm systems ranging from lowly abundant communities (i.e., two to three species) to highly abundant communities (100s to 1000s species). The arrows represent fluid flow in the environment.

and competition in synthetic, dual species model communities (Chandler et al. 2012, Hosni et al. 2011). These studies have not only provided fundamental understanding of the molecular mechanisms of QS, but also important insights into the possible roles of QS when socializing with other organism(s) in a community. Rather than existing as a member of a lowly abundant community, bacteria in nature often assemble into highly complex, multispecies biofilm communities (i.e., 100s to 1000s species), under dynamically changing environmental conditions (Tan et al. 2017). To define how QS behaves and functions in these highly diverse communities has been a real challenge of the field. Nonetheless, the increasing accessibility to sequencing technologies, coupled with the advancements in mass spectrometry, have made study of QS in complex communities *in situ* a possibility in recent years. In this chapter, we review the QS studies associated with multispecies communities, ranging from relatively low diversity, with defined community memberships, to significantly complex communities with undefined species (Figure 12.1). Specifically, we concentrate on recent findings on the functions of QS discovered in both natural (i.e., marine environments) and engineered (i.e., sludge) communities, and we conclude with a discussion on the applications of QS in the respective fields.

12.2 Quorum Sensing in Simple, Defined Multispecies Communities

The role of QS in simple synthetic communities with defined memberships can be studied both in the context of cooperation and as well as competition between multiple species that make up the community. Characteristics of QS that allow it to be important for cooperation are the diffusible nature of QS signals optimal for intraspecies communications, cell-density dependency, and control of public goods or extracellular virulence factors. Several studies have been done to elucidate the cooperative effect of QS in the context of monospecies planktonic cultures and biofilms (for detailed review refer Abisado et al. 2018). On the other hand, it is now believed that QS is important for competitive interspecies interactions because of the same public goods, eavesdropping on heterologous signals, broad ligand specificity of the LuxR homologs, as well as regulation of extracellular toxins or antimicrobials. More importantly, roles of QS in communities might differ depending on whether they exist as planktonic cultures or in structured form such as multispecies biofilms (Figure 12.1).

12.2.1 Quorum Sensing in Planktonic Co-Cultures

Several examples of co-cultures in planktonic conditions are available in which the importance of QS has been elucidated in interspecies interactions. In mixed-species cultures of *Burkholderia cepacia* and *Pseudomonas aeruginosa*, both bacterial species were found to eavesdrop on the QS signals of the other species and thereby regulate downstream virulence functions to enable co-survival (Lewenza et al. 2002). In the case of *Serratia plymuthica-Escherichia coli* planktonic cocultures it was shown that *E. coli* had a growth advantage depending on the presence or absence of *S. plymuthica* QS system (Moons et al. 2006). *E. coli* grew well with wild type *S. plymuthica* until log phase after which growth declined; only a small decrease in growth was observed in co-cultures of *E. coli* with QS deficient *S. plymuthica*. Similarly, it was demonstrated for the *Burkholderia thailandensis-Chromobacterium violaceum* dual-species model that the production of QS-dependent antimicrobials provides fitness advantage to the producing strain by killing the other strain (Chandler et al. 2012). Moreover, the signal receptor of *C. violaceum* can bind and sense the foreign signals produced by *B. thailandensis* and thereby provide it growth advantage in dual-species co-cultures. Studies on *Agrobacterium tumefaciens-P. aeruginosa* co-cultures showed that *A. tumefaciens* had a QS-dependent increase in growth in liquid co-cultures (An et al. 2006). In co-cultures of *P. aeruginosa* with *B. multivorans*, *P. aeruginosa* was found to have a growth advantage in a QS-dependent manner. Here, production of QS regulated antimicrobials, namely, rhamnolipids, phenazines, and hydrogen cyanide, together provided a competitive advantage to *P. aeruginosa* (Smalley et al. 2015).

12.2.2 Quorum Sensing in Mixed-Species Biofilms

Biofilms are surface-associated or aggregates of structured bacterial communities that remain embedded in self-produced extracellular matrix components (Burmølle et al. 2014, Hall-Stoodley et al. 2004). Bacteria in biofilms are protected from stress inducing conditions such as antimicrobial compounds, desiccation, predation, etc. Bacteria also communicate with one another using diffusible compounds in a cell density dependent manner through QS, which also depends on diffusion limitation of the signal (Darch et al. 2018). In case of multispecies biofilms, the outcome of interactions cannot be predicted because it depends on the spatial organization of individual species and nutritional factors in these biofilms. In addition to spatial organization of different bacterial species, the presence of signals that can be detected by multiple species as well as their role in regulating downstream functions are important for deciding the outcome of interactions in multispecies biofilms (Federle and Bassler 2003). The degradation of signals by neighboring bacteria can also modify the physiology and properties of a multispecies biofilm. As a result, bacterial species either cooperate with one another to improve their fitness or compete with one another for space and nutrients in multiple species biofilms. These are properties of a community rather than a population, and at present information on multispecies biofilm is lacking as most studies involve monospecies biofilms. Examples of studies on multispecies biofilms where QS has been shown to be important are described below.

One example of a dual-species co-cultivation biofilm model used well-studied bacterial species *P. aeruginosa* and *A. tumefaciens* (An et al. 2006). In dual-species biofilms, *P. aeruginosa* blanketed *A. tumefaciens* microcolonies in a surface-motility dependent manner. In the presence of wild type *P. aeruginosa*, *A. tumefaciens* formed biofilms that continued to decrease in biomass compared to monospecies biofilms. On the other hand, *A. tumefaciens* biofilm biomass remained constant in the presence of QS negative mutant of *P. aeruginosa*. This suggests a role for *P. aeruginosa* QS in providing competitive fitness to *P. aeruginosa* in the presence of *A. tumefaciens*. In the case of *A. tumefaciens-P. aeruginosa* dual species, planktonic co-cultures behaved quite similar to the dual-species biofilms made up of the same bacterial species.

QS-dependent antimicrobial production has been shown to be important for formation of dual species biofilms consisting of *S. plymuthica* and *E. coli* (Moons et al. 2006). In dual-species biofilms of *E. coli* with *S. plymuthica*, *E. coli* was capable of detecting *S. plymuthica* QS signals and at the same time was susceptible to the QS-dependent antimicrobials produced by *S. plymuthica*. In this study authors have shown that in the absence of production of a QS-dependent antimicrobial compound by *S. plymuthica*, *E. coli* biomass reached higher levels in mixed-species biofilms of these two species. The wild types of the two species also did not form mixed-species colonies containing both bacteria indicating competitive interaction. In this example of dual-species interactions, the antagonistic outcome was similar in both planktonic and biofilm conditions and it was dependent on *S. plymuthica* QS. Therefore, the production of an antibacterial compound drastically affected the existence of another sensitive species in mixed-species biofilms.

AI-2 (i.e., autoinducer-2) is another signal produced by several Gram-positive and Gram-negative strains often occurring in close proximity in nature or host-associated environments (McDougald et al. 2003). Other examples of mixed-species biofilms include *Streptococcus gordonii-Porphyromonas* (McNab et al. 2003) and *Streptococcus oralis-Actinomyces* (Rickard et al. 2006) biofilms in which AI-2 was shown to have an effect on biofilm growth. Both these bacteria are members of the oral biofilm community. *S. gordonii-P. gingivalis* are able to form biofilms of higher biomass in the presence of a heterologous wild type strain producing AI-2 (McNab et al. 2003). Similarly, *S. oralis-A. naeslundii* had a higher biovolume of the mixed-species biofilm due to higher production of AI-2 by *S. oralis*. A mixed species biofilm containing *A. naeslundii* wild type and *S. oralislux S* (coding for AI-2) mutant did not show synergy and could easily be dispersed (Rickard et al. 2006). *Streptococcus mutans* forms mixed species biofilms with *S. gordonii* as part of the dental plaque formation. Mutation of *luxS* in *S. gordonii* altered mixed species biofilm composition and resistance to chlorhexidine. On the other hand, the presence of AI-2 had an effect on biofilm formation by *S. mutans* suggesting an important role for AI-2 in dual species biofilm formation by these two bacterial species (Wang et al. 2017). These are examples of cooperative effect of QS in aggregates of bacterial species that mutually benefit one another.

The role of QS has also been determined for mixed-species biofilms associated with their plant or animal hosts. For example, QS plays an important role in the biofilms of *Pseudomonas syringae* associated with other epiphytic bacteria by regulating plant virulence (Dulla and Lindow 2009). The presence of large amounts 3OC6-HSL (i.e., *N*-(3-Oxohexanoyl)-L-homoserine lactone) produced by epiphytic bacteria led to premature induction of QS and suppression of swarming motility of *P. syringae*. This led to lesser disease severity on infected leaves compared to wild type *P. syringae*. A similar effect, i.e., lesser disease severity, was seen when a multispecies community consisting of epiphyte and *P. syringae* were inoculated together on susceptible bean plants. Importantly, inoculation with epiphytic bacteria in which QS had been quenched led to high disease severity similar to inoculation with *P. syringae* alone. Therefore, detection of AHLs (i.e., acylated homoserine lactone) produced by epiphytic bacteria decided the disease outcome when bean plants were inoculated with *P. syringae*. Another example is the bacterial consortium of *Erwinia toletana*, *Pantoea agglomerans*, and *Pseudomonas savastanoi* found in the olive knot disease where disease severity is dependent on the presence or absence of QS (Hosni et al. 2011). Two bacterial species in this multi-species community, namely *E. toletana* and *P. savastanoi*, produced identical AHLs and could detect AHLs produced by other species. *P. savastanoi* QS mutants were much reduced for causing disease on olive plants. Coinoculations of the wild type *E.*

toletana with *P. savastanoi* QS mutant restored full virulence showing a major role for interspecies communication in this community. In both cases bacterial eavesdropping on foreign QS signals forms the basis for interspecies interactions between these host-associated bacterial species.

P. aeruginosa and *Burkholderia cepacia* can coexist in the lungs of CF patients and form multispecies biofilms; here they produce AHLs to regulate virulence and biofilm formation (Huber et al. 2001, Van Delden and Iglewski 1998). Mixed biofilms formed *B. cepacia* are capable of sensing and responding to different AHLs produced by *P. aeruginosa*, suggesting cooperation between these two bacterial species (Riedel et al. 2001). Moreover, Sibley et al. have shown that interactions between *P. aeruginosa* and other non-pathogenic oropharyngeal isolates have the ability to modulate the severity of infection so that the outcome is virulent, avirulent, or synergistic depending on the isolates used (Sibley et al. 2008).

P. aeruginosa dominates in co-cultures with *Staphylococcus aureus* both in planktonic and biofilm mode of growth. Several different models have been used to understand the interaction between *S. aureus* and *P. aeruginosa* as they are known to occur together in disease conditions such as chronic wounds and cystic fibrosis lung infections but are difficult to grow together *in vitro*. These two bacterial species are now known to be involved in both antagonistic and cooperative interactions with each other, and QS plays an important role in these interactions (Hotterbeekx et al. 2017). *S. aureus* has been reported to dominate biofilms and non-attached aggregates acting as the early colonizer before promoting attachment by *P. aeruginosa* (Alves et al. 2018). *P. aeruginosa*, on the other hand, has been shown to promote invasive phenotype of *S. aureus*. Therefore, coinfection by both species is more virulent compared to single species (DeLeon et al. 2014). QS regulated exoproducts, namely LasB elastase, rhamnolipids, exotoxins, and phenazines, play an important role making *P. aeruginosa* the dominant species in CF (Hotterbeekx et al. 2017). LasB elastase has been shown to be a major antibiofilm protease causing both inhibition of biofilm formation as well as dispersal of mature biofilms of *S. aureus* without affecting growth (Park et al. 2012). Serum albumin, one of the components of the wound environment, binds and inhibits the *P. aeruginosa* QS signaling molecule. As a result, *P. aeruginosa* is unable to produce its array of QS-dependent virulence factors and therefore cannot compete with *S. aureus* in these polymicrobial infections (Smith et al. 2017). When together, *P. aeruginosa* usually has a growth advantage over *S. aureus* due to the production of pyocyanin and 4-hydroxy-2-heptylquinoline N-oxide (HQNO) (Biswas et al. 2009, Hoffman et al. 2006). Both of these compounds inhibit *S. aureus* respiration. This gives rise to growth inhibition and generates small colony variants of *S. aureus*. In addition to this, *S. aureus* also serves as a source of iron for *P. aeruginosa* during *in vivo* co-infection in a rat dialysis membrane peritoneal model (Mashburn et al. 2005). On a different note, *P. aeruginosa* is also able to sense the presence of AI-2in these polymicrobial biofilms and upregulate its QS-dependent virulence functions or exoproducts (Li et al. 2015). In these biofilms, N-acetyl glucosamine (GlcNAc), which is part of Gram-positive bacteria cell walls, is also sensed by *P. aeruginosa* to upregulate the production of Pseudomonas Quinolone signal (PQS), which in turn controls production of virulence factors (Williams and Cámara 2009, Korgaonkar et al. 2013, Déziel et al. 2004).

12.2.3 Cheating in Biofilms

Bacterial QS mediated cooperation in bacteria is also associated with secretion of QS regulated extracellular substances that enable nutrient acquisition, protection from the environment, and biofilm formation. These extracellular substances are referred to as public goods as they can also be used by non-producing cells. Occasionally, cheaters arise that avoid the cost of producing public goods but use them. While optimal use of public goods leads to enhanced fitness of the entire population, exploitation of public goods is associated with reduced fitness and collapse of the biofilm community (Mukherjee and Bassler 2019, Popat et al. 2012). Therefore, public good dependent cooperation needs to be maximized while preventing or managing the cheater population by various processes such as policing, diffusion limitation, spatial structure of biofilms, etc. For example, *P. aeruginosa lasI* mutants are unable to produce 3OC12-HSL (i.e., *N*-(3-Oxododecanoyl)-L-homoserine lactone) but at the same time can respond to wild type produced signals. These *lasI* mutants thus outcompete wild type in well-mixed cultures (Mund et al. 2017). When grown as monocultures on adenosine as the sole carbon source, *lasI* mutants are defective for growth as they are unable to use 3OC12-HSL to activate *nuh*, which encodes intracellular nucleoside hydrolase required for adenosine catabolism. *nuh* encodes a private good that cannot be shared between producers and non-producers. In mixed cultures, *lasI* mutants grow very well as they are able to activate *nuh* and use adenosine using 3OC12-HSL produced by wild type. Also, increasing viscosity of the medium reduces diffusion of the QS signal and minimizes cheating by the *lasI* mutant (Mund et al. 2017). Similarly, QS driven coregulation of production of a public good and a private good can reduce social cheating and collapse of wild type population. Here, *lasR* mutants are able to grow on a medium containing casein by using QS regulated extracellular proteases secreted by wild type to metabolize this substrate. However, these *lasR* mutants cannot grow on a medium containing adenosine as *nuh*, the enzyme required for catabolism of adenosine is a private good as mentioned previously. Both proteases and adenosine hydrolase are regulated by QS, but one is a public good while the other is a private good. Other examples of control of social cheating include the cyanide production and immunity to cyanide conferred by the *rhlR-rhlI* QS system of *P. aeruginosa* and the growth of *V. cholerae* cells on chitin (Drescher et al. 2014, Wang et al. 2015). Wild type *P. aeruginosa* is able to produce cyanide and at the same time is immune to its effects. In contrast, the *lasR* mutant is sensitive to cyanide due to lack of expression of the *rhlR* regulon, which confers immunity. In this way, the occurrence of *lasR* cheaters are minimized leading to population stability. *V. cholerae* produces chitin degrading enzymes (e.g. chitinases) to solubilize chitin to monomers that can be taken up easily. Chitinase-non-producers are also able to take up these *N*-acetyl glucosamine monomers. However, in thick biofilms, diffusion of these monomers is much less thus allowing the enzyme producers to take up the monomers efficiently before they reach the cheaters. In this manner, chitinase-producers are fitter compared to non-producers in biofilms but not in well-mixed cultures. Also, flow mediated removal of chitin degradation products also ensures that the proximally located producers obtain maximum growth benefit as opposed to distally located cheaters. Control of cheaters in multispecies biofilms has not been studied yet, and information regarding this is lacking. However, based on studies in monospecies biofilms, we can predict that cheaters would be important in stability of multispecies biofilms as well. Control of cheaters might involve diffusion limitation of public goods or segregation of co-operators and cheaters. So, the mechanisms involved in control of cheaters in monospecies biofilms would be also be applicable to multispecies biofilms.

The processes controlled by QS are varied, and therefore, QS plays an important role in interspecies interactions of bacteria; more studies are needed using different multispecies communities to understand this better. Important questions to find answers include how QS regulates multispecies community dynamics and affects physiology and survival of these polymicrobial communities. New multispecies experimental models need to incorporate both QS and quorum quenching (QQ) to better understand the ecology of signaling in multispecies communities as it occurs in nature.

12.3 Quorum Sensing in Complex, Undefined Multispecies Communities

While most QS studies in simple synthetic communities involved two to three species that were co-cultured in well-defined laboratory conditions, as mentioned above, microbial consortia consisted of hundreds to thousands of species, growing in dynamic natural or engineered ecosystems, and were typically examined in the context of undefined communities (Figure 12.1) (Tan et al. 2017, Abisado et al. 2018, Hmelo 2017). With the increasing number of species, the number and types of interactions within communities grow exponentially (Røder et al. 2015), making studies of QS challenging. Nevertheless, despite the complexity, it was first reported in 2008 that AHL-mediated QS was involved in the formation of a multispecies biofilm on a membrane bioreactor (Yeon et al. 2009a). Using a signal interference approach or enzymatic QQ, Yeon et al. were able to inhibit native QS processes, including AHL signaling, extracellular polymeric substances (EPS) production, and biofilm development by a sludge community, thus alleviating the biofouling issue (Yeon et al. 2009a). Such QS-inspired anti-fouling technology has not only generated tremendous interest among the water/wastewater treatment communities (Oh and Lee 2018), but these hallmark experiments have also instilled many to study the roles of QS in other environments. We have now begun to appreciate that QS not only regulates the architecture and stability of biofilm communities (Ma et al. 2018a, 2018b, Tan et al. 2014), but also community functions that govern important biogeochemical cycling of carbon, phosphorus, and nitrogen (Hmelo et al. 2011, Krupke et al. 2016, Van Mooy et al. 2011, Tang et al. 2015, 2018b, Yeon et al. 2009a), as well as influencing the health of coral communities in general (Golberg et al. 2011, Zimmer et al. 2014, Certner and Vollmer 2015, 2018) (Table 12.1). Examples of studies where different roles of QS were revealed in both natural and engineered multispecies communities will be discussed below.

12.3.1 Biofilm Development and EPS Formation

Biofilm formation is one of the most common social phenotypes that has been linked to QS regulation in many bacterial species. More often than not, QS modulates the progression of biofilm growth through the control of cellular motility, EPS, and hydrolytic enzymes expression. The outcomes of such regulation can be rather controversial, depending on the species, the developmental stage of biofilm, and the type of QS system involved. While QS encourages biofilm maturation in species, such as *Aeromonas hydrophila* (Talagrand-Reboul et al. 2017), *P. aeruginosa* (Davies et al. 1998), *B. cepacia* (Huber et al. 2001), and *S. mutans* (Senadheera and Cvitkovitch 2008), it promotes biofilm dispersal, on the other hand, among *Vibrio cholera* (Waters et al. 2008), *Xanthomonas campestris* (Dow et al. 2003), *Pseudomonas putida* (Gjermansen et al. 2005), *S. aureus* (Boles and Horswill 2008), and *P. aeruginosa* (Dong et al. 2008). Maturation of *P. aeruginosa* biofilm is driven by AHL-based QS, but expression of PQS (i.e., another QS system) by the same host can disperse biofilm cells at a later stage (Kaplan 2010). Similarly, QS has also been found to be essential in multispecies biofilm development, particularly among communities of floccular sludge. Besides membrane biofouling (Lee et al. 2016c, Yeon et al. 2009a), transformation of fluffy floccular sludge into compact microbial granules (i.e., suspended biofilm aggregates) was also reported to be QS-associated (Ma et al. 2018a, Tan et al. 2014). Using a sequencing batch reactor system, an activated sludge community that consisted of more than 500 different microbial species was converted into a suspended, highly structured granular biofilm after 105 days of reactor operation (Tan et al. 2014). Biomass conversion from flocs to granules that initiated between day 35 and 56 was accompanied by 100-fold increase in specific AHLs expression and EPS production (Tan et al. 2014). One possible mechanism underlying the process could be that AHL expression (i.e., QS) induced EPS synthesis that eventually led to granule formation. Contradictory to such hypothesis, however, Li et al. failed to demonstrate a clear effect of QS on granulation by continuous spiking of either AHL-producing or -quenching strains to an activated sludge community (Li et al. 2017). In fact, AHL modulation (i.e., supplementation of AHL-producing or -quenching strains) was found to suppress biofilm development, resulting in granules with smaller sizes compared to the community without modulation (Li et al. 2017). While these findings may highlight the complexity of QS regulation in undefined communities, the results should be deciphered more carefully given that the genetic differences between the supplemented strains apart from their QS traits, and the type and amount of AHLs expressed in the reactor at appropriate timings may play a crucial role in the granulation outcomes.

QS is not exclusive to aerobic microbial communities. AHLs and DSF (diffusible signaling factor), for instance, were found to accumulate in anaerobic microbial consortia including different anaerobic granules used to treat industrial wastewaters from pharmaceutical, brewery, alcohol, starch, or paper-making

TABLE 12.1

Quorum Sensing Studies in Natural or Engineered Multispecies Communities

QS Role/ Function	Multispecies Community	QS Signal	QS Signal Detection	Level of Evidence	References
Biofilm development and/or EPS formation	Activated sludge (i.e., membrane bioreactor)	C6, C8-HSL	TLC/Biosensor	**Correlation study** (i.e., biofouling positively correlates with native signal production); Causation study (i.e., signal quenching prevents biofouling)	Yeon et al. (2009a)
	Activated sludge (i.e., microbial granulation); > 500 species	C6, C8, C10-HSL, 3OC6, 3OC8-HSL	LC-MS/MS	**Correlation study** (i.e., granulation positively correlates with native signal production); **Causation study** (i.e., signal addback induces EPS production)	Tan et al. (2014)
	Anaerobic sludge (i.e., microbial granulation)	C8, C10-HSL	LC-MS/MS	**Correlation study** (i.e., granulation positively correlates with native signal production); **Causation study** (i.e., signal addback induces EPS production)	Ma et al. (2018a)
	ANAMMOX (i.e., flocculation/ aggregation); Dominated by *Candidatus* Jettenia caeni	C6, C8, C12-HSL 3OC6, 3OC8-HSL	LC-MS/MS	**Causation study** (i.e., signal addback induces specific EPS production as revealed by metabolomics, as well as enhances flocculation)	Tang et al. (2018b)
Biological nitrogen removal	ANAMMOX; Dominated by *Candidatus* Brocadia fulgida	C6, C8, C12-HSL	LC-MS/MS	**Causation study** (i.e., signal addback [C6, C8-HSL] induces specific ammonia removal rate and/or growth rate of *Candidatus* Brocadia fulgida while C12-HSL addition decreases specific ammonia removal rate and growth rate of *Candidatus* Brocadia fulgida)	Tang et al. (2015)
	ANAMMOX; Dominated by *Candidatus* Jettenia caeni	C6, C8, C12-HSL 3OC6, 3OC8-HSL	LC-MS/MS	**Causation study** (i.e., signal addback [C6, C8-HSL, 3OC6, 3OC8-HSL] induces specific ammonia removal rate and/or growth rate of *Candidatus* Jettenia caeni, while C12-HSL addition decreases specific ammonia removal rate and growth rate of *Candidatus* Jettenia caeni)	Tang et al. (2018b)
	Aerobic/ ANAMMOX; Dominated by *Candidatus* Kuenenia	C12-HSL	LC-MS/MS	**Correlation study** (i.e., high ammonia removal rate correlates with high biomass and C12-HSL production) **Causation study** (i.e., signal addback [C10, C12-HSL but not C4, C6-HSL] induces specific ammonia removal rate)	De Clippeleir et al. (2011)
Extracellular hydrolytic enzyme production	Particulate organic carbon (POC) associated community	C8, C12-HSL	LC-MS/MS	**Causation study** (i.e., signal addback [3OC6, 3OC8-HSL, C10, C12, C14-HSL] induces differential expressions of enzymes including aminopeptidase, lipase, and phosphatase)	Hmelo et al. (2011), Krupke et al. (2016)
	Trichodesmium colony	*In situ* signal detection was not performed. However, AHL producers were isolated from the *Trichodesmium* colony.	N.A.	**Causation study** (i.e., signal addback [mixture of C10, C12, C14-HSL] induces APase activity of *Trichodesmium* colony)	Van Mooy et al. (2011)
Host infection	Coral halobiont	*In situ* signal detection was not performed.	N.A.	**Causation study** (i.e., pre-incubation of healthy coral microbiota with C6-HSL can transform the microbiota to induce white-band disease on the coral *Acropora cervicornis*. The disease development can be arrested using QS inhibitor, furanone)	Certner et al. (2015, 2018)

Abbreviations: QS, quorum sensing; TLC, thin layer chromatography; LC-MS/MS, liquid chromatography-mass spectrometry; ANAMMOX, anaerobic ammonium oxidation.

factories (Feng et al. 2014, Ma et al. 2018b). Similar to the role of QS proposed in aerobic granulation (Tan et al. 2014), floc-to-granule transition in an anaerobic reactor was found positively correlated with elevated levels of AHL and EPS, and that the synthesis of EPS was inducible by the native AHLs (i.e., C8-HSLor *N*-Octanoyl-L-homoserine lactone and C10-HSLor *N*-Decanoyl-L-homoserine lactone) when supplemented exogenously (Ma et al. 2018a). Consistent with these findings, a methanogenic archaeon *Methanosaeta harundinacea* 6Ac, isolated from an up-flow anaerobic sludge bed granule, was characterized to use carboxylated AHLs to coordinate transition from a short cell to filamentous growth (Zhang et al. 2012). Given that filamentation is key to anaerobic granulation and that *M. harundinacea* is ubiquitous in anaerobic digestors (Zhou et al. 2015), it is speculated that *M. harundinacea* may facilitate the granulation process via QS. Likewise, the aggregation of ANAMMOX (i.e., anaerobic ammonium oxidation) consortia was recently found to be enhanced by AHL signaling (Tang et al. 2018b). Ametabolomic analysis revealed that AHL-induced accumulation of specific intracellular amino acids, such as alanine, valine and glutamate, and amino sugars, e.g., ManNAc(N-acetylmannosamine) and UDP-GlcNAc (uridinediphosphate-*N*-acetylgalactosamine), were positively correlated with extracellular proteins and polysaccharides expression as well as the aggregate size (Tang et al. 2018b). Importantly, specific AHLs were required to induce different combinations of amino acids and amino sugars expression from microbes with diverse metabolic pathways (Tang et al. 2018b). This suggests that multiple microorganisms/species may collectively involve in the synthesis of EPS to promote ANAMMOX aggregation as a ramification of the communal QS activities.

12.3.2 Biological Nitrogen Removal

Biological nitrogen removal (BNR) through the conventional nitrification-denitrification processes and the ANAMMOX activity is key to the nitrogen cycle. Many species implicated in the BNR processes have been documented to contain single or multiple AHL-based QS genetic modules and/or be capable of expressing a variety of AHLs in a cell density-dependent manner (Mellbye et al. 2017). These microbes include the ammonium-oxidizing bacteria (AOB; e.g., *Nitrosomonas europeae* strain Schmidt and *Nitrosospira multiformis*) (Batchelor et al. 1997, Burton et al. 2005, Gao et al. 2014, Norton et al. 2008), the nitrite-oxidizing bacteria (NOB; e.g., *Nitrospira moscoviensis*, *Nitrobacter winogradskyi*, and *Nitrobacter humburgensis*) (Mellbye et al. 2015, Nasuno et al. 2012, Starkenburg et al. 2006), the complete ammonium-oxidizing bacteria (COMAMMOX; e.g., *Nitrospira inopinate*, *Candidatus Nitrospira nitrificans*, and *Candidatus* Nitrospira nitrosa) and the ANAMMOX bacteria (e.g., *Candidatus* Jettenia caeni) (Tang et al. 2019). Although the importance of QS has been unveiled in certain nitrifiers, such as growth resumption of the starved *N. europaea* cell suspensions (Batchelor et al. 1997) and modulation of nitrogen oxide(s) fluxes in *N. winogradskyi* (Mellbye et al. 2016), QS regulations in many other nitrifying species remain unexplored. This in part could be due to the slow growth and the lack of genetic manipulation tools for the nitrifying bacteria, despite being able to grow as an isogenic culture. By contrast, experiments using culture-independent methods, including metagenomics and metabolomics, have generated important insights into the various roles and features of QS in the ANAMMOX communities (Tang et al. 2015, 2018a, 2018b, 2019). For example, it was found that the ANAMMOX activity that removes ammonium by coupling nitrite into nitrogen gas is futile unless the cell density is higher than 10^{10}–10^{11} cells mL^{-1}, and that corresponded to a marked increase in the AHL production by the ANAMMOX culture (Tang et al. 2015), consistent with the conventional notion of cell density-dependent QS activation and gene expression (Mukherjee and Bassler 2019). Also, it was shown that exogenous AHLs could induce AHL production by the ANAMMOX biomass (Tang et al. 2015), indicating the engagement of the canonical autoinduction loop that typically occurs in a QS-proficient organism (Mukherjee and Bassler 2019). Importantly, such AHL-mediated QS is rather common among the ANAMMOX communities (De Clippeleir et al. 2011, Tang et al. 2015, 2018b). Not only that, AHLs were produced by ANAMMOX communities dominated either by *Candidatus* Brocadia fulgida (Tang et al. 2015) or *Candidatus* Jettenia caeni (Tang et al. 2018b), the AHL species synthesized by these very different ANAMMOX bacteria were surprisingly similar. In fact, a core group of AHLs consisted of C6-HSL (i.e., *N*-Hexanoyl-L-homoserine lactone) and/or 3OC6-HSL, C8-HSL and/or 3OC8-HSL (i.e., *N*-(3-Oxooctanoyl-L-homoserine lactone), and C12-HSL (i.e., *N*-Dodecanoyl-L-homoserine lactone) was commonly produced by the different ANAMMOX communities (Tang et al. 2015, 2018b). Accordingly, two *hdtS* AHL synthases, i.e., *jqsl-1* and *jqsl-2*, have also been successfully identified from the *Candidatus* Jettenia caeni genome and characterized to synthesize AHLs mentioned above, except for the C12-HSL (Tang et al. 2019).

Despite the high similarity in the AHL profiles, individual AHL species may exhibit a distinct, community-specific QS role to optimize the ANAMMOX process. As examples, only C6-HSL was capable of inducing the growth of *Candidatus* Brocadia fulgida (Tang et al. 2015) while the amplification of *Candidatus* Jettenia caeni could be encouraged solely by the 3OC6-HSL (Tang et al. 2018b). On the other hand, addition of C6-HSL or C8-HSL, with and without oxo-substitutions, to the ANAMMOX communities was found to increase the specific ammonium removal rate in contrast to a reduction effect mediated by the C12-HSL (Tang et al. 2015, 2018b). In fact, C12-HSL supplementation was found to enhance nitrite removal disproportionally to the rate of ammonium removal, implying the growth of other competing microbes that utilized nitrite for purposes other than for ammonium removal via the ANAMMOX process (Tang et al. 2015). This was further confirmed by the qPCR and metagenomic analyses showing that C12-HSL enhanced the growth of heterotrophs while suppressed the numbers of ANAMMOX bacteria, suggesting a QS-mediated competition (Tang et al. 2015, 2018b). This finding is also consistent with the fact that both AHL synthases identified in *Candidatus* Jettenia caeni were not able to produce C12-HSL (Tang et al. 2019). Therefore, it is most likely that the C12-HSL found in the ANAMMOX communities was synthesized by the competing heterotrophs at the physiological conditions for them to thrive in an ANAMMOX-dominant consortium. It is important to note, however, that an opposite effect of C12-HSL (i.e., supporting the ANAMMOX activity) has also been observed in microbial communities enriched with *Candidatus* Kuenenia, yet another ANAMMOX bacterium (De Clippeleir et al. 2011). This, again, highlights the huge disparity in the AHL-mediated QS behaviors among the very different bacteria, as similarly exemplified in cases of QS-controlled biofilm development. This may also imply that any QS-augmentation of the communal behaviors should be customized based on the signal typing of the particular community to order to achieve a desired outcome.

12.3.3 Extracellular Hydrolytic Enzyme Expression

Bacteria can secrete a suite of hydrolytic enzymes to break down complex organic substrates in the immediate environment into soluble monomers for internalization and used as a nutrient for growth. The deployment of these extracellular enzymes can benefit both producing and nonproducing cells, and the enzymes are thus considered to be public goods (Drescher et al. 2014). Production of the public goods is metabolically expensive; thus, in some bacteria, QS is often engaged to optimize the expression of the secreted enzymes such that each cell in the population can contribute to its share of goods for the benefit of all (Mukherjee and Bassler 2019). Consistent with this, increasing evidence suggests that QS coordination of extracellular enzymatic activity is prevalent among marine microbial communities, and may directly influence the carbon flux and phosphorus cycling in the ocean (Decho et al. 2009, Hmelo et al. 2011, Krupke et al. 2016, Van Mooy et al. 2011). Enzymatic dissolution and remineralization of the sinking particulate organic carbon (POC) (Eppley and Peterson 1979), for instance, has been linked to the QS among heterotrophic bacteria associated with POC (Krupke et al. 2016, Hmelo et al. 2011). A diversity of AHLs was identified *in situ* within POC, and exogenous AHLs appeared to regulate microbial activities, including the expression of aminopeptidase, lipase, and phosphatase, of POC communities sampled at different biogeographical locations or seasons (Hmelo et al. 2011, Krupke et al. 2016). However, the QS response of each POC community was rather random, and could vary substantially depending on the biogeography, the type and concentration of AHLs supplemented, incubation time, and the enzymatic activities involved (Krupke et al. 2016). For example, AHL augmentation could lead to the increased expression of phosphatases and suppression of other enzymes, while an opposite effect may be true for other instances (Krupke et al. 2016). Microspatial heterogeneity, with respect to cellular activities, community memberships, chemical gradients, etc., has been proposed to lead to the observed variations (Krupke et al. 2016). Despite the variability, it is apparent that QS has a clear impact on the rate of POC disaggregation, which is known to influence the balance of atmospheric/oceanic CO_2 concentrations, as POC plays an important role in sequestering CO_2 from the atmosphere to the deep ocean (i.e., the marine biological carbon pump) (Eppley and Peterson 1979) and thus regulating global climate. A comprehensive understanding for how QS regulates the varied enzymatic responses of POC may therefore be critical, especially since we are facing the crisis of global warming.

Nutrients are very scarce in the oligotrophic regions of the ocean. While most phytoplankton, including the filamentous cyanobacterium *Trichodesmium*, are capable of fixing nitrogen and carbon efficiently from the atmosphere (Capone et al. 1997), their growth is often limited by the supply of phosphorus (Mills et al. 2004, Sañudo-Wilhelmy et al. 2001). To meet their phosphorus demand, *Trichodesmium* and other heterotrophs associated with the colonies express alkaline phosphatases to hydrolyze bioavailable phosphate from dissolved organic phosphorus molecules (Orchard et al. 2010). In particular, the expression of extracellular alkaline phosphatases (Luo et al. 2009), as confirmed by the use of fluorogenic substrates (i.e., 6,8-difluoro-4-methylumbelliferyl phosphate and 4-methylumbelliferyl phosphate) that cannot be transported across the cell membrane (Mahaffey et al. 2014) by heterotrophic epibionts in the *Trichodesmium* colonies were found to be stimulated by AHL signaling. Unlike the irregular QS response observed during POC degradation, AHL-induced alkaline phosphatase expression was highly specific and reproducible across multiple *Trichodesmium* colonies sampled from different oceans (Van Mooy et al. 2011). For example, the alkaline phosphatase activity of *Trichodesmium* colonies harvested from various locations in the North Atlantic and North Pacific were doubled consistently in response to amendment with a cocktail of saturated, but not the oxo-substituted, AHLs. The saturated AHLs were affirmed to be present *in situ* by mass spectrometry, and such AHL species are also highly abundant in the natural cyanobacterial mat communities (Decho et al. 2009), although it is not known if AHL-mediated QS is similarly involved in regulating the expression of alkaline phosphatases in those habitats. Intriguingly, as opposed to the stimulatory effect of AHLs, the alkaline phosphatase activity of the *Trichodesmium* colonies was attenuated by exogenous AI-2 (Van Mooy et al. 2011), suggesting a multi-signal regulation event. Supporting this, recent metagenomic analyses reported that the genomes of the epibionts, rather than *Trichodesmium*, are enriched with diverse QS genetic modules (Frischkorn et al. 2017). Regardless of the complexity of QS regulation involved, it is clear that QS has a strong impact on the microbial physiology by affecting how the epibionts of *Trichodesmium* colonies acquire limited phosphorus from the environment. As *Trichodesmium* colonies are important nutrient sources to the marine food webs (Hmelo 2017), it is essential to further understand how QS regulation in the heterotrophic epibionts may eventually affect the growth of the host *Trichodesmium*, and if *Trichodesmium* could indeed be benefited from this seemingly mutualistic relationship with their heterotrophic occupants.

12.3.4 Host Infection

Successful infection of a host often necessitates the expression of multiple virulence factors. In many pathogenic bacteria, including diverse species of marine *Vibrios*, the virulence genes are controlled by QS (Henke and Bassler 2004, Higgins et al. 2007, Tait et al. 2005, 2010, Kimes et al. 2011). In fact, QS has been associated with the escalation of coral epizootic incidence and severity in the last 30 years, but the specific role of QS in coral disease progression remains unclear (Hmelo 2017). Recent studies on the coral white band disease (WBD) suggest that AHL signaling has an ability to transform a benign coral microbiota into a disease-causing vector (Certner and Vollmer 2015). Specifically, the authors noted that pre-incubating homogenates (i.e., containing bacterial cells) derived from healthy corals with C6-HSL enabled the homogenates to induce the development of WBD-like symptoms in previously healthy corals (Certner and Vollmer 2015). This was similar to the effects when homogenates extracted from WBD corals were used for dosing healthy corals (Certner and Vollmer 2018). Importantly, it was shown that the development of WBD could be arrested by adding a QS inhibitor (QSI, i.e., halogenated furanone) to the WBD-derived homogenates (Certner and Vollmer 2018). While all corals (n = 15) incubated with WBD microbiota (i.e., the WBD derived homogenates; +WBD/-QSI) developed WBD signs within 24 h, only one of the fifteen corals dosed with QSI-amended WBD microbiota (+WBD/+QSI) contracted disease (Certner and Vollmer 2018), highlighting the crucial role of QS in coral disease development. The disease prevention by QSI could be partly explained by the modulation of the community composition. Microbiota analyses of the coral microbial communities suggest that QSI prevents the establishment of a WBD-causing microbiota through targeted eradication of disease-causing species, such as *Vibrionaceae* and *Flavobacteriaceae*, while preserving beneficial coral mutualists including putative symbionts *Endozoicomonas* and *Halomonadaceae* (Certner and Vollmer 2018). Thus, it is likely that QS enables WBD pathogen(s) to elicit disease signs in healthy corals by timely activation of virulence-related genes, which can be attenuated by QSI addition. By preventing QS activation, the growth of disease-causing species can be suppressed, and this is possible either directly by QSI, or outcompeted by other coral symbionts (Sharp and Ritchie 2012). These findings are congruent with the emerging dysbiosis hypothesis where microbiota imbalance or corrupted host-microbe mutualism can lead to coral diseases (Bourne et al. 2009, Egan and Gardiner 2016, Munn 2015). In healthy corals, QS-dependent virulence expression by opportunistic pathogens is likely to be suppressed by QSI or QQ enzymes produced by other coral symbionts (Skindersoe et al. 2008, Golberg et al. 2013) or the coral hosts (Manefield et al. 1999). However, at times, when coral-microbe balance is perturbed, such as a consequence to increased water temperature, the QS inhibition mechanism may be lost. This allows the pathogen(s) to express QS-dependent virulence and competition factors to overcome the host-symbiont defense, causing rapid disease progression. Future work will be critical to determine how QS initiates the coral disease development, and if the QSI technology could be applied to prevent coral disease initiation and spread, or even to recover corals from a diseased state.

It is apparent that QS-controlled behaviors are prevalent among biofilm communities in both natural and engineered ecosystems, despite the high species diversity and environmental complexity. Not only do these collective behaviors affect the survival and fitness of the communicating organism(s), but they have significant impacts on the global nutrient cycling and the health of coral ecosystems. Although the mechanistic details remain to be defined in most cases, the prominent QS-regulated phenotypes promise the future applications of QS or QSI technologies to manipulate community activities in efforts whether to prevent disease progression or to increase microbial productivity.

12.4 The Applications of Quorum Sensing in Microbial Community of Environmental Engineering Systems

Since the concept of using QS inhibitors to control some diseases was presented in 2003 (Greenberg 2003), various QS inhibitors and QQ enzymes have been discovered (Table 12.2). However, there are still a number of obstacles left in applying this concept to actual clinical techniques, such as toxicity, stability, side effects, etc. of a drug (Whiteley et al. 2017). It is worth noting that QS studies in the field of membrane water filtration (Yeon et al. 2009a), where studies began years later than in medicine, are now fairly close to application to the real world (Lee et al. 2016c). In the environmental engineering systems of complex microbial communities, the barriers to identifying the mechanism are high, but the concerns about the side effects of actual application are thought to be small compared to the medical applications to humans. This section will introduce examples of QS based applications in environmental engineering systems of the complex microbial community.

12.4.1 Quorum Quenching for the Inhibition of Membrane Biofouling

Membrane biofouling is a process whereby the surface of a membrane is covered by microbial cells and their products, thus deteriorating the water filtration performance in membrane systems such as membrane bioreactor (MBR), ultrafiltration (UF), nanofiltration (NF), and reverse osmosis (RO) (Flemming et al. 1997). In the process of membrane biofouling, the deposition of microbial cells and the growth of biofilm occur simultaneously. While antimicrobial treatment for removing biofouling layer can generate resistance by mutational processes of microorganisms (Woodford and Ellington 2007), QQ strategies to inhibit biofilm formation can minimize the possibility of causing resistance because its mechanism is not related to bacterial growth (Bjarnsholt et al. 2010, Cady et al. 2012). In 2009, the correlation between AHL QS and membrane biofouling was demonstrated, presenting the possibility of membrane biofouling control by interfering intercellular communication via degradation of AHL signal molecules (i.e., QQ) in MBRs for wastewater treatment (Yeon et al. 2009a). In constant flux operation of MBRs, the trans-membrane pressure (TMP) increases as membrane fouling progresses, and TMP increases are usually observed in two phases: a gradual increase in TMP (1st phase) followed by an abrupt TMP increase (2nd phase) that occurs with massive formation of biofilm on the membrane surface (Zhang et al. 2006, Hwang et al. 2012, Oh and Lee 2018). Yeon et al. proved that the increase of AHL levels in biofouling layer and that of TMP showed very similar patterns in MBRs (Yeon et al. 2009a). They also demonstrated that the injection of QQ enzymes (porcine kidney acylase I, EC 3.5.1.14) delayed the start of the 2nd phase TMP increase in a lab scale MBR without sacrificing biodegradation of organics while the addition of AHL signals put the sudden TMP increase forward.

In order to prevent the loss of free enzymes, the porcine kidney acylase I was immobilized either on magnetic particles (Yeon et al. 2009b) or nanofiltration membrane surfaces (Kim et al. 2011). However, practical issues regarding the cost and stability of enzyme applications have led to the question of whether this strategy is a valid approach for the MBR process of wastewater treatment. As an alternative for QQ enzymes, whole cell application of QQ bacteria has been tried to control biofouling in an MBR (Oh et al. 2012). Despite the lower catalytic activity of whole cell than the isolated enzymes, the application of whole cell is preferred for various industrial purposes owing to the following reasons: (i) it can be prepared more easily and cheaply, (ii) it has

TABLE 12.2

Quorum Sensing Inhibitors and Their Applications

QS Inhibitors	Source	Applications	References
Enzymes			
AHL acylase	Porcine kidney	Membrane biofouling	Yeon et al. (2009)
	Pseudomonas sp. 1A1	Membrane biofouling	Cheong et al. (2013)
	Delftia sp. T6	Membrane biofouling	Yavuztürk Gül et al. (2017)
	Aspergillus melleus	Medical devices	Ivanova et al. (2015)
AHL lactonase	*Rhodococcus* sp. BH4	Membrane biofouling	Oh et al. (2012)
	Bacillus methylotrophicus sp. WY	Membrane biofouling	Khan et al. (2016)
	Bacillus sp. T5	Membrane biofouling	Yavuztürk Gül et al. (2017)
	Enterococcus sp. HEMM-1	Membrane biofouling	Ham et al. (2018)
	Bacillus sp. 240B1	Plant disease	Dong et al. (2001)
	A. tumefaciens C58	Plant disease	D'Angelo-Picard et al. (2011)
	Bacillus sp. AI96	Infection in fish	Cao et al. (2012)
	Bacillus sp. B546	Infection in fish	Chen et al. (2010)
	Bacillus licheniformis DAHB1	Infection in shrimp	Vinoj et al. (2014)
	Sulfolobus solfataricus	Medical devices	Ng et al. (2011)
	Sulfolobus solfataricus	Therapeutic application	Hraiech et al. (2014)
	Bacillus sp. ZA1	Therapeutic application	Gupta et al. (2015)
QS inhibiting compounds			
Vanillin	vanilla beans	Membrane biofouling	Kappachery et al. (2010)
Piper betle extract	*Piper betle*	Membrane biofouling	Siddiqui et al. (2012)
2(5H)-furanone	*Delisea pulchra*	Membrane biofouling	Ponnusamy et al. (2010)
Kojic acid	*Aspergillus oryzae*	Marine biofouling	Dobretsov et al. (2011)
Farnesol (AI-2 inhibition)	*Candida albicans*	Membrane biofouling	Lee et al. (2016)

better stability than free enzyme in long-term applications, (iii) it is easier to handle compared to enzymes (Ishige et al. 2005). Moreover, the MBR process does not require an external nutritional supply for the survival of QQ bacterial cells because they can acquire sufficient resources (e.g., carbon, nutrients, etc.) from wastewater. The first QQ strain that was used in an MBR for biofouling control was a recombinant *Escherichia coli* (pMBP-His-aiiA), which can produce AHL-lactonase, AiiA (Oh et al. 2012). The recombinant QQ bacteria was encapsulated in a medium consisting of porous hollow-fiber membranes named QQ-vessel. In this experiment, a control strain *E. coli*, which harbors the empty vector (pMBP-His-parallel1), was encapsulated in the vessel of the control reactor to exclude effects other than the QQ effect of adding exogenous bacteria to the MBR. After the anti-biofouling effect of QQ-vessel was proved by showing significant delay of TMP increase in MBRs, the recombinant QQ strain was replaced with an indigenous QQ bacteria strain, *Rhodococcus* sp. BH4, which was isolated from a real MBR plant. The experimental results that showed the reduction of biofouling by applying the indigenous QQ bacteria raised the practical applicability of QQ strategy to control biofouling in MBRs to a higher level. Various studies were subsequently conducted to find more efficient QQ-media such as QQ-bead (Kim et al. 2013b, Lee et al. 2016a, 2016c, Khan et al. 2016, Xiao et al. 2018), QQ-cylinder (Lee et al. 2016b), QQ-hollow cylinder (Lee et al. 2016d), QQ-sheet (Nahm et al. 2017), rotating microbial carrier frame (Köse-Mutlu et al. 2016, Ergön-Can et al. 2017), etc., which could more efficiently interfere with the intercellular QS communication in MBRs and consequently be more effective in biofouling inhibition. The anti-biofouling effect of QQ in MBRs was tested not only in a lab-scale (5 L/day) but also in pilot-scales (1 ~ 10 m^3/day) treating real municipal wastewater (Lee 2017, Lee et al. 2016c).

To date, inhibition of EPS production has been reported as the main mechanism of biofouling reduction in MBRs by QQ application (Jiang et al. 2013, Weerasekara et al. 2014, Kim et al. 2011, 2013a, 2013b, Yeon et al. 2009a, 2009b, Cheong et al. 2014, Maqbool et al. 2015, Lee et al. 2016a, 2016c, Ergön-Can et al. 2017, Lade et al. 2017, Hasnain et al. 2017, Waheed et al. 2018). The reduction in EPS weakens the structure of biofilm, making it easier to detach bacterial cells by shear stress (Oh and Lee 2018). According to the aforementioned studies, QQ did not adversely affect wastewater treatment performance, such as the removal rates of chemical oxygen demand (COD), total nitrogen (TN), ammonia nitrogen (NH_4-N), and nitrate nitrogen (NO_3-N).

12.4.2 Quorum Sensing in Bioeletrochemical Systems to Enhance Energy Recovery

Microbial fuel cells (MFCs) are bioelectrochemical systems (BES) that use biofilms as catalysts for converting chemical energy directly into electrical energy (Rabaey and Verstraete 2005, Logan et al. 2006, Lovley 2006). MFC has a distinct advantage in that it can utilize low-grade biomass or even wastewater to produce bioelectricity (Huang and Logan 2008, Zhou et al.

2013). In a typical two-chamber MFC, biofilms in the anaerobic anode chamber oxidize organic or some inorganic substrates, then the electrons released during the biodegradation process are delivered to the electrode directly or by electron mediator (Wang et al. 2013c). The electrical power is produced when the electrons flow through an external circuit across an electrical load.

The role of QS in electricity generation had been first studied in the *P. aeruginosa*-inoculated MFCs. It is demonstrated that QS and current generation in MFC inoculated with *P. aeruginosa* are closely associated with the production of phenazine, which act as electron mediators to promote respiration of *P. aeruginosa* with the electrode (Venkataraman et al. 2010). Yong et al. found that the wild-type *P. aeruginosa* strain CGMCC 1.860 used an electron mediator compound having high mid-point potential as its major electron shuttle rather than using low mid-point potential phenazines (Yong et al. 2011). They transformed multi-copy broad-host plasmid pYC-rhlIR into the strain CGMCC 1.860 to overexpress *rhl* QS system for the elimination of that redox compound and the overproduction of phenazine, which resulted in 1.6 times higher current output. On the other hand, Wang et al. constructed a PQS defective but phenazine overproducing *P. aeruginosa* strain, which synthesized higher concentrations of phenazines under anaerobic conditions and consequently showed around five-fold higher maximum current density compared to the parent strain (Wang et al. 2013c).

Recently, several studies were conducted to elucidate the role of QS in mixed culture BES. Chen et al. reported that adding exogenous AHLs to the anode chamber increased the energy recovery of MFC, while adding acylase, which degrades the endogenous AHLs secreted by electrochemically active bacteria (EAB) and decreases the energy recovery (Chen et al. 2017). The endogenous or exogenous AHLs increased the electrode-associated biomass, the proportion of EABs, the biofilm compactness, and the ratio of live/dead cells, which led to higher energy recovery. Meanwhile, microbial electrolysis cell (MEC) is a different type of BES, a system that shares basic principles with MFC but produces energy sources such as hydrogen. Liu et al. reported that the addition of short chain AHLs (3OC6-HSL) to the anode chamber enhanced the electron transfer activities, resulting in a higher hydrogen yield and energy recovery in an MEC system inoculated with a mixed community of wastewater (Liu et al. 2015). Analysis of the microbial community demonstrated that adding short chain AHLs altered the communities of both anodic and cathodic biofilms toward a more favorable environment to transfer electrons to the electrodes by increasing the population of EAB and reducing that of hydrogen scavengers, such as homo-acetogens or methanogens (Cai et al. 2016). The role of QS in bioelectrochemical systems is illustrated in Figure 12.2a.

12.4.3 Quorum Sensing in Bioaugmentation to Improve Pollutant Removal in Wastewater Treatment

Bioaugmentation in wastewater treatment is a process which attempts to enhance treatment by increasing diversity and/or activity of microorganisms via direct introduction of either selected naturally occurring microorganisms or genetically altered microorganisms to the wastewater treatment plant (Stephenson and Stephenson 1992). Practically, the technical core of bioaugmentation is colonization of introduced bacteria in the activated sludge system, so the method should be able to maintain persistent survival rates and the activity of the introduced bacteria (Teng et al. 2010). Bacteria communicate with each other by using signal molecules with the aim of acclimatizing themselves to the environment according to the QS theory (Darch et al. 2012). Therefore, communication between native bacterial community and introduced bacteria might play an essential role in colonization of introduced bacteria and the performance of bioaugmentation.

A series of studies was carried out on the role of QS signaling in bioaugmentation with a nicotine-degrading bacterium *Pseudomonas* sp. HF-1 (Wang et al. 2012, 2014a, 2014b). The roles of AHLs and their release patterns were studied under different operating conditions in strain HF-1 bioaugmented system for tobacco treatment (Wang et al. 2012). The results showed that swarming was induced by C6-HSL and 3OC6-HSL; the formation of EPS by C6-HSL, 3OC6-HSL, and 3OC8-HSL; and biofilm formation by C6-HSL and 3OC8-HSL. This nicotine-degrading bacterium strain HF-1 was inoculated at different concentrations into activated sludge to better construct a bioaugmented system for tobacco wastewater treatment, and the level of AI-2 and AHL was monitored (Wang et al. 2014a). During

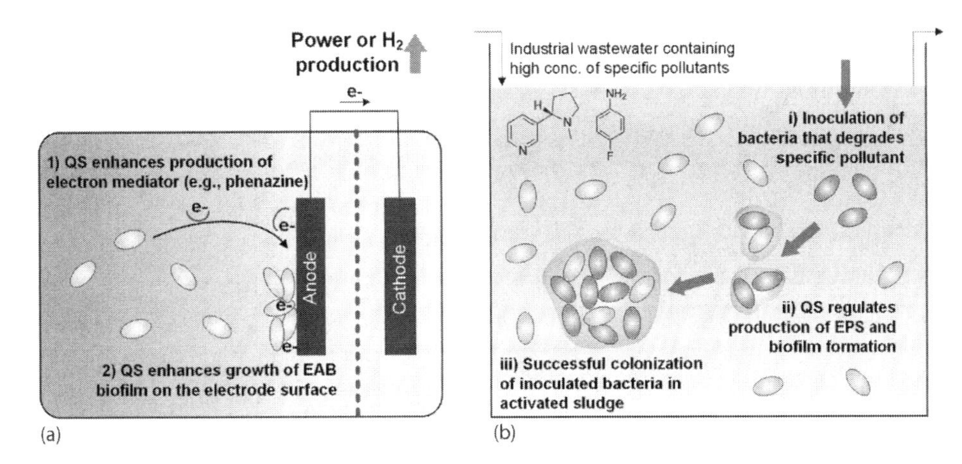

FIGURE 12.2 The applications of quorum sensing in (a) bioelectrochemical systems to enhance energy recovery and (b) bioaugmentation to improve pollutant removal in wastewater treatment.

the period of HF-1 inoculation, when compared with failed bioaugmented systems, AI-2 was significantly increased in the successful systems, suggesting that AI-2-mediated bacterial communication played an essential role in the colonization of HF-1, while constant level of short-chain AHL was detected only in the successful systems. When the HF-1 bioaugmented system was operated under pH 5.5 for three cycles and pH 8.0 for the rest, pH control increased the level of C6-HSL, 3OC6-HSL, and 3OC8-HSL, which resulted in the increased EPS secretion and the formation of HF-1 biofilm (Wang et al. 2014b).

Acinetobacter sp. TW can degrade fluorochemicals as well as nicotine (Wang et al. 2011, 2013b). Wang et al. demonstrated that AHL signaling was involved in modulating EPS production and biofilm formation of strain TW, but not in 4-fluoroaniline degradation in TW bioaugmented system for treating fluorine-containing industrial wastewater (Wang et al. 2013b). In detail, under optimal conditions for autoinducer release (25°C, pH 5, 800 mg/L 4-fluoroaniline, and 0% NaCl), the level of 3OC6-HSL and C6-HSL were significantly increased, thereby inducing EPS secretion and biofilm formation to facilitate bacterial colonization. Similarly, short chain AHLs were found to be increased to assist colonization of strain TW in bioaugmentation systems treating tobacco wastewater (Wang et al. 2013a). They demonstrated that long chain AHLs were also secreted in the bioaugmentation system and helped bacteria to resist the nicotine toxicity. The role of QS in bioaugmentation systems for treating industrial wastewater is illustrated in Figure 12.2b.

In summary, since the correlation between QS and biofouling in MBRs consisting of complex microbial community was discovered, a number of studies have been attempted over the past decade to identify the roles of QS in complex engineered systems and further develop QS or QQ techniques to improve system performance. Signal interference by QQ have been applied to control biofouling of membrane filtration systems, while QS signal induction has been tried to improve the performance of granular sludge systems, bioelectrochemical systems, and bioaugmented systems for pollutant removal. Furthermore, this QS-based approach is about to expand the area to a wider variety of environmental engineered systems, including ANAMMOX process for nitrogen removal (Sun et al. 2018, Tang et al. 2015, 2018b) and biofilm reactors (Hu et al. 2016a, 2016b). It is certain that this QS-based strategy will be the main focus of future research to improve the performance of complex microbial engineered systems, and to assess the validity of each technique closely and bring it from the bench to the real plant, a number of further studies are required.

12.5 Concluding Remarks

Quorum sensing is prevalent and functional in both natural and engineered ecosystems and has a central role in enhancing the fitness of microbial communities. The species composition and function and their spatial organization in biofilms are expected to influence bacterial communication, leading to heterogenous QS response. The extent to which these biological elements may affect the ultimate structure and function of the communities, and how this may vary in response to other environmental factors such as fluid flow and pH fluctuations, remain to be addressed, however. Future investigations using synthetic model communities with increased complexity coupled with computational simulations are likely to provide more mechanistic insights into how QS works in species-rich communities in dynamically changing environments. As the accessibility to advanced technologies increase, we anticipate to yield new insights, beyond QS functions, into the QS regulatory networks and the communicating organism(s) by studying the natural or engineered communities directly. Together, these novel understandings of QS will aid to better predict and control of the community behaviors, whether these pertain to increased nitrogen removal by ANAMMOX, enhanced energy recovery by MFCs, biofouling prevention in MBR, or conservation of coral ecosystems.

ACKNOWLEDGMENTS

This research was supported by the Singapore Centre for Environmental Life Science Engineering (SCELSE), which is funded by the National Research Foundation Singapore, Ministry of Education, Nanyang Technological University and National University of Singapore, under its Research Centre of Excellence Programme.

REFERENCES

Abisado, R. G., S. Benomar, J. R. Klaus, A. A. Dandekar, and J. R. Chandler. 2018. Bacterial quorum sensing and microbial community interactions. *mBio* 9 (3):e02331–17.

Alves, P. M., E. Al-Badi, C. Withycombe, P. M. Jones, K. J. Purdy, and S. E. Maddocks. 2018. Interaction between *Staphylococcus aureus* and *Pseudomonas aeruginosa* is beneficial for colonisation and pathogenicity in a mixed biofilm. *Pathog. Dis.* 76 (1):fty003.

An, D., T. Danhorn, C. Fuqua, and M. R. Parsek. 2006. Quorum sensing and motility mediate interactions between *Pseudomonas aeruginosa* and *Agrobacterium tumefaciens* in biofilm cocultures. *Proc. Natl. Acad. Sci. USA* 103 (10):3828–3833.

Barnard, A. M. L., S. D. Bowden, T. Burr, S. J. Coulthurst, R. E. Monson, and G. P. C. Salmond. 2007. Quorum sensing, virulence and secondary metabolite production in plant soft-rotting bacteria. *Philos. Trans. R. Soc. Lond. B Biol. Sci.* 362 (1483):1165–1183.

Batchelor, S., M. Cooper, S. Chhabra et al. 1997. Cell density-regulated recovery of starved biofilm populations of ammonia-oxidizing bacteria. *Appl. Environ. Microbiol.* 63 (6):2281–2286.

Biswas, L., R. Biswas, M. Schlag, R. Bertram, and F. Götz. 2009. Small-colony variant selection as a survival strategy for *Staphylococcus aureus* in the presence of *Pseudomonas aeruginosa*. *Appl. Environ. Microbiol.* 75 (21):6910–6912.

Bjarnsholt, T., T. Tolker-Nielsen, N. Hoiby, and M. Givskov. 2010. Interference of *Pseudomonas aeruginosa* signalling and biofilm formation for infection control. *Expert Rev. Mol. Med.* 12:e11.

Boles, B. R., and A. R. Horswill. 2008. Agr-mediated dispersal of *Staphylococcus aureus* biofilms. *PLoS Path.* 4 (4):e1000052.

Bourne, D. G., M. Garren, T. M. Work, E. Rosenberg, G. W. Smith, and C. D. Harvell. 2009. Microbial disease and the coral holobiont. *Trends Microbiol.* 17 (12):554–562.

Burmølle, M., D. Ren, T. Bjarnsholt, and S. J. Sørensen. 2014. Interactions in multispecies biofilms: Do they actually matter? *Trends Microbiol.* 22 (2):84–91.

Burton, E., H. Read, M. Pellitteri, and W. Hickey. 2005. Identification of acyl-homoserine lactone signal molecules produced by *Nitrosomonas europaea* strain Schmidt. *Appl. Environ. Microbiol.* 71 (8):4906–4909.

Cady, N. C., K. A. McKean, J. Behnke et al. 2012. Inhibition of biofilm formation, quorum sensing and infection in *Pseudomonas aeruginosa* by natural products-inspired organosulfur compounds. *PLoS One* 7 (6):e38492.

Cai, W., Z. Zhang, G. Ren et al. 2016. Quorum sensing alters the microbial community of electrode-respiring bacteria and hydrogen scavengers toward improving hydrogen yield in microbial electrolysis cells. *Appl. Energy.* 183:1133–1141.

Cao, Y., S. He, Z. Zhou et al. 2012. Orally administered thermostable *N*-acyl homoserine lactonase from *Bacillus* sp. strain A196 attenuates *Aeromonas hydrophila* infection in zebrafish. *Appl. Environ. Microbiol.* 78 (6):1899–1908.

Capone, D. G., J. P. Zehr, H. W. Paerl, B. Bergman, and E. J. Carpenter. 1997. *Trichodesmium*, a globally significant marine cyanobacterium. *Science* 276 (5316):1221.

Certner, R. H., and S. V. Vollmer. 2015. Evidence for autoinduction and quorum sensing in white band disease-causing microbes on *Acropora cervicornis. Sci. Rep.* 5:11134.

Certner, R. H., and S. V. Vollmer. 2018. Inhibiting bacterial quorum sensing arrests coral disease development and disease-associated microbes. *Environ. Microbiol.* 20 (2):645–657.

Chandler, J. R., S. Heilmann, J. E. Mittler, and E. P. Greenberg. 2012. Acyl-homoserine lactone-dependent eavesdropping promotes competition in a laboratory co-culture model. *ISME J.* 6:2219.

Chen, R., Z. Zhou, Y. Cao, Y. Bai, and B. Yao. 2010. High yield expression of an AHL-lactonase from *Bacillus* sp. B546 in Pichia pastoris and its application to reduce *Aeromonas hydrophila* mortality in aquaculture. *Microb. Cell Fact.* 9 (1):39.

Chen, S., X. Jing, J. Tang, Y. Fang, and S. Zhou. 2017. Quorum sensing signals enhance the electrochemical activity and energy recovery of mixed-culture electroactive biofilms. *Biosens. Bioelectron.* 97:369–376.

Cheong, W. S., S. R. Kim, H. S. Oh et al. 2014. Design of quorum quenching microbial vessel to enhance cell viability for biofouling control in membrane bioreactor. *J. Microbiol. Biotechnol.* 24 (1):97–105.

Cheong, W. S., C. H. Lee, Y. H. Moon et al. 2013. Isolation and identification of indigenous quorum quenching bacteria, *Pseudomonas* sp. 1A1, for biofouling control in MBR. *Ind. Eng. Chem. Res.* 52 (31):10554–10560.

Déziel, E., F. Lépine, S. Milot et al. 2004. Analysis of *Pseudomonas aeruginosa* 4-hydroxy-2-alkylquinolines (HAQs) reveals a role for 4-hydroxy-2-heptylquinoline in cell-to-cell communication. *Proc. Natl. Acad. Sci. USA* 101 (5):1339–1344.

D'Angelo-Picard, C., E. Chapelle, P. Ratet, D. Faure, and Y. Dessaux. 2011. Transgenic plants expressing the quorum quenching lactonase AttM do not significantly alter root-associated bacterial populations. *Res. Microbiol.* 162 (9):951–958.

Darch, S. E., O. Simoska, M. Fitzpatrick et al. 2018. Spatial determinants of quorum signaling in a *Pseudomonas aeruginosa* infection model. *Proc. Natl. Acad. Sci. USA* 115 (18):4779–4784.

Darch, S. E., S. A. West, K. Winzer, and S. P. Diggle. 2012. Density-dependent fitness benefits in quorum-sensing bacterial populations. *Proc. Natl. Acad. Sci.* 109 (21):8259.

Davies, D. G., M. R. Parsek, J. P. Pearson, B. H. Iglewski, J. W. Costerton, and E. P. Greenberg. 1998. The involvement of cell-to-cell signals in the development of a bacterial biofilm. *Science* 280 (5361):295.

De Clippeleir, H., T. Defoirdt, L. Vanhaecke et al. 2011. Long-chain acylhomoserine lactones increase the anoxic ammonium oxidation rate in an OLAND biofilm. *Appl. Microbiol. Biotechnol.* 90 (4):1511–1519.

Decho, A. W., P. T. Visscher, J. Ferry et al. 2009. Autoinducers extracted from microbial mats reveal a surprising diversity of *N*-acylhomoserine lactones (AHLs) and abundance changes that may relate to diel pH. *Environ. Microbiol.* 11 (2):409–420.

DeLeon, S., A. Clinton, H. Fowler, J. Everett, A. R. Horswill, and K. P. Rumbaugh. 2014. Synergistic interactions of *Pseudomonas aeruginosa* and *Staphylococcus aureus* in an in vitro wound model. *Infect. Immun.* 82 (11):4718–4728.

Dobretsov, S., M. Teplitski, M. Bayer, S. Gunasekera, P. Proksch, and V. J. Paul. 2011. Inhibition of marine biofouling by bacterial quorum sensing inhibitors. *Biofouling* 27 (8):893–905.

Dong, Y.-H., L.-H. Wang, J.-L. Xu, H.-B. Zhang, X.-F. Zhang, and L.-H. Zhang. 2001. Quenching quorum-sensing-dependent bacterial infection by an *N*-acyl homoserine lactonase. *Nature* 411 (6839):813–817.

Dong, Y.-H., X.-F. Zhang, S.-W. An, J.-L. Xu, and L.-H. Zhang. 2008. A novel two-component system BqsS-BqsR modulates quorum sensing-dependent biofilm decay in *Pseudomonas aeruginosa. Commun. Integr. Biol.* 1 (1):88–96.

Dow, J. M., L. Crossman, K. Findlay, Y.-Q. He, J.-X. Feng, and J.-L. Tang. 2003. Biofilm dispersal in *Xanthomonas campestris* is controlled by cell–cell signaling and is required for full virulence to plants. *Proc. Natl. Acad. Sci. USA* 100 (19):10995.

Drescher, K., C. D. Nadell, H. A. Stone, N. S. Wingreen, and B. L. Bassler. 2014. Solutions to the public goods Dilemma in bacterial biofilms. *Curr. Biol.* 24 (1):50–55.

Dulla, G. F., and S. E. Lindow. 2009. Acyl-homoserine lactone-mediated cross talk among epiphytic bacteria modulates behavior of *Pseudomonas syringae* on leaves. *ISME J.* 3 (7):825.

Egan, S., and M. Gardiner. 2016. Microbial dysbiosis: Rethinking disease in marine ecosystems. *Front. Microbiol.* 7:991–991.

Elias, S., and E. Banin. 2012. Multi-species biofilms: Living with friendly neighbors. *FEMS Microbiol. Rev.* 36 (5):990–1004.

Engebrecht, J., K. Nealson, and M. Silverman. 1983. Bacterial bioluminescence: Isolation and genetic analysis of functions from *Vibrio fischeri. Cell* 32 (3):773–781.

Eppley, R. W., and B. J. Peterson. 1979. Particulate organic matter flux and planktonic new production in the deep ocean. *Nature* 282 (5740):677.

Ergön-Can, T., B. Köse-Mutlu, İ. Koyuncu, and C. H. Lee. 2017. Biofouling control based on bacterial quorum quenching with a new application: Rotary microbial carrier frame. *J. Membr. Sci.* 525 (Supplement C):116–124.

Federle, M. J., and B. L. Bassler. 2003. Interspecies communication in bacteria. *J. Clin. Invest.* 112 (9):1291–1299.

Feng, H., Y. Ding, M. Wang et al. 2014. Where are signal molecules likely to be located in anaerobic granular sludge? *Water Res.* 50:1–9.

Flemming, H.-C., and J. Wingender. 2010. The biofilm matrix. *Nat. Rev. Microbiol.* 8:623.

Flemming, H.-C., and S. Wuertz. 2019. Bacteria and archaea on Earth and their abundance in biofilms. *Nat. Rev. Microbiol.* 17 (4):247–260.

Flemming, H. C., G. Schaule, T. Griebe, J. Schmitt, and A. Tamachkiarowa. 1997. Biofouling—The Achilles heel of membrane processes. *Desalination* 113 (2–3):215–225.

Frischkorn, K. R., M. Rouco, B. A. S. Van Mooy, and S. T. Dyhrman. 2017. Epibionts dominate metabolic functional potential of *Trichodesmium* colonies from the oligotrophic ocean. *ISME J.* 11:2090.

Gao, J., A. Ma, X. Zhuang, and G. Zhuang. 2014. An *N*-acyl homoserine lactone synthase in the ammonia-oxidizing bacterium *Nitrosospira multiformis*. *Appl. Environ. Microbiol.* 80 (3):951–958.

Gjermansen, M., P. Ragas, C. Sternberg, S. Molin, and T. Tolker-Nielsen. 2005. Characterization of starvation-induced dispersion in *Pseudomonas putida* biofilms. *Environ. Microbiol.* 7 (6):894–904.

Golberg, K., E. Eltzov, M. Shnit-Orland, R. S. Marks, and A. Kushmaro. 2011. Characterization of quorum sensing signals in coral-associated bacteria. *Microb. Ecol.* 61 (4):783–792.

Golberg, K., V. Pavlov, R. S. Marks, and A. Kushmaro. 2013. Coral-associated bacteria, quorum sensing disrupters, and the regulation of biofouling. *Biofouling* 29 (6):669–682.

Goo, E., C. D. Majerczyk, J. H. An et al. 2012. Bacterial quorum sensing, cooperativity, and anticipation of stationary-phase stress. *Proc. Natl. Acad. Sci. USA* 109 (48):19775.

Greenberg, E. P. 2003. Bacterial communication and group behavior. *J. Clin. Invest.* 112 (9):1288–1290.

Gupta, P., S. Chhibber, and K. Harjai. 2015. Efficacy of purified lactonase and ciprofloxacin in preventing systemic spread of *Pseudomonas aeruginosa* in murine burn wound model. *Burns* 41 (1):153–162.

Hall-Stoodley, L., J. W. Costerton, and P. Stoodley. 2004. Bacterial biofilms: From the natural environment to infectious diseases. *Nat. Rev. Microbiol.* 2 (2):95.

Ham, S. Y., H. S. Kim, E. Cha, J. H. Park, and H. D. Park. 2018. Mitigation of membrane biofouling by a quorum quenching bacterium for membrane bioreactors. *Bioresour. Technol.* 258:220–226.

Hasnain, G., S. J. Khan, M. Z. Arshad, and H. Y. Abdullah. 2017. Combined impact of quorum quenching and backwashing on biofouling control in a semi-pilot scale MBR treating real wastewater. *J. Chem. Soc. Pak.* 39 (02):215–223.

Henke, J. M., and B. L. Bassler. 2004. Three parallel quorum-sensing systems regulate gene expression in *Vibrio harveyi*. *J. Bacteriol.* 186 (20):6902.

Higgins, D. A., M. E. Pomianek, C. M. Kraml, R. K. Taylor, M. F. Semmelhack, and B. L. Bassler. 2007. The major *Vibrio cholerae* autoinducer and its role in virulence factor production. *Nature* 450:883.

Hmelo, L. R. 2017. Quorum sensing in marine microbial environments. *Ann. Rev. Mar. Sci.* 9 (1):257–281.

Hmelo, L. R., T. J. Mincer, and B. A. S. Van Mooy. 2011. Possible influence of bacterial quorum sensing on the hydrolysis of sinking particulate organic carbon in marine environments. *Environ. Microbiol. Rep.* 3 (6):682–688.

Hoffman, L. R., E. Déziel, D. A. D'Argenio et al. 2006. Selection for *Staphylococcus aureus* small-colony variants due to growth in the presence of *Pseudomonas aeruginosa*. *Proc. Natl. Acad. Sci. USA* 103 (52):19890–19895.

Hosni, T., C. Moretti, G. Devescovi et al. 2011. Sharing of quorum-sensing signals and role of interspecies communities in a bacterial plant disease. *ISME J.* 5:1857.

Hotterbeekx, A., S. Kumar-Singh, H. Goossens, and S. Malhotra-Kumar. 2017. In vivo and in vitro interactions between *Pseudomonas aeruginosa* and *Staphylococcus* spp. *Front. Cell. Infect. Microbiol.* 7:106.

Hraiech, S., J. Hiblot, J. Lafleur et al. 2014. Inhaled lactonase reduces *Pseudomonas aeruginosa* quorum sensing and mortality in rat pneumonia. *PLoS One* 9 (10):e107125.

Hu, H., J. He, J. Liu, H. Yu, and J. Zhang. 2016a. Biofilm activity and sludge characteristics affected by exogenous *N*-acyl homoserine lactones in biofilm reactors. *Bioresour. Technol.* 211:339–347.

Hu, H. Z., J. G. He, J. Liu, H. R. Yu, J. Tang, and J. Zhang. 2016b. Role of *N*-acyl-homoserine lactone (AHL) based quorum sensing on biofilm formation on packing media in wastewater treatment process. *RSC Adv.* 6 (14):11128–11139.

Huang, L., and B. E. Logan. 2008. Electricity generation and treatment of paper recycling wastewater using a microbial fuel cell. *Appl. Microbiol. Biotechnol.* 80 (2):349–355.

Huber, B., K. Riedel, M. Hentzer et al. 2001. The cep quorum-sensing system of *Burkholderia cepacia* H111 controls biofilm formation and swarming motility. *Microbiology* 147 (9):2517–2528.

Hwang, B. K., C. H. Lee, I. S. Chang, A. Drews, and R. Field. 2012. Membrane bioreactor: TMP rise and characterization of bio-cake structure using CLSM-image analysis. *J. Membr. Sci.* 419:33–41.

Ishige, T., K. Honda, and S. Shimizu. 2005. Whole organism biocatalysis. *Curr. Opin. Chem. Biol.* 9 (2):174–180.

Ivanova, K., M. M. Fernandes, A. Francesko et al. 2015. Quorum-quenching and matrix-degrading enzymes in multilayer coatings synergistically prevent bacterial biofilm formation on urinary catheters. *ACS Appl. Mater. Interfaces* 7 (49):27066–27077.

Jiang, W., S. Xia, J. Liang, Z. Zhang, and S. W. Hermanowicz. 2013. Effect of quorum quenching on the reactor performance, biofouling and biomass characteristics in membrane bioreactors. *Water Res.* 47 (1):187–196.

Köse-Mutlu, B., T. Ergön-Can, İ. Koyuncu, and C.-H. Lee. 2016. Quorum quenching MBR operations for biofouling control under different operation conditions and using different immobilization media. *Desal. Water Treat.* 57 (38):17696–17706.

Kaplan, J. B. 2010. Biofilm dispersal: Mechanisms, clinical implications, and potential therapeutic uses. *J. Dent. Res.* 89 (3):205–218.

Kappachery, S., D. Paul, J. Yoon, and J. H. Kweon. 2010. Vanillin, a potential agent to prevent biofouling of reverse osmosis membrane. *Biofouling* 26 (6):667–672.

Khan, R., F. Shen, K. Khan et al. 2016. Biofouling control in a membrane filtration system by a newly isolated novel quorum quenching bacterium, *Bacillus methylotrophicus* sp. WY. *RSC Adv.* 6 (34):28895–28903.

Kim, H. W., H. S. Oh, S. R. Kim et al. 2013a. Microbial population dynamics and proteomics in membrane bioreactors with enzymatic quorum quenching. *Appl. Microbiol. Biotechnol.* 97 (10):4665–4675.

Kim, J. H., D. C. Choi, K. M. Yeon, S. R. Kim, and C. H. Lee. 2011. Enzyme-immobilized nanofiltration membrane to mitigate biofouling based on quorum quenching. *Environ. Sci. Technol.* 45 (4):1601–1607.

Kim, S. R., H. S. Oh, S. J. Jo et al. 2013b. Biofouling control with bead-entrapped quorum quenching bacteria in membrane bioreactors: Physical and biological effects. *Environ. Sci. Technol.* 47 (2):836–842.

Kimes, N. E., C. J. Grim, W. R. Johnson et al. 2011. Temperature regulation of virulence factors in the pathogen *Vibrio coralliilyticus*. *ISME J.* 6:835.

Korgaonkar, A., U. Trivedi, K. P. Rumbaugh, and M. Whiteley. 2013. Community surveillance enhances *Pseudomonas aeruginosa* virulence during polymicrobial infection. *Proc. Natl. Acad. Sci. USA* 110 (3):1059–1064.

Krupke, A., L. R. Hmelo, J. E. Ossolinski, T. J. Mincer, and B. A. S. Van Mooy. 2016. Quorum sensing plays a complex role in regulating the enzyme hydrolysis activity of microbes associated with sinking particles in the ocean. *Front. Mar. Sci.* 3:55.

Lade, H., W. J. Song, Y. J. Yu, J. H. Ryu, G. Arthanareeswaran, and J. H. Kweon. 2017. Exploring the potential of curcumin for control of *N*-acyl homoserine lactone-mediated biofouling in membrane bioreactors for wastewater treatment. *RSC Adv.* 7 (27):16392–16400.

Lee, C. H. 2017. "Unfinished voyage of quorum quenching MBR for biofouling control & energy savings: Let microorganisms handle their business." 8th IWA Membrane Technology Conference & Exhibition for Water and Wastewater Treatment and Reuse, Singapore.

Lee, K., S. Lee, S. H. Lee et al. 2016a. Fungal quorum quenching: A paradigm shift for energy savings in membrane bioreactor (MBR) for wastewater treatment. *Environ. Sci. Technol.* 50 (20):10914–10922.

Lee, S., S. H. Lee, K. Lee et al. 2016b. Effect of the shape and size of quorum-quenching media on biofouling control in membrane bioreactors for wastewater treatment. *J. Microbiol. Biotechnol.* 26 (10):1746–1754.

Lee, S., S.-K. Park, H. Kwon et al. 2016c. Crossing the border between laboratory and field: bacterial quorum quenching for anti-biofouling strategy in an MBR. *Environ. Sci. Technol.* 50 (4):1788–1795.

Lee, S. H., S. Lee, K. Lee et al. 2016d. More efficient media design for enhanced biofouling control in a membrane bioreactor: Quorum quenching bacteria entrapping hollow cylinder. *Environ. Sci. Technol.* 50 (16):8596–8604.

Leung, V., and C. M. Lévesque. 2012. A stress-inducible quorum-sensing peptide mediates the formation of persister cells with noninherited multidrug tolerance. *J. Bacteriol.* 194 (9):2265–2274.

Lewenza, S., M. B. Visser, and P. A. Sokol. 2002. Interspecies communication between *Burkholderia cepacia* and *Pseudomonas aeruginosa*. *Can. J. Microbiol.* 48 (8):707–716.

Li, H., X. Li, Z. Wang et al. 2015. Autoinducer-2 regulates *Pseudomonas aeruginosa* PAO1 biofilm formation and virulence production in a dose-dependent manner. *BMC Microbiol.* 15 (1):192.

Li, Y.-S., X.-R. Pan, J.-S. Cao et al. 2017. Augmentation of acyl homoserine lactones-producing and -quenching bacterium into activated sludge for its granulation. *Water Res.* 125:309–317.

Liu, W., W. Cai, A. Ma et al. 2015. Improvement of bioelectrochemical property and energy recovery by acylhomoserine lactones (AHLs) in microbial electrolysis cells (MECs). *J. Power Sources* 284:56–59.

Logan, B. E., B. Hamelers, R. Rozendal et al. 2006. Microbial fuel cells: Methodology and technology. *Environ. Sci. Technol.* 40 (17):5181–5192.

Lovley, D. R. 2006. Microbial fuel cells: Novel microbial physiologies and engineering approaches. *Curr. Opin. Biotechnol.* 17 (3):327–332.

Luo, H., R. Benner, R. A. Long, and J. Hu. 2009. Subcellular localization of marine bacterial alkaline phosphatases. *Proc. Natl. Acad. Sci. USA* 106 (50):21219–21223.

Ma, H., S. Ma, H. Hu, L. Ding, and H. Ren. 2018a. The biological role of *N*-acyl-homoserine lactone-based quorum sensing (QS) in EPS production and microbial community assembly during anaerobic granulation process. *Sci. Rep.* 8 (1):15793.

Ma, H., X. Wang, Y. Zhang et al. 2018b. The diversity, distribution and function of *N*-acyl-homoserine lactone (AHL) in industrial anaerobic granular sludge. *Bioresour. Technol.* 247:116–124.

Mahaffey, C., S. Reynolds, C. E. Davis, and M. C. Lohan. 2014. Alkaline phosphatase activity in the subtropical ocean: Insights from nutrient, dust and trace metal addition experiments. *Front. Mar. Sci.* 1:73.

Manefield, M., R. de Nys, K. Naresh et al. 1999. Evidence that halogenated furanones from *Delisea pulchra* inhibit acylated homoserine lactone (AHL)-mediated gene expression by displacing the AHL signal from its receptor protein. *Microbiology* 145 (2):283–291.

Maqbool, T., S. J. Khan, H. Waheed, C.-H. Lee, I. Hashmi, and H. Iqbal. 2015. Membrane biofouling retardation and improved sludge characteristics using quorum quenching bacteria in submerged membrane bioreactor. *J. Membr. Sci.* 483:75–93.

Mashburn, L. M., A. M. Jett, D. R. Akins, and M. Whiteley. 2005. *Staphylococcus aureus* Serves as an Iron Source for *Pseudomonas aeruginosa* during In Vivo Coculture. *J. Bacteriol.* 187 (2):554–566.

McDougald, D., S. Srinivasan, S. A. Rice, and S. Kjelleberg. 2003. Signal-mediated cross-talk regulates stress adaptation in Vibrio species. *Microbiology* 149 (7):1923–1933.

McNab, R., S. K. Ford, A. El-Sabaeny, B. Barbieri, G. S. Cook, and R. J. Lamont. 2003. LuxS-based signaling in *Streptococcus gordonii*: Autoinducer 2 controls carbohydrate metabolism and biofilm formation with *Porphyromonas gingivalis*. *J. Bacteriol.* 185 (1):274–284.

Mellbye, B. L., P. J. Bottomley, and L. A. Sayavedra-Soto. 2015. Nitrite-oxidizing bacterium *Nitrobacter winogradskyi* produces *N*-acyl-homoserine lactone autoinducers. *Appl. Environ. Microbiol.* 81 (17):5917–5926.

Mellbye, B. L., A. T. Giguere, P. J. Bottomley, and L. A. Sayavedra-Soto. 2016. Quorum quenching of *Nitrobacter winogradskyi* suggests that quorum sensing regulates fluxes of nitrogen oxide(s) during nitrification. *mBio* 7 (5):e01753–16.

Mellbye, B. L., E. Spieck, P. J. Bottomley, and L. A. Sayavedra-Soto. 2017. Acyl-homoserine lactone production in nitrifying bacteria of the genera *Nitrosospira*, *Nitrobacter*, and *Nitrospira* identified via a survey of putative quorum-sensing genes. *Appl. Environ. Microbiol.* 83 (22):e01540–17.

Mills, M. M., C. Ridame, M. Davey, J. La Roche, and R. J. Geider. 2004. Iron and phosphorus co-limit nitrogen fixation in the eastern tropical North Atlantic. *Nature* 429 (6989):292–294.

Moons, P., R. Van Houdt, A. Aertsen, K. Vanoirbeek, Y. Engelborghs, and C. W. Michiels. 2006. Role of quorum sensing and antimicrobial component production by *Serratia plymuthica* in formation of biofilms, including mixed biofilms with *Escherichia coli*. *Appl. Environ. Microbiol.* 72 (11):7294–7300.

Mukherjee, S., and B. L. Bassler. 2019. Bacterial quorum sensing in complex and dynamically changing environments. *Nat. Rev. Microbiol.* 187 (2):554–566.

Mund, A., S. P. Diggle, and F. Harrison. 2017. The fitness of *Pseudomonas aeruginosa* quorum sensing signal cheats is influenced by the diffusivity of the environment. *mBio* 8 (3):e00353–17.

Munn, C. B. 2015. The role of vibrios in diseases of corals. *Microbiol. Spectr.* 3 (4). doi:10.1128/microbiolspec.VE-0006-2014.

Nahm, C. H., D. C. Choi, H. Kwon et al. 2017. Application of quorum quenching bacteria entrapping sheets to enhance biofouling control in a membrane bioreactor with a hollow fiber module. *J. Membr. Sci.* 526:264–271.

Nasuno, E., N. Kimura, M. J. Fujita, C. H. Nakatsu, Y. Kamagata, and S. Hanada. 2012. Phylogenetically novel LuxI/LuxR-type quorum sensing systems isolated using a metagenomic approach. *Appl. Environ. Microbiol.* 78 (22):8067.

Ng, F. S. W., D. M. Wright, and S. Y. K. Seah. 2011. Characterization of a phosphotriesterase-like lactonase from *Sulfolobus solfataricus* and its immobilization for disruption of quorum sensing. *Appl. Environ. Microbiol.* 77 (4):1181–1186.

Norton, J. M., M. G. Klotz, L. Y. Stein et al. 2008. Complete genome sequence of *Nitrosospira multiformis*, an ammonia-oxidizing bacterium from the soil environment. *Appl. Environ. Microbiol.* 74 (11):3559–3572.

Oh, H.-S., and C.-H. Lee. 2018. Origin and evolution of quorum quenching technology for biofouling control in MBRs for wastewater treatment. *J. Membr. Sci.* 554:331–345.

Oh, H. S., K. M. Yeon, C. S. Yang et al. 2012. Control of membrane biofouling in MBR for wastewater treatment by quorum quenching bacteria encapsulated in microporous membrane. *Environ. Sci. Technol.* 46 (9):4877–4884.

Orchard, E. D., J. W. Ammerman, M. W. Lomas, and S. T. Dyhrman. 2010. Dissolved inorganic and organic phosphorus uptake in *Trichodesmium* and the microbial community: The importance of phosphorus ester in the Sargasso Sea. *Limnol. Oceanogr.* 55 (3):1390–1399.

Park, J.-H., J.-H. Lee, M. H. Cho, M. Herzberg, and J. Lee. 2012. Acceleration of protease effect on *Staphylococcus aureus* biofilm dispersal. *FEMS Microbiol. Lett.* 335 (1):31–38.

Ponnusamy, K., D. Paul, Y. S. Kim, and J. H. Kweon. 2010. 2(5H)-Furanone: A prospective strategy for biofouling-control in membrane biofilm bacteria by quorum sensing inhibition. *Braz. J. Microbiol.* 41 (1):227–234.

Popat, R., S. A. Crusz, M. Messina, P. Williams, S. A. West, and S. P. Diggle. 2012. Quorum-sensing and cheating in bacterial biofilms. *Proc. R. Soc. B: Biol. Sci.* 279 (1748):4765–4771.

Qazi, S. N. A., E. Counil, J. Morrissey et al. 2001. *agr* expression precedes escape of internalized *Staphylococcus aureus* from the host endosome. *Infect. Immun.* 69 (11):7074.

Røder, H. L., P. K. Raghupathi, J. Herschend et al. 2015. Interspecies interactions result in enhanced biofilm formation by co-cultures of bacteria isolated from a food processing environment. *Food Microbiol.* 51:18–24.

Rabaey, K., and W. Verstraete. 2005. Microbial fuel cells: Novel biotechnology for energy generation. *Trends Biotechnol.* 23 (6):291–298.

Rendueles, O., and J.-M. Ghigo. 2012. Multi-species biofilms: How to avoid unfriendly neighbors. *FEMS Microbiol. Rev.* 36 (5):972–989.

Rickard, A. H., R. J. Palmer Jr, D. S. Blehert et al. 2006. Autoinducer 2: A concentration-dependent signal for mutualistic bacterial biofilm growth. *Mol. Microbiol.* 60 (6):1446–1456.

Riedel, K., M. Hentzer, O. Geisenberger et al. 2001. *N*-acylhomoserine-lactone-mediated communication between *Pseudomonas aeruginosa* and *Burkholderia cepacia* in mixed biofilms. *Microbiology* 147 (12):3249–3262.

Rutherford, S. T., and B. L. Bassler. 2012. Bacterial quorum sensing: Its role in virulence and possibilities for its control. *Cold Spring Harb. Perspect. Med.* 2 (11):a012427.

Sañudo-Wilhelmy, S. A., A. B. Kustka, C. J. Gobler et al. 2001. Phosphorus limitation of nitrogen fixation by *Trichodesmium* in the central Atlantic Ocean. *Nature* 411 (6833):66–69.

Senadheera, D., and D. G. Cvitkovitch. 2008. Quorum sensing and biofilm formation by *Streptococcus mutans*. In *Bacterial Signal Transduction: Networks and Drug Targets*, ed. Ryutaro Utsumi, 178–188. New York: Springer.

Sharp, K. H., and K. B. Ritchie. 2012. Multi-partner interactions in corals in the face of climate change. *Biol. Bull.* 223 (1):66–77.

Sibley, C. D., K. Duan, C. Fischer et al. 2008. Discerning the complexity of community interactions using a *Drosophila* model of polymicrobial infections. *PLoS Path* 4 (10):e1000184.

Siddiqui, M. F., M. Sakinah, L. Singh, and A. W. Zularisam. 2012. Targeting *N*-acyl-homoserine-lactones to mitigate membrane biofouling based on quorum sensing using a biofouling reducer. *J. Biotechnol.* 161 (3):190–197.

Skindersoe, M. E., P. Ettinger-Epstein, T. B. Rasmussen, T. Bjarnsholt, R. de Nys, and M. Givskov. 2008. Quorum sensing antagonism from marine organisms. *Mar. Biotechnol.* 10 (1):56–63.

Smalley, N. E., D. An, M. R. Parsek, J. R. Chandler, and A. A. Dandekar. 2015. Quorum sensing protects *Pseudomonas aeruginosa* against cheating by other species in a laboratory coculture model. *J. Bacteriol.* 197 (19):3154–3159.

Smith, A. C., A. Rice, B. Sutton et al. 2017. Albumin inhibits *Pseudomonas aeruginosa* quorum sensing and alters polymicrobial interactions. *Infect. Immun.* 85 (9):e00116–17.

Smith, R. S., and B. H. Iglewski. 2003. *P. aeruginosa* quorum-sensing systems and virulence. *Curr. Opin. Microbiol.* 6 (1):56–60.

Solano, C., M. Echeverz, and I. Lasa. 2014. Biofilm dispersion and quorum sensing. *Curr. Opin. Microbiol.* 18:96–104.

Starkenburg, S. R., P. S. Chain, L. A. Sayavedra-Soto et al. 2006. Genome sequence of the chemolithoautotrophic nitrite-oxidizing bacterium *Nitrobacter winogradskyi* Nb-255. *Appl. Environ. Microbiol.* 72 (3):2050–2063.

Stephenson, D., and T. Stephenson. 1992. Bioaugmentation for enhancing biological wastewater treatment. *Biotechnol. Adv.* 10 (4):549–559.

Sully, E. K., N. Malachowa, B. O. Elmore et al. 2014. Selective chemical inhibition of agr quorum sensing in *Staphylococcus aureus* promotes host defense with minimal impact on resistance. *PLoS Path.* 10 (6):e1004174.

Sun, Y., Y. Guan, D. Zeng, K. He, and G. Wu. 2018. Metagenomics-based interpretation of AHLs-mediated quorum sensing in *Anammox* biofilm reactors for low-strength wastewater treatment. *Chem. Eng. J.* 344:42–52.

Tait, K., Z. Hutchison, F. L. Thompson, and C. B. Munn. 2010. Quorum sensing signal production and inhibition by coral-associated vibrios. *Environ. Microbiol. Rep.* 2 (1):145–150.

Tait, K., I. Joint, M. Daykin, D. L. Milton, P. Williams, and M. Cámara. 2005. Disruption of quorum sensing in seawater abolishes attraction of zoospores of the green alga Ulva to bacterial biofilms. *Environ. Microbiol.* 7 (2):229–240.

Talagrand-Reboul, E., E. Jumas-Bilak, and B. Lamy. 2017. The social life of aeromonas through biofilm and quorum sensing systems. *Front. Microbiol.* 8:37-37.

Tan, C. H., K. S. Koh, C. Xie et al. 2014. The role of quorum sensing signalling in EPS production and the assembly of a sludge community into aerobic granules. *ISME J.* 8 (6):1186–1197.

Tan, C. H., K. W. K. Lee, M. Burmølle, S. Kjelleberg, and S. A. Rice. 2017. All together now: Experimental multispecies biofilm model systems. *Environ. Microbiol.* 19 (1):42–53.

Tang, X., Y. Guo, B. Jiang, and S. Liu. 2018a. Metagenomic approaches to understanding bacterial communication during the anammox reactor start-up. *Water Res.* 136:95–103.

Tang, X., Y. Guo, S. Wu, L. Chen, H. Tao, and S. Liu. 2018b. Metabolomics uncovers the regulatory pathway of acyl-homoserine lactones based quorum sensing in *Anammox* consortia. *Environ. Sci. Technol.* 52 (4):2206–2216.

Tang, X., Y. Guo, T. Zhu, H. Tao, and S. Liu. 2019. Identification of quorum sensing signal AHLs synthases in *Candidatus* Jettenia caeni and their roles in *Anammox* activity. *Chemosphere* 225:608–617.

Tang, X., S. Liu, Z. Zhang, and G. Zhuang. 2015. Identification of the release and effects of AHLs in *Anammox* culture for bacteria communication. *Chem. Eng. J.* 273:184–191.

Teng, Y., Y. Luo, M. Sun, Z. Liu, Z. Li, and P. Christie. 2010. Effect of bioaugmentation by *Paracoccus* sp. strain HPD-2 on the soil microbial community and removal of polycyclic aromatic hydrocarbons from an aged contaminated soil. *Bioresour. Technol.* 101 (10):3437–3443.

Van Delden, C., and B. H. Iglewski. 1998. Cell-to-cell signaling and *Pseudomonas aeruginosa* infections. *Emerg. Infect. Dis.* 4 (4):551.

Van Mooy, B. A. S., L. R. Hmelo, L. E. Sofen et al. 2011. Quorum sensing control of phosphorus acquisition in *Trichodesmium* consortia. *ISME J.* 6:422.

Venkataraman, A., M. Rosenbaum, J. B. A. Arends, R. Halitschke, and L. T. Angenent. 2010. Quorum sensing regulates electric current generation of *Pseudomonas aeruginosa* PA14 in bioelectrochemical systems. *Electrochem. Commun.* 12 (3):459–462.

Vinoj, G., B. Vaseeharan, S. Thomas, A. J. Spiers, and S. Shanthi. 2014. Quorum-quenching activity of the AHL-lactonase from *Bacillus licheniformis* DAHB1 inhibits *Vibrio* biofilm formation in vitro and reduces shrimp intestinal colonisation and mortality. *Mar. Biotechnol. (NY)* 16 (6):707–715.

Waheed, H., S. Pervez, I. Hashmi, S. J. Khan, and S. R. Kim. 2018. High-performing antifouling bacterial consortium for submerged membrane bioreactor treating synthetic wastewater. *Int. J. Environ. Sci. Technol.* 15 (2):395–404.

Wang, J. H., H. Z. He, M. Z. Wang et al. 2013a. Bioaugmentation of activated sludge with *Acinetobacter* sp. TW enhances nicotine degradation in a synthetic tobacco wastewater treatment system. *Bioresour. Technol.* 142:445–453.

Wang, M., A. L. Schaefer, A. A. Dandekar, and E. P. Greenberg. 2015. Quorum sensing and policing of *Pseudomonas aeruginosa* social cheaters. *Proc. Natl. Acad. Sci.* 112 (7):2187–2191.

Wang, M., J. Xu, J. Wang et al. 2013b. Differences between 4-fluoroaniline degradation and autoinducer release by *Acinetobacter* sp TW: Implications for operating conditions in bacterial bioaugmentation. *Environ. Sci. Pollut. R.* 20 (9):6201–6209.

Wang, M. Z., H. Z. He, X. Zheng, H. J. Feng, Z. M. Lv, and D. S. Shen. 2014a. Effect of *Pseudomonas* sp. HF-1 inoculum on construction of a bioaugmented system for tobacco wastewater treatment: Analysis from quorum sensing. *Environ. Sci. Pollut. Res. Int.* 21 (13):7945–7955.

Wang, M. Z., G. Q. Yang, X. Wang, Y. L. Yao, H. Min, and Z. M. Lu. 2011. Nicotine degradation by two novel bacterial isolates of *Acinetobacter* sp. TW and *Sphingomonas* sp. TY and their responses in the presence of neonicotinoid insecticides. *World J. Microbiol. Biotechnol.* 27 (7):1633–1640.

Wang, M. Z., X. Zheng, H. Z. He, D. S. Shen, and H. J. Feng. 2012. Ecological roles and release patterns of acylated homoserine lactones in *Pseudomonas* sp. HF-1 and their implications in bacterial bioaugmentation. *Bioresour. Technol.* 125:119–126.

Wang, M. Z., X. Zheng, K. Zhang et al. 2014b. A new method for rapid construction of a *Pseudomonas* sp. HF-1 bioaugmented system: Accelerating acylated homoserine lactones secretion by pH regulation. *Bioresour. Technol.* 169:229–235.

Wang, V. B., S. L. Chua, B. Cao et al. 2013c. Engineering PQS biosynthesis pathway for enhancement of bioelectricity production in *Pseudomonas aeruginosa* microbial fuel cells. *PLoS One* 8 (5):e63129.

Wang, X., X. Li, and J. Ling. 2017. *Streptococcus gordonii* LuxS/autoinducer-2 quorum-sensing system modulates the dual-species biofilm formation with *Streptococcus mutans*. *J. Basic Microbiol.* 57 (7):605–616.

Waters, C. M., W. Lu, J. D. Rabinowitz, and B. L. Bassler. 2008. Quorum sensing controls biofilm formation in *Vibrio cholerae* through modulation of cyclic di-GMP levels and repression of *vpsT*. *J. Bacteriol.* 190 (7):2527.

Weerasekara, N. A., K. H. Choo, and C. H. Lee. 2014. Hybridization of physical cleaning and quorum quenching to minimize membrane biofouling and energy consumption in a membrane bioreactor. *Water Res.* 67:1–10.

Whiteley, M., S. P. Diggle, and E. P. Greenberg. 2017. Progress in and promise of bacterial quorum sensing research. *Nature* 551:313.

Williams, P., and M. Cámara. 2009. Quorum sensing and environmental adaptation in *Pseudomonas aeruginosa*: A tale of regulatory networks and multifunctional signal molecules. *Curr. Opin. Microbiol.* 12 (2):182–191.

Woodford, N., and M. J. Ellington. 2007. The emergence of antibiotic resistance by mutation. *Clin. Microbiol. Infect.* 13 (1):5–18.

Xiao, Y., H. Waheed, K. Xiao, I. Hashmi, and Y. Zhou. 2018. In tandem effects of activated carbon and quorum quenching on fouling control and simultaneous removal of pharmaceutical compounds in membrane bioreactors. *Chem. Eng. J.* 341:610–617.

Yavuztürk Gül, B., and I. Koyuncu. 2017. Assessment of new environmental quorum quenching bacteria as a solution for membrane biofouling. *Proc. Biochem.* 61:137–146.

Yeon, K. M., W. S. Cheong, H. S. Oh et al. 2009a. Quorum sensing: A new biofouling control paradigm in a membrane bioreactor for advanced wastewater treatment. *Environ. Sci. Technol.* 43 (2):380–385.

Yeon, K. M., C. H. Lee, and J. Kim. 2009b. Magnetic enzyme carrier for effective biofouling control in the membrane bioreactor based on enzymatic quorum quenching. *Environ. Sci. Technol.* 43 (19):7403–7409.

Yong, Y. C., Y. Y. Yu, C. M. Li, J. J. Zhong, and H. Song. 2011. Bioelectricity enhancement via overexpression of quorum sensing system in *Pseudomonas aeruginosa*-inoculated microbial fuel cells. *Biosens. Bioelectron.* 30 (1):87–92.

Zhang, G., F. Zhang, G. Ding et al. 2012. Acyl homoserine lactone-based quorum sensing in a methanogenic archaeon. *ISME J.* 6:1336.

Zhang, J., H. C. Chua, J. Zhou, and A. G. Fane. 2006. Factors affecting the membrane performance in submerged membrane bioreactors. *J. Membr. Sci.* 284 (1–2):54–66.

Zhou, L., H. Yu, G. Ai, B. Zhang, S. Hu, and X. Dong. 2015. Transcriptomic and physiological insights into the robustness of long filamentous cells of *Methanosaeta harundinacea* prevalent in upflow anaerobic sludge blanket granules. *Appl. Environ. Microbiol.* 81 (3):831.

Zhou, M., H. Wang, D. J. Hassett, and T. Gu. 2013. Recent advances in microbial fuel cells (MFCs) and microbial electrolysis cells (MECs) for wastewater treatment, bioenergy and bioproducts. *J. Chem. Technol. Biotechnol.* 88 (4):508–518.

Zimmer, B. L., A. L. May, C. D. Bhedi et al. 2014. Quorum sensing signal production and microbial interactions in a polymicrobial disease of corals and the coral surface mucopolysaccharide layer. *PLoS One* 9 (9):e108541.

13

Breaking Bad: Understanding How Bacterial Communication Regulates Biofilm-Related Oral Diseases

Andrea Muras and Ana Otero

CONTENTS

13.1 Introduction: Quorum Sensing and Biofilm Formation

Biofilms are complex microbial communities attached to a surface or to each other, embedded in a matrix of self-produced extracellular polymers and exhibiting an altered phenotype with respect to growth rate and gene transcription (Donlan and Costerton 2002). Biofilms are involved in many different dynamic processes, exhibiting a high number of metabolic activities and taking part in the production and degradation of organic matter, the remediation of environmental contaminants, and cycling of different nutrients such as nitrogen and sulfur and many metals (Gupta et al. 2016). Besides their environmental relevance, biofilms are crucial in disease development, since it is estimated that up to 80% of all microbial infections are biofilm-based; biofilm infections on medical devices are the most important health hurdles nowadays. Device-related biofilm infections can be developed on or within multiple surfaces such as contact lenses, mechanical heart valves, and central or urinary catheters, but other, non-device related biofilm infections are responsible for common diseases such as caries, periodontitis, or osteomyelitis (Jamal et al. 2018). Nowadays, biofilm formation is a major problem in human health, increasing morbidity and mortality.

The biofilm-mode of living usually has a key role in virulence and promotes resistance to external aggressions, including antimicrobial substances, and also acts as a reservoir in persisting infections, which can be fatal, such as endocarditis or cystic fibrosis. Additionally, our own body, trying to fight against the biofilm, produces an overreaction of the body tissues through the secretion of inflammatory cytokines causing wounds at the surrounding tissues (Wilson 2001). Bacteria with a biofilm-mode of living present an altered growth rate and gene transcription compared to planktonic cells (Donlan and Costerton 2002), with a large group of genes being differentially regulated. Some of the altered phenotypes found in biofilm-associated bacteria include antibiotic resistance, reduced growth rate, secretion of different molecules, and virulence factors (Hall-Stoodley and Stoodley 2005). Additionally, biofilms protect the bacteria from UV radiation, pH stress, chemical exposure, phagocytosis, dehydration, and antibiotics (Gupta et al. 2016). In this sense, biofilms are thought to resist antibacterial substances by the physical barrier made up of the matrix of polysaccharides, protein, lipids, and nucleic acids. Moreover, bacterial biofilms facilitate gene horizontal transfer among bacteria (Lewis 2001), which can increase the appearance of antibiotic resistance.

Since biofilms are formed in very different environmental conditions, the transition from a planktonic lifestyle to the attachment on a surface can be triggered by nutrient content, pH, iron, and oxygen, and these external signals are different among microorganisms (O'Toole, Kaplan, and Kolter 2000). Since many sensing systems can integrate the environmental stimuli into signaling pathways, multiple routes are involved in the bacterial adhesion and biofilm formation. In these complex, evolved microbial communities, the physiological and chemical interactions, both mutualistic and antagonistic, follow similar strategies to those of higher forms of life (O'Toole, Kaplan, and Kolter 2000). In this sense, the bacterial communication processes are crucial for the microbe-microbe interactions required for multicellular behavior. In numerous bacterial species, biofilm development is regulated by a chemical communication system, referred to as **quorum sensing (QS)**, which can control the attachment, maturation, dispersion, or cellular aggregation of related genes

(Grandclément et al. 2016). QS systems are thought to enable bacteria to assess population size and initiate social behaviors in a coordinated action (Fuqua, Winans, and Greenberg 1994). Numerous important pathogens such as the Gram-negatives *Pseudomonas aeruginosa* and *Acinetobacter baumannii* and the Gram-positives *Staphylococcus aureus* and *Streptococcus mutans* regulate different stages of biofilm development using QS systems. The bacterial biofilm microenvironment is usually sensed by the majority of the bacteria inside the biofilm through QS systems that regulate the biofilm-mode lifestyle. QS may be involved in different stages of biofilm development, including (1) attachment, (2) proliferation and maturation, and (3) dispersion (Joo and Otto 2012). In some cases, QS signals induce changes in the production of exopolysaccharides, lipids, nucleic acids, and proteases that are needed to build the biofilm matrix. In fact, in some pathogenic bacteria, QS signals are produced only in the conditions that allow cell attachment (Mayer et al. 2018). In other cases, such as in *Pseudomonas aeruginosa*, *Staphylococcus aureus*, and *Vibrio cholera*, QS activates the dispersion mechanisms inducing surfactant production and cell death (Solano, Echeverz, and Lasa 2014). QS signals, as indicators of cell population density, are integrated by bacteria with other environmental signals (temperature, pH, osmolarity, oxidative stress, nutrient deprivation) to improve their adaptation to ecosystem stress in order to optimize their survival probabilities (Williams et al. 2007). The high cell density achieved within the biofilm matrix is an optimal environment for the occurrence of QS processes. In fact, the acyl-homoserine-lactone (AHLs) QS signals produced by Gram-negative bacteria can reach 1000-fold higher concentrations inside biofilms than in open environments (Flemming et al. 2016). Although each bacterium can exhibit a variety of QS systems, producing different types of signaling molecules, these cell-to-cell communication systems are highly specific and accurate. One limitation regarding the study of the role of QS processes in biofilm formation is that in most studies they are usually performed using monospecific biofilms. Therefore, the knowledge about the possible synergies and interactions between the different microorganisms affecting the production of QS molecules is limited. QS signals can also be sensed by bacteria that do not produce the signals. For example, the addition of AHL at 1 μM promotes the biofilm formation in the pathogenic bacteria *Escherichia coli*, and *Salmonella enterica* serovar Typhimurium. These bacteria do not produce AHL but have AHL sensors, increasing not only the bacterial adhesion but also the biomass and the production of exopolymeric substances in response to the presence of QS signals produced by other bacteria (Bai and Rai 2016). Therefore, QS constitutes a sophisticated three-dimensional network of multi-level and large-scale signaling systems in which the different implicated molecules can affect various signaling pathways also influencing each other (Jayaraman and Wood 2008).

13.2 Oral Biofilm Formation

Dental plaque is a complex biofilm of a high clinical relevance resulting from the accumulation and interaction of oral microorganisms attached to a tooth surface. Although the oral cavity is an extremely diverse environment, dental plaque is a very well-studied polymicrobial biofilm that presents a high taxonomic diversity, with 500–1000 bacterial species being identified so far (Aas et al. 2005), and whose complexity is beginning to be understood thanks to the application of metagenomic techniques. Dental plaque formation consists of a series of physical, chemical, and biological processes. Initially, saliva provides a wide range of complex molecules such as active proteins and glycoproteins that are adsorbed by the tooth surface forming the acquired pellicle. During the first stage of dental plaque development, the planktonic bacteria attach to this acquired pellicle by physical forces or by bacterial appendages connecting the tooth and the early colonizers (Marsh et al. 2016). This first layer of biofilm, containing the initial and early colonizers such as *Streptococcus*, *Veillonela*, or *Actinomyces*, adheres to receptors of the acquired pellicle through surface molecules (adhesins). In a second phase, other microorganisms attach to the first colonizers, such as the intermediate colonizer *Fusobacterium nucleatum*, followed by secondary colonizers, like *Lactobacillus* spp., using a specific cell-to-cell recognition mechanism between distinct genetic bacteria known as coaggregation. Primary colonizers can coaggregate with each other but usually not with secondary colonizers, which normally coaggregate with *F. nucleatum*, but not with each other. Because *F. nucleatum* is able to coaggragate with almost all oral bacteria, it is considered a key component of dental plaque (Kolenbrander et al. 2002, 2006). During maturation, the attached microorganisms become irreversibly adhered and begin to communicate and coordinate with each other, expressing the specific genes of biofilm growth mode. At the end of the maturation stage, the biofilms usually present a multi-layer structure and their thickness can reach 100 μm. Finally, some biofilm cells return to planktonic life in order to spread and colonize new surfaces (Gupta et al. 2016).

Dental plaque is omnipresent and, in most cases, is composed of commensal bacteria, creating a harmless or even beneficial microbial community co-existing with the host in symbiosis. The ecological equilibrium of the microbial habitants of the dental plaque, which have co-evolved with each other and with their host, is maintained through competitive and cooperative interactions. Therefore, an imbalance of the resident microbiota and associated functions, referred as dysbiosis, resulting from changes in local environment conditions, is responsible for the most common oral human diseases: dental caries and periodontal disease (Marsh 1994, 2003; Belda-Ferre et al. 2012; Hajishengallis, Darveau, and Curtis 2012). Additionally, other pathologies such as Alzheimer's disease and oral cancer are thought to be related to oral diseases (Sampaio-Maia et al. 2016; Pritchard et al. 2017). In this sense, the entire community and its functional activities are responsible for the progression of these oral diseases that cannot be attributed to a single or a few oral pathogens (Duran-Pinedo and Frias-Lopez 2015; Yost et al. 2015). The interaction between environmental conditions and the oral microbiota is also crucial for the development of oral diseases. For example, the risk of suffering caries depends on the high and repetitive sugar intake, which causes a decrease in the oral pH derived from the acid production by the acidogenic oral bacteria. In periodontal disease, an increase of the production of crevicular fluid due to an overreaction of the host immune system induces the dysbiosis of the oral microbiota, allowing the predominance of the periodontopathogens.

13.3 Quorum Sensing and Oral Biofilms

Despite it is generally accepted that QS related genes are required for the successful establishment of biofilm in oral pathogens, knowledge of microbial interactions and signaling processes within dental plaque is still limited, and little is known about the influence of these processes on commensal microbiota and the establishment of dysbiosis. Among the three most studied QS signals, the role of the autoinducer peptides and AI-2 signals is generally accepted. On the contrary and despite increasing evidence being accumulated indicating a role of acyl-homoserine-lactones (AHLs) QS signals in dental plaque formation, the current paradigm excludes a role of these Gram-negative QS signals in oral pathogenic processes. The study of QS processes in dental plaque is of paramount importance, since the interference with bacterial coaggregation and/or the reduction of the dental plaque is supposed to have beneficial effects on dental health. Therefore, the interference with the signals involved in these processes or the inhibition of functions related to disease initiation and progression could be potentially used in future oral diseases preventive strategies (Simón-Soro and Mira 2015; Fteita et al., 2017). Moreover, some authors suggest the possibility to identify and quantify AHL molecules produced by pathogens as biomarkers and to monitor the development of bacterial diseases (Kumari et al. 2006).

13.3.1 Autoinducer Peptides (AIPs)

The QS molecules used by Gram-positives are **autoinducer peptides (AIPs)**, which consist of post-translationally modified peptides that range from 5 to 34 amino acids in length. Depending on their structure, three different AIP families have been identified: (1) oligopeptide lantibiotics (Van der Meer et al. 1993; Quadri 2002) with exceptionally potent bactericidal activities; (2) the 16-membered thiolactone peptides (Chan, Coyle, and Williams 2004), which are the most extensively studied; and (3) isoprenylated tryptophan peptides (Ansaldi et al. 2002), which are the principal chemical architecture used by Gram-positives to mediate QS.

Gram-positive oral streptococci pathogens respond to AIPs through a two-component signal transduction system, in which the signal molecule binds to a membrane-bound histidine kinase sensor that leads to phosphorylation of response regulator proteins regulating gene expression. AIPs are usually highly species-specific or even strain-specific, but could also be sensed or induced by other bacteria. Although the competence-stimulating peptide (CSP) was identified in the oral streptococci *S. mutans*, *S. gordonii*, and *S. intermedius* (Table 13.1), the majority of the studies focus on the role of this QS system in *S. mutans* (Table 13.1). CSP is detected in the external environment by the two-component signal system ComCDE and is involved in the control of biofilm formation, bacteriocin synthesis, stress resistance, and autolysis (Petersen, Pecharki, and Scheie 2004; Senadheera and Cvitkovitch 2008; Perry, Cvitkovitch, and Levesque 2009; Senadheera et al. 2009).

TABLE 13.1

Phenotypes Regulated by the QS Signals CSP and XIP in Oral Gram-Positive Pathogens and the Method Used for the Study. The Effect of the Method Used on the Phenotype Was Indicated as an Increase (+) or a Decrease (–) in Expression and/or Production.

Pathogen	Signal	Genes Involved	Method	Phenotype	References
S. intermedius	CSP	Operon comCDE	Adding synthetic CSP	Biofilm formation (+) and genetic competence (+)	Petersen, Pecharki, and Scheie (2004)
S. gordonii	CSP	Operon comCDE	Mutation by transposon insertion	Biofilm formation (–)	Loo, Corliss, and Ganesjkumar (2000)
S. mutans	CSP	Operon comCDE	Genomic analyses	CSP gene identification	Li et al. (2001b)
			Adding CSP	Competence (+)	Li et al. (2001b)
			Mutation by insertion-duplication	Competence (–)	Li et al. (2001b)
			Knockout mutants	Acid tolerance (–)	Li et al. (2001a)
			Restoring by CSP	Acid tolerance (+)	Li et al. (2001a)
			Knockout and deletion mutants	Biofilm formation (–)	Li et al. (2002a)
			Restoring by CSP	Biofilm formation (+)	Li et al. (2002a)
			Luciferase reporter gene expression	Mutacines (–)	Kreth et al. (2006)
			Mutation by double cross-over recombination	Mutacines (–)	Kreth et al. (2006)
			Gene expression in response to addition of CSP	Mutacines (+)	Kreth et al. (2006)
			Adding synthetic CSP	Biofilm formation (+)	Petersen, Pecharki, and Scheie (2004)
			Transcriptome CSP-regulated genes (microarray based)	Bacteriocin production (–)	Perry et al. (2009)
			Adding CSP	Release of DNAe and cell death (+)	Perry, Cvitkovitch, and Levesque (2009)
S. mutans	XIP	ComSR	Adding synthetic XIP and deletions mutants	Competence (±)	Mashburn-Warren, Morrison, and Federle (2010)

The addition of synthetic CSP 11325 at 0.2–100 nM to the culture media promotes biofilm formation on polystyrene surfaces by *S. intermedius* without affecting the growth rate (Petersen, Pecharki, and Scheie 2004). In *S. mutans* the role of CSP-based QS systems on biofilm formation was described using mutants of the different QS-related genes ComC (precursor of CSP), ComD (membrane-bound histidine kinase), ComE (cognate response regulator), and ComX (now called sigX-inducing peptide, XIP). Mutants showed structural changes in the biofilm or reduced biomass in mucin-coated polystyrene surfaces (Li et al. 2002; Senadheera and Cvitkovitch 2008). Another QS system referred to as sigX-inducing peptide (XIP), previously termed ComX, was identified in all sequenced members of genus *Streptococcus*. The pre-peptide ComS is processed and secreted to produce XIP, which is detected intracellularly by an Rgg-type intracellular transcriptional regulator (ComR), and it is related to competence development (Mashburn-Warren, Morrison, and Federle 2010). The *sigX* gene is stimulated directly by XIP or indirectly by CSP (Reck, Tomash, and Wagner-Döbler 2015). It has been suggested that CSP in *S. mutans* should be renamed mutacin inducing peptide (MIP), since Reck, Tomash, and Wagner-Döbler (2015) have shown that in *S. mutans* this signal only controls mutacin transcription. Additionally, three different regulatory pathways have been described to control *sigX* in different groups of streptococci (Mashburn-Warren, Morrison, and Federle 2010). Interestingly, *Aggregatibacter actinomycetemcomitans* is able to induce the complete QS regulon of *S. mutans* in co-culture, although the mechanism is still unknown (Szafranski et al. 2017), indicating a complex network of intercellular signaling.

13.3.2 Autoinducer-2 (AI-2)

Both Gram-negative and Gram-positive bacteria are able to synthetize the interconvertible group of signaling molecules named **autoinducer-2 (AI-2)**. The AI-2 synthase, encoded by *luxS* gene, presents a wide distribution among prokaryotes (Schauder et al. 2001; Federle and Bassler 2003). These QS molecules were suggested to be the basis of a common language used for interspecies communication. Although a key role of AI-2 in biofilm formation has been proposed in several studies, the effects observed when the gene *luxS* is mutated have to be interpreted carefully since this gene also plays a crucial role in methionine metabolism, recycling of homocysteine, or accumulation of intermediates of SAM metabolism (Pereira, Thompson, and Xavier 2013) and participates in protein, RNA, and DNA synthesis (Sztajer et al. 2008).

Previous studies have identified *LuxS* among many species of oral bacteria, including *A. actinomycetemcomitans*, *Streptococcus mutans*, *S. gordonii*, *S. oralis*, *Porphyromonas gingivalis*, and other oral pathogens (Fong et al. 2001; Burgess et al. 2002; Merritt et al. 2003; McNab et al. 2003; Rickard et al. 2006). The production of this QS signal was also detected in different oral pathogenic bacteria (Table 13.2), including both collection strains and strains isolated from patients with periodontitis (Whittaker, Klier and Kolenbrander 1996; Burgess et al. 2002; Frias, Olle, and Alsina 2001; Kolenbrander et al. 2006). In this sense, several reports point to an important role of AI-2 signal in dental plaque formation. The external addition of partially purified AI-2 from *F. nucleatum* affected biofilm formation in monospecific cultures and multispecies mixtures including the pathogens *P. gingivalis*, *Treponema denticola* and *Tannerella forsythia* (Jang et al. 2013). Additionally, some bacteria need to detect the QS signals produced by others to form mixed biofilms; for instance, the mutation of *luxS* in *S. gordonii* showed that AI-2 is needed in order to form mixed-biofilms with *P. gingivalis* (McNab et al. 2003). The presence of AI-2 is essential for mutualistic biofilm growth in co-cultures of *S. oralis* and *Actinomyces naeslundi* on saliva as a sole source of nutrients (Rickard et al. 2006), both of which do not grow well in monoculture. The quantity of AI-2 needed for this mutualistic growth (0.8 nM–8 nM) (Rickard et al. 2006) was much lower than the AI-2 concentration measured in whole saliva (200 nM–1000 nM) (Campagna, Gooding, and May 2009). This could indicate that the AI-2 concentration is low at early stages of biofilm development and increases with maturation of the dental plaque. Furthermore, commensal species produce and respond to AI-2 in lower concentrations than periodontopathogens (Kolenbrander et al. 2006).

13.3.3 Acyl-Homoserine Lactones (AHLs)

Among the different QS signal molecules described so far (LaSarre and Federle 2013) the **N-acyl homoserine lactones (AHLs)** produced by Gram-negative bacteria are the best studied and characterized. AHLs are constituted by a homoserine lactone ring (HSL) linked by an amide bond to a fatty acid (between 4 and 20 carbons). Although these diffusible molecules are typical from Gram-negative bacteria, the production of AHL by the Gram-positive bacterium *Exiguobacterium* MPO has been described (Biswa and Doble 2013). The most common AHL-based QS system comprises a LuxI-type signal generator and a LuxR-type intracellular receptor (Ng and Bassler 2009), but other families of synthases, named LuxM/AinS and HdtS have also been described, not sharing homology with the LuxI synthase family (Gilson, Kuo, and Dunlap 1995; Laue et al. 2000). Additionally, alternative synthases such as RpaI, PpyS, and DarABC, regarded as "dialect" synthases (Brameyer, Bode and Heermann 2015), were found to produce signaling molecules different from AHLs, which are also involved in QS processes due to their ability to activate the LuxR-type receptors. Moreover, some bacteria can interact with the autoinducers synthetized by other bacteria because they possess one or more LuxR homologs (LuxR orphan) but do not produce AHLs or have a recognizable LuxI autoinducer synthase (Patankar and Gonzalez 2009). Traditionally AHLs are considered non-relevant in the oral cavity due to the unsuccessful identification of these molecules in pure cultures of oral pathogenic bacteria (Whittaker, Klier, and Kolenbrander 1996; Burgess et al. 2002; Frias, Olle, and Alsina 2001; Kolenbrander et al. 2006). However, these studies present some limitations concerning the low number of strains tested and the biosensors used (Shao and Demuth 2010). Beyond the limitations of the use of biosensors for the detection of AHLs (Frias, Olle, and Alsina 2001; Burgess et al. 2002), it should be considered that production of AHLs may change depending on culture medium and conditions (Mayer et al. 2018; Muras et al. 2018b). Despite this generally accepted paradigm, evidence points to a role of AHLs in dental plaque formation. The detection of AHLs has been described in saliva from healthy and unhealthy individuals (Kumari et al. 2006; Kumari, Pasini, and Daunert 2008),

TABLE 13.2

Phenotypes Regulated by the QS signal AI-2 in Oral Pathogens and Method Used for the Study. The Effect of the Method Used on the Phenotype Was Indicated as an Increase (+) or a Decrease (–).

Pathogen	Method	Phenotype	References
A. actinomycetemcomitans	Biosensor and gene expression	Growth in iron-limiting conditions (+)	Fong, Gao, and Demuth (2003)
	Mutation and gene expression	Biofilm formation (–) and virulence (–)	Novak et al. (2010)
A. naeslundi	Mutations and supplementation	Biofilm formation in co-culture with *S. oralis* (–)	Rickard et al. (2006)
F. nucleatum	Biosensor	AI-2 activity	Frias, Olla, and Alsina (2001)
P. gingivalis	Mutations and biosensor	Hemin uptake (+), hemin-regulated outer membrane protein (–)	Chun et al. (2001)
	Mutations and biosensor	Protease and haemagglutinating activity (–),	Burgess et al. (2002)
	Adding AI-2	Biofilm formation in co-culture with *luxS*-null *S. gordonii* (+)	McNab et al. (2003)
P. intermedia	Biosensor	AI-2 activity	Frias, Olla, and Alsina (2001)
S. anginosus	Mutation	Biofilm formation (–)	Petersen et al. (2006)
	Mutation and supplementation	Antibiotic susceptibility (+)	Ahmed, Petersen, and Scheie (2007)
S. intermedius	Mutation	Biofilm formation (–) and hemolytic activity (–)	Ahmed, Petersen, and Scheie (2008a, 2008b)
	Mutation and supplementation	Biofilm formation (–) and antibiotic susceptibility (+)	Ahmed, Petersen, and Scheie (2009)
S. gordonii	Biosensor	AI-2 activity	Frias, Olla, and Alsina (2001)
	Identification and mutation of *luxS*	Biofilm formation in co-culture with *luxS*-null *P. gingivalis* (–)	McNab et al. (2003)
	Identification and mutation of *luxS*	Carbohydrate metabolisms (–)	McNab et al. (2003)
S. mutans	Biosensor	AI-2 activity	Frias, Olla, and Alsina (2001)
	Mutation and supplementation by co-culture	Biofilm formation (–) and glucosyltransferase genes gtfB and gtfC (+)	Yoshida et al. (2005)
S. oralis	Biosensor	AI-2 activity	Frias, Olla, and Alsina (2001)
	Mutations and supplementation	Biofilm formation in co-culture with *A. naeslundi* (–)	Rickard et al. (2006)
S. sanguinis	Biosensor	AI-2 activity	Frias, Olla, and Alsina (2001)

and several AHL-producing bacteria including *Enterobacter* sp., *Klebsiella pneumoniae*, *Pseudomonas putida*, *Citrobacter amalonaticus* (*Levinea amalonaticus*) L8A, and *Burkholderia* were isolated from the tongue and dental plaque (Chen et al. 2013; Goh et al. 2014, 2016; Yin 2012a, 2012b), although these species are not considered part of the resident microbiota of the mouth.

The addition of AHLs and AHL analogues affect growth and gene expression of oral pathogens. The addition of AHLs at 1 μM induced changes in lactic acid production and protease activity, two pathology-related phenotypes, in three different *in vitro* oral biofilm models without affecting bacterial growth significantly (Muras et al. 2020). The external addition of specific AHLs to *in vitro* oral biofilm models also generated a shift in the bacterial composition: the relative presence of the orange-complex bacteria *Peptostreptococcus* and *Prevotella* increased in a periodontal biofilm model when the QS signal C6-HSL was added (Muras et al. 2020). The addition of C14-HSL not only changed the protein expression but also slowed down the bacterial growth in *P. gingivalis* (Komiya-Ito et al. 2006). Moreover, the use of AHL analogues reduced the biofilm formation in this pathogen (Asahi et al. 2010). Despite the fact that no AHLs could be found in the supernatant of *P. gingivalis* cultures, an ORF with significant homology in amino acid sequence with the AHL synthase HdtS (25% identity and 48% amino acid similarity) was identified in *P. gingivalis* W83 (Komiya-Ito et al. 2006). Sequences similar to HdtS can be found in the genomes of *Actinomyces naeslundii* (43% cover and 33% identity), *A. actinomycetemcomitans* (32% cover and 33% identity), *Prevotella intermedia* (71% cover and 26% identity), and *Veillonella parvula* (25% cover and 48% identity) (Muras, unpublished results). Even though these pathogens would not produce AHL, we cannot exclude that they respond to the addition of AHLs or AHL analogues because of the presence of AHL receptor homologs which do not have an associated AHL synthase. In this sense, an AHL-receptor LuxR-type homologue was identified in *P. gingivalis* ATCC33277 (Wu, Lin and Xie 2009; Chawla et al. 2009). Although further investigation is needed, all these data suggest that this type of QS signal could play an important role in the development of oral biofilm, being implicated in the development of oral diseases.

13.4 Prevention of Oral Biofilm Formation by Silencing Quorum Sensing Systems

Microbial attachment to a surface is a universal phenomenon in nature and is essential for biofilm formation. Therefore, a promising strategy to control the biofilm formation is to prevent the microbial attachment and to reduce the development of the biofilm structure. Current intervention strategies attempt to prevent the initial microbial attachment through biofilm inhibitors or to facilitate the penetration of drugs into the biofilm matrix to kill the associated cells. The use of new biomaterials with different physical and chemical properties is also being explored (Ivanova, 2015). However, the future approaches to control bacterial biofilms will be based on inhibition of genes involved in biofilm formation (Donlan and Costerton 2002), in some cases through the disruption of the QS systems (Ivanova et al. 2015, Muras et al. 2018a).

The search for molecules with the ability to interfere with QS, a process generally known as quorum quenching (QQ), has been proposed as an alternative approach for controlling pathogenic bacterial behavior to treat or prevent infectious diseases, since many species of pathogenic bacteria control their virulence factors through QS systems (Romero et al. 2014, 2015; Mayer et al. 2015, 2018). Furthermore, since QQ strategies do not interfere directly with bacterial growth, the probabilities of inducing resistance or tolerance against these mechanisms are lower than for antibiotics (García-Contreras, Maeda, and Wood 2016; Guendouze et al. 2017) contributing to solving the problem of bacterial antibiotic resistance. In fact, it has been demonstrated that QQ strategies increase the susceptibility to antibiotics in biofilm-forming pathogens (Simonetti et al. 2016).

Since many oral bacteria coordinate their pathogenic traits and capacity to form biofilm though QS, the interception of QS signals in order to inhibit functions related to disease initiation and progression could be used in future preventive strategies for oral diseases (Simón-Soro and Mira 2015; Muras et al. 2018a). Previous studies trying to inactivate QS processes to prevent and treat oral diseases focused mainly in the interference with the general QS signal AI-2 and CSP, the QS signal typical of streptococci. Several studies have confirmed the QS signal AI-2 as a potential target for inhibiting biofilm formation in oral pathogenic bacteria (Cho et al. 2016; Jang et al. 2013; Kasper et al. 2014; Kolderman et al. 2015; Ryu et al. 2016; Muras et al. 2018a). Furanone, an inhibitor of both AHL and AI-2 mediated QS channels (Givskov et al. 1999; Zang et al. 2009), and D-ribose, an AI-2 receptor inhibitor, reduced the expression of adhesion molecules and the coagregation in co-cultures of *F. nucleatum* with *P. gingivalis*, *T. denticola*, or *T. forsythia* and in monocultures of *F. nucelatum* (Jang et al. 2013). These QS inhibitors also decreased bone destruction and decreased the number of bacteria, reducing the progression of periodontitis in comparison to the controls in a mice infection model with *P. gingivalis* (Cho et al. 2016). D-galactose was also used to reduce biofilm formation by *F. nucleatum*, *P. gingivalis*, and *T. forsythia*, blocking the AI-2 receptor (Ryu et al. 2016). Other molecules, such as the amino acid-L-arginine and S-Aryl-l-cysteine sulphoxides and related organosulfur compounds were also reported to decrease oral biofilm formation by interfering with the AI-2 QS system (Kasper et al. 2014; Kolderman et al. 2015). Some studies have reported the inhibition of biofilm formation by *S. mutans* through interference with its AI-2 QS system. The evaluation of the synergistic inhibitory effect of ribose and xylitol on streptococcal biofilm formation concluded that ribose can be used as a protective agent since the pentose would compete with AI-2 for the ABC transporter (Lee et al. 2015). Recently, the use of QQ against AI-2 to prevent biofilm formation by *S. mutans* was proposed as a promising strategy for the prevention of dental caries (Muras et al. 2018a).

QQ processes seem to be already present in dental plaque, since *S. gordonii*, *S. sanguinis*, *S. mitis*, *S. oralis*, *P. gingivalis*, and *T. denticola* were reported to be able to inactivate or degrade *S. mutans* CSP (Wang et al. 2011), and this ability inhibited bacteriocin production and biofilm formation in these caries-associated bacterium. The presence of QQ activity against AHL is also frequently found in cultivable bacteria isolated from dental plaque and saliva of a healthy and a periodontal donor. The frequency of QQ strains was significantly higher in the periodontal sample, indicating a possible role of these strategies in the establishment of a dysbiosis (Muras, unpublished results).

Despite the fact that the role of AHLs in oral biofilm formation and oral disease is not unequivocally demonstrated, several results point to the possible use of QQ strategies against AHL-mediated QS processes to fight bacterial diseases. This fact is especially interesting since several inhibitors and QQ enzymes that can efficiently interfere with AHL QS signals are available (Romero et al. 2015). The addition of an analogue of C12-HSL (10 μM) reduced biofilm formation in *P. gingivalis* by approximately 25% (Asahi et al. 2010). However, no structure-activity relationship could be found among the different AHL analogues with the ability to reduce the biofilm formation in this pathogen. In another study, the cariogenic potential of oral biofilms was reduced by the addition of a C12-HSL analogue by inducing changes in the biofilm bacterial composition, decreasing *Streptococcus* spp. (54%) while increasing *Actinobacillus* spp. (580%). However, QS inhibition as an explanation of these ecological changes was excluded by the authors because no competition between the analogue and the natural AHL was observed (Janus et al. 2016). Curiously, extracts from the plant *Achyranthes aspera* presenting QQ activity against AHLs showed anti-biofilm activity against a cariogenic *S. mutans* KMS07 isolate (Murugan et al. 2013).

Different types of enzymes capable of degradating AHLs have been described (reviewed by Romero et al. 2015). The genes that codified this type of QQ enzymes are classified in two main groups: lactonases and acylases, although other types of QQ enzymes have also been described. AHL lactonases hydrolyze the ester bond of the homoserine lactone ring (HSL) (Dong et al. 2001). On the contrary, the AHL acylases cleave the AHL amide bond generating the corresponding free fatty acid and a homoserine lactone ring (Leadbetter and Greenberg 2000). Experiments carried out in our laboratory with the AHL-lactonase Aii20J (Mayer et al. 2015), which was recently reported to reduce biofilm formation in the nosocomial pathogen *Acinetobacter baumannii* ATCC17978 (Mayer et al. 2018), demonstrated an important reduction of the oral biofilm formed in BHI or BHI supplemented with sucrose from a saliva sample obtained from a healthy donor (Unpublished results, Figure 13.1). The reduction of biofilm formation was measured using the xCELLigence® system (Muras et al. 2018a, Mira et al. 2019). Further analyses are required to elucidate the changes in microbial composition and biofilm structure generated by the QQ of AHL-mediated processes in oral biofilms. Altogether these results further support a role of AHL-type QS signals in dental plaque formation.

The use of QQ compounds in combination with antibiotic, antiseptic, and/or biofilm dispersing agents could be an interesting strategy to control and prevent bacterial diseases. In fact, the combination of analogues of AHLs with antibiotics caused a reduced viability of *P. gingivalis* biofilms, being more effective

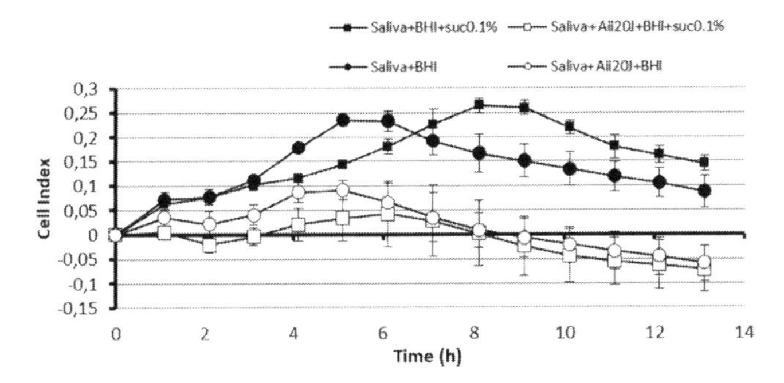

FIGURE 13.1 Effect of the AHL-lactonase Aii20J (20 µg/mL) on oral biofilm formation measured using the xCELLigence® system. The oral biofilm was generated with saliva from a healthy donor cultured with BHI and BHI supplemented with 0.1% sucrose at 37°C. Biofilm formation is expressed in Cell Index units.

than the single use of antibiotics (Asahi et al. 2011). These preliminary results obtained *in vitro* are extremely encouraging regarding the possibility of the use of QQ strategies for treating and preventing oral diseases. In fact, several dentistry products have been patented with the ability to inhibit growth and formation of oral biofilm such as a carbonate derivative with QSI activity registered by Colgate-Palmolive Company (Trivedi, Xu and Miksa 2009, Patent no US8617523B2). A garlic-based formulation with QSI activity, showing antibacterial and anti-inflammatory properties (Patent n US201201897) is currently distributed and in use in United Kingdom (Grandclément et al. 2016). Nevertheless, a deeper knowledge of the QS mechanisms that characterize the healthy dental plaque and the dysbiosis processes driving to oral disease establishment is required before QQ strategies can be applied as a preventive or curative approach for the treatment of these important pathologies.

13.5 Conclusions

Biofilms constitute a major problem in human health since they are crucial in disease development. Furthermore, the bacteria with a biofilm-mode of living present an altered phenotype with respect to growth rate and gene transcription compared to planktonic cells (Donlan and Costerton 2002), which contributes to its persistence and antibiotic resistance. It is generally accepted that QS-related genes are required for the successful establishment of biofilm in oral pathogens. The role of AI-2 and AIPs QS signals in oral pathogen biofilm formation has been confirmed for many important oral pathogens *in vitro*; however, the importance of the microbial interactions in mixed populations and the effect of signaling processes on commensal and pathogenic microbiota *in vivo* is largely unknown. The study of QS processes in dental plaque is of paramount importance, since its interference is supposed to have beneficial effects on dental health. The potential use of QQ strategies to prevent biofilm formation by oral pathogens is an attractive alternative since these approaches do not interfere with bacterial growth, and hence, the probabilities of inducing resistance or tolerance against these mechanisms are lower. Moreover, the potential role of AHLs in dental plaque formation and dysbiosis related to oral diseases would open the possibility of applying new QQ strategies to prevent and treat different bacterial oral diseases. Specifically, mono- and mixed-biofilm models are needed to improve our knowledge of dental plaque regulation and formation. However, further studies are required to confirm the role of the different QS molecules in dental plaque formation *in vivo* and the viability to apply QQ approaches to dental health practice.

ACKNOWLEDGMENTS

This work was supported by the grant "Axudas do Programa de Consolidación e Estructuración de Unidades de Investigación Competitivas (GPC)" from the Consellería de Cultura, Educación e Ordenación Universitaria, Xunta de Galicia (ED431B2017/53). A.M. was supported by a predoctoral fellowship from the Consellería de Cultura, Educación e Ordenación Universitaria, Xunta de Galicia (ED481A-2015/311).

REFERENCES

Aas, J.A., B.J. Paster, L.N. Stokes, I. Olsen, F.E Dewhirst. 2005. Defining the Normal Bacteria Flora of the Oral Cavity. *Journal of Clinical Microbiology* 43(11): 5721–5732.

Ahmed, N.A., F.C. Petersen, A.A. Scheie. 2007. AI-2 Quorum Sensing Affects Antibiotic Susceptibility in *Streptococcus anginosus*. *The Journal of Antimicrobial Chemotherapy* 60(1):49–53.

Ahmed, N.A., F.C. Petersen, A.A. Scheie. 2008a. Biofilm Formation and Autoinducer-2 Signaling in *Streptococcus intermedius*: Role of Thermal and pH Factors. *Oral Microbiology and Immunology* 23(6): 492–497.

Ahmed, N.A., F.C. Petersen, A.A. Scheie. 2008b. LuxS and Expression of Virulence Factors in *Streptococcus intermedius*. *Oral Microbiology and Immunology* 23(1): 79–83.

Ahmed, N.A., F.C. Petersen, A.A. Scheie. 2009. AI-2/LuxS is Involved in Increased Biofilm Formation by *Streptococcus intermedius* in the Presence of Antibiotics. *Antimicrobials Agents of Chemotheraphy* 53(10): 4258–4263.

Ansaldi, M., D. Marolt, T. Stebe, I. Mandic-Mulec, D. Dubnau. 2002. Specific Activation of the *Bacillus* Quorum-Sensing Systems by Isoprenylated Pheromone Variants. *Molecular Microbiology* 44(6): 1561–1573.

Asahi, Y., Noiri, Y, Igarashi, J, Asai, H, Suga, H, Ebisu, S. 2010. Effects of *N*-acyl Homoserine Lactone Analogues on *Porphyromonas gingivalis* Biofilm Formation. *Journal of Periodontal Research* 45(2): 255–261. doi:10.1111/j.1600-0765.2009.01228. x.

Asahi, Y., Y. Noiri, J. Igarashi, H. Suga, H. Azakami, S Ebisu. 2011. Synergistic Effects of Antibiotics and an *N*-acyl Homoserine Lactone Analog on *Porphyromonas gingivalis* Biofilms. *Journal Applied Microbiology* 112(2): 404–411. doi:10.1111/j.1365-2672.2011.05194. x.

Bai, J.A. and R.V. Rai. 2016. Effect of Small N Acyl Homoserine Lactone Quorum Sensing Signals on Biofilms of Food-borne Pathogens. *Journal of Food Science and Technology* 53(9): 3609–3614.

Belda-Ferre, P., L.D. Alcaraz, R. Cabrera-Rubio, H. Romero, A. Simón-Soro, M. Pignatelli, A. Mira. 2012. The Oral Metagenome in Health and Disease. *The ISME Journal* 6(1):46–56. doi:10.1038/ismej.2011.85.

Biswa, P. and M. Doble. 2013. Production of Acylated Homoserine Lactone by Gram-Positive Bacteria Isolated from Marine Water. *FEMS Microbiology Letters* 343(1): 34–41. doi: 10.1111/1574-6968.12123.

Brameyer, S., H.B. Bode and R. Heermann. 2015. Languages and Dialects: Bacterial Communication Beyond Homoserine Lactones. *Trends Microbiology* 23: 521–523.

Burgess, N.A., D.F. Kirke, P. Williams, K Winzer, K.R. Hardie, N.L. Meyers, J. Aduse-Opoku, A.M. Curtis, M. Cámara. 2002. LuxS-dependent Quorum Sensing in *Porphyromonas gingivalis* Modulates Protease and Haemagglutinin Activities but is not Essential for Virulence. *Microbiology* 148: 763–772.

Campagna, S.R., R.J. Gooding, A.L. May. 2009. Direct Quantification of the Quorum Sensing Signal, Autoinducer-2, in Clinically Relevant Samples by Liquid Chromatography-Tandem Mass Spectrometry. *Analytical Chemistry* 81(15): 6374–6381. doi:10.1021/ac900824j.

Chan, W.C., B.J. Coyle, P. Williams. 2004. Virulence Regulation and Quorum Sensing in Staphylococcal Indections: Competitive AgrC Antagonists as Quorum Sensing Inhibitors. *Journal of Medicinal Chemistry* 47(19): 4633–4641.

Chawla, A., T. Hirano, B.W. Bainbridge, D.R. Demuth, H. Xie, R.J. Lamont. 2011. Community signaling between *Stretococcus gordonii* and *Porphyromonas gingivalis* is controlled by the transcriptional regulator CdhR. *Molecular Microbiology* 78: 1510–1522.

Chen, J.W., S. Chin, K.K. Tee, W.F. Yin, Y.M. Choo, K.G. Chan. 2013. *N*-Acyl Homoserine Lactone-Producing *Pseudomonas putida* Strain T2-2 From Human Tongue Surface. *Sensors* 13(10): 13192–13203. doi:10.3390/s131013192.

Cho, Y.J., H.Y. Song, H. Ben Amara, B.K. Choi, R. Eunju, Y.A Cho, Y. Seol et al. 2016. In Vivo Inhibition of *Porphyromonas gingivalis* Growth and Prevention of Periodontitis with Quorum-Sensing Inhibitors. *Journal of Periodontology* 87(9): 1075–1082. doi:10.1902/jop.2016.160070.

Chun, W.O., P. Yoonsuk, R.J. Lamont, R. McNab, B. Barbieri, D.R. Demuth. 2001. Signaling System in *Porphyromonas gingivalis* Based on a LuxS Protein. *Journal of Bacteriology* 183(13): 3903–3909. doi:10.1128/JB.183.13.3903-3909.2001.

Dong, Y.H., L.H. Wang, J.L. Xu, H.B. Zhang, X.F. Zhang, H. Zhang. 2001. Quenching Quorum-Sensing-Dependent Bacterial Infection by an *N*-Acyl Homoserine Lactonase. *Nature* 411, 813–817.

Donlan, R.M. and J.W. Costerton. 2002. Biofilms: Survival Mechanisms of Clinically Relevant Microorganisms. *Clinical Microbiology Review* 15(2): 167–193.

Duran-Pinedo, A.E. and J. Frias-Lopez. 2015. Beyond Microbial Community Composition: Functional Activities of the Oral Microbiome in Health and Disease. *Microbes and Infection* 17(7): 505–516. doi:10.1016/j.micinf.2015.03.014.

Federle, M.J. and B.L. Bassler. 2003. Interspecies Communication in Bacteria. *The Journal of Clinical Investigation* 112(9): 1291–1299.

Flemming, H.C., J. Wingender, U. Szewzyk, P. Steinberg, S.A. Rice, S. Kjelleberg. 2016. Biofilms: An Emergent Form of Bacterial Life. *Nature Reviews Microbiology* 14(9): 563–575. doi: 10.1038/nrmicro.2016.94.

Fong, K.P., L. Gao, D.R. Demuth. 2003. luxS and arcB Control Aerobic Growth of *Actinobacillus actinomycetemcomitans* Under Iron Limitation. *Infection and Immunology* 71(1): 298–308.

Fong, K.P., W.O. Chung, R.J. Lamont, D.R. Demuth. 2001. Intra- and Interspecies Regulation of Gene Expression by *Actinobacillus actinomycetemcomitans* LuxS. *Infection and Immunity* 69(12): 7625–7634.

Frias, J., E. Olle, M. Alsina. 2001. Periodontal Pathogens Produce Quorum Sensing Signal Molecules. *Infection and Immunity* 69(2): 3431–3434.

Fteita, D., A.A. Musrati, E. Könönen, X. Ma, M. Gürsoy, M. Peurla, E. Söderling, H.O. Sintim, U.K. Gürsoy. 2017. Dipeptidyl Peptidase IV and Quorum Sensing Signaling in Biofilm-Related Virulence of *Prevotella aurantiaca*. *Anaerobe* 48: 152–159. doi:10.1016/j.anaerobe.2017.08.009.

Fuqua, W.C., S.C. Winans, E.P. Greenberg. 1994. Quorum Sensing in Bacteria: The LuxR-LuxI Family of Cell Density-Responsive Transcriptional Regulators. *Journal of Bacteriology* 176(2): 269–275.

García-Contreras, R., T. Maeda, T.K. Wood. 2016. Can Resistance Against Quorum-Sensing Interference Be Selected? *The ISME Journal* 10(1): 4–10. doi:10.1038/ismej.2015.84.

Gilson, L., A. Kuo, P.V. Dunlap. 1995. AinS and a New Family of Autoinducer Synthesis Proteins. *Journal of Bacteriology* 177(23): 6946–6951.

Givskov, M., R. de Nys, M. Manefield, L. Gram, R. Maximilien, L. Eberl, S. Molin, P.D. Steinberg, S. Kjelleberg. 1999. Eukaryotic Interference with Homoserine Lactone-mediated Prokaryotic Signalling. *Journal of Bacteriology* 178: 6618–6622.

Goh, S.Y., S.A. Khan, K.K. Tee, N.H.A Kasim, W.F. Yin, K.G. Chan. 2016. Quorum Sensing Activity of *Citrobacter amalonaticus* L8A, a Bacterium Isolated from Dental Plaque. *Scientific Reports* 6: 20702. doi:10.1038/srep20702.

Goh, S.Y., W.S. Tan, S.A. Khan, H.P. Chew, N.H.A. Kasim, W.F. Yin, K.G. Chan. 2014. Unusual Multiple Production of N-acylhomoserine Lactones a by *Burkholderia* sp. strain C10B Isolated from Dentine Caries. *Sensors* 14(5):8940–8949. doi:10.3390/s140508940.

Grandclément, C., M. Tannières, S. Moréra, Y. Dessaux, D. Faure. 2016. Quorum Quenching: Role in Nature and Applied Developments. *FEMS Microbiology Reviews* 40(1): 86–116. doi:10.1093/femsre/fuv038.

Guendouze, A., L. Plener, J. Bzdrenga, P. Jacquet, B. Rémy, M. Elias, J.P. Lavigne, D. Daudé, E. Chabrière. 2017. Effect of Quorum Quenching Lactonase in Clinical Isolates of *Pseudomonas aeruginosa* and Comparison with Quorum Sensing Inhibitors. *Frontiers in Microbiology* 8: 227. doi:10.3389/fmicb.2017.00227.

Gupta, P., S. Sarkar, B. Das, S. Bhattacharjee, P. Tribedi. 2016. Biofilm, pathogenesis and prevention—A Journey to Break the Wall: A Review. *Archives of Microbiology* 198(1): 1–15. doi:10.1007/s00203-015-1148-6.

Hajishengallis, G., R.P. Darveau, M.A. Curtis. 2012. The Keystone-Pathogen Hypothesis. *Nature Reviews in Microbiology* 10(10):717–725. doi:10.1038/nrmicro2873.

Hall-Stoodley, L., P. Stoodley. 2005. Biofilm Formation and Dispersal and the Transmission of Human Pathogens. *Trends in Microbiology* 13(1): 7–10.

Ivanova, K., M.M. Fernandes, A. Francesko, E. Mendoza, J. Guezquez, M. Burnet, T. Tzanov, 2015. Quorum-Quenching and Matrix-Degrading Enzymes in Multilayer Coatings Synergistically Prevent Bacterial Biofilm Formation on Urinary Catheters. *ACS Applied Materials and Interfaces* 7(49): 27066–27077. doi:10.1021/acsami.5b09489.

Jamal, M., W. Ahmad, S. Andleeb, F. Jalil, M. Imran, M.A. Nawaz, T. Hussain, M. Ali, M. Rafiq, M.A. Kamil. 2018. Bacterial Biofilm and Associated Infections. *Journal of the Chinese Medical Association* 81(1): 71–77. doi:10.1016/j.jcma.2017.07.012.

Jang, Y.J., Y.J. Choi, S.H. Lee, H.K. Jun, B.K. Choi. 2013. Autoinducer 2 of *Fusobacterium nucleatum* as a Target Molecule to Inhibit Biofilm Formation of Periodontopathogens. *Archives of Oral Biology* 58: 17–21. doi:10.1016/j.archoralbio.2013.08.006.

Janus, M.M., W. Crielaard, E. Zaura, B.J. Keijser, B.W. Brandt, B.P. Krom. 2016. A Novel Compound to Maintain a Healthy Oral Plaque Ecology in Vitro. *Journal of Oral Microbiology* 8: 32513. doi:10.3402/jom. v8.32513.

Jayaraman, A., T.K. Wood. 2008. Bacterial Quorum Sensing: Signals, Circuits, and Implications for Biofilms and Disease. *Annual Review of Biomedical Engineering* 10: 145–167. doi:10.1146/annurev.bioeng.10.061807.160536.

Joo, H.S., M. Otto. 2012. Molecular Basis of in Vivo Biofilm Formation by Bacterial Pathogens. *Chemistry and Biology* 19(12): 1503–1513. doi:10.1016/j.chembiol.2012.10.022.

Kasper, S.H., D. Samarian, A.P. Jadhav, A.H. Rickard, R.A. Musah, N.C. Cady. 2014. S-Aryl-L-cysteine Sulphoxides and Related Organosulphur Compounds Alter Oral Biofilm Development and AI-2-based Cell-Cell Communication. *Journal Applied Microbiology* 117: 1472–1486. doi:10.1111/jam.12616.

Kolderman, E., D. Bettampadi, D. Samarian, S.E. Dowd, B. Foxman, N.S. Jakubovics, A.H. Rickard. 2015. L-arginine Destabilizes Oral Multi-Species Biofilm Communities Developed in Human Saliva. *PLoS One* 10: e0121835.

Kolenbrander, P.E., R.J. Palmer, A.H. Rickard, N.S. Jakubovics, N.I. Chalmers, P.I. Diaz. 2006. Bacterial Interactions and Successions During Plaque Development. *Periodontology 2000* 42: 47–49.

Kolenbrander, P.E., R.N. Andersen, D.S. Blehert, P.S. Egland, J.S. Foster, R.J. Jr. Palmer. 2002. Communication Among Oral Bacteria. *Microbiology and Molecular Biology Reviews* 66(3): 486–505.

Komiya-Ito, A., T. Ito, A. Yamanaka, K. Okuda, S. Yamada, T. Kato. 2006. *N*-tetradecanoyl Homoserine Lactone, Signaling Compound for Quorum Sensing, Inhibits *Porphyromonas gingivalis* Growth. *Research Journal of Microbiology* 1(12): 353–359.

Kreth, J., J. Merritt, L. Zhu, W. Shi, F. Qi. 2006. Cell-Density- and ComE-Dependent Expression of a Group of Mutacin and Mutacin-like Genes in *Streptococcus mutans*. *FEMS Microbiology Letters* 265(1): 11–17.

Kumari, A., P. Pasini, S. Daunert. 2008. Detection of Bacterial Quorum Sensing *N*-acyl Homoserine Lactones in Clinical Samples. *Analytical and Bioanalytical Chemistry* 391(5): 1619–1927. doi:10.1007/s00216-008-2002-3.

Kumari, A., P. Pasini, S.K. Deo, D. Flomenhoft, H. Shashidhar, S. Daunert. 2006. Biosensing Systems for the Detection of Bacterial Quorum Signaling Molecules. *Analytical Chemistry* 78(22): 7603–7609.

LaSarre, B., M.J. Federle. 2013. Exploiting Quorum Sensing to Confuse Bacterial Pathogens. *Microbiology and Molecular Biology Reviews* 77(1): 73–111. doi:10.1128/MMBR.00046-12.

Laue, B.E., Y. Jiang, S.R. Chhabra, S. Jacob, G.S.A Stewart, A. Hardam, J.A. Downie, F. O'Gara, P. Williams. 2000. The Biocontrol Strain *Pseudomonas fluorescens* F113 Produces the Rhizobium Small Bacteriocin, *N*-(3-hydroxy-7-cis-tetradecenoyl) Homoserine Lactone, Via HdtS, A Putative Novel *N*-acylhomoserine Lactone Synthase. *Microbiology* 146(10): 2469–2480.

Leadbetter, J.R., E.P. Greenberg. 2000. Metabolism of Acyl-homoserine Lactone Quorum-Sensing Signals by *Variovorax paradoxus*. *Journal of Bacterioloy* 182(24): 6921–6926.

Lee, H.J., S.C. Kim, J. Kim, A. Do, S.Y. Han, B.D. Lee, H.H. Lee, M.C. Lee, S.H. Lee, T. Oh, S. Park, S.H Hong. 2015. Synergistic Inhibition of Streptococcal Biofilm by Ribose and Xylitol. *Archives of Oral Biology* 60:304–312. doi:10.1016/j.archoralbio.2014.11.004.

Lewis, K. 2001. Riddle of Biofilm Resistance. *Antimicrobial Agents and Chemotherapy* 45(4): 999–1007.

Li, Y.H., M.N. Hanna, G. Svensäter, R.P. Ellen, D.G. Cvitkovich. 2001a. Cell Density Modulates Acid Adaptation in *Streptococcus mutans*: Implications for Survival in Biofilms. *Journal of Bacteriology* 183(23): 6875–6884.

Li, Y.H., P.C. Lau, J.H. Lee, R.P. Ellen, D.G. Cvitkovitch. 2001b. Natural Genetic Transformation of *Streptococcus mutans* Growing in Biofilms. *Journal of Bacteriology* 183(3): 897–908.

Li, Y.Y., N. Tang, M.B. Aspiras, P.P. Lau, J.H. Lee, R.P. Ellen, D.G. Cvitkovitch. 2002. A Quorum-Sensing Signalling System Essential for Genetic Competence in *Streptococcus mutans* is Involved in biofilm formation. *Journal of Bacteriology.* 184(10): 2699–2708.

Loo, C.Y., D.A. Corliss, N. Ganeshkumar. 2000. *Streptococcus gordonii* Biofilm Formation: Identification of Gene that Code for Biofilm Phenotype. *Journal of Bacteriology* 182 (5): 1374–1382.

Marsh, P.D. 1994. Microbial Ecology of Dental Plaque and Its Significance in Health and Disease. *Advances in Dental Research* 8(2):263–271.

Marsh, P.D. 2003. Are Dental Diseases Examples of Ecological Catastrophes? *Microbiology* 149(2): 279–294.

Marsh, P.D., M.A.O. Lewis, H. Rogers, D.W. Williams, M. Wilson. 2016. *Marsh & Martin's Oral Microbiology*. 6th ed. Elsevier, Edinburgh.

Mashburn-Warren, L., D.A. Morrison, M.J. Federle. 2010. A Novel Double-Tryptophan Peptide Pheromone Controls Competence in *Streptococcus* spp. Via Rgg regulator. *Molecular Microbiology* 78(3): 589–606. doi:10.1111/j.1365-2958.2010.07361. x.

Mayer, C., A. Muras, M. Romero, M. López Díaz, M. Tomás, A. Otero. 2018. Multiple Quorum Quenching Enzymes Are Active in the Nosocomial Pathogen *Acinetobacter baumannii* ATCC17978. *Frontiers Cell Infection and Microbiology* 8: 310. doi:10.3389/fcimb.2018.00310.

Mayer, C., M. Romero, A. Muras, A. Otero. 2015. Aii20J, a Wide Spectrum Thermo-Stable *N*-Acylhomoserine Lactonase from the Marine Bacterium *Tenacibaculum* sp. 20J Can Quench AHL-Mediated Acid Resistance in *Escherichia coli*. *Applied Microbiology and Biotechnology* 99(22): 9523–9539. doi:10.1007/s00253-015-6741-8.

McNab, R., S.K. Ford, A. El-Sabaeny, B. Barbieri, G.S Cook, R.J. Lamont. 2003. Lux-S Based Signaling in *Streptococcus gordonii*: Autoinducer 2 Controls Carbohydrate Metabolism and Biofilm Formation with *Porphyromonas gingivalis*. *Journal of Bacteriology* 185(1): 274–284.

Merritt, J., F. Qi, S.D. Goodman, M.H. Anderson, W. Shi. 2003. Mutation of *luxS* Affects Biofilm Formation in *Streptococcus mutans*. *Infection and Immunology* 71(4): 1972–1979.

Mira, A., E. Buetas, B. Rosier, D. Mazurel, A. Villanueva-Castellote, C. Llena, M.D. Ferrer. 2019. Development of an in Vitro System to Study Oral Biofilms in Real Time through Impedance Technology: Validation and Potential Applications. *Journal of Oral Microbiology* 11(1): 1609838. doi:10.1080/20002297.2019.1609838.

Muras, A., C. Mayer, P. Otero-Casal, R.A.M. Exterkate, B.W. Brandt, W. Crielaard, A. Otero, B.P. Krom. 2020. Short chain N-acylhomoserine lactone quorum sensing molecules promote periodontal pathogens in in vitro oral biofilms. *Applied and Environmental Microbiology* 86: e01941–19. doi:10.1128/AEM.01941-19.

Muras, A., C. Mayer, M. Romero, T. Camino, M.D. Ferrer, A. Mira, A. Otero. 2018a. Inhibition of *Streptococcus mutans* Biofilm Formation by Extracts of *Tenacibaculum* sp. 20J, a Bacterium with Wide-Spectrum Quorum Quenching Activity. *Journal of Oral Microbiology* 10(1): e1429788. doi:10.1080/20002297.2018.1429788.

Muras, A., M. López-Pérez, C. Mayer, A. Parga, J. Amaro-Blanco, A. Otero. 2018b. High Prevalence of Quorum Sensing and Quorum Quenching Activity Among Cultivable Bacteria and Metagenomic Sequences in the Mediterranean Sea. *Genes* 9(2): 100. doi:10.3390/genes9020100.

Murugan, K., K. Sekar, S. Sangeetha, S. Ranjitha, S.A Sohaibani. 2013. Antibiofilm and Quorum Sensing Inhibitory Activity of *Achryranthes aspera* on Cariogenic *Streptococcus mutans*: An In Vitro and In Silico Study. *Pharmaceutical Biology*. 51: 728–736. doi:10.3109/13880209.2013.764330.

Ng, W.L. and B.L. Bassler. 2009. Bacterial Quorum-Sensing Network Architectures. *Annual Review of Genetics* 43: 197–222. doi:10.1146/annurev-genet-102108-134304.

Novak, E.A., H. Shao, C.A., Daep, D.R. Demuth. 2010. Autoinducer-2 and QseC Control Biofilm Formation and in Vivo Virulence of *Aggregatibacter actinomycetemcomitans*. *Infection and Immunology* 78(7): 2919–2926. doi:10.1128/IAI.01376-09.

O'Toole, G., H.B. Kaplan, R. Kolter. 2000. Biofilm Formation as Microbial Development. *Annual Review of Microbiology* 54:49–79.

Patankar, A.V., J.E. González. 2009. Orphan LuxR Regulators of Quorum Sensing. *FEMS Microbiology Review* 33(4): 739–756. doi:10.1111/j.1574-6976.2009.00163.x.

Pereira, C.S., J.A. Thompson, K.B. Xavier. 2013. AI-2-Mediated Signalling in Bacteria. *FEMS Microbiology Review* 37(2): 156–181. doi:10.1111/j.1574-6976.2012. 00345.x.

Perry, J.A., D.G. Cvitkovitch, C.M. Levesque. 2009. Cell Death in *Streptococcus mutans* Biofilms: A Link Between CSP and Extracellular DNA. *FEMS Microbiology Letters* 299(2): 261–266. doi:10.1111/j.1574-6968.2009.01758.x.

Perry, J.A., M.B. Jones, S.N. Peterson, D.G. Cvitkovitch, C.M. Levasque. 2009. Peptide Alarmone Signaling Triggers and Auto-Active Bacteriocin Necessary for Genetic Competence. *Molecular Microbiology* 72: 905–917.

Petersen, F.C., A.A.N. Ahmed, A. Naemi, A.A. Scheie. 2006. LuxS-mediated Signalling in *Streptococcus anginosus* and its Role in Biofilm Formation. *Antoine van Leeuwenhoek* 90(2): 109–112.

Petersen, F.C., D. Pecharki, A.A. Scheie. 2004. Biofilm Mode of Growth of *Streptococcus intermedius* Favored by a Competence-Stimulating Signaling Peptide. *Journal of Bacteriology* 186(18): 6327–6331.

Pritchard, A.B., S. Crean, I. Olsen, S.K. Singhrao. 2017. Periodontitis, Microbiomes and Their Role in Alzheimer's Disease. *Frontiers in Aging Neuroscience* 9: 336. doi:10.3389/fnagi.2017.00336.

Quadri, L.E. 2002. Regulation of Antimicrobial Peptide Production by Autoinducer-Mediated Quorum Sensing in Lactic Acid Bacteria. *Antoine Van Leewenhoek* 82: 133–145.

Reck, M., J. Tomash, I. Wagner-Döbler. 2015. The Alternative Sigma Factor sigX Controls Bacteriocin Synthesis and Competence, the Two Quorum Sensing Regulated Traits in *Streptococcus mutans*. *PLoS Genetics* 11(7): E1005353. doi:10.1371/journal.pgen.1005353.

Rickard, A.H., R.J.J. Palmer, D.S. Blehert, S.R. Campagna, M.F. Semmelhack, P.G. Egland, B.L. Bassler, P.E. Kolenbrander. 2006. Autoinducer 2: A Concentration-Dependent Signal for Mutualistic Bacterial Biofilm Growth. *Molecular Microbiology* 60: 1446–1456.

Romero, M., A. Muras, C. Mayer, N. Buján, B. Magariños, A. Otero. 2014. Quenching of AHLs production by the fish pathogen *Edwardsiella tarda* in Vitro Using Cell Extracts of the Marine Bacterium *Tenacibaculum* sp. Strain 20J. *Disease of Aquatic Organismes* 108(3): 217–225. doi:10.3354/dao02697.

Romero, M., C. Mayer, A. Muras, A. Otero. 2015. "Silencing bacterial communication through enzymatic quorum sensing inhibition" in *Quorum Sensing vs Quorum Quenching: A Battle with No End in Sight* edited by V.C. Kalia, Springer, India. pp. 219–236.

Ryu, E.J., J. Sim, J. Sim, J. Lee, B.K. Choi. 2016. D-Galactose as an Autoinducer 2 Inhibitor to Control the Biofilm Formation of Periodontopathogens. *Journal of Microbiology* 54(9): 632–637. doi:10.1007/s12275-016-6345-8.

Sampaio-Maia, B., I.M. Caldas, M.L. Pereira, D. Pérez-Mongiovi, R. Araujo. 2016. The Oral Microbiome in Health and Its Implication in Oral and Systemic Diseases. *Advances in Applied Microbiology* 97: 171–210. doi:10.1016/bs.aambs.2016.08.002.

Schauder, S., K. Shokat, M.G. Surrette, B.L. Bassler. 2001. The LuxS Family of Bacterial Autoinducers: Biosynthesis of a Novel Quorum-Sensing Molecule. *Molecular Microbiology* 41: 463–476.

Senadheera, D., D.G. Cvitkovitch. 2008. Quorum Sensing and Biofilm Formation by *Streptococcus mutans*. *Advances in Experimental Medicina and Biology* 631. 178–188. doi:10.1007/978-0-387-78885-2_12.

Senadheera, D., K. Krastel, R. Mair, A. Persadmeht, J. Abranches, R.A. Burne, D.G. Cvitkovitch. 2009. Inactivation of VicK Affects Acid Production and Acid Survival of *Streptococcus mutans*. *Journal of Bacteriology* 191(20): 6415–6424. doi:10.1128/JB.00793-09.

Shao, H., D.R. Demuth. 2010. Quorum Sensing Regulation of Biofilm Growth and Gene Expression by Oral Bacterial and Periodontal Pathogens. *Periodontology 2000* 52(1): 53–67. doi:10.1111/j.1600-0757.2009.00318. x.

Simonetti, O., O. Cirioni, I. Cacciatore, L. Baldassarre, O. Orlando, E. Pierpaoli, G. Lucarini, et al. 2016. Efficacy of the Quorum Sensing Inhibitor FS10 Alone and in Combination with

Tigecycline in an Animal Model of Staphylococcal Infected Wound. *PLoS One* 11(6): e0151956. doi:10.1371/journal.pone.0151956.

Simón-Soro, A., A. Mira, 2015. Solving the Etiology of Dental Caries. *Trends in Microbiology* 23(2):76–82. doi:10.1016/j.tim.2014.10.010.

Solano, C., M. Echeverz, I. Lasa. 2014. Biofilm Dispersion and Quorum Sensing. *Current Opinion in Microbiology* 18: 96–104. doi:10.1016/j.mib.2014.02.008.

Szafranski, S.P., Z.L. Deng, J. Tomasch, M. Jarek, S. Bhuju, M. Rohde, H. Sztajer, I. Wagner-Döbler. 2017. Quorum Sensing of *Streptococcus mutans* Is Activated by *Aggregatibacter actinomycetemcomitans* and by the Periodontal Microbiome. *BMC Genomics* 18(1): 238. doi:10.1186/s12864-017-3618-5.

Sztajer, H., A. Lemme, R. Viñchez, S. Schulz, R. Geffers, C.Y.Y. Yip, C.M. Levesque, D.G. Cvitkovitch, I. Wagner-Döbler. 2008. Autoinducer-2-Regulated Genes in *Streptococcus mutans* UA159 and Global Metabolic Effect of the *luxS* Mutation. *Journal of Bacteriology* 190(1): 401–415.

Trivedi, H.M., T. Xu, D. Miksa. 2009. Anti-biofilm Carbonate Compounds for Use in Oral Care Compositions. *Patent No.* US8617523B2.

Van der Meer, J.R., J. Polman, M.M. Beerthuyzen, R.J. Siezen, O.P. Kuipers, W.M. De Vos. 1993. Characterization of the *Lactococcus lactis* nisin A Operon Genes nisP, Encoding a Subtilisin-like Serine Protease Involved in Precursor Processing, and nisR, Encoding a Regulatory Protein Involved in Nisin Biosynthesis. *Journal of Bacteriology* 175(9): 2578–2588.

Wang, B.Y., A. Deuthc, J. Hong, H.K. Kuramitsu. 2011. Proteases of an Early Colonizer Can Hinder *Streptococcus mutans* Colonization In Vitro. *Journal of Dental Research* 90(4): 501–505. doi:10.1177/0022034510388808.

Whittaker, C.J., C.M. Klier, P.E. Kolenbrander. 1996. Mechanisms of Adhesion by Oral Bacteria. *Annual Review in Microbiology* 50:513–552.

Williams, P., K. Winzer, W.C Chan, M. Cámara, 2007. Look Who's Talking: Communication and Quorum Sensing in the Bacterial World. *Philosophical Transactions of Royal Society of London Series B. Biological Sciences* 362(1483): 1119–1134.

Wilson, M. 2001. Bacterial Biofilms and Human Diseases. *Science Progress* 84(3): 235–254.

Wu, J., X. Lin, H. Xie. 2009. Regulation of hemin binding proteins by a novel transcriptional activator in *Porphyromonas gingivalis*. *Journal of Bacteriology* 191: 115–122.

Yin, W.F., K. Purmal, S. Chin, K.Y. Chan, C.L. Koh, C.K. Sam, K.G. Chan. 2012a. *N*-Acyl Homoserine Lactone Production by *Klebsiella pneumoniae* Isolated from Human Tongue Surface. *Sensors* 12(3): 3472–3483. doi:10.3390/s120303472.

Yin, W.F., K. Purmal, S. Chin, K.Y. Chan, K.G. Chan. 2012b. Long Chain *N*-Acyl Homoserine Lactone Production by *Enterobacter* sp. Isolated from Human Tongue Surfaces. *Sensors* 12(11): 14307–14314. doi:10.3390/s121114307.

Yoshida, A., T. Ansai, T. Takehara, H.K. Kuramitsu. 2005. LuxS-based Signaling Affects *Streptococcus mutans* Biofilm Formation. *Applied and Environmental Microbiology* 71: 2372–2380.

Yost, S., A.E. Duran-Pinedo, R. Teles, K. Krishnan, J. Frias-Lopez. 2015. Functional Signatures of Oral Dysbiosis Duting Periodontitis Progression Revealed by Microbial Metatranscriptome Analysis. *Genome Medicine* 7(1): 27. doi:10.1186/s13073-015-0153-3.

Zang, T., B.W. Lee, L.M. Cannon, K.A. Ritter, S. Dai, D. Ren, T.K. Wood, Z.S. Zhou. 2009. A Naturally Occurring Brominated Furanone Covalently Modifies and Inactives LuxS. *Biorganic and Medicinal Chemistry Letters* 19(21): 6200–6204. doi:10.1016/j.bmcl.2009.08.095.

14

Quorum Sensing in Autotrophic Nitrogen Removal Systems for Wastewater Treatment

Guangxue Wu, Zhaolu Feng, Yuepeng Sun, and Tianle Li

CONTENTS

14.1 The Autotrophic Nitrogen Removal System and Its Possible Quorum Sensing

Wastewater treatment helps the control of disease spreading and contributes to the protection of human health and ecological safety, which was considered as an important sanitation technology in the previous century (Wu et al., 2018). Usually, activated sludge or biofilm wastewater treatment systems are applied to remove organic carbon, nitrogen, and phosphorus from wastewater. Since 2014, the 100-year anniversary of the activated sludge process, experts began to rethink about future development of wastewater treatment. The concept has been proposed that wastewater should be considered as a resource rather than a waste through the recovery of water, organic carbon, nitrogen, and phosphorus. For example, recovery of energy by converting organic carbon to methane and phosphorus through precipitation should be pursued. For nitrogen, the removal from wastewater should be the target, and biological processes should be applied for its removal. Among all nitrogen removal technologies, the autotrophic nitrogen removal process is becoming a promising technology as energy neutral or even positive in wastewater treatment (Wang et al., 2018).

In wastewater treatment processes, conventional biological nitrogen removal is mainly achieved through full nitrification and denitrification processes, while autotrophic nitrogen removal is mainly achieved through the partial nitrification and anaerobic ammonia oxidation (anammox) processes. Generally, for full nitrification, ammonia nitrogen (NH_4-N) in wastewater is oxidized to nitrite nitrogen (NO_2-N) by ammonia oxidizing bacteria (AOB) and further to nitrate nitrogen (NO_3-N) by nitrite oxidizing bacteria (NOB) under aerobic conditions, with oxygen as the electron acceptor. For partial nitrification (or nitritation), only the first step carried out by AOB would occur, and the second step by NOB would be inhibited. Usually, for conventional nitrogen removal, the produced NO_3-N is reduced to nitrogen gas (N_2) by denitrification with organic carbon as the electron donor. In the autotrophic nitrogen removal process, after nitritation, in the following anammox process, NH_4-N will react with the produced NO_2-N, and both will be converted to N_2. By this means, nitrogen in wastewater will be removed and released to the atmosphere in the form of N_2. Autotrophic nitrogen removal through nitritation and anammox processes can be achieved within one combined reactor with the two processes occurring simultaneously or within sequential two stage systems with the two processes occurring separately in two reactors (Figure 14.1). However, challenges are also faced for autotrophic nitrogen removal from wastewater. On the one hand, the suppression of NOB should be achieved, because NOB can compete with AOB and anammox for substrates (i.e., oxygen and NO_2-N). On the other hand, successful application of these autotrophic nitrogen removal processes requires balancing activities among nitrifiers, anammox, and heterotrophs. Especially, in the combined nitritation and anammox process, temperature, dissolved oxygen concentration, and organics may break the balance and further deteriorate nitrogen removal (Jenni et al., 2014).

Quorum sensing (QS) is a density-dependent mechanism for microbial interactions in microbial ecology systems, which depends on the production, secretion, and reception of signal substances. Once the microorganisms accumulate to certain concentrations with high density, signal substances could be produced and induce cascade gene responses (Wongsuk et al., 2016). Quorum sensing is an approach responsible for microbial communication, which is very important for carrying out group behaviors such as bioluminescence production, biofilm formation, extracellular polymeric substances (EPS) production, genetic exchange, and virulence factor expression (Ng and Bassler, 2009;

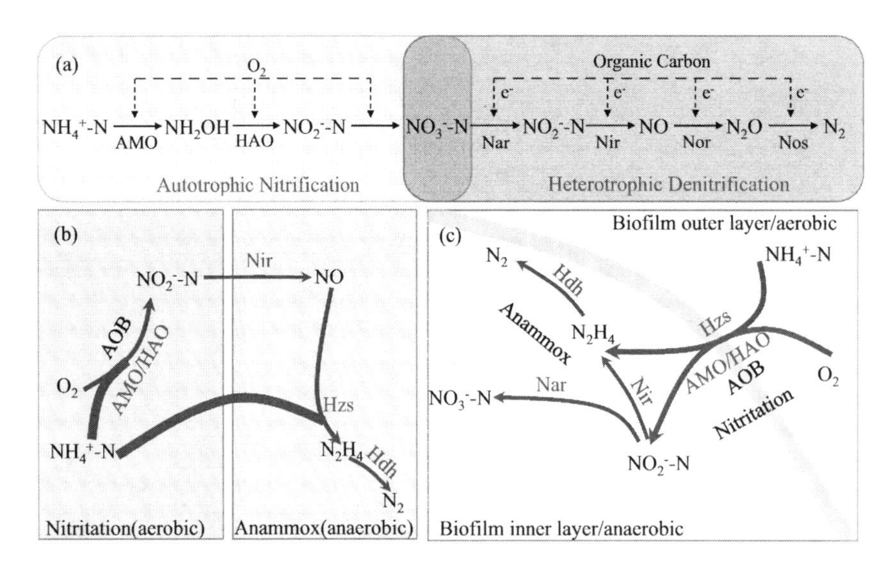

FIGURE 14.1 Autotrophic nitrification and heterotrophic denitrification (a), and autotrophic partial nitrification and anammox nitrogen removal diagram in two-stage systems (b) or in one combined reactor (c). AMO, HAO, Nir, Nar, Nor, Nos, Hdh, and Hzs represent ammonia monooxygenase, hydroxylamine dehydrogenase, nitrite reductase, nitrate reductase, nitric oxide reductase, nitrous oxide reductase, hydrazine dehydrogenase, and hydrazine synthase, respectively.

Hawver et al., 2016). To date, researchers in the field of biology have identified the QS in pure-cultivating Gram-negative and Gram-positive bacteria, and the functional microorganisms with identified QS in biological wastewater treatment systems have received increasing attention (Huang et al., 2016; Zhang and Li, 2016; Chen et al., 2018).

In autotrophic nitrogen removal processes, the functional microorganisms of AOB, NOB, and anammox bacteria are all autotrophic bacteria with slow growing rates. Therefore, usually they prefer the form of flocs or granules. In addition, the genomes of many nitrifying bacteria contain genes associated with microbial cell-cell signaling or quorum sensing (Mellbye et al., 2017). Therefore, QS may play important roles in the autotrophic nitrogen removal system (Table 14.1).

For example, for nitrifiers, Mellbye et al. (2017) found that in all *N*-acyl homoserine lactone (AHL)-positive nitrifiers except *Nitrospira moscoviensis*, there were statistically significant increases as the optical density at 600 nm (OD_{600}) approached 0.01, while *N. moscoviensis* showed a statistically significant increase in the AHL concentration at an extremely low OD_{600} of 0.005; in addition, AHL concentrations also increased significantly at the OD_{600} of 0.01. For anammox bacteria, only when the cell is higher than 10^{10}–10^{11} cells/mL do they begin to take on activity (Ding et al., 2013; Strous et al., 1999). In addition, when the anammox bacteria concentration (represented as volatile suspended solids, VSS) was above 0.4 g/L, the specific anoxic ammonium oxidation of 91 ± 12 mg/g·d was significantly ($p < 0.05$) higher than at low biomass concentrations of 26 ± 15 mg/g·d (De Clippeleir et al., 2011). All these evidence support the biomass dependence phenomenon similar to QS, indicating that QS could exist in the nitrifying and anammox systems.

TABLE 14.1

Common QS Bacteria/System, Signal Molecules and Behaviors

Autotrophic QS Bacteria/System	Signal Molecule	Application	QS Behavior	References
Nitrosomonas europaea	C_6-HSL C_8-HSL C_{10}-HSL	Pure culture	Nitrification: Ammonia oxidation	Burton et al. (2005)
Nitrosospira multiformis; *Nitrosospira briensis*	C_{14}-HSL 3-oxo-C_{14}-HSL	Pure culture	Nitrification: Ammonia oxidation	Mellbye et al. (2017) Gao et al. (2014)
Nitrobacter winogradskyi	C_{10}-HSL $C_{10:1}$-HSL	Pure culture	Nitrification: Nitrite oxidation	Mellbye et al. (2017) Shen et al. (2016)
Anammox	C_{12}-HSL	Anammox granular sludge reactor (AGSR)	Anammox biomass activity	Zhao et al. (2018a)
	C_6-HSL C_8-HSL C_{12}-HSL	Lab-scale SBR	Anammox biomass activity	Tang et al. (2015)
	C_6-HSL C_8-HSL	Biofilm	Biofilm formation and nitrifiers activity	Sun et al. (2018a)

For each type of microbial community, usually, two types of living strategies can be adopted. The r-strategy microorganism possesses a high growth rate, a low substrate affinity, and a relatively low yield coefficient, while the K-strategy microorganism has a low growth rate, a high affinity for substrate, and a relatively high yield coefficient (Pianka, 1970; Andrews and Harris, 1986). Mellbye et al. (2016) speculated that since *Nitrobacter* are *r*-strategists exploiting high substrate concentrations and sporadically growing with high biomass densities, they may make better use of cell-density-dependent QS genetic regulation. Therefore, the growth physiology of microorganisms should be paid attention during the investigation of QS in the autotrophic nitrogen removal communities.

For wastewater treatment, usually activated sludge systems or biofilm systems would be applied. In activated sludge systems, microorganisms are in the form of suspended solids and flocs, while for biofilm systems, microorganisms attach to the carriers for better growth. The high cell densities of 10^{10} cell/mL would occur only in biofilms and cell aggregates, where a high biomass density would be possible (Batchelor et al., 1997). Therefore, QS would occur better in the biofilm systems rather than the conventional activated sludge systems. On the other hand, one should remember that the function of QS would be controlled by signal substances, and the diffusion of signal substances would be reduced within biofilms (Batchelor et al., 1997). Therefore, the types, concentrations, and distribution of signal substances in the wastewater treatment process should be clarified comprehensively.

14.2 Detection of Signal Substances and Genes in Autotrophs for Nitrogen Removal

QS is achieved through producing signal substances. Usually, the QS phenomenon can be measured through biological reporter strains, thin-layer chromatography and by detecting signal substances produced by QS microorganisms (Fuqua and Winans, 1996; Shaw et al., 1997; Sun et al., 2018c). QS signal substances are often referred to as auto-inducers, which can be classified based on their molecular structures, such as AHLs, autoinducer 2 (AI-2), and diffusible signal factor (DSF) (Huang et al., 2016; Zhang and Li, 2016). Some testing examples are summarized in Table 14.2. All AHLs contain an acyl-homoserine lactonic ring, with different acylation branched chains in terms of the lengths of side chains and β position substituent groups (i.e., hydrogen, carbonyl, or hydroxyl) (Fuqua et al., 1996; Li et al., 2015). To date, some methods for the detection of signal substances in the liquid and biomass phases have been developed, including AHLs, bis-(3′-5′)-cyclic dimeric guanosine monophosphate (c-di-GMP),

TABLE 14.2

Common Quorum Sensing Signals and Their Detection Techniques

Signals	Function	Detection	References
N-acyl-homoserine lactones (AHLs)	1. Intraspecies communication 2. Biofilm formation EPS production 3. Aerobic granulation 4. Bacterial activity	Liquid chromatography/tandem mass Spectrometry (HPLC-MS/MS); High-pressure liquid chromatography (HPLC); Thin-layer chromatograph (TLC); SPE-ultra performance liquid chromatography/tandem mass spectrometry (UPLC-MS/MS)	Tan et al. (2014) Ishizaki et al. (2017) Gao et al. (2014) Hu et al. (2016)
Autoinducer 2 (AI-2)	1. Inter-species interactions 2. Biofilm formation 3. Pathogenicity factors and toxins 4. Bioluminescence 5. Swarming motility	Biosensor strains; HPLC-MS/MS High-performance liquid chromatography with UV detector (HPLC–UV); Gas chromatography–mass spectrometry (GC–MS); HPLC-fluorescence detector (HPLC–FLD)	Zhao et al. (2018a) Campagna et al. (2009) Song et al. (2014)
Autoinducing peptides (AIP)	1. Intercellular communication 2. Virulence and antimicrobial compounds	Ultrahigh performance liquid chromatography (UHPLC)	Junio et al. (2013)
Diffusible signal factors (DSF)	1. Production of exoenzymes 2. The presence of DSF leads to bacterial dispersion, whereas the absence of DSF leads to EPS production, aggregation, and biofilm formation 3. Govern the synthesis or turnover of cyclic di-GMP	HPLC; LC-MS	Ryan and Dow (2011) Zhou et al. (2016)
Diguanylate monophosphate (cyclic-di-GMP)	1. Coordinate biofilm formation and pathogenicity 2. Regulate virulence and motility factors 3. Regulate the synthesis of exopolysaccharide 4. Determine granular stability	HPLC; LC-MS/MS	Chua et al. (2015) Bouffartigues et al. (2015) Yang et al. (2018) Guo et al. (2017b)

AI-2, and DSF (Guo et al., 2017b; He et al., 2010; Sun et al., 2018c; Xu et al., 2017). However, usually, only one type of signal substance has been focused on in the specific study, which would be fine for isolated microorganisms or specific types of bacteria. However, for mixed culture systems, such as wastewater treatment processes, the result would be limited for detecting only specific types of signal substances.

Burton et al. (2005) reported the types and levels of AHLs produced by *Nitrosomonas europaea* strain Schmidt, with the detected AHLs of *N*-hexanoyl-L-homoserine lactone (C_6-HSL), *N*-octanoyl-L-homoserine lactone (C_8-HSL), and *N*-decanoyl-L-homoserine lactone (C_{10}-HSL) from extracts of chemostat culture effluent by independent methods of reporter strains (*Chromobacterium violaceum* and *Agrobacterium tumefaciens* NTLR) and gas chromatography-mass spectrometry (GC-MS), and the detected AHLs concentrations of 0.4–2.2 nM. Following, Mellbye et al. (2017) identified the AHLs produced by the AOB *Nitrosospira multiformis* and *Nitrosospirabriensis* and the NOB *Nitrobacter vulgaris* and *N. moscoviensis* as C_{10}-HSL, *N*-3-hydroxy-tetradecanoyl-L-homoserine lactone (3-OH-C_{14}-HSL), a monounsaturated AHL of 7, 8-*trans*-*N*-(decanoyl) homoserine lactone ($C_{10:1}$-HSL), and C_8-HSL, respectively.

By using the reporter strain of *Escherichia coli*, in the study of Batchelor et al. (1997), the signal substances of AHL for QS could be detected when AOB of *N. europaea* were cultured with the dosage of ammonia after concentration to 10^{10} cells/mL, indicating the density-dependent property. Furthermore, Mellbye et al. (2015) found that *Nitrobacter winogradskyi* produced two distinct acyl-HSLs, C_{10}-HSL and a monounsaturated acyl-HSL ($C_{10:1}$-HSL), in a cell-density- and growth phase-dependent manner. For example, the chemostat experiments suggest that acyl-HSL production was highest at a medium cell density (OD_{600} of 0.0119) corresponding with a doubling time (T_d) of 1.93 days, and a similar result was observed during batch culture on day 4, when the T_d of the culture was around 1.75 days; in addition, equally intriguing was the sharp decrease in production and concentration of acyl-HSL when cells entered deep stationary phase (Mellbye et al., 2015).

Beside the density or growth status effects, the metabolic pathway also affects QS and the type of produced signal substances. Shen et al. (2016) found that the accumulated acyl-HSL in the autotrophic culture of *N. winogradskyi* exhibited a five-fold increase in Miller units/OD_{600} compared with the heterotrophic culture. In addition, *N*-heptanoyl-L-homoserine lactone (C_7-HSL) and $C_{10:1}$-HSL were present in the extract of the cells cultured in the mixotrophic medium, whereas only $C_{10:1}$-HSL was detected in the extracts of the cells cultured in the heterotrophic and autotrophic media. *N. winogradskyi* can utilize nitrate, nitrite, carbon dioxide, and organic compounds, which may result in the release of different AHLs in response to various survival conditions.

In the wastewater treatment systems for nitrogen removal, signal substances have also been detected. In the biofilm supernatant, *N*-dodecanoyl-L-homoserine lactone (C_{12}-HSL) was detected by the reporter strain *Chromobacterium violaceum* CV026 and liquid chromatography with tandem mass spectrometry (LC-MS-MS), which was further detected in an anaerobic anammox-enriched community, but not in an aerobic AOB-enriched community (De Clippeleir et al., 2011), showing the enriched microbial community affecting the detected signal substances. Also, Zhao et al. (2018b) detected C_{12}-HSL in the supernatant of an anammox granular sludge reactor (AGSR). In addition, Tang et al. (2015) confirmed the production of more types of AHLs of C_6-HSL, C_8-HSL, and C_{12}-HSL in the anammox culture. The operational conditions and the existence of flocs may all affect the types of produced AHLs, which might cause the different results from various studies. The detection method should be also referred for the efficient detection of signal substances. In the study of Sun et al. (2018a, 2018b, 2019), signal substances in nitrification, anammox, and combined nitritation/anammox systems have been investigated by differentiating signal substances in the liquid and solid phases. For example, C_6-HSL and C_8-HSL were dominant in biomass and EPS phases of the nitrification and anammox biofilm systems. C_6-HSL, C_8-HSL, and C_{10}-HSL were confirmed to be universal signal substances and distributed mainly in EPS and biomass phases in the combined nitritation/anammox system. The results showed that C_6-HSL and C_8-HSL mainly existed in the liquid and solid phases (EPS and biomass phases), respectively (Figure 14.2).

Functional genes responsible for signal substance production are also different from various types of microorganisms, which may be also affected by growth conditions. Mellbye et al. (2016) found that only *Nitrobacter* species contained both *ncgABC* and clearly annotated autoinducer synthase and receptor genes associated with QS. Shen et al. (2016) confirmed the *nwiI* gene of *Nitrobacter winogradskyi* to be a homoserine lactone synthase, and several acyl-homoserine lactone signals with 7–11 carbon acyl groups, including a novel signal, $C_{10:1}$-HSL, could be identified in *N. winogradskyi*. Gao et al. (2014) identified one gene (*nmuI*) similar to AHL synthase in *N. multiformis*, and when was introduced into *Escherichia coli*, *N*-tetradecanoy-L-homoserine

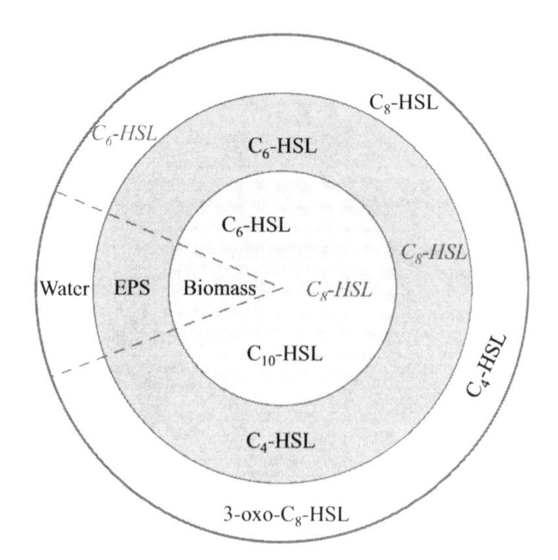

FIGURE 14.2 The profile of AHLs signal substances in nitrification, anammox, and combined nitritation/anammox biofilm systems. The italic AHLs represent dominant signal substances. C_{10}-HSL and 3-oxo-C_8-HSL were only identified in combined nitritation/anammox systems.

lactone (C_{14}-HSL) and *N*-3-oxo-tetradecanoyl-homoserine lactone (3-oxo-C_{14}-HSL) were detected using an AHL biosensor and liquid chromatography-mass spectrometry (LC-MS); however, by extracting *N. multiformis* culture supernatants, no AHL product was obtained. Burton et al. (2005) found that *Nitroso monaseuropaea* strain Schmidt could produce AHLs, while these bacteria lacked LuxI and LuxM homologs, suggesting that AHL synthesis occurs through other pathways, possibly mediated by an HdtS homolog.

However, Mellbye et al. (2017) found that cell density-dependent AHL production requires an AHL synthase gene, but the presence of putative AHL synthase and receptor genes does not guarantee the production of detectable AHLs under the applied conditions. In addition, the expression of the functional gene is also dependent on different conditions even for the coupled genes responsible for the same signal substance production. Mellbye et al. (2015) found that the *N. winogradskyi* genome contains genes encoding a putative acyl-HSL autoinducer synthase (nwi0626, *nwiI*) and a putative acyl-HSL autoinducer receptor (nwi0627, *nwiR*). *nwiI* was induced in a cell density-dependent manner. In the chemostat, expression of *nwiI* was induced greater than 10-fold at medium (OD_{600} of 0.012) and high (OD_{600} of 0.069) cell densities compared to the level at a low (OD_{600} of 0.0056) cell density. However, *nwiR* showed no change in gene expression in response to cell density under these conditions.

14.3 Autotrophic Biomass Production and Activities Affected by Quorum Sensing

During the batch attachment experiment, Li et al. (2015) found that AHLs enhanced biofilm growth and the type of AHLs had a significant effect. Dosing *N*-3-oxo-hexanoyl-homoserine lactone (3-oxo-C_6-HSL) yielded the highest biomass density on the plastic plate, and the attached cells were 2.56 times that of the control, and 2 times for C_{10}-HSL, *N*-3-oxo-octanoyl-homoserine lactone (3-oxo-C_8-HSL), and N-3-oxo-decanoyl-homoserine lactone (3-oxo-C_{10}-HSL), while 1.47 times for C_8-HSL and 1.55 times for C_6-HSL. This indicated that for AHL molecules without β-position substituent groups of C_6-HSL, C_8-HSL, and C_{10}-HSL, adhesion growth was enhanced with the length of the N-group side chain, while for AHL with the β-position substituent group of a carbonyl group (including 3-oxo-C_6-HSL, 3-oxo-C_8-HSL, and 3-oxo-C_{10}-HSL), the adhesion growth ability increased with decreasing the length of the N-group side chain (Li et al., 2015). During the long-term study, compared to the mean size of 168.5 μm for the loose seed sludge flocs, the size of the sludge without the addition of the signal molecule reached 208.1 μm, while it was 309.3 and 356.7 μm with C_6-HSL or 3-oxo-C_6-HSL added (Li et al., 2015). In addition, Wu et al. (2017) found when cellular extract of AHLs from the nitrifying granular sludge (NGS) was added into the nitrifying activated sludge, the cell adhesion and aggregation could be improved, and the microbial activity, the concentration of extracellular proteins, and the biomass were increased, eventually contributing to the nitrifying sludge granulation. Therefore, AHLs can enhance biofilm growth.

With the addition of long-chain AHLs for 16 days, Gao et al. (2019) found that the specific ammonia-oxidizing-rate based on biomass represented by suspended solids (SS) increased from 1.6 to 2.8 mg/g·h, the abundance of archaeal *amoA* genes increased 1.9 times faster than that of bacterial *amoA* genes in spite of the fact that the total quantity of archaeal *amoA* genes (max 1.08×10^6 copies/L sludge) did not exceed the bacterial *amoA* genes (max 7.39×10^7 copies/L sludge). The possible reason could be due to the fact that large signal molecules with relative stability were more capable of regulating sludge phenotypes and changes in community composition and function. In addition, for the oxygen-limited autotrophic nitrification/denitrification (OLAND) biomass, De Clippeleir et al. (2011) found that the addition of 67 mg/L C_{10}-HSL and 30 mg/L C_{12}-HSL significantly ($p < 0.05$) increased the anoxic ammonium oxidation rate with a factor of 1.5 and 1.4, respectively, while the addition of C_{12}-HSL (30 mg/L) did not have a significant effect on the aerobic ammonium oxidation rate. On the contrary, Li et al. (2015) showed that AHL could improve the nitritation (ammonia oxidation to nitrite) efficiency, while it did not affect nitratation (nitrite oxidation to nitrate) of the autotrophic nitrifying sludge, with the obtained maximum ammonia degradation rate coefficients of 0.704/h (C_6-HSL), 0.652/h (3-oxo-C_6-HSL), 0.568/h (C_8-HSL), 0.549/h (3-oxo-C_8-HSL), 0.539/h (C_{10}-HSL), 0.510/h (control), and 0.476/h (3-oxo-C_{10}-HSL). As the length of the AHL N side chain decreased, its contribution to improve ammonia oxidation increased. The AHLs without the β-position substituent group had a greater effect on the promotion of ammonia removal compared to the AHLs of the same side chain length but with a carbonyl as the substituent group.

In the anammox system, by adding anammox granular sludge reactor (AGSR) supernatant (S1) and C_{12}-HSL (S3), the specific anammox activity (SAA) increased from 94.84 mg N/g VSS·d (control) to 128.78 mg N/g VSS·d (S1) and 139.27 mg N/g VSS·d (S3), and also the start-up time of the anammox process was reduced from 80 to 66 d (Zhao et al., 2018b), which could be because AHLs retained more anammox by increasing the secretion of EPS. In the study of Tang et al. (2015), C_6-HSL and C_8-HSL could significantly increase activity of anammox bacteria, resulting in the increased ammonium nitrogen removal percentage by 35% and 20%, respectively, whereas C_{12}-HSL apparently promoted the growth of heterotrophic bacteria.

From the gene aspect, by adding a recombinant AiiAl actonase to *N. winogradskyi* cultures to degrade AHLs and mRNA sequencing (mRNA-Seq) to identify putative QS-controlled genes, Mellbye et al. (2016) found that the expression of *nirK* and *nirK* cluster genes (*ncgABC*) increased up to 19.9-fold under QS-proficient conditions (minus active lactonase).

For practical application, wastewater treatment systems may experience a starvation period due to limited wastewater or nutrient conditions. QS may affect the recovery of autotrophic microorganisms. Batchelor et al. (1997) found that for suspended *N. europaea* starved for 28 days in ammonium-free medium, the lag period was 53.4 h in the absence of 3-oxo-C_6-HSL, and the lag phase dramatically decreased to 22.2 h with the added 3-oxo-C_6-HSL of 0.01 μg/mL and further to 10.8 h at 0.1 μg/mL, while starved biofilms exhibited no lag phase prior to nitrite production, even after starvation for 43.2 days, although there was evidence of cell loss during starvation. The time taken to establish new steady states of ammonium and nitrite increased with the starvation period, from 60 h after starvation for 7.7 days to 170 h after starvation for 43.2 days, suggesting that the decrease in population size occurred continuously during the starvation period (Batchelor et al., 1997). Therefore, for autotrophic nitrogen removal systems, biofilm would be a better choice for stable operation. In addition, the addition of exogenous C_6-HSL could enhance the production of endogenous C_6-HSL by biomass, which could further accelerate nitrogen removal (Tang et al., 2015), which can be adopted for the start-up of systems with potential QS effect.

14.4 EPS in Autotrophic Nitrogen Removal Systems Affected by Quorum Sensing

EPS is produced by microorganisms, especially in biofilm systems, responding to stress conditions, such as limited nutrients or disinfection conditions. In addition, EPS can be used as the stored organic carbon for microbial utilization under stress conditions. Usually, QS can induce the production of more EPS under high biomass density conditions. For example, Paggi et al. (2003) observed the existence of potential AHL-type autoinducers in the archaeon *Natronococcus occultus*, and these autoinducers might participate in the production/activation of extracellular proteases. In nitrifying granular sludge, the extracellular proteins increased from 15.9 to 31.2 (control), 74.6 (dosing C_6-HSL), and 63.4 (dosing 3-oxo-C_6-HSL) mg/g SS, while the extracellular polysaccharides decreased from 47.2 mg/g SS to a similar level of approximately 20 mg/g SS (Li et al., 2015). In addition, by Fourier transform infrared spectroscopy, a new band at 1547 cm^{-1} (the N-H bending and C-N stretching (Amide II) of proteins) was observed only in the granular sludge dosed with 3-oxo-C_6-HSL, demonstrating that 3-oxo-C_6-HSL dosage had a direct interaction with the extracellular proteins of autotrophic nitrifying sludge (Li et al., 2015). Therefore, AHL quorum sensing favored the production of more extracellular proteins, which could enhance the formation and structure stability of autotrophic granules.

Operational conditions applied for wastewater treatment can affect QS production and its functions. By operating reactors at conventional stepwise nitrogen loading rates (R1) and alternating weekly to be higher or lower than that of R1 (R2), Guo et al. (2017a) found that the aggregation process of anammox was significantly faster under alternating loadings and was significantly correlated with higher levels of c-di-GMP and EPS production. The increased production of fructose 6-phosphate and UDP-*N*-acetyl-D-glucosamine, the precursor substances for the synthesis of exopolysaccharides, could be induced by higher levels of c-di-GMP. Higher levels of intracellular hydrophobic amino acids were also positively correlated with increased extracellular protein levels, considering the significant increase in peptides under alternating loadings. EPS induction was likely to be partly a consequence of the regulation of c-di-GMP, confirmed from the fact that c-di-GMP produced at the 1 h time point, while EPS production did not increase until the 2 h time point.

Under stress conditions, EPS production regulated by QS also plays very important roles. Guo et al. (2017b) found that under unfavorable conditions, anammox bacteria would synthesize high levels of c-di-GMP first and then produced more EPS, which would enhance the bacteria granulation to against the unfavorable conditions. For example, under the stress conditions, the anammox bacteria had more intracellular c-di-GMP than the control (aerobic condition: 12.96 vs. 10.05 pmol/mg bacterial protein; low pH: 13.76 to 10.05 pmol/mg bacterial protein), extracellular proteins increased more obviously and more statistically than polysaccharides (proteins: aerobic condition vs control: 69.73 vs. 62.50 mg/gVSS; low pH condition vs control: 72.73 vs. 62.50 mg/gVSS), and the anammox bacteria grew in the form of bigger granules than the control, especially for the condition with low pH (108.60 vs. 100.33 μm).

By applying the metabolic analysis, Tang et al. (2018a) found AHLs mainly regulated the synthesis of the amino acids Ala, Val, and Glu and selectively regulated Asp and Leu to affect extracellular proteins. Simultaneously, all AHLs regulated the ManNAc biosynthetic pathways, while 3-oxo-C_6-HSL, 3-oxo-C_8-HSL, and C_6-HSL particularly enriched the UDP-GlcNAc pathway to promote exopolysaccharides, resulting in different aggregation levels of the anammox consortia. In detail, 3-oxo-C_6-HSL increased the amount of Asp, 3-oxo-C_8-HSL exclusively increased the amounts of Asp, Leu, C_6-HSL and C_{12}-HSL increased the amount of Leu, and all AHLs enhanced ManNAc biosynthesis. Thus, AHLs promoted the production of extracellular proteins mainly by increasing the amounts of Ala, Val, and Glu and by promoting the enrichment of Asp and Leu.

AHLs might affect nitrifying activity by regulating biofilm formation in terms of EPS production. The higher protein contents and adhesion protein genes abundances in the tight EPS of nitrification and anammox systems was likely a result of high AHLs contents in the EPS phase (Sun et al., 2018a, 2018b). Besides, high contents of hydrophobic amino acids (histidine, alanine, valine, etc.) in the EPS phase might attribute to high contents of AHLs (mainly C_6-HSL and C_8-HSL) in anammox systems (Sun et al., 2018b). Recently, a metagenomics-based analysis revealed that abundances of exopolysaccharides, fructose-6-phosphate, and UDP-GlcNAc genes might be correlated with the contents of AHLs and abundances of AHLs-QS genes (Sun et al., 2019) (Figure 14.3). In addition, nitrifiers (*Nitrosomonas* and *Nitrospira*) and anammox (*Candidatus* Kuenenia) harbored both EPS and AHLs synthesis genes, indicating that AHLs might exert important roles in EPS production in the nitrification and anammox systems (Sun et al., 2019). These studies may provide the genetic evidences that AHLs QS favored the production of more extracellular proteins.

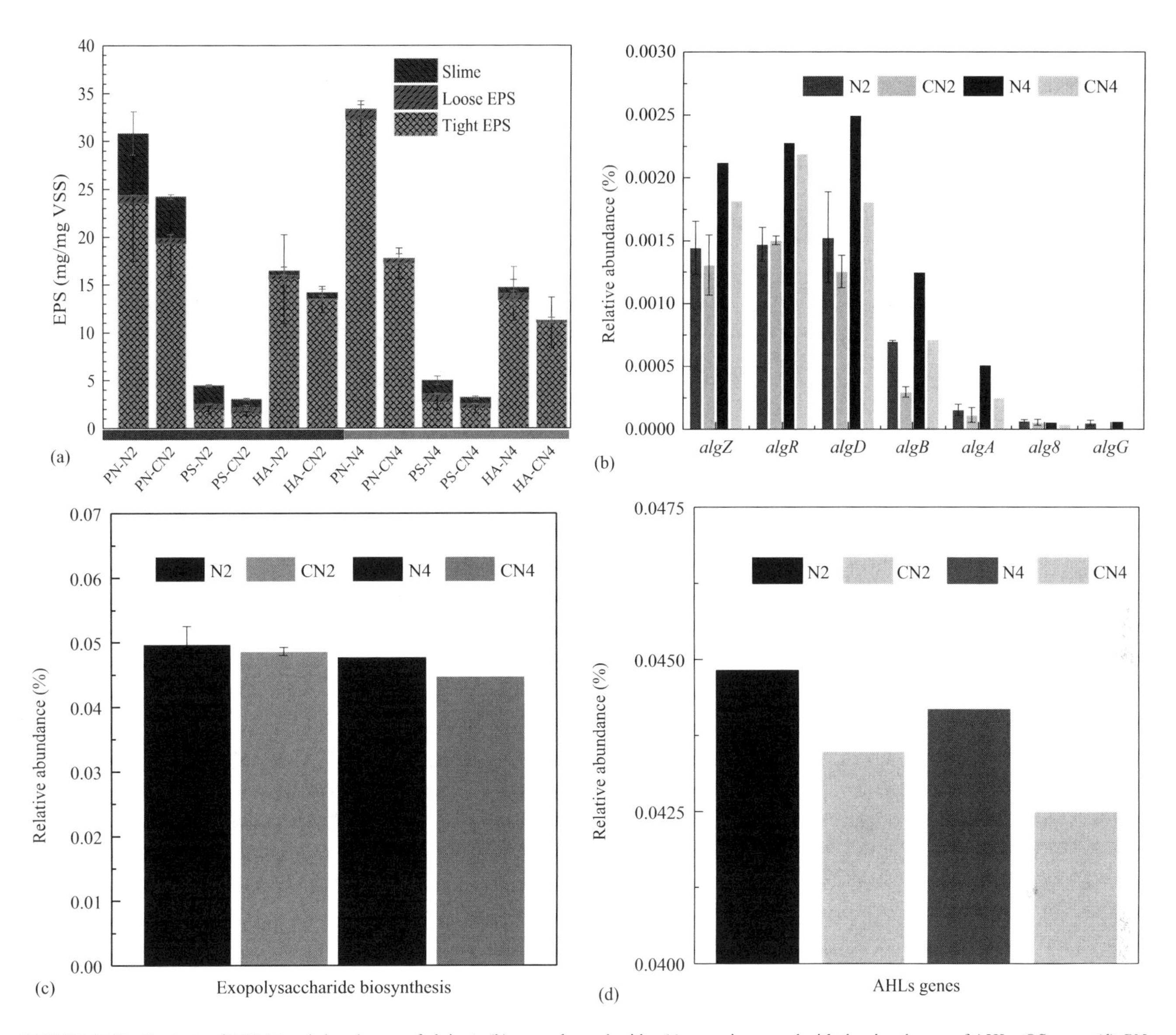

FIGURE 14.3 Contents of EPS (a) and abundances of alginate (b), exopolysaccharides (c) genes increased with the abundances of AHLs-QS genes (d). PN, PS, and HA represent protein, polysaccharides, and humic acids, respectively. N and CN represent combined nitrification and anammox systems without and with the addition of organics. (From Sun, Y. et al., *Bioresour. Technol.*, 272, 146–55, 2019.) *algA*, *algB*, *alg8*, *algD*, *algR*, *algG*, and *algZ* are genes encoding alginate production.

14.5 Quorum Sensing Induced Microbial Interactions During Autotrophic Nitrogen Removal

In the autotrophic nitrogen removal systems, AOB, NOB, anammox bacteria, and heterotrophs co-exist. Therefore, the interactions among all these microorganisms are very important for the clarification of their functions and also proposing the regulation strategies for improving system performance. Recently, there have been several papers published in this field that enhance our understanding of the ecology of this system.

For microbial interactions, due to the production of similar types of AHLs or other signal substances, microbial communication through QS could be possible. Mellbye et al. (2017) proposed that AHL cross talk between the AOB *N. multiformis* and the NOB *N. winogradskyi* and *N. vulgaris* might be possible, as they all utilize a similar AHL with a 10-carbon acyl tail and can be found in soil environments; in addition, AHL receptor homologs vary in signal specificity, so possible cross talk with a *Nitrospira* species that produces C_8-HSL, similar to *N. moscoviensis*, may also be possible. Therefore, through regulation of signal substances, microbial ecology could be managed.

In the anammox system, Lawson et al. (2017) found that in the anammox granules, genomes affiliated with *Brocadia* and *Chlorobi* dominated the abundance and gene expression, with the relative abundance and gene expression of *Brocadia* of 62% and 54% and *Chlorobi* of 21% and 26%. When respiring nitrate (produced during anammox) to nitrite, *Chlorobi* in anammox granules could degrade and catabolize extracellular peptides bounded in the EPS matrix, with less costly amino acids (e.g., glutamine) being preferentially used as carbon and energy sources. Heterotrophic bacteria *Chlorobi* and *Chloroflexi*-affiliated bacteria were missing pathways for the synthesis of many hydrophobic amino acids, which could be synthesized by *Brocadia* and other heterotrophs (*Bacteroidetes*-affiliated bacteria and *Proteobacteria*-affiliated bacteria). Also, several heterotrophs were missing key genes involved in B-vitamin biosynthesis, which were expressed in the *Brocadia* genome, suggesting that anammox bacteria might support B-vitamin requirements for the community. Furthermore, Zhao et al. (2018c) found that (i) *Armatimonadetes* and *Proteobacteria* could contribute to the secondary metabolites molybdopterin cofactor and folate for anammox bacteria to benefit their activity and growth; (ii) partial nucleotide sugars could be biosynthesized by *Brocadia* and then transported inside *Chloroflexi* for exopolysaccharide production; (iii) *Chloroflexi*-affiliated bacteria encoded the function of biosynthesizing exopolysaccharides for anammox consortium aggregation, based on the partial nucleotide sugars produced by anammox bacteria; (iv) *Chlorobi*-affiliated bacteria had the ability to degrade extracellular proteins produced by anammox bacteria to amino acids to affect consortium aggregation sequestered in the extracellular matrix for self-use and as carbon and energy sources for other organisms in the anammox consortia; (v) the *Chloroflexi*-affiliated bacteria harbored genes for a nitrite loop and could have a dual role in anammox performance during reactor start-up, where the nitrate produced from the anammox reaction can be reduced to nitrite by partial denitrifiers in the anammox consortia. Subsequently, the nitrite can be reused by the anammox bacteria, and the nitrite loop increases the nitrogen removal efficiency. For the possible QS types, Tang et al. (2018b) found that bacterial communication was more active at the initial start-up than in the high loading-rate phase, and *hdts* (key gene for producing AHL) and *rpfF* (key gene for producing DSF) were the primary communication engines in the anammox community. Among these two genes, anammox bacteria mainly used *hdts* genes to communicate with others.

In the partial nitrification and anammox systems, Speth et al. (2016) confirmed the existence of the nitrite loop. Most of the organisms capable of nitrate respiration and the hydrogenase-encoding organisms were present in the anaerobic fraction, while none of these organisms encode a known nitrite reductase, suggesting they extrude the formed nitrite, potentially allowing cyclic feeding. Formed nitrate can be reduced to nitrite, using either

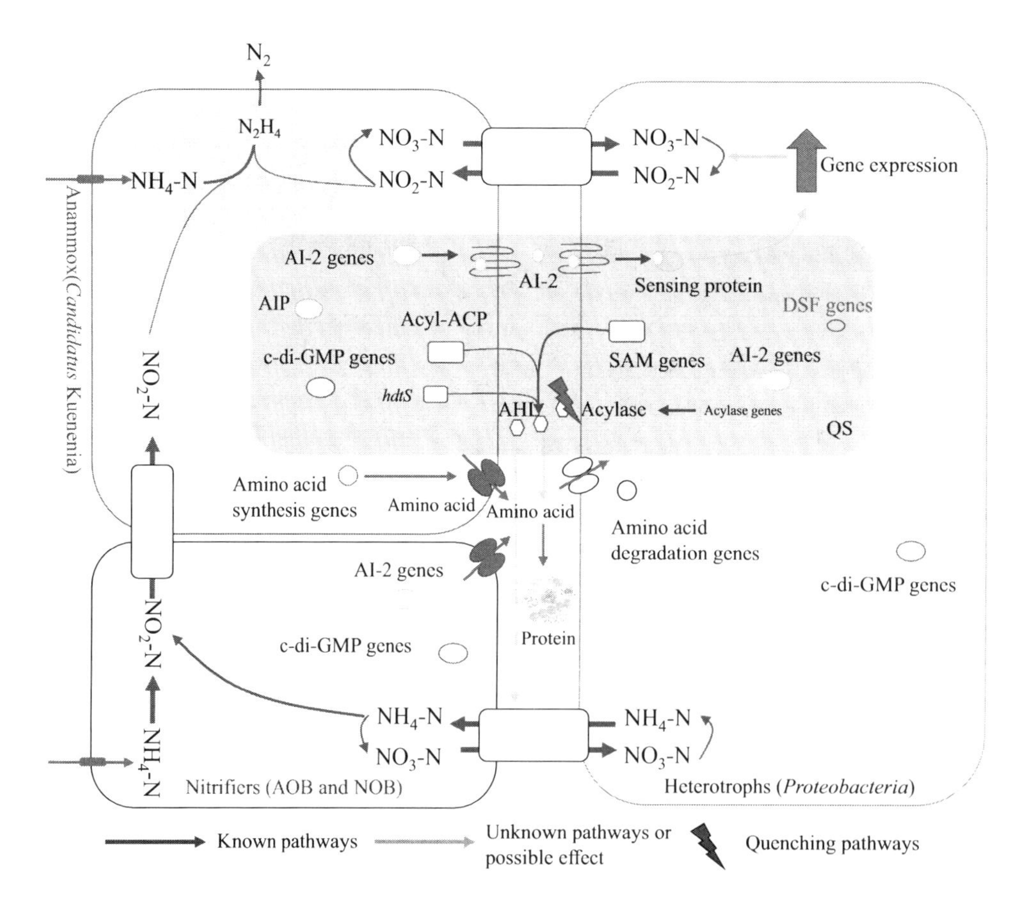

FIGURE 14.4 Proposed material and information cycles among *Candidatus* Kuenenia, nitrifiers, and heterotrophs (*Proteobacteria*). NO_2-N oxidation and reduction by *Candidatus* Kuenenia and heterotrophs, respectively, result in the distributed NO_2-N loop. NH_4-N oxidation and reduction by AOB and heterotrophs (NO_3-N assimilation/dissimilation reduction), respectively, result in the NH_4-N loop. Signal substances (such as AHLs and AI-2) are synthesized and sensed/degraded by *Candidatus* Kuenenia and heterotrophs, respectively. (From Sun, Y. et al., *Bioresour. Technol.*, 272, 146–55, 2019.)

organic matter or hydrogen as the electron donor, by *Chlorobi*-affiliated bacteria, *Acidobacteria*, and *Omnitrophica*, making additional nitrite available for *Candidatus* Brocadia-affiliated bacteria. The hydrogen required for autotrophic nitrate reduction can be formed through fermentation of organic matters by *Chlorobi*-affiliated bacteria and *Chloroflexi*-affiliated bacteria. The organic matter required for both nitrate reduction and fermentation can come from the substantial amount of electron donor in the influent, dead organic matter in the granule or extracellular granule matrix synthesized by autotrophs.

Based on the above results, the possible microbial interactions in the autotrophic nitrogen removal systems could be summarized in Figure 14.4 as modified from Sun et al. (2019). For the conversion of nitrogen, this includes the cooperation of AOB and anammox for converting ammonia and nitrite to N_2, with the production of nitrate. The heterotrophs would convert nitrate to nitrite, which can be further utilized by anammox bacteria. While for the carbon sources, the organic carbon produced by autotrophs will release to the solution, which can be utilized by heterotrophs, and the CO_2 produced by heterotrophs may be utilized by autotrophs. For microbial communication, the signal substances produced will be used as the language for cross talking among all microorganisms. Through the above mechanisms, the material flow and information communication could be completed.

14.6 Conclusions and Future Development Perspectives

In the autotrophic nitrogen removal systems, some QS related studies have been reported. Especially in recently years, some novel results have been obtained by combining LC-MS based chemical methods and also the metagenomics based molecular methods. The main conclusions could be summarized:

- QS exists in the autotrophic nitrogen removal systems, with different types of signal substances and also different functions under various conditions.
- QS can enhance biofilm growth, microbial activities, and EPS production. However, different types of signal substances may have different effects.
- Microbial communities within autotrophic nitrogen removal systems are very important for system improvement and also for system control.

However, the following aspects should be paid further attention in future studies.

- The detection methods should be further developed, especially with the rapid development of chromatography techniques. For signal substance detection, not only the liquid phase signal substances should be measured, but also those inside the biomass or EPS. In addition, the database for metagenomics-based analysis should be further built for better comparison from different studies.
- Through intensive studies, the signal substances distribution and function in various types of systems would be obtained, which would benefit the clarification of their function and control. Through the advanced analytical methods, the possible QS and the types of signal substances could be enlarged, which may enhance understanding of their potential roles in autotrophic nitrogen removal systems.
- The system operational conditions, the physiology status of microorganisms, and the microbial interactions should be investigated intensively for their effects on QS and the production of signal substances.

ACKNOWLEDGMENT

This research was supported by the National Natural Science Foundation of China (No: 51738005).

REFERENCES

Andrews, J. H. and R. F. Harris. 1986. r and K-selection and microbial ecology. In *Advances in Microbial Ecology*, ed. K. C. Marshall, Vols. 9, pp. 99–147. New York: Springer US.

Batchelor, S. E, M. Cooper, S. R. Chhabra, L. A. Glover, G. S. A. B. Stewart, P. Williams and J. I. Prosser. 1997. Cell density-regulated recovery of starved biofilm populations of ammonia-oxidizing bacteria. *Applied and Environmental Microbiology* 63(6): 2281–6.

Bouffartigues, E., J. A. Moscoso, R. Duchesne, T. Rosay, L. Fito-Boncompte, G. Gicquel, O. Maillot et al. 2015. The absence of the *Pseudomonas aeruginosa* OprF protein leads to increased biofilm formation through variation in c-di-GMP level. *Frontiers in Microbiology* 6: 630.

Burton, E. O., H. W. Read, M. C. Pellitteri and W. J. Hickey. 2005. Identification of Acyl-homoserine lactone signal molecules produced by *Nitrosomonas europaea* strain Schmidt. *Applied and Environmental Microbiology* 71(8): 4906–9.

Campagna, S. R., J. R. Gooding and A. L. May. 2009. Direct quantitation of the quorum sensing signal, autoinducer-2, in clinically relevant samples by liquid chromatography-tandem mass spectrometry. *Analytical Chemistry* 81(15): 6374–81.

Chen, H., A. Li, Q. Wang, D. Wu, C. Cui and F. Ma. 2018. *N*-Acyl-homoserine lactones and autoinducer-2-mediated quorum sensing during wastewater treatment. *Applied Microbiology and Biotechnology* 102: 1119–30.

Chua, S., K. Sivakumar, M. Rybtke, M. Yuan, J. B. Andersen, T. E. Nielsen, M. Givskov et al. 2015. C-di-GMP regulates *Pseudomonas aeruginosa* stress response to tellurite during both planktonic and biofilm modes of growth. *Scientific Reports* 5: 10052.

De Clippeleir, H., T. Defoirdt, L. Vanhaecke, S. E. Vlaeminck, M. Carballa, W. Verstraete and N. Boon. 2011. Long-chain acylhomoserine lactones increase the anoxic ammonium oxidation rate in an OLAND biofilm. *Applied Microbiology and Biotechnology* 90(4): 1511–9.

Ding, S., P. Zheng, H. Lu, J. Chen, Q. Mahmood and G. Abbas. 2013. Ecological characteristics of anaerobic ammonia oxidizing bacteria. *Applied Microbiology and Biotechnology* 97: 1841–9.

Fuqua, C. and S. C. Winans. 1996. Conserved cis-acting promoter elements are required for density-dependent transcription of *Agrobacterium tumefaciens* conjugal transfer genes. *Journal of Bacteriology* 178(2): 435–40.

Fuqua, C., S. C. Winans and E. P. Greenberg. 1996. Census and consensus in bacterial ecosystems: The LuxR–LuxI family of quorum-sensing transcriptional regulators. *Annual Review of Microbiology* 50: 727–51.

Gao, J., A. Ma, X. Zhuang and G. Zhuang. 2014. An *N*-acyl homoserine lactone synthase in the ammonia-oxidizing bacterium *Nitrosospira multiformis*. *Applied and Environmental Microbiology* 80(3): 951–8.

Gao, J., Y. Duan, Y. Liu, X. Zhuang, Y. Liu, Z. Bai, W. Ma and G. Zhuang. 2019. Long- and short-chain AHLs affect AOA and AOB microbial community composition and ammonia oxidation rate in activated sludge. *Journal of Environmental Sciences* 78: 53–62.

Guo, Y., S. Liu, X. Tang and F. Yang. 2017b. Role of c-di-GMP in anammox aggregation and systematic analysis of its turnover protein in *Candidatus Jetteniacaeni*. *Water Research* 113: 181–90.

Guo, Y., S. Liu, X. Tang, C. Wang, Z. Niu and Y. Feng. 2017a. Insight into c-di-GMP regulation in anammox aggregation in response to alternating feed loadings. *Environmental Science and Technology* 51(16): 9155–64.

Hawver, L. A., S. A. Jung and W.-L. Ng. 2016. Specificity and complexity in bacterial quorum-sensing systems. *FEMS Microbiology Reviews* 40: 738–52.

He, Y.-W., J. Wu, J.-S. Cha and L.-H. Zhang. 2010. Rice bacterial blight pathogen *Xanthomonas oryzae* pv. *Oryzae* produces multiple DSF-family signals in regulation of virulence factor production. *BMC Microbiology* 10: 187–96.

Hu, H., J. He, J. Liu, H. Yu, J. Tang and J. Zhang. 2016. Role of *N*-acyl-homoserine lactone (AHL) based quorum sensing on biofilm formation on packing media in wastewater treatment process. *RSC Advances* 6(14): 11128–39.

Huang, J., Y. Shi, G. Zeng, Y. Gu, G. Chen, L. Shi, Y. Hu, B. Tang and J. Zhou. 2016. Acyl-homoserine lactone-based quorum sensing and quorum quenching hold promise to determine the performance of biological wastewater treatments: An overview. *Chemosphere* 157: 137–51.

Ishizaki, S., R. Sugiyama and S. Okabe. 2017. Membrane fouling induced by AHL-mediated soluble microbial product (SMP) formation by fouling-causing bacteria co-cultured with fouling-enhancing bacteria. *Scientific Reports* 7: 8482.

Jenni, S., S. E. Vlaeminck, E. Morgenroth and K. M. Udert. 2014. Successful application of nitritation/anammox to wastewater with elevated organic carbon to ammonia ratios. *Water Research* 49: 316–26.

Junio, H. A., D. A. Todd, K. A. Ettefagh, B. M. Ehrmann, J. S. Kavanaugh, A. R. Horswill and N. B. Cech. 2013. Quantitative analysis of autoinducing peptide I (AIP-I) from *Staphylococcus aureus* cultures using ultrahigh performance liquid chromatography-high resolving power mass spectrometry. *Journal of Chromatography B* 930(1): 7–12.

Lawson, C. E., S. Wu, A. S. Bhattacharjee, J. J. Hamilton, K. D. McMahon, R. Goel and D. R. Noguera. 2017. Metabolic network analysis reveals microbial community interactions in anammox granules. *Nature Communication* 8: 15416.

Li, A.-J., B.-L. Hou and M.-X. Li. 2015. Cell adhesion, ammonia removal and granulation of autotrophic nitrifying sludge facilitated by *N*-acyl-homoserine lactones. *Bioresource Technology* 196: 550–8.

Mellbye, B. L., A. T. Giguere, P. J. Bottomley and L. A. Sayavedra-Soto. 2016. Quorum Quenching of *Nitrobacter winogradskyi* suggests that quorum sensing regulates fluxes of nitrogen oxide(s) during nitrification. *mBio* 7(5): e01753–16.

Mellbye, B. L., E. Spieck, P. J. Bottomley and L. A. Sayavedra-Soto. 2017. Acyl-Homoserine lactone production in nitrifying bacteria of the genera *Nitrosospira*, *Nitrobacter*, and *Nitrospira* identified via a survey of putative quorum-sensing genes. *Applied and Environmental Microbiology* 83(22): pii e01540-17.

Mellbye, B. L., P. J. Bottomley and L. A. Sayavedra-Soto. 2015. Nitrite-oxidizing bacterium *Nitrobacter winogradskyi* produces *N*-Acyl-Homoserine lactone autoinducers. *Applied and Environmental Microbiology* 81(17): 5917–26.

Ng, W. L. and B. L. Bassler. 2009. Bacterial quorum-sensing network architectures. *Annual Review of Genetics* 43: 197–222.

Paggi, R. A., C. B. Martone, C. Fuqua and R. E. Castro. 2003. Detection of quorum sensing signals in the haloalkaliphilic archaeon *Natronococcus occultus*. *FEMS Microbiology Letter* 221: 49–52.

Pianka, E. R. 1970. On r- and K- selection. *The American Naturalist* 104(940): 592–7.

Ryan, R. P. and J. M. Dow. 2011. Communication with a growing family: Diffusible signal factor (DSF) signaling in bacteria. *Trends in Microbiology* 19(3): 145–52.

Shaw, P. D., G. Ping, S. L. Daly, C. Cha, J. E. Cronan, K. L. Rinehart and S. K. Farrand. 1997. Detecting and characterizing *N*-acyl-homoserine lactone signal molecules by thin-layer chromatography. *Proceedings of the National Academy of Sciences of the United States of America* 94(12): 6036–41.

Shen, Q., J. Gao, J. Liu, S. Liu, Z. Liu, Y. Wang, B. Guo, X. Zhuang and G. Zhuang. 2016. A new acyl-homoserine lactone molecule generated by *Nitrobacter winogradskyi*. *Scientific Report* 6: 22903.

Song, X.-N, H.-B. Qiu, X. Xiao, Y.-Y. Cheng, W.-W. Li, G.-P. Sheng, X.-Y. Li and H.-Q. Yu. 2014. Determination of autoinducer-2 in biological samples by high-performance liquid chromatography with fluorescence detection using pre-column derivatization. *Journal of Chromatography A* 1361: 162–8.

Speth, D. R., M. H. in't Zandt, S. Guerrero-Cruz, B. E. Dutilh and M. S. M. Jetten. 2016. Genome-based microbial ecology of anammox granules in a full-scale wastewater treatment system. *Nature Communication* 7: 11172.

Strous, M., J. A. Fuerst, E. H. M. Kramer, S. Logemann, G. Muyzer, K. T. van de Pas-Schoonen, R. Webb, J. G. Kuenen and M. S. M. Jetten. 1999. Missing lithotroph identified as new planctomycete. *Nature* 400: 446–9.

Sun, Y., K. He, Q. Yin, S. Echigo, G. Wu and Y. Guan. 2018c. Determination of quorum-sensing signal substances in water and solid phases of activated sludge systems using liquid chromatography-mass spectrometry. *Journal of Environmental Science* 69: 85–94.

Sun, Y., Y. Guan, D. Wang, K. Liang and G. Wu. 2018a. Potential roles of acyl homoserine lactone based quorum sensing in sequencing batch nitrifying biofilm reactors with or without the addition of organic carbon. *Bioresource Technology* 259: 136–45.

Sun, Y., Y. Guan, D. Zeng, K. He and G. Wu. 2018b. Metagenomics-based interpretation of AHLs-mediated quorum sensing in Anammox biofilm reactors for low-strength wastewater treatment. *Chemical Engineering Journal* 344: 42–52.

Sun, Y., Y. Guan, H. Wang and G. Wu. 2019. Autotrophic nitrogen removal in combined nitritation and Anammox systems through intermittent aeration and possible microbial interactions by quorum sensing analysis. *Bioresource Technology* 272: 146–55.

Tan, C.,K. S. Koh, C. Xie, M. Tay, Y. Zhou, R. Williams, W. J. Ng, S. A. Rice and S. Kjelleberg. 2014. The role of quorum sensing signalling in EPS production and the assembly of a sludge community into aerobic granules. *ISME Journal* 8(6): 1186–97.

Tang, X., S. Liu, Z. Zhang and G. Zhuang. 2015. Identification of the release and effects of AHLs in anammox culture for bacteria communication. *Chemical Engineering Journal* 273: 184–91.

Tang, X., Y. Guo, B. Jiang and S. Liu. 2018b. Metagenomic approaches to understanding bacterial communication during the anammox reactor start-up. *Water Research* 136: 95–103.

Tang, X., Y. Guo, S. Wu, L. Chen, H. Tao and S. Liu. 2018a. Metabolomics uncovers the regulatory pathway of Acyl-homoserine lactones based quorum sensing in anammox consortia. *Environmental Science and Technology* 52(4): 2206–16.

Wang, C., S. Liu, X. Xu, C. Zhang, D. Wang and F. Yang. 2018. Achieving mainstream nitrogen removal through simultaneous partial nitrification, anammox and denitrification process in an integrated fixed film activated sludge reactor. *Chemosphere* 203: 457–66.

Wongsuk, T., P. Pumeesat and N. Luplertlop. 2016. Fungal quorum sensing molecules: Role in fungal morphogenesis and pathogenicity. *Journal of Basic Microbiology* 56(5): 440–7.

Wu, G., K. He, J. Miao, Q. Yin and Y. Zhao. 2018. Strategies for sustainable wastewater treatment based on energy recovery and emerging compounds control: A mini-review. *Desalination and Water Treatment* 127: 26–31.

Wu, L.-J., A.-J. Li, B.-J. Hou and M.-X. Li. 2017. Exogenous addition of cellular extract *N*-acyl-homoserine-lactones accelerated the granulation of autotrophic nitrifying sludge. *International Biodeterioration & Biodegradation*, 118: 119–25.

Xu, F., X. Song, P. Cai, G. Sheng and H. Yu. 2017. Quantitative determination of AI-2 quorum-sensing signal of bacteria using high performance liquid chromatography-tandem mass spectrometry. *Journal of Environmental Sciences* 52: 204–9.

Yang, Y., Y. Li, T. Gao, Y. Zhang and Q. Wang. C-di-GMP turnover influences motility and biofilm formation in *Bacillus amyloliquefaciens* PG12. 2018. *Research in Microbiology* 169: 205–13.

Zhang, W. and C. Li. 2016. Exploiting quorum sensing interfering strategies in Gram-negative bacteria for the enhancement of environmental applications. *Frontiers in Microbiology* 6: 1535.

Zhao, J., C. Quan, L. Jin and M. Chen. 2018a. Production, detection and application perspectives of quorum sensing autoinducer-2 in bacteria. *Journal of Biotechnology* 268: 53–60.

Zhao, R., H. Zhang, F. Zhang and F. Yang. 2018b. Fast start-up anammox process using Acyl-homoserine lactones (AHLs) containing supernatant. *Journal of Environmental Sciences* 65: 127–32.

Zhao, Y., S. Liu, B. Jiang, Y. Feng, T. Zhu, H. Tao, X. Tang and S. Liu. 2018c. Genome-centered metagenomics analysis reveals the symbiotic organisms possessing ability to cross-feed with anammox bacteria in anammox consortia. *Environmental Science and Technology* 52(19): 11285–96.

Zhou, L., L.-H. Zhang, M. Cámara and Y.-W. He. 2016. The DSF family of quorum sensing signals: Diversity, biosynthesis, and turnover. *Trends in Microbiology* 25(4): 293.

15

Mechanism and Types of Quorum Sensing Inhibitors

Christiane Chbib

CONTENTS

15.1 Introduction

Bacterial quorum sensing (QS) is defined as the bacterial ability to control gene expression. This process is stimulated through the recognition of a minimal threshold stimulatory concentration of chemicals known as autoinducers.[1] Quorum sensing is implicated in bacterial bioluminescence, first discovered in *Vibrio fischeri*,[2] where light emission is controlled by the secretion and detection of autoinducer molecules after a certain threshold is reached. Other implications of QS include virulence factors, biofilm formation, conjugation, sporulation, and swarming-motility.[3–6] New QS systems and signal molecules are identified. Autoinducers (AI) are conventionally divided into three groups: acyl homoserine lactones (AHL, AI-1)[7] used in Gram negative bacteria, oligopeptides used in Gram positive bacteria, and a universal group of autoinducer-2 (AI-2) found in both Gram positive and negative bacteria.[8–12] This chapter will discuss the types and mechanisms of quorum sensing inhibitors.

15.2 Quorum Sensing Inhibition Pathways

Studies have shown that cooperative genes regulated by the quorum sensing of bacteria are sensitive to nutrient availability. This fact proposes the integration of the metabolism process in bacterial QS.[13] Boyle and coworkers provided new insights into the metabolic prudence and its stabilization of the cooperation in multicellular groups. They found that when the conditions are rich in nutrients, *Pseudomonas aeruginosa* uses population density to modulate the amount of *rhlAB* expression, whereas when there is starvation, the bacteria adapt by adjusting the secretion of rhamnolipids biosurfactant that helps in the swarming process.[13]

At least six QS pathways have been identified in multiple bacteria (Table 15.1).

Quorum quenching (QQ) is defined as the approach of reducing or destroying the QS signal.[15] The strategies currently used for QQ include (1) inhibition of the signal generation, (2) interfering with the signal sequestration and degradation,

TABLE 15.1

Different Quorum Sensing Pathways in Bacteria[14]

Pathway	Signal Molecule	Bacteria
AI-1	Various AHLs	Gram negative
PQS and HHL	*Pseudomonas* quinolone signal 4-hydroxy-2-heptylquinoline	Gram negative
AI-3	Epinephrine/Norepinephrine	Gram negative
AI-2	Furanosyl borate esters of DPD	Gram negative and positive
AIP	Various oligopeptides	Gram positive
CAI-1	(*S*)-3-hydroxytricedan-4-one	Gram negative (*V. cholerae*)

and (3) interfering with signal reception and transduction.[16] Some of the major enzymes involved in quorum quenching include lactonase that breaks the lactone ring, acylase that attacks the acyl group, and oxidoreductase that reduces the carbonyl moieties.

15.3 Techniques and Tools to Study QS Inhibition

To test the QS inhibition of the compounds, different techniques were applied including the use of LuxS inhibition assay defined as a reaction initiated with LuxS and DTNB (5,5′-dithiobis (2-nitrobenzoic acid) addition. The reaction is monitored at 412 nm with a spectrophotometer.[74] Detection of homocysteine release, due to the cleavage *S*-ribosyl-homocysteine by LuxS enzyme, is also used to determine the activity of the QS inhibitor. The use of either Ellman reagent or mercaptide can detect the release of homocysteine by interacting chemically with the homocysteine released. The product of the chemical reaction can be detected at 412 nm and 245 nm, respectively. The QS inhibition selector strain contains a gene that encodes for a lethal protein fused into the promoter and results in the growth block and cell death. An example of reporter systems used in QS is *E. coli*-based reporter strains, which identify AHL and AI-2 molecules; they carry the *ccdB* lethal gene.[17] Other methods to determine QS activity include OD600 to measure bacterial growth.

15.3.1 Autoinducing Peptide (AIP)-Mediated Quorum Sensing Pathway

This type of autoinducer is an oligopeptide that is primarily functional in Gram positive bacteria. It can include 5–17 amino acids and is post-translationally modified when there is the incorporation of lactone, thiolactone, lanthionine, or isoprenyl groups.[14] AIP binds to the receptors on the bacterial surface formed by the two-component adaptive system which includes a sensing component (receptor) and a response-regulatory protein which controls the activation of the gene. Upon binding of the AIP to the receptor protein, the synthesis and release of AIP occurs. The histidine sensor kinase protein (H) is activated leading to the autophosphorylation of histidine and the transfer of a phosphoryl group to a cognate response regulator, which leads to the activation of the QS.[18,19]

The AIP inhibition pathway can be targeted by multiple inhibition of (1) AIP receptor; (2) histidine kinase; (3) phosphoryl transfer from histidine to aspartate; (4) enzymes responsible for steps related to synthesis, post-translational modification, and cleavage of AIP precursor; and (5) efflux transporter, which will block the export of AIP.[17,18]

15.3.1.1 Inhibitors That Target Histidine Kinase

The evaluation and screening of potential inhibitors is based on *in vitro* inhibition of the autophosphorylation step in the pathway and bacterial-cell based assay with downstream reporter genes to monitor the two-component signaling activity.[17,20,21]

Compound **1** (Figure 15.1) was found to inhibit AlgR1-AlgR2 phosphorylation in *Pseudomonas aeruginosa* at 50 μg/mL (217 μM).[21] The hydrophobic tyramine inhibitor compound **2** has IC_{50} = 1.6 μM by inhibiting the transfer of phosphate from ATP to KinA.[22] The 2 salicylanilides Closantel **3** and tetrachlorosalicylanilide **4**, have shown to inhibit the two-component system by inhibiting the autophosphorylation of KinA with IC_{50} = 3.8 μM for **3** and IC_{50} = 45 μM for **4**.[23,24] Compounds **5** and **6** have shown to inhibit the phosphorylation in CheA, NRn, and KinA with 112 μM and 124 μM, respectively.[22] Compound **7** inhibits the autophosphorylation of histidine protein kinase YycG in Gram positive bacteria with IC_{50} = 64 μM.[25] The tryptophan derivative **8** has shown to inhibit histidine kinase YycG with IC_{50} = 44 μM comparable to **7**.[26]

15.3.1.2 Inhibitors That Target AIP Receptors

A truncated derivative of the AIP-II thiolactone peptide trAIP-II (**9**; Figure 15.2) was found by Muir and coworkers to have a broad spectrum of inhibition of Staphylococcus *aureus*. When tested against *agr* Group I-IV strains, the IC_{50} exhibited the following inhibition: 272 ± 67, 209 ± 39, 10 ± 1, and 188 ± 50 nM. This finding shows that trAIP-II inhibits *agr* signaling and bacterial QS.[27] Some of the reported inhibition of the trAIP-II analogues are stated in Table 15.2.[28] Based on the results below, the authors concluded the essential moieties in trAIP-II for the activity of the compound (Figure 15.3).[28]

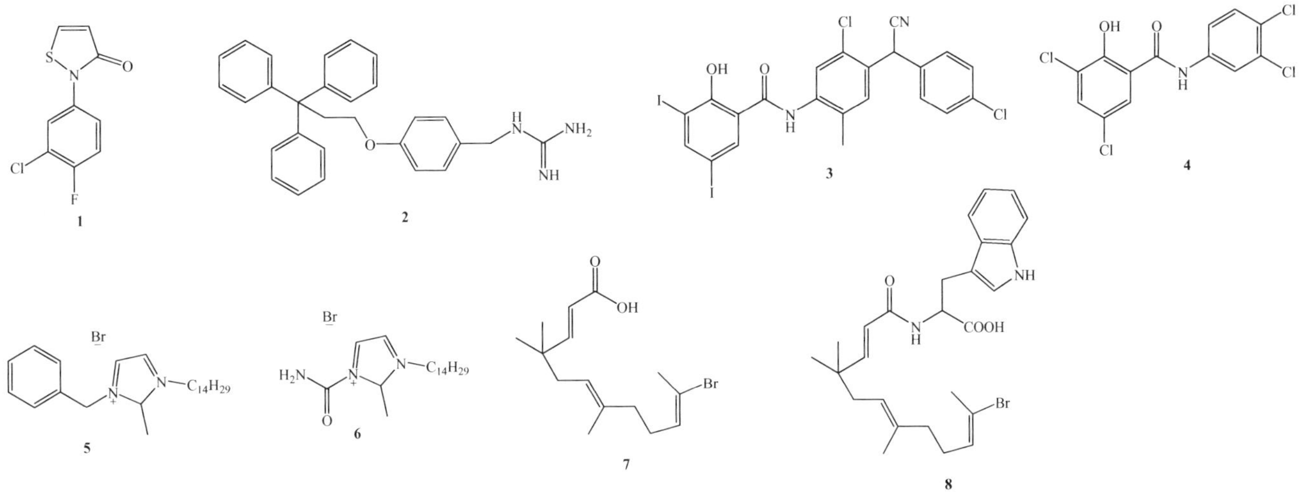

FIGURE 15.1 Histidine kinase inhibitors.[21–26]

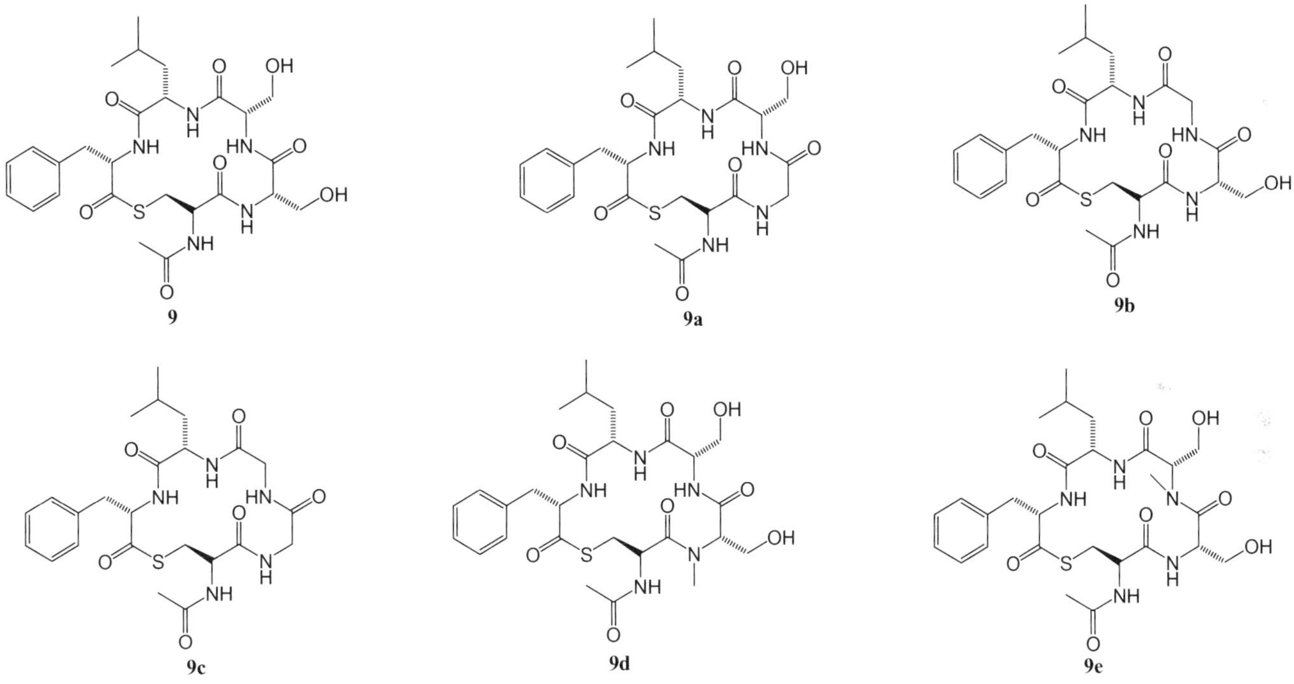

FIGURE 15.2 Structure of trAIP-II and its analogues.[27,28]

TABLE 15.2

IC_{50} Value of trAIP-II Analogues for AgrC-I and AgrC-II Inhibition[28]

Compound Number	IC_{50} Value for AgrC-I Inhibition (nM)	IC_{50} Value for AgrC-II Inhibition (nM)
9	90	293
9a	189	1160
9b	40	194
9c	77	238
9d	639	7880
9e	319	1050

Substitution of aspartate (Asp) in AIP1 by alanine (Ala) led to the potent inhibition of all *S. aureus* AgrCs with IC_{50} between 0.-21nM for different AgrCs.[29–31] Wright and coworkers linked the IC_{50} results of the AgrC inhibitors (Figure 15.4 and Table 15.3) with the hydrophobic interactions of one or both C-terminal non-polar residues in the peptide to a better ligand-receptor recognition, allowing for either activation or inhibition of QS.[32]

To determine the critical functional group in *S. aureus* group I AIP, William and coworkers studied the inhibition of the *agr* locus. Synthesized altered AIP analogues showed the ability to convert from activators to inhibitors *in vivo* against *S. aureus agr P3* strains[33] (Figure 15.5).

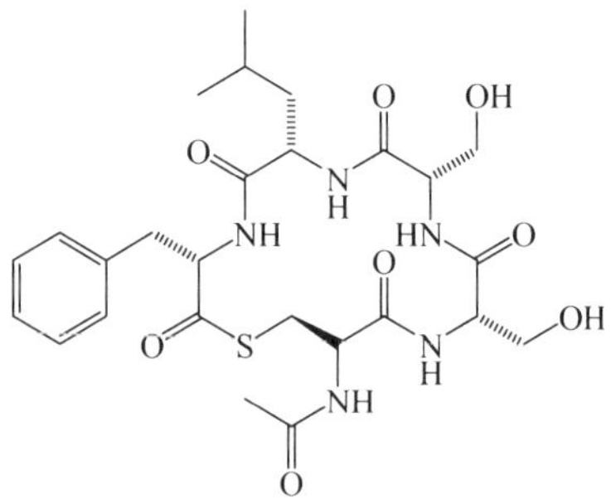

Benzyl and isopropy lgroups → Essential for the inhibitory activity
NH and OH groups → Non-essential for the inhibitory activity
Amide groups → Essential for cognate inhibitory activity

FIGURE 15.3 Essential moieties for optimal activity of trAIP-II.[28]

TABLE 15.3
Reported IC_{50} of Each AgrC Inhibitor[32]

Compound Number	IC_{50} Value Inhibition (nM)
10	AgrC-1: 5–21 AgrC-3: 0.3 AgrC-2: 4–8 AgrC-4: 3
11	AgrC-1: 137 AgrC-2: 2.8
12	AgrC-1: 295 AgrC-2: 19
13	AgrC-1: 303 AgrC-2: 18
14	AgrC-1: 10 AgrC-3: 0.3 AgrC-2: 230 AgrC-4: 12
15	AgrC-1: 5 AgrC-3: 0.1 AgrC-2: 5 AgrC-4: 5

15.3.2 The AHL Quorum Sensing Pathway

N-acyl-L-homoserine lactones AHL (**19**; Figure 15.6) is a signaling molecule used for Gram negative bacteria. It is composed of fatty acyl chains (4–16 carbons) linked to a lactonized homoserine through an amide bond.[34] LuxR-like proteins are the major regulatory proteins found in Gram negative bacteria. In Table 15.4, we discuss the type of organisms and their specific regulatory proteins in relation to their phenotypes.

The *N*-acyl-L-homoserine lactones (**19**, AHLs) are derived from S-adenosylmethionine (SAM, **20**, Scheme 15.1) with the assistance of the enzymatic action of LuxI as shown in Scheme 15.1.[48,49] Intramolecular cyclization of the homocysteine fragment with the elimination of the 5'-*S*-methyl-5'-thioadenosine and the concomitant acylation of the amino group with the common acyl carriers provided various AHL analogues.

The mechanism of QS of AHL starts with the diffusion of the latter in the cell. The interaction with the LuxR-type protein leads to acyl-HSL bond formation. It is divided into two domains: first, the conserved one which includes the *N*-terminal of LuxR from acyl-HSL-binding region; second, the C-terminus with a helix-turn-helix motif (HTH) required for DNA binding. The binding between *N*-terminal and acyl-HSL leads to conformation changes followed by a DNA binding.[50]

15.3.2.1 AHL Synthesis Inhibitors

Compounds *S*-adenosylhomocysteine **22** and sinefungin **23** are analogues of SAM (Figure 15.7). They inhibit the RhlI-catalyzed synthesis of C_4-HSL, a regulatory protein which therefore leads to the inhibition of quorum sensing. Because of the universal nature of SAM in living organisms, the use of this strategy for the inhibition of the QS leads to undesired side effects, which decreased further research on this mechanism.[51,52]

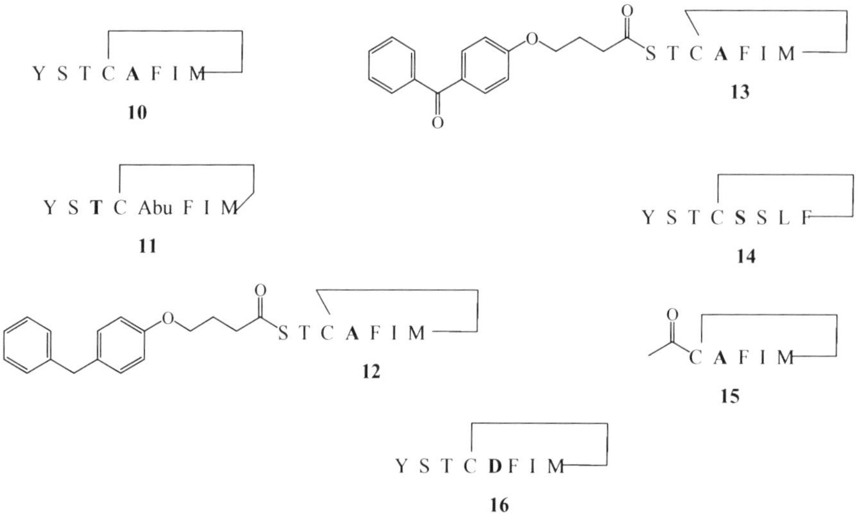

Y S T C A F I M
10
S T C A F I M
13
Y S T C Abu F I M
11
Y S T C S S L F
14
S T C A F I M
12
C A F I M
15
Y S T C D F I M
16

FIGURE 15.4 Inhibitors of AgrC receptors.[29–32]

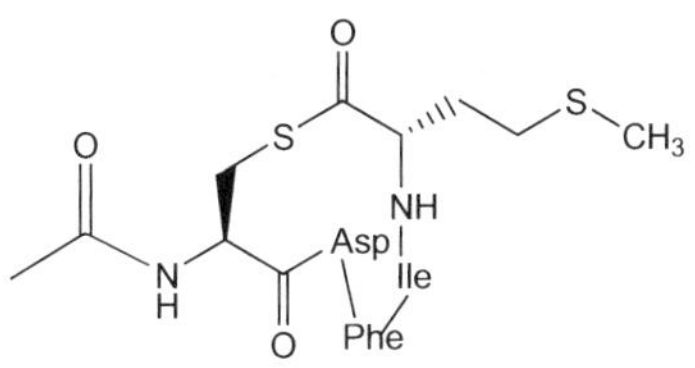

IC_{50} = 8 μM

16

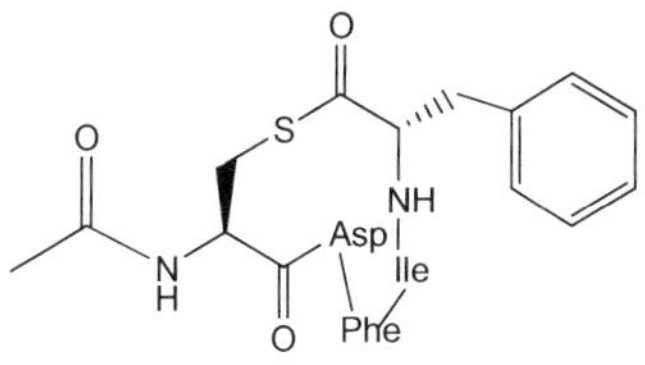

IC_{50} = 4 μM

17

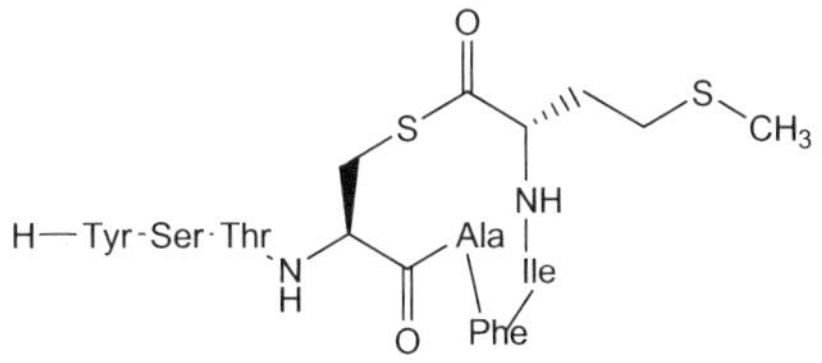

IC_{50} = 0.033 μM

18

FIGURE 15.5 Synthesized AIP analogues with their IC_{50} values.[33]

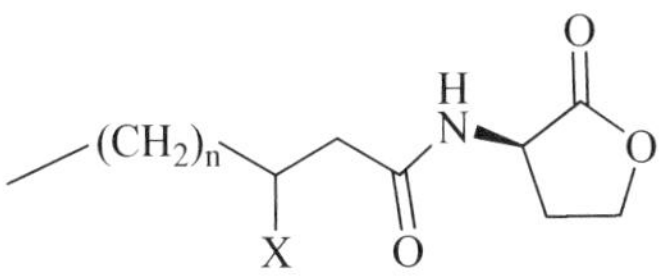

AHL

X= H, OH, or = O; n= 0-10

19

FIGURE 15.6 Chemical structure of AHL molecule.[34]

15.3.2.2 AHL Receptor Inhibitors

The synthesized analogues that act as AHL receptor inhibitors are grouped into three different structural modifications of the AHL molecule in its: (a) lactone ring, (b) acyl side chain, and (c) both lactone and acyl parts, in addition to molecules that inhibit AHL receptors and are unrelated structurally to AHL.

15.3.2.2.1 Modification of the Lactone Moiety

Modified lactones like compound **24** (Figure 15.8) have shown the ability to inhibit 70% fluorescent intensity at 100-fold

TABLE 15.4

The Different AHLs in Gram Negative Bacteria and Their Functions[35–47]

Organism	Signal	Regulatory Proteins	Phenotype
A. hydrophila	C_4-HSL	AhyI/AhyR	Exoprotease production
A. salmonicida	C_4-HSL	AsaI/AsaR	Extracellular protease
A. tumedaciens	3-oxo-C_8-HSL	TraI/TraR	Ti plasmid conjugation
B. cepacia	C_8-HSL	CepI/CepR	Protease siderophores
C. violaceum	C_6-HSL	CviI/CviR	Exoenzyme, antibiotics
E. carotovora	C_6-HSL	ExpI/ExpR	Exoenzyme, carbapenem antibiotic
E. chrysanthemi	3-oxo-C_6-HSL	ExpI/ExpR	Pectate lyase
P. aureofaciens	C_6-HSL	PhzI/PhzR	Phenazine antibiotics
P. aeruginosa	3-oxo-C_{12}-HSL	LasI/LasR RhlI/Rhlr	Multiple extracellular enzyme, biofilm formation
R. sphaeroides	C_{14}-HSL	CerI/RhlR	Dispersal from bacterial aggregate
S. liquefaciens	C_4-HSL	SwrI/SwrR	Extracellular protease
V. fischeri	C_6-HSL	LuxI/LuxR	Bioluminescence
V. harveyi	3-hydroxy-C_4-HSL	LuxM/N	Bioluminescence
Y. pseudoruberculosis	C_6-HSL	YpsR/I YthR/I	Regulation of clumping and motility
Y. pestis	C_{10}-HSL/C_6-HSL	YspI/YspR YpeI/YspR	Regulation of clumping and motility

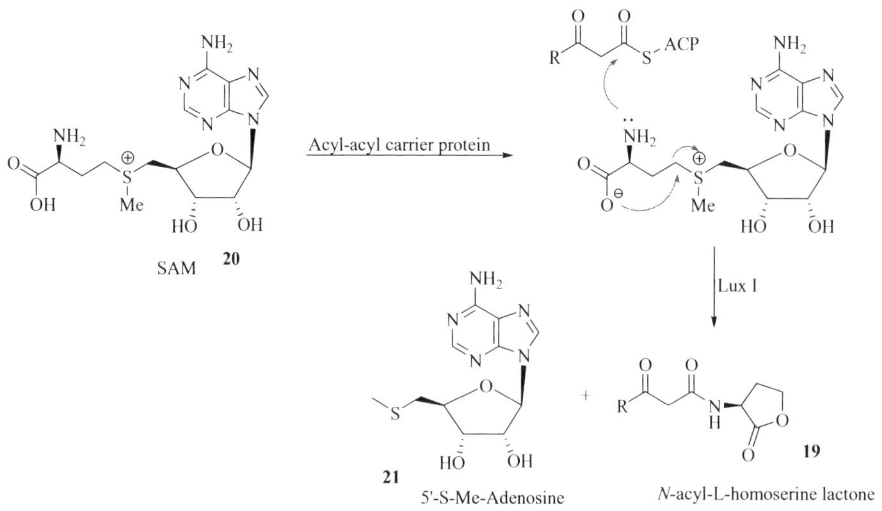

SCHEME 15.1 Biosynthesis of *N*-(acyl)-L-homoserine lactone.[48,49]

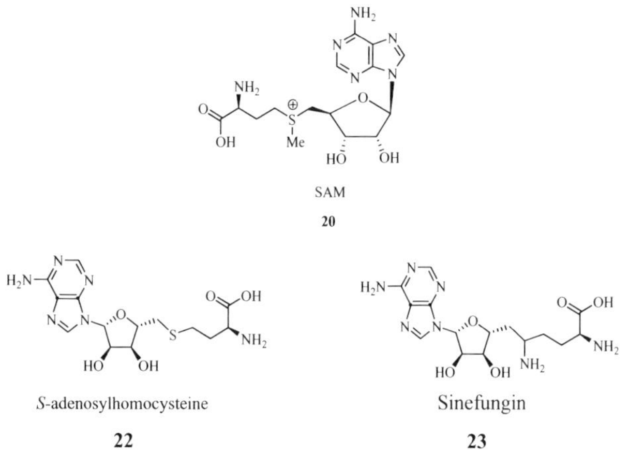

FIGURE 15.7 SAM analogues.[51,52]

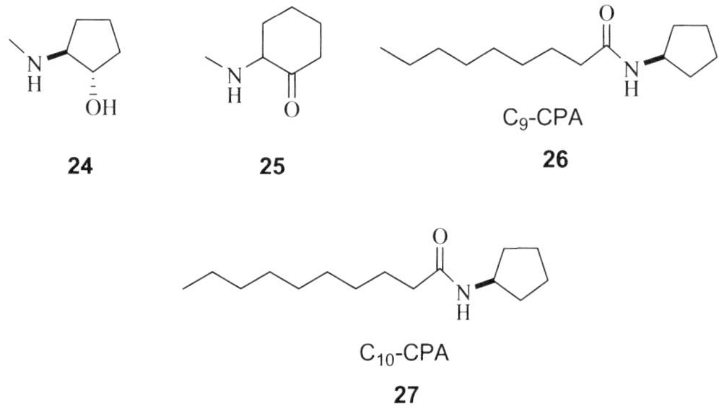

FIGURE 15.8 AHL lactone ring modification that inhibits AHL receptor.[53,54]

excess over that of PA-1 unlike compound **25** with only 35%.[53] Compound **26** has shown to inhibit prodigiosin production regulated by the production of *Serratia marcescens* AS-1 and the swarming motility and biofilm formation facilitated by *Serratia liquefaciens* MG1 (AS-1).[54] Compound **27** (Figure 15.8) has shown to inhibit *P. aeruginosa* quorum sensing *lasB-lacZ* expression by PA1 with IC_{50} = 80 μM and *rhlA-lacz* expression by PAI2.

15.3.2.2.2 Modification of the Acyl Side Chain

Some AHL derivatives can be agonists in certain bacterial species and strains but antagonists in others.[55] This can be caused by the ability of cross-communication among bacteria facilitated by these compounds.[56] The need to optimize the chemical structures to remove the agonistic activity but at the same time retain the antagonistic ones was studied. Multiple conclusions were drawn based on the activity of the compounds synthesized: (1) For the quorum sensing receptors to respond to the analogues, a difference of only two carbons is necessary. (2) Too many changes in the structure eliminates the agonist effect while maintains the inhibitory one. (3) Homoserine moiety has shown to be crucial for the retention of the activity of the compound.[57] Thus, if antagonists are desired then integration of bulky groups and unsaturated bonds in the structure is important. In addition, reducing the 3-oxo group can lead to receptor inhibition.[58,59] Compounds **28** and **29** that fit optimization conditions have shown to antagonize AhyR and CviR receptors.[60]

Reverchon and coworkers synthesized 22 different analogues (Figure 15.9) that were tested for agonistic or antagonistic activity on the *V. fisheri* LuxR model (Table 15.5). The introduction of the two-branching moiety at C5 with *tert*-butyl group like in **31** or adamentyl group like in **34** have shown to eliminate the agonistic activity of the compounds. Compounds **36–44** have all shown antagonistic activity solely. The addition of a bulky group might cause steric hindrance and alter the interaction with LuxR receptor as shown in compounds **34**, **45**, and **46**.

Doutheau and coworkers synthesized 11 new analogues of *N*-acylhomoserine lactone (Figure 15.10).[61] A major feature change in this family is the replacement of carboxamide bond by a sulfonamide. The activity testing was performed against *E. coli* strain NM522. Based on the IC_{50} values (Figure 15.10), **57**, **58**, **61**, and **62** did not show activity. This research shows

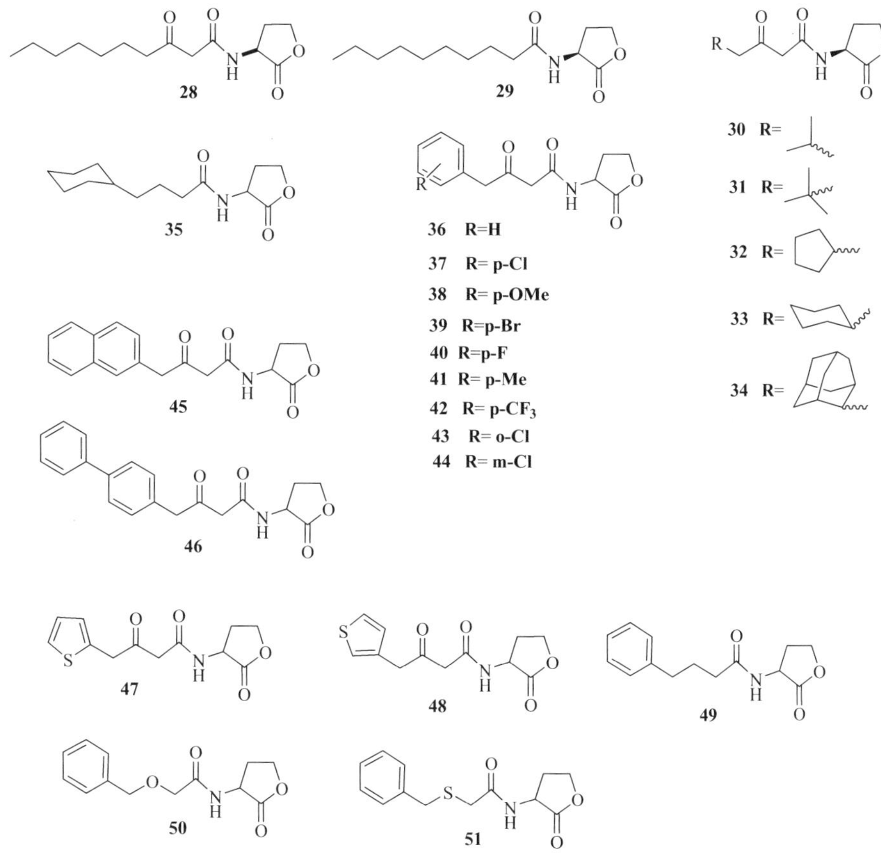

28 29 30 R= 31 R= 32 R= 33 R= 34 R= 35

36 R=H
37 R= p-Cl
38 R= p-OMe
39 R=p-Br
40 R=p-F
41 R= p-Me
42 R= p-CF_3
43 R= o-Cl
44 R= m-Cl

45 46 47 48 49 50 51

FIGURE 15.9 AHL acyl side chain modification that inhibits AHL receptor.[60]

TABLE 15.5
Activity of Compounds Synthesized by Reverchon and Coworkers

Compound Number	Agonistic Activity	Antagonistic Activity
30	$EC_{50} = 0.50\ \mu M$	—
31	No activity	No activity
32	$EC_{50} = 0.25\ \mu M$	—
33	$EC_{50} = 1.10\ \mu M$	—
34	No activity	No activity
35	EC_{50} = more than 6 μM	—
36	No activity	$IC_{50} = 2\ \mu M$
37	No activity	$IC_{50} = 3\ \mu M$
38	No activity	$IC_{50} = 3\ \mu M$
39	No activity	$IC_{50} = 2\ \mu M$
40	No activity	$IC_{50} = 4\ \mu M$
41	No activity	$IC_{50} = 4\ \mu M$
42	No activity	$IC_{50} = 10\ \mu M$
43	No activity	$IC_{50} = 6\ \mu M$
44	No activity	$IC_{50} = 4\ \mu M$
45	No activity	No activity
46	No activity	No activity
47	No activity	No activity
48	No activity	No activity
49	No activity	$IC_{50} = 2–3\ \mu M$
50	No activity	$IC_{50} = 2–3\ \mu M$
51	No activity	$IC_{50} = 2–3\ \mu M$

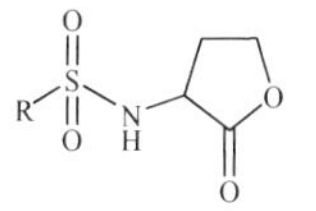

52 R= C_5H_{11} IC_{50} = 2 mM
53 R= C_4H_9 IC_{50} = 12 mM
54 R= C_6H_{13} IC_{50} = 6.5 mM
55 R= C_7H_{14} IC_{50} = 9 mM
56 R= C_8H_{15} IC_{50} = 10 mM
57 R= C_9H_{19} IC_{50} = 40 mM
58 R= $PhCH_2$ IC_{50} = 40 mM
59 R= $PhCH_2CH_2$ IC_{50} = 3 mM
60 R= $PhCH_2CH_2CH_2$ IC_{50} = 20 mM

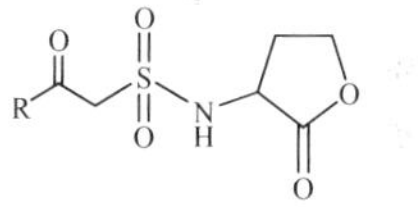

61 R= C_3H_7 IC_{50} = 40 mM
62 R= $PhCH_2$ IC_{50} = 40 mM

FIGURE 15.10 AHL acyl side chain modification that inhibits AHL receptor synthesized by Doutheau and coworkers.[61]

that the combination of 3-oxo and sulfonamide moieties results in the lack of inhibition. The length of the acyl chain plays a role in the activity of the compound where the shortest chain of 4 carbons in this family is **53** with a poor activity characteristic. Compound **52** with 5 carbon chain acyl groups shows to be the most potent with IC_{50} of 2 μM. The optimal structure-activity relationship appears to be at 5 carbon length, whereas the increase to 8 carbons will decrease the compound's inhibitory activity like in **54–56**. The addition of phenyl moiety and

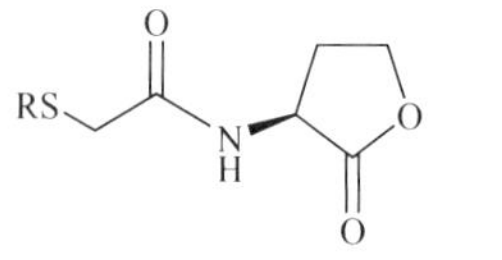

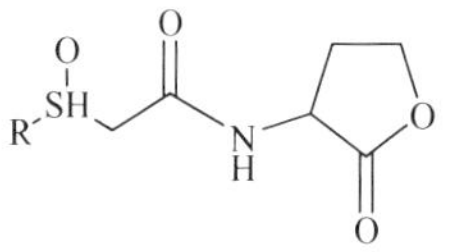

63 R= $(C_2H)_6CH_3$ IC_{50} = 6 mM

64 R= $(CH_2)_8CH_3$ IC_{50} = 50 mM

65 R= p-CH_3Ph IC_{50} = 50 mM

66 R= $(CH_2)_6CH_3$ IC_{50} = 50 mM

FIGURE 15.11 AHL acyl side chain modification that inhibits AHL receptor synthesized by Nielsen and coworkers.[62]

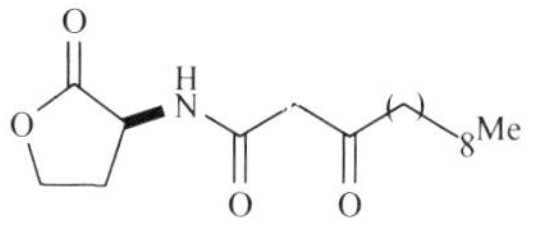

74 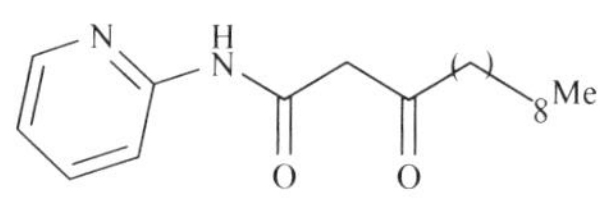 **75**

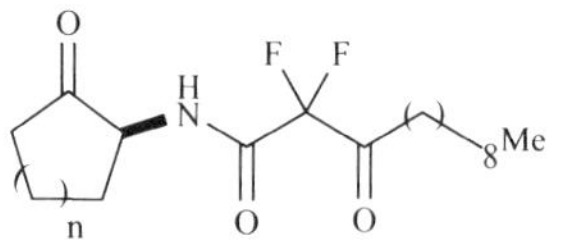

76 n=1
77 n=2

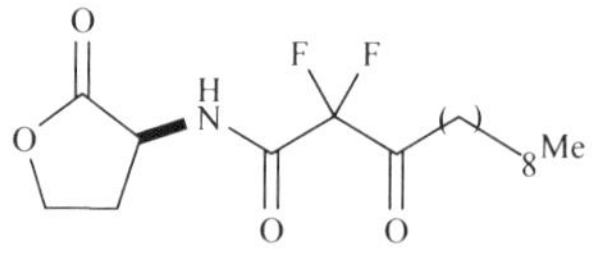

78

FIGURE 15.13 AHL analogues synthesized by Spring and coworkers.[64]

sulfonamide seems also to activate the inhibitory activity (IC_{50} of 3 μM) like with **58–60**.

Nielson and coworkers[62] incorporated at the acyl chain a sulfone or sulfoxide group (Figure 15.11). Compounds **63–66** showed antagonistic effects against the *las* system in *P. aeruginosa*.

Compounds **67–73** synthesized by Blackwell and coworkers[63] (Figure 15.12) have shown to have a strong antagonistic activity against the quorum sensing process in *V. fisheri*. The IC_{50}s of the compounds range from 0.6 to 4.1 μM.

Structure activity conclusions were drawn based on the IC_{50}/EC_{50} values: (1) acyl group of maximum 8 carbon number and containing either an aromatic ring with electron-withdrawing groups or a straight chain can antagonize LasR, TraR, and LuxR. (2) Phenylacetanoyl homoserine lactone (PHL) with electron withdrawing group and lipophilic group at 4-position exhibited antagonistic effects against TraR and LuxR. (3) PHL with electron withdrawing group and lipophilic group at 3-position exhibited antagonistic effects against LasR. (4) At least one carbon spacer between the lactone and aromatic ring grants an antagonistic effect. (5) The sulfonyl groups that replace carbonyl groups in TraR and LuxR antagonists have shown to keep their antagonistic activity.

15.3.2.2.3 Modification of Both Lactone and Acyl Moieties

Spring and coworkers synthesized four different analogues of AHL, modifying both lactone and acyl moieties (Figure 15.13).[64] The compounds were tested to find more about the immunomodulating

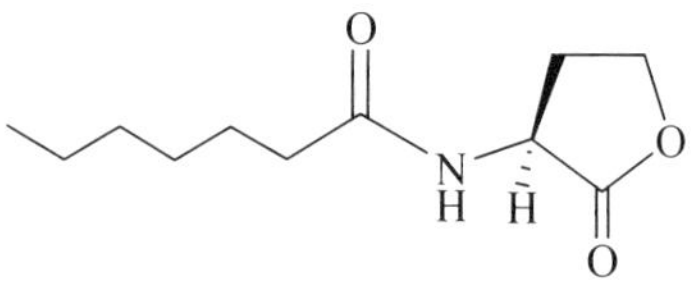

67 IC_{50} = 1.4 mM

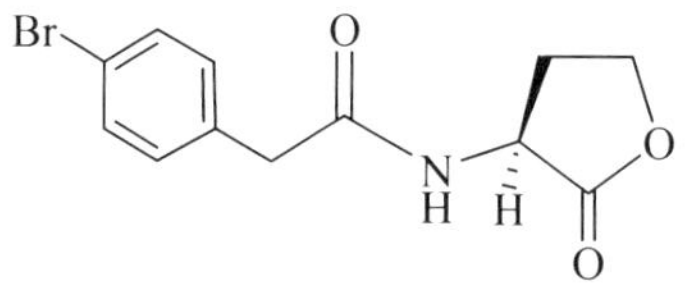

68 IC_{50} = 3.7 mM

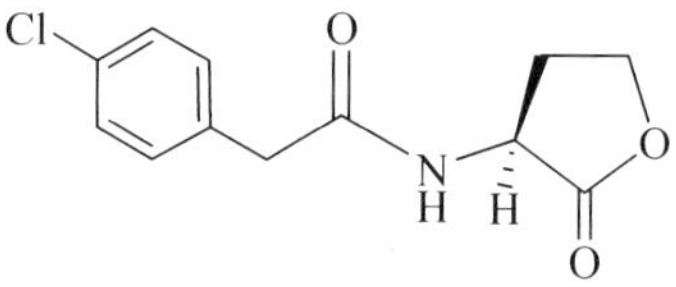

69 IC_{50} = 4.1 mM

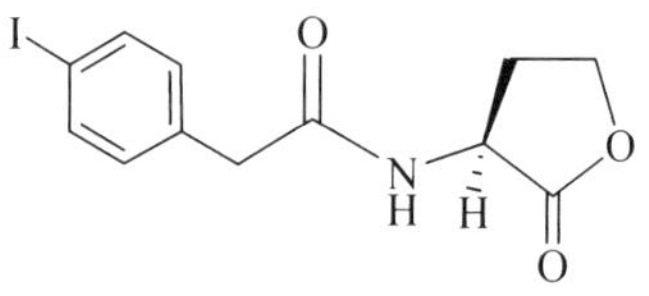

70 IC_{50} =0.9mM

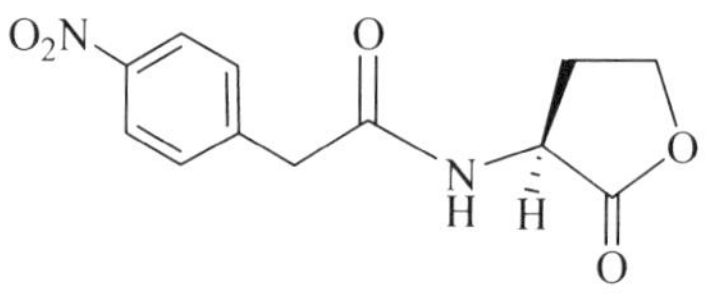

71 IC_{50} = 1.0 mM

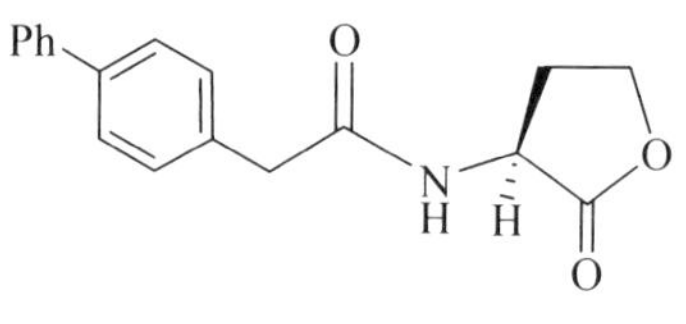

72 IC_{50} = 1.1 mM

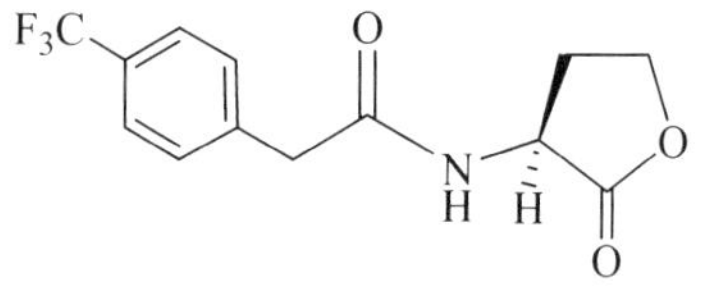

73 IC_{50} = 0.6 mM

FIGURE 15.12 AHL acyl side chain modification that inhibits AHL receptor synthesized by Blackwell and coworkers.[63]

effects of AHL analogues in *P. aeruginosa*. This bacterium has two quorum sensing systems; one of them, *las*, has the OdDHL molecule as its autoinducer (**74**; Figure 15.13). It was shown that OdDHL has effects on the immune system response of the host organism. When tested against the virulence factors pyoverdin and pyocanin, the analogues showed activating effects at 100 μM in the pigment production test in *P. aeruginosa* mutant. The potency of the compounds was **76**,**78** > **77** > **75**.

15.3.2.3 Inhibitors Not Structurally Related to AHL: Furanones

After the finding of inhibitory activity of **79** and **80** against quorum sensing (Figure 15.14), many furanones were synthesized to aim at optimizing the inhibition. Thus **81** and **82** were synthesized and showed inhibitory activity against quorum sensing by altering acyl-HSl signals. They have been shown to be potent in pulmonary infections caused by *P. aeruginosa* in mice. More data needs to be documented.[65–67]

15.3.3 Targeting the Autoinducer-2 Quorum Sensing Pathway

15.3.3.1 LuxS Mechanism of Catalysis

The formation of AI-2 (**87**) starts with S-adenosyl-L-homocysteine (**84**, SAH, Scheme 15.2), which is a product of S-adenosyl L-methionine (**83**, SAM)-dependent methylation. The SRH is cleaved by the LuxS enzyme into L-homocysteine (Hcy) and 4,5-dihydroxy-2,3-pentanedione (DPD), which is the precursor of AI-2[1] (**87**, Scheme 15.2). The formation of an AI-2 molecule is the result of a straightforward path that involves DPD (**86**) and boronic acid[68] (Scheme 15.2).

Pei and coworkers in 2003 proposed the pathway for the catalytic mechanism of *S*-ribosyl-homocysteine presented in Scheme 15.3. This scheme defines three major steps in the LuxS catalysis pathway: First the migration of carbonyl group from C1–C2 position (**88** to **90**), then the shift of the carbonyl to C3 position (**90–91**), and finally the β-elimination Hcy and formation of DPD (**92** to **94**).[69,70]

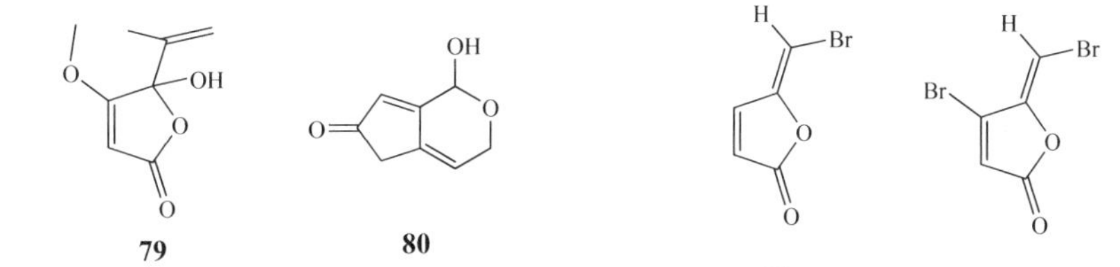

79
Penicillic acid
80
Patulin
Natural product
81
82
Synthetic bromo-furanones

FIGURE 15.14 Brominated furanones.[65–67]

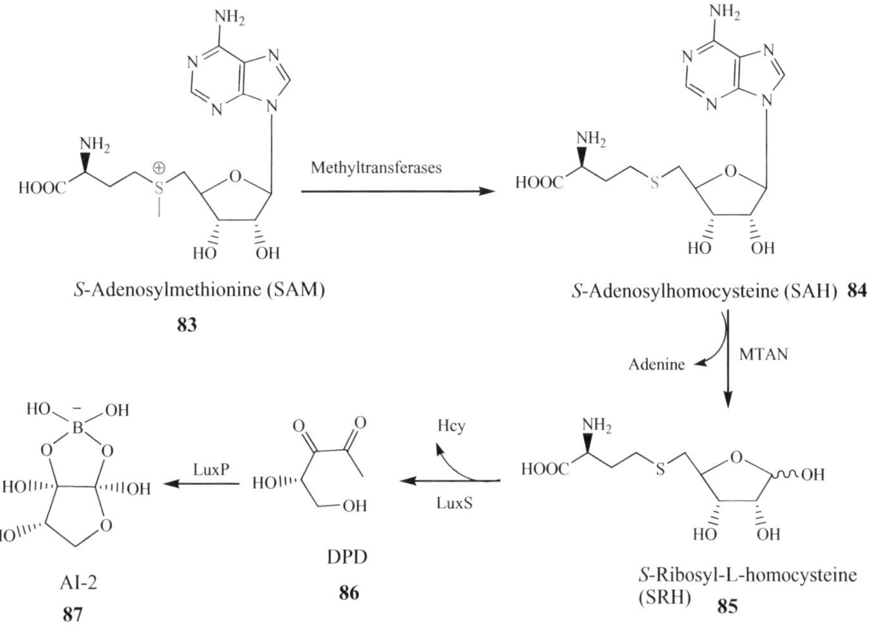

S-Adenosylmethionine (SAM) **83**
Methyltransferases
S-Adenosylhomocysteine (SAH) **84**
Adenine
MTAN
S-Ribosyl-L-homocysteine (SRH) **85**
Hcy
LuxS
DPD **86**
LuxP
AI-2 **87**

SCHEME 15.2 Pathway for the conversion of SAM to AI.[68]

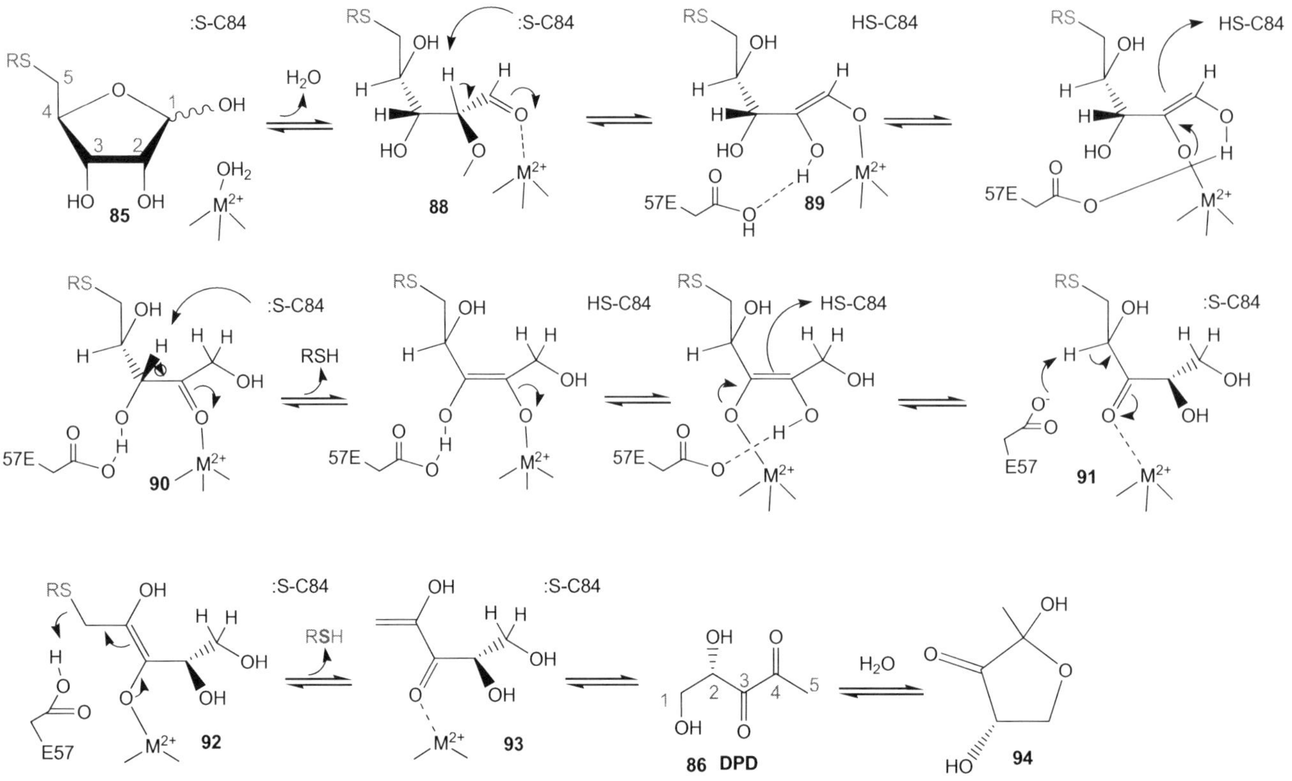

SCHEME 15.3 Proposed mechanism of LuxS-catalyzed reaction.[69,70]

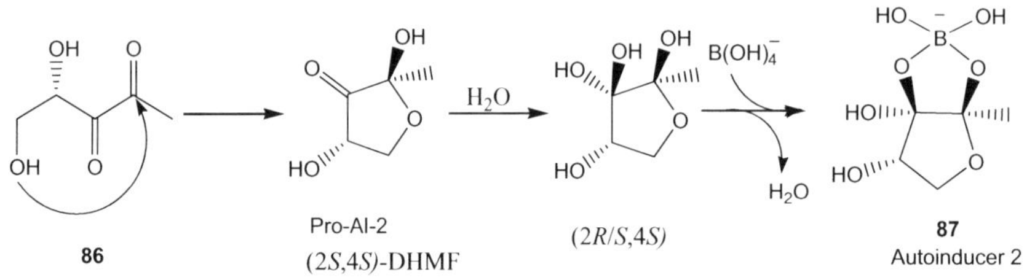

SCHEME 15.4 Cyclization of DPD to AI-2.[71]

15.3.3.2 Cyclization of DPD to AI-2

DPD **86** undergoes spontaneous cyclization to the corresponding tetrahydrofuranose derivative known as pro-AI-2. Complexation with boronic acid leads to the formation of AI-2 (**87**; Scheme 15.4).[71]

15.3.3.2.1 S-Ribosyl-homocysteine (SRH) Analogues

Zhou and coworkers[72,73] are the first to develop synthetic analogues of *S*-ribosyl-homocysteine (SRH), which act by competing with the natural substrate SRH (Figure 15.15). When assayed as LuxS inhibitors, compounds **95** and **96** have shown activity against *Bacillus subtilis* Co(II)-LuxS, but their inhibition constant Ki was not documented. The lack of the anomeric hydroxyl group in **95** prevents the first step in the catalysis of LuxS enzyme known as the aldo-ketose isomerization. The addition of one carbon in the homocysteine part in **96** prevents further β-elimination of the homocysteine. Pei and coworkers[74] have designed and synthesized open ribonic acid moiety analogues. The hydroxamate group in **97** is envisioned to produce a stable isostere with high affinity to LuxS. The metal ion in LuxS that gives the stability of the enediolate intermediate (e.g. **90**, Scheme 15.3) is expected to bind to the planar hydroxamate group in **97**. Compound **98** is the stereoisomer of **97** at the β-carbon of the hydroxamate and has a similar mechanism of inhibition. Both **97** and **98** showed inhibition with respect to Ki of 0.72 and 0.37 μM. The finding of Pei and Zhou et. al. confirms the importance of the homocysteine group for the activity as LuxS inhibition. Wnuk *et al* synthesized compounds where the hemiacetal is substituted by a hemiaminal moiety **99–100** (Figure 15.15).[75] It is expected that the binding strengths and rate of production of the open chain aldehyde would be different. When tested against *P. aeruginosa*, compounds **100** and **102** have shown to decrease *las* activity to 43% but exhibited an increase of *rhl* activity by 104%. Compound **101** displayed *las* activity of 0% and *rhl* activity of 0%.

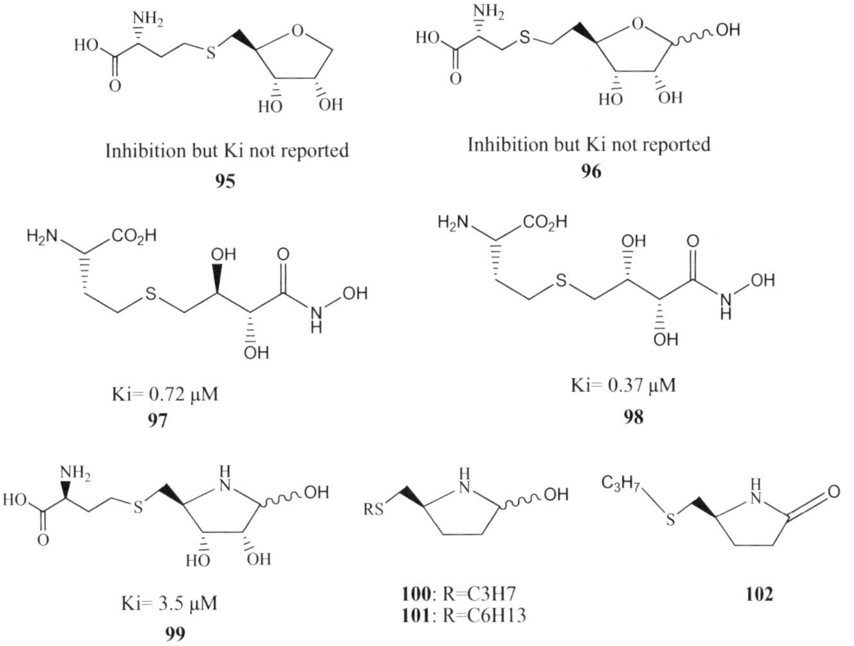

Inhibition but Ki not reported
95

Inhibition but Ki not reported
96

Ki= 0.72 μM
97

Ki= 0.37 μM
98

Ki= 3.5 μM
99

100: R=C3H7
101: R=C6H13

102

FIGURE 15.15 SRH analogues.[72,73]

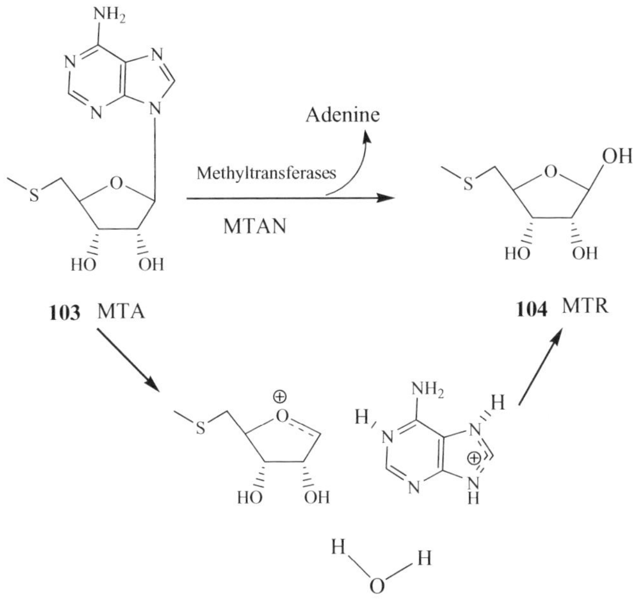

Adenine

Methyltransferases

MTAN

103 MTA

104 MTR

SCHEME 15.5 MTAN-catalyzed pathway.[76]

15.3.3.2.2 MTAN Transition State Analogues

MTAN is important as an enzyme during the synthesis process of AI-2. By losing its adenine, MTAN **103** becomes MTR (**104**; Scheme 15.5). Early transition state has been reported where the bond between the ribose and adenine moieties is partially broken. Late transition state is described by the dissociation of the ribosyl cation from the adenine leaving group fully (Scheme 15.5).[76]

Schramm and coworkers designed and synthesized different analogues (**105–110**; Figure 15.16). Compounds **105–108** specifically target the initial transition state of MTA resembling early SN1 transition state whereas compounds **109** and **110** are intended to target the late transition state.[76]

15.3.3.2.3 AI-2 Antagonists/Inhibitors

Wang and coworkers[77,78] screened and identified three different groups of AI-2 inhibitors (Figure 15.17). *Para*-substituted phenylboronic acids with no additional ionizable functional groups have shown to have the highest activity (**111–115**; Figure 15.17). Among aromatic diols series, pyrogallols showed to have better activity when compared to catechols (**116–200**; Figure 15.17). Wang and coworkers also identified the sulfones **201** and **202** (Figure 15.17) to have some AI-2 antagonistic activity with respect to IC_{50} of 35 ± 3 μM and 55 ± 7 μM.[79,80]

15.3.3.2.4 Sources of QS Inhibitors

By screening plant samples, Wood and coworkers[79] discovered the presence of natural products with QS inhibition activity: ursolic acid and 7-hydroxyindole. Some fatty acids like linoleic acid, oleic acid, stearic acid, and palmitic acid have proved to have inhibitory activity against the AI-2 pathway.[80] Gilbert and coworkers reported the anti-QS activity of cinnamaldehyde, a product used as a flavor in the food industry.[81]

15.3.4 Summary of the Phenotypic Effects of QS Inhibitors

Emergence of Resistance: Reports of bacterial isolates in cystic fibrosis patients detected bacteria carrying efflux-enhancing mutations that showed resistance to QS inhibitors (Table 15.6).[82]

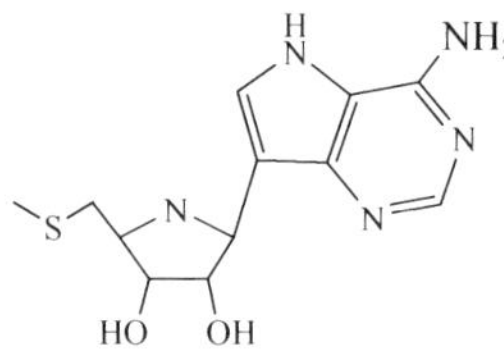

Ki= 1.0 ± 0.05 μM

105

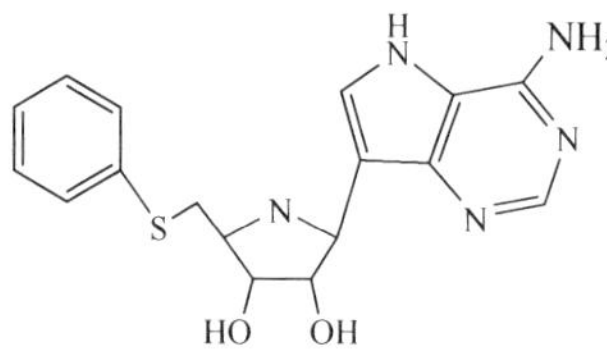

Ki= 335 ± 20 nM

106

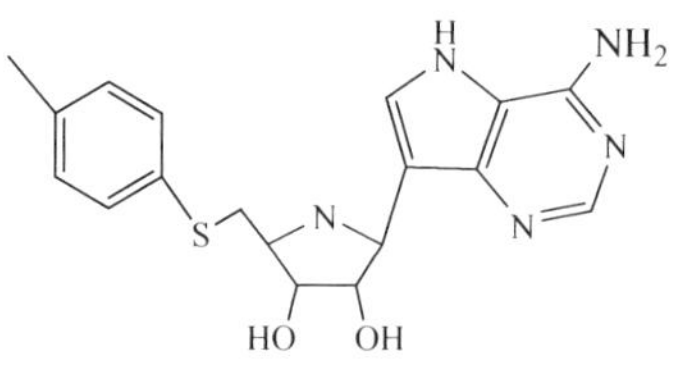

Ki= 60 ± 12 nM

107

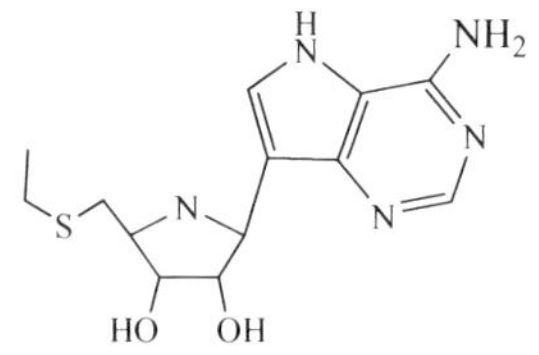

Ki= 40 ± 0.1 nM

108

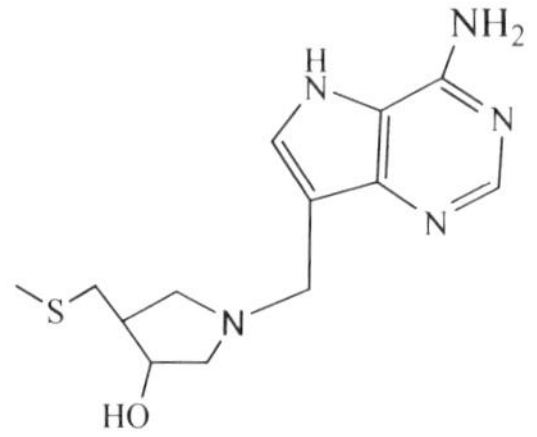

Ki= 24 ± 10 nM

109

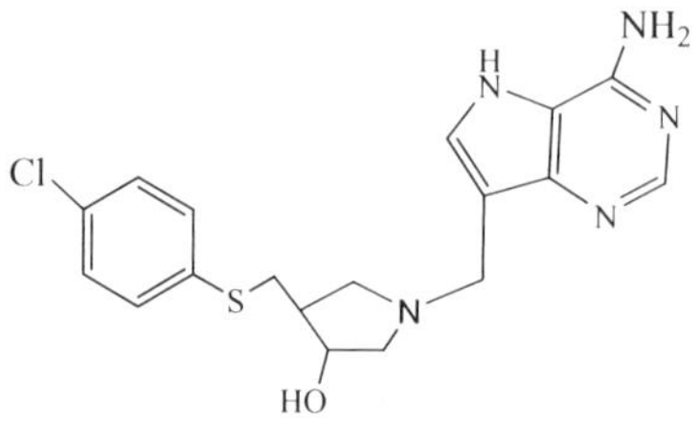

Ki= 2.3 ± 0.1 nM

Ki*= 360 ± 22 pM

110

FIGURE 15.16 MTAN transition state analogues.[76]

111 IC_{50} = 9 ± 5 mM

112 IC_{50} = 5 ± 2 mM

113 IC_{50} = 4 ± 1 mM

(a)

114 IC_{50} =6 ± 4 mM

115 IC_{50} = 4 ± 1 mM

116 IC_{50} = 2 ± 1 mM

117 IC_{50} = 4 ± 1 mM

118 IC_{50} = 3 ± 1 mM

(b)

119 IC_{50} =3 ± 1 mM

200 IC_{50} = 4 ± 2 mM

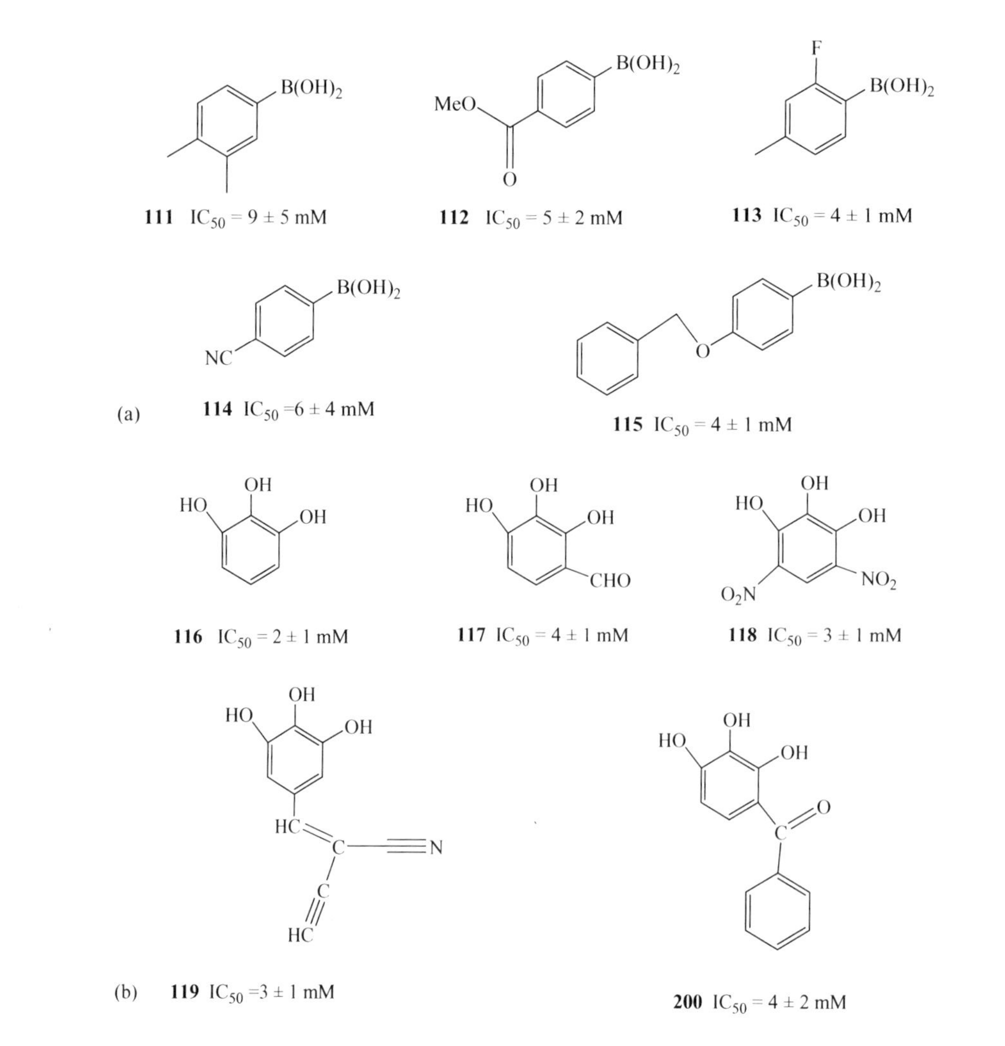

FIGURE 15.17 AI-2 inhibitors. (a) Para-substituted phenyl boronic acid. (b) Pyrogallols. (*Continued*)

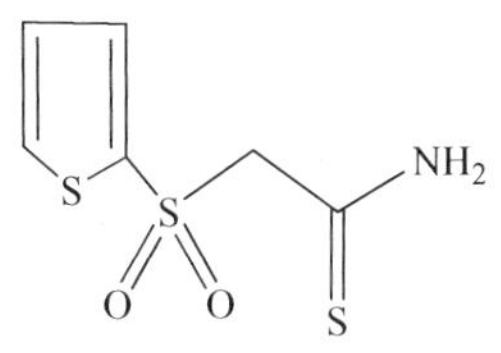

(c) **201** IC_{50} = 35 ± 3 mM 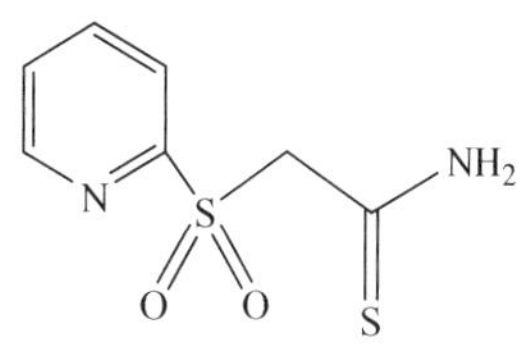 **202** IC_{50} = 55 ± 7 mM

FIGURE 15.17 (Continued) AI-2 inhibitors. (c) Sulfone.[77,78]

TABLE 15.6

Effects of QS Inhibitors on QS Regulated Phenotypes[21,28,53,54,61–64,72,73]

Compound Number	QS Inhibitors	Phenotypes
1	Histidine kinase inhibitors	Targets AlgR1-AgR2 phosphorylation
5 and **6**	Histidine kinase inhibitors	• Targets autophosphorylation activity of histidine kinases ChA, NRII, and KinA • Inhibits sporulation, bacterial chemotaxis, nitrogen assimilation
3 and **4**	Histidine kinase inhibitors	• Targets carboxyl-terminal catalytic domain of kinases • Causes protein aggregation or disruption of membrane integrity
9	AIP receptor antagonist	• Blocks *agr* • Inhibits virulence
26 and **27**	AHL receptor antagonist	Inhibits prodigiosin products, swarming motility, and biofilm formation
52 to **62**	AHL receptor antagonist	Targets luminescence with 3-oxo-C6HSL
63 to **66**	AHL receptor antagonist	Blocks LuxR in the presence of *N*-(3-oxohexanoyl)-L-homoserine lactone
67 to **73**	AHL receptor antagonist	• Targets TraR, lasR, and LuxR • Targets luminescence
74 to **78**	AHL analogues	Affects pyoverdine and pyocanin concentration
81 and **82**	Brominated furanones	Affects swarming velocity
100 and **102**	Lux S inhibitors	Affects bioluminescence and decrease in *las* activity

15.4 Challenges and Future Work

Quorum sensing inhibitors have proven to be a new strategy for bypassing bacterial resistant strains. They are unique in interfering with bacterial virulent gene expression and not growth. The major limitation is the lack of structure-activity relationship optimization.[83] Since AI-2 is not essential for cell growth or survival, interference with its synthesis and activity should not lead to resistance development.

15.5 Conclusion

In conclusion, there is a need for more data in regard to quorum sensing inhibitors. Multiple aspects need to be studied: (1) Optimizing the quorum sensing inhibitors structures, and (2) other biological activity data than measuring MIC and IC_{50}, which are the most reported. This includes the study of the effect of the inhibitors on bioluminescence in some bacteria and/or biofilm formation.

REFERENCES

1. Waters C.M., Bassler B.L. 2005. Quorum sensing: Cell-to-cell communication in bacteria. *Annu. Rev. Cell Dev. Biol.* 21:319–346.
2. Suga H., Smith K.M. 2003. Molecular mechanisms of bacterial quorum sensing as a new drug target. *Curr. Opin. Chem. Biol.* 7:586–591.
3. Praneenararat T., Geske G.D., Blackwell H.E. 2009. Efficient synthesis and evaluation of quorum-sensing modulators using small molecule macroarrays. *Org. Lett.* 20: 4600–4603.
4. Gonzalez, J.E., Keshavan N.D. 2006. Messing with bacterial quorum sensing. *Microbiol. Mol. Biol. Rev.* 70:859–870.
5. Williams P. 2007. Quorum sensing, communication and cross-kingdom signalling in the bacterial world. *Microbiology* 153:3923–3938.

6. Williams P., Winzer K., Chan W.C., Camara M. 2007. Look who's talking: Communication and quorum sensing in the bacterial world. *Philos. Trans. R. Soc.* 362:1119–1134.
7. Mattmann M.E., Blackwell H.E. 2010. Small molecules that modulate quorum sensing and control virulence in *Pseudomonas aeruginosa*. *J. Org. Chem.* 75:6737–6746.
8. Galloway W.R., Hodgkinson J.T., Bowden S.D., Welch., M., Spring D.R. 2011. Quorum sensing in Gram-negative bacteria: Small-molecule modulation ofAHL and AI-2 quorum sensing pathways. *Chem. Rev.* 111:28–67.
9. Vendeville A., Winzer K., Heurlier K., Tang C.M., Hardie K.R. 2005. Making 'sense' of metabolism: Autoinducer-2, LuxS and pathogenic bacteria. *Nat. Rev. Microbiol.* 3:383–396.
10. De Keersmaecker S.C., Sonck K., Vanderleyden J. 2006. Let LuxS speak up in AI-2 signaling. *Trends Microbiol.* 14: 114–119.
11. Xavier K.B., Bassler B.L. 2003. LuxS quorum sensing: More than just a numbers game. *Curr. Opin. Microbiol.* 6:191–197.
12. Duan K., Dammel C., Stein J., Rabin H., Surette M.G. 2003. Modulation of *Pseudomonas aeruginosa* gene expression by host microflora through interspecies communication. *Mol. Microbiolol.* 50:1477–1491.
13. Boyle K.E., Monaco H., Ditmarsch D., Deforet M., Xavier J.B. 2015. Integration of metabolic and quorum sensing signals governing the decision to cooperate in a bacterial social trait. *PLoS Comput. Biol.* 11:e1004279.
14. Ni N., Li M., Wang J., Wang B. 2009. Inhibitors and antagonists of bacterial quorum sensing. *Med. Res. Rev.* 29:65–124.
15. Roy V., Adams B.L., Bentley W.E. 2011. Developing next generation antimicrobials by intercepting AI-2 mediated quorum sensing. *Enzyme Microb. Technol.* 49:113–123.
16. Geske G.D., O'Neill J.C., Blackwell H.E. 2008. Expanding dialogues: From natural autoinducers to non-natural analogues that modulate quorum sensing in Gram-negative bacteria. *Chem. Soc. Rev.* 37:1432–1447.
17. Weiland-Brauer N., Pinnow N., Schmitz R. 2015. Novel reporter for identification of interference with acyl homoserine lactone and autoinducer-2 quorum sensing. *Appl. Env. Microbiol.* 81;1477–14589.
18. Lyon G.J., Muir T.W. 2003. Chemical signaling among bacteria and its inhibition. *Chem. Biol.* 10:1007–1021.
19. Kleerebezem M., Quadri L.E., Kuipers O.P., deVos W.M. 1997. Quorumsensing by peptide pheromones and two-component signal-transduction systems in Gram-positive bacteria. *Mol. Microbiol.* 24:895–904.
20. Hilliard J.J., Goldschmidt R.M., Licata L., Baum E.Z., Bush K. 1999. Multiple mechanisms of action for inhibitors of histidine protein kinases from bacterial two-component systems. *Antimicrob. Agents. Chemother.* 43:1693–1699.
21. Matsushita M., Janda K.D. 2002. Histidine kinases as targets for new antimicrobial agents. *Bioorg. Med. Chem.* 10:855–867.
22. Barrett J.F., Goldschmidt R.M., Lawrence L.E. et. al. 1998. Antibacterial agents that inhibit two-component signal transduction systems. *PROc Natl Acad Sci USA* 9:5317–5322.
23. Macielag M.J., Demers J.P., Fraga-Spano S.A. et al. 1998. Substituted salicylanilides as inhibitors of two-component regulatory systems in bacteria. *J. Med. Chem.* 41:2939–2945.
24. Stephenson K., Yamaguchi Y., Hoch J.A. 2000. The mechanism of action of inhibitors of bacterial two-component signal transduction systems. *J. Biol. Chem.* 275:38900–38904.
25. Kitayama T., Yamamoto K., Utsumi R. et. al. 2001. Chemistry of zerumbone. 2. Regulation of ring bond cleavage and unique antibacterial activities of zerumbone derivatives. *Biosci. Biotechnol. Biochem.* 65:2193–2199.
26. Kitayama T., Iwabuchi R., Minagawa S. et. al. 2007. Synthesis of a novel inhibitor against MRSA and VRE: Preparation from zerumbone ring opening material showing histidine-kinase inhibition. *Bioorg. Med. Chem. Lett.* 17:1098–1101.
27. Lyon G.J., Mayville P., Muir T.W., Novick R.P. 2000. Rational design of a global inhibitor of the virulence response in *Staphylococcus aureus*, based in part on localization of the site of inhibition of the receptor-histidine kinase, *AgrC*. *Proc. Natl. Acad. Sci. USA* 24:13330–13335.
28. George E.A., Novick R.P., Muir T.W. 2008. Cyclic peptide inhibitors of *staphylococcal* virulence prepared by Fmoc-based thiolactone peptide synthesis. *J. Am. Chem. Soc.* 14:4914–4924.
29. McDowell P., Affas Z., Reynolds C. et. al. 2001. Structure, activity and evolution of the group I Thiolactone peptide quorum-sensing system of staphylococcus aureus. *Mol. Microbiol.* 41: 503–512.
30. Lyon G.J., Wright J.S., Muir T.W., Novick R.P. 2002. Key determinants of receptor activation in the *Agr* autoinducing peptides of *Staphylococcus aureus*. *Biochemistry* 41:10095–10104.
31. Scott R.J., Lian L., Muharram S.H. et. al. 2003. Side-chain-to-tail thiolactone peptide inhibitors of the *Staphylococcal* quorum sensing system. *Bioorg. Med. Chem. Lett.* 13:2449–2453.
32. Wright J.S.III, Lyon G.J., George E.A., Muir T.W., Novick R.P. 2004. Hydrophobic interactions drive ligand-receptor recognition for activation and inhibition of *staphylococcal* quorum sensing. *Proc. Natl. Acad. Sci.* USA 46:16168–16173.
33. McDowell P., Affas Z., Reynolds C. et. al. 2001. Structure activity and evolution of the group I thiolactone peptide quorum-sensing system of *Staphylococcus aureus*. *Mol. Microbiol.* 41:503–512.
34. Hoang T.T. 2002. Beta-ketoacyl acyl carrier protein reductase (FabG) activity of the fatty acid biosynthetic pathway is a determining factor of the 3-oxo-homoserine lactone acyl chain lengths. *Microbiology* 148:3849–3856.
35. Swift S., Karlyshev A.V., Fish L. et. al. 1997. Quorum sensing in *Aeromonas hydrophilia* and *Aeromonas salmonicida*: Identification of the LuxRI homologs AhyRI and AsaRI and their cognate *N*-acylhomoserine lactone signal molecules. *J. Bacteriol.* 179:5271–5281.
36. Sokol P.A., Sajjan U., Visser M.B., Gingues S., Forstner J., Kooi C. 2003. The CepIR quorum-sensing system contributes to the virulence of *Burkholderia cenocepacia* respiratory infection. *Microbiology* 149:3649–3658.
37. McClean K.H., Winson M.K., Fish L. et. al. 1997. Quorum sensing and *Chromobacterium violaceum*: Exploitation of violacein production and inhibition for the detection of *N*-acylhomoserine lactones. *Microbiology* 143:3709–3711.
38. Welch M., Dutton J.M., Glansdorp F.G. et. al. 2005. Structure-activity relationships of *Erwinia carotovora* quorum sensing signaling molecules. *Bioorg. Med. Chem. Lett.* 19:4237–4238.
39. Nasser W., Bouillant M.L., Salmond G., Reverchon S. 1998. Characterization of the *Erwinia chrysanthemi* expI-expR

locus directing the synthesis of two *N*-acyl-homoserine lactone signal molecules. *Mol. Microbiol.* 29:1391–1405.
40. Zhang Z.G., Pierson L.S. 2001. A second quorum-sensing system regulates cell surface properties but not phenazine antibiotic production in *Pseudomonas aureofaciens*. *Appl. Environ. Microbiol.* 67:4305–4315.
41. Smith R.S., Harris S.G., Phipps R., Iglewski B. 2002. The *Pseudomonas aeruginosa* quorum-sensing molecule *N*-(3-oxododecanoyl)homoserine lactone contributes to virulence and induces inflammation in vivo. *J. Bacteriol.* 184:1132–1139.
42. Jiang Y., Camara M., Chhabra S.R. et. al. 1998. In vitro biosynthesis of the *Pseudomonas aeruginosa* quorum-sensing signal molecule *N*-butanoyl-L-homoserine lactone. *Mol. Microbiol.* 28:193–203.
43. Puskas A., Greenberg E.P., Kaplan S., Schaefer A.L. 1997. A quorum-sensing system in the free-living photosynthetic bacterium *Rhodobacter sphaeroides*. *J. Bacteriol.* 179:7530–7537.
44. Eberl L., Winson M.K., Sternberg C. et. al. 1996. Involvement of *N*-acyl-L-hormoserine lactone autoinducers in controlling the multicellular behaviour of *Serratia liquefaciens*. *Mol. Microbiol.* 20:127–136.
45. Callahan S.M., Dunlap P.V. 2000. LuxR- and acyl-homoserine-lactone-controlled non-lux genes define a quorum-sensing regulon in *Vibrio fischeri*. *J. Bacteriol.* 182:2811–2822.
46. Cao J.G., Meighen E.A. 1989. Purification and structural identification of an autoinducer for the luminescence system of *Vibrio harveyi*. *J. Biol. Chem.* 264:21670–21676.
47. Kirwan J.P., Gould T.A., Schweizer H.P., Bearden S.W., Murphy R.C., Churchill M.E., Quorum-sensing signal synthesis by the *Yersinia pestis* acyl-homoserine lactone synthase YspI. *J. Bacteriol.* 188:784–788.
48. Lyon G.J., Muir T.W. 2003. Chemical signaling among bacteria and its inhibition. *Chem. Biol.* 10:1007–1021.
49. Roy V., Fernandes R., Tsao C.Y., Bentley W.E. 2009. Cross species quorum quenching using a native AI-2 processing enzyme. *Chem. Biol.* 5:223–232.
50. Ni N., Li M., Wang J., Wang B. 2009. Inhibitors and antagonists of bacterial quorum sensing. *Med. Res. Rev.* 29:65–124.
51. Parsek M.R., Val D.L., Hanzelka B.L., Cronan J.E. Jr, Greenberg E.P. 1999. Acyl homoserine-lactone quorum sensing signal transduction. *Proc. Natl. Acad. Sci.* USA 96:4360–4365.
52. Musk D.J., Hergenrother P.J. 2006. Chemical countermeasures for the control of bacterial biofils: Effective compounds and promising targets. *Curr. Med. Chem.* 13:2163–2177.
53. Phillips G.J. 2001. Green fluorescent protein-A bright idea for the study of bacterial protein localization. *FEMS Microbiol. Lett.* 204:9–18.
54. Givskov M., de Nys R., Manefield M. et. al. 1996. Eukaryotic interference with homocysteine lactone-mediated prokaryotic signaling. *J. Bacteriol.* 178:6618–6622.
55. Eberhard A., Widrig C.A., McBath P., Schineller J.B. 1986. Analogs of the autoinducer of bioluminescence in *Vibrio fischeri*. *Arch Microbiol.* 146:35–40.
56. Hentzer M., Eberl L., Nielsen J., Givskov M. 2003. Quorum sensing-A novel target for the treatment of biofilm infections. *Biodrugs* 17:241–250.
57. Schaefer A.L., Hanzelka B.L., Eberhard A., Greenberg E.P. 1996. Quorum sensing in *Vibrio fischeri*: Probing autoinducer-LuxR interactions with autoinducer analogs. *J. Bacteriol.* 178: 2897–2901.
58. Zhu J., Beaber J.W., More M.I., Fuque C., Eberhard A., Winans S.C. 1998. Analogs of the autoinducer 3-oxooctanoyl-homoserine lactone strongly inhibit activity of the TraR protein of *Argobacterium tumefaciens*. *J. bacterial.* 180:5398–5405.
59. Reverchon S., Chantegrel B., Deshayes C., Doutheau A., Cotte-Pattat N. 2002. New synthetic analogues of *N*-acyl homoserine lactones as agonists or antagonists of transcriptional regulators involved in bacterial quorum sensing. *Bioorg. Med. Chem. Lett.* 12:1153–1157.
60. Swift S., Lynch M.J., Fish L. et. al. 1999. Quorum sensing-dependent regulation and blockade of exoprotease production in *Aeromonas hydrophila*. *Infect. Immun.* 10:5192–5199.
61. Castang S., Chantegrel B., Deshayes C. et. al. 2004. *N*-sulfonyl homoserine lactones as antagonists of bacterial quorum sensing. *Bioorg. Med. Chem. Lett.* 14:5145–5149.
62. Persson T., Hansen T.H., Rasmussen T.B., Skinderso M.E., Givskov M., Nielsen J. 2005. Rational design and synthesis of new quorum sensing inhibitors derives from acylated homoserine lactones and natural products from garlic. *Org. Biomol. Chem.* 3:253–262.
63. Geske G.D., O'neill J.C., Blackwell H.E. 2007. *N*-phenylacetanoyl-L-homoserine lactones can strongly antagonize or superagonize quorum sensing in *Vibrio fisheri*. *ACS Chem. Biol.* 5:315–319.
64. Thomas G.L., Bohner C.M., Williams H.E. et. al. 2006. Immunomodulatory effects of *Pseudomonas aeruginosa* quorum sensing small molecules probes on mammalian macrophages. *Mol. Biosyst.* 2:132–137.
65. Hentzer M., Wu H., Andersen J.B. et. al. 2003. Attenuation of *Pseudomonas aeruginosa* virulence by quorum sensing inhibitors. *EMBO J.* 22:3803–3815.
66. Wu H., Song Z., Hentzer M. et. al. 2004. Synthetic furanones inhibit quorum-sensing and enhance bacterial clearance in *Pseudomonas aeruginosa* lung infection in mice. *J. Antimicrob. Chemother.* 53:1054–1061.
67. Hentzer M., Riedel K., Rasmussen T.B. et. al. 2002. Inhibition of quorum sensing in *Pseudomonas aeruginosa* biofilm bacteria by a halogenated furanone compound. *Microbiol-Sgm* 148:87–102.
68. Roy V., Adams B.L., Bentley W.E. 2011. Developing next generation antimicrobials by intercepting AI-2 mediated quorum sensing. *Enzyme Microb. Technol.* 49:113–123.
69. Shen G., Rajan R., Zhu J.G., Bell C.E., Pei D.H. 2006. Design and synthesis of substrate and intermediate analogue inhibitors of S-ribosylhomocysteinase. *J. Med. Chem.* 49:3003–3011.
70. Zhu J.G., Dizin E., Hu X.B., Wavreille A.S., Park J., Pei D.H. 2003. S-ribosylhomocysteinase (LuxS) is a mononuclear iron protein. *Biochem.* 42:4717–4726.
71. Kadirvel M., Stimpson W.T., Moumene-Afifi S. et. al. 2010. Synthesis and bioluminescence-inducing properties of autoinducer (S)-4,5-dihydroxypentane-2,3-dione and its enantiomer. *Bioorg. Med. Chem. Lett.* 20:2625–2628.
72. Alfaro J.F., Zhang T., Wynn D.P., Karschner E.L., Zhou Z.S. 2004. Synthesis of LuxS inhibitors targeting bacterial cell-cell communication. *Org. Lett.* 6:3043–3046.
73. Zhao G., Wan W., Mansouri S., Alfaro J.F. et al. 2003. Chemical synthesis of S-ribosyl-L-homocysteine and activity assay as a LuxS substrate. *Bioorg. Med. Chem. Lett.* 13:3897–3900.

74. Shen G., Rajan R., Zhu J.G., Bell C.E., Pei D.H. 2006. Design and synthesis of substrate and intermediate analogue inhibitors of S-ribosylhomocysteinase. *J. Med. Chem.* 49:3003–3011.
75. Malladi V.L., Sobczak A.J., Meyer T.M. et. al. 2011. Inhibition ofLuxS by S-ribosylhomocysteine analogues containing a [4-aza]ribose ring. *Bioorg. Med. Chem.* 19:5507–5519.
76. Singh V., Shi W., Almo S.C. et. al. 2006. Structure and inhibition of a quorum sensing target from *Streptococcus pneumoniae. Biochemistry* 43:12929–12941.
77. Ni N., Chou H.T., Wang J.*et. al.* 2008. Identification of boronic acids as antagonists of bacterial quorum sensing in *Vibrio harveyi. Biochem. Biophy. Res. Commun.* 369:590–594.
78. Ni N., Choudhary G., Li M., Wang B. 2008. Pyrogallol and its analogs can antagonize bacterial quorum sensing in *Vibrio harveyi. Bioorg. Med. Chem. Lett.* 18:1597–1572.
79. Lee J., Bansal T., Jayaraman A., Bentley W.E., Wood T.K. 2007. Enterohemmorrhagic *Eschericia coli* biofilms are inhibited by 7-hydroxyindole and stimulated by isatin. *Appl. Environ. Microbiol.* 73:4100–4109.
80. Widmer K.W., Soni K.A., Hume M.E. et. al. 2007. Identification of poultry meat-derived fatty-acids functioning as quorum sensing signal inhibitors to autoinducer-2 (AI-2). *J. Food. Sci.* 72:363–368.
81. Niu C., Afre S., Gilbert E.S. 2006. Subinhibitory concentrations of cinnamaldehyde interfere with quorum sensing. *Lett. Appl. Microbiol.* 43:48–494.
82. Garcia-Contretas R., Nunez-Lopez L., Jasso-Chavez., R., Kwan B.W., Belmont J.A. Rangel-Vega A., Maeda T., Wood R.K. 2015. Quorum sensing enhancement of the stress response promotes resistance to quorum quenching and prevents social cheating. *ISME J.* 9:115–125.
83. Defoirdt T., Boon N. Bossier P., 2010.Can bacteria evolve resistance to quorum sensing disruption? *PLoS Pathog.* 6, e1000989.

16

Role of Small Volatile Signaling Molecules in the Regulation of Bacterial Antibiotic Resistance and Quorum Sensing Systems

José E. Belizário, Marcos Sulca-Lopez, Marcelo Sircili, and Joel Faintuch

CONTENTS

16.1 Introduction

The human body is co-habited by viruses, bacteria, protozoa, and yeasts that make up our microbiomes (Belizario and Napolitano, 2015). Together, these microbial communities living in symbioses produce a great variety of nutrients and small metabolites, ribosomal and non-ribosomal oligopeptides, and peptides such as antimicrobial peptides (AMPs), bacteriocins, and pigments. AMPs are natural antibiotics produced by microorganisms, plants, and animals that function as important components of the innate immune system and defensive response against exogenous pathogens (Galán et al., 2013). Thus, the richness and diversity of bacterial strains and their multiple functional metabolic and catabolic activities correlate with healthy and disease states (Tremaroli and Backhed, 2012, Belizario and Napolitano, 2015). Recent advances in next-generation sequencing and metagenomic approaches have helped to estimate the variation of antibiotic-like biosynthetic gene clusters (Donia et al., 2014) and antibiotic resistance gene apparatus in chromosomal and plasmid DNA in thousands of bacterial species (Yarygin et al., 2017). For instance, *P. aeruginosa* intrinsic resistome revealed a set of AmpC type metallo-β-lactamases (MBLs) genes such as IMP (imipenemase), VIM (Verona integron-encoded metallo-β-lactamase), and SPM (novel metallo-β-lactamase), conferring resistance to β-lactam-β-lactamase inhibitor combinations (D'Costa et al., 2006). Resistomes to aminoglycosides, quinones, and macrolides were unrevealed, confirming the cross-resistance among antibiotics (Lanza et al., 2018). Pathogenic microorganisms employ several mechanisms in attaining antibiotic and multi-drug resistance under prevailing conditions such as the biofilms (Vu et al., 2009). Recent studies demonstrated that a large set of biologically active small molecules can regulate the translation of genes in response to local changes in cell number and density, a phenomenon known as quorum sensing (QS) (Rutherford and Bassler, 2012, Schuster et al., 2013). Two major QS processes have been well described: type AI-1, which is involved in intraspecies communication, and type AI-2, which is related to interspecies communication (Rutherford and Bassler, 2012). It is not yet known how and when the intra- and interspecies communication and interactions within microorganisms can contribute to the alarming rate and spread of antibiotic resistance among patients undergoing infection treatment. In this review, we first update on metabolic and oxidative biochemical processes bacteria and human body that produce and release the organic volatile compounds (VOCs) associated with several diseases. We discuss the question of whether there exists a role of VOCs in the regulation of antibiotic resistance and quorum sensing communication within microorganisms that live in our own body and other ecological niches.

16.2 Bacterial VOCs Signatures

Hippocrates (c. 460–370 BC) was the first to claim that the odor emitted by humans could serve as classical sign for early diagnosis of human diseases. Dogs, after conditional training, can sniff out clinical diseases by smelling patient's specimens. The mixed-expired breath (99.995%) consists of nitrogen (78%), oxygen (13%), carbon dioxide (5%), water vapor (4%), and inert gases, and the remainder (<50 ppmv) is a mixture of volatile compounds with low molecular weight (<1 kDa) (Boots et al., 2012, Amann et al., 2014, de Lacy Costello et al., 2014, Broza et al., 2014, Ahmed et al., 2017). Volatomics or Breathomics refers to repertory of volatile organic compounds (VOCs) derived from specific cellular and tissue metabolism of the host cells and microbiota (Ahmed et al., 2019). Systemic VOCs pass efficiently from the blood into the alveolar air and then to the respiratory tract consisting of lungs, pharynx, larynx, nose, oral cavity, sinuses, and finally to the environment. Breath biopsy is a term that refers to the VOCs sampling in exhaled air or released in a headspace of aqueous liquids or solid biological specimens (Ahmed et al., 2019). Numerous VOCs, proteins, and peptides detected in water condensates (or aerosols) have been advocated as biological markers for the diagnosis of oxidative stress, inflammation, carcinogens, and microbial infection (Boots et al., 2012, Amann et al., 2014, de Lacy Costello et al., 2014, Ahmed et al., 2017, Timm et al., 2018). The ^{13}C-urea or ammonia breath test is the most sensitive and specific technique for diagnosis of *H. pylori*, while the nitric oxide test is used to diagnose asthma, acetone to diabetes mellitus and ketonemia, and isoprene and ammonia to renal disorders (Ulanowska et al. 2011). VOCs are also emitted from biological fluids such as breath, blood, saliva, skin, milk, and feces and can be collected in headspace, inert polymer bags, or directly onto sorbent tubes for chemical analysis (Herbig and Beauchamp, 2014, Amann et al., 2014). Advanced technologies such as gas chromatography (GC), gas chromatography and quadrupole mass spectrometry (GC-MS), selected ion flux tube mass spectrometry (SIFT-MS), mass spectrometry with ionic molecule reaction (IMR-MS), ion mobility spectrometry (IMS), field asymmetric ion mobility spectrometry (FAIMS), and proton transfer reaction mass spectrometry (PTR-MS) are among the technologies for detection and quantification of physiologically relevant VOCs. Diverse kinds of gas sensors or electronic noses (eNose) are in development for direct detection and identification of VOCs (Jansson and Larsson, 1969, Wilson, 2015, Amann et al., 2014). New apparatus and devices for collecting, concentrating, separating, and identifying breath proteins, metabolites, and VOCs have been progressively improved (Phillips et al., 2013, Ahmed et al., 2018). GC-GC-TOF-MS (two dimensional analyses) is the gold standard method for profiling, chemical identification, and quantification of VOCs and primary and secondary metabolites (Phillips et al., 2013). Real-time measurement of VOCs in healthy volunteers versus a diseased cohort may allow diagnostic and therapeutic monitoring in the future.

The gut human microbiome harbors a rich and diverse array of biosynthetic and biochemical pathways that is mostly distinct in source and composition from that of human cells. The end products of anaerobic bacteria fermentation of oligosaccharides include succinate and a mixture of volatile fatty acids, mainly acetate, propionate, *cis*-2-methylcrotonate, 2-methylbutyrate and 2-methylvalerate, short chain fatty acids (SCFAs), as well as alcohols, propanols, hydrocarbons, aldehydes, ketones, and sulfated organic compounds (Pereira et al., 2015, Audrain et al., 2015, Rees et al., 2018). The main sulfur-containing flatus components in healthy individuals are hydrogen sulfide (H_2S), followed by methanethiol and dimethyl sulfide. VOCs can discriminate pathogen interaction commonly observed in clinical infections (Filipiak et al., 2012). For example, high levels of 2-methylbutylacetate and methyl 2-methylbutyrate, two known antimicrobial VOCs, were found only in co-culture of *E. cloacae* and *P. aeruginosa* (Lawal et al., 2018). Bos and colleagues revised 51 articles in which they identified 161 VOCs produced during sepsis in children infected by Gram-positive and -negative bacteria (Bos et al., 2013). Palma and colleagues (2018) constructed a database with 792 VOCs emitted by diverse species of bacteria, protozoa, and fungi during their growth in various *in vitro* and *in vivo* conditions. Over 2000 VOCs emitted by human body fluids have been identified and catalogued according to key classes and chemical structures (de Lacy Costello et al., 2014, Vinaixa et al., 2016, Vizcaino and Crawford, 2016, Lemfack et al., 2017). The standardized repositories such as MVOC (http://bioinformatics.charite.de/mvoc) have allowed both quantitative and qualitative analyses of VOCs in their biochemical pathways and biological roles in health and disease (Abdullah et al., 2015, de Lacy Costello et al., 2014).

Figure 16.1 displays graphically a compilation of VOCs and their signatures for seven clinically relevant Gram-positive and -negative bacterial strains. In each balloon are the most representative chemical structures of VOCs for a bacterial strain. In the central balloon are included the VOCs: isopentanol, formaldehyde, methyl mercaptan, and trimethylamine, which are VOCs produced by all bacteria and not by the host. Some sets of VOCs are unique and considered potential candidate biomarkers for a bacterial strain. They are as follows: isovaleric acid and 2-methyl-methylbutanol for *Staphylococcus aureus* (SA); 1-undecane, 2,4-dimethyl-1-heptane, 2-butanone, 2-propanol, ammonia, 2-acetophenone, hydrogen cyanide, and methylthiocyanide for *Pseudomonas aeruginosa* (PA); methanol, pentanol, ethyl acetate, indole, 1-octanol, and hexanol for *Escherichia coli* (EC); 2,2,4,4,tetramethyloxolane, 3Z-octenyl acetate, and 3-methylcyclohexene for *Klebsiella penumoniae*; and methyl 4-methylpentanoate, 4-methylpentanoic acid, and 1-methyl-2-(1-methylethyl)-benzene for *Clostridium difficile* (CD) (Bos et al., 2013, Palma et al., 2018).

H_2S, nitric oxide (NO) (nitric oxide), hydrogen cyanide (HCN), carbon dioxide (CO_2), and carbon monoxide (CO) are gaseous inorganic volatiles involved in the regulation of many biological responses (Shatalin et al. 2011). When produced in low concentrations, they produce physiological effects; however, at high concentrations they can cause high toxicity (Clemente et al., 2013, de Lacy Costello et al., 2014). H_2S is oxidized into thiosulfate by the colonic epithelium and then into tetrathionate (Shatalin et al. 2011). Tetrathionate is a terminal electron acceptor during anaerobic respiration (Ribet and Cossart, 2015). It serves as a substrate for methane synthesis, one of the most abundant gases in the environment. Gram-positive and Gram-negative bacteria produce large quantities of indole, a microbial metabolite of tryptophan, that serve as an intercellular and extracellular signal in microbial communication. Indole produced by *E. coli* increases resistance of epithelial cells to colonization by *Salmonella* spp. (Kohli et al., 2018). It is important to mention that bacterial metabolic responses to antibiotics and compounds that interfere with folate, cell wall, or protein synthesis resemble the bacterial metabolic responses induced by exposure to hydrogen peroxide, e.g., an oxidative stress response (Zampieri et al., 2017). Many putative enzymes involved in VOCs synthesis and degradation are still unknown, indicating that novel metabolic pathways for production of secondary metabolites exist. Finally, we still do not know exactly how some VOCs are internalized and transduced, and if they use their owner receptors to trigger specific biochemical processes.

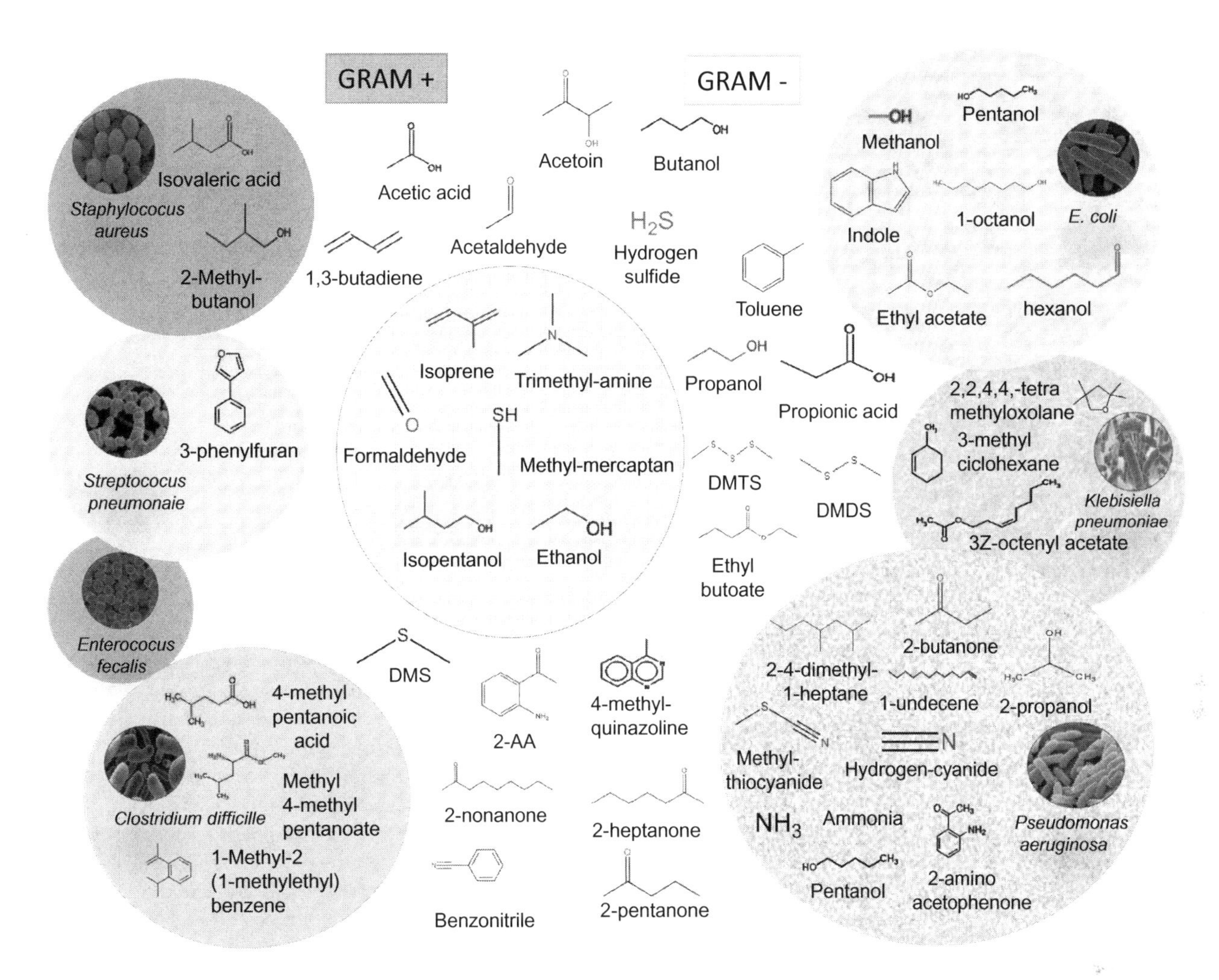

FIGURE 16.1 Chemical classes of volatile organic compounds (VOCs) released by four Gram-positive (*Staphylococcus aureus, Enterococcus faecalis, Streptococcus pneumoniae, Clostridium difficile*) and three Gram negative (*E. coli, Klebsiella pneumoniae, Pseudomonas aeruginosa*) bacteria strains. Each bacterial strain inside of circle has a set of VOCs that serve a potential biomarker for their presence. The central circle shows the VOCs produced by all bacteria. Isopentanol, formaldehyde, methyl mercaptan, and trimethylamine are produced only by bacteria and not by the host. Metabolites outside of colored circles are produced by the host and bacteria and considered as sub-products or intermediates of the metabolic pathways. Acetaldehyde, ethanol, and isoprene are found in large amounts in human exhaled breath. The inorganic compounds and sulfur-containing compounds associated with an inflammatory process include ammonia, nitric oxide, hydrogen cyanide, H_2S (hydrogen sulfide), DMS (dimethyl sulfide), DMDS (dimethyl disulfide), and DMTS (dimethyl trisulfide). Abbreviations: 2-AA, aminoacetophenone. (Adapted from Bos, L.D.J. et al., *PLOS Pathog.*, 9, e1003311, 2013; Palma, S.I.C.J. et al., *Sci. Rep.*, 8, 3360, 2018.)

16.3 Natural Antibiotics and Resistance Mechanisms

Antibiotics are low molecular weight natural molecules generated from ribosome and non-ribosome synthesis that inhibit bacterial growth (Galán et al., 2013). Most bacteria are protected from their own antibiotic through expression of intrinsic antibiotic resistance genes (Galán et al., 2013). Antibiotic exposure can rapidly provoke global metabolic changes that reflect antibiotic mechanisms of action (Zampieri et al., 2017, Paharik et al., 2017). The total burden of antibiotic resistance genes within the gastrointestinal, skin, oral, and vagina microbiomes is not yet known (Forslund et al., 2013, Donia et al., 2014, Dan and Hughes, 2014, Lanza et al., 2018). Antibiotic resistance to practically all antibiotic families was detected in both antibiotic- and non-antibiotic-producing bacteria (Galán et al., 2013, Paharik et al., 2017). Various intrinsic resistomes developed by multiple mechanisms within and among pathogens have been found using next generation DNA sequencing methods (Forslund et al., 2013, Hunter et al., 2014, Donia et al., 2014, Yarygin et al., 2017). Antibiotic resistance can be a consequence of errors or mutations, which are acquired via intrinsic (spontaneously), acquired, and adaptive resistance mechanisms (Galán et al., 2013, Culyba et al., 2015, Paharik et al., 2017). Application of sublethal bactericidal pressure can accelerate horizontal gene transfer (HGT) of plasmidial resistance genes, mutagenesis, and transient hypermutator phenotype or hypermutable state in microorganisms (Dan and Hughes, 2014). Bacterial resistant colonies form heterogeneous populations that co-evolve through stochastic events to give rise to drug-resistant mutants with great phenotypic variability (Nikaido, 2009, Andersson and Hughes, 2010, Baumler and Sparandio, 2016).

Most bacterial strains, as exemplified by *E. coli* and *Salmonella*, evolved multiple DNA damage (SOS) and DNA repair pathways for fixing various types of DNA damage. The homologous recombination, nucleotide excision repair, and translesion synthesis (TLS) are the classical systems of DNA repair (Baharoglu and Mazel, 2013). DNA lesion can be triggered by numerous stressors, including UV light, oxygen radicals, and antibiotics. They are recognized by either the damage sensor, RecA, or the SOS regulator, LexA. RecA binds to damaged ssDNA and induces the autocatalytic cleavage of LexA repressor. LexA regulates the expression of 15 genes, including genes involved in DNA repair or the DNA polymerases: Pol II (encoded by the gene polB), Pol IV (encoded by dinB), and Pol V (encoded by umuD and umuC) in *E. coli* (Dan and Hughes, 2014). Together, DNA polymerases IV and V catalyze translesion error-prone DNA synthesis by replacing the replicative Pol III DNA polymerase. However, these polymerases lack 3′-5′ exonuclease proofreading activity, in particular, Pol V, and this is the most important event for induction of mutation during SOS-induced response (Baharoglu and Mazel, 2013, Culyba et al., 2015). These enzymes can cause simple and double strand breaks, which facilitate the integration of mobile genetic elements (MGEs), such as transposable elements, plasmids, and viruses into the bacterial genome (Nordmann, 2014, Culyba et al., 2015). Together, these mechanisms play a major role in genomic diversity and evolution as well as pathogen dissemination and treatment failure.

Bactericidal antibiotics induce a common mechanism of cell death by stimulating the formation of lethal amounts of oxidative radicals. The changes in the oxidation-reduction cellular balance promote the formation of active oxygen species, including hydrogen peroxide (H_2O_2) and superoxide radical (O_2•), that cause damage to proteins, lipids, and DNA (Dwyer et al., 2014). These high reactive compounds are produced by respiratory chain protein complexes or via interaction between Fe^{2+} and free O_2 molecules in a reaction named the Fenton and Harber Weiss reaction (Fang, 2013, Dwyer et al., 2014). Oxygen, O_2•, and H_2O_2 can produce oxidative damage by converting mitochondrial proteins containing $[4Fe\text{-}4S]^{2+}$ iron and sulfur clusters into $[3Fe\text{-}4S]^{1+}$ and then $[2Fe\text{-}2S]^{2+}$ thereby promoting their quick degradation (Kohanski et al., 2010, Fang, 2013, Dwyer et al., 2014). Bacteria have developed antioxidant defense systems for free radical detoxification. These systems are under the control of three transcription factors: OxyR, SoxR, and SoxS (Kohanski et al., 2010, Dwyer et al., 2014). SoxR and SoxS regulons control the expression of antioxidant enzyme's catalase and superoxide dismutase (Mn-SOD), which are responsible for controlling H_2O_2 toxicity. OxyR regulon control the expression of the enzyme's catalase, glutaredoxin, and glutathione reductases, which control superoxide radical toxicity (Kohanski et al., 2010, Dwyer et al., 2014).

Fe^{2+} is an essential nutrient source for bacterial metabolism, survival, and virulence (Kehl-Fie and Skaar, 2010). Fe^{2+} acquisition in most bacterial species are mediated by siderophores (pyoverdine, pyochelin), which are low-molecular-weight and high-affinity iron-chelating compounds. A powerful strategy used by the host to combat invasive pathogens is the restriction of essential metals, such as Fe^{2+}, manganese (Mn^{2+}), Cu^{2+} (cupper), magnesium (Mg^{2+}), and zinc (Zn^{2+}). These transition metals are mainly incorporated into metalloproteins that include metalloenzymes, storage proteins, and transcription factors. Nutritional immunity is a process in which humans and other mammals confiscate and control host transition metals, limiting the amount available to invasive pathogenic bacteria (Hood and Skaar, 2012). To circumvent host Fe^{2+} restriction, many bacteria species pass to overproduce Fur (Ferric uptake regulator)-controlled regulatory genes that include the high affinity metal transporters, bacterioferritins (sequestration of iron), and the enzymes superoxide dismutase, ferredoxin, ferredoxin reductase, oxidoreductases, and dehydrogenases. These enzymes are vital to defense against iron catalyzed oxidative stress. It is interesting to note that many VOCs produced after tissue oxidative stress, such as nitric oxide, ethanes, and pentanes, may have a role in antibiotic resistance.

16.4 Quorum Sensing Systems and VOCs

Bacteria grow in nature as biofilms or as isolated free-living cell (planktonic) cultures. Biofilms are heterogeneous matrices, in which one or more bacterial species adhere and colonize in a cell density-dependent manner (Vu et al., 2009). Biofilms protect the microbial community from phagocytosis by naïve macrophages and antibiotic attack, thus over 65% of bacterial chronic infections in clinical settings exhibit antibiotic tolerance (Vu et al., 2009, Høibya et al., 2010). The molecular mechanism of antimicrobial resistance in biofilms is not fully understood (Vu et al., 2009, Høibya et al., 2010). The microbial cells growing in a biofilm are slow (sessile form) and physiologically distinct from planktonic cultures, i.e., free-floating cells. Bacterial species growing in planktonic state, as exemplified by the marine *Vibrio fischeri* and plant pathogen *Agrobacterium tumefaciens*, produce a large variety of small natural metabolites that act as signaling molecules to monitor changes in their cell number and density (Ruby, 1996). This mechanism, named quorum sensing (QS), is able to regulate bacterial population density and trigger adequate (transcriptional) responses that affect the group-living bacteria in synchrony (collectively) when the population reaches a certain threshold (Bassler, 1999, Rutherford and Bassler, 2012). This phenomenon involves the production, detection, and response to extracellular signaling molecules named autoinducers (AI) and autoinducing peptides (AIPs) by Gram negative and Gram positive, respectively. The type AI-1 QS regulatory system is formed by four elements: (i) a transcriptional regulator (protein family R); (ii) a *cis*-acting DNA palindromic sequence; (iii) an acyl-homoserine lactone (AHL), which is the signaling molecule or autoinducer (AI-1); and (iv) the AHL synthase protein (protein family I), which synthesizes the AI. These autoinducers facilitate the specific binding to DNA promoters and transcriptional activation of QS-controlled genes. Gram-negative release mainly *N*-acyl-homoserine lactones (AHLs). These molecules bind to cytoplasmic receptors that also act as transcription factors. Gram-positive bacteria release cyclic small peptides (AIPs) that are detected by two-component signal transduction systems composed of a sensor kinase and a response regulator (Håvarstein et al., 1995, Bassler, 1999, Rutherford and Bassler, 2012). AIPs bind to extracellular domains of specific membrane histidine kinase receptors, and this interaction activates the kinase activity of the receptor, which phosphorylates itself and a cognate cytoplasmic response regulator allowing the expression of QS genes. A wide range of biological processes, including bioluminescence, gene

TABLE 16.1

Name, Structure, Regulatory System, Receptor, and Potential Functions of Representative Quorum Sensing Molecules

Class	Structure	Producing Bacteria	Quorum Sensing System	Receptor	Potential Functions
N-Acyl-homoserine Lactone (AHLs)	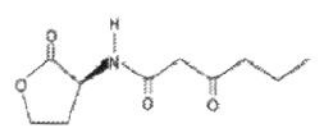	Gram Negative	AI-1 System LuxI and LuxR synthases	LuxI and LuxR	Bioluminescence symbiosis
4-Hydroxy-2-alkyl Quinolines (HAQs)	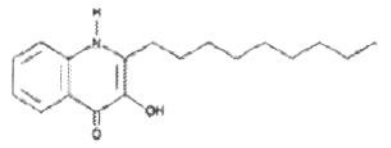	Gram Negative	AI-1 system LasI and RhlI synthases	LasR and RhIR	Biofilm Elastase Antibiotic tolerance Virulence factors
Furanosyl- borate	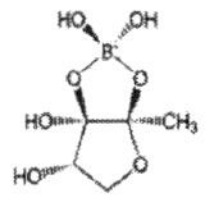	Gram Positive Gram Negative	AI-2 system LuxS synthase	LuxP and Lsr	Bioluminescence Metabolic process Biofilm formation
Cyclic Dipeptides	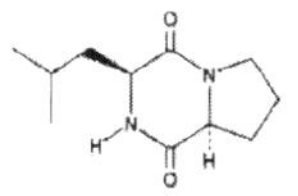	Gram Negative	Cyclic di-GMP	Ycgr and Yhjh	Motility Biofilm formation
Oligopeptides	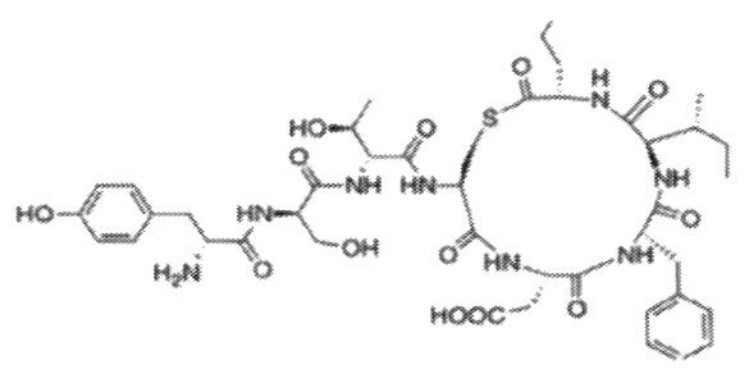	Gram Positive	AgrC/AgrA	AgrC/AgrA (histidine kinase)	Biofilm formation Virulence factors Conjugation
Indole (VOC)	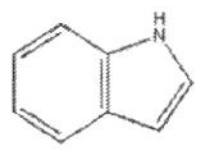	Gram Negative	AI-3 system SdiA	SdiA (?)	Biofilm formation Antibiotic tolerance Persister cell formation
2-aminoacetophene (2-AA, VOC)	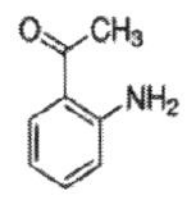	Gram Negative	MvfR	MvfR (?)	Biofilm formation Antibiotic tolerance Persister cell formation

transfer, sporulation, antibiotic production and resistance, biofilm formation, pathogen/host interaction, and virulence factor secretion are regulated by QS (Bassler, 1999, Rutherford and Bassler, 2012). Table 16.1 shows a short list of representative QS system, chemical name, structure, and potential functions of these signaling molecules.

A great diversity of quorum sensing signaling molecules and operons have been discovered to operate in prokaryotic microorganisms (Falcão et al., 2004, Schuster et al., 2013). The LuxI/LuxR system was first characterized in the marine Gram-negative bacterium *V. fisheri* (Ruby, 1996, Schuster et al., 2013). LuxI AHL synthase catalyzes the production of *N*-(β-ketocaproyl)-L-homoserine lactone utilizing *S*-adenosylmethionine and hexanoyl-acyl carrier protein. *N*-acyl-homoserine lactone (acyl-HSL) signal molecules vary in acyl group length (C4-C18), in the substitution of C3 (hydrogen, hydroxyl, or oxo group), and in the presence or absence of one or more carbon-carbon double bonds in the fatty acid chain. For instance, *N*-(3-oxo-hexanoyl)-l-homoserine lactone (oxoC6) binds to LuxR activating transcription of the luxICDABEGH operon, which encodes the enzyme luciferase. When luciferase protein is transcribed and activated, bioluminescence light is produced in a synchronized manner by the entire colony of the marine bioluminescence bacterium *V. fischeri* during certain nocturnal behaviors of animal hosts like *E. scolopes* squids (Ruby, 1996).

Autoinducer-2 (AI-2), a furanosyl borate diester or tetrahydroxy furan, is produced by the luxS-like genes encoded by (2S,4S)-2-methyl-2,3, 3,4-tetrahydroxy tetrahydrofuran-borate synthase. These genes were identified in Gram-negative and Gram-positive bacteria, including *Pasteurella*, *Photorhabdus*, *Haemophilus*, and *Bacillus* (Falcão et al., 2004, Rutherford and Bassler, 2012, Schuster et al., 2013). Furanosyl borates bind to intracellular receptors named LuxP soluble periplasmic AI-2 binding proteins (Hardie and Heurlier, 2008).

Various other species, including *P. aeruginosa*, *E. coli*, *Salmonella*, *Klebsiella*, *Shigella*, and *Enterobacter* use QS as a regulatory system for communication, biofilm formation, and virulence factor production (Falcão et al., 2004, Rutherford and Bassler, 2012, Hense and Schuster, 2015, Lee and Zhang, 2015). *P. aeruginosa* QS consists of a few sets of connected systems. *P. aeruginosa* species carry at least three QS systems that control more than 10% of its genome (Lee and Zhang, 2015). Two of them are controlled by the regulators LasR and RhIR, which are induced by *N*-acyl-homoserine lactone (acyl-HSL) signaling molecules *N*-3-oxododecanoyl homoserine lactone (3-oxo-C12-HSL) and *N*-3-butanoyl-Dl-homoserine lactone (C4-HSL). The third QS is under the control of *las*, *iqs*, *pqs*, and *rhl* genes (Hense and Schuster, 2015, Lee and Zhang, 2015). Hydroxy-2-heptylquinolone (HHQ) and 3,4-dihydroxy-2-heptyquinoline (PQS) peptide synthesis is under the control of the regulator PqsR (or MvfR), which serve as their own receptor. *PqsR* is regulated by the pqsABCD cluster and requires the synthesis of *pqsA* and *pqsD* genes (Que et al., 2013). *P. aeruginosa* growing in culture media is characterized by a sweet, grapelike odor of 2′-aminoacetophenone (2-AA), a single benzene secondary volatile organic chemical (Whiteley et al., 1999). This odor increases greatly in response to antibiotic treatment (Que et al., 2013). 2-AA is involved in antibiotic tolerance and accumulation of persister cells in bacterial biofilms (Cabral et al., 2018). 2-AA, together with PQS and HHQ 2-AA, are synthesized by pqsABCD operon enzymes, under the transcriptional control of MvfR (Que et al., 2013). However, 2-AA is a natural inhibitor of MvfR (Que et al., 2013). The most interesting feature of these three molecules is that they also act as inflammatory mediators through NF-κB and extracellular regulated kinase (ERK)1/2 pathways (Kim et al., 2010, Que et al., 2013). 2-AA limits inflammation by dampening pro-inflammatory cytokine activation, allowing successful bacterial persistence in host tissues (Bandyopadhaya et al., 2012).

Escherichia, *Salmonella*, *Klebsiella*, and *Enterobacter* genera codes to the LuxR homolog SdiA (suppressor of division inhibitor), which is considered QS system 3 (Michael et al., 2001, Ahmer, 2004). However, these bacteria do not encode an AHL synthase, but they are capable of sensing AHL molecules produced by other species, for example, *Yersinia enterocolitica*. SdiA control the transcription of genes involved in different stages of biofilm formation, such as *bcsA*, *csgA*, *csgD*, *fliC*, and *fimA* induced by enteropathogenic *Escherichia coli* (Culler et al., 2018). SdiA, like other LuxR-type receptors, is capable of recognizing a large variety of AIs as well as indole, a volatile nontoxic compound (Michael et al., 2001). SdiA transcription factor activity is induced by indole, and this inhibits *E. coli* biofilm formation (Lee et al., 2007). Indole at concentrations of 1 mM can inhibit the ability of SdiA to respond to acyl-HSLs. Indole also inhibits *Salmonella* virulence (Kohli et al., 2018). A recent study suggested that 1-octanoyl-rac-glycerol (OCL) interacts with SdiA, and the complex initiates the signaling cascade for the synthesis several response genes (Nguyen et al., 2015).

The Gram-positive *S. aureus* genome codes for the agr QS system for controlling bacterial virulence (Thoendel et al., 2011). AI in *agr* systems is an autoinducer AI peptide (AIP), encoded by the *agrBDCA* operon (Novick and Geisinger, 2008, Thodendel et al., 2011). The four *S. aureus* cyclic AIP peptides, termed AIP-I AIP-II, AIP-III, and AIP-IV, each contains a five-residue thiolactone ring and are involved on linkage of the cysteine sulfhydryl to the C-terminal carboxylate. The AIPs structure and amino acid sequence are conserved, thus a working model by their activation and inhibition within the ligand binding pocket of each receptor has been investigated (Lyon et al., 2002). AgrD is the precursor peptide of AIP, and AgrB is an integral membrane endopeptidase essential to biosynthesis of AIPs. AgrC and AgrA form a two-component pair in which AgrC is the membrane histidine kinase and AgrA is a response regulator. Extracellular AIP binding leads to its auto-phosphorylation and subsequent activation of membrane-anchored sensor kinase, AgrC, which in turn activates the P2 and P3 promoters. These promoters increase the transcription of RNA II and III regulatory RNAs. The RNA III hairpins by forming two loop-loop interactions with precursor mRNAs regulating positively and negatively various target genes, including *agr* system genes and the repressor of toxins (Rot) gene. Rot acting as a regulatory factor downregulates the expression of secreted virulence factors (alpha and delta toxins, proteases, lipases) and upregulates the expression the surface protein A (Lyon et al., 2002). The role of the Arg QS system in controlling an extracellular polysaccharide intercellular adhesin (PIA) responsible for biofilm formation may be indirect, and more investigation is needed (Novick and Geisinger, 2008, Thoendel et al., 2011).

The quorum quenching enzymes AHL acylases and AHL lactonases are involved in the cleavage of amide linkage of AHLs and inactivation of the lactone ring, respectively. A large series of natural and synthetic halogenated and brominated furanones has been shown to interfere with AHL and AI-2 QS pathways in Gram-negative and Gram-positive bacteria (Kayumov et al., 2014). These quorum quenching enzymes (QQ) and QS inhibitors as well as modified peptides and oligopeptides can attenuate the production of bacterial toxins, biofilm formation, and pathogenicity, and represent a new generation of antibacterial drugs (Kalia, 2013, Fong et al., 2018, Haque et al., 2019). Nonetheless, there is no compelling evidence for and against the possibility of acyl-HSLs production to have a specific role in the control of commensal or pathogenic intestinal microbial communities (Swearingen et al., 2013).

Plant-growth-promoting rhizobacteria are symbiotic root-colonizing bacteria that promote the growth of plants living at distant sites by releasing a variety of airborne chemical signals or pheromones (Audrain et al., 2015, Farag et al., 2013). Molecular analyses of the soil and rhizosphere identified thousands of unique bacterial species specialized in the production of bioactive secondary metabolites. These include the monoterpenes (γ-terpinene, α-pinene, β-pinene, and β-myrcene), pyrazines, indoles, and the sulfur-derivative compounds DMS, DMDS, and DMTS (Tyc et al., 2017). These compounds act as biological agents for regulation of different biological processes, such as competence and sporulation, plant protection against pathogens and herbivores, as well as attracting pollinators and seed dispersers, such bees and birds (Tyc et al., 2017). The plant-microbe and inter- and intra-kingdom microbe-microbe communications work in a similar way of quorum quenching mechanisms (Kim et al., 2013). Various conditions trigger the production of different VOCs in plants including nutrients, temperature, pH, moisture, light, as well as their interaction with other microorganisms. Linalool, α-terpineol, 4-terpineol, (Z) and (E)-linalool oxides are significantly over-emitted by tomato plants as defense against plant pathogenic bacteria such as *Pseudomonas syringae* and *Xanthomonas* spp. (López-Gresa et al., 2017). Therefore, over-expressing enzymes involved in the biosynthesis of these volatile compounds may open a new biotechnological strategy to bacterial resistance in plants.

VOCs emitted by rhizobacteria can modify antibiotic bacterial resistance and tolerance in various experimental settings (Audrain et al., 2015, Tyc et al., 2017). For example, exposure of *E. coli* to volatiles emitted by *Burkholderia ambifaria* increased its resistance to gentamicin and kanamycin. Furthermore, exposure to trimethylamine increased resistance to tetracycline aminoglycoside antibiotics to human pathogenic Gram-positive and Gram-negative bacteria (Audrain et al., 2015, Tyc et al., 2017). Overall, these studies confirm the multiple and distinct roles of VOCs as putative antibiotics and chemical modulators of antibiotic resistance to bacterial pathogens and the plant, animal, and human body's physiological processes.

16.5 Concluding Remarks

Long-term antibiotic treatment is known to decrease the bacterial diversity and number of ecological niches, favoring microbiome dysbiosis associated with many human diseases. Various investigations on bacterial evolutionary pathways to antibiotic mechanism of action have confirmed the roles of intrinsic oxidative stress, SOS response, and DNA repair systems for early stages of the infection process and antibiotic resistance. Clinical studies of resistomes for different classes of natural and synthetic antibacterial compounds suggest that various bioactive secondary metabolites, small volatile compounds, oligopeptides, and peptides are involved in imitating antibiotic resistance as well as quorum sensing cascades. The combined systematic analysis of metabolic pathways of VOCs, QS-interfering compounds, and secondary metabolites originated from inter- and intraspecies bacterial communication and surviving will pave future advances to bacteria biology. This could also open new horizons to development of strategic therapies for human infection pathogens as well as animal and plant pathogens.

Conflicts of Interest

The authors declare no conflicts of interest.

Funding

The authors are funded by Conselho Nacional de Desenvolvimento Científico e Tecnológico (CNPq, proc 486048/2011 and 312206/2016-0) and Fundação de Amparo a Pesquisa do Estado de São Paulo (2010/06707-0, 2012/02497-7, 2014/20847-0).

Authors' Contributions

MSL, MS, JF, and JEB participated in designing the study and the critical review of the content of all manuscripts cited here. All authors have approved the final version of the manuscript and are responsible for all aspects, ensuring its accuracy and completeness.

REFERENCES

Abdullah AA, Altaf-Ul-Amin M, Ono N, Sato T, Sugiura T, Morita AH, Katsuragi T, Muto A, Nishioka T, Kanaya S (2015). Development and mining of a volatile organic compound database. *Biomed Res Int* 2015: 139254.

Ahmed W, Lawal O, Nijsen TM, Goodacre R, Fowler SJ (2017). Exhaled volatile organic compounds of infection: A systematic review. *ACS Infect Dis* 3(10): 695–710.

Ahmed WM, Brinkman P, Weda H, Knobel HH, Xu Y et al. (2019). Methodological considerations for large-scale breath analysis studies: Lessons from the U-BIOPRED severe asthma project. *J Breath Res* 13(2019): 016001.

Ahmed WM, Geranios P, White IR, Lawal O, Nijsen TM et al. (2018). Development of an adaptable headspace sampling method for metabolic profiling of the fungal volatome. *Analyst* 143(17): 4155–4162.

Ahmer BM (2004). Cell-to-cell signalling in *Escherichia coli* and *Salmonella enterica*. *Mol Microbiol* 52: 933–945.

Amann A, Costello Bde L, Miekisch W, Schubert J, Buszewski B, Pleil J, Ratcliffe N, Risby T (2014). The human volatilome: Volatile organic compounds (VOCs) in exhaled breath, skin emanations, urine, feces and saliva. *J Breath Res* 8: 034001.

Andersson D, Hughes D (2010). Antibiotic resistance and its cost: Is it possible to reverse resistance? *Nat Rev Microbiol* 8(4): 260–271.

Audrain B, Farag MA, Ryu C-M, Ghigo J-M (2015). Role of bacterial volatile compounds in bacterial biology. *FEMS Microbiol Rev* 39(2): 222–233.

Baharoglu Z, Mazel D (2013). SOS, the formidable strategy of bacteria against aggressions. *FEMS Microbiol Rev* 38: 1126–1145.

Bandyopadhaya A, Kesarwani M, Que Y-A, He J, Padfield K et al. (2012). The Quorum Sensing volatile molecule 2-amino acetophenon modulates host immune responses in a manner that promotes life with unwanted guests. *PLoS Pathog* 8(11): e1003024.

Bassler BL (1999). How bacteria talk to each other: Regulation of gene expression by quorum sensing. *Curr Opin Microbiol* 2: 582–587.

Baumler AJ, Sparandio V (2016). Interactions between the microbiota and pathogenic bacteria in the gut. *Nature* 535(7610): 85–93.

Belizario JE, Napolitano M (2015). Microbiomes and their roles in dysbiosis, common diseases and novel therapeutic approaches. *Front Microbiol* 6: 1050.

Boots AW, van Berkel JJ, Dallinga JW, Smolinska A, Wouters EF et al. (2012). The versatile use of exhaled volatile organic compounds in human health and disease. *J Breath Res* 6: 027108.

Bos LDJ, Sterk PJ, Schultz MJ (2013). Volatile metabolites of pathogens: A systematic review. *PLOS Pathog* 9(5): e1003311.

Broza YY, Zuri L, Haick H (2014). Combined volatolomics for monitoring of human body chemistry. *Sci Rep* 4: 4611.

Cabral DJ, Wurster JI, Belenky P (2018). Antibiotic persistence as a metabolic adaptation: Stress, metabolism, the host, and new directions. *Pharmaceuticals* 11: 14.

Clemente JC, Ursell LK, Parfrey LW, Knight R (2013). The impact of the gut microbiota on human health: An integrative view. *Cell* 148(6): 1258–1270.

Culler HF, Couto SCF, Higa JS, Ruiz RM, Yang MJ, Bueris V, Franzolin MR, Sircili MP (2018). Role of SdiA on biofilm formation by atypical enteropathogenic *Escherichia coli*. *Genes* 9: 253.

Culyba MJ, Mo CY, Kohli RM (2015). Targets for Combating the Evolution of Acquired Antibiotic Resistance. *Biochemistry* 54(23): 3573–3582.

D'Costa VM, McGrann KM, Hughes DW, Wright GD (2006). Sampling the antibiotic resistome. *Science* 311(5759): 374–377.

Dan IA, Hughes D (2014). Microbiological effects of sublethal levels of antibiotics. *Nat Rev Microbiol* 12(7): 465–478.

de Lacy Costello B, Amann A, Al-Kateb H, Flynn C, Filipiak W, Khalid T, Osborne D, Ratcliffe NM (2014). A review of the volatiles from the healthy human body. *J Breath Res* 8(1): 014001.

Donia MS, Cimermancic P, Schulze CJ, Wieland Brown LC, Martin J, Mitreva M, Clardy J, Linington RG, Fischbach MA (2014). A systematic analysis of biosynthetic gene clusters in the human microbiome reveals a common family of antibiotics. *Cell* 158(6):1402–1414.

Dwyer DJ, Belenky PA, Yang JH, MacDonald C, Martell, JD et al. (2014). Antibiotics induce redox-related physiological alterations as part of their lethality. *Proc Natl Acad Sci U S A* 111(20): E2100–E2109.

Falcão JP, Sharp F, Sperandio V (2004). Cell-to-cell signaling in intestinal pathogens. *Curr Issues Intest Microbiol* 5(1): 9–17.

Fang FC (2013). Antibiotic and ROS linkage questioned. *Nat Biotechnol* 31(5): 415–416.

Farag MA, Zhang H, Ryu CM (2013). Dynamic chemical communication between plants and bacteria through airborne signals: induced resistance by bacterial volatiles. *J Chem Ecol* 39: 1007–1018.

Filipiak W, Sponring A, Bauer M, Filipiak A, Ager C et al. (2012). Molecular analysis of volatile metabolites released specifically by *Staphylococcus aureus* and *Pseudomonas aeruginosa*. *BMC Microbiol* 12: 113.

Fong J, Zhang C, Yang R, Boo ZZ, Tan SK et al. (2018). Combination therapy strategy of quorum quenching enzyme and quorum sensing inhibitor in suppressing multiple quorum sensing pathways of *P. aeruginosa*. *Sci Rep* 8(1): 1155. doi: 10.1038/s41598-018-19504-w.

Forslund K, Sunagawa S, Kultima JR, Mende DR, Arumugam M et al. (2013). Country-specific antibiotic use practices impact the human gut resistome. *Genome Res* 23: 1163–1169.

Galán J-C, González-Candelas F, Rolain J-M, Cantón R (2013). Antibiotics as selectors and accelerators of diversity in the mechanisms of resistance: From the resistome to genetic plasticity in the β-lactamases world. *Front Microbiol* 4: 9.

Haque S, Yadav DK, Bisht SC, Yadav N, Singh V, Dubey KK, Jawed A, Wahid M, Dar SA (2019). Quorum sensing pathways in Gram-positive and -negative bacteria: Potential of their interruption in abating drug resistance. *J Chemother* 22: 1–27.

Hardie KR, Heurlier K (2008). Establishing bacterial communities by "word of mouth": LuxS and autoinducer 2 in biofilm development. *Nat Rev Microbiol* 6: 635–643.

Håvarstein LS, Coomaraswami G, Morrison DA (1995). An unmodified heptadecapeptide pheromone induces competence for genetic transformation in Streptococcus pneumonia. *Proc Natl Acad Sci USA* 92: 11140–11144.

Hense BA, Schuster M (2015). Core principles of bacterial autoinducer systems. Microbiol Mol Biol Rev 79 (1):153–169.

Herbig J, Beauchamp J (2014). Towards standardization in the analysis of breath gas volatiles. *J Breath Res* 8(2014): 37101.

Høibya N, Bjarnsholta T, Givskov M, Molin S, Ciofu O (2010). Antibiotic resistance of bacterial biofilms. *Int J Antimicrob Ag* 35(4): 322–332.

Hood MI, Skaar EP (2012). Nutritional immunity: Transition metals at the pathogen-host interface. *Nat Rev Microbiol* 10(8): 525–537.

Hunter S, Corbett M, Denise H, Fraser M, Gonzalez-Beltran A et al. (2014). EBI metagenomics – A new resource for the analysis and archiving of metagenomic data. *Nucleic Acids Res* 42: D600–D606.

Jansson BO, Larsson BT (1969). Analysis of organic compounds in human breath by gas chromatography-mass spectrometry. *J Lab Clin Med* 74: 961–966.

Kalia VC (2013). Quorum sensing inhibitors: An overview. *Biotechnol Adv* 31: 224–245.

Kayumov AR, Khakimullina, EN, Sharafutdinov IS, Trizna EY, Latypova LZ et al. (2014). Inhibition of biofilm formation in *Bacillus subtilis* by new halogenated furanones. *J Antibiot* 68: 297–301.

Kehl-Fie TE, Skaar EP (2010). Nutritional immunity beyond iron: A role for manganese and zinc. *Curr Opin Chem Biol* 14(2): 218–224.

Kim K, Kim YU, Koh BH, Hwang SS, Kim SH et al. (2010) HHQ and PQS, two *Pseudomonas aeruginosa* quorum-sensing molecules, down-regulate the innate immune responses through the nuclear factor-kappaB pathway. *Immunology* 129: 578–588.

Kim K-S, Lee S, Ryu C-M (2013). Interspecific bacterial sensing through airborne signals modulates locomotion and drug resistance. *Nat Commun* 4: 1809.

Kohanski MA, DePristo MA, Collins JJ (2010). Sublethal antibiotic treatment leads to multidrug resistance via radical-induced mutagenesis. *Cell* 37(3): 311–320.

Kohli N, Crisp Z, Riordan R, Li M, Alaniz RC, Jayaraman A (2018). The microbiota metabolite indole inhibits *Salmonella* virulence: Involvement of the PhoPQ two-component system. *PLoS One* 13(1): e0190613.

Lanza VF, Baquero F, Martínez JL, Ramos-Ruíz R, González-Zorn B et al. (2018). In-depth resistome analysis by targeted metagenomics. *Microbiome* 6(1): 11.

Lawal O, Knobel H, Weda H, Nijsen TME, Goodacre R, Fowler SJ, BreathDx consortium (2018). TD/GC-MS analysis of volatile markers emitted from mono- and cocultures of *Enterobacter cloacae* and *Pseudomonas aeruginosa* in artificial sputum. *Metabolomics* 14: 66.

Lee J, Jayaraman A, Wood TK (2007). Indole is an inter-species biofilm signal mediated by SdiA. *BMC Microbiol* 7: 42.

Lee J, Zhang L (2015). The hierarchy quorum sensing network in *Pseudomonas aeruginosa*. *Protein Cell* 6(1): 26–41.

Lemfack MC, Gohlke BO, Toguem SMT, Preissner S, Piechulla B, Preissner R (2017). mVOC 2.0: A database of microbial volatiles. *Nucleic Acids Res* 46(D1): D1261–D1265.

López-Gresa MP, Lisón P, Campos L, Rodrigo I, Rambla JL, Granell A, Conejero V, Bellés JM (2017). A non-targeted metabolomics approach unravels the VOCs associated with the tomato immune response against *Pseudomonas syringae*. *Front Plant Sci* 8: 1188.

Lyon GJ, Wright JS, Muir TW, Novick RP (2002). Key determinants of receptor activation in the agr autoinducing peptides of *Staphylococcus aureus*. *Biochem* 41(31): 10095–10104.

Michael B, Smith JN, Swift S, Heffron F, Ahmer BM (2001). SdiA of *Salmonella enterica* is a LuxR homolog that detects mixed microbial communities. *J Bacteriol* 183: 5733–5742.

Nguyen Y, Nguyen NX, Rogers JL, Liao J, MacMillan JB, Jiang Y, Sperandio V (2015). Structural and mechanistic roles of novel chemical ligands on the SdiA quorum-sensing transcription regulator. *mBio* 6: 1–10.

Nikaido H (2009). Multidrug resistance in bacteria. *Annu Rev Biochem* 78: 119–146.

Nordmann P (2014). Carbapenemase-producing *Enterobacteriaceae*: Overview of a major public health challenge. *Med Mal Infect* 44: 51–56.

Novick RP, Geisinger E (2008). Quorum sensing in *Staphylococci*. *Annu Rev Genet* 42: 541–564.

Paharik EA, Schreiber HL, Spaulding CN, Dodson KW, Hultgren SJ (2017). Narrowing the spectrum: The new frontier of precision antimicrobials. *Genome Med* 9(1): 110.

Palma SICJ, Traguedo AP, Porteira AR, Frias MJ, Gamboa H, Roque ACA (2018). Machine learning for the meta analyses of microbial pathogens' volatile signatures. *Sci Rep* 8: 3360.

Pereira J, Porto-Figueira P, Cavaco C, Taunk K, Rapole S, Dhakne R, Nagarajaram H. Camara JS (2015). Breath analysis as a potential and non-invasive frontier in disease diagnosis: An overview. *Metabolites* 5: 3–55.

Phillips M, Cataneo RN, Chaturvedi A, Kaplan PD, Libardoni M, Mundada M, Patel U, Zhang X. (2013). Detection of an extended human volatome with comprehensive two-dimensional gas chromatography time-of-flight mass spectrometry. *PLoS One* 8(9): e75274.

Que Y-A, Hazan R, Strobel B, Maura D, He J et al. (2013). A quorum sensing small volatile molecule promotes antibiotic tolerance in bacteria. *PLoS One* 8(12): e80140.

Rees CA, Burklund A, Stefanuto PH, Schwartzman JD, Hill JE (2018). Comprehensive volatile metabolic fingerprinting of bacterial and fungal pathogen groups. *J Breath Res* 12(2): 026001.

Ribet D, Cossart P (2015). How bacterial pathogens colonize their hosts and invade deeper tissues. *Microbes Infection* 17(3): 173–183.

Ruby EG (1996). Lessons from a cooperative, bacterial-animal association: The Vibrio fischeri-Euprymnascolopes light organ symbiosis. *Annu Rev Microbiol* 50: 591–624.

Rutherford ST, Bassler BL (2012). Bacterial quorum sensing: Its role in virulence and possibilities for its control. *Cold Spring Harb Perspect Med* 2(11): a012427.

Schuster M, Sexton DJ, Diggle SP, Greenberg EP (2013). Acyl-homoserine lactone quorum sensing: From evolution to application. *Ann Rev of Microbiol* 67(1): 43–63.

Shatalin K, Shatalina E, Mironov A, Nudler E (2011). H_2S: A universal defense against antibiotics in bacteria. *Science* 334(6058): 986–990.

Swearingen MC, Sabag-Daigle A, Ahmer BMM (2013). Are there acyl-homoserine lactones within mammalian intestines? *J Bacterol* 195(2): 173–179.

Thoendel M, Kavanaugh JS, Flack CE, Horswill AR (2011). Peptide signaling in the staphylococci. *Chem Rev* 111 (1):117–151.

Timm CM, Lloyd EP, Egan A, Mariner R, Karig D (2018). Direct Growth of Bacteria in Headspace Vials Allows for Screening of Volatiles by Gas Chromatography Mass Spectrometry. *Front Microbiol* 9: 491.

Tremaroli V, Bäckhed F (2012). Functional interactions between the gut microbiota and host metabolism. *Nature* 489: 242–249.

Tyc O, Song C, Dickschat JS, Vos M, Garbeva P (2017). The ecological role of volatile and soluble secondary metabolites produced by soil bacteria. *Trends Microbiol* 25(4): 280–292.

Ulanowska A., Kowalkowski T, Hrynkiewicz K, Jackowski M, Buszewski B (2011). Determination of volatile organic compounds in human breath for *Helicobacter pylori* detection by SPMEGC/MS. *Biomed Chromatogr* 25(3): 391–397.

Vinaixa M, Schymanski EL, Neumann S, Navarro M, Salek RM, Yanes O (2016). Mass spectral databases for LC/MS and GC/MS-based metabolomics: State of the field and future prospects. *Trends Analyt Chem* 78: 23–35.

Vizcaino MI, Crawford JM (2016). Secondary metabolic pathway-targeted metabolomics. *Meth Mol Biol* 1401: 175–195.

Vu B, Chen M, Crawford RJ, Ivanova EP (2009). Bacterial extracellular polysaccharides involved in biofilm formation. *Molecules* 14(7): 2535–2554.

Whiteley M, Lee KM, Greenberg EP (1999). Identification of genes controlled by quorum sensing in *Pseudomonas aeruginosa*. *Proc Natl Acad Sci U S A* 96: 13904–13909.

Wilson AD (2015). Advances in electronic-nose technologies for the detection of volatile biomarker metabolites in the human breath. *Metabolites* 5(1): 140–163.

Yarygin KS, Kovarsky BA, Bibikova TS, Melnikov DS, Tyakht AV, Alexeev DG (2017). ResistoMap – Online visualization of human gut microbiota antibiotic resistome. *Bioinformatics* 33(14): 2205–2206.

Zampieri M, Zimmermann M, Claassen M, Sauer U (2017). Nontargeted metabolomics reveals the multilevel response to antibiotic perturbations. *Cell Rep* 19: 1214–1228.

17

Recent Advances in Science of Quorum Sensing: An Overview of Natural Product Inhibitors

Ana Ćirić, Petrović Jovana, Ivanov Marija, Kostić Marina, and Soković Marina

CONTENTS

17.1 Introduction

Even though the discovery of antibiotics at the beginning of 20th century drastically improved quality of life and life expectancy worldwide, uncontrolled and inadequate use of these agents has brought civilization back almost to the same starting point when these agents were not in use. To this contributes the fact that even today, various bacterial infections cause nearly 16 million deaths each year, while numerous strains of pathogenic microorganisms became resistant to the once effective, first-choice antibiotics. This situation forced the scientific community to search for new, alternative ways to cope with the rising problem. Aside from the pharmaceutical breakthroughs in the development of synthetic and natural sourced agents to eradicate pathogenic microorganisms, discovery of quorum sensing (QS) phenomenon turned out to be one of the most promising areas when it comes to inhibition of microorganisms' growth.

The definition of QS depends on the literature source dealing with the issue, but general scientific consensus defines it as a selective gene expression of bacteria coordinated with the population density in given environmental conditions, with expression occurring only when bacterial products/structures are beneficial for the entire colony rather than individuals [1,2]. Some of the QS regulated processes include: production of various virulence factors, bioluminescence, sporulation, competitiveness, and biofilm formation—though it is still questionable to what extent [3]. Regardless of the trait that is under QS regulation, this mechanism ensures the same—the rational utilization of available resources (mainly nutrients), enabling biological response only and exclusively in the presence of a critical number of individuals of the given species, so as to minimize dissipation of energy sources which are later on used only when it is in the interest of the entire colony [2]. This practically means that, when in low concentrations, bacteria behave as single cell organisms, whereas, when their concentration reaches threshold level, they start behaving like one "multicellular" organism [4]. Competitiveness is yet another QS regulated feature which enables one species to produce inhibitors so as to reduce the number of competitive species in the same environment [2]. Early on, Ji et al. [5] demonstrated that *Staphylococcus aureus* produces signaling molecules which may alter gene expression of other strains in multicellular communities and suggest that this particular system *S. aureus* uses to exclude its opponents.

QS regulation is also rather important when it comes to the development of disease symptoms reflected through several factors: the number of present bacteria, the time and the entrance route of the infection into the blood system, the effects of host defense mechanisms, as well as production of virulence factors such as proteins or other molecules harmful for the host, capsules which allow survival of pathogen under unfavorable conditions, adhesion proteins that enable the pathogen adherence to the host surface, etc. [6,7]. The role of each of these factors is more than obvious—they give a certain type of advantage to its producent. For example, a bacterium that synthesizes a toxin as a virulence factor has no effect as an individual cell, so accordingly, production of toxins by that cell would be a loss of resources. Conversely, if a sufficiently large number of cells are present, host colonization (i.e. disease onset) can be successfully initiated through coordinated gene expression which ends up in toxin synthesis [8].

Regardless to the trait that is under QS regulation, all activation mechanisms follow a certain order: bacteria produce signaling molecules autoinducers (AIs), which at low cell density diffuse away through the membrane so their concentration inside cells approximates the concentration in the environment; after bacterial multiplication when cell density reaches the threshold level, accumulation of a fair amount of these signaling molecules becomes detectable by all the present members of community. This recognition results in gene expression triggering, which is eventually responsible for the social behavior of a community. In more detail, the process unrolls as follows: when a signal molecule attaches to a receptor, it activates enzyme kinase, which undergoes autophosphorylation, after which it delivers phosphate to the gene transcription regulator, which ends up in its activation [4,9].

On more than one occasion it has been demonstrated that both types of bacteria—Gram-positive and Gram-negative—use QS strategy to adequately determine and respond to the cell density with respect to available resources. In Gram-negative bacteria, this phenomenon is widespread; species using this type of regulation synthesize a specific signal molecule—acyl homoserine lactones (AHL)—that regulate QS, whereas the presence of a peptide signaling pathway coupled with the production of small cyclic or linear peptides (oligopeptides from 5 to 10 amino acid) was confirmed in Gram-positive species [1]. AHLs attach to cytoplasmic receptor proteins and activate them, and upon signal detection bind to promoter regions of target genes to activate or repress their transcription [10]. Peptides use a two-component system family and interact with proteins embedded in the membrane. On some occasions they can be moved to cytoplasm, after which they interact with appropriate receptors [11,12]. But, since there is no strictly black/white situation in nature, this division into only two regulatory pathways should also include some transitional forms. Thus, a transitional signaling pathway, containing features of both types of bacteria, has already been described in *Vibrio harveyi*. This species uses AHLs as a signaling molecule as do Gram-negative bacteria, but as opposed to them, and similar to Gram-positive ones, it uses two component signal transductions. The mentioned species *V. harveyi* is also characterized by a newly described signaling molecule as well—AI-2 (furanosyl borate). According to Antunes et al. [4], structure of AI-2 molecules still largely remains in the dark, though it is known that aside from being a signaling molecule, it is also a metabolism by-product and comprehension of its metabolic pathway has actually questioned this molecule's role as a signaling molecule. In addition to AI-2 molecules, several authors described additional ones, such as hydroxyl-palmitic acid methylester and methyl dodecanoic acid [13,14], but nevertheless, AHLs and peptide-based signaling molecules remained the most studied [4,15]. Possible mechanisms of anti-QS activity in Gram-negative bacteria are presented in Figure 17.1.

The very possibility to inhibit growth of pathogenic microorganisms via interruption of bacterial signaling pathways proved to have encouraging results, so consequently, QS has been intensively studied in the field of microbiology in recent years [3,16]. This resulted in data expansion on QS regulation in numerous species that nowadays serve as appropriate model systems, including *Vibrio fischeri*, *V. harveyi*, *Pseudomonas aeruginosa*, *Erwinia carotovora*, *Agrobacterium tumefaciens*, *Chromobacterium violaceum*, etc. Furthermore, new facts suggest that bacteria do not only communicate within, but also between different species as well, and this may be the key feature responsible for their multimillennial successful survival in harsh environmental conditions [2]. *P. aeruginosa* has been extensively

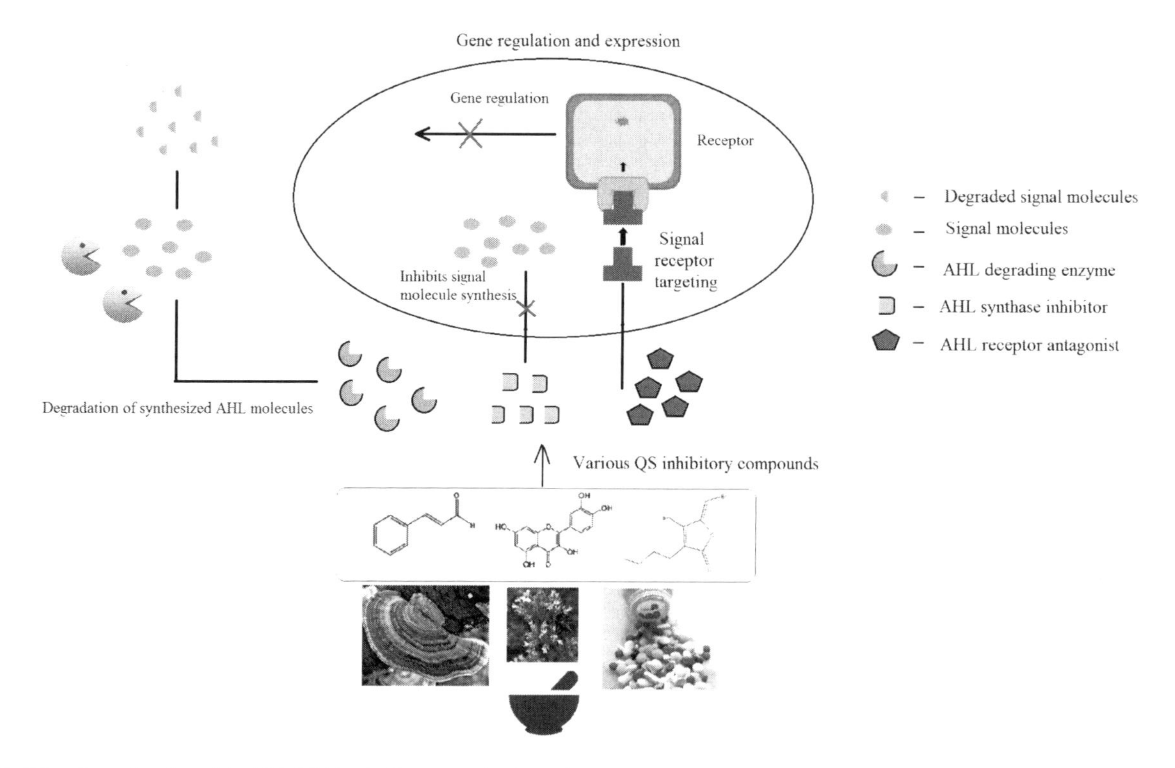

FIGURE 17.1 Example of possible QS inhibition in Gram-negative bacteria.

studied as a reporter strain, so detailed features will be presented below. According to Hall et al. [17], a Gram-negative bacterium *P. aeruginosa* is a very well elaborated model system for studying different aspects of QS regulation. It is a frequent opportunistic pathogen in patients with cystic fibrosis (CF) because it is easily adapted to the conditions present in the lungs and at a high degree causes fatal outcomes [4,18]. After colonization of the lungs, this species exhibits the so-called mucoid phenotype and produces significant amounts of mucus [19]. In this phase, *P. aeruginosa* increased the rate of mutations throughout the entire genome, especially in the regions that are responsible for pathogenicity. The virulent factors that are necessary for acute infection have been negatively selected in isolates that are the causes of chronic infections, and the strains isolated from CF turned out to be different than the original strains after 8 years of infection [19]. Numerous organ systems may be infected by *P. aeruginosa*, including most often the respiratory and urinary tracts, as well as the vascular and central nervous systems. Organ damage, as well the chronic nature of pseudomonal infections, is most often the result of virulence factors which *P. aeruginosa* produces, such as phenazine pigments, pyocyanin, or pyoverdin, to mention a few [20]. Aside from being able to produce pigments, *P. aeruginosa* is also able to increase its tolerance to antimicrobial agents after formation of cell aggregates—biofilm. According to Sakuragi and Kolter [21] and Van Baarlen et al. [19], biofilms are communities comprised of one or multiple bacterial species immersed in a matrix made up from polysaccharides, proteins, nucleic acids, and lipids, with high resistance rate to the applied antibiotics (up to 1000 times more than in planktonic phase) [16]. When the population density becomes optimal, the QS system activates the expression of a large number of unrelated genes, and the products of these genes allow the transition of free-living bacteria to biofilm [22]. After nutrients become limiting factor, bacterial cells are released from biofilm and once again, they enter the planktonic phase, which allows them to subsequently colonize new substrates [23]. This points to the fact that biofilm formation is closely related to the density of the population, i.e., QS regulation plays an important role in the formation of biofilms though it is still not determined to what extent [24]. Even though there is a high probability that biofilm formation is at least partially regulated through the QS signaling pathway, it is considered that other types of regulation under certain set of conditions may also have a particular effect in forming this structure [25].

A crucial step for colonization is the bacteria's ability to move, so the species uses different types of motilities depending on the surrounding conditions: "swarming," "swimming," "twitching," "gliding," and "sliding" [26]. "Swarming" is the rapid movement of a large number of bacteria on the substrate, carried out by using one rotating flagellum. Many bacteria that use swarming synthesize surfactants (surface active agents—amphipathic molecules) that reduce the surface tension between the surface and the bacterial cell in order to enable surface movement. "Swimming" is enabled by the rotating force of multiple flagella, but unlike swarming, it refers to the movement of an individual organism in a fluid medium. "Twitching" is achieved by the extension and retraction of the type IV pili, which results in a slow movement of cells that look like twitching. "Gliding" is an active movement over the surface that runs along the longitudinal axis of the cell without the assistance of additional structures (flagella for example). It is believed that this type of movement evolved several times independently in different lines. "Sliding" is a passive form of propagation over a substrate that does not require the presence of an active trigger, but this movement relies on surfactants that reduce surface tension by allowing the colony to expand [26,27].

According to Abisado et al. [2], the QS type of regulation has been confirmed in other organisms as well—viruses, eukaryotic microorganisms, even in ants. This points to the fact that the density of population becomes increasingly important in regulation of communities, as well as in matters of competition with other cohabitating organisms. The main advantage that the existence of QS regulation gives to populations possessing it is the increase of fitness and overall success when it comes to reproduction in nutrient deprived environments.

Numerous efforts have been made toward finding new natural and synthetic molecules that may be used for the development of novel therapeutic approaches in treating microorganisms [28–31]. The potential of this approach is actually driven by the fact that natural substances can inhibit bacterial growth via several mechanisms—for example, at the same time, they may act on both virulence factors and biofilm formation without fear of developing evolutionary pressure, which eventually favors only strains resistant to the applied molecule. After a lot of research, scientists today are somewhat close when describing the ideal quorum sensing inhibitor. The molecule in question should meet the following criteria: it should be small, chemically stable, highly specific, and a non-toxic agent for eukaryotic cells [32]. Low molecular weight substances of natural origin as well as selected synthetic compounds with certain chemical stability represent ideal potential inhibitors of QS [33,16]. Due to the fact that in the last fifty years scientists accumulated a significant amount of knowledge with respect to QS regulatory mechanisms, the focus of this chapter will include the most recent update on QS inhibitors research, including the most promising results in the last ten years.

17.2 Plants as Anti-Quorum Sensing Agents

Plants have long tradition of use as food and remedies in different parts of the world. Due to the astonishing number of species which produce a wide range of biologically active compounds, they represent a good basis for various types of research, among which their potential as potential quorum sensing inhibitors has already been described and appreciated. Many plants possess the opportunity to interfere with bacterial signaling systems (LuxR/LasR receptors) via secretion of signal-degrading enzymes, signal blockers, or signal mimics receptors [34]. Some of them promote signal AHL mimics that may activate or prevent quorum sensing [35,36]. Accordingly, researchers are focused on defining a bioactive principle from plants' extracts, fractions, or essential oils in order to make them applicable in the pharmaceutical industry. Data on medicinal plants and their corresponding activities as potential QS inhibitors, summarized in Table 17.1, indicate this has been an intensive field of research. Thus, in the short screening, Zolboo et al. [37] demonstrated that alcohol extracts of 66 medicinal plants collected in Mongolia inhibited of violacein production by the *C. violaceum* CV026,

TABLE 17.1
List of Medicinal Plants Capable of Interlocking QS

Plant Species	Type of Extract	Inhibition Against	QS Activity	Mechanism of Action	References
Adenanthera pavonina L.	Ethanolic Ethyl acetate fraction	*C. violaceum* CV026 and ATCC 12472 *P. aeruginosa* PAO1	Violacein and pyocyanin production Proteolytic and elastolytic activity Swarming motility Biofilm formation	–	[60]
Amomum tsaoko Crevost & Lemarié	Methanolic	*S. aureus* ATCC 6538 *S.* typhimurium ATCC 50013 *P. aeruginosa* ATCC 9027 *C. violaceum* ATCC 12472	Biofilm formation Violacein production	–	[61]
Angelica pancicii Vandas	Methanolic Ethanolic, Aqueous	*P. aeruginosa* PAO1	Pyocyanin production Biofilm formation Twitching and flagella motility	–	[50]
Arctium lappa L.	Methanolic	*E. coli, P. mirabilis, S. marcescens*—clinical isolates	Violacein and prodigiosin production Swarming motility Cell surface hydrophobicity Biofilm formation	–	[62]
Camellia nitidissima C.W. Chi	Dichloromethane fraction	*P. aeruginosa* PAO1	Pyocyanin production Swarming and swimming motility	–	[63]
Cassia alata L.	Ethanolic	*C. violaceum* CV026 *P. aeruginosa* PAO1	Violacein and pyocyanin production Proteolytic and elastolytic activities Swarming motility Biofilm formation	–	[64]
Centella asiatica (L.) Urban	Ethanolic	*C. violaceum* CV026 ATCC 12472, ATCC 31532 *P. aeruginosa* PAO1	Violacein and pyocyanin production Proteolytic activity Elastolytic activity Swarming motility Biofilm formation	–	[65]
Citrus maxima Merr.	Flavonoid extract	*V. anguillarum* *C. violaceum* CV026	Violacein production Biofilm formation Swimming and swarming motility	Inhibited AI-2 production, inhibited signaling compound toward the LasR receptor	[66]
Coriandrum sativum L.	Essential oil	*C. coli* ATCC 33559 and 873 *C. jejuni* ATCC 33560 and 225421 *C. violaceum* ATCC 12472	Biofilm formation Promoted biofilm dispersion Violacein production	–	[67]
Dorycnium herbaceum Vill.	Acetone Ethyl acetate Ethanolic	*P. aeruginosa* ATCC 27853 and clinic isolate	Biofilm formation	–	[48]
Echinophora sibthorpiana Guss.	Essential oil Methanolic Ethanolic Aqueous	*P. aeruginosa* PAO1	Pyocyanin production Biofilm formation Twitching and flagella motility	–	[51]
Eucalyptus globulus Labill. *Eucalyptus radiata* A.Cunn. ex DC.	Essential oil	*C. violaceum* ATCC 12472	Violacein production	– –	[68]
Ferulago macedonica Micevski & E. Mayer	Essential oil Methanolic Ethanolic Aqueous	*P. aeruginosa* PAO1	Pyocyanin production Biofilm formation Twitching and flagella motility	–	[51]

(Continued)

TABLE 17.1 (*Continued*)

List of Medicinal Plants Capable of Interlocking QS

Plant Species	Type of Extract	Inhibition Against	QS Activity	Mechanism of Action	References
Heracleum orphanidis Boiss.	Essential oil	*P. aeruginosa* PAO1	Pyocyanin production Biofilm formation Twitching and flagella motility	–	[49]
Hypericum perforatum L.	Methanolic Ethanolic Aqueous	*P. aeruginosa* PAO1 lasB-gfp and *rhlA-gfp* strains	Biofilm formation	Blocked the LasIR pathway	[56]
Hyptis suaveolens (L.) Poit.	Hexane extract	*C. violaceum* ATCC 12472 *E. coli, P. vulgaris, P. mirabilis, K. pneumoniae, S. marcescens*—clinical isolates	Reduction—Protease and hemolysin Violacein production Swarming and swimming motility Biofilm formation	–	[69]
Laserpitium ochridanum Micevski	Methanolic Ethanolic Aqueous	*P. aeruginosa* PAO1	Pyocyanin production Biofilm formation Twitching and flagella motility	–	[52]
Mangifera indica L.	Methanolic Acetone Ethyl acetate	*C. violaceum* ATCC 12472 *P. aeruginosa* PAO1 *A. hydrophila* WAF38	Violacein and pyocyanin production Reduction—Elastase, protease, chitinase Exopolysaccharide production Swarming motility Biofilm formation	–	[70]
Melilotus albus Medik.	Acetone Ethyl acetate Ethanolic	*P. aeruginosa* ATCC 27853 and clinic isolate	Biofilm formation	–	[48]
Micromeria thymifolia (Scop.) Fritsch	Essential oil	*P. aeruginosa* PAO1	Pyocyanin production Biofilm formation Twitching and flagella motility	–	[59]
Murraya koenigii (L.) Sprengel	Essential oil	*P. aeruginosa* PAO1	Elastase reduction Pyocyanin production	–	[58]
Myrmecodia tuberosa Jack	Methanolic Ethanolic Aqueous Ethyl acetate Hexane fraction	*P. aeruginosa* PAO1 *S. aureus* PCA	Swimming, swarming, and twitching motility Biofilm formation	–	[71]
Rosmarinus officinalis L.	Aqueous	*P. chlororaphis* (aureofaciens)30-84, *S. aureus* agr P3::blaZ pRN8826 *C. albicans* 10231	Quantification of phenazine β-Lactamase production Tyrosol production	–	[72]
Rubus rosaefolius Sm.	Phenolic	*C. violaceum* ATCC 6357 *A. hydrophila* IOC/FDA110-36 *S. marcescens* UFOP-001	Violacein production Swarming motility Biofilm formation	–	[73]
Rumex alveolatus Los.-Losinsk.	Methanolic	*P. aeruginosa* ATCC 1310 *S. aureus* ATCC 25923	Pyocyanin production Proteases production Biofilm formation	–	[45]
Sesbania grandiflora (L.) Poiret	Ethanolic	*V. cholerae* ITDI0063	Biofilm formation	–	[47]
Syzygium cumini L.	Methanolic Ethanolic	*C. violaceum* CV026 *K. pneumoniae*	Violacein production Biofilm formation Exopolysaccharyde production	Antagonist of the autoinducer-binding receptors LasR	[74]
Trapa natans L	Methanolic Acetone Ethyl acetate	*P. aeruginosa* PAO1, DSMZ 22644, PA14, PA14-R3 ("lasI P*rsal::lux*), PAO1*pqsA* (CTX*lux::pqsA*) *C. violaceum* CV026	Biofilm formation Swarming motility Pyocyanin, violacein, and elastase production	Inhibitory effects on LasR and PqsR system	[54]

(Continued)

TABLE 17.1 (*Continued*)
List of Medicinal Plants Capable of Interlocking QS

Plant Species	Type of Extract	Inhibition Against	QS Activity	Mechanism of Action	References
Thymus vulgaris L.	Essential oil	*P. fluorescens* KM121 *C. violaceum* CV026	Production of autoinductors Swimming motility Flagella gene expression Biofilm formation Violacein production	Effects on the level of mRNA of the *flgA* gene	[75]
Valeriana officinalis L. & Maillefer	Aqueous	*P. chlororaphis* (aureofaciens) 30–84 *C. albicans* 10231	Quantification of phenazine β-lactamase production Tyrosol production	–	[72]
Vernonia blumeoides Hook	Ethyl acetate Dichloromethane Hexane Methanolic	*C. violaceum* ATCC 12472, CV026, ATCC 31532 *A. tumefaciens* A136	Violacein production β-galactosidase expression	–	[76]
Viscum album L.	Ethanolic Chlorofom Dichloromethane Methanolic	*C. violaceum* CV026	Biofilm formation Violacein production	Inhibitory effects on signal molecule 3-oxo-C6-HSL	[53]

Note: Not tested mechanism of action.

using the standard disc-diffusion method. Of the tested species, the most potent proved to be extracts of *Hedysarum alpinum*, *Spongiocarpella grubovii*, and *Goniolimon speciosum*, which reduced violacein production in the range of 23.5 to 36.5 mm in zone inhibition. Furthermore, Choi et al. [38] screened extracts prepared from 388 medicinal plants originated from South Korea. The aforementioned extracts exhibited anti-QS activity toward several bacterial strains as model systems: *C. violaceum* CV017, *P. aeruginosa* PAO1, *Yersinia enterocolitica* ATCC 9610, and *Agrobacterium tumefaciens*. The extracts of *Cornus controversa* and *Cynanchum wilfordii* showed the most promising anti-QS activity. Similarly, anti-QS potential was also achieved with extracts prepared from the leaves of 61 medicinal plants from India; production of pigments violacein and pyocyanin in *C. violaceum* MTCC 2656 and *P. aeruginosa* MTCC 2297, respectively, was strongly inhibited with the extracts, as well as swarming motility in *P. aeruginosa*. The most promising inhibitory activity in terms of violacein production was demonstrated for the extracts of *Astilbe rivularis*, *Fragaria nubicola*, and *Osbeckia nepalensis*. Phytochemical analyses of these extracts revealed that triterpenes and flavonoid compounds could be responsible for the mentioned anti-QS potential [39]. According to the study by Yüzbaşıoğlu et al. [40], anti-QS effects have also been shown with 36 extracts from 26 plant species belonging to the families Compositaceae, Onograceae, and Fagaceae. Furthermore, this was the first report on anti-QS potential of the crude extracts from *Tanacetum balsamita* subsp. *balsamitoides*, *Epilobium angustifolium*, *Quercus frainetto*, and *Q. robur* which were capable to completely suppress violacein production in *C. violaceum*, without affecting bacterial growth.

On the other hand, numerous studies indicate that extracts may affect bacterial growth, but not via affecting QS regulation. For example, ethanolic extracts of 25 medicinal plants from southwestern Kenya did not exhibit QS effect against a transformed *E. coli* top 10; extracts of only two species, *Elaeodendron buchananii* and *Acacia gerrardii*, poorly inhibited QS regulated functions, without affecting its biofilm formation [41]. Furthermore, extracts prepared from *Mentha longifolia* subsp. *longifolia* and *Hypericum orientale* stimulated the production of pigment violacein rather than inhibited it [40].

Great anti-QS potential of methanolic extracts prepared from *Citrus sinensis*, *Laurus nobilis*, *Elettaria cardamomum*, *Allium cepa*, and *Coriandrum sativum* was reported by Al-Haidari et al. [42]. (Table 17.1). With *C. violaceum* ATCC 12472 and *P. aeruginosa* PA14 used as reporter strains, strong violacein and pyocyanin inhibition, twitching reduction, as well as biofilm formation was achieved. This study showed great anti-QS potential of several medicinal plants especially toward *P. aeruginosa*, which may have practical significance when formulating new antimicrobial agents. Aside from terpenoids and flavonoids, hydrolysable tannins identified in Indian plants—*Mangifera indica*, *Phyllanthus emblica*, *Terminalia chebula*, *Punica granatum*, *Syzygium cumini*, and *Terminalia bellirica*—were also able to inhibit QS-regulated virulence in *C. violaceum* 12472 [43]. On the contrary, condensed tannins could not inhibit AHL production, thus not affecting bacterial communication.

Furthermore, four fractions and the methanolic extract obtained from *Pistacia atlantica* proved to be promising quorum-sensing inhibitors against *P. aeruginosa* PAO1. Treatment with 0.06, 0.10, 0.15, 0.16, and 0.25 mg/mL of fractions and extract showed significant reduction in biofilm formation and pyocyanin production. This high activity of tested fractions and extract against *P. aeruginosa* was ascribed to bioactive flavonoids identified in fractions, which had strong degrading activity of R protein [44]. Aside from methanolic and ethanolic extracts, mixed aqueous extracts turned to have anti-QS effects as well. For example, the methanolic and aqueous extracts of leaves and root of *Rumex alveolatus* inhibited pyocyanin and protease production, as well as biofilm formation, but the aqueous extract itself did not possess any anti-quorum sensing activity. At concentration 62.5 mg/mL of methanol, extracts effectively dislodged biofilm by 60%–80% and 40% in *P. aeruginosa* and

S. aureus, respectively. The same extracts inhibited protease activity up to 42% and 27%, respectively, and inhibited pyocyanin production by 66% in *P. aeruginosa* [45]. Good antimicrobial and QS activity of methanol extracts of *R. alveolatus* was attributed to high concentration of polyphenols including tannins, anthraquinones, and flavonoids. As for the other reporter strains, studies indicate that various extracts could inhibit their QS regulated functions as well. Certain virulence aspects (anti-adhesion and antibiofilm potential) of multidrug resistance strain Carbapenem Resistant *E. coli* could be reduced with extracts of *Berberis aristata*, *Camellia sinensis*, and *Holarrhena antidysenterica* [46]. The qualitative and quantitative analysis showed that all three extracts had low content of tannin compounds, moderate content of flavonoid compounds, and high content of alkaloid compounds, to which the authors attributed high anti-QS effects. Virulence factors of *V. cholerae* could be reduced with ethanolic extracts of *Sesbania grandiflora*; high contents of alkaloids and tannins were held responsible for promising antibiofilm activity at a low concentration of 0.98 mg/mL [47].

Numerous other studies indicate that various plant extracts prepared from different solvents may act as QS inhibitors. The acetone, ethyl acetate, and ethanol extract of *Melilotus albus* and *Dorycnium herbaceum* showed an inhibitory effect toward formation of biofilm in *P. aeruginosa* ATCC 27853 and clinic isolate as reporter strains in a range of concentrations, from 5 mg/mL to 20 mg/mL [48]. The great antibiofilm activity of tested extracts was attributed to higher concentration of phenolic compounds. The investigation of Mileski et al. [49–52] evaluated anti-QS activity of methanolic, ethanolic, and aqueous extracts of various plants belonging to the Apiaceae family (*Heracleum orphanidis*, *Angelica pancicii*, *Echinophora sibthorpiana*, *Ferulago macedonica*, and *Laserpitium ochridanum*) using *P. aeruginosa* PAO1 strain as a model system. The best effect on biofilm formation, twitching, and flagella motility as well as pyocyanin production was achieved with ethanolic and aqueous extracts of *A. pancicii* at concentration 0.50 mg/mL. Other plants showed QS inhibitory activity in the following order: *H. orphanidis* > *L. ochridanum* > *E. sibthorpiana* > *F. macedonica*. The anti-quorum properties of the ethanol, chloroform, and dichloromethane:methanol extracts of leaf, stem, and fruits of *Viscum album* were evaluated using *C. violaceum* CV026 by [53]. In the mentioned study, all the tested extracts inhibited biofilm formation, but only dichloromethane:methanol extracts produced remarkable inhibitory effects on violacein production. The mechanism behind this activity turned out to be inhibition of signal molecule production (3-oxo-C6-HSL), which is responsible for bacterial communication in most Gram-negative bacteria. The study of Aleksic et al. [54] showed that methanolic, acetone, and ethyl acetate leaf extracts of *Trapa natans* exhibited significant inhibition toward biofilm formation, swarming motility, elastase production, pyocyanin, and violacein production of *P. aeruginosa* (PAO1, PA14) and *C. violaceum* CV026. Incubation of tested strains with 50–100 mg/mL of acetate extracts showed reduction in biofilm formation by 15%, while methanolic extract stimulated formation of this highly organized structure by 20%. All the extracts showed promising decrease in swarming motility at subinhibitory levels (0.2 of the previously determined minimal inhibitory concentration), but the methanolic extract exhibited the best reduction. The same concentrations of extracts inhibited elastase production by nearly 60% and pyocyanin production by 50% [54]. The mechanism behind this activity included inhibitory effects of the extracts on LasR and PqsR systems in *P. aeruginosa*, but also stimulating activity onto the RhlR system which had as a consequence increase in biofilm formation. Chong et al. [55] demonstrated inhibitory effects on several QS regulated traits with four different Chinese herbal plant extracts prepared from *Poria cum Radix pini*, *Angelica dahurica*, *Rhizoma cibotii*, and *Schizonepeta tenuifolia*. Hexane, chloroform, and methanolic extracts of plants showed drastic reduction in swarming motility and production of pyocyanin in *P. aeruginosa* at a concentration of 1 mg/mL, as well as decrease by 50% in violacein production in *C. violaceum*. All the tested extracts, except *S. tenuifolia*, possessed bioluminescence activities, inhibiting about 57% lux-based *E. coli* biosensors, pSB401 and pSB1075. It is assumed that mechanisms of action included prevention of AHL binding to the transcriptional regulator, as well as inhibitory effects of compounds present in extract onto the RhlR system in *P. aeruginosa*. The ethanolic, methanolic, and acetone extract and ultra-sonicated extracts of *Hypericum perforatum* proved to be effective quorum sensing inhibitors against *P. aeruginosa* PAO1, asB-gfp and rhlA-gfp strains. Treatment of 250 μg/mL of all tested extracts was able to significantly block the LasIR pathway of *P. aeruginosa*, which is an important quorum-sensing signal receptor [56]. On the other hand, tested extracts didn't successfully inhibit biofilm formation, which indicates that disturbance of RhlR system was not sufficient to prevent the inhibition.

Essential oils produced by aromatic and medicinal plants also showed promising anti-QS effects and the literature contains many data supporting this. Even though the QS mechanism of essential oils is still poorly understood, what we do know is that this activity is directly dependent on the chemical components identified in the oil as well as their abundance. One of the studies that points to this fact was performed by Cervantes-Ceballos et al. [57] in which the authors evaluated the anti-QS potential of six essential oils distilled from aromatic and medicinal plants cultivated in Colombia: *Aloysia triphylla*, *Cymbopogon nardus*, *Lippia origanoides*, *Hyptis suaveolens*, *Swinglea glutinosa*, and *Eucalyptus globulus*. The results achieved with 25 μg/mL showed that all tested oils possessed low, but not negligible, QS inhibition against *E. coli* (pJBA132) in the following order: *S. glutinosa* (16.3%) > *H. suaveolens* (14.5%) > *E. globulus* (13.2%) > *A. triphylla* (11.6%) > *L. origanoides* (10.5%) > *C. nardus* (9.1%). Efficiency of essential oils as QS inhibitors was demonstrated also with *Murraya koenigii* and *Micromeria thymifolia* against *P aeruginosa* PAO1. *M. koenigii* significantly reduced pyocyanin production at a concentration of 0.3% (v/v) with 64.2%, whereas *M. thymifolia* efficiently inhibited twitching motility, biofilm formation (22.8%–26.5%), pyocyanin production (74.4%) [58,59]. Recently, Mileski et al. [49–51] evaluated the anti-quorum sensing properties of essential oils of *Heracleum orphanidis*, *Ferulago macedonica*, and *Echinophora sibthorpiana* against *P. aeruginosa* PAO1. The oils in question showed antibiofilm activity in 18.33%– 19.08%, 28.22%, and 7.84% respectively. The essential oil of *E. sibthorpiana* better reduced structures involved with twitching and swarming of bacterial cultures, in comparison to the other tested species.

17.3 Macromycetes as Anti-Quorum Sensing Agents

Mushrooms are widely known for their nutritional value, but they also produce various metabolites that have been shown to have different biological activities: antimicrobial, antioxidant, immunostimulatory effects, etc. Due to this fact, fruiting bodies of some species, aside from being rich source of nutritionally desirable compounds, also produce compounds highly appreciated for medical purposes [77]. Various extracts prepared using different solvents proved to have anti-QS activity; furthermore, their activity toward inhibition on biofilm formation, virulence factors production, as well as changes in structures used for bacterial movements—twitching and flagella motility—have been described on more than one occasion [77,78] (Table 17.2).

According to Soković et al. [79], hot water extract of *Agaricus blazei* demonstrated anti-QS activity using *P. aeruginosa* PAO1 strain as a model system by acting on several QS regulated features. The aforementioned extract inhibited formation of biofilm by (98.37 %), even higher than commercial antibiotics streptomycine and ampicillin used as positive controls (88.36% and 92.16, respectively) and also reduced production of pigment pyocyanin, at subinhibitory concentration. Furthermore, Glamočlija et al. [80], tested anti-QS potential of four wild growing *Agaricus* species: *A. macrosporus*, *A. campestris*, *A. bitorquis*, and *A. bisporus* collected in Serbia. Comparative analysis of the tested ethanolic extracts showed that different extracts showed different anti-QS potential: *A. macrosporus* sample had better antibiofilm potential than the other tested species in the study, whereas *A. campestris* sample most efficiently reduced flagella and *A. bitorquis* sample had the best twitching effect (Table 17.2) [80]. Methanolic extract prepared from *Agrocybe aegerita* fruiting body showed very good activity against pyocyanin production and antibiofilm activity according to Petrović et al. [79]. In studies by Kostić et al. [81] and Fernandes et al. [78] similar results were obtained as well. Methanolic extracts obtained from wild growing fruiting bodies of *Armillaria mellea* and *Polyporus squamosus* showed good activity against formation of biofilm using *P. aeruginosa* as model system (Table 17.2). Additionally, methanolic extract of the honey mushroom strongly inhibited pyocyanin production, which indicates its potential use as an inhibitory agent toward virulence factor secretion, whereas Dryad's saddle mushroom demonstrated effectiveness in twitching activity inhibition but had no antibiofilm potential [78,81]. Results regarding the anti-QS potential of a very well-known medicinal *Inonotus obliquus* mushroom are in accordance with the aforementioned, indicating the mushroom's potential in reduction of twitching and flagella motility, as well as pyocyanin production [82].

A comprehensive study by Zhu et al. [83] screened anti-QS potential of 103 basidiomycetes using *C. violaceum* CV026 as a model system. Of the tested species, only 14 samples, *A. bisporus*, *A. aegerita*, *Antrodia camphorata*, *Auricularia auricula*, *A. polytricha*, *Cordyceps sinensis*, *Coriolus versicolor*, *Flammulina velutipes*, *Ganoderma lucidum*, *I. obliquus*, *Lentinus edodes*, *Phellinus igniarius*, *Pleurotus ostreatus*, and *Sparassis crispa*, inhibited production of virulence factor—pigment violacein. Furthermore, methanolic extracts of the *Tremella fuciformis* and *G. lucidum* mushroom also have been shown to inhibit violacein production in a concentration-dependent manner [84,85]. Also, ethanolic extract of the *Amanita rubescens* showed anti-QS

TABLE 17.2

Macromycetes Extracts as Quorum Sensing Inhibitors

Macromycetes	Type of Extract	Inhibition Against	QS Activity	Mechanism of Action	References
Agaricus bitorquis (Quèlet) Sacc.	Ethanolic	*P. aeruginosa* PAO1	Twitching	–	[80]
Agaricus blazei Murill	Aqueous		Biofilm formation	–	[79]
			Pyocyanin production	–	
Agaricus campestris L.	Ethanolic		Flagella motility	–	[80]
Agaricus macrosporus (F.H.Møller & Jul. Schff.) Pilát	Ethanolic		Biofilm formation	–	[80]
Agrocybe aegerita (Brig.)	Methanolic		Biofilm formation	–	[79]
			Pyocyanin production	–	
Amanita rubescens (Pers. Ex Fr.) Gray	Ethanolic	*C. violaceum* CV026	Violacein production	–	[86]
Auricularia auricular (Bull.) J. Schröt	Pigments		Violacein production	–	[87]
Armillaria mellea (Vahl: Fr.) Kummer	Methanolic	*P. aeruginosa* PAO1	Biofilm formation	–	[81]
			Pyocyanin production	–	
Ganoderma lucidum (W.Curt.:Fr.) P. Karst	Methanolic	*C. violaceum* CV026	Violacein production	–	[85]
Inonotus obliquus (Ach. Ex Pers.) Pilát	Aqueous	*P. aeruginosa* PAO1	Twitching activity	–	[82]
	Ethanolic		Pyocyanin production	–	
Polyporus squamosus (Huds.) Fr	Methanolic		Biofilm formation	–	[78]
			Twitching activity	–	
Tremella fuciformis Berk.	Methanolic	*C. violaceum* CV026	Violacein production	–	[84]

Note: Not tested mechanism of action.

activity against *C. violaceum* (Table 17.2) [86]. Pigments isolated from *Auricularia auricular* fruiting bodies showed promising inhibitory activity toward QS regulation, since they inhibited violacein production in the CV026 strain [87].

17.4 Natural Compounds as Novel Anti-Quorum Sensing Agents

Nature is considered an immense source of compounds with diverse structures associated with various health-beneficial effects and bioactive properties: antimicrobial, antioxidant, antiproliferative/cytotoxic, anti-neurodegenerative, and many others [88]. It is estimated that around 50% of different medications available at market are of natural origin. Among them, a few groups of compounds emerged as possible "bioactive leaders," such as flavonoids and terpenes, which stand out due to their biodiversity, abundance, and plethora of biological activities already described in the literature [89–91]. Due to this fact, the selected groups will be elaborated in detail through the following section of the chapter.

Flavonoids are polyphenolic compounds, derivatives of 2-phenyl-benzo-γ-pyrone, abundantly present in various plants, fruits, and vegetables. They exhibit a range of biological activities, including inhibition of certain features regulated by QS in some microorganisms besides bacteria, like fungi *Candida albicans* [92]. To the importance of flavonoids as potential anti-QS agents also contributes the fact that few very comprehensive studies investigated quenching potential of extensive libraries of compounds. For example, Skogman et al. [93] screened a total of 465 natural and synthetic flavonoids and found 3 main compounds able to interfere with virulence factors—pigment production of *C. violaceum* and biofilm growth of *P. aeruginosa* and *E. coli* (Table 17.3). Similar research continued later by Manner and Fallarero [94] screened anti QS potential of as many as 3040 natural compounds. Among the tested compounds, the best activity has been demonstrated for a group of flavonoid compounds, with two natural product derivatives belonging to this group as "lead" bioactive compounds. Potent activity was shown toward production of pigment violacein in *C. violaceum* as well as several virulence factors in *P. aeruginosa*- biofilm formation and swarming and swimming motility (Table 17.3) [94]. The potential of natural flavonoids as "quorum quenchers" was demonstrated also after their extraction from *Piper delineatum*; compounds in question were able to interfere with *V. harveyi* QS probably by altering downstream elements LuxO in the QS pathway of *V. harveyi*. They also inhibited formation of biofilm in a dose dependent manner (15.6–500 μM) [95]. Several aspects of QS regulated features in methicillin and vancomycin resistant strains of *S. aureus* have been evaluated with the flavonoid morin. Its antibiofilm potential has been confirmed on different biofilm formation stages: the compound in question inhibited not only the adhesion of cells (early stages of biofilm formation), but also disrupted already established structures of *S. aureus* biofilm (later stages). The elaboration of morin's mechanism of action demonstrated that the compound reduced production of extracellular polymeric substances (EPS) which are included in adherence and cell communication, as well as motility and expansion of bacterial colonies, factors also included in its virulence ability [96]. The similar effect against *S. aureus* was observed also by Silva et al. [97] with myrcetin, which had a profound impact on biofilm formation as well as pigment production. Quercetin, on the contrary, showed only strong potential against formation of biofilm in *P. aeruginosa*, which may be of clinical significance since it has been demonstrated that *P. aeruginosa* significantly increases its tolerance to antibiotics by forming highly organized structure of biofilm [98]. Flavonoids tested in the study by Paczkowski et al. [99] also showed QS inhibitory activity toward *P. aeruginosa* via inhibition of the autoinducer-receptors binding, LasR and RhlR, as the most probable mode of action. Phloretin and 7,8-dihydroxyflavone were also able to reduce expression of *P. aeruginosa* QS regulated genes and suppress its pyocyanin production [99].

Terpenes are hydrocarbons with an isoprene skeleton, abundantly identified in naturally sourced substances with numerous biological activities, including anti-QS [100]. For example, three terpenoids obtained from *Platostoma rotundifolium*—cassipourol, β-sitosterol, and α-amyrin—have shown anti-QS properties toward *P. aeruginosa.* All identified terpenoids could prevent biofilm formation, while only cassipourol and β-sitosterol could also alter gene expression associated with QS in *P. aeruginosa* [101]. Another terpenoid compound, carvacrol, which is abundantly identified in various essential oils, showed potential in interfering with QS related traits of both *P. aeruginosa* and *C. violaceum* (pigment production and biofilm formation) [102] (Table 17.3). Furthermore, citral and citronellal—glycomonoterpenes identified in numerous natural sources—were able to inhibit various bacterial QS traits as is shown in Table 17.3 [103]. Citral inhibited biofilm formation as well as motility and endotoxin production in *Cronobacter sakazakii* and pigment production in *C. violaceum* [104]; linalool disrupted biofilm formation of *Acinetobacter baumannii* at different growth stages, and inhibited violacein production using *C. violaceum* as the reporter strain [105]. As opposed to this, linalool and α-terpineol did not show any anti-QS potential regarding inhibition of violacein production, in the study by Mukherji and Prabhune [106]. Furthermore, in the same study, glycomonoterpenols were able to interfere with 8 hours old biofilm of both *V. cholerae* and *P. aeruginosa*, which suggests that production of violacein and inhibition of biofilm formation have different regulatory pathways [106]. Among different sesquiterpenoids tested in the study by Gilabert et al. [107], viridiflorol had the most prominent activity toward inhibition of *P. aeruginosa* and *S. aureus* biofilm formation. Production of elastase is associated with host tissue damage and is one of many QS regulated virulence factors in *P. aeruginosa* [108]. Triterpenoids ursolic and betulinic acid were able to reduce its activity for more than 90%, which seems to be important [107]. Gilabert et al. [107] also suggest that selected sesqui- and triterpenoids identified in liverwort *Lepidozia chordulifera* were able to interfere with some of the QS regulated processes like biofilm formation and elastase activity (Table 17.3), but none of them was able to reduce production of signaling molecules—autoinducers known to be regulators of QS.

TABLE 17.3

Flavonoids and Terpenes as Quorum Sensing Inhibitors

Group	Compound	Inhibition Against	QS Activity	Mechanism of Action	References
Flavonoids	(−)-(2S)-7,5′-Dihydroxy-5,3′-dimethoxyflavanone	*V. harveyi* BB886 and BB170	QS-mediated bioluminescence Biofilm formation	Interaction with elements downstream LuxO in the QS pathway	[95]
	2-(2-chlorophenyl)-4-oxochromen-3-yl propanoate	*C. violaceum* ATCC 31532	Violacein production	–	[94]
		P. aeruginosa PAO1, ATCC 9027 and 15422	Biofilm formation	–	
			Swarming and swimming motility	–	
		E. coli K-12	Biofilm formation	–	
	2-(4-methoxyphenyl)-4-oxochromen-3-yl decanoate	*C. violaceum* ATCC 31532	Violacein production	–	
		P. aeruginosa PA01, ATCC 9027, and 15422	Biofilm formation	–	
			Swarming and swimming motility	–	
		E. coli K-12	Biofilm formation	–	
	2′,4′,4-trihydroxy-3,6′-dimethoxychalchone	*V. harveyi* BB886 and BB170	QS-mediated bioluminescence Biofilm formation	Interaction with elements downstream LuxO in the QS pathway	[95]
	2′,5-dimethoxyflavone	*C. violaceum* CV026	Violacein production	Disruption of the LuxI/LuxR system	[93]
		P. aeruginosa PAO1, ATCC 700829, ATCC 9027, and ATCC 15442	Biofilm formation and viability	–	
		E. coli ATCC 10536 and ATCC 700928	Biofilm formation and viability	–	
	6-methylflavone	*C. violaceum* CV026	Violacein production	Disruption of the LuxI/LuxR system	
		P. aeruginosa PAO1, ATCC 700829, ATCC 9027, and ATCC 15442	Biofilm formation and viability	–	
		E. coli ATCC 10536 and ATCC 700928	Biofilm formation and viability	–	
	7,8-dihydroxyflavone	*P. aeruginosa* PA14	Inhibitors of the QS receptor LasR	Antagonist of the autoinducer-binding receptors LasR and RhlR	[99]
	Baicalein	*P. aeruginosa* PA14	Inhibitors of the QS receptor LasR		
	Chrysin	*P. aeruginosa* PA14	Inhibitors of the QS receptor LasR		
	Kaempferide	*C. violaceum* CV026	Violacein production	Disruption of the LuxI/LuxR system	[93]
		P. aeruginosa PAO1, ATCC 700829, ATCC 9027, and ATCC 15442	Biofilm formation and viability	–	
		E. coli ATCC 10536 and ATCC 700928		–	
	Morin	Methicillin and vancomycin resistant *S. aureus*	Motility and colony spreading Biofilm and EPS production	Inhibits the DNA binding activity of SarA (Staphylococcal accessory regulator A)	[96]
	Myricetin	*S. aureus* ATCC 6538 and ATCC Newman 25904	Adhesion and biofilm formation	Binding to the Sortase A, alteration in expression of cell-wall proteins involved in adhesion, down regulation of the global regulator saeR	[97]
		S. aureus ATCC 6538	Staphyloxanthin production	–	
	Phloretin	*P. aeruginosa* PA14	Inhibitors of the QS receptor LasR	Antagonist of the autoinducer-binding receptors LasR and RhlR	[99]
	Quercetin	*P. aeruginosa* PAO1	Biofilm formation	–	[98]
		P. aeruginosa PAO1	Twitching motility	–	
		P. aeruginosa PA14	Inhibitors of the QS receptor LasR	Antagonist of the autoinducer-binding receptors LasR and RhlR	[99]

(*Continued*)

TABLE 17.3 (*Continued*)

Flavonoids and Terpenes as Quorum Sensing Inhibitors

Group	Compound	Inhibition Against	QS Activity	Mechanism of Action	References
Terpenes	Carvacrol	*P. aeruginosa* ATCC 10154	Biofilm formation	–	[102]
			Pyocyanin production	–	
		C. violaceum ATCC 12472	Violacein production	–	
	Cassipourol	*P. aeruginosa* PAO1	Biofilm formation	Inhibition of quorum sensing-associated genes expression in las and rhl systems	[101]
			Pyocyanin production		
	Citral	*C. sakazakii* ATCC 29544	Swimming and swarming ability	–	[104]
			Biofilm formation	–	
			Adhesion and invasion	–	
		C. violaceum ATCC 12472	Violacein production	–	
	G–citral	*C. violaceum* CV026	Violacein production	–	[103]
		A. tumefaciens NTL4 (pZLR4)	Pigment inhibition	–	
		P. aeruginosa NCIM 5029	Biofilm formation	–	
			Pyoverdine production	–	
		V. cholerae MTCC 0139	Biofilm formation	–	
		C. sakazakii	Biofilm formation	–	
	G-citron	*C. violaceum* CV026	Violacein production	–	
		A. tumefaciens NTL4 (pZLR4)	Pigment inhibition	–	
		P. aeruginosa NCIM 5029	Biofilm formation	–	
			Pyoverdine production	–	
		V. cholerae MTCC 0139	Biofilm formation	–	
		C. sakazakii	Biofilm formation	–	
	Glyco-linalool Glyco-terpineol	*V. cholerae* MTCC 0139 and *P. aeruginosa* NCIM 5029	Inhibition of established biofilm	–	[106]
	Linalool	*A. baumannii* LMG 1025, LMG 1041, AcB 10/10, AcB 23/10, AcB 24/10	Biofilm formation	–	[105]
		C. violaceum ATCC 12472	Violacein production	–	
	Viridiflorol	*P. aeruginosa* ATCC 27853	Biofilm formation	–	[107]
		S. aureus ATCC 6538		–	
	α-amyrin	*P. aeruginosa* PAO1	Biofilm formation	–	[101]
	β-sitosterol	*P. aeruginosa* PAO1	Biofilm formation	Inhibition of quorum sensing-associated genes expression in las and rhl systems	
			Pyocyanin production		
	Ursolic and betulinic acids	*P. aeruginosa* ATCC 27853	Elastase activity	–	[107]

Note: Not tested mechanism of action.

17.5 Secondary Metabolites Isolated from Microorganisms as QS Inhibitors

Microorganisms coexist in nature with different inter- or intraspecies interactions among each other. They are known to synthesize various molecules in order to help other microorganisms establish the infection (e.g. *Candida albicans* and *S. aureus* coinfections), but they also produce compounds in order to eliminate competition from the site of infection [108].

A series of studies points to this fact as well. For example, from approximately 200 bacterial strains collected from the coral *Pocillopora damicornis*, *Vibrio alginolyticus* distinguished thanks to its high anti-QS activity toward both *C. violaceum* and *P. aeruginosa*, with rhodamine isothiocyanate probably responsible for these activities (Table 17.4) [109].

TABLE 17.4

Microorganisms and Their Secondary Metabolites as QS Inhibitors

Microorganism	Compound Responsible for Activity	Inhibition Against	QS Activity	Mechanism of Action	References
Rhizobium sp. NAO1	N-butyryl homoserine lactone	*P. aeruginosa* PAO1	Biofilm formation Elastase activity and siderophore production	Competition with the auto-inducers produced by *P. aeruginosa*	[110]
Roseofilum reptotaenium D. Casamatta, D. Stanic, M. Gantar & L.L. Richardson 2012	Lyngbic acid	*V. harveyi* JMH626	Luminescence inhibition	Inhibitor of CqsS-mediated signaling	[111]
Vibrio alginolyticus (Miyamoto et al. 1961) Sakazaki 1968	Rhodamine isothiocyanate	*C. violaceum* 12472	Violacein production	–	[109]
		P. aeruginosa PAO1	Biofilm formation	Disruption of the las and/or rhl system of PAO1	
			Swarming and swimming motility	Disruption of the las and/or rhl system of PAO1	
			Elastase and rhamnolipid production	Disruption of the las and/or rhl system of PAO1	

Note: Not tested mechanism of action.

Furthermore, as described by Chang et al. [110], *Rhizobium* sp. NAO1 produced molecules that could compete with auto-inducers responsible for the QS of *P. aeruginosa*. This enabled *Rhizobium* sp. NAO1 to interfere with numerous QS regulated traits of *P. aeruginosa*, such as disruption of biofilm formation and down regulation of AHL-mediated virulence factors. Furthermore, lyngbic acid identified from cyanobacterium *Roseofilum reptotaenium* was able to reduce luminescence in *V. harveyi* strain probably via interference with CAI-1 (cholerae autoinducer 1) receptor CqsS (cholerae quorum-sensing sensor) (Table 17.4) [111].

17.6 Natural Products as QS Inhibitors—Perspective and Application

The extensive and prolonged use of antibiotics caused bacteria to develop resistance to these agents via several mechanisms [112]. This is considered to be the consequence of evolutionary pressure by which only the fittest, aka the resistant strains, survive and adapt. Along with the fact that there were no major breakthroughs in the field of antibiotic development, the phenomenon of QS has become rather important when planning new strategies to eliminate diseases caused by pathogenic microorganisms. It is well known that plants, mushrooms, and their products have been used by ancient people as remedies and for prevention of infectious diseases. One of the perspective aspects when evaluating antimicrobial activity turned out to be estimation of their anti-QS activity as well. The extensive search for any potential QS inhibitors revealed that among naturally sourced substances, many have exerted promising anti-QS potential [33]. The downside of these research includes the fact that most of the studies have been performed using *in vitro* systems, which indicates the necessity to expand the results using *in vivo* systems, as well as to perform clinical trials to estimate their eventual potential as treatments. Furthermore, few limitations regarding the use of natural substances as QS inhibitors have already been acknowledged. For instance, the chemical nature of compounds may restrict their use due possible toxic effects on the consumer. This has already been demonstrated in the case of patulin, a very efficient and perspective anti-QS agent, unsuitable for human consumption due to its toxic nature. Moreover, another major obstacle refers to the amount of bioactive substance that has to be ingested so as to exert equivalent effect as in *in vitro* systems. Thus, for garlic extract to be an efficient QS inhibitor, one must consume an equivalent of 50 garlic bulbs per day, which would most certainly have secondary adverse effects on the consumer. Nevertheless, the possible leverage when administering natural QS inhibitors should not be omitted as well, since true potential may lie in combined therapy of natural QS inhibitors and antibiotics. This was previously validated in the case of biofilm disruption: applied natural QS inhibitor "softened" the rigid structure of biofilm, making it more vulnerable for the antibiotic treatment [33].

Nevertheless, even in an ideal scenario, a potential medicine has to pass a series of steps to reach the stage of commercial distribution, among which clinical trials are the most important for safe human consumption. For example, the antibiofilm effect of natural products as mouthwash ingredients has been evaluated on patients with some sort of dental inflammation [113]. Several authors also demonstrated this potential with 10% *Ricinus communis* oil solution in patients with stomatitis, thus indicating the possibility of this natural product to be passed on to the next phase of trial [114,115]. Furthermore, consummation of oral cranberry extract (proanthocyanidin-A, PAC-A) and a formula composed of cranberry extract, solidago, orthosiphon, and birch (CISTIMEV PLUS®) reduced bacterial biofilms in patients with indwelling urinary catheters [116].

On the other hand, many authors have protected their research by making patents; for example, Tufenkji et al. [117] demonstrated that phenolic-rich maple syrup extract, combined with antibiotics exerts good anti-biofilm activity against four pathogenic strains (*E. coli* CFT073, *P. mirabilis* HI4320, and *P. aeruginosa* PAO1 and PA14). Furthermore, another patent is based on strong inhibitory effect of polyacetylene compound falcarindiol (isolated from Umbelliferae and Araliaceae plants) toward Gram-positive and Gram-negative bacteria [118]; extracts prepared from *Carex dimorpholepis* [119], *Cercis chinensis* [120], *Citrus junos* [121], *Myristica fragrans* [122], *Schinus terebinthifolia* [123], *Vaccinium macrocarpon* [124], *Styrax paralleoneurus*, and *Styrax tonkinensis* [125] have shown significant reduction in biofilm formation. All the aforementioned results indicate that natural substances could potentially be used as a new source of safe and beneficial QS inhibitory substances.

Given the demonstrated potential of anti-QS compounds, as well as the accumulation of enormous amounts of data, numerous efficient and cost-effective methods have been developed for measuring various aspects of QS [126]. To make a long story short, virtual screening combined with molecular docking turned out to be convenient in terms of saving money and time, since out of several thousand compounds, software found only several to be suitable for further studies. Even though this approach is useful, the long path when developing new, effective therapeutics which includes *in vitro* studies followed by *in vivo* studies and assessments of the compound's safety, toxicity, biodegradability, etc. should not be neglected [127].

17.7 Conclusions

From the presented results, it could be concluded that certain natural sourced substances are suitable QS inhibitors, probably due to their molecular weight and structure. With this in mind, it is safe to say that some plant extracts, fractions, oils, pure compounds, and macromycetes are of key importance when managing bacterial growth and pathogenesis since they have been demonstrated to alter regulation of their virulence genes. Still, assessment of anti-QS potential toward several reporter strains is a time-consuming and rather expensive process, while the gained benefits do not surpass the invested. The fact that only a few clinical trials regarding anti-QS activity have been performed so far indicates that pharmaceutical companies lack the interest for the subject, probably due to insufficient profit. This is a problem that we should all strive to solve and surpass not only for gaining new knowledge and insights into bacterial social behavior, but also for the benefit of all humankind.

ACKNOWLEDGMENT

The authors are grateful to the Ministry of Education, Science and Technological Development of Serbia for financial support (Grant number 173032).

REFERENCES

1. Rutherford, S.T., Bassler, B.L. 2012. Bacterial quorum sensing: Its role in virulence and possibilities for its control. *Cold Spring Harbor Perspectives in Medicine*, 2 (11): 1–26.
2. Abisado, R.G., Benomar, S., Klaus, J.R., Dandekar, A.A., Chandler, J.R. 2018. Minireview bacterial quorum sensing and microbial community interactions. *mBio*, 9:e02331–17. https://doi.org/ 10.1128/mBio02331-17, pp. 1–14.
3. Miller, M.B., Bassler, B.L. 2001. Quorum sensing in bacteria. *Annual Review of Microbiology*, 55: 165–199.
4. Antunes, L.C.M., Ferreira, R.B.R., Buckner, M.M.C., Finlay, B.B. 2010. Quorum sensing in bacterial virulence. *Microbiology*, 156(8): 2271–2282.
5. Ji, G., Beavis, R., Novick, R.P. 1997. Bacterial interference caused by autoinducing peptide variants. *Science*, 276: 2027–2030.
6. Peterson, J.W. 1996. Bacterial pathogenesis. *In*: Baron, S. (Eds.). *Medical Microbiology*, 4th edition, Chapter 7. University of Texas Medical Branch at Galveston, Galveston, TX, pp. 1–20.
7. Wilson, J.W., Schurr, M.J., LeBlanc, C.L., Ramamurthy, R., Buchanan, K.L., Nickerson, C.A. 2002. Mechanisms of bacterial pathogenicity. *Postgraduate Medical Journal*, 78: 216–224.
8. Madigan, M.T., Martinko, J.M., Parker, J. 2006. Metabolic diversity. *In*: *Brock Biology of Microorganisms*, Carlson, G. (Eds.), Pearson Education, Upper Saddle River, NJ, pp. 549–611.
9. Defoirdt, T., Brackman, G., Coenye, T. 2013. Quorum sensing inhibitors: How strong is the evidence? *Trends in Microbiology*, 21(12): 619–624.
10. Fuqua, W.C., Winans, S.C., Greenberg, E.P. 1994. Quorum sensing in bacteria: The LuxR-LuxI family of cell density-responsive transcriptional regulators. *Journal of Bacteriology*, 176(2): 269–275.
11. Novick, R.P., Geisinger, E. 2008. Quorum sensing in staphylococci. *Annual Review of Genetics*, 42: 541–564.
12. Pottathil, M., Lazzazzera, B.A. 2003. The extracellular Phr peptide-Rap phosphatase signaling circuit of *Bacillus subtilis*. *Frontiers in Bioscience*, 8: 32–45.
13. Dong, Y.H., Zhang, L.H. 2005. Quorum sensing and quorum-quenching enzymes. *Journal of Microbiology*, 43: 101–109.
14. McDougald, D., Rice, S.A., Kjelleberg, S. 2007. Bacterial quorum sensing and interference by naturally occurring biomimics. *Analytical and Bioanalytical Chemistry*, 387: 445–453.
15. Amara, N., Krom, B.P., Kaufmann, G.F., Meijler, M.M. 2011. Macromolecular inhibition of quorum sensing: Enzymes, antibodies, and beyond. *Chemical Reviews*, 111: 195–208.
16. Kalia, V.C. 2013. Quorum sensing inhibitors: An overview. *Biotechnology Advances*, 31(2): 224–245.
17. Hall, S., McDermott, C., Anoopkumar-Dukie, S., McFarland, A.J., Forbes, A., Perkins, A.V., Grant, G.D. 2016. Cellular effects of pyocyanin, a secreted virulence factor of *Pseudomonas aeruginosa*. *Toxins*, 8(8): 1–14.
18. Karatuna, O., Yagci, A. 2010. Analysis of quorum sensing-dependent virulence factor production and its relationship with antimicrobial susceptibility in *Pseudomonas aeruginosa* respiratory isolates. *Clinical Microbiology and Infection*, 16(12): 1770–1775.
19. Van Baarlen, P., Van Belkum, A., Summerbell, R.C., Crous, P.W., Thomma, B.P.H.J. 2007. Molecular mechanisms of pathogenicity: How do pathogenic microorganisms develop cross-kingdom host jumps? *FEMS Microbiology reviews*, 31(3): 239–277.
20. El-Fouly, M.Z., Sharaf, A.M., Shahin, A.A.M., El-Bialy, H.A., Omara, A.M.A. 2015. Biosynthesis of pyocyanin pigment by *Pseudomonas aeruginosa*. *Journal of Radiation Research and Applied Sciences*, 8(1): 36–48.
21. Sakuragi, Y., Kolter, R. 2007. Quorum-sensing regulation of the biofilm matrix genes (pel) of *Pseudomonas aeruginosa*. *Journal of Bacteriology*, 189(14): 5383–5386.

22. Toole, G.O., Kaplan, H.B., Kolter, R. 2000. Biofilm formation as microbial development. *Annual Review of Microbiology*, 54: 49–79.
23. Abdallah, M., Benoliel, C., Drider, D., Dhulster, P., Chihib, N.E. 2014. Biofilm formation and persistence on abiotic surfaces in the context of food and medical environments. *Archives of Microbiology*, 196(7): 453–472.
24. Chow, S., Gu, K., Jiang, L., Nassour, A. 2011. Salicylic acid affects swimming, twitching and swarming motility in *Pseudomonas aeruginosa*, resulting in decreased biofilm formation. *Journal of Experimental Microbiology and Immunology*, 15: 22–29.
25. Kjelleberg, S., Molin, S. 2002. Is there a role for quorum sensing signals in bacterial biofilms? *Current Opinion in Microbiology*, 5(3), 254–258.
26. Kearns, D.B. 2010. A field guide to bacterial swarming motility. *Nature Reviews Microbiology*, 8(9): 634–644.
27. Melville, S., Craig, L. 2013. Type IV pili in Gram-positive bacteria. *Microbiology and Molecular Biology Reviews*, 77(3): 323–341.
28. Geske, G.D., Wezeman, R.J., Siegel, A.P., Blackwell, H.E. 2005. Small molecule inhibitors of bacterial quorum sensing and biofilm formation. *Journal of the American Chemical Society*, 127: 12762–12763.
29. Geske, G.D., Mattmann, M.E., Blackwell, H.E. 2008. Evaluation of a focused library of *N*-aryl L-homoserine lactones reveals a new set of potent quorum sensing modulators. *Bioorganic and Medicinal Chemistry Letters*, 18: 5978–5981.
30. de Nys, R., Givskov, M., Kumar, N., Kjelleberg, S., Steinberg, P.D. 2006. Furanones, *Progress in Molecular and Subcellular Biology*, 42: 55–86.
31. Kalia, V.C., Purhoit, H.J. 2011. Quenching the quorum sensing system: Potential antibacterial drug targets. *Critical Reviews in Microbiology*, 37(2): 121–140.
32. Scutera, S, Zucca, M., Savoia, D. 2014. Novel approaches for the design and discovery of quorum-sensing inhibitors. *Expert Opinion on Drug Discovery*, 9(4): 353–366.
33. Rasmussen, T.B., Givskov, M. 2006. Quorum sensing inhibitors: A bargain of effects. *Microbiology*, 152(4): 895–904.
34. Degrassi, G., Devescovi, G., Solis, R., Steindler, L., Venturi, V. 2007. *Oryza sativa* rice plants contain molecules that activate different quorum-sensing *N*-acyl homoserine lactone biosensors and are sensitive to the specific AiiA lactonase. *FEMS Microbiology Letter*, 269: 213–220.
35. Gao, M., Teplitski, M., Robinson, J.B., Bauer, W.D. 2003. Production of substances by *Medicago truncatula* that affect bacterial quorum sensing. *Molecular Plant Microbe Interaction*, 16: 827–834.
36. Teplitski, M., Chen, H., Rajamani, S., Gao, M., Merighi, M., Sayre, R.T., Robinson, J.B., Rolfe, B.G., Bauer, W.D. 2004. *Chlamydomonas reinhardtii* secretes compounds that mimic bacterial signals and interfere with quorum sensing regulation in bacteria. *Plant Physiology*, 134: 137–146.
37. Zolboo, B., Tsenguunmaa, L., Undram, O., Batkhuu, J. 2015. Quorum sensing screening of some medicinal plants from Mongolia. *Journal of Agricultural Sciences*, 13: 63–65.
38. Choi, O., Kang, D-W., Kyung Cho, S., Lee, Y., Kang, B., Bae. J., Seunghoe Kim, S., Lee, H.J., Lee, E.S., Kim, J. 2018. Anti-quorum sensing and anti-biofilm formation activities of plant extracts from South Korea. *Basic Research*, 8: 411–417.
39. Tiwary, K.B., Ghosh, R., Moktan, S., Ranjan, K.V., Dey, P., Choudhury, D., Dutta, S., Deb, D., Das, P.A., Chakraborty, R. 2017. Prospective bacterial quorum sensing inhibitors from Indian medicinal plant extracts. *Letters in Applied Microbiology*, 65: 2–10.
40. Çepnİ Yüzbaşıoğlu, E., Bona, M., Şerbetçi, T., Güre, F. 2018. Evaluation of quorum sensing modulation by plant extracts originating from Turkey, plant biosystems. *An International Journal Dealing with All Aspects of Plant Biology*, 152: 376–385.
41. Omwenga, E.O., Hensel, A., Pereira, S., Shitandi, A.A., Goycoolea, F.M. 2017. Antiquorum sensing, antibiofilm formation and cytotoxicity activity of commonly used medicinal plants by inhabitants of Borabu sub-county, Nyamira County, Kenya. *PLoS ONE*, 12:e0185722.
42. Al-Haidari, A.R., Shaaban, I.M., Ibrahim, S., Mohamed, A.G. 2016. Anti-quorum sensing activity of some medicinal plants. *Journal of Traditional, Complementary, and Alternative Medicines*, 13: 67–71.
43. Shukla, V., Bhathena, Z. 2016. Broad spectrum anti-quorum sensing activity of tannin-rich crude extracts of Indian medicinal plants. *Scientifica*, 2016: 8.
44. Kordbacheh, H., Eftekhar, F., Ebrahimi, N.S. 2017. Anti-quorum sensing activity of *Pistacia atlantica* against *Pseudomonas aeruginosa* PAO1 and identification of its bioactive compounds. *Microbial Pathogenesis*, 110: 390–398.
45. Korkorian, N., Mohammadi-Sichani, M. 2017. Anti-quorum sensing and antibacterial activity of *Rumex alveolatus*. *Zahedan Journal of Research in Medical Sciences*, 19: e56009.
46. Thakur, P., Chawla, R., Tanwar, A., Chakotiya, S.A., Narula, A., Goel, R., Arora, R., Sharma, K.R. 2016. Attenuation of adhesion, quorum sensing and biofilm mediated virulence of carbapenem resistant *Escherichia coli* by selected natural plant products. *Microbial Pathogenesis*, 92: 76–85.
47. Guzman, D.M.P.J., Cortes, D.A., Neri, D.K., Cortez, E.C., Lim De las Alas, P.T. 2018. Antibacterial and antibiofilm activities of *Sesbania grandiflora* against foodborne pathogen *Vibrio cholera*. *Journal of Applied Pharmaceutical Science*, 8: 067–071.
48. Stefanović, D.O., Tešić, D.J., Čomić, R.L. 2015. *Melilotus albus* and *Dorycnium herbaceum* extracts as source of phenolic compounds and their antimicrobial, antibiofilm, and antioxidant potentials. *Journal of Food Drug Analysis*, 23: 417–424.
49. Mileski, K., Ćirić, A., Trifunovic, S.S., Ristić, M., Sokovic, M., Matevski, V., Tesevic, V.V., Jadranin, M.B., Marin D.P., Džamić, A. 2016. *Heracleum orphanidis*: Chemical characterisation, and comparative evaluation of antioxidant and antimicrobial activities with specific interest in the influence on Pseudomonas aeruginosa PAO1. *Food and Function*, 7: 4061–4074.
50. Mileski, S.K., Trifunović, S.S., Ćirić, D.A., Šakić, M.Ž., Ristić, S.M., Todorović, M.N., Matevski, S.V., Marin, D.M., Tešević, V.V., Džamić, M.A. 2017a. Research on chemical composition and biological properties including antiquorum sensing activity of *Angelica pancicii* vandas aerial parts and roots. *Journal of Agricultural and Food Chemistry*, 65: 10933–10949.
51. Mileski, K., Ćirić, A., Matevski, V., Marin, P., Soković, M., Džamić, A. 2017b. Inhibition of quorum sensing virulente factors of *Pseudomonas aeruginosa* PAO1 by *Ferulago macedonica* and *Echinophora sibthorpiana* extracts and essential oils. *Lekovite Sirovine*, 37: 33–40.

52. Mileski, S.K., Ćirić, D.A., Petrović, D.J., Ristić, S.M., Matevski, S.M., Marin, D.M., Džamić, M.A. 2017c. *Laserpitium ochridanum*: Antioxidant, antimicrobial and anti-quorum sensing activities against *Pseudomonas aeruginosa*. *Journal of Applied Botany and Food Quality*, 90: 330–338.
53. Erdönmez, D., Kenar, N., Erkan, T.K. 2018. Screening for anti-quorum sensing and anti-biofilm activity in *Viscum album* L. extracts and its biochemical composition. *Trakya University Journal of Natural Sciences*, 19: 175–186.
54. Aleksic, I., Ristivojevic, P., Pavic, A., Radojević, I., Čomić, L.R., Vasiljevic, B., Opsenica, D., Milojković-Opsenica, D., Senerovic, L. 2018. Anti-quorum sensing activity, toxicity in zebrafish (*Danio rerio*) embryos and phytochemical characterization of *Trapa natans* leaf extracts. *Journal of Ethnopharmacology*, 10: 148–158.
55. Chong, M.Y., How, Y.K., Yin, F.W., Chan, G.K. 2018. The effects of Chinese herbal medicines on the quorum sensing-regulated virulence in *Pseudomonas aeruginosa* PAO1. *Molecules*, 23,972: 2–14.
56. Doğan, Ş., Gökalsın, B., Şenkardeş, İ., Doğan, A., Sesal, N.C. 2019. Anti-quorum sensing and anti-biofilm activities of *Hypericum perforatum* extracts against *Pseudomonas aeruginosa*. *Journal of Ethnopharmacology*, 10: 293–300.
57. Cervantes-Ceballos, L., Caballero-Gallardo, K., Olivero-Verbel, J. 2015. Repellent and anti-quorum sensing activity of six aromatic plants occurring in Colombia. *Natural Product Communications*, 10: 1753–1757.
58. Ganesh, S.P., Rai, R.V. 2018. Attenuation of quorum-sensing-dependent virulence factors and biofilm formation by medicinal plants against antibiotic resistant *Pseudomonas aeruginosa*. *Journal of Traditional and Complementary Medicine*, 8(1): 170–177.
59. Bukvicki, R.D., Ćirić, A., Sokovic, D.M., Vannini, L., Nissen, L., Novakovic, M.M., Vujisic, Lj., Asakawa, Y., Marin D.M. 2016. *Micromeria thymifolia* essential oil suppresses quorum-sensing signaling in *Pseudomonas aeruginosa*. *Natural Product Communications*, 11: 1903–1906.
60. Vasavi, H., Arun, A., Rekha, P.D. 2015. Anti-quorum sensing potential of *Adenanthera pavonina*. *Pharmacognosy Research*, 7: 105–109.
61. Rahman, T.R.Md., Lou, Z., Yu, F., Wang, P., Wang, H. 2017. Anti-quorum sensing and anti-biofilm activity of *Amomum tsaoko* (Amommum tsao-ko Crevost et Lemarie) on foodborne pathogens. *Saudi Journal of Biological Sciences*, 24: 324–330.
62. Rajasekharan, S.K., Ramesh, S., Bakkiyaraj, D., Elangomathavan, R., Kamalanathan, C. 2015. Burdock root extracts limit quorum-sensing-controlled phenotypes and biofilm architecture in major urinary tract pathogens. *Urolithiasis*, 43: 29–40.
63. Yang, R., Guan, Y., Zhou, J., Sun, B., Wang, Z., Chen, H., He, Z., Jia, A., 2018. Phytochemicals from *Camellia nitidissima* chi flowers reduce the pyocyanin production and motility of *Pseudomonas aeruginosa* PAO1. *Frontiers in Microbiology*, 8: 1–13.
64. Rekha, P.D., Vasavi, S.H., Vipin, C., Saptami, K., Arun, B.A. 2016.A medicinal herb *Cassia alata* attenuates quorum sensing in *Chromobacterium violaceum* and *Pseudomonas aeruginosa*. *Letters in Applied Microbiology*, 64: 231–238.
65. Vasavi, S.H., Arun, B.A., Rekha, D.P. 2016. Anti-quorum sensing activity of flavonoidrich fraction from *Centella asiatica* L. against *Pseudomonas aeruginosa* PAO1. *Journal of Microbiology Immunology and Infection*, 49: 8–15.
66. Liu, Z., Pan, Y., Li, X., Jie, J., Zeng, M. 2017. Chemical composition, antimicrobial and antiquorum sensing activities of pummelo peel flavonoid extract. *Industrial Crops and Products*, 109: 862–868.
67. Duarte, A., Luís, A., Oleastro, M., Domingues, C.F. 2016. Antioxidant properties of coriander essential oil and linalool and their potential to control *Campylobacter* spp. *Food Control*, 61: 115–122.
68. Luís, Â., Duarte, A., Gominho, J., Domingues, C.F. Duarte, P.A. 2016. Chemical composition, antioxidant, antibacterial and anti-quorum sensing activities of *Eucalyptus globulus* and *Eucalyptus radiate* essential oils. *Industrial Crops and Products*, 79: 274–282.
69. Salini, R., Sindhulakshmi, M., Poongothai, T., Pandian, K.S. 2015. Inhibition of quorum sensing mediated biofilm development and virulence in uropathogens by *Hyptis suaveolens*. *Antonie Van Leeuwenhoek*, 107: 1095–1106.
70. Husain, F.M., Ahmad, I., Al-thubiani, A.S., Abulreesh, H.H., AlHazza, I.M., Aqil, F. 2017. Leaf extracts of *Mangifera indica* L. inhibit quorum sensing–regulated production of virulence factors and biofilm in test bacteria. *Frontiers in Microbiology*, 8: 727.
71. Hertiani, T., Utami, S., Pratiwi, T., Rihardini, I.M., Khaerani Cahyaningrum, P. 2018. Investigation on inhibitory potential of *Myrmecodia tuberosa* on quorum sensing-related pathogenicity in *Pseudomonas aeruginosa* PAO1 and *Staphylococcus aureus* Cowan I Strains. *Pakistan Journal of Biological Sciences*, 21: 101–109.
72. Biswas, P., Lokur, A. 2017. Detection of anti-quorum sensing activity of *Rosmarinus officinalis* and *Valeriana officinalis* using microbial biosensor strain. *International Journal of Pharmaceutical Sciences and Research*, 8: 5205–5214.
73. Oliveira, B.D., Rodrigues, C.A., Cardoso, I.M.B., Ramos, C.C.L.A., Bertoldi, C.M., Jason Guy Taylor, G.J., Rodrigues da Cunha, L., Uelinton Manoel Pinto, M.U. 2016. Antioxidant, antimicrobial and anti-quorum sensing activities of *Rubus rosaefolius* phenolic extract. *Industrial Crops and Products*, 84: 59–66.
74. Gopu, V., Kothandapani, S., Shetty, P.H. 2015. Quorum quenching activity of *Syzygium cumini* (L.) Skeels and its anthocyanin malvidin against *Klebsiella pneumoniae*. *Microbial Pathogenesis*, 79: 61–69.
75. Myszka, K., Schmidt, T.M., Majcher, M., Juzwa, W., Olkowicz, M., Czaczyk, K. 2106. Inhibition of quorum sensing-related biofilm of *Pseudomonas Fluorescens* KM121 by *Thymus vulgare* essential oil and its major bioactive compounds. *International Biodeterioration and Biodegradation*, 114: 252–259.
76. Aliyu, B.A., Koorbanally, A.N., Moodley, B., Singh, P., Yousuf Chenia, Y.H. 2016. Quorum sensing inhibitory potential and molecular docking studies of sesquiterpene lactones from *Vernonia blumeoides*. *Phytochemistry*, 126: 23–33.
77. Petrović, J., Glamočlija, J., Stojković, D., Nikolić, M., Ćirić, A., Fernandes, A., Ferreira, I.C.F.R., Soković, M. 2014. Bioactive composition, antimicrobial activities and the influence of *Agrocybe aegerita* (Brig.) Sing on certain quorum-sensing-regulated functions and biofilm formation by *Pseudomonas aeruginosa*. *Food and Function*, 5: 3296–3303.
78. Fernandes, Â., Petrović, J., Stojković, D., Barros, L., Glamočlija, J., Soković, M., Martins, A., Ferreira, I.C.F.R. 2016. *Polyporus squamosus* (Huds.) Fr from different origins:

Chemical characterization, screening of the bioactive properties and specific antimicrobial effects against *Pseudomonas aeruginosa*. *LWT* (Lebensmittel—Wissenschaft und Technologie)—*Food Science and Technology*, 69: 91–97.
79. Soković, M., Ćirić, A., Glamočlija, J., Nikolić, M., van Griensven, L.J.L.D. 2014. *Agaricus blazei* hot water extract shows anti quorum sensing activity in the nosocomial human pathogen *Pseudomonas aeruginosa*. *Molecules*, 19: 4189–4199.
80. Glamočlija, J., Stojković, D., Nikolić, M., Ćirić, A., Barros, L., Ferreira, C.F.R.I., Soković M. 2015a. Comparative study on edible *Agaricus* mushrooms as functional foods. *Food and Function*, 6: 1900–1910.
81. Kostić, M., Smiljković, M., Petrović, J., Glamočlija, J., Barros, L., Ferreira, C.F.R.I., Ćirić, A., Soković, M. 2017. Chemical, nutritive composition and wide-broad bioactive properties of honey mushroom *Armillaria mellea* (Vahl: Fr.) Kummer. *Food and Function*, 8: 3239–3249.
82. Glamočlija, J., Ćirić, A., Nikolić, M., Fernandes, Â., Barros, L., Calhelha, C.R., Ferreira, C.F.R.I., Soković, M., van Griensven, L.J.L.D. 2015b. Chemical characterization and biological activity of Chaga (*Inonotus obliquus*), a medicinal 'mushroom'. *Journal of Ethnopharmacology*, 162: 232–332.
83. Zhu, H., Wang, S., Zhang, S., Cao, C. 2011a. Inhibiting effect of bioactive metabolites produced by mushroom cultivation on bacterial quorum sensing-regulated behaviors. *Chemotherapy*, 57(2): 292–297.
84. Zhu, H., Sun, S.J. 2008. Inhibition of bacterial quorum sensing-regulated behaviours by *Tremella fuciformis* extract. *Current Microbiology*, 57: 418–422.
85. Zhu, H., Liu, W., Tian, B., Liu, H., Ning, S. 2011b. Inhibition of quorum sensing in the opportunistic pathogenic bacterium *Chromobacterium violaceum* by an extract from fruiting bodies of Lingzhi or Reishi medicinal mushroom, *Ganoderma lucidum* (W.Curt.:Fr.) P. Karst. (higher Basidiomycetes). *International Journal of Medicinal Mushrooms*, 13(6): 559–564.
86. Tabbouche, S.A., Gürgen, A., Yildiz, S., Kiliç, A.O., Sökmen, M. 2017. Antimicrobial and anti-quorum sensing activity of some wild mushrooms collected from Turkey. *Journal of Science and Technology MSU* , 5(2): 453–457.
87. Zhu, H., He, C.C., Chu, Q.H. 2011c. Inhibition of quorum sensing in *Chromobacterium violaceum* by pigments extracted from *Auricularia auricular*. *Letters in Applied Microbiology*, 52(3): 269–274.
88. Mushtaq, S., Abbasi, B.H., Uzair, B., Abbasi, R. 2018. Natural products as reservoirs of novel therapeutic agents. *EXCLI Journal*, 4(17): 420–451.
89. González-Burgos E, Gómez-Serranillos M.P. 2012. Terpene compounds in nature: A review of their potential antioxidant activity. Current Medicinal Chemistry, 19: 5319–5341.
90. Sülsen, V., Lizarraga, E., Mamadalieva, N., Lago, J.H. 2017. Potential of terpenoids and flavonoids from *Asteraceae* as anti-inflammatory, antitumor, and antiparasitic agents. *Evidence-Based Complementary and Alternative Medicine*, Vol. 2017, Article ID 6196198, pp. 1–2.
91. Panche, A.N., Diwan, A.D., Chandra, S.R. 2016. Flavonoids: An overview. *Journal of Nutritional Science*, 29(5): e47.
92. Smiljkovic, M., Kostic, M., Stojkovic, D., Glamoclija, J., Sokovic, M. 2018. Could flavonoids compete with synthetic azoles in diminishing *Candida albicans* infections? *Current Medicinal Chemistry*, doi: 10.2174/0929867325666180629133218.
93. Skogman, M.E., Kanerva, S., Manner, S., Vuorela, P.M., Fallarero, A. 2016. Flavones as quorum sensing inhibitors identified by a newly optimized screening platform using *Chromobacterium violaceum* as reporter bacteria. *Molecules*, 21(9), pii: E1211.
94. Manner, S., Fallarero, A. 2018. Screening of natural product derivatives identifies two structurally related flavonoids as potent quorum sensing inhibitors against Gram-negative bacteria. *International Journal of Molecular Sciences*, 19(5), pii: E1346.
95. Martín-Rodríguez, A.J., Ticona, J.C., Jiménez, I.A., Flores, N., Fernández, J.J., Bazzocchi, I.L. 2015. Flavonoids from *Piper delineatum* modulate quorum-sensing-regulated phenotypes in *Vibrio harveyi*. *Phytochemistry*, 117: 98–106.
96. Chemmugil, P., Lakshmi, P.T.V., Annamalai, A. 2019. Exploring morin as an anti-quorum sensing agent (anti-QSA) against resistant strains of *Staphylococcus aureus*. *Microbial Pathogenesis*, 127: 304–315.
97. Silva, L.N., Da Hora, G.C.A., Soares, T.A., Bojer, M.S., Ingmer, H., Macedo, A.J., Trentin, D.S. 2017. Myricetin protects *Galleria mellonella* against *Staphylococcus aureus* infection and inhibits multiple virulence factors. *Scientific Reports*, 7(1): 2823.
98. Pejin, B., Ciric, A., Markovic, J.D., Glamoclija, J., Nikolic, M., Stanimirovic, B., Sokovic, M. 2015. Quercetin potently reduces biofilm formation of the strain *Pseudomonas aeruginosa* PAO1 *in vitro*. *Current Pharmaceutical Biotechnology*, 16(8): 733–737.
99. Paczkowski, J.E., Mukherjee, S., McCready, A.R., Cong, J.P., Aquino, C.J., Kim, H., Henke, B.R., Smith, C.D., Bassler, B.L. 2017. Flavonoids suppress *Pseudomonas aeruginosa* virulence through allosteric inhibition of quorum-sensing receptors. *The Journal of Biological Chemistry*, 292(10): 4064–4076.
100. Bouyahya, A., Dakka, N., Et-Touys, A., Abrini, J., Bakri, Y. 2017. Medicinal plant products targeting quorum sensing for combating bacterial infections. *Asian Pacific Journal of Tropical Medicine*, 10(8): 729–743.
101. Rasamiravaka, T., Ngezahayo, J., Pottier, L., Ribeiro, S. O., Souard, F., Hari, L., Stévigny, C., Jaziri, M.E. and Duez, P. 2017. Terpenoids from *Platostoma rotundifolium* (Briq.). A. J. Paton alters the expression of quorum sensing-related virulence factors and the formation of biofilm in *Pseudomonas aeruginosa* PAO1. *International Journal of Molecular Science*, 18: 1270.
102. Tapia-Rodriguez, M.R., Hernandez-Mendoza, A., Gonzalez-Aguilar, G.A., Martinez-Tellez, M.A., Martins, C.M., Ayala-Zavala J.F. 2017. Carvacrol as potential quorum sensing inhibitor of *Pseudomonas aeruginosa* and biofilm production on stainless steel surfaces. *Food Control*, 75: 255–261.
103. Patil, A., Joshi-Navre, K., Mukherji, R., Prabhune, A. 2017. Biosynthesis of glycomonoterpenes to attenuate quorum sensing associated virulence in bacteria. *Applied Biochemistry and Biotechnology*, 181(4): 1533–1548.

104. Shi, C., Sun, Y., Liu, Z., Guo, D., Sun, H., Sun, Z., Chen, S., Zhang, W., Wen, Q., Peng, X., Xia X. 2017. Inhibition of *Cronobacter sakazakii* virulence factors by citral. *Scientific Reports*, 24(7): 43243.
105. Alves, S., Duarte, A., Sousa, S., Domingues, F.C. 2016. Study of the major essential oil compounds of *Coriandrum sativum* against *Acinetobacter baumannii* and the effect of linalool on adhesion, biofilms and quorum sensing. *Biofouling*, 32(2): 155–165.
106. Mukherji, R., Prabhune, A. 2015. A new class of bacterial quorum sensing antagonists: Glycomonoterpenols synthesized using linalool and alpha terpineol. *World Journal of Microbiology and Biotechnology*, 31(6): 841–849.
107. Gilabert, M., Marcinkevicius, K., Andujar, S., Schiavone, M., Arena, M.E., Bardón, A. 2015. Sesqui- and triterpenoids from the liverwort *Lepidozia chordulifera* inhibitors of bacterial biofilm and elastase activity of human pathogenic bacteria. *Phytomedicine*, 22(1): 77–85.
108. Dufour, N., Rao, R.P. 2011. Secondary metabolites and other small molecules as intercellular pathogenic signals. *FEMS Microbiology Letters*, 314(1): 10–17.
109. Song, Y., Cai, Z.H., Lao, Y.M., Jin, H., Ying, K.Z., Lin, G.H., Zhou, J. 2018. Antibiofilm activity substances derived from coral symbiotic bacterial extract inhibit biofouling by the model strain *Pseudomonas aeruginosa* PAO1. *Microbial Biotechnology*, 11(6): 1090–1105.
110. Chang, H., Zhou, J., Zhu, X., Yu, S., Chen, L., Jin, H., Cai, Z. 2017. Strain identification and quorum sensing inhibition characterization of marine-derived *Rhizobium* sp. NAO1. *Royal Society Open Science*, 4(3): 170025.
111. Meyer, J.L., Gunasekera, S.P., Scott, R.M., Paul, V.J., Teplitski, M. 2016. Microbiome shifts and the inhibition of quorum sensing by black band disease cyanobacteria. *The ISME Journal*, 10(5): 1204–1216.
112. Kalia, V.C., Wood, T.K., Kumar, P. 2014. Evolution of resistance to quorum-sensing inhibitors. *Microbial Ecology*, 68(1): 13–23.
113. Lu, L., Hu, W., Tian, Z., Yuan, D., Yi, G., Zhou, Y., Cheng, Q., Zhu, J., Li, M. 2019. Developing natural products as potential anti-biofilm agents. *Chinese Medicine*, 14(11): 1–17.
114. Salles, M.M., Badaro, M.M., Arruda, C.N., Leite, V.M., Silva, C.H., Watanabe, E., Oliveira, Vde C., Paranhos, Hde F. 2015. Antimicrobial activity of complete denture cleanser solutions based on sodium hypochlorite and *Ricinus communis*-a randomized clinical study. *Journal of Applied Oral Science*, 23: 637–642.
115. Arruda, C.N.F., Salles, M.M., Badaro, M.M., de Cassia Oliveira, V., Macedo, A.P., Silva-Lovato, C.H., de Freitas Oliveira Paranhos, H. 2017. Effect of sodium hypochlorite and *Ricinus communis* solutions on control of denture biofilm: A randomized crossover clinical trial. *The Journal of Prosthetic Dentistry*, 117: 729–734.
116. Cai, T., Caola, I., Tessarolo, F., Piccoli, F., D'Elia, C., Caciagli, P., Nollo, G., Malossini, G., Nesi, G., Mazzoli, S., Bartoletti, R. 2014. Solidago, orthosiphon, birch and cranberry extracts can decrease microbial colonization and biofilm development in indwelling urinary catheter: A microbiologic and ultrastructural pilot study. *World Journal of Urology*, 32: 1007–1014.
117. Tufenkji, N., Maisuria, V., inventors. 2016. Synergistic combination of a phenolic-rich maple syrup extract and an antibiotic. Patent US2016339071(A1).
118. Yu, W., Gong, Q., Zheng, H., Song, Y, inventors. 2015. Bacterial quorum sensing inhibitor and antibacterial application thereof. Patent CN104784160(A).
119. Lee, J.H., Lee, J.T., Cho, M.H., Kim, J.A., inventors. 2015. Composition for inhibiting biofilm containing *Carex dimorpholepis* extract. Patent KR101517716(B1).
120. Kim, S.K., Kuk, M., inventors. 2016. Quorum sensing inhibitor using plant extracts for preventing the proliferation of Gram-negative bacteria. Patent KR101656875(B1).
121. Lee, N.R., Oh, S.K., Chang, H.J., Choi, S.W., Kim, S.M., Kim, H.J., inventors. 2017. Composition for inhibiting quorum sensing comprising *Citrus junos* extract as effective component. Patent KR20170037909(A).
122. Kim, J., Cvitkovitch, D.G., inventors. 2016.Compositions and methods for reducing caries-causing oral bacteria. Patent US2016213604(A1).
123. Quave, C.L., Lyles, J., inventors. 2017. Botanical extracts and compounds from schinus plants and methods of use. Patent US2017007652(A1).
124. Tufenkji, N., Maisuria, V.B., inventors. 2017. Use of cranberry derived phenolic compounds as antibiotic synergizing agent against pathogenic bacteria. Patent WO2017096484(A1).
125. Silberstein, T., Lutz, R., Cohen, E., Sandori-Kazaz, H., inventors. 2017. Compositions of saponins and plant extracts. Patent WO2017191629(A1).
126. Turovskiy, Y., Kashtanov, D., Paskhover, B., Chikindas, M.L. 2007. Quorum sensing: Fact, fiction, and everything in between. *Advances in Applied Microbiology*, 62: 191–234.
127. Ding, T., Li, T., Li, J. 2018. Identification of natural product compounds as quorum sensing inhibitors in *Pseudomonas fluorescens* P07 through virtual screening. *Bioorganic and Medicinal Chemistry*, 26(14): 4088–4099.

18

Nanomaterials as Quorum Sensing Inhibitors

Jamuna A Bai and V. Ravishankar Rai

CONTENTS

18.1 Introduction

Quorum sensing (QS) or bacterial communication is the population density dependent regulation of expression of certain phenotypes. It includes synthesis, secretion, and detection of signaling molecules also known as auto-inducers. The Gram-negative bacteria produce acyl homoserine lactones as signaling molecules, while the Gram-positive bacteria use autoinducing peptides as QS signals (Parsek et al., 1999; Defoirdt et al., 2007). QS has an important role in regulating various phenotypes in bacteria including virulence and biofilm formation. The disruption of QS attenuates bacterial virulence; hence, it is a promising anti-infective strategy (Defoirdt, 2017). The inhibition of QS mechanism is also known as quorum quenching (QQ) (Grandclement et al., 2016). QS signaling can be disrupted by inhibition of autoinducer synthesis, degradation of the autoinducer, and interception of its interaction with the receptor (Kalia, 2013; Tang and Zhang, 2014).

The first isolated natural QS inhibitory compound was a halogenated furanone from marine red alga *Delisea pulchra* (Gram et al., 1996). Since then a number of natural and synthetic bioactive compounds have been identified with QS inhibitory activity. Medicinal plants such as garlic and essential oils of cinnamon and clove have been reported for their QS inhibition potential. The phytochemicals including cinnamaldehdye, eugenol, baicalin, quercetin, naringenin, and kempferol have shown significant bacterial QS (Paul et al., 2018). Most of these compounds either degrade the signaling molecules or are antagonists and show competitive inhibition (Rampioni et al., 2014). Similarly, synthetic QS inhibitors including small molecules have been explored for their use as next generation anti-infectives (Sintim et al., 2010). Inactivation or degradation of QS signals, mainly AHLs using enzymes, have been reported. These QQ enzymes have been isolated from diverse bacteria, and those that degrade AHLs are classified into lactonases, acylases, and oxidoreductases (Dong et al., 2001; Lin et al., 2003).

A number of QS inhibitory compounds have been investigated for their potential as anti-infective therapeutics. Some examples of QS inhibitors studied as prophylactics are against *P. aeruginosa*, *Staphylococcus aureus*, *Salmonella typhimurium*, and *Serratia marcescens* (Quave et al., 2015; Rasko et al., 2008; Srinivasan et al., 2017).

Though the QS inhibitors have shown potential as anti-infective therapeutics, their clinical application has been limited due to no knowledge of their long- and medium-term effects, inability to prepare pharmacological formulations due to the inherent property of these molecules, low stability, specific targeting, and limited availability. To overcome these limitations, nanotechnological approaches can be used to develop nanostructured materials and nanoparticles which can themselves act as anti-QS agents or improve the efficacy of the QS inhibitors by ensuring their targeted delivery and controlled release (Holban et al., 2016; Colino et al., 2018). Some metal nanoparticles and nanoformulations have shown promising *in vitro* and *in vivo* anti-QS activity (Qais et al., 2018; Bai and Rai et al., 2018).

The unique size-dependent properties of nanomaterials make them indispensible for applications in biology and medicine (Salata, 2004; Ramos et al., 2017). This chapter focuses on the QS inhibitory activity of nanomaterials and their ability to reduce pathogenicity and biofilm formation in bacteria. The potential

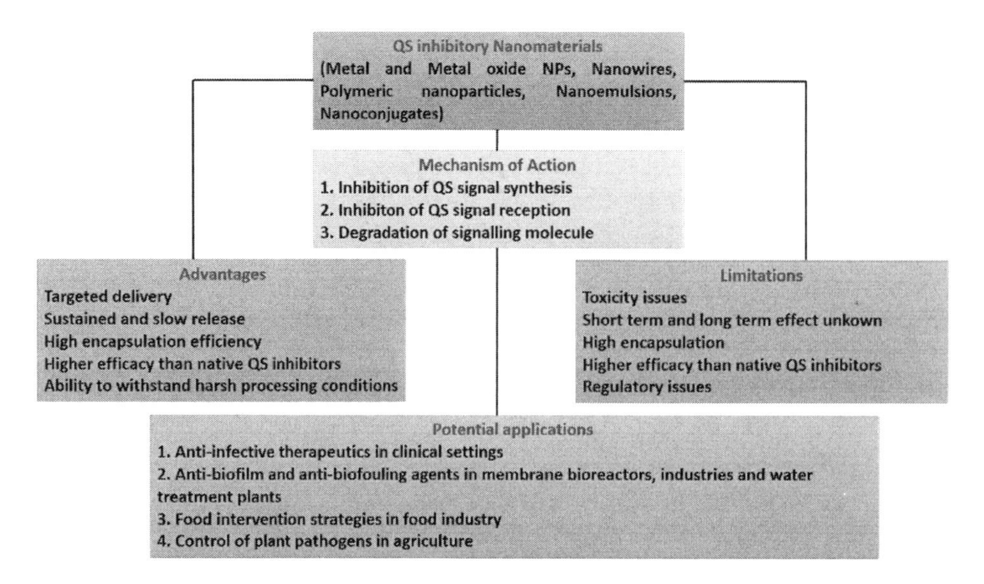

FIGURE 18.1 An overview of nanoparticles as potential quorum sensing inhibitors.

application of the QS inhibitory nanomaterials as novel therapeutics, food intervention techniques, for plant protection, and as antibiofouling agents are discussed (Figure 18.1).

18.2 Nanoparticles as Quorum Sensing Inhibitors

The broad-spectrum antimicrobial activity of nanoparticles is well documented (Vittal and Bai, 2011; Wang et al., 2017). The studies on the quorum sensing inhibitory activity of nanoparticles is limited and only recently it is been increasingly investigated. In the following section, we review the mechanism of action of some of the nanomaterials with quorum sensing inhibitory properties (Table 18.1).

18.2.1 Inhibition of Signal Synthesis

The antimicrobial activity of silver and silver composites and their mechanism of action has been well studied (Naik et al., 2013). However, there are few studies on the antibiofilm activity of silver and its compounds (Babapour et al., 2011; Naik and Kowshik 2014a), and the anti-QS activity has not yet been explored. Silver nanoparticles (AgNPs) and single-wall carbon nanotubes (SWCNT) at sub-inhibitory concentration were able to inhibit the AHL signal synthesis in the phytopathogens *Pseudomonas syringae* and *Pantoea stewartii*. *P. syringae* synthesizes 6 short chain AHLs (C4-HSL, C6-HSL, 3OC6-HSL, C7-HSL, C8-HSL, and 3OC8-HSL) with C6-HSL and 3OC6-HSL being the most predominant AHLs produced. The pathogen *P. stewartii* produces 11 AHLs including both short chain and long chain AHLs (C4-HSL, C6-HSL, 3OC6-HSL, C7-HSL, C8-HSL, 3OC8-HSL, 3OC10-HSL, C12-HSL, 3OC12-HSL, C14-HSL, and 3OC14-HSL). The MIC of nanoparticles AgNP (20 nm) and SWCNT (1.4 nm) on the phytopathogens was determined to be 1 mg/L and 200 mg/L, respectively. In *P. syringae*, the AGNPs at 0.5 mg/L significantly inhibited the production of all AHLs except 3OC8-HSL, and SWCNT at 20 mg/L was able to slightly inhibit C4-HSL and C8-HSL synthesis. The AgNPs could inhibit only C4-HSL production and had no effect on other AHLs in *P. stewartii*. The SWCNTs had no effect on AHLs synthesis in *P. stewartii* at the tested concentration (Mohanty et al., 2016). The nanomaterials had differential effects on the production of different AHLs at sub MIC in the phytopathogens. The study shows that the QS modulatory effect of nanomaterials is dependent on bacterial species.

The fungi and bacteria have been increasingly used for synthesizing nanoparticles and are considered as green nano-factories (Duran et al., 2005). Fungal cell mass and the secreted metabolites can reduce $Ag+$ ions to AgNPs by oxidoreductases reactions (Verma et al., 2010). The fungi are preferred over bacteria for nanoparticle synthesis due to high metal tolerance, metal uptake capability, and easy handling for mass cultivation (Dias et al., 2002). Mycosynthesized silver nanoparticles inhibited AHL synthesis in the clinical pathogen *P. aeruginosa*. The silver nanoparticles synthesized from *Rhizopus arrhizus* BRS-07 had an average particle size of 28 nm. The anti-QS activity of these nanoparticles was confirmed by their inhibition of violacein production in *C. violaceum*. The silver nanoparticles at 25 μg/mL inhibited the production of C12-AHL and C4-AHL in *P. aeruginosa* as observed by HPTLC studies. The silver nanoparticles significantly downregulated the expression of virulence genes *lasA*, *lasB*, *phzA1*, and *rhlA*. This resulted in inhibition in production of virulence factors including LasA protease, LasB elastase, pyocyanin, pyochelin, rhamanolipid, and alginate. Further, the silver nanoparticles also inhibited biofilm formation and maturation. The study clearly showed that the QS inhibitory activities in *P. aeruginosa* were due to downregulation of LasIR-RhlIR transcriptional activity by silver nanoparticles (Singh et al., 2015).

18.2.2 Inhibition of Signal Reception

Silicon dioxide nanoparticles have high biocompatibility, and their particle size can be controlled. This has led to their immense application in the biomedical field (Wang and Benicewicz, 2013). Cyclodextrins (CD) are cyclic oligosaccharides having seven glucopyranose units bound by α-1,4-glycosidic linkages.

TABLE 18.1

Quorum Sensing Inhibitory Activity of Nanomaterials and Their Potential Biological Applications

Nanomaterials	Target Pathogens	Molecular Target and Mechanism of Action	Potential Application	References
Silver nanoparticles (20 nm)	*Pseudomonas syringae*	Inhibition of signal (C4-HSL, C6-HSL, 3OC6-HSL, C7-HSL, C8-HSL) synthesis	Control of phytopathogens	Mohanty et al. (2016)
Silver nanoparticles (20 nm)	*Pantoea stewartii*	Inhibition of C4-HSL synthesis	Control of phytopathogens	Mohanty et al. (2016)
Single-wall carbon nanotubes (1.4 nm)	*P. syringae*	Inhibition of C4-HSL and C8-HSL synthesis	Control of phytopathogens	Mohanty et al. (2016)
Silver nanoparticles (28 nm)	*P. aeruginosa*	Inhibition of C12-AHL and C4-AHL synthesis, downregulation of expression of virulence genes *lasA*, *lasB*, *phzA1*, and *rhlA*	Anti-infective therapeutics	Singh et al. (2015)
Silicon dioxide nanoparticles-cyclodextrin nanoconjugates (50 nm)	*Vibrio fischeri*	Inhibiting signal reception by binding to AHLs and reducing the expression of both *luxA* and *luxR*	Anti-infectives	Miller et al. (2015)
Selenium nanoparticles-honey conjugates (12–15 nm)	*Chromobacterium violaceum* and *P. aeruginosa*	Inhibition of signal reception to LasR site; inhibition of violacein of *C. violaceum*; inhibition of elastase, erotease, pyocyanin, biofilm, motility, and rhamnolipid in *P. aeruginosa*	Anti-infectives and anti-biofilm agents	Prateeksha et al. (2017)
Magnetic nanoparticles (200 nm) immobilized acylhomoserine lactonase	*E. coli* (PJN105Q)	Quorum quenching hydrolysis of 3O-C10 HSL	Nanobiocatalyst	Beladiya et al. (2015)
Gold nanoparticles coated with AHL lactonase	Multidrug-resistant *Proteus* sp.	Degraded C6-HSL; Inhibited exopolysaccharide production and biofilm formation	Nanobiocatalyst, anti-infectives; anti-biofilm agents	Vinoj et al. (2015)
Ultra-small solid lipid nanoparticles (<100 nm) incorporated with 2-heptyl-6-nitro-4-oxo-1,4-dihydroquinoline-3-carboxamide	*P. aeruginosa*	PqsR antagonist; Inhibition of virulence factor production in animal infection model	Nanocarriers for anti-infectives	Nafee et al. (2015)
CAI-1 NPs (112 nm)	*V. cholerae*	Reduced biofilms and virulence	Anti-infectives and anti-biofilm agents	Lu et al. (2015)
Chitosan-quercetin conjugates (250–400 nm)	*E. coli* Top 10	Inhibition of QS-regulated phenotypes	Nanocarriers for QS inhibitors	Nguyen et al. (2017)
Kaempferol-chitosan nanoparticles (192 nm)	*C. violaceum*	Inhibition of violacein production	Nanoencapsulation of QS inhibitors	Ilk et al. (2017)
Nanochitosan (114–155 nm)	*E. coli*	Decrease in QS-regulated GFP expression	Anti-infectives	Qin et al. (2017); Qin et al. (2018)
PEG-coated LasR inhibitor V-06-018 nanocarriers (70–400 nm)	*P. aeruginosa*	Inhibition of LasR-regulated pyocyanin pigment production	Anti-infectives	Lu et al. (2018)
Cinnamaldehyde-chitosan nanoparticles (208.12 nm)	*P. aeruginosa*	Inhibition of pyocyanin, protease, rhamnolipid production, biofilm formation and motility	Anti-infectives	Pattanaik et al. (2018a)
Cinnamaldehyde-chitosan nanoparticles (149 nm)	*E. coli* TOP10	Decrease in QS-regulated GFP expression	Nanoencapsulation of QS inhibitors	Qin et al. (2018)
Ferulic acid chitosan-tripolyphosphate nanoparticles (215.55 nm)	*P. aeruginosa*	Inhibition of pyocyanin, protease, rhamnolipid production, biofilm formation and motility	Nanocarriers for QS inhibitors	Pattanaik et al. (2018b)

(Continued)

TABLE 18.1 (*Continued*)

Quorum Sensing Inhibitory Activity of Nanomaterials and Their Potential Biological Applications

Nanomaterials	Target Pathogens	Molecular Target and Mechanism of Action	Potential Application	References
Ajoene-chitosan NPs (685.7 nm)	*P. aeruginosa*	Inhibition of protease, elastase, pyochelin, haemolysin, alginate, rhamnolipids, and biofilms	Anti-infectives for biofilm-associated pyelonephritis	Vadekeetil et al. (2017)
Silver nanowires (200–250 nm)	*C. violaceum* and *P. aeruginosa*	Inhibition of violacein production in *C. violaceum*. inhibition of biofilm formation in *P. aeruginosa*	Anti-biofilm agents	Wagh et al. (2013)
ZnO nanoparticles (<50 nm)	*P. aeruginosa*	Inhibition of biofilms, pyocyanin and pyochelin production, hemolytic activity, and PQS signaling in *P. aeruginosa*	Anti-infectives	Lee et al. (2014a)
ZnO nanoparticles (<50 nm)	*P. aeruginosa* PA01 and PA14	Inhibition of elastase and biofilms in *P. aeruginosa*	Anti-infectives	Garcia-Lara et al. (2015)
Niclosamide nanocrystals (450 nm)	*P. aeruginosa*	Inhibition of QS-regulated virulence factors in *P. aeruginosa*	Anti-infectives	Costabile et al. (2015)
Silver nanoparticles (12 nm) Silver nanoparticles (70 nm)	*P. aeruginosa*	Inhibition of pyocyanin and hemolysin production and biofilm formation	Anti-infectives	Ali et al. (2017); Shah et al. (2019)
ZnO nanoparticles (<50 nm and <100 nm)	*P. aeruginosa*	Inhibition of pyocyanin production and biofilm formation	Anti-infectives	Aswathanarayan and Vittal (2017)
Tin dioxide nanoflowers (50 nm)	*P. aeruginosa* and *Serratia marcescens*	Inhibition of synthesis of pyocyanin in *P. aeruginosa* and prodiogsin in *S. marcescens*; Anti-biofilm activity on both pathogens	Anti-infectives	Al-Shabib et al. (2018)
Gold nanoparticles (50 nm)	*P. aeruginosa*	Inhibition of synthesis of pyocyanin and motility in *P. aeruginosa*	Anti-infectives	San Diego et al. (2018); Khan et al. (2019)
Chitosan-quercetin (190 nm) and chitosan-baicalein (187 nm) nanocapsules	*E. coli* Top 10	Inhibition of QS-regulated gene expression	Nanoencapsulation of QS inhibitors	Omwenga et al. (2018)
Selenium NPs (90 nm)	*C. violaceum* and *P. aeruginosa*	Inhibition of violacein production in *C. violaceum* and pyocyanin in *P. aeruginosa*	Anti-infectives	Gomez et al. (2019)
Tellurium NPs (125 nm)	*C. violaceum* and *P. aeruginosa*	Inhibition of violacein production in *C. violaceum* and pyocyaninin *P. aeruginosa*	Anti-infectives	Gomez et al. (2019)
Silver NPs (20–40 nm)	*S. marcescens*	Inhibition of prodigiosin, EPS productions and biofilms; downregulation in the expression of virulence and biofilm formation genes *fimA*, *fimC*, *flhD*, and *bsmB*	Anti-infectives	Ravindran et al. (2019)
AgCl-TiO_2 nanoparticles (6–7 nm)	*C. violaceum*	Inhibition of violacein production in *C. violaceum*	Food intervention	Naik and Kowshik (2014)
Nanoemulsions of cumin (52 nm), pepper (82 nm), and fennel (59 nm) essential oils	*C. violaceum*, *E. coli*, and *Klebsiella pneumoniae*	Inhibition of violacein production in *C. violaceum*; inhibition of EPS production and biofilm formation in the food borne-pathogen.	Food intervention	Venkadesaperumal et al. (2016)
Zinc nanostructures (24 nm)	*C. violaceum*, *P. aeruginosa*, *Listeria monocytogenes*, and *E. coli*.	Inhibition of violacein production in *C. violaceum*; inhibition ofelastase, protease, pyocyanin, and alginate in *P. aeruginosa* PAO1; reduction of *lasB* and *pqsA* transcriptional activity; inhibition of EPS production and biofilm formation in the food borne-pathogen	Food intervention	Al-Shabib et al. (2016)

(*Continued*)

TABLE 18.1 (*Continued*)

Quorum Sensing Inhibitory Activity of Nanomaterials and Their Potential Biological Applications

Nanomaterials	Target Pathogens	Molecular Target and Mechanism of Action	Potential Application	References
Nanoemulsions of eugenol and methyl salicylate	*E. coli* O157:H7	Downregulated expression of curli, type I fimbriae, Shiga-like toxins (*stx1* and *stx2*), quorum sensing (*luxS*, *luxR*, and *tnaA*), LPS biosynthesis (*waaL*, *waaP*, *waaD*, and *waaJ*), and *ler*-controlled toxins (*espD*, *escJ*, *escR*, and *tir*) genes	Food intervention	Prateeksha et al. (2019)
Gold nanoemulsions (75 nm)	*Pectobacterium carotovorum* subsp. *Carotovorum* and *Agrobacterium tumefaciens* NT1	Inhibition of EPS and hydrolytic enzymes (cellulose and pectinase), biofilms and AHL production	Control of phytopathogens	Joe et al. (2015)
Nanofilters of acylase	*P. aeruginosa* PA01	Quorum quenching activity: degradation of N-acetyl-L-methionine and inhibition of biofilms	Anti-biofouling membrane	Kim et al. (2011)
Acylase immobilized on mesoporous silica and magnetic nanoparticles	*P. aeruginosa* PA01	Quorum quenching activity: degradation of AHLs and inhibition of biofilms	Membrane filtration for water reclamation	Lee et al. (2014b)
Acylase has been immobilized on carboxylated polyaniline nanofibers	*P. aeruginosa* PA01	Quorum quenching activity: degradation of AHLs and inhibition of biofilms	Anti-biofouling activity	Lee et al. (2017)

CD can form a truncated cone shape with a hydrophilic exterior and a hydrophobic interior. β-CD can be detected when it forms a complex with the guest molecule. It can also alter reaction pathways, and increase accelerates the guest molecule involved reactions (Kato et al., 2006). Silicon dioxide nanoparticles (SiNP) of 50 nm size loaded with β-cyclodextrin inhibited AHL signal reception in *Vibrio fischeri*. The marine bacteria *V. fischeri* synthesizes and responds to N-3-oxo-hexanoyl-L-homoserine lactone (3OC6-HSL) and N-octanoyl-L-homoserine lactone (C8-HSL) using the LuxA/R QS system. Bioluminescence in *V. fischeri* is an AHL-mediated QS process regulated by *lux* operon. SiNPs and SiNP loaded with β-cyclodextrin decreased bioluminescence in *V. fischeri* by reducing the expression of both *luxA* and *luxR*. The β-cyclodextrin are able to bind to AHLs and disrupt QS signaling. However, on engineering β-cyclodextrin with SiNPs, an increased quorum quenching activity could be observed. The SiNPs on functionalizing target AHLs and prevent their binding to receptors resulting in downregulation of *luxA* and *luxR* and thus disrupt QS (Miller et al., 2015). Thus, QS inhibition can be achieved by binding to AHLs and reducing their concentration and preventing cell to cell communication. Similarly, selenium nanoparticles conjugated with honey (12–15 nm) inhibited QS in *C. violaceum* and *P. aeruginosa* PA01 (Prateeksha et al., 2017). Honey, a polyphenolic compound, has been studied for its QS inhibitory activity (Truchado et al., 2009). However, its bioavailability is a limitation for its therapeutic application (Jadaun et al., 2015). Selenium nanoparticles (SeNPs) have been used as carriers for pharmaceutical agents to increase bio-efficacy (Liu et al., 2012; Yu et al., 2014). The selenium nano-scaffold of honey at 4.5 μg/mL inhibited violacein production in *C. violaceum* wildtype and biofilm formation in *P. aeruginosa* PA01 by >90%. It significantly reduced the synthesis of virulence factors including pyocyanin and elastase activity. The nanoconjugates repressed the expression of genes involved in the production of virulence factors and synthesis of rhamnolipids and biofilms. *In vivo* studies showed that nanoconjugates induced wound healing and reduced pathogenicity of *P. aeruginosa* PAO1 in mouse models. The mechanism of action of the nanoconjugates was revealed by molecular docking studies. *In silico* analysis showed that hydrogen bonding and hydrophobic interaction of honey to AHL binding site of LasR disrupted QS and thus inhibited virulence factor production in *P. aeruginosa* PAO1. Thus, selenium nanoparticles improve the efficacy of anti-virulence agents by improving its delivery (Prateeksha et al., 2017).

18.2.3 Degradation of Signals

Magnetic nanoparticles (200 nm) immobilized with the acylhomoserine lactonase enzymes can hydrolyze and degrade AHLs. Purified recombinant AiiA (r-AiiA) were immobilized on magnetic nanoparticles, and it was observed that immobilization improved the AHL degrading activity. The r-AiiA-MNP nanobiocatalyst effectively hydrolyzed 3O-C10 AHL and inhibited QS in the biosensor strain *E. coli* (PJN105Q). The QQ nanocatalyst can be recovered using an external magnetic field and used multiple times for AHL hydrolysis (Beladiya et al., 2015). Gold nanoparticles coated with acyl homoserine lactonase proteins (AiiA AuNPs) ranging in size from 10 to 30 nm inhibited QS-regulated phenotypes in multidrug-resistant *Proteus* species. On coating with gold nanoparticles, the AHL degradation activity of AiiA doubled and was 1.8 ± 0.12 mg liter^{-1} h^{-1} in comparison to uncoated AiiA. AiiA AuNPs degraded C6-HSL significantly within 2 h. Similarly, the AiiA AuNPs at 2 μM inhibited exopolysaccharide production and biofilm formation by 85%, whereas AiiA enzymes decreased biofilm formation by only 45%. The synergistic activity of the AHL lactonase and the gold nanoparticles as AiiA AuNPs can find application as antibiofilm activity against pathogenic *Proteus* species (Vinoj et al., 2015).

18.3 Nanomaterials as Vehicles for Quorum Sensing Inhibitors

The nanomaterials, due to their physicochemical properties and small particle size, have been used as carriers for compounds with anti-quorum sensing activity to improve the efficacy of their delivery.

Ultra-small solid lipid nanoparticles (<100 nm) have been designed to incorporate novel QS inhibitors. A PqsR antagonist (2-heptyl-6-nitro-4-oxo-1,4-dihydroquinoline-3-carboxamide), was discovered that strongly inhibited the virulence in *P. aeruginosa in cellulo* (IC50 of 2 μM toward pyocyanin) and *in vivo* animal infection models. The lipophilicity of the QS inhibitor limited its application as a therapeutic. However, the use of biocompatible and biodegradable nanoparticles was found to be an effective strategy for the controlled delivery of QS inhibitors at mucosal sites. On encapsulation in the solid lipid nanoparticles, the quorum sensing inhibitors exhibited a sevenfold superior anti-virulence activity. The nanoparticles had a high encapsulation efficiency of 68%–95% and penetrated effectively into artificial sputum. The nanocarriers enabled the effective delivery of QS inhibitory based anti-infectives due to high loading efficiency, prolonged release, and mucus penetrating ability (Nafee et al., 2014).

In *Vibrio cholerae*, the accumulation of the signaling molecule CAI-1 autoinducer results in decreased virulence factor production and biofilm formation, and induces the production of proteases which detach the pathogen from the intestine (Zhu et al., 2002). Therefore, CAI-1 autoinducer can be used for QS based treatment of cholera, based on its mode of action of transition to non-virulence (Wei et al., 2011). However, the inherent hydrophobicity of CAI-1 limits its oral delivery. The use of nanocarriers has the potential to successfully deliver CAI-1 to *V. cholerae* in intestinal crypts. Therefore, the flash NanoPrecipitation method was used to develop CAI-1 nanoparticles (CAI-1 NPs) with average diameter of 112 ± 1.7 nm. Using this method, nanocarriers coated with water soluble PEGs capable of encapsulating hydrophobic compounds were prepared. The CAI-1 NPs at 50 μM CAI-1 reduced biofilm production in *V. cholerae* by 71% ± 2%, in comparison to 32% ± 4% reduction with only free CAI-1. The nanoparticles were able to activate QS response by 5 orders of higher magnitude than the free signaling molecule. The CAI-1 NPs had an effective penetration through intestinal mucus layer and biofilms and thereby reduced *V. cholera* virulence (Lu et al., 2015).

Quercetin, a hydrophobic flavonoid, has been incorporated into chitosan nanoparticles and studied for its QS inhibitory activity. Ionotropic gelation technique was used to prepare nanoparticles of chitosan, sulfo-butyl-ether-β-cyclodextrin (Captisol®), and pentasodium tripolyphosphate. The nanoparticles had a size range of 250–400 nm with high zeta potential ranging from +31 to +40 mV. By increasing amounts of Captisol®, Quercetin, could be incorporated in the NPs formulation, without altering significantly its physicochemical properties. These NPs could inhibit QS in *E. coli* top 10 biosensor by 61.12%, and the use of Captisol® prolonged the delivery of quercetin. Studies showed that the interaction between nanoparticles and the bacterial membrane resulted in the inhibition of QS-regulated activities (Thanh and Goycoolea, 2017).

Kaempferol, a flavonoid compound, is known for its antimicrobial, anti-inflammatory, anticancer, and antioxidant applications. However, the low water solubility of flavonoids affects their chemical stability and bioavailability (Gupta et al., 2016). The use of nanoparticles can increase the solubility and bioavailability of kaempferol in a controlled manner (Yadav and Sawant, 2010). Chitosan nanoparticles were loaded with kaempferol and tested for QS inhibitory activity. The nanoparticles had an average size of 192.27 ± 13.6 nm with zeta potential of +35 mV. The loading and encapsulation efficiency of the phytochemical in the chitosan nanoparticles was 78% and 93%, respectively. In comparison to the pure compound, kaempferol-chitosan nanoparticles inhibited violacein production in *C. violaceum* CV026 by 76%. Even after 30 days of storage, the nanoformulation of kaempferol significantly inhibited production of violacein pigment in the biosensor. Encapsulation of kaempferol in chitosan nanoparticles significantly improved its anti-quorum sensing activity (Ilk et al., 2017).

Nanocapsules are nano-vesicular systems having a lipid core stabilized by a surfactant and coated by a polymeric shell, and are prepared by spontaneous emulsification (Lopez-montilla et al., 2002). Chitosan polymer based nanocapsules are used as nanocarriers for lipophilic and hydrophilic drugs (Goycoolea et al., 2012). The effect of nanocapsules prepared from chitosan (114–155 nm size with zeta potential of +50 mV) on *E. coli* quorum sensing reporter strains has been studied. The nanochitosan capsules promoted aggregation of *E. coli* cells by influencing the zeta potential of the cells. The nanocapsules inhibited quorum sensing in *E. coli* as observed by the decrease in GFP expression which is under QS regulon in the reporter strain. It was inferred that QS was disrupted as nanochitosan mediated bacterial aggregation prevented diffusion of the AHL signaling molecules (Qin et al., 2017). Follow-up studies showed that non-toxic chitosan with medium degree of polymerization and molecular weight of 115 kDa at low concentrations was able to inhibit QS in *E. coli* biosensors. It was observed that only at concentration ≤30.6 μg/mL, chitosan induced aggregation and disrupted QS. It was confirmed that bacterial aggregation prevented diffusion of AHLs resulting in QS inhibition (Qin et al., 2018a).

Similarly, the flash NanoPrecipitation method has been used to prepare nanoformulations against *P. aeruginosa* infection. A LasR inhibitor, V-06-018, with limited water solubility was discovered. Therefore, PEG-coated nanocarriers were developed to encapsulate V-06-018 and deliver the QS inhibitor through pulmonary mucus. The inhibitor V-06-018 was dissolved in organic solvent and later mixed in water in presence of excipients and polyethylene glycol-based stabilizers. The resulting PEG-coated V-06-018 nanocarriers ranging in particle size from 70 to 400 nm were water dispersible and stable. The V-06-018 encapsulated nanocarrier was able to reduce the LasR-regulated pyocyanin pigment production proving that its QS inhibitory potential was not reduced even on nanoencapsulation. Further, the nano V-06-018 was able to effectively penetrate through pulmonary mucus to reach *P. aeruginosa* infection (Lu et al., 2018).

Cinnamaldehyde disrupts QS activity by decreasing the DNA binding ability of LuxR and inhibiting AHL production. Chitosan nanoparticles have been used to encapsulate cinnamaldehyde to increase its efficacy as a QS inhibitory agent. Cinnamaldehyde encapsulated chitosan nanoparticles of average diameter of 208.12 nm size were prepared using ionic gelation method. The nanoparticles significantly inhibited the expression of virulence factors including pyocyanin, protease, rhamnolipid production, biofilm formation, and motility in *P. aeruginosa* PAO1. The encapsulation efficiency of cinnamaldehyde was 65.04% ± 3.14% and *in vitro* release study showed its sustained release. The nanoencapsulation of cinnamaldehyde improved its QS inhibitory activity and also aided its slow and sustained release. Thus, biocompatible nanoparticles can be used as carriers for effective targeting of QS systems in pathogenic bacteria (Pattnaik et al., 2018a). Similarly, chitosan encapsulated cinnamaldehyde has been studied for its QS inhibition property in *E. coli* TOP biosensor strain. It was observed that nanoencapsulated cinnamaldehyde showed greater QS inhibition than free cinnamaldehyde and nanoemulsions of cinnamldehyde. It was inferred that nanocapsules interact with *E. coli* cells by electrostatic interaction and thereby effectively deliver the QS inhibitors to the bacteria. The electrostatic adsorption of the chitosan nanocapsules to the bacterial cell envelope results in improved QS inhibitory activity of cinnamaldehdye (Qin et al., 2018b).

Ferulic acid was encapsulated with chitosan-tripolyphosphate to prepare nanoparticles with mean diameter of 215.55 nm. The nanoencapsulated particles showed significant QS inhibition against *P. aeruginosa* PAO1 in comparison to ferulic acid. The NPs were able to downregulate the synthesis of QS-regulated virulence factors. At 500 μg/mL, ferulic acid nanoparticles inhibited pyocyanin production by 89.47%, whereas ferulic acid could inhibit the pigment production by only 70%. A similar effect was observed in the case of inhibition of protease activity and motilities in *P. aeruginosa* PAO1 by the ferulic acid NPs. On treatment at sub-MIC of 500 μg/mL, biofilm formation, EPS and rhamnolipid production was reduced by 84%, 74%, and 73%, respectively. The NPs also significantly reduced cell surface hydrophobicity by 16%. Microscopy studies showed that the NPs altered the biofilm architecture and reduced biofilm formation. The improved efficacy of ferulic acid NPs in comparison to ferulic acid in attenuating QS and inhibiting biofilm formation was attributed to the use of nanocarriers which resulted in slow and sustained release of the QS inhibitor ferulic acid at the target sites (Pattnaik et al., 2018b).

Nanoformulation of ajoene, a QS inhibitor was developed using chitosan and dextran sulphate. The quorum sensing inhibitor-loaded nanoparticles (685.7 nm) had a high zeta potential of 37.9. The nanoformulation was developed to intravesically target *P. aeruginosa* biofilm-associated murine pyelonephritis. Under *in vitro* and *in vivo* conditions, the NPs combined with ciprofloxacin exhibited significant antivirulence activity by reducing the virulence factors and biofilm in *P. aeruginosa*. The nanoformulation exhibited twofold increase in antivirulence activity in comparison to only ajoene. It significantly reduced the virulence factors including protease, elastase, pyochelin, haemolysin, alginate, and rhamnolipids. The significant biofilm inhibitory activity was attributed to increased solubility, enhanced uptake, and sustained release of ajoene. And in combination with ciprofloxacin, there was a reduction in virulence factor production by 73%–97%. The nanoformulations showed higher anti-biofilm activity than ajoene. It was also observed that complete eradication, i.e., >99%, of biofilm was possible using the combination of nanoparticles and ciprofloxacin. The mechanism for biofilm inhibition was the capability of nanoformulation to maintain a high concentration of ajoene near the biofilms. The chitosans present in the nanoformulation can depolarize the cell membrane of cells in biofilms and increase their permeability to NPs. Further, the nanoformulation was intravesically administered to target *P. aeruginosa* associated murine pyelonephritis. *In vivo* tracking using fluorescent labeling confirmed the targeted delivery of NPs. Bacteriological and histopathological studies confirmed the therapeutic efficacy of nanoformulation of ajoene. Thus, the nanoparticle formulations of QS inhibitors by targeted administration can be used as potential therapeutics for treating pyelonephritis caused by *P. aeruginosa*. Hence, we propose a novel next-generation therapeutic and its appropriate targeting route against biofilm-associated pyelonephritis (Vadekeetil et al., 2019).

18.4 Potential Application of Nano-QS Inhibitors

18.4.1 Therapeutics and Anti-Biofilm Agents

In *P. aeruginosa*, the QS mechanism regulates virulence factor production and in respiratory infections and cystic fibrosis. Hence, targeting QS machinery can help in reducing *P. aeruginosa* pathogenicity (Deziel et al., 2005; Flickinger et al., 2011; Klein et al., 2012). Silver nanoparticles have been extensively studied for antimicrobial activity (Kim et al., 2011a). However, silver nanowires (SNWs) on stabilizing with PVP have shown no inhibitory activity. The stabilization of SNWs with ethylene glycol and PVP as stabilizers may prevent its interaction and thereby cause a noninhibitory effect (Slistan-Grijalva et al., 2005). Silver nanowires (SNWs) of 200–250 nm sizes were screened for inhibition of violacein production in *Chromobacterium violaceum*. The silver nanowires at sub inhibitory concentrations of 0.5 mg/mL inhibited violacein production by 60% in the biosensor. Further, SNWs significantly inhibited biofilm formation in the pathogen *P. aeruginosa* in a dose dependent manner and at 4 mg/mL showed maximum anti-biofilm activity without effecting cell viability. Thus, QS mediated inhibition of biofilms of *P. aeruginosa* by SNWs is a promising therapeutic approach (Wagh et al., 2013).

The toxic effects of heavy metal ions and metal resistance have been studied in *P. aeruginosa* (Teitzel and Parsek, 2003). It has also been observed that the heavy metal ions of silver and copper have antibiofilm activity in *P. aeruginosa* due to their toxic effects (Bjarnsholt et al., 2007; Harrison et al., 2008). A number of metal ions and metal nanoparticles were tested for QS inhibitory activity in *P. aeruginosa*. ZnO nanoparticles (<50 nm) significantly inhibited biofilms, pyocyanin and pyochelin production, hemolytic activity, and PQS signaling in *P. aeruginosa*.

at concentrations as low as 1 mM; the ZnO nanoparticles were able to inhibit biofilm formation by more than 95% on polystyrene surfaces and completely inhibited biofilm formation on glass surfaces. Though, the ZnO nanoparticles inhibited pyocyanin production, they induced the production of pyoverdine and had no effect on rhamnolipid synthesis. DNA microarray studies showed that the ZnO nanoparticles repressed pyocyanin synthesis gene (*phz* operon) and induced synthesis of pyoverdine gene (*pvd*S). Studies using *P. aeruginosa* PAO1 transposon mutants showed that the NPs require CzcR (*czc* two-component response regulator) to inhibit pyocyanin production. Similarly, it was revealed that CzcR and RhlR were involved in the inhibition of biofilms by the nanoparticles. Hydrophobicity studies showed that ZnO nanoparticles increased the hydrophilicity of *P. aeruginosa* cells, which influenced the inhibition of *P. aeruginosa* biofilm formation. The study revealed that ZnO nanoparticles could inhibit QS-regulated virulence determinants in pathogens by a mechanism involving the upregulation of czc efflux pump (Lee et al., 2014a).

ZnO nanoparticles of 65.17 nm have been observed to inhibit quorum sensing in the laboratory, clinical, and environmental isolates of *P. aeruginosa*. The ZnO nanoparticles at 1 mmol l^{-1} strongly inhibited virulence factors in the laboratory PAO1 and PA14. The nanoparticles inhibited elastase production in both the lab strains by 90%. In PA14, pycoyanin production was inhibited by 75%. The ZnO nanoparticles strongly inhibited biofilm formation in PAO1 and PA14 by 97% and 94%, respectively. Further, the ZnO nanoparticles were also able to inhibit the production of elastase in all clinical strains. Similarly, the nanoparticles inhibited biofilm formation in the clinical and environmental strains by 26%–100%. Thus, ZnO nanoparticles were able to inhibit quorum sensing in the various strains of *P. aeruginosa* at sub-MIC, though there was some variation in the levels of inhibition of virulence factors in different strains (Garcia-Lara et al., 2015).

The major problem associated with QS inhibitors for their *in vivo* application against *P. aeruginosa* is their penetration through mucus and bacterial biofilms (Suk et al., 2011). Bio biodegradable nanocarriers can be used for effective delivery of QS inhibitors. One such example is use of anti-infective inhalation therapy for nanocarriers (Forier et al., 2014; Weber et al., 2014). The anthelmintic drug niclosamide was repurposed as an antivirulence agent and tested for its QS inhibitory activity against *P. aeruginosa*. The niclosamide nanocrystals were prepared with high pressure homogenization and stabilized using polysorbates. These nanocrystals had a diameter of 450 nm and a zeta potential of −20 mV. The nanocrystals at 2.5–10 µM were able to inhibit QS-regulated traits in *P. aeruginosa*. Under *in vitro* conditions, the nanoformulations (≤10 µM) did not affect the viability of bronchial epithelial cells. On intratracheal administration in rats at 100-fold higher than the QS inhibitory concentration, no toxicity was observed. Thus, dry powders of niclosamide nanoparticles reconstituted in saline solution could be used as inhalable nanosuspensions. These inhaled antivirulence drugs are a promising therapeutic option to treat cystic fibrosis caused by *P. aeruginosa* (Costabile et al., 2015).

Crataeva nurvala is a medicinal plant having extensive usage in Ayurveda medicinal systems (Walia et al., 2007). Biosynthesized silver nanoparticles from the bark extract of the medicinal plant *Crataeva nurvala* were found to have average diameter of 15.2 nm. These silver nanoparticles at 15 µg mL^{-1} efficiently inhibited biofilm formation in drug-resistant clinical isolates of *P. aeruginosa* by 79.70%. The nanoparticles were internalized in the cell and prevented bacterial colonization and prevented biofilm initiation. Further, the silver nanoparticles inhibited production of virulence factors pyocyanin by 74.64% and hemolysin by 47.7% in *P. aeruginosa* (Ali et al., 2017).

ZnO Nps of varying size also have been investigated for antibiofilm activity based on the mechanism of QS inhibition. The ZnO NPs <100 nm showed significant antibiofilm activity against *P. aeruginosa* PA01 in comparison to NPs <50 nm in size. The ZnO NPs at sub-MIC of 3.125 µg/m were also able to inhibit the production of the virulence factor (Aswathanarayan and Vittal, 2017). In a similar study, phyto-synthesized ZnO NPs at slightly higher concentrations (50–100 µg/mL) have significantly reduced the synthesis of the virulence factor pyocyanin (Alavi et al., 2019).

The QS inhibitory activity of tin dioxide nanoflowers (50 nm) has also been confirmed by its ability to inhibit violacein production in *C. violaceum*. At a concentration of 16 µg/mL, the nanoflowers were able to inhibit violacein production by 74%. The nanoflowers also inhibited virulence factor production and disrupted biofilm formation in the pathogens *P. aeruginosa* and *Serratia marcescens*. In *P. aeruginosa* PAO1, the NPs at sub-MICs reduced elastase and protease activity by 59% and 64%, respectively. The NPs showed dose dependent activity against inhibition of pyocyanin and alginate production, and maximum inhibitory activity was observed at 64 µg/mL of the NPs. Similarly, the NPs also inhibited the QS-regulated virulence factor prodigiosin production in *S. marcescens* by 54% at 64 µg/mL concentration. The monophasic nanoflowers showed significant inhibitory activity against both biofilm formation and in preformed biofilms of the pathogens as observed by SEM and confocal studies. The NPs were able to inhibit EPS, swarming motility, and cell surface hydrophobicity and thus disrupt biofilm formation. The study clearly shows that tin dioxide nanoflowers are promising QS inhibitor agents against pathogenic bacteria (Al-Shabib et al., 2018).

Gold nanoparticles biosynthesized from *Lysinibacillus* sp. and *P. stutzeri* isolated from hyperalkaline were studied for QS inhibitory activity. The gold NPs showed concentration dependent inhibition of pyocyanin production in *P. aeruginosa* PA01 without affecting the growth of the organism. The gold NPs produced by *Lysinibacillus* sp. and *P. stutzeri* at 50 mL colloidal suspension inhibited pyocyanin production by 65.47% and 80.81%, respectively. Thus, the gold nanoparticles bioynthesized from bacteria adapted to alkaline conditions have potential anti-infective activity as seen by their ability to inhibit virulence factor pyocyanin production (San Diego et al., 2018).

The flavonoids quercetin and baicalein have been encapsulated in chitosan, and these nanocapsules have been tested for QS inhibition and antibiofilm activity in *E. coli* top 10 biosensor. The nancapsules of quercetin and baicalein had a diameter of 190 nm and 187 nm and a high zeta (ζ) potential of +48.1 and +48.4 mV, respectively. It was observed that the encapsulation with chitosan improved the QS inhibitory activity of both the flavonoids without affecting the growth of the biosensor. At low concentrations, the nanocapsules were able to inhibit biofilm formation in the biosensor strain. The free flavonoids had a cytotoxic effect on the MDCK-C7 cell lines at higher concentration, whereas nanoencapsulated flavonoids were conferred with a cytoprotective effect. Thus, nanocapsulation increased the QS inhibitory activity of the flavonoids by its sustained release and also reduced its toxicity to mammalian cells (Omwenga et al., 2018).

Selenium NPs (90 nm) and tellurium NPs (125 nm) have also shown QS inhibitory activity. Selenium NPs were able to inhibit violacein production in *C. violaceum* and biofilm formation in *P. aeruginosa* by 80%. The selenium NPs were able to disrupt the biofilm architecture and structure of *P. aeruginosa*, resulting in formation of discrete micro-aggregates. The tellurium NPs also inhibited violacein production and biofilm formation in *P. aeruginosa* and at much lower concentrations. When the biofilms were treated with selenium NPs, the percentage of viable cells was constant, whereas on treatment with tellerium NPs, cell viability was reduced. In case of selenium nanoparticles, the mechanism of action is interference in QS signal biosynthesis by NPs, while tellurium NPs may have disrupted QS signal reception and response. Thus, based on the QS inhibitory behavior, both selenium and tellurium NPs are promising as anti-biofilm agents against pathogenic bacteria (Gomez et al., 2019).

Gold nanoparticles (53 nm) were synthesized and stabilized with fucoidan. The NPs had an inhibitory effect on the QS-regulated traits in *P. aeruginosa*. The gold NPs had a minimum inhibitory concentration of 512 µg/mL and at sub-inhibitory concentration they inhibited biofilm formation and motility in *P. aeruginosa*. The NPs at 128 µg/mL inhibited biofilm formation by 86%. Microscopic studies showed that the NPs inhibited the attachment of the bacterial cells to substrates such as nylon and glass surfaces. Thus, the gold NPs inhibited attachment of sessile cells and prevented biofilm formation in *P. aeruginosa*. At higher concentrations of 128–256 µg/mL, the gold NPs were able to disrupt and disperse preformed mature biofilms. The gold NPs strongly inhibited the production of virulence factors including pyocyanin, alginate, and pyoverdin at 256 µg/mL by 87.7%, 53%, and 95%, respectively. Similarly, the NPs at highest concentration of 256 µg/mL inhibited swimming motility completely and swarming and twitching motility by 53% and 72%, respectively. Thus, fucoidan stabilized gold nanoparticles can be a potential anti-biofilm and anti-virulent agent against *P. aeruginosa* infections (Khan et al., 2019).

Silver nanoparticles were synthesized using the aqueous extracts of leaves *Piper betle*. The silver NPs synthesized from the plant extract were polydispers in nature and 20–70 nm in size. The silver NPs at 20 µg/mL inhibited violacein production in *C. violaceum* by 97%. The NPs at sub-lethal concentration of 8 µg/mL inhibited virulence factors pyocyanin, elastase production, and biofilm formation in *P. aeruginosa* PA01 by 82%, 77%, and 78%, respectively. The NPs were able to disrupt the structures of biofilms in *P. aeruginosa*. Thus, eco-friendly, rapid, and cost-effective synthesized NPs could be used as anti-infectives against *P. aeruginosa* infections (Shah et al., 2019).

Similarly, silver nanoparticles in the size range of 20–40 nm were synthesized using *Vetiveria zizanioides* root extract. These NPs were able to inhibit QS-regulated virulence factors including prodigiosin, exopolysaccharide productions, and biofilm formation in *S. marcescens* at sub-MICs. The NPs at 2 µg/mL inhibited prodigiosin, EPS production, and biofilm formation by 78%, 60%, and 84%, respectively. The NPs treated *S. marcescens* showed irregularly shaped biofilms due to cell destruction, reduced microcolonies, and EPS content. The transcriptomic studies showed downregulation in the expression of QS-regulated virulence and biofilm formation genes including *fimA, fimC, flhD*, and *bsmB* gene. The biosynthesized silver NPs based on its QS inhibitory and antibiofilm could be used as anti-infectives against hospital-acquired *S. marcescens* infections (Ravindran et al., 2018).

A nanoformulation containing both antibiotic and QS inhibitors has been tested for its efficacy as an anti-biofilm agent in *P. aeruginosa* biofilms. A bioresponsive polymer formulation was prepared using modified alginate nanoparticles. The nanoformulation was used to deliver together both the antibiotic ciprofloxacin and the quorum sensing inhibitor 3-amino-7-chloro-2-nonylquinazolin-4(3H)-one to mature biofilms of *P. aeruginosa*. The polymer contained a pH-responsive linker between the polysaccharide backbone and the QS inhibitor. It encapsulated ciprofloxacin by charge-charge interactions, i.e., the positively-charged drug interacted with the carboxyl residues of the alginate matrix. In the low pH regions of biofilms, there was a dual-action release of both antibiotic and QS inhibitors due to cleaving of the QSI-linker to the alginate matrix and reduced charge-charge interactions between antibiotic and the alginate matrix as a result of carboxyl side-chain protonation. In a biofilm model, the release of antibiotic and QS inhibitors from the pH-responsive nanoparticles significantly reduced the viability of the biofilm compared to only the antibiotic treatment. The alginate NPs penetrated deeply into *P. aeruginosa* biofilms due to the charges of the NPs and the release of the QS inhibitor. In both a 2D keratinocyte and a 3D *ex vivo* skin infection model, the dual-action bio-responsive release nanoparticles effectively cleared the infection. Thus, the novel nanoformulation is a promising combination therapeutic to prevent biofilm as well as kill matured *P. aeruginosa* biofilms (Singh et al., 2019).

18.4.2 Food Industry

The anti-quorum sensing activity of AgCl-TiO_2 nanoparticles (6–7 nm) was observed in the *C. violaceum*. The NPs at 100 and 200 µg/mL inhibited violacein production in *C. violaceum* by 82% and 100%, respectively. The NP coated glass slides containing 100 µg/mL were able to completely inhibit the biofilm formation of the bacteria. Mass spectrometry studies of NP treated cell fractions showed no peaks corresponding to the QS signal oxo-octanoyl homoserine lactone but presence of additional small peaks which could be the degradation products or precursors of the QS signal. The inorganic materials used in the study can withstand harsh processing conditions and are not toxic. Hence, these NPs can find potential application in the food packaging industry (Naik and Kowshik, 2014b).

Essential oils (EOs) are complex mixtures of volatile compounds extracted from plants. The essential oils of *Cinnamomum verum*, *Syzygium aromaticum*, *Salvia sclarea*, *Juniperus communis*, Citrus lemon, and *Origanum majorana* have shown quorum sensing inhibitory activity (Niu et al., 2006; Khan et al., 2009; Kerekes et al., 2013). The essential oils have antimicrobial activity at high concentrations. The EOs on dilution due to formation of droplets have low antimicrobial efficacy (Valero and Giner, 2006). However, by using nanoemulsion, the concentration of EO in the nanoformulation can be lowered. Nanoemulsions have proven to be efficient delivery systems for active ingredients. The large surface area of nanoemulsion also facilitates rapid diffusion of bioactives (Teixeria et al., 2007). Nanoemulsions of cumin, pepper, and fennel essential oils were prepared by ultrasonication, and these had a size of 52.89 nm, 82.08 nm, and 59.52 nm, respectively. Fennel and cumin nanoemulsions significantly inhibited violacein production in *C. violaceum* than pepper nanoemulsions. The fennel nanoemulsions at 50 μL/mL inhibited violacein production by 75%, whereas pepper emulsion at 30 μL/mL inhibited pigment production by only 32%. The nanoemulsions were able to inhibit EPS production and biofilm formation in the food-borne pathogens *E. coli* and *K. pneumoniae*. Thus, the essential oil nanoemulsions can find potential application as antibiofilm agents against food-borne pathogens based on their QS inhibitory activity (Venkadesaperumal et al., 2016).

Zinc nanostructures with a particle size of 24 nm were synthesized using *Nigella sativa* seed extract. The nanostructures inhibited QS phenotypes in *C. violaceum* and *P. aeruginosa* biosensors. At sub-MIC of 400 μg/mL, the NPs significantly inhibited violacein synthesis in *C. violaceum* CVO26 by 91%. At the highest sub-MIC concentration, the NPs reduced virulence factor production, i.e., 82% of elastase, 77% of protease, 93% of pyocyanin, and 73% of alginate in *P. aeruginosa* PAO1. The NPs at 10–40 μg/mL were able to reduce *lasB* and *pqsA* transcriptional activity by 35%–85% and 41%–84%, respectively. Similarly, the NPs were also able to significantly affect motility and biofilm formation in *P. aeruginosa* PAO1. The zinc NPs at sub-inhibitory concentrations also inhibited biofilm formation in food borne pathogens *Listeria monocytogenes* and *E. coli*. The treatment with NPs reduced the number of microcolonies during the biofilm formation resulting in formation of weak biofilms with reduced biomass. The NPs also affected preformed biofilms in both the pathogens. The Zn NPs treated biofilms had poor and weakened architecture and reduced thickness. The *in vivo* efficacy of the nanostructures was tested in food models such as lettuce. Thus, the ZnOs based on their anti-QS and biofilm inhibitory activity could be used as potential food packaging material and/or as food preservatives (Al-Shabib et al., 2016).

Nanoemulsions of two bioactive compounds, eugenol and methyl salicylate, derived from essential oils of *G. fragrantissima* were evaluated for antivirulence and biofilm inhibitory activity in *Escherichia coli* O157:H7. The nanoemulsions of the bioactive compounds inhibited biofilms and virulence factor production without affecting the growth of the pathogen. Transcriptomic studies showed that the nanoemulsions downregulated the expression of curli, type I fimbriae, Shiga-like toxins (*stx1* and *stx2*), quorum sensing (*luxS*, *luxR* and *tnaA*), LPS biosynthesis (*waaL*, *waaP*, *waaD* and *waaJ*), and *ler*-controlled toxins (*espD*, *escJ*, *escR* and *tir*) genes. This resulted in decreased attachment and biofilm formation and reduced pathogenicity. The nanoemulsions were also able to significantly inhibit biofilm formation of *E. coli* on glass, plastic, and meat surfaces in comparison to eugenol and methyl salicylate coatings. Thus, the nanoemulsification approach enhanced the QS inhibitory activity of eugenol and methyl salicylate (Prateeksha et al., 2019).

18.4.3 Control of Phytopathogens

Nanoemulsions are water-in-oil formulations of oil, water, surfactant, and cosurfactant prepared by mechanical extrusion techniques. The droplets are of uniform size of 20 to 200 nm in diameter. Due to their size, the nanoemulsions destabilize the lipid envelope and fuse with bacterial cell walls (Aboofazeli, 2010). Nanoemulsions of gold (75 nm) were studied for QS inhibitory activity against the plant pathogen *Pectobacterium carotovorum* subsp. *Carotovorum*, a causal agent of soft rot disease in horticultural crops. The gold nanoemulsions significantly reduced biofilms of *P. carotovorum*. Similarly, the gold nanoemulsions strongly inhibited swarming and swimming motility exhibited by the pathogen. The nanoemulsions were able to inhibit AHL mediated β-galactosidase activity in the biosensor strain *Agrobacterium tumefaciens* NT1. The nanoemulsions were able to reduce EPS and hydrolytic enzymes (cellulose and pectinase) and AHL production in *P. carotovorum*. The gold nanoemulsions completely attenuated virulence of *P. carotovorum* as observed by absence of lesion and any other signs of infection in potato tubers. Thus, nanoemulsions of gold NPs can be used as potential biocontrol agents to prevent soft root infections in potato tubers (Joe et al., 2015).

18.4.4 Anti-Biofouling Agents

Nanostructured materials which interfere with the QS signal molecules involved in biofilm growth have been developed for the mitigation of biofouling.

Acylase quorum quenching enzyme was immobilized on a nanofiltration membrane to prevent QS signaling between bacteria in the biocake and thereby mitigate biofouling. Characterization of the acylase-immobilized membrane showed that acylase with added chitosan formed a matrix on the membrane acting as an additional barrier and decreased water permeability. The quorum quenching activity of the membrane was confirmed by its degradation of its substrate N-acetyl-L-methionine. The acylase nanofiltration membrane was highly stable and retained 90% of its enzymic activity even after 20 cycles of usage. The biofilm formation by *P. aeruginosa* PA01 on immobilized membranes was only 24% in comparison to the control. It was also able to inhibit biofilm maturation and 3D structure formation resulting in biofilms of 20 μm thickness, whereas the biofilms in control were 45 μm thick. Thus, the acylase on the nanofiltration membrane effectively degraded AHLs produced by PAO1 and prevented the maturation of the PAO1 biofilm. The quantity of EPS secreted was only 30% of that control. Thus, the nanofilters containing acylase have high antibiofouling activity as these effectively degrade QS signaling and suppress EPS secretion and biofilm maturation (Kim et al., 2011b).

Acylase enzymes have been immobilized into spherical mesoporous silica and magnetic nanoparticles. The nanoscale enzyme reactors of acylase showed highly effective antifouling activity. The immobilized acylase were highly stable even after rigorous shaking at 200 rpm for 1 month, whereas the free acylase lost more than 90% of its activity within a day. On its application in membrane filtration for advanced water treatment, nanoscale enzyme reactors efficiently reduced maturation of *P. aeruginosa* PAO1 biofilms on the membrane surface and prevented membrane fouling. Thus, these have a great potential to be used in the membrane filtration for water reclamation (Lee et al., 2014b).

Acylase have been immobilized on carboxylated polyaniline nanofibers as these can act as nanobiocatalysts for antifouling applications because they have high enzyme loading capacity and stability. The immobilized acylase showed anti-biofouling activity against *Pseudomonas aeruginosa* under both static and continuous flow conditions. At 40 μg/mL of immobilized acylase, biofilm formation under static condition was 5 times lower than that of control. The addition of immobilized acylase reduced the synthesis of AHL signals. Thus, the nanofibers containing acylase have potential as antifouling agents as these significantly reduced biofilm formation and membrane biofouling based on its AHL quenching ability (Lee et al., 2017).

18.5 Conclusion

Nanomaterials have been evaluated and found highly effective as anti-QS agents. The nanoparticles can also be useful in delivery of known QS inhibitors for controlling bacterial infection and other QS mediated phenotypes such as biofilms, biofouling, etc. The formulations of nanoparticles can enhance efficacy and availability of QS inhibitors in *in vivo* conditions due to their stability, sustained release, and targeted delivery. Therefore, future investigations on nanoparticles as QS inhibitors should focus on mechanistic studies and *in vivo* studies to assess their therapeutic efficacy. Further, long-term effects and toxicity studies on nanomaterials used as either QS inhibitors or as delivery systems of anti-QS agents should be performed prior to their pharmacological applications. The QS inhibitory nanomaterials apart from anti-infective activity should also be explored for food intervention, plant protective applications, and mitigating biofouling.

REFERENCES

Aboofazeli R (2010) Nanometric-scaled emulsions (nanoemulsions). *Iran. J. Pharm. Res. IJPR* 9:325.

Alavi M, Karimi N, Salimikia I (2019) Phytosynthesis of zinc oxide nanoparticles and its antibacterial, antiquorum sensing, antimotility, and antioxidant capacities against multidrug resistant bacteria. *J. Ind. Eng. Chem. Res.* 72:457–473.

Ali SG, Ansari MA, Khan HM, Jalal M, Mahdi AA, Cameotra SS (2017) *Crataeva nurvala* nanoparticles inhibit virulence factors and biofilm formation in clinical isolates of *Pseudomonas aeruginosa*. *J. Basic Microbiol.* 57(3):193–203. doi:10.1002/jobm.201600175.

Al-Shabib NA, Husain FM, Ahmed F, Khan RA, Ahmad I, Alsharaeh E, Khan MS et al. (2016) Biogenic synthesis of Zinc oxide nanostructures from *Nigella sativa* seed: Prospective role as food packaging material inhibiting broad-spectrum quorum sensing and biofilm. *Sci. Rep.* 6:36761.

Al-Shabib NA, Husain FM, Ahmad N, Qais FA, Khan A, Khan A, Khan MS, Khan JM, Shahzad SA, Ahmad I (2018) Facile synthesis of tin oxide hollow nanoflowers interfering with quorum sensing-regulated functions and bacterial biofilms. *J. Nanomater.* doi:10.1155/2018/6845026.

Aswathanarayan JB, Vittal RR (2017) Antimicrobial, biofilm inhibitory and anti-infective activity of metallic nanoparticles against pathogens MRSA and *Pseudomonas aeruginosa* PA01. *Pharm. Nanotechnol.* 5(2):148–153. doi:10.2174/2211738505666170424121944.

Babapour A, Yang B, Bahang S, Cao W (2011) Low-temperature sol-gel-derived nanosilver-embedded silane coating as biofilm inhibitor. *Nanotech* 22:155602.

Bai JA, Rai RV (2018) Nanotechnological approaches in quorum sensing inhibition. In V. C. Kalia (Ed.), *Biotechnological Applications of Quorum Sensing Inhibitors*. Springer, Singapore, pp. 245–261. doi:10.1007/978-981-10-9026-4_12.

Barik SK, Singh BN (2019) Nanoemulsion-loaded hydrogel coatings for inhibition of bacterial virulence and biofilm formation on solid surfaces. *Sci. Rep.* 9(1):6520. doi:10.1038/s41598-019-43016-w.

Beladiya C, Tripathy RK, Bajaj P, Aggarwal G, Pande AH (2015) Expression, purification and immobilization of recombinant AiiA enzyme onto magnetic nanoparticles. *Prot. Exp. Purif.* 113:56–62. doi:10.1016/j.pep.2015.04.014.

Bjarnsholt T, Kirketerp-Moller K, Kristiansen S, Phipps R, Nielsen AK, Jensen PO (2007) Silver against *Pseudomonas aeruginosa* biofilms. *APMIS* 115:921–928.

Colino CI, Millán CG, Lanao JM (2018) Nanoparticles for signaling in biodiagnosis and treatment of infectious diseases. *Int. J. Mol. Sci.* 19(6):1627. doi:10.3390/ijms19061627.

Costabile G, d'Angelo I, Rampioni G, Bondì R, Pompili B, Ascenzioni F, Mitidieri E et al. (2015) Toward repositioning niclosamide for antivirulence therapy of *Pseudomonas aeruginosa* lung infections: Development of inhalable formulations through nanosuspension technology. *Mol. Pharm.* 12(8):2604–2617. doi:10.1021/acs.molpharmaceut.5b00098.

Defoirdt T (2017) Quorum-sensing systems as targets for antivirulence therapy. *Trends Microbiol.* 26:313–328. doi:10.1016/j.tim.2017.10.005.

Defoirdt T, Miyamoto CM, Wood TK, Meighen EA, Sorgeloos P, Verstraete W, Bossier P (2007) The natural furanone (5Z)-4-bromo-5-(bromomethylene)-3-butyl-2(5H)-furanone disrupts quorum sensing-regulated gene expression in *Vibrio harveyi* by decreasing the DNA-binding activity of the transcriptional regulator protein luxR. *Environ. Microbiol.* 9:2486–2495. doi:10.1111/j.1462-2920.2007.01367.x.

Deziel E, Gopalan S, Tampakaki AP, Lepine F, Padfield KE, Saucier M, Xiao G, Rahme LG (2005) The contribution of MvfR to Pseudomonas aeruginosa pathogenesis and quorum sensing circuitry regulation: Multiple quorum sensing-regulated genes are modulated without affecting lasRI, rhlRI or the production of N-acyl-L-homoserine lactones. *Mol. Microbiol.* 55:998–1014.

Dias MA, Lacerda IC, Pimentel PF, de Castro HF, Rosa CA (2002) Removal of heavy metals by an *Aspergillus terreus* strain immobilized in a polyurethane matrix. *Lett. Appl. Microbiol.* 34:46–50.

Dong YH, Wang LH, Xu JL, Zhang HB, Zhang XF, Zhang LH (2001) Quenching quorum sensing dependent bacterial infection by an *N*-acyl homoserine lactonase. *Nature* 411, 813–817.

Duran N, Marcato PD, Alves OL, Souza GI, Esposito E (2005) Mechanistic aspects of biosynthesis of silver nanoparticles by several *Fusarium oxysporum* strains. *J. Nanobiotechnol.* 3:8. doi:10.1186/1477-3155-3-8.

Flickinger ST, Copeland MF, Downes EM, Braasch AT, Tuson HH, Eun YJ, Weibel DB (2011) Quorum sensing between *Pseudomonas aeruginosa* biofilms accelerates cell growth. *J. Am. Chem. Soc.* 133:5966–5975.

Forier K, Raemdonck K, De Smedt SC, Demeester J, Braeckmans K, Coenye T (2014) Lipid and polymer nanoparticles for drug delivery to bacterial biofilms. *J. Control Release* 190:607–623. doi:10.1016/j.jconrel.2014.03.055.

Garcia-Lara B, Saucedo Mora MA, Roldan Sanchez JA, Perez-Eretza B, Ramasamy M, Lee J, Coria-Jimenez R, Tapia M, Varela-Guerrero V, Garcia-Contreras R (2015) Inhibition of quorum-sensing-dependent virulence factors and biofilm formation of clinical and environmental *Pseudomonas aeruginosa* strains by ZnO nanoparticles. *Lett. Appl. Microbiol.* 61(3):299–305. doi:10.1111/lam.12456.

Gomez BG, Arregui L, Serrano S, Santos A, Coronaa TP, Madrid Y (2019) Selenium and tellurium-based nanoparticles as interfering factors in quorum sensing-regulated processes: Violacein production and bacterial biofilm formation. *Metallomics.* 11:1104–1114. doi:10.1039/C9MT00044E.

Goycoolea FM, Valle-gallego A, Stefani R, Menchicchi B, David L, Rochas C, Santander-ortega MJ, Alonso MJ (2012) Chitosan-based nanocapsules: Physical characterization, stability in biological media and capsaicin encapsulation. *Colloids Surf. B: Biointerfaces* 2:1423–1434. doi:10.1007/s00396–012-2669-z.

Gram L, de Nys R, Maximilien R, Givskov M, Steinberg P, Kjelleberg S (1996) Inhibitory effects of secondary metabolites from the red alga *Delisea pulchra* on swarming motility of *Pro. mirabilis. Appl. Environ. Microbiol.* 62(11):4284–4287.

Grandclement C, Tannieres M, Morera S, Dessaux Y, Faure D (2016) Quorum quenching: Role in nature and applied developments. *FEMS Microbiol. Rev.* 40:86–116. doi:10.1093/femsre/fuv038.

Gupta A, Kaur CD, Saraf S (2016) Formulation, characterization, and evaluation of ligand-conjugated biodegradable quercetin nanoparticles for active targeting. *Artif Cells Nanomed Biotechnol.* 44:1–11.

Harrison JJ, Turner RJ, Joo DA, Stan MA, Chan CS, Allan ND (2008) Copper and quaternary ammonium cations exert synergistic bactericidal and antibiofilm activity against *Pseudomonas aeruginosa. Antimicrob. Agents Chemother.* 52 (2008):2870–2881.

Holban AM, Gestal MC, Grumezescu AM (2016) Control of biofilm-associated infections by signaling molecules and nanoparticles. *Int. J. Pharm.* 510:409–418.

Ilk S, Saglam N, Ozgen M, Korkusuzda F (2017) Chitosan nanoparticles enhances the anti-quorum sensing activity of kaempferol. *Int. J. Biol. Macromol.* 94:653–662. doi:10.1016/j.ijbiomac.2016.10.068.

Jadaun V, Prateeksha Singh BR, Paliya BS, Upreti DK, Rao CV, Rawat AKS (2015) Honey enhances the anti-quorum sensing activity and anti-biofilm potential of curcumin. *RSC Adv.* 5:71060–71070. doi:10.1039/C5RA14427B.

Joe MM, Benson A, Sarvanan VS, Tongmin S (2015) In vitro antibacterial activity of nanoemulsion formulation on biofilm, AHL production, hydrolytic enzyme activity, and pathogenicity of *Pectobacterium carotovorum sub sp. Carotovorum. Physiol. Mol. Plant Path.* 91:46–55. doi.org/10.1016/j.pmpp.2015.05.009.

Kalia VC (2013) Quorum sensing inhibitors: An overview. *Biotechnol. Adv.* 31:224–245. doi:10.1016/j.biotechadv.2012.10.004.

Kato N, Morohoshi T, Nozawa T, Matsumoto H, Ikeda T (2006) Control of gram-negative bacterial quorum sensing with cyclodextrin immobilized cellulose ether gel. *J. Incl. Phenom. Macrocycl. Chem.* doi:56:55–59 10.1007/s10847-006-9060-y.

Kerekes EB, Deak E, Tako M, Tserennadmid R, Petkovits T (2013) Anti-biofilm forming and anti-quorum sensing activity of selected essential oils and their main components on food-related microorganisms. *J. Appl. Microbiol.* 115:933–942.

Khan F, Manivasagan P, Lee JW, Pham DTN Oh J, Kim YM (2019) Fucoidan-stabilized gold nanoparticle-mediated biofilm inhibition, attenuation of virulence and motility properties in *Pseudomonas aeruginosa* PAO1. *Mar. Drugs* 17:208. doi:10.3390/md17040208.

Khan MSA, Zahin M, Hasan S, Husain FM, Ahmad I (2009) Inhibition of quorum sensing regulated bacterial functions by plant essential oils with special reference to clove oil. *Lett. Appl. Microbiol.* 49:354–359.

Kim JH, Choi DC, Yeon KM, Kim SR, Lee CH (2011b) Enzyme-immobilized nanofiltration membrane to mitigate biofouling based on quorum quenching. *Environ. Sci. Technol.* 45:1601–1607. doi:10.1021/es103483j.

Kim S, Baek Y-W, An Y-J (2011a) Assay-dependent effect of silver nanoparticles to *Escherichia coli* and *Bacillus subtilis. Appl. Microbiol. Biotechnol.* 92:1045–1052.

Klein T, Henn C, de Jong JC, Zimmer C, Kirsch B, Maurer CK, Pistorius D, Müller R, Steinbach A, Hartmann RW (2012) Identification of small-molecule antagonists of the *Pseudomonas aeruginosa* transcriptional regulator PqsR: Biophysically guided hit discovery and optimization. *ACS Chem. Biol.* 7:1496–1501.

Lee B, Yeon KM, Shim J, Kim SR, Lee CH, Lee J, Kim J (2014b) Effective antifouling using quorum-quenching acylase stabilized in magnetically-separable mesoporous silica. *Biomacromolecules* 15:1153–1159. doi:10.1016/j.jconrel.2014.06.055.

Lee J, Lee I, Nam J, Hwang DS, Yeon KM, Kim J (2017) Immobilization and Stabilization of acylase on carboxylated polyaniline nanofibers for highly effective antifouling application via quorum quenching. *ACS Appl. Mater. Interfaces* 9:15424–15432. doi:10.1021/bm401595q.

Lee JH, Kim YG, Cho MH, Lee J (2014a) ZnO nanoparticles inhibit *Pseudomonas aeruginosa* biofilm formation and virulence factor production. *Microbiol. Res.* 169:888–896. doi:10.1016/j.micres.2014.05.005.

Lin YH, Xu JL, Hu J, Wang LH, Ong SL, Leadbetter JR, Zhang LH (2003) Acyl-homoserine lactone acylase from *Ralstonia* strain XJ12B represents a novel and potent class of quorum-quenching enzymes. *Mol. Microbiol.* 47:849–860. doi:10.1046/j.1365-2958.2003.03351.x.

Liu W, Li XL, Wong YS, Zheng WJ, Zhang YB., Cao WQ, et al. (2012) Selenium nanoparticles as a carrier of 5-fluorouracil to achieve anticancer synergism. *ACS Nano* 6, 6578–6591. doi:10.1021/nn202452c.

Lopez-montilla JC, Herrera-Morales PE, Pandey S, Shah DO (2002) Spontaneous emulsification: Mechanisms, physicochemical aspects, modeling, and applications. *Dispers. Sci. Technol.* 23:219–268. doi:10.1080/01932690208984202.

Lu HD, Pearson E, Ristroph KD, Duncan GA, Prudhomme RK (2018) *Pseudomonas aeruginosa* pyocyanin production reduced by quorum-sensing inhibiting nanocarriers. *Int. J. Pharm.* 544(1):75–82. doi:10.1016/j.ijpharm.2018.03.058.

Lu HD, Spiegel A, Hurley A, Perez LJ, Maisel K, Ensign LM, Hanes J, Bassler BL, Semmelhack MF, Prud'homme RK (2015) Modulating *Vibrio cholerae* quorum sensing controlled communication using autoinducer loaded nanoparticles. *Nano Lett* 15:2235–2241. doi:10.1021/acs.nanolett.5b00151.

Miller KP, Wang L, Chen Y, Pellechia PJ, Benicewicz BC, Decho AW (2015) Engineering nanoparticles to silence bacterial communication. *Front Microbiol.* 6:189. doi:10.3389/fmicb.2015.00189.

Mohanty A, Tan CH, Cao B (2016) Impacts of nanomaterials on bacterial quorum sensing: Differential effects on different signals. *Environ. Sci. Nano.* 3:351–356. doi:10.1039/C5EN00273G.

Nafee N, Husari A, Maurer CK, Lu C, de Rossi C, Steinbach A, Hartmann RW, Lehr CM (2014) Schneider M. Antibiotic-free nanotherapeutics: Ultra-small, mucus-penetrating solid lipid nanoparticles enhance the pulmonary delivery and anti-virulence efficacy of novel quorum sensing inhibitors. *J. Control. Release.* 192:131–140. doi:10.1016/j.jconrel.2014.06.055.

Naik K, Chatterjee A, Prakash H, Kowshik M (2013) Mesoporous TiO_2 nanoparticles containing Ag ion with excellent antimicrobial activity at remarkable low silver concentrations. *J. Biomed. Nanotechnol.* 9:664–673.

Naik K, Kowshik M (2014a) Anti-biofilm efficacy of low temperature processed AgCl-TiO_2 nanocomposite coating. *Mater. Sci. Eng. C. Mater. Biol. Appl.* 34:62–68.

Naik K, Kowshik M (2014b) Anti-quorum sensing activity of AgCl-TiO_2 nanoparticles with potential use as active food packaging material. *J. Appl. Microbiol.* 117:972–983. doi:10.1111/jam.12589.

Niu C, Alfre S, Gilbert ES (2006) Subinhibitory concentrations of cinnamaldehyde interfere with quorum sensing. *Lett. Appl. Microbiol.* 43:489–494.

Omwenga EO, Hensel A, Shitandi A, Goycoolea FM (2018) Chitosan nanoencapsulation of flavonoids enhances their quorum sensing and biofilm formation inhibitory activities against an *E. coli* Top 10 biosensor. *Colloids Surf. B Biointerfaces* 164:125–133.

Parsek MR, Val DL, Hanzelka BL, Cronan JE, Greenberg EP (1999) Acyl homoserine-lactone quorum-sensing signal generation. *Proc. Natl. Acad. Sci. USA* 96:4360–4365. doi:10.1073/pnas.96.8.4360.

Pattnaik S, Barik S, Macha C, Chiranjeevi PV, Siddhardha B (2018a) Anti quorum sensing and antibiofilm efficacy of cinnamaldehyde encapsulated chitosan nanoparticles against *Pseudomonas aeruginosa* PAO1. *LWT* 97:752–759.

Pattnaik S, Barik S, Muralitharan G, Busi S (2018b) Ferulic acid encapsulated chitosan-tripolyphosphate nanoparticles attenuate quorum sensing regulated virulence and biofilm formation in *Pseudomonas aeruginosa* PAO1. *IET Nanobiotechnol.* 12(8):1056–1061. doi:10.1049/iet-nbt.2018.5114.

Paul D, Gopal J, Kumar M, Manikandan M (2018) Nature to the natural rescue: Silencing microbial chats. *Chem. Biol. Interact.* 25(280):86–98. doi:10.1016/j.cbi.2017.12.018.

Prateeksha, Singh BR, Shoeb M, Sharma S, Naqvi AH, Gupta VK, Singh BN (2017) Scaffold of selenium nanovectors and honey phytochemicals for inhibition of *Pseudomonas aeruginosa* quorum sensing and biofilm formation. *Front Cell Infect. Microbiol.* 7:93. doi:10.3389/fcimb.2017.00093.

Qais FA, Khan MS, Ahmad I (2018) Nanoparticles as quorum sensing inhibitor: Prospects and limitations. In V.C. Kalia (Ed.), *Biotechnological Applications of Quorum Sensing Inhibitors*. Springer, pp. 227–248. doi:10.1007/978-981-10-9026-4_11.

Qin X, Emich J, Goycoolea FM (2018a) Assessment of the quorum sensing inhibition activity of a non-toxic chitosan in an *N*-Acyl Homoserine Lactone (AHL)-based *Escherichia coli* biosensor. *Biomolecules* 8(3) pii: E87. doi:10.3390/biom8030087.

Qin X, Engwer C, Desai S, Vila-Sanjurjo C, Goycoole FM (2017) An investigation of the interactions between an *E. coli* bacterial quorum sensing biosensor and chitosan-based nanocapsules. *Colloids Surf. B: Biointerfaces* 149:358–368. doi:10.1016/j.colsurfb.2016.10.031.

Qin X, Kraft T, Goycoolea FM (2018b) Chitosan encapsulation modulates the effect of trans -cinnamaldehyde on AHL-regulated quorum sensing activity. *Colloids Surf. B: Biointerfaces* 169:453. doi:10.1016/j.colsurfb.2018.05.054.

Quave CL, Lyles JT, Kavanaugh JS, Nelson K, Parlet CP, Crosby HA, Heilmann KP, Horswill AR (2015) *Castanea sativa* (European Chestnut) leaf extracts rich in ursene and oleanene derivatives block *Staphylococcus aureus* virulence and Pathogenesis without detectable resistance. *PLoS One* 10(8):e0136486. doi:10.1371/journal.pone.0136486.

Ramos, AP, Cruz MAE, Tovani CB, Ciancaglini P (2017) Biomedical applications of nanotechnology. *Biophys. Rev.* 9(2):79–89. doi:10.1007/s12551-016-0246-2.

Rampioni G, Leoni L, Williams P (2014) The art of antibacterial warfare: Deception through interference with quorum sensing-mediated communication. *Bioorg. Chem.* 55:60–68. doi:10.1016/j.bioorg.2014.04.005.

Rasko DA, Moreira CG, Li de R, Reading NC, Ritchie JM, Waldor MK, Williams N, et al. (2008) Targeting QseC signaling and virulence for antibiotic development. *Science* 321(5892):1078–1080. doi:10.1126/science.1160354.

Ravindran D, Ramanathan S, Arunachalam K, Jeyaraj GP, Shunmugiah KP, Arumugam VR (2018) Phytosynthesized silver nanoparticles as antiquorum sensing and antibiofilm agent against the nosocomial pathogen Serratia marcescens: An in vitro study. *J. Appl. Microbiol.* 124(6):1425–1440. doi:10.1111/jam.13728.

Salata OV (2004) Applications of nanoparticles in biology and medicine. *J. Nanobiotechnol.* 2:3. doi:10.1186/1477-3155-2-3.

San Diego KDG, Alindayu JIA, Baculi RQ (2018) Biosynthesis of gold nanoparticles by bacteria from hyperalkaline spring and evaluation of their inhibitory activity against pyocyanin production. *J. Microbiol. Biotechnol. Food Sci.* 8(2):781–787. doi:10.15414/jmbfs.2018.8.2.781-787.

Shah S, Gaikwad S, Nagar S, Kulshrestha S, Vaidya V, Nawani N, Pawar S (2019) Biofilm inhibition and anti-quorum sensing activity of phytosynthesized silver nanoparticles against the nosocomial pathogen *Pseudomonas aeruginosa. Biofouling* 35(1):34–49. doi:10.1080/08927014.2018.1563686.

Singh BR, Singh A, Khan W, Naqvi AH, Singh HB (2015) Mycofabricated biosilver nanoparticles interrupt *Pseudomonas aeruginosa* quorum sensing systems. *Sci. Rep.* 5:13719. doi:10.1038/srep13719.

Singh N, Romero M, Travanut A, Monteiro PF, Jordana-Lluch E, Hardie KR, Williams P, Alexander MR, Alexander C (2019) Dual bioresponsive antibiotic and quorum sensing inhibitor combination nanoparticles for treatment of *Pseudomonas aeruginosa* biofilms *in vitro* and *ex vivo*. *Biomater. Sci.* 29. doi:10.1039/c9bm00773c.

Sintim HO, Smith JA, Wang J, Nakayama S, Yan L (2010) Paradigm shift in discovering next-generation anti-infective agents: Targeting quorum sensing, c-di-GMP signaling and biofilm formation in bacteria with small molecules. *Future Med Chem.* 2:1005–1035. doi:10.4155/fmc.10.185.

Slistan-Grijalva A, Herrera-Urbina R, Rivas-Silva JF, Ávalos-Borja M, Castillón-Barraza FF, Posada-Amarillas A (2005) Assessment of growth of silver nanoparticles synthesized from an ethylene glycol–silver nitrate–polyvinylpyrrolidone solution. *Physica E* 25:438–448.

Srinivasan R, Mohankumar R, Kannappan A, Karthick Raja V, Archunan G, Karutha Pandian S, Ruckmani K, Veera Ravi A (2017) Exploring the anti-quorum sensing and antibiofilm efficacy of phytol against *Serratia marcescens* associated acute pyelonephritis infection in wistar rats. *Front Cell Infect. Microbiol.* 7:498. doi:10.3389/fcimb.2017.00498.

Suk JS, Lai SK, Boylan NJ, Dawson MR, Boyle MP, Hanes J (2011) Rapid transport of muco-inert nanoparticles in cystic fibrosis sputum treated with *N*-acetyl cysteine. *Nanomedicine* 6:365–375.

Tang K, Zhang XH (2014) Quorum quenching agents: Resources for antivirulence therapy. *Mar. Drugs* 12(6):3245–3282. doi:10.3390/md12063245.

Teitzel GM, Parsek MR (2003) Heavy metal resistance of biofilm and planktonic *Pseudomonas aeruginosa. Appl. Environ. Microbiol.* 69:2313–2320.

Teixeira PC, Leite GM, Domingues RJ, Silva J, Gibbs PA, Ferreira JP (2007) Antimicrobial effects of a microemulsion and a nanoemulsion on enteric and other pathogens and biofilms. *Int J Food Microbiol*, 118:15–19. doi:10.1016/j.ijfoodmicro.2007.05.008.

Thanh NH, Goycoolea FM (2017) Chitosan/Cyclodextrin/TPP nanoparticles loaded with quercetin as novel bacterial quorum sensing inhibitors. *Molecules* 22(11) pii: E1975. doi:10.3390/molecules22111975.

Truchado P, Lopez-Galvez F, Gil MI, Tomas-Barberan FA, Allende A. (2009) Quorum sensing inhibitory and antimicrobial activities of honeys and the relationship with individual phenolics. *Food Chem.* 115:1337–1344. doi:10.1016/j.foodchem.2009.01.065.

Vadekeetil A, Chibber S, Harjai K (2019) Efficacy of intravesical targeting of novel quorum sensing inhibitor nanoparticles against *Pseudomonas aeruginosa* biofilm-associated murine pyelonephritis. *J. Drug Target.* 27:995–1003. doi:10.1080/1061186X.2019.1574802.

Valero M, Giner MJ (2006) Effects of antimicrobial components of essential oils on growth of Bacillus cereus INRA L2104 in and the sensory qualities of carrot broth. *Int. J. Food Microbiol.* 106:90–94.

Venkadesaperumal G, Rucha S, Sundar K, Shetty PH (2016) Anti-quorum sensing activity of spice oil nanoemulsions against food borne pathogens. *LWT—Food Sci. Tech* 66:225–231. doi:10.1016/j.lwt.2015.10.044.

Verma VC, Kharwar RN, Gange AC (2010) Biosynthesis of antimicrobial silver nanoparticles by the endophytic fungus *Aspergillus clavatus. Nanomedicine* 5(1):33–40.

Vinoj G, Pati R, Sonawane A, Vaseeharan B (2015) In vitro cytotoxic effects of gold nanoparticles coated with functional acyl homoserine lactone lactonase protein from *Bacillus licheniformis* and their antibiofilm activity against *Proteus* species. *Antimicrob. Agents Chemother.* 59:763–771. doi:10.1128/AAC.03047-14.

Vittal RR, Bai AJ (2011) Nanoparticles and their potential application as antimicrobials. *Science Against Microbial Pathogens: Communicating Current Research and Technological Advances*:197–209.

Wagh (nee Jagtap) MS, Patil RH, Thombre DK, Kulkarni MV (2013) Evaluation of anti-quorum sensing activity of silver nanowires. *Appl. Microbiol. Biotechnol.* 97:3593. doi:10.1007/s00253-012-4603-1.

Walia N, Kaur A, Babbar SB (2007) An efficient, in vitro cyclic production of shoots from adult trees of *Crataeva nurvala. Buch Ham. Plant. Cell. Rep.* 26:277–284.

Wang L, Benicewicz BC (2013) Synthesis and characterization of dye-labeled poly(methacrylic acid) grafted silica nanoparticles. *ACS Macro. Lett.* 2:173–176 doi:10.1021/mz3006507.

Wang L, Hu C, Shao L (2017) The antimicrobial activity of nanoparticles: Present situation and prospects for the future. *Int. J. Nanomed.* 12:1227–1249. doi:10.2147/IJN.S121956.

Weber S, Zimmer A, Pardeike J (2014) Solid lipid nanoparticles (SLN) and nanostructured lipid carriers (NLC) for pulmonary application: A review of the state of the art. *Eur. J. Pharm. Biopharm.* 86:7–22.

Wei Y, Perez LJ, Ng W-L, Semmelhack MF, Bassler BL (2011) Mechanism of *Vibrio cholerae* Autoinducer-1 Biosynthesis. *ACS Chem. Bio.* 6(4):356–365. doi:10.1021/cb1003652.

Yadav KS, Sawant KK (2010) Modified nanoprecipitation method for preparation of cytarabine-loaded PLGA nanoparticles. *AAPS Pharm. Sci Tech.* 11:1456–1465. doi:10.1208/s12249-010-s9519-4.

Yu B, Li XL, Zheng WJ, Feng YX, Wong YS, Chen TF (2014) pH-responsive cancer-targeted selenium nanoparticles: A transformable drug carrier with enhanced theranostic effects. *J. Mat. Chem. B* 2:5409–5418. doi:10.1039/C4TB00399C.

Zhu J, Miller MB, Vance RE, Dziejman M, Bassler BL, Mekalanos JJ (2002) Quorum-sensing regulators control virulence gene expression in *Vibrio cholera. PNAS* 99:3129–3134. doi:10.1073/pnas.052694299.

19

Phytochemical Compounds Targeting the Quorum Sensing System as a Tool to Reduce the Virulence Factors of Food Pathogenic Bacteria

Jesus M. Luna-Solorza, Francisco J. Vazquez-Armenta, A. Thalia Bernal-Mercado, M. Melissa Gutierrez-Pacheco, Filomena Nazzaro, and J. Fernando Ayala-Zavala

CONTENTS

19.1 Introduction

In recent years, advances in antibiotics, disinfectants, and hygiene assurances programs have increased to assure food safety around the world. Nevertheless, it is estimated that just in the United States, 48 million people get sick from foodborne illness, 128,000 are hospitalized, and 3,000 die each year (CDC 2019a). This means that the available solutions are not enough to fight pathogenic bacteria and it evidences the need for more effective alternatives. The current treatments for this problem are antibiotics, and they have lost efficacy against bacteria due to their incorrect use, generating resistance of food pathogens that are challenging the entire world. Moreover, the pharmaceutical companies have reduced antibiotics' manufacture for profitability issues, decreasing their production and the available options to attack this problem (Bartlett et al. 2013). Therefore, new strategies are necessary to fight foodborne infections and bacterial resistance.

Bacterial infections are mediated by virulence factors to survive in abiotic environments, allowing their survival and the host invasion (Cross 2008). These virulence factors include cell surface proteins to mediate bacterial adhesion; besides, bacteria produces and secretes carbohydrates and proteins to offer protection to the embedded cells. Bacteria developed a complex network called *quorum sensing*, to sense and respond to environmental changes to increase their energetic efficiency during these processes (Reverchon and Nasser 2013). *Quorum sensing* is an intercellular communication process that involves the production, secretion, and perception of specific molecules (Winzer and Williams 2001). Gram-negative bacteria use mostly acyl homoserine lactones, and Gram-positive bacteria use peptide signals in their communication, and both bacteria can use autoinducer-2 molecules (Heilmann and Götz 2010, Stacy et al. 2013). There are LuxI/LuxR, LuxS/LuxP based *quorum sensing* systems; the first uses Acyl-homoserine lactones, and the second detects autoinducer-2 (Schaefer et al. 1996, Schauder et al. 2001). On the other hand, the *agr* system uses the autoinducer peptide, and it is exclusive for Gram-positive bacteria (Novick and Geisinger 2008). The knowledge of *quorum sensing* can be helpful to design anti-virulence treatments focused on potential targets within these systems.

Plant extracts and their constituents have been recognized as *quorum sensing* inhibitors due to their similarity with signal molecules, blocking receptor proteins (Nazzaro et al. 2013). Phytochemicals are part of a broad group with different structures that include phenolic compounds, alkaloids, terpenes, and sulfur compounds (Truchado et al. 2012). Phenolic compounds like vanillic acid, cinnamic acid, and methyl gallate (Joshi et al. 2016, Hossain et al. 2017, Sethupathy et al. 2017); alkaloids as solenopsin A and berberine (Park et al. 2008, Aswathanarayan and Vittal 2018); terpenes and terpenoids like carvacrol, limonene, and trans-cinnamaldehyde (Kerekes et al. 2013, Chang et al. 2014, Tapia-Rodriguez et al. 2017); and sulfur compounds such as iberin (Jakobsen et al. 2012) have been able to disturb the *quorum sensing* system of pathogenic bacteria. For all this, phytochemicals are an alternative to reduce the virulence of food pathogenic bacteria due to the effect in the *quorum sensing* systems.

19.2 Incidences of Infectious Diseases Caused by Food Pathogenic Bacteria

Foodborne diseases caused by pathogenic bacteria are common around the world despite the advances in antibiotics, disinfectants, and hygiene programs, the. Due to this, the Food and Drug Administration (FDA) published a rule in 2001 titled Hazard Analysis and Critical Control Points (HACCP), which is a program of procedures that control and verify the biological risks during the sanitary processing in the food industry (Minor and Parrett 2017). However, 18 years later it is estimated that in United States (CDC 2019a) each year around 48 million people get sick from a foodborne illness, while in Canada 4 million people are infected (Singh et al. 2019). In the last six years, several foodborne disease outbreaks occurred in the United States due to the contamination of a variety of food (Figure 19.1). The number of cases decreased from 2015 to 2016, but an increase was recorded in 2018; these events can be related to deficient food safety assurance programs. Anybody can get a foodborne illness, but pregnant women, young children, older adults, and people with a weak immune system could be in a life-threatening situation. Figure 19.2 shows different foods like vegetables, poultry, fruits, and beef among the most likely to be contaminated (CDC 2019b). With this data, it can be seen that the presence of pathogenic bacteria in food has not been eliminated. For this reason, considering all of the foodborne illness statistics, it is necessary to use new strategies to decrease the number of people vulnerable to pathogenic bacteria.

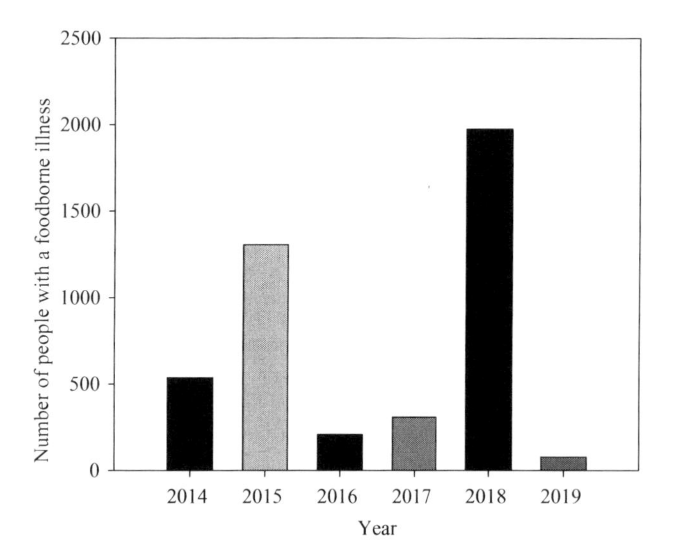

FIGURE 19.1 Total cases of people with a foodborne disease in the last six years.

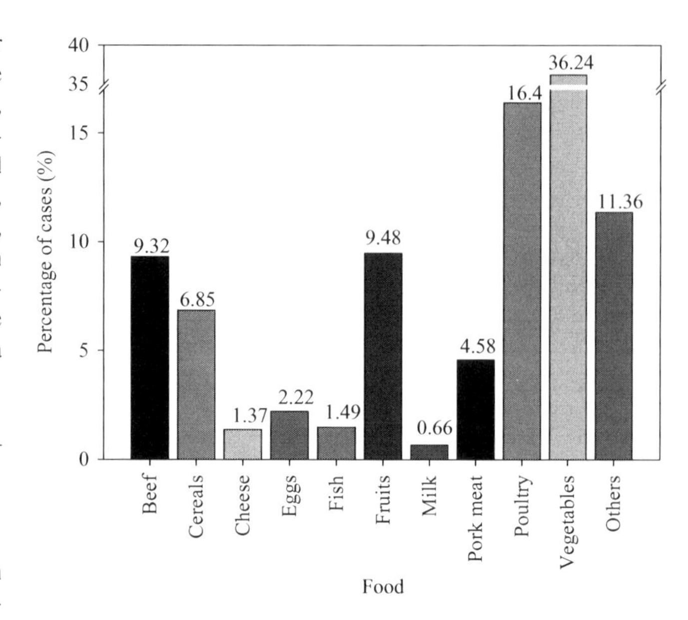

FIGURE 19.2 Percentage of cases of foodborne diseases due to each type of food. (Adapted from CDC, 2019b.)

Many pathogenic bacteria can cause foodborne illnesses, and these microorganisms can be practically in all types of food and get cross contamination during their processing or distribution. Among these bacteria, *Campylobacter jejuni*, *Clostridium botulinum*, *Escherichia coli*, *Listeria monocytogenes*, *Staphylococcus aureus*, and *Salmonella* Typhimurium are the most important. *Campylobacter jejuni* can be in the water, raw milk, or in undercooked meat; this bacteria cause diarrhea, cramping, abdominal pain, and fever (Maktabi et al. 2019). *Clostridium botulinum* can be in badly sterilized canned food; this bacterium produces different types of toxins, which affect the nervous system often causing death (Xin et al. 2019). *Escherichia coli* can be present in uncooked ground beef, unpasteurized juices, raw fruits, and vegetables; some symptoms of infection by this bacteria are similar to *Campylobacter* but *E. coli* could also cause acute kidney failure (Najafi et al. 2019). *Listeria monocytogenes* can be in ready-to-eat foods, cheeses made with raw milk, seafood, and salads; this pathogen causes fever and muscle aches and can cause problems with pregnancy such as a miscarriage (Scobie et al. 2019). *Staphylococcus aureus* is in the skin and hair of healthy people, and it has been reported in dairy products like raw milk or cheeses due to improper food handling. *S. aureus* produces inflammations, toxic shock syndrome, scalded skin syndrome, and even death (Han et al. 2019). *Salmonella* can be founded in eggs, poultry, and meat. *Salmonella* infections inflict diarrhea, fever, and abdominal cramps (CDC 2018b). Currently, different serotypes of *Salmonella* and *Escherichia coli*, as well as *Listeria monocytogenes*, are the predominant bacteria (Figure 19.3). *Salmonella* and *Listeria monocytogenes* were the most important bacteria

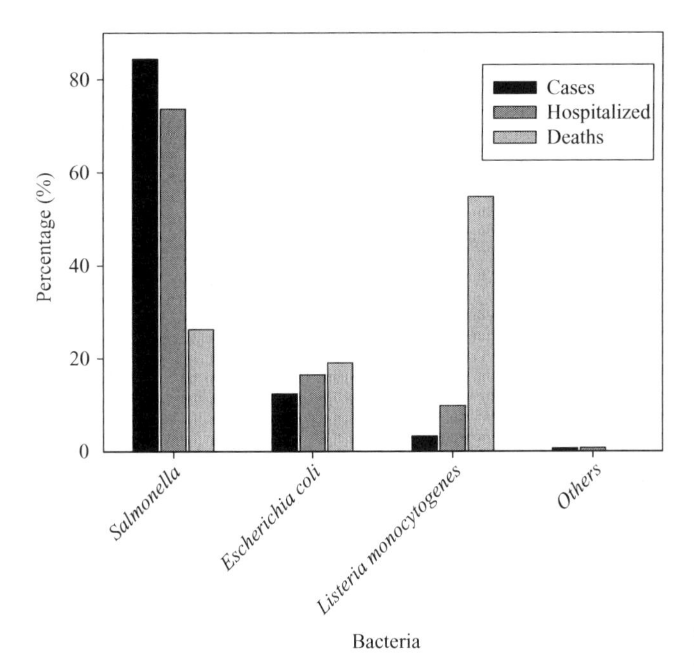

FIGURE 19.3 Percentage of total cases, hospitalizations, and deaths due to the presence of bacteria in food. (Adapted from CDC, Centers for Disease Control and Prevention, List of selected multistate foodborne outbreak investigations, https://www.cdc.gov/foodsafety/outbreaks/multistate-outbreaks/outbreaks-list.html, 2019b.)

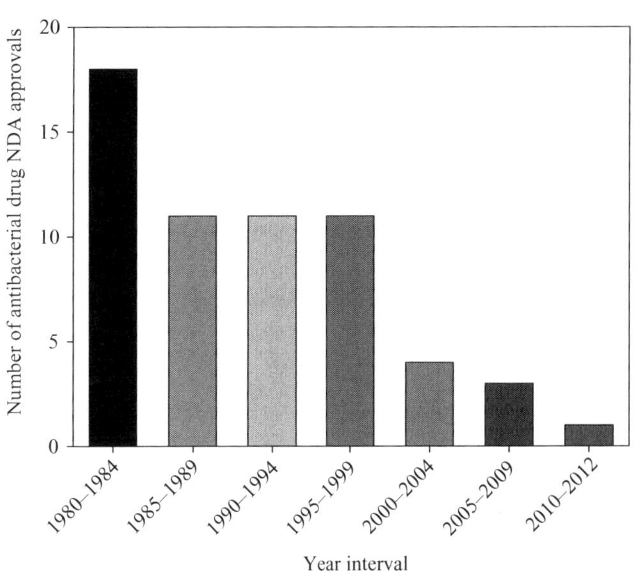

FIGURE 19.4 Number of antibacterial new drug application (NDA) approvals by year intervals. (Adapted from CDC, Centers for Disease Control and Prevention, Biggest threats and data, https://www.cdc.gov/drugresistance/biggest_threats.html?CDC_AA_refVal=https%3A%2F%2Fwww.cdc.gov%2Fdrugresistance%2Fthreat-report-2013%2Findex.html, 2018a.)

causing the most illness cases and the most number of hospitalization and fatal incidents, respectively.

The risk of contracting a foodborne infection is not the only thing that must matter to the scientific community. Also, it is necessary to pay more attention to the efficacy of the conventional treatments used in these diseases and their possible long-term risks. Such is the case of antibiotics, which are the most common therapies to alleviate these diseases. However, they are becoming less effective due to the resistance that bacteria acquire against them. This resistance is occurring around the world and is a crisis (Ventola 2015). Even the World Health Organization (2014) says that resistance to antibiotic threatens the affectivity of the treatments against a variety of infections caused by bacteria in the twenty-first century. Infections caused by resistant strains to multiple drugs are associated with higher mortality compared to susceptible bacteria. Also, expenses of more than 20 billion dollars per year are estimated in the United States (Munita and Arias 2016). Also it is estimated that 2 million people get sick from infection by bacteria resistant to antibiotics and at least 23,000 people die (CDC 2018a).

Each year the battle against bacterial diseases is harder due to the developed resistance to antibiotics. It has been reported that *E. coli* is resistant to some of the most frequently used antibiotics such as ciprofloxacin (Bernal-Mercado et al. 2018). Also, Torres et al. (2013) found that enterobacteria like *E. coli, Klebsiella pneumonia*, and *Candida* spp. were resistant to trimethoprim-sulfamethoxazole, third-generation cephalosporins, and ciprofloxacin. They also observed that some Gram-negative bacilli like *Pseudomonas aeruginosa* and *Acinetobacter baumannii* were resistant to imipenem, piperacillin with tazobactam, meropenem, nitrofurantoin, ciprofloxacin, and amikacin. Besides, they noticed that enterococcus was resistant to ciprofloxacin and ampicillin. This increase in the resistance to antibiotics is just the consequence for their non-appropriated use and the complexity of the bacterial biochemical machinery. Therefore, the use of new alternatives is of vital importance in the fight against infections and bacterial resistance.

The problem with the strategies nowadays is that the amount of new antibiotics available has decreased (Figure 19.4) (CDC 2018a). The low production of antibiotics is in large part because the pharmaceutical companies has stopped their manufacture for profitability issues (Bartlett et al. 2013). Because if the price of an antibiotic is compared with the price of treatments for chronic conditions, antibiotics are immeasurably lower (Gould and Bal 2013, Wright 2014). Even a cost-benefit study conducted by the Office of Health Economics in London calculated that the value of a new antibiotic is around 50 million US dollars, while a drug to treat a neuromuscular disease has a value of more than one billion US dollars (Ventola 2015). However, the social impact of these economic decisions has to be considered. For all the above, it is demonstrated that it is necessary to create new strategies to combat the diseases transmitted by pathogenic bacteria and minimize the resistance of them.

19.3 Targeting the Quorum Sensing to Control Bacterial Virulence Factors

The establishment of bacterial infection is mediated by virulence factors that enable bacteria to replicate and disseminate within the host manly allowing its invasion and helping to elude host defenses (Cross 2008). The concept of the virulence factor includes any bacterial product or strategies that contribute to the ability of the bacterium to cause disease (Sui et al. 2009). Classifications are commonly made according to function during a bacterial infection such as adhesion factors, invasion factors, capsules, endotoxins, exotoxins, or siderophores (Peterson 1996). They also can be classified depending their role during the infection as offensive (e.g. involved in active invasion or that directly cause damage to the host), defensive (e.g. involved evasion of the host defenses), regulation (e.g. involved in coordination of their expression), and non-specific (e.g. neither offensive nor defensive, or both depending on the context) (Sui et al. 2009). Virulence factors include bacterial toxins, cell surface proteins that mediate the bacterial attachment, cell surface carbohydrates and proteins that protect a bacterium, hydrolytic enzymes that may contribute to the pathogenicity of the bacterium, and also biofilm formation to survive. However, to achieve a successful infection, the regulation of expression of virulence factors must be tightly coordinated (Camejo et al. 2011).

Most of the pathogenic bacteria can survive and replicate in the external environment and specialized host niches. So, to avoid the unnecessary waste of energy due to the production of virulence factors outside the host, most bacteria develop complex regulatory networks that aid to sense and respond to environmental changes (Reverchon and Nasser 2013). One of these is the *quorum sensing* system (QS) that detects bacterial population to coordinate behaviors when bacterial density is high. QS is a communication process that involves the production, secretion, and perception of low-molecular-weight signals called autoinducers and enables individual cells to detect (*sense*) when the minimal number of bacteria (*quorum*) has been achieved to initiate a specific response (Winzer and Williams 2001). It has been assumed that the purpose of QS is to stimulate the production of "public goods," mainly extracellular enzymes or specific metabolites that could be better exploited by the bacterial populations at high cell densities (Darch et al. 2012). This also applies to bacterial pathogenesis, when high cell densities are necessary to successfully establish an infection or survive to host defenses (Ciofu et al. 2015).

QS systems are classified according to the nature of the used signal molecules. Acyl-homoserine lactones or autoinducer-1 (AHL or AI-1) are used by Gram-negative bacteria, while Gram-positive uses a polypeptide signal named autoinducer peptide (AIP) (Heilmann and Götz 2010, Stacy et al. 2013). The third system, using furanone derivatives named autoinducer-2 (AI-2), is found in both Gram-positive and Gram-negative bacteria. Also, autoinducer-3 (AI-3) have been described in Gram-negative bacteria; however, the structure and nature of the signal molecule remain to be identified.

19.4 LuxI/LuxR Based QS System

This system has been first described in the marine bacterium *Vibrio fischeri*, where cell density-dependent bioluminescence was observed, and it has been used as a model to generate fundamental knowledge on the QS. Two regulatory proteins constitute this system, the Acyl-homoserine-lactone synthase LuxI and the receptor protein LuxR, encoded by *luxI* and *luxR* genes, respectively. LuxI catalyzes AHL synthesis that transfers a fatty acid chain from an acylated acyl carrier protein (ACP) to S-adenosylmethionine (SAM). The result is the release of AHL and methylthioadenosine as products (Schaefer et al. 1996). In *V. fischeri*, the AHL synthetized by LuxI is 3-oxohexanoyl HSL; however, depending on the length of the fatty acid moieties of ACPs recognized by LuxI homologs, different bacteria can produce their AHLs resulting in the genus- and species-specific signals (Fuqua and Greenberg 2002).

AHLs are synthesized constitutively at low cell densities where they can diffuse through the bacterial membrane into the extracellular environment. When the bacterial population increases (high cell densities), the AHL concentration in the environment is high enough to diffuse back into the cell and to achieve their critical concentration required to bind to LuxR. LuxR is a transcriptional factor that is allosterically activated by AHL forming LuxR-AHL complex, which can form dimers or multimers and bind to the promoter region (Lux box) of the *luxCDABEGH* operon creating an autoinduction circuit, but also LuxR can bind other genes containing Lux boxes in the promoter region (Parker and Sperandio 2009). When the operon has been activated, the expression of virulence factors occurs; in the case of this strain, this results in the production of luminescence.

In *E. coli*, the LuxRhomolog is named SdiA; however, in its genome, there are no homologs for the AHL synthase LuxI, so *E. coli* does not produce AHL. Despite this, *E. coli* can sense and respond to external AHL produced by other bacteria (Kim et al. 2014). Also, SdiA plays an important role in the control of biofilm development, since it has been observed that disruption of gene *sdiA* decreased the expression of the repressor CrsB, enhancing biofilm formation (Suzuki et al. 2002). While in *Salmonella* spp., SdiA senses exogenous AHL and activates the *rck* operon and *srgE* gene (Smith and Ahmer 2003, Dyszel et al. 2010). It has been reported that Rck promotes the adhesion of *Salmonella* to epithelial cells during pathogenesis and other two genes of the *rck* operon, *pefI*, and *srgA* to affect the expression of fimbriae (Crago and Koronakis 1999, Smith et al. 2011). Although *E. coli* and *Salmonella* spp. do not synthesize AHL, it is possible that the QS receptor SdiA modulates virulence of both pathogens in the response of AHL synthetized by neighbor bacteria from other genera.

19.5 LuxS/LuxP Based Systems

These QS systems are present in both Gram-negative and Gram-positive bacteria and detect AI-2 molecules. These molecules are different from the AHLs described above and are synthesized from SAM metabolism. LuxS converts ribose-homocysteine into homocysteine and 4,5-dihydroxy-2,3-pentanedione (DPD), a compound that, in the presence of water, cyclizes into several furanones (Schauder et al. 2001). In *V. fisheri* the detection of AI-2 is via binding to receptor protein LuxP; this AI-2/LuxP complex interacts with LuxQ, a sensor kinase that triggers a signaling cascade that results in deactivation of the negative response regulator LuxO and luciferase production and bioluminescence (Parker and Sperandio 2009).

In *Vibrio parahaemolyticus*, an invasive pathogen acquired by eating marine products with poor sanitary quality, the QS system consists of the sensor histidine kinase that is active when autoinducer molecules are low and phosphorylate the LuxO regulator via the small histidinephosphorelay protein LuxU. Phosphorylated LuxO (LuxO-P) induces the expression of small quorum-regulatory RNAs (Qrrs) that interfere with the mRNA of the central output regulator of the system. At high cell density and in the presence of a high concentration of autoinducers LuxO becomes inactive, so Qrrs are no longer transcribed, and the QS outputs are translated (Gode-Potratz and McCarter 2011). The virulence factors repressed at high cell density are surface sensing regulon, protein secretion systems (T3SS1, T6SS1), and iron transporters, among others (Gode-Potratz and McCarter 2011). *Vibrio cholerae,* a toxigenic pathogen, employs parallel QS circuits that converge into a shared signaling pathway. The two systems are the CqsA/CqsS system, which synthesizes and senses CAI-1 (*S*-3-hydroxytridecan-4-one), and the LuxS/LuxPQ system, which produces and detects AI-2 (Jung et al. 2015). *Similar to V. parahaemolyticus, V. cholerae* uses QS to regulate virulence factor production, biofilm formation, secretion systems, and competence development (Jung et al. 2015). It has been proposed that the repression of virulence factors and biofilm formations at high cell density allows *Vibrio* to leave the host and reincorporates in the environment in large numbers and initiates a new cycle of infection.

In *E. coli,* the LuxS system uses AI-2 that is phosphorylated in the cytosol by LsrK after its uptake. Activated AI-2 binds and depress the *lsr* repressor LsrR. It has been reported that *lsrR* mutation affected the expression of virulence factors related to attachment, defense, and pathogenicity in *E. coli.* These genes include *csgE,* a curli production assembly/transport component, *htrE* involved in type II pilus assembly, and another fimbria-like protein. *lsrK* mutations affect *ppdD,* which encodes a putative major type IV pilin and two putative fimbrial proteins, *yadK,* and *yadN* (Li et al. 2007). Also, both mutations impair biofilm formation and alter its architecture (Li et al. 2007). In the other hand, Barrios et al. (2006) proposed that AI-2 stimulates biofilm formation and alters its architecture by stimulating flagellar motion and motility. Since the first step in *E. coli* infection is the adhesion of bacteria to microvilli of the epithelial cells in the intestine mediated by extracellular appendages such as fimbriae, type IV pilus, and flagella (McDaniel et al. 1995), the interruption of the LuxS type QS system could help to attenuate virulence.

In the other hand, *Salmonella* spp. also possess a LuxS type QS system that regulates virulence factors. In the first step of the infection, the fimbrial adhesins include Fim (fimbriae), Lpf (long polar fimbriae), Tafi (thin aggregative fimbriae), Pef (plasmid-encoded fimbriae), and curli (thin aggregative fimbriae) that play essential roles in the bacterial attachment to abiotic surfaces and the host cells. The curli synthesis is regulated by QS via the LuxR type regulator, CsgD (Römling et al. 2000, Brown et al. 2001). CsgD induces curli production by the transcriptional activation of *csgBAC* operon, while the cellulose synthesis is activated indirectly by ArdA regulator (Römling et al. 2000, Zogaj et al. 2003). Cellulose is also involved in the attachment to abiotic surfaces, cellular agglomeration, and biofilm formation that helps in the survival in non-enteric habits (Zogaj et al. 2003).

19.6 *Agr* System

This QS system is used exclusively by Gram-positive bacteria and was first described in *S. aureus* (Abdelnour et al. 1993). In this system, a polypeptide referred to as AIP is used as the signal molecule instead of the small diffusible molecules used by the other QS systems. The *agr* system consists of the operon *agrB-DCA* (Novick and Geisinger 2008). AgrB is a membrane-bound peptidase that cleaves and processes the propeptide derived from AgrD at the C-terminal end; besides this, AgrB catalyzes the formation of a thiolactone ring with a central cysteine and exports the autoinducer peptide in its active form. After the accumulation of the peptide in the extracellular space, the AIP is recognized by the receptor histidine kinase AgrC, which is part of a two-component system where ArgA is the response regulator. In the presence of the correct AIP, AgrC phosphorylates AgrA, which activates the transcription of selected genes by binding to the promoter regions P2 that include *agrBDCA* operon and P3 that transcript RNA III. It is the effector molecule that activates secretion of enterotoxins and other proteins in *S. aureus* (Wang and Muir 2016, Zetzmann et al. 2016).

S. aureus is naturally present in human and animal skin, nares, and respiratory and genital tracts; however, as an opportunistic pathogen, it can cause severe infection that affects many organs. This bacterium produces diverse virulence factors such as adhesins and many toxins (enterotoxins, toxic shock syndrome toxins, pore-forming hemolysins, and exfoliative toxins). It also produces coagulase that aids in the bacterial evasion of phagocytosis, collagenase, which hydrolyzes collagen and hyaluronidase, to hydrolyze hyaluronic acid that causes severe tissue damage (Drăgulescu and Codita 2015). QS regulation of virulence factors is carried out by an AgrA-dependent P3 operon, located back-to-back with P2, which encodes RNAIII, the mRNA for δ-toxin. RNAIII binds to virulence-factor mRNAs, suppressing the synthesis of proteins required for the adhesion step while repressing those involved in the invasion stage. Also, an essential role of the *agr* system provides benefits at the group level since agr mutants that do not perform QS increased infectivity when they are co-cultured with QS wild type strains (Pollitt et al. 2014).

An *agr* system is also present in *L. monocytogenes,* the etiological agent of listeriosis, a foodborne disease with a high mortality rate (20%–30%) presented as gastrointestinal or invasive forms (Vera et al. 2013). Invasive listeriosis is characterized by the adhesion and invasion to the epithelial cells with subsequent translocation to different organs, causing systemic infection in the host (Vera et al. 2013). The main difference with *agr* systems of *S. aureus* is the lack of RNAIII (Mellin and Cossart 2012), which suggests the participation of other transcriptional regulators affected by AgrA-dependent regulation (Zetzmann et al. 2016). It has been observed that *ΔagrD* mutant displayed significantly reduced biofilm formation, reduced invasion of Caco-2 intestinal epithelial, and also exhibited attenuated virulence in mice (Riedel et al. 2009). Also in *ΔagrD* mutant, the expression of virulence genes (*hlyA, actA, plcA, prfA,* and *inlA)* was downregulated (Riedel et al. 2009). In this pathogen, InlA modulates the bacterial entry to host cells by binding to the cadherin and inducing phagocytosis (Chen et al. 2011). Then, *L. monocytogenes* escapes from the phagocytic vacuoles with the help of the secreted pore-forming toxin Listeriolysin O (LLO) and phospholipases A (PlcA) and B (PlcB), to reach the cytosol where the bacteria can grow and multiply (Meyer-Morse et al. 2010, Hamon et al. 2012). Finally, the translocation of *L. monocytogenes* from cell to cell occurs through actin assembly inducing protein (ActA) that allows intracellular motility by the polymerization of actin fibers (Artola and Herrejón 2010, Travier et al. 2013). This highlights the importance of *agr* systems in the regulation of virulence of Gram-positive bacteria. All the explained systems have to be analyzed in genetic and enzymatic perspectives to visualize the potential targets to be attacked for emerging treatments.

19.7 Phytochemicals as Quenchers of the Quorum Sensing System

Phytochemicals have been explored in many research studies as agents capable of inhibiting biofilm formation or causing its eradication (Tapia-Rodriguez et al. 2017; Gutierrez-Pacheco et al. 2018). Biofilms are bacterial communities attached to a surface and covered by an adhesive and protective matrix composed of extracellular polymeric substances (EPS) excreted by themselves (Flemming et al. 2016). The biofilm is the most common way of life of most bacteria, and it is continuously in alteration. For the development and establishment of biofilms, many chemical and physical processes are involved such as growth, motility, adhesion, EPS production, among others (Bouyahya et al. 2017). This multifactorial process is produced by a combination of environmental conditions and a series of gene behavioral responses operating under high-cell density and controlled by QS (Kjelleberg and Molin 2002). To date, it has been suggested that QS regulation is important in different phases of biofilm formation for several bacteria under certain situations (Bouyahya et al. 2017). For example, QS in some pathogens regulates motility that favors bacteria to overcome the hydrodynamic forces of the media and to counteract the repulsion between bacteria and the surface to start the attachment (Yang and Defoirdt 2015, Flemming et al. 2016, Yang et al. 2018). In addition, QS regulates the production of exopolysaccharides which are necessary to maintain the surface biofilm adhesion and to give the structure of biofilms (Flemming et al. 2016, Bouyahya et al. 2017). With this in mind, phytochemicals have been used as QS inhibitors to affect biofilms due to an interference in this system that results in the affectation of essentials factors related to the development of biofilms such as motility and EPS production.

The disruption of bacterial communication can be achieved by interfering with the production or reception of signal molecules such as inhibition of signal synthesis, signal degradation, signal sequestration, and signal competition (Figure 19.5) (Vattem et al. 2007b, Truchado et al. 2015). Plant extracts and phytochemicals have been used since ancient times as treatment to combat various infections, which has been attributed to their antimicrobial properties. Phytochemicals have been recognized to act as inhibitors of intra- and inter-species communication systems because of their similarity with QS signals molecule and due to their ability to interact with signal producers or receptors (Nazzaro et al. 2013). Some of the benefits of using plant extracts and their main compounds could be that most of them are harmless for human health and they are rarely associated with side effects and bacterial resistance. Phytochemicals are part of a wide group of chemical compounds with a very different structure synthesized by the secondary metabolism of plants in the response to environmental stress (Truchado et al. 2012). The main phytochemicals with antimicrobial properties are phenolics, terpenes, terpenoids, sulfur, and alkaloids compounds and recently, it has been reported that these compounds possess anti-QS activity toward different pathogenic bacteria.

In this sense, phytochemicals have been proposed as an excellent alternative to counteract the virulence of pathogenic bacteria.

19.8 Phenolic Compounds

Phenolic compounds are secondary metabolites found in plants characterized by the presence of one or more aromatic rings with at least one free hydroxyl group. This big group includes simple phenols, phenolic acids, hydroxycinnamic acid derivatives, coumarins, and flavonoids (Oliveira et al. 2017). Phenolic compounds exert a wide range of biological activities and a high potential as antioxidants, anti-inflammatories, anti-allergics, anti-carcinogenics, antimicrobials, anti-QS, among others

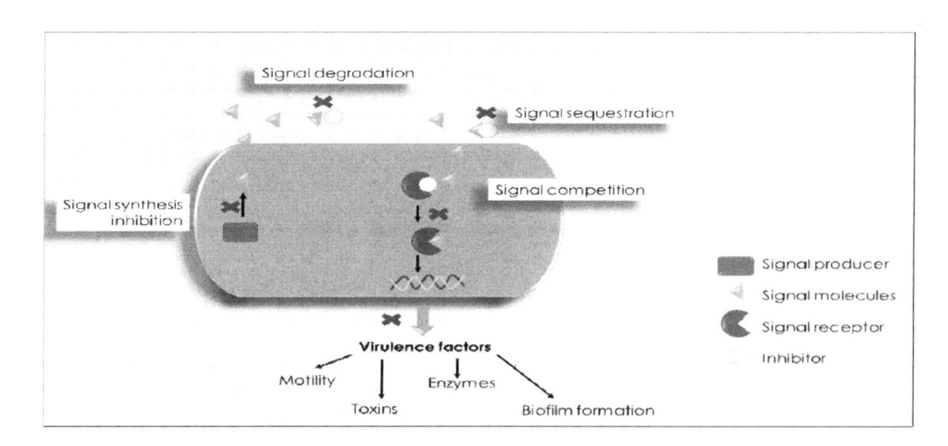

FIGURE 19.5 Potential mode of action of phytochemicals against QS of pathogenic bacteria. Phytochemicals can inhibit the production of QS signal molecules or their reception by degrading, sequestration, or imitating the QS signal molecules.

(Heleno et al. 2015, Truchado et al. 2015, Martins et al. 2016). Phenolics are the compounds most associated with anti-QS effects and anti-virulence factors compared to other phytochemicals (Oliveira et al. 2016). Several studies have suggested that phenolic compounds disturb the QS system, mainly of Gram-negative bacteria, and as a consequence, many virulence factors regulated by this system could be affected such as the motility, biofilm formation, toxin production, and enzyme activity.

Phenolic and cinnamic acids have been proposed as anti-QS in pathogenic bacteria. Vanillic acid, found as the main active constituent of *Actinidia deliciosa*, significantly affected in a dose-dependent manner biofilm and virulence factors regulated by QS-regulated in *Serratia marcescens*. Furthermore, vanillic acid increased the survival of *Caenorhabditis elegans* upon *S. marcescens* infection and modulated the expression of proteins involved in flagellin and fatty acid production (Sethupathy et al. 2017). As an anti-quorum strategy, the inhibition of acyl-homoserine lactones (AHL) production may attenuate the biofilm formation of *Pseudomonas* spp. With this in mind, ρ-coumaric and gallic acids at 120 and 240 μmol/L were used to inhibit the 3-oxo-C6-homoserinelactone (HSL) synthesis in *P. fluorescens* KM120. At these doses, phenolic compounds decreased the expression of *flgA* and diminished the rate of bacterial colonization on the stainless steel surface (Myszka et al. 2016a).

Joshi et al. (2016) reported that cinnamic acid and salicylic acid interfere with the QS system of *Pectobacterium aroidearum* and *P. carotovorum*, two phytopathogens. As a consequence, the expression of some virulence factors was also affected. In addition, the exposition of these compounds inhibited the expression of QS genes *expI*, *expR*, *PC11442* (*luxR* transcriptional regulator), and *luxS* (a component of the AI-2 system); therefore, the amount of AHLs was reduced, and other virulence genes (*pecS*, *pel*, *peh*, and *yheO*) controlled by QS were also down-regulated. This study proposed that the virulence of *Pectobacterium*, the causal agent of soft rot in many crops, could be treated with cinnamic and salicylic acids by the inhibition of production and accumulation of AHLs.

Methyl gallate, another phenolic acid, affected the QS system of the biosensor strain *Chromobacterium violaceum* affecting the synthesis and activity of AHLs. Also, this compound affected QS-associated virulence factors of *P. aeruginosa* PAO1 such as biofilm, motility, proteolytic, elastase, pyocyanin, and rhamnolipid activity. The possible molecular mechanism of this could be due to the downregulation of the expression of the autoinducer synthase (*lasI* and *rhlI*) and the receptor (*lasR* and *rhlR*) genes. Methyl gallate could be used as a potent anti-virulence agent to reduce the pathogenicity of *P. aeruginosa* and its infections (Hossain et al. 2017). In another study, it was observed that rosmarinic acid could be bound to the QS regulator RhlR of *P. aeruginosa* PAO1 and competed with the bacterial ligand *N*-butanoyl-homoserine lactone (C4-HSL). On the contrary, this compound stimulated the QS-dependent gene expression and increased the biofilm formation and production of pyocyanin and elastase (Corral-Lugo et al. 2016).

Flavonoids are one of the most active phenolic compounds with a wide variety of biological and antimicrobial properties. The flavonoids commonly found in citrus plants, quercetin, kaempferol, apigenin, and naringenin have been known to interfere with the QS system being effective to reduce AI-2-mediated bioluminescence production in the reporter strains *Vibrio harveyi* BB886 and MM32. The inhibition of AI-2 may influence biofilm formation; therefore, it was observed that all these compounds inhibited *V. harveyi* BB120 and *E. coli* O157:H7 biofilm formation in a dose-dependent manner (Vikram et al. 2010). Also, the flavonoid phloretin down-regulated genes of AI-2 synthesis and reduced *E. coli* O157:H7 biofilm formation (Lee et al. 2011).

Similarly, the flavonoid quercetin showed to act as a competitive inhibitor, since it can bind rigidly to LasR receptor protein as it was observed by molecular docking tests. The molecular dynamics simulation predicted that QS activity of this flavonoid occurs through the conformational changes between the receptor-quercetin complex. Moreover, quercetin at a range of 40–80 μg/mL reduced QS-dependent virulence factors such as violacein production of *C. violaceum*, extracellular polymeric substances amount, and biofilm biomass of *Klebsiella pneumoniae*, *P. aeruginosa*, and *Yersinia enterocolitica* (Gopu et al. 2015). Another study supports that quercetin had anti-QS activities (Ouyang et al. 2016). This flavonoid at 16 μg/mL inhibited the gene expression levels of *lasI*, *lasR*, *rhlI*, and *rhlR*, reducing the QS-regulated virulence factors of *P. aeruginosa* including pyocyanin, protease, elastase, and biofilm formation. In the other hand, naringenin, a flavonoid found in citrus fruits, reduced the production of 3-oxo-C12-HSL and C4-HSL by the reduction in QS-controlled genes such as *lasI* and *rhlI*, respectively (Vandeputte et al. 2011).

Coumarin is a potent QS inhibitor with anti-virulence efficacy against pathogenic bacteria with a broad spectrum of action. Gutiérrez-Barranquero et al. (2015) showed that coumarin at 100 μg inhibited the pigmentation of the reporter strains *S. marcescens* SP19 and *C. violaceum* CV026, suggesting that it could be blocking short and medium chain AHLs. Also, coumarin at 100 μg and 10 μM of exogenous AHLs achieved the higher pigmentation inhibition of *A. tumefaciens* NTL4, a reporter strain used to evaluate the interference with long chain AHLs. Besides, coumarin affected virulence factors under the QS control; the phenolic compound at 1.36 mM inhibited biofilm formation, reduced swarming motility, and inhibited phenazine production in a concentration-dependent manner. This could be associated to the decrement of the *pqsA* (*Pseudomonas* quinolone signal) that interfaces with AHL systems and *rhlI* genes expression caused by coumarin at 1.36 mM.

Stilbenes and hydrolyzable tannins such as resveratrol and ellagic acid, respectively, have demonstrated great potential as antimicrobials. It has been observed that the concentration of AHL of *Y. enterocolitica* and *P. carotovorum* incubated for 24 h has been considerably reduced with the use of resveratrol (20 μg/mL) and ellagic acid (4 μg/mL) (Truchado et al. 2012). The authors proposed that these phytochemicals interfered with the QS system due to a degradation-transformation of AHLs and the inhibition of AHL synthesis. Specifically in *Y. enterocoloitica*, 3-oxo-C6-HSL was reduced 85% by resveratrol and 44% by ellagic acid; while C6-HSL was reduced by 61% and 45%, respectively. The amount of 3-ox-C6-HSL and C6-HSL in *P. carotovorum* was reduced by approximately 50% by both compounds (Truchado et al. 2012). Also, these authors indicated that ellagic acid at 4 μg/mL and resveratrol at 20 μg/mL inhibited violacein production of *C. violaceum* by 79% and 75%, respectively.

Complex tannins such as punicalagin were evaluated against virulence factors and QS-related genes of *Salmonella*. This flavonoid at sub-inhibitory doses significantly decreased bacterial motility, which is in correspondence to the downregulation of the motility-related genes (*fliA*, *fliY*, *fljB*, *flhC*, and *fimD*) in quantitative reverse transcription. Also, punicalagin inhibited the production of violacein by *C. violaceum*, repressed the expression of QS-related genes (*sdiA* and *srgE*), and reduced *Salmonella* invasion of colonic cells (Li et al. 2014).

Several studies demonstrated the positive impact of plant extracts rich in phenolic compounds on anti-QS and anti-virulence activities. The extract of *Lagerstroemia speciosa* at 10 mg/mL significantly reduced QS-regulated virulence factors such as LasA protease, LasB elastase, and pyoverdin production of *P. aeruginosa* PAO1. In addition, the plant extract inhibits 83% of *P. aeruginosa* biofilm (Singh et al. 2012). Other study reported the efficacy of mango leaf extract on QS-regulated virulence factors in Gram-negative pathogens. The extract exerted a promising dose-dependent interference with violacein production of *C. violaceum*, which indicated the inhibition of QS. Also, the addition of mango leaf extract at 800 µg/mL achieved a reduction in swarming motility, exopolysaccharide production, biofilm formation, and elastase, total protease, and chitinase enzymes of *P. aeruginosa* PAO1 regulated by QS. The authors also demonstrated the efficacy of mango leaf extract to reduce mortality of *C. elegans* pre-infected with *P. aeruginosa* PAO1 (Husain et al. 2017). Similarly, the extract of *Rubus rosaefolius* showed antimicrobial and anti-QS activities due to the presence of phenolic compounds. At sub-inhibitory concentrations, the phenolic extract inhibited the QS assessed by the violacein production of *C. violaceum* and reduced the swarming and biofilm formation of *S. marcescens* and *Aeromonas hydrophila* (Oliveira et al. 2016).

Also, phenolic compounds have been proposed to act as inhibitors of Gram-positive QS systems and their virulence factors. The polyphenols extracted from Mexican avocado leaves cultivar Maria Elena decreased the expression level of *agrA* and *rnaIII* genes involved in the QS system and virulence factors expression of *S. aureus*, compared with the 16S rDNA gene (García-Moreno et al. 2017). Another study demonstrated that the stilbene resveratrol inhibited the expression of α-hemolysin (Hla) by the downregulation of *hla*, a gene that encodes for HLA, and *RNAIII*, the effector molecule of the *agr* system in *S. aureus*. Moreover, it was observed that this compound attenuated the bacterial virulence observing a relief in A549 cellular injury and protection in mice from *S. aureus* pneumonia (Tang et al. 2019). On the other hand, it was demonstrated that luteolin significantly inhibited α-toxin production of the sensitive and methicillin-resistant *S. aureus*. Also, the transcriptional levels of *agr* (accessory gene regulator) were inhibited by luteolin. It is known that the expression of α-toxin is controlled by several global regulatory systems such as Agr QS (Khan et al. 2015).

19.9 Alkaloids

Alkaloids are part of secondary plant metabolites acting as defense compounds against pathogens and predators. They are a class of organic compounds that contain basic nitrogen atoms, carbon, and hydrogen, and some of them may also contain oxygen and sulfur. Alkaloids are a focus of research due to their biological action as pharmacological and antimicrobial drugs (Cushnie et al. 2014, Matsuura and Fett-Neto 2017). However, little exploration has been conducted in the field of QS quenching using alkaloids.

Some studies have reported the efficacy of alkaloids to interfere with Gram-negative and Gram-positive QS systems. The alkaloid solenopsin A, a venom alkaloid found in the fire ant, disrupted QS signaling with the consequent control of virulence factors and biofilm formation of *P. aeruginosa*. The alkaloid showed activity on the *rhl* QS signaling system due to a competition of C4-HSL. Moreover, the presence of solenopsin decreased the amount of pyocyanin (>50%), elastase B production (30% decrease by 50 µmol/L solenopsin), and biofilm formation in *P. aeruginosa* via disruption of *rhl* QS system. The transcription of some QS-regulated genes *lasB*, *lasI*, and *rhl* were decreased ~2-fold by the alkaloid solenopsin (Park et al. 2008).

Aswathanarayan and Vittal (2018) reported that berberine (1.6 mg/mL), an isoquinoline alkaloid with pharmacological properties, inhibited the QS system observing a decrement of 62.67% in violacein production of *C. violaceum*. Furthermore, berberine at sub-inhibitory concentrations affected biofilm biomass by 71.70%, prevented biofilm formation, and inactivated biofilm maturation in *P. aeruginosa* PAO1, while in *S.* Typhimurium, berberine reduced biofilm formation by 31.20%, prevented invasion and adhesion by 55.37% in colonic cells and 54.68% in HT29, and reduced *in vivo* virulence in *C. elegans* by 65.38%. Also, the authors demonstrated by *in silico* studies that this alkaloid could interact with LasR and RhlR, the QS signal receptors. In another study, tomatidine, a steroidal alkaloid found in the skins and leaves of tomatoes, showed to be able to reduce the virulence of *S. aureus* regulated by the *agr* QS system. The addition of this compound reduced the hemolytic activity of *S. aureus* and reduced in a dose-depended manner the expression of *hld*, a gene that encodes for the δ-toxin, and a fragment of RNAIII, which is the effector of the *agr* system. Also, this compound at 12.8 mg/L repressed the expression of *hld, geh, nunc, plc*, and *splC*, *agr*-upregulated genes (Mitchell et al. 2011).

19.10 Terpenes and Terpenoids

Terpenes are naturally occurring hydrocarbons that are secondary plant metabolites consisting of the simple hydrocarbon molecule isoprene, whereas terpenoid refers to a terpene that has been modified by oxygen addition (Zwenger and Basu 2008). These compounds have shown to reduce the production of virulence factors regulated by QS such as motility, EPS production, biofilm formation, toxin, and pigment synthesis, among others. Recently, carvacrol, the main terpenoid of oregano essential oil, has been designated as potential QS inhibitor. Tapia-Rodriguez et al. (2017) reported that carvacrol affects QS and QS-regulated virulence factors of *P. aeruginosa*. Specifically, carvacrol reduced the expression of *lasR* and affected the synthesis of AHL, suggesting a post-transcriptional affectation of the LasI synthase; also, pyocyanin and biofilm formation were affected.

Gutierrez-Pacheco et al. (2018) reported that carvacrol (0.66 mM) reduced motility (48.2 mm reduction), EPS synthesis, and biofilm formation (1.1 log CFU/cm^2) of *P. carotovorum*. Similarly, Joshi et al. (2016) evaluated the effect of carvacrol and eugenol (250µM) on *P. carotovorum* subsp. *brasilense*, and both terpenoids interfered with the QS, resulting in an inhibition of *expI* and *expR*, biofilm formation, gene expression of plant cell wall degrading enzymes (PCWDEs), and the infection of plant tissues. On the other hand, it has been reported that carvacrol, citral, carvone, citral, geraniol, thymol, and eugenol inhibited the motility, EPS synthesis, and biofilm formation of the *P. carotovorum* and *P. fluorescens* at 0.1 mg/mL (Zhang et al. 2018b). Thymol also has shown to be effective in reducing the virulence factors Hla and the enterotoxins SEA and SEB of *S. aureus* and therefore, the hemolytic and TNF-inducing activities (Qiu et al. 2010).

Limonene, α-pinene, terpinen-4-ol, and linalool (main components of lemon, juniper, marjoram, and clary sage essential oils, respectively) significantly reduced the biofilm formation of *Bacillus cereus* and *E. coli* at concentrations below the minimal inhibitory concentrations. Among these, terpinen-4-ol showed higher QS inhibitory activity, reducing violacein production by the biosensor strain *C. violaceum* (Kerekes et al. 2013). Similarly, Khan et al. (2009) reported that clove essential oil, rich in terpenes such as eugenol, reduced 78.4% the QS-controlled violacein production of *C. violaceum* and up to 78% the swarming motility of *P. aeruginosa*.

Niu et al. (2006) reported that cinnamaldehyde (200 µmol) reduced up to 70% the LuxR transcription. In addition, this compound at concentrations of 60 and 100 µmol reduced by 55% and 60% the bioluminescence of the *V. harveyi* BB886 (mediated by 3-hydroxy-C4-AHL) and *V. harveyi* BB1170 (mediated by the autoinducer-2) respectively. Also, the synthesis of C6-AHL in *P. carotovorum* was affected by the presence of cinnamaldehyde at 0.05 mg/mL. In another study, methyl eugenol, a terpenoid of clove essential oil, showed anti-QS activity inhibiting violacein production in *C. violaceum*; also, this compound reduced by 87% and 93% the biofilm formation and bioluminescence of *P. aeruginosa* and *V. harveyi*, respectively (Sybiya Vasantha Packiavathy et al. 2012).

Diterpenes such as phytol showed anti-virulence activity, reducing QS-regulated virulence factors. Specifically, this compound reduced the biofilm formation of *P. aeruginosa* by 74%–84%, twitching and flagella motility, as well as pyocyanin synthesis (Pejin et al. 2015). Similarly, casbenediterpene, a diterpenoid from plants of *Euphorbiaceae* family, significantly reduced the biofilm formation of *Streptococcus mutans* at 250 µg/mL (Sá et al. 2012). On the other hand, Rasamiravaka et al. (2017) reported that the main terpenoids of *Platostoma rotundifolium*, inhibited QS-regulated and -regulatory genes expression in *las* and *rhl* systems of *P. aeruginosa*. These terpenoids disrupted the formation of biofilms at concentrations down to 12.5, 50 and 50 µM for cassipourol, β-sitosterol, and α-amyrin, respectively, and also reduced the production of total exopolysaccharides.

It has been reported that trans-cinnamaldehyde inhibited AHL synthesis by RhlI of *P. aeruginosa* with the concomitant reduction of pyocyanin production. Molecular docking analysis suggested that this compound binds LasI and EsaI, AHL synthases of *P. aeruginosa* and *Pantoea stewartti*, respectively. Trans-cinnamaldehyde interacted with their substrate binding sites, specifically with Arg30 by hydrogen bonding and with Phe105 by hydrophobic and pi-pi interactions in the hydrophobic tunnel of acyl acyl-carrier protein (ACP) union site in LasI (Chang et al. 2014).

19.11 Sulfur Compounds

Sulfur-containing compounds have shown anti-QS activity against Gram-positive and Gram-negative pathogens. Iberin, an isothiocyanate, produced by many members of the *Brassicaceae* family, inhibited QS of *P. aeruginosa* interfering with the LasI/R and RhlI/R systems. Iberin (32 µg/mL) down-regulated genes under QS control such as *rhlA* and *rhlB* (involved in rhamnolipid synthesis), *piv* (that codify to protease IV), *lecA* (the cytotoxic galactophiliclectin), *lasA*, and *lasB* (involved in elastase production), *cbpD* and *chiC* (encoding a chitin binding protein precursor and chitinase), *phzM* (encoding a putative phenazine-specific methyltransferase), *phzS* (encoding a flavin-containing monoxygenase, involved in pyocyanin synthesis), *phzC-phzG* operon, and *phzAB* (encoding phenazine biosynthesis proteins). All these genes contribute significantly to the virulence and pathogenicity of *P. aeruginosa* (Jakobsen et al. 2012).

Other sulfur-containing compounds such as sulforaphane and erucin, found in broccoli and other crucifers, showed anti-QS activity toward *P. aeruginosa*. It was evidenced that these two isothiocyanates effectively binds the transcriptional activator LasR, resulting in an inhibition of the QS activation with the concomitant reduction of biofilm formation and pyocyanin synthesis at concentrations of 50 µM and above.

Several mechanisms have been proposed to explain the anti-QS activity of phytochemicals; most of these agree that the structural similarities of these compounds with the substrates (used to synthesize QS signals) such as SAM and acyl-ACP in the case of the Gram-negative AHL type QS and with the signal molecule, causing an interruption of the cell-cell communication (Gutierrez-Pacheco et al. 2019). In addition, it is important to highlight that some phytochemicals also showed structural similarities with known QS inhibitors such as the well-documented furanones from *Deliseapulchra* and the synthetic inhibitor N-3-oxocyclohex-1-enyl octanamide (J8-C8) (Figure 19.6). Mainly, J8-C8 is an AHL synthase inhibitor; this binds to TofI (AHL synthase of the plant pathogen *Burkholderia glumae*), occupying the binding site of the acyl chain of the acyl-ACP (Chung et al. 2011), whereas halogenated furanones from *D. pulchra* caused destabilization of LuxR type protein. On the other hand, compounds like ambuic acid inhibited the QS of Gram-positive bacteria (Nakayama et al. 2009). Even when the structures of substrates/inhibitors and phytochemicals are not equal, their chemical characteristics (polarity, size, among others) could allow them to react with active protein sites or cause structure conformation changes, abolishing the binding of protein ligands (Gutierrez-Pacheco et al. 2019).

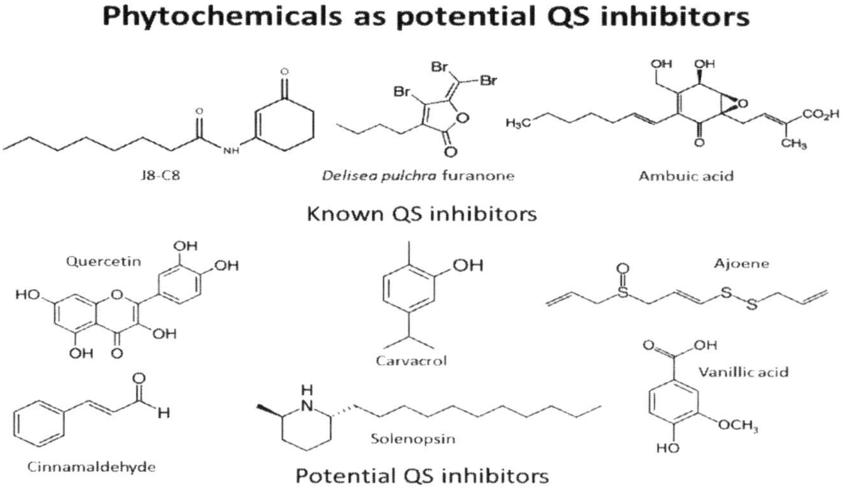

FIGURE 19.6 Structure comparison of known QS inhibitors and phytochemicals as potential anti-QS compounds.

Table 19.1 gives an overview of some studies reporting the effectiveness of phytochemicals to inhibit the QS system of some pathogens. It is important to highlight that most of the studies are conducted with biosensors bacteria because they produce some metabolites that can be measured easily compared to most food pathogens. The chemical analysis and quantification of signal molecules in some bacteria are difficult to evaluate especially in Gram-positive; hence, there is a need for new methodologies to evaluate the QS and their signal molecules. However, the use of bacterial biosensors could be an approach of the phytochemicals potential since they could have a similar effect in food bacteria with similar QS system. Although there are various studies evaluating phytochemicals as QS quenching, there is a lack of information about their mechanism of action.

Most of the studies hypothesize that the possible mechanisms by which phytochemicals exert an anti-QS activity is throughout an inhibition of signal synthesis and reception, gene expression, or a direct interaction with the signal molecule (Gutierrez-Pacheco et al. 2019). Their structural characteristics such as their size, amphipathicity, and hydrophobicity, could allow phytochemicals to interact throughout hydrogen bonding and hydrophobic interactions with the QS proteins; in fact, the recent studies in the field have hypothesized as the primary mechanism the possible affectation of these proteins. Tapia-Rodriguez et al. (2017) reported that the possible anti-QS mechanism of action of carvacrol against *P. aeruginosa* is throughout an affectation of *lasR* gene expression and a post-translational affectation of LasI, affecting AHL synthesis. In addition, these authors determined by molecular docking analysis that this phytochemical interacts with aminoacids of the active site pocket of LasI and within the binding pocket of LasR.

Similarly, Joshi et al. (2016) reported that eugenol and carvacrol inhibit the QS of the plant pathogen *Pectobacterium carotovorum* subsp. *brasilense* by a possible affectation of QS proteins ExpI/ExpR. These compounds inhibited AHL synthesis and gene expression and formed Π-Π interactions with Phe123 and Phe102, respectively, in the active site of a structural model protein. It has been reported that some phytochemicals like phenolic compounds and EO components could mimic AHLs and be sensed by the LuxR proteins affecting gene translation of virulence factors (Tapia-Rodriguez et al. 2017). Besides the advances in the study of the anti-QS activity of phytochemicals, it is necessary to evaluate the direct interaction of the inhibitors and their target, establish a more specific relation between the decreasing in the QS and the attenuation of bacterial virulence, and to determine the anti-QS activity of other phytochemical compounds.

19.12 Methods for Screening the Anti-QS Activity of Phytochemicals

The increase in the cases of bacterial infections caused by pathogens as well as the resistance to antibiotics has led the researchers to look for compounds that block the QS and therefore the synthesis of QS-regulated virulence factors (Vattem et al. 2007a). It has been reported that some synthetic compounds, as well as phytochemicals present in plants, herbs, and spices, showed anti-QS activity (Adonizio et al. 2006, Koh et al. 2013). Considering the variety of plants in nature, there are still many compounds that have the potential to act as QS inhibitors. There are different methods to determine the anti-QS activity of phytochemicals, biosensor strains being the most used.

Biosensor strains include some bacteria that have a transcriptional response regulator (*luxR* homolog) coupled with a reporter gene allowing a readily observable phenotype such as bioluminescence, green fluorescence protein, pigment synthesis, and β-galactosidase activity in response to QS activation (Steindler and Venturi 2007, Thornhill and McLean 2018). Principal of the assays using biosensor strains are performed to evaluate if phytochemicals reduce the synthesis of signal molecules. A commonly used biosensor is based on *C. violaceum*, a Gram-negative bacteria that produces a purple pigment called violacein, regulated via the CviI/CviR QS system, which produces and responds to C6-AHL (Thornhill and McLean 2018). Most of the studies evaluate the anti-QS activity of phytochemicals growing *C. violaceum* in the presence of the test compounds and

TABLE 19.1

Overview of Various Class of Phytochemicals as Quorum Sensing Inhibitors in Bacterial Pathogens

Phytochemical	Dose	Bacteria	Biological Action	Interference with QS	References
Phenolic Compounds					
Vanillic acid 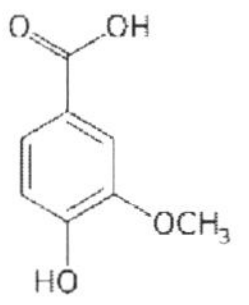	0.125–0.5 mg/mL	*Serratia marcescens*	Inhibition of protease, biofilm, hemolysin, prodigiosin, and lipase Modulation of the expression of proteins involved in S-layers, histidine, flagellin, and fatty acid production Increment of the survival of *Caenorhabditis elegans* upon the infection	Not specified	Sethupathy et al. (2017)
Cinnamic acid 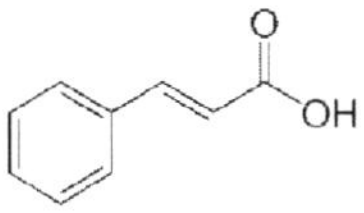	0.1 mg/mL	*Pseudomonas aeruginosa*	Inhibition of pyocyanin, protease, LasB, elastase, chitinase, and biofilm formation	*In silico* competitive inhibition for LasR and RhlR toward the ligand binding domain of the transcriptional activators	Rajkumari et al. (2018)
Chlorogenic acid 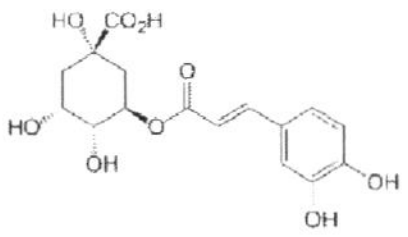	0.04–2.56 mg/mL	*Chromobacterium violaceum* *P. aeruginosa*	Inhibition of biofilm formation, swarming motility, and chitinolytic activity and violacein production Inhibition of biofilm, formation, swarming motility, protease, and elastase activities and rhamnolipid and pyocyanin production	Downregulation of QS relate genes *pqsA, pqsR, lasI, lasR, rhlI, rhlR* Interaction through hydrogen bonds with QS receptors in computational modeling	Wang et al. (2019)
Methyl gallate 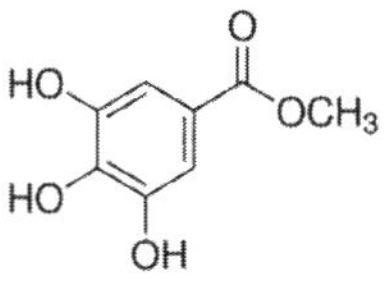	0.5 mg/mL	*C. violaceum*	Reduction of violacein production	Not specified	Tan et al. (2015)
Caffeic acid 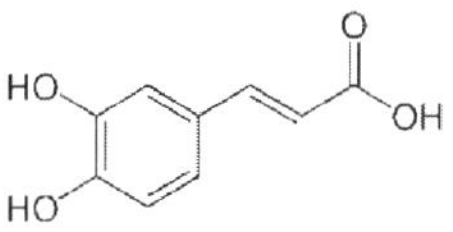 Ferulic acid 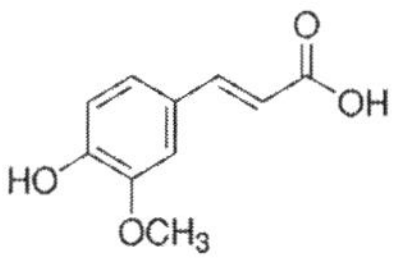	4 mM	*P. aeruginosa*	Reduction of pyocyanin, biofilm formation, and sw motility	Not specified	Ugurlu et al. (2016)
Catechin 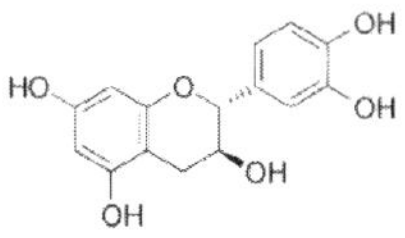	0.25–4 mM	*C. violaceum* CV026 *P. aeruginosa* PAO1	Inhibition of violacein production Inhibition of pyocyanin and elastase production, biofilm formation	Downregulation of QS relate genes expression	Vandeputte et al. (2010)
Quercetin 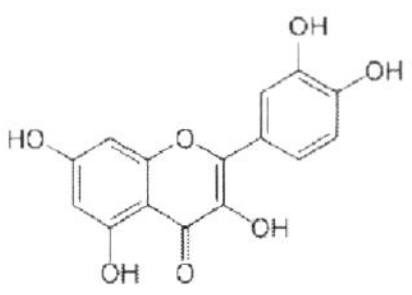	0.016 mg/mL	*P. aeruginosa* PAO1	Inhibition of biofilm formation and pyocyanin, protease, and elastase production	Reduction of the expression levels of *lasI, lasR, rhlI*, and *rhlR*	Ouyang et al. (2016)

(Continued)

TABLE 19.1 (*Continued*)

Overview of Various Class of Phytochemicals as Quorum Sensing Inhibitors in Bacterial Pathogens

Phytochemical	Dose	Bacteria	Biological Action	Interference with QS	References
Phloretin Naringenin Apigenin Quercetin	100 µM	*P. aeruginosa*	Reduction of pyocyanin production and swarming motility	Antagonism of the autoinducer binding receptors, LasR, and RhlR	Paczkowski et al. (2017)
Epigallocatechingallate 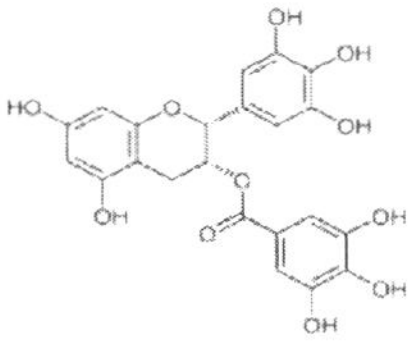	2–16 mM	*P. aeruginosa*	Reduction of biofilm, pyocyanin, elastase, and protease	Not specified	Tran et al. (2013)
7-hydroxycoumarin 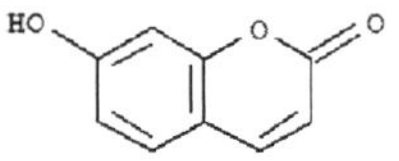	2 mg/mL	*C. violaceum*	Reduction of violacein pigment	Not specified	Monte et al. (2014)
Morin 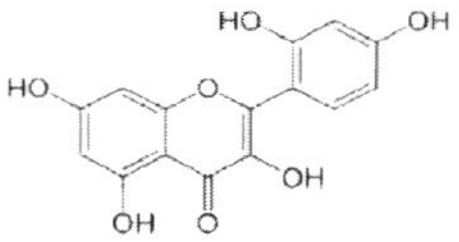	0.05–0.4 mg/mL	*Staphylococcus aureus*	Inhibition of biofilm, motility, spreading, and EPS production	Interaction through hydrogen bonds with global regulatory protein (SarA)	Chemmugil et al. (2019)
Curcumin 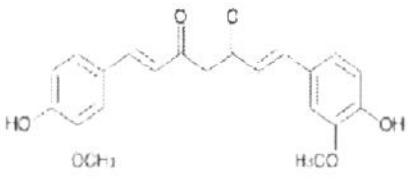	0.01 mg/mL	*Pectobacterium wasabiae SCC3193,* *P. carotovorum subsp. Carotovorum Pcc21* *P. carotovorum subsp. Carotovorum Pcc*	Inhibition of pectatelyase and polygalacturonase enzymes production and swimming and swarming motility	Inhibition of QS signal molecules, 3-oxo-C6-HSL, and 3-oxo-C8-HSL synthesis	Sivaranjani et al. (2016)
Cinnamaldehyde 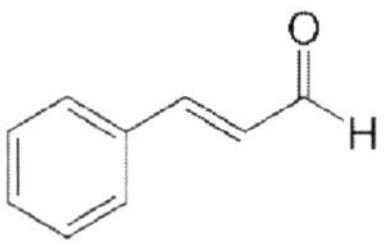	0.05–0.1 mg/mL	*C. violaceum* CV026 *P. aeruginosa*	Reduction of violacein pigment Reduction of biofilm, motility, and down-regulation of gene expression of *csgAB* (curli formation)	Not specified	Kim et al. (2015)
Phillyrin	0.25 mg/mL	*P. aeruginosa* PAO1 *C. violaceum*	Reduction of swimming motility, rhamnolipid, pyocyanin, and elastase production Inhibition of violacein pigment	Not specified	Zhou et al. (2019b)
Zingerone 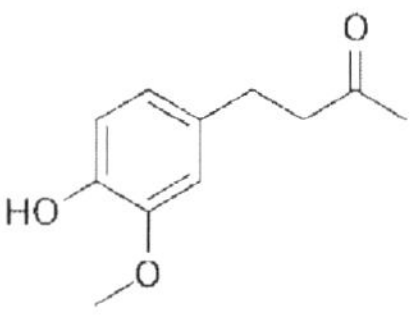	0.5 mg/mL	*C. violaceum* *P. aeruginosa*	Inhibition of pyocyanin and violacein production	Not specified	Kumar et al. (2014)
Cranberry extract rich in proanthocyanidins	0.2 mg/mL	*P. aeruginosa*	Protection of *D. melanogaster* from the infection Inhibition of staphylolytic, elastolytic, and alkaline proteolytic activities	Reduction of AHLs levels Inhibition of LasI/RhlI synthases and QS transcriptional regulators LasR/RhlR genes Binding to QS transcriptional regulator through their ligand binding sites in molecular docking	Maisuria et al. (2016)

(*Continued*)

TABLE 19.1 (*Continued*)

Overview of Various Class of Phytochemicals as Quorum Sensing Inhibitors in Bacterial Pathogens

Phytochemical	Dose	Bacteria	Biological Action	Interference with QS	References
Alkaloids					
Caffeine 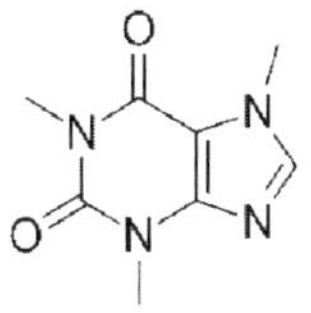	0.1–1.0 mg/mL	*C. violaceum* CV026 *P. aeruginosa* PA01	Inhibition of violacein production Inhibition of swarming motility	Inhibition of acyl homoserine lactone production	Norizan et al. (2013)
Hordenine 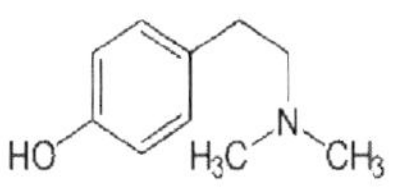	0.25, 0.5 and 0.1 mg/mL	*S. marcenscens*	Inhibition of biofilm formation and destruction of biofilm architecture Improvement the susceptibility of the preformed biofilms to antibiotic Change of membrane permeability Inhibition in protease activity and extracellular polysaccharides, prodigiosin, and hemolysin production Change in bacterial metabolism Attenuation of tomato plant infection model	Inhibition of acyl-homoserine lactones production Downregulation in the expressions of *fimA*, *fimC*, *flhD*, *bmsA*, *pigA*, *pigB*, *sodB*, and *zwf*	Zhou et al. (2019a)
Capsaicin	16 μg/mL	*S. aureus*	Reduction of α-toxin production	Reduction of relative expression levels of *hla* and *RNAIII* involved in QS *agr* system	Qiu et al. (2012)
Reserpine	0.1–0.8 mg/mL	*P. aeruginosa* PA01	Reduced biofilm formation, cell motility, extracellular polymer substances, pyocyanin, rhamnolipid, protease, and elastase activity	Reduced QS-controlled gene expression *lasRlasI, rhlR, rhlI*, and *mvfR* Interaction with synthase LasI and QS transcriptional regulators LasR/MvfR in molecular docking analysis	Parai et al. (2018)
Terpenes					
Carvacrol 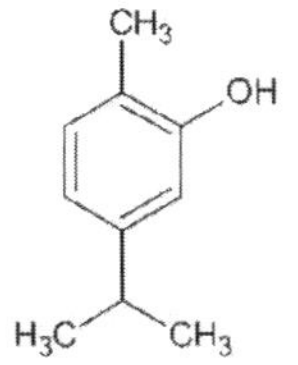	<0.5 mM	*C. violaceum* *Salmonella enterica* subsp. Typhimurium *Staphylococcus aureus* 0074 *C. violaceum* *P. aeruginosa* *Pectobacterium carotovorum* subsp. *brasilense*	Inhibition of biofilm formation Reduction in violacein production and chitinase activity Reduction in biofilm formation Inhibition of biofilm formation, gene expression of plant cell wall degrading enzymes, and reduction in the infection of plant tissues	Not specified Reduction of *cviI* gene expression (coding AHL synthase) Reduction of the expression of *lasR* and inhibition of AHL synthesis Downregulation of *expI* and *expr* genes	Burt et al. (2014), Tapia-Rodriguez et al. (2017), Joshi et al. (2016)

(*Continued*)

TABLE 19.1 (*Continued*)

Overview of Various Class of Phytochemicals as Quorum Sensing Inhibitors in Bacterial Pathogens

Phytochemical	Dose	Bacteria	Biological Action	Interference with QS	References
Thymol	<0.004 mL/mL	*Pseudomonas fluorescens* KM121	Suppression of bacterial motility and reduction of the mRNA level of flagella gene Inhibition of biofilm formation	Reduction of AHLs production	Myszka et al. (2016b)
Eugenol	0.012–0.096 mg/mL	*P. aeruginosa* PAO1 *C. violaceum* CV026 *S. aureus*	Inhibition of QS-regulated production of elastase, protease, chitinase, pyocyanin, and exopolysaccharide Inhibition of violacein production Reduction of biofilm biomass	Not specified	Al-Shabib et al. (2017)
Sulphur-Containing Compounds					
Sulforaphane	100 μM	*P. aeruginosa* *E. coli DH5a-lacZ*	Inhibited biofilm formation and pyocyanin production	Bind to LasR and inhibition of its activation	Ganin et al. (2013)
Allylisothiocyanate	0.005 mg/mL	*C. violaceum*	Inhibited violacein production	Not specified	Borges et al. (2014)

measuring the reduction of violacein by different methods such as inhibition halos or extracting them with solvents like butanol, ethanol, and acetone, for their spectrophotometric quantification at 585 nm (Alvarez et al. 2014).

C. violaceum CV026 is a biosensor that lacks the *luxI* gene and therefore is unable to produce AHL, but it responds to exogenous C6-AHL and C4-AHL. This strain is used to determine if the phytochemical inhibits AHL synthesis. Briefly, a bacterium (AHL producer) or a *C. violaceum* AHL over-producer strain (such as CV31532) is grown in the presence of the compound, and then, the AHLs are extracted and put in contact with CV026 cultures. The magnitude of pigment induction after 24 h was related to a reduction in AHL synthesis (Zhang et al. 2018a). Otherwise, if the compound (at non-inhibitory concentrations) is put on a plate where an AHL-producing strain is planted as well as CV026, the response is also possibly related to the interference in AHL recognition. These bacteria mainly recognize short-length AHL such as C6-AHL and C4-AHL, and for that reason, these experiments are complemented with the use of other biosensors (Thornhill and McLean 2018).

A. tumefasciens strains can recognize a broader range of AHLs such as C6-C12-AHL, although C14-C18-AHL have also been detected. *A. tumefasciens* A136 is an avirulent derivative of this microbe that lacks the Ti plasmid, and hence cannot synthesize AHL (McLean et al. 2004). This strain has a *traI-lacZ* fusion, does not contain a *traI* gene, and contains a plasmid that over-expresses TraR (pCF218) and a second plasmid carrying TraR-regulated promoter (*traI-lacZ*; pCF372). Similarly, the NTL4 strain contains the reporter gene *lacZ*, fused to a QS-dependant promoter. The activation of this promoter by exogenous AHLs causes concomitant expression of β-galactosidase, producing a blue color by X-gal degradation (Thornhill and McLean 2018). Anti-QS compounds inhibit the QS-dependent promoter and subsequent *lacZ* expression, thus limiting X-gal hydrolysis and the appearance of blue color. For these experiments, the strain KYC55, a 3-oxo-C8-AHL over-producer is used as a positive control.

E. coli pJBA132 (*luxRPluxI::gfp*) and *P. aeruginosalasB-gfp* (*PlasB::gfp*) *and rhlA-gfp* (*PrhlA::gfp*) are biosensor strains and carry a *gfp* gene that codifies to the GFP protein. In the presence of exogenous AHL, LuxR positively affects the expression of the *luxI* promoter (*PluxI*), which then controls the expression of the *gfp* reporter gene. The QS-inhibitory effect of the potential QS inhibitors is determined by their capacity to reduce the GFP fluorescence emission at 515 nm (without affecting bacterial growth) attributed to a lower *gfp* expression caused by the low AHL concentration in the local environment (Andersen et al. 2001).

Other biosensor strains can be used to detect the anti-QS activity of phytochemicals, particularly affecting the AI-2 QS system. This is the case of *V. harveyi* BB170 (defective in the synthesis of AI-1 but capable of synthesizes basal levels of AI-2) and *V. harveyi* MM32 (not able to synthesizes AI-1 and AI-2). In the presence of AI-2 molecules, the expression of QS genes related to bioluminescence is activated. To determine the anti-QS activity of phytochemicals, the bacteria (AI-2 producer) grows in the presence of the compound, and after incubation, the cell-free supernatant (containing AI-2 molecules) is put in contact with the biosensors *V. harveyi* BB170 or *V. harveyi* MM32 and incubated. The reduction of bioluminescence (at 490 nm) is related to AI-2 synthesis inhibition and anti-QS activity (Taga and Xavier 2011).

There are more sophisticated methodologies to determine the anti-QS activity of phytochemicals such as the analysis of gene expression of specific QS genes after growth of the bacteria in the presence of the phytochemical. On the other hand, techniques such as thin-layer chromatography or even the detection and quantification of the signal molecules by liquid chromatography can be indicative of the anti-QS activity of phytochemicals (Thornhill and McLean 2018). On the other hand, microarray-based transcriptome analysis has been a useful tool because it gives information about the transcripts under QS regulation that are reduced or abolished in the presence of phytochemical compounds, and it is possible to associate their effect with alterations of genes implicated in specific QS-regulated virulence factors (Jakobsen et al. 2011). For example, using this technology, it was possible to determine that guava leaf extract significantly downregulated 816 genes of the *C. violaceum* genome, many of which are QS-regulated in this and other Gram-negative bacteria (Ghosh et al. 2014).

19.13 Patents on Phytochemicals as Quorum Sensing Inhibitors and Commercial Applications

The use of phytochemicals or natural extracts rich in these compounds with anti-QS activity also has been patented. Mathee et al. (2009) proposed the use of ellagitannins for the inhibition of QS, providing methods of treating bacterial infections in mammalians subjects. Gong et al. (2015) patented the extraction and purification of falcarindiol, a polyacetylene compound that showed anti-QS properties against *P. aeruginosa* affecting biofilm formation and virulence. Also, the method of preparation and application of QS inhibitors from camphor tree leaves has been reported by Ge et al. (2018). In the invention, the effect of camphor essential oil is tested against *C. violaceum* showing inhibition of QS-dependent violacein production. Also, this essential oil exerted antimicrobial activity against *E. coli, S. aureus*, and *P. aeruginosa*. Other patented plant extracts with anti-QS activity include *Carexdimorpholepis* extract (Cho et al. 2015) and Baktaegi tree extract (Kim and Kuk 2016). Despite the existence of patents, there are no natural products on the market destined for the inhibition of QS. Several companies commercialize natural extracts of plants, fungi, and algae rich in compounds that have been reported as QS inhibitors. However, the effectiveness of the extracts has not been evaluated, and therefore manufacturers do not make these statements.

19.14 Conclusions

The current knowledge about *QS* systems can allow giving new alternatives. Phytochemical compounds could be that alternative due to their multiple functions against infections of pathogenic bacteria; the *QS* system is the main factor in the presence of its virulence factors and its pathogenicity. Besides, these compounds have a wide variety of mechanisms to eliminate bacteria, which could decrease the likelihood that the bacteria will become resistant to these compounds. There are patents of phytochemical compounds as *QS* inhibitors, but there are no commercial products available which inhibit quorum sensing. For all this, attacking the system of quorum sensing using phytochemical compounds to control pathogenic bacteria is an alternative with a full future in the commercial sector due to their efficacy and the lack of this type of products.

REFERENCES

Abdelnour, Arturo, Staffan Arvidson, Tomas Bremell, C. Ryden, and Andrzej Tarkowski. 1993. The Accessory Gene Regulator (Agr) Controls *Staphylococcus aureus* Virulence in a Murine Arthritis Model. *Infection and Immunity* 61:3879–85.

Adonizio, Allison L., Kelsey Downum, Bradley C. Bennett, and Kalai Mathee. 2006. Anti-Quorum Sensing Activity of Medicinal Plants in Southern Florida. *Journal of Ethnopharmacology* 105:427–35.

Al-Shabib, Nasser Abdulatif, Fohad M. Husain, Iqbal Ahmad, and Mohammad Hassan Baig. 2017. Eugenol Inhibits Quorum Sensing and Biofilm of Toxigenic Mrsa Strains Isolated from Food Handlers Employed in Saudi Arabia. *Biotechnology & Biotechnological Equipment* 31:387–96.

Alvarez, Maria V., Luis A. Ortega-Ramirez, M. Melissa Gutierrez-Pacheco, et al. 2014. Oregano Essential Oil-Pectin Edible Films as Anti-Quorum Sensing and Food Ántimicrobial Agents. *Frontiers in Microbiology* 5:699.

Andersen, Jens Bo, Arne Heydorn, Morten Hentzer, et al. 2001. Gfp-Based *N*-Acyl Homoserine-Lactone Sensor Systems for Detection of Bacterial Communication. *Applied and Environmental Microbiology* 67:575–85.

Artola, B. Sánchez, and E. Palencia Herrejón. 2010. Infecciones Por Listeria. *Medicine-Programa de Formación Médica Continuada Acreditado* 10:3368–72.

Aswathanarayan, Jamuna Bai, and Ravishankar Rai Vittal. 2018. Inhibition of Biofilm Formation and Quorum Sensing Mediated Phenotypes by Berberine in *Pseudomonas aeruginosa* and *Salmonella* Typhimurium. *RSC Advances* 8:36133–41.

Barrios, Andrés F. González, Rongjun Zuo, Yoshifumi Hashimoto, Li Yang, William E. Bentley, and Thomas K. Wood. 2006. Autoinducer 2 Controls Biofilm Formation in *Escherichia coli* through a Novel Motility Quorum-Sensing Regulator (Mqsr, B3022). *Journal of Bacteriology* 188:305–16.

Bartlett, John G., David N. Gilbert, and Brad Spellberg. 2013. Seven Ways to Preserve the Miracle of Antibiotics. *Clinical Infectious Diseases* 56:1445–50.

Bernal-Mercado, Ariadna, Francisco Vazquez-Armenta, Melvin Tapia-Rodriguez, et al. 2018. Comparison of Single and Combined Use of Catechin, Protocatechuic, and Vanillic Acids as Antioxidant and Antibacterial Agents against Uropathogenic *Escherichia coli* at Planktonic and Biofilm Levels. *Molecules* 23:2813.

Borges, Anabela, Sofia Serra, Ana Cristina Abreu, Maria J. Saavedra, Antonio Salgado, and Manuel Simões. 2014. Evaluation of the Effects of Selected Phytochemicals on Quorum Sensing Inhibition and in Vitro Cytotoxicity. *Biofouling* 30:183–95.

Bouyahya, Abdelhakim, Nadia Dakka, Abdeslam Et-Touys, Jamal Abrini, and Youssef Bakri. 2017. Medicinal Plant Products Targeting Quorum Sensing for Combating Bacterial Infections. *Asian Pacific Journal of Tropical Medicine* 10:729–43.

Brown, Peter K., Charles M. Dozois, Cheryl A. Nickerson, Amy Zuppardo, Jackie Terlonge, and Roy Curtiss. 2001. Mlra, a Novel Regulator of Curli (Agf) and Extracellular Matrix Synthesis by *Escherichia coli* and *Salmonella enterica* Serovar Typhimurium. *Molecular Microbiology* 41:349–63.

Burt, S. A., V. T. Ojo-Fakunle, J. Woertman, and E. J. Veldhuizen. 2014. The Natural Antimicrobial Carvacrol Inhibits Quorum Sensing in *Chromobacterium violaceum* and Reduces Bacterial Biofilm Formation at Sub-Lethal Concentrations. *PLoS One* 9:e93414.

Camejo, Ana, Filipe Carvalho, Olga Reis, Elsa Leitão, Sandra Sousa, and Didier Cabanes. 2011. The Arsenal of Virulence Factors Deployed by *Listeria monocytogenes* to Promote Its Cell Infection Cycle. *Virulence* 2:379–94.

CDC, Centers for Disease Control and Prevention. 2018a. Biggest ThreatsandData. https://www.cdc.gov/drugresistance/biggest_threats.html?CDC_AA_refVal=https%3A%2F%2Fwww.cdc.gov%2Fdrugresistance%2Fthreat-report-2013%2Findex.html (Accessed April 10, 2019).

CDC, Centers for Disease Control and Prevention. 2018b. *E. coli (Escherichia coli).* https://www.cdc.gov/salmonella/general/index.html (Accessed March 16, 2018).

CDC, Centers for Disease Control and Prevention. 2019a. Foodborne Illnesses and Germs. https://www.cdc.gov/foodsafety/foodborne-germs.html (Accessed April 5, 2019).

CDC, Centers for Disease Control and Prevention. 2019b. List of Selected Multistate Foodborne Outbreak Investigations https://www.cdc.gov/foodsafety/outbreaks/multistate-outbreaks/outbreaks-list.html (Accessed April 3, 2019).

Ciofu, Oana, Tim Tolker-Nielsen, Peter Østrup Jensen, Hengzhuang Wang, and Niels Høiby. 2015. Antimicrobial Resistance, Respiratory Tract Infections and Role of Biofilms in Lung Infections in Cystic Fibrosis Patients. *Advanced Drug Delivery Reviews* 85:7–23.

Corral-Lugo, Andrés, Abdelali Daddaoua, Alvaro Ortega, Manuel Espinosa-Urgel, and Tino Krell. 2016. Rosmarinic Acid Is a Homoserine Lactone Mimic Produced by Plants That Activates a Bacterial Quorum-Sensing Regulator. *Science Signaling* 9:ra1.

Crago, Aimee M., and Vassilis Koronakis. 1999. Binding of Extracellular Matrix Laminin to *Escherichia coli* Expressing the *Salmonella* Outer Membrane Proteins Rck and Pagc. *FEMS Microbiology Letters* 176:495–501.

Cross, Alan S. 2008. What Is a Virulence Factor? *Critical Care (London, England)* 12:196–96.

Cushnie, T. P. Tim, Benjamart Cushnie, and Andrew J. Lamb. 2014. Alkaloids: An Overview of Their Antibacterial, Antibiotic-Enhancing and Antivirulence Activities. *International Journal of Antimicrobial Agents* 44:377–86.

Chang, Chien-Yi, Thiba Krishnan, Hao Wang, et al. 2014. Non-Antibiotic Quorum Sensing Inhibitors Acting against *N*-Acyl Homoserine Lactone Synthase as Druggable Target. *Scientific Reports* 4:7245.

Chemmugil, P., P. T. V. Lakshmi, and Arunachalam Annamalai. 2019. Exploring Morin as an Anti-Quorum Sensing Agent (Anti-Qsa) against Resistant Strains of *Staphylococcus aureus. Microbial Pathogenesis* 127:304–15.

Chen, Yuhuan, William H. Ross, Richard C. Whiting, et al. 2011. Variation in *Listeria monocytogenes* Dose Responses in Relation to Subtypes Encoding a Full-Length or Truncated Internalin A. *Applied and Environmental Microbiology* 77:1171–80.

Cho, Moo Hwan, Jung Ae Kim, Jin Hyung Lee, Jin Tae Lee. "Composition for Inhibiting Biofilm Containing Carex Dimorpholepis Extract." Patent number KR101517716B1, Industry-Academic Cooperation Foundation, Yeungnam University, 4 May 2015.

Chung, Jiwoung, Eunhye Goo, Sangheon Yu, et al. 2011. Small-Molecule Inhibitor Binding to an *N*-Acyl-Homoserine Lactone Synthase. *Proceedings of the National Academy of Sciences* 108:12089–94.

Darch, Sophie E., Stuart A. West, Klaus Winzer, and Stephen P. Diggle. 2012. Density-Dependent Fitness Benefits in Quorum-Sensing Bacterial Populations. *Proceedings of the National Academy of Sciences* 109:8259–63.

Drăgulescu, Elena-Carmina, and Irina Codita. 2015. Host-Pathogen Interaction in Infections Due to *Staphylococcus aureus. Staphylococcus aureus* Virulence Factors. *Romanian Archives of Microbiology and Immunology* 74:46–64.

Dyszel, Jessica L., Jenee N. Smith, Darren E. Lucas, et al. 2010. *Salmonella enterica* Serovar Typhimurium Can Detect Acyl Homoserine Lactone Production by *Yersinia enterocolitica* in Mice. *Journal of Bacteriology* 192:29–37.

Flemming, Hans-Curt, Jost Wingender, Ulrich Szewzyk, Peter Steinberg, Scott A. Rice, and Staffan Kjelleberg. 2016. Biofilms: An Emergent Form of Bacterial Life. *Nature Reviews Microbiology* 14:563.

Fuqua, Clay, and E. Peter Greenberg. 2002. Signalling: Listening in on Bacteria: Acyl-Homoserine Lactone Signalling. *Nature Reviews Molecular Cell Biology* 3:685.

Ganin, Hadas, Josep Rayo, Neri Amara, Niva Levy, Pnina Krief, and Michael M. Meijler. 2013. Sulforaphane and Erucin, Natural Isothiocyanates from Broccoli, Inhibit Bacterial Quorum Sensing. *MedChemComm* 4:175–79.

García-Moreno, Miguel A., Myriam A. de la Garza-Ramos, Guillermo C. G. Martínez-Ávila, Adriana Gutiérrez-Díez, Maria Ojeda-Zacarías, and Victor E. Aguirre-Arzola. 2017. Inhibición De La Expresión Del Sistema Agr De *Staphylococcus aureus* Resistente a Meticilina Mediante El Uso De Polifenoles Totales De Hojas De Aguacate Mexicano (*Persea Americana* Var. *Drymifolia*). *Nova Scientia* 9:200–21.

Ge, Yifei, Yongyu Li, Jingyuan Liu, Wenting Wang, Shaohua Wu, Yingxiang Wu, Liaoyuan Zhang, and Yanhu Zhang. "Preparation Method and Application of Camphor Essential Oil-Based Bacterial Quorum Sensing Inhibitor." Patent number CN109463402A, 30 November 2018.

Ghosh, Runu, Bipransh Kumar Tiwary, Anoop Kumar, and Ranadhir Chakraborty. 2014. Guava Leaf Extract Inhibits Quorum-Sensing and *Chromobacterium violaceum* Induced Lysis of Human Hepatoma Cells: Whole Transcriptome Analysis Reveals Differential Gene Expression. *PLoS One* 9:e107703.

Gode-Potratz, Cindy J., and Linda L. McCarter. 2011. Quorum Sensing and Silencing in *Vibrio parahaemolyticus. Journal of Bacteriology* 193:4224–37.

Gong, Qianhong, Yang Song, Wengong Yu, and Hongda Zheng. "Bacterial Quorum Sensing Inhibitor and Antibacterial Application Thereof." Patent number CN104784160A, Ocean University of China, 22 July 2015.

Gopu, Venkadesaperumal, Chetan Kumar Meena, and Prathapkumar Halady Shetty. 2015. Quercetin Influences Quorum Sensing in Food Borne Bacteria: In-Vitro and In-Silico Evidence. *PLoS One* 10:e0134684.

Gould, Ian M., and Abhijit M. Bal. 2013. New Antibiotic Agents in the Pipeline and How They Can Help Overcome Microbial Resistance. *Virulence* 4:185–91.

Gutiérrez-Barranquero, José A., F. Jerry Reen, Ronan R. McCarthy, and Fergal O'Gara. 2015. Deciphering the Role of Coumarin as a Novel Quorum Sensing Inhibitor Suppressing Virulence Phenotypes in Bacterial Pathogens. *Applied Microbiology and Biotechnology* 99:3303–16.

Gutierrez-Pacheco, Maria Melissa, Ariadna Thalia Bernal-Mercado, Francisco Javier Vazquez-Armenta, et al. 2019. Quorum Sensing Interruption as a Tool to Control Virulence of Plant Pathogenic Bacteria. *Physiological and Molecular Plant Pathology*

Gutierrez-Pacheco, M. M., G. A. Gonzalez-Aguilar, M. A. Martinez-Tellez, et al. 2018. Carvacrol Inhibits Biofilm Formation and Production of Extracellular Polymeric Substances of *Pectobacterium carotovorum* Subsp. *Carotovorum. Food Control* 89:210–18.

Hamon, Mélanie Anne, David Ribet, Fabrizia Stavru, and Pascale Cossart. 2012. Listeriolysin O: The Swiss Army Knife of *Listeria. Trends in Microbiology* 20:360–68.

Han, Daobin B., Yurong R.Yan, Jianmin M. Wang, et al. 2019. An Enzyme-Free Electrochemiluminesce Aptasensor for the Rapid Detection of *Staphylococcus aureus* by the Quenching Effect of Mos2-Ptnps-Vancomycin to S2o82-/O-2 System. *Sensors and Actuators B-Chemical* 288:586–93.

Heilmann, Christine, and Friedrich Götz. 2010. Cell–Cell Communication and Biofilm Formation in Gram-Positive Bacteria. *Bacterial Signaling* 1:7–22.

Heleno, Sandrina A., Anabela Martins, Maria João R. P. Queiroz, and Isabel C. F. R. Ferreira. 2015. Bioactivity of Phenolic Acids: Metabolites Versus Parent Compounds: A Review. *Food Chemistry* 173:501–13.

Hossain, Md Akil, Seung-Jin Lee, Na-Hye Park, et al. 2017. Impact of Phenolic Compounds in the Acyl Homoserine Lactone-Mediated Quorum Sensing Regulatory Pathways. *Scientific Reports* 7:10618.

Husain, Fohad M., Iqbal Ahmad, Abdullah S. Al-thubiani, Hussein H. Abulreesh, Ibrahim M. AlHazza, and Farrukh Aqil. 2017. Leaf Extracts of *Mangifera indica* L. Inhibit Quorum Sensing–Regulated Production of Virulence Factors and Biofilm in Test Bacteria. *Frontiers in Microbiology* 8:727.

Jakobsen, Tim Holm, Steinn Kristinn Bragason, Richard Kerry Phipps, et al. 2012. Food as a Source for Quorum Sensing Inhibitors: Iberin from Horseradish Revealed as a Quorum Sensing Inhibitor of *Pseudomonas aeruginosa. Applied and Environmental Microbiology* 78:2410–21.

Jakobsen, Tim Holm, Maria van Gennip, Louise Dahl Christensen, Thomas Bjarnsholt, and Michael Givskov. "Qualitative and Quantitative Determination of Quorum Sensing Inhibition in Vitro." In *Quorum Sensing: Methods and Protocols*, edited by Kendra P. Rumbaugh, 253–63. Totowa, NJ: Humana Press, 2011.

Joshi, Janak Raj, Netaly Khazanov, Hanoch Senderowitz, Saul Burdman, Alexander Lipsky, and Iris Yedidia. 2016. Plant Phenolic Volatiles Inhibit Quorum Sensing in Pectobacteria and Reduce Their Virulence by Potential Binding to Expi and Expr Proteins. *Scientific Reports* 6:38126.

Jung, Sarah A., Christine A. Chapman, and Wai-Leung Ng. 2015. Quadruple Quorum-Sensing Inputs Control *Vibrio cholerae* Virulence and Maintain System Robustness. *PLoS Pathogens* 11:e1004837.

Kerekes, E.-B., Éva Deák, Miklós Takó, et al. 2013. Anti-Biofilm Forming and Anti-Quorum Sensing Activity of Selected Essential Oils and Their Main Components on Food-Related Micro-Organisms. *Journal of Applied Microbiology* 115:933–42.

Khan, Burhan A., Anthony J. Yeh, Gordon Yc Cheung, and Michael Otto. 2015. Investigational Therapies Targeting Quorum-Sensing for the Treatment of *Staphylococcus aureus* Infections. *Expert Opinion on Investigational Drugs* 24:689–704.

Khan, Mohd Sajjad Ahmad, Maryam Zahin, Sameena Hasan, Fohad Mabood Husain, and Iqbal Ahmad. 2009. Inhibition of Quorum Sensing Regulated Bacterial Functions by Plant Essential Oils with Special Reference to Clove Oil. *Letters in Applied Microbiology* 49:354–60.

Kim, Soo Ki and Min Kuk. "Quorum Sensing Inhibitor Using Plant Extracts for Preventing the Proliferation of Gram-Negative Bacteria." Korean Patent number KR101656875B1, Konkuk University Industrial Cooperation Corp., 12 September. 2016.

Kim, Truc, Thao Duong, Chun-ai Wu, et al. 2014. Structural Insights into the Molecular Mechanism of *Escherichia coli* Sdia, a Quorum-Sensing Receptor. *Acta Crystallographica Section D: Biological Crystallography* 70:694–707.

Kim, Yong-Guy, Jin-Hyung Lee, Soon-Il Kim, Kwang-Hyun Baek, and Jintae Lee. 2015. Cinnamon Bark Oil and Its Components Inhibit Biofilm Formation and Toxin Production. *International Journal of Food Microbiology* 195:30–39.

Kjelleberg, Staffan, and Soeren Molin. 2002. Is There a Role for Quorum Sensing Signals in Bacterial Biofilms? *Current Opinion in Microbiology* 5:254–58.

Koh, Chong-Lek, Choon-Kook Sam, Wai-Fong Yin, et al. 2013. Plant-Derived Natural Products as Sources of Anti-Quorum Sensing Compounds. *Sensors* 13:6217–28.

Kumar, N. Vijendra, Pushpa Srinivas Murthy, Javagal R. Manjunatha, and Bettadaiah Kempaiah 2014. Synthesis and Quorum Sensing Inhibitory Activity of Key Phenolic Compounds of Ginger and Their Derivatives. *Food Chemistry* 159:451–57.

Lee, Jin-Hyung, Sushil Chandra Regmi, Jung-Ae Kim, et al. 2011. Apple Flavonoid Phloretin Inhibits *Escherichia coli* O157: H7 Biofilm Formation and Ameliorates Colon Inflammation in Rats. *Infection and Immunity* 79:4819–27.

Li, Guanghui, Chunhong Yan, Yunfeng Xu, et al. 2014. Punicalagin Inhibits *Salmonella* Virulence Factors and Has Anti-Quorum-Sensing Potential. *Applied and Environmental Microbiology* 80:6204–11.

Li, Jun, Can Attila, Liang Wang, Thomas K. Wood, James J. Valdes, and William E. Bentley. 2007. Quorum Sensing in *Escherichia coli* Is Signaled by Ai-2/Lsrr: Effects on Small RNA and Biofilm Architecture. *Journal of Bacteriology* 189:6011–20.

Maisuria, Vimal B., Yossef Lopez-de Los Santos, Nathalie Tufenkji, and Eric Déziel. 2016. Cranberry-Derived Proanthocyanidins Impair Virulence and Inhibit Quorum Sensing of *Pseudomonas aeruginosa. Scientific Reports* 6:30169.

Maktabi, Slavash, Masoud Ghorbanpoor, Masomeh Hossaini, and Amirabbas Motavalibashi. 2019. Detection of Multi-Antibiotic Resistant *Campylobacter coli* and *Campylobacter jejuni* in Beef, Mutton, Chicken and Water Buffalo Meat in Ahvaz, Iran. *Veterinary Research Forum* 10:37–42.

Martins, Natália, Lillian Barros, and Isabel C. F. R. Ferreira. 2016. In Vivo Antioxidant Activity of Phenolic Compounds: Facts and Gaps. *Trends in Food Science & Technology* 48:1–12.

Mathee, Kalai, Allison L. Adonizio, Frederick Ausubel, Jon Clardy, Bradley Bennett, and Kelsey Downum. "Ellagitannins as Inhibitors of Bacterial Quorum Sensing." Patent number WO2009114810A2, The Florida International University Board of Trustees, 23 December. 2009.

Matsuura, Hélio Nitta, and Arthur Germano Fett-Neto. 2017. Plant Alkaloids: Main Features, Toxicity, and Mechanisms of Action, In: Gopalakrishnakone P., Carlini C., Ligabue-Braun R. (eds) *Plant Toxins. Toxinology*. Springer, Dordrecht, Netherlands.

McDaniel, Timothy K., Karen G. Jarvis, Michael S. Donnenberg, and James B. Kaper. 1995. A Genetic Locus of Enterocyte Effacement Conserved among Diverse Enterobacterial Pathogens. *Proceedings of the National Academy of Sciences* 92:1664–68.

McLean, Robert J. C., Leland S. Pierson, and Clay Fuqua. 2004. A Simple Screening Protocol for the Identification of Quorum Signal Antagonists. *Journal of Microbiological Methods* 58:351–60.

Mellin, Jeffery R., and Pascale Cossart. 2012. The Non-Coding RNA World of the Bacterial Pathogen *Listeria monocytogenes*. *RNA Biology* 9:372–78.

Meyer-Morse, Nicole, Jennifer R. Robbins, Chris S. Rae, et al. 2010. Listeriolysin O Is Necessary and Sufficient to Induce Autophagy During *Listeria monocytogenes* Infection. *PLoS One* 5:e8610.

Minor, Travis, and Matt Parrett. 2017. The Economic Impact of the Food and Drug Administration's Final Juice Haccp Rule. *Food Policy* 68:206–13.

Mitchell, Gabriel, Myriame Lafrance, Simon Boulanger, et al. 2011. Tomatidine Acts in Synergy with Aminoglycoside Antibiotics against Multiresistant *Staphylococcus aureus* and Prevents Virulence Gene Expression. *Journal of Antimicrobial Chemotherapy* 67:559–68.

Monte, Joana, Ana Abreu, Anabela Borges, Lúcia Simões, and Manuel Simões. 2014. Antimicrobial Activity of Selected Phytochemicals against *Escherichia coli* and *Staphylococcus aureus* and Their Biofilms. *Pathogens* 3:473–98.

Munita, Jose M., and Cesar A. Arias. 2016. Mechanisms of Antibiotic Resistance. *Microbiology Spectrum* 4.

Myszka, Kamila, Marcin T. Schmidt, Wojciech Białas, Mariola Olkowicz, Katarzyna Leja, and Katarzyna Czaczyk. 2016a. Role of Gallic and P-Coumaric Acids in the Ahl-Dependent Expression of Flga Gene and in the Process of Biofilm Formation in Food-Associated *Pseudomonas fluorescens* Km120. *Journal of the Science of Food and Agriculture* 96:4037–47.

Myszka, Kamila, Marcin T. Schmidt, Małgorzata Majcher, Wojciech Juzwa, Mariola Olkowicz, and Katarzyna Czaczyk. 2016b. Inhibition of Quorum Sensing-Related Biofilm of *Pseudomonas Fluorescens* Km121 by *Thymus vulgare* Essential Oil and Its Major Bioactive Compounds. *International Biodeterioration & Biodegradation* 114:252–59.

Najafi, Sohelia, Morad Rahimi, and Zahra Nikousefat. 2019. Extra-Intestinal Pathogenic *Escherichia coli* from Human and Avian Origin: Detection of the Most Common Virulence-Encoding Genes. *Veterinary Research Forum* 10:43–49.

Nakayama, Jiro, Yumi Uemura, Kenzo Nishiguchi, Norito Yoshimura, Yasuhiro Igarashi, and Kenji Sonomoto. 2009. Ambuic Acid Inhibits the Biosynthesis of Cyclic Peptide Quormones in Gram-Positive Bacteria. *Antimicrobial Agents and Chemotherapy* 53:580–86.

Nazzaro, Filomena, Florinda Fratianni, and Raffaele Coppola. 2013. Quorum Sensing and Phytochemicals. *International Journal of Molecular Sciences* 14:12607–19.

Niu, Cyr, S. Afre, and Eric S. Gilbert. 2006. Subinhibitory Concentrations of Cinnamaldehyde Interfere with Quorum Sensing. *Letters in Applied Microbiology* 43:489–94.

Norizan, Siti Nur Maisarah, Wai-Fong Yin, and Kok-Gan Chan. 2013. Caffeine as a Potential Quorum Sensing Inhibitor. *Sensors* 13:5117–29.

Novick, Richard P., and Edward Geisinger. 2008. Quorum Sensing in *Staphylococci*. *Annual Review of Genetics* 42:541–64.

Oliveira, B. D. A., A. C. Rodrigues, M. C. Bertoldi, J. G. Taylor, and Uelinton Manoel Pinto. 2017. Microbial Control and Quorum Sensing Inhibition by Phenolic Compounds of Acerola (*Malpighia emarginata*). *International Food Research Journal* 24:2228–37.

Oliveira, Brígida D'Ávila, Adeline Conceição Rodrigues, Bárbara Moreira Inácio Cardoso, et al. 2016. Antioxidant, Antimicrobial and Anti-Quorum Sensing Activities of *Rubus rosaefolius* Phenolic Extract. *Industrial Crops and Products* 84:59–66.

Ouyang, J., Fengjun Sun, Wenli Feng, et al. 2016. Quercetin Is an Effective Inhibitor of Quorum Sensing, Biofilm Formation and Virulence Factors in *Pseudomonas aeruginosa*. *Journal of Applied Microbiology* 120:966–74.

Paczkowski, Jon E., Sampriti Mukherjee, Amelia R. McCready, et al. 2017. Flavonoids Suppress *Pseudomonas aeruginosa* Virulence through Allosteric Inhibition of Quorum-Sensing Receptors. *Journal of Biological Chemistry* 292:4064–76.

Parai, Debaprasad, Malabika Banerjee, Pia Dey, Arindam Chakraborty, Ekramul Islam, and Samir Kumar Mukherjee. 2018. Effect of Reserpine on *Pseudomonas aeruginosa* Quorum Sensing Mediated Virulence Factors and Biofilm Formation. *Biofouling* 34:320–34.

Park, Junguk, Gunnar F. Kaufmann, J. Phillip Bowen, Jack L. Arbiser, and Kim D. Janda. 2008. Solenopsin a, a Venom Alkaloid from the Fire Ant *Solenopsis invicta*, Inhibits Quorum-Sensing Signaling in *Pseudomonas aeruginosa*. *Journal of Infectious Diseases* 198:1198–201.

Parker, Chris T., and Vanessa Sperandio. 2009. Cell-to-Cell Signalling During Pathogenesis. *Cell Microbiology* 11:363–9.

Pejin, Boris, Ana Ciric, Jasmina Glamoclija, Milos Nikolic, and Marina Sokovic. 2015. In Vitro Anti-Quorum Sensing Activity of Phytol. *Natural Product Research* 29:374–77.

Peterson, Johnny W. 1996. Bacterial Pathogenesis. In: Baron S., (ed) *Source Medical Microbiology*. 4th ed. University of Texas Medical Branch at Galveston, Galveston, TX.

Pollitt, Eric J. G., Stuart A. West, Shanika A. Crusz, Maxwell N. Burton-Chellew, and Stephen P. Diggle. 2014. Cooperation, Quorum Sensing, and Evolution of Virulence in *Staphylococcus aureus*. *Infection and Immunity* 82:1045–51.

Qiu, Jiazhang, Xiaodi Niu, Jianfeng Wang, et al. 2012. Capsaicin Protects Mice from Community-Associated Methicillin-Resistant *Staphylococcus aureus* Pneumonia. *Plos One* 7:e33032.

Qiu, Jiazhang, Dacheng Wang, Hua Xiang, et al. 2010. Subinhibitory Concentrations of Thymol Reduce Enterotoxins a and B and A-Hemolysin Production in *Staphylococcus aureus* Isolates. *PLoS One* 5:e9736.

Rajkumari, Jobina, Subhomoi Borkotoky, Ayaluru Murali, Kitlangki Suchiang, Saswat Kumar Mohanty, and Siddhardha Busi. 2018. Cinnamic Acid Attenuates Quorum Sensing Associated Virulence Factors and Biofilm Formation in *Pseudomonas aeruginosa* Pao1. *Biotechnology Letters* 40:1087–100.

Rasamiravaka, Tsiry, Jérémie Ngezahayo, Laurent Pottier, et al. 2017. Terpenoids from Platostoma Rotundifolium (Briq.) Aj Paton Alter the Expression of Quorum Sensing-Related Virulence Factors and the Formation of Biofilm in *Pseudomonas aeruginosa* Pao1. *International Journal of Molecular Sciences* 18:1270.

Reverchon, Sylvie, and William Nasser. 2013. Dickeya Ecology, Environment Sensing and Regulation of Virulence Programme. *Environmental Microbiology Reports* 5:622–36.

Riedel, Christian U., Ian R. Monk, Pat G. Casey, Mark S. Waidmann, Cormac G. M. Gahan, and Colin Hill. 2009. Agrd-Dependent Quorum Sensing Affects Biofilm Formation, Invasion, Virulence and Global Gene Expression Profiles in *Listeria monocytogenes*. *Molecular Microbiology* 71:1177–89.

Römling, Ute, Manfred Rohde, Arne Olsen, Staffan Normark, and Jürgen Reinköster. 2000. Agfd, the Checkpoint of Multicellular and Aggregative Behaviour in *Salmonella* Typhimurium Regulates at Least Two Independent Pathways. *Molecular Microbiology* 36:10–23.

Sá, Nairley Cardoso, Theodora Thays Arruda Cavalcante, Amanda Ximenes Araújo, et al. 2012. Antimicrobial and Antibiofilm Action of Casbane Diterpene from *Croton nepetaefolius* against Oral Bacteria. *Archives of Oral Biology* 57:550–55.

Scobie, Antonia, Sanch Kanagarajah, Ross J. Harris, et al. 2019. Mortality Risk Factors for Listeriosis—A 10 Year Review of Non-Pregnancy Associated Cases in England 2006–2015. *Journal of Infection* 78:208–14.

Schaefer, Amy L., Dale L. Val, Brian L. Hanzelka, John E. Cronan Jr, and E. Peter Greenberg. 1996. Generation of Cell-to-Cell Signals in Quorum Sensing: Acyl Homoserine Lactone Synthase Activity of a Purified *Vibrio fischeri* Luxi Protein. *Proceedings of the National Academy of Sciences of the United States of America* 93:9505.

Schauder, Stephan, Kevan Shokat, Michael G. Surette, and Bonnie L. Bassler. 2001. The Luxs Family of Bacterial Autoinducers: Biosynthesis of a Novel Quorum-Sensing Signal Molecule. *Molecular Microbiology* 41:463–76.

Sethupathy, Sivasamy, Sivagnanam Ananthi, Anthonymuthu Selvaraj, et al. 2017. Vanillic Acid from *Actinidia deliciosa* Impedes Virulence in *Serratia marcescens* by Affecting S-Layer, Flagellin and Fatty Acid Biosynthesis Proteins. *Scientific Reports* 7:16328.

Singh, Brahma N., H. B. Singh, Akanksha Singh, Braj R. Singh, Aradhana Mishra, and C. S. Nautiyal. 2012. *Lagerstroemia speciosa* Fruit Extract Modulates Quorum Sensing-Controlled Virulence Factor Production and Biofilm Formation in *Pseudomonas aeruginosa*. *Microbiology* 158:529–38.

Singh, Maleeka, Kavita Walia, and Jeffrey M Farber. 2019. The Household Kitchen as the “Last Line of Defense” in the Prevention of Foodborne Illness: A Review and Analysis of Meat and Seafood Recipes in 30 Popular Canadian Cookbooks. *Food Control* 100:122–29.

Sivaranjani, Murugesan, Subramanian Radhesh Krishnan, Arunachalam Kannappan, Manikandan Ramesh, and Arumugam Veera Ravi. 2016. Curcumin from *Curcuma longa* Affects the Virulence of *Pectobacterium wasabiae* and *P. Carotovorum* Subsp. *Carotovorum* Via Quorum Sensing Regulation. *European Journal of Plant Pathology* 146:793–806.

Smith, James L., Pina M. Fratamico, and Xianghe Yan. 2011. Eavesdropping by Bacteria: The Role of Sdia in *Escherichia coli* and *Salmonella enterica* Serovar Typhimurium Quorum Sensing. *Foodborne Pathogens and Disease* 8:169–78.

Smith, Jenée N., and Brian M. M. Ahmer. 2003. Detection of Other Microbial Species by *Salmonella*: Expression of the Sdia Regulon. *Journal of Bacteriology* 185:1357–66.

Stacy, Danielle M., Sebastian T. Le Quement, Casper L. Hansen, et al. 2013. Synthesis and Biological Evaluation of Triazole-Containing *N*-Acyl Homoserine Lactones as Quorum Sensing Modulators. *Organic & Biomolecular Chemistry* 11:938–54.

Steindler, Laura, and Vittorio Venturi. 2007. Detection of Quorum-Sensing *N*-Acyl Homoserine Lactone Signal Molecules by Bacterial Biosensors. *FEMS Microbiology Letters* 266:1–9.

Sui, Shannan J. Ho, Amber Fedynak, William W. L. Hsiao, Morgan G. I. Langille, and Fiona S. L. Brinkman. 2009. The Association of Virulence Factors with Genomic Islands. *PloS One* 4:e8094.

Suzuki, Kazushi, Xin Wang, Thomas Weilbacher, et al. 2002. Regulatory Circuitry of the Csra/Csrb and Bara/Uvry Systems of *Escherichia coli*. *Journal of Bacteriology* 184:5130–40.

Sybiya Vasantha Packiavathy, Issac Abraham, Palani Agilandeswari, Khadar Syed Musthafa, Shunmugiah Karutha Pandian, and Arumugam Veera Ravi. 2012. Antibiofilm and Quorum Sensing Inhibitory Potential of *Cuminum cyminum* and Its Secondary Metabolite Methyl Eugenol against Gram Negative Bacterial Pathogens. *Food Research International* 45:85–92.

Taga, Michiko E., and Karina B. Xavier. 2011. Methods for Analysis of Bacterial Autoinducer-2 Production. *Current Protocols in Microbiology* 23:1C. 1.1–1C. 1.15.

Tan, Yuen Ping, Eric Wei Chiang Chan, and Crystale Siew Ying Lim. 2015. Potent Quorum Sensing Inhibition by Methyl Gallate Isolated from Leaves of *Anacardium occidentale* L.(Cashew). *Chiang Mai Journal of Science* 42:650–56.

Tang, Feng, Li Li, Xian-Mei Meng, et al. 2019. Inhibition of Alpha-Hemolysin Expression by Resveratrol Attenuates *Staphylococcus aureus* Virulence. *Microbial Pathogenesis* 127:85–90.

Tapia-Rodriguez, Melvin R., Adrian Hernandez-Mendoza, Gustavo A. Gonzalez-Aguilar, Miguel Angel Martinez-Tellez, Claudia Miranda Martins, and J. Fernando Ayala-Zavala. 2017. Carvacrol as Potential Quorum Sensing Inhibitor of *Pseudomonas aeruginosa* and Biofilm Production on Stainless Steel Surfaces. *Food Control* 75:255–61.

Thornhill, Starla G., and Robert J. C. McLean. Surfaces. Gustavo A. Gonzalez-Aguilar, Miguel Angel Martinez-Tellez, Claudia Miranda Martins, and J Fernando Ay *Quorum Sensing*, 3–24: Springer, 2018.

Torres, M. A. De Lira, Flores Santos, L. E. Fragoso Morales, et al. 2013. Infecciones Del Tracto Urinario Asociado a Catéter Vesical. Áreas De Cirugía Y Medicina Interna De Dos Hospitales Del Sector Público. *Enfermedades Infecciosas y Microbiología* 33:13–18.

Tran, Linda, Arjun Naik, Brian Koronkiewicz, and Bela Peethambaran. 2013. Epigallocatechin Gallate Inhibits Biofilm Production and Attenuates Virulent Factors of *Pseudomonas aeruginosa* and *Psuedomonas fluorescence. Journal of Natural Remedies* 14:106–11.

Travier, Laetitia, Stéphanie Guadagnini, Edith Gouin, et al. 2013. Acta Promotes *Listeria monocytogenes* Aggregation, Intestinal Colonization and Carriage. *PLoS Pathog* 9:e1003131.

Truchado, Pilar, Mar Larrosa, Irene Castro-Ibáñez, and Ana Allende. 2015. Plant Food Extracts and Phytochemicals: Their Role as Quorum Sensing Inhibitors. *Trends in Food Science & Technology* 43:189–204.

Truchado, Pilar, Francisco A. Tomás-Barberán, Mar Larrosa, and Ana Allende. 2012. Food Phytochemicals Act as Quorum Sensing Inhibitors Reducing Production and/or Degrading Autoinducers of *Yersinia enterocolitica* and *Erwinia carotovora. Food Control* 24:78–85.

Ugurlu, Aylin, Aysegul Karahasan Yagci, Seyhan Ulusoy, Burak Aksu, and Gulgun Bosgelmez-Tinaz. 2016. Phenolic Compounds Affect Production of Pyocyanin, Swarming Motility and Biofilm Formation of *Pseudomonas aeruginosa. Asian Pacific Journal of Tropical Biomedicine* 6:698–701.

Vandeputte, Olivier M., Martin Kiendrebeogo, Sanda Rajaonson, et al. 2010. Identification of Catechin as One of the Flavonoids from *Combretum albiflorum* Bark Extract That Reduces the Production of Quorum-Sensing-Controlled Virulence Factors in *Pseudomonas aeruginosa* Pao1. *Applied and Environment Microbiology* 76:243–53.

Vandeputte, Olivier M., Martin Kiendrebeogo, Tsiry Rasamiravaka, et al. 2011. The Flavanone Naringenin Reduces the Production of Quorum Sensing-Controlled Virulence Factors in *Pseudomonas aeruginosa* Pao1. *Microbiology* 157:2120–32.

Vattem, Dhiraj A., K. Mihalik, Sylvia H. Crixell, and Robert J.C. McLean. 2007b. Dietary Phytochemicals as Quorum Sensing Inhibitors. *Fitoterapia* 78:302–10.

Ventola, C. Lee. 2015. The Antibiotic Resistance Crisis: Part 1: Causes and Threats. *Pharmacy and Therapeutics* 40:277.

Vera, Alejandra, Gerardo González, Mariana Domínguez, and Helia Bello. 2013. Principales Factores De Virulencia De Listeria Monocytogenesy Su Regulación. *Revista chilena de infectología* 30:407–16.

Vikram, Amit, Guddadarangavvanahally K. Jayaprakasha, Palmy Jesudhasan, S. D. Pillai, and B. S. Patil. 2010. Suppression of Bacterial Cell–Cell Signalling, Biofilm Formation and Type Iii Secretion System by Citrus Flavonoids. *Journal of Applied Microbiology* 109:515–27.

Wang, Boyuan, and Tom W. Muir. 2016. Regulation of Virulence in *Staphylococcus aureus*: Molecular Mechanisms and Remaining Puzzles. *Cell Chemical Biology* 23:214–24.

Wang, Hong, Weihua Chu, Chao Ye, et al. 2019. Chlorogenic Acid Attenuates Virulence Factors and Pathogenicity of *Pseudomonas aeruginosa* by Regulating Quorum Sensing. *Applied Microbiology and Biotechnology* 103:903–15.

Winzer, Klaus, and Paul Williams. 2001. Quorum Sensing and the Regulation of Virulence Gene Expression in Pathogenic Bacteria. *International Journal of Medical Microbiology* 291:131–43.

World Health Organization. 2014. *Antimicrobial Resistance: Global Report on Surveillance*. World Health Organization, Geneva, Switzerland.

Wright, Gerard D. 2014. Something Old, Something New: Revisiting Natural Products in Antibiotic Drug Discovery. *Canadian Journal of Microbiology* 60:147–54.

Xin, Wenwen W., Yong Huang, Bin Ji, et al. 2019. Identification and Characterization of *Clostridium botulinum* Strains Associated with an Infant Botulism Case in China. *Anaerobe* 55:1–7.

Yang, Qian, and Tom Defoirdt. 2015. Quorum Sensing Positively Regulates Flagellar Motility in Pathogenic *Vibrio harveyi. Environmental Microbiology* 17:960–68.

Yang, Yang, Yun Liu, Mingxu Zhou, and Guoqiang Zhu. 2018. Both Quorum Sensing (Qs)-I and Ii Systems Regulate *Escherichia coli* Flagellin Expression. *Pakistan Journal of Zoology* 50:1807–13.

Zetzmann, Marion, Andrés Sánchez-Kopper, Mark S. Waidmann, Bastian Blombach, and Christian U. Riedel. 2016. Identification of the Agr Peptide of *Listeria monocytogenes. Frontiers in Microbiology* 7:989.

Zhang, Ying, Jie Kong, Fei Huang, et al. 2018a. Hexanal as a Qs Inhibitor of Extracellular Enzyme Activity of Erwinia Carotovora and Pseudomonas Fluorescens and Its Application in Vegetables. *Food Chemistry* 255:1–7.

Zhang, Ying, Jie Kong, Yunfei Xie, et al. 2018b. Essential Oil Components Inhibit Biofilm Formation in *Erwinia carotovora* and *Pseudomonas fluorescens* Via Anti-Quorum Sensing Activity. *LWT* 92:133–39.

Zhou, Jin-Wei, Ling-Yu Ruan, Hong-Juan Chen, et al. 2019a. Inhibition of Quorum Sensing and Virulence in *Serratia marcescens* by Hordenine. *Journal of Agricultural and Food Chemistry* 67:784–95.

Zhou, Shuxin, An Zhang, and Weihua Chu. 2019b. Phillyrin Is an Effective Inhibitor of Quorum Sensing with Potential as an Anti-*Pseudomonas aeruginosa* Infection Therapy. *Journal of Veterinary Medical Science*:18–0523.

Zogaj, Xhavit, Werner Bokranz, Manfred Nimtz, and Ute Römling. 2003. Production of Cellulose and Curli Fimbriae by Members of the Family Enterobacteriaceae Isolated from the Human Gastrointestinal Tract. *Infection and Immunity* 71:4151–58.

Zwenger, Sam, and Chhandak Basu. 2008. Plant Terpenoids: Applications and Future Potentials. *Biotechnology and Molecular Biology Reviews* 3:1.

20

Targeting Bacterial Communication to Improve Bacterial Infections Therapy: Implications for Phage-Based Treatments—A Mathematical Perspective

Adrián Cazares, Rodolfo García-Contreras, Christina Kuttler, and Judith Pérez-Velázquez

CONTENTS

20.1 Introduction

A revival in the interest for phage-based therapeutics has brought new hopes to the war against antibiotic resistance. However, recent studies exploring the interlink between bacteriophages and bacterial QS, which regulates virulence in many bacterial pathogens, raise important considerations about the effect of this relationship toward outcomes. In principle, these findings could be exploited for therapeutic purposes. Nevertheless, under certain circumstances, they also suggest phage therapy could promote an increase in virulence, potentially threatening the safe use of phages as antimicrobials. The aim of this chapter is to revise this evidence and place it in a broader evolutionary and ecological context. We discuss how mathematical modeling can be helpful in developing safer ways to treat bacterial infections which take into account long-term development of these known interactions. Overall, we believe this discussion is pivotal for the development and evaluation of new phage-based treatment strategies, which can be both safe and evolutionary stable.

20.2 Phage Therapy

With the discovery of phages came the idea of employing them to treat bacterial infections (Breederveld 2019). In spite of the clear potential of using phages as antimicrobial agents, a range of factors prevented PT of winning broad acceptance apart from some East European countries (Loc-Carrillo and Abedon 2011; Roach and Debarbieux 2017). However, in the face of the ongoing antibiotic crisis, PT has regained notorious popularity (Matsuzaki et al. 2014; Breederveld 2019; Reardon 2014). PT offers many advantages; it is claimed that phages are species-specific, thus eluding unwanted removal of beneficial bacteria (Breederveld 2019; Loc-Carrillo and Abedon 2011). Phages also are believed to be especially suitable for infections which are difficult to treat, e.g. chronic infections (Torres-Barceló 2018). Phage-based therapies are not limited to the application of natural viral particles but include the use of engineered phages (Breederveld 2019) and diverse products of phage origin (e.g., lytic proteins) (Torres-Barceló 2018). As an example, engineered phages can even be used to deliver agents against diseases like cancer (Korolev, Xavier, and Gore 2014). By PT we will refer to the application of natural or synthetic phages to kill bacteria via lysis.

Disadvantages of PT have been previously revised but deemed as limited as compared to its benefits (Loc-Carrillo and Abedon 2011). However, published work revising a range of PT drawbacks has been accumulating over recent years (Torres-Barceló 2018; Roach and Debarbieux 2017; Breederveld 2019). Factors on detriment of PT can be broadly categorized as determinants limiting phage infection, potential incompatibility with other therapies, difficulties on phage stocks production and stability, and safety and side effects issues (see Torres-Barceló 2018; Breederveld 2019; Roach and Debarbieux 2017 for details). Selection of resistance to phage infection and interaction of phages with the immune system stand as some of the major concerns about PT, although now more research is being dedicated on to addressing at least the former problem (Sweere et al. 2019; Wright et al. 2018). A clear example is the work by Wright et al. (2018) who comprehensively characterized the phage-bacteria interaction networks leading to cross-resistance against a large set of *P. aeruginosa* virulent phages.

Engineered phages have caused reasons to be optimistic. Pires et al. (2016) reviewed how phages may be engineered to kill only antibiotic resistant cells. Previous work has shown that engineered phages increased survival in waxworm larvae infected with resistant *E. coli*.

In addition, combinations of PT and antibiotics are currently on trial. The Phagoburn study (Phagoburn 2018), recruiting victims with wounds infected by *Escherichia coli* or *Pseudomonas aeruginosa*, involves the application of phage preparations, in some cases together with standard antibiotics. A number of patient-specific cases have also added to the hopeful wave (Chan et al. 2018; Dedrick et al. 2019; Duplessis et al. 2018; Furr et al. 2018; LaVergne et al. 2018). Relatively recently, the FDA has approved PT trials against *P. aeruginosa* (AmpliPhi 2018). Furthermore, there are also now private biotech companies aiming to apply PT (PhagoMed 2018; Adaptive Phage Therapeutics 2018), and specialized PT research centers (Center for Innovative Phage Applications and Therapeutics 2018), along with the continuous generation of PT-oriented phage collections (Merril 2010).

In light of all these advances, we would like to highlight and discuss bacteriophage-QS system (P-QS) interactions. To this end, we first briefly talk about the complex social interactions which appear as result of QS; we later present known evidence regarding P-QS interactions. We then discuss how mathematical modeling can be helpful in developing safer ways to treat bacterial infections which take into account long-term development of these known interactions.

20.3 Micro-Socio-Biology

Quorum sensing is a cell to cell signaling mechanism that enables bacteria to collectively control gene expression. *Pseudomonas aeruginosa* is involved in expression of cooperative behaviors including the production of costly exoproducts like exoenzymes and siderophores (Popat et al. 2015). Individuals that take advantage of such public goods (PG) without contributing to their production are considered social cheaters (Diggle et al. 2007) and can potentially invade the population (Sandoz, Mitzimberg, and Schuster 2007). However, QS systems regulating PG are widespread in nature and conserved among several bacterial species. Mechanisms counteracting the effects of social cheating include growth of bacteria in environments that promote the physical separation of cooperators and cheaters, thus leading to decreasing the cheater's fitness (Hense et al. 2007; Mund et al. 2016), e.g. growth of bacterial populations in highly viscous medium that limits the diffusion of the PG (Rolf et al. 2009); a high degree of relatedness between individuals, also called kin selection (Popat et al. 2015); growth utilizing a combination of QS controlled public and private goods as carbon sources (Dandekar, Chugani, and Greenberg 2012); and growth in the presence of toxic compounds such as H2O2 (García-Contreras et al. 2014) and toxic compounds like HCN (Wang et al. 2015) and pyocyanin produced by the cooperators as a policing mechanism (Castañeda-Tamez et al. 2018), which preferentially decrease the growth of cheaters.

20.4 Bacteriophage-QS System (P-QS) Interactions

Although phage-bacteria interactions are known to play a pivotal part in disease development, relatively few studies exist exploring the key role of bacterial QS in this context. This is surprising, as QS in many pathogens such as *Pseudomonas aeruginosa* regulates virulence. These studies suggest that bacteria use QS not only to induce an infection in their host but also to regulate the defense in their role of host, i.e. from phages. Our research group has shown, using an integrative approach—experimental and mathematical—that bacteriophages select active QS systems in *Pseudomonas aeruginosa* (Saucedo-Mora et al. 2017). We proposed that this phenomenon preserves bacterial QS in the environment and during infections, and may potentially select virulent bacterial strains. This is not an isolated case.

The earliest evidence we found is for *Pseudomonas aeruginosa* (Glessner et al. 1999), where reports indicate that both the *las* and *rhl* QS systems are required for type 4 pilus-dependent twitching motility and infection by the pilus-specific phage D3112cts.

How phages interact with QS signals has been reported in other bacteria such as *Vibrio anguillarum* (Tan, Svenningsen, and Middelboe 2015), since mutants that are permanently locked in a high-cell density state are almost completely immune to the phage KVP40, due to QS-mediated down regulation of the OmpK receptor used by the phage. This phenomenon contributes to a higher phage attachment and a higher death rate by the KVP40 phage toward the mutants locked in a low-cell density state.

Escherichia coli (Høyland-Kroghsbo, Maerkedahl, and Svenningsen 2013) has the ability to sense QS signals produced by other bacterial species like *N*-acyl-L-homoserine lactones (AHL), reducing the numbers of the λ phage receptor LamB; in contrast a *sdiA* mutant which is unable to sense AHL expresses high levels of LamB and is more susceptible to the λ phage. Nevertheless, no direct link between the sensing of its own population density and phage protection is reported yet for *E. coli*.

Davies et al. (2016) observed that temperate phages related to D3112 promote evolution (adaptation to biofilm environment) in cystic fibrosis (CF) sputum, including the accumulation of QS mutants. This is interesting, as it seems to contradict our findings: D3112-like phages acting as transposable mutagenic agents represent a source of mutations that are later selected, particularly on QS genes within a CF-like environment, versus D3112-like viruses acting as infective agents contributing to selection of QS active systems in competitions between wild-type (WT) and mutant. It is worth noticing, however, their work deals with mutations and selection over generations.

Hoque et al. (2016) found that QS can protect *V. cholera* against predatory phages. *V. cholera* mutants (inactivated AI synthase genes) were more susceptible to multiple phages compared to the parent bacteria. Lim et al. (2016) speculated that resistance of a *P. aeruginosa* strain to the lytic bacteriophage PB1 might be due to *las*-deficiency (Taj et al. 2014) and showed that AHL can increase T4 phage production and lysis against *Escherichia coli* and that indole reduced the lysis activity and production of T4 bacteriophage.

Qin et al. (2017) reported that QS is involved in the defense of phage K5 infection in *Pseudomonas aeruginosa*. Mumford and Friman (2017) studied QS signaling using a QS-deficient *lasR Pseudomonas aeruginosa* PAO1 and the lytic phage PT7, together with competitors *Staphylococcus aureus* and *Stenotrophomonas maltophilia*. They found that phage selection decreased the total bacterial densities in the QS-deficient lasR pathogen communities, whereas an increase was observed in pathogen communities including the QS signaling PAO1 strain. Remarkably, Silpe and Bassler (2019) recently showed that phages and *V. cholera* rely on the same signal molecule for pathogenesis and Igler and Abedon (2019) discussed some ecological explanations of these findings.

The whole scale of interaction becomes quite complex when we consider CRISPR-Cas systems. Although this is not the subject of our interest, we still discuss briefly some known results as we consider that an integrative approach should ideally not ignore this perspective.

20.5 CRISPR-Cas Systems and QS

Many *P. aeruginosa* strains, including PA14, have CRISPR-Cas systems, usually of the I-F type (van Belkum et al. 2015). CRISPR-Cas in *P. aeruginosa* confers adaptive immunity against phages and other mobile genetic elements (Cady et al. 2012). Additionally, CRISPR-Cas systems can play a role in intracellular gene regulation (Ratner, Sampson, and Weiss 2015). This also seems to be the case in *P. aeruginosa* (Cady et al. 2012) as several recent publications indicate a possible crosslinking between CRISPR-Cas and QS in PA14. CRISPR-Cas in *P. aeruginosa* affects biofilm development (Zegans et al. 2009), although the mechanisms behind this are not fully understood. In line with this finding, Wu and Li (2016) reported that the QS receptor *LasR* is modulated by CRISPR-Cas in PA14. Furthermore, it was shown that the expression of cas genes in PA14 is positively controlled by *las* and *rhl* QS, which promotes the interference with foreign DNA (Høyland-Kroghsbo et al. 2017).

CRISPR-Cas defense acts against both temperate and lytic phages. It is not fully understood how spacers for defense against new lytic phages in naïve bacteria are generated, as these phages can have fast intracellular cycles. Infected naïve cells thus may be lysed before the process of spacers acquisition from such new phages is finished. It has been shown that bacteria can solve this problem by generating spacers from defective phages mutants, which are impeded in an effective lytic cycle (Hynes, Villion, and Moineau 2014). PA14 can switch between two modes of anti-phage defense strategies depending on the environmental conditions. Constitutive defense by down regulation of surface proteins to avoid attachment is associated to constitutive costs. It dominates in nutrient rich-media, where the risk of infection is generally high. In contrast, inducible CRISPR-Cas mediated defense dominates under nutrient poor conditions, where QS may allow for an additional fine tuning. Interestingly, it has been reported that starvation in certain nutrients can upregulate QS systems in *P. aeruginosa* (Mellbye and Schuster 2011).

Høyland-Kroghsbo et al. (2017) showed that *P. aeruginosa* uses either surface protein reduction or CRISPR-Cas system activity to gain immunity against certain phages. The choice of one response over the other seems to be highly dependent on the environmental conditions (namely nutrients). They did not explore the possible role of QS in this context. Gao et al. (2015) showed QS regulation of CRISPR-Cas system in *B. glumae* PG1. However, they did not investigate whether this is connected with phage resistance. Leung et al. (2015) investigated the role of LexA in the formation of persisters induced by the CSP-ComDE QS regulatory system in *Streptococcus*. This, however, did not include phages.

Faruque et al. (2005) and Hoque et al. (2016) reported that cholera phages may play a role in emergence of new *V. cholera* pandemic serogroups or clones.

20.6 Implications for Phage-Based Therapeutics

All the discussed studies are evidence that P-QS interactions may be widespread. It is then clear that they may have important implications for phage-based therapies. We elaborate here further with an example.

Saucedo-Mora et al. (2017) found, by screening temperate and lytic bacteriophages (82 and 62, respectively), that several bacteriophages show a preference for lysing QS defective *P. aeruginosa* mutants, and specifically explored this phenomenon using the D3112 and JBD30 temperate viruses. Both phages preferentially kill the *lasR rlhR* mutant over the parental wild-type strain both in single cultures and in mixed ones in different mediums *in vitro* and *in vivo* inside *Galleria mellonella* larvae. In addition, JBD30 showed preferential attachment to the mutant when stationary phase cultures were tested (while D3112 virus was not yet tested). Nevertheless, functional phage units are significantly produced at higher rates in wild-type cultures. In addition, further experimental tests with *G. mellonella* and mathematical modeling showed that the selection of the wild type (QS proficient phenotype) by JBD30 virus increased the virulence of WT and mutant mixtures, since the WT strain shows individually 100-fold higher virulence than the mutant toward the host. Our results revealed that phages allow the selection of QS systems and counteract the exploitation of the PG by cheaters.

It was recently shown that temperate phages increase the fitness of their *P. aeruginosa* hosts *in vivo* during lung infections in competitions with the isogenic PA01 host lacking the phages. Given that we used temperate phages, we explored the impact of the lysogeny on the QS selection. Our findings suggest that during infection, temperate bacteriophages may maintain cooperative behavior by eliminating QS-deficient social cheaters that lack the phages. In contrast, if the QS-deficient mutant is carrying the temperate phage, the proportions do not significantly change relative to conditions without phage, and when both strains carried the phage from the beginning of the competition, the mutant proportions only decreased slightly.

Using a mathematical model, we showed that the net outcome with respect to the balance between *lasR rhlR* mutant and wild-type depends not only on costs of PG production and the level of

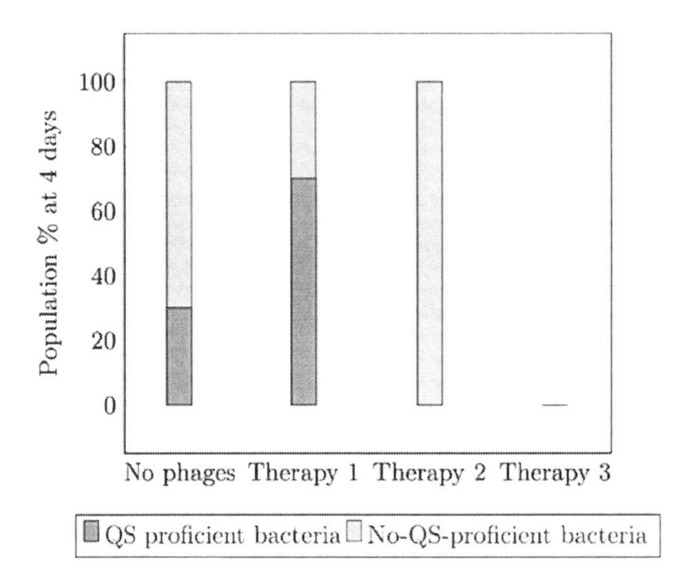

FIGURE 20.1 Mathematical modeling is used to predict outcomes of PT.

protection against phages but also on the phages concentration which dynamically changes over time, and may vary depending on the environmental conditions. We further developed a model to include explicitly the dynamics of phages, which were also measured. We explored possible outcomes to PT (see Figure 20.1) in light of the reported results regarding P-QS interactions. We found that, if not taken into consideration, there can be scenarios which lead to selection of the virulent strains.

We proposed similar approaches that can be used to test other known P-QS interactions for other bacteria and other phages, as mathematical models are generic enough to allow specifying other bacterial and phage species.

20.7 Mathematical Modeling

Mathematical models offer a generic way to investigate interactions in cellular communities' subject to stress (drugs, virus, nutrient starvation) and the possibility of testing numerically a variety of therapies and therapy protocols. This can in principle save costs and allows predictions beyond experimental possibilities. In Jalalimanesh et al. (2018) and Castañeda-Tamez et al. (2018) using a mathematical agent-based model, we were able to test how the bacterial environment diffusion properties may be contributing to stabilize QS in certain growth conditions.

There are a numbers of approaches which have been used to model QS; see our review on this topic (Pérez-Velázquez, Gölgeli, and García-Contreras 2016; Pérez-Velázquez and Hense 2018).

A series of mathematical approaches have been used in addressing issues relevant to PT, ranging from practical to complex. In Kasman et al. (2002), the so-called measure of multiplicity of infection (MOI) is deemed as inappropriate for low cell concentrations, and an alternative measure is proposed which takes into account the cell concentration and adsorption time. Ali and Wahl (2017) presented a series of mathematical models which were used to study CRISPR-Cas system effects on biofilm formation. They predict that CRISPR-immune bacteria in a biofilm will be difficult to eradicate by PT but also suggest that lysogens without a functioning CRISPR system may invade and dominate the biofilm., which has implication for therapy. If CRISPR bacteria were removed from the biofilm, PT has a much higher chance of success. In Bonsma-Fisher, Soutière, and Goyal (2018), a stochastic model of individual spacers under QS control is developed. According to their results, spacer abundance variability arises mostly due to population dynamics rather than due to individual parameters of spacers. Li et al. (2017) studied two key members of the microbiota of the freshwater metazoan *Hydra vulgaris*, *Curvibacter* sp., and *Duganella* sp. under the presence of an inducible prophage in the *Curvibacter* sp. genome; using a mathematical model, they showed that the interplay of the lysogenic life cycle of the *Curvibacter* phage and the lytic life cycle of *Duganella* sp. may be responsible for the complex competitive link between the two bacteria.

To address population interactions (such as those of phages and bacteria) using mathematical models, one can consider population dynamics. Population dynamics models are often comprised of a set of ordinary differential equations. This type of model requires measurements of variables which change in time, such as bacterial and phage amount and autoinducer concentration. We have developed a variety of models for other bacterial species (such as *P. syringae* (Pérez-Velázquez et al. 2015) and *P. putida* (Barbarossa and Kuttler 2016). Population dynamics models, however, may require to include evolutionary aspects. Tools such as adaptive dynamics, a mathematical framework for studying evolution which extends evolutionary game theory to account for more realistic ecological dynamics, and it can incorporate both frequency- and density-dependent selection, which may be considered depending on the data sets. This approach allows consideration of standard ecological models and puts them into an evolutionary perspective. Our work in this area comprises Mund et al. (2016) and Mund, Kuttler, and Pérez-Velázquez (2019).

There are also mathematical models which take into consideration spatial structure, already mentioned as playing a pivotal role in stability of cooperation. Experiments which culture the bacteria varying structure or viscosity of the environment can be used to feed these models. They can explicitly describe spatial structure or can consider compartments. They typically involve partial (which describe changes in more than one variable, for example time and space) differential equations as opposed to ordinary differential equations. Spatial structured models such as agent based models can explicitly account for cell location and include gradients of oxygen and nutrients (Jalalimanesh et al. 2018). This modeling approach allows to include bacterial cells properties such as QS state or whether a cell is lysogenic.

Modeling complex environments such as bacterial habitats, adding interactions with phages, and accounting for QS may a require a combination of tools, often leading to hybrid models, which are composed of both spatially explicit models and differential equations which track changes in both time and space.

Overall, mathematical models can be used toward exploring the long-term dynamics of the interactions of the WT and QS mutants. For example, to further explore whether phages could promote the expression of QS controlled virulence factors, rather than attenuate them. Different therapy protocols can be tested and simulated with the associated mathematical models, for example including QS inhibition. Fozard et al. (2012) developed a computational model (individual-based) of QQ for a developing biofilm. Anand, Rai, and Thattai (2013) used an experimentally validated computational model of LuxI/LuxR QS to study the effects of using inhibitors, both when used individually and combined.

We now briefly discuss how the other evidence of P-QS may be revised in a similar light to our results. According to the results of Tan, Svenningsen, and Middelboe (2015) in a mixture of WT bacteria and QS-defective mutants, PT would select the WT phenotype since the expression of the phage receptor in the QS mutants will be higher. This outcome can potentially increase the populations virulence. This can be tested with a mathematical model.

The results of Mumford and Friman (2017) seem to imply that in the absence of other bacteria, phage will select the less virulent population (QS mutants). However, we do not know for sure as they did not do the associated experiments; therefore, a mathematical model can help simulate this scenario.

20.8 Discussion

This current antibiotic crisis can be largely blamed on our lack our knowledge of bacterial life-styles and survival strategies at the time we started to use antibiotics widely. It was unknown to us that bacteria had millions of years of evolution behind them; we also ignored how virulence in bacteria was regulated, and that they dispose of a series of sophisticated mechanisms to survive (such as efflux pumps and antibiotic resistance enzymes). Now we possess enough evidence that may prevent us from committing similar mistakes and there exist a variety of aspects one can target to better understand bacterial infection development and therefore improve the efficacy of already developed therapies or propose complementary ones.

PT features prominently as an alternative to antibiotics to fight bacterial infections. Solid evidence may help us ensure that PT is used safely. An interdisciplinary approach which includes mathematical models can be useful here.

Several sources of stress, either biotic or abiotic, which bacteria are commonly exposed to in their natural environments, potentially contribute to the selection of functional QS systems. Phages are an omnipresent instance of biotic stress for bacteria. Our aim was to bring to the discussion of how phages are involved in the evolutionary stability of bacterial QS systems and what are the implications of this for PT. More importantly, how this interlinks may impact the success or failure of PT. Mathematical models can help evaluate a wide range of possibilities.

P-QS interactions must be to be thoroughly explored and understood. Interestingly, until now, this has not been widely investigated. Our research group has shown, using an integrative approach—experimental and mathematical—that bacteriophages select active QS systems in *Pseudomonas aeruginosa*. Under some circumstances, our results imply the promotion of increased virulence by bacteriophages, potentially threatening its safety of use in humans.

Bruchmann, Kirchen, and Schwartz (2013) suggested that subinhibitory concentrations of antibiotics may promote QS (and by this probably resistance to phages) if they do not kill the bacteria. It is likely that in chronic infections not all cells are reached by sufficient high concentrations of antibiotics to kill them. Thus, a combination of QQ drugs (which might also be antibiotic active) and PT represents an alternative to test.

Of pivotal importance, it is to also minimize human body's beneficial microbes' annihilation through the use of antibiotics. Phage-based therapeutics has been put forward as a promising alternative (Matsuzaki et al. 2014), listed as one of seven prongs to combat antibiotic resistance (Reardon 2014). In a recent opinion paper, Young and Gill (2015) called for a better understanding of the phage-host interactions, in order for phage-based therapeutics to become a reality. Phage therapy has not prospered, according to them, due to the poor understanding of how phages function.

Although QS has been shown to play a role in phage response in *Vibrio anguillarum* and in *Escherichia coli*, the QS-phage tolerance is so far under explored in *P. aeruginosa*, an archetypical Gram negative bacterial model for QS studies.

For *E. coli* and *V. anguillarum*, it had been suggested that a possible advantage for therapeutics would be the simultaneous application of QS disruption therapies and bacteriophages (which will be more active against QS deficient bacteria), and since we demonstrated the same phenomenon for *P. aeruginosa*, we suggest this possibility should be explored both experimentally and theoretically with our model bacterial model, in order to optimize treatment schemes and hence propose strategies that increase chances of archiving bacterial clearance and hence the elimination of infections.

Some phages have developed strategies to overcome CRSPR–Cas defense. We intuitively hypothesized that such phages may not promote high virulent WT PA14 compared to QS mutants, as PA14 should have no fitness advantage in the presence of these phages. Surprisingly, we find the opposite. We currently do not know whether this is an effect of an unknown mechanism in the CRISPR-Cas phage interplay, or caused by non-CRISPR-Cas mediated defense.

Our research group is not only interested in further exploring these interactions but also in developing tools and methods which may allow to test and compare possible scenarios with the help of an integrated approach. In particular, the combination of QQ and phages, QS-Phages and antibiotics and other therapies which take into consideration the complex interactions between bacteria and phages. More importantly, the use of complementary methods such as mathematical modeling can help simulate a range of scenarios which cannot be tested experimentally.

Bottom line is that a form of resistance to therapy, a fight for survival and perseverance seems inevitable, as this is an intrinsic part of life. Recent studies have called for a change in the strategy toward improving therapy efficiency: using ecology and evolutionary principles to predict and prevent resistance development (Baquero, Coque, and de la Cruz 2011; Korolev, Xavier, and Gore 2014). The aim of this work was to show that this approach is indeed possible when one considers a wider context of interactions and a broad set of tools to tackle the complexity of phage-bacteria-QS interaction and its key impact for infection development.

REFERENCES

"Adaptive Phage Therapeutics." 2018. http://www.aphage.com, Merril and Greg Merril (eds.).

Ali, Qasim, and Lindi M. Wahl. 2017. "Mathematical Modelling of CRISPR-Cas System Effects on Biofilm Formation." *Journal of Biological Dynamics* 11 (sup2): 264–84. https://doi.org/10.1080/17513758.2017.1314025.

AmpliPhi. 2018. "AmpliPhi Biosciences Announces Positive FDA Feedback for Its Clinical Stage Bacteriophage Product Candidate AB-PA01 Targeting *Pseudomonas aeruginosa* Infections." 2018. https://www.businesswire.com/news/home/20180918005339/en/AmpliPhi-Biosciences-Announces-Positive-FDA-Feedback-Clinical.

Anand, Rajat, Navneet Rai, and Mukund Thattai. 2013. "Interactions among Quorum Sensing Inhibitors." *PLoS One* 8 (4): e62254. https://doi.org/10.1371/journal.pone.0062254.

Chan BK, Turner PE, Kim S, Mojibian HR, Elefteriades JA, Narayan D. Phage treatment of an aortic graft infected with Pseudomonas aeruginosa. Evol Med Public Health. 2018;2018(1):60–66. Published 2018 Mar 8. doi:10.1093/emph/eoy005

Baquero, Fernando, Teresa M. Coque, and Fernando de la Cruz. 2011. "Ecology and Evolution as Targets: The Need for Novel Eco-Evo Drugs and Strategies to Fight Antibiotic Resistance." *Antimicrobial Agents and Chemotherapy* 55 (8): 3649–60. https://doi.org/10.1128/AAC.00013-11.

Barbarossa, V. Maria, and Christina Kuttler. 2016. "Mathematical Modeling of Bacteria Communication in Continuous Cultures." *Applied Sciences*. https://doi.org/10.3390/app6050149.

Belkum, Alex van, Leah B. Soriaga, Matthew C. LaFave, Srividya Akella, Jean-Baptiste Veyrieras, E. Magda Barbu, Dee Shortridge, et al. 2015. "Phylogenetic Distribution of CRISPR-Cas Systems in Antibiotic-Resistant *Pseudomonas aeruginosa*." *MBio* 6 (6): e01796–15. https://doi.org/10.1128/mBio.01796-15.

Bonsma-Fisher, Madeleine, Dominique Soutière, and Sidhartha Goyal. 2018. "How Adaptive Immunity Constrains the Composition and Fate of Large Bacterial Populations." *Proceedings of the National Academy of Sciences* 115 (32): E7462–E7468. https://doi.org/10.1073/pnas.1802887115.

Breederveld, Roelf S. 2019. "Phage Therapy 2.0: Where Do We Stand?" *Lancet Infectious Diseases* 19 (1): 2–3.

Bruchmann, Julia, Silke Kirchen, and Thomas Schwartz. 2013. "Sub-Inhibitory Concentrations of Antibiotics and Wastewater Influencing Biofilm Formation and Gene Expression of Multi-Resistant *Pseudomonas aeruginosa* Wastewater Isolates." *Environmental Science and Pollution Research* 20 (6): 3539–49. https://doi.org/10.1007/s11356-013-1521-4.

Cady, Kyle C., Joe Bondy-Denomy, Gary E. Heussler, Alan R. Davidson, and George A. Toole. 2012. "The CRISPR/Cas Adaptive Immune System of *Pseudomonas aeruginosa* Mediates Resistance to Naturally Occurring and Engineered Phages." *Journal of Bacteriology* 194 (21): 5728LP–5738. https://doi.org/10.1128/JB.01184-12.

Castañeda-Tamez, Paulina, Jimena Ramírez-Peris, Judith Pérez-Velázquez, Christina Kuttler, Ammar Jalalimanesh, Miguel Á Saucedo-Mora, J. Guillermo Jiménez-Cortés, et al. 2018. "Pyocyanin Restricts Social Cheating in *Pseudomonas aeruginosa*." *Frontiers in Microbiology* 9: 1348. https://doi.org/10.3389/fmicb.2018.01348.

Center for Innovative Phage Applications and Therapeutics. 2018. "Center for Innovative Phage Applications and Therapeutics." UC San Diego School of Medicine, Steffanie Strathdee and Robert "Chip" Schooley (eds.).

Chan Benjamin K., Paul E. Turner, Samuel Kim, Hamid R. Mojibian, John A. Elefteriades, and Deepak Narayan. 2018. Phage treatment of an aortic graft infected with *Pseudomonas aeruginosa. Evolution, Medicine & Public Health* 2018 (1): 60–66. doi:10.1093/emph/eoy005

Dandekar, Ajai A., Sudha Chugani, and E. Peter Greenberg. 2012. "Bacterial Quorum Sensing and Metabolic Incentives to Cooperate." *Science* 338 (6104): 264LP–266. https://doi.org/10.1126/science.1227289.

Davies, Emily V., Chloe E. James, Irena Kukavica-Ibrulj, Roger C. Levesque, Michael A. Brockhurst, and Craig Winstanley. 2016. "Temperate Phages Enhance Pathogen Fitness in Chronic Lung Infection." *The ISME Journal* 10 (10): 2553–2555. https://doi.org/10.1038/ismej.2016.51.

Dedrick, Rebekah M., Carlos A. Guerrero-Bustamante, Rebecca A. Garlena, Daniel A. Russell, Katrina Ford, Kathryn Harris, Kimberly C. Gilmour, et al. 2019. "Engineered Bacteriophages for Treatment of a Patient with a Disseminated Drug-Resistant Mycobacterium Abscessus." *Nature Medicine* 25 (5): 730–33. https://doi.org/10.1038/s41591-019-0437-z.

Diggle, Stephen P., Ashleigh S. Griffin, Genevieve S. Campbell, and Stuart A. West. 2007. "Cooperation and Conflict in Quorum-Sensing Bacterial Populations." *Nature* 450: 411. https://doi.org/10.1038/nature06279.

Duplessis, C., B. Biswas, B. Hanisch, M. Perkins, M. Henry, J. Quinones, D. Wolfe, L. Estrella, and T. Hamilton. 2018. "Refractory *Pseudomonas* Bacteremia in a 2-Year-Old Sterilized by Bacteriophage Therapy." *Journal of the Pediatric Infectious Diseases Society* 7 (3): 253–56. https://doi.org/10.1093/jpids/pix056.

Faruque, Shah M., Iftekhar Bin Naser, M. Johirul Islam, A. S. G. Faruque, A. N. Ghosh, G. Balakrish Nair, David A. Sack, and John J. Mekalanos. 2005. "Seasonal Epidemics of Cholera Inversely Correlate with the Prevalence of Environmental Cholera Phages." *Proceedings of the National Academy of Sciences* 102 (5): 1702–7. https://doi.org/10.1073/pnas.0408992102.

Fozard, John A., Michael Lees, J. R. King, and B. S. Logan. 2012. "Inhibition of Quorum Sensing in a Computational Biofilm Simulation." *Biosystems* 109 (2): 105–14. https://doi.org/10.1016/j.biosystems.2012.02.002.

Furr, C.-L. L., S. M. Lehman, S. P. Morales, F. X. Rosas, A. Gaidamaka, I. P. Bilinsky, P. C. Grint, R. T. Schooley, and S. Aslam. 2018. "P084 Bacteriophage Treatment of Multidrug-Resistant *Pseudomonas aeruginosa* Pneumonia in a Cystic Fibrosis Patient." *Journal of Cystic Fibrosis* 17: S83. https://doi.org/10.1016/S1569-1993(18)30381-3.

Gao, Rong, Dagmar Krysciak, Katrin Petersen, Christian Utpatel, Andreas Knapp, Christel Schmeisser, Rolf Daniel, Sonja Voget, Karl-Erich Jaeger, and Wolfgang R. Streit. 2015. "Genome-Wide RNA Sequencing Analysis of Quorum Sensing-Controlled Regulons in the Plant-Associated *Burkholderia glumae* PG1 Strain." *Applied and Environmental Microbiology* 81 (23): 7993–8007. https://doi.org/10.1128/AEM.01043-15.

García-Contreras, Rodolfo, Leslie Nuñez-López, Ricardo Jasso-Chávez, Brian W. Kwan, Javier A. Belmont, Adrián Rangel-Vega, Toshinari Maeda, and Thomas K. Wood. 2014. "Quorum Sensing Enhancement of the Stress Response Promotes Resistance to Quorum Quenching and Prevents Social Cheating." *The Isme Journal* 9: 115. https://doi.org/10.1038/ismej.2014.98.

Glessner, Alex, Roger S. Smith, Barbara H. Iglewski, and Jayne B. Robinson. 1999. "Roles of *Pseudomonas aeruginosa* Las and Rhl Quorum-Sensing Systems in Control of Twitching Motility." *Journal of Bacteriology* 181 (5): 1623–29. https://www.ncbi.nlm.nih.gov/pubmed/10049396.

Hense, Burkhard A., Christina Kuttler, Johannes Müller, Michael Rothballer, Anton Hartmann, and Jan-Ulrich Kreft. 2007. "Does Efficiency Sensing Unify Diffusion and Quorum Sensing?" *Nature Reviews Microbiology* 5: 230. https://doi.org/10.1038/nrmicro1600.

Hoque, M. Mozammel, Iftekhar Bin Naser, S. M. Nayeemul Bari, Jun Zhu, John J. Mekalanos, and Shah M. Faruque. 2016. "Quorum Regulated Resistance of *Vibrio cholerae* against Environmental Bacteriophages." *Scientific Reports* 6: 37956. https://doi.org/10.1038/srep37956.

Høyland-Kroghsbo, Nina M., Jon Paczkowski, Sampriti Mukherjee, Jenny Broniewski, Edze Westra, Joseph Bondy-Denomy, and Bonnie L. Bassler. 2017. "Quorum Sensing Controls the *Pseudomonas aeruginosa* CRISPR-Cas Adaptive Immune System." *Proceedings of the National Academy of Sciences* 114 (1): 131–35. https://doi.org/10.1073/pnas.1617415113.

Høyland-Kroghsbo, Nina Molin, Rasmus Baadsgaard Maerkedahl, and Sine Lo Svenningsen. 2013. "A Quorum-Sensing-Induced Bacteriophage Defense Mechanism." *mBio* 4 (1): e00362; e00362–12. https://doi.org/10.1128/mBio.00362-12.

Hynes, Alexander P., Manuela Villion, and Sylvain Moineau. 2014. "Adaptation in Bacterial CRISPR-Cas Immunity Can Be Driven by Defective Phages." *Nature Communications* 5: 4399. https://doi.org/10.1038/ncomms5399.

Igler, Claudia, and Stephen T. Abedon. 2019. "Commentary: A Host-Produced Quorum-Sensing Autoinducer Controls a Phage Lysis-Lysogeny Decision." *Frontiers in Microbiology* 10: 1171. https://doi.org/10.3389/fmicb.2019.01171.

Jalalimanesh, Ammar, Christina Kuttler, Rodolfo García-Contreras, and Judith Pérez-Velázquez. 2018. "An Agent-Based Model to Study Selection of *Pseudomonas aeruginosa* Quorum Sensing by Pyocyanin: A Multidisciplinary Perspective on Bacterial Communication BT." In, *Quantitative Models for Microscopic to Macroscopic Biological Macromolecules and Tissues* edited by Luis Olivares-Quiroz and Osbaldo Resendis-Antonio, 133–47. Cham, Switzerland: Springer International Publishing. https://doi.org/10.1007/978-3-319-73975-5_7.

Kasman, Laura M., Alex Kasman, Caroline Westwater, Joseph Dolan, Michael G. Schmidt, and James S. Norris. 2002. "Overcoming the Phage Replication Threshold: A Mathematical Model with Implications for Phage Therapy." *Journal of Virology* 76 (11): 5557–64. https://doi.org/10.1128/JVI.76.11.5557-5564.2002.

Korolev, Kirill S., Joao B. Xavier, and Jeff Gore. 2014. "Turning Ecology and Evolution against Cancer." *Nature Reviews Cancer* 14 (5): 371–80. http://dx.doi.org/10.1038/nrc3712.

LaVergne, Stephanie, Theron Hamilton, Biswajit Biswas, M. Kumaraswamy, R. T. Schooley, and Darcy Wooten. 2018. "Phage Therapy for a Multidrug-Resistant Acinetobacter Baumannii Craniectomy Site Infection." *Open Forum Infectious Diseases* 5 (4): ofy064. https://doi.org/10.1093/ofid/ofy064.

Leung, Vincent, Dragana Ajdic, Stephanie Koyanagi, and Céline M Lévesque. 2015. "The Formation of *Streptococcus mutans* Persisters Induced by the Quorum-Sensing Peptide Pheromone Is Affected by the LexA Regulator." *Journal of Bacteriology* 197 (6): 1083–94. https://doi.org/10.1128/JB.02496-14.

Li, Xiang-Yi, Tim Lachnit, Sebastian Fraune, Thomas C. G. Bosch, Arne Traulsen, and Michael Sieber. 2017. "Temperate Phages as Self-Replicating Weapons in Bacterial Competition." *Journal of The Royal Society Interface* 14 (137). https://doi.org/10.1098/rsif.2017.0563.

Lim, Wee S., Kevin K. S. Phang, Andy Tan, Sam F.-Y. Li, and Dave Ow. 2016. "Small Colony Variants and Single Nucleotide Variations in Pf1 Region of PB1 Phage-Resistant *Pseudomonas aeruginosa*." *Frontiers in Microbiology* 7. https://doi.org/10.3389/fmicb.2016.00282.

Loc-Carrillo, Catherine, and Stephen T. Abedon. 2011. "Pros and Cons of Phage Therapy." *Bacteriophage* 1 (2): 111–14. https://doi.org/10.4161/bact.1.2.14590.

Matsuzaki, Shigenobu, Jumpei Uchiyama, Iyo Takemura-Uchiyama, and Masanori Daibata. 2014. "Perspective: The Age of the Phage." *Nature* 509: S9. https://doi.org/10.1038/509S9a.

Mellbye, Brett, and Martin Schuster. 2011. "The Sociomicrobiology of Antivirulence Drug Resistance: A Proof of Concept." *mBio* 2 (5): e00131–11. https://doi.org/10.1128/mBio.00131-11.

Merril, Carl R. 2010. "PhageBank\texttrademark. Biological Defense Research Directorate (BDRD) of the U.S. Naval Medical Research Center (NMRC)."

Mumford, Rachel, and Ville Petri Friman. 2017. "Bacterial Competition and Quorum-Sensing Signalling Shape the Eco-Evolutionary Outcomes of Model In Vitro Phage Therapy." *Evolutionary Applications* 10 (2): 161–69. https://doi.org/10.1111/eva.12435.

Mund, Anne, C. Kuttler, Judith Pérez-Velázquez, and Burkhard A. Hense. 2016. "An Age-Dependent Model to Analyse the Evolutionary Stability of Bacterial Quorum Sensing." *Journal of Theoretical Biology* 405: 104–15. https://doi.org/10.1016/j.jtbi.2015.12.021.

Mund, Anne, Christina Kuttler, and Judith Pérez-Velázquez. 2019. "Existence and Uniqueness of Solutions to a Family of Semi-Linear Parabolic Systems Using Coupled Upper-Lower Solutions." *Discrete & Continuous Dynamical Systems—B* 22 (11): 1–13. https://doi.org/10.3934/dcdsb.2019102.

Pérez-Velázquez, Judith, Meltem Gölgeli, and Rodolfo García-Contreras. 2016. "Mathematical Modelling of Bacterial Quorum Sensing: A Review." *Bulletin of Mathematical Biology* 78 (8): 1585–639. https://doi.org/10.1007/s11538-016-0160-6.

Pérez-Velázquez, Judith, and Burkhard A. Hense. 2018. "Differential Equations Models to Study Quorum Sensing BT—Quorum Sensing: Methods and Protocols." In, edited by Livia Leoni and Giordano Rampioni, 253–71. New York: Springer New York. https://doi.org/10.1007/978-1-4939-7309-5_20.

Pérez-Velázquez, Judith, Beatriz Quiñones, Burkhard A. Hense, and Christina Kuttler. 2015. "A Mathematical Model to Investigate Quorum Sensing Regulation and Its Heterogeneity in *Pseudomonas syringae* on Leaves." *Ecological Complexity* 21: 128–41. https://doi.org/10.1016/j.ecocom.2014.12.003.

Phagoburn. 2018. "Phagoburn." European Commission under the 7th Framework Programme for Research and Development, http://www.phagoburn.eu.

PhagoMed. 2018. "PhagoMed." https://www.phagomed.com/

Pires, Diana P., Sara Cleto, Sanna Sillankorva, Joana Azeredo, and Timothy K. Lu. 2016. "Genetically Engineered Phages: A Review of Advances over the Last Decade." *Microbiology and Molecular Biology Reviews* 80 (3): 523LP–543. https://doi.org/10.1128/MMBR.00069-15.

Popat, R., D. M. Cornforth, L. McNally, and S. P. Brown. 2015. "Collective Sensing and Collective Responses in Quorum-Sensing Bacteria." *Journal of The Royal Society Interface* 12 (103): 20140882. https://doi.org/10.1098/rsif.2014.0882.

Qin, Xuying, Qinghui Sun, Baixue Yang, Xuewei Pan, Yang He, and Hongjiang Yang. 2017. "Quorum Sensing Influences Phage Infection Efficiency via Affecting Cell Population and Physiological State." *Journal of Basic Microbiology* 57 (2): 162–70. https://doi.org/10.1002/jobm.201600510.

Ratner, Hannah K., Timothy R. Sampson, and David S. Weiss. 2015. "I Can See CRISPR Now, Even When Phage Are Gone: A View on Alternative CRISPR-Cas Functions from the Prokaryotic Envelope." *Current Opinion in Infectious Diseases* 28 (3): 267–74. https://doi.org/10.1097/QCO.0000000000000154.

Reardon, Sara. 2014. "Phage Therapy Gets Revitalized." *Nature* 510: 15. https://doi.org/10.1038/510015a.

Roach, Dwayne R., and Laurent Debarbieux. 2017. "Phage Therapy: Awakening a Sleeping Giant." *Emerging Topics in Life Sciences* 1 (1): 93–103. https://doi.org/10.1042/ETLS20170002.

Rolf, Kümmerli, Griffin Ashleigh S., West Stuart A., Buckling Angus, and Harrison Freya. 2009. "Viscous Medium Promotes Cooperation in the Pathogenic Bacterium *Pseudomonas aeruginosa*." *Proceedings of the Royal Society B: Biological Sciences* 276 (1672): 3531–38. https://doi.org/10.1098/rspb.2009.0861.

Sandoz, Kelsi M., Shelby M. Mitzimberg, and Martin Schuster. 2007. "Social Cheating in *Pseudomonas aeruginosa* Quorum Sensing." *Proceedings of the National Academy of Sciences* 104 (40): 15876–81. https://doi.org/10.1073/pnas.0705653104.

Saucedo-Mora, Miguel A., Paulina Castañeda-Tamez, Adrián Cazares, Judith Pérez-Velázquez, Burkhard A. Hense, Daniel Cazares, Wendy Figueroa, et al. 2017. "Selection of Functional Quorum Sensing Systems by Lysogenic Bacteriophages in *Pseudomonas aeruginosa*." *Frontiers in Microbiology* 8: 1669. https://doi.org/10.3389/fmicb.2017.01669.

Silpe, Justin E., and Bonnie L. Bassler. 2019. "A Host-Produced Quorum-Sensing Autoinducer Controls a Phage Lysis-Lysogeny Decision." *Cell* 176 (1): 268–80.e13. https://doi.org/10.1016/j.cell.2018.10.059.

Sweere, Johanna M., Jonas D. Van Belleghem, Heather Ishak, Michelle S. Bach, Medeea Popescu, Vivekananda Sunkari, Gernot Kaber, et al. 2019. "Bacteriophage Trigger Antiviral Immunity and Prevent Clearance of Bacterial Infection." *Science* 363 (6434): eaat9691. https://doi.org/10.1126/science.aat9691.

Taj, Muhammad, Lin Lian Bing, Qi Zhang, Ji Ling, Imran Hassani, Taj Hassani, Zohra Samreen, Aslam Mangle, and Wei Yunlin. 2014. "Quorum Sensing Molecules Acyl-Homoserine Lactones and Indole Effect on T4 Bacteriophage Production and Lysis Activity." *Pakistan Veterinary Journal* 34: 397–99.

Tan, Demeng, Sine Lo Svenningsen, and Mathias Middelboe. 2015. "Quorum Sensing Determines the Choice of Antiphage Defense Strategy in *Vibrio anguillarum*." *mBio* 6 (3). https://doi.org/10.1128/mBio.00627-15.

Torres-Barceló, Clara. 2018. "Phage Therapy Faces Evolutionary Challenges." *Viruses* 10 (6). https://doi.org/10.3390/v10060323.

Wang, Meizhen, Amy L. Schaefer, Ajai A. Dandekar, and E. Peter Greenberg. 2015. "Quorum Sensing and Policing of *Pseudomonas aeruginosa* Social Cheaters." *Proceedings of the National Academy of Sciences* 112 (7): 2187 LP–2191. https://doi.org/10.1073/pnas.1500704112.

Wright, Rosanna C. T., Ville-Petri Friman, Margaret C. M. Smith, and Michael A. Brockhurst. 2018. "Cross-Resistance Is Modular in Bacteria–Phage Interactions." *PLoS Biology* 16 (10): 1–22. https://doi.org/10.1371/journal.pbio.2006057.

Wu, Min, and Rongpeng Li. 2016. "A Novel Role of the Type I CRISPR-Cas System in Impairing Host Immunity by Targeting Endogenous Genes." *The Journal of Immunology* 196 (1 Supplement): 200.14. http://www.jimmunol.org/content/196/1_Supplement/200.14.

Young, Ry, and Jason J. Gill. 2015. "MICROBIOLOGY. Phage Therapy Redux—What Is to Be Done?" *Science* 350 (6265): 1163–64.

Zegans, Michael E., Jeffrey C. Wagner, Kyle C. Cady, Daniel M. Murphy, John H. Hammond, and George A. O'Toole. 2009. "Interaction between Bacteriophage DMS3 and Host CRISPR Region Inhibits Group Behaviors of *Pseudomonas aeruginosa*." *Journal of Bacteriology* 191 (1): 210LP–219. https://doi.org/10.1128/JB.00797-08.

21

Quorum Quenching Monoclonal Antibodies for the Detection and Treatment of Gram-Negative Bacterial Infections

Soumya Palliyil

CONTENTS

21.1 *Pseudomonas aeruginosa* Pathogenesis and Quorum Sensing

The human pathogen *Pseudomonas aeruginosa* causes life threatening acute and chronic infections in the lungs of cystic fibrosis patients and is a leading cause of ventilator associated pneumonia, bacteraemia, urinary catheter infections, and wound and surgical/transplantation infections. It is intrinsically resistant to multiple classes of antimicrobial agents used in clinics and expresses an array of virulence factors which can cause extensive damage to the host tissue during infection. These factors along with the ability of the organism to adapt to environmental changes make it a challenging pathogen to treat in patients. According to the Center for Disease Control (CDC) figures, 13% of the estimated 51,000 healthcare associated *P. aeruginosa* infections in the United States are caused by multidrug-resistant strains attributing to the death of roughly 400 individuals each year. Following initial colonization, along with its intrinsic ability to develop resistance, *Pseudomonas* expresses a large number of virulence factors which mediate its survival in the host. However following infection, the bacteria is exposed to inflammatory responses including oxidative stress mediated by the host immune system and chemotoxic stress induced by antibiotic treatment. These environmental stress factors result in the switching to a persistent, alginate overproducing mucoid phenotype which exhibits enhanced resistance to multiple antibiotics and host mediated immune response (Malhotra et al. 2018, Pedersen et al. 1992). In addition to expression of alginates, *Pseudomonas* also produces an array of virulence factors that cause host tissue damage and evasion of host immune response. Extensive studies have shown the role of *P. aeruginosa* quorum sensing (QS) circuits in orchestrating pathogenesis by regulating the expression of extracellular virulence factors (and also biofilm formation) (Miller and Bassler 2001). QS is generally controlled by low molecular weight autoinducer compounds which accumulate and reach a critical threshold concentration as bacterial numbers increase in the surrounding environment and interact with cognate receptors, triggering the expression of a set of target genes.

21.2 Quorum Sensing Networks in *P. aeruginosa*

QS circuits in *Pseudomonas* are highly interconnected and organized in a hierarchical order to respond to external stress cues and control of virulence genes expression. Four sets have been reported so far, namely *las*, *rhl*, *pqs*, and *iqs*, with the *las* system at the top of this signaling hierarchy (de Kievit and Iglewski 2000, Gambello and Iglewski 1991, Pesci et al. 1999, Lee et al. 2013). The Las and Rhl systems are acyl homoserine lactone (AHL) dependent circuits where *N*-3-oxododecanoyl-homoserine lactone (3-oxo-C12-HSL) and *N*-butyryl-homoserine lactone (C4-HSL) act as respective signaling molecules. Extracellular virulence factors such as flagellum, adhesion factors including alginate, exotoxin A, exoenzyme S, elastases (LasA and LasB), alkaline protease, rhamnolipids, hydrogen cyanide, and phospholipase C are known to be controlled by the *las* and *rhl* systems and the bacteria produce these factors in a cell density dependent manner (de Kievit and Iglewski 2000, Whiteley, Lee, and Greenberg 1999). Apart from controlling the expression of virulence factors, 3-oxo-C12-HSL itself exerts significant immuno modulatory effects on mammalian immune responses. Telford et al. (1998, 36–42) showed that this compound (at 10–30 μM concentrations) reduced lymphocyte responses, especially T cell activation and proliferation in mouse spleens, suppressed TNF-α and IL-12 cytokine production by murine peritoneal macrophages, and at higher concentrations (80 μM) suppressed antibody production. A third QS signal called 2-heptyl-3-hydroxy-4-quinolone (PQS), which is chemically different from AHL autoinducers, also regulates the production of elastase, 3-oxo-C12-HSL, phospholipase, and pyocyanin (McGrath, Wade, and Pesci 2004). PQS synthesis is controlled by the Las and Rhl systems, and its bioactivity is dependent on the Rhl system (Pesci et al. 1999). In addition, PQS has been shown to induce the expression of RhlI autoinducer synthase which directs the synthesis of C4-HSL, thereby serving as a connecting link between the las and rhl signaling systems (McKnight, Iglewski, and Pesci 2000). Higher levels of PQS expression are observed at late stationary phase, suggesting that they are not involved in sensing cell density. PQS and its precursor compounds were found in the sputum of CF patients and positively correlated with the density of *P. aeruginosa* in these samples (Collier et al. 2002). Most recently a fourth intercellular signaling molecule called 2-(2-hydroxyphenyl)-thiazole-4-carbaldehyde has been identified and the circuit it controls is named integrated quorum sensing (IQS) system (Lee et al. 2013). The IQS system expression is tightly controlled by the *las* system and controls the production of PQS, C4-HSL, and virulence factors such as pyocyanin, rhamnolipids, and elastase. Whole genome studies have revealed up to 10% *Pseudomonas* genes regulated by QS and broadened its role beyond controlling the expression of virulence factors to a global regulatory system mediating cellular functions as well (Schuster and Greenberg 2006).

21.3 Quorum Quenching Monoclonal Antibodies

The ability of microbes to develop resistance against the drugs targeting them is increasing at an alarming speed, overtaking the rate of discovery of novel antimicrobial agents. This combined with the lack of financial incentives for developing new compounds and hurdles associated with regulatory approvals have slowed down big pharma's interest in anti-infectives therapeutics resulting in the decline of a robust drug discovery pipeline combating multiple pathogens. Antimicrobial resistance (AMR), if left unchecked, is predicted to kill 10 million people each year by 2050, more than the number of deaths caused by cancer (The review on antimicrobial resistance, O'Neill 2016). There is a real need to develop alternative therapies with novel mechanisms of action such as anti-virulence strategies, drugs that can prevent resistance development, and the use of biologics therapies such as monoclonal antibodies in combination with existing antibiotics to target unconventional pathways and activating the body's immune system for effective clearance of drug-resistant bacteria. Along with novel therapies, the battle against AMR should be supported by better diagnostics including rapid point of care (POC) tests differentiating bacterial and viral infections, tests to identify resistance, narrow spectrum diagnostics that are based on pathogen specific biomarkers, and tests that allow patient stratification and recruitment into clinical trials. Our laboratory investigated the immuno modulation of QS molecules by monoclonal antibodies (mAbs) as a novel approach to prevent *P. aeruginosa* infections and as tools to detect these compounds in bodily fluids as a possible first clue to an undiagnosed Gram-negative infection. *Pseudomonas* isolates from the lungs of cystic fibrosis patients with chronic infection were found to produce 3-oxo-C12 HSL and C4-HSL QS molecules, with a high number of bacteria switching to mucoid strains maintaining C4-HSL and rhamnolipids expression throughout the late stages of infection (Bjarnsholt et al. 2010). Isolates from patients with urinary tract infections associated with catherization were shown to produce several AHL molecules including C4-HSL, 3-oxo-C12 HSL (Kumar et al. 2011). Studies reported by Barr et al. (2015) showed the presence of QS signaling molecules (including PQS, HHQ, 3-oxo-C12 HSL, C4-HSL) in the sputum, plasma, and urine samples of CF patients with chronic *P. aeruginosa* infections and the values positively correlated with pulmonary bacterial load. The presence of QS compounds in bodily fluids makes them attractive biomarkers for developing rapid noninvasive POC diagnostic tests for detecting *Pseudomonas* in early pulmonary infections. The extracellular distribution of QS signals makes them ideal targets for anti-infective therapy since the evolutionary pressure on bacteria to develop resistance will be limited (Figure 21.1). Several laboratories have reported the development of anti-QS agents and demonstrated their ability to reduce bacterial virulence and drug resistance and their

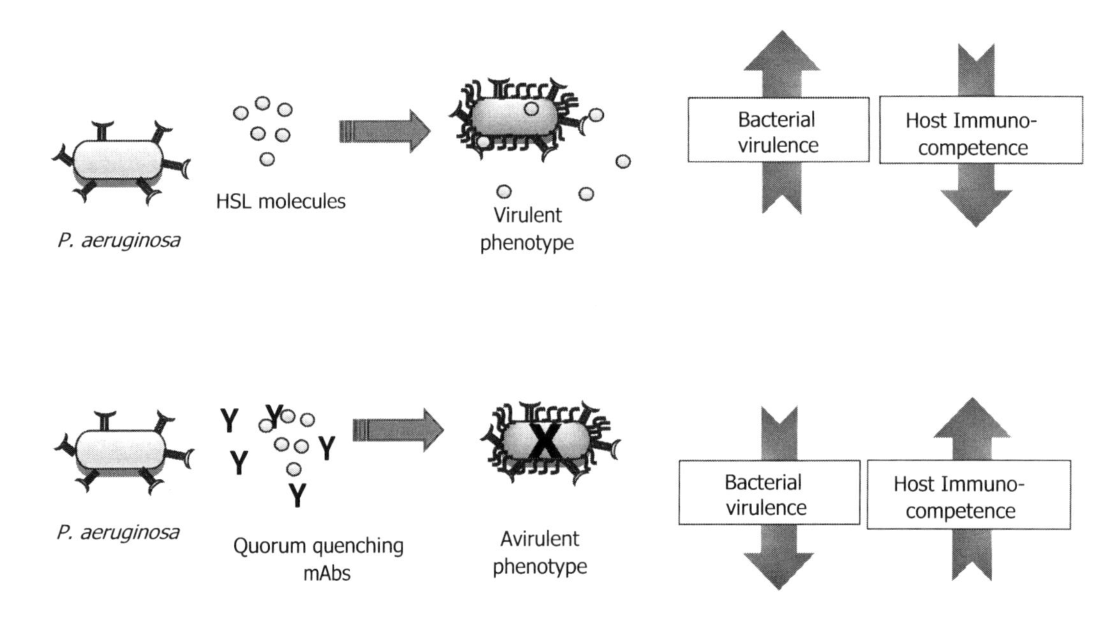

FIGURE 21.1 Diagrammatic representation of antibody mediated quorum quenching and virulence control in *Pseudomonas aeruginosa* infections.

synergistic effects in combination therapies with existing antibiotics (Jiang et al. 2019). Monoclonal antibody mediated inhibition of the QS process may offer exquisite target specificity and less off-target cytotoxicity. Quorum quenching using antibodies would not only block the signaling cascade and virulence factors production, but also neutralize 3-oxo-C_{12}-HSL, a molecule that beyond its action as a QS signal has immunomodulatory effects important for *P. aeruginosa* infections.

In our laboratory, a sheep immunization strategy was adopted to develop antibodies with high affinity and sensitivity toward HSL compounds. Sheep immunization was chosen for this HSL application as this approach has previously been shown to generate high affinity antibodies against haptenic targets (Charlton, Harris, and Porter 2001), and sheep polyclonal antibodies have been used as specific high affinity immunologic probes for analytical and clinical purposes for many years (Li, Kilpatrick, and Whitelam 2000). In *P. aeruginosa* the autoinducers 3-oxo-C12-HSL and C4-HSL share a similar lactone ring structure but have different subgroups at the third carbon position and also vary in the length of their acyl side chain. For this project we synthesized three HSL antigens *N*-acyl-C12-HSL, 3-oxo-C12-HSL, and 3-OH-C12-HSL with the same lactone ring structure and varying functional groups at C3 position in order to drive the immune response toward cross-reactive antibodies (Figure 21.2). These antigens were conjugated to the carrier protein thyroglobulin (TG) and used as immunogen to enhance the chance of eliciting an antibody response in sheep. Analysis of immunized sheep polyclonal serum using a series of binding and competition ELISAs confirmed successful immunization and presence of neutralizing antibodies that recognized native HSL compounds with high sensitivity and affinity. When *P. aeruginosa* culture was grown in the presence of sheep polyclonal serum, a significant reduction in the expression of elastase was observed revealing the neutralizing effect of these polyclonal antibodies against homoserine lactones (Palliyil et al. 2014). Using PBLs from the immunized sheep as starting material, an HSL antibody phage display library was constructed in a single chain Fv (scFv) format which allows rapid selection of specific binders through a process called biopanning. Forced epitope selection (FES) strategy was designed to drive selection toward the enrichment of high sensitivity and cross-reactivity clones using three HSL subgroups conjugated to bovine serum albumin (BSA) and TG carrier proteins. A different subgroup of HSL conjugate in decreasing concentration was used for each round of biopanning and an additional free antigen elution step was incorporated in the third round of selection to recover binders which preferentially recognized native HSL compounds (Figure 21.3). A number of cross-reactive phage clones were identified after three rounds of library selection, and their ability to recognize free HSL compounds was analyzed through monoclonal phage competition ELISA. A panel of unique native HSL binders was reformatted into sheep-mouse chimeric IgGs and had picomolar sensitivities (IC_{50} values as determined by competition ELISA) for 3-oxo-C12-HSL signaling molecule. These values are quite impressive, considering the chemical nature of these lipid-like compounds (average molecular weight 300 Da) which possess only a small head like structure and lack critical antigenic features such as aromaticity or charge. Modeling of these sensitive anti-HSL antibodies indicated that the level of sensitivity observed was achieved through the generation of a deep and negatively charged binding pocket (Al Qaraghuli et al. 2015).

The protective effects of the sheep derived, anti-quorum sensing antibodies were demonstrated in a slow killing model of the nematode worm *Caenorhabditis elegans*. *P. aeruginosa* "slow killing" is mediated by the accumulation of bacterial load within the intestinal lumen of *C. elegans*, and *lasR* mutants are less pathogenic as they fail to produce virulence factors such as elastase and protease, which suggests that the QS system is an

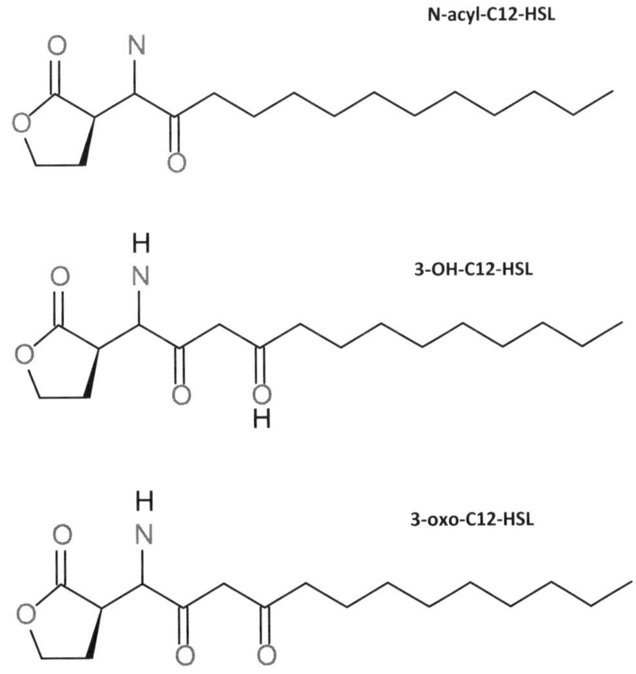

FIGURE 21.2 Chemical structures of three HSL subgroups used for sheep immunization.

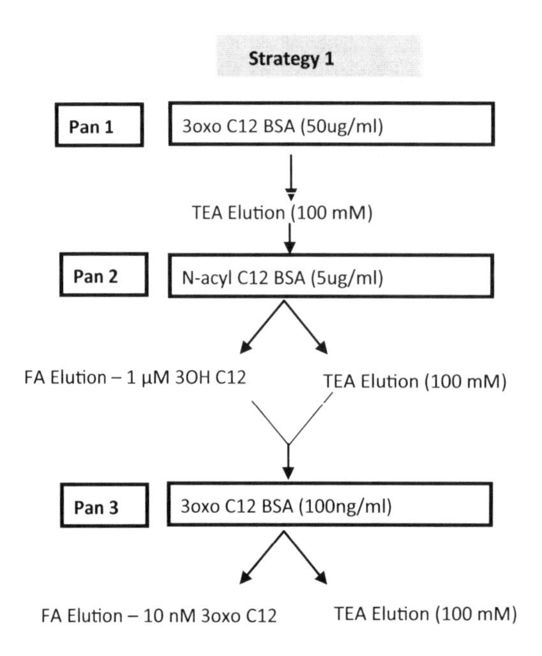

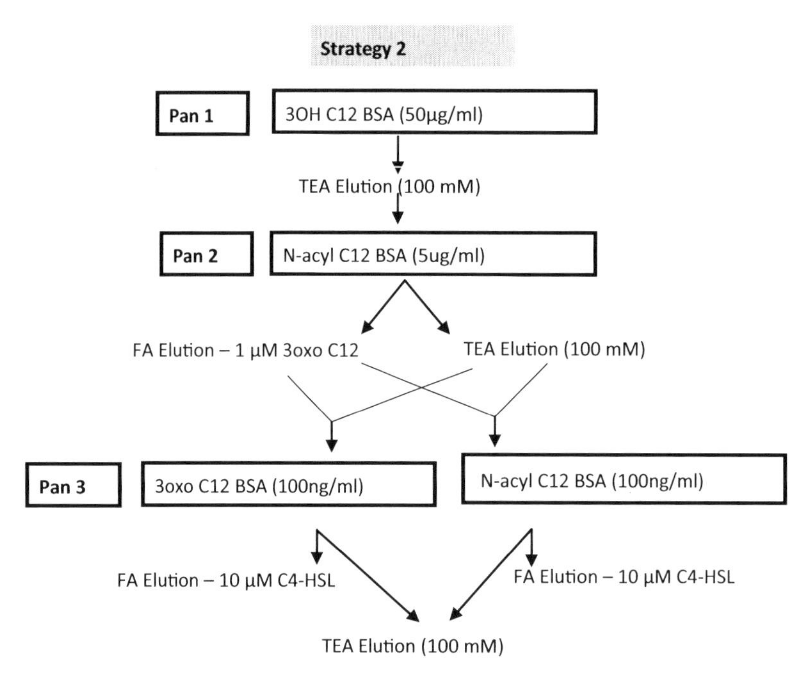

FIGURE 21.3 Outline of the forced epitope selection (FES) strategy adopted for the generation of cross-reactive anti-HSL clones.

essential regulatory factor for the establishment and/or proliferation of the bacteria within the nematodes (Tan, Mahajan-Miklos, and Ausubel 1999a). In a slow killing model, treatment with HSL mAbs significantly increased the survival rate of *C. elegans*, which was similar to the defective slow killing observed in *lasR* mutant strains (Tan, Mahajan-Miklos, and Ausubel 1999b). We believe that the mode of action is based around antibody mediated scavenging of HSL compounds, suppression of virulence factors, and inflammatory responses associated with *P. aeruginosa* infections and reduction of the immunomodulatory effects of 3-oxo-C12-HSL. The therapeutic potential of our HSL mAbs was tested in a non-neutropenic lung model of mice infected with *P. aeruginosa* PA058 where antibody monotherapy demonstrated significant efficacy, prolonging survival by up to 83%. Since no significant reduction in bacterial counts was observed in the lungs of infected mice, it is proposed that HSL specific antibodies protect mice possibly through antibody mediated scavenging of HSL compounds. This mode of action does not necessarily affect bacterial numbers but probably prevents a switch to a more pathogenic phenotype (Palliyil et al. 2014). HSL scavenging may disrupt bacterial communication to such an extent that the bacteria are "held" as a benign or nonpathogenic phenotype. A similar protective effect of mAbs against the cytotoxicity of 3-oxo-C12-HSL compound was validated in murine macrophage-like cell line by Kaufmann and group (2008), where the antibody sustained cell viability in a concentration dependent manner. In this case, the antibody was also shown to prevent the phosphorylation of the mitogen-activated protein (MAP) kinase p38, which is another biochemical marker for the cytotoxic effects induced by 3-oxo-C12-HSL compound. In addition, reduction of 3-oxo-C12-HSL induced cleavage of the enzyme Poly (ADP-ribose) polymerase (PARP), which is an apoptotic marker, was observed in macrophages treated with the mAb RS2-1G9 (Kaufmann et al. 2008). However, the protective effect of this monoclonal antibody in an *in vivo* animal model study has yet to be reported. It has been widely accepted that antibodies with sub-nanomolar affinity exhibit improved therapeutic efficacy, if the targets to be bound are present in low concentrations. Our study demonstrates the power of animal immunization and phage display based selection strategies to isolate high affinity monoclonal antibodies toward hapten targets like homoserine lactones which inherently lack antigenic properties such as aromaticity and charge.

Competition ELISA performed using *P. aeruginosa* culture filtrates grown in the presence of urine demonstrated the ability of HSL-2 mAb to detect native autoinducer compounds in this bodily fluid without significant loss of sensitivity (Figure 21.4). The limit of detection of HSL mAb for 3-oxo-C12-HSL is higher than the extractable levels of HSL compounds found in many *P. aeruginosa* cultures and within the range of concentrations reported in the sputum of CF patients (Pearson et al. 1995, Erickson et al. 2002, Barr et al. 2015). HSL-2 mAb retained functional recognition of its antigen in complex matrices such as urine, which is crucial as this would avoid the need for sample pre-treatment in a diagnostic setting. Rapid and accurate detection and confirmation of *P. aeruginosa* colonization in CF lungs and wound infections would enable clinicians to initiate early targeted antibiotic therapy for optimizing patient outcomes. Using HSL mAbs, simple and reliable point of care diagnostic tests based on a lateral flow immunoassay format can be developed where homoserine lactone compounds serve as early biomarkers of infection.

21.4 Conclusions

Monoclonal antibodies targeting acylhomoserine lactones have the potential to block quorum sensing signaling in *P. aeruginosa* and thereby regulate the expression of virulence factors involved in bacterial pathogenesis. We have successfully generated high sensitivity mAbs toward the predominant cell signaling compound 3-oxo-C12 -HSL from an immunized sheep phage display antibody library and employing recombinant antibody engineering tools. In a non-neutropenic lung model of mice infected with *P. aeruginosa* PA058, HSL mAb monotherapy demonstrated significant efficacy and prolonged their survival without reduction of bacterial load in the lungs. The absence of inherent bactericidal activity may prove highly advantageous, as this therapeutic approach may not apply the same selection pressures on bacteria to develop resistance thereby enhancing the effects of co-therapy approaches. In addition, the high sensitivity of HSL mAbs could be exploited to develop an immunoassay-based diagnostic system to detect the presence of specific markers of infection (homoserine lactones) in bodily fluids for an early therapeutic intervention. Further work in this area could lead to the development of

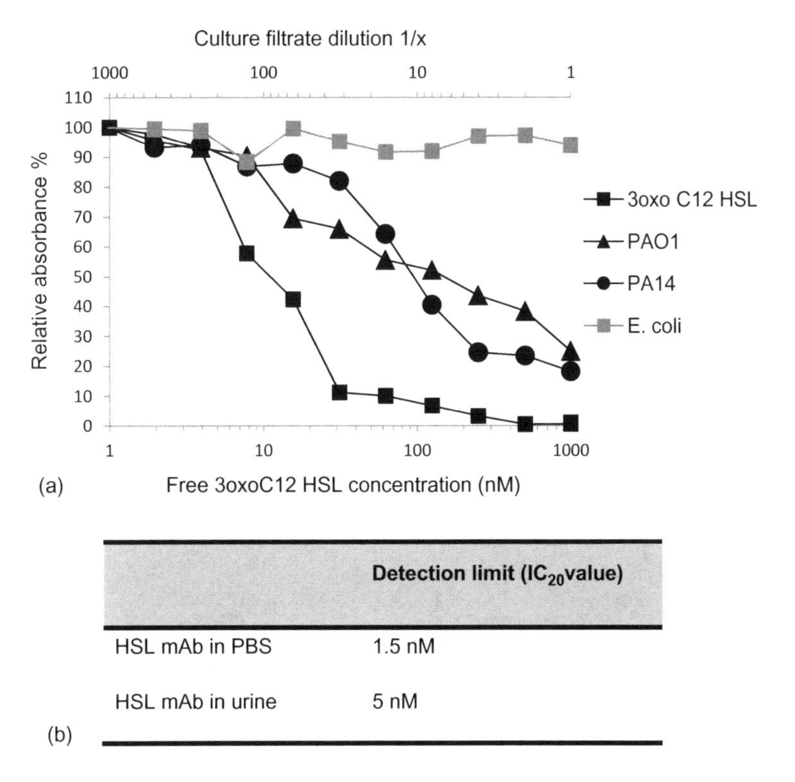

	Detection limit (IC_{20}value)
HSL mAb in PBS	1.5 nM
HSL mAb in urine	5 nM

FIGURE 21.4 Sensitivity and limit of detection of HSL mAb. (a) Competitive inhibition of HSL-2 mAb binding to 3-oxo-C12-BSA conjugate in *P. aeruginosa* PA01 and PA14 culture filtrates. Purified free 3-oxo-C12-HSL was used as a positive control and *Escherichia coli* OP50 as a negative control. (b) Limit of detection of HSL mAb in *P. aeruginosa* culture filtrates grown in urine and PBS.

additional mAbs which can block other QS circuits operating in *P. aeruginosa* and related Gram-negative bacteria as novel anti-virulence therapeutic strategies for controlling the emergence of multidrug-resistant pathogens.

REFERENCES

Al Qaraghuli, M. M., S. Palliyil, G. Broadbent, D. C. Cullen, K. A. Charlton, and A. J. Porter. 2015. "Defining the Complementarities between Antibodies and Haptens to Refine our Understanding and Aid the Prediction of a Successful Binding Interaction." *BMC Biotechnology* 15: 99.

Barr, H. L., N. Halliday, M. Camara, D. A. Barrett, P. Williams, D. L. Forrester, R. Simms, et al. 2015. "*Pseudomonas aeruginosa* Quorum Sensing Molecules Correlate with Clinical Status in Cystic Fibrosis." *The European Respiratory Journal* 46 (4): 1046–1054.

Bjarnsholt, T., P. O. Jensen, T. H. Jakobsen, R. Phipps, A. K. Nielsen, M. T. Rybtke, T. Tolker-Nielsen, et al. 2010. "Quorum Sensing and Virulence of *Pseudomonas aeruginosa* during Lung Infection of Cystic Fibrosis Patients." *PloS One* 5 (4): e10115.

Charlton, K., W. J. Harris, and A. J. Porter. 2001. "The Isolation of Super-Sensitive Anti-Hapten Antibodies from Combinatorial Antibody Libraries Derived from Sheep." *Biosensors & Bioelectronics* 16 (9–12): 639–646.

Collier, D. N., L. Anderson, S. L. McKnight, T. L. Noah, M. Knowles, R. Boucher, U. Schwab, P. Gilligan, and E. C. Pesci. 2002. "A Bacterial Cell to Cell Signal in the Lungs of Cystic Fibrosis Patients." *FEMS Microbiology Letters* 215 (1): 41–46.

de Kievit, T. R. and B. H. Iglewski. 2000. "Bacterial Quorum Sensing in Pathogenic Relationships." *Infection and Immunity* 68 (9): 4839–4849.

Erickson, D. L., R. Endersby, A. Kirkham, K. Stuber, D. D. Vollman, H. R. Rabin, I. Mitchell, and D. G. Storey. 2002. "*Pseudomonas aeruginosa* Quorum-Sensing Systems may Control Virulence Factor Expression in the Lungs of Patients with Cystic Fibrosis." *Infection and Immunity* 70 (4): 1783–1790.

Gambello, M. J. and B. H. Iglewski. 1991. "Cloning and Characterization of the *Pseudomonas aeruginosa lasR* Gene, a Transcriptional Activator of Elastase Expression." *Journal of Bacteriology* 173 (9): 3000–3009.

Jiang, Q., J. Chen, C. Yang, Y. Yin, and K. Yao. 2019. "Quorum Sensing: A Prospective Therapeutic Target for Bacterial Diseases." *BioMed Research International* 2019: 2015978.

Kaufmann, G. F., J. Park, J. M. Mee, R. J. Ulevitch, and K. D. Janda. 2008. "The Quorum Quenching Antibody RS2-1G9 Protects Macrophages from the Cytotoxic Effects of the *Pseudomonas aeruginosa* Quorum Sensing Signalling Molecule *N*-3-Oxo-Dodecanoyl-Homoserine Lactone." *Molecular Immunology* 45 (9): 2710–2714.

Kumar, R., S. Chhibber, V. Gupta, and K. Harjai. 2011. "Screening & Profiling of Quorum Sensing Signal Molecules in *Pseudomonas aeruginosa* Isolates from Catheterized Urinary Tract Infection Patients." *The Indian Journal of Medical Research* 134: 208–213.

Lee, J., J. Wu, Y. Deng, J. Wang, C. Wang, J. Wang, C. Chang, Y. Dong, P. Williams, and L. H. Zhang. 2013. "A Cell-Cell Communication Signal Integrates Quorum Sensing and Stress Response." *Nature Chemical Biology* 9 (5): 339–343.

Li, Y., J. Kilpatrick, and G. C. Whitelam. 2000. "Sheep Monoclonal Antibody Fragments Generated using a Phage Display System." *Journal of Immunological Methods* 236 (1–2): 133–146.

Malhotra, S., D. H. Limoli, A. E. English, M. R. Parsek, and D. J. Wozniak. 2018. "Mixed Communities of Mucoid and Nonmucoid *Pseudomonas aeruginosa* Exhibit Enhanced Resistance to Host Antimicrobials." *mBio* 9 (2). doi:10.1128/mBio.00275-18.

McGrath, S., D. S. Wade, and E. C. Pesci. 2004. "Dueling Quorum Sensing Systems in *Pseudomonas aeruginosa* Control the Production of the *Pseudomonas* Quinolone Signal (PQS)." *FEMS Microbiology Letters* 230 (1): 27–34.

McKnight, S. L., B. H. Iglewski, and E. C. Pesci. 2000. "The *Pseudomonas* Quinolone Signal Regulates Rhl Quorum Sensing in *Pseudomonas aeruginosa*." *Journal of Bacteriology* 182 (10): 2702–2708.

Miller, M. B. and B. L. Bassler. 2001. "Quorum Sensing in Bacteria." *Annual Review of Microbiology* 55: 165–199.

O'Neill, J. 2016. "Tacking drug-resistant infections globally – Final report and recommendations." The review on antimicrobial resistance. https://amr-review.org/.

Palliyil, S., C. Downham, I. Broadbent, K. Charlton, and A. J. Porter. 2014. "High-Sensitivity Monoclonal Antibodies Specific for Homoserine Lactones Protect Mice from Lethal *Pseudomonas aeruginosa* Infections." *Applied and Environmental Microbiology* 80 (2): 462–469.

Pearson, J. P., L. Passador, B. H. Iglewski, and E. P. Greenberg. 1995. "A Second *N*-Acylhomoserine Lactone Signal Produced by *Pseudomonas aeruginosa*." *Proceedings of the National Academy of Sciences of the United States of America* 92 (5): 1490–1494.

Pedersen, S. S., N. Hoiby, F. Espersen, and C. Koch. 1992. "Role of Alginate in Infection with Mucoid *Pseudomonas aeruginosa* in Cystic Fibrosis." *Thorax* 47 (1): 6–13.

Pesci, E. C., J. B. Milbank, J. P. Pearson, S. McKnight, A. S. Kende, E. P. Greenberg, and B. H. Iglewski. 1999. "Quinolone Signaling in the Cell-to-Cell Communication System of *Pseudomonas aeruginosa*." *Proceedings of the National Academy of Sciences of the United States of America* 96 (20): 11229–11234.

Schuster, M. and E. P. Greenberg. 2006. "A Network of Networks: Quorum-Sensing Gene Regulation in *Pseudomonas aeruginosa*." *International Journal of Medical Microbiology: IJMM* 296 (2–3): 73–81.

Tan, M. W., S. Mahajan-Miklos, and F. M. Ausubel. 1999a. "Killing of *Caenorhabditis elegans* by *Pseudomonas aeruginosa* used to Model Mammalian Bacterial Pathogenesis." *Proceedings of the National Academy of Sciences of the United States of America* 96 (2): 715–720.

Tan, M. W., S. Mahajan-Miklos, and F. M. Ausubel. 1999b. "Killing of *Caenorhabditis elegans* by *Pseudomonas aeruginosa* used to Model Mammalian Bacterial Pathogenesis." *Proceedings of the National Academy of Sciences of the United States of America* 96 (2): 715–720.

Telford, G., D. Wheeler, P. Williams, P. T. Tomkins, P. Appleby, H. Sewell, G. S. Stewart, B. W. Bycroft, and D. I. Pritchard. 1998. "The *Pseudomonas aeruginosa* Quorum-Sensing Signal Molecule *N*-(3-Oxododecanoyl)-L-Homoserine Lactone has Immunomodulatory Activity." *Infection and Immunity* 66 (1): 36–42.

Whiteley, M., K. M. Lee, and E. P. Greenberg. 1999. "Identification of Genes Controlled by Quorum Sensing in *Pseudomonas aeruginosa*." *Proceedings of the National Academy of Sciences of the United States of America* 96 (24): 13904–13909.

22

Novel Intervention Techniques in the Food Industry

Carolina Ripolles-Avila and José Juan Rodríguez-Jerez

CONTENTS

22.1 Introduction

Finding ways to control production chains is a constant task on the part of governments and health institutions nowadays, as food safety comes into play. This is due to the threat of foodborne diseases, which are considered as both a worldwide public health problem and a financial issue, making food safety and hygiene control in the food industry so relevant (Rivera et al. 2018). Among the factors that can directly affect food safety and are constantly changing are emerging pathogens that cause foodborne diseases, in addition to changes associated with lifestyle, eating habits, age of the population, and socioeconomic status (Hussain and Dawson 2013). To this effect, foodborne diseases can be defined as a broad group of pathologies caused by ingesting water or food contaminated with pathogenic microorganisms, chemical contaminants, or biotoxins, and dependent on the individual reactions of the human body to microorganisms (WHO 2006).

Outbreaks can occur when two or more people suffer similar symptoms after ingesting the same food product. The contamination of food products can originate at any point during the time lapse between preparing and consuming the food, leading to the appearance of foodborne diseases which are responsible for a wide range of symptoms of varying severity, such as diarrhea, vomiting, other gastrointestinal complications, and even immunological problems, multiorgan failure, and death (WHO 2015). Effective hygiene control is therefore hugely important to prevent these adverse health effects (FAO-WHO 2003). Millions of people become ill every year because the food supplied through the food chain is contaminated by pathogenic microorganisms. In fact, in 2015 the World Health Organization reported that a foodborne disease is acquired by 1 in 10 people every year, representing a total of approximately 600 million cases causing 420,000 deaths (WHO 2015). Currently, the most common microbial hazards that produce human zoonoses and are therefore relevant to food safety are the microbial pathogens *Campylobacter* spp., *Salmonella* spp., Shiga toxin producing *Escherichia coli*, *Yersinia enterocolitica*, *Listeria monocytogenes*, *Coxiella burnetii*, *Brucella* spp., *Francisella tularensis*, and viruses such as norovirus and West Nile virus (EFSA-ECDC 2019) (ordered by reported numbers and notification rates). The foodborne diseases produced by these pathogens are not only manifest in outbreaks, which is the most visible aspect, but they create a much more extensive and persistent problem (Rivera et al. 2018). Studies that have evaluated and approximately quantified the incidence of foodborne diseases in the population have reached different conclusions, and the differences between the estimates made in the evaluations and the lack of accuracy in terms of actual incidences are understood when we realize that a high percentage of foodborne diseases are not reported, recorded, or diagnosed by health authorities. This implies that the percentage of cases registered by health authorities is very small in relation to the real incidence among the population (Arendt et al. 2013). Consequently, food safety is a problem related not only to declarable diseases, but also to small episodes that entail a reduced quality of life for consumers, despite this large second group not being clearly reflected in the official epidemiological data.

Food safety and quality assurance programs in food industries require food microbiology testing to guarantee the absence of pathogens in food products. The impact of foodborne diseases on food companies can otherwise have severe consequences, ranging from loss of reputation to bankruptcy. Hence, research and development into analytical methods to detect food product contamination by foodborne pathogens has experienced rapid growth in recent years, with the market expected to reach $15 billion by 2019 (Mangal et al. 2015).

Several causes of outbreaks related to foodborne diseases have been observed, although their origin is most often associated with poor handling, cross-contamination, and insufficient hygiene or problems related to environmental factors (Callejón et al. 2015; Ripolles-Avila, Ríos-Castillo, and Rodríguez-Jerez 2018). In all these cases, the origin is insufficient cleaning or disinfection due to an inadequate microbiological contamination control systems, which is why efficient control during all stages from processing to transportation is vitally important. If good microbiological controls are applied, the sanitation practices applicable to the facilities must be modified. A high level of contamination by microorganisms normally implies that the cleaning products need to be reviewed because if the organic matter residues are not completely removed, the surfaces can easily become recontaminated.

22.2 Cross-Contamination: A Leading Factor of Food Contamination

Cross-contamination during food processing has been identified as an important factor associated with the transmission of foodborne diseases (Brown et al. 2017). This concept is defined as the direct or indirect contamination by microorganisms or other chemical or physical hazards of an uncontaminated matrix by a matrix that is contaminated, be it via food, food-contact surfaces, or operators. All this is made possible by the ability of microorganisms to adhere to surfaces and subsequently trigger the formation of biofilms (González-Rivas et al. 2018).

To date, several studies have been conducted to find the origin of foodborne diseases. Although identifying the origin of these diseases is often complicated, due to both the lack of precise data provided by governments and industries and incomplete investigations, it is important to make every effort to do so. Among these studies, one carried out at the European level by the WHO in 1995 estimated that 25% of outbreaks were associated with cross-contamination. Factors such as inefficient hygiene practices, contaminated surfaces, and a lack of hygiene in the manipulators were highlighted as being decisive for cross-contamination to occur (Tirado and Schmidt 2001). Laufer et al. (2015) conducted a follow-up evaluation of all outbreaks of infections transmitted by *Salmonella* spp. attributed to beef in the United States during the period 1973–2011. Among the different factors involved, cross-contamination was identified as the origin of the outbreaks analysed in 45% of the cases. Controlling surfaces at the microbiological contamination level is therefore of great interest, since it is these (i.e., food-contact surfaces, equipment, and utensils) that cause pathogen transmission when in contact with food when they have previously been microbiologically contaminated.

Surfaces can become a very important reservoir of microorganisms, which can reach food along different pathways. One of them is through food handlers, who can have pathogens present at low infective doses on their hands that can easily be transferred to food-contact surfaces, where they can stay for a long time and even remain viable after applying cleaning and disinfection treatments (Todd et al. 2009). This exponentially increases the danger of cross-contamination with the consequent safety risk to public health. The different surfaces of the food industry can also act as reservoirs of spoilage microorganisms (Bagge-Ravn et al. 2003) which, when transferred to a food product, can cause deterioration problems due to the enzymatic activities of the microorganism itself, with the consequent loss of food quality. It is therefore essential that cleaning and disinfection strategies are in place to achieve significant reductions in the microbial load and improve hygiene conditions. Regular monitoring of the effectiveness of cleaning and disinfection practices could serve as a crucial preventive measure. It is often the norm to insist on analyzing the surfaces that are in direct contact with food, but when talking about cross-contamination, a more global view is required that takes all the areas of the installation into account. This entails a more complex approach than the one considered in routine controls.

Cross-contamination and the biological agents that cause food spoilage are clearly leading factors in the transmission of foodborne diseases. The impact and current importance of predictive microbiology, which seeks microbial responses to environmental factors with the aim of controlling them, has led to the generation of predictive models that allow us to characterize the variability associated with microbial transfer from food-contact surfaces to other surfaces or food products. In recent years, the understanding of the dynamics of transfer in predictive models has enabled us to identify quantifiable links between the parameters of processing control and microbial levels, simplifying the complexity of these relationships to implement them in risk assessment models. However, nowadays these have certain limitations due mainly to their lack of reproducibility (Possas et al. 2017).

22.3 Biofilms, a High-Level Problem

In nature, microorganisms can be found in two different states. The first is floating freely, a state that is called planktonic and is especially relevant in the propagation of microorganisms. The second is adhered to a surface, a state called sessile, which is essential to allow microorganisms to persist and resist adverse environmental conditions.

Microorganisms have the ability to adhere to living or inert surfaces, forming cellular aggregates that are held together and embedded in a self-produced matrix formed by a conglomerate of different types of biopolymers composed of extracellular polymeric substances (EPS), proteins, and extracellular DNA (eDNA) (Costerton et al. 1995; Donlan 2001; González-Rivas et al. 2018). The EPS form the skeleton of the generated three-dimensional structure, and they play a key role in adhesion to surfaces and cohesion (Flemming and Wingender 2010). These microbial communities attached to surfaces are known as biofilms, the formation of which is a dynamic process that occurs sequentially and includes five main stages (Costerton et al. 1995):

(i) reversible fixation of individual cells to a surface; (ii) irreversible binding to the surface; (iii) formation of microcolonies; (iv) maturation of the biofilm; and last, (v) dissociation of the cells and the colonization of a new surface.

Biofilms have been known about since the seventeenth century when Anton van Leeuwenhoek (1632–1723) first observed microorganisms on the surface of teeth using a constructed microscope (Donlan and Costerton 2002). It was not until the mid-twentieth century that the structures were described in more detail and their importance for the survival of microorganisms identified, linking them to the transmission of foodborne diseases (Heukelekian and Heller 1940; Jones, Roth, and Sanders 1969; Zobell 1943). Nutman et al. (2016) pointed out, however, that this way of life can be traced back 3,700 million years, since fossil remains have been found with successive accumulations of calcified biofilms.

Microbial cells making up biofilms can acquire new characteristics through horizontal gene transfer, can be protected from environmental fluctuations, and can cooperate among them to access more food and increase their metabolic efficiency (Davey and O'Toole 2000; González-Rivas et al. 2018). These structures confer resistance on the cells against antimicrobial agents, ultraviolet light, desiccation, and treatments with disinfectants, compared with the same cells in a planktonic state (Borucki et al. 2003). Additionally, the ability of bacteria to survive environmental stress related to food processing environments, such as refrigeration, disinfection, acidity, and salinity increases when these structures are formed (Giaouris et al. 2014). The communities of microorganisms that make up biofilms can be composed of one or multiple species (Davey and O'Toole 2000). However, in the food industry biofilms are not usually formed by simple three-dimensional groupings of identical cells, but by heterogeneous sub-populations with different behaviors that contribute to the global success of the biofilm (Bridier et al. 2015). Multispecies biofilms have been shown to be more resistant to biocides than monospecies. This is due, in part, to their more complex structure (Sanchez-Vizuete et al. 2015). For example, *E. coli* O157:H7 can produce 400 times more biomass when it coexists in biofilms together with *Acinetobacter calcoaceticus* compared to a monoculture under the same environmental conditions (Habimana et al. 2010). One of the explanations as to why multispecies biofilms have a higher resistance to biocides compared to monospecies is the specific nature and composition of the matrix. Interactions between microorganisms enhance the formation of these structures so that they can efficiently protect themselves from the environment in general.

Effective hygienization is required to control the microbial load of food industry surfaces. To this effect, as an essential part of the HACCP system, which guarantees the safety of the products elaborated in the food industries, there are the cleaning and disinfection programs. Sanitation has different objectives: (i) to eliminate microbiological, chemical, and physical residues that contaminate a food product and negatively affect the health of the consumer, leading, for example, to foodborne diseases; (ii) to eliminate microorganisms that may cause an alteration of the organoleptic characteristics and reduce the product shelf life; (iii) to help maintain machinery, utensils, and facilities in proper condition; (iv) to transmit an image of hygiene and order in the food industry to both customers and workers. This involves a series of operations based on an action plan, which consists of several stages including the use of a cleaning product and a subsequent disinfecting agent, followed by maintaining the food-contact surfaces and the work environment in suitable hygienic conditions. However, the entire procedure begins with the ability to detect the existence of surface-active biofilms. A mature biofilm can be hundreds of microns in diameter, allowing it to spread to key areas of the facility. Nonetheless, hundreds of microns in size is difficult to detect with the naked eye, and some products have recently been introduced to detect these structures (Ripolles-Avila, Ríos-Castillo, and Rodríguez-Jerez 2018), helping to control facilities and guarantee the safety of the food products.

One property that stands out in terms of the protection afforded by the matrix on the cells that form the biofilm is resistance to antimicrobials. This could be due to the biocide diffusing slowly or incompletely inside the biofilm, a different physiology of the cells that form it, a differentiation of part of the population to persistent cells, an expression of different stress adaptation responses (Jefferson 2004), or perhaps a combination of all these factors. These aspects mean that conventional cleaning and disinfection protocols are not usually effective to eradicate biofilms (Álvarez-Ordóñez and Briandet 2016), which is why investigating new methods or substances for sanitizing them has become an important research area that seeks to contribute new approaches. Consumers' recent and growing negative perception of synthetic chemical substances and resistance to antimicrobials has prompted a shift in the research related to these procedures toward alternative solutions (Witte et al. 2017; Gabriel et al. 2018). These include the use of enzymes, bacteriophages, physical technologies, chitosan, and substances for interfering with cell-to-cell communication, the latter being discussed in-depth in the following sections. Additionally, new ways of researching the control of pathogenic and spoilage biofilms have recently been proposed that do not consist of eliminating the entire microbial load on the food-contact surface but leave the microbiota that does not damage the food product, causing a displacement of the total growth of pathogenic and spoilage microorganisms (Ripolles-Avila, Hascoët et al. 2019). Integrating these recent findings will be very useful as a control to prevent biofilms, which have repercussions both at a public health level and at a product quality level.

22.4 Alternative Strategies to Eradicate Biofilms Based on Quorum Quenching for Use in the Food Industry

Microorganisms have the ability to communicate with each other to coordinate and control their behavior depending on cell density (González-Rivas et al. 2018), a process known as *quorum sensing* (QS). This communication takes place because microorganisms synthesize small diffusible signaling molecules called bacterial pheromones or autoinducers (AI) (Skandamis and Nychas 2012), which can further detect and respond to these molecules (Figure 22.1). As described, a high concentration of cells is found in microbial biofilms, so as essential components of biofilm physiology, AI activity and the regulation of gene expression dependent on the density of biofilm cells have been proposed as relevant (Parsek and Greenberg 2005).

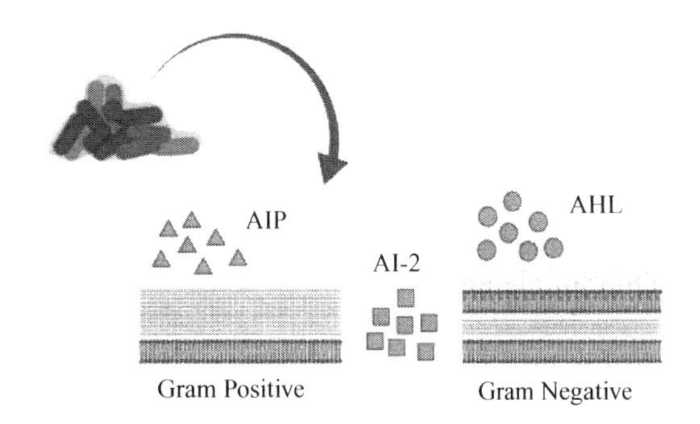

FIGURE 22.1 Quorum sensing systems that can be distinguished in bacteria. Some of the most relevant are the autoinducing peptide system (AIP), acylhomoserin lactone (AHL), and autoinducer-2 (AI-2).

Although each microorganism uses its own QS system, the operating mechanism is the same and is comprised of three steps: (i) the production of signaling molecules, (ii) their accumulation, and (iii) their detection by nearby microorganisms (LaSarre and Federle 2013). Therefore, now that strategies to control QS are becoming increasingly relevant, it is very important to focus on at least one of these three steps when characterizing the compounds that inhibit this communication. Furthermore, three different QS systems can generally be distinguished in bacteria (Figure 22.1): the autoinducing peptide system (AIP) associated with Gram-positive bacteria, the acylhomoserin lactone (AHL) QS system, which is common in Gram-negative bacteria, and the autoinducer-2 (AI-2) QS system in both Gram-negative and Gram-positive bacteria (Brackman and Coenye 2015). It has recently been reported that an autoinducer-3 (AI-3) in *E. coli* O157:H7 and other intestinal bacteria (Nahar et al. 2018) have been found.

The regulation of biofilm formation, maturation, and detachment is controlled by QS systems. For example, the biofilm maturation of *Serratia liquefaciens* has been shown to be influenced by quorum sensing production and detection based on AHL. These species generate heterogeneous biofilms consisting of long filaments of aggregated cells. It has been observed that mutants in the *swrI* gene, which is involved in the production of AHL signaling molecules, result in very thin biofilms that lack filaments and have poorly formed aggregates (Labbate et al. 2004). At a laboratory level, most research on QS systems has focused on discovering the different chemical substances produced as signal molecules and the genes involved in their production, but it has been limited to the communication of a single species. However, the bacteria that naturally foment biofilms are found living in multispecies communities. Because the same QS signal molecules, be they chemically related or distinct, can be produced by different bacterial species, it may be given that a species responds to a QS molecule generated by a different species such that they can share information, a phenomenon that has been called bacterial "cross-talk." In this context, for example, bacteria that do not have AHL production genes such as *Salmonella* spp. or *E. coli* do have an AHL response regulatory protein that enables them to respond to the AHL molecules produced by other bacteria present in the local environment by detecting the signal (Diggle and Williams 2017). Furthermore, Almeida et al. (2017) showed that although *Salmonella* spp. does not produce AI-1, it was able to respond to it when it is synthesized by other bacterial species, triggering biofilms formation. This implies that a specific bacterial species may be able to communicate with other species and respond to distinct multispecies stimuli.

22.4.1 Quorum Sensing Inhibitors (QSI)

While much remains to be learned about the involvement of this controlling system in biofilms, it is considered as a key aspect and has become another alternative focus for their control. Brackman and Coenye (2015) indicated that these systems can be blocked in four different ways: (i) inhibition of signal molecule synthesis; (ii) inhibition of the binding between the signal and the receptor; (iii) degradation of the signals; and (iv) inhibition of the signal translation cascade.

Regarding the first case, to form AHL molecules for microbial communication, a transfer of an acyl derivative to the amino group of adenosylmethionine S (SAM) by a protein of the Luxi family is required. On the other hand, the formation of AI-2 is a bit more complex, requiring two completely different enzymatic reactions. Thus, SAM inhibitors or fatty acid biosynthesis can be used as strategies for the inhibition of AHL production, the same thing happening with the AI-2 signaling molecules (Brackman and Coenye, 2015). For that reason, mutations in certain genes affect the synthesis of QS chemical signals, negatively influencing the formation of biofilms by causing the bacteria to form these structures defectively, immaturely, and with little structure. In fact, numerous studies have indicated that there is a direct relationship between the non-production of QS signals and a negative effect on the formation of biofilms, and therefore, a mutation in the gene that codes for producing this chemical signal triggers a lower formation of biofilms. For example, McCarthy et al. (2010) reported that suppressing the BCAM0227 gene in *Burkholderia cenocepaciammi* led to the non-production of the QS signaling molecule cis-2-dodecenoic acid (BDSF), implying that complex biofilm structures did not occur. In fact, the bacteria did not colonize the surface extensively and did not go beyond microcolonies. In addition, Udine et al. (2013) indicated that mutants of *B. cenocepaciammi* in the genes that produce the signaling molecules of BDSF and AHLs produced very deficient biofilm structures. To confirm the role of these signaling molecules, they were both introduced into the growth culture medium, resulting in the production of more complex biofilm structures. Smooth, unstructured biofilms are formed when AHL synthesis is blocked in *Pseudomonas aeruginosa* by the gene *lasI*, as these mutant strains cannot synthesize 3-oxo C12-HSL, thus preventing the production of the signal molecule (Hentzer and Givskov 2003). In the case of the production of AI-2, it has been described that elimination of the *luxS*, a gene that is key to the production of these molecules, directly affects the synthesis of this molecules, directly lowering *Streptococcus intermedius* biofilms formation, and this can be extrapolated to other species of this genera (Ahmed, Petersen, and Scheie 2009).

In the second case, less biofilm formation may occur when the junction between the signal and the receiver is compromised due to inhibition. On this regard, the mechanism focuses on the generation of analog signal molecules that block the junctions with the receptor and, in this way, there is a blockage of the communication between microorganisms. Some studies have evaluated the modification of the signal molecules by the side of the chain; others have focused on more central parts (Brackman and Coenye, 2015). To this effect, altering the AHL to analogous molecules that interfere in the direct binding between the AHL and its receptor impacts negatively on biofilm formation. Biofilms formed by *P. aureginosa* were also significantly reduced when an analogue of the AHL was obtained by replacing the lactone ring with a cyclopentyl ring (Ishida et al. 2007). Although in this case the analogue was synthesized artificially, there are other types found in nature and are especially associated with plant food extracts. Different patents have indicated analogs of AI-2 that can interrupt the QS system in, for example, *Vibrio harveyi* (Pan and Ren 2009). Moreover, some garlic extracts contained anti-QS compounds, in particular penicillin and patulinic acid, which proved to have an effect of preventing communication between species of *Penicillium* spp., compromising its development (Rasmussen et al. 2005). However, this point will be discussed further in the following sections.

The third option consists of eliminating the QS chemical signals by enzymatic degradation once they have been produced, thus preventing their accumulation, a technique called *quorum quenching* which has become very relevant in recent times (Oh et al. 2017). To this effect, several enzymes, including acetylases, lactonases, reductases, and oxidases, have been shown to trigger a modification or degradation of the structure with a consequent loss of QS signal activity. However, no specific enzyme has been found to degrade QS signal molecules as they can be of a varying chemical nature. These enzymes can come from different sources, highlighting among them the capacity that some microorganisms have to produce anti-QS enzymes (Rehman and Leiknes 2018; Okano et al. 2019) that could promote the degradation of certain chemical signals employed in the QS system of other species and thus use them as an alternative to prevent the formation of biofilms. This point will be discussed in greater depth in following sections.

Last, a blocking can be produced by some compounds such as furanones and cinnamaldehyde at the QS system activation cascade level. Blocking the signal at this level is also a way of acting on the interruption of the communication between microorganisms, having consequences in regard of compromising the development of certain microorganisms. It has been reported that furanones present an anti-QS effect by inhibiting the compound *N*-dodecanoyl-DL-homoserine lactone (C12-HSL) on *Salmonella* Enteritidis, causing an inhibition of biofilm formation (Almeida et al. 2017). In this sense, it has been described in a patent that furarone 8 generates a high inhibition mediated by AHL, producing reductions of around 90% in microbial communication and subsequent inhibition of the formation of biofilms of *P. aeruginosa* (Parsek and Greenberg 2005). Furthermore, the two mentioned compounds (i.e., furanones and cinnamaldehyde) have been shown to be able to interfere with different QS systems, directly influencing the biofilm formation of many different species such as *E. coli* and *Vibrio* spp. (Brackman et al. 2011). The use of furanones is restricted due to their toxicity, while cinnamaldehyde is widely used as a flavoring additive in foods and beverages.

Most studies have been conducted with biofilms formed by microbial monospecies, while in real conditions biofilms are made up of tens or even hundreds of different species. Therefore, the control of a simple signal may involve a specific microorganism in the process of biofilm construction and not the total number of species involved. Consequently, tackling the matter of controlling QS signals requires a global vision and research that can address the real situation of food industry surfaces.

22.4.2 Types of QSI That Can Be Employed in the Food Industry

As described in the previous section, microorganisms can communicate with each other in natural ecosystems for two different reasons, either to cooperate with each other or to compete with others (Kalia 2013). In this last scenario, there is considerable interest in strategies that selectively eliminate QS systems through producing a wide range of compounds to control microbial biofilms. All these control systems use substances employed to block signals between microorganisms. It is important to consider that microorganisms are able to create alternative resistance mechanisms and it should not be ruled out that resistance to their action could be formed, as has happened with antibiotics or some antimicrobials. Therefore, their application should not reduce control of the surface microbiota. Some of these compounds are discussed below.

22.4.2.1 Anti-QS Enzymes

Since the twentieth century, the application of enzymes in the food industry as cleaning agents has gradually increased in scope in developed countries (Meireles et al. 2016). This technology enables a high degree of personalization, and operation protocols can be established to adapt to the specific industry, thereby improving the integral and specific cleaning of the processing environment. An interesting point about enzymatic technology in the food industry is that enzymes are able to disintegrate biofilm matrices, helping to improve surface hygiene. They can also exhibit other properties apart from the biochemical decomposition of the biofilm matrix, such as the capacity to degrade surface cell adhesion, promote cell lysis, deactivate other necessary enzymes that can be important for microbial growth, and inhibit QS signaling systems (Nahar et al. 2018).

The first time it was demonstrated that certain enzymes have the ability to degrade QS signals was in microorganisms isolated from the soil, or more precisely bacteria from the genus *Variovorax* spp. and *Bacillus* spp. (Dong et al. 2000; Lead better and Greenberg 2000). Since then, numerous enzymes involved in the degradation or modification of QS signals, and therefore considered as potential QS inhibitors, have been reported (Table 22.1). The most important of these are (Grandclément et al. 2016): lactonases, involved in the ring opening of the homoserin lactone; amidases, also called acylases, which cleave the amide bond of the AHL and consequently release fatty acids and homoserine lactone; reductases, which convert the 3-oxo-substituted AHL into its 3-hydroxyl-substituted cognate AHL; oxidases, which mostly catalyze the oxidation of the acyl chain; enzymes active toward 2-alkyl-4(1H)-quinolones (AQs); and enzymes acting on AI-2 such as AI-2 kinase. The mechanisms regulated by QS are repressed by the action of these enzymes so that they can no longer act as signals (Giaouris and Simões 2018). It has been shown that for multidrug-resistant *P. aeruginosa* (MDRPA) biofilms, a lactonase treatment decreases the production of virulence factors and increases susceptibility to antibiotics (Kiran et al. 2011). However, it must be noted that

TABLE 22.1

Studies on Quorum Quenching Enzymes for Biofilm Control and Their Key Findings

Anti-QS Enzyme	Produced By	Surface	Key Finding	References
AiiA lactonase	Engineered T7 phage	Polyvinyl chloride plastic	Inhibition of biofilm formation of mixed-species containing *P. aeruginosa* and *E. coli*.	Pei and Lamas-Samanamud (2014)
OHL-lactonase	*Enterobacter ludwigii*	Polyvinyl chloride plastic	Significant reduction in biofilm formation and the exopolisaccharide production of *Y. enterocolitica* by 66.15% and 70.18%, respectively.	Gopu et al. (2016)
Dehydrogenase	*Planktomarina temperata*	Polystyrene	Inhibition of *E. coli* K12, *B. subtilis*, and *S. aureus* biofilm formation. Not effective against *P. aeruginosa*.	Weiland-Bräuer et al. (2016)
SsoPox lactonase	SsoPox enzyme added (0.5 mg mL^{-1})	Polyurethane	Inhibition of two virulence factors, proteases and pyocyanin, and biofilm formation in 51 clinical *P. aeruginosa* isolates.	Guendouze et al. (2017)
AHL lactonase (MomL)	*E. coli* BL21 harboring MomL expression plasmid	Polystyrene	Reduction of biofilm formation and increased biofilm susceptibility to different antibiotics in *P. aeruginosa* and *A. baumannii*.	Zhang, Brackman, and Coenye (2017)
AHL-acylase and lactonase	Proteobacteria from marine origin (*Erythrobacter*, *Labrenzia*, and *Bacterioplanes*)	Polycarbonate membrane	Inhibition of *P. auregionasa* biofilm formation and kinetics.	Rehman and Leiknes (2018)
AiiM lactonase	*Microbacterium testaceum*	Polyvinyl alcohol (PVA)	Inhibition of *S. marcescens* AS-1 attachment and subsequent biofilm formation.	Okano et al. (2019)

unlike quorum quenching compounds that penetrate the target cell, anti-QS enzymes act in an extracellular way, preventing or diminishing the selective pressure on the development of microbial resistances (Fetzner 2015).

On this regard, some advantages could be observed regarding the use of this type of enzymes as an anti-QS strategy. Among them, it would be highlighted its clear effectiveness in terms of the direct elimination of signals, without causing resistance to antimicrobials and being a feasible alternative in terms of the prevention of microbial development and consequent formation of biofilms. Nevertheless, this alternative could also present some disadvantages, highlighting among them the possible cost to massively produce these compounds and then purify them to include them in the development of products for food industry applications.

As has been reported, biofilm formation prevention on food industry surfaces can occur through the inclusion of anti-QS enzymes (Thallinger et al. 2013), which have demonstrated to be one of the most encouraging molecules for QS signal degradation, thereby preventing biofilm formation by controlling the stages of bacterial granule formation (Nahar et al. 2018). It is important to emphasize that this type of treatment leads to prevention of the formation of these structures and not their direct elimination, especially when they are in a mature state, for which other strategies must be considered (reviewed by, e.g., González-Rivas et al. 2018). Furthermore, it is vital that studies under real conditions are carried out since inhibitory effects have only been indicated in *in vitro* conditions. Additionally, research studies should include other types of materials, such as stainless steel, widely used as a food-contact surface.

Interestingly, the approach proposed until now regarding anti-QS enzymes has been to use them as an integral part of a product to be used for cleaning operations in the food industry. However, what could be truly relevant but not previously considered is the use of microorganisms with the ability to produce anti-QS enzymes as an alternative to conventional surface sanitization. To this effect, a very recent study proposed that the hygienic theory of surfaces used traditionally by the food industry could be reconsidered to incorporate microorganisms that compete with food pathogens and do not have any type of spoilage effect on the product (Ripolles-Avila, Hascoët et al. 2019). If the microorganisms employed do not allow pathogenic microorganisms to grow into biofilms because there is an ecological displacement due to competition, and in turn these microorganisms do not cause any repercussions at the level of product quality and safety, this could be a really useful new hygienic alternative to apply in the food industry. This approach could also be applied if bacteria that produce anti-QS enzymes are found that are able to inhibit biofilms of microorganisms on surfaces in the food industry by both competition and the production of these enzymes. For example, in a study conducted by Torabi Delshad et al. (2018) different microorganisms capable of producing anti-QS enzymes were used to reduce the swimming motility and biofilm formation of *Yersinia ruckeri*. Furthermore, this was tested *in vivo* for trout, and the results indicated that the microorganisms inoculated into the fish were able to control enteric redmouth disease. This idea could be extrapolated to other contexts, including the food industry, and applied as an alternative. In another study, Balabanova et al. (2017) reported that the bacteria culture isolated from meat products (*P. aeruginosa*, *S. aureus*, *Bacillus subtilis*, and *Salmonella enterica*) together with a marine bacterium called *Cobetia amphilecti* resulted in the inhibition of cell growth and the complete degradation of the *P. aeruginosa* and *B. subtilis* biofilms. In this case, the significant anti-biofilm factor would be an extracellular nucleolytic enzyme with the ability to produce QS inhibition. The aforementioned drawback on the cost of producing anti-QS enzymes could be overcome if this strategy were used, since enzymes would not be needed to be produced, but just to grow this type of microorganisms, which by themselves would produce the enzymes.

22.4.2.2 Anti-QS Food Plant Extracts

Products of natural origin have long been extremely important in terms of finding alternatives to promoting health, and plant, mineral, and animal products have particularly constituted a relevant therapeutic arsenal. It was after the discovery of penicillin that the search for antimicrobial substances in higher plants gained momentum (Araujo Nogueira et al. 2017). To this effect, a group of compounds that have been described as strong QSIs are plant extracts from different botanical groups, including fruits, vegetables, fresh herbs, and spices (Kalia 2013). It seems that the interruption of QS systems by plant compounds could serve as a defense mechanism to fight against bacterial infections. Among the keys to the success of plant food extracts and phytocemics which may contribute to their effectiveness are: (i) chemically stable molecules, (ii) highly effective low mass molecules, and (iii) not harmful to human health (Rasmussen and Givskov 2006). Furthermore, it has been observed over time that including extracts of plant foods and phytochemicals is interesting as a food preservation technique, although their use must be stipulated in the specific system in which they will function, since they cannot easily be transferred from one QS system to another (Truchadoa et al. 2015). All these can be considered important and interesting advantages on these type of compounds for being used as potential inhibitors of the communication between microorganisms, preventing their development and further biofilm formation. In addition, natural resources may be alternatives to be implemented in real conditions because they present a low environmental impact. However, it could be pointed out as a disadvantage that they can have bacteriostatic and bactericidal effects, and thus present some resistance over time.

Many research studies related to the ability of plant extracts to interfere in QS communication systems between microorganisms have been carried out (Table 22.2). Among the mechanisms of action of the plant food extracts and phytochemicals that have been highlighted so far are the similarity of the compounds in the chemical structure and the QS signals (especially the AHL) and their ability to degrade signal receptors (Kalia 2013). Additionally, in the case of AHLs, the LuxR family of transcriptional regulators act as receptors or regulators, while the LuxP/Q type family of proteins regulate the QS type AI-2. In the latter case, LuxP is a periplasmic binding protein that binds to AI-2. Biofilm formation of carbapenem-resistant *E. coli* was significantly reduced due to the anti-QS effect of *Camellia sinensis*, *Holarrhenaantidys enterica*, and *Berberis aristata*, the last one presenting the highest activity (Thakur et al. 2016). Last, it has been reported that extracts of plant foods and phytochemicals have some other mechanisms of action that can be used to inhibit bacterial QS. However, only a few of the studies available focus on the role of plant food extracts and phytochemicals, since QSI deals with the possible mechanisms of action (Truchadoa et al. 2015). Furthermore, a hurdle effect can be observed by the combined effect of some natural compounds with organic acids. It has been observed on *E. coli* O157:H7 and *S. Typhimurium* that lactic and malic acids have an inhibition effect of AI-2, diminishing biofilm production of these microorganisms (Almasoud et al. 2016).

One challenging task for the food industry is the elimination of mature biofilms (Ripolles-Avila et al. 2018; Ripolles-Avila, Cervantes-Huaman et al. 2019). A currently emerging alternative is the use of substances such as natural plant extracts, which generate an inhibition of the adherence of

TABLE 22.2

Studies on Anti-QS Compounds Derived from Plant Extracts for Biofilm Control and Their Key Findings

Anti-QS Compound	Derived From	Target Microorganism	Key Finding	References
Vanillin	*Vanilla planifolia*	*Aeromonas hydrophila*	Strong biofilm reduction in different membranes, including microfiltration, ultrafiltration, and reverse osmosis membranes.	Kappachery et al. (2010)
Pomegranate peel extract	*Punica granatum*	*Y. enterocolitica*	Decreased QS-associated biofilm formation and swimming motility.	Oh et al. (2015)
Hexane extract of *H. suaveolens* (HEHS)	*Hyptis suaveolens*	*E. coli*, *Proteus vulgaris*, *Proteus mirabilis*, *Klebsiella pneumoniae*, and *Serratia marcescens*	HEHS promotes the loosening of the biofilm architecture and strongly inhibits *in vitro* biofilm formation.	Salini et al. (2015)
Star anise	*Illicium verum*	*S. aureus*, *S. Typhimurium*, and *P. aeruginosa*	The extract inhibited the formation of pathogenic biofilms by up to 87% in a dose-dependent manner.	Rahman et al. (2017)
Cinnamaldehyde	*Cinnamomum verum*	*Pseudomonas fluorescens*	Cinnamaldehyde interacted with the LuxR-type protein of the bacteria, which constituted the molecular basis of the QS inhibition observed.	Li et al. (2018)
Essential oils (EOs) extracted from dried leaves	*Thymus daenensis* and *Satureja hortensis*	*S. aureus*	Significant inhibitory effect of the EOs on biofilm formation and disruption at sub-MIC concentrations.	Sharifi et al. (2018)
Skeels leaf extract	*Syzygium cumini*	*S. aureus* and *P. aeruginosa*	Biofilm inhibition up to 86% and 86.40% in *P. aeruginosa* and *S. aureus*, respectively. Moreover, the extract also disrupted some virulence phenotypes of both pathogens.	Gupta et al. (2019)
Methanolic extract	*Plectranthus tenuiflorus*	*P. aeruginosa* PAO1	The production of biofilm-aggravating phenotypes such as exopolysaccharides, alginate, and rhamnolipid were significantly reduced by the plant extract.	Hnamte et al. (2019)

microorganisms and do not involve bacterial death. The components of the extracts can hinder the fixation of microorganisms, keeping cells in a planktonic state and triggering their greater susceptibility to other antimicrobial agents such as the disinfectants used in the food industry. This has similarly been proposed as a strategy for antiviral therapy (Araujo Nogueira et al. 2017). It is, nonetheless, a strategy to prevent the formation of structures.

In a recent study conducted by Salmi-Mani et al. (2018), antibacterial PET surfaces with a base of vanillin derivatives, which have been shown to have an important anti-QS effect, were developed. Significant reductions were observed in the formation of biofilms of *Rhodococcus wratislaviensis* and *S. aureus* as Gram-positive bacteria, and *E. coli* and *P. aeruginosa* as Gram-negative bacteria, in the microbial tests performed on the material. In another study conducted by Nielsen et al. (2018), it was indicated that materials such as stainless steel and polyethylene coated with isoeugenol proved to be effective in preventing the formation of biofilms of three pathogenic microorganisms, including *S. aureus*, *L. monocytogenes*, and *P. fluorescens*, and nonviable cells were even detected on the surfaces coated with this antimicrobial. Similarly, extracts from Red Globe and Carignan grape stems containing polyphenols inhibited the adhesion of *L. monocytogenes* to stainless steel and polypropylene surfaces due to the ability of this compound to inhibit motility and modify its adhesion (Vazquez-Armenta et al. 2018). Thus, the development of new materials that include compounds that can prevent biofilm formation may be a relevant alternative to control these structures in the food industry. However, it should not be forgotten that this would be a preventive strategy for biofilms and would in no way substitute the cleaning and disinfection protocols applied in the food industry.

22.4.2.3 Anti-QS Nanoparticles

New expectations for researchers have emerged in recent times thanks to advances in nanotechnology, which is gaining increasing importance and attracting huge attention due to its application at different levels, including medicine, bioremediation, diagnostics, and agriculture, among others (Valcárcel and López-Lorente 2016). On comparing the bulk form with the form of a nanoparticle, it has been indicated that the latter has exceptional chemical and physical properties that are far superior to those of bulk materials, making them able to interact differently with biological systems and contribute to antimicrobial activity (Wagh et al. 2013). This finding has led to the research and development of new antibacterial agents in the form of nanomaterials for use as an alternative to disinfectants.

The different physicochemical and functional properties of nanoparticles make them a valuable antimicrobial agent (Table 22.3). To this effect, they can be used as broad-spectrum antimicrobial agents since they exhibit a wide range of actions and include strains that are resistant to multiple drugs (Qais, Khan, and Ahmad 2018). This makes them a viable alternative to conventional antimicrobials employed in the food industry. There are also metallic nanoparticles such as calcium oxide or magnesium oxide, an important characteristic that helps antimicrobial activity due to their being alkaline (Sawai, Himizu, and Yamamoto 2005). These types of nanoparticles composed of alkaline metals are slightly more soluble and contribute to the alkalinity of the medium, which is not achieved with semiconductor metal nanoparticles such as zinc oxide (Zhang et al. 2007). Another aspect that determines the bacteriostatic and bactericidal condition is the electrostatic nature of the positively charged nanoparticles, as in the case of cerium oxide nanoparticles (Thill et al. 2006).

TABLE 22.3

Studies on Anti-QS Nanoparticles (NPs) for Biofilm Control and Their Key Findings

Anti-QS Nanoparticle	Surface	Target Microorganism	Key Finding	References
AgCl-TiO_2 NPs	Glass	*Chrobacterium violaceum*	Anti-QS activity related to the absence of the signaling molecule AHL. Inhibition of biofilm formation.	Naik and Kowshik (2014)
Cu NPs	Polyvinyl chloride	*P. aeruginosa*	CuNP treatments at 100 ng mL^{-1} resulting in a 94%, 89%, and 92% reduction in the biofilm, cell surface hydrophobicity, and exopolysaccharides, respectively.	Lewis Oscar et al. (2015)
Ag NPs	*In silico* (computational study)	*P. aeruginosa*	LasI/RhlI synthase was inhibited by Ag, blocking the biosynthesis of AHLs. Therefore, no QS occurred and biofilm formation was prevented.	Ali et al. (2017)
Honey polyphenol carrying Ag NPs	Polystyrene	*P. aeruginosa* PAO1	Inhibition of elastin- degrading elastase, exoprotease, pyocyanin, swarming motility, rhamnolipid and biofilm formation.	Prateeksha et al. (2017)
Ag–TiO_2, TiO_2–Ag, Ag–Cu and Cu–Ag nanocomposites (NCs)	Polystyrene	*P. aeruginosa*, *S. aureus*, and *E. coli*	The effectiveness on biofilm reduction was higher on: Ag–TiO_2> TiO_2–Ag > Cu–Ag > Ag–Cu.	Alavi and Karimi (2018)
Ag NPs	Polystyrene	*S. marcescens*	The Ag NPs were found to attenuate the production of QS-dependent virulence factors (prodigiosin, protease, lipase, EPS) and biofilm formation.	Ravindran et al. (2018)
AgWPA NPs	Polystyrene	*S. aureus*	AgWPA NPs were remarkably efficient at suppressing the formation of *S. aureus* biofilms.	Liang et al. (2019)
Ag NPs	Polystyrene	*S. aureus*	Genes encoding capsular polysaccharides, intercellular adhesion and virulence were downregulated with the silver NPs.	Singh, Rajwade, and Paknikar (2019)

There are other types of nanoparticles that are photoactivable such as titanium oxide, which when activated by UV light can trigger an inhibition of the growth of bacteria resistant to drying (Sadiq, Chandrasekaran, and Mukherjee 2010).

There are three important parameters that influence the antimicrobial effectiveness of nanoparticles. These are (Raghunath and Perumal 2017): (i) particle size, (ii) solubility in aqueous medium, and (iii) the release of metal ions. In general, the mode of action of nanoparticles can be considered as completely different from that attributed to conventional antibiotics. The mode of action of the latter usually includes alterations of both the cellular wall and the route of the nucleic acids, and enzyme destruction (Zhu et al. 2013). Nanoparticles, by contrast, have a completely different mode of action, including alteration of the cellular processes at both the molecular and the biochemical levels, membrane damage, oxidative and non-oxidative stress induction, and the release of metal ions (Qais, Khan, and Ahmad 2018). It has been suggested that because nanoparticles possess so many antibacterial actions at once, bacterial cells must produce multiple simultaneous genetic changes to develop resistance, which makes it very difficult for them to develop these resistances early on (Zaidi, Misba, and Khan 2017).

Given the potential of nanotechnology, different studies have recently been carried out in which this technology is combined with plant extracts to enhance their action. To this effect, a study conducted by Subhaswaraj et al. (2018) encapsulated cinnamaldehyde in chitosan nanoparticles with the aim of observing the attenuation of the virulence regulated by QS in *P. aeruginosa* PAO1. The study revealed that a slow, sustained release of cinnamaldehyde influenced an inhibition of the QS, which affected the decrease in the production of virulence factors and the formation of associated biofilms. Similarly, in another study conducted by Pattnaik et al. (2018) the anti-QS and anti-biofilm potential of ferulic acid encapsulated chitosan-tripolyphosphate nanoparticles was evaluated, concluding that the slow, sustained release of ferulic acid by the NPs at the target sites increased its effectivity to produce anti-QS activity, significantly attenuating the swimming and swarming motility of *P. aeruginosa* PAO and reducing biofilm formation. In another new development, a group of researchers found that materials containing poly (3-hydroxybutyrate) nanocomposites and biosynthesized silver nanoparticles for active coatings and packaging applications showed strong antimicrobial activity against the foodborne pathogens *S. enterica* and *L. monocytogenes* (Castro-Mayorga et al. 2018). All these results suggest new avenues to develop novel food-contact materials to prevent biofilm formation and new disinfectant products that include these compounds in the post-antibiotic era.

However, disinfection requires good cleaning. Any system that seeks to disinfect or control microbial growth requires the regular use of cleaners. These products should preferably have an anti-biofilm action to facilitate the penetrability of these substances inside the biofilm. Solutions that are not accompanied by rigorous, compatible cleaning will have a low efficiency. This part will further be discussed in the following section.

For all those reasons, current research has begun to focus on the exploitation of nanotechnology for the generation of nanoantimicrobial products, including among them, these nano-anti-QS inhibitors. It has been observed that these can present a series of advantages such as having a greater solubility in the matrix of the biofilms, a greater penetration in the mucus of the biofilms, effective administration, and maintenance of the activity of the QS inhibitors (Singh et al. 2017). However, some drawbacks have been described in the use of this type of substances, especially when they contain antimicrobials that can cause some resistance. Among them can be highlighted the low stability and the possible cytotoxicity, which makes them limited for being used in real conditions for the moment, despite presenting very promising results in *in vitro* conditions (Piras et al. 2015).

It must be taken into consideration that new alternatives may produce benefits associated with the improvement in food safety; however, as just mentioned, nanomaterials have been indicated because they can present multiple risks to human and environmental health. For this reason, it is important that rigorous risk assessments are carried out and a clear and concise regulation is established about it. Although there are some European Union regulations providing specific provisions for nanomaterials, such as Regulation (EC) No. 1333/2008 on food additives, Regulation (EC) No. 450/2009 on active and intelligent materials intended to come into contact with food, or Regulation (EU) No.10/2011 on plastic materials intended to come into contact with food, it is difficult for the private sectors and food industries to have a clear orientation on the applicable regulatory framework since there are no specific regulations for all food categories (Jain et al. 2016).

22.4.3 Cleaning and Disinfection in the Food Industry and the Integration of These QS Inhibitors into Prevention Protocols for Biofilm Formation

In the food industry, hygiene should not only be considered as a basic stage of the production process but must also be viewed from a global perspective and as an important prerequisite of the HACCP system. As is known, the main objective of implementing a comprehensive hygiene system is to eliminate the pathogenic and spoilage microorganisms present on food-contact surfaces (González-Rivas et al. 2018).

Cleaning is the first part of the food industry's sanitation program and is the operation whose main objective is the detachment and elimination of the organic and inorganic dirt attached to surfaces, objects, and utensils (Holah 1995). The surfaces are thereby conditioned so that the disinfectant action is optimal. Another objective of cleaning should be to break or dissolve the extracellular matrix associated with biofilms to enable the subsequent penetration of disinfectants into microbial cells (Simões et al. 2006). Disinfection is the procedure whose main objective is to eliminate or inactivate the microorganisms present on work surfaces to achieve safe levels of microbiological contamination and ensure that the product does not represent a risk to public health or the quality of the food products. Preventing the development of microorganisms on surfaces during production should indirectly avoid transfer to the product (Holah 1995). The development of novel strategies that include new effective cleaning products to reduce adhering organic matter and therefore eliminate the possible focus of microbial adhesion is of enormous interest to the food industry (Guerrero-Navarro et al. 2019).

Although much research has been done so far on QS and anti-QS, additional studies are needed that focus on the transfer of the integral mechanism of anti-QS action and its relationship with microbial biofilms to achieve real, feasible applications. The identification, characterization, and applicability of anti-QS compounds could be truly relevant for reducing both the virulence and pathogenicity of bacteria, including the bacteria resistant to antimicrobials, and the formation of biofilms on contact surfaces with food, thus preventing cross-contamination (Tiwari et al. 2016). To this effect, on observing the results of the different studies carried out up to now, the use of compounds with an anti-QS effect could be suitable for the food industry to prevent the formation of biofilms once the surfaces have been cleaned and disinfected. The application of these substances could prevent the few cells in the sessile state remaining on the surface from communicating with each other, thus reducing the chance of them generating the structures. However, this is something that should be investigated at an industrial level to observe the percentage of efficiency and feasibility it may have for the food industry.

22.5 Application of Anti-QS Inhibitors on Food Products

The deterioration of food products is the result of the enzymatic activities of different microorganisms found in the food itself. Some research has shown that QS plays a crucial role in food quality because the different signaling molecules produced by microorganisms are able to regulate various enzymatic activities such as lipolysis and proteolysis (Ammor, Michaelidis, and Nychas 2008). This suggests that communication between the cells is very relevant in the deterioration of food products, although studies related to the role triggered by QS in the alteration of the foods are scarce (González-Rivas et al. 2018).

Furthermore, in food spoilage it is important to highlight the concept of specific spoilage organisms (SSO). For example, in meat these microorganisms are found mainly in the skin and on the surface of the muscle, as well as on the working food surfaces. After slaughter, these microorganisms increase in concentration and can spread to the internal tissue. Some researchers have suggested that the signaling molecules that trigger QS are produced by these microorganisms. These organisms are related to the family Enterobacteriaceae, *Pseudomonas* spp., *Aeromonas* spp., *Brochothrix thermosphacta*, and lactic acid bacteria. Moreover, there is another important concept, the ephemeral spoilage organism (ESO). This becomes dominant in the product or on the surface due to the selection that occurs during processing or storage. For example, conditions of cold aerobic storage and high relative humidity are selective for *Pseudomonas* spp. (Ammor, Michaelidis, and Nychas 2008). Therefore, an understanding of both microbial ecology and the product and its environment are essential in QS and deterioration.

Nowadays, with the significant increase in resistance to antibiotics and antimicrobials among pathogens and other bacteria found in the environment, we must evolve and find escape strategies against traditional antibacterial agents and conventional techniques. Recent studies have revealed possible strategies to improve the quality of products by inhibiting the QS systems of spoilage microorganisms in food products. These anti-QS compounds could be interesting new natural biopreservatives for the food industry. For example, in a study conducted by Zhang et al. (2018) it was indicated that hexanal was able to control the deterioration caused by *Erwinia carotovora* and *P. fluorescens* by inhibiting the production of AHLs and the extracellular enzymes regulated by QS. This significantly inhibited soft rot in Chinese cabbage and the burning of lettuce leaves, making the development of a new conservation technique possible. In another study conducted by S. Zhu et al. (2014), it was shown how the compound (Z)-5-(bromomethylene) furan-2(5H)-one (BMF), an inhibitor of AI-2, significantly reduced the expression of luxS and the activity of AI-2, causing the extracellular proteolytic activities of *Shewanella putrefaciens* and *Shewanella baltica*, microorganisms involved in the deterioration of refrigerated shrimps, to be inhibited. The study showed the potential application of AI-2 inhibitors to prolong the shelf life of fresh shrimp meat.

22.6 Economic Feasibility

This is one of the main problems involved when applying new alternatives to the usual sanitization processes. Clearly the use of new substances could lead to an increase in production costs, and there is also a need for research and the technological development of the different substances discussed. Furthermore, once how they function is understood and their efficiency verified, they must be further tested on an industrial scale with the subsequent extra costs involved. However, non-quality and lack of food security always have enormous negative consequences. According to Vysochynska and Intxaurburu (2016) the non-quality costs can be classified as:

- Prevention costs
- Appraisal costs
- Internal failure costs
- External fail

With an effective control system, the prevention and internal failure costs would be restrained, while any action means an increased initial investment. However, if the procedure is applied correctly, final production costs would necessarily be reduced after a short time.

In many cases the wrong approach is to focus excessively on the price of the chemical and forget what really matters, which is the quality of the product and the safety of consumers. In this chapter we have highlighted the essential points in the control of biofilm formation, the mechanisms that allow their adhesion, and how to avoid the signals that allow the transmission of information between microorganisms. However, we must not forget the ultimate goal of any of these actions: In food safety, it is about preventing pathogens from reaching consumers or preventing them from being dangerous to their health. This is more important than the costs involved in controlling the hazards, otherwise foodborne diseases will appear, leading to the closure of the company or huge economic losses.

REFERENCES

Ahmed, Nibras A., Fernanda C. Petersen, and Anne A. Scheie. 2009. "AI-2/LuxS Is Involved in Increased Biofilm Formation by *Streptococcus intermedius* in the Presence of Antibiotics." *Antimicrobial Agents and Chemotherapy* 53 (10): 4258–63. doi:10.1128/AAC.00546-09.

Alavi, Mehran, and Naser Karimi. 2018. "Antiplanktonic, Antibiofilm, Antiswarming Motility and Antiquorum Sensing Activities of Green Synthesized Ag–TiO_2, TiO_2–Ag, Ag–Cu and Cu–Ag Nanocomposites against Multi-Drug-Resistant Bacteria." *Artificial Cells, Nanomedicine, and Biotechnology* 46 (3): 399–413. doi:10.1080/21691401.2018.1496923.

Ali, Syed Ghazanfar, Mohammad Azam Ansari, Qazi Mohd. Sajid Jamal, Haris M. Khan, Mohammad Jalal, Hilal Ahmad, and Abbas Ali Mahdi. 2017. "Antiquorum Sensing Activity of Silver Nanoparticles in *Pseudomonas aeruginosa*: An *in Silico* Study." *In Silico Pharmacology* 5 (12): 1–7. doi:10.1007/s40203-017-0031-3.

Almasoud, Ahmad, Navam Hettiarachchy, Srinivas Rayaprolu, Dinesh Babu, Young Min Kwon, and Andy Mauromoustakos. 2016. "Inhibitory Effects of Lactic and Malic Organic Acids on Autoinducer Type 2 (AI-2) Quorum Sensing of *Escherichia coli* O157: H7 and *Salmonella* Typhimurium." *LWT – Food Science and Technology* 66: 560–64. doi:10.1016/j.lwt.2015.11.013.

Almeida, Felipe Alves de, Natan de Jesus Pimentel-Filho, Uelinton Manoel Pinto, Hilário Cuquetto Mantovani, Leandro Licursi de Oliveira, and Maria Cristina Dantas Vanetti. 2017. "Acyl Homoserine Lactone-Based Quorum Sensing Stimulates Biofilm Formation by *Salmonella* Enteritidis in Anaerobic Conditions." *Archives of Microbiology* 199 (3): 475–86. doi:10.1007/s00203-016-1313-6.

Álvarez-Ordóñez, Avelino, and Romain Briandet. 2016. "Editorial: Biofilms from a Food Microbiology Perspective: Structures, Functions, and Control Strategies." *Frontiers in Microbiology* 7 (November): 1–3. doi:10.3389/fmicb.2016.01938.

Ammor, Mohammed Salim, Christos Michaelidis, and George-John E. Nychas. 2008. "Insights into the Role of *Quorum Sensing* in Food Spoilage." *Journal of Food Protection* 71 (7): 1510–25. doi:10.4315/0362-028X-71.7.1510.

Araujo Nogueira, Jose Walter, Renata Albuquerque Costa, Magda Turini da Cunha, and Theodora Thays Arruda Cavalcante. 2017. "Antibiofilm Activity of Natural Substances Derived from Plants." *African Journal of Microbiology Research* 11 (26): 1051–60. doi:10.5897/ajmr2016.8180.

Arendt, Susan, Lakshman Rajagopal, Catherine Strohbehn, Nathan Stokes, Janell Meyer, and Steven Mandernach. 2013. "Reporting of Foodborne Illness by U.S. Consumers and Healthcare Professionals." *International Journal of Environmental Research and Public Health* 10: 3684–3714. doi:10.3390/ijerph10083684.

Bagge-Ravn, Dorthe, Kelna Gardshodn, Lone Gram, and Birte Fonnesbech Vogel. 2003. "Comparison of Sodium Hypochlorite–Based Foam and Peroxyacetic Acid–Based Fog Sanitizing Procedures in a Salmon Smokehouse: Survival of the General Microflora and *Listeria monocytogenes*." *Journal of Food Protection* 66 (April): 592–98. doi:10.4315/0362-028X-66.4.592.

Balabanova, Larissa, Oksana Son, Valery Rasskazov, Yulia Noskova, Liudmila Tekutyeva, Marina Eliseikina, Lubov Slepchenko, Olga Nedashkovskaya, and Anna Podvolotskaya. 2017. "Nucleolytic Enzymes from the Marine Bacterium Cobetia Amphilecti KMM 296 with Antibiofilm Activity and Biopreservative Effect on Meat Products." *Food Control* 78: 270–78. doi:10.1016/j.foodcont.2017.02.029.

Borucki, Monica K., Jason D. Peppin, David White, Frank Loge, and Douglas R. Call. 2003. "Variation in Biofilm Formation among Strains of *Listeria monocytogenes*." *Applied and Environmental Microbiology* 69 (12): 7336–42. doi:10.1128/AEM.69.12.7336.

Brackman, Gilles, and Tom Coenye. 2015. "Quorum Sensing Inhibitors as Anti-Biofilm Agents." *Current Pharmaceutical Design* 21 (1): 5–11. doi:10.2174/1381612820666140905114627.

Brackman, Gilles, Shari Celen, Ulrik Hillaert, Serge van Calenbergh, Paul Cos, Louis Maes, Hans J. Nelis, and Tom Coenye. 2011. "Structure-Activity Relationship of Cinnamaldehyde Analogs as Inhibitors of AI-2 Based Quorum Sensing and Their Effect on Virulence of *Vibrio* Spp." *PLoS ONE* 6 (1): 1–10. doi:10.1371/journal.pone.0016084.

Bridier, Arnaud, Pilar Sanchez-Vizucte, Morgan Guilbaud, J-C. Piard, Muriel Naïtali, and Romain Briandet. 2015. "Biofilm-Associated Persistence of Food-Borne Pathogens." *Food Microbiology* 45: 167–78. doi:10.1016/j.fm.2014.04.015.

Brown, Laura G., E. Rickamer Hoover, Carol A. Selman, Erik Coleman, and Helen Schurz Rogers. 2017. "Outbreak Characteristics Associated with Identification of Contributing Factors to Foodborne Illness Outbreaks." *Epidemiology and Infection* 145 (11): 2254–62. doi:10.1017/S0950268817001406.

Callejón, Raquel M., M. Isabel Rodríguez-Naranjo, Cristina Ubeda, Ruth Hornedo-Ortega, María Carmen Garcia-Parrilla, and Ana M. Troncoso. 2015. "Reported Foodborne Outbreaks Due to Fresh Produce in the United States and European Union: Trends and Causes." *Foodborne Pathogens and Disease* 12 (1): 32–38. doi:10.1089/fpd.2014.1821.

Castro-Mayorga, J. L., M. A. M. Reis, F. Freitas, J. M. Lagaron, and M. A. Prieto. 2018. "Biosynthesis of Silver Nanoparticles and Polyhydroxybutyrate Nanocomposites of Interest in Antimicrobial Applications." *International Journal of Biological Macromolecules* 108: 426–35. doi:10.1016/j.ijbiomac.2017.12.007.

Costerton, J. W., Zbigniew Lewandowski, Douglas E. Caldwell, Darren R. Korber, and Hilary Lappin-Scott. 1995. "Microbial Biofilms." *Annual Review of Microbiology* 49 (1): 711–45. doi:10.1146/annurev.mi.49.100195.003431.

Davey, Mary Ellen, and George A. O'Toole. 2000. "Microbial Biofilms: From Ecology to Molecular Genetics." *Microbiology and Molecular Biology Reviews* 64 (4): 847–67. doi:10.1128/MMBR.64.4.847-867.2000.

Diggle, S. P., and P. Williams. 2017. "Quorum Sensing": *Brenner's Encyclopedia of Genetics:* 2nd edn., pp. 1–4. doi:10.1016/B978-0-12-374984-0.01252-3.

Dong, Y.-H., J.-L. Xu, X.-Z. Li, and L.-H. Zhang. 2000. "AiiA, an Enzyme That Inactivates the Acylhomoserine Lactone Quorum-Sensing Signal and Attenuates the Virulence of *Erwinia carotovora*." *Proceedings of the National Academy of Sciences* 97 (7): 3526–31. doi:10.1073/pnas.97.7.3526.

Donlan, Rodney M. 2001. "Biofilm Formation: A Clinically Relevant Microbiological Process." *Clinical Infectious Diseases* 33 (8): 1387–92. doi:10.1086/322972.

Donlan, Rodney M., and J. William Costerton. 2002. "Biofilms: Survivalmechanisms of Clinically Relevant Microorganisms." *Clinical Microbiology Reviews* 15 (2): 167–19. doi:10.1128/CMR.15.2.167.

EFSA-ECDC. 2019. "The European Union Summary Report on Trends and Sources of Zoonoses, Zoonotic Agents and Food-Borne Outbreaks in 2018." doi:10.2903/j.efsa.2019.5926.

FAO-WHO. 2003. *Food Hygiene*. Fourth edition. ISBN 978-92-5-105913-5.

Fetzner, Susanne. 2015. "Quorum Quenching Enzymes." *Journal of Biotechnology* 201: 2–14. doi:10.1016/j.jbiotec.2014.09.001.

Flemming, Hans Curt, and Jost Wingender. 2010. "The Biofilm Matrix." *Nature Reviews Microbiology* 8 (9): 623–33. doi:10.1038/nrmicro2415.

Gabriel, Alonzo A., Ma Luisa P. Ballesteros, Leo Mendel D. Rosario, Roy B. Tumlos, and Henry J. Ramos. 2018. "Elimination of *Salmonella enterica* on Common Stainless Steel Food Contact Surfaces Using UV-C and Atmospheric Pressure Plasma Jet." *Food Control* 86: 90–100. doi:10.1016/j.foodcont.2017.11.011.

Giaouris, Efstathios, and Manuel V, Simões. 2018. In the Food Industry and Alternative Control Strategies. *Foodborne Diseases*. doi:10.1016/B978-0-12-811444-5/00011-7.

Giaouris, Efstathios, Even Heir, Michel Hébraud, Nikos Chorianopoulos, Solveig Langsrud, Trond Møretrø, Olivier Habimana, Mickaël Desvaux, Sandra Renier, and George John Nychas. 2014. "Attachment and Biofilm Formation by Foodborne Bacteria in Meat Processing Environments: Causes, Implications, Role of Bacterial Interactions and Control by Alternative Novel Methods." *Meat Science* 97 (3): 289–309. doi:10.1016/j.meatsci.2013.05.023.

González-Rivas, Fabián, Carolina Ripolles-Avila, Fabio Fontecha-Umaña, Abel Guillermo Ríos-Castillo, and José Juan Rodríguez-Jerez. 2018. "Biofilms in the Spotlight: Detection, Quantification, and Removal Methods." *Comprehensive Reviews in Food Science and Food Safety* 17 (5): 1261–76. doi:10.1111/1541-4337.12378.

Gopu, Venkadesaperumal, Chetan Kumar Meena, Ayaluru Murali, and Prathapkumar Halady Shetty. 2016. "Quorum Quenching Activity in the Cell-Free Lysate of *Enterobacter ludwigii* Isolated from Beef and Its Effect on Quorum Sensing Regulation in *Yersinia enterocolitica*." *RSC Advances* 6 (25): 21277–84. doi:10.1039/c5ra25440j.

Grandclément, Catherine, Mélanie Tannières, Solange Moréra, Yves Dessaux, and Denis Faure. 2016. "Quorum Quenching: Role in Nature and Applied Developments." *FEMS Microbiology Reviews* 40 (1): 86–116. doi:10.1093/femsre/fuv038.

Guendouze, Assia, Laure Plener, Janek Bzdrenga, Pauline Jacquet, Benjamin Rémy, Mikael Elias, Jean Philippe Lavigne, David Daudé, and Eric Chabrière. 2017. "Effect of Quorum Quenching Lactonase in Clinical Isolates of *Pseudomonas aeruginosa* and Comparison with *Quorum Sensing* Inhibitors." *Frontiers in Microbiology* 8 (February): 1–10. doi:10.3389/fmicb.2017.00227.

Guerrero-Navarro, A. E., A. G. Ríos-Castillo, C. Ripolles-Avila, A. S. Hascoët, X. Felipe, and J. J. Rodriguez-Jerez. 2019. "Development of a Dairy Fouling Model to Assess the Efficacy of Cleaning Procedures Using Alkaline and Enzymatic Products." *LWT – Food Science and Technology* 106 (September): 44–49. doi:10.1016/j.lwt.2019.02.057.

Gupta, Kuldeep, Salam Pradeep Singh, Ajay Kumar Manhar, Devabrata Saikia, Nima D. Namsa, Bolin Kumar Konwar, and Manabendra Mandal. 2019. "Inhibition of *Staphylococcus aureus* and *Pseudomonas aeruginosa* Biofilm and Virulence by Active Fraction of Syzygium Cumini (L.) Skeels Leaf Extract: *In-Vitro* and *In Silico* Studies." *Indian Journal of Microbiology* 59 (1): 13–21. doi:10.1007/s12088-018-0770-9.

Habimana, Olivier, Even Heir, Solveig Langsrud, Anette Wold Åsli, and Trond Møretrø. 2010. "Enhanced Surface Colonization by *Escherichia coli* 0157:H7 in Biofilms Formed by an *Acinetobacter calcoaceticus* Isolate from Meat-Processing Environments." *Applied and Environmental Microbiology* 76 (13): 4557–59. doi:10.1128/AEM.02707-09.

Hentzer, Morten, and Michael Givskov. 2003. "Pharmacological Inhibition of *Quorum Sensing* for the Treatment of Chronic Bacterial Infections." *Journal of Clinical Investigation* 112 (9): 1300–1307. doi:10.1172/JCI20074.

Heukelekian, H, and A Heller. 1940. "Relation between Food Concentration and Surface for Bacterial Growth." *Journal of Bacteriology* 40 (4): 547–58.

Hnamte, Sairengpuii, Pattnaik Subhaswaraj, Sampath Kumar Ranganathan, Dinakara Rao Ampasala, Gangatharan Muralitharan, and Busi Siddhardha. 2019. "Methanolic Extract of Plectranthus Tenuiflorus Attenuates Quorum Sensing Mediated Virulence and Biofilm Formation in *Pseudomonas aeruginosa* PAO1." *Journal of Pure and Applied Microbiology* 12 (4): 1985–96. doi:10.22207/jpam.12.4.35.

Holah, John T. 1995. "Disinfection of Food Production Areas." *Revue Scientifique et Technique* 14 (2): 343–63.

Hussain, Malik Altaf, and Christopher O Dawson. 2013. "Economic Impact of Food Safety Outbreaks on Food Businesses." *Foods* 2: 585–89. doi:10.3390/foods2040585.

Ishida, Takenori, Tsukasa Ikeda, Noboru Takiguchi, Akio Kuroda, Hisao Ohtake, and Junichi Kato. 2007. "Inhibition of Quorum Sensing in *Pseudomonas aeruginosa* by *N*-Acyl Cyclopentylamides." *Applied and Environmental Microbiology* 73 (10): 3183–88. doi:10.1128/AEM.02233-06.

Jain, Aditi, Shivendu Ranjan, Nandita Dasgupta, and Chidambaram Ramalingam. 2016. "Nanomaterials in Food and Agriculture: An Overview on Their Safety Concerns and Regulatory Issues." *Critical Reviews in Food Science and Nutrition* 58 (2): 297–317. doi:10.1080/10408398.2016.1160363.

Jefferson, Kimberly K. 2004. "What Drives Bacteria to Produce a Biofilm?" *FEMS Microbiology Letters* 236 (2): 163–73. doi:10.1016/j.femsle.2004.06.005.

Jones, H. C., I. L. Roth, and W. M. Sanders. 1969. "Electron Microscopic Study of a Slime Layer." *Journal of Bacteriology* 99 (1): 316–25.

Kalia, Vipin Chandra. 2013. "Quorum Sensing Inhibitors: An Overview." *Biotechnology Advances* 31 (2): 224–45. doi:10.1016/j.biotechadv.2012.10.004.

Kappachery, Sajeesh, Diby Paul, Jeyong Yoon, and Ji Hyang Kweon. 2010. "Vanillin, a Potential Agent to Prevent Biofouling of Reverse Osmosis Membrane." *Biofouling* 26 (6): 667–72. doi:10.1080/08927014.2010.506573.

Kiran, S., P. Sharma, K. Harjai, and N. Capalash. 2011. "Enzymatic Quorum Quenching Increases Antibiotic Susceptibility of Multidrug Resistant *Pseudomonas aeruginosa*." *Iranian Journal of Microbiology* 3 (1): 1–12.

Labbate, Maurizio, Shu Yeong Queck, Kai Shyang Koh, Scott A. Rice, Michael Givskov, and Staffan Kjelleberg. 2004. "Quorum Sensing-Controlled Biofilm Development in *Serratia liquefaciens* MG1." *Journal of Bacteriology* 186 (3): 692–98. doi:10.1128/JB.186.3.692-698.2004.

LaSarre, B., and M. J. Federle. 2013. "Exploiting Quorum Sensing to Confuse Bacterial Pathogens." *Microbiology and Molecular Biology Reviews* 77 (1): 73–111. doi:10.1128/MMBR.00046-12.

Laufer, A. S., J. Grass, K. Holt, J. M. Whichard, P. M. Griffin, and L. H. Gould. 2015. "Outbreaks of *Salmonella* Infections Attributed to Beef – United States, 1973–2011." *Epidemiology and Infection* 143 (9): 2003–13. doi:10.1007/s11910-013-0409-5.Depression.

Leadbetter, Jared R., and E. P. Greenberg. 2000. "Metabolism of Acyl-Homoserine Lactone Quorum-Sensing Signals by *Variovorax paradoxus*." *Journal of Bacteriology* 182 (24): 6921–26. doi:10.1128/JB.182.24.6921-6926.2000.Updated.

Lewis Oscar, Felix, Davoodbasha MubarakAli, Chari Nithya, Rajendran Priyanka, Venkatraman Gopinath, Naiyf S. Alharbi, and Nooruddin Thajuddin. 2015. "One Pot Synthesis and Anti-Biofilm Potential of Copper Nanoparticles (CuNPs) against Clinical Strains of *Pseudomonas aeruginosa*." *Biofouling* 31 (4): 379–91. doi:10.1080/08927014.2015.1048686.

Li, Tingting, Dangfeng Wang, Nan Liu, Yan Ma, Ting Ding, Yongchao Mei, and Jianrong Li. 2018. "Inhibition of Quorum Sensing-Controlled Virulence Factors and Biofilm Formation in *Pseudomonas fluorescens* by Cinnamaldehyde." *International Journal of Food Microbiology* 269 (February): 98–106. doi:10.1016/j.ijfoodmicro.2018.01.023.

Liang, Ziwei, Yanfei Qi, Shuanli Guo, Kun Hao, Mingming Zhao, and Na Guo. 2019. "Effect of AgWPA Nanoparticles on the Inhibition of *Staphylococcus aureus* Growth in Biofilms." *Food Control* 100: 240–46. doi:10.1016/j.foodcont.2019.01.030.

Mangal, Manisha, Sangita Bansal, Satish K. Sharma, and Ram K. Gupta. 2015. "Molecular Detection of Foodborne Pathogens: A Rapid and Accurate Answer to Food Safety." *Critical Reviews in Food Science and Nutrition* 56 (9): 1568–84. doi:10.1080/10408398.2013.782483.

McCarthy, Yvonne, Liang Yang, Kate B. Twomey, Andrea Sass, Tim Tolker-Nielsen, Eshwar Mahenthiralingam, J. Maxwell Dow, and Robert P. Ryan. 2010. "A Sensor Kinase Recognizing the Cell-Cell Signal BDSF (*Cis*-2-Dodecenoic Acid) Regulates Virulence in *Burkholderia cenocepacia*." *Molecular Microbiology* 77 (5): 1220–36. doi:10.1111/j.1365-2958.2010.07285.x.

Meireles, Ana, Anabela Borges, Efstathios Giaouris, and Manuel Simões. 2016. "The Current Knowledge on the Application of Anti-Biofilm Enzymes in the Food Industry." *Food Research International* 86: 140–46. doi:10.1016/j.foodres.2016.06.006.

Nahar, Shamsun, Md Furkanur Rahaman Mizan, Angela Jie won Ha, and Sang Do Ha. 2018. "Advances and Future Prospects of Enzyme-Based Biofilm Prevention Approaches in the Food Industry." *Comprehensive Reviews in Food Science and Food Safety* 17 (6): 1484–1502. doi:10.1111/1541-4337.12382.

Naik, K., and M. Kowshik. 2014. "Anti-Quorum Sensing Activity of AgCl-TiO_2 Nanoparticles with Potential Use as Active Food Packaging Material." *Journal of Applied Microbiology* 117 (4): 972–83. doi:10.1111/jam.12589.

Nielsen, C. K., G. Subbiahdoss, G. Zeng, Z. Salmi, J. Kjems, T. Mygind, T. Snabe, and R. L. Meyer. 2018. "Antibacterial Isoeugenol Coating on Stainless Steel and Polyethylene Surfaces Prevents Biofilm Growth." *Journal of Applied Microbiology* 124 (1): 179–87. doi:10.1111/jam.13634.

Nutman, Allen P., Vickie C. Bennett, Clark R. L. Friend, Martin J. Van Kranendonk, and Allan R. Chivas. 2016. "Rapid Emergence of Life Shown by Discovery of 3,700-Million-Year-Old Microbial Structures." *Nature* 537 (7621): 535–38. doi:10.1038/nature19355.

Oh, Hyun Suk, Chuan Hao Tan, Jiun Hui Low, Miles Rzechowicz, Muhammad Faisal Siddiqui, Harvey Winters, Staffan Kjelleberg, Anthony G. Fane, and Scott A. Rice. 2017. "Quorum Quenching Bacteria Can Be Used to Inhibit the Biofouling of Reverse Osmosis Membranes." *Water Research* 112: 29–37. doi:10.1016/j.watres.2017.01.028.

Oh, Soo Kyung, Hyun Joo Chang, Hyang Sook Chun, Hyun Jin Kim, and Nari Lee. 2015. "Pomegranate (*Punica Granatum L.*) Peel Extract Inhibits Quorum Sensing and Biofilm Formation Potential in *Yersinia enterocolitica*." *Korean Journal of Microbiology and Biotechnology* 43 (4): 357–66. doi:10.4014/mbl.1510.10004.

Okano, Chigusa, Daichi Murota, Eri Nasuno, Ken ichi Iimura, and Norihiro Kato. 2019. "Effective Quorum Quenching with a Conformation-Stable Recombinant Lactonase Possessing a Hydrophilic Polymeric Shell Fabricated via Electrospinning." *Materials Science and Engineering C* 98: 437–44. doi:10.1016/j.msec.2019.01.007.

Pan, Jiachuan, and Dacheng Ren. 2009. "Quorum Sensing Inhibitors: A Patent Overview." *Expert Opinion on Therapeutic Patents* 19 (11): 1581–1601. doi:10.1517/13543770903222293.

Parsek, Matthew R., and E. P. Greenberg. 2005. "Sociomicrobiology: The Connections between Quorum Sensing and Biofilms." *Trends in Microbiology*. doi:10.1016/j.tim.2004.11.007.

Pattnaik, Subhaswaraj, Subhashree Barik, Gangatharan Muralitharan, and Siddhardha Busi. 2018. "Ferulic Acid Encapsulated Chitosan-Tripolyphosphate Nanoparticles Attenuate Quorum Sensing Regulated Virulence and Biofilm Formation in *Pseudomonas aeruginosa* PAO1." *IET Nanobiotechnology* 12 (8): 1056–61. doi:10.1049/iet-nbt.2018.5114.

Pei, Ruoting, and Gisella R. Lamas-Samanamud. 2014. "Inhibition of Biofilm Formation by T7 Bacteriophages Producing Quorum-Quenching Enzymes." *Applied and Environmental Microbiology* 80 (17): 5340–48. doi:10.1128/aem.01434-14.

Piras, Anna M., Giuseppantonio Maisetta, Stefania Sandreschi, Matteo Gazzarri, Cristina Bartoli, Lucia Grassi, Semih Esin, Federica Chiellini, and Giovanna Batoni. 2015. "Chitosan Nanoparticles Loaded with the Antimicrobial Peptide Temporin B Exert a Long-Term Antibacterial Activity *in Vitro* against Clinical Isolates of *Staphylococcus epidermidis*." *Frontiers in Microbiology* 6 (April): 1–10. doi:10.3389/fmicb.2015.00372.

Possas, Arícia, Elena Carrasco, R. M. García-Gimeno, and Antonio Valero. 2017. "Models of Microbial Cross-Contamination Dynamics." *Current Opinion in Food Science* 14: 43–49. doi:10.1016/j.cofs.2017.01.006.

Prateeksha, Braj R. Singh, M. Shoeb, S. Sharma, A. H. Naqvi, Vijai K. Gupta, and Brahma N. Singh. 2017. "Scaffold of Selenium Nanovectors and Honey Phytochemicals for Inhibition of *Pseudomonas aeruginosa* Quorum Sensing and Biofilm Formation." *Frontiers in Cellular and Infection Microbiology* 7 (March): 1–14. doi:10.3389/fcimb.2017.00093.

Qais, Faizan Abul, Mohammad Shavez Khan, and Iqbal Ahmad. 2018. "Nanoparticles as Quorum Sensing Inhibitor: Prospects and Limitations BT." In *Biotechnological Applications of Quorum Sensing Inhibitors*, edited by Vipin Chandra Kalia, pp. 227–44. Singapore: Springer Singapore. doi:10.1007/978-981-10-9026-4_11.

Raghunath, Azhwar, and Ekambaram Perumal. 2017. "Metal Oxide Nanoparticles as Antimicrobial Agents: A Promise for the Future." *International Journal of Antimicrobial Agents* 49 (2): 137–52. doi:10.1016/j.ijantimicag.2016.11.011.

Rahman, M. D. Ramim Tanver, Zaixiang Lou, Jun Zhang, Fuhao Yu, Yakindra Prasad Timilsena, Caili Zhang, Yi Zhang, and Amr M. Bakry. 2017. "Star Anise (*Illicium Verum Hook.* f.) as Quorum Sensing and Biofilm Formation Inhibitor on Foodborne Bacteria: Study in Milk." *Journal of Food Protection* 80 (4): 645–53. doi:10.4315/0362-028x.jfp-16-294.

Rasmussen, Thomas B., Thomas Bjarnsholt, Mette Elena Skindersoe, Morten Hentzer, Peter Kristoffersen, Manuela Ko, John Nielsen, Leo Eberl, and Michael Givskov. 2005. "Screening for Quorum-Sensing Inhibitors (QSI) by Use of a Novel Genetic System, the QSI Selector." *Journal of Bacteriology* 187 (5): 1799–1814. doi:10.1128/JB.187.5.1799.

Rasmussen, Thomas B., and Michael Givskov. 2006. "Quorum Sensing Inhibitors: A Bargain of Effects." *Microbiology* 152 (4): 895–904. doi:10.1099/mic.0.28601-0.

Ravindran, D., S. Ramanathan, K. Arunachalam, G. P. Jeyaraj, K. P. Shunmugiah, and V. R. Arumugam. 2018. "Phytosynthesized Silver Nanoparticles as Antiquorum Sensing and Antibiofilm Agent against the Nosocomial Pathogen *Serratia marcescens*: An *in Vitro* Study." *Journal of Applied Microbiology* 124 (6): 1425–40. doi:10.1111/jam.13728.

Rehman, Zahid Ur, and Tor Ove Leiknes. 2018. "Quorum-Quenching Bacteria Isolated from Red Sea Sediments Reduce Biofilm Formation by *Pseudomonas aeruginosa*." *Frontiers in Microbiology* 9 (July): 1–13. doi:10.3389/fmicb.2018.01354.

Ripolles-Avila, C., Abel Guillermo Ríos-Castillo, and José Juan Rodríguez-Jerez. 2018. "Development of a Peroxide Biodetector for a Direct Detection of Biofilms Produced by Catalase-Positive Bacteria on Food-Contact Surfaces." *CyTA – Journal of Food* 16 (1): 506–15. doi:10.1080/19476337.2017.1418434.

Ripolles-Avila, C., A. S. Hascoët, A. E. Guerrero-Navarro, and J. J. Rodríguez-Jerez. 2018. "Establishment of Incubation Conditions to Optimize the *in Vitro* Formation of Mature *Listeria monocytogenes* Biofilms on Food-Contact Surfaces." *Food Control* 92: 240–48. doi:10.1016/j.foodcont.2018.04.054.

Ripolles-Avila, C., A. S. Hascoët, J. V. Martínez-Suárez, R. Capita, and J. J. Rodríguez-Jerez. 2019. "Evaluation of the Microbiological Contamination of Food Processing Environments through Implementing Surface Sensors in an Iberian Pork Processing Plant: An Approach towards the Control of *Listeria monocytogenes*." *Food Control* 99 (November 2018): 40–47. doi:10.1016/j.foodcont.2018.12.013.

Ripolles-Avila, C., B. H. Cervantes-Huaman, A. S. Hascoët, J. Yuste, and J. J. Rodríguez-Jerez. 2019. "Quantification of Mature *Listeria monocytogenes* Biofilm Cells Formed by an *in Vitro* Model: A Comparison of Different Methods." *International Journal of Food Microbiology* 289 (October 2018): 209–14. doi:10.1016/j.ijfoodmicro.2018.10.020.

Rivera, Dácil, Viviana Toledo, Angélica Reyes-Jara, Paola Navarrete, Mark Tamplin, Bon Kimura, Martin Wiedmann, Primal Silva, and Andrea I. Moreno Switt. 2018. "Approaches to Empower the Implementation of New Tools to Detect and Prevent Foodborne Pathogens in Food Processing." *Food Microbiology* 75: 126–32. doi:10.1016/j.fm.2017.07.009.

Sadiq, I. Mohammed, N. Chandrasekaran, and A. Mukherjee. 2010. "Studies on Effect of TiO_2 Nanoparticles on Growth and Membrane Permeability of *Escherichia coli*, *Pseudomonas aeruginosa*, and *Bacillus subtilis*." *Current Nanoscience* 6 (4): 381–87. doi:10.2174/157341310791658973.

Salini, Ramesh, Muthukrishnan Sindhulakshmi, Thirumaran Poongothai, and Shunmugiah Karutha Pandian. 2015. "Inhibition of Quorum Sensing Mediated Biofilm Development and Virulence in Uropathogens by *Hyptis suaveolens*." *Antonie van Leeuwenhoek, International Journal of General and Molecular Microbiology* 107 (4): 1095–1106. doi:10.1007/s10482-015-0402-x.

Salmi-Mani, Hanène, Gabriel Terreros, Nadine Barroca-Aubry, Caroline Aymes-Chodur, Christophe Regeard, and Philippe Roger. 2018. "Poly(Ethylene Terephthalate) Films Modified by UV-Induced Surface Graft Polymerization of Vanillin Derived Monomer for Antibacterial Activity." *European Polymer Journal* 103 (November 2017): 51–58. doi:10.1016/j.eurpolymj.2018.03.038.

Sanchez-Vizuete, P., B. Orgaz, S. Aymerich, D. Le Coq, and R. Briandet. 2015. "Pathogens Protection against the Action of Disinfectants in Multispecies Biofilms." *Frontiers in Microbiology* 6 (June): 1–12. doi:10.3389/fmicb.2015.00705.

Sawai, Jun, Kyoko Himizu, and Osamu Yamamoto. 2005. "Kinetics of Bacterial Death by Heated Dolomite Powder Slurry." *Soil Biology and Biochemistry* 37 (8): 1484–89. doi:10.1016/j.soilbio.2005.01.011.

Sharifi, A., A. Mohammadzadeh, T. Zahraei Salehi, and P. Mahmoodi. 2018. "Antibacterial, Antibiofilm and Antiquorum Sensing Effects of *Thymus daenensis* and *Satureja hortensis* Essential Oils against *Staphylococcus aureus* Isolates." *Journal of Applied Microbiology* 124 (2): 379–88. doi:10.1111/jam.13639.

Simões, M., L. C. Simões, I. Machado, M. O. Pereira, and M. João Vieira. 2006. "Control of Flow-Generated Biofilms with Surfactants: Evidence of Resistance and Recovery." *Food and Bioproducts Processing* 84 (C4): 338–45. doi:10.1205/fbp06022.

Singh, Brahma N., Prateeksha, Dalip K. Upreti, Braj Raj Singh, Tom Defoirdt, Vijai K. Gupta, Ana Olivia De Souza, et al. 2017. "Bactericidal, Quorum Quenching and Anti-Biofilm Nanofactories: A New Niche for Nanotechnologists." *Critical Reviews in Biotechnology* 37 (4): 525–40. doi:10.1080/07388551.2016.1199010.

Singh, Nimisha, Jyutika Rajwade, and K. M. Paknikar. 2019. "Transcriptome Analysis of Silver Nanoparticles Treated *Staphylococcus aureus* Reveals Potential Targets for Biofilm Inhibition." *Colloids and Surfaces B: Biointerfaces* 175 (November 2018): 487–97. doi:10.1016/j.colsurfb.2018.12.032.

Skandamis, Panagiotis N., and George-John E. Nychas. 2012. "Quorum Sensing in the Context of Food Microbiology." *Applied and Environmental Microbiology* 78 (16): 5473–82. doi:10.1128/aem.00468-12.

Subhaswaraj, Pattnaik, Subhashree Barik, Chandrasekhar Macha, Potu Venkata Chiranjeevi, and Busi Siddhardha. 2018. "Anti Quorum Sensing and Anti Biofilm Efficacy of Cinnamaldehyde Encapsulated Chitosan Nanoparticles against *Pseudomonas aeruginosa* PAO1." *LWT – Food Science and Technology* 97 (June): 752–59. doi:10.1016/j.lwt.2018.08.011.

Thakur, Pallavi, Raman Chawla, Ankit Tanwar, Ankita Singh Chakotiya, Alka Narula, Rajeev Goel, Rajesh Arora, and Rakesh Kumar Sharma. 2016. "Attenuation of Adhesion, Quorum Sensing and Biofilm Mediated Virulence of Carbapenem Resistant *Escherichia coli* by Selected Natural Plant Products." *Microbial Pathogenesis* 92: 76–85. doi:10.1016/j.micpath.2016.01.001.

Thallinger, Barbara, Endry N. Prasetyo, Gibson S. Nyanhongo, and Georg M. Guebitz. 2013. "Antimicrobial Enzymes: An Emerging Strategy to Fight Microbes and Microbial Biofilms." *Biotechnology Journal* 8 (1): 97–109. doi:10.1002/biot.201200313.

Thill, Antoine, Ophélie Zeyons, Olivier Spalla, Franck Chauvat, Jerôme Rose, Mélanie Auffan, and Anne Marie Flank. 2006. "Cytotoxicity of CeO_2 Nanoparticles for *Escherichia coli*: Physico-Chemical Insight of the Cytotoxicity Mechanism." *Environmental Science and Technology* 40 (19): 6151–56. doi:10.1021/es060999b.

Tirado, C., and K. Schmidt. 2001. "WHO Surveillance Programme for Control of Foodborne Infections and Intoxications: Preliminary Results and Trends across Greater Europe." *Journal of Infection* 43 (1): 80–84. doi:10.1053/jinf.2001.0861.

Tiwari, Ruchi, Kumaragurubaran Karthik, Rajneesh Rana, Yashpal Singh Mali, Kuldeep Dhama, and Sunil Kumar Joshi. 2016. "Quorum Sensing Inhibitors/Antagonists Countering Food Spoilage Bacteria-Need Molecular and Pharmaceutical Intervention for Protecting Current Issues of Food Safety." *International Journal of Pharmacology* 12 (3): 262–71. doi:10.3923/ijp.2016.262.271.

Todd, Ewen C. D., Judy D. Greig, Charles A. Bartleson, and Barry S Michaels. 2009. "Outbreaks Where Food Workers Have Been Implicated in the Spread of Foodborne Disease. Part 9. Washing and Drying of Hands To Reduce Microbial Contamination." *Journal of Food Protection* 73 (10): 1937–55. doi:10.4315/0362-028X-73.10.1937.

Torabi Delshad, Somayeh, Siyavash Soltanian, Hassan Sharifiyazdi, and Peter Bossier. 2018. "Effect of Quorum Quenching Bacteria on Growth, Virulence Factors and Biofilm Formation of *Yersinia ruckeri in Vitro* and an *in Vivo* Evaluation of Their Probiotic Effect in Rainbow Trout." *Journal of Fish Diseases* 41 (9): 1429–38. doi:10.1111/jfd.12840.

Truchadoa, Pilar, Mar Larrosaa, Irene Castro-Ibáñez, and Ana Allende. 2015. "Plant Food Extracts and Phytochemicals: Their Role as Quorum Sensing Inhibitors." *Trends in Food Science & Technology* 43: 189–204. doi:10.1016/j.tifs.2015.02.009.

Udine, Claudia, Gilles Brackman, Silvia Bazzini, Silvia Buroni, Heleen van Acker, Maria Rosalia Pasca, Giovanna Riccardi, and Tom Coenye. 2013. "Phenotypic and Genotypic Characterisation of *Burkholderia cenocepacia* J2315 Mutants Affected in Homoserine Lactone and Diffusible Signal Factor-Based Quorum Sensing Systems Suggests Interplay between Both Types of Systems." *PLoS ONE* 8 (1): 7–10. doi:10.1371/journal.pone.0055112.

Valcárcel, Miguel, and Ángela I. López-Lorente. 2016. "Recent Advances and Trends in Analytical Nanoscience and Nanotechnology." *Trends in Analytical Chemistry* 84: 1–2. doi:10.1016/j.trac.2016.05.010.

Vazquez-Armenta, F. J., A. T. Bernal-Mercado, J. Lizardi-Mendoza, B. A. Silva-Espinoza, M. R. Cruz-Valenzuela, G. A. Gonzalez-Aguilar, F. Nazzaro, F. Fratianni, and J. F. Ayala-Zavala. 2018. "Phenolic Extracts from Grape Stems Inhibit *Listeria monocytogenes* Motility and Adhesion to Food Contact Surfaces." *Journal of Adhesion Science and Technology* 32 (8): 889–907. doi:10.1080/01694243.2017.1387093.

Vysochynska, Oksana, and Gurutze Intxaurburu. 2016. "Total Cost of Poor Quality." https://addi.ehu.es/bitstream/handle/10810/20676/Total Cost of Poor Quality.pdf?sequence=1.

Wagh, Mohini S., Rajendra H. Patil, Deepali K. Thombre, Milind V. Kulkarni, Wasudev N. Gade, and Bharat B. Kale. 2013. "Evaluation of Anti-Quorum Sensing Activity of Silver Nanowires." *Applied Microbiology and Biotechnology* 97 (8): 3593–3601. doi:10.1007/s00253-012-4603-1.

Weiland-Bräuer, Nancy, Martin J. Kisch, Nicole Pinnow, Andreas Liese, and Ruth A. Schmitz. 2016. "Highly Effective Inhibition of Biofilm Formation by the First Metagenome-Derived AI-2 Quenching Enzyme." *Frontiers in Microbiology* 7 (July): 1–19. doi:10.3389/fmicb.2016.01098.

WHO. 2006. "*WHO Consultation to Develop a Strategy to Estimate the Global Burden of Foodborne Diseases*." World Health Organization, Geneva, Switzerland.

WHO. 2015. "*Who Estimates of the Global Burden of Foodborne Diseases*." ISBN: 978-92-4-156516-5.

Witte, Anna Kristina, Martin Bobal, Roland David, Beat Blättler, Dagmar Schoder, and Peter Rossmanith. 2017. "Investigation of the Potential of Dry Ice Blasting for Cleaning and Disinfection in the Food Production Environment." *LWT – Food Science and Technology* 75: 735–41. doi:10.1016/j.lwt.2016.10.024.

Zaidi, Sahar, Lama Misba, and Asad U. Khan. 2017. "Nano-Therapeutics: A Revolution in Infection Control in Post Antibiotic Era." *Nanomedicine: Nanotechnology, Biology, and Medicine* 13 (7): 2281–2301. doi:10.1016/j.nano.2017.06.015.

Zhang, Lingling, Yunhong Jiang, Yulong Ding, Malcolm Povey, and David York. 2007. "Investigation into the Antibacterial Behaviour of Suspensions of ZnO Nanoparticles (ZnO Nanofluids)." *Journal of Nanoparticle Research* 9 (3): 479–89. doi:10.1007/s11051-006-9150-1.

Zhang, Ying, Jie Kong, Fei Huang, Yunfei Xie, Yahui Guo, Yuliang Cheng, He Qian, and Weirong Yao. 2018. "Hexanal as a QS Inhibitor of Extracellular Enzyme Activity of *Erwinia carotovora* and *Pseudomonas fluorescens* and Its Application in Vegetables." *Food Chemistry* 255 (February): 1–7. doi:10.1016/j.foodchem.2018.02.038.

Zhang, Yunhui, Gilles Brackman, and Tom Coenye. 2017. "Pitfalls Associated with Evaluating Enzymatic Quorum Quenching Activity: The Case of MomL and Its Effect on *Pseudomonas aeruginosa* and *Acinetobacter baumannii* Biofilms." *PeerJ* 5: e3251. doi:10.7717/peerj.3251.

Zhu, Suqin, Haohao Wu, Mingyong Zeng, Liu Zunying, Yuanhui Zhao, and Shiyuan Dong. 2014. "Regulation of Spoilage-Related Activities of *Shewanella putrefaciens* and *Shewanella baltica* by an Autoinducer-2 Analogue, (Z)-5-(Bromomethylene) Furan-2(5H)-One." *Journal of Food Processing and Preservation* 39 (6): 719–28. doi:10.1111/jfpp.12281.

Zhu, Xuena, Evangelia Hondroulis, Wenjun Liu, and Chen Zhong Li. 2013. "Biosensing Approaches for Rapid Genotoxicity and Cytotoxicity Assays upon Nanomaterial Exposure." *Small* 9 (9–10): 1821–30. doi:10.1002/smll.201201593.

Zobell, Claude E. 1943. "The Effect of Solid Surfaces upon Bacterial Activity." *Journal of Bacteriology* 46 (1): 39–56. doi:10.1146/annurev.arplant.58.032806.103848.

23

Quorum Sensing and Quorum Quenching in Food-Related Bacteria

Barbara Tomadoni and Alejandra Ponce

CONTENTS

23.1 Introduction

23.1.1 Food-Related Bacteria and Biofilm Formation

An enormous amount of food is wasted because of food spoilage. Food spoilage can be caused by either physical or biochemical changes that occur naturally in food, or it can be produced by microbial activity (Ammor, Michaelidis, and Nychas 2008). The latter is the most common reason of food spoilage and it is of prior concern to the food industry, not only because of the economic loss that it implies, but also because it constitutes a health hazard (Aswathanarayan and Vittal 2011).

Food spoilage by microorganisms is manifested by visible growth and changes in the food characteristics, mainly appearance, texture, and flavor. Microorganisms produce different types of enzymes (i.e. saccharolytic, proteolytic, pectinolytic, and lipolytic enzymes) whose metabolic end products can be associated with food spoilage. These metabolic products can cause off-flavor that results in rejection of the product (Ammor, Michaelidis, and Nychas 2008). Most common food spoilage bacteria are *Pseudomonas* spp., lactic acid bacteria (LAB), spore-forming bacteria (such as, *Bacillus* and *Clostridium*) and Enterobacteriaceae. Other important food spoilage bacteria include the Gram-negative *Acinetobacter, Achromobacter, Alcaligenes, Flavobacterium, Moraxella, Photobacterium*, and *Psychrobacter*, and the Gram-positive genera *Brevibacterium, Brochothrix, Corynebacterium*, and *Micrococcus*. These spoilage organisms tend to be associated mainly with chilled proteinaceous foods (i.e. meat, fish, and their derivative products) and dairy products.

Some microorganisms in food products can cause diseases and are considered a threat to human health. These are called foodborne pathogens, and may occur in all types of foods, from fruits and vegetables, to dairy, fish, and meat products. Foodborne pathogens do not necessarily cause food spoilage but can cause foodborne illness to consumers. Among them, *Listeria monocytogenes, Campylobacter jejuni, Salmonella* spp., *Yersinia enterocolitica, Clostridium* spp., *Staphylococcus aureus, Bacillus cereus*, and *Escherichia coli* have been responsible for most of the food-related outbreaks in the last decade. Foodborne pathogens associated with fresh produce are one of the major concerns in foodborne outbreaks recorded worldwide (Amrutha, Sundar, and Shetty 2017).

On the other hand, there are some food-related bacteria that are beneficial or desirable in the food industry, i.e. that produce desirable changes and intentionally alter food products' characteristics (Figure 23.1). Some beneficial bacteria can preserve foods through products of fermentation. LAB (such as *Carnobacterium, Enterococcus, Lactobacillus, Lactococcus, Leuconostoc, Oenococcus, Pediococcus, Streptococcus, Tetragenococcus, Vagococcus*, and *Weissella*) are one of the most important food-grade microorganisms used in food fermentation. These bacteria can degrade a variety of carbohydrates, with lactic acid as the most predominant end product. Many lactic acid bacteria also produce bacteriocins that have antimicrobial activity that is antagonistic to other bacteria, especially toward bacteria closely related to the bacteriocin producing strain (Doyle, Steenson, and Meng 2013).

Figure 23.1 shows the main classification of food-related bacteria, as previously described.

Undesired food-related bacteria, both spoilage and pathogenic, have the capacity to attach themselves to processing surfaces and form biofilms. Bacterial biofilm, also called biofouling, is a matrix formed by bacterial populations clusters adhering either cell-to-cell, or to surfaces or interfaces. This population of bacterial cells is embedded in hydrated polymeric matrices, and in this way have an increased resistance to environmental attacks, like cleaning agents and antibiotics (Gopu, Chandran, and Shetty 2018).

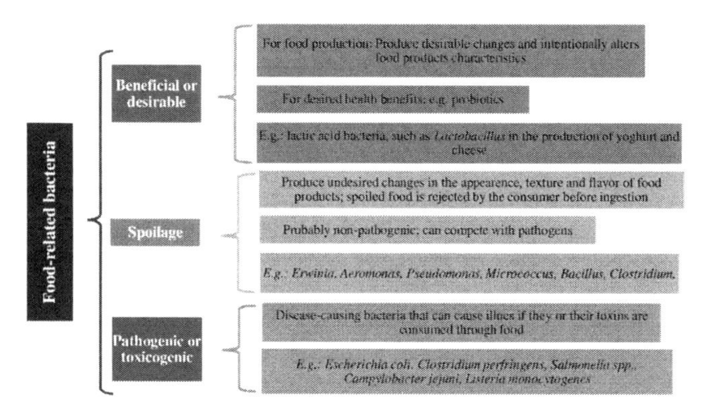

FIGURE 23.1 Food-related bacteria.

Therefore, foodborne pathogens are more harmful when they are in the form of biofilms, because they are highly prevalent and difficult to completely eliminate. Furthermore, bacteria can form biofilms in different surfaces present in the food industry, such as, glass, plastic, metal, and wood (Aswathanarayan and Vittal 2011).

Figure 23.2 portrays a schematic representation of the biofilm formation process. These stages slightly differ between Gram-positive and Gram-negative bacteria (Vasudevan 2014), though we can simplify them into the following:

1. Initial contact with the surface
2. Primary attachment/reversible attachment
3. Accumulation and intracellular aggregation/Irreversible attachment
4. Biofilm maturation
5. Detachment and dispersion.

Briefly, biofilm formation begins with the initial contact of the bacteria cell with the surface or interface. In Gram-negative bacteria (e.g. *Pseudomonas aeruginosa*, *Vibrio cholerae*, *Escherichia coli*, and *Salmonella enterica*), motility is an important characteristic for the primary attachment (Lasa et al. 2005). Motility seems to help bacteria reach the surface and counteract hydrophobic repulsions. However, even though it favors the process of biofilm formation, it is not an essential requirement, since many non-motile Gram-positive bacteria (e.g. *Staphylococcus* and *Streptococcus*) are capable of forming biofilms. In the case of Gram-positive bacteria, surface proteins have been described to participate in the primary attachment. Once the bacteria have attached to the surface, cellular division begins, and they form microcolonies. Subsequently, bacteria begin to secrete an exopolysaccharide (EPS) that constitutes the biofilm matrix and forms mushroom-like structures (Figure 23.2), where canals can be found. The EPS composition is different among bacteria and varies from alginate in *P. aeruginosa*, cellulose in *Salmonella typhimurium*, glucose and galactose rich polymer in *V. cholerae*, and poly-*N*-acetylglucosamine in *S. aureus*, among others. Furthermore, recent studies have demonstrated that even the same bacteria, when exposed to different environmental conditions, can produce different exopolysaccharides. For example, some *P. aeruginosa* strains are capable of producing a glucose-rich polymer that forms a film in the interphase with air. Finally, some bacteria are released or detached from within the biofilm matrix to colonize new surfaces, closing the cycle of biofilm forming process (Figure 23.2) (Vlamakis et al. 2013; Vasudevan 2014; Meliani and Bensoltane 2015).

In the food industry, the presence of biofilms is very common, and they can be found in a variety of places because they can be formed on many surfaces and different materials, as it was previously explained. Since these formations can contain pathogenic microorganisms and have a greater resistance to disinfection, the chances of contamination of the product and of provoking food infections are increased, which is why the presence of biofilms in the contact surfaces of the food industry is considered a risk to human health (Gopu, Chandran, and Shetty 2018).

Even though the majority of bacteria species have the capacity to form biofilms, some genera can form them more easily and rapidly than others; that is the case of *Pseudomonas*, *Listeria*, *Enterobacter*, *Flavobacterium*, *Alcaligenes*, *Staphylococcus*, and *Bacillus* (Lee Wong 1998).

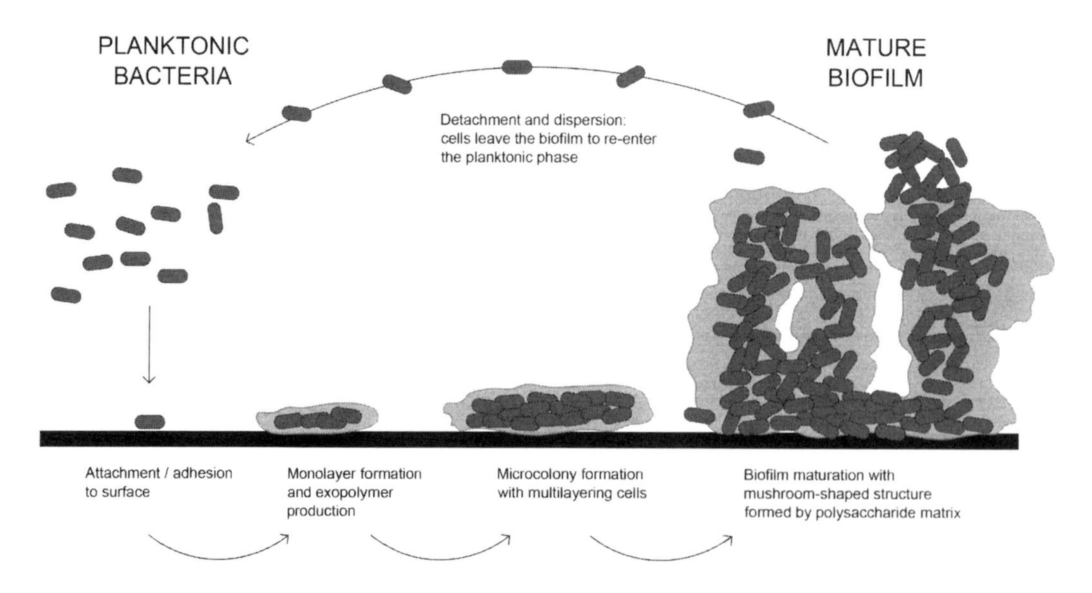

FIGURE 23.2 Biofilm formation process. (Adapted from Vasudevan, R., *J. Microbiol. Exp.*, 1, 1–16, 2014.)

23.1.2 Quorum Sensing and Its Role in Food Spoilage and Food Safety

Antimicrobial agents and disinfection reactants are often misused, with the possible risk of emergence and spread of resistant microorganisms. This phenomenon is a major problem worldwide that generates growing concern in public health. The resistance and potential virulence of the bacteria could be related to their ability to detect and respond to population density. This ability is called "cell-to-cell communication" or "quorum sensing" (QS), which regulates gene expression in response to cell population density. The cells involved produce and excrete substances, called autoinducers (AIs), that serve as a chemical signal to induce the collective genetic expression; i.e. it is a form of cellular communication (Alvarez, Moreira, and Ponce 2012). In general, acyl-homoserine lactones (AHLs) have been identified as inducers in Gram-negative bacteria, and oligopeptides in the case of Gram-positive bacteria.

The different quorum sensing signaling systems are usually classified into four main groups. Two of them are found in Gram-negative bacteria, which use autoinducer type 1 and type 3 (AI-1 and AI-3, respectively). The third type is found in Gram-positive bacteria, which uses an autoinducing polypeptide system (AIP), normally involved in intraspecies communication (Figure 23.3). Finally, a fourth system is used for interspecies communication, found in both Gram-positive and Gram-negative bacteria, which uses autoinducer type 2 (AI-2). All of these systems begin with bacteria production and autoinducer release into the environment. The detection of these chemically distinct autoinducers, and the resulting alteration in gene expression, is specific to each system (Aswathanarayan and Vittal 2011).

QS mechanisms allow bacteria to regulate a series of physiological activities. These processes include symbiosis, virulence, competition, conjugation, antibiotic production, motility, sporulation, and film or biofilm formation (Nazzaro, Fratianni, and Coppola 2013). In recent years, QS inhibition has been a research area of growing interest due to its possible applications in medicine, industry, and biotechnology. QS mechanism has been widely studied to explain bacterial pathogenicity, and it has also been discovered that food spoilage by microbial activity is influenced by phenotypes regulated by this same system. For example, some reports have associated milk spoilage to QS systems in psychrotrophic bacteria *Pseudomonas* spp., *Serratia* spp., *Enterobacter* spp., and *Hafnia alvei* (Wang et al. 2019). These spoilage bacteria produce enzymes, such as proteases and lipases, that can cause severe problems in the dairy industry such as milk protein hydrolysis, development of off-flavors, shelf-life reduction, decrease of yield during cheese production, milk heat-stability loss, and gelation of UHT milk (Tapia-Rodriguez et al. 2017). The production of these enzymes could be regulated by QS. Other QS-regulated enzyme activities have been reported associated with seafood products. AHL-modulated exoenzyme activities have been described in *Serratia proteamaculans* B5a isolated from cold-smoked salmon (Christensen et al. 2003); also, C4-HSL increased exoenzyme production in *Pseudomonas psychrophila* PSPF19 and promoted the spoilage of refrigerated freshwater fish (Aswathanarayan and Vittal 2014).

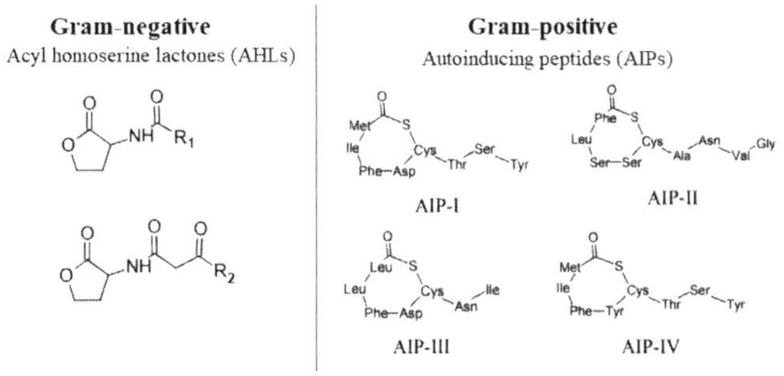

The discovery of quorum sensing signals in damaged food products has brought a new perspective to study the process of food spoilage. In this sense, inhibition of cellular communication or QS could be a good strategy to prevent or reduce both bacterial pathogenicity and food spoilage.

Furthermore, biofilm formation has also been associated with quorum sensing. There has been much experimental evidence suggesting that QS regulates the biofilm formation process (Aswathanarayan and Vittal 2011). QS systems seem to be involved in every phase of the process. They regulate the population density and the metabolic activity within the mature biofilm to adapt the nutritional demands to the available resources. The release of bacteria from within the extracellular matrix of the biofilm also appears to be induced by QS mechanisms.

The adhesion of different pathogenic bacteria including *Bacillus*, *Salmonella*, *Listeria*, *Staphylococcus*, and *Escherichia* with subsequent biofilm formation in food processing surfaces can cause food spoilage and/or foodborne diseases (Amrutha, Sundar, and Shetty 2017). The outbreaks associated with food-related bacteria show that there has not been a proper control by the use of chemicals, such as chlorine and hydrogen peroxide, normally used in food industries on a regular basis or that more resistant strains to these chemicals have aroused.

With the increasing occurrence of foodborne outbreaks and the increasing amount of food waste caused by spoilage microorganisms, there is a clear necessity for the research of novel strategies to prevent or reduce microbial contamination in food industries, especially those that aim at inhibiting the quorum sensing system. Hence, the aim of this chapter is to summarize the latest findings in quorum sensing systems on food-related bacteria and its inhibition.

FIGURE 23.3 Natural autoinducers in Gram-positive and Gram-negative bacteria.

23.2 Quorum Quenching (QQ) on Food-Related Bacteria

Currently, interference with the QS systems is primarily achieved by inhibiting the synthesis of autoinducers, degrading autoinducers, interfering with autoinducer receptors, or inhibiting the autoinducer/receptor complex formation (Zhang and Li 2016). The molecules responsible for inhibition of autoinducer-induced QS systems or the autoinducer-regulated phenotype are called quorum sensing inhibitors (QSIs). An interesting feature of QSIs is that they operate at lower than minimum inhibitory concentration (MIC). Thus, bacteria do not perceive threats to their survival and continue to grow without causing disease (Kalia 2013;

Kalia and Purohit 2011). The inactivation or degradation of QS signal molecules is known as QS inhibition or quorum quenching (QQ). QSIs or QQ compounds can be used as novel biopreservatives which can inhibit food spoilage or pathogenic bacteria virulence, maintaining the organoleptic, physical, and nutritional quality of fresh, processed, packaged, and ready-to-eat food products for safer consumer health (Nychas, Marshall, and Sofos 2007).

There are different types of QSIs, ranging from synthetic to natural ones of different origin (i.e. microbial, fungal, plant, or animal). QSIs include a wide variety of chemicals, such as furanones and their related structural analogs (Zang et al. 2009; Liu et al. 2011; Steenackers et al. 2010), bismuth porphyrin complexes (Galkin et al. 2015), glycosylation reagents of glycosylated flavonoids (Brango-Vanegas et al. 2014), glycomonoterpenols (Mukherji and Prabhune 2015), heavy metals (Vega et al. 2014), and nanomaterials (Wagh Nee Jagtap et al. 2013; Miller et al. 2015; Singh et al. 2015). Other chemically synthesized compounds, antibodies, and natural bioactive molecules have also been reported to act as QSIs (Certner and Vollmer 2018; Defoirdt et al. 2012; Delago, Mandabi, and Meijler 2016; Kalia 2013; Kalia and Purohit 2011; Ding et al. 2017; Ding, Li, and Li 2018; Joshi et al. 2016; Reuter, Steinbach, and Helms 2016).

The ideal QS inhibitors have been defined as chemically stable and highly effective low molecular-mass molecules, with a high degree of specificity for the QS regulator without toxic side effects on the bacteria or an eventual eukaryotic host (Asfour 2018). There is an increasing demand for safer food with acceptable shelf-life, and a growing tendency of rejection toward the use of chemical additives. Therefore, biocontrol approaches are gaining interest, where the use of extracts or pure compounds obtained from natural sources are perceived as "greener" and offer some specific beneficial features compared to more conventional treatments. For example, plant food extracts and phytochemicals have been proven to be effective against food-related bacteria with the additional advantage of antioxidant and other beneficial properties for human health (Tomadoni, Moreira, and Ponce 2016).

Plants produce many and varied compounds, such as simple phenolics, flavonoids, alkaloids, and terpenoids (Asfour 2018), and there is a great interest in the biological and therapeutic activities of these natural products in inhibiting QS on foodborne spoilage and pathogenic bacteria. Hence, there is a clear demand for anti-QS agents to overcome bacterial resistance to antibiotics and cleaning agents, and there is a necessity to examine and identify alternative greener and safer approaches for controlling pathogens. The study of plants, plant products, and their purified components could open up the possibility of using these compounds as new anti-QS agents.

Enzymes have also been tested as QSIs, usually called quorum quenching or anti-QS enzymes (Meireles et al. 2016). N-acyl homoserine lactonases and acylases are a few examples of such enzymes. Lactonases can hydrolyze the bond in the homoserine ring, avoiding the binding of AHLs to transcriptional regulators, hence, disrupting the QS system (Thallinger et al. 2013). Kiran et al. (2011) studied the anti-QS activity of a lactonase against *Pseudomonas aeruginosa* and achieved 69%–77% of biofilm reduction, as well as a decrease in the production of virulence factors. In a recent study, using the latter, Singh et al. (2015) demonstrated that *Aspergillus clavatus* produced a mixture of amylases, proteases, and pectinases that reduced biofilms of *Pseudomonas aeruginosa* and *Bacillus subtilis* by 82% and 75%, respectively. An engineered T7 bacteriophage expressing a lactonase with a wide range of anti-QS activity was developed by Pei and Lamas-Samanamud (2014). This engineered phage resulted in inhibition of biofilm formation in polyvinyl chloride microtiterplates when applied to biofilms containing both species of *P. aeruginosa* and *Escherichia coli*.

The most important food spoilage and pathogenic bacteria QS systems and the latest findings on their inhibition will be thoroughly discussed. Table 23.1 gives an overview on the QS signals and QS systems reported in food-related bacteria, while Table 23.2 shows a summary of the latest studies on QSIs.

QSIs are emerging as a new form of antimicrobial agents with potential application in many different fields, including medicine (both human and veterinary), food science and agriculture, and aquaculture, among others. A good number of biotechnology firms have recently emerged aiming specifically at the development of formulations for disrupting or inhibiting QS systems (e.g. QSI Pharma A/S, Denmark; Quorex Pharmaceuticals Inc., Carlsbad, USA; 4SC AG, Germany), which evidence that

TABLE 23.1

QS Systems and QS Signals on Food-Related Bacteria

Food-Related Bacteria	QS System	QS Signals	References
Aeromonas hydrophila	lux-like	C4-HSL	Garde et al. (2010)
Aeromonas veronii	lux-like	C4-HSL, C6-HSL, C7-HSL, C8-HSL	Zhao et al. (2018)
Bacillus cereus	PlcRa	Heptapeptide PapRa7	Huillet et al. (2012)
Bacillus subtilis	Rap-Phr	Pentapeptide or hexapeptide signals	Omer Bendori et al. (2015)
Clostridium difficile	lux-like	AI-2	Carter et al. (2005)
Clostridium perfringens	lux-like agr-like	Furanosyl borate diester AI-2 AIP	Yu et al. (2017)
Escherichia coli O157:H7	luxS	AI-2	Anand and Griffiths (2003)
Pseudomonas aeruginosa	las system rhl system	*N*-(3-oxododecanoyl) homoserine lactone (OdDHL, 3-oxo-C12-HSL), *N*-butyryl-L-homoserine lactone (C4-HSL)	Curutiu et al. (2018)
Salmonella enterica	luxS	AI-2	Choi, Shin, and Ryu (2007)
Salmonella typhimurium	luxS	AI-2	Choi et al. (2012)
Staphylococcus aureus	agr-like	AIP	Xu et al. (2017)

TABLE 23.2

Quorum Sensing Inhibitors (QSIs) on Food-Related Bacteria

Food-Related Bacteria	QSI	References
Bacillus subtilis, *Bacillus cereus*, *Listeria monocytogenes*, *Staphylococcus aureus*, and methicillin-resistant *Staphylococcus aureus* (MRSA)	Grape, apple, and pitahaya extracts and individual phenolic compounds	Zambrano et al. (2019)
Chromobacterium violaceum and *Pseudomonas aeruginosa*	Onion extract	Al-Yousef et al. (2017)
Clostridium difficile	*Leptospermum scoparium* honey	Hammond, Donkor, and Brown (2014)
Erwinia carotovora and *Yersinia enterocolitica*	Cinnamaldehyde, ellagic acid, pomegranate extract, resveratrol, and rutin	Truchado et al. (2012)
Escherichia coli reporter strain AI1-QQ	Extract of *Albizia schimperiana* (root) and extract of *Justicia schimperiana* (seed)	Bacha et al. (2016)
Listeria spp.	Antifungal metabolite produced from *Pestalotiopsis* spp. and *Monochaetia* spp.	Nakayama et al. (2009)
MRSA	Metal oxide nanoparticles	Agarwala, Choudhury, and Yadav (2014)
Pseudomonas aeruginosa	Carvacrol	Tapia-Rodriguez et al. (2017)
Pseudomonas aeruginosa	Mandarin essential oil	Luciardi et al. (2016)
Pseudomonas aeruginosa	Sodium ascorbate	El-Mowafy, Shaaban, and Abd El Galil (2014)
Pseudomonas fluorescens	Thymus vulgare essential oil	Myszka et al. (2016)
P. aeruginosa PA01 and *Salmonella typhimurium*	Berberine	Aswathanarayan and Vittal (2018)
Salmonella typhimurium	Extracts of Amomum tsaoko	Rahman et al. (2017)
Staphylococcus strains	FS8 and tigecycline	Simonetti et al. (2013)
S. aureus	Azithromycin	Gui et al. (2014)
S. aureus	*Leptospermum scoparium* honey	Lu et al. (2014)
S. aureus	Coumarin	Gutierrez-Barranquero et al. (2015)

commercial interests associated with these fields are massive. As a result, many patents regarding quorum sensing inhibitors are being released. Table 23.3 shows the latest patents regarding QSIs for food-related bacteria.

23.2.1 QS and QQ in Gram-Negative Food-Related Bacteria

As previously stated, in Gram-negative bacteria AHLs are the most common class of autoinducers. They have a core *N*-acylated homoserine-lactone ring and a 4–18 carbon acyl chain that can contain modifications. Hundreds of bacterial species contain LuxI-type synthases that produce these AHLs. The length of the acyl chain can affect stability, which may have consequences on signaling dynamics (Papenfort and Bassler 2016).

Figure 23.4 shows a schematic representation of typical quorum sensing mechanism in Gram-negative bacteria. AHLs are produced in the cell and freely diffuse across the inner and outer membranes. When the concentration of AHLs is sufficiently high, which occurs at high cell density, they bind to cytoplasmic receptors. The AI-bound receptors regulate expression of the genes in the QS regulon (Czajkowski and Jafra 2009) (Figure 23.4). In some cases of Gram-negative bacterial QS, AIs are detected by two-component histidine kinase receptors that function analogously to those soon described for Gram-positive QS bacteria.

In addition to AHLs, many more signal molecules with different chemical structures have been discovered in Gram-negative pathogens and spoilage bacteria, including autoinducer-2 (AI-2), quinolones, indole, pyrones, and dialkylresorcinols. Furthermore, it is becoming evident that bacteria usually do not rely on only one signal molecule, and different quorum sensing system architectures have been described, including both hierarchical and parallel configurations (Defoirdt 2018).

Gram-negative foodborne pathogens, such as *Escherichia coli* O157:H7, *Pseudomonas*, *Salmonella*, *Yersinia enterocolitica*, and *Campylobacter jejuni*, form biofilms on food surfaces and equipment used in the food industry, leading to serious health problems, and QS systems appear to be crucial in every step of biofilm formation (Williams et al. 2007). Milk and dairy products are easily susceptible to spoilage by psychrotrophic bacteria, such as *Serratia* spp. and *Pseudomonas* spp. These bacteria are also responsible for the slime formed on meat surfaces and spoilage of meat and meat products (Dunstall et al. 2005; Yang et al. 2009) stored under aerobic chill conditions (3°C–8°C) where QS through AHL signals, such as C4-HSL, 3-oxo-C6-HSL, C6-HSL, C8-HSL, and C12-HSL, have been detected (Steindler and Venturi 2007).

Understanding the mechanism of QS systems in food environments can help in finding a solution to the problem of food spoilage and foodborne diseases. QSIs can be evaluated and promoted as food preservatives to enhance food safety and therefore, to improve consumers' health. Different QSIs have been studied for inhibition of QS systems in Gram-negative bacteria. For example, enzymes produced by a wide range of prokaryotes with an ability to degrade QS signal molecules (particularly, AHLs) include lactonases, oxidoreductases, acylases, and phosphotriesterase-like lactonases (Fetzner 2015), whereas oxidoreductases target AI-2 (Weiland-Bräuer et al. 2016). A variety of small molecules including certain intermediates of the AHL biosynthesis route, non-cognate AHLs, and dicyclic peptides have also been reported to act as QSIs (Bauer and Robinson 2002). However, for

TABLE 23.3

Patents on Quorum Sensing Inhibitors

Patent Title	Inventor(s)	Reference Number and Date of Publication
Compounds and methods for modulating communication and virulence in quorum sensing bacteria	Helen E. Blackwell Grant D. Geske Rachel W. Wezeman	US7642285B2 2010-01-05
Ellagitannins as inhibitors of bacterial quorum sensing	Kalai Mathee Allison L. Adonizio Frederick Ausubel Jon Clardy Bradley Bennett Kelsey Downum	US20110105421A1 2011-05-05
Modulation of bacterial quorum sensing with synthetic ligands	Helen E. Blackwell Grant D. Geske Jennifer C. O'Neill	WO2008116029A1 2008-09-25
Non-lactone carbocyclic and heterocyclic antagonists and agonists of bacterial quorum sensing	Helen E. Blackwell Christine E. McInnis	US20140142156A1 2014-05-22
Compositions for regulating or modulating quorum sensing in bacteria, methods of using the compounds, and methods of regulating or modulating quorum sensing in bacteria	Binghe Wang Nanting Ni Junfeng Wang Chung-Dar Lu Han-Ting Chou Minyong Li Shilong Zheng Yunfeng Cheng Hanjing Peng	US8653258B2 2014-02-18
Antibody-mediated disruption of quorum sensing in bacteria	Kim D. Janda Gunnar F. Kaufmann Junguk Park	EP2211889B1 2014-08-20
Blockers of the quorum sensing system of Gram-negative bacteria	Dr. Aldo Ammendola Dr. Katharina Aulinger-Fuchs Dr. Astrid Gotschlich Dr. Martin Lang Dr. Wael Saeb Dr. Udo Sinks Dr. Andreas Wuzik	EP1475092A1 2004-11-10
Quorum sensing inhibitors	Michael Givskov Liang Yang Yang Yi Sean Tan	US9988380B2 2018-06-05
Peptide with quorum-sensing inhibitory activity, polynucleotide that encodes said peptide, and the uses thereof	Ana Maria Otero Casal Manuel Romero Bernárdez Celia Mayer Mayer	US20160244735A1 2016-08-25

Note: This table was generated by performing a search using the keywords "quorum sensing inhibitors" + "food" on the website of Google Patents: https://patents.google.com. This search yielded >350 results, of which the most relevant examples are here listed.

food-related bacteria the latest trend has aimed at more natural QSI, obtained mainly from essential oils, and fruit and vegetable extracts, among many others.

Extracts of various plant parts have been found to inhibit QS in Gram-negative bacteria because of the similarity in their chemical structure to those of AHLs, and also because of their ability to degrade signal receptors LuxR/LasR. Extracts of *Phyllanthus emblica*, *Terminalia bellirica*, *T. chebula*, *P. granatum*, *Syzygium cumini*, and *M. indica* (flower) were found to effectively interfere with the QS system of Gram-negative bacteria (*Chromobacterium violaceum*) over a wide range of subinhibitory concentrations (Shukla and Bhathena 2016).

Ethyl acetate fraction of onion peel extract (ONPE) inhibited the production of QS-mediated virulence factors, such as violacein in *C. violaceum* and pyocyanin in *P. aeruginosa*. ONPE also reduced QS-controlled biofilm formation, exopolysaccharide production, and swarming motility in *P. aeruginosa* (Al-Yousef et al. 2017). A recent study reports that *Amomum tsaoko* (Black Cardamom) has QSI properties. At a concentration of 4 mg/mL, extracts from *Amomum tsaoko* showed the maximum QSI property against *Salmonella typhimurium* (Rahman et al. 2017). Bacha et al. (2016) demonstrated that methanol extract of *Albizia schimperiana* (root) and petroleum ether extract of *Justicia schimperiana* (seed) inhibited QS in *E. coli* reporter strain AI1-QQ.

Some studies evaluated individual compounds as QSIs. For example, carvacrol or cymophenol ($C_6H_3(CH_3)(OH)C_3H_7$, a monoterpenoid phenol) inhibited *Pseudomonas aeruginosa* biofilms at 0.9–7.9 mM, compared to nontreated bacteria on a stainless steel surface. Pyocyanin production by *P. aeruginosa* was reduced up to 60% at 3.9 mM of carvacrol. Higher doses of carvacrol affected *P. aeruginosa* viability (Tapia-Rodriguez et al. 2017). Other studies demonstrated that phytochemical compounds such as cinnamaldehyde, ellagic acid, pomegranate extract, resveratrol, and rutin interfere with the QS system of *Yersinia enterocolitica* and *Erwinia carotovora* (Truchado et al. 2012).

Sodium ascorbate is another chemical molecule which has been studied against *Pseudomonas aeruginosa* with anti-QS activity. Sodium ascorbate concentrations lower than MIC resulted in significant reduction of the signaling molecules C4-HSL and 3-oxo-C12-HSL (El-Mowafy, Shaaban, and Abd El Galil 2014). In addition, inhibition/attenuation of pyocyanin production, biofilm formation, and motility of *Pseudomonas* was observed. The effect of ascorbate on the expression of QS regulatory genes was also analyzed. QS regulatory genes expression (LasI, LasR, RhlI, RhlR, PqsR, and PqsA) was repressed compared to untreated *Pseudomonas aeruginosa* PAO1 (analyzed by RT-PCR), confirming that sodium ascorbate QS inhibition was achieved on gene expression at the molecular level (El-Mowafy, Shaaban, and Abd El Galil 2014).

Berberine, a quaternary ammonium salt from the protoberberine group of benzylisoquinoline alkaloids found in such plants as *Berberis*, showed significant antibiofilm activity against *P. aeruginosa* PA01 at sub-MIC concentrations. It prevented biofilm formation and inactivated biofilm maturation in pathogenic bacteria at concentrations ranging from 0.019 to 1.25 mg/mL. Berberine showed an interaction with the quorum sensing signal receptors LasR and RhlR. Furthermore, its anti-infective properties in *Salmonella typhimurium* were studied. At sub-MIC concentrations of 0.019 mg/mL, it reduced biofilm formation in *S. typhimurium* by 31.20%. It significantly prevented invasion and adhesion of *Salmonella* in a human colonic cell (HT-29) by 55.37% and 54.68%, respectively (Aswathanarayan and Vittal 2018). These results suggest that berberine, previously recognized for its antimicrobial activity, could find potential application as an anti-biofilm and anti-infective agent based on its quorum sensing inhibitory activity.

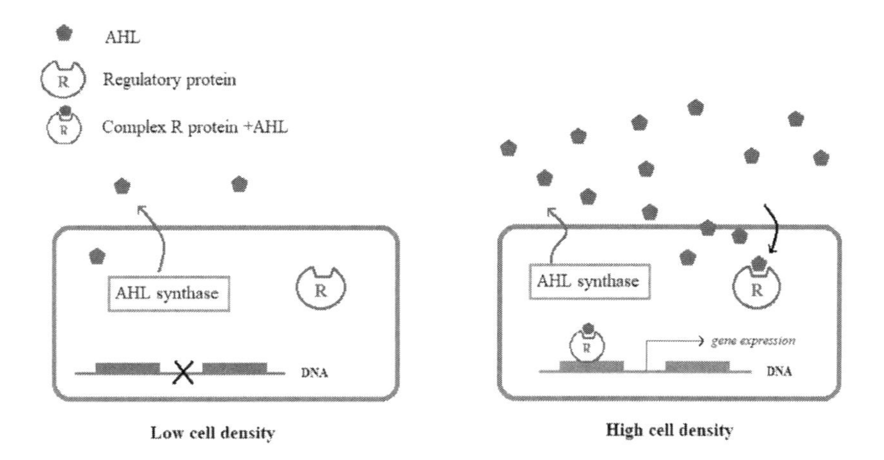

FIGURE 23.4 Gram-negative bacteria quorum sensing mechanism. Bacterial cells produce signal molecules *N*-acyl-homoserine lactones (AHLs) using a specific enzyme (AHL synthase). AHLs are transferred into the environment and back to the cells by either passive diffusion or active ATP-dependent transport systems (low cell density). With an increasing bacterial population density, the AHLs accumulate in the environment and in the bacterial cells. When the concentration of AHLs reaches a threshold level (quorum state), the signal molecules interact with the regulatory protein (R). The regulatory protein (R), in most cases, acts as a positive transcription regulator. The complex of R protein and AHL binds to promoters of target genes and initiates their expressions in a cell density-dependent manner (high cell density). (Adapted from Czajkowski, R., and Jafra, S., *Acta Biochim. Polon.*, 56, 1–16, 2009.)

23.2.2 QS and QQ in Gram-Positive Food-Related Bacteria

In the case of Gram-positive bacteria, autoinducers are usually oligopeptides, also called autoinducing peptides (AIPs). These molecules, unlike AHLs, are very specific, and they provide the strain the capacity to communicate in an intraspecific manner. Oligopeptide signals cannot diffuse freely through the cell membrane like AHLs do; instead they need to be transported through ATP-binding protein exporters. Furthermore, AIPs do not directly bind to a transcriptional activator. When the AIP extracellular concentration reaches a certain threshold value, it stimulates the autophosphorylation of the transmembrane protein, which then phosphorylates the response regulator (Sturme et al. 2002; Ng and Bassler 2009; Ivanova, Fernandes, and Tzanov 2013). Once the response regulator is modified, it can eventually regulate gene expression. Hence, the quorum sensing system of Gram-positive bacteria is much more complex than the Gram-negative one. Figure 23.5 shows a schematic representation of a Gram-positive QS mechanism.

A quorum sensing peptide system is used to regulate the development of bacterial competence in *Bacillus subtilis* and *Streptococcus pneumoniae*, conjugation in *Enterococcus faecalis*, and virulence and biofilm formation in *Staphylococcus aureus* (Kievit and Iglewski 2000). Autoinducing peptides (AIPs) have also been shown to regulate QS in other food-related Gram-positive

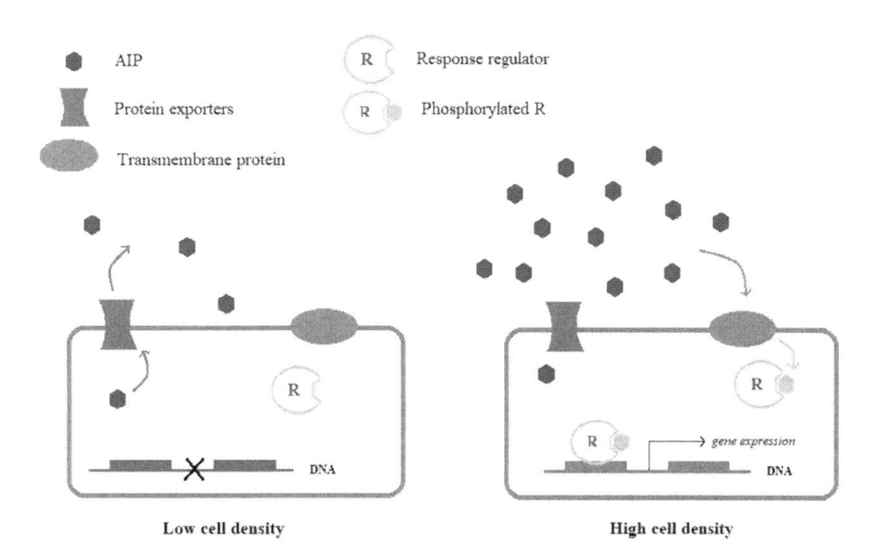

FIGURE 23.5 Gram-positive bacteria quorum sensing mechanism. Oligopeptide signals are transferred into the environment through ATP-binding protein exporters (low cell density). With an increasing bacterial population density, the AIPs accumulate in the environment. Once the extracellular concentration of AIPs reaches a threshold level, it stimulates autophosphorylation of the transmembrane protein, which in turn phosphorylates the response regulator (R). In its phosphorylated state, R activates targeted genes and initiates their expressions in a cell density-dependent manner (high cell density). (Adapted from Vasudevan, R., *J. Microbiol. Exp.*, 1, 1–16, 2014.)

bacteria, such as various species of *Clostridium* and *Enterococcus* (Carter et al. 2005; Yu et al. 2017; Ali et al. 2017).

S. aureus is one of the most studied Gram-positive bacteria because it causes a wide range of disease states that range from mild to life-threatening, and several outbreaks have been associated with contaminated food products (Gallina et al. 2013; Fetsch and Johler 2018). It is a human opportunistic pathogen, mainly because of its capacity to produce enterotoxins at temperatures between 10°C and 46°C. This species is able to grow and multiply on the mucous membranes and skin of food handlers, which makes it a major concern for food industries, because staphylococcal enterotoxins are heat-stable and are secreted during its growth in the food matrix, eventually contaminated by either animals or food handlers. Moreover, the emergence of methicillin-resistant *S. aureus* (MRSA) in farm animals has become a major issue of concern because animal-derived foods are a primary contamination source for this resistant pathogen which has the ability to form biofilms on many different types of animal surfaces (Galié et al. 2018).

The virulence of this *S. aureus* depends on the temporal expression of a varied array of virulence factors, and a good strategy to control it would be to inhibit its quorum sensing system. In this foodborne pathogen, the *agr* quorum sensing system controls expression of many virulence factors. This system is comprised of four genes: *agrB*, *agrD*, *agrC*, and *agrA*. The autoinducer propeptide, encoded by *agrD*, is first processed to the active AIP by the *agrB* transporter and then released into the extracellular environment. Once it has reached a sufficient concentration, it activates the two-component system (*agrC–agrA*) to subsequently modify gene expression (Yu et al. 2017).

Listeria monocytogenes is also an important Gram-positive food-borne pathogen, mainly because it has the ability to survive in a varied range of environmental conditions. Its accessory gene regulator *agr* system was shown to impact on biofilm formation and virulence and has been proposed as one of the regulatory mechanisms involved in adaptation to these changing environments (Zetzmann et al. 2016).

Among the *Clostridium* genera, *C. perfringens* is a spore-forming anaerobic bacterium that is widely distributed in soil, feces, and foods, and it is also part of the normal intestinal microbiota of both animals and humans. *C. perfringens* causes a number of human and animal diseases, because it produces at least 16 different extracellular toxins, including food poisoning, gas gangrene in humans, enterotoxemia in sheeps and goats, lamb dysentery, and necrotic enteritis in poultry (Yu et al. 2017). *C. perfringens* virulence is explained by at least two different QS systems: the LuxS and the *agr*-like systems. The LuxS QS system, common to both Gram-negative and Gram-positive species, uses a signaling molecule called furanosyl borate diester (AI-2), which is synthesized from S-adenosylmethionine by the LuxS enzyme (Yu et al. 2017). On the other hand, the *agr*-like system, is found only in Gram-positive species, as previously explained for *S. aureus*. *AgrB* and *agrD* orthologues were recently identified in the *C. perfringens* genome (Vidal et al. 2009; Ohtani et al. 2009) and found to be positively involved in the regulation of several toxins production.

Bacillus species also present a major concern in the food industry, particularly in dairy products, as they can form biofilms in pipelines and on equipment surfaces used in the entire line of production. These biofilms represent a continuous hygienic problem and can lead to serious economic losses due to food spoilage and equipment impairment. Biofilm formation by *Bacillus subtilis* is apparently dependent on LuxS quorum sensing (QS) by Autoinducer-2 (AI-2) (Duanis-Assaf et al. 2016).

Many synthetic compounds were found to be important in the inhibition of pathogenesis and/or biofilm formation of Gram-positive bacteria through QQ. For example, metal oxide nanoparticles were proved to be active against methicillin resistant *S. aureus* (MRSA) (Agarwala, Choudhury, and Yadav 2014); also, azithromycin showed activity against α-hemolysin and biofilm formation by *S. aureus* (Gui et al. 2014), and FS8 and tigecycline, which prevented biofilm formation by *Staphylococcus* strains (Simonetti et al. 2013).

However, the new trend in QSIs aims at more natural compounds from varied origins. For example, ambuic acid (an antifungal metabolite produced from *Pestalotiopsis* spp. and *Monochaetia* spp.) has shown quorum sensing inhibition in *Listeria* spp. through repression of peptide biosynthesis (Nakayama et al. 2009). Some studies carried out on New Zealand manuka (*Leptospermum scoparium*) honey revealed that this honey can inhibit biofilm formation of important food-related pathogenic bacteria such as *S. aureus* (Lu et al. 2014) and *Clostridium difficile* (Hammond, Donkor, and Brown 2014). Coumarin (an organic chemical compound found in many plants) was also studied and proved to inhibit the QS-regulated biofilm formation in the Gram-positive pathogen *S. aureus* (Gutierrez-Barranquero et al. 2015).

In a recent study, Zambrano et al. (2019) studied the effects of fruit extracts against different food pathogenic and spoilage bacteria. Particularly they evaluated grape, apple, and pitahaya (dragon fruit) extracts against five Gram-positive bacteria of importance in the food industry: *Bacillus subtilis*, *Bacillus cereus*, *Listeria monocytogenes*, *Staphylococcus aureus*, and methicillin-resistant *Staphylococcus aureus* (MRSA). The antimicrobial and anti-biofilm forming activities against the food-borne pathogen and spoilage bacteria, and the anti-quorum sensing capacity of the extracts were investigated with encouraging results. Furthermore, these authors evaluated individual phenolic compounds (i.e. gallic acid, vanillic acid, syringic acid, p-coumaric acid, 4-hydroxybenzoic acid, cinnamic acid, (+)-catechin, (–)-epicatechin, polydatin, quercetin, and resveratrol). MICs found for every phenolic compound against *L. monocytogenes*, *S. aureus*, and MRSA were >500 μg/mL. However, with sublethal doses of 100 μg/mL, the individual phenolic compounds showed significant anti-biofilm forming capacities in the Gram-positive strains. Among the tested bacteria, the most sensitive was *S. aureus*. In this context, earlier studies had verified that phenolic extracts and certain phenolic compounds, especially flavonoids, can efficiently prevent the biofilm formation in *S. aureus* (Pratiwi et al. 2015; Lopes et al. 2017). These extracts and individual phenolic compounds that proved to be effective can be used as a source of antimicrobials, natural food preservatives, and/or sanitizers to control food-related bacteria.

23.3 Future Trends and Challenges

The various processes modulated by quorum sensing are primarily concerned with the regulation of virulence, determination of bacterial pathogenicity, development of genetic competence, transfer of conjugative plasmids, sporulation, biofilm formation, antimicrobial peptide synthesis, and symbiosis for their survival under adverse environmental conditions. The enormous amount of food waste by microbial spoilage added to the outbreaks caused by foodborne pathogens and draws attention to the fact that food-related bacteria are becoming more and more resistant to today's preservation measurements; the necessity to develop new strategies for their control is clear.

The identification and characterization of anti-QS compounds as new antipathogenic and antibiofilm forming candidates might play an important role in reducing the virulence and pathogenicity of bacteria, including drug-resistant bacteria, in reducing the biofilm forming hazard for safe infection-free food products, and it will contribute to the silence of food-borne pathogens for safer food production. The consumer trends also aim at greener preservation methods and greener products; hence, quorum quenching compounds from natural origin may be a feasible alternative.

This chapter will help in understanding the concept of quorum sensing, in promoting its application, and in exploring the role of signaling molecules and potential of QQ compounds, to use them as a biocontrol tool to avoid food spoilage and foodborne outbreaks, and to reduce economic loss in the food industry.

ACKNOWLEDGMENTS

This work was supported by Consejo Nacional de Investigaciones Científicas y Técnicas (CONICET), Agencia Nacional de Promoción Científica y Tecnológica (ANCyT), and Universidad Nacional de Mar del Plata (UNMDP).

REFERENCES

Agarwala, Munin, Bula Choudhury, and R. N. S. Yadav. 2014. "Comparative Study of Antibiofilm Activity of Copper Oxide and Iron Oxide Nanoparticles against Multidrug Resistant Biofilm Forming Uropathogens." *Indian Journal of Microbiology* 54 (3): 365–68. doi:10.1007/s12088-014-0462-z.

Al-Yousef, Hanan M., Atallah F. Ahmed, Nasser A. Al-Shabib, Sameen Laeeq, Rais A. Khan, Md T. Rehman, Ali Alsalme, Mohamed F. Al-Ajmi, Mohammad S. Khan, and Fohad M. Husain. 2017. "Onion Peel Ethylacetate Fraction and Its Derived Constituent Quercetin 4'-O-Beta-D Glucopyranoside Attenuates Quorum Sensing Regulated Virulence and Biofilm Formation." *Frontiers in Microbiology* 8: 1675. doi:10.3389/fmicb.2017.01675.

Ali, Liaqat, Mohsan Ullah Goraya, Yasir Arafat, Muhammad Ajmal, Ji-Long Chen, and Daojin Yu. 2017. "Molecular Mechanism of Quorum-Sensing in *Enterococcus faecalis:* Its Role in Virulence and Therapeutic Approaches." *International Journal of Molecular Sciences* 18 (5): 960. doi:10.3390/ijms18050960.

Alvarez, M. V., María R. Moreira, and Alejandra Ponce. 2012. "Antiquorum Sensing and Antimicrobial Activity of Natural Agents with Potential Use in Food." *Journal of Food Safety* 32 (3): 379–87. doi:10.1111/j.1745-4565.2012.00390.x.

Ammor, M. S., C. Michaelidis, and G.-J. E. Nychas. 2008. "Insights into the Role of Quorum Sensing in Food Spoilage." *Journal of Food Protection* 71 (7): 1510–25. doi:10.4315/0362-028X-71.7.1510.

Amrutha, Balagopal, Kothandapani Sundar, and Prathapkumar Halady Shetty. 2017. "Effect of Organic Acids on Biofilm Formation and Quorum Signaling of Pathogens from Fresh Fruits and Vegetables." *Microbial Pathogenesis* 111: 156–62. doi:10.1016/j.micpath.2017.08.042.

Anand, S. K., and Mansel W. Griffiths. 2003. "Quorum Sensing and Expression of Virulence in *Escherichia coli* O157:H7." *International Journal of Food Microbiology* 85 (1–2): 1–9.

Asfour, Hani Z. 2018. "Anti-Quorum Sensing Natural Compounds." *Journal of Microscopy and Ultrastructure* 6 (1): 1–10. doi:10.4103/JMAU.JMAU_10_18.

Bacha, Ketema, Yinebeb Tariku, Fisseha Gebreyesus, Shibru Zerihun, Ali Mohammed, Nancy Weiland-Brauer, Ruth A. Schmitz, and Mulugeta Mulat. 2016. "Antimicrobial and Anti-Quorum Sensing Activities of Selected Medicinal Plants of Ethiopia: Implication for Development of Potent Antimicrobial Agents." *BMC Microbiology* 16 (1): 139. doi:10.1186/s12866-016-0765-9.

Aswathanarayan, Jamuna Bai, and Ravishankar Rai Vittal. 2018. "Inhibition of Biofilm Formation and Quorum Sensing Mediated Phenotypes by Berberine in *Pseudomonas aeruginosa* and *Salmonella typhimurium*." *RSC Advances* 8 (63): 36133–41. doi:10.1039/C8RA06413J.

Aswathanarayan, Jamuna Bai, and Ravishankar Rai Vittal. 2011. "Bacterial Quorum Sensing and Food Industry." *Comprehensive Reviews in Food Science and Food Safety* 10 (3): 183–93. doi:10.1111/j.1541-4337.2011.00150.x.

Aswathanarayan, Jamuna Bai, and Ravishankar Rai Vittal. 2014. "Quorum Sensing Regulation and Inhibition of Exoenzyme Production and Biofilm Formation in the Food Spoilage Bacteria *Pseudomonas psychrophila* PSPF19." *Food Biotechnology* 28 (4): 293–308. doi:10.1080/08905436.2014.963601.

Bauer, Wolfgang D., and Jayne B. Robinson. 2002. "Disruption of Bacterial Quorum Sensing by Other Organisms." *Current Opinion in Biotechnology* 13 (3): 234–37.

Brango-Vanegas, J., G. M. Costa, C. F. Ortmann, E. P. Schenkel, F. H. Reginatto, F. A. Ramos, C. Arevalo-Ferro, and L. Castellanos. 2014. "Glycosylflavonoids from Cecropia Pachystachya Trecul Are Quorum Sensing Inhibitors." *Phytomedicine: International Journal of Phytotherapy and Phytopharmacology* 21 (5): 670–75. doi:10.1016/j.phymed.2014.01.001.

Carter, Glen P., Des Purdy, Paul Williams, and Nigel P. Minton. 2005. "Quorum Sensing in *Clostridium difficile:* Analysis of a LuxS-Type Signalling System." *Journal of Medical Microbiology* 54 (Pt 2): 119–27. doi:10.1099/jmm.0.45817-0.

Certner, Rebecca H., and Steven V. Vollmer. 2018. "Inhibiting Bacterial Quorum Sensing Arrests Coral Disease Development and Disease-Associated Microbes." *Environmental Microbiology* 20 (2): 645–57. doi:10.1111/1462-2920.13991.

Choi, Jeongjoon, Dongwoo Shin, Minjeong Kim, Joowon Park, Sangyong Lim, and Sangryeol Ryu. 2012. "LsrR-Mediated Quorum Sensing Controls Invasiveness of *Salmonella typhimurium* by Regulating SPI-1 and Flagella Genes." *PLoS One* 7 (5): e37059. doi:10.1371/journal.pone.0037059.

Choi, Jeongjoon, Dongwoo Shin, and Sangryeol Ryu. 2007. "Implication of Quorum Sensing in *Salmonella enterica* Serovar Typhimurium Virulence: The LuxS Gene Is Necessary for Expression of Genes in Pathogenicity Island 1." *Infection and Immunity* 75 (10): 4885–90. doi:10.1128/IAI.01942-06.

Christensen, Allan B., Kathrin Riedel, Leo Eberl, Lars R. Flodgaard, Soren Molin, Lone Gram, and Michael Givskov. 2003. "Quorum-Sensing-Directed Protein Expression in *Serratia proteamaculans* B5a." *Microbiology (Reading, England)* 149 (Pt 2): 471–83. doi:10.1099/mic.0.25575-0.

Curutiu, Carmen, Florin Iordache, Veronica Lazar, Aurelia Magdalena Pisoschi, Aneta Pop, Mariana Carmen Chifiriuc, and Alina Maria Hoban. 2018. "Impact of *Pseudomonas aeruginosa* Quorum Sensing Signaling Molecules on Adhesion and Inflammatory Markers in Endothelial Cells." *Beilstein Journal of Organic Chemistry* 14: 2580–88.

Czajkowski, Robert, and Sylwia Jafra. 2009. "Quenching of Acyl-Homoserine Lactone-Dependent Quorum Sensing by Enzymatic Disruption of Signal Molecules." *Acta Biochimica Polonica* 56 (1): 1–16.

Defoirdt, Tom. 2018. "Quorum-Sensing Systems as Targets for Antivirulence Therapy." *Trends in Microbiology* 26 (4): 313–28. doi:10.1016/j.tim.2017.10.005.

Defoirdt, Tom, Tore Benneche, Gilles Brackman, Tom Coenye, Patrick Sorgeloos, and Anne Aamdal Scheie. 2012. "A Quorum Sensing-Disrupting Brominated Thiophenone with a Promising Therapeutic Potential to Treat Luminescent Vibriosis." *PloS One* 7 (7): e41788. doi:10.1371/journal.pone.0041788.

Delago, Antonia, Aviad Mandabi, and Michael M. Meijler. 2016. "Natural Quorum Sensing Inhibitors – Small Molecules, Big Messages." *Israel Journal of Chemistry* 56 (5): 310–20. doi:10.1002/ijch.201500052.

Ding, Ting, Tingting Li, and Jianrong Li. 2018. "Identification of Natural Product Compounds as Quorum Sensing Inhibitors in *Pseudomonas fluorescens* P07 through Virtual Screening." *Bioorganic and Medicinal Chemistry* 26 (14): 4088–99. doi:10.1016/j.bmc.2018.06.039.

Ding, Ting, Tingting Li, Zhi Wang, and Jianrong Li. 2017. "Curcumin Liposomes Interfere with Quorum Sensing System of *Aeromonas sobria* and in Silico Analysis." *Scientific Reports* 7 (1): 8612. doi:10.1038/s41598-017-08986-9.

Doyle, Michael P., Larry R. Steenson, and Jianghong Meng. 2013. "8 Bacteria in Food and Beverage Production." doi:10.1007/978-3-642-31331-8.

Duanis-Assaf, Danielle, Doron Steinberg, Yunrong Chai, and Moshe Shemesh. 2016. "The LuxS Based Quorum Sensing Governs Lactose Induced Biofilm Formation by *Bacillus subtilis*." *Frontiers in Microbiology*. https://www.frontiersin.org/article/10.3389/fmicb.2015.01517.

Dunstall, George, Michael T. Rowe, G. Brian Wisdom, and David Kilpatrick. 2005. "Effect of Quorum Sensing Agents on the Growth Kinetics of *Pseudomonas* spp. of Raw Milk Origin." *Journal of Dairy Research* 72 (3): 276–80. doi:10.1017/S0022029905000713.

El-Mowafy, S. A., M. I. Shaaban, and K. H. Abd El Galil. 2014. "Sodium Ascorbate as a Quorum Sensing Inhibitor of *Pseudomonas aeruginosa*." *Journal of Applied Microbiology* 117 (5): 1388–99. doi:10.1111/jam.12631.

Fetsch, Alexandra, and Sophia Johler. 2018. "*Staphylococcus aureus* as a Foodborne Pathogen." *Current Clinical Microbiology Reports* 5 (2): 88–96. doi:10.1007/s40588-018-0094-x.

Fetzner, Susanne. 2015. "Quorum Quenching Enzymes." *Journal of Biotechnology* 201: 2–14. doi:10.1016/j.jbiotec.2014.09.001.

Galié, Serena, Coral García-gutiérrez, Elisa M. Miguélez, Claudio J. Villar, Felipe Lombó, and Giovanni Di Bonaventura. 2018. "Biofilms in the Food Industry: Health Aspects and Control Methods." *Frontiers in Microbiology* 9: 1–18. doi:10.3389/fmicb.2018.00898.

Galkin, Mycola, Volodimir Ivanitsia, Yuriy Ishkov, Boris Galkin, and Tetiana Filipova. 2015. "Characteristics of the *Pseudomonas aeruginosa* PA01 Intercellular Signaling Pathway (Quorum Sensing) Functioning in Presence of Porphyrins Bismuth Complexes." *Polish Journal of Microbiology* 64 (2). Poland: 101–6.

Gallina, S., D. M. Bianchi, A. Bellio, C. Nogarol, G. Macori, T. Zaccaria, F. Biorci, E. Carraro, and L. Decastelli. 2013. "Staphylococcal Poisoning Foodborne Outbreak: Epidemiological Investigation and Strain Genotyping." *Journal of Food Protection* 76 (12): 2093–98. doi:10.4315/0362-028x.jfp-13-190.

Garde, Christian, Thomas Bjarnsholt, Michael Givskov, Tim Holm Jakobsen, Morten Hentzer, Anetta Claussen, Kim Sneppen, Jesper Ferkinghoff-Borg, and Thomas Sams. 2010. "Quorum Sensing Regulation in *Aeromonas hydrophila*." *Journal of Molecular Biology* 396 (4): 849–57. doi:10.1016/J.JMB.2010.01.002.

Gopu, Venkadesaperumal, Sivasankar Chandran, and Prathapkumar Halady Shetty. 2018. "Significance and Application of Quorum Sensing in Food Microbiology." In *Quorum Sensing and Its Biotechnological Applications*, 193–219. Springer, Singapore. doi:10.1007/978-981-13-0848-2_13.

Gui, Zhihong, Huafu Wang, Ting Ding, Wei Zhu, Xiyi Zhuang, and Weihua Chu. 2014. "Azithromycin Reduces the Production of α-Hemolysin and Biofilm Formation in *Staphylococcus aureus*." *Indian Journal of Microbiology* 54 (1): 114–17. doi:10.1007/s12088-013-0438-4.

Gutierrez-Barranquero, Jose A., F. Jerry Reen, Ronan R. McCarthy, and Fergal O'Gara. 2015. "Deciphering the Role of Coumarin as a Novel Quorum Sensing Inhibitor Suppressing Virulence Phenotypes in Bacterial Pathogens." *Applied Microbiology and Biotechnology* 99 (7): 3303–16. doi:10.1007/s00253-015-6436-1.

Hammond, Eric N., Eric S. Donkor, and Charles A. Brown. 2014. "Biofilm Formation of *Clostridium difficile* and Susceptibility to Manuka Honey." *BMC Complementary and Alternative Medicine* 14: 329. doi:10.1186/1472-6882-14-329.

Huillet, Eugénie, Marcel H. Tempelaars, Gwenaëlle André-Leroux, Pagakrong Wanapaisan, Ludovic Bridoux, Samira Makhzami, Watanalai Panbangred, Isabelle Martin-Verstraete, Tjakko Abee, and Didier Lereclus. 2012. "PlcRa, a New Quorum-Sensing Regulator from *Bacillus cereus*, Plays a Role in Oxidative Stress Responses and Cysteine Metabolism in Stationary Phase." *PLoS One* 7 (12): e51047. doi:10.1371/journal.pone.0051047.

Ivanova, K., M. M. Fernandes, and T. Tzanov. 2013. "Current Advances on Bacterial Pathogenesis Inhibition and Treatment Strategies." In *Microbial Pathogens and Strategies for Combating Them: Science, Technology and Education*, 322–36. Formatex Research Center, Spain. doi:10.13140/RG.2.1.3988.7840.

Joshi, Janak Raj, Netaly Khazanov, Hanoch Senderowitz, Saul Burdman, Alexander Lipsky, and Iris Yedidia. 2016. "Plant Phenolic Volatiles Inhibit Quorum Sensing in Pectobacteria and Reduce Their Virulence by Potential Binding to ExpI and ExpR Proteins." *Scientific Reports* 6: 38126. doi:10.1038/srep38126.

Kalia, Vipin Chandra. 2013. "Quorum Sensing Inhibitors: An Overview." *Biotechnology Advances* 31 (2): 224–45. doi:10.1016/j.biotechadv.2012.10.004.

Kalia, Vipin Chandra, and Hemant J. Purohit. 2011. "Quenching the Quorum Sensing System: Potential Antibacterial Drug Targets." *Critical Reviews in Microbiology* 37 (2): 121–40. doi:10.3109/1040841X.2010.532479.

Kievit, T. R. de, and B. H. Iglewski. 2000. "Bacterial Quorum Sensing in Pathogenic Relationships." *Infection and Immunity* 68 (9): 4839–49. https://www.ncbi.nlm.nih.gov/pubmed/10948095.

Kiran, S., P. Sharma, K. Harjai, and N. Capalash. 2011. "Enzymatic Quorum Quenching Increases Antibiotic Susceptibility of Multidrug Resistant *Pseudomonas aeruginosa*." *Iranian Journal of Microbiology* 3 (1): 1–12. https://www.ncbi.nlm.nih.gov/pubmed/22347576.

Lasa, I., J. L. del Pozo, J. R. Penadés, and J. Leiva. 2005. "Biofilms Bacterianos e Infección." *Anales Del Sistema Sanitario de Navarra*. scieloes.

Lee Wong, Amy C. 1998. "Biofilms in Food Processing Environments." *Journal of Dairy Science* 81 (10): 2765–70. doi:10.3168/JDS.S0022-0302(98)75834-5.

Liu, Xiaoguang, Jinli Jia, Roman Popat, Catherine A. Ortori, Jun Li, Stephen P. Diggle, Kexiang Gao, and Miguel Camara. 2011. "Characterisation of Two Quorum Sensing Systems in the Endophytic *Serratia plymuthica* Strain G3: Differential Control of Motility and Biofilm Formation According to Life-Style." *BMC Microbiology* 11 (1): 26. doi:10.1186/1471-2180-11-26.

Lopes, Laenia Angelica Andrade, Jessica Bezerra Dos Santos Rodrigues, Marciane Magnani, Evandro Leite de Souza, and Jose P. de Siqueira-Junior. 2017. "Inhibitory Effects of Flavonoids on Biofilm Formation by *Staphylococcus aureus* that Overexpresses Efflux Protein Genes." *Microbial Pathogenesis* 107: 193–97. doi:10.1016/j.micpath.2017.03.033.

Lu, Jing, Lynne Turnbull, Catherine M. Burke, Michael Liu, Dee A. Carter, Ralf C. Schlothauer, Cynthia B. Whitchurch, and Elizabeth J. Harry. 2014. "Manuka-Type Honeys can Eradicate Biofilms Produced by *Staphylococcus aureus* Strains with Different Biofilm-Forming Abilities." *PeerJ* 2: e326. doi:10.7717/peerj.326.

Luciardi, María Constanza, María Amparo Blázquez, Elena Cartagena, Alicia Bardón, and Mario Eduardo Arena. 2016. "Mandarin Essential Oils Inhibit Quorum Sensing and Virulence Factors of *Pseudomonas aeruginosa*." *LWT – Food Science and Technology* 68: 373–80. doi:10.1016/J.LWT.2015.12.056.

Meireles, Ana, Anabela Borges, Efstathios Giaouris, and Manuel Simões. 2016. "The Current Knowledge on the Application of Anti-Biofilm Enzymes in the Food Industry." *Food Research International* 86: 140–46. doi:10.1016/j.foodres.2016.06.006.

Meliani, Amina, and Ahmed Bensoltane. 2015. "Review of Pseudomonas Attachment and Biofilm Formation in Food Industry." *Poultry, Fisheries & Wildlife Sciences* 3 (1): 1–7. doi:10.4172/2375-446X.1000126.

Miller, Kristen P., Lei Wang, Yung-Pin Chen, Perry J. Pellechia, Brian C. Benicewicz, and Alan W. Decho. 2015. "Engineering Nanoparticles to Silence Bacterial Communication." *Frontiers in Microbiology* 6: 189. doi:10.3389/fmicb.2015.00189.

Mukherji, Ruchira, and Asmita Prabhune. 2015. "A New Class of Bacterial Quorum Sensing Antagonists: Glycomonoterpenols Synthesized Using Linalool and Alpha Terpineol." *World Journal of Microbiology & Biotechnology* 31 (6): 841–49. doi:10.1007/s11274-015-1822-5.

Myszka, Kamila, Marcin T. Schmidt, Małgorzata Majcher, Wojciech Juzwa, Mariola Olkowicz, and Katarzyna Czaczyk. 2016. "Inhibition of Quorum Sensing-Related Biofilm of *Pseudomonas fluorescens* KM121 by Thymus Vulgare Essential Oil and Its Major Bioactive Compounds." *International Biodeterioration & Biodegradation* 114: 252–59. doi:10.1016/J.IBIOD.2016.07.006.

Nakayama, Jiro, Yumi Uemura, Kenzo Nishiguchi, Norito Yoshimura, Yasuhiro Igarashi, and Kenji Sonomoto. 2009. "Ambuic Acid Inhibits the Biosynthesis of Cyclic Peptide Quormones in Gram-Positive Bacteria." *Antimicrobial Agents and Chemotherapy* 53 (2): 580–86. doi:10.1128/AAC.00995-08.

Nazzaro, Filomena, Florinda Fratianni, and Raffaele Coppola. 2013. "Quorum Sensing and Phytochemicals." *International Journal of Molecular Sciences* 14 (6): 12607–19. doi:10.3390/ijms140612607.

Ng, Wai-Leung, and Bonnie L. Bassler. 2009. "Bacterial Quorum-Sensing Network Architectures." *Annual Review of Genetics* 43: 197–222. doi:10.1146/annurev-genet-102108-134304.

Nychas, G. J. E., D. L. Marshall, and J. N. Sofos. 2007. "Meat, Poultry, and Seafood." In *Food Microbiology: Fundamentals and Frontiers*, 3rd ed., 105–40. American Society of Microbiology, Washington, DC.

Ohtani, Kaori, Yonghui Yuan, Sufi Hassan, Ruoyu Wang, Yun Wang, and Tohru Shimizu. 2009. "Virulence Gene Regulation by the Agr System in *Clostridium perfringens*." *Journal of Bacteriology* 191 (12): 3919–27. doi:10.1128/JB.01455-08.

Omer Bendori, Shira, Shaul Pollak, Dorit Hizi, and Avigdor Eldar. 2015. "The RapP-PhrP Quorum-Sensing System of *Bacillus subtilis* Strain NCIB3610 Affects Biofilm Formation through Multiple Targets, Due to an Atypical Signal-Insensitive Allele of RapP." Edited by G. A. O'Toole. *Journal of Bacteriology* 197 (3): 592–602. doi:10.1128/JB.02382-14.

Papenfort, Kai, and Bonnie L. Bassler. 2016. "Quorum Sensing Signal-Response Systems in Gram-Negative Bacteria." *Nature Reviews. Microbiology* 14 (9): 576–88. doi:10.1038/nrmicro.2016.89.

Pei, Ruoting, and Gisella R. Lamas-Samanamud. 2014. "Inhibition of Biofilm Formation by T7 Bacteriophages Producing Quorum-Quenching Enzymes." *Applied and Environmental Microbiology* 80 (17): 5340–48. doi:10.1128/AEM.01434-14.

Pratiwi, Sylvia U. T., Ellen L. Lagendijk, Triana Hertiani, Sandra De Weert, A. M. Cornellius, and J. J. Van Den Hondel. 2015. "Antimicrobial Effects of Indonesian Medicinal Plants Extracts on Planktonic and Biofilm Growth of *Pseudomonas aeruginosa* and *Staphylococcus aureus*." *International Journal of Pharmacy and Pharmaceutical Sciences* 7 (4 SE-Original Article(s)). https://innovareacademics.in/journals/index.php/ijpps/article/view/4021.

Rahman, Md Ramim Tanver, Zaixiang Lou, Fuhao Yu, Peng Wang, and Hongxin Wang. 2017. "Anti-Quorum Sensing and Anti-Biofilm Activity of Amomum Tsaoko (Amommum

Tsao-Ko Crevost et Lemarie) on Foodborne Pathogens." *Saudi Journal of Biological Sciences* 24 (2): 324–30. doi:10.1016/j.sjbs.2015.09.034.

Reuter, Kerstin, Anke Steinbach, and Volkhard Helms. 2016. "Interfering with Bacterial Quorum Sensing." *Perspectives in Medicinal Chemistry* 8: 1–15. doi:10.4137/PMC.S13209.

Shukla, Varsha, and Zarine Bhathena. 2016. "Broad Spectrum Anti-Quorum Sensing Activity of Tannin-Rich Crude Extracts of Indian Medicinal Plants." *Scientifica* 2016: 1–8. doi:10.1155/2016/5823013.

Simonetti, Oriana, Oscar Cirioni, Federico Mocchegiani, Ivana Cacciatore, Carmela Silvestri, Leonardo Baldassarre, Fiorenza Orlando, et al. 2013. "The Efficacy of the Quorum Sensing Inhibitor FS8 and Tigecycline in Preventing Prosthesis Biofilm in an Animal Model of Staphylococcal Infection." *International Journal of Molecular Sciences* 14 (8): 16321–32. doi:10.3390/ijms140816321.

Singh, Braj R., Brahma N. Singh, Akanksha Singh, Wasi Khan, Alim H. Naqvi, and Harikesh B. Singh. 2015. "Mycofabricated Biosilver Nanoparticles Interrupt *Pseudomonas aeruginosa* Quorum Sensing Systems." *Scientific Reports* 5: 13719. doi:10.1038/srep13719.

Steenackers, Hans P., Jeremy Levin, Joost C. Janssens, Ami De Weerdt, Jan Balzarini, Jos Vanderleyden, Dirk E. De Vos, and Sigrid C. De Keersmaecker. 2010. "Structure-Activity Relationship of Brominated 3-Alkyl-5-Methylene-2(5H)-Furanones and Alkylmaleic Anhydrides as Inhibitors of *Salmonella* Biofilm Formation and Quorum Sensing Regulated Bioluminescence in *Vibrio harveyi*." *Bioorganic & Medicinal Chemistry* 18 (14): 5224–33. doi:10.1016/j.bmc.2010.05.055.

Steindler, Laura, and Vittorio Venturi. 2007. "Detection of Quorum-Sensing *N*-Acyl Homoserine Lactone Signal Molecules by Bacterial Biosensors." *FEMS Microbiology Letters* 266 (1): 1–9. doi:10.1111/j.1574-6968.2006.00501.x.

Sturme, Mark H. J., Michiel Kleerebezem, Jiro Nakayama, Antoon D. L. Akkermans, Elaine E. Vaughan, and Willem M. De Vos. 2002. "Cell to Cell Communication by Autoinducing Peptides in Gram-Positive Bacteria." *Antonie Van Leeuwenhoek* 81: 233–43.

Tapia-Rodriguez, Melvin R., Adrian Hernandez-Mendoza, Gustavo A. Gonzalez-Aguilar, Miguel Angel Martinez-Tellez, Claudia Miranda Martins, and J. Fernando Ayala-Zavala. 2017. "Carvacrol as Potential Quorum Sensing Inhibitor of *Pseudomonas aeruginosa* and Biofilm Production on Stainless Steel Surfaces." *Food Control* 75: 255–61. doi:10.1016/J.FOODCONT.2016.12.014.

Thallinger, Barbara, Endry N. Prasetyo, Gibson S. Nyanhongo, and Georg M. Guebitz. 2013. "Antimicrobial Enzymes: An Emerging Strategy to Fight Microbes and Microbial Biofilms." *Biotechnology Journal* 8 (1): 97–109. doi:10.1002/biot.201200313.

Tomadoni, B., M. R. Moreira, and A. Ponce. 2016. "Anti-Quorum Sensing Activity of Natural Compounds against Chromobacterium Violaceum." *Annals of Food Science and Nutraceuticals* 1 (1): 43–48.

Truchado, Pilar, Francisco A. Tomás-Barberán, Mar Larrosa, and Ana Allende. 2012. "Food Phytochemicals Act as Quorum Sensing Inhibitors Reducing Production and/or Degrading Autoinducers of *Yersinia enterocolitica* and Erwinia Carotovora." *Food Control* 24 (1–2): 78–85. doi:10.1016/j.foodcont.2011.09.006.

Vasudevan, R. 2014. "Biofilms: Microbial Cities of Scientific Significance." *Journal of Microbiology & Experimentation* 1 (3): 1–16. doi:10.15406/jmen.2014.01.00014.

Vega, Leticia M., Jacques Mathieu, Yu Yang, Barry H. Pyle, Robert J. C. McLean, and Pedro J. J. Alvarez. 2014. "Nickel and Cadmium Ions Inhibit Quorum Sensing and Biofilm Formation without Affecting Viability in *Burkholderia multivorans*." *International Biodeterioration & Biodegradation* 91: 82–87. doi:10.1016/J.IBIOD.2014.03.013.

Vidal, Jorge E., Jianming Chen, Jihong Li, and Bruce A. McClane. 2009. "Use of an EZ-Tn5-Based Random Mutagenesis System to Identify a Novel Toxin Regulatory Locus in *Clostridium perfringens* Strain 13." *PLoS One* 4 (7): e6232. doi:10.1371/journal.pone.0006232.

Vlamakis, Hera, Yunrong Chai, Pascale Beauregard, Richard Losick, and Roberto Kolter. 2013. "Sticking Together: Building a Biofilm." *Nature Reviews Microbiology* 11 (3): 157–68. doi:10.1038/nrmicro2960.

Wagh Nee Jagtap, Mohini S., Rajendra H. Patil, Deepali K. Thombre, Milind V. Kulkarni, Wasudev N. Gade, and Bharat B. Kale. 2013. "Evaluation of Anti-Quorum Sensing Activity of Silver Nanowires." *Applied Microbiology and Biotechnology* 97 (8): 3593–601. doi:10.1007/s00253-012-4603-1.

Wang, Yanbo, Feifei Wang, Chong Wang, Xiuting Li, and Linglin Fu. 2019. "Positive Regulation of Spoilage Potential and Biofilm Formation in *Shewanella baltica* OS155 via Quorum Sensing System Composed of DKP and Orphan LuxRs." *Frontiers in Microbiology*. https://www.frontiersin.org/article/10.3389/fmicb.2019.00135.

Weiland-Bräuer, Nancy, Martin J. Kisch, Nicole Pinnow, Andreas Liese, and Ruth A. Schmitz. 2016. "Highly Effective Inhibition of Biofilm Formation by the First Metagenome-Derived AI-2 Quenching Enzyme." *Frontiers in Microbiology* 7: 1098. doi:10.3389/fmicb.2016.01098.

Williams, Paul, Klaus Winzer, Weng C. Chan, and Miguel Camara. 2007. "Look Who's Talking: Communication and Quorum Sensing in the Bacterial World." *Philosophical Transactions of the Royal Society of London. Series B, Biological Sciences* 362 (1483): 1119–34. doi:10.1098/rstb.2007.2039.

Xu, Tao, Xu-Yang Wang, Peng Cui, Yu-Meng Zhang, Wen-Hong Zhang, and Ying Zhang. 2017. "The Agr Quorum Sensing System Represses Persister Formation through Regulation of Phenol Soluble Modulins in *Staphylococcus aureus*." *Frontiers in Microbiology*. https://www.frontiersin.org/article/10.3389/fmicb.2017.02189.

Yang, Liang, Morten Theil Rybtke, Tim Holm Jakobsen, Morten Hentzer, Thomas Bjarnsholt, Michael Givskov, and Tim Tolker-Nielsen. 2009. "Computer-Aided Identification of Recognized Drugs as *Pseudomonas aeruginosa* Quorum-Sensing Inhibitors." *Antimicrobial Agents and Chemotherapy* 53 (6): 2432–43. doi:10.1128/AAC.01283-08.

Yu, Qiang, Dion Lepp, Iman Mehdizadeh Gohari, Tao Wu, Hongzhuan Zhou, Xianhua Yin, Hai Yu, et al. 2017. "The Agr-Like Quorum Sensing System Is Required for Pathogenesis of Necrotic Enteritis Caused by *Clostridium perfringens* in Poultry." *Infection and Immunity* 85 (6). doi:10.1128/IAI.00975-16.

Zambrano, Carolina, Erika Beáta Kerekes, Alexandra Kotogán, Tamás Papp, Csaba Vágvölgyi, Judit Krisch, and Miklós Takó. 2019. "Antimicrobial Activity of Grape, Apple and Pitahaya Residue Extracts after Carbohydrase Treatment against Food-Related Bacteria." *LWT* 100: 416–25. doi:10.1016/j.lwt.2018.10.044.

Zang, Tianzhu, Bobby W. K. Lee, Lisa M. Cannon, Kathryn A. Ritter, Shujia Dai, Dacheng Ren, Thomas K. Wood, and Zhaohui Sunny Zhou. 2009. "A Naturally Occurring Brominated Furanone Covalently Modifies and Inactivates LuxS." *Bioorganic & Medicinal Chemistry Letters* 19 (21): 6200–4. doi:10.1016/j.bmcl.2009.08.095.

Zetzmann, Marion, Andrés Sánchez-Kopper, Mark S. Waidmann, Bastian Blombach, and Christian U. Riedel. 2016. "Identification of the Agr Peptide of Listeria Monocytogenes." *Frontiers in Microbiology* 7: 989. doi:10.3389/fmicb.2016.00989.

Zhang, Weiwei, and Chenghua Li. 2016. "Exploiting Quorum Sensing Interfering Strategies in Gram-Negative Bacteria for the Enhancement of Environmental Applications." *Frontiers in Microbiology* 6: 1535. doi:10.3389/fmicb.2015.01535.

Zhao, Dandan, Fei Lyu, Shulai Liu, Jianyou Zhang, Yuting Ding, Wenxuan Chen, and Xuxia Zhou. 2018. "Involvement of Bacterial Quorum Sensing Signals in Spoilage Potential of Aeromonas Veronii Bv. Veronii Isolated from Fermented Surimi." *Journal of Food Biochemistry* 42 (2): e12487. doi:10.1111/jfbc.12487.

24

Quorum Quenching as an Anti-biofouling Strategy for Wastewater Reuse and Biofouling Affected Industries

Ioannis D. Kampouris, Dimitra C. Banti, and Petros Samaras

CONTENTS

24.1 Introduction

The majority of environmental bacteria are harmless and non-pathogenic; however, they are able to impact negatively the human activities by causing impediments on certain technologies and infrastructures, leading to severe damage, destruction, or increased costs of operation, cleaning, and restoration (Klahre and Flemming 2000, Le-Clech et al. 2006, Schultz et al. 2011, Fitridge et al. 2012, Polman et al. 2013). The main issue that is attributed to bacteria is known as "biofouling" and arises from the bacterial growth on solid surfaces. The origin of biofouling is associated with the production of extracellular polymeric substances (EPS) and the formation of biofilm on those surfaces (Le-Clech et al. 2006). The biofilm matrix allows a completely different lifestyle, in comparison with the planktonic state (Flemming and Wingender 2010). It immobilizes cells and increases their survival, along with their interactions in the complex environmental microbial communities. However, there are several technologies and activities that are critically affected by biofilm production, such as membrane bioreactors (MBRs), ships (marine biofouling), and aquacultures. In addition, biofouling affects the common daily life and industry, since it leads to clogging of waterlines, reduction of heat transfer in heat exchangers, and numerous problems in water-processing systems. Therefore, biofouling is a phenomenon that is commonly present in many types of industrial water-related processes.

The formation of biofilm and the production of EPS is regulated by a type of intercellular signaling in bacteria, known as quorum sensing (QS) (Shrout and Nerenberg 2012). This specific type of communication allows the bacterial cells to sense the size of their population and regulate the expression of specific genes by utilizing small auto-inducer molecules. Up to now, several QS-systems have been discovered. Those systems are enabling intra-species and inter-species communication to regulate several phenotypes, such as biofilm formation, EPS production, virulence, and competence (Williams et al. 2000, Shrout and Nerenberg 2012, Papenfort and Bassler 2016). The bacteria utilize a broad spectrum of auto-inducer molecules, and there are general groups of common types of auto-inducers that are shared among several bacterial species (Williams et al. 2007). The most common QS auto-inducers groups are the *N*-acyl-homoserine lactones (AHLs), the auto-inducing peptides (AIPs), and the auto-inducer type-2 (AI-2). The AHLs are amphipathic molecules, which are used mainly by Gram-negative bacteria, while the AIPs are being utilized mainly by Gram-positive bacteria (Firmicutes, Tenericutes, and Actinobacteria). The AI-2 signaling systems are present to both Gram-positive and Gram-negative bacteria. The most common AI-2 auto-inducer is the signaling molecule 4,5-dihydroxy-2,3-pentanedione (DPD) (Lee et al. 2018a). Since quorum sensing regulates the expression of biofilm phenotype, the inhibition of quorum sensing, also known as quorum quenching (QQ), is able to disrupt and reduce the expression of EPS along with biofilm formation (Hentzer et al. 2002), as it is clearly depicted in Figure 24.1. Therefore, the establishment of a strategy aiming to the inhibition of quorum sensing may result to the utilization of quorum quenching as an anti-biofouling strategy that can reduce the costs and the damages caused by biofouling in the infrastructures (Malaeb et al. 2013). As mentioned before,

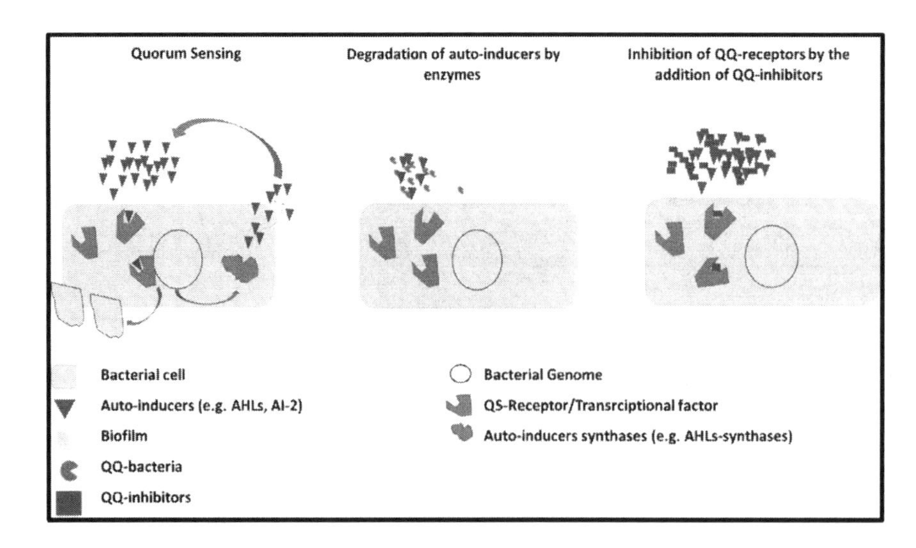

FIGURE 24.1 Description of quorum sensing mechanisms and quorum quenching inhibition with the use of enzymes and inhibitors. Usually bacteria use auto-inducer molecules as signals for the production of biofilm. The auto-inducers bind several receptors/transcriptional factors and regulate the production of EPS and more signaling molecules. The degradation of these compounds by enzymes can lead to elimination of the signal and reduction of EPS production. Another option is the use of QQ inhibitors that have competitive activity to the receptor/transcriptional factor.

the mechanism of QS is based on the production, secretion, and sensing of signaling molecules. When these molecules reach a threshold concentration, they change the gene expression in the population. At low population density or high diffuse rates, acyl-homoserine lactones (AHL) are decreased and the LuxR receptor (a transcriptional activator of the Lux operon that is activated when bacterial cell density is high) is degraded (Figure 24.1). When the AHL concentration reaches a specific concentration, the AHL signaling molecules bind LuxR to make an AHL/LuxR complex, hence activating the receptor. AHL-based signaling is predominantly found in bacteria belonging to Proteobacteria (Gram-negative), although there are some exceptions. The QS mechanism of Gram-positive bacteria has an analogous function, with the difference that the specific signal is an autoinducer peptide (AIP). In this system, the AIP precursors are produced, modified post-transcriptionally, and secreted via specific transporters. When mature AIPs are in a high concentration, they bind to a transmembrane histidine kinase (HK) and the HK receptor is activated, which activates the downstream response regulator (DRR). This activated RR initiates transcription of specific genes (Siddiqui et al. 2015).

The QS mechanism of AHL interspecies communication, can be disrupted by the following QQ strategies: (a) inhibiting the production of QS signaling molecules, (b) degradation of AHL, (c) reducing the activity of AHL cognate receptor protein or AHL synthase, and (d) mimicking the signal molecules primarily by synthetic compounds as analogues of signal molecules (Rasmussen and Givskov 2006, Kalia et al. 2013, Siddiqui et al. 2015) (Figure 24.1). The free enzymes (or the enzymes that QQ microorganisms produce) are divided in three categories (acylases, lactonases, and oxidoreductases) (Figure 24.2).

The MBR systems used for wastewater treatment are considered as a promising technology, due to their various advantages. The biggest advantage of this technology is the direct separation of biomass from the effluent, eliminating the need for disinfection (Le-Clech et al. 2006). MBRs are also able to eliminate the

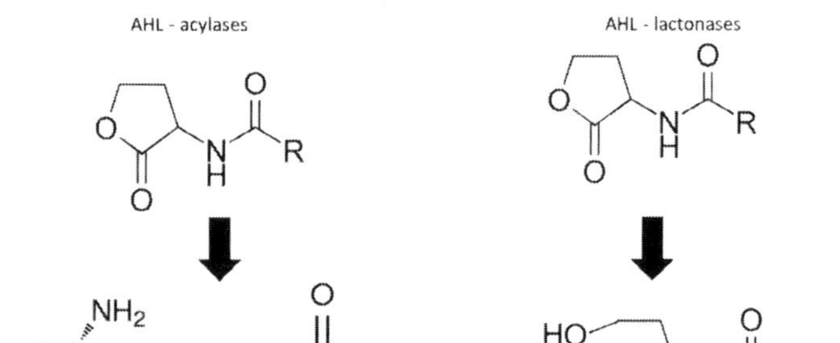

FIGURE 24.2 Description of mechanisms of the two main categories of QQ enzymes that have been used or are located in QQ microorganisms for biofouling mitigation: acylases and lactonases. The acylases are able to catalyze the degradation of amide bonds while the lactonases are able to degrade the lactone ring.

release of bacteria in the environment, including antibiotic resistant bacteria and pathogens from wastewater treatment plants, which are considered as emergent contaminants (Manaia et al. 2018). However, membrane biofouling represents a primary drawback to the broader application of this technology, increasing the operational costs of the system due to frequent membrane cleaning and replacement costs. Since the activated sludge is an ideal environment for bacterial growth and biofilm formation, with required nutrients and constant supply of oxygen, the MBRs represent suitable model systems to observe the EPS production and biofilm formation (Daims et al. 2016). Those factors influenced the research on biofilm formation in membrane bioreactors and resulted in the establishment of quorum quenching as a promising strategy to mitigate biofouling mainly in MBRs. Therefore, this chapter will focus mainly on the status of research on QQ methods associated to MBR systems and the future challenges of quorum quenching as an anti-biofouling strategy. In addition, the last part of the chapter is focused on other current applications of quorum quenching and the possibility of applying this approach as an antifouling strategy to other industrial sectors impacted by biofilm formation.

24.2 Quorum Quenching Methods and Applications in Membrane Bioreactors

24.2.1 The Origin of Membrane Biofouling

The MBR technology combines activated sludge treatment with membrane filtration. Daims et al. (2006) referred to wastewater treatment as the ideal ecosystem than can serve as a model to explore the interaction of complex microbial communities. Considering this, it should be noted that the MBR is one of the best systems (along with several other wastewater treatment approaches) to explore the complexity of biofilm formed by bacterial communities and biofouling. It has been mentioned that biofouling in MBRs is mainly attributed to two parameters: (a) the adhesion of microbial products from bacteria (soluble microbial products-SMP and EPS) and (b) the attachment of bacterial cells on the filtration membrane, where they can form a biofilm cake layer (Gkotsis et al. 2014, Banti et al. 2017, 2018). Usually the process of biofilm formation starts with the colonization of solid surfaces by plactonic bacteria, which after attachment starts to produce EPS and form the biolfim matrix. However, in activated sludge the intensive mixing usually causes the formation of aggregates (flocs). Those aggregates are composed of EPS, the same material as biofilm, and they have higher adhesive properties than separate plactonic bacteria.

While membrane fouling is a problem of microbiological nature, the first studies considered the microbial part as a "black box." This "black box" concept may assist the chemical characterization of foulants and the understanding of membrane fouling. However, the "black box" concept did not provide efficient solutions for fouling mitigation in MBRs (Vanysacker et al. 2014). The initial studies were focused on physicochemical methods, often neglecting the complexity of microbial interactions and failing to properly mitigate biofouling, or they ended up by elevating the operational costs to very high levels. For example, the physical cleaning with high aeration resulted in increased costs due to high energy demand. In addition, the frequent cleaning of the membrane by chemical methods resulted in the fast destruction of the membrane while the added chemicals (biocides or antibiotics) contributed to the deterioration of the quality of the treatment and the treated effluent (Malaeb et al. 2013). While the modification of membrane properties may lead to an improvement, the various SMPs produced by bacteria are able to change the membrane proteins and create a conditioning film that can strengthen the flocs adhesion (Vanysacker et al. 2014, Malaeb et al. 2013). Furthermore, those approaches did not prevent the occurrence of biofouling. Taking this into account, several research groups started to focus on biological solutions as an advanced method to mitigate biofouling.

In recent years, several biological applications have been proposed as antifouling strategies, such as the use of predators (rotifiers) or the enzymatic degradation of biofilm (Malaeb et al. 2013). Nevertheless, there is one specific biological method, the disruption of quorum sensing and specifically the degradation of AHLs, on which most of the current published studies have been focused (Malaeb et al. 2013). Before the establishment of this method, there were phylogenetic studies aiming to elucidate the bacterial population in activated sludge and biofilm cake layer. Those studies showed a dominance of *Alphaproteobacteria*, *Bettaproteobacteria*, and *Gammaproteobacteria* (Miura et al. 2007, Xia et al. 2010). Secondly, lab-scale studies that focused on the biofilm formation and EPS production from strains that belong to Proteobacteria showed that biofilm and EPS production are regulated by AHLs mediated communication, using the *luxI*/*luxR* dependent systems (Case et al. 2008). The AHLs are a group of various compounds; however, they have similar chemical structure, a lactone ring and an acyl-chain, which may vary in the number of carbon atoms and substitution (Shrout and Nerenberg 2012). Different species of bacteria produce various types of AHLs, although an extraordinary characteristic of this specific system is that various (different in acyl-chain number) AHL molecules are able to activate the LuxR receptors of most of the species of Proteobacteria. Therefore, the estimation of population density is crossing the species boundaries and enables bacteria from different species to coordinate their actions. The breakthrough for understanding the biofouling ecology in MBR systems was achieved through the connection of this interspecies cell-to-cell communication with biofilm development on the membrane (Yeon et al. 2009a). The above-mentioned research team analyzed sludge biofilm structures with Gram staining, and it was found that there was a predominance of Gram-negative bacteria (Proteobacteria). Then, the team tried to confirm (with success) the hypothesis that AHLs concentration in mixed liquor affects biofouling. Following this confirmation, a plethora of studies were published focused on quorum sensing inhibition, specifically in MBRs. The published studies can be categorized in three main groups: (1) enzymatic applications of QQ, (2) applications of QS inhibitors, and (3) application of QQ microorganisms (Figures 24.3 and 24.4).

24.2.2 Enzymatic Applications of Quorum Quenching in MBRs

As mentioned before, the enzymes that degrade AHLs are divided generally in three categories: acylases, lactonases, and AHL-oxidoreductases. The acylases are present to many types of organisms; for example, mammals (e.g., porcine acylase) are able to degrade amino-acyl compounds such as AHLs. Since AHLs are amino-acyl compounds, the acylases are really useful as quorum quenching agents. Yeon et al. (2009a) used this type of enzyme to prove that degradation of AHLs could reduce biofouling. However, they used the MBR in recycle mode to maintain the concentration of the enzyme stable in the bioreactor. It was clear that without a proper encapsulation or stabilization of the QQ-enzymes, their implementation as technology would not be able to overpass the up scaling from lab to full-scale plants. To overcome this specific issue, Yeon et al. (2009b) developed magnetic carriers, and they included magnetic ion-exchange resin as a magnetic core and enzyme immobilization agent. While the activity of the immobilized enzyme was slightly lower than the activity of free acylase, the stability of the enzyme was increased to high levels, sustaining its activity

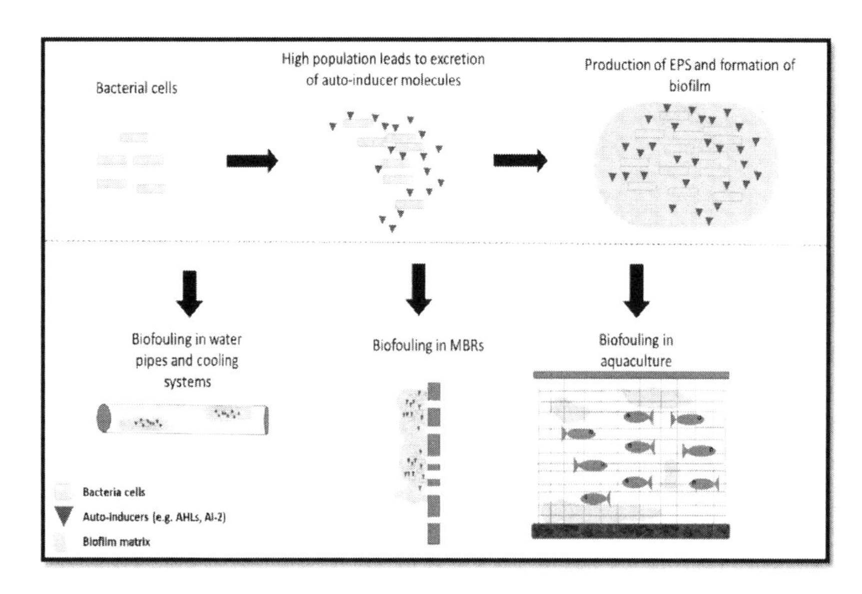

FIGURE 24.3 The upper part is a schematic explanation of quorum sensing mediated biofilm formation in bacterial populations. The lower part describes the technologies/infrastructures that are impacted by biofilm formation (biofouling).

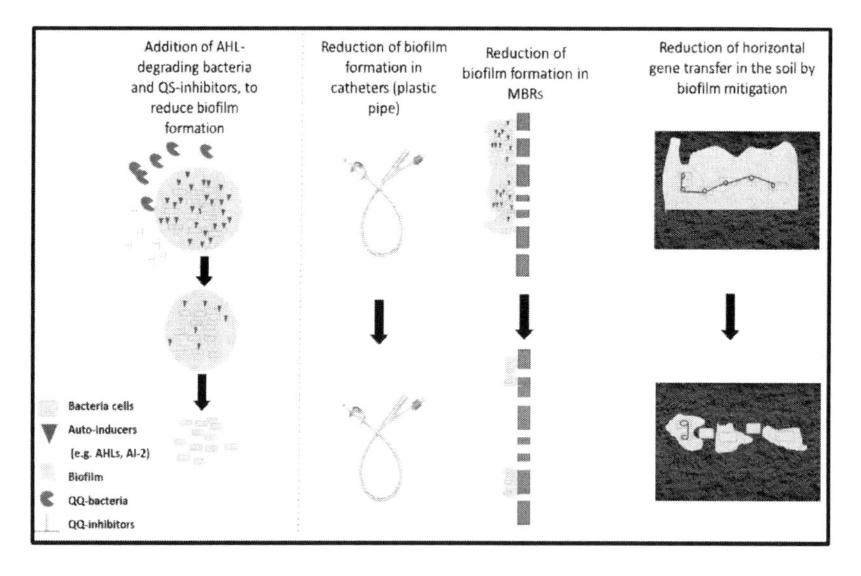

FIGURE 24.4 Schematic explanation of quorum quenching methods and their current applications in medicine, wastewater treatment, and agriculture.

for further cycles of operation. Kim et al. (2013a) utilized again the same approach of magnetic carriers; however, in that study they focused on the impact of the degradation of AHLs in biofilm, EPS composition, and the effect of quorum quenching in microbial communities. While a protein and a polysaccharide fraction of EPS were reduced in biofilms, there was also a change to the relative abundance of *Gammaproteobacteria* in the biofilm of the QQ MBR in comparison with the biofilm of the control MBR. Several other alternative applications have been investigated, such as the immobilization of the enzymes in the membrane (Kim et al. 2011), while Jiang et al. (2013) performed encapsulation of the enzyme in alginate beads. The immobilization of enzymes in alginate provided several advantages, such as the optimum mixing of enzymes. In addition, in this study it was demonstrated that acylase not only reduced the EPS biocake, but also caused the reduction of SMP and EPS (mg/L) in the mixed liquor of the QQ MBRs. Furthermore, they also observed a change in the hydrophobicity profile of EPS. However, all the previous studies anticipated a strong issue regarding the use of enzymes: free or immobilized enzymes should constantly need to be renewed, which means that they are not suitable for long run operation. In conclusion, while the enzymatic degradation of AHLs was the fastest, simplest, and optimum way to show quorum quenching potential as antifouling strategy, the application of enzymatic quorum sensing inhibition to real-scale MBR is not a cost-efficient option for commercial applications. Aiming to gain the great advantages of enzymatic QQ for biofouling mitigation in wastewater treatment and simultaneously overtake their drawbacks of relatively high cost and low stability of these enzymes, QQ bacteria, which can produce AHL-degrading QQ enzymes, have occasionally been successfully isolated from activated sludge and applied in MBRs. A summary of them is

TABLE 24.1
QQ Enzymes or Active Ingredients Isolated from Activated Sludge Used for Biofouling Mitigation

Enzyme or Active Ingredient	Source	Substrate	References
AHL QQ			
Porcine acylase (EC 3.5.1.14)	Porcine kidney	*N*-acyl-aliphatic-L-amino acid	Yeon et al. (2009b)
AHL lactonase	*Rhodococcus* sp. BH4	C6, C7, C8, C10, 3OC6, 3OC8	Oh et al. (2012)
AHL lactonase	*Bacillus methylotrophicus* sp. WY	C8, C10, C12, C14, 3OC6, 3OC12	Khan et al. (2016)
AHL lactonase	*Bacillus* sp. T5	C8	Gul et al. (2018)
AHL lactonase	*Enterococcus* sp. HEMM-1	C4, C6	Ham et al. (2018)
AHL acylase	*Delftia* sp. T6	C8	Gul et al. (2018)
AHL acylase	*Acinetobacter* sp. Ooi24	C10	Ochiai et al. (2013)
Culture supernatant	*Pseudomonas* sp. 1A1	C6, C8, C10, C12, 3OC12	Cheong et al. (2013)
Culture supernatant	*Staphylococcus* sp. ML-1	3OC6	Kim et al. (2015)
Culture supernatant	*Micrococcus* sp. SY-1	C6	Kim et al. (2015)
Culture supernatant	*Pseudomonas* sp. SS-1	C6	Kim et al. (2015)
Cell lysate	*Afipia* sp. MS-1	3OC6	Kim et al. (2015)
Cell lysate	*Acinetobacter* sp. HMS-1	C6	Kim et al. (2015)
Cell lysate	*Streptococcus* sp. SL-2	3OC6	Kim et al. (2015)
Whole cell	*Rhodococcus* sp. B5	C6	Kim et al. (2015)
Whole cell	*Microbacterium* sp. HSY-2	C6	Kim et al. (2015)
AI-2 QQ			
Culture supernatant (hydrophilic, <400 Da)	*Acinetobacter* sp. DKY-1	AI-2 (DPD, 4,5-dihydroxy-2,3-pentandione)	Lee et al. (2018a)
Farnesol	*Candida albicans* Tup1/Tup1	Blockage of LuxS synthesis	Lee et al. (2016a)

Source: Lee, K. et al., *Bioresour. Technol.*, 270, 656–668, 2018b.
Abbreviations: 3OC12, *N*-(3-oxododecanoyl) HSL; AHLs, *N*-acyl homoserine lactones; AI-2, autoinducer-2; C10:1, 7,8-*trans*-*N*-(decanoyl) HSL; C6, *N*-hexanoyl HSL. The number after C in AHLs indicates the number of carbons in the acyl group.

presented in Table 24.1. Additional information regarding the sources/microorganisms that produce AHL-degrading enzymes are presented and analyzed in Section 2.4.

24.2.3 Utilizing Quorum Sensing Inhibitors for MBR Biofouling Mitigation

Another alternative method that was explored by a few research groups was the addition of compounds that can block AHLs binding to LuxR receptors. The origin of those specific approaches lies in medical field studies. For example, Givskov et al. (1996) suggested the use of furanone compounds as biofilm disruptors, since they have the ability to interfere and reduce the production of AHL in *luxI*/*luxR* systems. Those halogenated furanone compounds could be used to minimize the harmful activities of certain plant and animal pathogens. However, since furanone compounds showed a few limitations in their use, due to their linkage with halogen elements, there was a research interest for natural products with quorum sensing inhibition activity. This led to the discovery of extracts from certain plants that were able to exhibit quorum quenching properties. Therefore, extracts from various plants have been examined, such as garlic (Bjarnsholt et al. 2005), *Vanilla planifolia* (Choo et al. 2006), and *Piper betle* (Siddiqui et al. 2012). From the above extracts, only the *Piper betle* extract had been investigated for MBR biofouling mitigation (Siddiqui et al. 2012, 2013), until now. The addition of the extract to a lab-scale MBR was able to mitigate biofouling and reduce the production of EPS in the biocake.

However, except for the use of crude extracts, few studies have been focused on the potential of specific metabolites that are produced from certain plant species. For example, Lade et al. (2014) investigated the potential of active compounds from particular plant extracts, specifically curcumin (tumeric) and (–)-epigallocatechin gallate (green tea) to inhibit AHL-mediated signaling and biofilm formation in Gram-negative bacteria that were isolated from MBR activated sludge samples. The biofilm formation indicated that each of the compounds inhibited respectively 52%–68% and 59%–78% of biofilm formation among all the tested bacteria. However, when the researchers combined the extracts, they observed an effect around 95%–99% of biofilm reduction. Another compound that has been investigated for its anti-biofouling potential is vanillin (4-hydroxy-3-methoxy benzaldehyde), which is the one of the main constituents of vanilla beans (and their extracts), and it is used as a food ingredient in ice cream and chocolate products. Vanillin has also been proposed as a potential biofouling control agent in membrane systems applied for water treatment (Ponnusamy et al. 2009; Kappachery et al. 2010). Those studies reported that addition of vanillin, in concentrations ranging from 0.063 to 0.25 mg/mL, reduced biofilm formation by *Aeromonas hydrophila* on polystyrene surface without impeding the growth of planktonic cells.

Nam et al. (2015) further explored the application of vanillin as a biofouling agent in MBR. The authors added vanillin in different concentration doses in an MBR. They observed that doses greater than 125 mg/L in 100 mL of sludge that was sampled from MBRs exhibited reduction of biofilm formed on the membrane surfaces around 25%. In addition, they tested the potential of vanillin for biofouling mitigation in MBR, with a working volume 3 L. They observed that the transmembrane pressure (TMP) of the MBR with the vanillin dose of 250 mg/L showed a delay in comparison with the control MBR. Furthermore, they measured the EPS concentrations at the end of the MBR operation, where they observed a significantly less amount of EPS in the biocake of the MBR operating with vanillin addition than the control MBR. Therefore, they concluded that vanillin could be used as an anti-biofouling agent for MBRs. While the addition of those commercial compounds (or extracts) is cheaper than quorum quenching enzymes, there is a critical issue regarding their use: they tend to wash off in permeate, and therefore their continuous addition in the mixed liquor will be necessary to be able to obtain the same effect. Tse et al. (2018) tried to overcome this issue by developing an anti-biofouling membrane by incorporating chlorinated polyethylene onto the surface of a microfiltration membrane using Williamson ether synthesis reaction. They reported that the modified membrane had significantly less biofouling compared to the unmodified membrane. Moreover, the incorporation of vanillin decreased further the amount of live cells, dead cells, and polysaccharides that were covering the surface by 82%, 77%, and 51%, respectively, compared to the unmodified membrane. Therefore, vanillin-modified membranes may result in an alternative option that improves MBR performance.

24.2.4 Application of QQ Microorganisms for Biofouling Mitigation

The application of QQ microorganisms for the mitigation of biofouling is the third and most promising application of the QQ method. The enzymes and quorum sensing inhibitors are efficient in biofouling mitigation media, but they have a relatively small life, even if they are fixed in the membrane; therefore, they do not represent sustainable options for real-scale MBRs. The QQ microorganisms are able to produce such degrading enzymes (bacteria) or quorum sensing inhibitors (fungi and bacteria), with the proper supplementation of nutrients. Usually there are two choices for the application of microorganisms: (1) the transformation of a commercial bacterial strain (such as *E. coli*) with a gene for an enzyme which degrades AHLs and (2) the isolation of environmental strains that are able to degrade AHLs or producing quorum sensing inhibitors. Therefore, their activity may last longer than enzymes and inhibitors, while they will need fewer cycles of renewal. Nevertheless, due to intensive competition in aquatic, soil, or gut ecosystems, many bacteria present quorum quenching potential that could be used to compete with other bacterial species (Dong et al. 2004). As mentioned before, there are usually three types of enzymes that induce quorum quenching: lactonases, acylases, and oxidoreductases (Uroz et al. 2009). Apart from the fact that even one single bacterial strain can possess several types of QQ enzymes from this category, bacteria have also enzymes, for their own metabolic activities that will enable them to use the reaction products as a nutrient source. This removal of products shifts the equilibrium and optimizes the reaction conditions of AHLs degradation while enabling higher degradation than the addition of one type of enzyme at high concentration.

Studies focused on the isolation of bacterial strain able to degrade AHLs have been published in the area of probiotics many years before the research of anti-biofouling in MBRs, even from activated sludge. Oh et al. (2012) published for the first time the use of QQ bacteria as anti-biofouling agents in MBR. They used two types of bacteria, one strain of *E. coli*, transformed with an AHL-lactonase gene, and one strain that was isolated from activated sludge of an urban wastewater treatment plan (*Rhodococcus* sp. BH4). The strain was isolated with a minimal medium that was containing AHLs as the only carbon source. They tested both the transformed and the indigenous strain by placing them in a microbial vessel and adding it in a lab-scale MBR with a working volume of 800 mL. Both of the strains showed a delay of TMP increase. However, the use of *Rhodococcus* sp. BH4 was considered a better choice, since the transformed *E. coli* would require the addition of antibiotics to retain the plasmid. *Rhodococcus* sp. BH4 was able to degrade AHLs intracellularly. Nevertheless, the recombinant *E. coli* was used to show the possibility that quorum quenching could mitigate biofouling of certain strains in reverse osmosis membranes (Oh et al. 2017).

During recent years, several studies were published related to the use of quorum quenching microorganisms in MBR, focusing on the application of *Rhodococcus* sp. BH4 as QQ microorganisms. Jahangir et al. (2012) published two important findings regarding the encapsulation of QQ strains. The first one was that when the encapsulation medium was not located in the same tank as the membrane, the effect of QQ would be low. However, when the recirculation rate was high, they showed that it is possible to solve this problem. In addition, the addition of the encapsulated medium in the membrane tank and the application of high recirculation rate result in a much longer delay of TMP increase. Therefore, the configuration of the membrane, along with the good mixing and distribution of mixed liquor through recirculation could play a crucial role in the effectiveness of the encapsulated QQ strain. These results showed that the application of such a method should consider the complexity of AHLs mediated signaling. The AHLs are auto-inducers, and the gene *luxI* is one of the genes that are expressed when AHLs are binding to LuxR receptors. This means that there is a positive feedback of the over-production of AHLs, and unless constant degradation is taking place, there will be no effect of quorum quenching in activated sludge. Therefore, this study showed that if the total bacterial community is not quenched efficiently in activated sludge, they will soon start producing AHLs and the mitigation of biofouling will not be effective.

Cheong et al. (2013) showed the isolation of another bacterial strain *Pseudomonas* sp. 1A1 for biofouling mitigation. The specific strain was also able to degrade AHLs, but in comparison with *Rhodococcus* sp. BH4, this strain was able to produce extracellular enzymes. *Pseudomonas* sp. 1A1 was also encapsulated in a porous medium and exhibited the same anti-biofouling activity. To ensure the uniform degradation of the AHLs through activated sludge, Kim et al. (2013b) encapsulated the strain *Rhodococcus* sp. BH4 in alginate beads. The alginate beads provided also physical washing of the biocake. The effect of anti-biofouling activity was similar as the other porous media. However, alginate is hydrophilic polysaccharide and prone to degradation by the microorganisms. Another option (Kim et al. 2015) was the coating of alginate with polysulfonate coating. This technique may increase the mechanical strength and stability of the alginate beads. However, this process uses phase inversion with an organic solvent that is toxic to microorganisms and may reduce the amount of encapsulated bacteria. As a result, it requires further activation of the encapsulated bacteria with a nutrient source.

The method of immobilized *Rhodococcus* sp. BH4 in alginate beads has been further explored (Lee et al. 2016a). These authors tested the effect of this process in a real-scale MBR that had an anoxic tank and was fed with urban wastewater. The specific strain exhibited similar results in real-scale too, minimizing biofouling. However, several other efforts have been made to find an even better optimum medium for encapsulation of the QQ stains (specifically *Rhodococcus* sp. BH4). One example is the use of QQ cylinders (Lee et al. 2016b) or the use of rotary microbial carrier frames (Ergön-Can et al. 2017).

Other bacterial stains have also been proved as potential efficient quorum quenchers, able to mitigate biofouling. Gül et al. (2018) isolated bacterial strains from several different environments. They took samples from a coastal environment, a highly polluted pond, and a saltern in Turkey. The strains that were isolated belong to the genera *Shewanella*, *Klebsiella*, *Acinetobacter*, *Bacillus*, *Deftia*, *Vibrio*, *Comamonas*, *Microbacterium*, and *Pseudomonas*. The researchers selected the strain *Bacillus* sp. T5, which was encapsulated in hollow fibers. The strain was also able to mitigate biofouling in an MBR with working volume of 4.5 L. Kampouris et al. (2018) isolated four bacterial stains from the activated sludge of a municipal wastewater treatment plant in Greece, using a medium which contained hexanoyl-homoserine lactone as the sole carbon source. Only one strain (*Lactobacillus* sp. SBR04MA) showed a high degradation potential of AHLs. The authors examined the activity of this strain in a lab-scale MBR with 25 L working volume, which consisted of two tanks: one aeration tank and one membrane tank. The strain also was immobilized in alginate beads. They operated an MBR in two runs: one run with empty beads and a second with beads containing the strain. In both runs, the alginate beads were trapped in the aeration tank, leaving the membrane tank without control or QQ beads. However, the beads degraded very fast, in seven days. Both of the runs started with the same critical flux and the authors did not observe any direct effect after the addition of the beads. After the degradation of beads and the release of the strain, they observed an increase of the critical flux and reduction of the TMP rate at high fluxes only in the QQ mode of operation. Despite the release of *Lactobacillus* sp. SBR04MA cells, due to degradation of encapsulation medium, the cells of the strain did not lose their activity. In contrast, the sludge biocommunity was affected by the strains, resulting in increased critical flux and minimization of TMP rate at high fluxes. Furthermore, the results showed that the addition of the strain reduced the overall EPS and SMP concentration in the mixed liquor and the corresponding SMP amounts passing through the membrane pores. Therefore, the strain *Lactobacillus* sp. SBR04MA showed that there is also a potential use of novel QQ strains, with the form of free cells. The use of free cells could release the efforts for encapsulation and lower the costs of the commercial application of QQ in membrane bioreactors. However, further research is necessary to establish an appropriate addition mode of free cells in a real-scale MBR, along with the isolation of novel strains that may better perform in the form of free cells.

A few studies have investigated the effect of interference in AI-2 mediated communication. Lee et al. (2018a) showed the disruption of AI-2 by an indigenous activated sludge bacterium *Acinetobacter* sp. DKY-1. This specific strain was able to degrade an AI-2 signaling molecule DPD. However, according to Lee et al. (2018a) the mechanism behind AI-2 inactivation from *Acinetobacter* sp. DKY-1 was the secretion of a hydrophilic metabolite with small molecular weight that was inhibiting AI-2 mediated communication. When beads with the encapsulated strain were added into an MBR, a significantly decreased DPD concentration was observed together with reduced membrane biofouling. Apart from bacteria, fungal strains have been used to mitigate biofouling in the bioreactor. The use of *Candida albicans* strain that was able to interfere with AI-2 mediated communication and prevent membrane biofouling has also been reported (Lee et al. 2018b). The specific strain was producing farnesoyl, which is a product that interferes with AI-2 signaling and therefore, farsenoyl can reduce biofilm formation in species belonging to Gram-negative and Gram-positive bacteria. Utilization of alginate beads containing the specific fungal strain results in lower TMP increase rate. These results showed clearly that fungal QQ strains could also be used for mitigation of biofouling.

Another approach of quorum quenching is the combination of inhibitors with QQ bacteria, such as biostimulation with gamma-caprolactone (GCL), a molecule with similar structure as AHLs. Yu et al. (2016) explored the use of the enrichment of a QQ bacterial consortium from activated sludge by GCL, along with the effect of biostimulation on biofouling control in MBR treating urban wastewater. In this study, quantitative PCR was carried out to monitor the concentration of the QQ gene *qsdA*. The results demonstrated that after enrichment with GCL, the bacterial community of activated sludge could effectively degrade AHLs, while the concentration of the gene *qsdA* in the mixed liquor increased. In addition, lower rate of TMP increase and lower concentration of EPS (mg/g VSS) were reported during biostimulation.

In general, the use of bacterial or fungal strains to inhibit QS mediated signaling is a promising anti-biofouling method, mainly because the cells can act as a factory and produce enzymes or metabolites, while the feed with untreated wastewater in MBR generally provides the necessary nutrients for their growth support. However, further research is necessary to optimize biofouling mitigation. There are many unknown quorum quenching microorganisms in the environment, which potentially may exhibit characteristics such as faster degradation rate or higher potential of survival in MBRs with urban or industrial wastewater.

24.2.5 Methods for Screening QQ-Activity Identifying Microorganisms and Biofouling Mitigation

The isolation of quorum quenching bacteria can be done by the use of selective media. The selective media contain AHLs as the only carbon source (Oh et al. 2012, Gül et al. 2017, Kampouris et al. 2018). A very detailed medium for isolation of QQ strains has been described by Chan et al. 2009. This medium is supplemented with N-3-oxo-hexanoylhomoserine lactone (3-oxo-C6-HSL) that constitutes an acylhomoserine lactone (AHL) used as signaling molecules in Gram-negative bacterial cell-to-cell communication, as the sole source of carbon and nitrogen along with other inorganic nutrients that are necessary for bacterial growth.

The screening for the detection of AHLs can be performed by using biosensor strains such as *Chromobacterium violaceum* CV026 or *Agrobacterium tumefaciens* A136. *Chromobacterium violaceum* is a Gram-negative bacterium commonly found in soil and water that produces the characteristic purple pigment violacein. The biosensor strain cannot produce AHLs by itself; however, when cultured in a medium containing AHLs, a violet pigment (violacein) is produced. *Agrobacterium tumefaciens* A136 (Zhu et al. 1998) is also a Gram-negative bacterium which is used as biosensor for AHLs. The specific transformants of this strain, instead of producing pigment, produce fluorescent proteins that enable the AHLs quantification and QQ activity evaluation (Jahangir et al. 2012). Usually the strains are identified by the use of molecular markers, especially the gene 16Sr RNA (Oh et al. 2012, Gül, Koyuncu 2017, Kampouris et al. 2018).

Various immobilization and coating techniques of QQ enzymes and QQ microorganisms have been developed aiming to improve biofouling mitigation. Direct fixation with coating or embedding of QQ enzymes (acylase or lactonase) onto magnetic enzyme carriers has been performed to avoid the loss of free enzymes (Yeon et al. 2009b). Another solution for QQ enzymes (acylase) immobilization is alginate beads that constitute a highly biocompatible material (Jiang et al. 2013). Moreover, according to Kim et al. (2011), QQ enzymes (porcine kidney acylase I) have been directly immobilized onto the membrane surface. Some years later Kim et al. (2018) developed novel biocatalytic membranes for enzymatic inhibition of bacterial QS, achieving effective control of membrane biofouling in MBRs. Yang (2011) extracted lactonase aiiA from recombinant *E. coli* XL1-Blue and immobilized it to mesoporous silica (Lee et al. 2018a). Several different types of vessels for carrying QQ microorganisms have also been developed. Ergön-Can et al. (2017) created a rectangular QQ media structured in a frame with two faces made of microfilters, and it was rotated on a vertical axle placed in the center of an MBR. *Rhodococcus* sp. BH4 was entrapped in this vessel. QQ beads have also been used for entrapping QQ bacteria (*Rhodococcus* sp. BH4). Kim et al (2013b) developed a spherical bead made of a porous microstructure with a large inner void space in which QQ bacteria could be entrapped. Alginate QQ beads successfully mitigated biofouling in an MBR fed with synthetic wastewater but were not sufficient for applications in MBRs fed with real wastewater. This may be attributed to the fact that alginate beads are easily degraded into real wastewater that typically contains a wide variety of components including salts. Therefore, the alginate beads were reinforced with a coating of synthetic polymers, known as microcapsulation by Kim et al. (2015).

Lee et al (2016b) searched and further improved the QQ bacteria immobilization by using biocompatible hydrogel polymers (e.g., polyvinyl alcohol), in which QQ bacteria were entrapped by polymer coagulation and crosslinking. However, as QQ beads have an inactive inner core region, more efficient designs of QQ media, such as cylinder and hollow cylinders, have also been developed (Lee et al., 2016b). Finally, mobile QQ sheets (1 × 1 cm), with a thickness of ~0.5 mm, were found to be more effective, because QQ sheets are able to slide between membrane fibers and thus can wash the membrane surfaces effectively (Nahm et al. 2017, Lee et al. 2018b).

Following the screening for QQ activities, most of studies verified by potential of biofouling mitigation have been investigated by using several physical parameters, chemical methods, or microscopy in lab-scale or pilot-scale MBRs. The main physical parameter that most of the studies evaluate is the reduction of TMP increase rate during the addition of a QQ-strain/enzyme/inhibitor (Oh et al. 2012, Cheong et al. 2014, Kim et al. 2018) or the alteration of the critical flux (Kampouris et al. 2018). Apart from the physical parameters, several chemical parameters are reported, such as the reduction of concentration of EPS and SMP in the mixed liquor. The concertation is measured chemically by protein and polysaccharide quantification (Jiang et al. 2013, Lee et al. 2016c). The use of microscopy techniques has been reported, as evaluation for biofouling mitigation potential, by measuring the size of the flocs (Kampouris et al. 2018) or the biofilm's biomass (Oh et al. 2012).

24.2.6 Submitted Patents for Biofouling Mitigation Using Quorum Quenching

Extensive research has been carried out on the development of QQ enzymes, inhibitors, or microorganisms to reduce biofouling. At the same time, several patents have been registered based on the QQ mechanism in order to scale them up in real-scale wastewater treatment plants. Some of the most promising patents are presented below.

According to the international patent WO2018101888A1, an innovative bio-bead has been created for biofouling control in aerobic and anaerobic membrane bioreactors. In more detail, a composite material for use in aerobic membrane bioreactors to reduce and/or prevent membrane biofouling has been created comprised of:

- A crosslinked polymeric matrix material compatible with bacterial growth
- A bacterial population and
- A biocompatible adsorbent material, wherein
- The bacterial population and the adsorbent material are homogeneously mixed throughout the polymeric matrix material, and the bacterial population substantially consists of one or more bacterial species selected from *Microbacterium* sp. QQ1 (KR058848), *Enterobacter cloaca* QQ13 (KR058854), *Pseudomonas* sp. QQ3 (KR058846), and *Rhodococcus* sp. BH4.

Additionally, a composite material for use in anaerobic membrane bioreactors to reduce and/or prevent membrane biofouling has been created comprised of:

- A crosslinked polymeric matrix material compatible with bacterial growth and
- A bacterial population, wherein
- The bacterial population is homogeneously mixed throughout the polymeric matrix material; the bacterial population substantially consists of *Microbacterium* sp. QQ1 (KR058848) and
- The homogeneously mixed bacterial population and polymeric matrix material are coated with a coating

material selected from one or more of the groups consisting of polyvinylchloride, polyvinylidene fluoride, polytetrafluoroethylene, polysulfone, polyethersulfone, polyacrylonitrile, polyethylene, and polypropylene.

The U.S. patent US20150237870A1 registered the creation of genetically engineered bacteriophages that inhibit the biofilm formation. Embodiments are directed to the patented engineered bacteriophages that produce polypeptides that interfere with or quench quorum sensing. In certain aspects, the quorum quenching enzyme AiiA has a broad specificity for degrading substrates, small signal molecule acyl homoserine factories (AHL) that initiate the quorum sensing pathway with global impact for diverse bacteria in biofilm. Certain embodiments are directed to the engineered T7 aiiA phage that effectively degrades AHL from diverse bacteria and inhibits formation of the mixed species biofilms containing *Pseudomonas aeruginosa* and *E. coli*. Engineered phages can be used as antifouling and antibacterial agents in industrial as well as in clinical settings by lysing the host bacteria and stably expressing quorum quenching enzyme having a broad substrate specificity and impacting diverse bacteria in the community.

The Chinese patent CN104909450A registered a preparation and application method of QQ inhibitors in order to mitigate efficiently biofouling in aerobic membrane bioreactors for wastewater treatment. The patented fouling inhibitor has been prepared according to the following procedure:

- Preparation of 600 mL solution of 2% ~ 4% mass fraction of sodium alginate.
- Preparation of 250 mL solution of 30% ~ 45% mass fraction of allicin solution.
- Preparation of 150 mL solution of ethanol solution with 5% ~ 8% mass fraction of brominated furanones.
- Mixing of the prepared allicin solution with the ethanol solution containing brominated furanones, and then mixing with the sodium alginate solution and allowed to stand for aging for 2–4 h.
- Calcium chloride solution is then added dropwise, and it is freeze-dried in order to form the QQ inhibitor beads.

Finally, alginate microspheres are created containing allicin and brominated furanones as inhibitors of membrane biofouling.

24.2.7 The Challenges for Application of QQ Methods in Real-Scale Urban or Industrial Wastewater Treatment Plants

From the previous discussion regarding the research and patents registered on QQ methods, the use of QQ strain seems to be the most promising one for real-scale application. However, there are several challenges, which are not exclusively related to quorum quenching but in general to wastewater treatment and reuse process, and it is necessary to considered them before real-scale application. For example, it is unknown whether a strain will perform adequately under extreme conditions. In some cases, the temperature may be too low in the winter or too high in the summer for the survival, sustainability, or performance of the strains. Furthermore, a load of chemicals such as dyes, disinfectants, drug metabolites, and antibiotic residues are reaching daily the WWTP. The composition of the influent may vary and in cases where the concentration of toxic compounds is high, it may have an impact on the QQ strains and affect their survival rate. Another possible effect of QQ that needs further research is related to the potential impact on the quality of the treatment and the effluent. Since bacteria may organize several behaviors via quorum sensing, such as nitrification, there is a risk that quorum quenching may affect those processes. For example, Yu et al. (2018) reported that QQ suppressed the nitrification process in extreme conditions. Therefore, further research is necessary to those processes before the establishment of QQ methods. In addition to the effect of QQ strains on the conventional parameters of quality in the effluent, a consideration should be placed regarding the impact of quorum quenching on the removal of contaminants of emergent concern (CECs), as CECs include a large number of potentially harmful chemical compounds, such as antibiotics, disinfectants, and drug metabolites (Deblonde et al. 2011). The impact of QQ in the removal of those compounds should be necessary to be investigated. In addition, it is necessary to further look for novel QQ microorganisms, which may perform better or have better survival rates in MBRs. Especially in the AI-2 mediated signaling, there is only one bacterial species isolated as quencher and one fungal strain that has been tested. It is still unknown if there are several AI-2 quenchers with the ability to improve quenching. Even if several challenges remain, quorum quenching is one of the few methods that effectively mitigates biofouling and has attracted the interest of several research groups from several countries. While the research in this field for the current isolated strains and for novel unknown potential QQ strains is still in progress, the improvement of the QQ strategies will surely keep expanding.

The development of QQ-based strategies aiming to mitigate biofouling in membrane bioreactors and other industry applications is a relatively recent field of applied research, as most studies have been published within the last years. These strategies are only in a preliminary phase, and several questions on the exact nature of biofilms both in terms of implicated microbes and composition of the extracellular matrix remain, but undoubtedly, this field bears high hopes for future applications.

24.3 Current and Potential Applications of QQ in Other Industrial Sectors and Infrastructures

Even though MBR is not the only technology or infrastructure that is impacted by biofouling, it is the one that has attracted the scientific interest for the application of QQ methods. This is due to the need for improvement of wastewater reuse, and due to the fact that MBRs provide ideal ecosystems to study the effect of quorum sensing. Nevertheless, few studies explored applications of quorum quenching in topics other than wastewater treatment. Ivanova et al. (2015) used acylase coatings in the catheters to prevent biofilm formation. Lai et al. (2017) showed a promising application of quorum quenching, which was not related directly to biofilm formation. They reported that the use of quorum sensing inhibitors, 4-nitropyridine N-oxide (4-NPO) and 3,4-dibromo-2H-furan-5-one (DBF) had an effect in the abundance of the tested antibiotic resistance genes. Specifically, the application of quorum sensing inhibitors led to a significant decrease of tetracycline resistance genes and the gene *intl1*, which is associated with mobile genetic elements and resistance genes.

Biofouling in aquacultures is a problem affecting the target culture species, while infrastructures are exposed to biofilm formatting microorganisms, with significant production impacts. Biofouling can cause increased weight of stock and equipment (e.g., panels, nets, ropes, and floats) due to biomass deposition, as a result of biofouling and cage deformation and structural fatigue due to the extra weight imposed by fouling (Fitridge et al. 2012). Several approaches to control and mitigate aquaculture biofouling have been recommended such as antifouling coatings or the use of biocides. The antifouling coatings may include copper or other metals which may prove harmful to ecosystems and along with biocides, predators and humans may ingest the fish and shellfish that have accumulated these contaminants. In cases where antibiotics are used as biocides, there is an increased risk from the development of antibiotic resistance in bacteria (Guardiola et al. 2012). Therefore, quorum quenching could represent an alternative strategy to improve aquaculture performance. Dobretsov et al. (2011) showed that quorum sensing inhibitors could reduce biofouling by bacteria and diatoms. Specifically, they incubated glass slides in containers filled with seawater for one week. On glass slides exposed to inhibitors (330 mM and 1 mM kojic acid) they observed less bacteria and diatoms adhesion. In addition, Golberg et al. (2013) reported the isolation of bacterial strains from corrals that were able to interfere with AHLs-mediated signaling. Despite the abundance of quorum quenching strains isolated from marine environments, there is lack of studies showing the performance of quorum quenching in aquaculture biofouling. This is probably due to the difficulty in establishing a proper quorum quenching method in an open system such as fisheries, in contrast with relatively closed systems such as MBR systems. However, by the use of a proper carrier or immobilization method, such as for the QQ strains in MBR reactors, there is a great potential for the utilization of quorum quenching as an efficient technique to mitigate biofouling.

Biofouling can occur in metal pipes and cooling systems, which are part of many industrial plants. In water treatment and transportation networks, biofouling can lead to reduction in flow capacity of water-transfer systems up to 10% (Perkins et al. 2014). In cooling systems, biofouling is a serious problem: it causes clogging and increases energy consumption, through decreasing heat transfer. Furthermore, mainly in wastewater systems, biofilm formation is a prior step to a more destructive process, the microbial induced corrosion (MIC) (Esnault et al. 2011). The main bacteria in the MIC process are the sulphate-reducing bacteria (SRB). The SRB are able to reduce sulphite, thiosulphate, and sulfur to hydrogen sulphide (H_2S). Phylogenetically the most abundant group of SRB belongs to *Delta proteobacteria* (Scarascia et al. 2012). Regarding the SRB biofilm, the rate of corrosion increases up to 10,000 times, in comparison with the planktonic state. Therefore, MIC is a process influenced by biofouling. Since *Delta proteobacteria* represent the most abundant group of SRB, they use mainly AHLs to regulate biofilm formation. It is worth mentioning that with the use of QQ enzymes, inhibitors, or QQ strain, the biofilm formation of those bacteria could be prevented, and prevention of biofilm could lead to minimization of MIC. However, similar to the aquaculture fouling, there is not a reported application of quorum quenching in such systems. This is probably mainly because the establishment of this method as an efficient method, in those systems, is more difficult than a membrane bioreactor. For example, there is no uniform mixing as activated sludge in most of the systems (metal pipes, cooling systems), so there is a requirement for different methods of immobilization of bacteria with other types of encapsulation media. However, there is potential that in the future, quorum quenching could be established as a potential strategy to prevent biofouling and corrosion caused by the corresponding inducing microorganisms in metal pipes and cooling systems. Since those infrastructures are common in all industrial facilities, quorum quenching could be applied and reduce the costs of operation and maintenance in these industries.

24.4 Final Remarks

The use of enzymes, inhibitors, and microorganisms that interfere with quorum sensing has been studied mainly in activated sludge membrane bioreactor systems treating municipal wastewaters, concluding that quorum quenching is a promising method to reduce membrane biofouling. Several bacterial strains degrading AHLs and one strain that interferes with AI-2 communication have been isolated during recent years. While most of the studies' results show successful mitigation of biofouling, further research is necessary to elucidate the impact on the quality parameters or focused on finding novel QQ strains that would show greater performance. In addition to MBRs, the QQ-methods have been applied only in a few but critical cases; however, they could be potentially utilized in several other applications such as in aquacultures, metal pipes, and cooling systems, minimizing the biofilm formation and preventing other effects caused by biofilm formation, microbial induced corrosion.

REFERENCES

Banti, D.C., Karayannakidis, P.D., Samaras, P., Mitrakas, M.G., 2017. "An Innovative Bioreactor Set-Up That Reduces Membrane Fouling by Adjusting the Filamentous Bacterial Population." *Journal of Membrane Science*, *542*: 430–438. doi:10.1016/j.memsci.2017.08.034.

Banti, D.C., Samaras, P., Tsioptsias, C., Zouboulis, A., Mitrakas, M., 2018. "Mechanism of SMP Aggregation within the Pores of Hydrophilic and Hydrophobic MBR Membranes and Aggregates Detachment." *Separation and Purification Technology*, 202: 119–129. doi:10.1016/j.seppur.2018.03.045.

Bjarnsholt, T., Jensen, P.Ø., Rasmussen, T.B., Christophersen, L., Calum, H., Hentzer, M., Hougen, H.P., 2005. "Garlic Blocks Quorum Sensing and Promotes Rapid Clearing of Pulmonary Pseudomonas Aeruginosa Infections." *Microbiology* 151 (12): 3873–3880. doi:10.1099/mic.0.27955-0.

Case, R.J., Labbate, M., Kjelleberg, S., 2008. "AHL-Driven Quorum-Sensing Circuits: Their Frequency and Function among the Proteobacteria." *ISME Journal* 2 (4): 345–349. doi:10.1038/ismej.2008.13.

Chan, K.G., Yin, W.F., Sam, C.K., Koh, C.L., 2009. A Novel Medium for the Isolation of N-acylhomoserine Lactone-Degrading Bacteria. *Journal of Industrial Microbiology and Biotechnology* 36 (2): 247–251. doi:10.1007/s10295-008-0491-x.x.

Cheong, W.S., Kim, S.R., Oh, H.S., Lee, S.H., Yeon, K.M., Lee, C.H., Lee, J.K., 2014. "Design of Quorum Quenching Microbial Vessel to Enhance Cell Viability for Biofouling Control in Membrane Bioreactor." *Journal of Microbiology and Biotechnology* 24 (1): 97–105. doi:10.4014/jmb.1311.11008.

Cheong, W.-S., Lee, C.-H., Moon, Y.-H., et al., 2013. "Isolation and Identification of Indigenous Quorum Quenching Bacteria, *Pseudomonas* sp. 1A1, for Biofouling Control in MBR." *Industrial and. Engineering Chemical Research.* 52 (31): 10554–10560. doi: 10.1021/ie303146f.

Choo, J.H., Rukayadi, Y., Hwang, J.K., 2006. "Inhibition of Bacterial Quorum Sensing by Vanilla Extract." *Letters in Applied Microbiology* 42 (6): 637–641. doi:10.1111/j.1472-765X.2006.01928.x.

Daims, H., Taylor, M.W., Wagner, M., 2006. "Wastewater Treatment: A Model System for Microbial Ecology." *Trends in Biotechnology* 24 (11): 483–489. doi:10.1093/ejcts/ezr101.

Deblonde, T., Cossu-Leguille, C., Hartemann, P., 2011. "Emerging Pollutants in Wastewater: A Review of the Literature." *International Journal of Hygiene and Environmental Health* 214 (6): 442–448. doi:10.1016/j.ijheh.2011.08.002.

Dobretsov, S., Teplitski, M., Bayer, M., Gunasekera, S., Proksch, P., Paul, V.J., 2011. "Inhibition of Marine Biofouling by Bacterial Quorum Sensing Inhibitors." *Biofouling* 27 (8): 893–905. doi: 10.1080/08927014.2011.609616.

Dong, Y.H., Zhang, X.F., Xu, J.L., Zhang, L.H., 2004. "Insecticidal *Bacillus thuringiensis* Silences Erwinia Carotovora Virulence by a New Form of Microbial Antagonism, Signal Interference." *Applied and Environmental Microbiology* 70 (2): 954–960. doi:10.1128/AEM.70.2.954-960.2004.

Ergön-Can, T., Köse-Mutlu, B., Koyuncu, I., Lee, C.H., 2017. "Biofouling Control Based on Bacterial Quorum Quenching with a New Application: Rotary Microbial Carrier Frame." *Journal of Membrane Science* 525: 116–124. doi:10.1016/j.memsci.2016.10.036.

Fitridge, I., Dempster, T., Guenther, J., de Nys, R., 2012. "The Impact and Control of Biofouling in Marine Aquaculture: A Review." *Biofouling* 28 (7): 649–669. doi:10.1080/08927014.2012.700478.

Flemming, H.C., Wingender, J., 2010. "The Biofilm Matrix." *Nature Reviews Microbiology* 8 (9): 623–633. doi:10.1038/nrmicro2415.

Givskov, M., Nys, R.De., Manefield, M., Gram, L., Maximilien, R.I.A., Eberl, L.E.O., Molin, S., Steinberg, P.D., Kjelleberg, S., 1996. "Eukaryotic Interference with Homoserine Lactone-Mediated Prokaryotic Signalling." *Journal of Bacteriology* 178 (22): 6618–6622. doi:10.1128/jb.178.22.6618-6622.1996.

Gkotsis, P., Banti, D., Peleka, E., Zouboulis, A., Samaras, P., 2014. "Fouling Issues in Membrane Bioreactors (MBRs) for Wastewater Treatment: Major Mechanisms, Prevention and Control Strategies." *Processes* 2 (4): 795–866. doi:10.3390/pr2040795.

Golberg, K., Pavlov, V., Marks, R.S., Kushmaro, A., 2013. "Coral-Associated Bacteria, Quorum Sensing Disrupters, and the Regulation of Biofouling." *Biofouling* 29 (6): 669–682. doi:10.1080/08927014.2013.796939.

Guardiola, F.A., Cuesta, A., Meseguer, J., Esteban, M.A., 2012. "Risks of Using Antifouling Biocides in Aquaculture." *International Journal of Molecular Sciences* 13 (2): 1541–1560. doi:10.3390/ijms13021541.

Gül, B.Y., Koyuncu, I., 2017. "Assessment of New Environmental Quorum Quenching Bacteria as a Solution for Membrane Biofouling." *Process Biochemistry* 61: 137–146. doi:10.1016/j.procbio.2017.05.030.

Gül, B.Y., Imer, D.Y., Park, P.-K., Koyuncu, I., 2018. "Selection of quorum quenching (qq) bacteria for membrane biofouling control: effect of different gram-staining qq bacteria, *bacillus sp. T5* and *delftia sp. T6*, on microbial population in membrane bioreactors." *Water Science Technology* 78 (2): 358–366. doi:10.2166/wst.2018.305.

Hentzer, M., Riedel, K., Rasmussen, T.B., Heydorn, A., Andersen, J.B., Parsek, M.R., Rice, S.A., et al. 2002. "Inhibition of Quorum Sensing in *Pseudomonas aeruginosa* Biofilm Bacteria by a Halogenated Furanone Compound." *Microbiology* 148: 87–102. doi:10.1099/00221287-148-1-87.

Ivanova, K., Fernandes, M.M., Francesko, A., Mendoza, E., Guezguez, J., Burnet, M., Tzanov, T., 2015. "Quorum-Quenching and Matrix-Degrading Enzymes in Multilayer Coatings Synergistically Prevent Bacterial Biofilm Formation on Urinary Catheters." *Applied Materials and Interfaces* 7 (49): 27066–27077. doi:10.1021/acsami.5b09489.

Jahangir, D., Oh, H.S., Kim, S.R., Park, P.K., Lee, C.H., Lee, J.K., 2012. "Specific Location of Encapsulated Quorum Quenching Bacteria for Biofouling Control in an External Submerged Membrane Bioreactor." *Journal of Membrane Science* 411–412: 130–136. doi:10.1016/j.memsci.2012.04.022.

Jiang, W., Xia, S., Liang, J., Zhang, Z., Hermanowicz, S.W., 2013. "Effect of Quorum Quenching on the Reactor Performance, Biofouling and Biomass Characteristics in Membrane Bioreactors." *Water Research* 47 (1): 187–196. doi:10.1016/j.watres.2012.09.050.

Kalia, V.C., Wood, T.K., Kumar, P., 2013. "Evolution of Resistance to Quorum-Sensing Inhibitors." *Microbial Ecology* 68(1): 13–23. doi: 10.1007/s00248-013-0316-y.

Kampouris, I.D., Karayannakidis, P.D., Banti, D.C., Sakoula, D., Konstantinidis, D., Yiangou, M., Samaras, P.E., 2018. "Evaluation of a Novel Quorum Quenching Strain for MBR Biofouling Mitigation." *Water Research* 143: 56–65. doi:10.1016/j.watres.2018.06.030.

Kappachery, S., Paul, D., Yoon, J., Kweon, J.H., 2010. "Vanillin, a Potential Agent to Prevent Biofouling of Reverse Osmosis Membrane." *Biofouling* 26 (6): 667–672. doi:10.1080/08927014.2010.506573.

Khan, R., Shen, F., Khan, K., 2016. "Biofouling Control in a Membrane Filtration System by a Newly Isolated Novel Quorum Quenching Bacterium, *Bacillus methylotrophicus* sp." Wy. RSC *Advances* 6 (34), 28895–28903. doi:10.1039/x0xx00000x.

Kim, H.W., Hyun, S.O., Kim, S.R., Lee, K.B., Yeon, K.M., Lee, C.H., Kim, S., Lee, J.K., 2013a. "Microbial Population Dynamics and Proteomics in Membrane Bioreactors with Enzymatic Quorum Quenching." *Applied Microbiology and Biotechnology* 97 (10): 4665–4675. doi:10.1007/s00253-012-4272-0.

Kim, J.H., Choi, D.C., Yeon, K.M., Kim, S.R., Lee, C.H., 2011. "Enzyme-Immobilized Nanofiltration Membrane to Mitigate Biofouling Based on Quorum Quenching." *Environmental Science and Technology* 45 (4): 1601–1607. doi:10.1021/es103483j.

Kim, S.R., Lee, K.B., Kim, J.E., Won, Y.J., Yeon, K.M., Lee, C.H., Lim, D.J., 2015. "Macroencapsulation of Quorum Quenching Bacteria by Polymeric Membrane Layer and Its Application to MBR for Biofouling Control." *Journal of Membrane Science* 473: 109–117. doi:10.1016/j.memsci.2014.09.009.

Kim, S.R., Oh, H.S., Jo, S.J., Yeon, K.M., Lee, C.H., Lim, D.J., Lee, C.H., Lee, J.K., 2013b. "Biofouling Control with Bead-Entrapped Quorum Quenching Bacteria in Membrane Bioreactors: Physical and Biological Effects." *Environmental Science and Technology* 47 (2): 836–842. doi:10.1021/es303995s.

Kim, T.H., Lee, I., Yeon, K.-M., Kim J., 2018. "Biocatalytic Membrane with Acylase Stabilized on Intact Carbon Nanotubes for Effective Antifouling via Quorum Quenching." *Journal of Membrane Science* 554: 357–365. doi: 10.1016/j.memsci.2018.03.020

Klahre, J., Flemming, H.C., 2000. "Monitoring of Biofouling in Papermill Process Waters." *Water Research* 34 (14): 3657–3665. doi:10.1016/S0043-1354(00)00094-4.

Lade, H., Paul, D., Kweon, J.H., 2014. "Quorum Quenching Mediated Approaches for Control of Membrane Biofouling," *International Journal of Biological Science* 10: 547. doi:10.7150/ijbs.9028

Lai, B.M., Zhang, K., Shen, D.S., Wang, M.Z., Shentu, J.L., Li, N., 2017. "Control of the Pollution of Antibiotic Resistance Genes in Soils by Quorum Sensing Inhibition." *Environmental Science and Pollution Research* 24 (6): 5259–5267. doi:10.1007/s11356-016-8260-2.

Le-Clech, P., Chen, V., Fane, T.A.G., 2006. Fouling in Membrane Bioreactors Used in Wastewater Treatment." *Journal of Membrane Science* 284 (1–2): 17–53. doi:10.1016/j.memsci.2006.08.019.

Lee, K., Kim, Y.W., Lee, S., Lee, S.H., Nahm, C.H., Kwon, H., Park, P.K., Choo, K.H., Koyuncu, I., Drews, A., Lee, C.H., Lee, J.K., 2018a. "Stopping Autoinducer-2 Chatter by Means of an Indigenous Bacterium (Acinetobacter Sp. DKY-1): A New Antibiofouling Strategy in a Membrane Bioreactor for Wastewater Treatment." *Environmental Science and Technology* 52 (11): 6237–6245. doi:10.1021/acs.est.7b05824.

Lee, K., Yu, H., Zhang, X., Choo, K.H., 2018b. "Quorum Sensing and Quenching in Membrane Bioreactors: Opportunities and Challenges for Biofouling Control." *Bioresource technology* 270: 656–668. doi: 10.1016/j.biortech.2018.09.019.

Lee, K., Lee, S., Lee, S.H., Kim, S.R., Oh, H.S., Park, P.K., Choo, K.H., Kim, Y.W., Lee, J.K., Lee, C.H., 2016a. "Fungal Quorum Quenching: A Paradigm Shift for Energy Savings in Membrane Bioreactor (MBR) for Wastewater Treatment." *Environmental Science and Technology* 50 (20): 10914–10922. doi:10.1021/acs.est.6b00313.

Lee, S., Lee, S.H., Lee, K., Kwon, H., Nahm, C.H., Lee, C.H., Park, P.K., Choo, K.H., Lee, J.K., Oh, H.S., 2016b. "Effect of the Shape and Size of Quorum-Quenching Media on Biofouling Control in Membrane Bioreactors for Wastewater Treatment." *Journal of Microbiology and Biotechnology* 26 (10): 1746–1754. doi:10.4014/jmb.1605.05021.

Lee, S., Park, S.K., Kwon, H., Lee, S.H., Lee, K., Nahm, C.H., Jo, S.J., Oh, H.S., Park, P.K., Choo, K.H., Lee, C.H., Yi, T., 2016c. "Crossing the Border between Laboratory and Field: Bacterial Quorum Quenching for Anti-Biofouling Strategy in an MBR." *Environmental Science and Technology* 50 (4): 1788–1795. doi:10.1021/acs.est.5b04795.

Malaeb, L., Le-Clech, P., Vrouwenvelder, J.S., Ayoub, G.M., Saikaly, P.E., 2013. "Do Biological-Based Strategies Hold Promise to Biofouling Control in MBRs?" *Water Research* 47 (15): 5447–5463. doi:10.1016/j.watres.2013.06.033.

Manaia, Célia M., Jaqueline Rocha, NazarenoScaccia, Roberto Marano, Elena Radu, Francesco Biancullo, Francisco Cerqueira, et al. 2018. "Antibiotic Resistance in Wastewater Treatment Plants: Tackling the Black Box." *Environment International* 115: 312–324. doi:10.1016/J.ENVINT.2018.03.044.

Miura, Y., Watanabe, Y., Okabe, S., 2007. "Membrane biofouling in pilot-scale membrane bioreactors (MBRs) treating municipal wastewater: impact of biofilm formation." *Environmental Science & Technology* 41 (2): 632–638. doi:10.1021/es0615371.

Nahm, C.H., Choi, D.-C., Kwon, H., et al., 2017. "Application of Quorum Quenching Bacteria Entrapping Sheets to Enhance Biofouling Control in a Membrane Bioreactor with a Hollow Fiber Module." *Journal of Membrane Science* 526: 264–271. doi: 10.1016/j.memsci.2016.12.046

Nam, A., Kweon, J., Ryu, J., Lade, H., Lee, C., 2015. "Reduction of Biofouling Using Vanillin as a Quorum Sensing Inhibitory Agent in Membrane Bioreactors for Wastewater Treatment." *Membrane Water Treatment* 6 (3): 189–203. doi:10.1109/OCEANSKOBE.2008.4531093.

Ochiai, S., Morohoshi, T., Kurabeishi, A., Shinozaki, M., Fujita, H., SAWADA, I., Ikeda, T. 2013. "Production and Degradation of Nacylhomoserine Lactone Quorum Sensing Signal Molecules in Bacteria Isolated from Activated Sludge." *Biosci. Biotechnol. Biochem.* 77 (12): 2436–2440.

Oh, H.S., Tan, C.H., Low, J.H., Rzechowicz, M., Siddiqui, M.F., Winters, H., Kjelleberg, S., Fane, A.G., Rice, S.A., 2017. "Quorum Quenching Bacteria Can Be Used to Inhibit the Biofouling of Reverse Osmosis Membranes." *Water Research* 112. doi:10.1016/j.watres.2017.01.028.

Oh, H.S., Yeon, K.M., Yang, C.S., Kim, S.R., Lee, C.H., Park, S.Y., Han, J.Y., Lee, J.K., 2012. "Control of Membrane Biofouling in MBR for Wastewater Treatment by Quorum Quenching Bacteria Encapsulated in Microporous Membrane." *Environmental Science and Technology* 46 (9): 4877–4884. doi:10.1021/es204312u.

Papenfort, K., Bassler, B.L., 2016. "Quorum Sensing Signal-Response Systems in Gram-Negative Bacteria." *Nature Reviews Microbiology* 14 (9): 576–588. doi:10.1038/nrmicro.2016.89.

Perkins, S.C.T., Henderson, A.D., Walker, J.M., Sargison, J.E., Li, X.L., 2014. "The Influence of Bacteria-Based Biofouling on the Wall Friction and Velocity Distribution of Hydropower Pipes." *Australian Journal of Mechanical Engineering* 12 (1): 77–88. doi:10.7158/M12-087.2014.12.1.

Polman, H., Verhaart, F., Bruijs, M., 2013. "Impact of Biofouling in Intake Pipes on the Hydraulics and Efficiency of Pumping Capacity." *Desalination and Water Treatment* 51 (4–6): 997–1003. doi:10.1080/19443994.2012.707371.

Ponnusamy, K., Paul, D., Kweon, J.H., 2009. "Inhibition of Quorum Sensing Mechanism." *Environmental Engineering Science* 26 (8): 1359–1363. doi:10.1371/journal.pone.0173179.

Rasmussen, T.B., Givskov, M., 2006. "Quorum Sensing Inhibitors: A Bargain of Effects." *Microbiology* 152: 895–904. doi:10.1099/mic.0.28601-0.

Scarascia, G., Wang, T., Hong, P.-Y., 2016. "Quorum Sensing and the Use of Quorum Quenchers as Natural Biocides to Inhibit Sulfate-Reducing Bacteria." *Antibiotics* 5 (4): 39. doi:10.3390/antibiotics5040039.

Schultz, M.P., Bendick, J.A., Holm, E.R., Hertel, W.M., 2011. "Inhibition of Marine Biofouling by Bacterial Quorum Sensing Inhibitors." *Biofouling* 27 (8): 893–905. doi:10.1080/08927014.2011.609616.

Shrout, J.D., Nerenberg, R., 2012. "Monitoring Bacterial Twitter: Does Quorum Sensing Determine the Behavior of Water and Wastewater Treatment Biofilms?" *Environmental Science and Technology* 46 (4): 1995–2005. doi:10.1021/es203933h.

Siddiqui, M.F., L. Singh, A.W. Zularisam, M. Sakinah. 2013. "Biofouling Mitigation Using Piper Betle Extract in Ultrafiltration MBR." *Desalination and Water Treatment* 51 (37–39): 6940–6951. doi:10.1080/19443994.2013.793477.

Siddiqui, M.F., Rzechowicz, M., Harvey, W., Zularisam, A.W., Anthony, G.F. 2015. "Quorum Sensing Based Membrane Biofouling Control for Water Treatment: A Review." *Journal of Water Process Engineering*, 7: 112–122. doi: 10.1016/j.jwpe.2015.06.003.

Siddiqui, Muhammad Faisal, Mimi Sakinah, Lakhveer Singh, A.W. Zularisam. 2012. "Targeting N-Acyl-Homoserine-Lactones to Mitigate Membrane Biofouling Based on Quorum Sensing Using a Biofouling Reducer." *Journal of Biotechnology* 161: 190–197. doi:10.1016/j.jbiotec.2012.06.029.

Tse, L.K., Takada, K., Jiang, S.C., 2018. "Surface Modification of a Microfiltration Membrane for Enhanced Anti-Biofouling Capability in Wastewater Treatment Process." *Journal of Water Process Engineering* 26: 55–61. doi:10.1016/j.jwpe.2018.09.006.

Uroz, S., Dessaux, Y., Oger, P., 2009. "Quorum Sensing and Quorum Quenching: The Yin and Yang of Bacterial Communication." *ChemBioChem* 10 (2): 205–216. doi:10.1002/cbic.200800521.

Vanysacker, Louise, Bart Boerjan, Priscilla Declerck, Ivo F.J. Vankelecom. 2014. "Biofouling Ecology as a Means to Better Understand Membrane Biofouling." *Applied Microbiology and Biotechnology* 98 (19): 8047–8072. doi:10.1007/s00253-014-5921-2.

Williams, P., Camara, M., Hardman, A., Swift, S., Milton, D., Hope, V.J., Winzer, K., Middleton, B., Pritchard, D.I., Bycroft, B.W., 2000. "Quorum Sensing and the Population-Dependent Control of Virulence." *Philosophical Transactions of the Royal Society B: Biological Sciences* 355 (1397): 667–680. doi:10.1098/rstb.2000.0607.

Williams, P., Winzer, K., Chan, W.C., Cámara, M., 2007. "Look Who's Talking: Communication and Quorum Sensing in the Bacterial World." *Philosophical Transactions of the Royal Society B: Biological Sciences* 362 (1483): 1119–1134. doi:10.1098/rstb.2007.2039.

Xia, S., Li, J., He, S., Xie, K., Wang, X., Zhang, Y., Duan, L., Zhang, Z., 2010. "The effect of organic loading on bacterial community composition of membrane biofilms in a submerged polyvinyl chloride membrane bioreactor." *Bioresource Technology* 101 (17): 6601–6609. doi:10.1016/j.biortech.2010.03.082.

Yang, F., Wang, L.H., Wang, J., Dong, Y.H., Hu, J.Y., Zhang, L.H., 2005 "Quorum Quenching Enzyme Activity Is Widely Conserved in the Sera of Mammalian Species." *FEBS Letters* 579 (17): 3713–3717. doi:10.1016/j.jpcs.2010.01.013.

Yeon, K.M., Lee, C.H., Kim, J., 2009a. "Quorum Sensing: A New Biofouling Control Paradigm in a Membrane Bioreactor for Advanced Wastewater Treatment." *Environmental Science and Technology* 43 (2): 380–385. doi:10.1021/es8019275.

Yeon, K.M., Lee, C.H., Kim, J., 2009b. "Magnetic enzyme carrier for effective biofouling control in the membrane bioreactor based on enzymatic quorum quenching." *Environmental science and Technology* 43 (19): 7403–7409. doi:10.1021/es901323k.

Yu, H., Liang, H., Qu, F., He, J., Xu, G., Hu, H., Li, G., 2016. "Biofouling Control by Biostimulation of Quorum-Quenching Bacteria in a Membrane Bioreactor for Wastewater Treatment." *Biotechnology and Bioengineering* 113 (12): 2624–2632. doi:10.1002/bit.26039.

Yu, H., Qu, F., Zhang, X., Wang, P., Li, G., Liang, H., 2018. "Effect of quorum quenching on biofouling and ammonia removal in membrane bioreactor under stressful conditions." *Chemosphere* 199: 114–121. doi:10.1016/j.chemosphere.2018.02.022.

Zhu, J., Winans, S.C., 1998. "Activity of the quorum-sensing regulator *TraR* of Agrobacterium *tumefaciens* is inhibited by a truncated, dominant defective TraR-like protein." *Molecular Microbiology* 27 (2): 289–297. doi:10.1046/j.1365-2958.1998.00672.x.

25

Application of Quorum Sensing Inhibitors in Anti-biofouling Membranes

Sunny C. Jiang and Leda K. Tse

CONTENTS

25.1 Introduction

Rapid advancements of membrane technology have enabled the production of clean water from nontraditional water resources, including seawater and municipal sewage, to meet the ever-increasing human demand of water. Desalination of seawater for potable uses has a long history and has been regarded as an effective way to alleviate freshwater scarcity and ease water stress in arid regions [1]. Desalination processes separate dissolved salts and minerals from seawater and brackish water to obtain water suitable for human and animal consumption, irrigation, or other industrial uses [2]. Various desalination technologies have been developed over the past three decades including distillation-based thermal processes such as multistage flash/multi-effect distillation and vapor compression distillation, and membrane-based processes including electrodialysis, reverse osmosis (RO), and nanofiltration. RO desalination is, at present, the most commonly used method for the production of freshwater from seawater and other saline water sources due to the advancement of thin-film composite membranes. To date, over 15,000 desalination plants have been installed worldwide with approximately 50% based on RO membrane technology [1]. It is likely that the use of RO desalination will continue to expand, due to its relatively lower energy cost and simplicity.

Beyond desalination of seawater, municipal wastewater is now reclaimed for agricultural, industrial, and even potable reuse in arid regions. For example, Israel uses ~84% treated wastewater in agricultural production [3]. Orange County Sanitation District in Southern California reclaims 100 million gallons per day of municipal wastewater to replenish the groundwater reservoir as a potable water resource [4]. Reclamation of wastewater can serve as a key long-term component of regional water resources because wastewater quantity is reliable and increases with population growth. Membrane-based wastewater treatment technologies, including microfiltration and RO membrane, are increasingly used to treat wastewater for production of high-quality water for reuse [5].

25.2 Membrane Biofouling Is a Key Challenge of Water Purification

One of the critical challenges limiting the effective application of membrane technology for water purification is membrane fouling [6]. Although all water distribution surfaces are subjected to fouling, membrane fouling has the largest impact on water production rate and energy consumption rate. Paul and coworkers surveyed 70 RO membrane installations in the United States [7] and reported "above average" problems with membrane fouling, with biofouling representing the most common operational problem experienced. In the Middle East, where the largest amount of desalted water is produced in the world, it was noted that around 70% of the seawater RO (SWRO) membrane installations suffer from biofouling problems [6].

Membrane fouling is a process where solutes or particles including bacteria in feed water deposit and grow on the membrane surface, which reduces the flux of water and deteriorates the quality of product water. Therefore, in the desalination and water reclamation industry, membrane cleaning is necessary to recover permeate flux or improve salt rejection from fouling [8]. Cleaning processes often use undefined commercial cleaning products, including alkaline solutions, acids, metal chelating agents, and surfactants, which partially restore the performance of the fouled RO membranes. However, cleaning processes can increase operation difficulty, decrease membrane life span, and increase membrane replacement costs [9]. Energy loss due to membrane fouling has been estimated at up to 50% of the total energy required for RO membrane desalination [10].

Many pretreatment technologies have been implemented to reduce membrane fouling. For example, the advances in pretreatment technologies (i.e., microfiltration and ultrafiltration) have significantly reduced the turbidity of pretreated water and inorganic fouling, yet, organic and biofouling continue to plague the desalination industry [10]. Moreover, pretreatment membranes are also subjected to fouling and result in increase of operational cost and reduction of water productivities.

Biofouling is attributed to the attachment, growth of deposited microbial cells, and secretion of bacterial metabolic byproducts (often called extracellular polymeric substance, or EPS) on the membrane surface to form a biofilm [10]. Previous research has found that even when 99.99% (4-log) of all bacteria were eliminated by microfiltration, the surviving cells still adhered to surfaces, multiplied, and created biofilm composing of live and dead cells and EPS to cause significant fouling and deterioration of RO membrane performance. Other pretreatment methods such as oxidant-based disinfection, ultraviolet irradiation, and coagulation followed by granular media filtration can reduce the number of bacteria in the RO feed seawater significantly, but would typically not eliminate biofilm formation on the RO membranes [11,12]. Many studies have concluded that disinfection of feed waters does not necessarily avoid the occurrence of biofouling [13]. In addition to the inefficiency, chemical biocides such as chlorine that have been used for decades to control biofouling in other systems are not applicable to polyamide thin film composite RO membranes due to the membranes' sensitivity to oxidation [14]. Chlorination of feed water followed by de-chlorination before contacting with RO membrane was ineffective in desalination plants [12,15,16]. The chlorination process was suspected to break down large organic molecules to smaller organic molecules that are more bioavailable to support biofilm growth [16].

The focus of biofouling control has more recently moved toward membrane surface modification [17]. The incorporation of antimicrobial nanomaterials (e.g., silver or titanium oxide [TiO_2]) has been studied by several groups. However, several significant drawbacks of such approaches have retarded the development of the novel membranes. One key challenge is to ensure long-term retention of nanoparticles in the membrane structure to preserve the antibacterial activity after the RO membrane is used in operation. By using TiO_2 as the antimicrobial nanomaterial in membranes, there is a need for a continuous UV light source for photocatalysis [18]; meanwhile, a possible degradation of RO membranes (e.g., polyamide layer) may occur due to photocatalysis [17]. Moreover, the unstable incorporation of antimicrobial nanomaterial into RO membranes results in the leakage of nanoparticles to the bulk fluid, which are hazardous to the environment when discharged with the brine reject [17].

25.3 Biofilm Formation Is Due to Quorum Sensing among Bacteria

Much has been discussed in this book regarding quorum sensing. Biofilm development and maturation for both Gram-negative (G−) and Gram-positive (G+) bacteria are regulated by quorum sensing (QS) pathways. During QS, bacteria communicate with each other by synthesizing and secreting signaling molecules that accumulate to a threshold level based on the cell population density [19]. After the threshold is reached, the signaling molecule binds to the appropriate transcription regulator and either activates or represses target genes to trigger biofilm development [19]. Acyl-homoserine lactone (AHL) QS molecules, known as autoinducer 1 (AI-1), are commonly found in G− proteobacteria (Table 25.1). In this system, the LuxI is believed to initiate acyl-HSL synthesis and the LuxR acts as the transcriptional activator, which controls the lux operon directly. With few exceptions, the proteins that synthesize these autoinducers constitute an evolutionarily conserved family of homologues known as the LuxI family of autoinducer synthases [20].

For G+ bacteria, amino acid peptides are the most common QS molecule. These active signaling molecules are generated by cleaving translated proteins to form short peptide chains and are further modified depending on the type of bacteria [30–33]. In addition, a universal QS pathway responsible for inter-species communication in both G− and G+ bacteria has been discovered and named the autoinducer 2 (AI-2). The AI-2 was first identified from rearrangement and spontaneous reaction of 4,5- dihydroxy-2,3-pentanedione (DPD) synthesized by LuxS [33,34]. However, the AI-2 QS pathway needs to be further studied to identify all the AI-2 molecules and the luxS genes responsible for biofilm development in specific bacteria. AI-2 QS is only identified in the limited numbers of bacteria (Table 25.2). Other signaling molecules implicated in bacterial

TABLE 25.1

AHL Signaling System for Biofilm Regulation among Bacterial Isolates

Bacteria Genus	Specific AHL Signaling Molecule	References
Alteromonas	C4-HSL; C6-HSL	[21,22]
Aeromonas	C4-HSL; C6-HSL	[23,24]
Nitrosomonas	C6-HSL; C8-HSL; C10-HSL; 3OH-C6-HSL	[25,26]
Ochrobactrum	C4-HSL; C6-HSL	[21,22]
Pseudomonas	C4-HSL	[27]
Pseudoalteromonas	—	[21]
Shewanella	C4-HSL; C6-HSL	[21,22]
Thalassomonas	—	[21,28]
Vibrio	C4-HSL; C6-HSL; 3OH-C4-HSL	[22,27,29]

TABLE 25.2

AI-2 Signaling System in Marine and Wastewater Bacterial Isolates

Bacteria Genus	References
Bacillus	[38]
Escherichia	[39]
Salmonella	[40]
Vibrio	[41,42]

quorum sensing include oligopeptides and nitric oxide, but they have not been well studied. Among all known QS molecules, the auto-inducer type 1 (AI-1) signaling system is thought to play a significant role in seawater RO membrane biofouling because the dominant bacteria present in the seawater intake and on RO membrane surfaces belong to the G− γ-Proteobacteria class [35–37].

25.4 Quorum Sensing Inhibition as a Strategy for Anti-biofouling

Understanding QS signaling pathways in the formation of biofilm has led to attempts to develop QS inhibition molecules for control of biofouling at a cellular level. Since up to 60% of bacterial species that were actively involved in biofilm formation on RO membranes have been shown to produce QS molecules [43], disruption or inhibition of the QS systems appears to be a promising approach for controlling microbial biofilm, thus reducing biofouling.

Wide ranges of QS inhibitors (QSI) or quorum quenchers that target specific G− or G+ or both types of bacteria have been investigated using model microbial organisms.

QS signal competitors, often called QS quenchers or QSI, consist of a range of natural and synthetic compounds, known to suppress QS pathways by either degrading the signaling molecule, blocking the signal generation (luxI inhibitor), or blocking signal reception (luxR inhibitor).

Laboratory studies have applied QSI to control biofouling by model bacterial strains in bench-scale membrane fouling experiments. For example, Ponnusamy and coworkers [44] found that a furanone, AI-2 antagonist, could suppress the formation of biofilm caused by strains of *Aeromonas hydrophila* isolated from a fouled RO membrane system. A commercially available vanillin (4-hydroxy-3-methoxybenzaldehyde), which is believed to interfere or modify the AHL's structure and to hinder the ability of AHL's ability to bind to the receptor protein [45,46], was shown to be an effective biofilm inhibitor [47]. Other AHL inhibitors, such as acylase I, have been used to effectively inhibit biofouling by a mixed culture of G+ and G− bacteria. Acylase I was also effective to reduce biofouling when immobilized on beads in a submerged membrane bioreactor in a nanofiltration membrane [48,49]. In these experiments, the formation of mature biofilm was prevented due to the QSI activity in reduction of bacterial extracellular polymeric substances (EPS) secretion. The acylase-immobilized nanofiltration membrane maintained more than 90% of its initial enzyme activity for more than 20 repetitive cycles of reaction and washing procedure. In a lab-scale continuous cross flow nanofiltration system operated at a constant pressure, the flux with the QSI-immobilized nanofiltration membrane was maintained over 90% of its initial flux, whereas the unmodified nanofiltration membrane decreased to 60% due to the observed severe biofouling [50]. Gule et al. [51] reported the first investigation on the inhibition of initial adhesion of bacteria using a furanone-modified polymer. They have shown the modified polymers had good antimicrobial and cell-adhesion inhibition for *P. aeruginosa*, *E. coli*, *S. typhimurium*, *S. aureus*, and *K. pneumoniae*, individually and in mixed cultures.

25.5 Application of QS Inhibition in Seawater Desalination

The applicability of QS inhibition for anti-biofouling in SWRO membrane desalination was first investigated by Katebian et al. [52]. Different from previous research that used laboratory strains of bacteria as model organisms, this study isolated a diverse group of bacteria from fouled desalination cartridge filters and SWRO membranes at two desalination plants in California and in Western Australia. The production of QS molecules by these biofouling bacteria were first characterized. They were shown to be low molecular weight AHL, similar to known auto-inducer 1 (AI-1) signaling molecule. AI-1 production in the mixed culture of the four different biofilm-forming marine bacteria was greater than in individual bacterial cultures, suggesting interspecies interactions.

Four potential QSIs were tested for reduction of biofilm produced by the marine bacterial isolates and the indigenous microbial community in natural seawater. The potential QSIs, cinnamaldehyde (CNMA), kojic acid (KJ), vanillin (VA), and a brominated furanone compound (F-30), were selected based on: (1) previous work demonstrating the target QS molecule for biofilm inhibition (AI-1 or AI-2); (2) prior demonstration of biofilm reduction in model bacterial system; (3) commercial availability; (4) nontoxic to humans at lower doses; (5) inexpensive to manufacture. Each QSI was added to the bulk artificial seawater inoculated with marine biofouling bacteria or natural seawater with indigenous bacterial community. The experiments were carried out at accelerated fouling condition, where organic nutrients were added to stimulate biofilm production. The results showed two of the four QSIs, VA and CNMA, significantly reduced biofilm formed by marine bacteria on SWRO using mixed bacterial cultures. Confocal microscopy examination of membrane surfaces further confirmed that VA significantly reduced biofilm EPS and dead cells on the RO membrane surface. The decrease in EPS was further evident using quantitative analysis (by COMTSTAT2), where biomass and thickness were reduced by half in the experiments with QSI treatments. Based on these results, they concluded VA, the AI-1 system inhibitor, consistently and effectively reduced SWRO membrane EPS production and dead cells for both the mixed isolates and native bacterial communities in natural seawater.

The suppressed EPS secretion on SWRO membrane surfaces is likely to offer long-term benefits in preventing permeate flux decline in membrane desalination since EPS has been attributed to reducing efficiency of convectional process [53,54]. Moreover, VA treatment did not significantly change the live cell biomass on RO membrane surfaces, indicating the reduction of EPS was not due the lethal effect of VA on bacteria. The effectiveness of VA was attributed to its ability to interfere with bacterial AHL without the need to diffuse into the cell membrane. The scavenging of AHL in cell suspension effectively reduced the level of AHL and its binding to the transcription regulator to trigger biofilm production. This mechanism was more effective in comparison with QSI that requires diffusion into the cell membrane as a competitor to the auto-inducer binding site [45,46].

CNMA, the second most effective QSI identified in the study using bacterial isolates, lacked the consistency in SWRO membrane biofilm reduction for mixed bacterial culture and native bacterial communities. The efficiency of CNMA to inhibit biofilm formation may rely on its diffusion through the cell membrane to inhibit the AI-1 and AI-2 molecules from binding to the appropriate receptor protein [55,56]. The interactions between CNMA, marine bacteria, and membrane surface are complex and are not yet well understood.

25.6 Incorporation of QSI on SWRO Membranes

The promising outcomes of QSI in biofouling reduction for seawater desalination have motivated the research of SWRO membrane surface modification. A stable incorporation of QSI molecule on membrane polymer has the potential to create *in situ* anti-biofouling membranes. Studies were carried out to attach QSI molecules onto SWRO membrane surface.

The common SWRO membrane is a type of thin-film composite (TFC) polyamide membrane. TFC membranes consist of a top thin-film polyamide layer (0.2 µm thick) supported by a micro-porous polysulfone intermediate layer (40–50 µm thick) and a polyester structural support (120–150 µm) [57]. TFC membranes achieve high permeate flux and salt rejection rates due to the high water permeability of the thin-film layer and strong support of a micro-porous intermediate layer, which withstands high-pressure compaction [57–60]. A stable adsorption of QSI molecules onto polyamide layer is required to maintain the anti-biofouling activity during desalination operation. Moreover, membrane surface properties have significant influences on surface modification.

Two types of commercially available and commonly used seawater grade polyamide RO membranes, Dow Filmtec™ SW30XLE (Midland, Michigan) and Hydranautics SWC5 (Oceanside, California) were tested for incorporation of QSI using physical deposition methods [61–63]. CNMA and VA, the two QSIs that have shown effectiveness against seawater bacterial biofilm in bulk fluid were applied in the physical adsorption experiments using cross-flow membrane cassettes [64]. The incorporation of CNMA and VA onto polyamide surface was investigated using Raman *in vivo* microscopy (Renishaw) to assess QSI attachment and retention on the modified membrane surface. The research showed that both types of SWRO membrane could incorporate CNMA onto polyamide surfaces as indicated by the significant peaks at Raman shift belonged to C-H, C=O, and C=C bonds on the aldehyde and benzene functional groups of CNMA [65,66]. Similarly, VA-SWC5 and VA-SW30XLE membranes shared the same Raman spectra with VA in its natural state (powder) [65,67,68]. A new peak was also observed for VA-RO membranes, suggesting formation of a new bond between VA and the polyamide. Moreover, both VA and CNMA signal peaks were stable during membrane washing and compaction although reduction of the signal peak at the end of the high-pressure RO biofouling experiment was observed on the modified membrane surfaces.

QSI surface-modified SWRO membranes compared well with original virgin membranes in pure water permeate fluxes and the salt rejections. However, there were significant changes in surface hydrophobicity. Before QSI incorporation, the virgin SW30XLE membrane was significantly more hydrophilic than the SWC5 membrane [69,70]. After QSI incorporation, the contact angle of modified SWC5 membranes decreased to become more hydrophilic, while the modified SW30XLE membranes increased to become less hydrophilic. Since CNMA and VA have limited water solubility due to the presence of hydrophobic (i.e., benzene) and hydrophilic (i.e., aldehyde) functional groups, the QSI layer on the RO membrane surface significantly altered the contact angle to reflect the property of either CNMA or VA.

Anti-biofouling experiments using mixed culture of biofilm producing marine bacteria indicated QSI incorporation on both types of SWRO membranes significantly reduced bacterial cell colonization and bacterial EPS. Permeate flux improved in QSI-modified SW30XLE but was not obvious for modified SWC5 membranes. The initial flux rates of both modified membranes were slightly lower in comparison with the original virgin membranes. This is likely because the QSI acts as a foulant to block water transport through the membrane as the QSI adsorbs onto the polyamide SWRO surface. However, QSI modification did not impact the overall membrane performance during the fouling period.

In a high-pressure RO system subjected to harsh biofouling conditions, QSI-modified RO membranes experienced a significant reduction in biofilm coverage on membrane surfaces. QSI-modified RO membranes also exhibited a slow decline of permeate flux compared to virgin membranes.

The mechanism of biofilm reduction for QSI-modified RO membranes could be due to either the QSI ability of CNMA or VA to block bacterial communication or other not yet understood mechanisms that deserve future research. It was also unclear if the reduction of VA and CNMA Raman spectral peak areas on modified membranes after the biofouling experiment was due to detachment of QSI from the polyamide surface or because of the consumption of QSI by bacteria in serving the function of QS inhibition.

Future research should improve methods for QSI attachment on RO surfaces. It is essential to ensure that the QSI activity remains intact. The anti-biofilm capabilities of VA and CNMA rely on the compounds' ability to disrupt bacterial communication by either altering the bacterial QS molecule or reducing the ability of QS molecules to bind to the receptor protein, respectively. A deeper understanding on the QSI's anti-biofilm capabilities and the chemical structure of the QSI is necessary to successfully modify the RO surface.

25.7 QSI in Membrane Bioreactor for Wastewater Treatment

Membrane biofouling is not only limited to RO membrane for seawater desalination. Membrane bioreactors (MBR) that are now widely used in municipal wastewater treatment plants are also plagued by membrane biofouling. MBR technology integrates two processes, activated sludge and low-pressure membrane filtration, into a single system that degrades organic matter and clarifies the effluent [71]. A microfiltration or ultrafiltration membrane is typically used in the low-pressure membrane filtration. MBR removes the need of a secondary sedimentation basin, which greatly decreases the space needed in a densely populated urban area. MBR technology offers a stable effluent water quality, shorter sludge retention times, and potential 30%–50%

smaller energy footprint than the conventional process [72,73]. The chlorinated polyethylene (CPE) microfiltration membrane is a common type of membrane used in MBR. CPE membrane is resistant to acid, alkali, ozone, and chlorine. Physical and chemical processes are commonly used for regeneration of fouled CPE membrane. However, repeated cleanings increase both energy and chemical cost of the water treatment operation.

Attempts were made to chemically link VA, the AI-1 QSI, onto CPE membranes in order to create an *in-situ* anti-biofouling membrane for MBR. However, the chemical stability of CPE prevents direct physical adsorption or chemical reaction between VA and CPE. Intermediate chemical reactions are needed to reactive CPE. Tse et al. (2019) adopted a Williamson ether synthesis reaction to replace chlorine on CPE with VA (Masuda et al., 1993). Potassium carbonate (PC) and N, N-dimethylformamide (DMF) were used as intermediates in the reaction. In this reaction, VA first reacted with PC to produce an alkoxide ion. DMF was used to slow down the reaction rate of the highly reactive alkoxide ion so that it could replace the organohalide bond on CPE [74] to form an ether bond between VA and CPE. The formation of the new ether bound and the attachment of VA structure to CPE membrane were confirmed using Raman *in vivo* microscopy and FTIR spectroscopy analysis. The stability of the VA-modified CPE membrane was assessed under cross-flow conditions and after membrane cleaning using sodium hypochlorite. After 1 hour of sodium hypochlorite cleaning, no significant change of chemical peaks was observed on membrane surfaces based on peak area revealed by Raman spectroscopy, suggesting chemical linked VA on CPE were stable.

The anti-biofouling ability of the VA-CPE membrane was compared with original CPE membrane using activated sludge samples collected from a local wastewater treatment plant. After 48 hours of accelerated fouling experiment, VA-CPE membrane experienced a reduction in live and dead bacteria covering the surface compared to the unmodified. The EPS formation on the membrane surface also followed a similar trend. The COMSTAT 2 analysis further verified the significant reduction ($p < 0.05$ for all) in live and dead cells' biomass and biofilm thickness compared to the unmodified CPE membrane. Overall, the addition of VA on the CPE membrane surface led to a greater resistance toward bacterial colonization and biofilm production from activated sludge compared to unmodified CPE membranes, indicating QSI's potential to reduce membrane biofouling in wastewater treatment plants.

25.8 Future Research Directions

As shown in the few examples of the recent research, understanding and controlling bacterial QS communication pathways at a biological level holds the potential to inhibit biofilm formation and membrane biofouling. However, the practical values of the current research in industrial water purification applications will require future research. For example, administering QSI in the bulk fluid has been shown to be effective at reducing membrane biofilm. During desalination operation, the QSI chemicals could be dosed to the desalination feed water using a side stream at specified intervals during high membrane biofouling seasons (i.e., during spring algal blooms or after heavy rainfall associated with land runoffs). A pilot-scale field study would be necessary to test the dosing rate and timing of the dosing application in coordination with the study of seasonal variability of water quality of the intake water. The dosing of commercially available QSI agent (i.e., VA) into feed water is a relatively simply installation in desalination treatment systems. However, the key drawback of QSI dosing in bulk fluid is the maintenance and the cost associated with the practice. This approach will most likely be a Band-Aid solution to temporarily mitigate biofouling.

Surface modification of existing TFC membrane or developing new membrane materials that directly incorporate QSIs on membrane surface may have the best potential to prevent membrane biofouling in the long-term. Previous studies have demonstrated such potential but future work is needed to incorporate QSI on the TFC layer of the membrane and to expose the active QSI sites to interact with QS molecules. There are two main challenges in QSI membrane modification: (1) the position of the chemical bond between polyamide TFC layer and QSI should not prevent the inhibition reaction of QSI with QS molecules; (2) the QSIs selected should not require diffusion into the cell in order to block QS signaling because once bonded to a membrane surface the diffusion would not be possible. Therefore, an improved understanding of QS pathway and QSI active sites during inhibition reaction is a key for selection of effective QSI for membrane modification. Such knowledge will be critical to design a better approach for chemical modification of membrane surfaces.

Many QSIs have been found in nature and are produced commercially. Only a few QSIs have been explored in membrane biofouling research. More work will be needed to establish the best QSI to use together with membrane materials. The selection of QSI should also be considered for the source of intended feed water quality and potential microbial composition.

REFERENCES

1. Greenlee, L.F., et al., Reverse osmosis desalination: Water sources, technology, and today's challenges. *Water Research*, 2009. **43**(9): p. 2317–2348.
2. Li, D. and H.T. Wang, Recent developments in reverse osmosis desalination membranes. *Journal of Materials Chemistry*, 2010. **20**(22): p. 4551–4566.
3. PDIWA, *The State of Israel: National Water Efficiency Report*. 2011, The State of Israel: Planning Department of the Israeli Water Authority.
4. OCSD, GWRS final expansion—final implementation plan. Project No. SP-173, Effluent reuse study. 2016: https://www.ocwd.com/media/5119/sp-173-vol1.pdf.
5. Tse, L.K., K. Takada, and S.C. Jiang, Surface modification of a microfiltration membrane for enhanced anti-biofouling capability in wastewater treatment process. *Journal of Water Process Engineering*, 2018. **26**: p. 51–61.
6. Gamal Khedr, M., Membrane fouling problems in reverse osmosis desalination plants. *Desalination & Water Reuse*, 2011. **10**(3): p. 8–17.
7. Paul, D.H., Reverse osmosis: Scaling, fouling and chemical attack. *Desalination & Water Reuse*, 1991. **1**: p. 8–11.
8. Ang, W.S., S.Y. Lee, and M. Elimelech, Chemical and physical aspects of cleaning of organic-fouled reverse osmosis membranes. *Journal of Membrane Science*, 2006. **272**(1–2): p. 198–210.

9. Kang, G.D. and Y.M. Cao, Development of antifouling reverse osmosis membranes for water treatment: A review. *Water Research*, 2012. **46**(3): p. 584–600.
10. Matin, A., et al., Biofouling in reverse osmosis membranes for seawater desalination: Phenomena and prevention. *Desalination*, 2011. **281**: p. 1–16.
11. AWWA Research Foundation, Lyonnaise des eaux-Dumez, and S.A.W.R. Commission., *Water Treatment Membrane Processes*. 1996, New York: McGraw-Hill.
12. Al-Abri, M., et al., Chlorination disadvantages and alternative routes for biofouling control in reverse osmosis desalination. *npj Clean Water*, 2019. 2: p. 1–16.
13. Hassan, A.M., et al., Performance evaluation of SWCC SWRO plants. *Desalination*, 1989. **74**(1–3): p. 37–50.
14. Avlonitis, S., W. Hanbury, and T. Hodgkiess, Chlorine degradation of aromatic polyamides. *Desalination*, 1992. **85**(3): p. 321–334.
15. Khan, M., et al., Does chlorination of seawater reverse osmosis membranes control biofouling? *Water Research*, 2015. **78**: p. 84–97.
16. Jiang, S.C. and N. Voutchkov, *Investigation of Desalination Membrane Biofouling*. 2014, Alexandria: Water Reuse Foundation.
17. Lee, K.P., T.C. Arnot, and D. Mattia, A review of reverse osmosis membrane materials for desalination—development to date and future potential. *Journal of Membrane Science*, 2011. **370**(1): p. 1–22.
18. Zhang, H., et al., Tuning photoelectrochemical performances of Ag-TiO_2 nanocomposites via reduction/oxidation of Ag. *Chemistry of Materials*, 2008. **20**(20): p. 6543–6549.
19. Dobretsov, S., et al., The effect of quorum-sensing blockers on the formation of marine microbial communities and larval attachment. *FEMS Microbiology Ecology*, 2007. **60**(2): p. 177–188.
20. Gray, K.M. and J.R. Garey, The evolution of bacterial LuxI and LuxR quorum sensing regulators. *Microbiology*, 2001. **147**(Pt 8): p. 2379–2387.
21. Huang, Y.L., et al., Diversity and acyl-homoserine lactone production among subtidal biofilm-forming bacteria. *Aquatic Microbial Ecology*, 2008. **52**(2): p. 185.
22. Cuadrado-Silva, C.T., et al., Detection of quorum sensing systems of bacteria isolated from fouled marine organisms. *Biochemical Systematics and Ecology*, 2013. **46**(0): p. 101–107.
23. Lynch, M.J., et al., The regulation of biofilm development by quorum sensing in *Aeromonas hydrophila*. *Environmental Microbiology*, 2002. **4**(1): p. 18–28.
24. Swift, S., et al., Quorum sensing in *Aeromonas hydrophila* and *Aeromonas salmonicida*: Identification of the LuxRI homologs AhyRI and AsaRI and their cognate N-acylhomoserine lactone signal molecules. *Journal of Bacteriology*, 1997. **179**(17): p. 5271–5281.
25. Burton, E., et al., Identification of acyl-homoserine lactone signal molecules produced by Nitrosomonas europaea strain Schmidt. *Applied and Environmental Microbiology*, 2005. **71**: p. 4906–4909.
26. Batchelor, S., et al., Cell density-regulated recovery of starved biofilm populations of ammoniaoxidizing bacteria. *Applied and Environmental Microbiology*, 1997. **63**: p. 2281–2286.
27. Waters, C.M. and B.L. Bassler, Quorum sensing: Cell-to-cell communication in bacteria. *Annual Review of Cell and Developmental Biology*, 2005. **21**: p. 319–346.
28. Mohamed, N.M., et al., Diversity and quorum—Sensing signal production of Proteobacteria associated with marine sponges. *Environmental Microbiology*, 2008. **10**(1): p. 75–86.
29. Hammer, B.K. and B.L. Bassler, Quorum sensing controls biofilm formation in *Vibrio cholerae*. *Molecular Microbiology*, 2003. **50**: p. 101–104.
30. Miller, M.B. and B.L. Bassler, Quorum sensing in bacteria. *Annual Review of Microbiology*, 2001. **55**: p. 165–199.
31. Xavier, K.B., et al., Phosphorylation and processing of the quorum-sensing molecule autoinducer-2 in enteric bacteria. *ACS Chemical Biology*, 2007. **2**(2): p. 128–136.
32. Dunny, G.M. and B.A.B. Leonard, Cell-cell communication in gram-positive bacteria. *Annual Review of Microbiology*, 1997. **51**: p. 527–564.
33. Shrout, J.D. and R. Nerenberg, Monitoring bacterial twitter: Does quorum sensing determine the behavior of water and wastewater treatment biofilms? *Environmental Science & Technology*, 2012. **46**(4): p. 1995–2005.
34. Xavier, K.B., et al., Phosphorylation and processing of the quorum-sensing molecule autoinducer-2 in enteric bacteria. *ACS Chemical Biology*, 2007. **2**: p. 128.
35. Lee, J. and I.S. Kim, Microbial community in seawater reverse osmosis and rapid diagnosis of membrane biofouling. *Desalination*, 2011. **273**(1): p. 118–126.
36. Bae, H., et al., Changes in the relative abundance of biofilm-forming bacteria by conventional sand-filtration and microfiltration as pretreatments for seawater reverse osmosis desalination. *Desalination*, 2011. **273**(2–3): p. 258–266.
37. Zhang, M., et al., Composition and variability of biofouling organisms in seawater reverse osmosis desalination plants. *Applied and Environmental Microbiology*, 2011. **77**(13): p. 4390–4398.
38. Pereira, C.S., et al., Identification of functional LsrB-like autoinducer-2 receptors. *Journal of Bacteriology*, 2009. **191**(22): p. 6975–6987.
39. Xavier, K.B. and B.L. Bassler, Regulation of uptake and processing of the quorum-sensing autoinducer AI-2 in *Escherichia coli*. *Journal of Bacteriology*, 2005. **187**(1): p. 238–248.
40. Miller, S.T., et al., *Salmonella typhimurium* recognizes a chemically distinct form of the bacterial quorum-sensing signal AI-2. *Molecular Cell*, 2004. **15**(5): p. 677–687.
41. Barrios, A.F.G., et al., Autoinducer 2 Controls Biofilm Formation in *Escherichia coli* through a novel motility quorum-sensing regulator (MqsR, B3022). *Journal of Bacteriology*, 2006. **188**(1): p. 305–316.
42. Beutler, B., et al., Genetic analysis of host resistance: Toll-like receptor signaling and immunity at large. *Annual Review of Immunology*, 2006. **24**: p. 353–389.
43. Kim, S., et al., Biofouling of reverse osmosis membranes: Microbial quorum sensing and fouling propensity. *Desalination*, 2009. **247**(1–3): p. 303–315.
44. Ponnusamy, K., et al., 2(5h)-Furanone: A prospective strategy for biofouling-control in membrane biofilm bacteria by quorum sensing inhibition. *Brazilian Journal of Microbiology*, 2010. **41**(1): p. 227–234.

45. Ponnusamy, K., D. Paul, and J.H. Kweon, Inhibition of quorum sensing mechanism and *Aeromonas hydrophila* biofilm formation by vanillin. *Environmental Engineering Science*, 2009. **26**(8): p. 1359–1363.
46. Ponnusamy, K., et al., Anti-biofouling property of vanillin on *Aeromonas hydrophila* initial biofilm on various membrane surfaces. *World Journal of Microbiology and Biotechnology*, 2013: p. 1–9.
47. Kappachery, S., et al., Vanillin, a potential agent to prevent biofouling of reverse osmosis membrane. *Biofouling*, 2010. **26**(6): p. 667–672.
48. Yeon, K.M., et al., Quorum sensing: A new biofouling control paradigm in a membrane bioreactor for advanced wastewater treatment. *Environmental Science & Technology*, 2009. **43**(2): p. 380–385.
49. Jiang, W., et al., Effect of quorum quenching on the reactor performance, biofouling and biomass characteristics in membrane bioreactors. *Water Research*, 2013. **47**(1): p. 187–196.
50. Kim, J.H., et al., Enzyme-immobilized nanofiltration membrane to mitigate biofouling based on quorum quenching. *Environmental Science & Technology*, 2011. **45**(4): p. 1601–1607.
51. Gule, N.P., et al., Immobilized Furanone derivatives as inhibitors for adhesion of bacteria on modified poly(styrene-co-maleic anhydride). *Biomacromolecules*, 2012. **13**(10): p. 3138–3150.
52. Katebian, L., et al., Inhibiting quorum sensing pathways to mitigate seawater desalination RO membrane biofouling. *Desalination*, 2016. **393**: p. 135–143.
53. Flemming, H.C., Reverse osmosis membrane biofouling *Experimental Thermal and Fluid Science*, 1997. **14**: p. 382.
54. Fonseca, A.C., et al., Extra-cellular polysaccharides, soluble microbial products, and natural organic matter impact on nanofiltration membranes flux decline. *Environmental Science & Technology*, 2007. **41**(7): p. 2491–2497.
55. Brackman, G., et al., Cinnamaldehyde and cinnamaldehyde derivatives reduce virulence in *Vibrio* spp. by decreasing the DNA-binding activity of the quorum sensing response regulator LuxR. *BMC Microbiology*, 2008. **8**(1): p. 149.
56. Niu, C., S. Afre, and E. Gilbert, Subinhibitory concentrations of cinnamaldehyde interfere with quorum sensing. *Letters in Applied Microbiology*, 2006. **43**(5): p. 489–494.
57. Cadotte, J.E., *Reverse osmosis membrane*. U.S. Patent No. 4,039,4400, 1977.
58. Wang, X.-L., W.-N. Wang, and D.-X. Wang, Experimental investigation on separation performance of nanofiltration membranes for inorganic electrolyte solutions. *Desalination*, 2002. **145**(1): p. 115–122.
59. Abdul Azis, P.K., I. Al-Tisan, and N. Sasikumar, Biofouling potential and environmental factors of seawater at a desalination plant intake. *Desalination*, 2001. **135**(1–3): p. 69–82.
60. Belfort, G., R.H. Davis, and A.L. Zydney, The behavior of suspensions and macromolecular solutions in crossflow microfiltration. *Journal of Membrane Science*, 1994. **96**(1): p. 1–58.
61. Azari, S. and L. Zou, Using zwitterionic amino acid l-DOPA to modify the surface of thin film composite polyamide reverse osmosis membranes to increase their fouling resistance. *Journal of Membrane Science*, 2012. **401**: p. 68–75.
62. Wang, Z.-G., J.-Q. Wang, and Z.-K. Xu, Immobilization of lipase from *Candida rugosa* on electrospun polysulfone nanofibrous membranes by adsorption. *Journal of Molecular Catalysis B: Enzymatic*, 2006. **42**(1): p. 45–51.
63. Hilal, N., M. Khayet, and C.J. Wright, *Membrane Modification: Technology and Applications*. 2012, Boca Raton: CRC Press.
64. Katebian, L., M.R. Hoffmann, and S.C. Jiang, Incorporation of quorum sensing inhibitors onto reverse osmosis membranes for biofouling prevention in seawater desalination. *Environmental Engineering Science*, 2017. **35**: p. 261–269.
65. da Silva, A.M.M., et al., β-Cyclodextrin Complexes of Benzaldehyde, Vanillin and Cinnamaldehyde: A Raman Spectroscopic Study1. *Carbohydrate Chemistry*, 1995. **14**(4–5): p. 677–684.
66. Stewart, D., et al., Fourier-transform infrared and Raman spectroscopic evidence for the incorporation of cinnamaldehydes into the lignin of transgenic tobacco (*Nicotiana tabacum L.*) plants with reduced expression of cinnamyl alcohol dehydrogenase. *Planta*, 1997. **201**(3): p. 311–318.
67. Binoy, J., I.H. Joe, and V. Jayakumar, Changes in the vibrational spectral modes by the nonbonded interactions in the NLO crystal vanillin. *Journal of Raman Spectroscopy*, 2005. **36**(12): p. 1091–1100.
68. Aggarwal, R., et al., Measurement of the Absolute Raman Cross Sections of Diethyl Phthalate, Dimethyl Phthalate, Ethyl Cinnamate, Propylene Carbonate, Tripropyl Phosphate, 1, 3-Cyclohexanedione, 3'-Aminoacetophenone, 3'-Hydroxyacetophenone, Diethyl Acetamidomalonate, Isovanillin, Lactide, Meldrum's Acid, p-Tolyl Sulfoxide, and Vanillin. DTIC Document, 2013.
69. Lee, S., et al., Membrane characterization by dynamic hysteresis: Measurements, mechanisms, and implications for membrane fouling. *Journal of Membrane Science*, 2011. **366**(1): p. 17–24.
70. McCutcheon, J.R. and M. Elimelech, Influence of membrane support layer hydrophobicity on water flux in osmotically driven membrane processes. *Journal of Membrane Science*, 2008. **318**(1): p. 458–466.
71. Radjenović, J., et al., Membrane bioreactor (MBR) as an advanced wastewater treatment technology, in *Emerging Contaminants from Industrial and Municipal Waste*, D. Barcelo, M. Petrović, and M.L. de Alda (eds.). 2008, Berlin: Springer. p. 37–101.
72. Visvanathan, C., R.B. Aim, and K. Parameshwaran, Membrane separation bioreactors for wastewater treatment. *Critical Reviews in Environmental Science and Technology*, 2000. **30**(1): p. 1–48.
73. Ng, H.Y. and S.W. Hermanowicz, Membrane bioreactor operation at short solids retention times: Performance and biomass characteristics. *Water Research*, 2005. **39**(6): p. 981–992.
74. Li, J.J., *Name Reactions: A Collection of Detailed Mechanisms and Synthetic Applications Fifth Edition*. 2014, Berlin: Springer Science & Business Media.

26

Quorum Sensing and Quorum Quenching Based Antifouling Mechanism: A Paradigm Shift for Biofouling Mitigation in a Membrane Bioreactor (MBR)

Shabila Parveen, Sher Jamal Khan, and Imran Hashmi

CONTENTS

26.1 Introduction

Worldwide 2.1 billion people have no access to safely managed drinking water and live in countries experiencing high water stress. Wastewater that flows back untreated into the ecosystem is estimated at 80%. To allow the human and industrial effluent's safe disposal, avoiding danger to human health and the natural environment, wastewater treatment is required. Therefore, there is a pressing need to innovate and improve wastewater treatment techniques for safe wastewater disposal and reuse (United Nations 2018).

Among the wastewater treatment technologies, biological treatment options are cost-effective, which harness the action of microorganisms to decompose the organic contaminants and uptake the nutrients. Among these, membrane bioreactors (MBRs) are particularly appropriate systems to obtain higher quality effluent with less hydrated sludge (Chang et al. 2011, Cicek 2003). MBRs have an extraordinary biodegradation capacity with higher biomass concentration. Besides the several advantages of MBRs, the high cost of membranes and associated membrane fouling hinders the application of MBR for wastewater treatment to a higher extent; it also requires regular cleaning (Lade et al. 2014, Meng et al. 2009).

Biomass and the microbial byproducts are major membrane foulants compared to the other organic and inorganic substances. The substrate in the influent is consumed by microorganisms for growth, maintenance, and generation of extracellular polymeric substances (EPS). These extracellular substances help to stick the microbial flocs and formation of biofilm on the membrane surface and cause membrane biofouling (Lin et al. 2014).

In literature, several operational parameters including hydraulic and sludge retention times (HRT and SRT), and intensity of aeration or concentration of dissolved oxygen (DO) are optimized to reduce the membrane fouling. Even with various available conventional control strategies, the membrane biofouling remains undesirable in MBR application for wastewater treatment. This area, therefore, remains the main concern in the development of the MBRs and their wide-spreading application (Lade et al. 2014, Lewandowski and Beyenal 2005, Zhang et al. 2015a).

Upon the relationship found between quorum sensing (QS) and EPS/biofilm formation, an antifouling strategy called quorum quenching (QQ) has been studied in the last two decades for its potential application in MBR. To control the phenotypic expressions such as biofilm and EPS formation, virulence, motility, and luminescence, a competence that is regulated by QS, several anti-QS techniques are being explored including the addition of enzymes in the solution or immobilized form, and inoculating the QQ species in the MBRs. Controlling QS is considered to be a cost-effective and sustainable option which reduced the membrane cleaning or replacement cost (Siddiqui et al. 2015).

In this chapter, an overview of membrane biofouling and its impact on flux declination and its correlation with QS has been provided. Moreover, it comprehensively reviews the quorum sensing and quenching-based antifouling in MBRs, compared with conventional strategies including operational, physical, and chemical approaches. The opportunities and challenges associated with the implications of QQ-based antifouling in full-scale MBRs are also discussed.

26.2 Membrane Bioreactor

MBR is a state-of-the-art, promising biological wastewater treatment technology to obtain higher quality effluent with less production of waste sludge and smaller footprints (Drews 2010, Lade et al. 2014). Advantages of using MBRs are listed in Table 26.1.

MBR is a compact technology which combines the biological degradation process with membrane filtration. Membranes used for MBR range from microfiltration to nanofiltration. Application of MBRs eliminates the use of clarifier and reduces the hydraulic retention time (HRT) to obtain the effluent, matching the stringent standards. It has indicated higher retention of microbial flocs and suspended solids, demonstrating effectiveness for water reclamation or reuse. Hence, MBR has the potential to replace other conventional treatment units such as sedimentation, sludge thickening, filtration, and flocculation and disinfection (Judd 2010, Le-Clech et al. 2006, Liu and Fang 2003).

26.2.1 Types and Configuration of MBRs

26.2.1.1 Aerobic versus Anaerobic MBRs

The two major groups of MBR are aerobic and anaerobic MBRs. In the aerobic MBR, a membrane filtration unit is coupled with an aerobic bioreactor while in the anaerobic MBR, a membrane filtration unit is integrated with an anaerobic bioreactor (Benitez et al. 1995, Yamamoto and Muang Win 1991).

Aerobic MBR requires aeration systems to supply oxygen to microorganisms in the bioreactor and to scour the membrane surface. In anaerobic MBR, the bioreactor and filtration are both closed units. Therefore, the prime advantages of anaerobic MBR over aerobic MBR is the aeration-energy savings, lower sludge production, and potential biogas recovery (Baek and Pagilla 2006). The disadvantages of anaerobic MBR are the difficulty in retaining slow-growth anaerobic microorganisms with shorter

TABLE 26.1
Advantages of MBR Technology

	Advantages
Permeate quality	High and stable effluent quality, absolute bacterial removal, independent from influent quality due to the buffering effect of high MLSS values, high rate of nitrification
Biodegradation efficiency	Highly efficient, the possibility of growth of specific microorganisms and integration nutrient removal processes
Footprint	Small footprint and reactor requirement, compact design
Operation	Easy and flexible operation, decentralization and satellite technologies are adaptable with MBR, achievable automation
Biomass retention	High
Load volume	High volumetric loading
Sludge production	Low sludge production
Disinfection capacity	Suspended solids concentration is lower, larger particles removal leads to effective disinfection
Water reclamation	High potential

HRT during low-strength wastewater treatment and permeate quality rarely meeting discharge standards (Lin et al. 2013).

26.2.1.2 Submerged versus Side-Stream MBRs

The MBR process configuration is the way in which the membrane unit is combined with the bioreactor. There are two MBR process configurations: submerged or immersed (iMBR) and side-stream MBR (sMBR). There are also various membrane configurations; in practice, five membrane configurations are used: Hollow fiber, spiral-wound, plate-and-frame (flat sheet), pleated filter cartridge, and tubular. The first three types are widely used in MBR while plated filter cartridge and tubular modules are not used widely (Baek and Pagilla 2006) (Figure 26.1).

Side-stream MBR was commercialized in the 1970s, and the submerged MBR appeared in the 1990s. With the emergence of immersed system, MBR installation and total installed flow capacity grew exponentially (Judd 2010).

In submerged MBR, the membrane is present inside the bioreactor. The negative pressure on the permeate side of the membrane is the driving force across the membrane. It simplifies the previous side-stream MBR and reduces power consumption. Lower permeation flux in submerged MBR results from operation at a lower transmembrane pressure (TMP) (Chiemchaisri et al. 1993, Yamamoto et al. 1988).

In side-stream MBR, the mixed liquor is recirculated through the membrane module situated outside the bioreactor. The pressure created by high-cross-flow velocity through the membrane unit is the driving force for the system. The cost associated with recirculation of mixed liquor is high while the benefits include higher effluent fluxes, easier membrane maintenance, and less complicated scale-up (Baek and Pagilla 2006, Trouve et al. 1994).

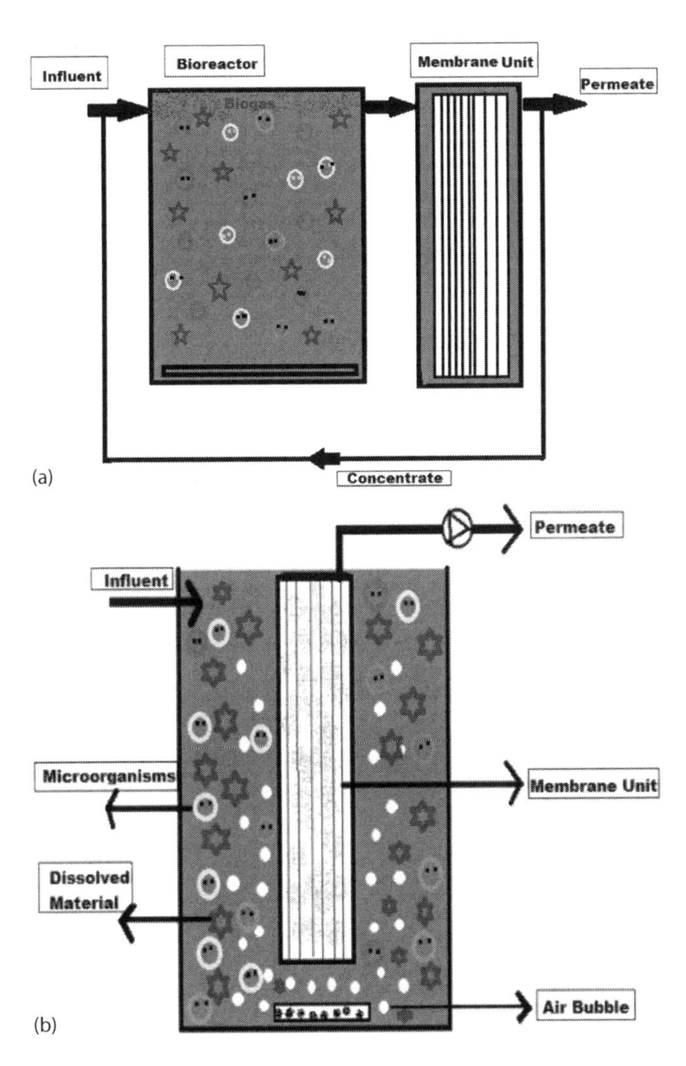

FIGURE 26.1 The interaction of bioreactor and membrane in (a) a side-steam MBRs and (b) submerged MBR.

26.2.2 Application of MBRs

Aerobic MBR is used majorly for treatment of municipal (low organic load), industrial, and textile (intermediate level organic load) wastewater. However, the anaerobic MBR process has focused more on industrial wastewater of high-organic load, for example in alcohol distillery and brewery wastewater treatment (Choo and Lee 1996).

The current application areas of MBRs are municipal and industrial wastewater treatment. However, it has potentially wider application areas including groundwater and drinking water abatement, solid waste digestion, and odor control (Cicek 2003).

Although there are various advantages and potential benefits of MBR, the limitations that hamper MBR's performance and the widespread application include associated cost and membrane fouling.

26.3 Membrane Fouling

A major limitation preventing the widespread use of MBR is membrane fouling. Membrane fouling is the process by which constituents and species present in the water deposit on the membrane surface, clog the pores, and hamper the water flux (Akamatsu et al. 2010, Meng et al. 2009).

26.3.1 Categories of Membrane Fouling

Membrane fouling is classified in various ways. It had been categorized on the basis of nature and types of foulants, and their interaction with the membrane. Membrane fouling is usually determined in terms of different types of resistance.

26.3.1.1 Organic, Inorganic, and Biofouling

Organic, inorganic, and biofouling are the three major categories of membrane fouling based on the chemical nature and type of foulants. Inorganic fouling occurs due to the deposits of inorganic materials like salts, clay, and metal oxides. The deposition of salts form scales on the membrane surface. Organic fouling is the adsorption of organic materials like proteins, oil, grease, polysaccharides, surfactants, humic substances, and biopolymers on the membrane surface. Biofouling involves biofilm formation on the membrane surface upon deposition of compounds, attachment, and growth of microorganisms (Basile et al. 2015, Jiang 2007, Meng et al. 2009).

Among these three major categories, it is the membrane biofouling that persists and is presently an active field of research.

26.3.1.2 Reversible and Irreversible Membrane Biofouling

Reversible membrane biofouling refers to the fouling of the membrane surface that requires physical cleaning such as scrubbing the surface or backwashing through permeate; on the other hand, irreversible membrane fouling is the internal fouling of membrane pores that require chemical cleaning (Basile et al. 2015).

26.4 Membrane Biofouling

Among the various types of membrane fouling, biofouling is most complicated and associated with deposition, growth, and metabolism of the microbial cells on the surface of the membrane and insides of the pores. Microbial metabolites including EPS and SMP are also components of the biofouling. These metabolites, either in bound form (EPS) or soluble form (SMP), produced by microorganisms and cells are considered the drivers of biofouling in MBR. The microbial metabolites are majorly constituted by proteins, polysaccharides, and humic substances. The spatial and temporal characteristics of microbial growth make biofouling a complex process (Afriat et al. 2006, Aslam et al. 2017, Gao et al. 2013, Kochkodan and Hilal 2015, Krzeminski et al. 2017, Laspidou and Rittmann 2002, Le-Clech et al. 2006, Lin et al. 2013, Malaeb et al. 2013, Rosenberger et al. 2006).

26.4.1 Mechanism of Biofouling

Understanding the mechanisms and identifying the responsible components can help to control the fouling. Factors that are commonly reported to affect fouling include the membrane material and its architect, sludge characteristics, and parameters of MBR operation Hence it includes the biological, membrane operation, and design parameters altogether. Constituents of EPS including polysaccharides and proteins are reported to be the major contributors in biofouling. Moreover, components of EPS also form effluent organic matters, potential pollutants in receiving water bodies.

Biofouling starts with the adhesion of microorganism and other biofoulants on the membrane surface. The microorganisms utilize the nutrients and substrate in the wastewater for growth and sustenance. The colonization and aggregate buildup in the pores and on the surface of the membrane forms a biofilm, enclosed in extracellular polymeric substances (EPS) (Flemming 2011).

Biofouling takes place in five stages as shown in Figure 26.2. Following are the stages of biofouling: (1) formation of conditioning film on the surface of the membrane; it comprises organic macromolecules and inorganic compounds, (2) transport of microorganisms to the condition films through diffusion, convection, turbulent, eddy, and chemotaxis; microorganisms

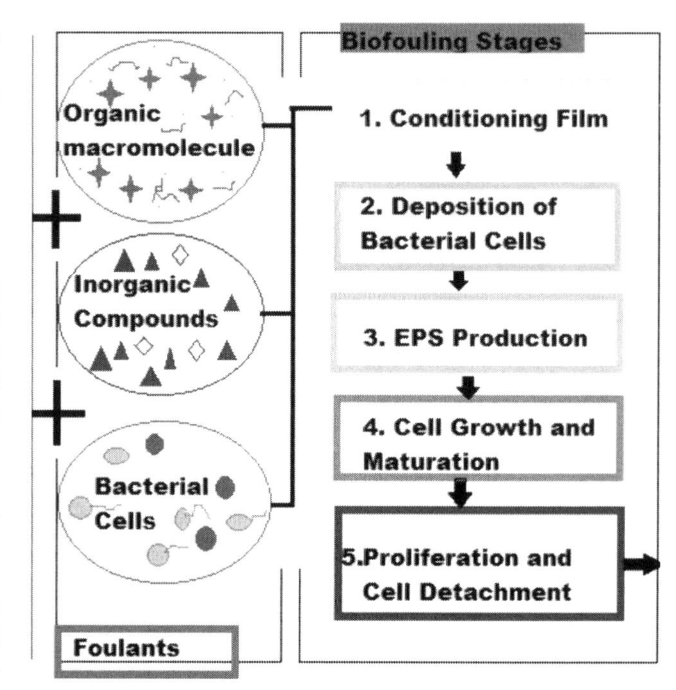

FIGURE 26.2 Mechanism of biofouling in membrane bioreactor.

adhere as results of electrostatic and van der Waals forces, and hydrophobic interactions, (3) continuous attachment of microorganisms to the surface of membrane; EPS adhere them together and form biofilm, (4) maturation of biofilm and cell growth, and (5) three dimensional proliferation of biofilm with few cell detachment (Deng et al. 2016b).

The constituents of conditioning film determine the surface properties of the membrane, playing a vital role in the interaction among microorganisms, biofoulants, and membrane surface. The conditioning film is constituted by organic macromolecules and biofoulants (proteins, polysaccharides, humic substances) and other inorganic compounds. EPS, soluble microbial products (SMP) molecules, and other foulants interact with the membrane surface through adsorption, covalent bond, electrostatic attraction, and thermodynamic and hydrodynamic interactions (Al-Juboori and Yusaf 2012, Hong et al. 2014, Nguyen et al. 2012, Pasmore et al. 2001).

26.4.2 Role of EPS in Membrane Biofouling

26.4.2.1 EPS Matrix

In biological wastewater treatment systems, microorganisms agglomerate and form flocs. These flocs are surrounded by the EPS matrix. The physical functions attributed to EPS are adhesion, aggregation of microbial cells, the cohesion of biofilm or flocs, and retention of water. It provides the stages for the emergence of the properties of microbial flocs or the biofilm. Several other ongoing phenomena in the EPS matrix are (i) provide a habitat, (ii) capture the organic or inorganic compounds from influent, (iii) act as external digestion system by retaining several enzymes, (iv) provide environment for inter-cellular interaction, and (v) tolerance to antimicrobial agents (Flemming et al. 2016).

26.4.2.2 Composition of EPS

The biofilm formation on the membrane surface that causes membrane biofouling is composed of microbes and EPS matrix. It is reported that biofilm is constituted of 15% microbial communities and 85% extracellular polymers (Pandit et al. 2018).

EPSs are extracellular polymeric substances of biological origin that help form to microbial aggregates. These biopolymers are composed of various organic macromolecules present around cells and in microbial aggregates. The composition includes mostly polysaccharides and proteins along with certain amounts of humic acids, nucleic acids, lipids, uronic acids, acetate, pyruvate, phosphate, etc. These components are held together through hydrogen bonds, linkage and multivalent cation (Conrad et al. 2003, Liu and Fang 2003, Nielsen et al. 1996, Tian 2008). Kim et al. (2013b) reported that microorganisms produce more protein and polysaccharides per unit mass of microorganism with biofilm maturation.

26.4.2.3 Classification of EPS

EPSs are categorized as soluble EPS and bound EPS. The bound EPS are further divided into loosely bound (LB) EPS and tightly bound (TB) EPS. Soluble EPSs are considered the same as the soluble microbial products (SMP). LB-EPSs diffuses in the outer layer of the cell while TB-EPSs forms the inner layer. The correlation between LB-EPSs and membrane biofouling is reported to be significant (Jorand et al. 1995, Laspidou and Rittmann 2002, Poxon and Darby 1997, Ramesh et al. 2007) (Figure 26.3).

EPSs are termed as membrane foulants; they adhere to the membrane surface, block membrane pores, contribute to the formation of cake layers on the membrane surface, and induce osmotic effects. They provide structural and functional unity to bacterial biofilm. They facilitate biofilm adherence on the membrane surface, enhance bacterial growth, and protect the biofilm itself. Not only the bound EPS, but SMP also play a major role in membrane biofouling. Composition of both origins depends upon the influent's constituents and operating parameters of MBRs. In literature, MBRs are tried to optimize in a way to reduce the production of extracellular substances. Conventional membrane fouling is insufficient to obtain flux improvement up to the mark.

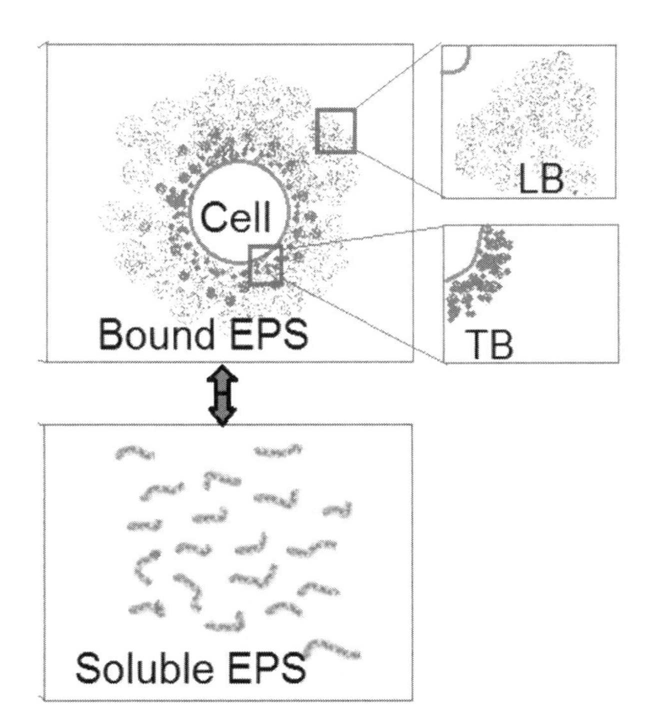

FIGURE 26.3 Extracellular polymeric substances (EPS) structure in terms of loosely bound (LB), tightly bound (TB), and soluble EPS.

26.4.2.4 Formation of EPS

The microbial aggregates or biofilm consist of microbial cells enveloped by EPS. EPS are located at or outside of the microbial cells and help in sticking together these cells (Laspidou and Rittmann 2002). Microorganisms release extracellular substances during all stages of microbial growth, but the composition and extent of concentration vary with the growth phases. Previous studies have reported contradictory results regarding the relationship between the EPS production and growth stages (More et al. 2014).

Carbon and nitrogen are primary elements of EPS. The C/N ratio in the influent significant impact the composition and concentration, but again not any consistently favorable condition to limit the membrane biofouling have been reported in the literature (Ye et al. 2011).

For the past few years, the mechanism of EPS formation have been explored, and a biological phenomenon of cells-cells communications has been found to be responsible for such extracellular secretion (Lin et al. 2014). The working parameters, severe environmental conditions, and presence of toxic substances in the influent substantially disturb the level of communication and led to variate the composition and concentration of extracellular secretion (Hou et al. 2015, Tan et al. 2014). Such dynamics of EPS production are linked to the membrane biofouling (Trinh et al. 2016). This discovery has shifted the paradigm from conventional to biological ways to control the membrane biofouling in MBRs.

26.4.3 Role of Quorum Sensing in Membrane Biofouling

Quorum sensing (QS) has acquired consideration in the past decade due to its role in membrane biofouling. QS is used for the environmental sensing system in which the bacterial species monitor their own population density. The cells-cells communication regulates the collective behavior in bacteria, such as biofilm formation, bioluminescence, and virulence, among others.

The mechanism is driven by signal molecules called autoinducers, such as (1) oligopeptides (5–10 amino acid cyclic thiolactone), (2) *N*-acyl homoserine lactones (AHLs), (3) furanosyl borate (Autoinducer-2, AI-2), (4) hydroxyl-palmitic acid methylester, and (5) methyl dodecanoic acid. Not all microbials use similar signal molecules (Kalia 2013). AHLs and peptide-based are most common in Gram-negative and -positive strains, respectively. The signal molecules concentration is associated to change in gene expression, which link to the production of extracellular secretion (Camilli and Bassler 2006, Shao and Bassler 2012, Waters and Bassler 2005).

There are several reported biofilm-specific taxa that play a role in QS and are involved in biofouling the membrane, including *Enterobacteriales*, *Pseudomonadales*, and *Acinetobacter*. Lim et al. (2012) identified *Enterobacter cancerogenus* ATCC 35316 abundantly in biofilm formed over the membrane surface of MBR.

26.4.3.1 QS Mechanism in Gram-Negative Bacterial Species

AHLs-based QS system is prevalent among Gram-negative bacterial species. AHL produced by each Gram-negative bacteria species is specific to it. The AHL QS system is a single synthase-regulator complex. The mechanism involves synthesizing signal molecules by synthase gene (luxl), which is distributed inside and around the cells; these signal molecules bind to the receptor and activate the transcriptional gene (LuxR). At a low population density or high diffusion rate, AHL are at low concentration and the LuxR (activated at high cell density) receptor is degraded. The receptor is activated when the AHL concentration reaches a specific concentration by forming AHL/LuxR complex (Siddiqui et al. 2015).

A few other signal molecules involved in bacterial QS of Gram-negative species have been discovered, namely 2-(2-hydroxyphenyl)-thiazole-4-carbaldehyde (IQS), cholera autoinducer 1 (CAI-1), and 3,5-dimethylpyrazin-2-ol (DPO). Pseudomonas quinolone signal (PQS) is used by *P. aeruginosa* for QS. It is reported as having a connection with AHL-based QS (Heeb et al. 2011, Lee and Zhang 2015b).

AHLs-based QS is reported to be the leading mechanisms of production of EPS, which could be linked to the biofouling and attracted the researchers to limit the AHL production lesser the critical concentration for gene expression (Mukherjee et al. 2018, Oh et al. 2012, Siddiqui et al. 2012).

26.4.3.2 QS Mechanism in Gram-Positive Bacterial Species

Autoinducing peptide (AIP)-based QS is utilized by Gram-positive bacterial species for cell-to-cell communication. The system in Gram-positive QS bacterial species acts in an analogous fashion; the signal involves AIP. The mechanism involves the production of AIP precursors, post-transcription modification, and secretion through specific transporters. At maturity, AIP concentration increases, and they bind to transmembrane histidine kinases, which get activated and in turn activate the downstream response regulator. This process of activated regulators initiates specific gene transcription (Siddiqui et al. 2015).

26.4.3.3 Other QS Mechanisms in Bacterial Species

Autoinducer-2 (AI-2) signal molecules are a universal language involved in the communication of both Gram-negative and Gram-positive bacterial species. AI-2 are interconverting molecules by nature and are derived from 4,5- dihydroxy-2,3-pentanedione (DPD) (Schauder et al. 2001).

The mechanism involves signal recognition and transduction through the two-component pathway. LuxP and LuxQ constitute the first component and phosphotransferase protein LuxU and LuxO constitute the second component. The two LuxPQ in tetramer form binds with AI-2 and undergo conformational change preventing phosphorylation of the cytoplasmic proteins LuxU and LuxO. The production of LuxR leads to QS (Konai et al. 2018, Wang et al. 2004).

26.4.3.4 QS by Archaeal Species

QS signal molecules regulate the archaeal biofilm formation and the way it coordinates biofilm formation in bacterial communities (Orell et al. 2013). Luo et al. (2015) attempted the characterization of the archaeal community involved in membrane biofouling of MBR. They found that the archaeal community in biofilm was constituted by *Thermoprotei*, *Thermoplasmata*, *Thermococci*, *Methanopyri*, *Methanomicrobia*, and *Halobacteria*.

Fröls et al. (2012) and Koerdt et al. (2010) also demonstrated the potential of Haloarchaea and Crenarchaea species in adherence to surfaces and biofilm formation.

26.5 Membrane Biofouling Control Strategies

Removal of biofilm is difficult; neither natural nor biocide-based microbial control in biofilm has eradicated biofilms from surfaces sustainably. The approaches adopted for biofouling control are synthesizing new membrane materials and variations in operational parameters such as HRT, SRT, and aeration. Several physical and chemical ways are also reported to delay the membrane biofouling. On the whole, these methods only delay biofouling. Biofouling persists, lowering the membrane efficiency and causing high energy consumption for filtration. A new biological approach is emerging as a more sustainable and cost-effective option. After the role of EPS, SMP, and other excretions by microorganisms were proven, the focus has been expanded to more biological processes involved in membrane fouling.

Along with traditional methods, design parameters also determine fouling while new strategies are emerging like inhibition of quorum sensing, an antifouling strategy (Lee et al. 2007).

26.5.1 Conventional Biofouling Control Strategies

26.5.1.1 Material Methods

Membranes are being modified by several ways including the new membrane materials, incorporation of nanoparticles while synthesizing, and change in surface characteristics and architect. One of the popular approaches is the hydrophilic modification of the membrane surface. It enhances water permeability and delays the membrane cleaning requirement. The drawback of this method is that the membrane efficiency is not recovered fully once the membrane is fouled (Shen et al. 2017, Zhang et al. 2015b).

Materials for the synthesis of composite membranes have been investigated in the past to enhance membrane permeability. This includes carbon nanotubes, titania nanotubes, and aquaporin proteins (de Groot and Grubmüller 2001, Hinds et al. 2004, Zhang et al. 2006). But these materials are not extensively evaluated for real application in membrane technologies (Lee et al. 2018).

26.5.1.2 Physical Methods

Several physical methods are reported in the literature to control the biocake formation on membranes such as shear induced by aeration, modification in the membrane module design, variation in hydrodynamics parameters, application of electrical pulses, relaxation, and backwashing.

The physical methods are applicable for reversible membrane fouling while the irreversible membrane fouling persists with the application of only physical cleaning methods.

26.5.1.3 Chemical Methods

Unlike the biocake layer formation on the membrane surface, the internal membrane pore plugged with fine particles and organics constitutes irreversible fouling. The irreversibly fouled membrane requires chemical cleaning (Weerasekara et al. 2016). Chemical approaches include chemical cleaning and chemical additives addition in MBR (Oh and Lee 2018).

Chlorine effectively removes biofilms by working like an oxidizing biocide, therefore suppressing bacterial growth and eliminating the polymeric components of the biofilm. It physically removes the biofilm from surfaces. Bleach treatment (with 5% NaClO) also shows such an impact on biofilms (Pandit et al. 2018).

Another measure is the addition of flux enhancers. Adding cationic polymers to the mixed liquor of MBR enhances the filterability. Addition of nanomaterials, Fullerence C_{60} inhibits the respiratory activity and attachment of the bacterial species *Escherichia coli* (Chae et al. 2009, Yoon et al. 2005).

Mitigating membrane biofouling in MBR by powdered activated carbon (PAC) has also been studied. It provides a large surface area, support medium, and habitat for bacterial activities at low temperature (Khan et al. 2012, Ma et al. 2012, Seo et al. 2002).

Various studies show that wastewater composition, HRT, and SRT impact the characteristics of biological mixed liquor. Changing the characteristics of biological mixed liquor by the addition of coagulants, carbon adsorbents, or media in MBR indicated lower membrane biofouling (Deng et al. 2016a, Inaba et al. 2017).

There are certain drawbacks of using chemical methods for biofouling control. Polymeric membranes used in majority MBR have low chemical membrane stability, especially with strong oxidants, including chlorine (Ahmad et al. 2016, Kochkodan and Hilal 2015, Mansouri et al. 2010). Moreover, a membrane's chemical resistance to chlorine, other cleaning agents, and pH conditions requires the use of chemically durable membrane materials due to the destructive effect of chemicals on certain membrane material by residual concentration (Wang et al. 2010a). Furthermore, the addition of additives causes another problem, which is excessive chemical sludge production (Lee et al. 2018).

26.5.1.4 Operational Methods

Traditionally, correlation of mixed liquor suspended solids (MLSS) concentration with fouling remained majorly under focus. The biomass retained in the MBR during sludge retention time (SRT) undergoes endogenous decay (autolysis) releasing dissolved organic matter (DOM), eventually contributing to membrane biofouling. Optimization of operating conditions, such as HRT, SRT, MLSS, and aeration rate is important for decreased DOM that resulted in reduced biofouling and better effluent quality (Miura et al. 2007, Xiong and Liu 2010).

Longer HRT and SRT reduces membrane fouling (Jiang et al. 2018). However, this method is not sustainable and changes in these parameters affect biological treatment performance (Lee et al. 2018).

Other operational strategies include air scouring, backwashing, and cyclic operation (De Temmerman et al. 2015, Tabraiz et al. 2017, Zheng et al. 2018). Air scouring, or enhanced shear, is an energy-intensive process.

26.5.2 Biological Methods

The conventional membrane biofouling control strategies targeted the accumulated biomass on the membrane surface. With the discovery of the role of QS in biofilm development on the membrane surface, there is a new paradigm in biofouling control in MBRs (Lee et al. 2016c).

The biological anti-biofouling strategies focus upon targeting the sources that initiate or accelerate biofilm formation at its developmental stage. The biofouling process is majorly biological in nature. The biological strategies have the potential to provide a sustainable solution to a problem of biological origin. They include the use of microbial predation, bacteriophages, and the quorum quenching.

26.5.2.1 Bacterial Quorum Quenching Process

QS plays a key role in biofilm formation on membrane surfaces via signal molecules mediating extracellular secretions. Mitigation of biofouling by disrupting signal molecules, called quorum quenching (QQ), is an emerging and innovative technique to delay biofouling naturally (Bouayed et al. 2016, Bzdrenga et al. 2017, Yeon et al. 2008, Zhu et al. 2018).

Previously, different QQ approaches, including QS inhibitor molecules and QS signal degrading enzymes, have been used to degrade QS signal molecules. Later, bacterial species producing these degrading enzymes have been employed in MBR. The three-major QQ enzymes reported controlling membrane biofouling are AHL acylases, AHL lactonases, and AHL oxidoreductase (Maqbool et al. 2015, Xiao et al. 2018).

The AHL-based QS in Gram-negative bacterial species has three points of target: the signal generating point, the signal molecules, and the signal receptor. The QQ enzymes degrade signal molecules into by-products. These by-products are unable to induce QS and are most probably utilized as sources of carbon, nitrogen, or energy (Amara et al. 2010). These points of the target are illustrated in Figure 26.4.

What makes the QQ-based biofouling control technique more attractive is its nominal or no negative impact on treatment efficiency and no production of by-products that pose a threat to health or the environment.

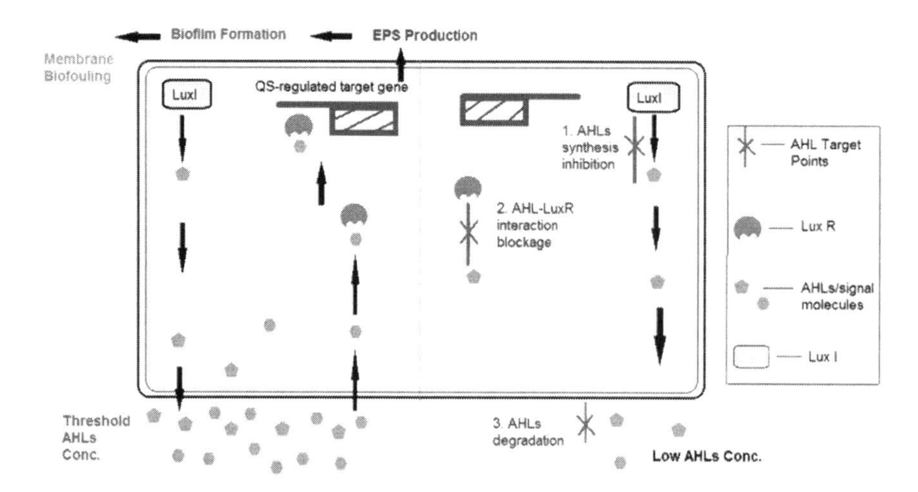

FIGURE 26.4 Bacterial quorum sensing and quorum quenching mechanism.

26.5.2.1.1 QQ Mechanism

Different mechanisms have been studied for the inhibition of QS activities, including:

1. QS signal production control: This strategy involves techniques to disrupt the signal molecules production. For instance, the LuxI genes in Gram-negative bacterial species produce AHLs. The target in this mechanism is, therefore, the LuxI genes. It aims at the complete disruption of signal molecules production (Chen et al. 2011). Most of the studies on the inhibition of AHL synthesis has centered on the utilization of analogs of S-adenosyl-L-methionine (SAM). SAM is an amino acid donor for homoserine lactone ring moieties (Huang et al. 2016).
2. QS signal (AHL) degradation: The targets in this strategy are the signal molecules. The production is not stopped, while after production the density is controlled by the degradation of signal molecules. This technique is heavily practiced (Sio et al. 2006). There are three options reported including chemical, metabolic, and enzymatic degradation (Huang et al. 2016). Among these, the metabolic pathway has been found more effective in the case of MBRs (Ham et al. 2018, Maqbool et al. 2015, Oh et al. 2012).
3. QS signal activity (AHL cognate receptor protein or AHL synthase) control: This method involves the control at QS gene expression site; that is, the signal molecule interaction with genes that are expressed is targeted. In the case of AHL-based QS, the LuxR-AHL complex triggers the biofilm formation, hence this the technique avoids the complex from forming (Parveen and Cornell 2011).
4. QS signal mimicking by synthetic compounds as signal molecule's analogs: This is a rare technique in which such compounds are introduced in the system, which have more affinity toward the group of genes otherwise expressed by signal molecules attachment (Chen et al. 2011).

26.5.2.2 Enzymatic Disruption of EPS, Cell Wall Hydrolases, and Using D-amino Acids

The degradation of EPS by application of enzymes disrupts the physical integrity of EPS, denatures the proteins, and negatively impacts the carbohydrate and lipids. These enzymes include protease, protein K, Bacterial Amylase Novo, umamizyme, papain, Amano Protease A, amyloglucosidase, polarzyme, everlase, trysin, savinase, and subtilisin (Loiselle and Anderson 2003, Molobela et al. 2010). The drawback of this method is enzyme instability and associated cost.

The cell wall hydrolases method utilizes enzymes via cell degradation to disrupt fouling layers on the membrane surface. The enzymes having affinities toward macromolecules are used in this method to improve membrane performance (Wong et al. 2015).

Lysozyme, protease, lipases, and amylases, hydrolytic enzymes by nature, have shown biofouling retardation capacity (Conte et al. 2006, Wong et al. 2015, Xiong and Liu 2010). However, the enzyme efficiency is negatively impacted due to the hydrophobic attraction between the layer and hydrolysis products (Porcelli and Judd 2010).

D-amino acids in trace concentrations have indicated biofilm disruption ability. Various bacterial species, for example, *Vibrio cholera* and *Bacillus subtilis*, produce different D-amino acids. They change the peptidoglycan layer structure, interfere with protein synthesis, and inhibit cell growth (Cava et al. 2011, Kolodkin-Gal et al. 2010, Meng et al. 2017). D-amino acids may be species-specific (Yu et al. 2016), limiting their wider application for biofilm control.

26.5.2.3 Bacteriophage, Protozoan, and Metazoan Predation

Bacteriophages are constituted by proteins encapsulating deoxyribonucleic acid (DNA) or ribonucleic acid (RNA) genomes. The antifouling mechanism of bacteriophage involves bacterial lysis or extrusion and replication of genomes inside the bacterial cell before lysis. Apart from its application in virulence control, it finds application in membrane biofouling control in wastewater treatment technologies (Branch et al. 2016, Purnell et al. 2015).

However, the bacterial composition in practice is complex and bacteriophages display high specificity against target bacteria. Furthermore, its application on a large scale is challenging (Bagheri and Mirbagheri 2018, Wu et al. 2017).

Protozoan and metazoan species are utilized as predators on the wastewater treatment plant. They modify the physical integrity of the biofilm layer on the membrane surface. The predators move across the biofilm and leave behind pores, lowering the impermeability due to membrane biofouling (Derlon et al. 2013). However, introducing non-indigenous worms in a wastewater treatment system is challenging (Bagheri and Mirbagheri 2018).

26.5.3 Hybrid Methods

Several methods in combination further mitigate the membrane biofouling issue. While the problem still exists, membrane biofouling is a natural process and it involves a complex web of cellular communications named QS (Yeon et al. 2008).

Combining two methods has shown improved membrane biofouling control in MBR. Such attempts include hybridizing quorum quenching with other antifouling methods. A couple of examples include QQ with back pulse method and relaxation (Weerasekara et al. 2014) and QQ with chlorination technique. A combination with adsorbent has also been tried in previous studies. This hybrid control measure has shown substantial performance, and multiple times increase of flux has been reported.

Biostimulation of QQ by addition of gamma-caprolactone (GCL), which is structurally similar to AHL, showed an increase in QQ activity in MBR (Yu et al. 2016).

26.6 Quorum Quenching: The New Biofouling Control Paradigm

Conventional biofouling control methods were largely the physical and chemical strategies with a focus on removal of accumulated biomass on the membrane surface. The establishment of the role of QS in membrane biofouling in MBR opened a new paradigm in biofouling control via QQ. QQ aimed at inhibition of biofilm formation at its developmental stage (Le-Clech et al. 2006, Lee et al. 2016a, Yeon et al. 2008) (Figure 26.5).

26.6.1 Historical Development of QQ in MBR

The involvement of QS in virulence caused by biofilm formation by *Pseudomonas aeruginosa* was elaborated by Davies et al. (1998). QS's role was found in swarming, EPS production, and biofilm formation through cell-to-cell communication and genes regulation. QQ was successfully applied to control biofilm formation on surfaces of medical devices and plant tissue (Baveja et al. 2004). These studies were the precursor of the introduction of QS and QQ concept in MBR for membrane biofouling control.

The idea of QS in MBR started developing in the year 2002, and it was hypothesized that QS plays a role in biofilm formation on the membrane surface through signal molecules, regulating the group behaviors. The hypothesis was proved, and a positive correlation was found between microbial QS and membrane biofouling (Oh and Lee 2018, Yeon et al. 2008). Various studies

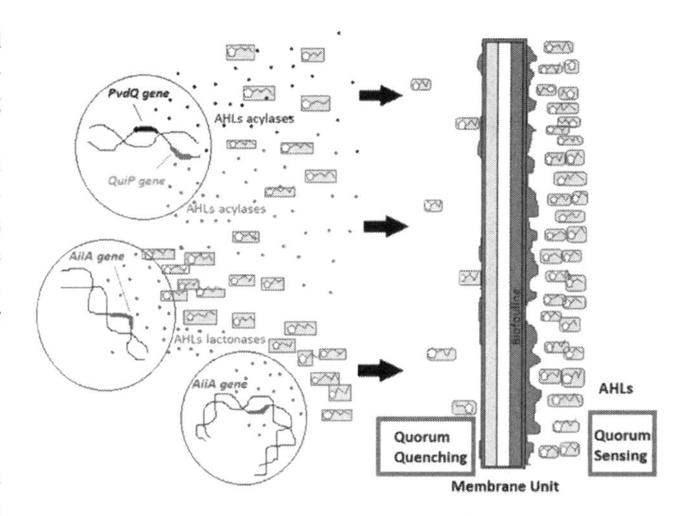

FIGURE 26.5 Effect of QQ enzymes producing genes on membrane biofouling.

attempted to elucidate the QQ in MBR and bring the technique closer to application in large scale MBR.

The initial studies before 2012 were dominated by enzymatic disruption of signal molecules in MBR and other membrane technologies. *Acylase I* reduced biofilm formation by *Aeromonas hydrophila* and *Pseudomonas putida* on borosilicate (36% and 23%), polystyrene (60% and 73%), and the reverse osmosis (RO) membrane (20% and 24%) (Paul et al. 2009). *Acylase* immobilized into sodium alginate capsules for enzymatic quorum quenching in MBRs indicated lower membrane biofouling (Jiang et al. 2012).

Later, bacterial QQ was introduced to overcome the cost associated with enzymes and their instability in MBR. Free-moving beads entrapped with quorum quenching bacteria were applied for the inhibition of biofouling in an MBR. Cell entrapping beads (CEBs) with a porous microstructure were prepared by entrapping QQ bacteria (*Rhodococcus* sp. BH4) into alginate beads (Kim et al. 2013b, Oh et al. 2012).

Further attempts have resulted in various media design, including cell entrapped micro-vessels (Jahangir et al. 2012, Oh et al. 2012), beads (Kim et al. 2014, Maqbool et al. 2015, Weerasekara et al. 2014), macrocapsules coated with polymer, hollow cylinders (Lee et al. 2016b), and QQ sheets (Nahm et al. 2017).

26.6.2 Enzymatic QQ in MBR

QQ enzymes were discovered in a wide range of bacteria and were classified into three major types according to their enzymatic mechanisms: AHL lactonase (lactone hydrolysis), AHL acylase (amidohydrolysis), and AHL oxidase and reductase (oxidoreduction). The metal centers at the active sites of these enzymes are considerably diverse (Lee et al. 2013).

AHL lactonases hydrolyze the lactone ring of AHL. They produce *N*-acyl-homoserine, which is unable to regulate bacterial behavior via the QS process. AHL acylases cleave the AHL amide bond and produce free fatty acid and lactone ring. Oxidoreductase targets the acyl side chain by oxidative or reducing activities (Romero et al. 2008, Zhang et al. 2002b).

The first enzyme introduced in MBR was porcine kidney acylase I. It indicated AHL signal molecule degradation and lowered membrane biofouling (Yeon et al. 2008). The short catalytic lifetime of porcine kidney acylase led to the enzyme immobilization on magnetic carriers. The study indicated a lower membrane biofouling. It was attributed to fewer EPS production due to enzymatic QQ activity. The enzymatic QQ neither significantly interfered with the entire microbial community nor seriously affected the general performance of MBR in organic content removal from wastewater (Yeon et al. 2008).

Jiang et al. (2012) using sodium alginate beads immobilized acylase I enzyme in MBR. The results indicated improved sludge characteristics and lower EPS production leading to delayed membrane biofouling; the filtration time was increased multiple times compared to the control. Lee et al. (2014) used QQ acylase stabilized in magnetically-separable mesoporous silica. It efficiently alleviated the biofilm maturation of *Pseudomonas aeruginosa* PAO1 on the membrane surface, thereby enhancing the filtration performance by membrane biofouling prevention. Application of acylase I on borosilicate, polystyrene, and reverse osmosis membrane showed lower biofilm formation in a study conducted by Paul et al. (2009).

Enzymatic QQ has indicated lower EPS production. Furthermore, it offers various benefits including insignificant impact on the overall microbial community in MBR, membrane biofouling control at the development stage of biofilm formation, the insignificant impact of MBR's general performance, and organics removal from wastewater (Yeon et al. 2008). Despite the advantages associated with enzyme-based QQ approaches, the application of can be technically challenging (Table 26.2).

TABLE 26.2

Advantages and Disadvantages of QQ Enzyme Inoculation in MBR

	Advantages	Disadvantages
Cost		High cost
Stability		Unstable without immobilizing media in MBR
Sustainability		Requires continuous addition in MBR
Application frequency		Continuous application for biofouling control
Application mode		Requires immobilizing media
Source	Various biological (prokaryotic and eukaryotic) sources	
Enzyme preparation		Requires extraction and preparation for final application in MBR
Enzyme purification		May require purification after extraction from source
By-product	Consumption in MBR does not produce harmful by-products	
Effect on MBR parameters	No negative impact on MBR parameters	
Effect on permeate quality	No negative impact on treated water quality	
Effect on biological ecology	No negative impact on microbial activity in MBR	

26.6.3 Bacterial QQ in MBR

Bacterial QQ was introduced as an alternative to enzymatic QQ. Instead of QQ enzyme introduction in MBR, QQ bacteria producing these QQ enzymes were introduced mainly immobilized in media. The objectives behind this shift toward QQ bacteria were to minimize the cost associated with enzyme extraction and purification and introduce QQ bacteria as a more sustainable and stable source of QQ enzyme in MBR (Ishige et al. 2005, Oh and Lee 2018, Oh et al. 2012).

During the past decade, attempts have been made to isolate and identify QQ bacterial species in MBR for biofouling control. Ten species were initially found to produce QQ enzymes including four *Bacillus* species, *Agrobacterium tumefaciens*, *Arthrobacter* sp., *Klebsiella pneumoniae*, *P. aeruginosa*, *Rastonia* sp., and *V. paradoxus* (Dong and Zhang 2005).

Rhodococcus sp. BH4 isolated from an MBR treating real wastewater showed QQ activity. Recombinant *E. coli* demonstrated delayed biofouling in MBR (Oh et al. 2012), with a substantial reduction in AHL concentration and concurrent increase of filtration duration. Cheong et al. (2013) isolated *Pseudomonas* sp. 1A1 from laboratory scale MRB and identified extracellular QQ activity in the species. Kim et al. (2014) indicated that QQ enzymes are produced by *Afipia* sp., *Acinetobacter* sp., *Pseudomonas* sp., *Micrococcus* sp., *Microbacterium* sp., and *Rhodococcus* sp.

Gül and Koyuncu (2017) identified QQ bacterial species belonging to nine genera, namely *Shewanella*, *Acinetobacter*, *Klebsiella*, *Bacillus*, *Deftia*, *Vibrio*, *Comamonas*, *Microbacterium*, and *Pseudomonas*. All these species degraded *N*-octanoyl-homoserine lactone (C8-HSL) in MBR. Gu et al. (2018) isolated an indigenous QQ bacterium *Acinetobacter bereziniae* strain from laboratory scale MBR.

Yu et al. (2018) investigated QQ impact on ammonia removal and biofouling under stressful conditions. The study indicated QQ as an effective biofouling control strategy under varying physical conditions in the MBR. Results showed decreased biofouling under low SRT and temperature. QQ influences sludge characteristics but does not impact pollutant degradation.

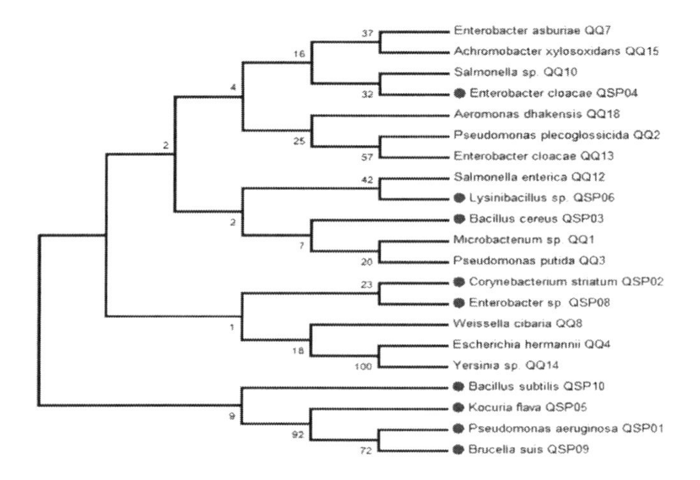

FIGURE 26.6 Neighbor-joining phylogenetic tree based on 16SrDNA gene sequences of quorum quenching (QQ) bacteria isolated from full-scale (with dots) and laboratory-scale MBR. (From Perveen, *Detection of quorum quenching enzymes producing genes in membrane bioreactor*, IESE NUST Islamabad, 2004.)

26.6.4 Fungal QQ in MBR

Fungal QQ to inhibit bacterial QS in MBR is a new paradigm among QQ methods. *Candida albicans*, a farnesol producing fungus species, was employed in MBR for membrane biofouling inhibition. *Candida albicans* was entrapped in polymer beads (AEBs) and applied in MBR fed by real wastewater. Lower membrane biofouling was achieved in AEB MBR. Further investigation of the mechanism involved in suppressed biofilm formation indicated that farnesol secreted from *Candida albicans* is able to mitigate biofilm formation via the suppression of AI-2 QS. Secondly, the energy saving in AEB MBR and MBR without AEBs was investigated. Results showed lower energy requirement for aeration due to the physical washing effect and biological QQ effect of AEBs (Lee et al. 2016a).

26.6.5 QQ Enzyme Producing Genes

Previous studies detected QQ enzymes encoding genes in QQ species. For an instant, Dong et al. (2002) conducted biochemical and molecular analyses on *Bacillus* species. The lactonase producing genes exhibited high levels of homology to *aiiA240B1* encoding AHL lactonase. Lactonase belongs to the metallo-lactamase superfamily. *aiiA* gene expression from *B. cereus* A24 and *Bacillus* sp. indicated a reduction in AHL concentrations (Ulrich 2004). Acylase activity has been studied for AHL-degradation in microbial species of *Variovorax paradoxus* and *Ralstonia* sp. XJ12B (Lin et al. 2003, Ma et al. 2013). The homologues of QQ enzymes have been found in complete bacterial genomes in various other species.

In a recent study, three bacterial species, namely *Bacillus cereus* QSP03, *Pseudomonas aeruginosa* QSP01, and *Bacillus subtilis QSP10*, were isolated from full-scale MBR treating real wastewater in Pakistan (Perveen 2018). DNA amplification with QQ gene-specific primers for AHL lactonase and acylase were carried out. The sequenced amplicons of *Pseudomonas aeruginosa* QSP01 and *Bacillus cereus* QSP03 after determination of their consensus sequence showed high similarity to coding sequences of already published *pvdQ*, *quiP*, and *aiiA* genes. All these sequences were aligned using MEGA 7 to infer their evolutionary relationship and construct a phylogenetic tree (Figure 26.7).

The molecular phylogenesis of QQ genes indicated existence in diverse bacterial species. The sequence similarity analysis has enabled division of lactonases and acylases into sub-groups. Lactonase can be divided into QsdAlactonase group and all other known lactonases are grouped together which includes AiiA enzyme. Further clustering of AiiA group of QQ enzyme resulted in AttM, AhlD, and AiiB lactonases. These are all groups of the MBL superfamily. Table 26.3 lists some of the QQ

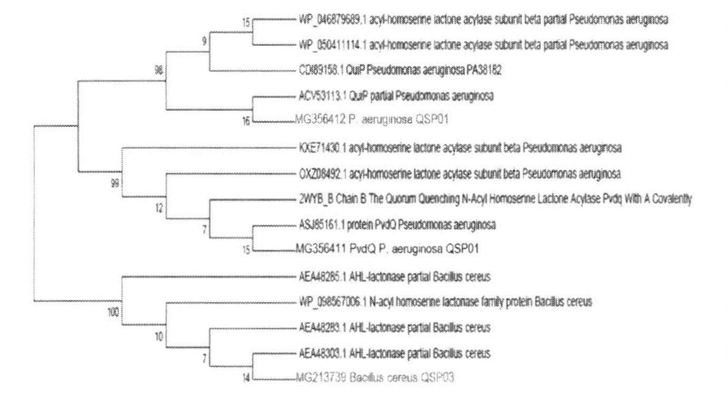

FIGURE 26.7 Neighbor-joining phylogenetic tree based on quorum quenching (QQ) gene sequences from literature and newly identified QQ genes in MBR (highlighted in gray).

TABLE 26.3

List of Some Known QQ Enzyme Producing Bacterial Species

QQ Enzyme/ Gene	Bacterial Strain	Reference
AHL-lactonase		
aiiA gene	*Bacillus* sp. *240B1* *Bacillus cereus A24* *Bacillus mycoides* *Bacillus thuringiensis* *Bacillus anthracis*	Dong et al. (2000, 2002), Lee et al. (2002), Ulrich (2004)
MomL	*Muricaudaolearia*	Tang et al. (2014)
attM gene	*Agrobacterium tumefaciens*	Zhang et al. (2002a)
aiiB gene	*Agrobacterium tumefaciens* C58	Carlier et al. (2003)
aiiS gene	*Agrobacterium radiobacter* K84	Uroz et al. (2009)
ahlD gene	*Arthrobacter* sp. IBN110	Park et al. (2003)
ahlK gene	*Klebsiella pneumoniae*	Park et al. (2003)
qlcA gene	*Acidobacteria*	Riaz et al. (2008)
aiiM gene	*Microbacterium testaceum* StLB037	Wang et al. (2010b)
qsdA gene	*Rhodococcus erythropolis* W2	Uroz et al. (2008)
aidH gene	*Ochrobactrum* sp. T63	Mei et al. (2010)
DlhR, QsdR1	*Rhizobium* sp. NGR234	Krysciak et al. (2011)
ahlS gene	*Solibacillus silvestris* StLB046	Morohoshi et al. (2012)
SsoPox	*Sulfolobus solfataricus* strain P2 *Rhodococcus* sp.	Elias et al. (2008), Merone et al. (2005)
GKL	*Geobacillus kaustophilus strain* HTA426	Park et al. (2006)
PPH	*Mycobacterium tuberculosis*	Chow et al. (2010)
MCP	*Mycobacterium avium subsp. paratuberculosis*	Afriat et al. (2006)
AHL-acylase		
aiiD gene	*Ralstonia eutropha*	Lin et al. (2003)
pvdQ gene	*Pseudomonas aeruginosa*	Sio et al. (2006), Huang et al. (2003)
quiP gene	*Pseudomonas aeruginosa*	Huang et al. (2006)
aiiC gene	*Anabaena* sp. PCC 7120	Romero et al. (2008)
ahlM gene	*Streptomyces* sp. M664	Park et al. (2005)
Aac	*Ralstonia solanacearum* *Shewanella* sp. MIB015	Chen et al. (2009), Morohoshi et al. (2005)
HacA	*Pseudomonas syringae*	Shepherd and Lindow (2009)
HacB	*Pseudomonas syringae* *Variovorax* sp. *Variovorax paradoxus Broad* *Tenacibaculum maritimum* *Comomonas* sp. D1 *Rhodococcus erythropolis* W2	Shepherd and Lindow (2009), Uroz et al. (2003, 2005, 2007), Leadbetter and Greenberg (2000), Romero et al. (2010)
Unidentified	*Pseudomonas* sp. 1A1	Cheong et al. (2013)
Oxidoreductase		
P450BM-3	*Bacillus megaterium* CYP102 A1	Chowdhary et al. (2007)
Unidentified	*Burkholderia* sp. strain GG4	Chan et al. (2011)
Unidentified	*Rho. erythropolis* W2	Uroz et al. (2005)

enzymes/genes from literature, adapted and modified from Chen et al. (2013), Lade et al. (2014), and Dong and Zhang (2005).

26.6.6 QQ Immobilizing Media

To utilize QQ bacteria in MBRs for long-term sustainable operations, one important strategy is cell immobilization to restrict the cells and prevent the gradual decrease of QQ bacteria population. Initially, AHL degrading enzymes were introduced by immobilizing the enzymes in nanofiltration membrane and magnetic carriers. Later to overcome the associated cost and enzyme loss through enzyme degradation, bacterial cells producing these enzymes are being introduced in MBR (Yeon et al. 2008).

The associated benefits include enhanced stability and sustainability of QQ in MBR. Immobilizing media protects QQ bacteria in a new and competing environment from other microbial species and unfavorable physical conditions. The microporous structure of the media allows for better mass transfer of dissolved oxygen, nutrients, substrates, and target signal molecules (Oh and Lee 2018) (Table 26.4).

26.6.7 QQ and QS Analysis Methods in MBR

Traditionally, various cultural-dependent techniques have been used in isolation of QQ and QS bacterial species in MBR.

Biosensor-based biochemical analysis is the most popular method to detect short and long chain AHL signal molecules produced by QS bacterial species and degraded by QQ bacterial species. Examples of biosensor species used in QQ and QS analysis, among others, are:

1. *Chromobacteriumviolaceum* CV026-*C. violaceum* via CviI/R AHL quorum sensing system regulates violacein production. It responds to C6-AHL, a short-chain AHL. CV026 is constructed as a violacein and AHL-negative double miniTn5 mutant of ATCC 31532. Exposure to external short chain AHL produces visual purple pigment (McClean et al. 1997).
2. *Agrobacterium tumefaciens* A136- the strain is constructed by eliminating the TraI/R QS system. It has two plasmids, pCF218 that produces the TraR response regulator, and pMV26, containing the traI promoter joined to the luxCDABE operon (Watson et al. 1975). The strain is sensitive to long-chain AHLs (C8-HSL, 3-O-C8-HSL, C10-HSL, C12-HSL, 3-O-C12-HSL and C14-HSL). In the presence of x-gal (β-galactosidase), blue pigment can be observed visually upon its interaction with exogenous AHL.
3. *Pseudomonas aeruginosa* QSIS2- The strain contains pLasB-SacB1 that encodes killing induced by exogenous AHLs. It is a *Pseudomanas aeruginosa* las IrhlI double mutant strain (Rasmussen et al. 2005).

Lade et al. (2014) have developed methods to analyze QS activity in bacterial species by well-diffusion assay. Other techniques for rapid analysis of QS in bacterial species include the parallel streak method and bioagar assay. In the parallel streak method, bacteria strain under study for QS activity is streaked on agar in parallel to the biosensor species. The signal molecule production by QS bacterial species produces color in the presence of biosensor species.

Recently, the biochemical characterization techniques are combined with molecular identification via 16SrRNA. The microbial community in MBR is also sampled and analyzed by pyrosequencing (Lim et al. 2012).

Furthermore, advanced analytical tools of gas chromatography–mass spectrometry (GC-MS) and high-performance liquid chromatography (HPLC) quantify AHLs molecules.

TABLE 26.4
Common QQ Enzymatic and Bacteria Entrapping Media Designs

QQ Media	Description	Reference
QQ sheets	QQ sheets were found more suitable with flat-sheet membrane module MBR. It indicated 2.5 times more QQ activity than QQ beads due to greater surface area provision by QQ sheets.	Nahm et al. (2017)
QQ microbial bag	For biofouling control in MBR, QQ bacteria was encapsulated in a dumpling shaped microbial bag.	Gu et al. (2018)
QQ hollow cylinder	Hollow cylinders have a greater surface area than beads. The design showed increased QQ activity as compared to QQ beads.	Lee et al. (2016b)
Polymer QQ beads	Alginate beads were coated with polymers to enhance their strength for long-term application in MBR. Water soluble and non-soluble polymer both indicated improved strength. Water-soluble polymer polyvinylalcohol is used currently in media preparation.	Kim et al. (2015)
Microbial vessel (MV)	Hydrophilic polyvinylidene fluoride (PVDF) hollow fiber microfiltration membrane was used to prepare microbial vessels entrapping *Rhodococcus* sp. BH4.	Köse-Mutlu et al. (2016)
Rotating microbial carrier frame (RMCF)	*Rhodococcus* sp. BH4 filled in four cubbyholes of a polycarbonate frame was covered with PVDF microfiltration membrane for application in laboratory scale MBR. It was found to be the most feasible option as per cost analysis.	
Ceramic microbial vessel	QQ microbial vessel was introduced in laboratory scale MBR. It enhanced the QQ bacterial cell viability and lowered biofouling in MBR.	Cheong et al. (2014)
QQ beads and cell entrapping beads	QQ bacteria were entrapped in alginate beads to introduce in MBR. They move freely in the bioreactor. They lowered membrane biofouling by signal molecule degradation and physical collision with cake layers on the membrane surface.	Kim et al. (2013b, 2014)
QQ microporous hollow fibers	QQ bacteria were encapsulated inside the lumen of microporous hollow fiber membranes. It indicated the anti-biofouling effect in MBR.	Oh et al. (2012)
Magnetic enzyme carriers (MECs)	Quorum quenching enzyme entrapped on MECs indicated lower biofouling in a laboratory scale MBR.	Kim et al. (2013a)

REFERENCES

Afriat, Livnat, Cintia Roodveldt, Giuseppe Manco, and Dan S Tawfik. 2006. "The latent promiscuity of newly identified microbial lactonases is linked to a recently diverged phosphotriesterase." *Biochemistry* no. 45 (46):13677–13686.

Ahmad, Rizwan, Zaki Ahmad, Asad Ullah Khan, Naila Riaz Mastoi, Muhammad Aslam, and Jeonghwan Kim. 2016. "Photocatalytic systems as an advanced environmental remediation: Recent developments, limitations and new avenues for applications." *Journal of Environmental Chemical Engineering* no. 4 (4):4143–4164.

Akamatsu, Kazuki, Wei Lu, Takashi Sugawara, and Shin-ichi Nakao. 2010. "Development of a novel fouling suppression system in membrane bioreactors using an intermittent electric field." *Water research* no. 44 (3):825–830.

Al-Juboori, Raed A, and Talal Yusaf. 2012. "Biofouling in RO system: Mechanisms, monitoring and controlling." *Desalination* no. 302:1–23.

Amara, Neri, Bastiaan P Krom, Gunnar F Kaufmann, and Michael M Meijler. 2010. "Macromolecular inhibition of quorum sensing: Enzymes, antibodies, and beyond." *Chemical Reviews* no. 111 (1):195–208.

Aslam, Muhammad, Amine Charfi, Geoffroy Lesage, Marc Heran, and Jeonghwan Kim. 2017. "Membrane bioreactors for wastewater treatment: A review of mechanical cleaning by scouring agents to control membrane fouling." *Chemical Engineering Journal* no. 307:897–913.

Baek, Seung H, and Krishna R Pagilla. 2006. "Aerobic and anaerobic membrane bioreactors for municipal wastewater treatment." *Water Environment Research* no. 78 (2):133–140.

Bagheri, Majid, and Sayed Ahmad Mirbagheri. 2018. "Critical review of fouling mitigation strategies in membrane bioreactors treating water and wastewater." *Bioresource Technology* no. 258:318–334.

Basile, Angelo, Alfredo Cassano, and Navin K Rastogi. 2015. *Advances in membrane technologies for water treatment: Materials, processes and applications*. Boston: Elsevier.

Baveja, Jasmina, MDP Willcox, EBH Hume, N Kumar, R Odell, and LA Poole-Warren. 2004. "Furanones as potential antibacterial coatings on biomaterials." *Biomaterials* no. 25 (20):5003–5012.

Benitez, Jaime, Abraham Rodríguez, and Roger Malaver. 1995. "Stabilization and dewatering of wastewater using hollow fiber membranes." *Water Research* no. 29 (10):2281–2286.

Bouayed, Naila, Nicolas Dietrich, Christine Lafforgue, Chung-Hak Lee, and Christelle Guigui. 2016. "Process-oriented review of bacterial quorum quenching for membrane biofouling mitigation in membrane bioreactors (MBRs)." *Membranes* no. 6 (4):52.

Branch, Amos, Trang Trinh, Guido Carvajal, Greg Leslie, Heather M Coleman, Richard M Stuetz, Jörg E Drewes, Stuart J Khan, and Pierre Le-Clech. 2016. "Hazardous events in membrane bioreactors–Part 3: Impacts on microorganism log removal efficiencies." *Journal of Membrane Science* no. 497:514–523.

Bzdrenga, Janek, David Daude, Benjamin Remy, Pauline Jacquet, Laure Plener, Mikael Elias, and Eric Chabriere. 2017. "Biotechnological applications of quorum quenching enzymes." *Chemico-Biological Interactions* no. 267:104–115.

Camilli, Andrew, and Bonnie L Bassler. 2006. "Bacterial small-molecule signaling pathways." *Science* no. 311 (5764):1113–1116.

Carlier, Aurélien, Stéphane Uroz, Bruno Smadja, Rupert Fray, Xavier Latour, Yves Dessaux, and Denis Faure. 2003. "The Ti plasmid of Agrobacterium tumefaciens harbors an attM-paralogous gene, aiiB, also encoding *N*-acyl homoserine lactonase activity." *Applied and Environmental Microbiology* no. 69 (8):4989–4993.

Cava, Felipe, Miguel A De Pedro, Hubert Lam, Brigid M Davis, and Matthew K Waldor. 2011. "Distinct pathways for modification of the bacterial cell wall by non-canonical D-amino acids." *The EMBO Journal* no. 30 (16):3442–3453.

Chae, So-Ryong, Shuyi Wang, Zachary D Hendren, Mark R Wiesner, Yoshimasa Watanabe, and Claudia K Gunsch. 2009. "Effects of fullerene nanoparticles on *Escherichia coli* K12 respiratory activity in aqueous suspension and potential use for membrane biofouling control." *Journal of Membrane Science* no. 329 (1–2):68–74.

Chan, Kok-Gan, Steve Atkinson, Kalai Mathee, Choon-Kook Sam, Siri Ram Chhabra, Miguel Camara, Chong-Lek Koh, and Paul Williams. 2011. "Characterization of *N*-acylhomoserine lactone-degrading bacteria associated with the *Zingiber officinale* (ginger) rhizosphere: Co-existence of quorum quenching and quorum sensing in *Acinetobacter* and *Burkholderia*." *BMC Microbiology* no. 11 (1):51.

Chang, Chia-Yuan, Kulchaya Tanong, Jia Xu, and Hokyong Shon. 2011. "Microbial community analysis of an aerobic nitrifying-denitrifying MBR treating ABS resin wastewater." *Bioresource Technology* no. 102 (9):5337–5344.

Chen, Chin-Nung, Chii-Jaan Chen, Chen-Ting Liao, and Chia-Yin Lee. 2009. "A probable aculeacin A acylase from the *Ralstonia solanacearum* GMI1000 is *N*-acyl-homoserine lactone acylase with quorum-quenching activity." *BMC Microbiology* no. 9 (1):89.

Chen, Fang, Yuxin Gao, Xiaoyi Chen, Zhimin Yu, and Xianzhen Li. 2013. "Quorum quenching enzymes and their application in degrading signal molecules to block quorum sensing-dependent infection." *International Journal of Molecular Sciences* no. 14 (9):17477–17500.

Chen, Guozhou, Lee R Swem, Danielle L Swem, Devin L Stauff, Colleen T O'Loughlin, Philip D Jeffrey, Bonnie L Bassler, and Frederick M Hughson. 2011. "A strategy for antagonizing quorum sensing." *Molecular Cell* no. 42 (2):199–209.

Cheong, Won-Suk, Sang-Ryoung Kim, Hyun-Suk Oh, Sang H Lee, Kyung-Min Yeon, Chung-Hak Lee, and Jung-Kee Lee. 2014. "Design of quorum quenching microbial vessel to enhance cell viability for biofouling control in membrane bioreactor." *Journal of Microbiology and Biotechnology* no. 24 (1):97–105.

Cheong, Won-Suk, Chi-Ho Lee, Yun-Hee Moon, Hyun-Suk Oh, Sang-Ryoung Kim, Sang H Lee, Chung-Hak Lee, and Jung-Kee Lee. 2013. "Isolation and identification of indigenous quorum quenching bacteria, *Pseudomonas* sp. 1A1, for biofouling control in MBR." *Industrial & Engineering Chemistry Research* no. 52 (31):10554–10560.

Chiemchaisri, Chart, Kazuo Yamamoto, and Saravanamut Vigneswaran. 1993. "Household membrane bioreactor in domestic wastewater treatment." *Water Science and Technology* no. 27 (1):171–178.

Choo, Kwang-Ho, and Chung-Hak Lee. 1996. "Membrane fouling mechanisms in the membrane-coupled anaerobic bioreactor." *Water Research* no. 30 (8):1771–1780.

Chow, Jeng Yeong, Bo Xue, Kang Hao Lee, Alvin Tung, Long Wu, Robert C Robinson, and Wen Shan Yew. 2010. "Directed evolution of a thermostable quorum-quenching lactonase from the amidohydrolase superfamily." *Journal of Biological Chemistry* no. 285 (52):40911–40920.

Chowdhary, Puneet K, Neela Keshavan, Hien Q Nguyen, Julian A Peterson, Juan E González, and Donovan C Haines. 2007. "*Bacillus megaterium* CYP102A1 oxidation of acyl homoserine lactones and acyl homoserines." *Biochemistry* no. 46 (50):14429–14437.

Cicek, N. 2003. "A review of membrane bioreactors and their potential application in the treatment of agricultural wastewater." *Canadian Biosystems Engineering* no. 45:6.37–649.

Conrad, Arnaud, Merja Kontro, Minna M Keinänen, Aurore Cadoret, Pierre Faure, Laurence Mansuy-Huault, and Jean-Claude Block. 2003. "Fatty acids of lipid fractions in extracellular polymeric substances of activated sludge flocs." *Lipids* no. 38 (10):1093–1105.

Conte, Amalia, Giovanna Giuliana Buonocore, Antonio Bevilacqua, M Sinigaglia, and Matteo Alessandro Del Nobile. 2006. "Immobilization of lysozyme on polyvinylalcohol films for active packaging applications." *Journal of Food Protection* no. 69 (4):866–870.

Davies, David G, Matthew R Parsek, James P Pearson, Barbara H Iglewski, J William Costerton, and E Peter Greenberg. 1998. "The involvement of cell-to-cell signals in the development of a bacterial biofilm." *Science* no. 280 (5361):295–298.

de Groot, Bert L, and Helmut Grubmüller. 2001. "Water permeation across biological membranes: Mechanism and dynamics of aquaporin-1 and GlpF." *Science* no. 294 (5550):2353–2357.

De Temmerman, Lieven, Thomas Maere, Hardy Temmink, Arie Zwijnenburg, and Ingmar Nopens. 2015. "The effect of fine bubble aeration intensity on membrane bioreactor sludge characteristics and fouling." *Water Research* no. 76:99–109.

Deng, Lijuan, Wenshan Guo, Huu Hao Ngo, Bing Du, Qin Wei, Ngoc Han Tran, Nguyen Cong Nguyen, Shiao-Shing Chen, and Jianxin Li. 2016a. "Effects of hydraulic retention time and bioflocculant addition on membrane fouling in a sponge-submerged membrane bioreactor." *Bioresource Technology* no. 210:11–17.

Deng, Lijuan, Wenshan Guo, Huu Hao Ngo, Hongwei Zhang, Jie Wang, Jianxin Li, Siqing Xia, and Yun Wu. 2016b. "Biofouling and control approaches in membrane bioreactors." *Bioresource Technology* no. 221:656–665.

Derlon, Nicolas, Nicolas Koch, Bettina Eugster, Thomas Posch, Jakob Pernthaler, Wouter Pronk, and Eberhard Morgenroth. 2013. "Activity of metazoa governs biofilm structure formation and enhances permeate flux during Gravity-Driven Membrane (GDM) filtration." *Water Research* no. 47 (6):2085–2095.

Dong, Yi-Hu, Andi R Gusti, Qiong Zhang, Jin-Ling Xu, and Lian-Hui Zhang. 2002. "Identification of quorum-quenching *N*-acyl homoserine lactonases from *Bacillus* species." *Applied and Environmental Microbiology* no. 68 (4):1754–1759.

Dong, Yi-Hu, Jin-Ling Xu, Xian-Zhen Li, and Lian-Hui Zhang. 2000. "AiiA, an enzyme that inactivates the acylhomoserine lactone quorum-sensing signal and attenuates the virulence of Erwinia carotovora." *Proceedings of the National Academy of Sciences* no. 97 (7):3526–3531.

Dong, Yi-Hu, and Lian-Hui Zhang. 2005. "Quorum sensing and quorum-quenching enzymes." *The Journal of Microbiology* no. 43 (1):101–109.

Drews, Anja. 2010. "Membrane fouling in membrane bioreactors—Characterisation, contradictions, cause and cures." *Journal of Membrane Science* no. 363 (1–2):1–28.

Elias, Mikael, Jérôme Dupuy, Luigia Merone, Luigi Mandrich, Elena Porzio, Sébastien Moniot, Daniel Rochu, Claude Lecomte, Mosè Rossi, and Patrick Masson. 2008. "Structural basis for natural lactonase and promiscuous phosphotriesterase activities." *Journal of Molecular Biology* no. 379 (5):1017–1028.

Flemming, Hans-Curt. 2011. "Microbial biofouling: Unsolved problems, insufficient approaches, and possible solutions." In *Biofilm Highlights*, Hans-C Flemming, Jost Wingender, and Ulrich Szewzyk (eds.), 81–109. Heidelberg: Springer.

Flemming, Hans-Curt, Thomas R Neu, and Jost Wingender. 2016. *The perfect slime: Microbialextracellular polymeric substances (EPS)*: IWA publishing, London, UK.

Fröls, Sabrina, Mike Dyall-Smith, and Felicitas Pfeifer. 2012. "Biofilm formation by haloarchaea." *Environmental Microbiology* no. 14 (12):3159–3174.

Gao, Da-Wen, Zhi-Dan Wen, Bao Li, and Hong Liang. 2013. "Membrane fouling related to microbial community and extracellular polymeric substances at different temperatures." *Bioresource Technology* no. 143:172–177.

Gu, Yanling, Jinhui Huang, Guangming Zeng, Yahui Shi, Yi Hu, Bi Tang, Jianxin Zhou, Weihua Xu, and Lixiu Shi. 2018. "Quorum quenching activity of indigenous quorum quenching bacteria and its potential application in mitigation of membrane biofouling." *Journal of Chemical Technology & Biotechnology* no. 93 (5):1394–1400.

Gül, Bahar Yavuztürk, and Ismail Koyuncu. 2017. "Assessment of new environmental quorum quenching bacteria as a solution for membrane biofouling." *Process Biochemistry* no. 61:137–146.

Ham, So-Young, Han-Shin Kim, Eunji Cha, Jeong-Hoon Park, and Hee-Deung Park. 2018. "Mitigation of membrane biofouling by a quorum quenching bacterium for membrane bioreactors." *Bioresource Technology* no. 258:220–226.

Heeb, Stephan, Matthew P Fletcher, Siri Ram Chhabra, Stephen P Diggle, Paul Williams, and Miguel Cámara. 2011. "Quinolones: From antibiotics to autoinducers." *FEMS Microbiology Reviews* no. 35 (2):247–274.

Hinds, Bruce J, Nitin Chopra, Terry Rantell, Rodney Andrews, Vasilis Gavalas, and Leonidas G Bachas. 2004. "Aligned multiwalled carbon nanotube membranes." *Science* no. 303 (5654):62–65.

Hong, Huachang, Meijia Zhang, Yiming He, Jianrong Chen, and Hongjun Lin. 2014. "Fouling mechanisms of gel layer in a submerged membrane bioreactor." *Bioresource Technology* no. 166:295–302.

Hou, Jun, Lingzhan Miao, Chao Wang, Peifang Wang, Yanhui Ao, and Bowen Lv. 2015. "Effect of CuO nanoparticles on the production and composition of extracellular polymeric substances and physicochemical stability of activated sludge flocs." *Bioresource Technology* no. 176:65–70.

Huang, Jean J, Jong-In Han, Lian-Hui Zhang, and Jared R Leadbetter. 2003. "Utilization of acyl-homoserine lactone quorum signals for growth by a soil pseudomonad and *Pseudomonas aeruginosa* PAO1." *Applied and Environmental Microbiology* no. 69 (10):5941–5949.

Huang, Jean J, Ashley Petersen, Marvin Whiteley, and Jared R Leadbetter. 2006. "Identification of QuiP, the product of gene PA1032, as the second acyl-homoserine lactone acylase of *Pseudomonas aeruginosa* PAO1." *Applied and Environmental Microbiology* no. 72 (2):1190–1197.

Huang, Jinhui, Yahui Shi, Guangming Zeng, Yanling Gu, Guiqiu Chen, Lixiu Shi, Yi Hu, Bi Tang, and Jianxin Zhou. 2016. "Acyl-homoserine lactone-based quorum sensing and quorum quenching hold promise to determine the performance of biological wastewater treatments: An overview." *Chemosphere* no. 157:137–151.

Inaba, Tomohiro, Tomoyuki Hori, Hidenobu Aizawa, Atsushi Ogata, and Hiroshi Habe. 2017. "Architecture, component, and microbiome of biofilm involved in the fouling of membrane bioreactors." *npj Biofilms and Microbiomes* no. 3 (1):5.

Ishige, Takeru, Kohsuke Honda, and Sakayu Shimizu. 2005. "Whole organism biocatalysis." *Current Opinion in Chemical Biology* no. 9 (2):174–180.

Jahangir, Daniyal, Hyun-Suk Oh, Sang-Ryoung Kim, Pyung-Kyu Park, Chung-Hak Lee, and Jung-Kee Lee. 2012. "Specific location of encapsulated quorum quenching bacteria for biofouling control in an external submerged membrane bioreactor." *Journal of Membrane Science* no. 411:130–136.

Jiang, Qi, Hao H Ngo, Long D Nghiem, Faisal I Hai, William E Price, Jian Zhang, Shuang Liang, Lijuan Deng, and Wenshan Guo. 2018. "Effect of hydraulic retention time on the performance of a hybrid moving bed biofilm reactor-membrane bioreactor system for micropollutants removal from municipal wastewater." *Bioresource Technology* no. 247:1228–1232.

Jiang, Tao. 2007. "Characterization and modelling of soluble microbial products in membrane bioreactors." Ghent: Ghent University.

Jiang, Tao, Hanmin Zhang, Dawen Gao, Feng Dong, Jifeng Gao, and Fenglin Yang. 2012. "Fouling characteristics of a novel rotating tubular membrane bioreactor." *Chemical Engineering and Processing: Process Intensification* no. 62:39–46.

Jorand, F, F Zartarian, F Thomas, JC Block, JY Bottero, G Villemin, V Urbain, and J Manem. 1995. "Chemical and structural (2D) linkage between bacteria within activated sludge flocs." *Water Research* no. 29 (7):1639–1647.

Judd, Simon. 2010. *The MBR book: Principles and applications of membrane bioreactors for water and wastewater treatment*. Amsterdam: Elsevier.

Kalia, Vipin Chandra. 2013. "Quorum sensing inhibitors: An overview." *Biotechnology Advances* no. 31 (2):224–245.

Khan, S Jamal, Chettiyappan Visvanathan, and Veeriah Jegatheesan. 2012. "Effect of powdered activated carbon (PAC) and cationic polymer on biofouling mitigation in hybrid MBRs." *Bioresource Technology* no. 113:165–168.

Kim, A Leum, Son-Young Park, Chi-Ho Lee, Chung-Hak Lee, and Jung-Kee Lee. 2014. "Quorum quenching bacteria isolated from the sludge of a wastewater treatment plant and their application for controlling biofilm formation." *Journal of Microbiology Biotechnology* no. 24 (11):1574–1582.

Kim, Hak-Woo, Hyun-Suk Oh, Sang-Ryoung Kim, Ki-Baek Lee, Kyung-Min Yeon, Chung-Hak Lee, Seil Kim, and Jung-Kee Lee. 2013a. "Microbial population dynamics and proteomics in membrane bioreactors with enzymatic quorum quenching." *Applied Microbiology and Biotechnology* no. 97 (10):4665–4675.

Kim, Sang-Ryoung, Ki-Baek Lee, Jeong-Eun Kim, Young-June Won, Kyung-Min Yeon, Chung-Hak Lee, and Dong-Joon Lim. 2015. "Macroencapsulation of quorum quenching bacteria by polymeric membrane layer and its application to MBR for biofouling control." *Journal of Membrane Science* no. 473:109–117.

Kim, Sang-Ryoung, Hyun-Suk Oh, Sung-Jun Jo, Kyung-Min Yeon, Chung-Hak Lee, Dong-Joon Lim, Chi-Ho Lee, and Jung-Kee Lee. 2013b. "Biofouling control with bead-entrapped quorum quenching bacteria in membrane bioreactors: Physical and biological effects." *Environmental Science & Technology* no. 47 (2):836–842.

Kochkodan, Victor, and Nidal Hilal. 2015. "A comprehensive review on surface modified polymer membranes for biofouling mitigation." *Desalination* no. 356:187–207.

Koerdt, Andrea, Julia Gödeke, Jürgen Berger, Kai M Thormann, and Sonja-Verena Albers. 2010. "Crenarchaeal biofilm formation under extreme conditions." *PloS one* no. 5 (11):e14104.

Kolodkin-Gal, Ilana, Diego Romero, Shugeng Cao, Jon Clardy, Roberto Kolter, and Richard Losick. 2010. "D-amino acids trigger biofilm disassembly." *Science* no. 328 (5978):627–629.

Konai, Mohini Mohan, Geetika Dhanda, and Jayanta Haldar. 2018. "Talking through chemical languages: Quorum sensing and bacterial communication." In *Quorum sensing and its biotechnological applications*, 17–42. Singapore: Springer.

Köse-Mutlu, Börte, Tülay Ergön-Can, İsmail Koyuncu, and Chung-Hak Lee. 2016. "Quorum quenching MBR operations for biofouling control under different operation conditions and using different immobilization media." *Desalination and Water Treatment* no. 57 (38):17696–17706.

Krysciak, D, C Schmeisser, S Preuss, J Riethausen, M Quitschau, S Grond, and WR Streit. 2011. "Involvement of multiple loci in quorum quenching of autoinducer I molecules in the nitrogen-fixing symbiont Rhizobium (*Sinorhizobium*) sp. strain NGR234." *Applied and Environmental Microbiology* no. 77 (15):5089–5099.

Krzeminski, Pawel, Lance Leverette, Simos Malamis, and Evina Katsou. 2017. "Membrane bioreactors–a review on recent developments in energy reduction, fouling control, novel configurations, LCA and market prospects." *Journal of Membrane Science* no. 527:207–227.

Lade, Harshad, Diby Paul, and Ji Hyang Kweon. 2014. "Quorum quenching mediated approaches for control of membrane biofouling." *International Journal of Biological Sciences* no. 10 (5):550.

Laspidou, Chrysi S, and Bruce E Rittmann. 2002. "A unified theory for extracellular polymeric substances, soluble microbial products, and active and inert biomass." *Water Research* no. 36 (11):2711–2720.

Le-Clech, Pierre, Vicki Chen, and Tony AG Fane. 2006. "Fouling in membrane bioreactors used in wastewater treatment." *Journal of Membrane Science* no. 284 (1–2):17–53.

Leadbetter, Jared R, and E Peter Greenberg. 2000. "Metabolism of acyl-homoserine lactone quorum-sensing signals by *Variovorax paradoxus*." *Journal of Bacteriology* no. 182 (24):6921–6926.

Lee, Byoungsoo, Kyung-Min Yeon, Jongmin Shim, Sang-Ryoung Kim, Chung-Hak Lee, Jinwoo Lee, and Jungbae Kim. 2014. "Effective antifouling using quorum-quenching acylase stabilized in magnetically-separable mesoporous silica." *Biomacromolecules* no. 15 (4):1153–1159.

Lee, Jasmine, Jien Wu, Yinyue Deng, Jing Wang, Chao Wang, Jianhe Wang, Changqing Chang, Yihu Dong, Paul Williams, and Lian-Hui Zhang. 2013. "A cell-cell communication signal integrates quorum sensing and stress response." *Nature Chemical Biology* no. 9 (5):339.

Lee, Jasmine, and Lianhui Zhang. 2015. "The hierarchy quorum sensing network in *Pseudomonas aeruginosa*." *Protein & Cell* no. 6 (1):26–41.

Lee, Kibaek, Seonki Lee, Sang Hyun Lee, Sang-Ryoung Kim, Hyun-Suk Oh, Pyung-Kyu Park, Kwang-Ho Choo, Yea-Won Kim, Jung-Kee Lee, and Chung-Hak Lee. 2016a. "Fungal quorum quenching: A paradigm shift for energy savings in membrane bioreactor (MBR) for wastewater treatment." *Environmental Science & Technology* no. 50 (20):10914–10922.

Lee, Kibaek, Huarong Yu, Xiaolei Zhang, and Kwang-Ho Choo. 2018. "Quorum sensing and quenching in membrane bioreactors: Opportunities and challenges for biofouling control." *Bioresource Technology* no. 270:656–668.

Lee, Sang H, Seonki Lee, Kibaek Lee, Chang H Nahm, Hyeokpil Kwon, Hyun-Suk Oh, Young-June Won, Kwang-Ho Choo, Chung-Hak Lee, and Pyung-Kyu Park. 2016b. "More efficient media design for enhanced biofouling control in a membrane bioreactor: Quorum quenching bacteria entrapping hollow cylinder." *Environmental Science & Technology* no. 50 (16):8596–8604.

Lee, Sang Jun, Sun-Yang Park, Jung-Ju Lee, Do-Young Yum, Bon-Tag Koo, and Jung-Kee Lee. 2002. "Genes encoding the *N*-acyl homoserine lactone-degrading enzyme are widespread in many subspecies of *Bacillus thuringiensis*." *Applied and Environmental Microbiology* no. 68 (8):3919–3924.

Lee, Seonki, Sang Hyun Lee, Kibaek Lee, Hyeokpil Kwon, Chang Hyun Nahm, Chung-Hak Lee, Pyung-Kyu Park, Kwang-Ho Choo, Jung-Kee Lee, and Hyun-Suk Oh. 2016c. "Effect of the shape and size of quorum-quenching media on biofouling control in membrane bioreactors for wastewater treatment." *Journal of Microbiology and Biotechnology* no. 26 (10):1746–1754.

Lee, Woo-Nyoung, In-Soung Chang, Byung-Kook Hwang, Pyung-Kyu Park, Chung-Hak Lee, and Xia Huang. 2007. "Changes in biofilm architecture with addition of membrane fouling reducer in a membrane bioreactor." *Process Biochemistry* no. 42 (4):655–661.

Lewandowski, Zbigniew, and Haluk Beyenal. 2005. "Biofilms: Their structure, activity, and effect on membrane filtration." *Water Science and Technology* no. 51 (6–7):181–192.

Lim, SooYeon, Seil Kim, Kyung-Min Yeon, Byoung-In Sang, Jongsik Chun, and Chung-Hak Lee. 2012. "Correlation between microbial community structure and biofouling in a laboratory scale membrane bioreactor with synthetic wastewater." *Desalination* no. 287:209–215.

Lin, Hongjun, Wei Peng, Meijia Zhang, Jianrong Chen, Huachang Hong, and Ye Zhang. 2013. "A review on anaerobic membrane bioreactors: Applications, membrane fouling and future perspectives." *Desalination* no. 314:169–188.

Lin, Hongjun, Meijia Zhang, Fangyuan Wang, Fangang Meng, Bao-Qiang Liao, Huachang Hong, Jianrong Chen, and Weijue Gao. 2014. "A critical review of extracellular polymeric substances (EPSs) in membrane bioreactors: Characteristics, roles in membrane fouling and control strategies." *Journal of Membrane Science* no. 460:110–125.

Lin, Yi-Han, Jin-Ling Xu, Jiangyong Hu, Lian-Hui Wang, Say Leong Ong, Jared Renton Leadbetter, and Lian-Hui Zhang. 2003. "Acyl-homoserine lactone acylase from Ralstonia strain XJ12B represents a novel and potent class of quorum-quenching enzymes." *Molecular Microbiology* no. 47 (3):849–860.

Liu, Yan, and Herbert HP Fang. 2003. "Influences of extracellular polymeric substances (EPS) on flocculation, settling, and dewatering of activated sludge." *Critical Reviews in Environmental Science and Technology* no. 33: 237–273.

Loiselle, Melanie, and Kimberly W Anderson. 2003. "The use of cellulase in inhibiting biofilm formation from organisms commonly found on medical implants." *Biofouling* no. 19 (2):77–85.

Luo, Jinxue, Jinsong Zhang, Xiaohui Tan, Diane McDougald, Guoqiang Zhuang, Anthony G Fane, Staffan Kjelleberg, Yehuda Cohen, and Scott A Rice. 2015. "Characterization of the archaeal community fouling a membrane bioreactor." *Journal of Environmental Sciences* no. 29:115–123.

Ma, Anzhou, Di Lv, Xuliang Zhuang, and Guoqiang Zhuang. 2013. "Quorum quenching in culturable phyllosphere bacteria from tobacco." *International Journal of Molecular Sciences* no. 14 (7):14607–14619.

Ma, Cong, Shuili Yu, Wenxin Shi, Wende Tian, Sebastiaan Gerard Jozef Heijman, and Louis Cornelis Rietveld. 2012. "High concentration powdered activated carbon-membrane bioreactor (PAC-MBR) for slightly polluted surface water treatment at low temperature." *Bioresource Technology* no. 113:136–142.

Malaeb, Lilian, Pierre Le-Clech, Johannes S Vrouwenvelder, George M Ayoub, and Pascal E Saikaly. 2013. "Do biological-based strategies hold promise to biofouling control in MBRs?" *Water Research* no. 47 (15):5447–5463.

Mansouri, Jaleh, Simon Harrisson, and Vicki Chen. 2010. "Strategies for controlling biofouling in membrane filtration systems: Challenges and opportunities." *Journal of Materials Chemistry* no. 20 (22):4567–4586.

Maqbool, Tahir, Sher Jamal Khan, Hira Waheed, Chung-Hak Lee, Imran Hashmi, and Hamid Iqbal. 2015. "Membrane biofouling retardation and improved sludge characteristics using quorum quenching bacteria in submerged membrane bioreactor." *Journal of Membrane Science* no. 483:75–83.

McClean, Kay H, Michael K Winson, Leigh Fish, Adrian Taylor, Siri Ram Chhabra, Miguel Camara, Mavis Daykin, John H Lamb, Simon Swift, and Barrie W Bycroft. 1997. "Quorum sensing and *Chromobacterium violaceum*: Exploitation of violacein production and inhibition for the detection of *N*-acylhomoserine lactones." *Microbiology* no. 143 (12):3703–3711.

Mei, Gui-Ying, Xiao-Xue Yan, Ali Turak, Zhao-Qing Luo, and Li-Qun Zhang. 2010. "AidH, an alpha/beta-hydrolase fold family member from an *Ochrobactrum* sp. strain, is a novel *N*-acylhomoserine lactonase." *Applied and Environmental Microbiology* no. 76 (15):4933–4942.

Meng, Fangang, So-Ryong Chae, Anja Drews, Matthias Kraume, Hang-Sik Shin, and Fenglin Yang. 2009. "Recent advances in membrane bioreactors (MBRs): Membrane fouling and membrane material." *Water Research* no. 43 (6):1489–1512.

Meng, Fangang, Shaoqing Zhang, Yoontaek Oh, Zhongbo Zhou, Hang-Sik Shin, and So-Ryong Chae. 2017. "Fouling in membrane bioreactors: An updated review." *Water Research* no. 114:151–180.

Merone, Luigia, Luigi Mandrich, Mosè Rossi, and Giuseppe Manco. 2005. "A thermostable phosphotriesterase from the archaeon *Sulfolobus solfataricus*: Cloning, overexpression and properties." *Extremophiles* no. 9 (4):297–305.

Miura, Yuki, Mirian Noriko Hiraiwa, Tsukasa Ito, Takanori Itonaga, Yoshimasa Watanabe, and Satoshi Okabe. 2007. "Bacterial community structures in MBRs treating municipal wastewater: Relationship between community stability and reactor performance." *Water Research* no. 41 (3):627–637.

Molobela, I Phyllis, T Eugene Cloete, and Mervyn Beukes. 2010. "Protease and amylase enzymes for biofilm removal and degradation of extracellular polymeric substances (EPS) produced by *Pseudomonas fluorescens* bacteria." *African Journal of Microbiology Research* no. 4 (14):1515–1524.

More, TT, Jay Shankar Singh Yadav, Song Yan, Rajeshwar Dayal Tyagi, and Rao Y Surampalli. 2014. "Extracellular polymeric substances of bacteria and their potential environmental applications." *Journal of Environmental Management* no. 144:1–25.

Morohoshi, Tomohiro, Atsushi Ebata, Shigehisa Nakazawa, Norihiro Kato, and Tsukasa Ikeda. 2005. "*N*-acyl homoserine lactone-producing or-degrading bacteria isolated from the intestinal microbial flora of ayu fish (*Plecoglossus altivelis*)." *Microbes and Environments* no. 20 (4):264–268.

Morohoshi, Tomohiro, Yoshiaki Tominaga, Nobutaka Someya, and Tsukasa Ikeda. 2012. "Complete genome sequence and characterization of the *N*-acylhomoserine lactone-degrading gene of the potato leaf-associated *Solibacillus silvestris*." *Journal of Bioscience and Bioengineering* no. 113 (1):20–25.

Mukherjee, Manisha, Yidan Hu, Chuan Hao Tan, Scott A Rice, and Bin Cao. 2018. "Engineering a light-responsive, quorum quenching biofilm to mitigate biofouling on water purification membranes." *Science Advances* no. 4 (12):eaau1459.

Nahm, Chang Hyun, Dong-Chan Choi, Hyeokpil Kwon, Seonki Lee, Sang Hyun Lee, Kibaek Lee, Kwang-Ho Choo, Jung-Kee Lee, Chung-Hak Lee, and Pyung-Kyu Park. 2017. "Application of quorum quenching bacteria entrapping sheets to enhance biofouling control in a membrane bioreactor with a hollow fiber module." *Journal of Membrane Science* no. 526:264–271.

Nguyen, Thang, Felicity Roddick, and Linhua Fan. 2012. "Biofouling of water treatment membranes: A review of the underlying causes, monitoring techniques and control measures." *Membranes* no. 2 (4):804–840.

Nielsen, Per Halkjær, Bente Frølund, and Kristian Keiding. 1996. "Changes in the composition of extracellular polymeric substances in activated sludge during anaerobic storage." *Applied Microbiology and Biotechnology* no. 44 (6):823–830.

Oh, Hyun-Suk, and Chung-Hak Lee. 2018. "Origin and evolution of quorum quenching technology for biofouling control in MBRs for wastewater treatment." *Journal of Membrane Science* no. 554:331–345.

Oh, Hyun-Suk, Kyung-Min Yeon, Cheon-Seok Yang, Sang-Ryoung Kim, Chung-Hak Lee, Son Young Park, Jong Yun Han, and Jung-Kee Lee. 2012. "Control of membrane biofouling in MBR for wastewater treatment by quorum quenching bacteria encapsulated in microporous membrane." *Environmental Science & Technology* no. 46 (9):4877–4884.

Orell, Alvaro, Sabrina Fröls, and Sonja-Verena Albers. 2013. "Archaeal biofilms: The great unexplored." *Annual Review of Microbiology* no. 67:337–354.

Pandit, Soumya, Shruti Sarode, and Kuppam Chandrasekhar. 2018. "Fundamentals of bacterial biofilm: Present state of art." In *Quorum sensing and Its biotechnological applications*, 43–60. Singapore: Springer.

Park, Sun-Yang, Byung-Joon Hwang, Min-Ho Shin, Jung-Ae Kim, Ha-Kun Kim, and Jung-Kee Lee. 2006. "*N*-acylhomoserine lactonase producing *Rhodococcus* spp. with different AHL-degrading activities." *FEMS Microbiology Letters* no. 261 (1):102–108.

Park, Sun-Yang, Hye-Ok Kang, Hak-Sun Jang, Jung-Kee Lee, Bon-Tag Koo, and Do-Young Yum. 2005. "Identification of extracellular *N*-acylhomoserine lactone acylase from a *Streptomyces* sp. and its application to quorum quenching." *Applied and Environmental Microbiology* no. 71 (5):2632–2641.

Park, Sun-Yang, Sang Jun Lee, Tae-Kwang Oh, Jong-Won Oh, Bon-Tag Koo, Do-Young Yum, and Jung-Kee Lee. 2003. "AhlD, an *N*-acylhomoserine lactonase in *Arthrobacter* sp., and predicted homologues in other bacteria." *Microbiology* no. 149 (6):1541–1550.

Parveen, Nikhat, and Kenneth A Cornell. 2011. "Methylthioadenosine/S-adenosylhomocysteine nucleosidase, a critical enzyme for bacterial metabolism." *Molecular Microbiology* no. 79 (1):7–20.

Pasmore, Mark, Paul Todd, Sara Smith, Dawn Baker, JoAnn Silverstein, Darrell Coons, and Christopher N Bowman. 2001. "Effects of ultrafiltration membrane surface properties on *Pseudomonas aeruginosa* biofilm initiation for the purpose of reducing biofouling." *Journal of Membrane Science* no. 194 (1):15–32.

Paul, Diby, Young Sam Kim, Kannan Ponnusamy, and Ji Hyang Kweon. 2009. "Application of quorum quenching to inhibit biofilm formation." *Environmental Engineering Science* no. 26 (8):1319–1324.

Perveen, Shabila. 2018. *Detection of quorum quenching enzymes producing genes in membrane bioreactor*, IESE NUST Islamabad.

Porcelli, Nicandro, and Simon Judd. 2010. "Chemical cleaning of potable water membranes: A review." *Separation and Purification Technology* no. 71 (2):137–143.

Poxon, Theresa L, and Jeannie L Darby. 1997. "Extracellular polyanions in digested sludge: Measurement and relationship to sludge dewaterability." *Water Research* no. 31 (4):749–758.

Purnell, Sarah, James Ebdon, Austen Buck, Martyn Tupper, and Huw Taylor. 2015. "Bacteriophage removal in a full-scale membrane bioreactor (MBR)–Implications for wastewater reuse." *Water Research* no. 73:109–117.

Ramesh, A, DJ Lee, and JY Lai. 2007. "Membrane biofouling by extracellular polymeric substances or soluble microbial products from membrane bioreactor sludge." *Applied Microbiology and Biotechnology* no. 74 (3):699–707.

Rasmussen, Thomas Bovbjerg, Thomas Bjarnsholt, Mette Elena Skindersoe, Morten Hentzer, Peter Kristoffersen, Manuela Köte, John Nielsen, Leo Eberl, and Michael Givskov. 2005. "Screening for quorum-sensing inhibitors (QSI) by use of a novel genetic system, the QSI selector." *Journal of Bacteriology* no. 187 (5):1799–1814.

Riaz, K, C Elmerich, A Raffoux, D Moreira, Y Dessaux, and D Faure. 2008. "Metagenomics revealed a quorum quenching lactonase QlcA from yet unculturable soil bacteria." *Communications in Agricultural and Applied Biological Sciences* no. 73 (2):3–6.

Romero, Manuel, Rubén Avendaño-Herrera, Beatriz Magariños, Miguel Cámara, and Ana Otero. 2010. "Acylhomoserine lactone production and degradation by the fish pathogen *Tenacibaculum maritimum*, a member of the Cytophaga–Flavobacterium–Bacteroides (CFB) group." *FEMS Microbiology Letters* no. 304 (2):131–139.

Romero, Manuel, Stephen P Diggle, Stephan Heeb, Miguel Camara, and Ana Otero. 2008. "Quorum quenching activity in *Anabaena* sp. PCC 7120: Identification of AiiC, a novel AHL-acylase." *FEMS Microbiology Letters* no. 280 (1):73–80.

Rosenberger, S, C Laabs, B Lesjean, R Gnirss, G Amy, M Jekel, and J-C Schrotter. 2006. "Impact of colloidal and soluble organic material on membrane performance in membrane bioreactors for municipal wastewater treatment." *Water Research* no. 40 (4):710–720.

Schauder, Stephan, Kevan Shokat, Michael G Surette, and Bonnie L Bassler. 2001. "The LuxS family of bacterial autoinducers: Biosynthesis of a novel quorum-sensing signal molecule." *Molecular Microbiology* no. 41 (2):463–476.

Seo, Gon, Satoshi Takizawa, and Shinichiro Ohgaki. 2002. "Ammonia oxidation at low temperature in a high concentration powdered activated carbon membrane bioreactor." *Water Science and Technology: Water Supply* no. 2 (2):169–176.

Shao, Yi, and Bonnie L Bassler. 2012. "Quorum-sensing non-coding small RNAs use unique pairing regions to differentially control mRNA targets." *Molecular Microbiology* no. 83 (3):599–611.

Shen, Liguo, Xuena Wang, Renjie Li, Haiying Yu, Huachang Hong, Hongjun Lin, Jianrong Chen, and Bao-Qiang Liao. 2017. "Physicochemical correlations between membrane surface hydrophilicity and adhesive fouling in membrane bioreactors." *Journal of Colloid and Interface Science* no. 505:900–909.

Shepherd, Ryan W, and Steven E Lindow. 2009. "Two dissimilar *N*-acyl-homoserine lactone acylases of *Pseudomonas syringae* influence colony and biofilm morphology." *Applied and Environmental Microbiology* no. 75 (1):45–53.

Siddiqui, Muhammad Faisal, Miles Rzechowicz, Harvey Winters, Zularisam Abd Wahid, and Anthony G Fane 2015. "Quorum sensing based membrane biofouling control for water treatment: A review." *Journal of Water Process Engineering* no. 7:112–122.

Siddiqui, Muhammad Faisal, Mimi Sakinah, Lakhveer Singh, and AW Zularisam. 2012. "Targeting *N*-acyl-homoserine-lactones to mitigate membrane biofouling based on quorum sensing using a biofouling reducer." *Journal of Biotechnology* no. 161 (3):190–197.

Sio, Charles F, Linda G Otten, Robbert H Cool, Stephen P Diggle, Peter G Braun, Rein Bos, Mavis Daykin, Miguel Cámara, Paul Williams, and Wim J Quax. 2006. "Quorum quenching by an *N*-acyl-homoserine lactone acylase from *Pseudomonas aeruginosa* PAO1." *Infection and immunity* no. 74 (3):1673–1682.

Tabraiz, Shamas, Sajjad Haydar, Paul Sallis, Sadia Nasreen, Qaisar Mahmood, Muhammad Awais, and Kishor Acharya. 2017. "Effect of cycle run time of backwash and relaxation on membrane fouling removal in submerged membrane bioreactor treating sewage at higher flux." *Water Science and Technology* no. 76 (4):963–975.

Tan, Chuan Hao, Kai Shyang Koh, Chao Xie, Martin Tay, Yan Zhou, Rohan Williams, Wun Jern Ng, Scott A Rice, and Staffan Kjelleberg. 2014. "The role of quorum sensing signalling in EPS production and the assembly of a sludge community into aerobic granules." *The ISME Journal* no. 8 (6):1186.

Tang, Kaihao, and Xiao-Hua Zhang. 2014. "Quorum quenching agents: resources for antivirulence therapy." *Marine drugs* no. 12 (6):3245–3282.

Tian, Yu. 2008. "Behaviour of bacterial extracellular polymeric substances from activated sludge: A review." *International Journal of Environment and Pollution* no. 32 (1):78–89.

Trinh, Trang, Amos Branch, Adam C Hambly, Guido Carvajal, Heather M Coleman, Richard M Stuetz, Jörg E Drewes, Pierre Le-Clech, and Stuart J Khan. 2016. "Hazardous events in membrane bioreactors–Part 1: Impacts on key operational and bulk water quality parameters." *Journal of Membrane Science* no. 497:494–503.

Trouve, E, V Urbain, and J Manem. 1994. "Treatment of municipal wastewater by a membrane bioreactor: Results of a semi-industrial pilot-scale study." *Water Science and Technology* no. 30 (4):151.

Ulrich, Ricky L. 2004. "Quorum quenching: Enzymatic disruption of *N*-acylhomoserine lactone-mediated bacterial communication in *Burkholderia thailandensis*." *Applied and Environmental Microbiology* no. 70 (10):6173–6180.

United Nations, UN. 2018. "Sustainable Development Goal 6 Synthesis Report on Water and Sanitation." Published by the United Nations New York, New York no. 10017.

Uroz, Stephane, Siri Ram Chhabra, Miguel Camara, Paul Williams, Phil Oger, and Yves Dessaux. 2005. "*N*-Acylhomoserine lactone quorum-sensing molecules are modified and degraded by *Rhodococcus erythropolis* W2 by both amidolytic and novel oxidoreductase activities." *Microbiology* no. 151 (10):3313–3322.

Uroz, Stephane, Cathy D'Angelo-Picard, Aurelien Carlier, Miena Elasri, Carine Sicot, Annik Petit, Phil Oger, Denis Faure, and Yves Dessaux. 2003. "Novel bacteria degrading *N*-acylhomoserine lactones and their use as quenchers of quorum-sensing-regulated functions of plant-pathogenic bacteria." *Microbiology* no. 149 (8):1981–1989.

Uroz, Stéphane, Yves Dessaux, and Phil Oger. 2009. "Quorum sensing and quorum quenching: The yin and yang of bacterial communication." *ChemBioChem* no. 10 (2):205–216.

Uroz, Stéphane, Phil Oger, Siri Ram Chhabra, Miguel Cámara, Paul Williams, and Yves Dessaux. 2007. "*N*-acyl homoserine lactones are degraded via an amidolytic activity in *Comamonas* sp. strain D1." *Archives of Microbiology* no. 187 (3):249–256.

Uroz, Stéphane, Phil M Oger, Emilie Chapelle, Marie-Thérèse Adeline, Denis Faure, and Yves Dessaux. 2008. "A *Rhodococcus* qsdA-encoded enzyme defines a novel class of large-spectrum quorum-quenching lactonases." *Applied and Environmental Microbiology* no. 74 (5):1357–1366.

Wang, Lian-Hui, Yawen He, Yunfeng Gao, Ji En Wu, Yi-Hu Dong, Chaozu He, Su Xing Wang, Li-Xing Weng, Jin-Ling Xu, and Leng Tay. 2004. "A bacterial cell–cell communication signal with cross-kingdom structural analogues." *Molecular Microbiology* no. 51 (3):903–912.

Wang, Pan, Zhiwei Wang, Zhichao Wu, Qi Zhou, and Dianhai Yang. 2010a. "Effect of hypochlorite cleaning on the physiochemical characteristics of polyvinylidene fluoride membranes." *Chemical Engineering Journal* no. 162 (3):1050–1056.

Wang, Wen-Zhao, Tomohiro Morohoshi, Masashi Ikenoya, Nobutaka Someya, and Tsukasa Ikeda. 2010b. "AiiM, a novel class of *N*-acylhomoserine lactonase from the leaf-associated bacterium *Microbacterium testaceum*." *Applied and Environmental Microbiology* no. 76 (8):2524–2530.

Waters, Christopher M, and Bonnie L Bassler. 2005. "Quorum sensing: Cell-to-cell communication in bacteria." *Annual Review of Cell and Developmental Biology* no. 21:319–346.

Watson, Bruce, Thomas C Currier, Milton P Gordon, Mary-Dell Chilton, and Eugene W Nester. 1975. "Plasmid required for virulence of Agrobacterium tumefaciens." *Journal of Bacteriology* no. 123 (1):255–264.

Weerasekara, Nuwan A, Kwang-Ho Choo, and Chung-Hak Lee. 2014. "Hybridization of physical cleaning and quorum quenching to minimize membrane biofouling and energy consumption in a membrane bioreactor." *Water Research* no. 67:1–10.

Weerasekara, Nuwan A, Kwang-Ho Choo, and Chung-Hak Lee. 2016. "Biofouling control: Bacterial quorum quenching versus chlorination in membrane bioreactors." *Water Research* no. 103:293–301.

Wong, Philip Chuen Yung, Jia Yi Lee, and Chee Wee Teo. 2015. "Application of dispersed and immobilized hydrolases for membrane fouling mitigation in anaerobic membrane bioreactors." *Journal of Membrane Science* no. 491:99–109.

Wu, Bing, Rong Wang, and Anthony G Fane. 2017. "The roles of bacteriophages in membrane-based water and wastewater treatment processes: A review." *Water Research* no. 110:120–132.

Xiao, Yeyuan, Hira Waheed, Keke Xiao, Imran Hashmi, and Yan Zhou. 2018. "In tandem effects of activated carbon and quorum quenching on fouling control and simultaneous removal of pharmaceutical compounds in membrane bioreactors." *Chemical Engineering Journal* no. 341:610–617.

Xiong, Yanghui, and Yu Liu. 2010. "Biological control of microbial attachment: A promising alternative for mitigating membrane biofouling." *Applied Microbiology and Biotechnology* no. 86 (3):825–837.

Yamamoto, Kazuo, Masami Hiasa, Talat Mahmood, and Tomonori Matsuo. 1988. "Direct solid-liquid separation using hollow fiber membrane in an activated sludge aeration tank." In *Water Pollution Research and Control Brighton*, 43–54. Pergamon Press, Brighton, UK.

Yamamoto, Kazuo, and Khin Muang Win. 1991. "Tannery wastewater treatment using a sequencing batch membrane reactor." *Water Science and Technology* no. 23 (7–9):1639–1648.

Ye, Fenxia, Yangfang Ye, and Ying Li. 2011. "Effect of C/N ratio on extracellular polymeric substances (EPS) and physicochemical properties of activated sludge flocs." *Journal of Hazardous Materials* no. 188 (1–3):37–43.

Yeon, Kyung-Min, Won-Seok Cheong, Hyun-Suk Oh, Woo-Nyoung Lee, Byung-Kook Hwang, Chung-Hak Lee, Haluk Beyenal, and Zbigniew Lewandowski. 2008. "Quorum sensing: A new biofouling control paradigm in a membrane bioreactor for advanced wastewater treatment." *Environmental Science & Technology* no. 43 (2):380–385.

Yoon, Seok-Hwan, John H Collins, Deepak Musale, Suma Sundararajan, Shih-Perng Tsai, G Anders Hallsby, Jian Feng Kong, Jeroen Koppes, and Philippe Cachia. 2005. "Effects of flux enhancing polymer on the characteristics of sludge in membrane bioreactor process." *Water Science and Technology* no. 51 (6–7):151–157.

Yu, Cong, Xuening Li, Nan Zhang, Donghui Wen, Charles Liu, and Qilin Li. 2016. "Inhibition of biofilm formation by D-tyrosine: Effect of bacterial type and D-tyrosine concentration." *Water Research* no. 92:173–179.

Yu, Huarong, Fangshu Qu, Xiaolei Zhang, Peng Wang, Guibai Li, and Heng Liang. 2018. "Effect of quorum quenching on biofouling and ammonia removal in membrane bioreactor under stressful conditions." *Chemosphere* no. 199:114–121.

Zhang, Hai-Bao, Lian-Hui Wang, and Lian-Hui Zhang. 2002a. "Genetic control of quorum-sensing signal turnover in Agrobacterium tumefaciens." *Proceedings of the National Academy of Sciences* no. 99 (7):4638–4643.

Zhang, Haimin, Xie Quan, Shuo Chen, and Huimin Zhao. 2006. "Fabrication and characterization of silica/titania nanotubes composite membrane with photocatalytic capability." *Environmental Science & Technology* no. 40 (19):6104–6109.

Zhang, Meijia, Bao-qiang Liao, Xiaoling Zhou, Yiming He, Huachang Hong, Hongjun Lin, and Jianrong Chen. 2015a. "Effects of hydrophilicity/hydrophobicity of membrane on membrane fouling in a submerged membrane bioreactor." *Bioresource Technology* no. 175:59–67.

Zhang, Rong-guang, Katherine M Pappas, Jennifer L Brace, Paula C Miller, Tim Oulmassov, John M Molyneaux, John C Anderson, James K Bashkin, Stephen C Winans, and Andrzej Joachimiak. 2002b. "Structure of a bacterial quorum-sensing transcription factor complexed with pheromone and DNA." *Nature* no. 417 (6892):971.

Zhang, Zhenghua, Yuan Wang, Greg L Leslie, and T David Waite. 2015b. "Effect of ferric and ferrous iron addition on phosphorus removal and fouling in submerged membrane bioreactors." *Water research* no. 69:210–222.

Zheng, Yi, Wenxiang Zhang, Bing Tang, Jie Ding, and Zhien Zhang. 2018. "Membrane fouling mechanism of biofilm-membrane bioreactor (BF-MBR): Pore blocking model and membrane cleaning." *Bioresource Technology* no. 250:398–405.

Zhu, Zhenya, Lei Wang, and Qingqing Li. 2018. "A bioactive poly (vinylidene fluoride)/graphene oxide@ acylase nanohybrid membrane: Enhanced anti-biofouling based on quorum quenching." *Journal of Membrane Science* no. 547:110–122.

27

Application of Quorum Quenching in the Control of Animal Bacterial Pathogens

Tsegay Teame, Rui Xia, Yalin Yang, Chenchen Gao, Fengli Zhang, Chao Ran, Zhen Zhang, Hongling Zhang, and Zhigang Zhou

CONTENTS

27.1 Introduction

Infectious diseases are the most serious economic or ecological problems in several species of terrestrial and aquatic animals. There are a number of triggering factors behind the emergence of each disease in animals, and each of them is unique. For instance, in terrestrial animals, infectious disease can introduce from other areas through transferring of an infected animal from one area to another via market channels. In the case of aquatic animals, infectious diseases may emerge through pathogen exchange with wild populations as in the case of Salmonid production, evolution from nonpathogenic microorganisms, and anthropogenic transfer of stocks (Murray and Peeler, 2005). Nowadays the intensification of animal production system frequently results in high population densities and causes stresses which increase the risk of infection diseases and their spreading. In the case of aquatic animals, expansion of the aquaculture sector requires introduction of a new species from the wild which may become a case of disease emerging and affect both wild and farmed fish adversely.

So far, the large number and type of antibiotic have been intensively used as preventive and curative measures in the production of terrestrial and aquatic animals, but such practices brought a particular concern because regular use of these chemical products has resulted in an increasing number of antibiotic-resistance strains of pathogenic microbes, rendered antibiotic treatments ineffective, and can also have harmful effects on the environment. Furthermore, the horizontal transfer of resistance determinants to human pathogens and the presence of antibiotics

Tsegay Teame and Rui Xia are contributed equally for this work.

residues in the aquaculture products for human consumption constitute important threats to public health (Defoirdt et al., 2007). Due to these reasons, scientists have been assessing new mechanisms and strategies to control the pathogenic microbes, in order to increase the animal production sustainably. Presently, several infectious diseases control mechanisms are developing and antimicrobial is one of the controlling methods. Mostly probiotic bacteria are used as microbial control agents. Verschuere et al. (2000) described that probiotics act to control disease in different modes of action including competing with harmful organisms for space and nutrients, producing inhibitor compounds for pathogens, enhancing the immune response of the host, and improving the quality of the water where the fish lived.

Currently the advancement of molecular and cellular technologies makes it easy to study the genetic and biochemical analysis of bacteria, bacterial pathogenesis, and the complex interaction between pathogenic bacteria and their hosts at cellular and molecular levels. The availability of a model animal to assess the bacterial pathogenicity factors *in vitro* contributes much to the study of bacterial diseases in animals. The genetic analysis study on bacterial virulence factors revealed that the pathogenic bacteria are different from the nonpathogenic by the presence of specific pathogenicity genes which deceptively have been acquired via horizontal genetic transfer through evolutionary processes. The pathogenicity potential of bacteria varies from species to species and even within species. Some can infect a wide variety of organisms, and some are specific to a single species. For any pathogenic organism, the most difficulty comes from the need to pass, survive, and overcome host defense lines. To overwhelm with these activities, pathogens use an extensive arsenal of virulence related factors. Moreover, the communication between bacteria cells is vital for the process of virulence and pathogenicity development.

Therefore, with the fast growing and expansion of the terrestrial as well as the aquatic animal farming, development of alternative methods to control infectious diseases is crucial. Pathogenic bacteria produce diverse virulence factors to infect animals. Donneberg (2000) revealed that motility, adhesion, host tissue degradation, iron acquisition, secretion of toxins, and the other chemicals that protect from the host defense are the main virulence factors regulated by the gene of pathogenic bacteria. Interference with the pathogenic bacteria genes involved in the production and activity of virulence factors is an alternative strategy to control animal pathogens (Defoirdt et al., 2013). Quorum quenching is important to control pathogenic bacteria via interfering with the quorum sensing (QS) system of the pathogenic bacteria or interferes with the host – pathogenic reaction system mechanisms in animals.

The classification of QQ agents in to small molecules (quorum sensing inhibitors, QSIs) and macromolecules (QQ enzymes) was based on their molecular weight. A diverse array of microorganisms communicate with each other in a cell density-dependent manner through the exchange of diffusible signal molecules or autoinducers, and this is termed a quorum sensing system. *N*-Acylhomoserine lactones (AHLs) are the major signaling molecules for QS in Gram-negative bacteria (Papenfort and Bassler, 2016). The AHL signal molecules produced by different organisms are highly conserved; they have the same homoserine lactone moiety, but may differ in the length and substitution of their acyl side chain. These structural features suggest that there may be at least four types of enzymes that could degrade AHL signals. Several opportunistic and pathogenic bacteria produce AHLs to regulate the expression of various genes related to their pathogenicity, biofilm formation, and antibiotic production (Burt et al., 2014; O'Loughlin et al., 2013). Bacteria that can recognize this QS communication have developed the capacity to interfere with it at different stages. Several studies showed that a wide range of QQ enzymes and QSIs have been reported to act on the QS system of microbes by various mechanisms including inhibiting the production of signal molecules, enzymatically degrading the signal molecules, competing for binding sites with signal molecules, inhibiting gene expression, and scavenging of AIs by antibodies and macromolecules such as cyclodextrins (Kalia, 2013; Morohoshi et al., 2013; Park et al., 2007a). Various QQ enzymes synthesized by several species of prokaryotes and eukaryotes including the lactonases, acylases, phosphotriesterase-like lactonases (PLLs), and oxidoreductases have an ability to degrade AHLs signal molecules, (Fetzner, 2015), and oxidoreductases enzymes also have the capability to degrade the AI-2 QS signal molecule (Dong et al., 2000; Weiland-Bräuer et al., 2016). A wide variety of plants are known to synthesize QSIs compounds (e.g., curcumin, coumarin, ajoene, citrus limonoids, flavonoids, and naringenin) that either degrade QS signals or compete for signal receptors (Brackman et al., 2008; Jakobsen et al., 2017; Vikram et al., 2011; Vandeputte et al., 2011; Teplitski et al., 2011; Paczkowski et al., 2017; Packiavathy et al., 2013; Gutiérrez-Barranquero et al., 2015), and other small molecule dicyclic peptides have also been reported to act as QSIs (Bauer and Robinson, 2002). Currently, quorum quenching enzymes and other natural or synthetic quorum sensing inhibitors have been discovered and evaluated as novel antimicrobial agents against different pathogenic bacteria in animals with promising diseases controlling results.

27.2 Quorum Sensing and Quorum Sensing Mediated Virulence Factor in Clinical Animal Bacterial Pathogens

27.2.1 Clinical Relevance of Animal Bacterial Pathogens and Their Quorum Sensing Systems

There are several pathogenic bacteria for terrestrial and aquatic animals, which are their virulence factor gene expression depending on their QS system. In aquatic animals, the common pathogenic bacterial species belong to the genera *Vibrio*, *Pseudomonas*, *Aeromonas*, *Flavobacterium*, *Yersinia*, *Edwardsiella*, *Streptococcus*, *Lactococcus*, *Renibacterium*, and *Mycobacterium* (Austin and Austin, 1999; Castillo-Juarez et al., 2015). These bacteria cause a great loss in the aquaculture sector. For instance, *V. harveyi*, a Gram-negative, pathogenic, marine fluorescence emitting bacteria commonly present in gut microflora of aquatic invertebrates, crustaceans, molluscs, and vertebrates including fishes. *V. harveyi*, an oligotropic pathogen reported to be associated with bright red syndrome (Soto-Rodriguez et al., 2012) and luminous vibriosis (Harikrishnan et al., 2011) to aquatic invertebrates and skin ulcers, eye lesions, gastro-enteritis (Cao and Meighen, 1989), and vasculitis to vertebrates, impedes the commercial development of aquaculture around the world. In the process of infection, the bacteria uses its QS system to express the virulence factor.

In animal production, the diseases caused by the species of bacteria, *Staphylococcus, Pseudomonas,* and *Salmonella*, play a great role in the reduction of animal production. *Staphylococcus aureus (S. aureus)* is a Gram-positive bacterium and use the *agr* (accessory gene regulator) QS system. The *S. aureus agr* are classified in to four various groups by sequence analysis method, and the autoinducers peptides of one *agr* group inhibit the expression of the *agr* expression of the other groups (Novick, 2003). Several haemolysins, enterotoxins, exfoliating toxins, enzymes, and surface proteins are regulated by *agr*. Therefore, *agr* has been assigned various functions in the pathogenesis of *S. aureus* infections in several animal species.

The virulence power of the bacteria strain is important to the level of infection in the animal, and it expresses in different levels in various animals. For instance, at low level of virulence, strains of *S. aurues* infects rabbits, causing skin infections, mastitis, and internal abscessation and occasionally infects an individual animal, while in the animal infected with highly virulent strains, it results chronic diseases and spreads rapidly throughout the body of the rabbit (Vancraeynest et al., 2006). In *S. aureus*, virulence factors are regulated by their QS system while the formation of biofilm is negatively regulated by the main QS system, and the *agr*-dependent expression plays a key role in the detachment of biofilms and subsequent colonization of new sites (Boles and Horswill, 2008). Therefore, although still speculative, the QS-dependent detachment from biofilms may render some *S. aureus* strains more capable of spreading through a herd and causing chronic problems at the herd level (e.g., highly virulent rabbit strains). This may also explain why agr+ and agr− variants of *S. aureus* might have a cooperative interaction in certain types of infections (Traber et al., 2008).

P. aeruginosa is also an important opportunistic pathogen in humans and other animals, and its infections are difficult to treat since this bacterium is resistant to several antimicrobial agents (Rubin et al., 2008), and its antimicrobial susceptibility is even lower when it is present in biofilms (Girard and Bloemberg, 2008). *P. aeruginosa* cause various tissue infections in different animal species, which indicates their adaptability of the bacterium. *P. aeruginosa* has two different QS systems, Las and Rhl, and a multitude of virulence factors are regulated by at least one of the two different QS systems (Girard and Bloemberg, 2008).

Biofilm formation is one of the characters of bacteria which is important to infect their host and the resist effectiveness of antimicrobials, which is regulated by the QS system of the bacteria. Pseudomonas is notorious for its capability to form biofilms. Through the formation of a biofilm, the bacteria are able to colonize all kinds of animal tissues, plants, and inert surfaces, thereby increasing environmental persistence. Moreover, bacteria inside a biofilm are difficult to kill either by antimicrobial substances or by host defense mechanisms, and therefore often result in chronic infections that do not respond to antimicrobial therapy. Since *P. aeruginosa* is found ubiquitously in water, wet surfaces are predisposed to be colonized with *P. aeruginosa* biofilm. *P. aeruginosa* is also a severe complicating factor in various animals including in the cases of equine ulcerative keratitis (Keller and Hendrix, 2005), in dermatitis, otitis externa and urinary tract infections in small animals (Rubin et al., 2008), in fleece rot in sheep (Kingsford and Raadsma, 1997), in bovine mastitis, which is often associated with contaminated udder washing water or contaminated intra-mammary dry-cow preparations (McLennan et al., 1997), and in hemorrhagic pneumonia in mink (Hammer et al., 2003). Even though QS often favors the expression of virulence genes, Tron et al. (2004) showed that *P. aeruginosa* canine otitis externa isolates share a deficiency of one or more components of the Rhl QS system, leading to the loss of the elastase virulence factor. The virulence factor that favors colonization of canine ears may be under the negative control of the QS system of the bacterium, or a factor that facilitates removal of the bacterium from the canine ears may be under the positive control of Rhl. The authors postulate that either this phenotype might be present before infection or else it might be acquired after infection in an effort to adapt to the novel environment.

Recent research suggests an alternative explanation for the presence of QS deficient strains on sites where it can be expected that biofilm formation should be favorable for colonization. Evolutionary biologists describe the phenomenon of "cheating" individuals in the population (Dunny et al., 2008). Such cheaters profit from the benefits of the cooperating population, without contributing to the community. Since these cheaters do not invest in the production of the biofilm, they have a competitive advantage relative to the biofilm producing individuals and, as a consequence, can become predominant in the community. However, when the proportion of cheaters becomes too large, this can affect the stability of the biofilm and the entire population may crash as a result (Dunny et al., 2008).

Salmonella species of bacteria are one of the most important terrestrial as well as aquatic animal pathogenic bacteria and have many diverse QS systems such as LuxR homologue, SdiA. The bacteria do not produce AHLs because of the absence of luxI homologue (Walters and Sperandio, 2006). Previous studies demonstrated that SdiA senses the production of AHLs, which is produced by other bacterial species (Michael et al., 2001) and in this way senses the presence of the intestinal environment. SdiA regulates the expression of virulence genes with different functions, ranging from resistance to complement killing to the expression of fimbriae. In murine, poultry, or bovine models, however, SdiA mutant strain is not attenuated and, very recently, SdiA dependent gene expression was shown to be activated in the gut of turtles, but not in the intestines of several homoiothermic animal species (Smith et al., 2006). The role of SdiA in Salmonella pathogenesis is so complex.

In *Salmonella enterica (S. enterica)* subspecies *enterica* serotype *Typhimurium* (*Salmonella typhimurium*), the luxS gene is directly involved in the AI-2 synthesis. It was recently demonstrated that luxS and the Salmonella AI-2, 4, 5 dihydroxy-2, 3-pentanedione, are necessary for Salmonella virulence (Choi et al., 2007). Deletion mutants lacking the luxS gene are severely attenuated in the expression of the virulence genes of the Salmonella pathogenicity island (SPI). These genes are required for the efficient invasion of intestinal epithelial cells and are therefore crucial in the pathogenesis of Salmonella infections in a wide variety of animal hosts, including poultry, pigs, and cattle (Boyen et al., 2006). Moreover, the contribution of the QS system in the development of infection in animals has been described in various other veterinary or zoonotic pathogens, such as *Clostridium perfringens* (Ohtani et al., 2002), *Yersinia pseudotuberculosis* (Atkinson et al., 2008), and *Campylobacter jejuni* (He et al., 2008). Therefore, these and other evidence showed that the QS system of the bacteria play a great role to develop infection in animals and resist the antimicrobial agents and the host defense mechanisms. Due to this, interfering on the QS system of these pathogenic bacteria is an alternative method to control infectious diseases in animals.

27.2.2 Quorum Sensing Dependent Virulence Factor Production in Animal Pathogenic Bacteria

Virulence factors are gene products that allow the pathogenic bacteria to infect and damage the host, including products involved in motility and adhesion of the pathogens to the host, protection from host defense mechanism, and host tissue degradation, iron acquisition, and toxins (Defoirdt, 2014). Virulence factors control the infectious cycles of pathogenic bacteria by promoting the pathogens entry, its growth and reproduction in the host's body, and exit from the host's body (Defoirdt, 2014; Donnenberg, 2000). These virulence factors are either secretory, membrane associated, or cytosolic in nature. The cytosolic factors facilitate the bacterium to undergo quick adaptive metabolic, physiological, and morphological shifts according to the different host conditions. There are several factors which enable pathogens to confer virulence activities, including swimming motility and chemotaxis, extracellular polysaccharide production, biofilm formation, and production of lytic enzymes such as hemolysin, caseinase, gelatinase, lipase, and phospholipase. A number of researchers identified different types of virulence factors from several bacteria species, and these virulence factors produced by different bacteria species are different in type and amount. For instance, the Gram-negative bacteria *P. aeruginosa* produces pyocyanin, elastase, rhamnolipids, lipase, lectin, biofilm, hydrogen cyanide, alkaline protease, and exotoxic A (Hauser, 2011; Karatuna and Yagci, 2010), whereas *Staphylococcus*, which are Gram-positive bacteria produces other types of virulence factors including fibronectin binding protein, hemolysin, protein A, lipase, and enterotoxin (Gallardo-Garcia et al., 2016). Studies have shown that the production of these virulence factors is regulated by the bacterial QS signaling systems (Aboushleib et al., 2015). In addition to this, in Gram-positive bacteria, such as *S. aureus* AIPs-based QS *agr* genes controls the expression of virulence genes (George and Muir, 2007). These genes are responsible for the synthesis of cell-surface proteins such as protein A and fibronectin binding proteins, secreted proteins that include enzymes (e.g., hemolysins and proteases), and toxins (e.g., toxic shock syndrome toxin and enterotoxin B).

During the process of infection, virulence factors of bacteria combat with defense mechanisms of the host. If virulence factors overcome the defense mechanisms of the host, infection is established; otherwise, the bacteria are eliminated by the host defense mechanisms from the host. Pathogenicity of microorganisms is determined by nature and type of its virulence factors. The production of virulence factors, which are importance molecules for the bacteria to invade the host, cause disease and evade the host immune defense and are crucial for the pathogenesis of infections in animals. Peterson (1996) classified virulence factors in bacteria into adherence factors, invasion factors, capsules, endotoxins, exotoxins, and siderophores. These various virulence factors play a significant role to break the host defense mechanism and establish the bacteria in the host. In extracellular pathogens, the secretory virulence factors act synergistically to kill the host cells.

Using the structure of one of the *S. aureus* AIPs, Park et al. (2007a) designed an analog of this cyclopeptide to raise antibodies upon coupling to two carrier proteins. These antibodies drastically reduced the production of α-hemolysin by *S. aureus* grown *in vitro* to a level where no hemolytic activity could be detected on blood agar plates. Disruption of QS to control the production of virulence factors seems to be an attractive broad-spectrum therapeutic strategy, and in some bacteria species, they produce more than one virulence factor. In this condition it is important to appropriate several QSIs to combat the effect of the bacteria on the host. Some of the QS mediated virulence factors of certain animal pathogenic bacteria are described (Table 27.1).

27.3 Quorum Quenching as a Control Strategy for Animal Pathogenic Bacteria

To reduce the risk caused by infectious diseases in animals requires alternative protection methods other than using the classical antibiotics. Interfering with the QS system of pathogenic bacteria has been proposed as a new anti-infective strategy. The connection between QS and pathogenic bacterial virulence generated excitement for a potentially new approach to fighting bacterial infections, and targeting and disrupting the QS system. Many pathogenic bacterial species relevant to terrestrial and aquatic animal production and veterinary medicine use the QS system to regulate production of virulence factors that are essential for bacterial infection. Preventing bacteria from producing virulence factors could be an important alternative strategy, known as antivirulence therapy or quorum quenching, for controlling bacterial diseases. Dong et al. (2000) conducted the first application of the QQ strategy in protecting against microbial infection by transforming aiiA gene into the phytopathogen *Erwinia carotovora* to reduce its decay phenotype in Chinese cabbage. The QQ system paradigm focuses on disarming bacteria by preventing virulence factor production or by neutralizing those factors.

Scientists have already identified three major paths by which to interfere with QS systems as part of QQ strategies, and these routes inhibit the synthesis of QS signal molecules, inhibit the interaction between a QS signal molecule and its related receptor, and quench extracellular QS signal molecules through neutralization or degradation. QQ enzymes or QSIs which can interfere with the synthesis of QS signal molecules or the binding to the receptors play a key role in the control of infectious bacterial diseases and their mechanisms are shown (Figure 27.1). QQ strains of bacteria interfere with QS, using quorum sensing inhibitors (QSIs) to block the action of AIs and quorum quenching (QQ) enzymes to degrade signaling molecules. Several studies demonstrated the effectiveness of QQ enzymes and QSIs in animal infectious disease controlling. QQ enzymes such as acylases and paraoxanases have been identified from animals such as

TABLE 27.1

QS Systems and QS-Mediated Virulence Factors of Some Common Economic Important Animal Bacterial Pathogens

Terrestrial Animal Pathogenic Bacteria	Autoinducer	Biosynthesis Protein	Sensing Protein	Virulence Factor	References
P. aeruginosa	AI-1, 2-heptyl-3-hydroxy-4(1H)-quinolone	LasI, pqsH, rh1I	LasR, pqsR, rh1R	Alkaline protease, elastase, lactin, flagella, LPS, pyocyanin, hemolysin, rhamnolipids	Christiaen et al. (2014); Lee and Zhang (2015); Schuster and Greenberg (2007)
B. cenocepacia	AI-1	CepI, CciI, RpfB	CepR, CciR, Clp	Lactin, elastase, protease, siderophores	Christiaen et al. (2014)
S. aureus	AIP	AgrB, AgrB1, FsrB	AgrA, BlpR, FsrA	Fibronectin binding protein, hemolysin, protein A, lipase and enterotoxin, exoprotease, lipase, urease, hemolysin, capsule	Gallardo-Garcia et al. (2016)
Salmonella enterica	AI-2	LuxS, RpfB	LsrR, QseB, QseF, LsrK	Spv, Mig-14, Agf, Lpf, RatB, BimA, Capsule I, BoaA, BoaB, Type IV pili, flagella	Choi et al. (2012)
Aquatic Animal Pathogenic Bacteria					
V. harveyi	AI-2, HAI-1, CAI-1	luxI, luxS, cqsA	luxR, luxP, cqsS	Protease, type III secretion, siderophore	Defoirdt et al. (2008)
V. parahaemolyticus	CAI-1, AI-2	cqsA	cqsS	Protease, amylase, aerolysin, RtxA, Flp type IV pili	Zhou et al. (2018)
V. anguillarum	AI-2, CAI-1 HHL, OH-HHL, ODHL	vanI, vanM, cqsA	vanR, vanN, cqsS	Toxin (rtx), flagellin A, haemolysins, protease, siderophore anguibactin	Li et al. (2008b); Norqvist et al. (1990); Ormonde et al. (2000); Rodkhum et al. (2005)
V. mediterranei	AI-2	LuxS, luxI	LuxR, LuxP	Protease, chitinase, caseinase and amylase	Torres et al. (2016, 2018)
V. campbellii	HAI-1, AI2, CAI-1	LuxM, luxS, cqsA	LuxN, luxR, luxP, cqsS	Toxin T1, metalloprotease, siderophore, chitinase A, phospholipase, type III secretion, extracellular polysaccharide	Defoirdt (2014); Natrah et al. (2012)
V. alginolyticus	AI-2, AI-1, CAI-1	LuxS, LuMAiS, CqsA	LuxU, LuxO, $LuxR_{val}$	Flagella, protease, and polysaccharide	Rui et al. (2008)
Aeromonas hydrophila	AI-1 AI-2 AI-3, BHL, HHL	LuxS, ahyI	LuxU, ahyR	Serine protease and metalloprotease, amylase, MaM7, VpadF, TDH, TRH	Chu et al. (2014); Natrah et al. (2012)
A. veronii	AI-1	acuI, luxI	acuR, luxR	Caseinase, esterase, amylase, lecithinase, hemolysin	Sun et al. (2016)
A. salmonicida	AI-2,*BHL, HHL*	asaI	asaR	Extracellular protease	Schwenteita et al. (2011)
Edwardsiella tarda	AI-2, BHL, HHL, OHHL, HeHL	LuxS, edwI, luxI	edwR, luxR	Protease, haemolytic, extracellular polysaccharide and siderophore	Rui et al. (2008)

Abbreviations: AI-2, autoinducer 2; BHL, *N*-butanoyl-L-homoserine lactone; CAI-1, (S)-3-hydroxytridecan-4-one; HAI-1, *N*-(3-hydroxybutanoyl)-L-homoserine lactone; HeHL, *N*-heptanoyl-L-homoserine lactone; HHL, *N*-hexanoyl-L-homoserine lactone; ODHL, *N*-(3-oxodecanoyl)-L-homoserine lactone; OH-HHL, *N*-(3-hydroxyhexanoyl)-L-homoserine lactone; OHHL, *N*-(3-oxohexanoyl)-L-homoserine lactone.

rats, mice, zebrafish, and humans. Acylase I from porcine kidney has been used in aquaculture as well as in the health care sector to inhibit AHL-mediated biofilm formation by *Aeromonas hydrophila* and *Pseudomonas putida* (Dong and Zhang, 2005; Paul et al., 2009). Paraoxanases from human epithelial cells and from the serum of mammals such as rats, goats, bovines, and horses inhibit AHL-mediated QS in *Pseudomonas aeruginosa (P. aeruginosa)* (Stoltz et al., 2007; Yang et al., 2005). Mayville et al. (1999) demonstrated that mice develop resistance against *Staphylococcus aureus* bacterial infection after treated with synthetic AIP-2, and Hentzer et al. (2003) showed that the treatment of mice with furanone develops resistance and decrease of virulence of *P. aeruginosa*. Paraoxonases 1 from human and murine act as host modulators of *P. aeruginosa* QS (Ozer et al., 2005). Moreover, competitive organisms are able to clear the signal molecule to quench QS (Kalia and Purohit, 2011). For example, *Escherichia coli* ingest AI-2s to influence the QS of *Vibrio harveyi (V. harveyi)* (Xavier and Bassler, 2003). Certain bacteria have an ability to degrade the AHLs of the QS signal molecules of other bacteria, and this activity protects several organisms including *Artemia* spp., rotifers, and larvae of turbot or prawn from bacterial infection (Nhan et al., 2010).

QQ strains such as *Stenotrophomonas maltophilia (S. moltophilia), Enterobacter hormaechei (E. hormaechei), Bacillus thuringiensis (B. thuringiensis), Citrobacter (C. gillenii),* and *Bacillus cereus (B. cereus)* were able to degrade AHL signal molecules and effectively reduce swimming motility, biofilm formation of *Yersinia ruckeri (Y. ruckeri)*, and finally able to control enteric red mouth disease in trout (Torabi Delshad et al., 2018). *Alteromonas stellipolaris* (*A. stellipolaris*) bacteria are able to

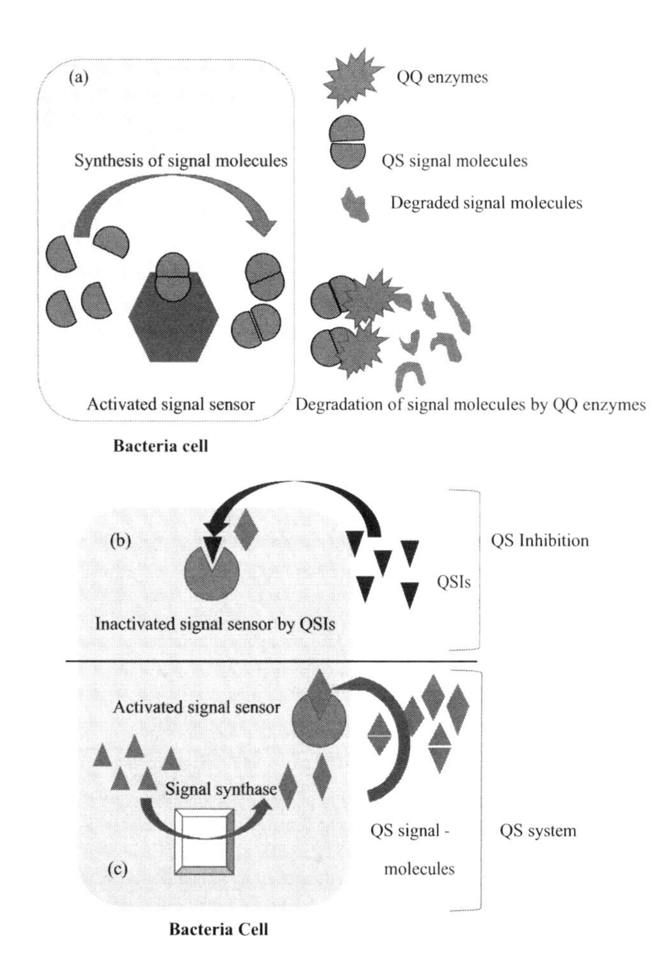

FIGURE 27.1 Simple diagrammatic representation of QQ agents and their mechanisms, during processes of (a) and (b), there is no cell-to-cell communication and no activation of gene expression related to virulence and biofilm formation, when process (c) undergoes activation of the virulence related gene of the pathogen and infects the animal.

degrade the accumulation of AHLs and reduce the production of protease and chitinase and swimming motility of a *Vibrio* species in cocultivation experiments *in vitro* from a bivalve hatchery site (Torres et al., 2016). Strains of *Klebsiella pneumonia* (JYQ_1 and JYQ_5), *Acinetobacter baumanni* JYQ_2, *Pseudomonas nitroeducens* JYQ_3, and *Pseudomonas* JYQ_4 are identified as quenchers using the *Chromobacterium violaceum* CV026 biosensor (Kaur and Yogalakshmi, 2019). Romero et al. (2014) isolated and demonstrated the marine bacterium *Tenacibaculum* sp. strain 20J as an effective quorum quencher bacterium and were important to treat or prevent infectious diseases such as *Edwardsiellosis* and other aquatic bacterial diseases. *Burkholderia glumae*, *Pectobacterium carotovora*, *P. aeruginosa*, or *Agrobacterium tumefaciens* are identified as potential AHL degrading enzymes and have been revealed to reduce or eliminate the accumulation of QS signal molecules in the environment and leads to decrease the capacity of the expression of genes involved in the QS process (Cho et al., 2007; Reimmann et al., 2002). Wang et al. (2019) isolated QQ enzyme from the coastal marine water and demonstrated that the biofilm formation associated with *P. aeruginosa* was significantly inhibited by QQ enzyme $AiiA_{S1\text{-}5}$ and also reduced the quorum sensing-mediated virulence traits of *A. hydrophila*, *P. aeruginosa*, and *Vibrio alginolyticus*. The effect of *Bacillus* sp. QSI-1 as an efficient quorum quencher on virulence factors production and biofilm formation of fish pathogen *A. hydrophila* was investigated in zebrafish gut, and QSI-1 reduced the accumulation of AHLs but did not affect the growth of *A. hydrophila* (Chu et al., 2014).

Fetzner (2015) described that AHL lactonases and acylases are the widely studied QQ enzymes that hydrolyze *N*-acylhomoserine lactone (AHL) signaling molecules. Since *A. hydrophila* is one of the pathogenic bacteria and have AHL-dependent quorum-sensing systems (Lynch et al., 2002), AHL lactonase could hydrolyze quorum-sensing signal molecules (butaryl-homoserine lactone and hexanoylhomoserine lactone) of the bacteria and reduce its virulence (Cao et al., 2012; He et al., 2013; Zhang et al., 2011). Moreover, the effectiveness of quenching enzymes such as AHL lactonases B565, AIO6, AI-96, and QsdA against *A. hydrophila* were confirmed by Cao et al. (2012) in zebrafish *in vivo* (AI-96) and Zhang et al. (2011) *in vitro*. AIO6 QQ enzyme was also effective against *A. hydropila* when tilapia were fed diets containing this enzyme (Cao et al., 2014). The effectiveness of QQ strains, QQ enzymes, and QSIs to control infectious diseases were verified in a small number of terrestrial animals (Table 27.2).

Antibiotics have been used for a long period of time for the treatment of various infectious bacterial diseases. The actions of antibiotics are either killing the bacteria or inhibiting their growth, and this also results in a strong selection pressure on the bacteria. However, the exhaustive applying of these compounds has resulted in the emerging of resistance strains of bacteria that reduce the effectiveness of the antimicrobial agents and is associated with treatment dose increase (Alanis, 2005). For instance, O'Neill (2016) reviewed the report of Wellcome Trust, which estimated that by 2050, antimicrobial resistance could cause 10 million additional deaths annually, and a cumulative loss to the world's GDP of $100 trillion. Inhibiting the secretion of virulence and disturbing the formation of biofilm is an important mechanism to affect the pathogenicity of bacteria, and QQ has been used as an alternative to antibiotics (Hentzer et al., 2003) but does not kill bacteria (Romero et al., 2012). The chance of the development of antibiotic-resistance bacteria during the application of the classical antibiotic for disease control is higher.

Diggle et al. (2007) suggested that the application of QQ on bacteria growth was dependent on the culture medium (whether it is rich or poor media) where the bacteria grow and might thereby introduce a selection pressure, although milder than the classical antibiotics strategy, and select for resistant bacteria (García-Contreras et al., 2015). Using QS-disrupted variants, a few studies have revealed that bacterial resistance to QS may arise. Maeda et al. (2012) observed mutations increasing efflux of C-30, an efficient QQ furanone, as a mechanism to overcome QS disruption. Sandoz et al. (2007) identified that the presence of some bacteria that ceased synthesis of quorum-regulated factors and these bacteria are termed "social cheaters." These types of mutant bacteria, which are insensitive to QS, might interfere with QQ mechanisms (Tay and Yew, 2013), but Gerdt and Blackwell (2014) demonstrated that the spreading power of these QQ-resistance bacteria were very slow and less fit than their normal strain. The rise of QQ-resistance bacteria depends on the usage strategy of the QQ enzyme, and mostly the QQ enzymes have minimum chance of development of resistance bacteria

TABLE 27.2

Applications of Quorum Quenching Strains, Quorum Quenching Enzymes, Inhibitors, and Bacteria Effective in Terrestrial Animal Pathogenic Infections

QQ Enzyme or QSIs	Source	Animal	Against	Effect on Bacteria	Dosage Used	Effect on the Animal	References
QQ Strains							
P. aeruginosa PAO1	*Pseudomonas, Pseudoalteromonas, Delftia, Arthrobacter*	*Caenorhabditis elegans*	*P. aeruginosa*	Inhibit biofilm formation, reduced elastase production	0.1 optical density (OD)	Increased survival rate	Christiaen et al. (2014)
B. cenocepacia LGM16656	Rhizosphere, water, mucus or intestines of flounders isolated bacteria	*C. elegans*	*Burkholderia cenocepacia* LGM16656	Inhibit biofilm formation, reduced elastase production	0.1 OD	Increased survival rate	Christiaen et al. (2014)
QQ Enzymes							
Lactonase AiiA	*Bacillus* spp.	Sheep erythrocytes	*Burkholderia thailandensis*	Abolished the accumulation of AHLs, reduces cell swarming and twitching motility	10 μL	Prevents the β-haemolysis of sheep erythrocytes	Ulrich (2004)
Lactonase AiiA	*Bacillus* sp. strain 240B1	*C. elegans*	*B. cenocepacia*	Abolished the accumulation of AHL	5 μL	Reduce Fast killing	Wopperer et al. (2006)
AHL acylase AiiD	*Ralstonia* spp.	*C. elegans*	*P. aeruginosa* PAO1	Decreasing its ability to swarm, reduce elastase, and pyocyanin production	40 μg	>80% survive	Lin et al. (2003)
AHL acylase $PvdQ^{L\alpha146W,F\beta24Y}$	*P. aeruginosa*	*Galleria mellonella*	*B. cenocepacia*	Reduced accumulation of AHLs	0.04 mg/mL	Increased survival rate	Koch et al. (2014)
Paraoxonase 1	Human	Mice	*P. aeruginosa*	Inactivating the 3OC12-HSL quorum-sensing signal, prevent biofilm formation	20 μg/mL	Increased survival	Ozer et al. (2005)
Paraoxidase PON1	Human	*Drosophila melanogaster*	*P. aeruginosa*	Inactivate 3OC12-HSL signal of the bacteria	—	Increased survival	Stoltz et al. (2008)
QSIs							
Phillyrin	Forsythia suspense	*C. elegans*	*P. aeruginosa*	Decrease biofilm formation, reduce swimming and twitching motility	0.5 mg/mL	Improved resistance	Zhou et al. (2019)
Hamamelitannin	*S. aureus* Mu50	Mice	*B. cenocepacia*	Reduce biofilm formation	50 μL	Reduced the microbial load in the lungs, increased survival, improve resistance	Brackman et al. (2011)

of all the QQ strategies due to the remote action nature of the enzyme and do not need to enter the bacterial cells. QQ enzymes are naturally broad-spectrum enzymes and can be engineered for altered specificity. Therefore, the use QQ strategy to control infectious animal diseases is with minimum probability of development of bacteria that can resist the effect.

27.4 The Advantage of Quorum Quenching Agents as Means of Treatment to Replace the Application of Antibiotics in Animals

Indeed, antibiotics play a major role for the treatment of various bacterial infectious diseases for aquatic as well as terrestrial animals. Moreover, these compounds have been used for growth promotion in animals for several years (Heuer et al., 2009). These days, the numbers of antibiotic-resistance bacteria are increasing and this makes scientists become more aware and should be used with more care and as a consequence, the development of alternative methods to control pathogenic bacteria in animal production is mandatory. In order to enter and damage their host animals, pathogenic bacteria produce several virulence factors and development of a protection mechanism from the host defense (Donnenberg, 2000). To infect their host, bacteria must produce their virulence factors. QQ therapy is an important method to reduce the effect of pathogenic bacteria on their host.

Quorum-quenching agents are inhibitors of bacterial virulence, rather than of bacterial growth (Cegelski et al., 2008). The QQ strategy has more advantages in contrast to classic antibiotics because QQ has less capacity to interfere with nontarget organisms, mostly the commensal microbiota, as it specifically targets virulence gene expression or gene regulation. In the process of gene regulation mechanism, QQ agents might interfere with the processes of regulation in nontarget organisms. In addition to this, mostly QQ agents result in selective pressure only under conditions in which the virulence genes are required; the tendency toward resistance development and spread is also lower (Defoirdt et al., 2010). QQ infectious bacterial disease controlling strategies are more effective when compared with the controlling strategy using antimicrobial, since the pressure of selectivity in the pathogenic bacteria is much less and weakens bacterial infections without decreasing growth, in contrast to antibiotics (Bjarnsholt et al., 2010; Rasko and Sperandio, 2010). Therefore, the applications of QQ strategies to control infectious diseases are more advantageous when compared with the classis antibiotic methods.

27.5 Screening, Identification and Characterization of Quorum Sensing Inhibitors for Animal Pathogenic Bacteria

Quorum sensing inhibitors are molecules which are responsible for inhibition of autoinducers, which induce QS systems or the autoinducer-regulated phenotype. The mechanism of interfering of the QSIs with the QS system is achieved through several modes of action, including by blocking of the synthesis of autoinducers, degrading autoinducers, interfering with autoinducer receptors, or inhibiting the formation of autoinducer/receptor complex (Lade et al., 2014). A number of qualitative and quantitative methods have been developed during the last decade to detect and characterize QSIs from different sources such as aquatic algae (Kwan et al., 2011), invertebrates (Skindersoe et al., 2008), and terrestrial plants (Chong et al., 2011; Koh et al., 2013). In addition to these, the enzyme which inactivates the QS signal molecules were also recorded from bacteria (Chen et al., 2013), mammalian cells (Yang et al., 2005), and plants (Koh et al., 2013).

Various methods have been implemented to identify strains that produce QSI compounds, followed by lead molecule purification. QS biosensors, which are genetically modified strains that express reporter genes such as lacZ, gfp, or luxCDABEG in response to the presence of specific QS signals (Steindler and Venturi, 2007), are one of the valuable tools for the identification of QSI molecules. Depending on the scale of the screening and targeted QS system, scientists apply several methods to demonstrate the efficiency of different QSIs. One of the methods is colorimetric tests via biosensor reporter bacteria. During the screening of compounds for QS inhibitory activity, appropriate biosensors need to be employed. These strains allow quantitative and qualitative detection of QS signals. Most of the biosensors are based on AHL and AI-2 reporters. This method uses many bacteria strains including *Chromobacterium violaceum* strain CV026 (McClean et al., 1997), *A. tumefaciens* strain NT1 (pZLR4) (Cha et al., 1998). In the latter system, sensitivity is higher when AHLs are oxidized at carbon-3 or with the luminescent substrate beta-Glo instead of the X-Gal substrate (Kawaguchi et al., 2008). Other series of tests target known QS-regulated functions such as the motility of *Yersinia enterocolitica* (Atkinson et al., 2006) or pyocyanin (Reimmann et al., 1997) or pyoverdin (Adonizio et al., 2008) production in *P. aeruginosa*. Some QSIs studies are focused on biofilms formation. They rely on qualitative methods such as Congo red agar (Nasr, 2012) or quantitative ones such as the Microtiter plate assay (Thenmozhi et al., 2009) to evaluate the characteristics of the biofilms. However, these tests cannot be used for high-throughput screening of QSIs.

To optimize large screenings, the QSI selector system was developed by Rasmussen et al. (2005) to screen QSIs from various food sources and herbal medicines. This technique involves a plasmid that carries a construct consisting of a luxR homolog gene and a killing function controlled by a QS-regulated promoter. Consequently, bacteria harboring this plasmid can grow on AHL-containing medium only in the presence of efficient QSI molecules. Desouky et al. (2013) also develop a high-throughput screen to highlight QSIs that target cyclic peptide-mediated QS in Gram-positive bacteria. It relies upon the use of both the *agr* and the *fsr* systems from *S. aureus* and *Enterococcus faecalis*. It combines *S. aureus agr* reporter strain that carries luciferase and green fluorescence protein genes under the control of the *agr* promoter, and a gelatinase-induction assay to examine the *E. faecalis fsr* QS system and look for QSIs. The main biosensor strains employed for screening and identification of QSIs from the marine environment based on their phenotypic expression are described in the following section. For the isolation of QSIs, the following methods are commonly used.

27.5.1 Pigment Producers Biosensor Strains

The bacterium *C. violaceum*, that produces the purple (violacein) pigment, which is regulated by the CviI/CviR of its QS systems, makes the bacterium one of the most widely used as biosensor for screening of several QSIs. An AHL-deficient non-pigment mutant strain (CV026) has been employed as biosensor to perceive the incidence of AHLs. In the presence of exogenous AHL, this nonpigmented biosensor produces violacein and forms purple colonies, and this bacterium is important to identify the QSIs, since the in the presence of QSIs the production of pigment from the bacterium become reduced. In addition to this mutant strain there are also other mutants, including CV017 and VIR24, which are important to isolate QSIs based on their bioactive compounds (Someya et al., 2009).

27.5.2 Bioluminescence-Based Biosensor Strains

Some strains of bacteria which produce bioluminescence are also important for the isolation of QS inhibitors. The inhibitory activity of bioluminescence strains of bacteria can be measured quantitatively using a luminometer or qualitatively by bioluminescence microscopy. *E. coli* bacteria strain based on its plasmid pSB401 is one of the most widely used bioluminescence-based QS reporter strain. This employs the luxCDABE operon from *Photorhabdus luminescens* under the control of PluxI gene and the *V. fischeri* luxR DNA fragment, and other reporters based on pSB403, pSB1075, pSB536, and pSU2007 plasmids were also used for screening of QS inhibitors (Rasmussen et al., 2005).

27.5.3 Beta Galactosidase-Based Biosensors

Agrobacterium tumefaciens strain NT1 bearing plasmid pZLR4 is a regularly applied biosensor in this QSIs isolation system.

The *A. tumefaciens* strain NT1 does not produce native AHLs and harbors a plasmid encoding β-galactosidase. β-galactosidase, also called lactase, is a glycoside hydrolase enzyme that catalyzes the hydrolysis of β-galactosides into monosaccharides through the breaking of a glycosidic bond. The colonies of this strain give different color based on the presence or absence of AHLs. For instance, the color becomes blue in the presence of exogenous AHLs whereas they become colorless in the presence of QS inhibitors. This strain has the ability to respond to a wide range of AHLs at very low concentration levels. In addition to this strain, *E. coli* strains are also a biosensor strain because these strains harbor the plasmid pKDT17 (Farrand et al., 2002).

27.5.4 Green Fluorescent Protein (Gfp)-Based Biosensor

The commonly used gfp-based AHL sensor plasmids are pKR-C12, pAS-C8, and QSIS3. pKR-C12 is based on lasB-gfp translational fusion wherein constitutive expression of lasR gene occurs under the influence Plac and is based on the plasmid pBBR1MCS-5. pAS-C8 biosensor depends on the quorum sensing method associated with *Burkholderia cepacia* and is responsive to the presence of *N*-octanoyl-L-homoserine lactine (C8-HSL). It contains Pcepl-gfp along with the cepR regulator gene, which is under the control of Plac. Qualitative and quantitative screening can be performed with these strains by using epifluorescence microscope (Andersen et al., 2001). The *E. coli* QSIS3-based system (derived from *Vibrio fischeri* LuxR QS gene) is another popular system in this category.

27.6 Anti-infectives Mechanisms of Quorum Sensing Agents

27.6.1 Anti-pathogenic Bacteria Mechanisms of the Quorum Sensing Inhibitors

Since the QS system of bacteria is vital for the development of virulence factor and pathogenesis of bacteria, interfering in the QS system is important to control the disease caused by the bacteria. So it is conceivable that inhibitors of bacteria QS system could have therapeutic application. Hence, it is important to study the mechanistic pathways of the QS inhibitors in order to effectively use them as anti-infectious agents. Even though there are different mechanisms for QS inhibitors in each pathway including inhibition of AIs synthesis, AIs receptor antagonism, inhibition of targets downstream of receptor binding, sequestration of AIs using (e.g., antibodies against AIs), the degradation of AIs using enzymes, inhibition of AI secretion/transport and antibodies that "cover" and therefore block AIs receptors (Asfour, 2018; d'Angelo-Picard et al., 2006; De Lamo Marin et al., 2007), the main QSIs are primarily performed by three mechanisms (Defoirdt et al., 2004). Understanding these processes of quorum sensing, the quorum disruption strategies generally adopt three common approaches including inhibition of quorum signal molecule synthesis, inhibition of quorum molecule/receptor interaction, and degradation of quorum molecules. Though many quorum inhibitors were developed for broad-spectrum effects, it is important to note that quorum sensing is largely unique to each bacterial species. Certain bacteria species have more than one QS system, and in this case, it is mandatory to apply a combination of inhibitors to control the pathogenicity of the bacteria effectively.

Since the pathogenic bacteria depends on its QS signaling to form biofilms and produce virulence factors, many studies demonstrated that blocking of the QS system or inactivating of the QS signal molecules by anti-quorum sensing agents makes pathogens more susceptible to host immune responses, antibiotics, and inducing the pathogenicity of the bacteria. Nowadays a number of QS disruption strategies, including receptor inactivation, signals synthesis inhibition, signals degradation, signaling blockage by antibody, and combining use with antibiotics, convey the potential of QS as the therapeutic target for bacterial diseases.

27.6.1.1 QSIs That Bind to Signaling Molecule Receptor Proteins in Animal Pathogens

Chemical analogues of signaling molecules compete for receptor protein binding sites because of their structural similarities. Previous studies demonstrated that a number of QSIs, most of which are chemical analogues of AHL and AI-2, have been shown to be effective to control bacterial diseases by interfering with their QS system. Swift et al. (1997) demonstrated that the exogenous addition of 10 μM *N*-(3-oxodecanoyl)-L-homoserine lactone effectively delayed the appearance of serine protease and reduced its production in *A. salmonicida*. An effective QSI, *N*-(heptylsulphanylacetyl)-L-homoserine lactone, competitively binds to LuxR and LasR and interferes with the QS system in *V. fischeri* and *P. aeruginosa* (Zhao et al., 2015). In addition to this, there are many other QS inhibitors, including plant extracts compounds (from garlic), homoserine lactone (HSL) derivatives, and acetamide, that positively inhibit the LuxR gene which is responsible for the expression of pathogenicity of bacteria (Persson et al., 2005). *N*-(propylsulphanylacetyl)-L-homoserine lactone and *N*-(pentylsulphanylacetyl)-L-homoserine lactone effectively reduce the expression of toxic proteases, such as AsaP1 by *A. salmonicida* at a dose level of 10 μM (Schwenteita et al., 2011). Natrah et al. (2012) found that *N*-tetradecanoyl-L-homoserine lactone effectively decreased the virulence of both *A. hydrophila* and *A. salmonicida* toward burbot larvae infections at a dose level of 10 μM.

AI-2-mediated QS can be blocked in a similar manner as described above about AHLs. In QSIs, an active 3-(methoxyphenylpropionamido) ribofuranosyl derivative, which is isolated from nucleoside analogues, was obtained that most likely interfered with signal transduction at the level of LuxPQ in *V. harveyi*. This QSI decreases protease activity in *V. anguillarum* by 23.3% and decreased biofilm formation in *V. anguillarum* and *V. vulnificus* by 35.11% and 17.15%, respectively (Zhao et al., 2015).

Cinnamaldehyde and its derivatives are potentially effective QSIs, as they decrease the binding of LuxR to promoter sequences and treatment at concentrations of 100–150 μM; this compound decreases protease activity in *V. anguillarum* and biofilm formation in *V. anguillarum* and *V. vulnificus* (Zhao et al., 2015). Cinnamaldehyde and 2-NO2-cinnamaldehyde completely protect Artemia against *V. harveyi* without affecting its growth. Treatment with 4-methoxy-cinnamaldehyde also decreases protease activity and biofilm formation in *V. anguillarum* (Brackman et al., 2008). As Natrah et al. (2012) reported, cinnamaldehyde effectively decreases the virulence of *A. hydrophila and A. salmonicida* toward burbot larvae at a dose of 0.01 μM. Furanone derivatives, which are structurally similar to AHL molecules, are potential QSIs that function similarly to cinnamaldehyde. Treatment with (5Z)-4-Bromo-5-(bromomethylene)-3-butyl 2 (5H)-furanone (100 μM) from the alga *Dellsea pulchra* inhibits toxin production in *V. harveyi* and protects prawns from *V. harveyi* infection (Manefield et al., 2000).

Zang et al. (2009) also identified two small peptides with a sequence similar to the C site of LuxS, and the peptides interact specifically with LuxS and inhibit AI-2 activity of the bacteria. They demonstrated that these small peptides control biofilm growth, decrease the expression of the T3SS genes esrA and orf26, and inhibit the dissemination and survival of *E. tarda* in infected fish, which provides new strategies for designing QSIs against pathogenic bacteria in aquaculture. Similarly, halogenated furanones covalently modify and inactivate and have been shown in several different microbes LuxS (Zang et al., 2009). Furanone (60–100 μg/mL) inhibits AI-2 signaling in *E. coli* and leads to the differential expression of many genes (Ren et al., 2005). Synthetic bromated furanone inhibits *Streptococci* biofilm formation via the AI-2-mediated QS pathway (Lonn-Stensrud et al., 2009). As a result, the application and mechanism of QSIs to control bacterial diseases in terrestrial and aquatic animals uses several mechanisms.

27.6.2 Interfering with the Host – Pathogenic Reaction System in Animal Pathogens

It is known that host stress has influenced the outcome of host microbe interactions, and this has been entirely associated with a direct effect on the intestinal barrier function and weakens the host immune system (Verbrugghe et al., 2012). Lyte et al. (2011) demonstrated that infectious bacteria have evolved specific detection systems for the stress hormones produced by the host and that detection of these stress hormones affects the growth and virulence of the pathogenic bacteria. Some bacteria such as *E. coli* and *Salmonella* spp. have the ability to sense and respond to the host signals including the catecholamine stress hormones adrenaline and noradrenaline (Defoirdt et al., 2013). These hormones are conserved among vertebrates and invertebrates and are important part of the acute stress responses in animals. The catecholamines hormones play a key role in the growth of the pathogenic bacteria by enhancing the availability of iron to the bacteria from the host when iron deficiency phenomenon has occurred (Lyte et al., 2011) and activating virulence gene expression of the pathogenic bacteria (Defoirdt et al., 2013).

Previous studies reported that catecholamines hormones in some pathogenic bacteria such as *E. coli* and *Salmonella* spp. disturb the production of virulence-related features such as motility and type III secretion (Bearson and Bearson, 2008; Rasko et al., 2008; Sperandio et al., 2003), hemolysin production (Karavolos et al., 2011), and expression of pilus and fimbrial adhesions (Lyte et al., 1997) and intestinal colonization in chicks, pigs, and calves (Methner et al., 2008; Pullinger et al., 2010). Karavolos et al. (2013) suggested that the presence of diverse bacterial adrenergic sensors such as QseC show different reactions from susceptibilities to inhibiting with eukaryotic cells α and β adrenergic receptors. Through these mechanisms, *Vibrios* and *Aeromonas* spp. also respond to catecholamines and QseC homologues (Defoirdt et al., 2013). QQ agents interfere with the QS system and affect the pathogens by inhibiting production of several enzymes related to virulence, affect the swimming motility of the bacteria, diminish biofilm formation, and reduce the accumulation of AHLs. These actions result in improvement of the immunity and survival rate of the host animal (Figure 27.2).

27.6.3 The Role of Quorum Quenching Enzymes and QSIs in Modulating the Gut Microbiota of Animals

Understanding how the bacterial QS system contributes to the development virulence factors is providing new insights into animal gut health and disease, insights that today's feed industry can influence now. Most animal diseases result from a pathophysiologic process that involves not only pathogens such as bacterium, viruses, or fungi, but also the host's microbiota and immune response. When a disease occurs in an animal, at the cause of enteric infections there is an imbalance in the intestinal immune-microbiota axis. The interaction between diet, microbiota and the host immune system is important for devising new strategies to struggle animal intestinal as well as physiological health. In animals, the normal intestinal microbiota consists of both commensal and disease-causing bacteria communities in a balanced relationship with each other, and many of the bacteria are important to the host (Wu et al., 2012b). The gut microbiota, mostly dominated by bacteria, are diverse and dynamic in composition and in both species and densities across the different parts of the gastrointestinal tract of the animal.

Intestinal microbiota plays a key role in protecting its host from enteric bacterial infection. However, several intestinal pathogens have developed ploys to outcompete the commensal intestinal bacteria community, resulting in the development of infection and disease in the host. Pathogenic bacteria use several mechanisms including utilization of microbiota derived sources of carbon and nitrogen to thwart the gut microbiota. Moreover, they also exploit the QS regulatory signals from the commensal microbiota and the host to promote their own growth and virulence development.

The gut microbiota and the host itself are the main sources of these metabolites, which are produced in the intestinal tract of vertebrates and participate in various physiological processes, including energy hemostasis, QS system, and host immune regulation. Some types of metabolites are only produced by the gut microbiota such as bacteriocins, short-chain fatty acids, and quorum-sensing autoinducers and are important to the host normal physiology. The fish gut possesses about 10^7–10^8 colony-forming units (CFU) g^{-1} bacteria (Pérez et al., 2010). There is some difference between the gut microbiota of freshwater and marine fish species. The most common bacteria found in the gut of freshwater fish species are *Aeromonas*, *Acinetobacter*, *Lactococcus*, *Flavobacterium*, and *Pseudomonas*, representatives of the family *Enterobacteriaceae*, and obligate anaerobic bacteria of the genera *Clostridium*, *Bacteroides*, and *Fusobacterium*. The population of these bacteria is influenced by the metabolites and the

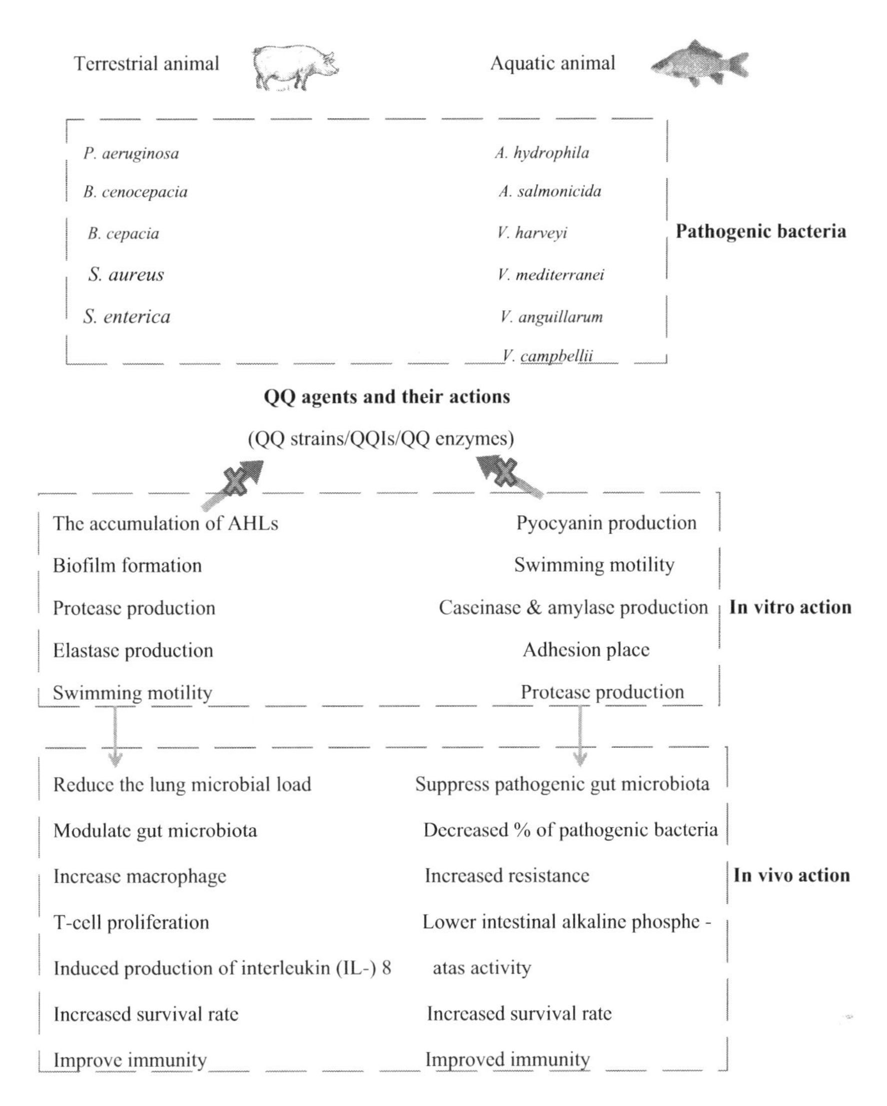

FIGURE 27.2 *In vitro* and *in vivo* effects of quorum quenching agents in terrestrial and aquatic animals.

effectors that promote species stability, equilibrium, adaptation, and survival in the fish gut (Kashinskaya et al., 2015). Bassler and Losick (2006) showed that QS play a key role in the stability of the bacteria in the fish gut, which provide bacteria to communicate intra- and interspecies and regulate cooperative behaviors and for the synthesis of extracellular digestive enzymes production, as well as modulating the population growth and survival in the fish gut.

Several pathogenic bacteria in fish including *Aeromonas* sp. *Edwardsiella tarda* and *Vibrio* spp. are opportunistic pathogens (Austin and Austin, 2007). The unbalanced growth of these pathogenic bacteria in the fish gut disrupts the core microbiome of the fish and results in disease development in the fish (Wu et al., 2012a). Thus, QQ is important for the host to maintain balance within gut ecosystem of microorganisms and inhibit the virulence gene expression of pathogenic bacteria in the intestinal environment. Even though more research is needed for better understanding, a few studies showed that QQ enzymes and QSIs participated in the modulation of the gut microbiota of animals. Cao et al. (2014) confirmed that addition of AHL lactonase AiiAB546 enzymes to the diet of zebrafish affects the gut microbiota diversity and composition. They also observed that the DGGE result of the gut microbiota analysis indicated that the fewest bands were recorded from the group of fish fed with a diet containing the QQ enzyme AiiAB546, and some bacteria species such as *Cetobacterium* sp. were only detected in the zebrafish fed AiiAB546 diet. *Weissella* sp. was undetectable in the presence of AiiAB546. Therefore, the investigators suggested that addition of QQ enzyme suppress mostly the pathogenic bacteria the diversity of gut microbiota. Zhou et al. (2016) demonstrated that the composition of gut microbiota of goldfish (*Carassius auratus*) was affected by quenching bacteria QSI-1, and the percentage of pathogenic bacteria *A. hydrophila* was decreased significantly. This result shows that QSI-1 protects fish from *A. hydrophila* infection by the quorum quenching pathway. It also suggests that the QQ enzyme produced by the probiotic bacteria plays a key role in modulating the gut microbiota structure and reduces the number of AHL-producing pathogenic bacteria in the fish gut.

Quorum-quenching probiotic bacteria significantly modulate the gut microbiota of swine, and these beneficial effects can contribute to the improvement of performance and gastrointestinal tract health of weaning pigs (Kim et al., 2018).

Thompson et al. (2015) revealed that AI-2 levels influence the abundance of the major phyla of the gut microbiota in mice. They showed that *E. coli* that increased intestinal AI-2 levels altered the composition of the antibiotic-treated mice gut microbiota, favoring the expansion of the *Firmicutes* phylum in the gut. Therefore, these studies verified that QQ enzymes and QSIs play a positive role in modulating the gut microbiota and improve the health of the gut. Most of these studies verified that QQ agents improve animal gut health by suppressing the growth of pathogenic bacteria.

27.6.4 The Role of Quorum Quenching in the Immune System of Animals

The ability of organisms to resist infection or disease caused by potential harmful microorganisms is termed immunity, and this system can be classified into innate and adaptive immunity. The development of an infection is the interaction between the pathogen and the immune system of the host. The immune system protects the body against the disease by recognizing and neutralizing the pathogen. The innate immune response includes both humoral and cellular defense such as the complement system and the processes played by granulocytes and macrophages. Immunity modulating agents are used for immunotherapy, which is defined as treatment of disease by inducing, enhancing, or suppressing an immune response. Ng and Bassler (2009) proposed that bacteria synthesize QQ enzymes not only as a self-modulatory system for their own QS system but also to interfere with the QS system of their competitors. For instance, *Pseudomonas* HSLs are not only important in the regulation of bacterial virulence genes but also interact with eukaryotic cells and modulate immune responses of the host (Saleh et al., 1999). Several studies demonstrated that many QQ enzymes and QSIs were identified as important to improve the immune system of organisms. Cao et al. (2010) demonstrated that the expression levels of some anti-infectious immunity-related factors, including interleukin 10 (IL-10), toll-like receptor 5b (TLR5b), and inducible nitric oxide synthase 2a (iNOS2a) in zebrafish, which were related to *A. hydrophila* virulent factors under regulation of quorum sensing, were affected by the addition of QQ enzyme AiiAB546. QS signal molecules of 3O-C12-HSL and C4-HSL strongly increased effectiveness of macrophage phagocytosis capacity in the innate immune system in mice during *P. aeruginosa* infections (Karlsson et al., 2011).

The signal molecules of *Pseudomonas*, 3-oxo-C12-HSL is an effective immunomodulator of various eukaryotic cells, and the production of this molecule greatly affects the capacity of pathogenicity bacterium (Tateda et al., 2003). They also showed that the signal molecule 3-oxo-C12-HSL was important to inactivate the interleukin-12 (IL-12) and the tumor necrosis factor alpha (TNFα) secretion by the lipopolysaccharide simulated macrophage and suppress T-cell proliferation. Imamura et al. (2005) observed a lower proportion of neutrophils in mice infected with mutant *P. aeruginosa* compared with the mice infected with the wild-type *P. aeruginosa*, and the histological examination revealed that mild inflammatory change occurred in the mice lungs infected with the mutant group.

The contribution of QSIs agents produced from *P. aeruginosa* in the immune system of the host was studied by several researchers. Smith et al. (2002) revealed that the induction of Cox-2 expression, cytokine, and chemokine mRNA was performed by 3-Oxo-C12-HSL compound in addition to gamma interferon secretion by T cells and NF-kβ activation in keratinocytes. Stoltz et al. (2008) also suggested that PON1 QQ enzyme play a key role in the innate immune response to quorum sensing dependent pathogenic infections in animals after tested in *Drosophila melanogaster* against *P. aeruginosa*. Cao et al. (2014) demonstrated that zebrafish feed with supplementation of AHL lactonase AiiAB546 enzyme influences the immunity of the fish after challenged with *A. hydrophila*. With *A. hydrophila* infection, co-injection or oral administration of AHL lactonases, AiiAB546 from *Bacillus* sp. B546 and AiiAAI96 from *Bacillus* sp. AI96 both isolated from the mud of a fish pond could successfully increase resistance and reduce the mortality of common carp and delay the death time of zebrafish (Cao et al., 2012; Chen et al., 2010). Intestinal alkaline phosphatase is a key enzyme in detoxifying the endotoxin component lipopolysaccharides (LPS) by dephosphorylation (Bates et al., 2007), and its activity was positively correlated with the number of *A. hydrophila* cells in the fish intestine challenge. Liu et al. (2016) confirmed that hybrid tilapia fed with a diet containing the QQ enzyme AHL lactonase AIO6 had lower intestinal alkaline phosphatase activity after challenge, indicating improved resistance of the fish against *A. hydropila*. Hence, these and other studies confirmed that QQ enzymes and QSIs are not only pivotal to degrade the AI signal molecules, but also interact with eukaryotic cells and modulate host immune response of aquatic and terrestrial animals.

27.7 Application and Limitations of Quorum Sensing to Control Animal Bacterial Pathogens

After understanding of the positive relationship between QS system and the virulence factors development in pathogenic bacteria, scientists identify and develop QQ enzymes and QS inhibitors for the interfering of the QS system. Recently, QQ gained more attention by scientists throughout the world, and many researchers have been describing the wide application of QQ to inhibit bacterial virulence, and limit the formation of biofilm and antibiotic resistance in several bacteria strains. Several QQ molecules have been identified and demonstrated as valuable tools for controlling infectious disease in aquatic as well as terrestrial animals. Presently, QQ agents have been widely used in various fields such as agriculture, ecology, the food industry, aquaculture, bioprocessing, and antifouling, in human and animal medicine, and in water treatment plants (Bzdrenga et al., 2017; Torres et al., 2016). Since the discovery of the first quorum-quenching enzyme encoded by aiiA (Dong et al., 2000), the prokaryotic-origin AHL-lactonases and AHL-acylases have been frequently used in investigations of the role of AHL signals owing to the convenience in cloning and expression. More recently, the importance of AHL quorum-sensing signaling in the regulation of virulence and other physiological functions in *Burkholderia thailandensis* and *Erwinia amylovora* has been demonstrated by expression of the AHL-lactonases encoded by the aiiA homologues in these two pathogens, respectively (Molina et al., 2005).

The QQ enzymes and other small natural and synthetic derivatives of QS inhibitors have been identified and valued as effective antimicrobial compounds against diverse pathogenic bacteria in animals. For instance, treatment of mice with synthetic furanones, the derivatives of the natural furanones produced by the seaweed *D. pulchra*, significantly decreased the cell number of *P. aeruginosa* in the infected lung tissues and the disease symptoms (Wu et al., 2004). Skin lesions on inoculated mice were reduced when AIP-II, the group-specific cell-to-cell communication signal produced by group-II *S. aureus*, was included in the inoculum mixture of group-I *S. aureus* bacterial cells (Mayville et al., 1999). The experiment identified the key structural features of the signals involved in activation and antagonism, and led to the design of a global inhibitor of the virulence response in *S. aureus* (Lyon et al., 2000). Even though there is some difficulty of the QQ enzymes and QSIs to produce in large scale with stability and tolerance to industrial processes or storage, still they have been used widely in various fields for biotechnological purposes, and in this review, we discuss the detailed applications of QQ strains, QQ enzymes, and QSIs in aquaculture.

27.7.1 The Application of Quorum Quenching in Aquaculture

For several pathogenic microorganisms in aquaculture, the virulence gene expression is directly related to their capacity to exchange information with other members of the population by the QS system (Bruhn et al., 2005). A pathogen with the higher capacity of detection of signaling molecules indicates the higher density of their population, which is the sign to become virulent. Increasing of the population density of the pathogens results in the accumulation of the signal molecules, and this in turn triggers the activation of gene expressions responsible for pathogenicity development, which helps for successful infection. According to the Food and Agriculture Organization (FAO), the world growth of aquaculture production of aquatic animals from 2001 to 2015 has been gradually reduced with annual growth rate was 5.9%, which is significantly lower than the growth rate recorded in the 1980s and 1990s (FAO, 2017). The major causes for the reduction of the growth rate of world aquaculture were bacterial infections, which comprise a significant constraint to the development of aquaculture in the world (Defoirdt et al., 2004). More importantly, massive use of antibiotics in aquaculture systems is leading to rapid evolution and spread of multiple antibiotic-resistant strains (Heuer et al., 2009), and this results in an increase in the number of resistance bacteria in aquatic animals.

Scientists have been developing several approaches to reduce these risks, including use of probiotics, bacteriophage therapies, or immunostimulants, and vaccines also play a great role to control the majority of fish pathogens (Adams et al., 2006). However, none of these methods seem to significantly solve the problems of bacterial infections in aquaculture. QQ is an appealing strategy that might reduce bacterial infections with a minimum possibility of development of resistance varieties of bacteria. Both Gram-positive and -negative bacteria have QS systems, including the major fish pathogenic Gram-negative bacteria such as *Aeromonas* or *Vibrio* spp. For those pathogenic bacteria, which depend on the QS system to coordinate their pathogenesis activities like *Vibrio* species, interfering with their QS signal molecules AHLs through a QQ mechanism is a likely an alternative for controlling the infectious disease development process instead of reducing the bacteria number directly by killing them (Bjarnsholt et al., 2010; Defoirdt et al., 2004; Natrah et al., 2012; Zhao et al., 2015). A number of application methods also have been explained.

Moreover, several studies have been described the pivotal application of QQ ploy to control infectious diseases caused by the bacteria in aquaculture species including turbot larvae (Tinh et al., 2008), giant freshwater prawns (Nhan et al., 2010), and bivalves (Torres et al., 2016). Torres et al. (2016) isolated four strains of QQ bacteria from the marine ecology for bivalves' aquaculture. *Acidovorax spp.* and *Ochrobactrum spp* bacteria, which were isolated from the intestines of sturgeon, sea bass, and prawns, improved the survival of *shrimp nauplii* challenged with the pathogen *V. campbellii* (Liu et al., 2007). *Pseudomonas* sp. strain FF16 and *Raoultella planticola* strain R5B1 were identified as effective inhibitors of the QS mediated virulence development in *Flavobacterium psychrophilum*, *Vibrio anguillarum*, and *A. hydrophila* of the fish pathogenic bacteria. These bacteria with QQ properties were proposed as a good probiotic candidate for the use of fish farming (e.g., salmon) (Fuente et al., 2015). Several fish pathogenic bacteria including *A. hydrophila* AH1, *V. harveyi* T4, and *Edwardsiella tarda* were also controlled by the small peptide inhibitors produced by the *E. coli* strain DH5α/p5906 (Sun and Zhang, 2016). AHL-degrading enzymes and QSIs were thus investigated for disrupting QS of fish pathogens. Purified lactonases were tested and in particular oral administration of AHL-lactonase from *Bacillus* sp. strain AI96 was shown to decrease *A. hydrophila* infection in zebrafish (Cao et al., 2012). Chen et al. (2010) observed similar results using AHL-lactonase (AiiAB546) from *Bacillus* sp. B546 produced by *Pichia pastoris* in common carp. Vinoj et al. (2014) also demonstrated that the pathogen colonization of Indian white shrimp was successfully reduced using AHL-Lactonase from *Bacillus licheniformis* DAHB1 strain.

Some studies also suggest that the application of QQ enzymes is more effective if we use them in combination with pre- and probiotics, immunostimulants, and vaccines to control and protect fish against a wide range of fish pathogenic infections. Accordingly, several studies confirmed that the QQ strategy can be an effective ploy to control aquatic animal infectious diseases that occur in the aquaculture system and reduce economic losses in the aquaculture industry. The detailed mechanisms of the QQ strategy on controlling of infectious diseases in aquatic animals were verified in a few aquatic animals as shown in Table 27.3.

27.7.2 *In Vivo* Application of QS Inhibitors to Control Animal Bacterial Pathogens and Their Limitations

Development of strategies that target the QS systems of bacteria are important anti-infectious strategies to control the production of virulence factors and biofilm formation of pathogenic bacteria. The development of novel nonantibiotic therapeutics aiming to suppress the virulent genes expression and prevent infection, without affecting bacterial cell viability, has attracted attention due to the decreased risk of creation of antibiotic-resistance bacteria in the environment. Due to the promising result of the

TABLE 27.3

Applications of Quorum Quenching Strains, Quorum Quenching Enzymes, and Quorum Sensing Inhibitors to Control Infectious Diseases in Aquatic Animals

QQ Agents	Source	Animal	Inhibition Against	Effect	Dosage Used	Percent Survival/ Damage Rate	References
QQ Strains							
Bacillus sp. YB1701	Coastal sediment	*Carassius auratus gibelio*	*A. hydrophila* and *V. parahemolyticus*	Reduce production of protease, amylase	5×10^{10} CFU/g	76.7%	Zhou et al. (2018)
Chlorella saccharophila CCAP211/48	Culture of algae and protozoa	Burbot	*A. salmonicida* HN-00	Interfere with QS system	10^6 cells/mL	About 93%	Natrah et al. (2012)
Chlamydomonas reinhardtii CCAP11/45	Culture of algae and protozoa	Burbot	*A. salmonicida* HN-00	Interfere with QS system	10^6 cells/mL	About 74%	Natrah et al. (2012)
Bacterial enrichment cultures	Enrichment cultures from European sea bass and Asian sea bass	*Macrobrachium rosenbergii*	*V. harveyi*	Interfere with QS and inhibit pathogenicity	10^6–10^7 CFU/mL	67%–75%	Nhan et al. (2010)
Alteromonas stellipolaris PQQ-42	Bivalve hatchery	*Oculina patagonica*	*V. mediterranei VibC*-Oc-097	Reduce the production of protease and chitinase and swimming motility	10^5 CFU/mL	Lower degree of tissue damage	Torres et al. (2016)
QQ Enzymes							
AHL-lactonase	*Oculina patagonica, Cladocora caespitosa* corals and sea water	*Artemia salina*	*V. mediterranei VibC-Oc-097, V. owensii VibC-Oc-106* and *V. coralliilyticus VibC-Oc-193*	Reduce swimming motility, caseinase and amylase activity was eliminated, chitinase and DNAse activities were diminished	10^6 CFU/mL	Significantly increased	Torres et al. (2018)
AHLs-lactonases	*Oculina patagonica, Cladocora caespitosa* corals and sea water	*Venerupis philippinarum*	*V. coralliilyticus VibC-Oc-193*	Reduce swimming motility	10^6 CFU/mL	Significantly increased	Torres et al. (2018)
Lactonase AiiA-B546	*Bacillus* sp. QSI-1	Zebrafish	*A. hydrophila*	Inhibit protease production and biofilm formation	10^8 CFU/g or 3.74 U/g	83.3%	Cao et al. (2012); Chu et al. (2014)
AHL lactonase $AiiA_{B546}$	*Bacillus* sp. B54	Common carp	*A. hydrophila* ATCC 7966	Decrease protease	10 µL	Increased survival rate by 25%, delayed mortality time	Chen et al. (2010)
AHL lactonase AIO6	*Lactobacillus plantarum* JCM 1149	Hybrid tilapia	*A. hydrophila* NJ-1	Compute for adhesion place in the intestine	4 U/g	Decreased lamina propria damaged, improve microvilli organization, improved resistance	Liu et al. (2016)
AHL lactonase$AiiA_{B546}$	*Bacillus* sp. B546	Zebrafish	*A. hydrophila*	Interfer with QS, influenced adhesion of *A. hydrophila* to the host gut	3.72 U/g	Increased resistance	Cao et al. (2014)

(*Continued*)

TABLE 27.3 (*Continued*)

Applications of Quorum Quenching Strains, Quorum Quenching Enzymes, and Quorum Sensing Inhibitors to Control Infectious Diseases in Aquatic Animals

QQ Agents	Source	Animal	Inhibition Against	Effect	Dosage Used	Percent Survival/ Damage Rate	References
HepS-AHL	Commercial	Arctic charr	*A. salmonicida* subsp. *achromogenes*, strain *Keldur265-87*	Inhibit protease production	10^3 CFU/fish	Increased survival	Schwenteita et al. (2011)
C14-HSL	Commercial	Burbot	*A. hydrophila* AH-1N	Interfere with QS system	10 µM	About 74% survived	Natrah et al. (2012)
C14-HSL	Commercial	Burbot	*A. salmonicida* HN-00	Interfere with QS system	10 µM	About 82% survived	Natrah et al. (2012)
QSIs							
3-hydroxybutyrate	Poly-β-hydroxybutyrate	*Artemia salina*	*V. campbellii*	Decrease phospholipase, protease, and motility activity	125 mM	About 80% survived	Defoirdt et al. (2018)
Polyhydroxybutyrates	Enrichment cultures	*Artemia salina*	*V. harveyi*, *V. campbellii* LMG21363	Degrade QS signal molecule – AHL	0.9%–9.4% of cell dry weight	Increased survival	Van Cam et al. (2009)
Coumarin	Plant	Salmon	*Vibrio anguillarum*	Inhibit biofilm formation	25–150 µg	Increased survival	Gutiérrez-Barranquero et al. (2015)
Curcumin	*Curcuma longa* plant	*Artemia salina* nauplii	*V. harveyi*	Reduce biofilm formation and motility	25–100 µg/mL	Increased survival up to 67%	Packiavathy et al. (2013)

biotechnological application of QSIs in controlling animal pathogenic bacteria, it attracts the attention of scientists. Presently, numerous studies have tried to apply QSIs to control bacterial infections in terrestrial as well as aquatic animals. Especially in aquatic animals, disease control with QSIs is very important, since the application of antibiotics is highly restricted and promising results have been obtained (Romero et al., 2012). Several *in vivo* trials including using bacterial consortia with AHL degradation activity were demonstrated to protect *Artemia* sp., rotifers, and larvae of turbot or prawn from infection (Nhan et al., 2010). Defoirdt et al. (2006) demonstrated that the application of furanones inhibits the pathogenicity of several *Vibio* species including *Vibrio harveyi*, *V. campbellii*, and *V. parahaemolyticus* against *Artemia* sp. The negative effect of *V. harveyi* also inhibited using furanones on rotifers, but furanones are toxic for both animals (Tinh et al., 2008). The synthetic furanone C-30 is not toxic at any concentration and effective against pathogenic *V. anguillarum* in rainbow trout (Rasch et al., 2004). The QQ enzyme AiiA was active to degrade the AHLs of *A. hyrophila* and was effective to control the disease caused by this bacterium in carp (Chen et al., 2010). In aquaculture, halogenated furanones have been holding great promise in protecting fishes (Benneche et al., 2011; Defoirdt et al., 2007). However, thiophenones are proving to be less toxic (Defoirdt et al., 2012; Yang et al., 2015). Supplementation of fish feed with variant of lactonases and peptides have been reported to inhibit QS mediated pathogenicity (Sun and Zhang, 2016).

In addition to these, the effectiveness of purified QQ enzyme PvdQ acylase against *P. aeruginosa* PAO1 in *C. elegans* resulted in reduction in pathogenicity and increased life span of nematodes (Papaioannou et al., 2009). In mice, the pathogenicity of *P. aeruginosa* PAO1 was reduced after pulmonary infections were treated with garlic extracts (Rumbaugh et al., 1999). These *in vivo* demonstrations showed that THE application of QSIs could be successfully used for blocking bacterial infections. Moreover, the interest of these mechanisms is that they do not affect directly the survival of the pathogen but the expression of virulence factors, and so they do not exert selective pressure, avoiding the appearance of resistance.

At present there are many patents related to strategies for interfering with QS systems as a method of controlling pathogenic bacterial diseases in terrestrial and aquatic animals. Most of the patents registered were focused on the mechanisms of inhibition of quorum sensing signal molecules AHLs and a few on the inhibition of AI-2. The patents registered for the inhibitor of the action of AHLs are based on compounds which have similar structure to the QS molecule, thereby acting as replacements for naturally occurring bacterial QS ligands and producing an antagonistic effect, or reducing the expression of QS gene expression. The sources of the compounds patented were isolated from various sources including plants, garlic extract (*S*-allyl-cysteine) (Steggles, 2011), and bacteria (Piericidin A1) acting as a competitive inhibitory activity against AHLs (Kim et al., 2007), red algae (*Rhodophyta* extracts) (Park et al., 2007b), a marine fungi, *Penicillium* (shown to inhibit QS mechanisms) (Qianhong et al., 2010), Chinese herbal extracts, *Epimedium* (diphylloside A) (Kangmin et al., 2009), and tropical mangrove shrubs like *Conocarpus erectus* (interfere with QS mechanisms) (Mathee et al., 2009).

Some of the patents registered, including the patent application WO2011154585 (A1), provide information on the use of a novel *alpha-proteobacteria* for quorum quenching, which is the invention (concerning the fields of biology, molecular biology, and aquaculture) specifically related to a new a-proteobacteria capable of degrading/AHLs for control of bacterial infectious diseases and prevention of biofilm formation. Patent application CN103275949 (A) provides data on quorum-quenching enzyme OLB-26 and coding genes, and the application thereof. For more information, see the review in Lopez et al. (2014).

27.7.3 Limitation of Quorum Sensing Inhibitors

In spite of the fact that mechanisms of QS system and QSIs have been widely studied and have a great potential to control numerous bacterial pathogenic phenotypes in animals, most of them are still at the preclinical stage and need to be translated to *in vivo* trials (Reuter et al., 2016). To date, few *in vivo* trials conducted with QS inhibitors are described. The practical use of the discovered anti-QS compounds is frequently limited because of their potential toxicity, instability, and reduced therapeutic effect compared to antibiotics (Ivanova et al., 2018). Nano-size transformation of the anti-virulent and antibacterial compounds have gained attention for engineering a new class of antimicrobials with unique properties, improved stability, and higher efficacy *in vitro* and *in vivo* compared to bulk solutions. Despite the numerous advantages of QSIs to control several bacterial diseases in animals, there are few reports on the disadvantages of QS inhibitors. QS inhibitors exert a negative effect on biofilm formation and related expression of virulence factors, but they can also exert a positive effect on the expression of virulence factors by the biofilm population. For instance, the type III secretion system of *P. aeruginosa*, which is a key determinant of virulence, is negatively regulated by the QS system. As a result, QS inhibition can promote virulence (Kong et al., 2009). Type VI secretion system of *P. aeruginosa* is negatively regulated by QS itself (Sana et al., 2012). Sometimes inhibition of QS accentuates the colonization of more virulent wild type along with the eradication of the less virulent mutated strains that give rise to the increased prevalence of virulent genotypes in nosocomial infections.

For certain QSIs, the concentration is vital to the effective action of the QS inhibitors against bacterial infection, and they work in a concentration dependent manner including furanones and synthetic HSLs and can activate the QS system at a particular concentration rather than inhibiting them. A few studies showed that some bacteria species could develop resistance against QSIs. For instance, *P. aeruginosa* has been reported to acquire resistance against QSI, furanone C-30, since the bacterium activates multidrug efflux pump MexAB-Opr, which is a result of mutation in the transcription repressors MexR and NalC that disrupt their regulatory activity (Das and Singh, 2018). Many such mutations can be found in clinical isolates as they are selected under immense antibiotic pressure. Another quorum quenching compound, 5-fluorouracil, has faced resistance in cellular uptake in the case of some clinical isolates and multidrug-resistant strains. Instead, these strains produce increased concentration of virulence factors in the presence of QS inhibitors like furanone. Moreover, a few studies have shown evidence of emergence of microbial resistance to QSI (García-Contreras et al., 2015; Kalia et al., 2014; Koul et al., 2016). Some bacteria strains under stress conditions act as "social cheaters" and stop responding to QS system. This action affects the efficacy of the QS inhibitors (Tay and Yew, 2013). However, since these mutants show reduced fitness, their chances of survival are expected to be low (Gerdt and Blackwell, 2014). The likelihood of the emergence of resistance to QSIs is also influenced by the mechanisms of action of the QSIs themselves. Hence, we may need to be more cautious and look for QSIs which are QS signal independent. QSIs may not be an effective method to control bacteria diseases when the animal has low immunity, since the immune system plays a great role in cleaning the bacteria from the body after it is inhibited by the QSIs.

In a few bacteria strains, the presence of more than one QS regulation system enhances the likelihood that herterodimers of transcriptional regulations will bind to promiscuous promoters and make the bacteria capable of expressing its genes in order to resist the effect caused by the environment (Koul and Kalia, 2017). The variability in the specific activities of AHL synthases in strains of *E. carotovora* and the multiplicity of LuxR signal receptor homologs in *Burkholderia mallei* are latent features which further support the bacterial potential to develop resistance to QSIs (Brader et al., 2005; Case et al., 2008). In *P. aeruginosa*, the QS system *-rhl* and *-las* are essential for biofilm formation, and their disruption depends upon the antibiotics used and the host immune system (Bjarnsholt et al., 2005).

The use of enzyme-based QSIs which act on QS signals outside the bacterial cell have been suggested to be ideal candidates (Defoirdt et al., 2010; Guendouze et al., 2017). The apprehension remains that bacteria may become provoked to produce larger quantities of QS signal molecules. In fact, it has been shown that QS-mediated swarming motility in *Serratia liquefaciens* and a LuxR-regulated PluxI-gfp (ASV) fusion in a mouse model inhibited by QSI-brominated furanone could be reversed by exogenous addition of AHLs (Hentzer et al., 2003). Hence, we must ensure that enzymes with high hydrolytic activity and stability be employed. Although these hydrolytic enzymes have a very broad spectrum of activity, the chance exists for bacteria to react by modifying their QS signal molecules. Bacteria also have the potential to modify their LuxR receptors to improve affinity for QS signal molecules (Hawkins et al., 2007), requiring that the QSI enzymes be modified to detect AIs even at very low concentrations. Another challenge in treating these infectious bacteria is that their QS-mediated pathogenicity occurs at high cell density; therefore, the ability to detect these bacteria and their QS signals at low densities (i.e., before the onset of disease) is crucial. With few limitations of QSIs, it can be expected to have numerous biotechnological applications in yet unexplored areas of economic importance (Bzdrenga et al., 2017). Therefore, QS inhibitors should be used judicially as their indiscriminate use may result in selection of QS inhibitor-resistant strains and worsen the biofilm mediated virulence.

27.8 Conclusions and Future Perspectives

Due to the increase in animal diseases, producers are looking for antibiotic alternatives to help maintain the health of their animals. The QQ strategy that works by disrupting QS systems is a valuable tool for promoting aquatic and terrestrial animal health. QS is important in controlling a variety of microbial cell activities, such as virulence, biofilm formation, and antibiotic resistance, which significantly impact animal health and production systems. Since the detection of QS systems in bacteria, researchers have been devoted to understanding the QS system and the development of strategies to disrupt and manipulate QS. QQ is a mechanism of interfering with the QS system. Quorum-sensing interfering agents can be either natural or synthetic compounds acting as inhibitors or agonists of signal molecule biosynthesis, signal molecule detection, or signal transduction, or enzymes that inactivate or degrade the signal molecules that sequester signal molecules and induce an immune response. In this review, many different QQ strains, QQ enzymes, and QSI agents that can inhibit bacterial QS with varying degrees of potency are described. Many of these QQ agents were shown to effectively inhibit QS, and the strong efficacy of several QQ enzymes and QSIs were demonstrated *in vitro* as well as *in vivo* using terrestrial and aquatic animals. These QQ agents can be QQ strains, QQ enzymes, and QSIs. The QQ strategy has more advantages than using the classic antibiotic to control infectious animal diseases with less probability of resistance bacteria development.

These studies point to a promising role for QQ strategy as quenchers of bacterial virulence and/or biofilm formation in various biotechnological settings without resulting in antibiotic-resistance. Although more research is needed to better understand different species, QS disruption is a QQ strategy that holds promise for reducing the effects of enteric bacterial diseases, preserving intestinal health, and promoting animal performance in antibiotic-free production systems. Even though the QQ mechanisms play a key role in controlling infectious diseases in animals, currently their application is limited to a few terrestrial and aquatic animals. Therefore, it is time to investigate the details of the application of QQ enzymes and QSIs in order to apply them for a wide range of animals. Detailed studies on the QS crosstalk and signal specificity, the effect of QQ system on the host's immune system, and QQ and host gut health are other interesting research areas that will impact the QQ strategy.

ACKNOWLEDGMENTS

This work was supported by the key project of Chinese National Programs for Fundamental Research and Development (973 Program, 2015CB150605), the National Natural Science Foundation of China (31872584, 3180131599, 31702354, 31602169, 31672294, 31572633), the Beijing earmarked fund for Modern Agro-industry Technology Research System (SCGWZJ 20201104-4), and Innovation Capability Support Program of Shaanxi (2018TD-021).

REFERENCES

Aboushleib HM, Omar HM, Abozahra R, Elsheredy A, and Baraka K, "Correlation of quorum sensing and virulence factors in *Pseudomonas aeruginosa* isolates in Egypt." *The Journal of Infection in Developing Countries* 9, no. 10 (2015): 1091–1099.

Adams A and Thompson KD, "Biotechnology offer revolution to fish health management." *Trends Biotechnology* 24 (2006): 201–205.

Adonizio A, Kong KF, and Mathee K, "Inhibition of quorum sensing controlled virulence factor production in *Pseudomonas aeruginosa* by south Florida plant extracts." *Antimicrobial Agents Chemistry* 52 (2008): 198–203.

Alanis AJ, "Resistance to antibiotics: Are we in the post-antibiotic era?" *Archives of Medical Research* 36 (2005): 697–705.

Andersen JB, Heydorn A, Hentzer M, Eberl L, Geisenberger O, Christensen BB, Molin S, and Givskov M, "gfp-based *N*-acyl homoserine-lactone sensor systems for detection of bacterial communication." *Applied and Environmental Microbiology* 67 (2001): 575–585.

Asfour HZ, "Anti-quorum sensing natural compounds." *Journal of Microscope Ultrastructure* 6, no. 1 (2018): 1–10.

Atkinson S, Chang CY, Patrick HL, Buckley CM, Wang Y, Sockett RE, Camara M, and Williams P, "Functional interplay between the *Yersinia pseudotuberculosis* YpsRI and YtbRI quorum sensing systems modulates swimming motility by controlling expression of flhDC and fliA." *Molecular Microbiology* 69 (2008): 137–151.

Atkinson S, Chang CY, Sockett RE, Cámara M, and Williams P, "Quorum sensing in Yersinia enterocolitica controls swimming and swarming motility." Journal of Bacteriology 188 (2006): 1451–1461. doi:10.1128/JB.188.4.1451-1461.2006.

Austin B and Austin DA, *Bacterial Fish Pathogens: Disease of Farmed and Wild Fish*. Springer, New York, 3rd ed., 1999.

Austin B and Austin DA, *Bacterial Fish Pathogens: Diseases of Farmed and Wild Fish*. Springer Publishers, Cham, Switzerland, 2007.

Bassler BL and Losick R, "Bacterially speaking." *Cell* 125, no. 2 (2006): 237–246.

Bates JM, Akerlund J, Mittge E, and Guillemin K, "Intestinal alkaline phosphatase detoxifies lipopolysaccharide and prevents inflammation in zebrafish in response to the gut microbiota." *Cell Host & Microbe* 2, no. 6 (2007): 371–382.

Bauer WD and Robinson JB, "Disruption of bacterial quorum sensing by other organisms." *Current Opinion in Biotechnology* 13, no. 3 (2002): 234–237.

Bearson BL and Bearson SMD, "The role of the QseC quorum-sensing sensor kinase in colonization and norepinephrine-enhanced motility of *Salmonella enterica* serovar Typhimurium." *Microbial Pathogenesis* 44, no. 4 (2008): 271–278.

Benneche T, Herstad G, Rosenberg M, Assev S, and Scheie AA, "Facile synthesis of 5-(alkylidene) thiophen-2 (5 H)-ones. A new class of antimicrobial agents." *RSC Advances* 1, no. 2 (2011): 323–332.

Bjarnsholt T, Jensen PO, Burmølle M, Hentzer M, Haagensen JAJ, Hougen HP, et al., "*Pseudomonas aeruginosa* tolerance to tobramycin, hydrogen peroxide and polymorphonuclear leukocytes is quorum-sensing dependent." *Microbiology* 151 (2005): 373–383.

Bjarnsholt T, Tolker-Nielsen T, Høiby N, and Givskov M, "Interference of *Pseudomonas aeruginosa* signalling and biofilm formation for infection control." *Expert Reviews in Molecular Medicine* 12 (2010): e11.

Boles BR and Horswill AR, "Agr-mediated dispersal of *Staphylococcus aureus* biofilms." *PLoS Pathogogy* 4 (2008): e1000052.

Boyen F, Pasmans F, Van Immerseel F, Morgan E, Adriaensen C, Hernalsteens JP, et al., "*Salmonella* Typhimurium SPI-1 genes promote intestinal but not tonsillar colonization in pigs." *Microbes Infection* 8 (2006): 2899–2907.

Brackman G, Cos P, Maes L, Nelis HJ, and Coenye T, "Quorum sensing inhibitors increase the susceptibility of bacterial biofilms to antibiotics in vitro and in vivo." *Antimicrobial Agents and Chemotherapy* 55 (2011): 2655–2661. doi:10.1128/AAC.00045-11

Brackman G, Defoirdt T, Miyamoto C, Bossier P, Van Calenbergh S, Nelis H, and Coenye T, "Cinnamaldehyde and cinnamaldehyde derivatives reduce virulence in *Vibrio* spp. by decreasing the DNA-binding activity of the quorum sensing response regulator LuxR." *BMC Microbiology* 8, no. 1 (2008): 149.

Brader G, Sjoblom S, Hyytiöinen H, Sims-Huopaniemi K, and Palva ET, "Altering substrate chain length specificity of an acylhomoserine lactone synthase in bacterial communication." *Journal of Biological Chemistry* 280 (2005): 10403–10409.

Bruhn JB, Dalsgaard I, Nielsen KF, Buchholtz C, Larsen JL, and Gram L, "Quorum sensing signal molecules (acylated homoserine lactones) in Gram-negative fish pathogenic bacteria." *Diseases of Aquatic Organisms* 65 (2005): 43–52.

Burt SA, Ojo-Fakunle VT, Woertman J, and Veldhuizen EJ, "The natural antimicrobial carvacrol inhibits quorum sensing in *Chromobacterium violaceum* and reduces bacterial biofilm formation at sub-lethal concentrations." *PLoS One* 9, no. 4 (2014): e93414.

Bzdrenga J, Daudé D, Rémy B, Jacquet P, Plener L, Elias M, and Chabrière E, "Biotechnological applications of quorum quenching enzymes." *Chemical and Biological Interactions* 267 (2017): 104–115.

Cao JG and Meighen EA, "Purification and structural identification of an autoinducer for the luminescence system of *Vibrio harveyi*." *Journal of Biological Chemistry* 264, no. 36 (1989): 21670–21676.

Cao Y, He S, Zhou Z, Zhang M, Mao W, Zhang H, and Yao B, "Orally administered thermostable *N*-acyl homoserine lactonase from *Bacillus* sp. strain Ai96 attenuates *Aeromonas hydrophila* infection in zebrafish." *Applied Environmental Microbiolology* 78, no. 6 (2012): 1899–1908.

Cao Y, Liu Y, Mao W, Chen R, He S, Gao X, Zhou Z, and Yao B. "Effect of dietary *N*-acyl homoserin lactonase on the immune response and the gut microbiota of zebrafish, *Danio rerio*, infected with *Aeromonas hydrophila*." *Journal of the World Aquaculture Society* 45, no. 2 (2014): 149–162.

Cao X, Wang Q, Liu Q, Liu H, He H, and Zhang Y, "Vibrio alginolyticus MviN is a LuxO-regulated protein and affects cytotoxicity toward EPC cell." *Journal of Microbiology and Biotechnology* 20 (2010): 271–280.

Case RJ, Labbate M, and Kjelleberg S, "AHL-driven quorum sensing circuits: Their frequency and function among the *Proteobacteria*." *ISME* 2 (2008): 345–349.

Castillo-Juarez I, Maeda T, Mandujano-Tinoco EA, Tomas M, Perez-Eretza B, Garcia-Contreras SJ, et al., "Role of quorum sensing in bacterial infections." *World Journal of Clinical Cases* 3 (2015): 575–598. doi:10.12998/wjcc.v3.i7.575.

Cegelski L, Marshall GR, Eldridge GR, and Hultgren SJ, "The biology and future prospects of antivirulence therapies." *Nature Reviews Microbiology* 6 (2008): 17–27.

Cha C, Gao P, Chen YC, Shaw PD, and Farrand SK, "Production of acyl-homoserine lactone quorum-sensing signals by Gram-negative plant associated bacteria." *Molecular Plant-Microbe* 11 (1998): 1119–1129.

Chen F, Gao Y, Chen X, Yu Z, and Li X, "Quorum quenching enzymes and their application in degrading signal molecules to block quorum sensing-dependent infection." *International Journal of Molecular Science* 14 (2013): 17477–17500.

Chen R, Zhou Z, Cao Y, Bai Y, and Yao B, "High yield expression of an AHL-lactonase from *Bacillus* sp. B546 in *Pichia pastoris* and its application to reduce *Aeromonas hydrophila* mortality in aquaculture." *Microbe Cell Fact* 9 (2010): 39–48.

Cho Y, Cramer RA, Kim KH, Davis J, Mitchell TK, Figuli P, et al., "The Fus3/Kss1 MAP kinase homolog Amk1 regulates the expression of genes encoding hydrolytic enzymes in *Alternaria brassicicola*." *Fungal Genetics and Biology* 44, no. 6 (2007): 543–553.

Choi J, Shin D, and Ryu S, "Implication of quorum sensing in *Salmonella enterica* serovar typhimurium virulence: The *luxS* gene is necessary for expression of genes in pathogenicity island 1." *Infection and Immunity* 75 (2007): 4885–4890.

Choi J, Shin D, Kim M, Park J, Lim S, and Ryu S. "LsrR-mediated quorum sensing controls invasiveness of *Salmonella typhimurium* by regulating SPI-1 and flagella genes." *PLoS One* 7, no. 5 (2012): e37059.

Chong YM, Yin WF, Ho CY, Mustafa MR, Hadi AHA, Awang K, et al., "Malabaricone C from *Myristica cinnamomea* exhibits anti-quorum sensing activity." *Journal of Natural Products* 74 (2011): 2261–2264.

Christiaen SEA, Matthijs N, Zhang X-H, Nelis HJ, Bossier P, and Coenye T, "Bacteria that inhibit quorum sensing decrease biofilm formation and virulence in *Pseudomonas aeruginosa* PAO1." *Pathogens and Disease* 70, no. 3 (2014): 271–279.

Chu, Weihua, Shuxin Zhou, Wei Zhu, and Xiyi Zhuang. "Quorum quenching bacteria *Bacillus* sp. QSI-1 protect zebrafish (*Danio rerio*) from *Aeromonas hydrophila* infection." *Scientific Reports* 4 (2014): 5446.

d'Angelo-Picard C, Haudecoeur E, Chevrot R, Dessaux Y, and Faure D, "The plant pathogen *Agrobacterium tumefaciens*: A model to study the roles of lactonases in the quorum-sensing regulatory network." *Biology of Plant-microbe Interactions* 5 (2006): 353–356.

Das L and Singh Y, "Quorum sensing inhibition: A target for treating chronic wounds." In Kalia VC (ed.), *Biotechnological Applications of Quorum Sensing Inhibitors*, pp. 111–126. Springer Nature, Singapore, 2018.

De Lamo Marin S, Xu Y, Meijler MM, and Janda KD, "Antibody catalyzed hydrolysis of a quorum sensing signal found in Gram-negative bacteria." *Bioorganic & Medicinal Chemistry Letters* 17 (2007): 1549–1552.

Defoirdt T, "Virulence mechanisms of bacterial aquaculture pathogens and antivirulence therapy for aquaculture." *Reviews in Aquaculture* 6, no. 2 (2014): 100–114.

Defoirdt T, Benneche T, Brackman G, Coenye T, Sorgeloos P, and Scheie AA, "A quorum sensing-disrupting brominated thiophenone with a promising therapeutic potential to treat luminescent vibriosis." *PloS one* 7, no. 7 (2012): e41788.

Defoirdt T, Boon N, Bossier P, and Verstraete W, "Disruption of bacterial quorum sensing: An unexplored strategy to fight infections in aquaculture." *Aquaculture* 240 (2004): 69–88.

Defoirdt T, Boon N, Sorgeloos P, Verstraete W, and Bossier P, "Quorum sensing and quorum quenching in *Vibrio harveyi*: lessons learned from in vivo work." *The ISME Journal* 2 (2008): 19–26.

Defoirdt T, Brackman G, and Coenye T, "Quorum sensing inhibitors: How strong is the evidence?" *Trends in Microbiology* 21, no. 12 (2013): 619–624.

Defoirdt T, Crab R, Wood TK, Sorgeloos P, Verstraete W, and Bossier P, "Quorum sensing-disrupting brominated furanones protect the gnotobiotic brine shrimp *Artemia franciscana* from pathogenic *Vibrio harveyi*, *Vibrio campbellii* and *Vibrio parahaemolyticus* isolates. *Applied and Environmental Microbiology* 72 (2006): 6419–6423.

Defoirdt T, Mai Anh NT, and De Schryver P, "Virulence-inhibitory activity of the degradation product 3-hydroxybutyrate explains the protective effect of poly-β-hydroxybutyrate against the major aquaculture pathogen *Vibrio campbellii*." *Scientific Reports* 8, no. 1 (2018): 7245.

Defoirdt T, Miyamoto CM, Wood TK, Meighen EA, Sorgeloos P, Verstraete W, and Bossier P, "The natural furanone (5Z)-4-bromo-5-(bromomethylene)-3-butyl-2 (5H)-furanone disrupts quorum sensing-regulated gene expression in *Vibrio harveyi* by decreasing the DNA-binding activity of the transcriptional regulator protein luxR." *Environmental Microbiology* 9 (2007): 2486–2495.

Defoirdt T, Ruwandeepika HAD, Karunasagar I, Boon N, and Bossier P, "Quorum sensing negatively regulates chitinase in *Vibrio harveyi*." *Environmental Microbiology Reports* 2 (2010): 44–49.

Defoirdt T, Sorgeloos P, and Bossier P, "Alternatives to antibiotics for the control of bacterial disease in aquaculture." *Current Opinion in Microbiology* 14 (2011): 251–258.

Desouky SE, Nishiguchi K, Zendo T, Lgarahi Y, Williams P, Sonomoto K, and Nakayama J, "High-throughput screening of inhibitors targeting Agr/Fsr quorum sensing in *Staphylococcus aureus* and *Enterococcus faecalis*." *Bioscience, Biotechnology, and Biochemistry* 77 (2013): 923–927.

Diggle SP, Gardner A, West SA, and Griffin AS, "Evolutionary theory of bacterial quorum sensing: When is a signal not a signal?" *Philosophical Transactions of the Royal Society B* 362 (2007): 1241–1249.

Dong Y-H and Zhang L-H, "Quorum sensing and quorum-quenching enzymes." *The Journal of Microbiology* 43, no. 1 (2005): 101–109.

Dong Y-H, Xu J-L, Li X-Z, and Zhang L-H, "AiiA, an enzyme that inactivates the acylhomoserine lactone quorum-sensing signal and attenuates the virulence of *Erwinia carotovora*." *Proceedings of the National Academy of Sciences* 97, no. 7 (2000): 3526–3531.

Donnenberg MS, "Pathogenic strategies of enteric bacteria." *Nature* 406, no. 6797 (2000): 768.

Dunny GM, Brickman TJ, and Dworkin M, "Multicellular behavior in bacteria: Communication, cooperation, competition and cheating." *Bioessays* 30 (2008): 296–298.

Farrand SK, Qin Y, and Oger P, "Quorum-sensing system of *Agrobacterium plasmids*: Analysis and utility." *Methods Enzymology* 358 (2002): 452–484.

Fetzner S, "Quorum quenching enzymes." *Journal of Biotechnology* 201 (2015): 2–14.

Food and Agriculture Organization (FAO). FAO Aquaculture Newsletter No. 56 (April). Rome, FAO, 2017.

Fuente MDL, Miranda CD, Jopia P, Gonzalez-Rocha G, Guiliani N, Sossa K, and Urrutia H, "Growth inhibition of bacterial fish pathogens and quorum-sensing blocking by bacteria recovered from Chilean salmonid farms." *Journal of Aquatic Animal Health* 27 (2015): 112–122.

Gallardo-Garcia MM, Sanchez-Espin G, Ivanova-Georgieva R, Ruiz-Morales J, Vinuela Gonzalez MM, and Garcia-Lopez MV, "Relationship between pathogenic, clinical, and virulence factors of *Staphylococcus aureus* in infective endocarditis versus uncomplicated bacteremia: A case-control study." *European Journal of Clinical Microbiology & Infectious Diseases* 35, no. 5 (2016): 821–828.

García-Contreras R, Maeda T, and Wood TK, "Can resistance against quorum-sensing interference be selected?" *The ISME Journal* 10 (2015): 4–10.

George EA and Muir TW, "Molecular mechanisms of agr quorum sensing in virulent *staphylococci*." *Chembiochemistry* 8 (2007): 847–855.

Gerdt JP and Blackwell HE, "Competition studies confirm two major barriers that can preclude the spread of resistance to quorum-sensing inhibitors in bacteria." *ACS Chemistry and Biology* 9 (2014): 2291–2299.

Girard G and Bloemberg GV, "Central role of quorum sensing in regulating the production of pathogenicity factors in *Pseudomonas aeruginosa*." *Future Microbiology* 3 (2008): 97–106.

Guendouze A, Plener L, Bzdrenga J, Jacquet P, Rémy B, Elias M, et al., "Effect of quorum quenching lactonase in clinical isolates of *Pseudomonas aeruginosa* and comparison with quorum sensing inhibitors." *Frontiers in Microbiology* 8 (2017): 227.

Gutiérrez-Barranquero JA, Reen FJ, McCarthy RR, and O'Gara F, "Deciphering the role of coumarin as a novel quorum sensing inhibitor suppressing virulence phenotypes in bacterial pathogens." *Applied Microbiology and Biotechnology* 99, no. 7 (2015): 3303–3316.

Hammer AS, Pedersen K, Andersen TH, Jørgensen JC, and Dietz HH, "Comparison of *Pseudomonas aeruginosa* isolates from mink by serotyping and pulsed-field gel electrophoresis." *Veterinary Microbiology* 94 (2003): 237–243.

Harikrishnan R, Balasundaram C, Jawahar S, and Heo MS, "Solanum nigrum enhancement of the immune response and disease resistance of tiger shrimp, *Penaeus monodon* against *Vibrio harveyi*." *Aquaculture* 318, no. 1–2 (2011): 67–73.

Hauser AR, "*Pseudomonas aeruginosa*: So many virulence factors, so little time." *Critical Care Medicine* 39, no. 9 (2011): 2193–2194.

Hawkins AC, Arnold FH, Stuermer R, Hauer B, and Leadbetter JR, "Directed evolution of *Vibrio fischeri* LuxR for improved response to butanoylhomoserine lactone." *Applied Environmental Microbiology* 73 (2007): 5775–5781.

He SX, Zhang MC, Xu L, Yang YL, Li Q, and Zhou ZG, "Effect of quorum-quenching enzyme from *Bacillus* sp. AI-96 on *Aeromonas hydrophila* NJ-1 bath challenge in zebrafish." *Journal of Fisheries of China* 1 (2013): 011.

He Y, Frye JG, Strobaugh TP, and Chen CY, "Analysis of AI-2/LuxS-dependent transcription in *Campylobacter jejuni* strain 81-176." *Foodborne Pathogens and Disease* 5 (2008): 399–415.

Hentzer M, Wu H, Andersen JB, Riedel K, Rasmussen TB, Bagge N, et al., "Attenuation of *Pseudomonas aeruginosa* virulence by quorum sensing inhibitors." *EMBO Journal* 22 (2003): 3803–3815.

Heuer OE, Kruse H, Grave K, Collignon P, Karunasagar I, and Angulo FJ, "Human health consequences of use of antimicrobial agents in aquaculture." *Clinical Infectious Diseases* 49 (2009): 1248–1253.

Imamura Y, Yanagihara K, Tomono K, Ohno H, Higashiyama Y, Miyazaki Y, et al., "Role of *Pseudomonas aeruginosa* quorum-sensing systems in a mouse model of chronic respiratory infection." *Journal of Medical Microbiology* 54, no. 6 (2005): 515–518.

Ivanova A, Ivanova K, and Tzanov T, "Inhibition of quorum-sensing: A new paradigm in controlling bacterial virulence and biofilm formation." In Kalia VC (ed.), *Biotechnological Applications of Quorum Sensing Inhibitors*, pp. 3–21. Springer Nature, Singapore, 2018.

Jakobsen TH, Warming AN, Vejborg RM, Moscoso JA, Stegger M, Lorenzen F, et al. "A broad range quorum sensing inhibitor working through sRNA inhibition." *Scientific Reports* 7, no. 1 (2017): 9857.

Kalia VC and Purohit HJ, "Quenching the quorum sensing system: Potential antibacterial drug targets." *Critical Reviews in Microbiology* 37 (2011): 121–140.

Kalia VC, "Quorum sensing inhibitors: An overview." *Biotechnology Advances* 31, no. 2 (2013): 224–245.

Kalia VC, Wood TK, and Kumar P, "Evolution of resistance to quorum-sensing inhibitors." *Microbial Ecology* 68 (2014): 13–23.

Kangmin D, Yuan W, and Yue W, "Antibiotic effective ingredient and use thereof." CN101385737A, 2009.

Karatuna O and Yagci A, "Analysis of quorum sensing-dependent virulence factor production and its relationship with antimicrobial susceptibility in *Pseudomonas aeruginosa* respiratory isolates." *Clinical Microbiology and Infection* 16, no. 12 (2010): 1770–1775.

Karavolos MH, Bulmer DM, Spencer H, Rampioni G, Schmalen I, Baker S, et al. "*Salmonella typhi* sense host neuroendocrine stress hormones and release the toxin haemolysin E." *EMBO Reports* 12, no. 3 (2011): 252–258.

Karavolos MH, Winzer K, Williams P, and Anjam Khan CM, "Pathogen espionage: Multiple bacterial adrenergic sensors eavesdrop on host communication systems." *Molecular Microbiology* 87, no. 3 (2013): 455–465.

Karlsson T, Glogauer M, Ellen RP, Loitto V-M, Magnusson K-E, and Magalhaes MAO, "Aquaporin 9 phosphorylation mediates membrane localization and neutrophil polarization." *Journal of Leukocyte Biology* 90, no. 5 (2011): 963–973.

Kashinskaya EN, Belkova NL, Izvekova GI, Simonov EP, Andree KB, Glupov VV, Baturina OA, Kabilov MR, and Solovyev MM, "A comparative study on microbiota from the intestine of *Prussian carp* (*Carassius gibelio*) and their aquatic environmental compartments, using different molecular methods." *Journal of Applied Microbiology* 119, no. 4 (2015): 948–961.

Kaur J and Yogalakshmi KN, "Screening of quorum quenching activity of the bacteria isolated from dairy industry waste activated sludge." *International Journal of Environmental Science and Technology* 16 (2019): 5421–5428.

Kawaguchi T, Chen YP, Norman RS, and Decho AW, "Rapid screening of quorum-sensing signal N-acyl homoserine lactones by an in vitro cell-free assay." *Applied and Environmental Microbiology* 74(12) (2008): 3667–3671. doi:10.1128/AEM.02869-07.

Keller RL and Hendrix DV, "Bacterial isolates and antimicrobial susceptibilities in equine bacterial ulcerative keratitis (1993–2004)." *Equine Veterinary Journal* 37 (2005): 207–211.

Kim CJ, Lee GS, Lee JK, Yun BS, Lee JC, and Park DJ, "Piericidin A1 with competitive inhibitory activity of acyl homoserin lactone." KR100743672B1, 2007.

Kim J, Kim J, Kim Y, Oh S, Song M, Choe JH, Whang K-Y, Kim KH, and Oh S, "Influences of quorum-quenching probiotic bacteria on the gut microbial community and immune function in weaning pigs." *Animal Science Journal* 89, no. 2 (2018): 412–422.

Kingsford NM and Raadsma HW, "The occurrence of *Pseudomonas aeruginosa* in fleece washings from sheep affected and unaffected with fleece rot." *Veterinary Microbiology* 54 (1997): 275–285.

Koch G, Nadal-Jimenez P, Reis CR, Muntendam R, Bokhove M, Melillo E, et al., "Reducing virulence of the human pathogen *Burkholderia* by altering the substrate specificity of the quorum-quenching acylase PvdQ." *Proceedings of the National Academy of Sciences* 111, no. 4 (2014): 1568–1573.

Koh CL, Sam CK, Yin WF, Tan LY, Krishnan T, Chong YM, and Chan KG, "Plant-derived natural products as sources of anti-quorum sensing compounds." *Sensors* 13 (2013): 6217–6228.

Kong W, Liang H, Shen L, Duan K, "Regulation of type III secretion system by Rhl and PQS quorum sensing systems in *Pseudomonas aeruginosa*." *Wei Sheng Wu Xue Bao* 49 (2009): 1158–1164.

Koul S and Kalia VC, "Multiplicity of quorum quenching enzymes: A potential mechanism to limit quorum sensing bacterial population." *Indian Journal of Microbiology* 57 (2017): 100–108.

Koul S, Prakash J, Mishra A, and Kalia VC, "Potential emergence of multi-quorum sensing inhibitor resistant (MQSIR) bacteria." *Indian Journal of Microbiology* 56 (2016): 1–18.

Kwan JC, Meickle T, Ladwa D, Teplitski M, Paul V, and Luesch H, "Lyngbyoic acid, a 'tagged' fatty acid from a marine *cyanobacterium*, disrupts quorum sensing in *Pseudomonas aeruginosa*." *Molecular Biosystem* 7 (2011): 1205–1216.

Lade H, Paul D, and Kweon JH, "Quorum quenching mediated approaches for control of membrane biofouling." *International Journal of Biological Sciences* 10 (2014): 550–565. doi:10.7150/ijbs.9028.

Lee J, and Zhang L, "The hierarchy quorum sensing network in *Pseudomonas aeruginosa*." *Protein Cell* 6 (2015): 26–41. doi:10.1007/s13238-014-0100-x.

Li L, Rock JL, and Nelson DR, "Identification and characterization of a repeat-in-toxin gene cluster in *Vibrio anguillarum*." *Infection and Immunity* 76 (2008a): 2620–2632.

Li X, Du G, and Chen J. "Use of enzymatic biodegradation for protection of plant against microbial disease." *Current Topics in Biotechnology* 4 (2008b): 1–12.

Lin Y-H, Xu J-L, Hu J, Wang L-H, Ong SL, Leadbetter JR, and Zhang L-H, "Acyl-homoserine lactone acylase from ralstonia strain XJ12B represents a novel and potent class of quorum-quenching enzymes." *Molecular Microbiology* 47, no. 3 (2003): 849–860.

Liu D, Thomas PW, Momb J, Hoang QQ, Petsko GA, Ringe D, and Fast W, "Structure and specificity of a quorum-quenching lactonase (AiiB) from *Agrobacterium tumefaciens*." *Biochemistry* 46, no. 42 (2007): 11789–11799.

Liu Z, Liu W, Ran C, Hu J, and Zhou Z, "Abrupt suspension of probiotics administration may increase host pathogen susceptibility by inducing gut dysbiosis." *Scientific Reports* 6 (2016): 23214.

Lonn-Stensrud J, Landin MA, Benneche T, Petersen FC, and Scheie AA, "Furanones, potential agents for preventing *Staphylococcus epidermidis* biofilm infections?" *Journal of Antimicrob Chemotheraphy* 63 (2009): 309–316.

Lynch MJ, Swift S, Kirke DF, Keevil CW, Dodd CE, and Williams P, "The regulation of biofilm development by quorum sensing in *Aeromonas hydrophila*." *Environmental Microbiology* 4 (2002): 18–28. doi:10.1046/j.1462-2920.2002.00264.x.

Lyon GJ, Mayville P, Muir TW, and Novick RP, "Rational design of a global inhibitor of the virulence response in *Staphylococcus aureus*, based in part on localization of the site of inhibition to the receptor-histidine kinase, agrc." *Proceedings of the National Academy of Sciences* 97, no. 24 (2000): 13330–13335.

Lyte M, Erickson AK, Arulanandam BP, Frank CD, Crawford MA, and Francis DH, "Norepinephrine-induced expression of the K99 pilus adhesin of enterotoxigenic *Escherichia coli*." *Biochemical and Biophysical Research Communications* 232, no. 3 (1997): 682–686.

Lyte M, Vulchanova L, and Brown DR, "Stress at the intestinal surface: Catecholamines and mucosa–bacteria interactions." *Cell and Tissue Research* 343, no. 1 (2011): 23–32.

Maeda T, García-Contreras R, Pu M, Sheng L, Garcia LR, Tomas M, et al., "Quorum quenching quandary: Resistance to antivirulence compounds." *ISME Journal* 6 (2012): 493–501.

Manefield M, Harris L, Rice SA, de Nys R, and Kjelleberg S, "Inhibition of luminescence and virulence in the black tiger prawn (*Penaeus monodon*) pathogen *Vibrio harveyi* by intercellular signal antagonists." *Applied and Environmental Microbiology* 66, no. 5 (2000): 2079–2084.

Mathee K, Adonizio AL, Ausubel F, Clardy J, Bennett B, and Downum K, "Use of ellagitannins as inhibitors of bacterial QS." WO2009114810A2, 2009.

Mayville P, Ji G, Beavis R, Yang H, Goger M, Novick RP, and Muir TW, "Structure-activity analysis of synthetic autoinducing thiolactone peptides from *Staphylococcus aureus* responsible for virulence." *Proceedings of the National Academy of Sciences* 96, no. 4 (1999): 1218–1223.

McClean KH, Winson MK, Fish L, et al., "Quorum sensing and *Chromobacterium violaceum*: Exploitation of violacein production and inhibition for the detection of N-acylhomoserine lactones." *Microbiology* 143 (1997): 3703–3711.

McLennan MW, Kelly WR, and O'Boyle D, "*Pseudomonas* mastitis in a dairy herd." *Australian Veterinary Journal* 75 (1997): 790–792.

Methner U, Rabsch W, Reissbrodt R, and Williams PH, "Effect of norepinephrine on colonisation and systemic spread of *Salmonella enterica* in infected animals: Role of catecholate siderophore precursors and degradation products." *International Journal of Medical Microbiology* 298, no. 5–6 (2008): 429–439.

Michael B, Smith JN, Swift S, Heffron F, and Ahmer BM, "SdiA of *Salmonella enterica* is a LuxR homolog that detects mixed microbial communities." *Journal of Bacteriology* 183 (2001): 5733–5742.

Molina L, Rezzonico F, Défago G, and Duffy B, "Autoinduction in *Erwinia amylovora*: Evidence of an acyl-homoserine lactone signal in the fire blight pathogen." *Journal of Bacteriology* 187, no. 9 (2005): 3206–3213.

Morohoshi T, Tokita K, Ito S, Saito Y, Maeda S, Kato N, and Ikeda T, "Inhibition of quorum sensing in Gram-negative bacteria by alkylamine-modified cyclodextrins." *Journal of Bioscience and Bioengineering* 116, no. 2 (2013): 175–179.

Murray AG and Peeler EJ, "A framework for understanding the potential for emerging diseases in aquaculture." *Preventive Veterinary Medicine* 67 (2005): 223–235.

Nasr RA, "Biofilm formation and presence of *icaAD* gene in clinical isolates of *staphylococci*." *Egyptian Journal of Medical Human Genetics* 13 (2012): 269–274.

Natrah FMI, Alam MI, Pawar S, Harzevili AS, Nevejan N, Boon N, Sorgeloos P, Bossier P, and Defoirdt T, "The impact of quorum sensing on the virulence of *Aeromonas hydrophila* and *Aeromonas salmonicida* towards burbot (*Lota lota* L.) larvae." *Veterinary Microbiology* 159, no. 1–2 (2012): 77–82.

Ng W-L and Bassler BL, "Bacterial quorum-sensing network architectures." *Annual Review of Genetics* 43 (2009): 197–222.

Nhan DT, Cam DTV, Wille M, Defoirdt T, Bossier P, and Sorgeloos P, "Quorum quenching bacteria protect *Macrobrachium rosenbergii* larvae from *Vibrio harveyi* infection." *Journal of Applied Microbiology* 109 (2010): 1007–1016.

Norqvist A, Norrman B, and Wolf-Watz H, "Identification and characterization of a zinc metalloprotease associated with invasion by the fish pathogen *Vibrio anguillarum*." *Infection and Immunity* 58 (1990): 3731–3736.

Novick RP, "Autoinduction and signal transduction in the regulation of staphylococcal virulence." *Molecular Microbiology* 48 (2003): 1429–1449.

O'Loughlin CT, Miller LC, Siryaporn A, Drescher K, Semmelhack MF, and Bassler BL, "A quorum-sensing inhibitor blocks *Pseudomonas aeruginosa* virulence and biofilm formation." *Proceedings of the National Academy of Sciences USA* 110 (2013): 17981–17986.

O'Neill J, "The review on antimicrobial resistance: Tracking drug resistant infections globally." Wellcome Trust and the Department of Health of UK Government, 2016.

Ohtani K, Hayashi H, and Shimizu T, "The *luxS* gene is involved in cell-cell signaling for toxin production in *Clostridium perfringens*." *Molecular Microbiology* 44 (2002): 171–179.

Ormonde P, Horstedt P, OToole R, and Milton DL. "Role of motility in adherence to and invasion of a fish cell line by *Vibrio anguillarum*." *Journal of Bacteriology* 182 (2000): 2326–2328.

Ozer EA, Pezzulo A, Shih DM, Chun C, Furlong C, Lusis AJ, Greenberg EP, and Zabner J, "Human and murine paraoxonase 1 are host modulators of *Pseudomonas aeruginosa* quorum-sensing." *FEMS Microbiology Letters* 253, no. 1 (2005): 29–37.

Packiavathy IASV, Sasikumar P, Pandian SK, and Ravi AV, "Prevention of quorum-sensing-mediated biofilm development and virulence factors production in *Vibrio* spp. by curcumin." *Applied Microbiology and Biotechnology* 97, no. 23 (2013): 10177–10187.

Paczkowski JE, Mukherjee S, McCready AR, Cong J-P, Aquino CJ, Kim H, Henke BR, Smith CD, and Bassler BL, "Flavonoids suppress *Pseudomonas aeruginosa* virulence through allosteric inhibition of quorum-sensing receptors." *Journal of Biological Chemistry* 292, no. 10 (2017): 4064–4076.

Papaioannou E, Wahjudi M, Nadal-Jimenez P, Koch G, Setroikromo R, and Quax WJ, "Quorum-quenching acylase reduces the virulence of *Pseudomonas aeruginosa* in a *Caenorhabditis elegans* infection model." *Antimicrobial Agents Chemistry* 53 (2009): 4891–4807.

Papenfort K and Bassler BL, "Quorumsensing signal-response systems in Gram-negative bacteria." *Nature Reviews Microbiology* 14, no. 9 (2016): 576–588.

Park J, Jagasia R, Kaufmann GF, Mathison JC, Ruiz DI, Moss JA, et al., "Infection control by antibody disruption of bacterial quorum sensing signaling." *Chemistry & Biology* 14 (2007a): 1119–1127.

Park SH, Kim JS, Seo YW, Kim YJ, and Kim MJ, "Antibacterial composition for inhibiting QS." KR100777780B1, 2007b.

Paul D, Youngsam K, Ponnusamy K, and Jihyang K, "Application of quorum quenching to inhibit biofilm formation." *Environmental Engineering Science* 26, no. 8 (2009): 1319–1324.

Pérez T, Balcázar JL, Ruiz-Zarzuela I, Halaihel N, Vendrell D, De Blas I, and Múzquiz JL, "Host–microbiota interactions within the fish intestinal ecosystem." *Mucosal Immunology* 3, no. 4 (2010): 355.

Persson T, Hansen TH, Rasmussen TB, Skinderso ME, Givskov M, and Nielsen J, "Rational design and synthesis of new quorum-sensing inhibitors derived from acylated homoserine lactones and natural products from garlic." *Organic and Biomolecular Chemistry* 3 (2005): 253–262.

Peterson JW, "Bacterial pathogenesis." In Baron S (ed.), *Medical Microbiology*. The University of Texas Medical Branch at Galveston, Galveston, TX, 4th ed., 1996.

Pullinger GD, Carnell SC, Sharaff FF, van Diemen PM, Dziva F, Morgan E, Lyte M, Freestone PPE, and Stevens MP, "Norepinephrine augments *Salmonella enterica*-induced enteritis in a manner associated with increased net replication but independent of the putative adrenergic sensor kinases QseC and QseE." *Infection and Immunity* 78, no. 1 (2010): 372–380.

Qianhong G, Hongbing L, Shouliang Y, Wengong Y, and Shanshan Z, "Compound separated and extracted from marine *Penicillium* and application thereof." CN101811959A, 2010.

Rasch M, Buch C, Austin B, Slierendrecht WJ, Ekmann KS, Larsen JL, et al., An inhibitor of bacterial quorum sensing reduces mortalities caused by vibriosis in rainbow trout (*Oncorhynchus mykiss* Walbaum). *System Applied Microbiology* 27 (2004): 350–359.

Rasko DA, Moreira CG, Li DR, Reading NC, Ritchie JM, Waldor MK, et al., "Targeting QseC signaling and virulence for antibiotic development." *Science* 321, no. 5892 (2008): 1078–1080.

Rasko DA, and Sperandio V, "Anti-virulence strategies to combat bacteria-mediated disease." *Nature Reviews Drug Discovery* 9 (2010): 117–128.

Rasmussen TB, Bjarnsholt T, Skindersoe ME, Hentzer M, Kristoffersen P, Kote M, Nielsen J, Eberl L, and Givskov M, "Screening for quorum-sensing inhibitors (QSI) by use of a novel genetic system, the QSI selector." *Journal of Bacteriology* 187 (2005): 1799–1814.

Reimmann C, Beyeler M, Latifi A, Winteler H, Foglino M, Lazdunski A, and Haas D, "The global activator GacA of *Pseudomonas aeruginosa* PAO positively controls the production of the autoinducer *N*-butyryl-homoserine lactone and the formation of the virulence factors pyocyanin, cyanide, and lipase." *Molecular Microbiology* 24 (1997): 309–319.

Reimmann C, Ginet N, Michel L, Keel C, Michaux P, Krishnapillai V, et al., Genetically programmed autoinducer destruction reduces virulence gene expression and swarming motility in *Pseudomonas aeruginosa* PAO1. *Microbiology* 148 (2002): 923–932.

Ren D, Zuo R, Gonzalez Barrios AF, Bedzyk LA, Eldridge GR, Pasmore ME, and Wood TK, "Differential gene expression for investigation of *Escherichia coli* biofilm inhibition by plant extract ursolic acid." *Applied and Environmental Microbiology* 71, no. 7 (2005): 4022–4034.

Reuter K, Steinbach A, and Helms V, "Interfering with bacterial quorum sensing." *Perspect Medicinal Chemistry* 8 (2016): 1–15.

Rodkhum C, Hirono I, Crosa JH, and Aoki T, "Four novel hemolysin genes of *Vibrio anguillarum* and their virulence to rainbow trout." *Microbial Pathogenesis* 39 (2005): 109–119.

Romero M, Acuna L, and Otero A, "Patents on quorum quenching: Interfering with bacterial communication as a strategy to fight infections." *Recent Patents on Biotechnology* 6 (2012): 2–12.

Romero M, Muras A, Mayer C, Buján N, Magariños B, and Otero A, "In vitro quenching of fish pathogen *Edwardsiella tarda* ahl production using marine bacterium *Tenacibaculum* sp. strain 20j cell extracts." *Diseases of Aquatic Organisms* 108, no. 3 (2014): 217–225.

Rubin J, Walker RD, Blickenstaff K, Bodeis-Jones S, and Zhao S, "Antimicrobial resistance and genetic characterization of fluoroquinolone resistance of *Pseudomonas aeruginosa* isolated from canine infections." *Veterinary Microbiology* 131 (2008): 164–172.

Rui H, Liu Q, Ma Y, Wang Q, and Zhang Y, "Roles of LuxR in regulating extracellular alkaline serine protease A, extracellular polysaccharide and mobility of *Vibrio alginolyticus*." *FEMS Microbiology Letters* 285, no. 2 (2008): 155–162.

Rumbaugh KP, Griswold JA, Iglewski BH, and Hamood AN, "Contribution of quorum sensing to the virulence of *Pseudomonas aeruginosa* in burn wound infections." *Infectious Immunology* 67 (1999): 5854–5862.

Saleh A, Figarella C, Kammouni W, Marchand-Pinatel S, Lazdunski A, Tubul A, Brun P, and Merten MD, "Pseudomonas aeruginosa quorum-sensing signal molecule N-(3-oxododecanoyl)-L-homoserine lactone inhibits expression of P2Y receptors in cystic fibrosis tracheal gland cells." *Infection and Immunity* 67, no. 10 (1999): 5076–5082.

Sana TG, Hachani A, Bucior I, Soscia C, Garvis S, Termine E, et al., "The second type VI secretion system of *Pseudomonas aeruginosa* strain PAO1 is regulated by quorum sensing and Fur and modulates internalization in epithelial cells." *Journal of Biological Chemistry* 287 (2012): 27095–27105.

Sandoz KM, Mitzimberg SM, Schuster M, "Social cheating in *Pseudomonas aeruginosa* quorum sensing." *Proceedings of the National Academy of Sciences of the United States of America* 104 (2007): 15876–15881.

Schuster M and Greenberg EP. "Early activation of quorum sensing in *Pseudomonas aeruginosa* reveals the architecture of a complex regulon." *BMC Genomics* 8 (2007): 287.

Schwenteita J, Gram L, Nielsen KF, Fridjonsson OH, Bornscheuer UT, Givskov M, and Gudmundsdottir BK, "Quorum sensing in *Aeromonas salmonicida* subsp. *achromogenes* and the effect of the autoinducer synthase AsaI on bacterial virulence." *Veterinary Microbiology* 147 (2011): 389–397.

Skindersoe ME, Ettinger-Epstein P, Rasmussen TB, Bjarnsholt T, de Nys R, and Givskov M, "Quorum sensing antagonism from marine organisms." *Marine Biotechnology* 10 (2008): 56–63.

Smith D, Wang JH, Swatton JE, Davenport P, Price B, Mikkelsen H, et al., "Variations on a theme: Diverse *N*-acyl homoserine lactone-mediated quorum sensing mechanisms in Gram negative bacteria." *Scientific Progress* 89 (2006): 167–211.

Smith RS, Harris SG, Phipps R, and Iglewski B, "The *Pseudomonas aeruginosa* quorum-sensing molecule N-(3-oxododecanoyl) homoserine lactone contributes to virulence and induces inflammation in vivo." *Journal of Bacteriology* 184, no. 4 (2002): 1132–1139.

Someya N, Morohoshi T, Okano N, Otsu E, Usuki K, Sayama M, Sekiguchi H, Ikeda T, and Ishida S, "Distribution of *N*-acylhomoserine lactone-producing fluorescent pseudomonads in the phyllosphere and rhizosphere of potato (*Solanum tuberosum* L.)." *Microbes Environment* 24 (2009): 305–314.

Soto-Rodriguez SA, Gomez-Gil B, Lozano R, del Rio-Rodríguez R, Diéguez AL, and Romalde JL, "Virulence of *Vibrio harveyi* responsible for the 'Bright-red' syndrome in the Pacific white shrimp *Litopenaeus vannamei*." *Journal of Invertebrate Pathology* 109, no. 3 (2012): 307–317.

Sperandio V, Torres AG, Jarvis B, Nataro JP, and Kaper JB, "Bacteria–host communication: The language of hormones." *Proceedings of the National Academy of Sciences* 100, no. 15 (2003): 8951–8956.

Steggles RS, "Composition comprising garlic extract and an antibiotic and/or antiseptic for use in the treatment of a multispecies bacterial infection." GB2472315A, 2011.

Steindler L and Venturi V, "Detection of quorum-sensing *N*-acyl homoserine lactone signal molecules by bacterial biosensors." *FEMS Microbiology Letters* 266 (2007): 1–9.

Stoltz DA, Ozer EA, Ng CJ, Yu JM, Reddy ST, Lusis AJ, et al., "Paraoxonase-2 deficiency enhances *Pseudomonas aeruginosa* quorum sensing in murine tracheal epithelia." *American Journal of Physiology-Lung Cellular and Molecular Physiology* 292, no. 4 (2007): L852–L860.

Stoltz DA, Ozer EA, Taft PJ, Barry M, Liu L, Kiss PJ, Moninger TO, Parsek MR, and Zabner J, "Drosophila are protected from *Pseudomonas aeruginosa* lethality by transgenic expression of paraoxonase-1." *The Journal of Clinical Investigation* 118, no. 9 (2008): 3123–3131.

Sun B and Zhang M, "Analysis of the antibacterial effect of an *Edwardsiella tarda* LuxS inhibitor." *Springer Plus* 5 (2016): 92–98.

Sun J, Zhang X, Gao X, Jiang Q, and Lin L, "Characterization of virulence properties of *Aeromonas veronii* isolated from Diseased Gibel Carp (*Carassius gibelio*)." *International Journal of Molecular Science* 17, no. 4 (2016): 496.

Swift S, Karlyshev AV, Fish L, Durant EL, et al., "Quorum sensing in *Aeromonas hydrophila* and *Aeromonas salmonicida*: Identification of the LuxRI homologs AhyRI and AsaRI and their cognate N-acylhomoserine lactone signal molecules." *Journal of Bacteriology* 179 (1997): 5271–5278.

Tateda, Kazuhiro, Yoshikazu Ishii, Manabu Horikawa, Tetsuya Matsumoto, Shinichi Miyairi, Jean Claude Pechere, Theodore J Standiford, Masaji Ishiguro, and Keizo Yamaguchi. "The *Pseudomonas aeruginosa* autoinducer *N*-3-oxododecanoyl homoserine lactone accelerates apoptosis in macrophages and neutrophils." *Infection and Immunity* 71, no. 10 (2003): 5785–5793.

Tay SB and Yew WS, "Development of quorum-based anti-virulence therapeutics targeting Gram-negative bacterial pathogens." *International Journal of Molecular Science* 14 (2013): 16570–16599.

Teplitski M, Merighi M, Gao M, and Robinson J, "Integration of cell-to-cell signals in soil bacterial communities." In Witzany G (ed.), *Biocommunication in Soil Microorganisms*, pp. 369–401. Springer, Berlin, Germany, 2011.

Thenmozhi R, Nithyanand P, Rathna J, and Pandian SK, "Antibiofilm activity of coral-associated bacteria against different clinical M serotypes of *Streptococcus pyogenes*." *FEMS Immunology & Medical Microbiology* 57 (2009): 284–294.

Thompson JA, Oliveira RA, Djukovic A, Ubeda C, and Xavier KB, "Manipulation of the quorum sensing signal Ai-2 affects the antibiotic-treated gut microbiota." *Cell Reports* 10, no. 11 (2015): 1861–1871.

Tinh NTN, Yen VHN, Dierckens K, Sorgeloos P, and Bossier P, "An acyl homoserine lactone-degrading microbial community improves the survival of first feeding turbot larvae (*Scophthalmus maximus* L.)." *Aquaculture* 285 (2008): 56–62.

Torabi Delshad S, Soltanian S, Sharifiyazdi H, and Bossier P, "effect of quorum quenching bacteria on growth, virulence factors and biofilm formation of *Yersinia ruckeri* in vitro and an in vivo evaluation of their probiotic effect in rainbow trout." *Journal of Fish Diseases* 41, no. 9 (2018): 1429–1438.

Torres M, Reina JC, Juan Carlos Fuentes-Monteverde, Gerardo Fernández, Jaime Rodríguez, Carlos Jiménez, and Inmaculada Llamas. "AHL-lactonase expression in three marine emerging pathogenic *Vibrio* spp. reduces virulence and mortality in brine shrimp (*Artemia salina*) and manila clam (*Venerupis philippinarum*)." *PloS one* 13, no. 4 (2018): e0195176.

Torres M, Rubio-Portillo E, Antón J, Ramos-Esplá AA, Quesada E, and Llamas I, "Selection of the *N*-acylhomoserine lactone-degrading bacterium *Alteromonas stellipolaris* PQQ-42 and of its potential for biocontrol in aquaculture." *Frontiers in Microbiology* 7 (2016): 646. doi:10.3389/fmicb.2016.00646

Traber KE, Lee E, Benson S, Corrigan R, Cantera M, Shopsin B, and Novick RP, "*agr* function in clinical *Staphylococcus aureus* isolates." *Microbiology* 154 (2008): 2265–2274.

Tron EA, Wilke HL, Petermann SR, and Rust L, "*Pseudomonas aeruginosa* from canine otitis externa exhibit a quorum sensing deficiency." *Veterinary Microbiology* 99 (2004): 121–129.

Ulrich RL, "Quorum quenching: Enzymatic disruption of *N*-acylhomoserine lactone-mediated bacterial communication in burkholderia thailandensis." *Applied and Environmental Microbiology* 70, no. 10 (2004): 6173–6180.

Van Cam DT, Hao NV, Dierckens K, Defoirdt T, Boon N, Sorgeloos P, and Bossier P, "Novel approach of using homoserine lactone-degrading and poly-β-hydroxybutyrate-accumulating bacteria to protect *Artemia* from the pathogenic effects of *Vibrio harveyi*." *Aquaculture* 291, no. 1–2 (2009): 23–30.

Vancraeynest D, Haesebrouck F, Deplano A, Denis O, Godard C, Wildemauwe C, and Hermans K, "International dissemination of a high virulence rabbit *Staphylococcus aureus* clone." *Journal of Veterinary Medicine B Infectious Diseases and Veterinary Public Health* 53 (2006): 418–422.

Vandeputte OM, Kiendrebeogo M, Rasamiravaka T, Stevigny C, Duez P, Rajaonson S, et al. "The flavanone naringenin reduces the production of quorum sensing-controlled virulence factors in *Pseudomonas aeruginosa* PAO1." *Microbiology* 157, no. 7 (2011): 2120–2132.

Verbrugghe E, Boyen F, Gaastra W, Bekhuis L, Leyman B, Van Parys A, Haesebrouck F, and Pasmans F, "The complex interplay between stress and bacterial infections in animals." *Veterinary Microbiology* 155, no. 2–4 (2012): 115–127.

Verschuere L, Rombaut G, Sorgeloos P, and Verstraete W, "Probiotic bacteria as biological control agents in aquaculture." *Microbiology and Molecular Biology Reviews* 64 (2000): 655–671.

Vikram A, Jesudhasan PR, Jayaprakasha GK, Pillai SD, and Patil BS, "Citrus limonoids interfere with *Vibrio harveyi* cell–cell signalling and biofilm formation by modulating the response regulator LuxO." *Microbiology* 157, no. 1 (2011): 99–110.

Vinoj G, Vaseeharan B, Thomas S, Spiers AJ, and Shanthi S, "Quorum-quenching activity of the AHL-lactonase from *Bacillus licheniformis* DAHB1 inhibits Vibrio biofilm formation in vitro and reduces shrimp intestinal colonisation and mortality." *Marine Biotechnology* 16, no. 6 (2014): 707–715.

Walters M and Sperandio V, "Quorum sensing in *Escherichia coli* and *Salmonella*." *International Journal of Medical Microbiology* 296 (2006): 125–131.

Wang T-N, Guan Q, Pain A, Kaksonen AH, and Hong P, "Discovering, characterizing, and applying acyl homoserine lactone-quenching enzymes to mitigate microbe-associated problems under saline conditions." *Frontiers in Microbiology* 10 (2019): 823.

Weiland-Bräuer N, Kisch MJ, Pinnow N, Liese A, and Schmitz RA, "Highly effective inhibition of biofilm formation by the first metagenome-derived AI-2 quenching enzyme." *Frontiers in Microbiology* 7 (2016): 1098.

Wopperer J, Cardona ST, Huber B, Jacobi CA, Valvano MA, and Eberl L, "A quorum-quenching approach to investigate the conservation of quorum-sensing-regulated functions within the *Burkholderia cepacia* complex." *Applied and Environmental Microbiology* 72, no. 2 (2006): 1579–1587.

Wu H, Song Z, Hentzer M, Andersen JB, Molin S, Givskov M, and Høiby N, "Synthetic furanones inhibit quorum-sensing and enhance bacterial clearance in *Pseudomonas aeruginosa* lung infection in mice." *Journal of Antimicrobial Chemotherapy* 53, no. 6 (2004): 1054–1061.

Wu S, Tian J, Wang G, Li W, and Zou H, "Characterization of bacterial community in the stomach of yellow catfish (*Pelteobagrus fulvidraco*)." *World Journal of Microbiology and Biotechnology* 5, no. 28 (2012a): 2165–2174.

Wu S, Wang G, Angert ER, Wang W, Li W, and Zou H, "Composition, diversity, and origin of the bacterial community in grass carp intestine." *PloS one* 7, no. 2 (2012b): e30440.

Xavier KB and Bassler BL, "LuxS quorum sensing: More than just a numbers game." *Current Opinion in Microbiology* 6 (2003): 191–197.

Yang F, Wang L-H, Wang J, Dong Y-H, Hu JY, and Zhang L-H, "Quorum quenching enzyme activity is widely conserved in the sera of mammalian species." *FEBS Letters* 579, no. 17 (2005): 3713–3717.

Yang Q, Yu Y, Lin Y, Ni Y, Sun J, Xu Y, et al., "Distribution and antimicrobial resistance profile of common pathogens isolated from respiratory secretion in CHINET Antimicrobial Resistance Surveillance Program, 2005–2014." *Chinese Journal of Infection and Chemotherapy* 16 (2016): 541–550.

Zang T, Lee BWK, Cannon LM, Ritter KA, Dai S, Ren D, et al., "A naturally occurring brominated furanone covalently modifies and inactivates LuxS." *Bioorganic & Medicinal Chemistry Letters* 19, no. 21 (2009): 6200–6204.

Zhang, MeiChao, YaNan Cao, Bin Yao, DongQing Bai, and ZhiGang Zhou. "Characteristics of quenching enzyme AiiO-AIO6 and its effect on *Aeromonas hydrophila* virulence factors expression." *Journal of Fisheries of China* 35, no. 11 (2011): 1720–1728.

Zhao J, Chen M, Quan C, and Fan S, "Mechanisms of quorum sensing and strategies for quorum sensing disruption in aquaculture pathogens." *Journal of Fish Diseases* 38, no. 9 (2015): 771–786.

Zhou H, Wang M, Smalley NE, Kostylev M, Schaefer AL, Greenberg EP, Dandekar AA, and Xu F, "Modulation of *Pseudomonas aeruginosa* quorum sensing by glutathione." *Journal of Bacteriology* 201 (9) (2019). doi:10.1128/JB.00685-18.

Zhou J, Lyu Y, Richlen M, Anderson DM, Cai, Z, "Quorum sensing is a language of chemical signals and plays an ecological role in algal-bacterial interactions." *Critical Reviews in Plant Sciences* 35(2) (2016): 81–105. doi:10.1080/07352689.2016.1172461.

Zhou S, Xia Y, Zhu C, and Chu W, "Isolation of marine *Bacillus* spp. with antagonistic and organic-substances-degrading activities and its potential application as a fish probiotic." *Marine Drugs* 16, no. 6 (2018): 196.

Index

Note: Page numbers in italic and bold refer to figures and tables, respectively.